Families of Functions

Linear Functions

$y = mx + b, m > 0$

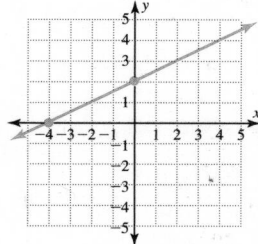

$y = mx + b, m < 0$

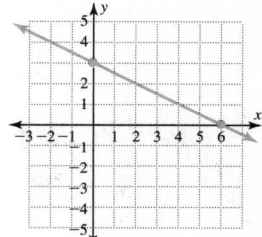

$y = mx + b, m = 0$
(a constant function)

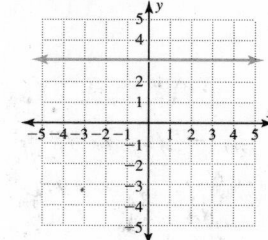

Absolute Value Function

$y = |x|$

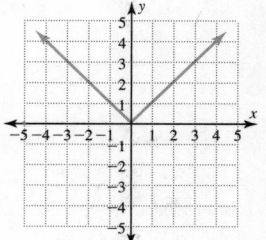

Quadratic Functions

$y = x^2$

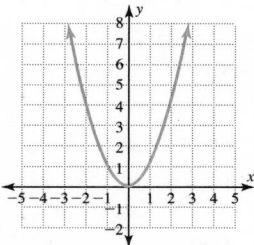

$y = ax^2 + bx + c, a > 0$

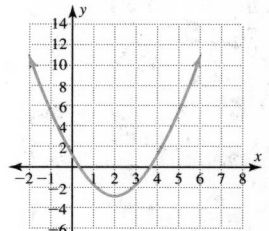

$y = ax^2 + bx + c, a < 0$

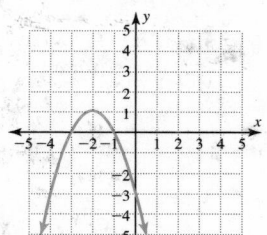

Square Root Function

$y = \sqrt{x}$

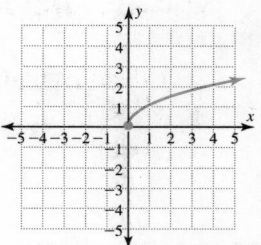

Cubic Functions

$y = x^3$

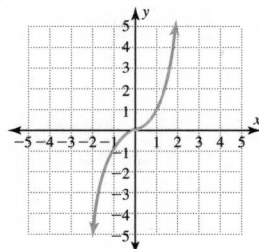

$y = ax^3 + bx^2 + cx + d,$
$a > 0$

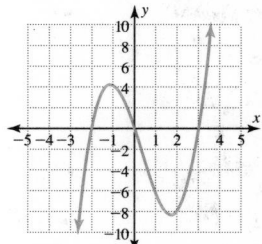

$y = ax^3 + bx^2 + cx + d,$
$a < 0$

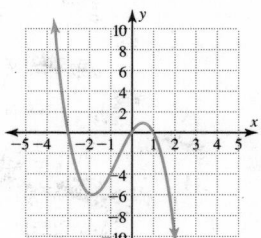

Cube Root Function

$y = \sqrt[3]{x}$

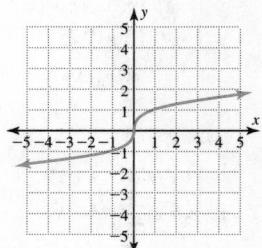

Rational Function

$y = \dfrac{1}{x}$

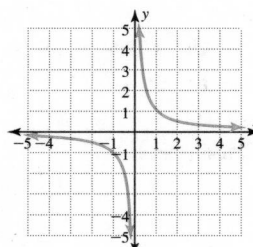

Exponential Growth Function

$y = b^x, b > 1$

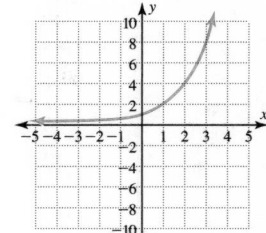

Exponential Decay Function

$y = b^x, 0 < b < 1$

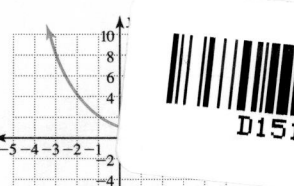

Logarithmic Function

$y = \log_b x, b > 0, b \neq 1$

Subsets of the Real Numbers

Natural numbers: $\{1, 2, 3, 4, 5, 6, \ldots\}$
Whole numbers: $\{0, 1, 2, 3, 4, 5, 6, \ldots\}$
Integers: $\{\ldots -3, -2, -1, 0, 1, 2, 3, \ldots\}$
Rational: $\left\{\dfrac{a}{b} : a, b \text{ are integers and } b \neq 0\right\}$
Irrational: The set of real numbers that are not rational.

CONSTANTS: $\pi \approx 3.14159$, $e \approx 2.71828$

Operations with Fractions

Addition: $\dfrac{a}{b} + \dfrac{c}{b} = \dfrac{a+c}{b}$ for $b \neq 0$

Subtraction: $\dfrac{a}{b} - \dfrac{c}{b} = \dfrac{a-c}{b}$ for $b \neq 0$

Reducing fractions: $\dfrac{ac}{bc} = \dfrac{a}{b}$ for $b \neq 0$ and $c \neq 0$

Multiplication: $\dfrac{a}{b} \cdot \dfrac{c}{d} = \dfrac{a \cdot c}{b \cdot d}$ for $b \neq 0$ and $d \neq 0$

Division: $\dfrac{a}{b} \div \dfrac{c}{d} = \dfrac{a}{b} \cdot \dfrac{d}{c} = \dfrac{a \cdot d}{b \cdot c}$ for $b \neq 0$, $c \neq 0$,
and $d \neq 0$

Properties of the Real Numbers

Commutative properties:

of addition $\qquad a + b = b + a$

of multiplication $\quad ab = ba$

Associative properties:

of addition $\qquad (a + b) + c = a + (b + c)$

of multiplication $\quad (ab)(c) = (a)(bc)$

Distributive properties of multiplication over addition:

$a(b + c) = ab + ac$

$(b + c)a = ba + ca$

Identities:

additive $\qquad\quad a + 0 = 0 + a = a$

multiplicative $\quad a \cdot 1 = 1 \cdot a = a$

Inverses:

additive $\qquad\quad a + (-a) = (-a) + a = 0$

multiplicative $\quad a \cdot \dfrac{1}{a} = \dfrac{1}{a} \cdot a = 1$

Order of Operations

Step 1. Start with the expression within the innermost pair of grouping symbols.
Step 2. Perform all exponentiations.
Step 3. Perform all multiplications and divisions as they appear from left to right.
Step 4. Perform all additions and subtractions as they appear from left to right.

Interval Notation

Inequality Notation	Verbal Meaning	Graph	Interval Notation
$x > a$	x is greater than a		$(a, +\infty)$
$x \geq a$	x is greater than or equal to a		$[a, +\infty)$
$x < a$	x is less than a		$(-\infty, a)$
$x \leq a$	x is less than or equal to a		$(-\infty, a]$
$a < x < b$	x is greater than a and less than b		(a, b)
$a < x \leq b$	x is greater than a and less than or equal to b		$(a, b]$
$a \leq x < b$	x is greater than or equal to a and less than b		$[a, b)$
$a \leq x \leq b$	x is greater than or equal to a and less than or equal to b		$[a, b]$
$-\infty < x < \infty$	x is any real number		$(-\infty, +\infty)$

IMPORTANT:

HERE IS YOUR REGISTRATION CODE TO ACCESS
YOUR PREMIUM McGRAW-HILL ONLINE RESOURCES.

For key premium online resources you need THIS CODE to gain access. Once the code is entered, you will be able to use the Web resources for the length of your course.

If your course is using **WebCT** or **Blackboard**, you'll be able to use this code to access the McGraw-Hill content within your instructor's online course.

Access is provided if you have purchased a new book. If the registration code is missing from this book, the registration screen on our Website, and within your WebCT or Blackboard course, will tell you how to obtain your new code.

Registering for McGraw-Hill Online Resources

TO gain access to your McGraw-Hill web resources simply follow the steps below:

1. USE YOUR WEB BROWSER TO GO TO: **http://www.mhhe.com/hallmercer/**
2. CLICK ON **FIRST TIME USER**.
3. ENTER THE REGISTRATION CODE* PRINTED ON THE TEAR-OFF BOOKMARK ON THE RIGHT.
4. AFTER YOU HAVE ENTERED YOUR REGISTRATION CODE, CLICK **REGISTER**.
5. FOLLOW THE INSTRUCTIONS TO SET-UP YOUR PERSONAL UserID AND PASSWORD.
6. WRITE YOUR UserID AND PASSWORD DOWN FOR FUTURE REFERENCE.
   KEEP IT IN A SAFE PLACE.

TO GAIN ACCESS to the McGraw-Hill content in your instructor's **WebCT** or **Blackboard** course simply log in to the course with the UserID and Password provided by your instructor. Enter the registration code exactly as it appears in the box to the right when prompted by the system. You will only need to use the code the first time you click on McGraw-Hill content.

Thank you, and welcome to your McGraw-Hill online Resources!

* YOUR REGISTRATION CODE CAN BE USED ONLY ONCE TO ESTABLISH ACCESS. IT IS NOT TRANSFERABLE.

0-07-249562-6 HALL/MERCER: BEGINNING & INTERMEDIATE ALGEBRA, 1/E

REGISTRATION CODE

prehistory-55027465

Beginning and Intermediate
ALGEBRA
THE LANGUAGE AND SYMBOLISM
OF MATHEMATICS

James W. Hall   ■   **Brian A. Mercer**
Parkland College   *Parkland College*

Boston   Burr Ridge, IL   Dubuque, IA   Madison, WI   New York   San Francisco   St. Louis
Bangkok   Bogotá   Caracas   Kuala Lumpur   Lisbon   London   Madrid   Mexico City
Milan   Montreal   New Delhi   Santiago   Seoul   Singapore   Sydney   Taipei   Toronto

McGraw-Hill Higher Education

A Division of The **McGraw-Hill** Companies

BEGINNING AND INTERMEDIATE ALGEBRA: THE LANGUAGE AND SYMBOLISM OF MATHEMATICS

1 2 3 4 5 6 7 8 9 0 VNH/VNH 0 9 8 7 6 5 4 3 2
1 2 3 4 5 6 7 8 9 0 VNH/VNH 0 9 8 7 6 5 4 3 2

ISBN 0–07–249562–6
ISBN 0–07–251344–6 (Instructor's Edition)

Publisher: *William K. Barter*
Senior sponsoring editor: *David Dietz*
Developmental editor: *Christien A. Shangraw*
Executive marketing manager: *Marianne C. P. Rutter*
Senior marketing manager: *Mary K. Kittell*
Project manager: *Jane Mohr*
Senior production supervisor: *Laura Fuller*
Lead media project manager: *Judi David*
Media technology producer: *Jeff Huettman*
Designer: *K. Wayne Harms*
Cover/interior designer: *Andrew Ogus*
Cover image: *Copyright 1992 by Photo Art Publishing Trust,*
Photo taken by Lou DeSerio, Sedona Arizona
Senior photo research coordinator: *John C. Leland*
Photo research: *Chris Hammond/PhotoFind LLC*
Supplement producer: *Brenda A. Ernzen*
Compositor: *The GTS Companies*
Typeface: *10/12 Times Roman*
Printer: *Von Hoffmann Press, Inc.*

The credits section for this book begins on page 865 and is considered an extension of the copyright page.

Library of Congress Cataloging-in-Publication Data

Hall, James W.
    Beginning and intermediate algebra : the language and symbolism of mathematics /
James W. Hall, Brian A. Mercer. — 1st ed.
        p   cm.
    Includes index.
    ISBN 0–07–249562–6
    1. Algebra. I. Mercer, Brian A. II. Title.

QA152.3 .H28   2003
512—dc21   2002023500
                                            CIP

www.mhhe.com

DEDICATION

To our families for their support and encouragement.

ABOUT THE AUTHORS

JAMES W. HALL is chair of the mathematics department and a professor of mathematics at Parkland College in Champaign, Illinois. He started teaching mathematics in 1969 at Northern Arizona University and also taught at Clayton State College in Georgia prior to joining Parkland College in 1975. From 1989 to 1990 he taught at Dandenong College in Victoria, Australia. He received a B.S. and an M.A. in mathematics from Eastern Illinois University and an Ed.D. from Oklahoma State University. He was Midwest Regional Vice President of AMATYC (American Mathematical Association of Two-Year Colleges) from 1987 to 1989 and President of IMACC (Illinois Mathematics Association of Community Colleges) from 1995 to 1996. In 1978 he edited the "Report on Microcomputers in the Classroom" for ICTM (Illinois Council of Teachers of Mathematics), and from 1991 to 1995 he was chairperson of the editorial review committee for AMATYC. He is currently writing team chair for Liberal Arts and Statistics for the AMATYC Crossroads Revision.

BRIAN A. MERCER is an assistant professor of mathematics at Parkland College in Champaign, Illinois. He started teaching in 1994 at Neoga High School in Neoga, Illinois. Prior to Parkland College in 1998 he also taught at Lakeland College in Mattoon, Illinois. He received a B.S. in mathematics from Eastern Illinois University and an M.S. in mathematics from Southern Illinois University. He is a member of AMATYC and is currently a board member of IMACC.

CONTENTS

PREFACE

The Universe is a grand book which cannot be read until one first learns to comprehend the language and become familiar with the characters in which it is composed. It is written in the language of mathematics.

—*Galileo*

Beginning and Intermediate Algebra: The Language and Symbolism of Mathematics covers the topics from both Beginning and Intermediate Algebra. It is fully integrated, rather than the combination of two separate texts.

Our primary goal is to implement the AMATYC standards, as outlined in *Crossroads in Mathematics,* and to give strong support to faculty members who teach this material. These standards were used as guiding principles to organize the topics. This organization is designed to work for students with a variety of learning styles and for teachers with a variety of experiences and backgrounds. Examples of this organization include an early presentation of function notation and graphing of linear equations in two variables.

The inclusion of multiple perspectives—verbal, numerical, algebraic, and graphical—has proven popular with a broad cross section of students. Calculator Perspectives help students see the relationship between mathematics and technology. The specific instructions provided in the Calculator Perspectives also eliminate the need for instructors to create separate keystroke handouts.

The Beginning Algebra portion of this text concentrates primarily on material related to linear equations. (Nonlinear material is reserved for the Intermediate Algebra portion of the text.) The review material in Chapter 1 is presented through the evaluation of algebraic expressions, the checking of solutions to equations, and other contexts. This non-traditional approach motivates students through real-life applications to review background concepts. It also gives students who have already had this material in high school a fresh approach and helps them to connect previously separated topics.

TEACHING APPROACH

Emphasis on the Rule of Four and Multiple Perspectives

The "rule of four" is a phrase that means concepts should be examined algebraically (symbolically), numerically, graphically, and verbally. Reviewers of the manuscript were very pleased that we integrated multiple perspectives throughout the book. We use multiple perspectives not only in examples and exercises, but also in definitions and exposition. Our experience leads us to believe that students who use the rule of four develop a deeper understanding of the concepts they study. They are less likely to memorize steps, they are more likely to retain the material they understand, and they are more likely to apply mathematics outside the classroom. (See AMATYC Standard P-4.)

Technology Is Built-In, Not Added-On

Topics that once were postponed until many manipulative skills had been developed can now be considered earlier by using technology to focus on concepts instead of computation. The use of a graphing calculator is demonstrated throughout the book. The students' use of technology enables them to examine realistic problems such as producing the payment schedule for a car loan. Together, realistic applications and the use of calculators facilitate the development of modeling skills by the students. Technology is woven throughout the text—it is not simply inserted into a standard presentation. (See AMATYC Standards I-2 and I-6.)

Functions in Beginning Algebra

The Beginning Algebra chapters give the student many opportunities to become familiar with function notation and with the input-output concept. This portion of the book concentrates primarily on material related to linear functions, with nonlinear material reserved for the Intermediate Algebra portion of the text. (See AMATYC Standard C-4.)

Functions in Intermediate Algebra

The Intermediate Algebra portion of the textbook contains an introduction to the definition of a function and various notations used to represent functions. This topic is intended to serve as a transition for students who are directly entering the book at this point. The use of function notation provides an opportunity to review the operations with algebraic expressions in a new context and to reexamine linear equations. Chapter 6 contains material on several families of functions, including linear, absolute value, quadratic, square root, cubic, and cube root. (See AMATYC Standard C-4.)

Mathematical Modeling and Word Equations

The residual value of mathematics—the mathematics that students can still use four or more years after taking a course—is not a collection of tricks or memorized steps. What endures is an understanding that allows students to see mathematics as useful in improving their daily lives. Most people encounter mathematics through words, either orally or in writing, not through equations that people want them to solve. Students need to model real problems in a course if we expect them to use mathematics on their own. Word equations help students bridge the gap between the statement of a word problem and the formation of an algebraic equation that models the problem. To that end, the text presents many realistic examples and exercises involving data (see the Index of Applications). (See AMATYC Standard C-2.)

Factoring

Factoring is developed gradually as multiplication is considered.

- In Section 1.6 we examine both multiplication and factoring when the distributive property is illustrated.

- In Section 5.5 we continue to stress the relationship between multiplication and factoring when the multiplication of polynomials is examined. The role of the distributive property is again emphasized.
- In Section 5.6 we use long division of polynomials to complete the factorization of a polynomial when one factor is known.
- Section 5.7 covers some special products and the corresponding factored forms.
- In Section 6.7, factoring out of the GCF (greatest common factor) of a polynomial is used to examine factoring from numerical and graphical perspectives. We then examine the relationship between linear factors of a polynomial, the zeros of a polynomial function, and the x-intercepts of a graph.
- A comprehensive algebraic approach to factoring polynomials is presented in Chapter 7.

Again, our goal is to develop connections among concepts, especially the relationship between the distributive property and multiplication and factoring. Chapter 7 also explores the relationship between the zeros of a function and the factors of a polynomial as introduced in Chapter 6.

Using Systems of Equations

Word problems that involve two unknowns are solved in Chapter 3 using two variables rather than one variable. This approach has been received well by the students who often have more trouble identifying two unknowns using one variable than using a separate variable for each unknown. This approach also received favorable feedback from teachers who class-tested the manuscript. Later in the book we examine alternate approaches that build on creating functional models and the relation of one variable to another.

Using Discrete Data

The book contains data and problems that give the students experience with discrete data. This will give students a better perspective on mathematical models, especially those who will use Intermediate Algebra as their prerequisite to an Introductory Statistics course. (See AMATYC Standards C-5 and C-6.)

Using the Language and Symbolism of Mathematics

Each exercise set starts with a few questions on using the language and symbolism of mathematics. One benefit of assigning these exercises that we have noted in our classes is that the students spend more time reading the book before starting the other exercises. (See AMATYC Standard I-5.)

NOTABLE FEATURES

A Different Kind of Chapter 1

The organization of Chapter 1, "Operations with Real Numbers," is intentionally different from that in most Beginning Algebra books. One benefit of this organization is that it stimulates the interest of students who may be reviewing this material. Many of the chapter's topics are presented either within a new context or in a nontraditional order. The material on operations with real numbers includes problems that evaluate algebraic expressions and problems that check solutions to equations. Also, the commutative, associative, and distributive properties are presented as needed within the arithmetic review rather than in an intimidating section that focuses only on terminology.

Calculator Perspectives in the Main Body of the Book

The TI-83 Plus™ is used to work sample problems in the book as concepts are developed. These Calculator Perspectives provide both students and teachers with calculator material right where it is needed without additional handouts or supplements. The TI-83 Plus™ was selected as the representative graphing calculator because it is the most popular model at the colleges we surveyed. Identical Calculator Perspectives, including screen shots and step-by-step keystroke instructions, are provided for other graphing calculator models at www.mhhe.com/hallmercer. (Some colleges require calculators rather than list them as optional because this is advantageous for students whose grants and scholarships are based on need.) (See AMATYC Standard P-1.)

Estimation Skills and Error Analysis Exercises

Estimation skills, concern for reasonable answers, and the ability to detect calculator errors should be developed by students at the same time they develop their calculator skills. Many examples and exercises in the book are specifically designed to help the students develop these skills. They are clearly labeled as such in the exercise sets. (See AMATYC Standard C-1.)

Mathematical Notes

Mathematical notes throughout the book give the students a sense of historical perspective and connect mathematics to other disciplines. These short vignettes give the origin of some of the symbols and terms that we now use and provide brief glimpses into the lives of some of the men and women of mathematics. (See AMATYC Standard P-3.)

Self-Checks

Self-checks and Self-check answers in each section help students become active learners and monitor their own progress. (See AMATYC Standard I-7.)

Example Format

The example format provides a clear model students can use to work the exercises. Sidebar explanations limit wordiness and allow students with different ability levels to use the examples in different ways. Examples are demonstrated from multiple perspectives so students can compare algebraic, numerical, graphical, and verbal approaches to a given problem. (See AMATYC Standard P-4.)

Geometrical-Based Problems

Examples and exercises based on geometric shapes are placed throughout the textbook. Many exercise sets have problems involving perimeter, area, and volume. (See AMATYC Standard C-3.)

Design of Exercises

Many exercises are composed of multiple parts to help the students face common misconceptions about the language and symbolism of mathematics. A few examples are:

Exercise 1.6, #13	**a.** Simplify $(6+4)^2$ **b.** Simplify $6^2 + 4^2$
Chapter 5, Review #71	**a.** Expand $2x(x+4y) - 3y(x+4y)$ **b.** Factor $2x(x+4y) - 3y(x+4y)$
Exercise 7.5, #55	**a.** Solve $(5m-3)(m-2) = 0$ **b.** Simplify $(5m-3)(m-2)$

Group Discussion Questions and Group Projects

Each exercise set has group discussion questions, and there is a group project at the end of each of the first ten chapters. Many of these exercises build bridges to past or future material, and many seek to engage students orally and in writing to enable interactive and collaborative learning. (See AMATYC Standard P-2.)

Key Concepts, Chapter Review, and Mastery Test

Each chapter ends with these features. The Key Concepts outline the main points covered in that chapter. The Chapter Review contains a selection of exercises designed to help students review material from the chapter and to gauge their readiness for an exam. The Chapter Review is longer than an hour exam, and the order of the questions may not parallel the order of each topic within the chapter. The Mastery Test is more directed in its purpose: each of its questions matches an objective stated at the beginning of one of the sections in the chapter. Students can use the Mastery Test diagnostically to determine which sections and objectives have been mastered and on which they need more work.

Diagnostic Review of Beginning Algebra

A comprehensive review of the first five chapters appears between Chapters 5 and 6. Every question is categorized to help the instructor assign questions for a specific purpose. For each problem, the student is provided with the correct answer and directed to specific examples in the first five chapters of the text for personal review. We have found this review especially useful for students entering directly into the Intermediate Algebra portion of the book.

SUPPLEMENTS FOR THE INSTRUCTOR

Instructor's Edition

This ancillary contains answers to problems and exercises in the text, including answers to all Language and Symbolism of Mathematics vocabulary questions, all end-of-section exercises, all end-of-chapter review exercises, and all end-of-chapter mastery tests.

Computerized Test Bank

The computerized test bank allows you to create well-formatted quizzes or tests using a large bank of algorithmically generated and static questions through an intuitive Windows or Macintosh interface. When creating a quiz or test, you can manually choose individual questions or have the software randomly select questions based on section, question type, difficulty level, and other criteria. Instructors also have the ability to add or edit test bank questions to create their own customized test bank. In addition to printed tests, the test generator can deliver tests over a local area network or the World Wide Web, with automatic grading.

Instructor's Solutions Manual

Prepared by Mark Smith of College of Lake County, this supplement contains detailed solutions to all the exercises in the text not included in the Student Solution's Manual (see below). The methods used to solve the problems in the manual are the same as those used to solve the examples in the textbook.

Student Study Guide

This supplement provides a great many student benefits (see below), but also has notable benefits for instructors. Novice instructors can use it as the framework for day-by-day class plans. Veteran teachers have testified to its helpfulness in making better use of class time and keeping students focused on active learning.

Online Learning Center

Web-based interactive learning is available for your students on the Online Learning Center, located at www.mhhe.com/hallmercer. Student resources are located in the Student Center, and include interactive applications, algorithmically generated practice exams and quizzes, audiovisual tutorials, and web links. Instructor resources are located in the Instructor Center and include links to PageOut, ALEKS®, and other recommended sites.

PageOut

PageOut is McGraw-Hill's unique point-and-click course website tool, enabling you to create a full-featured, professional-quality course website without knowing HTML coding. With PageOut you can post your course syllabus, assign McGraw-Hill Online Learning Center content, add links to important off-site resources, and maintain student results in the online grade book. You can send class announcements, copy your course site to share with colleagues, and upload original files. PageOut is free for every McGraw-Hill user, and if you're short on time, we even have a team ready to help you create your site!

SUPPLEMENTS FOR THE STUDENT

Student's Solutions Manual

Prepared by Mark Smith of College of Lake County, the Student's Solutions Manual contains complete worked-out solutions to all the odd-numbered exercises from the text, including all end-of-section exercises, all end-of-chapter review exercises, and all end-of-chapter mastery tests.

Student Study Guide

This supplement provides key language and symbolism, definitions, and procedures so students can avoid using valuable class time recopying these items. Its examples are consistent with those in the text, and its structure models the framework of a class lecture. When completed, it provides an organized set of notes for later study.

Hall/Mercer Video Series

The video series is composed of 11 videocassettes (one for each chapter of the text). An on-screen instructor introduces topics and works through examples using the methods presented in the text, including step-by-step instruction on graphing calculator operations. The video series is also available on video CDs.

Hall/Mercer Tutorial CD-ROM

This interactive CD-ROM is a self-paced tutorial linked directly to the text that reinforces topics through unlimited opportunities to review concepts and practice problem solving. The CD-ROM provides algorithmically generated "bookmarkable" practice exercises (including hints), section- and chapter-level testing with gradebook capabilities, and a built-in graphing calculator. This product requires virtually no computer training on the part of students and supports Windows and Macintosh systems.

Online Learning Center

Student resources are located in the Student Center on the Online Learning Center (OLC) and include interactive applications, algorithmically generated "bookmarkable" practice exercises (including hints), section- and chapter-level testing, a built-in graphing calculator, a glossary, audiovisual tutorials, and links to PageOut, NetTutor, and other fun and useful algebra websites. The OLC is located at www.mhhe.com/hallmercer, and a free password card is included with each new copy of this text.

ALEKS

ALEKS® (**A**ssessment and **LE**arning in **K**nowledge **S**paces) is an artificial intelligence-based system for individualized math learning, available over the World Wide Web. ALEKS® delivers precise, qualitative diagnostic assessments of students' math knowledge, guides them in the selection of appropriate new study material, and records their progress toward mastery of curricular goals in a robust classroom management system. It interacts with the student much as a skilled human tutor would, moving between explanation and practice as needed, correcting and analyzing errors, defining terms, and changing topics on request. By sophisticated modeling of a student's "knowledge state" for a given subject matter, ALEKS® can focus clearly on what the student is most ready to learn next, building a learning momentum that fuels success.

To learn more about ALEKS®, including purchasing information, visit the ALEKS® website at www.highed.aleks.com.

NetTutor

NetTutor is a revolutionary system that enables students to interact with a live tutor over the World Wide Web. Students can receive instruction from live tutors using NetTutor's Web-based, graphical chat capabilities. They can also submit questions and receive answers, browse previously answered questions, and view previous live chat sessions.

ACKNOWLEDGMENTS

We would like to thank our students who used the preliminary versions of this text for their suggestions and encouragement. We would also like to thank Kevin Hastings, Brenda Marshall, and Mark Smith who taught from the preliminary versions and who helped us to craft the final version. Jan Crutcher and Debra Blair, who typed the original manuscript, deserve a medal for their can-do attitude and for reading our writing.

We are also very appreciative of the support from McGraw-Hill. Bill Barter, Publisher, and David Dietz, Senior Sponsoring Editor, gave us the support to create a different book that incorporates technology and embraces the AMATYC standards. We are also very appreciative of the support of Christien Shangraw and Erin Brown, Developmental Editors, and Jane Mohr, Project Manager, for helping us accomplish all the miracles that put text, art, and answers into a quality textbook. Jeff Huettman, Media Technology Producer, did an outstanding job creating the Online Learning Center.

We also wish to thank Lou DeSerio for taking such a beautiful photo of the Sedona, Arizona, area, and Wayne Harms, Senior Designer, for incorporating it into the front cover. Mathematics is only one of the wonders of our world; another is the natural beauty exemplified by this photo of Cathedral Rock.

Special thanks are due to: John Hunt of Humboldt State University, and Frank Purcell and Elka Block of Twin Prime Editorial for their accuracy-check of the text; Michael J. Keller of Saint John's River Community College for providing the TI-83 Plus screenshots that appear throughout this book; LinkSystems International for their work on the Online Learning Center; Frank Pecchioni of Jefferson Community College for his authorship of the online glossary; John LaMaster of Indiana University–Purdue University for translating the text's TI-83 Plus™ Calculator Perspectives into identical Web-ready demonstrations for alternate calculator models; Ellen Sawyer of College of DuPage for her exhaustive accuracy and functionality check of the Online Learning Center material;

John Booze for his authorship of the computer test bank; VPG Integrated Media for their innovative work on the text's video series and James Tanton for his on-screen instruction; Mark Smith of College of Lake County, who not only authored the Instructor's and Student's Solutions Manuals for use with this text, but prepared the answers sections for the text itself, and used the manuscript of the text in his classroom and provided valuable feedback on its effectiveness.

Commentary from the following reviewers was indispensable during the development of this text:

Stan Barrick, *California State University Sacramento*
Charles Belair, *Medaille College*
Jackie L. Black, *Ashland Community College*
Arlene Blasius, *SUNY College at Old Westbury*
Philip Blau, *Shawnee State University*
Susan Bradley, *Angelina College*
Don Brown, *Georgia Southern University*
Debra Bryant, *Tennessee Technology College*
Barbara Burke, *Hawaii Pacific University*
Michael Butler, *Humboldt State University*
Tina Cannon, *Chattanooga State Technical Community College*
Debbie Cochener, *Austin Peay University*
Pat C. Cook, *Weatherford College*
Jaqueline R. Coomes, *Eastern Washington University*
Antonio S. David, *Del Mar College*
Yolanda Davis, *Central Texas College*
Dale Felkins, *Arkansas Tech University*
Joan Finney, *Dyersburg State Community College*
Judy Godwin, *Collin County Community College*
Kay Haralson, *Austin Peay University*
James Harris, *John A. Logan College*
Lonnie Hass, *North Dakota State University*
Bonnie M. Hodge, *Austin Peay State University*
Jeannie Hollar, *Lenoir-Rhyne College*
Steven Howard, *Rose State College*
Patricia Ann Jenkins, *South Carolina State University*
Steven Kahn, *Anne Arundel Community College*
David Kendall, *University of Massachusetts Amherst*
Monika Kurth, *Scott Community College*
Ann M. Loving, *J. Sargeant Reynolds Community College*
Doug Mace, *Kirtland Community College*
Kathleen Miranda, *SUNY College at Old Westbury*
Valerie Morgan-Krick, *Tacoma Community College*
Prinder Naidu, *Kennesaw State University*
Jon Odell, *Richland Community College*
Delene Perley, *Walsh University*
Nancy Ressler, *Oakton Community College*
Marianne Rey, *University of Alaska Anchorage*
Jacqueline H. Ruff, *Gordon College*
Jorge Sarmiento, *County College of Morris*
Martha Scarborough, *Motlow Community College*
Joyce Smith, *Chattanooga State Technical Community College*
Lee Ann Spahr, *Durham Tech Community College*
Brooks W. Spies, Jr., *Kansas City Community College*
Jane Theiling, *Dyersburg State Community College*
Peggy Tibbs, *Arkansas Technical University*
Lee Topham, *Kingwood College*
Patricia Widder, *William Rainey Harper College*
Jackie Wing, *Angelina College*
Judith B. Wood, *Central Florida Community College*

Walk-Through

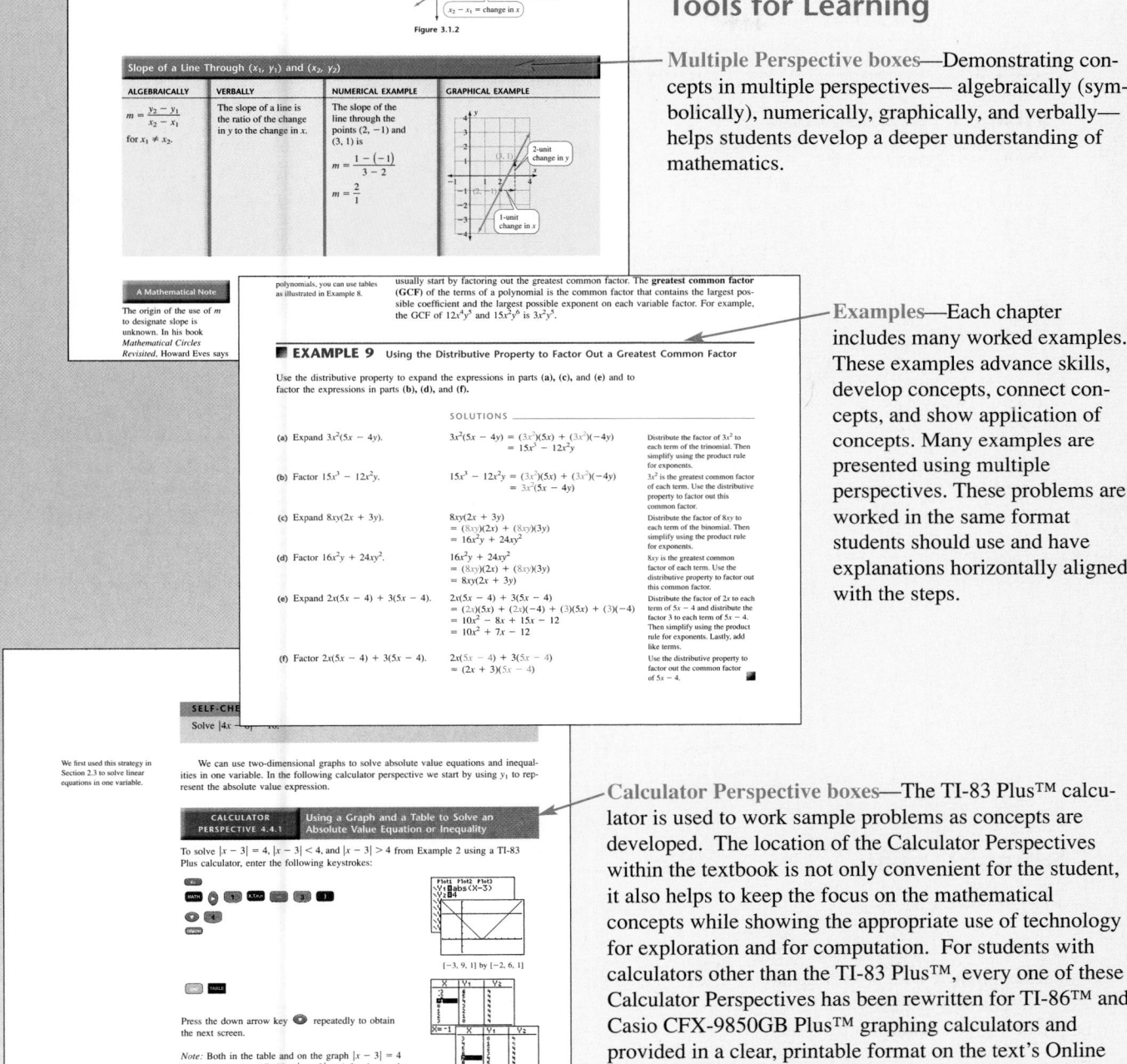

Tools for Learning

Multiple Perspective boxes—Demonstrating concepts in multiple perspectives— algebraically (symbolically), numerically, graphically, and verbally— helps students develop a deeper understanding of mathematics.

Examples—Each chapter includes many worked examples. These examples advance skills, develop concepts, connect concepts, and show application of concepts. Many examples are presented using multiple perspectives. These problems are worked in the same format students should use and have explanations horizontally aligned with the steps.

Calculator Perspective boxes—The TI-83 Plus™ calculator is used to work sample problems as concepts are developed. The location of the Calculator Perspectives within the textbook is not only convenient for the student, it also helps to keep the focus on the mathematical concepts while showing the appropriate use of technology for exploration and for computation. For students with calculators other than the TI-83 Plus™, every one of these Calculator Perspectives has been rewritten for TI-86™ and Casio CFX-9850GB Plus™ graphing calculators and provided in a clear, printable format on the text's Online Learning Center, complete with screenshots and step-by-step instructions.

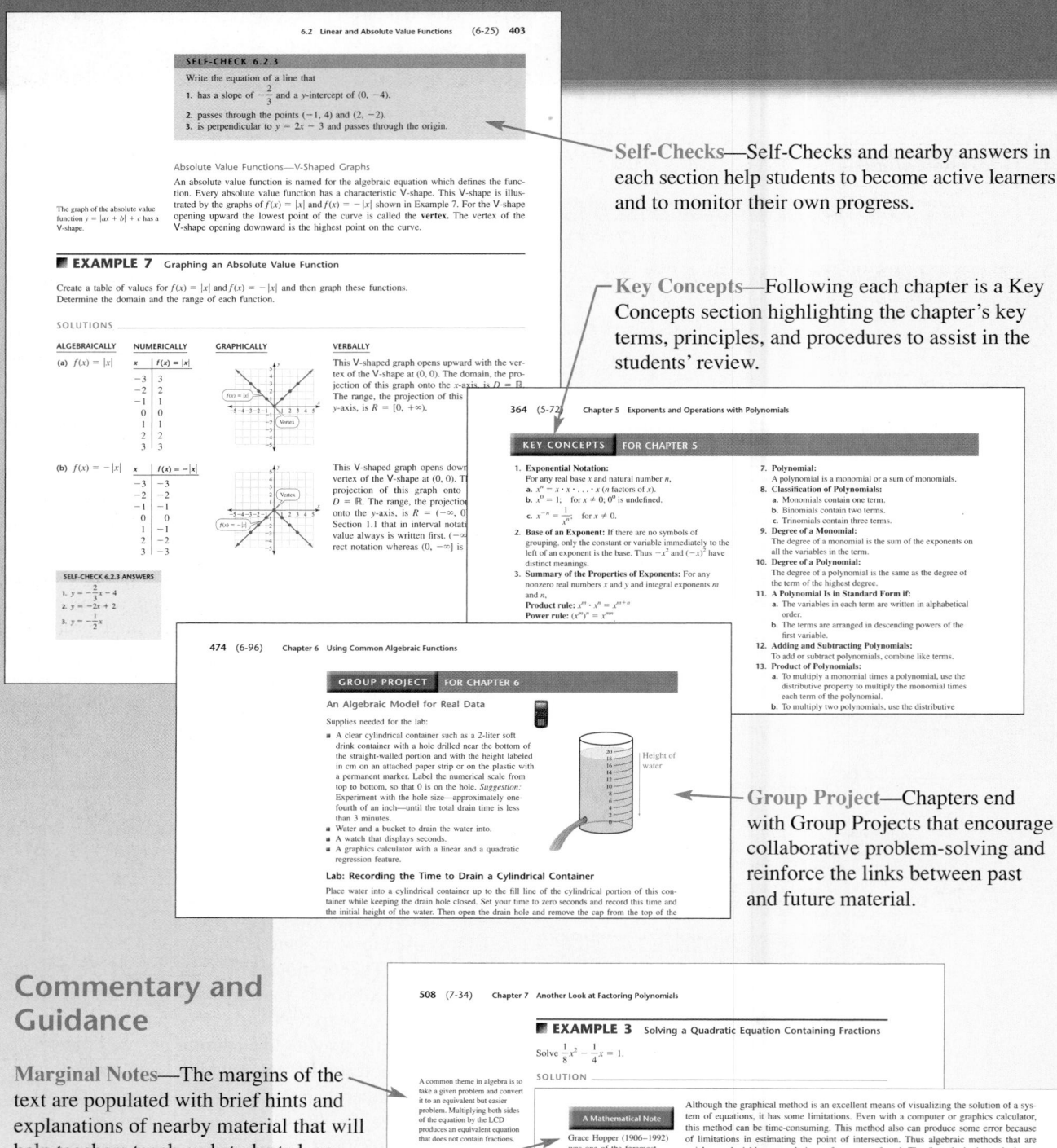

Self-Checks—Self-Checks and nearby answers in each section help students to become active learners and to monitor their own progress.

Key Concepts—Following each chapter is a Key Concepts section highlighting the chapter's key terms, principles, and procedures to assist in the students' review.

Group Project—Chapters end with Group Projects that encourage collaborative problem-solving and reinforce the links between past and future material.

Commentary and Guidance

Marginal Notes—The margins of the text are populated with brief hints and explanations of nearby material that will help teachers teach and students learn.

Mathematical Notes—These short vignettes describe the origin of some of the symbols and terms used in the text and provide brief glimpses into the lives of some of the men and women of mathematics.

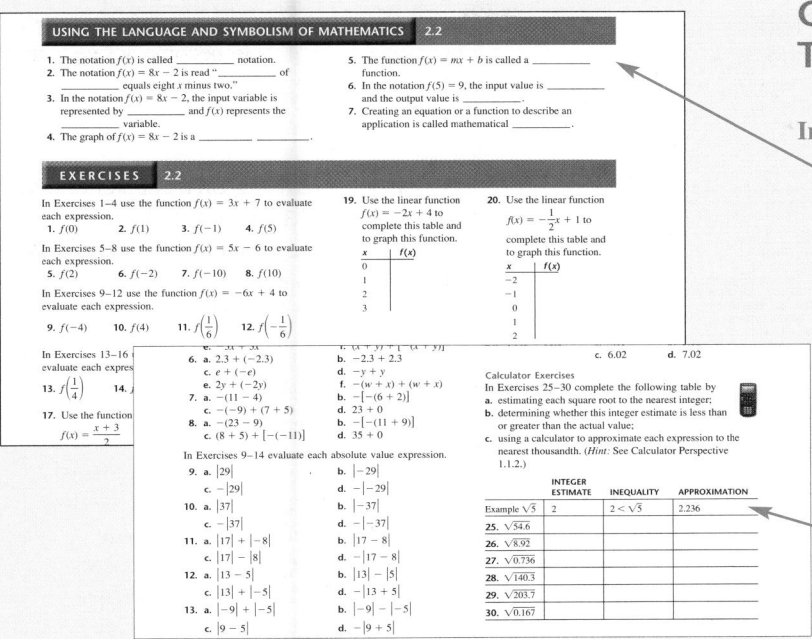

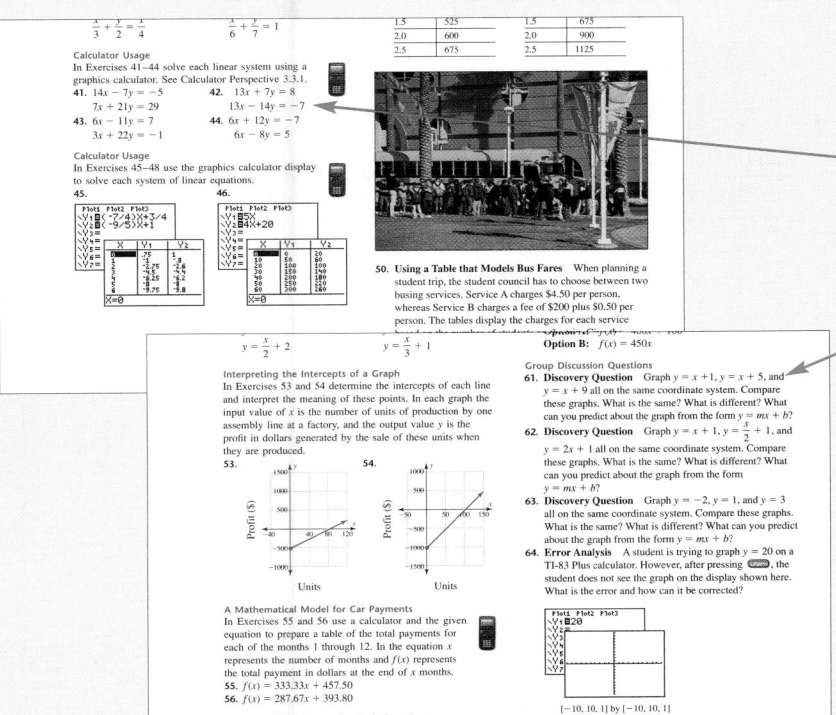

Conceptual Understanding Through Practice

In every section of the text:

Using the Language and Symbolism of Mathematics—Every section includes a series of fill-in-the-blank questions to help students gain fluency in the language and symbolism of mathematics.

Exercises—Each end-of-section exercise set is carefully constructed to develop and to reinforce the skills and concepts of algebra, and to provide an appropriate review of the section. Exercise sets include:

Estimation Skills and Error Analysis Exercises—Estimation skills, concern for reasonable answers, and the ability to detect calculator errors are critical elements of students' mathematical knowledge. There are examples and exercises in the book specifically designed to help the students develop these skills.

Calculator Exercises—When the use of a calculator is appropriate to the solution of an exercise, it is indicated in the text by a calculator icon, right next to the exercise.

Group Discussion Questions—These exercises involve students in interactive and collaborative learning and encourage them to communicate mathematics both orally and in writing.

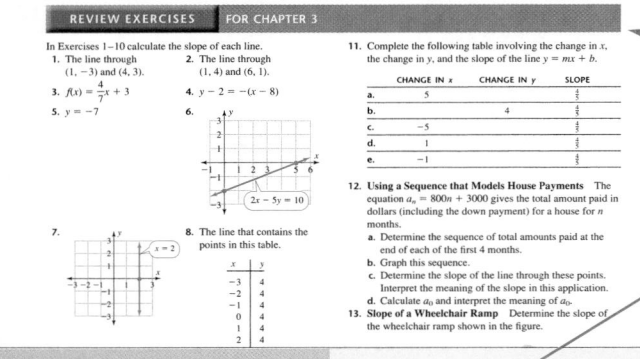

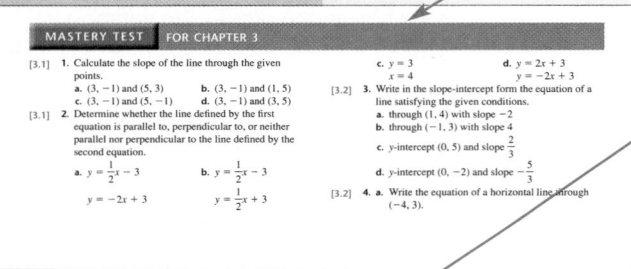

In every chapter of the text:

Chapter Review Exercises—These comprehensive exercise sets provide ample and well-distributed practice on the topics of the chapter.

Chapter Mastery Test—These tests are written specifically to cover each objective presented in the chapter.

Diagnostic Review—To ensure students have mastered the concepts of the previous chapters, a comprehensive diagnostic review, complete with answers, appears between Chapters 5 and 6. It will prove especially helpful to students who are entering the book at this intermediate junction and who need to assess their beginning algebra skills. Direct references to the relevant examples in Chapters 1 to 5 are provided alongside the answers for students who need to refresh or reinforce their mastery of beginning algebra concepts.

INDEX OF FEATURES

continued

Index of Calculator Perspectives
(continued)

Index of Mathematical Notes

Index of Applications

Agriculture, Biology, and Environment

continued

Index of Applications
(continued)

Education

Family and Consumer Finance

Geometry

continued

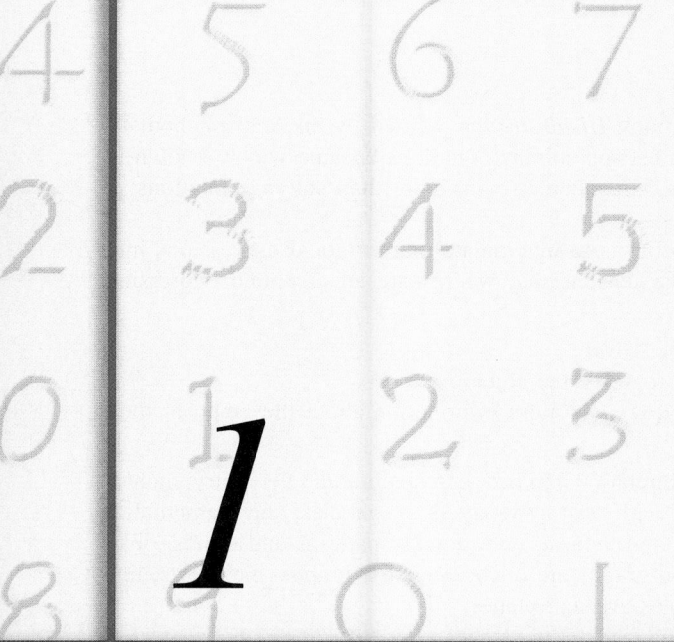

1

OPERATIONS WITH REAL NUMBERS

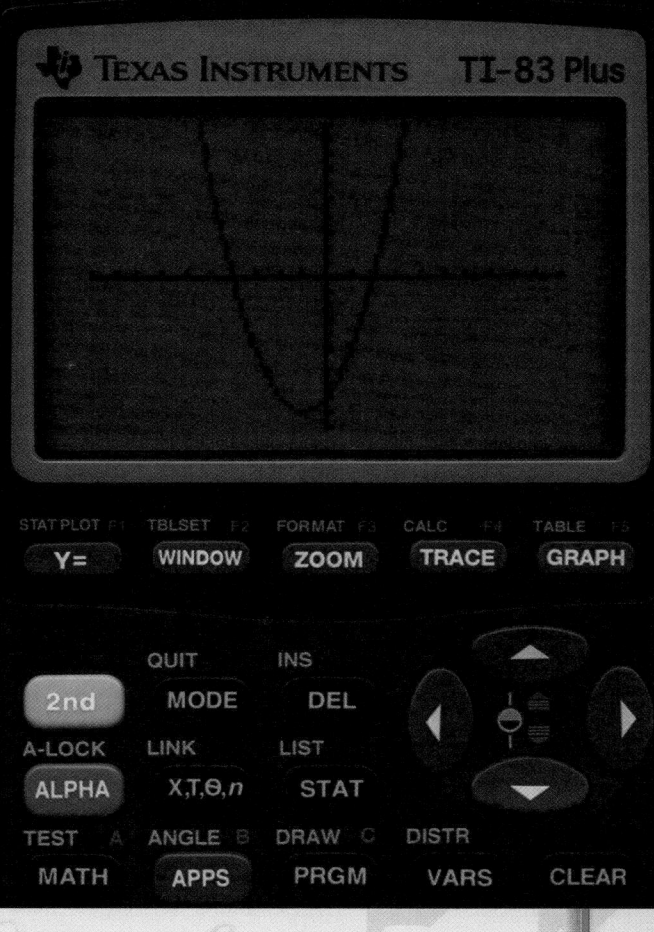

The word *algebra* comes from the book *Hisab al-jabr,* written by an Arab mathematician in A.D. 830; translations of this text on solving equations became widely known in Europe as *al-jabr.* Algebra, however, is concerned with more than solving equations; it is a generalization of arithmetic.

One of the primary uses of algebra is to manipulate mathematical expressions into a more desirable form. Three of the reasons that we rewrite an algebraic expression follow:

1. To put the expression in its simplest form.
2. To simplify an equation in order to solve the equation.
3. To place a function in a form that reveals more information about the graph of the function.

This book also will focus on the relationship between algebra and geometry, including various graphs. Graphs are used widely in newspapers, magazines, and mathematics texts. Graphs can help us to understand data, to note quickly patterns and trends, or to visualize abstract concepts. Common graphs are histograms, bar graphs, pie charts, line graphs, the real number line, and the Cartesian plane.

Section 1.1   The Real Number Line

Objectives:

1. Identify additive inverses.
2. Evaluate absolute value expressions.
3. Use interval notation.
4. Estimate and approximate square roots.
5. Identify natural numbers, whole numbers, integers, rational numbers, and irrational numbers.

To interpret or create graphs, we need to understand the terminology and concepts reviewed in this section. The number line is one of the most used graphs in mathematics. It provides a model of the real numbers and is used when students first are introduced to many concepts. In arithmetic we use only constants such as $-13, 0, \frac{2}{3}, \sqrt{2}$ and π, which have a fixed value. Algebra uses not only constants but also variables such as $a, b, P, W, x, y,$ and $z.$ A **variable** is a letter that can be used to represent different numbers. The constants and variables that we use most often will represent real numbers.

The Real Number Line

Origin

Negative Numbers ↓ Positive Numbers

$$\xleftarrow{\quad\;+\;\;+\;\;+\;\;+\;\;+\;\;+\;\;+\;\;+\;\;+\;\;+\;\;+\;\;}\rightarrow$$
$$-5\;-4\;-3\;-2\;-1\;\;0\;\;1\;\;2\;\;3\;\;4\;\;5$$

There are an infinite number of points on the number line and every point is associated with a **real number.** The set of all real numbers can be denoted by $\mathbb{R}.$ The numbers to the right of 0 are called the **positive numbers,** and those to the left of 0 are called the **negative numbers.** The number 0 is neither positive nor negative. The set of numbers $\{1, 2, 3, 4, \ldots\}$ associated with the points marked off to the right of the origin is called the set of **natural numbers.** The three dots inside this set notation indicate that these numbers continue without end and thus the set is infinite. The set $\{0, 1, 2, 3, 4, \ldots\}$ is called the set of **whole numbers.**

Real numbers that are the same distance from the origin but on opposite sides of the origin are called **opposites** of each other, or **additive inverses.** For example, -2 (negative two) and 2 are additive inverses of each other.

Each real number represents a directed distance from the origin. For example, $+2$ represents a distance 2 units to the right, whereas -2 represents a distance 2 units to the left of the origin.

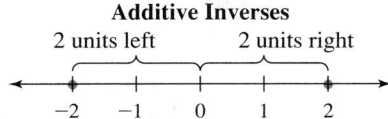

The sum of a number and its additive inverse is zero. For example, $-2 + 2 = 0$ and $5 + (-5) = 0$. The additive inverse of 0 is 0.

Zero is called the **additive identity** because 0 is the only real number with the property that $a + 0 = a$ and $0 + a = a$ for every real number a.

Opposites or Additive Inverses

ALGEBRAICALLY	VERBALLY	NUMERICAL EXAMPLE	GRAPHICAL EXAMPLE
1. If a is a real number, the *opposite* of a is $-a$.	The additive inverse of a real number is formed by changing the sign of the number.	-5 is the opposite of 5 5 is the opposite of -5	Opposites $-5 \quad 0 \quad 5$
2. $a + (-a) = 0$ $-a + a = 0$	The sum of a real number and its additive inverse is zero.	$5 + (-5) = 0$ $-5 + 5 = 0$	

Definitions and the solutions of problems will be presented using multiple representations throughout this book. The definition of additive inverses is given algebraically, verbally, numerically, and graphically. Comparing these representations can help you to increase your understanding of mathematics.

The set $\{\ldots, -4, -3, -2, -1, 0, 1, 2, 3, 4, \ldots\}$ is called the set of **integers.** The set of integers contains all the whole numbers and their opposites, or additive inverses. As illustrated in the figure, the integers are spaced one unit apart on the number line.

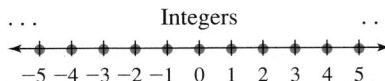

The opposite of any positive number is a negative number; for example, the opposite of 4 is -4. The opposite of any negative number is a positive number; for example, the opposite of -4 is 4. The opposite of x is $-x$. If x is -4, then $-x$ is 4. Thus $-(-4) = 4$, and in general $-(-x) = x$.

Double Negative Rule

ALGEBRAICALLY	VERBALLY	NUMERICAL EXAMPLE
For any real number a, $-(-a) = a$.	The opposite of the additive inverse of a is a.	$-(-5) = 5$

A Mathematical Note

π is defined as the ratio of the circumference of a circle to its diameter. $\pi \approx$ 3.14159265. ($\approx$ means approximately equal to.) However, the decimal form of π does not terminate or repeat. In 1897 House Bill #246 was introduced in the Indiana legislature to make 3.2 the value of π in that state. Fortunately, better judgment prevailed and this foolish bill was defeated.

EXAMPLE 1   Writing Inverse Additives

Write the additive inverse of each of the following real numbers:

SOLUTIONS

ADDITIVE INVERSE

(a) 7 -7 Except for zero, the additive inverse of a number is formed by changing the sign of the number.

(b) -3 $-(-3) = 3$ The opposite of negative three is three.

(c) 0   0   Zero is its own additive inverse.

(d) -27.89 27.89 $-(-27.89) = 27.89$

(e) $\dfrac{4}{17}$ $-\dfrac{4}{17}$

(f) π $-\pi$ The opposite of pi is negative pi.

EXAMPLE 2   Adding Additive Inverses and the Additive Identity

Evaluate each of the following sums:

SOLUTIONS

(a) $7 + (-7)$ $7 + (-7) = 0$ The sum of a number and its additive inverse is zero.

(b) $-7 + 7$ $-7 + 7 = 0$

(c) $-\pi + \pi$ $-\pi + \pi = 0$

(d) $3x + (-3x)$ $3x + (-3x) = 0$

(e) $19 + 0$ $19 + 0 = 19$ 0 is the additive identity; $a + 0 = a$ for all real numbers a.

Additive inverses are the same distance from the origin, so these numbers are said to have the same magnitude, or absolute value. The **absolute value** of a real number x, denoted by $|x|$, is the distance on the number line between 0 and x. Distance is never negative, so $|x|$ is never negative: $|0| = 0$, $|3| = 3$, and $|-3| = 3$. Both -3 and 3 are a distance of 3 units from the origin. If x is positive, $|x| = x$, as in $|3| = 3$. If x is negative, however, $|x| = -x$, as in $|-3| = -(-3) = 3$. Thus the definition of absolute value is often stated using the notation given in the following box.

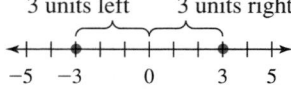

3 units left   3 units right

-5 -3 0   3   5

Absolute Value

ALGEBRAICALLY	VERBALLY	NUMERICAL EXAMPLE	GRAPHICAL EXAMPLE
$|x| = \begin{cases} x & \text{if } x \text{ is nonnegative} \\ -x & \text{if } x \text{ is negative} \end{cases}$	The absolute value of x is the distance between 0 and x.	$|2| = 2$ $|-2| = 2$	2 units left 2 units right -2 -1 0 1 2

■ EXAMPLE 3   Evaluating Absolute Value Expressions

Evaluate each of the following absolute value expressions:

SOLUTIONS

(a) $\|5.8\|$	$\|5.8\| = 5.8$	$\|x\| = x$ if x is nonnegative.
(b) $\|-4.29\|$	$\|-4.29\| = 4.29$	$\|x\|$ is the opposite of x if x is negative.
(c) $-\left\|\dfrac{5}{7}\right\|$	$-\left\|\dfrac{5}{7}\right\| = -\dfrac{5}{7}$	The additive inverse symbol is outside the absolute value symbols.
(d) $-\|-329\|$	$-\|-329\| = -(329)$ $= -329$	First $\|-329\|$ is replaced by 329, then the leading symbol in front denotes the opposite of this number.
(e) $\|9.7\| + \|-7.3\|$	$\|9.7\| + \|-7.3\| = 9.7 + 7.3$ $= 17$	First $\|9.7\|$ is replaced by 9.7 and $\|-7.3\|$ is replaced by 7.3. Then add 9.7 and 7.3.
(f) $\|9.7 - 7.3\|$	$\|9.7 - 7.3\| = \|2.4\|$ $= 2.4$	First subtract $9.7 - 7.3$. Then replace $\|2.4\|$ with 2.4.

Calculator perspectives will occur throughout the text and will integrate mathematical concepts with the appropriate use of technology.

CALCULATOR PERSPECTIVE 1.1.1   Evaluating Absolute Value Expressions

To enter $\|9.7 - 7.3\|$ from Example 3(f) on a TI-83 Plus calculator, enter the following keystrokes.

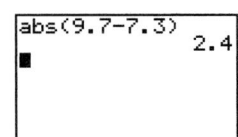

Note: The first option under NUM within the MATH menu is abs(. This includes the left parenthesis. The expression within the absolute value is completed by entering the right parenthesis.

SELF-CHECK 1.1.1

1. Write the additive inverse of -4.5, 0, and $\dfrac{17}{43}$.
2. Evaluate $\|7\|$, $\|-7\|$, $\|9 - 3\|$, $\|9\| + \|-3\|$, and $\|9\| - \|-3\|$.
3. Evaluate Example 3(e), $\|9.7\| + \|-7.3\|$, using a graphics calculator.

SELF-CHECK 1.1.1 ANSWERS

1. 4.5, 0, $-\dfrac{17}{43}$

2. $7, 7, 6, 12, 6$

3.
```
abs(9.7)+abs(-7.
3)
                17
```

The number line also provides an excellent model for examining the inequalities *less than* and *greater than*. Because these inequalities describe the order of numbers on the number line, they also are known as the **order relations**. Table 1.1.1 illustrates all possible orders for arranging two numbers x and y on the number line.

Table 1.1.1   Equality and Inequality Symbols

ALGEBRAIC NOTATION	VERBAL MEANING	GRAPHICAL RELATIONSHIP ON THE NUMBER LINE
$x = y$	x equals y	x and y are the same point.
$x \approx y$	x is approximately equal to y	x and y are "close" but are not the same point.
$x \neq y$	x is not equal to y	x and y are different points.
$x < y$	x is less than y	The point x is to the left of the point y.
$x \leq y$	x is less than or equal to y	The point x is on or to the left of the point y.
$x > y$	x is greater than y	The point x is to the right of the point y.
$x \geq y$	x is greater than or equal to y	The point x is on or to the right of the point y.

The statements $x < y$ and $y > x$ are equivalent because they both specify that the point x is to the left of the point y.

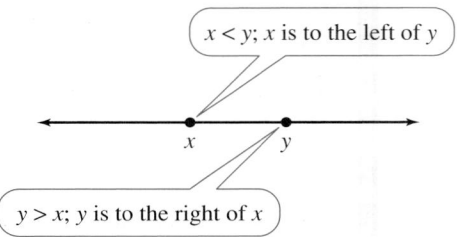

■ **EXAMPLE 4**   Showing Order Relationships on the Number Line

Plot each pair of numbers on the line, and determine the order relationship between the numbers.

SOLUTIONS _____

(a) -4 and -2

−5 −4 −3 −2 −1 0 1 2 3 4 5

-4 is to the left of -2; thus -4 is less than -2; $-4 < -2$

(b) π and 3

π

2.9   3   3.1   3.2   3.3   3.4

$\pi \approx 3.14159$; thus π is to the right of 3 and is greater than 3; $\pi > 3$

(c) 0 and -3

−5 −4 −3 −2 −1 0 1 2 3 4 5

0 is to the right of -3; thus $0 > -3$. ■

SELF-CHECK 1.1.2

Determine whether each statement is true (T) or false (F).

1. $4\frac{1}{3} > 4\frac{1}{2}$

2. $-7 \leq -3$

3. $\frac{1}{3} > 0.3$

4. $-5 \geq 0$

5. $9 > 9$

6. $9 \geq 9$

A Mathematical Note

The symbol "∞" was used to represent infinity by John Wallis in *Arithmetica Infinitorum* in 1655. The Romans had commonly used this symbol to represent 1000. Likewise, we now use the word *myriad* to mean any large number although the Greeks used it to mean 10,000.

The real number line extends infinitely to both the left and to the right. The infinity symbol "∞" is not a real number; rather, it signifies that the values continue through extremely large values without any end or bound. The symbol "$-\infty$" indicates values unbounded to the left and "$+\infty$" indicates values unbounded to the right. The interval notation $(3, +\infty)$ is a compact notation that represents all real numbers that are greater than 3 (see Table 1.1.2).

Table 1.1.2   Inequality Notation

INEQUALITY NOTATION	VERBAL MEANING	GRAPH	INTERVAL NOTATION
$x > 3$	x is greater than 3	(open circle at 3, extending right)	$(3, +\infty)$
$x \leq 5$	x is less than or equal to 5	(closed bracket at 5, extending left)	$(-\infty, 5]$

Note that in the interval notation in Table 1.1.3 a parenthesis indicates that a value is not included in the interval and a bracket indicates that a value is included in the interval. An infinity symbol at either end of an interval is enclosed within parentheses since this denotes an unbounded interval.

Table 1.1.3   Interval Notation

INEQUALITY NOTATION	VERBAL MEANING	GRAPH	INTERVAL NOTATION
$x > a$	x is greater than a	(open at a, extending right)	$(a, +\infty)$
$x \geq a$	x is greater than or equal to a	(closed at a, extending right)	$[a, +\infty)$
$x < a$	x is less than a	(open at a, extending left)	$(-\infty, a)$
$x \leq a$	x is less than or equal to a	(closed at a, extending left)	$(-\infty, a]$
$a < x < b$	x is greater than a and less than b	(open at a, open at b)	(a, b)
$a < x \leq b$	x is greater than a and less than or equal to b	(open at a, closed at b)	$(a, b]$
$a \leq x < b$	x is greater than or equal to a and less than b	(closed at a, open at b)	$[a, b)$
$a \leq x \leq b$	x is greater than or equal to a and less than or equal to b	(closed at a, closed at b)	$[a, b]$
$-\infty < x < \infty$	x is any real number	(extending both directions)	$(-\infty, +\infty)$

In the interval [a, b], a is called the left endpoint and b is called the right endpoint. The smaller value is always the left endpoint and the larger value is always the right endpoint. For example, [2, 5] is correct whereas [5, 2] is an incorrect notation because the larger value is listed first. The interval (−1, 3) contains all points between −1 and 3.

■ EXAMPLE 5   Writing Multiple Representations for Intervals

Express each interval in inequality notation, in verbal form, and graphically.

SOLUTIONS _____

INTERVAL NOTATION	INEQUALITY NOTATION	VERBALLY	GRAPHICALLY
(a) [2, +∞)	$x \geq 2$	x is greater than or equal to 2	←+++⊢+++++→ −1 0 1 2 3 4 5 6
(b) (−∞, π)	$x < \pi$	x is less than π	←+++++)+++→ −1 0 1 2 3 4 5 6 (π above)
(c) (−2, 1]	$-2 < x \leq 1$	x is greater than −2 and less than or equal to 1	←++(+++]++→ −4 −3 −2 −1 0 1 2 3
(d) [−3, 2]	$-3 \leq x \leq 2$	x is greater than or equal to −3 and less than or equal to 2	←+⊢+++++]+→ −4 −3 −2 −1 0 1 2 3

[1, −2) would not be correct. Interval notation is stated always with the smaller number to the left and the larger number to the right.

■

We use $-\sqrt{4}$ to represent −2, a negative number that is also a square root of 4. The key point is $\sqrt{x}$ denotes only one of the square roots of x and $-\sqrt{x}$ denotes the other.

Although calculators are used frequently to approximate square roots, it is important to be able to make rough mental estimates of square roots. For $x \geq 0$ the **principal square root of x,** denoted by $\sqrt{x}$, is a nonnegative real number r so that $r^2 = x$. For example, $\sqrt{4} = 2$ since $2^2 = 4$ and $\sqrt{9} = 3$ since $3^2 = 9$. To estimate $\sqrt{5}$, note that $\sqrt{4} < \sqrt{5} < \sqrt{9}$. Thus $2 < \sqrt{5} < 3$. A calculator approximation yields $\sqrt{5} \approx 2.2361$, a value consistent with our mental estimation.

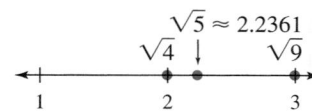

$$\sqrt{5} \approx 2.2361$$
$$\sqrt{4} \downarrow \qquad \sqrt{9}$$

SELF-CHECK 1.1.3

Express each of these intervals using interval notation.

1. All real numbers greater than or equal to −2.
2. $x < 4$
3. $-5 < x \leq 3$
4. ←+⊢+++++)++→ −π   1

EXAMPLE 6   Estimating and Approximating Square Roots

Use a calculator to approximate each of the square roots to the nearest thousandth and then mentally estimate this value to check the reasonableness of the calculation.

SOLUTIONS

(a) $\sqrt{47}$ Calculator: $\sqrt{47} \approx 6.856$
Because $47 \approx 49$, $\sqrt{47} \approx \sqrt{49} = 7$;
6.856 (which is a little less than 7) seems
to be a reasonable approximation of $\sqrt{47}$.

(b) $\sqrt{10{,}100}$ Calculator: $\sqrt{10{,}100} \approx 100.499$
Because, $10{,}100 > 10{,}000 = 100^2$,
100.499 seems to be a reasonable
approximation of $\sqrt{10{,}100}$.

CALCULATOR PERSPECTIVE 1.1.2   **Evaluating Square Roots**

To evaluate $\sqrt{25} + \sqrt{144}$ from Example 7(a) on a TI-83 Plus calculator, enter the following keystrokes.

The square root $\boxed{\sqrt{}}$ is the secondary feature of the $\boxed{x^2}$ key.

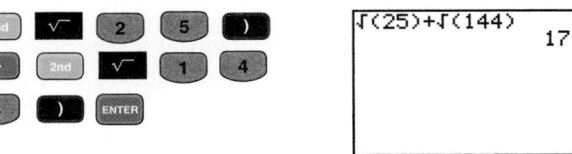

Note: Most graphics calculators include the left parenthesis when the square root button is pressed. Then the expression under the square root is entered and completed by pressing the right parenthesis.

EXAMPLE 7   Evaluating Expressions with Square Roots

Evaluate each of these expressions.

SOLUTIONS

(a) $\sqrt{25} + \sqrt{144}$ $\sqrt{25} + \sqrt{144} = 5 + 12$ $\sqrt{25} = 5$ since $5^2 = 25$, and
$\phantom{\sqrt{25} + \sqrt{144}} = 17$ $\sqrt{144} = 12$ since $12^2 = 144$.

(b) $\sqrt{25 + 144}$ $\sqrt{25 + 144} = \sqrt{169}$ First add the terms under the square root
$\phantom{\sqrt{25 + 144}} = 13$ symbol; then take the square root.
$\sqrt{169} = 13$ since $13^2 = 169$.

Note that in Example 7, $\sqrt{25} + \sqrt{144}$ is not equal to $\sqrt{25 + 144}$; you must be careful to interpret correctly this order of operations when working with square roots. More information on the order of operations is presented in Section 1.6.

SELF-CHECK 1.1.4

Evaluate each of these expressions without using a calculator.

1. $\sqrt{25 - 16}$
2. $\sqrt{25} - \sqrt{16}$

Evaluate each of these expressions using a graphics calculator.

3. $\sqrt{121}$
4. $\sqrt{625 - 576}$
5. $\sqrt{625} - \sqrt{576}$

The real numbers are classified as either rational or irrational. A real number which can be written as a ratio of two integers is called a **rational number.** (Notice that the first five letters of the word *rational* spell *ratio.*) The other real numbers, which cannot be written as the ratio of two integers, are called **irrational numbers.**

Rational and Irrational Numbers

Rational: A real number x is rational if $x = \frac{a}{b}$ for integers a and b, with $b \neq 0$. In decimal form a rational number is either a terminating decimal or an infinite repeating decimal.

Irrational: A real number is irrational if it is not rational. In decimal form an irrational number is an infinite nonrepeating decimal.

$b \neq 0$ is read *b is not equal to zero.*

The review of whole number arithmetic and the arithmetic of fractions is embedded in the arithmetic of real numbers in Sections 1.1–1.6.

All integers are also rational numbers since every integer can be written as the ratio of itself to 1. For example, -3, 0, and 7 can be written as $\frac{-3}{1}$, $\frac{0}{1}$, and $\frac{7}{1}$. Another form that can be interpreted as a ratio is percent notation. For example, 17% read "17 percent" means 17 parts out of 100; $17\% = \frac{17}{100}$. Other examples of rational numbers are $\frac{1}{7}$, $\frac{-2}{13}$, and $\frac{479}{-22}$. The numbers $4\frac{2}{5}$, 0.17, and 0.333... are also rational since each of these numbers can be written as a ratio of two integers: $4\frac{2}{5} = \frac{22}{5}$, $0.17 = \frac{17}{100}$, and $0.333\ldots = \frac{1}{3}$. Numbers such as π, $\sqrt{2}$, $\sqrt{3}$, and $-\sqrt{5}$ are irrational since they cannot be written as the ratio of two integers. The relationship of the important subsets of the real numbers that have been defined in this section is summarized by the tree diagram in the following figure:

SELF-CHECK 1.1.4 ANSWERS

1. 3 2. 1 3. 11
4. 7 5. 1

```
√(121)
                 11
√(625-576)
                  7
√(625)-√(576)
                  1
```

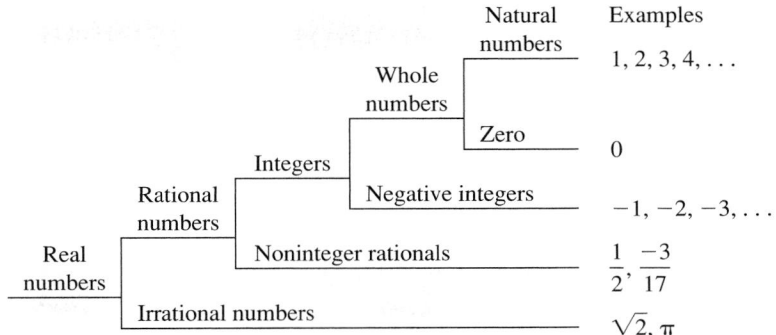

EXAMPLE 8   Classifying Real Numbers

Given the set $\left\{-5, -3.2, -\sqrt{3}, 0, \dfrac{2}{3}, 2, \pi, 4\dfrac{1}{3}, 6\right\}$, list all the elements in this set of real numbers that are:

SOLUTIONS

(a) Natural numbers   2, 6   The natural numbers consist of 1, 2, 3, ...

(b) Whole numbers   0, 2, 6   The whole numbers consist of 0, 1, 2, ...

(c) Integers   −5, 0, 2, 6   The integers consist of −2, −1, 0, 1, 2, ...

(d) Rational numbers $-5, -3.2, 0, \dfrac{2}{3}, 2, 4\dfrac{1}{3}, 6$ The rational numbers consist of the integers, fractions, and terminating or repeating decimals.

(e) Irrational numbers $-\sqrt{3}, \pi$ Irrational numbers consist of the real numbers that are not rational.

A number is not automatically irrational just because it has a square root symbol. For example, $\sqrt{4}$ is another notation for the rational number 2. All numbers with rational numbers as their square roots are called **perfect squares.** The integers 1, 4, 9, 16, and 25 are perfect squares since their square roots are 1, 2, 3, 4, and 5, respectively. The number $\dfrac{4}{81}$ is also a perfect square since $\sqrt{\dfrac{4}{81}} = \dfrac{2}{9}$. However, $\sqrt{2}$ is irrational since it can be shown that 2 is not a perfect square of any rational number.

EXAMPLE 9   Identifying Rational and Irrational Numbers

Classify each of the following real numbers as either rational or irrational:

SOLUTIONS

(a) $\sqrt{100}$ Rational $\sqrt{100} = 10$ since $10^2 = 100$

(b) $\sqrt{0.25}$ Rational $\sqrt{0.25} = 0.5$ since $0.5^2 = 0.25$

(c) $\sqrt{5}$ Irrational   5 is not the square of any rational number.

(d) $\sqrt{\dfrac{36}{49}}$ Rational $\sqrt{\dfrac{36}{49}} = \dfrac{6}{7}$ since $\left(\dfrac{6}{7}\right)^2 = \dfrac{36}{49}$

(e) $\sqrt{0}$ Rational

(f) 0.666...   Rational

$\sqrt{0} = 0$ since $0^2 = 0$

All repeating decimals are rational numbers. This number also can be written as $\frac{2}{3}$.

(g) 0.101101110...   Irrational

Although there is a pattern (the number of 1s is increasing between the 0s), this is not a repeating decimal.

SELF-CHECK 1.1.5

Given the set $\left\{ -4, -\sqrt{2}, -1, -\frac{3}{7}, 0, 4, 5\frac{3}{8}, 7 \right\}$, list all the elements in this set of real numbers that are

1. Natural numbers
2. Whole numbers
3. Integers
4. Rational numbers
5. Irrational numbers

SELF-CHECK 1.1.5 ANSWERS

1. 4, 7   2. 0, 4, 7
3. $-4, -1, 0, 4, 7$
4. $-4, -1, -\frac{3}{7}, 0, 4, 5\frac{3}{8}, 7$
5. $-\sqrt{2}$

USING THE LANGUAGE AND SYMBOLISM OF MATHEMATICS   1.1

1. A _____ is a letter that can be used to represent different numbers.
2. The notation $-a$ is read "the opposite of a" or "the _____ _____ of a."
3. The additive identity is _____ .
4. The additive inverse of $-a$ is _____ .
5. If one number is 7 units to the right of the origin, its additive inverse is _____ units to the _____ of the origin.
6. The notation $|x|$ is read "_____ _____ of x." $|x|$ represents the _____ between 0 and x on the number line.
7. The notation $x \geq y$ is read "x is _____ than or _____ to y."
8. The notation $x < y$ is read "x is _____ than y." On the number line x is to the _____ of y.
9. The notation $x \neq y$ is read "x is not _____ to y."
10. The notation $x \approx y$ is read "x is _____ _____ to y."
11. The notation $\sqrt{x}$ is read "the _____ _____ _____ of x."

12. A real number that is a terminating decimal is a _____ number.
13. A real number that is a repeating decimal is a _____ number.
14. A real number that is an infinite nonrepeating decimal is an _____ number.
15. The symbol π is the symbol for _____ .
16. The symbol ∞ is the symbol for _____ .
17. In interval notation a parenthesis indicates that a value **is/is not** included in the interval. (Select the correct choice.)
18. In interval notation a bracket indicates that a value **is/is not** included in the interval. (Select the correct choice.)
19. The set of all real numbers can be represented by $\mathbb{R}$ or by the interval notation _____ .
20. The notation $[2, -3]$ is incorrect notation because the _____ value is listed first. The correct notation for $-3 \leq x \leq 2$ is _____ .
21. The interval notation to represent $a < x \leq b$ is _____ .

Each exercise set will start with exercises on "Using the Language and Symbolism of Mathematics." It is highly recommended that you complete these before your instructor lectures on each section.

EXERCISES 1.1

In Exercises 1 and 2 write the additive inverse of each number.

1. a. 9   **b.** -13 **c.** $-\dfrac{3}{5}$

   d. 7.23   **e.** $-\sqrt{13}$ **f.** 0

2. a. 8   **b.** -30 **c.** $\dfrac{5}{7}$

   d. -8.94 **e.** $-\pi$ **f.** 0.03

3. Simplify each expression to either 7 or -7.
   a. $-(-7)$ **b.** $-[-(-7)]$ **c.** $-[-(-(-7))]$

4. Simplify each expression to either 9 or -9.
   a. $-(-9)$ **b.** $-[-(-9)]$ **c.** $-[-(-(-9))]$

In Exercises 5–8 simplify each expression.

5. a. $17 + (-17)$ **b.** $-17 + 17$
   c. $-\pi + \pi$ **d.** $x + (-x)$
   e. $-3x + 3x$ **f.** $(x + y) + [-(x + y)]$

6. a. $2.3 + (-2.3)$ **b.** $-2.3 + 2.3$
   c. $e + (-e)$ **d.** $-y + y$
   e. $2y + (-2y)$ **f.** $-(w + x) + (w + x)$

7. a. $-(11 - 4)$ **b.** $-[-(6 + 2)]$
   c. $-(-9) + (7 + 5)$ **d.** $23 + 0$

8. a. $-(23 - 9)$ **b.** $-[-(11 + 9)]$
   c. $(8 + 5) + [-(-11)]$ **d.** $35 + 0$

In Exercises 9–14 evaluate each absolute value expression.

9. a. $|29|$ **b.** $|-29|$
   c. $-|29|$ **d.** $-|-29|$

10. a. $|37|$ **b.** $|-37|$
   c. $-|37|$ **d.** $-|-37|$

11. a. $|17| + |-8|$ **b.** $|17 - 8|$
   c. $|17| - |8|$ **d.** $-|17 - 8|$

12. a. $|13 - 5|$ **b.** $|13| - |5|$
   c. $|13| + |-5|$ **d.** $-|13 + 5|$

13. a. $|-9| + |-5|$ **b.** $|-9| - |-5|$
   c. $|9 - 5|$ **d.** $-|9 + 5|$

14. a. $|-23| - |-11|$ **b.** $|-23| + |-11|$
   c. $|23 - 11|$ **d.** $-|23 + 11|$

In Exercises 15–18 evaluate each expression without using a calculator. Then use a calculator to check your result. (*Hint:* See Calculator Perspective 1.1.2.)

15. a. $\sqrt{9}$ **b.** $\sqrt{16}$
   c. $\sqrt{9} + \sqrt{16}$ **d.** $\sqrt{9 + 16}$

16. a. $\sqrt{25}$ **b.** $\sqrt{144}$
   c. $\sqrt{25} + \sqrt{144}$ **d.** $\sqrt{25 + 144}$

17. a. $\sqrt{1} + \sqrt{16} + \sqrt{64}$ **b.** $\sqrt{1 + 16 + 64}$

18. a. $\sqrt{9} + \sqrt{16} + \sqrt{144}$ **b.** $\sqrt{9 + 16 + 144}$

In Exercises 19–24 mentally estimate the value of each square root and then select the choice that is the closest to your estimate.

19. $\sqrt{17}$ **a.** 3.88   **b.** 4.12   **c.** 4.92   **d.** 9.24

20. $\sqrt{24}$ **a.** 4.90   **b.** 5.10   **c.** 12.01   **d.** 12.99

21. $\sqrt{62.41}$ **a.** 6.7   **b.** 7.1   **c.** 7.9   **d.** 8.9

22. $\sqrt{123.21}$ **a.** 61.6   **b.** 58.4   **c.** 12.3   **d.** 11.1

23. $\sqrt{9.1} + \sqrt{16.1}$ **a.** 4.03   **b.** 5.03   **c.** 6.03   **d.** 7.03

24. $\sqrt{9.1 + 16.1}$ **a.** 4.03   **b.** 5.02   **c.** 6.02   **d.** 7.02

Calculator Exercises

In Exercises 25–30 complete the following table by
a. estimating each square root to the nearest integer;
b. determining whether this integer estimate is less than or greater than the actual value;
c. using a calculator to approximate each expression to the nearest thousandth. (*Hint:* See Calculator Perspective 1.1.2.)

	INTEGER ESTIMATE	INEQUALITY	APPROXIMATION
Example $\sqrt{5}$	2	$2 < \sqrt{5}$	2.236
25. $\sqrt{54.6}$			
26. $\sqrt{8.92}$			
27. $\sqrt{0.736}$			
28. $\sqrt{140.3}$			
29. $\sqrt{203.7}$			
30. $\sqrt{0.167}$			

31. a. What value of x makes the statement $x < 7$ false but makes the statement $x \le 7$ true?
   b. What value of x makes both $x \le 3$ and $x \ge 3$ true statements?

32. a. What value of x makes the statement $x > 9$ false but makes the statement $x \ge 9$ true?
   b. What value of x makes both $x \le -5$ and $x \ge -5$ true statements?

In Exercises 33–38 insert $<$, $=$, or $>$ in the blank to make each statement true.

33. a. 83 _____ 38   **b.** -83 _____ -38
   c. -9.4 _____ 0   **d.** 0 _____ 0.009

34. a. -8.4 ___ -4.8 **b.** 8.4 ___ 4.8

   c. 0 ___ $\dfrac{1}{3}$ **d.** 0 ___ $-\dfrac{2}{5}$

35. a. $-\dfrac{1}{2}$ ___ $-\dfrac{1}{5}$ **b.** $8\dfrac{3}{4}$ ___ $8\dfrac{3}{5}$

   c. $\dfrac{1}{2}$ ___ 0.5 **d.** $|21|$ ___ $|-21|$

36. a. $-\dfrac{5}{8}$ ___ $-\dfrac{5}{9}$ **b.** $7\dfrac{1}{11}$ ___ $7\dfrac{1}{15}$

   c. $-\dfrac{1}{4}$ ___ -0.25 **d.** $|-37|$ ___ $|37|$

37. a. $|-5|$ ___ $-|5|$ **b.** $|-5|$ ___ $-|-5|$

   c. $\sqrt{97}$ ___ 10 **d.** $-|7-4|$ ___ $-|9-7|$

38. a. $-|7|$ ___ $-|-7|$ **b.** $-|-7|$ ___ $|7|$

   c. $\sqrt{27}$ ___ 5 **d.** $|14-12|$ ___ $|5-2|$

39. List all the elements of $\left\{-11, -4.8, -\sqrt{9}, 0, 1\dfrac{3}{5}, \sqrt{5}, 15\right\}$
   that are
   a. Natural numbers   **b.** Whole numbers
   c. Integers   **d.** Rational numbers
   e. Irrational numbers

40. List all the elements of $\left\{-9.3, -5, \sqrt{7}, 0, 1, 2, \sqrt{\dfrac{25}{36}}, 5\dfrac{1}{9}\right\}$
   that are
   a. Natural numbers   **b.** Whole numbers
   c. Integers   **d.** Rational numbers
   e. Irrational numbers

Multiple Representations

In Exercises 41–48 use the given information to complete each row as illustrated by the example in the first row.

	INEQUALITY NOTATION	VERBALLY	GRAPHICALLY	INTERVAL NOTATION
Example:	$x > 2$	x is greater than 2.	2	$(2, \infty)$
41.			-3 3	
42.			6	
43.	$x > 1$			
44.	$1 \le x \le 5$			
45.		x is greater than or equal to -3.		
46.		x is greater than or equal to 0 and less than 2.		
47.				$(-\infty, 4]$
48.				$(-3, 5)$

In Exercises 49–56 list the set or sets to which the given real number belongs. The choices are natural numbers, whole numbers, integers, rational numbers, and irrational numbers.

49. -18 **50.** $-\dfrac{1}{8}$ **51.** 81 **52.** $\sqrt{81}$

53. $\sqrt{8}$ **54.** 8 **55.** $5\dfrac{3}{7}$ **56.** 0

57. Plot these numbers on a real number line:

$$-5, -3.5, 0, 1\dfrac{3}{4}, 4.$$

58. Plot these numbers on a real number line:

$$-4, -2, -\dfrac{1}{2}, 0, 1, 2\dfrac{1}{2}, 5.$$

59. Plot these numbers on a real number line:

$$-2.5, \dfrac{5}{4}, |-3|, -|4|.$$

60. Plot these numbers on a real number line:

$$-\dfrac{3}{2}, -|-2|, |-2|, 2\dfrac{1}{3}.$$

In Exercises 61–66 identify each real number as either rational or irrational.

61. a. $\sqrt{9}$ **b.** $\sqrt{10}$

62. a. $\sqrt{20}$ **b.** $\sqrt{25}$

63. a. $0.171717...$ **b.** $0.171771777...$

64. a. $0.232232223...$ **b.** $-0.133133133...$

65. a. $\sqrt{0}$ **b.** $\sqrt{\dfrac{49}{16}}$

66. a. 0.13562 **b.** $0.101001000...$

Multiple Representations

In Exercises 67–72 write each verbal statement in algebraic form.

67. The absolute value of x is equal to y.

68. The square root of x is equal to y.

69. π is greater than 3.14 and less than 3.15.

70. The square root of 2 is greater than one and less than two.
71. The opposite of x is less than or equal to negative two.
72. The absolute value of negative eleven equals eleven.
73. a. Give an example of a number x so that
$$|x - 3| = x - 3$$
    b. Give an example of a number x so that
$$|x - 3| \neq x - 3$$
74. a. Give an example of a number x so that $\sqrt{x} < x$
    b. Give an example of a number x so that $\sqrt{x} > x$
    c. Give an example of a number x so that $\sqrt{x} = x$
75. a. What whole number is not a natural number?
    b. List two integers that are not natural numbers.
    c. List two negative rational numbers that are not integers.
    d. List two rational numbers between 1 and 2.
76. a. What number is its own additive inverse?
    b. List two positive rational numbers that are not integers.
    c. List two real numbers that are not rational numbers.
    d. List two rational numbers between -2 and -3.
77. **Home Runs** The following bar graph shows the total number of home runs hit by Mark McGwire of the St. Louis Cardinals for five consecutive seasons. Select the data that this bar graph best represents.
    a. 40, 50, 48, 78, 59
    b. 52, 58, 70, 65, 32

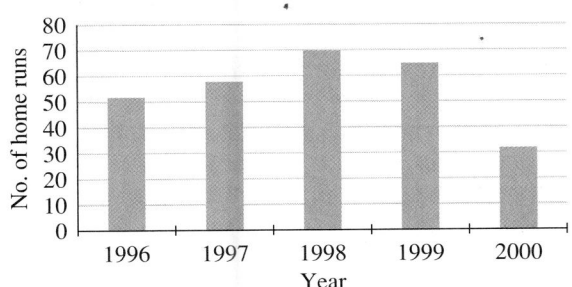

78. **Stock Price Change** The following bar graph shows the change in the price of a stock for a given week. Select the data that this bar graph best represents.
    a. $+\$1.55, -\$0.75, -\$0.89, +\$0.50, +\$1.85$
    b. $+\$1.50, -\$0.75, -\$0.89, +\$5, +\$1.45$

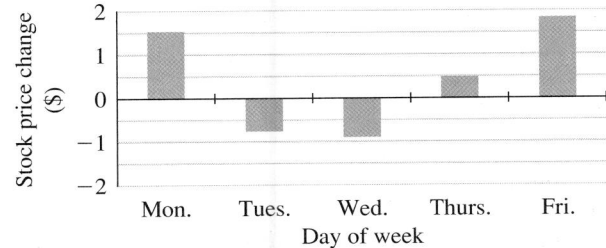

79. **Education Levels of Full-Time Workers** The following pie chart gives the education level breakdown

for full-time, year-round working women according to the U.S. Census Bureau (**http://www.census.gov/**).
Select the percents that this pie chart best represents.

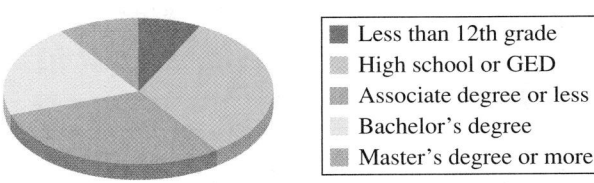

    a. Less than 12th grade: 7%, high school or GED: 33%, Associate degree: 30%, Bachelor's degree: 21%, Master's degree or more: 9%
    b. Less than 12th grade: 12%, high school or GED: 43%, Associate degree: 12%, Bachelor's degree: 20%, Master's degree or more: 13%

80. **Education Levels of Full-Time Workers** The following pie chart gives the education level breakdown for full-time, year-round working men according to the U.S. Census Bureau (**http://www.census.gov/**).
Select the percents that this pie chart best represents.

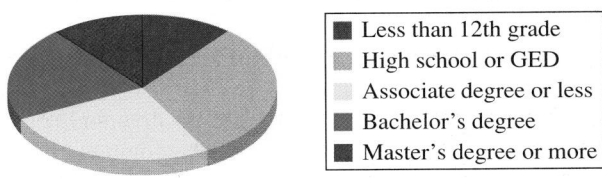

    a. Less than 12th grade: 5%, high school or GED: 50%, Associate degree: 15%, Bachelor's degree: 23%, Master's degree or more: 7%
    b. Less than 12th grade: 11%, high school or GED: 31%, Associate degree: 26%, Bachelor's degree: 21%, Master's degree or more: 11%

Group Discussion Questions

81. **Multiple Representations** There are a variety of notations that can be used to represent the same real number. For example, 0.25, $\frac{1}{4}$, and 25% all represent the same real number.
    a. Plot 0.25 on this number line:

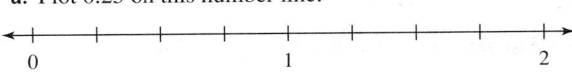

    b. Shade $\frac{1}{4}$ of this region:

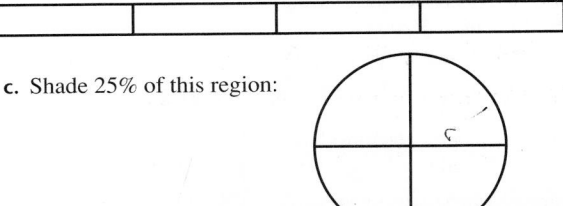

    c. Shade 25% of this region:

82. Discovery Question

a. Complete the following table, and compare the values of $\sqrt{a + b}$ and $\sqrt{a} + \sqrt{b}$:

a	b	$\sqrt{a + b}$	$\sqrt{a} + \sqrt{b}$
4	5	$\sqrt{9} = 3$	$\sqrt{4} + \sqrt{5} \approx 4.23$
9	16		
0	25		
4	0		
7	8		

b. Are $\sqrt{a + b}$ and $\sqrt{a} + \sqrt{b}$ never equal, sometimes equal, or always equal?

c. Complete the statement $\sqrt{a + b}$ _____ $\sqrt{a} + \sqrt{b}$ with either $<$, $\leq$, $>$, $\geq$, or $=$ so that the statement will be true for all nonnegative values of a and b.

83. Discovery Question

a. Complete the following table, and compare the values of $|x + y|$ and $|x| + |y|$.

x	y	$|x + y|$	$|x| + |y|$
-3	7	$|-3 + 7| = 4$	$|-3| + |7| = 10$
8	-6		
0	4		
-3	0		
-2	-5		
7	8		

b. Are $|x + y|$ and $|x| + |y|$ never equal, sometimes equal, or always equal?

c. Complete the statement $|x + y|$ _____ $|x| + |y|$ with either $<$, $\leq$, $>$, $\geq$, or $=$ so that the statement will be true for all values of x and y.

84. Writing Mathematically   Describe in your own words how to evaluate the absolute value of a real number. Is there any real number whose absolute value is not positive?

85. Writing Mathematically   Write a paragraph describing the relationship of the natural numbers, whole numbers, integers, rationals, irrationals, and real numbers.

Section 1.2   Addition of Real Numbers

Objectives: **6.** Add positive and negative real numbers.
7. Use the commutative and associative properties of addition.
8. Evaluate an algebraic expression for given values of the variables.

Positive and negative numbers both occur in many real-life applications: the price changes of a stock, the profit or loss of a business, the temperature on a cold day, the elevation of an object, the net yardage on a football play, and the directed distance as shown on a number line.

For today's quotes, check
http://quicken.com

SYMBOL	LAST	CHANGE
AOL	48.50	-3.07
DELL	42.38	$+2.23$
DJIA	10,299.24	-24.68
INTC	122.50	-0.33
NASDAQ	3205.11	$+25.24$
MSFT	61.44	$+1.85$

A Mathematical Note

The symbols "+" and "−", used to indicate addition and subtraction, were introduced by Johannes Widman, a Bohemian mathematician, in the fifteenth century. The words *plus* and *minus* are from the Latin for more and less, respectively.

In Example 1 we begin to consider operations with these signed numbers. We start by building on the skills that we already have with addition. Recall the terminology for the addition $5 + 8 = 13$; the numbers 5 and 8 are called **terms** or **addends,** and the result 13 is called the **sum.**

■ **EXAMPLE 1** Adding Real Numbers with Like Signs

Complete the following table using a positive number to indicate a profit and a negative number to indicate a loss:

QUARTERLY PROFIT/LOSS STATEMENT FOR H&M PRINTERS

	QUARTER 1 ($)	QUARTER 2 ($)	QUARTER 3 ($)	QUARTER 4 ($)
Branch A	+25,000 (profit)	+30,000 (profit)	−5,000 (loss)	−3,000 (loss)
Branch B	+40,000 (profit)	+50,000 (profit)	−7,000 (loss)	−4,000 (loss)
Total				

SOLUTIONS

The answer +$65,000 is usually just written as $65,000; and is understood to represent positive sixty-five thousand dollars.

Total for Quarter 1: $25,000 + $40,000 = +$65,000; a profit

Total for Quarter 2: $30,000 + $50,000 = +$80,000; a profit

Total for Quarter 3: −$5,000 + (−$7,000) = −$12,000; a loss

Total for Quarter 4: −$3,000 + (−$4,000) = −$7,000; a loss ■

The "+" symbol has two distinct meanings. One meaning indicates that the direction of a signed number on the number line is to the right of the origin. The second meaning indicates the operation of addition, which results in movement on the number line.

Example 1 illustrates that when both terms are positive, the sum is positive and its magnitude is found by adding the magnitudes or absolute values of the terms. When both terms are negative, the sum is negative and its magnitude is found by adding the magnitudes or absolute values of the terms. Another application of signed numbers is directed distance. The addition of real numbers can be represented by combining directed distances on the number line. View each term as a trip in a specific direction and addition as the combining of these individual trips. This is illustrated in the following box.

Addition of Real Numbers with Like Signs

VERBALLY	GRAPHICAL EXAMPLE	NUMERICAL EXAMPLE
To add numbers with like signs: 1. Add the absolute values of the terms 2. and use the same sign as the terms.		$(+3) + (+4) = +7$ $(-3) + (-4) = -7$

Example 2 illustrates the addition of terms with like signs, including the addition of fractions. The procedure for adding fractions is summarized in the following box.

Addition of Fractions

	VERBALLY	ALGEBRAICALLY	NUMERICAL EXAMPLE
	To add fractions with the same denominator, add the numerators and use the common denominator.	$\dfrac{a}{b} + \dfrac{c}{b} = \dfrac{a+c}{b}$ for $b \neq 0$.	$\dfrac{1}{5} + \dfrac{3}{5} = \dfrac{1+3}{5}$ $= \dfrac{4}{5}$
	To add fractions with unlike denominators, first express each fraction in terms of a common denominator and then add the numerators using this common denominator.	$\dfrac{a}{b} + \dfrac{c}{d} = \dfrac{ad}{bd} + \dfrac{bc}{bd} = \dfrac{ad+bc}{bd}$ for $b \neq 0$ and $d \neq 0$.	$\dfrac{1}{4} + \dfrac{2}{3} = \dfrac{3}{12} + \dfrac{8}{12}$ $= \dfrac{3+8}{12}$ $= \dfrac{11}{12}$

The group exercises at the end of this section contain a graphical perspective on adding fractions.

EXAMPLE 2 Adding Terms with Like Signs

Calculate the following sums:

SOLUTIONS

(a) $4.5 + 7.7$

$4.5 + 7.7 = 12.2$

Both terms and the sum are positive.

(b) $-2.9 + (-5.1)$

$-2.9 + (-5.1) = -8.0$

Both terms and the sum are negative. The absolute value of the sum is the sum of the absolute values of the terms.

(c) $\dfrac{2}{9} + \dfrac{5}{9}$

$\dfrac{2}{9} + \dfrac{5}{9} = \dfrac{2+5}{9}$

$= \dfrac{7}{9}$

$\dfrac{a}{b} + \dfrac{c}{b} = \dfrac{a+c}{b}$ for $b \neq 0$. To add fractions with the same denominator, add the numerators and use the common denominator.

(d) $-\dfrac{1}{4} + \left(-\dfrac{1}{3}\right)$

$-\dfrac{1}{4} + \left(-\dfrac{1}{3}\right) = -\dfrac{3}{12} + \left(-\dfrac{4}{12}\right)$

$= \dfrac{-3 + (-4)}{12}$

$= \dfrac{-7}{12}$

$= -\dfrac{7}{12}$

To add fractions with different denominators, first express each fraction in terms of a common denominator.

In this case express each term in terms of the least common denominator (LCD) of 12.

Then add these like terms. Both $\dfrac{-7}{12}$ and $-\dfrac{7}{12}$ represent the same number. This is discussed further in Section 1.5.

CALCULATOR PERSPECTIVE 1.2.1 — Converting Between Decimals and Fractions

To convert a fraction to a decimal, enter the fraction using the division symbol for the fraction bar. Pressing [ENTER] will then display the fraction in decimal form. To convert a decimal answer to a fraction, use the **Ans ▶ Frac** feature. To convert $\frac{3}{8}$ to a decimal and to enter $-\frac{1}{4} + \left(-\frac{1}{3}\right)$ from Example 2(d) on a TI-83 Plus calculator, enter the following keystrokes:

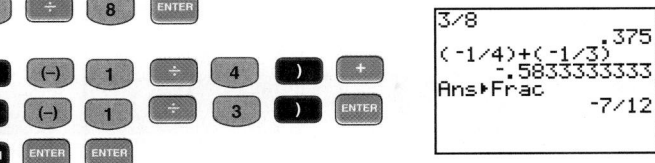

[3] [÷] [8] [ENTER]

[( ] [(-)] [1] [÷] [4] [ )] [+]

[( ] [(-)] [1] [÷] [3] [ )] [ENTER]

[MATH] [ENTER] [ENTER]

```
3/8
                 .375
(-1/4)+(-1/3)
          -.5833333333
Ans▶Frac
               -7/12
```

To form the opposite of a number, use the [(-)] key. The [−] key is used only for subtraction. Also note that by pressing [MATH] [ENTER] [ENTER] at the end of the expression with fractions, we can directly convert the answer to a fraction.

Note: Pressing the [MATH] key pulls up the **Frac** conversion as one option. Pressing [ENTER] once selects this option. Pressing [ENTER] a second time executes this conversion to a fraction. Although the parentheses enclosing the fractions are not required in this example, their usage makes the input more readable and ensures the correct interpretation.

SELF-CHECK 1.2.1

1. Use a calculator to check Example 2(b).

2. Use a calculator to add $\frac{2}{5} + \frac{3}{7}$.

3. Convert $\frac{5}{8}$ to a decimal.

4. Convert 0.875 to a fraction.

In Example 3 we will examine the addition of terms with unlike signs.

◼ EXAMPLE 3   Adding Real Numbers with Unlike Signs

Complete the following table using a positive number to indicate a profit and a negative number to indicate a loss.

BRANCH RESULTS FOR H&M PRODUCTIONS

	QUARTER 1 ($)	QUARTER 2 ($)	QUARTER 3 ($)	QUARTER 4 ($)
Texas branch	+35,000	−35,000	+5,000	−10,000
California branch	−10,000	+50,000	−10,000	+7,000
Total				

SELF-CHECK 1.2.1 ANSWERS

1. −8.0   2. $\frac{29}{35}$

3. 0.625   4. $\frac{7}{8}$

SOLUTION _____

Total for Quarter 1: $35,000 + (−$10,000) = +$25,000; a profit
Total for Quarter 2: −$35,000 + $50,000 = +$15,000; a profit
Total for Quarter 3: $5,000 + (−$10,000) = −$5,000; a loss
Total for Quarter 4: −$10,000 + $7,000 = −$3,000; a loss

Note that in Example 3 when the terms are of opposite signs, the net result is that one term tends to diminish the magnitude of the other. The sum has the same sign as the term with the larger magnitude. This observation is summarized in the following box as a rule for all real numbers.

Addition of Real Numbers with Unlike Signs

VERBALLY	GRAPHICAL EXAMPLE	NUMERICAL EXAMPLE
To add numbers with unlike signs: 1. Subtract the smaller absolute value from the larger absolute value	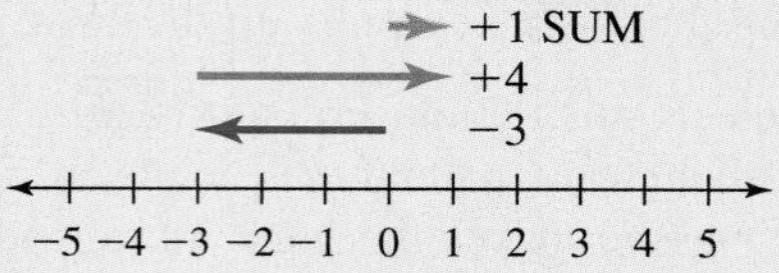 +1 SUM; +4; −3 (number line −5 to 5)	$-3 + (+4) = +1$
2. and use the sign of the term with larger absolute value.	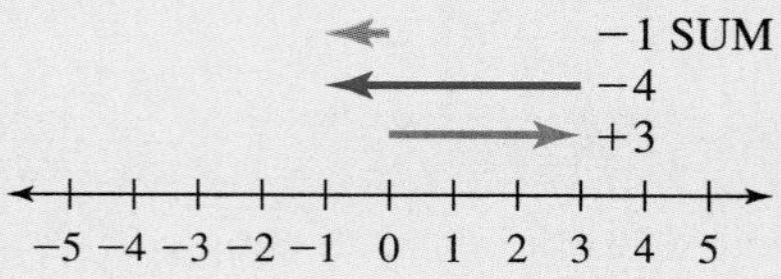 −1 SUM; −4; +3 (number line −5 to 5)	$(+3) + (-4) = -1$

■ EXAMPLE 4   Adding Terms with Unlike Signs

Calculate these sums:

SOLUTIONS _____

(a) $-4.5 + 7.7$ $-4.5 + 7.7 = 3.2$ The sum is positive, as is 7.7; and the difference of the absolute values is $7.7 - 4.5 = 3.2$.

(b) $\dfrac{1}{3} + \left(-\dfrac{1}{2}\right)$

$$\dfrac{1}{3} + \left(-\dfrac{1}{2}\right) = \dfrac{2}{6} + \left(-\dfrac{3}{6}\right)$$
$$= \dfrac{2 + (-3)}{6}$$
$$= \dfrac{-1}{6}$$
$$= -\dfrac{1}{6}$$

Use the least common denominator (LCD) of 6 to express each term in terms of a common denominator. The sum is negative since the term $-\dfrac{3}{6}$ has the larger absolute value.

(c) $2.9 + (-2.9)$ $2.9 + (-2.9) = 0$ The sum of a real number and its additive inverse is zero.

The procedure for adding real numbers is summarized in the following box. The key question to ask before adding two terms is, "Do the terms have like signs or unlike signs?"

Addition of Real Numbers

- Determine if the terms have like or unlike signs.
- If the terms have like signs, add the absolute values. The sum will have the same sign as both terms.
- If the terms have unlike signs, subtract the absolute values. The sum will have the same sign as the term with the larger absolute value.

The most commonly used grouping symbols for printed copy are parentheses "( )" and brackets "[ ]." Use only parentheses for calculators as brackets have a special meaning. This is covered in the calculator perspective in Section 1.6.

When grouping symbols are used, the implied order of operations is to perform first the operations within the innermost grouping symbols. We will examine order of operations in more detail in Section 1.6.

■ EXAMPLE 5   Adding Terms with Grouping Symbols

Determine the following sums:

SOLUTIONS

(a) $-7 + [-8 + (+5)]$

$$-7 + [-8 + (+5)] = -7 + (-3)$$
$$= -10$$

The indicated order of operations is first to add the terms inside the brackets.

(b) $[-9 + (-7 + 3)] + (-11)$

$$[-9 + (-7 + 3)] + (-11)$$
$$= [-9 + (-4)] + (-11)$$
$$= -13 + (-11)$$
$$= -24$$

Add the terms inside the inner parentheses first; then add the terms inside the brackets.

SELF-CHECK 1.2.2 ANSWERS

1. 3.2   2. $-\dfrac{1}{6}$ 3. 0

```
-4.5+7.7
              3.2
1/3+(-1/2)▶Frac
             -1/6
2.9+(-2.9)
               0
```

4. −24

```
(-9+(-7+3))+(-11
)
             -24
```

SELF-CHECK 1.2.2

Use a calculator to check the answers to Examples 4 and 5(b).

When we purchase two items at a store, the total will be the same no matter which item is charged first. (Can you imagine the confusion that would result if the order did affect the total price?) For example, $9 + 5 = 5 + 9$. We say that addition is **commutative** to describe the fact that the order of the terms does not change the sum. That is, $a + b = b + a$.

When we purchase several items at a store, we may cluster them together in different ways. For example, all canned goods and all fresh produce might be totaled separately. Again, if we subtotaled these groups separately, we are confident that the total price of all items will not change. We say that addition is **associative** to describe the fact that terms can be regrouped without changing the sum. For example, $(3 + 4) + 5$ and $3 + (4 + 5)$ both equal 12. In general, the associative property of addition says $(a + b) + c = a + (b + c)$.

Properties of Addition

	ALGEBRAICALLY	VERBALLY	NUMERICAL EXAMPLE
Commutative property	$a + b = b + a$	The sum of two terms in either order is the same.	$4 + 5 = 5 + 4$
Associative property	$(a + b) + c = a + (b + c)$	Terms can be regrouped without changing the sum.	$(3 + 4) + 5 = 3 + (4 + 5)$

The properties of addition, subtraction, multiplication, and division are presented as needed. By the end of Chapter 1 you will have reviewed all these operations.

Together the commutative and associative properties allow us to add terms in any order we desire. People who are very skilled at mental arithmetic often take advantage of this to add terms producing multiples of 10 or other convenient values. For expressions with several terms it is often convenient first to add all terms with like signs.

■ EXAMPLE 6   Using the Properties of Addition

Calculate these sums:

SOLUTIONS

(a) $47 + 19 + 53 + 31 + 7$

$47 + 19 + 53 + 31 + 7$
$= (47 + 53) + (19 + 31) + 7$
$= 100 + 50 + 7$
$= 157$

For convenience, we use the commutative and associative properties to regroup the terms in order to pair terms that produce multiples of 10. With practice you may be able to do this entire computation mentally. For most problems writing some of these steps is usually a good idea.

(b) $3 + (-4) + 5 + (-6) + 7$

$3 + (-4) + 5 + (-6) + 7$
$= (3 + 5 + 7) + [(-4) + (-6)]$
$= 15 + (-10)$
$= 5$

For convenience we first regroup terms with like signs together.

SELF-CHECK 1.2.3

Calculate these sums:
1. $-9 + (-7)$
2. $-1.9 + 3.7$
3. $\dfrac{3}{5} + \left(-\dfrac{1}{2}\right)$
4. $\left(-\dfrac{7}{11}\right) + \dfrac{7}{11}$
5. $-9 + 8 + (-13) + (-11) + 2$

SELF-CHECK 1.2.3 ANSWERS

1. -16 2. 1.8 3. $\dfrac{1}{10}$
4. 0 5. -23

Positive and negative results often are displayed by bar graphs. Sometimes it is useful to be able to take the displayed results and to calculate additional information.

■ **EXAMPLE 7**   Totaling the Results of a Bar Graph

Use the midyear results shown in this bar graph to determine the total profit of H&M Productions, Inc. for this 6-month period.

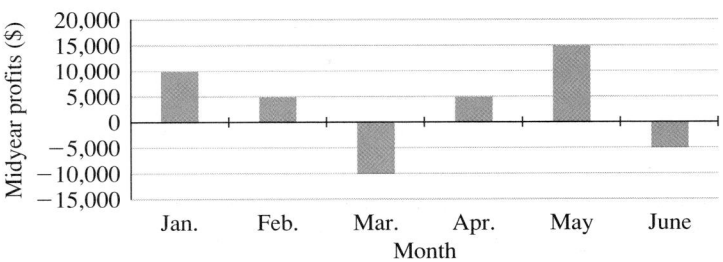

You will encounter algebraic concepts in a variety of contexts in the real world. Bar graphs are used frequently in newspapers and magazines to display numerical information that you should be able to interpret.

SOLUTIONS

$10,000 + 5,000 + (-10,000) + 5,000 + 15,000 + (-5,000)$
$= (10,000 + 5,000 + 5,000 + 15,000) + [(-10,000) + (-5,000)]$
$= 35,000 + (-15,000)$
$= 20,000$

Read each of the 6 monthly results from the table and then add these terms to obtain the 6-month total.

Answer: The total profit for the company for the first 6 months of the year was $20,000.   ■

Algebraic expressions can be as simple as a single constant or a variable or can be more involved when we combine constants and variables with operations. Examples of algebraic expressions are $-\dfrac{7}{12}$, π, x, $x + y$, $2x + y + (-11)$, $|x| + |y|$, and $\sqrt{x + y}$. Many different words and phrases are commonly used in word problems and applications to indicate addition. Table 1.2.1 lists some common phrases and corresponding algebraic expressions.

Table 1.2.1   Phrases Used to Indicate Addition

KEY PHRASE	VERBAL EXAMPLE	ALGEBRAIC EXAMPLE
Plus	"12 plus 8"	$12 + 8$
Total	"The total of $25 and $40"	$25 + 40
Sum	"The sum of x and y"	$x + y$
Increased by	"An interest rate r is increased by 0.5%"	$r + 0.005$
More than	"7 more than x"	$x + 7$

It is customary to convert percents to decimal form before performing operations; $0.5\% = 0.005$.

To **evaluate an algebraic expression** for given values of the variables means to replace each variable by the specific value given for that variable and then to simplify this expression. When a constant is substituted for a variable, it is often wise to use parentheses to avoid a careless error in sign or an incorrect order of operations.

EXAMPLE 8 Evaluating an Algebraic Expression

Evaluate these expressions for $x = 7$ and $y = -5$.

SOLUTIONS

(a) $x + y + 4$

$$\begin{aligned} x + y + 4 &= 7 + (-5) + 4 \\ &= [7 + 4] + (-5) \\ &= 11 + (-5) \\ &= 6 \end{aligned}$$

Replace x by 7 and y by -5 and then use the commutative property of addition to reorder the terms. Add the terms with like signs before combining the terms with unlike signs. Note the use of parentheses when -5 is substituted for y.

(b) $x + y + (-11)$

$$\begin{aligned} x + y + (-11) &= 7 + (-5) + (-11) \\ &= 7 + [(-5) + (-11)] \\ &= 7 + (-16) \\ &= -9 \end{aligned}$$

Replace x by 7 and y by -5 and then add the terms with like signs before combining the terms with unlike signs.

(c) $|x| + |y|$

$$\begin{aligned} |x| + |y| &= |7| + |-5| \\ &= 7 + 5 \\ &= 12 \end{aligned}$$

Replace x by 7 and y by -5 and then remove the absolute value symbols before adding the two terms.

SELF-CHECK 1.2.4

1. Write an algebraic expression for "11 more than y."
2. Write an algebraic expression for "x increased by 2."
3. Evaluate $x + y + (-5)$ for $x = 6$ and $y = -9$.

EXAMPLE 9 Using Estimation Skills to Check Answers

Mentally estimate the value of each expression and then calculate the exact value:

SOLUTIONS

		ESTIMATED ANSWER	ACTUAL ANSWER	
(a) $-24.937 + (-15.054)$	$-24.937 \approx -25$ $-15.054 \approx -15$	-25 (+) -15 -40	-24.937 (+) -15.054 -39.991	We rounded each term to the nearest integer to simplify the mental estimation.
(b) $-9.823 + 61.150$	$61.150 \approx 61$ $-9.823 \approx -10$	61 (+) -10 51	61.150 (+) -9.823 51.327	

SELF-CHECK 1.2.4 ANSWERS

1. $y + 11$ 2. $x + 2$
3. -8

USING THE LANGUAGE AND SYMBOLISM OF MATHEMATICS   1.2

1. In the addition $27 + (-13) = 14$, 27 and -13 are called _____ or _____ , and 14 is called the _____ .

2. The property that says $a + b = b + a$ for all real numbers a and b is called the _____ property of addition.

3. The property that says $(a + b) + c = a + (b + c)$ for all real numbers a, b, and c is called the _____ property of addition.

4. If x and y have the same sign, then the sum $x + y$ will have the same sign as _____ terms.

5. If x and y have unlike signs, then the sum $x + y$ will have the same sign as the term with the _____ absolute value.

6. Together the _____ and _____ properties of addition allow us to add terms in any order and to obtain the same sum.

7. The sum $\dfrac{w}{x} + \dfrac{y}{x} =$ _____ for $x \neq$ _____ .

8. To _____ an algebraic expression for given values of the variables means to replace each variable by the specific value given for that variable.

9. The phrase "x increased by 5" is represented algebraically by _____ .

10. The phrase "the sum of x, y, and z" is represented algebraically by _____ .

EXERCISES   1.2

In Exercises 1–22 calculate each sum without using a calculator.

1. a. $8 + (+5)$ b. $-8 + (+5)$
   c. $8 + (-5)$ d. $-8 + (-5)$

2. a. $9 + (+11)$ b. $-9 + (+11)$
   c. $9 + (-11)$ d. $-9 + (-11)$

3. a. $-45 + 45$ b. $31 + (-47)$
   c. $-58 + 67$ d. $-19 + (-24)$

4. a. $97 + (-97)$ b. $-50 + 23$
   c. $70 + (-58)$ d. $-31 + (-17)$

5. a. $-8 + (-7) + (-6)$ b. $-17 + [(-11) + 9]$

6. a. $-3 + (-9) + (-2)$ b. $[(-12) + (-9)] + 8$

7. a. $[11 + (-17)] + [(-8) + (-4)]$
   b. $[-19 + 15] + [17 + (-30)]$

8. a. $[-19 + 7] + [13 + (-19)]$
   b. $[23 + (-19)] + [41 + (-51)]$

9. $[-5 + (-6) + (-7)] + [8 + 4 + 3]$

10. $[11 + 12 + 13] + [(-10) + (-15) + (-16)]$

11. $-5 + 7 + (-4) + 8 + (-9) + 2$

12. $-3 + 5 + (-6) + (-7) + 4 + 11$

13. a. $\sqrt{25} + \left(-\sqrt{49}\right)$ b. $-\sqrt{81} + \sqrt{100}$

14. a. $-\sqrt{169} + \sqrt{64}$ b. $\sqrt{144} + \left(-\sqrt{400}\right)$

15. a. $-\left(\sqrt{16} + \sqrt{9}\right) + \sqrt{16 + 9}$
    b. $-\sqrt{9 + 16 + 144} + \sqrt{9} + \sqrt{16} + \sqrt{144}$

16. a. $-\left(\sqrt{25} + \sqrt{144}\right) + \sqrt{25 + 144}$
    b. $-\sqrt{1 + 16 + 64} + \sqrt{1} + \sqrt{16} + \sqrt{64}$

17. a. $|-17| + |17|$ b. $-|17| + |-17|$
    c. $|-23 + 19|$

18. a. $|19| + |-19|$ b. $-|19| + |-19|$
    c. $|-25 + 17|$

19. a. $\dfrac{7}{11} + \left(-\dfrac{3}{11}\right)$ b. $-\dfrac{2}{5} + \dfrac{3}{4}$
    c. $-\dfrac{1}{2} + \left(-\dfrac{1}{3}\right) + \left(-\dfrac{1}{4}\right)$

20. a. $-\dfrac{11}{17} + \dfrac{5}{17}$ b. $-\dfrac{2}{3} + \dfrac{3}{5}$
    c. $-\dfrac{2}{5} + \left(-\dfrac{1}{2}\right) + \dfrac{2}{3}$

21. a. $-85.6 + 85.6$ b. $1.2 + 0 + (-1.5)$
    c. $-4.8 + 0 + 6.09$

22. a. $-77.9 + 77.9$ b. $-5.4 + 0 + 8.6$
    c. $-8.6 + 0 + 6.09$

Multiple Representations

In Exercises 23–26 represent the addition numerically, verbally, or graphically as described in each problem. (*Hint:* See the following example.)

NUMERICALLY	VERBALLY	GRAPHICALLY*
$4 + 2 = 6$	The sum of four and two is six.	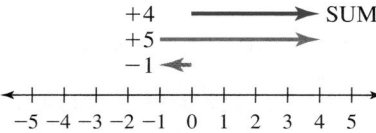

*An arrow 4 units long is drawn to the right, starting at the origin. From the tip of this arrow, draw another arrow 2 units to the right. The sum, 6, is represented by an arrow from the origin to the tip of the last arrow.

23. Represent the addition $5 + (-2) = 3$ both verbally and graphically.

24. Give the numerical and verbal representations for the addition shown here.

$$\begin{array}{c} +4 \longrightarrow \text{SUM} \\ +5 \longrightarrow \\ -1 \longleftarrow \\ \hline -5\ -4\ -3\ -2\ -1\ \ 0\ \ 1\ \ 2\ \ 3\ \ 4\ \ 5 \end{array}$$

25. Represent the addition $-3 + 5 = 2$ both verbally and graphically.

26. Represent the addition "The sum of two and three is five," both numerically and graphically.

In Exercises 27–34 evaluate each expression for $w = 7$, $x = -3$, and $y = -6$.

27. a. $|x|$ **b.** $-|x|$ **c.** $|-x|$

28. a. $|y|$ **b.** $-|y|$ **c.** $|-y|$

29. a. $w + x$ **b.** $\sqrt{w + x}$ **c.** $-\sqrt{w + x}$

30. a. $w + y$ **b.** $\sqrt{w + y}$ **c.** $-\sqrt{w + y}$

31. a. $w + x + y$ **b.** $-(w + x + y)$
    c. $-w + (-x) + y$

32. a. $w + [-x + (-y)]$ **b.** $-[w + (-x) + (-y)]$
    c. $-w + (-x + y)$

33. a. $w + x + (-y)$ **b.** $\sqrt{w + [(-x) + (-y)]}$
    c. $-\sqrt{w + [(-x) + (-y)]}$

34. a. $-(-x + y) + w$ **b.** $\sqrt{21 + w + x}$
    c. $-\sqrt{21 + w + x}$

Perimeter

The perimeter of a geometric figure is the distance around the figure. For a rectangle or triangle the perimeter is the sum of the lengths of all sides. In Exercises 35–38 calculate the perimeter of each figure. All units are in centimeters.

35.

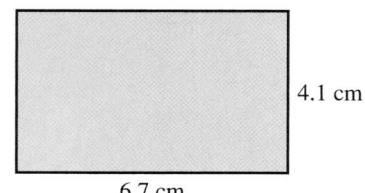

4.1 cm

6.7 cm

36.

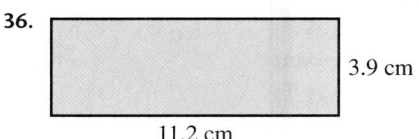

3.9 cm

11.2 cm

37.

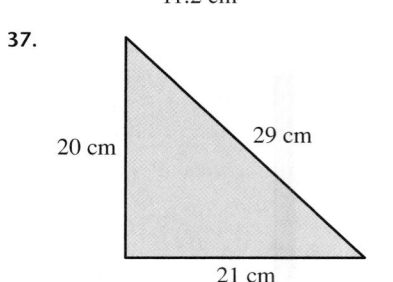

20 cm   29 cm

21 cm

38.

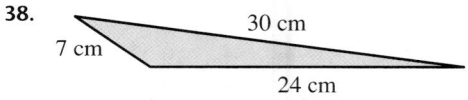

7 cm   30 cm

24 cm

39. The property that says $(5 + 7) + 9 = (7 + 5) + 9$ is the _____ property of addition.

40. The property that says $(5 + 7) + 9 = 5 + (7 + 9)$ is the _____ property of addition.

41. Use the associative property of addition to rewrite $11 + (12 + 13)$.

42. Use the commutative property of addition to rewrite $17 + 13$.

Tuition Bill

The following table gives the 2001–2002 tuition bill for a Parkland College Business Administration student.

	HOURS	TUITION ($)
Fall	12	660
Spring	14	770
Summer	6	330

43. Calculate the total number of semester hours taken by this student in 2001–2002.

44. Calculate the total tuition billed to this student for 2001–2002.

Earnings Reports

In Exercises 45 and 46 use the results shown in the bar graphs to determine the net income for MCI Worldcom and Intuit, Inc., respectively in this 4-year period. (**Source: http://infoseek.go.com**)

45.

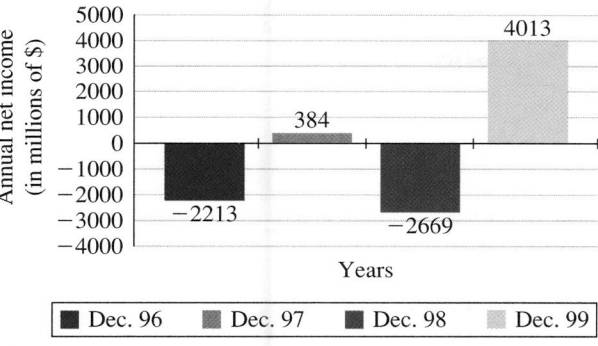

■ Dec. 96 ■ Dec. 97 ■ Dec. 98 ■ Dec. 99

46.

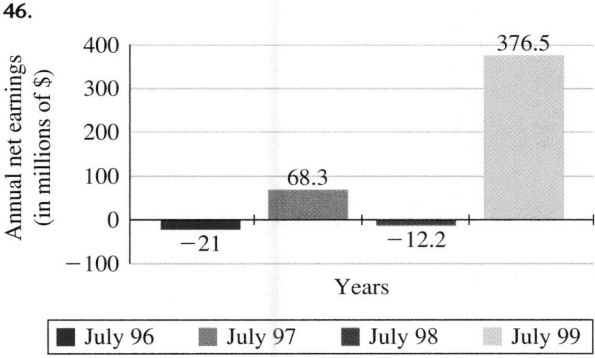

■ July 96 ■ July 97 ■ July 98 ■ July 99

Power Production

In Exercises 47–49 use the pie chart to answer the questions regarding electric power production in the United States in 1997. (**Source: http://www.census.gov**)

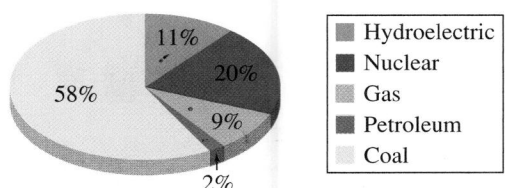

■ Hydroelectric
■ Nuclear
■ Gas
■ Petroleum
■ Coal

47. What percent of the electric power production do the combined sources of hydroelectric, gas, and petroleum generation produce?

48. What percent of the electric power production do the combined sources of coal and gas generation produce?

49. What percent of the electric power production is generated by sources other than coal and nuclear?

Education Attainment

In Exercises 50–52 use the pie chart to answer the questions regarding higher education in Illinois in 1999. (**Source: http://www.census.gov**)

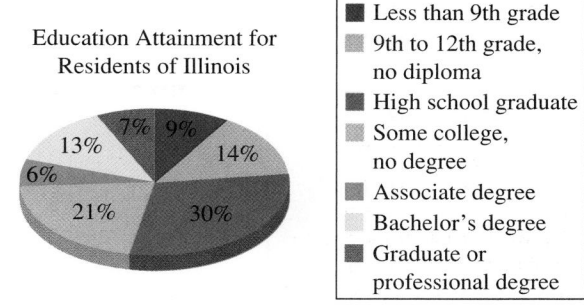

Education Attainment for Residents of Illinois

■ Less than 9th grade
▨ 9th to 12th grade, no diploma
■ High school graduate
▨ Some college, no degree
▨ Associate degree
▨ Bachelor's degree
■ Graduate or professional degree

50. What is the percent of Illinois residents that do not have a high school diploma?

51. What is the percent of Illinois residents that have some kind of college degree?

52. What is the percent of Illinois residents that do not have a college degree?

Thermometer Reading

53. A thermometer registers −32 degrees Fahrenheit (−32°F). The temperature rises 30°. What is the resulting temperature?

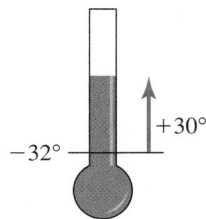

54. A thermometer registers −12°F. The temperature rises 21°. What is the resulting temperature?

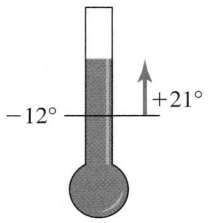

Perimeter of a Room

To order the crown molding trim for a basement recreation room, a contractor must calculate the total length of the trim needed. To do this the contractor calculates the perimeter of the room. In Exercises 55 and 56 calculate the perimeter of each room. All units are given in meters. (*Hint:* The perimeter can be calculated using the given measurements even though the measurements of some sides are not given.)

55.

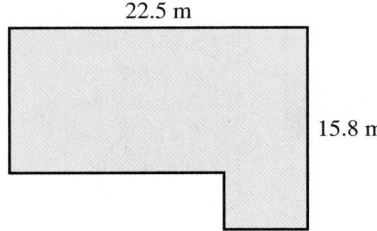

22.5 m

15.8 m

56.

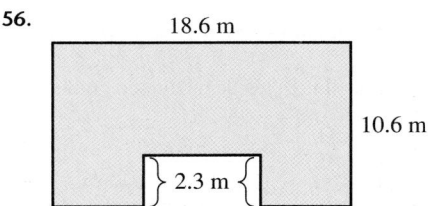

18.6 m

10.6 m

2.3 m

Perimeter of a Nickel Blank

57. Some of the United States coinage is based on the metric system, which was legalized in 1866. The nickel is 2 cm in diameter and has a mass of 5 grams. Nickels are stamped from a blank metal sheet. In the example shown here one nickel is stamped from a surrounding square. Calculate the perimeter of the square.

58. The blank metal sheet shown here has four nickels stamped from it. Calculate the perimeter of this rectangular sheet. (*Hint:* See Exercise 57.)

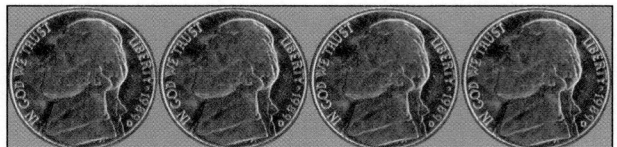

Business Profit or Loss

59. Complete the last row of the following table using a positive number to represent a total profit for the two stores and a negative number to represent a total that is a loss.

QUARTERLY PROFIT/LOSS STATEMENT FOR BRIAN'S BARBEQUE

QUARTER	1 ($)	2 ($)	3 ($)	4 ($)
Springfield store	+7000	+7000	−7000	−7000
Washington store	+5000	−5000	+5000	−5000
Total				

Football Yardage

60. A football team started a series on its own 20-yd line. The results of five plays are given in the following table. Complete this table by giving the position of the football at the end of each play.

PLAY	RESULT	NEW YARD LINE POSITION
a. 1	6-yd gain	
b. 2	3-yd loss	
c. 3	12-yd gain	
d. 4	0 yd	
e. 5	7-yd loss	

Estimation Skills and Calculator Skills

In Exercises 61–66 first mentally determine the correct sign of the sum and then use a calculator to calculate this sum. See Calculator Perspective 1.2.1 for Exercises 63 and 64. Approximate Exercises 65 and 66 to the nearest hundredth.

PROBLEM	SIGN OF SUM WITHOUT CALCULATOR	SUM WITH CALCULATOR
61. $-47.6 + 53.4$		
62. $47.6 + (-53.4)$		
63. $-\dfrac{8}{5} + \left(-\dfrac{3}{4}\right)$		
64. $\dfrac{8}{5} + \left(-\dfrac{3}{4}\right)$		
65. $\sqrt{3} + \sqrt{5}$		
66. $-\sqrt{7} + \sqrt{6}$		

Estimation Skills

In Exercises 67–74 mentally estimate each sum and then select the most appropriate answer.

67. $6.73296 + (-7.5186)$
  a. Less than -1 **b.** Between -1 and 0
  c. Between 0 and 1 **d.** Greater than 1

68. $-13.54462 + 14.92857$
  a. Less than -1 **b.** Between -1 and 0
  c. Between 0 and 1 **d.** Greater than 1

69. $-5.40073 + (-1.29875) + (-2.4463)$
  a. Less than -10 **b.** Between -10 and 0
  c. Between 0 and 10 **d.** Greater than 10

70. $12.438 + (-13.794) + (-6.214) + (-2.847)$
  a. Less than -10 **b.** Between -10 and 0
  c. Between 0 and 10 **d.** Greater than 10

71. $\pi + (-3.15)$
  a. Less than -1 **b.** Between -1 and 0
  c. Between 0 and 1 **d.** Greater than 1

72. $\dfrac{1}{3} + (-0.3333)$
  a. Less than -1 **b.** Between -1 and 0
  c. Between 0 and 1 **d.** Greater than 1

73. $-35.764 + 66.056$
  a. -100 **b.** $+100$
  c. -30 **d.** 30

74. $-35.764 + (-66.056)$
  a. -100 **b.** $+100$
  c. -30 **d.** 30

Calculator Usage

In Exercises 75–78 complete each row of the table to express each rational number in both fractional form and decimal form. See Calculator Perspective 1.2.1.

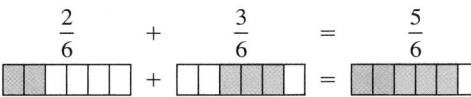

	FRACTIONAL FORM	DECIMAL FORM
Example	$\dfrac{1}{2}$	0.5
75.		0.4
76.		0.125
77.	$\dfrac{3}{8}$	
78.	$\dfrac{7}{16}$	

Multiple Representations

In Exercises 79–82 write each verbal statement in algebraic form.

79. a. The sum of a and b equals the sum of b and a.
  b. The sum of a real number x and its additive inverse is zero.

80. a. The sum of x and three is less than eleven.
  b. The sum of negative four and y is greater than or equal to twelve.

81. a. π is not equal to 3.14.
  b. π is approximately equal to 3.14.

82. a. The absolute value of the sum of five and negative three is not equal to the sum of the absolute value of five and the absolute value of negative three.
  b. The square root of the sum of thirty-six and sixty-four is less than the sum of the square root of thirty-six and the square root of sixty-four.

Multiple Representations

83. If r is the rate in mi/h (miles per hour) of one airplane, write an expression for the rate of an airplane that is traveling 50 mi/h faster.

84. An investor placed part of her portfolio in a savings bond and the rest in a CD (certificate of deposit). She invested $3000 more in the CD than in the savings bond. If P is the principal invested in the savings bond, write an expression for the amount invested in the CD.

85. If n represents an even integer, write an expression for the next integer.

86. If n represents an even integer, write an expression for the next even integer.

Group Discussion Questions

87. Challenge Question   Determine the following sums and explain your reasoning for part **c:**
  a. $1 + (-2) + 3 + (-4)$
  b. $1 + (-2) + 3 + (-4) + 5 + (-6)$
  c. $1 + (-2) + 3 + (-4) + \cdots + 99 + (-100)$

88. Multiple Representations   A fraction represents a part of something. The addition of fractions can be represented graphically by combining these parts as illustrated by the following two examples:

Example: $\dfrac{1}{5} + \dfrac{3}{5} = \dfrac{4}{5}$

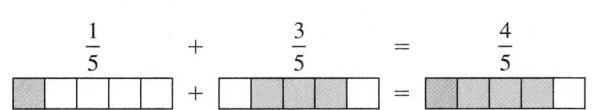

Example: $\dfrac{1}{3} + \dfrac{1}{2} = \dfrac{5}{6}$

$\dfrac{1}{3} + \dfrac{1}{2} =$

$\dfrac{2}{6} + \dfrac{3}{6} = \dfrac{5}{6}$

Using the preceding illustrated format, represent the addition of the fractions in parts **a–d:**

a. $\dfrac{2}{9} + \dfrac{5}{9}$

b. $\dfrac{1}{4} + \dfrac{1}{3}$

c. $\dfrac{1}{2} + \dfrac{2}{5}$

d. $\dfrac{3}{10} + \dfrac{7}{15}$

89. Error Analysis   If you use a TI-83 Plus calculator to evaluate the expression $-7 + (-8 + (+5))$ from Example 5(a) as shown here, this will produce an error. What is the source of the syntax error?

(*Hint:* Consider a similar expression, $-7 + (-8 + (-5))$, that has -5 instead of $+5$.)

90. Multiple Representations   A flowchart is a diagram that is used often by computer programmers to outline the strategy for a program, including all possible decisions that the program will need to make and the results of those decisions. Complete this flowchart for adding two real numbers.

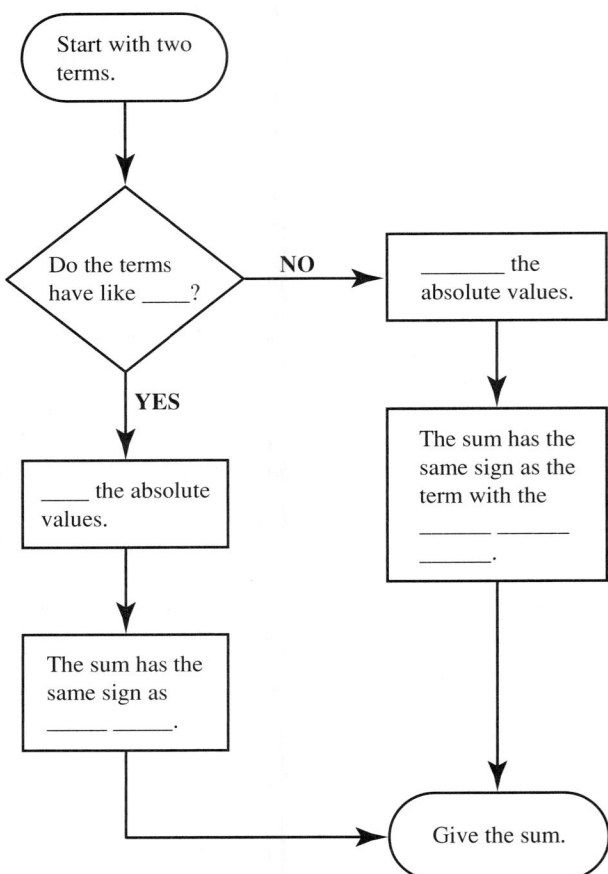

Use this flowchart to find each sum. Have one of your classmates read the steps of the flowchart as another student gives the input values.

a. $-12 + 5$

b. $17 + (-8)$

c. $-9 + (-13)$

d. $11 + 8$

Section 1.3   Subtraction of Real Numbers

Objectives:

9. Subtract positive and negative real numbers.

10. Calculate the terms of a sequence.

11. Check a possible solution of an equation.

Many mathematical concepts such as slope (Section 3.1) are an examination or comparison of different changes.

In this section we continue our presentation of operations with signed numbers by considering subtraction. Recall the terminology for the subtraction $13 - 5 = 8$; the number 13 is called the **minuend,** the number 5 is called the **subtrahend,** and 8 is called the **difference.** The **change** from a to b is determined by the subtraction $b - a$. In Example 1 we build on our previous skills to examine temperature changes and to develop the rules for subtracting signed numbers.

Before this example let us consider some of the words and phrases used in word problems and applications to indicate subtraction. Table 1.3.1 lists some common phrases and corresponding algebraic expressions.

Table 1.3.1 Phrases Used to Indicate Subtraction

KEY PHRASE	VERBAL EXAMPLE	ALGEBRAIC EXAMPLE
Minus	"x minus y"	$x - y$
Difference	"The difference between 12 and 8"	$12 - 8$
Decreased by	"An interest rate r is decreased by 0.5%"	$r - 0.005$
Less than	"7 less than x"	$x - 7$
Change	"The change from 70° to 96°"	$96° - 70°$

■ EXAMPLE 1 Subtracting Positive and Negative Real Numbers

The following table gives the Celsius temperatures at 8 A.M. and noon for four consecutive days in Champaign, IL. Use subtraction to calculate the change from the 8 A.M. temperature to the noon temperature.

Tuesday's Temperatures

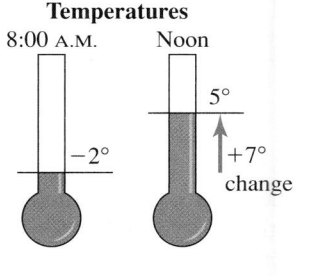

TEMPERATURES AT PARKLAND COLLEGE IN CHAMPAIGN, IL

	MONDAY	TUESDAY	WEDNESDAY	THURSDAY
Noon	+5°	+5°	−5°	−5°
8 A.M.	+2°	−2°	−2°	+2°
Change				

SOLUTION

The change from a to b is

$b - a$

The change from the 8:00 A.M. temperature to the noon temperature is

$\boxed{\text{Noon temperature}} - \boxed{\text{8:00 A.M. temperature}}$.

	MONDAY	TUESDAY	WEDNESDAY	THURSDAY
Change	$+5° - (+2°) = +3°$	$+5° - (-2°) = +7°$	$-5° - (-2°) = -3°$	$-5° - (+2°) = -7°$

We can check the answer to any subtraction problem by using addition. In Example 1:

$5 - 2 = 3$ checks because $3 + 2 = 5$
$5 - (-2) = 7$ checks because $7 + (-2) = 5$
$-5 - (-2) = -3$ checks because $-3 + (-2) = -5$
$-5 - 2 = -7$ checks because $-7 + 2 = -5$

We can say that subtraction is the inverse of addition and can define subtraction as the addition of an additive inverse.

Definition of Subtraction

ALGEBRAICALLY	VERBALLY	NUMERICAL EXAMPLE	GRAPHICAL EXAMPLE
For any real numbers x and y, $$x - y = x + (-y).$$	To subtract y from x, add the opposite of y to x.	$5 - 2 = 5 + (-2)$ $= 3$	$5 - 2 = 3$ DIFFERENCE $\longrightarrow$ -2 $\longleftarrow$ $+5 \longrightarrow$ $-5\ -4\ -3\ -2\ -1\ \ 0\ \ 1\ \ 2\ \ 3\ \ 4\ \ 5$

When using your calculator to perform a subtraction, you do not need to rewrite the problem as an addition problem (the calculator takes care of that for you). For pencil and paper calculations or mental computations you may find it wise either to write or think of the subtraction in terms of addition as illustrated in Example 2.

■ **EXAMPLE 2**   Subtracting Two Real Numbers

Determine the following differences:

SOLUTIONS

(a) $9 - 13$ $9 - 13 = 9 + (-13)$
$\qquad\qquad\qquad = -4$ Change the operation to addition, and change 13 to its additive inverse, -13. Do not change the sign of 9.

(b) $17 - (-5)$ $17 - (-5) = 17 + (5)$
$\qquad\qquad\qquad\quad = 22$ Change the operation to addition, and change -5 to its additive inverse, 5. Do not change the sign of 17.

(c) $-8 - 4$ $-8 - 4 = -8 + (-4)$
$\qquad\qquad\qquad = -12$ Change the operation to addition, and change 4 to its additive inverse, -4. Do not change the sign of -8.

(d) $-6 - (-11)$ $-6 - (-11) = -6 + (11)$
$\qquad\qquad\qquad\qquad = 5$ Change the operation to addition, and change -11 to its additive inverse, 11. Do not change the sign of -6. ■

For subtraction problems it is wise to write the conversion step until your proficiency level is very high. Nonetheless, it is natural to prefer using shortcuts. This can be done mentally as illustrated in Example 3.

EXAMPLE 3 Subtracting Two Real Numbers

Mentally calculate these differences:

SOLUTIONS

(a)	$-6 - 5$	$-6 - 5 = -11$	Think "−6 *plus* −5."
(b)	$-8 - (-3)$	$-8 - (-3) = -5$	Think "−8 *plus* +3."
(c)	$11 - (-5)$	$11 - (-5) = 16$	Think "11 *plus* +5."
(d)	$-7 - 0$	$-7 - 0 = -7$	Think "−7 *plus* 0."

SELF-CHECK 1.3.2

Determine the following differences:

1. $-7 - 8$
2. $-7 - (-8)$
3. $11 - 17$
4. $11 - (-17)$

The procedure for subtracting fractions is summarized in the following box.

Subtracting Fractions

VERBALLY	ALGEBRAICALLY	NUMERICAL EXAMPLE
To subtract fractions with the same denominator, subtract the numerators and use the common denominator.	$\frac{a}{b} - \frac{c}{b} = \frac{a - c}{b}$ for $b \neq 0$	$\frac{5}{7} - \frac{3}{7} = \frac{5 - 3}{7} = \frac{2}{7}$
To subtract fractions with different denominators, first express each fraction in terms of a common denominator and then subtract the numerators using this common denominator.	$\frac{a}{b} - \frac{c}{d} = \frac{ad}{bd} - \frac{bc}{bd} = \frac{ad - bc}{bd}$ for $b \neq 0$ and $d \neq 0$.	$\frac{3}{5} - \frac{1}{3} = \frac{9}{15} - \frac{5}{15} = \frac{9 - 5}{15} = \frac{4}{15}$

SELF-CHECK 1.3.2 ANSWERS

1. −15 2. 1 3. −6 4. 28

The group exercises at the end of this section contain a graphical perspective on subtracting fractions.

■ EXAMPLE 4   Subtracting Decimals and Fractions

Determine the following differences:

SOLUTIONS _____

(a) $7.7 - 4.05$ $7.7 - 4.05 = 3.65$

(b) $7.7 - (-4.05)$ $7.7 - (-4.05) = 7.7 + 4.05$ Change the operation to addition, and change -4.05 to its additive inverse
$= 11.75$ 4.05. Do not change the sign of 7.7.

(c) $-\dfrac{7}{13} - \dfrac{3}{13}$ $-\dfrac{7}{13} - \dfrac{3}{13} = \dfrac{-7-3}{13}$ $\dfrac{a}{b} - \dfrac{c}{b} = \dfrac{a-c}{b}$ for $b \neq 0$. To subtract fractions with the same denomina-

tor, subtract the numerators and use the common denominator.

$= \dfrac{-7+(-3)}{13}$

$= \dfrac{-10}{13}$

(d) $\dfrac{7}{10} - \dfrac{4}{15}$ $\dfrac{7}{10} - \dfrac{4}{15} = \dfrac{7}{10} \cdot \dfrac{3}{3} - \dfrac{4}{15} \cdot \dfrac{2}{2}$ To subtract fractions with different denominators, first express each fraction
in terms of a common denominator. In this case express each fraction in
terms of the least common denominator (LCD) of 30. Then subtract the
fractions using this LCD.

$= \dfrac{21}{30} - \dfrac{8}{30}$

$= \dfrac{21-8}{30}$

$= \dfrac{13}{30}$

■

SELF-CHECK 1.3.3

Use your calculator to calculate these differences:

1. $-37.4 - 87.8$ 2. $-37.4 - (-87.8)$

3. $\dfrac{5}{8} - \dfrac{7}{9}$ 4. $-\dfrac{3}{4} - \left(-\dfrac{7}{8}\right)$

SELF-CHECK 1.3.3 ANSWERS

1. -125.2 2. 50.4

3. $-\dfrac{11}{72}$ 4. $\dfrac{1}{8}$

Death Valley

Mt. Whitney

There are many words used in the English language that imply either addition or subtraction. Many of these words are antonyms since addition and subtraction are opposites. For example, we sometimes use the word *increase* to indicate addition and the word *decrease* to indicate subtraction. In the next example the word "change" implies that subtraction is the appropriate operation.

■ EXAMPLE 5   Determining Changes in Altitude

Death Valley, the lowest spot in the continental United States, is 282 ft below sea level. Mt. Whitney has an altitude of 14,495 ft above sea level. What is the change in altitude of a hiker who goes from Death Valley to the top of Mt. Whitney?

SOLUTION _____

$14,495 - (-282) = 14,495 + 282$ Change in altitude =
$= 14,777$ Mt. Whitney altitude − Death Valley altitude

Answer: The change in altitude from the bottom of Death Valley to the top of Mt. Whitney is 14,777 ft.

If the hiker then reversed his trip, going from the top of Mt. Whitney to the bottom of Death Valley, the subtraction would be $-282 - 14,495 = -14,777$, a drop in altitude.

CALCULATOR PERSPECTIVE 1.3.1	Comparing the Change in the Sign Key and the Subtraction Key

To subtract one number from another, use the ⊖ key. To form the opposite of a number, use the (-) key.

To enter $14,495 - (-282)$ from Example 5 on a TI-83 Plus calculator, enter the following keystrokes:

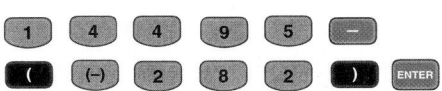

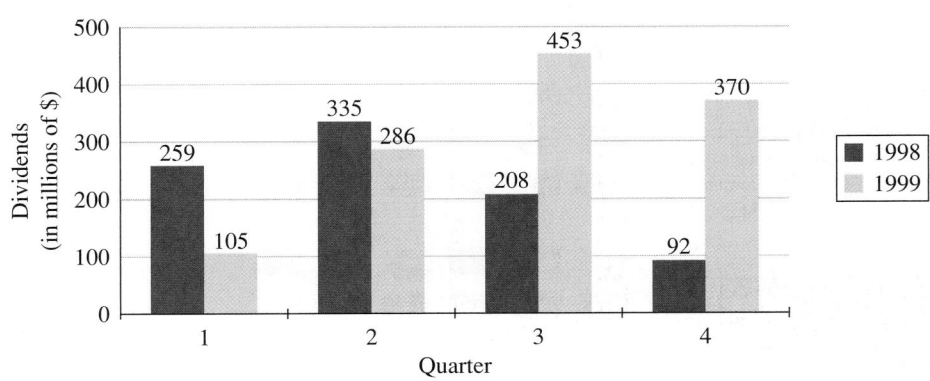

■ EXAMPLE 6   Determining Changes from a Bar Graph

The following bar graph displays the quarterly dividends for both 1998 and 1999 for *Texaco*. Use these values to determine the change in dividends for each quarter from 1998 to 1999.

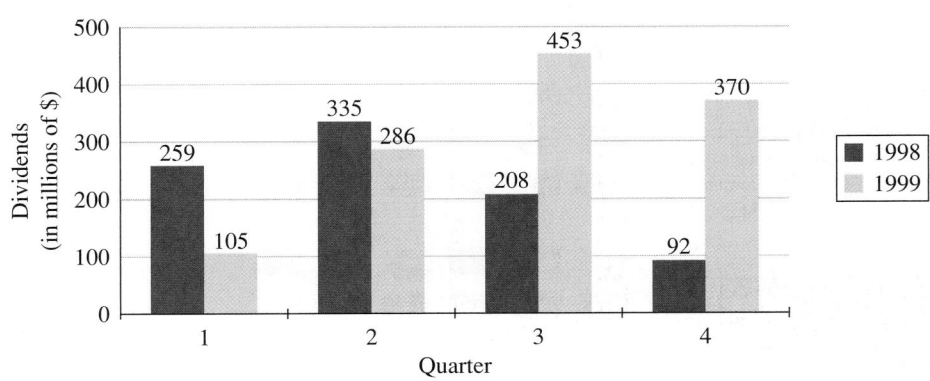

Source: http://www.infoseek.com

SOLUTIONS ─────────────────────────────

First Quarter: $105 - 259 = -154$ Change in dividends = 1999 dividends −
Second Quarter: $286 - 335 = -49$ 1998 dividends.
Third Quarter: $453 - 208 = 245$
Fourth Quarter: $370 - 92 = 278$

The dividends decreased \$154 million in the first quarter.
The dividends decreased \$49 million in the second quarter.
The dividends increased \$245 million in the third quarter.
The dividends increased \$278 million in the fourth quarter.

In the next example we first perform the operations inside the parentheses and then work from left to right.

■ EXAMPLE 7   Evaluating an Algebraic Expression

Evaluate these expressions for $w = -11$, $x = -9$, and $y = -5$.

SOLUTIONS _____

(a) $-w - (x + y)$

$$-w - (x + y) = -(-11) - [(-9) + (-5)]$$
$$= 11 - (-14)$$
$$= 11 + 14$$
$$= 25$$

Substitute for w, x, and y. Then simplify within the brackets before subtracting the result from 11.

(b) $w - (x - y)$

$$w - (x - y) = -11 - [-9 - (-5)]$$
$$= -11 - [-9 + 5]$$
$$= -11 - (-4)$$
$$= -11 + 4$$
$$= -7$$

Substitute for w, x, and y. Then simplify within the brackets before subtracting this result from -11.

CALCULATOR PERSPECTIVE 1.3.2   Storing Values Under Variable Names

To store values under any variable, use the STO and ALPHA keys, followed by any letter you need. Once the values have been stored under each variable, you may enter the expression to be evaluated.

To store $w = -11$, $x = -9$, and $y = -5$ and and to evaluate $-w - (x + y)$ from Example 7(a) on a TI-83 Plus calculator, enter the following keystrokes:

This whole sequence of keystrokes can be accomplished in one command by separating the individual store commands and the expression using the colon.

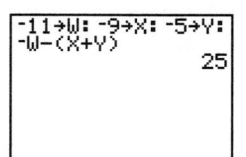

| (−) | 1 | 1 | STO▸ | ALPHA | W | ENTER |

| (−) | 9 | STO▸ | X,T,θ,n | ENTER |

| (−) | 5 | STO▸ | ALPHA | Y | ENTER |

| (−) | ALPHA | W | − | (| X,T,θ,n |

| + | ALPHA | Y |) | ENTER |

```
-11→W
              -11
-9→X
               -9
-5→Y
               -5
-W-(X+Y)
               25
```

Note: Because the variable x is used so frequently, the X,T,θ,n key is provided to enter the variable x with a single keystroke. ■

A basic human characteristic is curiosity and a search for patterns. In everyday language we use the term *sequence* to indicate a succession of events. Mathematically, a **sequence** is an ordered set of numbers with a first number, a second number, a third number, and so on. **Subscript notation** often is used to denote the terms of a sequence:

A subscripted 1 is used to denote the 1st term, and a subscripted 2 is used to denote the 2nd term.

a_1, a_2, and a_n. These terms are read *a sub one, a sub two,* and *a sub n,* respectively. If a sequence follows a predictable pattern, then we may be able to describe this pattern with a formula for a_n. This is illustrated in Example 8.

■ EXAMPLE 8 Calculating the Terms of a Sequence

Use the formula $a_n = n - 3$ to calculate the first five terms of this sequence: $a_1, a_2, a_3, a_4,$ and a_5.

SOLUTION

$a_n = n - 3$
$a_1 = 1 - 3 = -2$ Replace n by 1 to calculate a_1.
$a_2 = 2 - 3 = -1$ Replace n by 2 to calculate a_2.
$a_3 = 3 - 3 = 0$ Replace n by 3 to calculate a_3.
$a_4 = 4 - 3 = 1$ Replace n by 4 to calculate a_4.
$a_5 = 5 - 3 = 2$ Replace n by 5 to calculate a_5.

Answer: The first five terms are $-2, -1, 0, 1,$ and 2.

An equation that contains a variable may or may not be a true statement when a value is substituted for the variable. A value of a variable that makes the equation true is called a **solution** (or sometimes a **root**) of the equation, and this solution is said to **satisfy the equation.** The collection of all solutions is called the **solution set.**

For example, 3 is a solution of $x + 2 = 5$ because $3 + 2 = 5$; 3 satisfies the equation. To substitute a value into the equation to determine whether it satisfies the equation is referred to as a **check** of the value or as **checking a possible solution.**

■ EXAMPLE 9 Checking Possible Solutions of an Equation

Substitute 2 and 3 for x to determine whether either value is a solution of $x + 5 = x + (x + 2)$.

SOLUTIONS

Check $x = 2$:

$x + 5 = x + (x + 2)$ Evaluate each side of the equation for $x = 2$.
$2 + 5 \overset{?}{=} 2 + (2 + 2)$

$7 \overset{?}{=} 2 + 4$ Because the statement $7 = 6$ is false, 2 does not
$7 \overset{?}{=} 6$ does not check.   satisfy the equation.

Answer: 2 is not a solution.

Check $x = 3$:

$x + 5 = x + (x + 2)$ Evaluate each side of the equation for $x = 3$.
$3 + 5 \overset{?}{=} 3 + (3 + 2)$

$8 \overset{?}{=} 3 + 5$ Because the statement $8 = 8$ is true, 3 satisfies the
$8 \overset{?}{=} 8$ checks.   equation.

Answer: 3 is a solution.

A Mathematical Note

The first recorded use of "=" as the equal symbol was by Robert Recorde, an English author, in 1557. His justification for this notation was "No two things coulde be more equalle than two straight lines."

At this point we are not solving equations. We are only checking possible solutions and establishing what it means to be a solution.

We are using the notation $\overset{?}{=}$ as we check to determine whether the two sides of the equation are equal.

SELF-CHECK 1.3.4

1. Calculate the first five terms of the sequence with $a_n = n - 9$.
2. Substitute 5 and 6 for x to determine whether either value is a solution of $(x - 1) - x = x - 6$.

SELF-CHECK 1.3.4 ANSWERS

1. $-8, -7, -6, -5,$ and -4
2. 5 is a solution, and 6 is not a solution.

USING THE LANGUAGE AND SYMBOLISM OF MATHEMATICS   1.3

1. In the subtraction, $11 - 4 = 7$, 11 is called the
_____ , 4 is called the _____, and 7 is
called the _____.

2. The phrase "x decreased by 5" is represented algebraically
by _____.

3. The phrase "17 less than w" is represented algebraically
by _____.

4. The difference $\dfrac{w}{x} - \dfrac{y}{x} =$ _____ for $x \neq$
_____.

5. The change from x_1 to x_2 is represented algebraically by
_____.

6. The change from y_1 to y_2 is represented algebraically by
_____.

7. The notation a_1 is read _____ _____
_____ , and a_n is read _____ _____
_____ .

8. An ordered set of numbers with a first term, a second
term, and so on is called a _____.

9. A value of a variable that makes an equation a true
statement is called a _____ of an equation.

10. A solution of an equation is said to _____ the
equation.

11. To substitute a value into an equation to determine
whether it makes the equation true is referred to as
_____ a possible solution of the equation.

12. The collection of all solutions of an equation is called the
_____ _____ .

EXERCISES   1.3

In Exercises 1–22 calculate each difference without using
your calculator.

1. a. $7 - 13$ **b.** $7 - (-13)$
   c. $-7 - 13$ **d.** $-7 - (-13)$

2. a. $-15 - (-8)$ **b.** $15 - 8$
   c. $-15 - 8$ **d.** $15 - (-8)$

3. a. $-22 - (-15)$ **b.** $22 - 15$
   c. $-22 - 15$ **d.** $22 - (-15)$

4. a. $-11 - (-17)$ **b.** $-11 - 17$
   c. $11 - 17$ **d.** $11 - (-17)$

5. a. $-\dfrac{2}{3} - \left(-\dfrac{2}{3}\right)$ **b.** $-\dfrac{2}{3} - \dfrac{5}{3}$

   c. $\dfrac{2}{3} - \dfrac{3}{4}$ **d.** $-\dfrac{3}{5} - \left(-\dfrac{5}{6}\right)$

6. a. $-\dfrac{3}{5} - \left(-\dfrac{3}{5}\right)$ **b.** $-\dfrac{3}{7} - \dfrac{2}{7}$

   c. $\dfrac{3}{5} - \dfrac{1}{3}$ **d.** $-\dfrac{4}{7} - \left(-\dfrac{1}{2}\right)$

7. a. $-18 - (18 - 21)$ **b.** $(-18 - 18) - 21$
   c. $(-6 - 7) - (9 - 15)$

8. a. $-15 - [17 - (-4)]$ **b.** $(-15 - 17) - (-4)$
   c. $(-5 - 8) - (-4 - 19)$

9. a. $-7 - 8 - 9 - 10$ **b.** $-9 - [-6 - (14 - 23)]$
   c. $-17 + 31 - 7$

10. a. $-6 - 7 - 13 - 14$ **b.** $-4 - [-5 - (-11 - 3)]$
   c. $-4 + 25 - 6$

11. a. $46 - [(17 - 35) - (41 - 56)]$
   b. $-7 + 11 - 15 - 16 + 8 - 4$

12. a. $20 - [(-9 - 2) - (6 - 8)]$
   b. $23 - 16 - 4 + 7 + 5 - 3 + 11$

13. a. $-6 - (-9 + 5) + (-11 + 4)$
   b. $14 - [(9 - 12) - (-5 + 11)]$

14. a. $-7 - (18 - 33) - (-5 - 12)$
   b. $-17 + [(-5 - 7) - (-8 + 3)]$

15. a. $\sqrt{100} - \sqrt{64}$ **b.** $\sqrt{100 - 64}$

16. a. $\sqrt{169} - \sqrt{144}$ **b.** $\sqrt{169 - 144}$

17. a. $\sqrt{169 - 144 - 16}$
   b. $\sqrt{169} - (\sqrt{144} + \sqrt{16})$

18. a. $\sqrt{81 - 64 - 16}$ **b.** $\sqrt{81} - (\sqrt{64} + \sqrt{16})$

19. a. $|-7 - 3| - |-7 - (-3)|$
   b. $-|7 - 3| - |7 - (-3)|$

20. a. $|-9 - 14| - |-9 - (-14)|$
   b. $-|9 - 14| - |9 - (-14)|$

21. a. $43.8 - 43.8$ **b.** $-1.2 - 0 - 1.5$
   c. $-2.3 - (1.5 - 7.6)$

22. a. $27.9 - 27.9$ **b.** $-5.4 - 0 - 2.7$
   c. $-3.2 - (5.1 - 6.7)$

Multiple Representations

In Exercises 23–26 represent the subtraction numerically, verbally, or graphically as described in each problem. (*Hint:* See the following example.)

NUMERICALLY	VERBALLY	GRAPHICALLY

$2 - 5 = 2 + (-5)$ The difference
$\quad\quad = -3$ of two minus
five is negative
three.

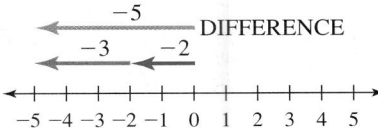

*An arrow 2 units long is drawn to the right, starting at the origin. From the tip of this arrow, draw another arrow 5 units to the left. The difference, -3, is represented by an arrow from the origin to the tip of the last arrow.

23. Represent the subtraction $3 - 1 = 2$ both verbally and graphically.

24. Give the numerical and verbal representations for the subtraction shown here.

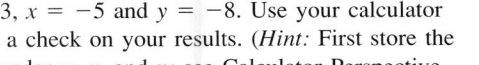

25. Represent the subtraction $-4 - 2 = -6$ both verbally and graphically.

26. Represent the subtraction "The difference of two minus three is negative one" both numerically and graphically.

In Exercises 27–32 evaluate each expression for $w = -3$, $x = -5$ and $y = -8$. Use your calculator only as a check on your results. (*Hint:* First store the values under w, x, and y; see Calculator Perspective 1.3.2.)

27. a. $w + x - y$ **b.** $w - x - y$
   c. $-w - (x - y)$

28. a. $w - (x - y)$ **b.** $-w - (x + y)$
   c. $w - x + y$

29. a. $(w - x) - (w - y)$ **b.** $(w - x) - (w + y)$
   c. $(w - y) - (x - y)$

30. a. $(y - x) + (x - w)$ **b.** $-(w + x + y)$
   c. $-(w - x + y)$

31. a. $|w - x + y|$ **b.** $|w| - |x| + |y|$
   c. $\sqrt{w + x - y}$

32. a. $|w + x - y|$ **b.** $|-w| - |x| - |y|$
   c. $\sqrt{-w - x - y}$

In Exercises 33–36 check both $x = 4$ and $x = 5$ to determine whether either value is a solution of the given equations.

33. a. $x + 7 = 12$ **b.** $-x - 7 = -11$

34. a. $x - 10 = 6$ **b.** $-x + 3 = -2$

35. a. $x - 1 = -(x - 7)$ **b.** $(x - 3) + x = x + 2$

36. a. $x - 2 = -(x - 8)$ **b.** $x - 2 = 6 - x$

In Exercises 37 and 38 calculate the first five terms of each sequence.

37. a. $a_n = n - 4$ **b.** $a_n = 3 - n$

38. a. $a_n = 11 - n$ **b.** $a_n = n - 7$

39. Length of a Fence A fencing company has been contracted to construct a fence on three sides of a rectangular lot. The contractor has 120 ft of chain-link fencing available for this job. If the width of the fenced area is 25 ft, determine the amount of fencing available for the length of this rectangle.

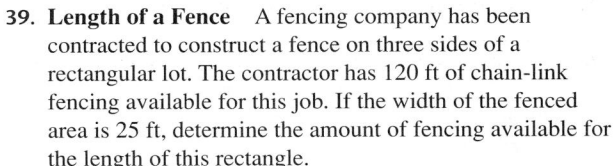

40. Material Wasted in Coin Production The area of a nickel is π square centimeters (cm^2). Two nickels are stamped from a blank sheet with an area of 8 cm^2 as shown in the following figure. Determine the exact area of the metal sheet that is wasted by this stamping process.

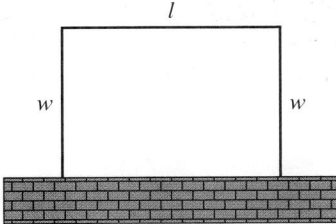

Tuition Bill

The following table gives 2001–2002 tuition comparisons for Austin Community College (ACC) and the University of Texas (UT).

SEMESTER HOURS	TUITION AT ACC ($)	TUITION AT UT ($)
6	192	504
12	384	1,008
18	576	1,512

41. How much can a student save by taking 12 semester hours at ACC rather than at UT?

42. How much can a student save by taking 18 semester hours at ACC rather than at UT?

Change in Sales

43. Determine the change in sales for the second quarter from 1997 to 1998, for Corel Corp.
(Source: http://infoseek.go.com)

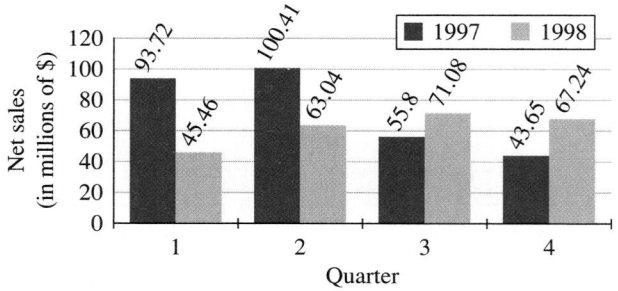

44. Determine the change in sales for the fourth quarter from 1997 to 1998 for Danaher Corp.
(Source: http://infoseek.go.com)

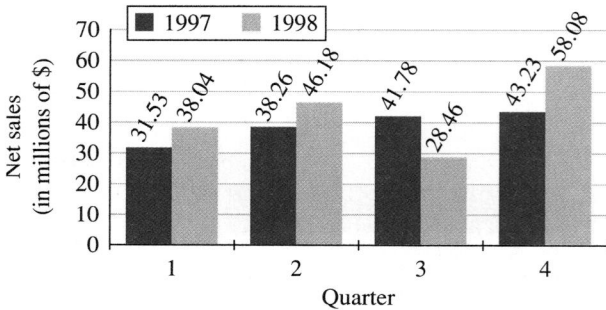

In Exercises 45 and 46 use the results shown in the pie charts to determine the value of the missing percent entry.
(Source: http://www.census.gov)

45.

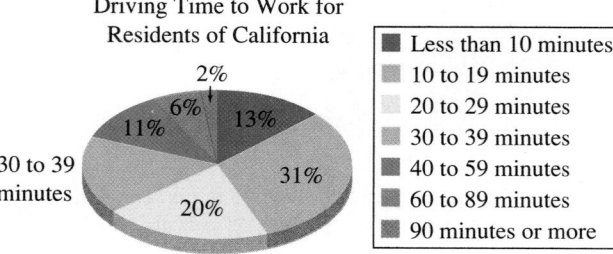

46.

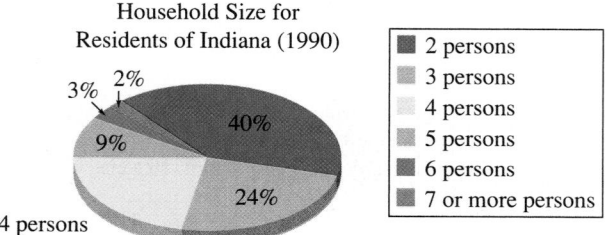

Change in Thermometer Readings

In Exercises 47–48 determine the change in Fahrenheit temperature from 8 A.M. to noon.

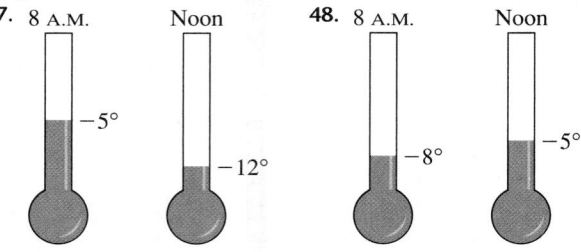

In Exercises 49–54 find each temperature change.

49. **a.** Find the change in temperature from 2° to 8°.
     b. Find the change in temperature from 8° to 2°.
50. **a.** Find the change in temperature from 7° to 3°.
     b. Find the change in temperature from 3° to 7°.
51. **a.** Find the change in temperature from −3° to 6°.
     b. Find the change in temperature from 6° to −3°.
52. **a.** Find the change in temperature from −2° to 3°.
     b. Find the change in temperature from 3° to −2°.
53. **a.** Find the change in temperature from −5° to −9°.
     b. Find the change in temperature from −9° to −5°.
54. **a.** Find the change in temperature from −7° to −1°.
     b. Find the change in temperature from −1° to −7°.

Change in Earnings per Share

In Exercises 55 and 56 complete the last row of the table to determine the change in earnings per share for each company from 1998 to 1999. *Source:* http://infoseek.go.com

55.

EARNINGS PER SHARE FOR BROADCOM CORP.

	QUARTER 1	QUARTER 2	QUARTER 3	QUARTER 4
1999	0.11	0.01	0.14	0.18
1998	0.07	0.05	0.05	0.08
Change				

56.

EARNINGS PER SHARE FOR INTER-TEL INC.

	QUARTER 1	QUARTER 2	QUARTER 3	QUARTER 4
1999	0.20	0.28	0.27	0.30
1998	0.20	−0.37	0.22	0.25
Change				

57. Football Yardage   A football team lost 7 yd on their first play, gained 10 yd on the second play, and gained $2\frac{1}{2}$ yd on the third play. What was the net yardage for the three plays?

58. Age Difference   Two of the greatest mathematicians in history were Archimedes, born in 287 B.C., and Newton, born in 1642 A.D. How many years separate their births?

59. Credit Card Balance   A credit card account had a balance of $143.12 at the start of July. On the July statement there was a credit of $37.16 for items returned. There were also purchases of $14.83, $17.89, and $34.11. The payment made in July was $75. What was the new balance prior to the finance charge?

60. Submarine Depth   A submarine was submerged to a depth of 85 ft. It ascended 40 ft, descended 25 ft, and then descended another 50 ft. What was its new depth?

61. Length of a Steel Rod   The length of a steel rod was 79.4506 m when first measured. As the temperature rose, the bar expanded by 0.0036 m. Then the temperature dropped, and the bar contracted by 0.0052 m. What is the resulting length of the bar?

62. Water Storage   The reservoir for Charleston started the month with 15,000 acre-feet of water. (An acre-foot of water is a volume of water determined by one acre covered to a depth of one foot.) Due to evaporation and usage, the water supply decreased 3000 acre-feet during the month. On the last day of the month the runoff from a 3-inch rain increased the supply by 1500 acre-feet. What is the volume of water in the reservoir at the end of the month?

Estimation and Calculator Skills

In Exercises 63–68 mentally determine the correct sign of the difference and then use your calculator to calculate this difference. Approximate Exercises 67 and 68 to the nearest thousandth. (*Hint:* See Calculator Perspective 1.3.1.)

PROBLEM	SIGN OF DIFFERENCE (WITHOUT CALCULATOR)	DIFFERENCE (WITH CALCULATOR)
63. $-78.43 - 97.5$		
64. $-78.43 - (-97.5)$		
65. $\dfrac{7}{3} - \dfrac{3}{8}$		
66. $-\dfrac{7}{3} - \left(-\dfrac{3}{8}\right)$		
67. $\sqrt{71} - \sqrt{17}$		
68. $-\sqrt{83} - (-\sqrt{77})$		

In Exercises 69–74 mentally estimate the value of each expression and then select the most appropriate answer.

69. $12.9854 - 13.0168$
  a. Less than -1 **b.** Between -1 and 0
  c. Between 0 and 1 **d.** Greater than 1

70. $-0.65438 - 0.360583$
  a. Less than -1 **b.** Between -1 and 0
  c. Between 0 and 1 **d.** Greater than 1

71. $0.666 - \dfrac{2}{3}$
  a. Less than -1 **b.** Between -1 and 0
  c. Between 0 and 1 **d.** Greater than 1

72. $3.14 - \pi$
  a. Less than -1 **b.** Between -1 and 0
  c. Between 0 and 1 **d.** Greater than 1

73. $1 - 17.89 - 82.416$
  a. -100 **b.** 100
  c. -65 **d.** 65

74. $-40.23 - (60.28 - 70.41)$
  a. -50 **b.** -30
  c. -10 **d.** 20

Multiple Representations

In Exercises 75–78 write each verbal statement in algebraic form.

75. a. x decreased by eight equals z.
  b. The change from y to z equals x.

76. a. The difference of x minus y equals seven.
  b. The difference of negative seven minus w equals x plus five.

77. a. The absolute value of x minus the absolute value of the opposite of y equals eleven.
  b. The absolute value of the quantity x minus y equals four.

78. a. The square root of the quantity x minus three is approximately equal to two and nine-tenths.
  b. The square root of x minus the square root of y is less than six.

Multiple Representations

79. If r is the rate in mi/h (miles per hour) of one airplane, write an expression for the rate of an airplane that is traveling 40 mi/h slower.

80. If r is the speed in mi/h of an airplane flying in still air, write an expression for the ground speed of this airplane when it encounters a 20 mi/h wind.

81. An investor placed part of her portfolio in a savings bond and the rest in a certificate of deposit (CD). She invested $2000 less in the CD than in the savings bond. If P is the principal invested in the savings bond, write an expression for the amount invested in the CD.

82. An investor has $10,000 to invest. If A represents the amount invested in stocks, write an expression for the amount remaining for other investments.

83. If n represents an even integer, write an expression for the previous even integer.

84. If n represents an even integer, write an expression for the previous integer.

Group Discussion Questions

85. Writing Mathematically
  a. Explain the difference between the meanings of the symbol "+" in the expressions $-7 - (+4)$ and $-7 + (-4)$.
  b. Explain the difference between the meanings of the symbol "−" in the expressions $8 - 3$ and $8 + (-3)$.
  c. Explain the difference between the ⊟ key and the ⊡ key on a calculator.

86. Writing Mathematically
  a. Make a list of words or phrases that are used often to indicate the operation of addition.
  b. Make a list of words or phrases that are used often to indicate the operation of subtraction.

87. Multiple Representations A fraction represents a part of something. The subtraction of fractions can be represented graphically by removing some of these parts as illustrated by the following two examples:

Example: $\dfrac{5}{7} - \dfrac{3}{7} = \dfrac{2}{7}$

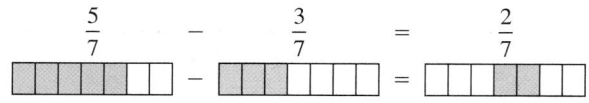

Example: $\dfrac{1}{2} - \dfrac{1}{3} = \dfrac{1}{6}$

$$\frac{3}{6} \qquad - \qquad \frac{2}{6} \qquad = \qquad \frac{1}{6}$$

Using the preceding format illustration, represent the subtraction of the fractions in parts **a–c**:

a. $\dfrac{7}{9} - \dfrac{2}{9}$ **b.** $\dfrac{7}{12} - \dfrac{1}{6}$ **c.** $\dfrac{2}{3} - \dfrac{1}{2}$

88. Error Analysis If you use a TI-83 Plus calculator to evaluate the expression $-7 + (-8 + 5)$ as shown here, this will produce an error. What is the source of the syntax error?

89. Multiple Representations Complete the following flowchart for subtracting two real numbers:

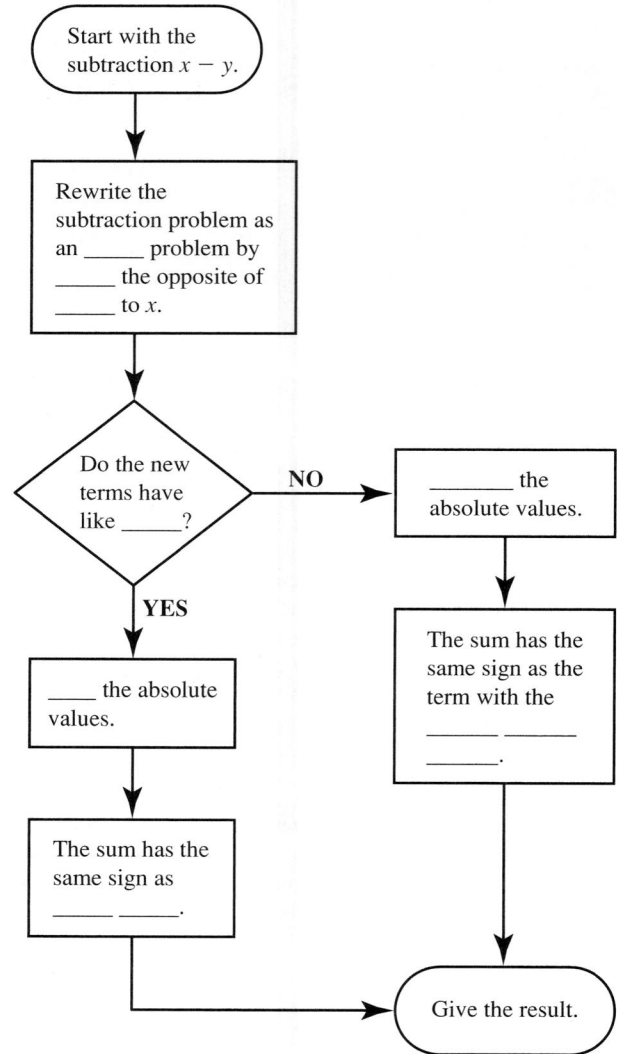

Use this flowchart to find each difference. Have one of your classmates read the steps of the flowchart while another student gives the input values.
  a. $-12 - 5$ **b.** $17 - (-8)$
  c. $-9 - (-13)$ **d.** $11 - 8$

Section 1.4   Multiplication of Real Numbers

Objectives: **12.** Multiply positive and negative real numbers.
13. Use the commutative and associative properties of multiplication.
14. Use algebraic formulas.

A Mathematical Note

The symbol "×" for multiplication was introduced by an English mathematician, William Oughtred (1574–1660).

This section examines the multiplication of signed numbers. Recall the terminology for the multiplication $5 \times 8 = 40$; both 5 and 8 are called **factors** of the **product** 40. The operation of multiplication is seldom represented by the times sign "×" in algebra, in part, due to the possible confusion with the variable x. Instead, we usually use implied multiplication, the dot, or parentheses as shown in the following box. The symbol "*" is used more on calculator displays and in computer programs than in written mathematics.

Notations for the Product of the Factors x and y					
xy	$x \cdot y$	$(x)(y)$	$x(y)$	$(x)y$	$x * y$

Some of the words and phrases used in word problems and applications to indicate multiplication are given in Table 1.4.1.

Table 1.4.1   Phrases Used to Indicate Multiplication

KEY PHRASE	VERBAL EXAMPLE	ALGEBRAIC EXAMPLE
Times	"x times y"	xy
Product	"The product of 5 and 7"	$5 \cdot 7$
Multiplied by	"The rate r is multiplied by the time t"	rt
Twenty percent of	"Twenty percent of x"	$0.20x$
Twice	"Twice y"	$2y$

■ EXAMPLE 1   Evaluating an Algebraic Expression

Evaluate each of these products for $x = 4$, $y = 5$, and $z = 6$.

SOLUTIONS

(a) xy $xy = (4)(5) = 20$ Substitute 4 for x and 5 for y in the implied product of x and y.

(b) yx $yx = (5)(4) = 20$ Compare the product of yx to xy in part (a).

(c) $(xy)(z)$ $(xy)(z) = (4 \cdot 5)(6)$ Substitute 4 for x, 5 for y, and 6 for z in this product of three
$= 20(6)$ factors.
$= 120$

(d) $(x)(yz)$ $x(yz) = (4)(5 \cdot 6)$ Compare this result to the result in part (c).
$= 4(30)$
$= 120$

Using the correct mathematical vocabulary gives you an advantage throughout your coursework. Remember that factors are multiplied and terms are added.

SELF-CHECK 1.4.1

1. Calculate the product of -5 and 17.
2. Calculate 25% of 80.
3. Write an algebraic expression for "four times w."
4. The radius r of a circle is tripled. Write an algebraic expression for the new radius.

Example 1 illustrates that multiplication is both commutative and associative. Together the commutative and associative properties of multiplication allow us to multiply factors in any order we desire.

Properties of Multiplication

	ALGEBRAICALLY	VERBALLY	NUMERICAL EXAMPLE
Commutative property	$ab = ba$	The product of two factors in either order is the same.	$4 \cdot 5 = 5 \cdot 4$
Associative property	$(ab)(c) = (a)(bc)$	Factors can be regrouped without changing the product.	$(4 \cdot 5)(6) = (4)(5 \cdot 6)$

The rules for multiplying signed numbers can be developed by considering the following four cases. Also recall that the product of 0 and any other factor is 0.

Case 1: A Positive Factor Times a Positive Factor

We are already familiar with this case. The product of two positive factors is positive. For example, $3 \cdot 4 = 12$.

Case 2: A Positive Factor Times a Negative Factor

Because multiplication can be interpreted as repeated addition, $3(-4)$ can be interpreted as $(-4) + (-4) + (-4) = -12$. Thus the product of a positive factor and a negative factor is negative.

Case 3: A Negative Factor Times a Positive Factor

Because multiplication is commutative, $(-4)(3)$ equals $3(-4) = -12$. Thus the product of a negative factor and a positive factor is negative.

Case 4: A Negative Factor Times a Negative Factor

Examine the pattern illustrated by the following products:

$$3(-3) = -9$$
$$2(-3) = -6$$
$$1(-3) = -3$$
$$0(-3) = 0$$
$$-1(-3) = ?$$

As the factor to the left decreases by 1, the product to the right increases by 3.

SELF-CHECK 1.4.1 ANSWERS

1. -85 2. 20 3. $4w$
4. $3r$

To continue this pattern, $(-1)(-3) = 3$. This suggests that the product of two negative factors is positive. For example, $(-3)(-4) = 12$.

These four cases can be summarized by examining the product of like signs and unlike signs.

Multiplication of Two Real Numbers

Like signs: Multiply the absolute values of the two factors and use a positive sign for the product.

Unlike signs: Multiply the absolute values of the two factors and use a negative sign for the product.

Zero factor: The product of 0 and any other factor is 0.

Example 2 illustrates the multiplication of two real numbers, including the multiplication of fractions. The procedure for multiplying fractions is summarized in the following box.

Multiplication of Fractions

VERBALLY	ALGEBRAICALLY	NUMERICAL EXAMPLE
To multiply two fractions, multiply the numerators and multiply the denominators.	$\dfrac{a}{b} \cdot \dfrac{c}{d} = \dfrac{ac}{bd}$ for $b \neq 0$ and $d \neq 0$.	$\dfrac{2}{7} \cdot \dfrac{3}{5} = \dfrac{2 \cdot 3}{7 \cdot 5}$ $= \dfrac{6}{35}$

◼ EXAMPLE 2   Multiplying Two Real Numbers

Determine the following products:

SOLUTIONS

(a) $5(-9)$ $5(-9) = -45$ The product of unlike signs is negative.

(b) $-6(7)$ $-6(7) = -42$ The product of unlike signs is negative.

(c) $\left(\dfrac{-2}{3}\right)\left(\dfrac{-4}{5}\right)$ $\left(\dfrac{-2}{3}\right)\left(\dfrac{-4}{5}\right) = \dfrac{(-2)(-4)}{3 \cdot 5}$ $\dfrac{a}{b} \cdot \dfrac{c}{d} = \dfrac{a \cdot c}{b \cdot d}$ for $b \neq 0$ and $d \neq 0$. To multiply two fractions, multiply the numerators and multiply the denominators. The product of like signs is positive.

$= \dfrac{8}{15}$

(d) $\left(\dfrac{-5}{6}\right)\left(\dfrac{7}{11}\right)$ $\left(\dfrac{-5}{6}\right)\left(\dfrac{7}{11}\right) = \dfrac{-5 \cdot 7}{6 \cdot 11}$

$= \dfrac{-35}{66}$ The product of unlike signs is negative.

(e) $(-17)(0)$ $(-17)(0) = 0$ The product of 0 and any other factor is 0.   ◼

SELF-CHECK 1.4.2

Mentally determine the sign of each product and then use a calculator to calculate the product.

1. $(-45.7)(0.13)$

2. $17.6(-4.8)$

3. $\left(-\dfrac{2}{3}\right)\left(-\dfrac{5}{7}\right)$

4. $0\left(-\dfrac{8}{19}\right)$

CALCULATOR PERSPECTIVE 1.4.1	Using the Multiplication Symbol on a Calculator

Two notations for indicating the product of two factors are illustrated in this calculator perspective.

To evaluate $-6(7)$ from Example 2(b) on a TI-83 Plus calculator, enter the following keystrokes:

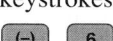  

or

```
-6*7
          -42
-6(7)
          -42
```

Note: In the first expression the "*" symbol indicates multiplication. In the second expression the multiplication is implied.

The product of several nonzero factors is either positive or negative as illustrated in Table 1.4.2 and as summarized in the box following the table.

Table 1.4.2 Sign Pattern for Negative Factors

NUMBER OF NEGATIVE FACTORS	PRODUCT	SIGN OF PRODUCT
2 (even)	$(-1)(-2) = +2$	Positive
3 (odd)	$(-1)(-2)(-3) = -6$	Negative
4 (even)	$(-1)(-2)(-3)(-4) = +24$	Positive
5 (odd)	$(-1)(-2)(-3)(-4)(-5) = -120$	Negative
6 (even)	$(-1)(-2)(-3)(-4)(-5)(-6) = +720$	Positive

When a product is positive, it is customary to eliminate the plus symbol. For example, 2 means $+2$.

The sign pattern of negative factors is reexamined when exponents are covered.

Product of Negative Factors

The product is positive if the number of negative factors is even.

The product is negative if the number of negative factors is odd.

Thus to multiply several signed factors, it may be more efficient to determine the sign of the product first and then to multiply the absolute values of the factors.

▮ **EXAMPLE 3**   Determining Products of Negative Factors

Determine the following products:

SOLUTIONS _____

(a) $(-2)(3)(-5)(-6)$

$(-2)(3)(-5)(-6) = -180$

The product is negative because the number of negative factors is odd (three).

(b) $(-17)(33)(0)(-45)$

$(-17)(33)(0)(-45) = 0$

Since one factor is 0, the product is 0.

(c) $(-8)(-5)(2)(-2)(-10)$

$(-8)(-5)(2)(-2)(-10) = 1600$

The product is positive because the number of negative factors is even (four). The product is written as 1600, which represents +1600. ▮

The following examples illustrate the effect of a factor of -1:

$$-1(-5) = +5 \qquad (-1)(+5) = -5$$

Thus the product of a negative one and a number results in the opposite of that number.

A Factor of -1		
ALGEBRAICALLY	**VERBALLY**	**NUMERICAL EXAMPLE**
For any real number a, $-1 \cdot a = -a$	The product of negative one and any real number is the opposite of that real number.	$-1 \cdot 3 = -3$ and $-1 \cdot (-4) = 4$

We will use this property in the next example to write $-x$ as $(-1)(x)$.

▮ **EXAMPLE 4**   Evaluating Algebraic Expressions

Evaluate these expressions for $x = -3$, $y = -4$, and $z = -5$.

SOLUTIONS _____

(a) $-x$

$$\begin{aligned} -x &= (-1)(x) \\ &= (-1)(-3) \\ &= 3 \end{aligned}$$

Replace $-x$ by $(-1)(x)$ and substitute -3 for x.

(b) $-xyz$

$$\begin{aligned} -xyz &= (-1)(x)(y)(z) \\ &= (-1)(-3)(-4)(-5) \\ &= (3)(-4)(-5) \\ &= (-12)(-5) \\ &= 60 \end{aligned}$$

Substitute -3 for x, -4 for y, and -5 for z. Note that the product of an even number of negative factors is positive. ▮

The **reciprocal** or **multiplicative inverse** of any nonzero real number a can be represented by $\frac{1}{a}$. The product of a number and its multiplicative inverse is positive one.

For example, $3\left(\frac{1}{3}\right) = 1$ and $-\frac{3}{5}\left(-\frac{5}{3}\right) = 1$.

One is called the **multiplicative identity** because 1 is the only real number with the property that $1 \cdot a = a$ and $a \cdot 1 = a$ for every real number a.

Reciprocals or Multiplicative Inverses

ALGEBRAICALLY	VERBALLY	NUMERICAL EXAMPLE
For any real number $a \neq 0$, $a \cdot \dfrac{1}{a} = 1$.	For any real number a other than zero, the product of the number a and its multiplicative inverse $\dfrac{1}{a}$ is one. Zero has no multiplicative inverse.	$-4 \cdot \left(-\dfrac{1}{4}\right) = 1$, and $\dfrac{4}{3} \cdot \dfrac{3}{4} = 1$

The product of zero and any real number is always zero; this product can never be one. Thus zero has no multiplicative inverse.

■ EXAMPLE 5   Determining Multiplicative Inverses

Determine the multiplicative inverse of each number and then multiply each number by its multiplicative inverse.

SOLUTIONS

(a) $\dfrac{2}{7}$ $\left(\dfrac{2}{7}\right)\left(\dfrac{7}{2}\right) = 1$ The reciprocal or multiplicative inverse of $\dfrac{2}{7}$ is $\dfrac{7}{2}$.

(b) $-\dfrac{1}{5}$ $\left(-\dfrac{1}{5}\right)(-5) = 1$

(c) y (for $y \neq 0$) $y\left(\dfrac{1}{y}\right) = 1$

(d) 0 0 has no multiplicative inverse. ■

SELF-CHECK 1.4.3
Write the multiplicative inverse of each number.

1. 8   2. $-\dfrac{3}{8}$ 3. 1   4. -1

SELF-CHECK 1.4.3 ANSWERS

1. $\dfrac{1}{8}$ 2. $-\dfrac{8}{3}$ 3. 1   4. -1

CALCULATOR PERSPECTIVE 1.4.2	Finding the Multiplicative Inverse

The multiplicative inverse of a number can be found by pressing the [x^{-1}] key after the variable or number. To determine the multiplicative inverse of $-\dfrac{1}{5}$ from Example 5(b) on a TI-83 Plus calculator, enter the following keystrokes:

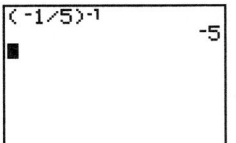

Note: The notation x^{-1} will be discussed in Section 5.3.

■ EXAMPLE 6   Checking a Possible Solution of an Equation

Check -5 to determine whether it is a solution of $-3x = 10 - x$.

SOLUTION _____

$$-3x = 10 - x \qquad \text{Evaluate each side of the equation for } x = -5.$$
$$-3(-5) \overset{?}{=} 10 - (-5)$$
$$15 \overset{?}{=} 10 + 5$$
$$15 \overset{?}{=} 15 \text{ checks.} \qquad \text{Because this statement is true, } -5 \text{ satisfies the equation.}$$

Answer: -5 is a solution of $-3x = 10 - x$ ■

One important use of equations is to state formulas. A **formula** is an equation that states the relationship among different quantities. For example, the formula $A = l \cdot w$ relates the area of a rectangle, A, to its length, l, and its width, w. Each variable used in a formula has a selected meaning and represents every possible value to which it applies. It is helpful to use representative letters for variables, such as V for volume, P for perimeter, r for radius, and t for time. A list of common formulas, including many formulas from geometry, is given on the inside back cover of this book.

■ EXAMPLE 7   Calculating the Interest on an Investment

Use the formula $I = PRT$ to calculate the interest on an investment of $5000 at a rate of 8% per year for one year.

SOLUTION _____

$$I = PRT \qquad \text{Substitute into the interest formula \$5000}$$
$$I = (5000)(0.08)(1) \qquad \text{for the principal } P, 0.08 \text{ for the interest rate}$$
$$I = 400 \qquad R \text{ of 8\%, and 1 for the time } T \text{ of one year.}$$

Answer: The interest on this loan for one year is $400. ■

The formula for a sequence uses subscript notation and matches each input value of n with an output value a_n that is calculated using the formula.

■ EXAMPLE 8   Calculating the Terms of a Sequence

Use the formula $a_n = -5n$ to calculate the first five terms of this sequence.

SOLUTION _____

$a_n = -5n$
$a_1 = -5(1) = -5$ The input is 1, the output is -5.
$a_2 = -5(2) = -10$ The input is 2, the output is -10.
$a_3 = -5(3) = -15$ The input is 3, the output is -15.
$a_4 = -5(4) = -20$ The input is 4, the output is -20.
$a_5 = -5(5) = -25$ The input is 5, the output is -25.

Answer: The first five terms are $-5, -10, -15, -20$, and -25.

SELF-CHECK 1.4.4

Use a calculator to answer these exercises.
1. The multiplicative inverse of -100.
2. The multiplicative inverse of 0.
3. Use the formula $I = PRT$ to determine the interest on an investment of $13,500 at 8.5% for one year.
4. Determine the first five terms of the sequence defined by $a_n = -6n$.

SELF-CHECK 1.4.4 ANSWERS
1. -0.01 or $-\dfrac{1}{100}$
2. An error message; 0 has no multiplicative inverse.
3. $1147.50
4. $-6, -12, -18, -24$, and -30

Note that there is an implied multiplication when two values are separated by parentheses.

■ EXAMPLE 9   Estimating a Product

Estimate $-25.32(-3.89)$ and then use a calculator to perform this multiplication.

SOLUTION _____

ESTIMATED ANSWER

$-25.32 \approx -25$ -25
$-3.89 \approx -4$ $\times$ $\underline{(-4)}$
                         $+100$

CALCULATOR VALUE

```
-25.32(-3.89)
          98.4948
■
```

Answer: 98.4948 seems reasonable based on the estimated value of 100.

USING THE LANGUAGE AND SYMBOLISM OF MATHEMATICS   **1.4**

1. In the multiplication $9 \cdot 11 = 99$, 9 and 11 are called _____ and 99 is called the _____.
2. The phrase "The product of P times R times T" is represented algebraically by _____.
3. The phrase "Fourteen percent of C" is represented algebraically by _____.
4. The property that says $ab = ba$ for all real numbers is called the _____ property of multiplication.

5. The property that says $(ab)(c) = (a)(bc)$ for all real numbers is called the _____ property of multiplication.

6. If either x or y is zero, the product $xy =$ _____.

7. If x and y have the same sign, then the sign of the product xy is _____.

8. If x and y have unlike signs, then the sign of the product xy is _____.

9. If 12 * 5 is shown on a calculator display, then the result will be _____.

10. The product $\dfrac{w}{x} \cdot \dfrac{y}{z} =$ _____ for $x \neq$ _____ and $z \neq$ _____.

11. The real numbers $-\dfrac{4}{7}$ and $-\dfrac{7}{4}$ are _____ or multiplicative _____ of each other.

12. The product of a number and its reciprocal is _____.

13. The only real number that does not have a reciprocal is _____.

14. The multiplicative identity is _____.

EXERCISES 1.4

In Exercises 1–16 calculate each product without using your calculator.

1. a. $7(-11)$ **b.** $-7(11)$
 c. $(-7)(-11)$ **d.** $-(-7)(-11)$

2. a. $-5(12)$ **b.** $-(5)(12)$
 c. $(5)(-12)$ **d.** $(-5)(-12)$

3. a. $2(-3)(10)$ **b.** $2(-3)(-10)$
 c. $(-2)(-3)(-10)$ **d.** $-(-2)(-3)(-10)$

4. a. $2(5)(-20)$ **b.** $2(-5)(-20)$
 c. $(-2)(-5)(-20)$ **d.** $-(-2)(-5)(-20)$

5. a. $(-1)(-1)(-1)(-8)$ **b.** $(-9)(-17)(0)(83)$
 c. $(-1)(2)(-3)(5)$ **d.** $(-10)(-10)(-10)$

6. a. $(-1)(1)(-1)(-8)$ **b.** $(12)(-43)(0)(-97)$
 c. $(-1)(-2)(3)(-5)$ **d.** $(-10)(10)(-10)$

7. a. $-0.1(1234)$ **b.** $-100(1234)$
 c. $-0.001(1234)$ **d.** $-1000(1234)$

8. a. $-0.01(-45.96)$ **b.** $-10(-45.96)$
 c. $-100(45.96)$ **d.** $-0.001(45.96)$

9. a. $\dfrac{1}{2} \cdot \dfrac{3}{5}$ **b.** $-\dfrac{1}{3} \cdot \dfrac{2}{5}$
 c. $\left(-\dfrac{5}{11}\right)\left(-\dfrac{3}{8}\right)$ **d.** $-\left(-\dfrac{5}{3}\right)\left(\dfrac{2}{13}\right)$

10. a. $\dfrac{1}{3} \cdot \dfrac{4}{5}$ **b.** $-\dfrac{1}{2} \cdot \dfrac{3}{7}$
 c. $\left(-\dfrac{3}{4}\right)\left(-\dfrac{5}{7}\right)$ **d.** $-\left(-\dfrac{4}{5}\right)\left(-\dfrac{3}{13}\right)$

11. a. $|-5||-4|$ **b.** $|(-5)(-4)|$
 c. $\sqrt{4}\sqrt{9}$ **d.** $\sqrt{(4)(9)}$

12. a. $|-3||-2||-1|$ **b.** $|(-3)(-2)(-1)|$
 c. $\sqrt{(4)(25)}$ **d.** $\sqrt{4}\sqrt{25}$

13. a. $\left(\dfrac{1}{5}\right)(5)$ **b.** $(5)\left(\dfrac{1}{5}\right)$
 c. $(-5)\left(-\dfrac{1}{5}\right)$ **d.** $1(-5)$

14. a. $\left(\dfrac{2}{3}\right)\left(\dfrac{3}{2}\right)$ **b.** $\left(\dfrac{3}{2}\right)\left(\dfrac{2}{3}\right)$
 c. $\left(-\dfrac{3}{2}\right)\left(-\dfrac{2}{3}\right)$ **d.** $\left(-\dfrac{3}{2}\right)1$

15. a. $\left(\dfrac{1}{6.3}\right)(6.3)$ **b.** $\left(-\dfrac{1}{6.3}\right)(-6.3)$
 c. $\left(-\dfrac{1}{1.7}\right)(1.7)$ **d.** $(1.7)1$

16. a. $(5.9)\left(\dfrac{1}{5.9}\right)$ **b.** $\left(-\dfrac{1}{5.9}\right)(-5.9)$
 c. $\left(-\dfrac{1}{5.9}\right)(5.9)$ **d.** $1(5.9)$

In Exercises 17–20 evaluate each expression for $w = -2$, $x = -5$, and $y = -30$.

17. a. $(wx)(y)$ **b.** $w(xy)$ **c.** $(-wx)(y)$
18. a. $(wx)(-y)$ **b.** $w[(x)(-y)]$ **c.** $-(wx)(-y)$
19. a. $-3w(xy)$ **b.** $4w(xy)$ **c.** $(-2w)(3x)(-y)$
20. a. $-10w(xy)$ **b.** $(-10wx)(-y)$ **c.** $(-5w)(-2x)(-y)$

In Exercises 21–24 check both $x = -2$ and $x = 2$ to determine whether either value is a solution of these equations.

21. a. $-4x = x + 10$ **b.** $-5x = x - 12$
22. a. $-x = x - 4$ **b.** $-x = x + 4$
23. a. $xx = 4$ **b.** $-xx = 10$
24. a. $xx = 9$ **b.** $-xx = -4$

In Exercises 25 and 26 calculate the first five terms of each sequence.

25. a. $a_n = 2n$ **b.** $a_n = n \cdot n$
26. a. $a_n = 3n$ **b.** $a_n = n \cdot n \cdot n$

Comparing Addition and Multiplication
In Exercises 27–32 first compute the sum of x and y and then compute the product of x and y.

	x	y	$x + y$	xy
Example	3	4	7	12
27.	5	8		
28.	-5	8		
29.	5	-8		
30.	-5	-8		
31.	0	π		
32.	π	-1		

Multiple Representations
In Exercises 33 and 34 write each verbal statement in algebraic form.

33. a. The product of eight and x.
   b. The product of seven and y equals eleven.
34. a. Twice the value of y.
   b. The value of x is twice the value of y.

Multiple Representations
35. If P is the initial value of an investment that triples in value after 10 years, write an expression for the new value of this investment.
36. A board is cut into two pieces so that the longer piece is 4 times as long as the shorter piece. If s is the length of the shorter piece, write an expression for the length of the longer piece.

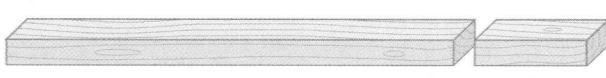

s

In Exercises 37–42 fill in each blank.
37. a. The result of the addition of a and b is called their _____.
   b. The result of the subtraction "a minus b" is called their _____.
   c. The result of the multiplication of a times b is called their _____.
38. a. The sum of a number and its additive inverse is _____.
   b. The product of a number and its multiplicative inverse is _____.
39. a. The property that says $5 \cdot (7 \cdot 9) = 5 \cdot (9 \cdot 7)$ is the _____ property of multiplication.
   b. The property that says $5 \cdot (7 \cdot 9) = (5 \cdot 7) \cdot 9$ is the _____ property of multiplication.
40. a. The property that says $wx + yz = yz + wx$ is the commutative property of _____.
   b. The property that says $wx + yz = wx + zy$ is the commutative property of _____.
41. a. The property that says $w + x(yz) = w + (xy)(z)$ is the associative property of _____.
   b. The property that says $w + (x + yz) = (w + x) + yz$ is the associative property of _____.

42. _____ is the only real number that does not have a multiplicative inverse.

Area of a Triangle
In Exercises 43–46 use the formula $A = \dfrac{1}{2} bh$ to calculate the area of each triangle.

43.
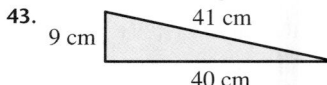
9 cm   41 cm   40 cm

44.
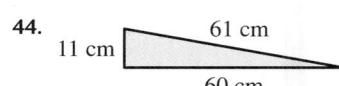
11 cm   61 cm   60 cm

45.
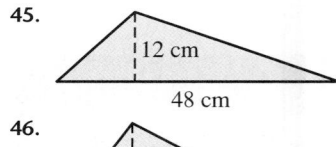
12 cm   48 cm

46.
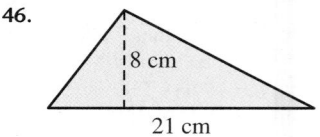
8 cm   21 cm

Volume of a Box
In Exercises 47 and 48 use the formula $V = l \cdot w \cdot h$ to calculate the volume of each object.

47. The mini-refrigerator shown here has interior dimensions of 14 in by 15 in by 16 in. Determine the interior volume of this refrigerator in cubic inches (in^3).

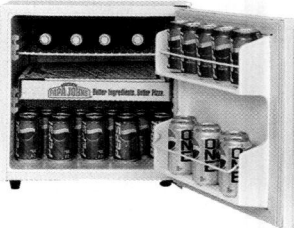

48. A box that contains 10 reams of printer paper has dimensions of 9 in by 11.5 in by 18 in. Determine the interior volume of this box in cubic inches (in^3).

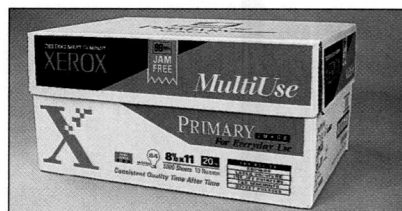

49. Perimeter of a Hexagon   Each side of the regular hexagon (six equal sides) shown here is 2.47 cm long. Determine the perimeter of this hexagon

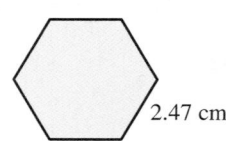

2.47 cm

50. Perimeter of a Pentagon
Each side of the regular pentagon (five equal sides) shown is 11.56 cm long. Determine the perimeter of this pentagon.

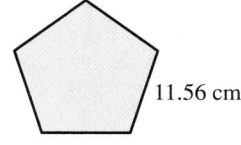

11.56 cm

Simple Interest
In Exercises 51–54 use a calculator and the formula $I = P \cdot R \cdot T$ to calculate the interest for each investment.

51. An investment of $4000 at a rate of 7% for one year.
52. An investment of $7000 at a rate of 8% for one year.
53. An investment of $1250 at a rate of 7.25% for one year.
54. An investment of $3725 at a rate of 8.5% for one year.

Active Ingredient
The amount of the active ingredient in a solution can be calculated by multiplying the percent of concentration of the active ingredient times the volume of the solution. In 4 gallons (gal) of 65% antifreeze solution there would be (0.65)(4) or 2.6 gal of pure antifreeze and the rest is presumed to be water. In Exercises 55–58 use a calculator to calculate the amount of the active ingredient in each solution.

55. 500 milliliters of a 15% juice solution
56. 200 gal of a 3.8% insecticide solution
57. 3 liters of a 22.5% hydrochloric acid solution
58. 1.5 gal of a 70% antifreeze solution

Distance Traveled
The distance an object travels can be calculated by finding the product of its rate and the time it travels. In Exercises 59–62, use the formula $D = R \cdot T$ to compute the distance each object travels.

59. A plane flies 420 mi/h for 3 h.
60. An explorer walks 3 days and averages 15 mi per day.
61. An ant crawls 2 m per minute for 6.5 min.
62. A spacecraft flies 6200 ft per second for 120 sec.

Circumference of a Circle
In Exercises 63 and 64 use the formula $C = 2\pi r$ to approximate the circumference of each circle to the nearest tenth of a centimeter.

63.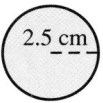
2.5 cm

64.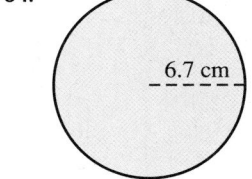
6.7 cm

65. Mass of Four Nickels   A nickel is 2 cm in diameter and has a mass of 5 grams. Four nickels are stamped from a

rectangular sheet as shown here. Determine the total mass of the four nickels.

66. Area of a Nickel Blank   A nickel is 2 cm in diameter and has a mass of 5 grams. Four nickels are stamped from a rectangular sheet as previously shown. Determine the area of this rectangular sheet.

67. Sales Tax on a Television   Determine the Texas sales tax on a television set that costs $499.99. The sales tax rate in Texas is 6.25%.

68. Sales Tax on a Microwave Oven   Determine the Minnesota sales tax on a microwave oven that costs $229.99. The sales tax rate in Minnesota is 6.5%.

69. Weekly Wages   A student worked 52 hours during one week. If she earned $5.25 per hour for the first 40 hours and $7.88 per hour for overtime, how much did she earn that week?

70. Cost of Bolts   The wholesale cost of a bolt was reduced from $0.027 to $0.023. How much will be saved if 1000 bolts are bought at the lower price rather than at the higher price?

71. Basketball Ticket Payments   The following table is a partially completed order form for college basketball tickets. Complete the last column of this table by calculating the cost for each game. Then determine the total cost for all these tickets.

GAME	NUMBER OF TICKETS	PRICE PER TICKET ($)	COST ($)
1	4	16	
2	2	16	
3	3	25	
4	6	25	

72. Concert Ticket Payments   The following table is a partially completed order form for concert tickets. Complete the last column of this table by calculating the cost for each concert and then determining the total cost for all these tickets.

CONCERT	NUMBER OF TICKETS	PRICE PER TICKET ($)	COST ($)
1	5	40	
2	3	35	
3	2	50	
4	1	80	

73. **Income from Sales** The first bar graph gives the sales price per item for each quarter for a business. The second bar graph gives the number of items sold each quarter. Calculate the sales revenue for this item for each quarter.

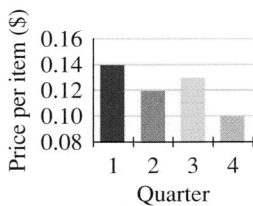 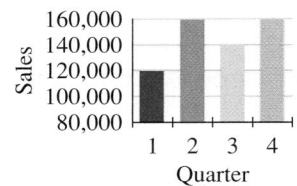

74. **Profit from Sales** The first bar graph gives the profit per item for each quarter for a business. The second bar graph gives the number of items sold each quarter. Calculate the sales profit for this item for each quarter.

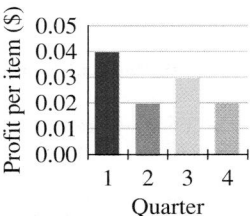 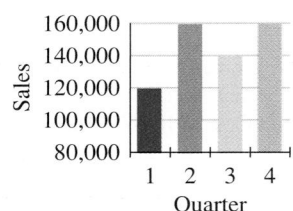

Estimation and Calculator Skills

In Exercises 75–80 mentally determine the correct sign of the product and then use a calculator to calculate the product. Approximate the product in Exercises 79 and 80 to the nearest thousandth. (*Hint:* See Calculator Perspective 1.4.1.)

PROBLEM	SIGN OF PRODUCT (WITHOUT CALCULATOR)	PRODUCT (WITH CALCULATOR)
75. $-8.7(4.5)(-3.2)$		
76. $-5.6(-7.8)(-9.3)$		
77. $\left(-\dfrac{4}{5}\right)\left(-\dfrac{7}{13}\right)$		
78. $\left(-\dfrac{3}{5}\right)\left(\dfrac{4}{7}\right)\left(-\dfrac{6}{11}\right)$		
79. $\left(-\sqrt{2}\right)\left(-\sqrt{3}\right)\left(-\sqrt{5}\right)$		
80. $\sqrt{7}\left(-\sqrt{11}\right)\left(-\sqrt{13}\right)$		

Estimation Skills

In Exercises 81–84 mentally estimate the value of each expression and then select the most appropriate answer.

81. $(-9.9)(15.12)$
   - **a.** -149.7 **b.** -249.12 **c.** 0
   - **d.** 149.7 **e.** 249.12

82. $(-159.63)(-5.1)$
   - **a.** -814.113 **b.** -514.113 **c.** 51.413
   - **d.** 81.4113 **e.** 814.113

83. $(93.45)(187.8)(0)(-35.3)$
   - **a.** $-620,000$ **b.** $-260,000$ **c.** 0
   - **d.** $260,000$ **e.** $620,000$

84. $(-0.09)(-311.83)$
   - **a.** -28.0647 **b.** -58.0647 **c.** 0
   - **d.** 28.0647 **e.** 58.0647

Group Discussion Questions

85. **Risks and Choices** Being a good citizen involves making personal decisions based on facts. Consider the following situation, which involves alcohol consumption.

   Three workers each had a drink after work. One drank a 12-ounce (oz) bottle of beer that was 5% alcohol; the second had a 4-oz glass of wine that was 15% alcohol; and the third had a 1.5-oz shot that was 40% alcohol. Which worker consumed the most alcohol?

86. **Writing Mathematically** The cost of one medium soft drink at a fast-food service is $1.13. However, if you buy two at the same time, the cost is $2.27. Can you give a mathematical explanation for this?

87. **Writing Mathematically** Discuss the reason that the "*" symbol is used to denote multiplication on computers rather than using the symbols "×" and "·", which were already common conventions for multiplication.

88. **Multiple Representations** Create a flowchart for multiplying two real numbers. (*Hint:* See the group discussion questions in Sections 1.2 and 1.3.)

Section 1.5 Division of Real Numbers

Objectives: **15.** Divide positive and negative real numbers.
   16. Express ratios in lowest terms.

This section examines the division of signed numbers. Recall the terminology for the division $16 \div 8 = 2$; 16 is called the **dividend,** 8 is called the **divisor,** and 2 is called the **quotient.**

The notation x/y can be easily misinterpreted, especially when you enter expressions such as $\dfrac{x}{y+z}$ on a calculator. We will consider this again in Section 1.6 when the order of operations is covered.

Notations for the Quotient of x Divided by y for $y \neq 0$			
$x \div y$	$\dfrac{x}{y}$	x/y	$x : y$

Some of the words and phrases used in word problems and applications to indicate division are given in Table 1.5.1.

Caution: x divided by y is written as $\dfrac{x}{y}$, but x divided into y is written as $\dfrac{y}{x}$.

Table 1.5.1   **Phrases Used to Indicate Division**

KEY PHRASE	VERBAL EXAMPLE	ALGEBRAIC EXAMPLE
Divided by	"x divided by y"	$\dfrac{x}{y}$
Quotient	"The quotient of 5 and 3"	$5 \div 3$
Ratio	"The ratio of x to 2"	$x : 2$

A Mathematical Note

The symbol "$\div$" for division is an imitation of fractional division; the dots indicate the numerator and the denominator of a fraction. This symbol was invented by an English mathematician John Pell (1611–1685).

If two people agree to share a \$50 restaurant bill equally, we can divide the bill by 2; \$50 $\div$ 2 = \$25 for each person. Equivalently, we could say that each person gets one-half the bill; $\dfrac{1}{2}(\$50) = \$25$ for each person. Dividing by 2 produces the same result as multiplying by $\dfrac{1}{2}$.

The division $50 \div 2 = 25$ checks because $2 \cdot 25 = 50$. We can say that division is the inverse of multiplication, and we can define division as the multiplication of a multiplicative inverse.

Definition of Division		
ALGEBRAICALLY	**VERBALLY**	**NUMERICAL EXAMPLE**
For any real numbers x and y with $y \neq 0$ $$x \div y = x\left(\dfrac{1}{y}\right).$$	Dividing two real numbers is the same as multiplying the first number by the multiplicative inverse of the second number.	$3 \div 2 = 3\left(\dfrac{1}{2}\right)$

Since division is defined in terms of multiplication, the rules for dividing signed numbers are derived from the rules for multiplying signed numbers.

Division of Two Real Numbers

Like signs: Divide the absolute values of the two numbers and use a positive sign for the quotient.

Unlike signs: Divide the absolute values of the two numbers and use a negative sign for the quotient.

Zero dividend: $\dfrac{0}{x} = 0$ for $x \neq 0$.

Zero divisor: $\dfrac{x}{0}$ is undefined for every real number x.

Example 1 illustrates the division of two real numbers, including the division of fractions. The procedure for dividing fractions is summarized in the following box.

The reciprocal or the multiplicative inverse of $\dfrac{c}{d}$ is $\dfrac{d}{c}$ for c and d not equal to zero.

Division of Fractions

VERBALLY	ALGEBRAICALLY	NUMERICAL EXAMPLE
To divide two fractions, multiply the first fraction by the reciprocal of the second fraction.	$\dfrac{a}{b} \div \dfrac{c}{d} = \dfrac{a}{b} \cdot \dfrac{d}{c}$ $= \dfrac{ad}{bc}$ for $b \neq 0$, $c \neq 0$, and $d \neq 0$.	$\dfrac{3}{4} \div \dfrac{4}{5} = \dfrac{3}{4} \cdot \dfrac{5}{4}$ $= \dfrac{15}{16}$

■ EXAMPLE 1   Dividing Real Numbers

Determine the following quotients:

SOLUTIONS _____

(a) $21 \div (-7)$

(b) $-36 \div (+4)$

(c) $-45 \div (-9)$

$21 \div (-7) = -3$ The quotient of unlike signs is negative.

$-36 \div (+4) = -9$ The quotient of unlike signs is negative.

$-45 \div (-9) = 5$ The quotient of like signs is positive.

(d) $\dfrac{1}{2} \div \dfrac{1}{3}$

$\dfrac{1}{2} \div \dfrac{1}{3} = \dfrac{1}{2} \times \dfrac{3}{1}$

$= \dfrac{3}{2}$

$\dfrac{a}{b} \div \dfrac{c}{d} = \dfrac{a}{b} \cdot \dfrac{d}{c} = \dfrac{a \cdot d}{b \cdot c}$ for $b \neq 0$, $c \neq 0$, and $d \neq 0$.

To divide by $\dfrac{1}{3}$, multiply by its multiplicative inverse $\dfrac{3}{1}$.

The product of like signs is positive.

(e) $\dfrac{-3}{5} \div \dfrac{4}{7}$

$\dfrac{-3}{5} \div \dfrac{4}{7} = \dfrac{-3}{5} \cdot \dfrac{7}{4}$ To divide by $\dfrac{4}{7}$, multiply by its multiplicative inverse $\dfrac{7}{4}$.

$= \dfrac{-3 \cdot 7}{5 \cdot 4}$

$= \dfrac{-21}{20}$ The product of unlike signs is negative.

SELF-CHECK 1.5.1

1. Calculate the quotient of 46 and -23
2. Write the ratio of 11 to 17 in fractional form.
3. Write an equation for the ratio C to d is π.
4. Calculate the quotient $\left(\dfrac{2}{3}\right) \div \left(-\dfrac{7}{11}\right)$.

Fractions are generally expressed in **reduced form;** a form where the numerator and denominator have no common factor other than -1 or 1. To multiply fractions, it is usually easier to divide both the numerator and the denominator first by any reducing factor and then to simplify the numerator and the denominator.

■ EXAMPLE 2   Simplifying Fractions

Express each fraction in reduced form.

SOLUTIONS _____

(a) $\dfrac{30}{66}$

$\dfrac{30}{66} = \dfrac{5 \cdot \overset{1}{6}}{6 \cdot 11}$ For $b \ne 0$ and $c \ne 0$, $\dfrac{ac}{bc} = \dfrac{a}{b}$ with c the common reducing factor.

$= \dfrac{5}{11}$ In this example divide both the numerator and the denominator by the reducing factor 6.

(b) $\dfrac{-33}{35} \cdot \dfrac{49}{22}$

$\dfrac{-33}{35} \cdot \dfrac{49}{22} = \dfrac{-3 \cdot \overset{1}{11}}{5 \cdot 7} \cdot \dfrac{\overset{1}{7} \cdot 7}{2 \cdot 11}$ Factor the numerator and the denominator and divide them by their common factors, 7 and 11.

$= \dfrac{-3 \cdot 7}{5 \cdot 2}$

$= \dfrac{-21}{10}$ The product of unlike signs is negative.

(c) $\dfrac{27}{50} \div \dfrac{12}{10}$

$\dfrac{27}{50} \div \dfrac{12}{10} = \dfrac{\overset{9}{27}}{\underset{5}{50}} \cdot \dfrac{\overset{1}{10}}{\underset{4}{12}}$ Invert the divisor and multiply. The reducing factors are 3 and 10.

$= \dfrac{9}{20}$

Division by Zero

The division $50 \div 2 = 25$ checks because $2 \cdot 25 = 50$. Likewise, if $3 \div 0 = a$, then $0 \cdot a = 3$ must check. However, this is impossible because $0 \cdot a = 0$ for all values of a. Thus this division by zero is undefined. Next we examine $0 \div 0 = b$, which will check if $0 \cdot b = 0$. However, this is true for all real numbers b. We say that $\dfrac{0}{0}$ is **indeterminate** since there is no reason to select or determine one value of b as preferable to any other value of b. Thus this division by zero also is undefined.

■ **EXAMPLE 3** Dividing Expressions Involving Zero

Determine the following quotients:

SOLUTIONS

(a) $\dfrac{10}{5}$ $\dfrac{10}{5} = 2$ Check: $2 \cdot 5 = 10$.

(b) $\dfrac{0}{5}$ $\dfrac{0}{5} = 0$ Check: $0 \cdot 5 = 0$.

(c) $\dfrac{5}{0}$ $\dfrac{5}{0}$ is undefined Division by zero is undefined. There is no value to multiply times 0 that will produce 5.

(d) $\dfrac{0}{0}$ $\dfrac{0}{0}$ is undefined Division by zero is undefined. All values multiplied times 0 will yield 0, thus there is no unique quotient. ■

| CALCULATOR PERSPECTIVE 1.5.1 | Using the Division Symbol on a Calculator |

To evaluate $\dfrac{10}{5}, \dfrac{0}{5}$, and $\dfrac{5}{0}$ from Example 3 on a TI-83 Plus calculator, enter the following keystrokes:

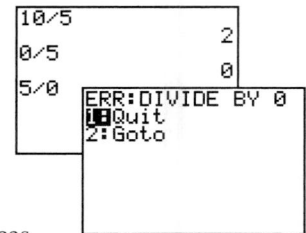

Note: Entering ÷ on a calculator produces a slash symbol on the calculator screen.
Dividing any real number by zero is undefined. ■

SELF-CHECK 1.5.2

Mentally determine the sign of each quotient and then use a calculator to calculate the quotient.

1. $-8.05 \div 2.3$ **2.** $0.728 \div (-0.014)$ **3.** $\left(-\dfrac{2}{3}\right) \div \left(-\dfrac{5}{7}\right)$ **4.** $-\dfrac{3}{7} \div 0$

Since $1 \div (-2) = -\dfrac{1}{2}$ and $(-1) \div 2 = -\dfrac{1}{2}$, we can write $\dfrac{1}{-2} = \dfrac{-1}{2} = -\dfrac{1}{2}$. The relationship of the sign of the numerator, the sign of the denominator, and the sign of a fraction are given in the following box.

Three Signs of a Fraction

ALGEBRAICALLY	VERBALLY	NUMERICAL EXAMPLE
For all real numbers a and b with $b \neq 0$ $-\dfrac{a}{b} = \dfrac{-a}{b} = \dfrac{a}{-b} = -\dfrac{-a}{-b}$ and $\dfrac{a}{b} = \dfrac{-a}{-b} = -\dfrac{-a}{b} = -\dfrac{a}{-b}$	Each fraction has three signs associated with it. Any two of these signs can be changed and the value of the fraction will stay the same.	$-\dfrac{2}{3} = \dfrac{-2}{3} = \dfrac{2}{-3} = -\dfrac{-2}{-3}$ and $\dfrac{2}{3} = \dfrac{-2}{-3} = -\dfrac{-2}{3} = -\dfrac{2}{-3}$

SELF-CHECK 1.5.3

Enter three different forms of the fraction negative two-thirds into a graphics calculator and compare the results.

■ **EXAMPLE 4**   Evaluating Algebraic Expressions

Evaluate the following expressions for $x = -24$, $y = -4$, and $z = -2$:

SOLUTIONS

(a) $\dfrac{x}{y}$

$\dfrac{x}{y} = \dfrac{-24}{-4}$ Substitute -24 for x and -4 for y.

$= \dfrac{6(\overset{1}{\cancel{-4}})}{1(\underset{1}{\cancel{-4}})}$ Divide both the numerator and the denominator by the reducing factor -4. $\dfrac{-4}{-4}$ equals 1. Often this is understood and is not written.

$= 6$ The quotient of like signs is positive.

SELF-CHECK 1.5.3 ANSWER

```
-(2/3)
        -.6666666667
-2/3
        -.6666666667
2/(-3)
        -.6666666667
■
```

All three values are equal.

(b) $\dfrac{-x}{z}$

$\dfrac{-x}{z} = \dfrac{-(-24)}{-2}$ Substitute for x and z.

$= \dfrac{24}{-2}$ Simplify the numerator.

$= \dfrac{\overset{1}{\cancel{2}}(12)}{\underset{1}{\cancel{2}}(-1)}$ Divide both the numerator and the denominator by the reducing factor 2.

$= -12$ The quotient of unlike signs is negative.

■ **EXAMPLE 5** Estimating a Quotient

Estimate 682.5 ÷ 2.1 and then use a calculator to perform this division.

SOLUTION _____

ESTIMATED ANSWER

$682.5 \approx 680$

$2.1 \approx 2$ $680 \div 2 = 340$

Answer: 325 seems reasonable based on the estimated value of 340.

CALCULATOR VALUE

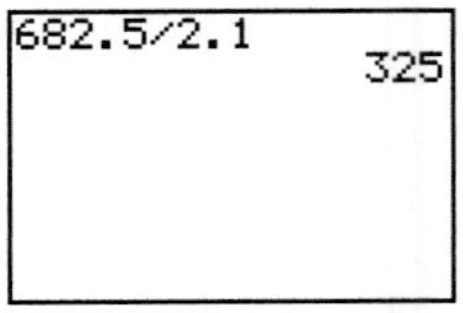

```
682.5/2.1
              325
```

In Example 5 the estimated value of 340 is 15 more than the actual value of 325. To judge how good this estimate is, we should examine not only our error of estimate, +15, but also the relative size of this error compared to the magnitude of the number being estimated. **Relative error** is the actual error divided by the actual value. In this

*The **error**, or error of estimate, is the estimated value minus the actual value.*

case $\frac{15}{325} \approx 0.0462$ or as a percent 4.62%. Thus our mental estimate in Example 5 was within 5% of the actual answer. Many calculator errors are the result of keystrokes which produce incorrectly placed decimal points, missing digits, and so on. These errors generate large relative errors and thus often can be detected by simply thinking about our results and asking ourselves if the results seem reasonable based on our mental estimate.

■ **EXAMPLE 6** Calculating the Relative Error

A contractor estimated the dimensions of the region shown to the right as 45 m by 60 m and calculated the approximate area. What was the relative error of this estimate?

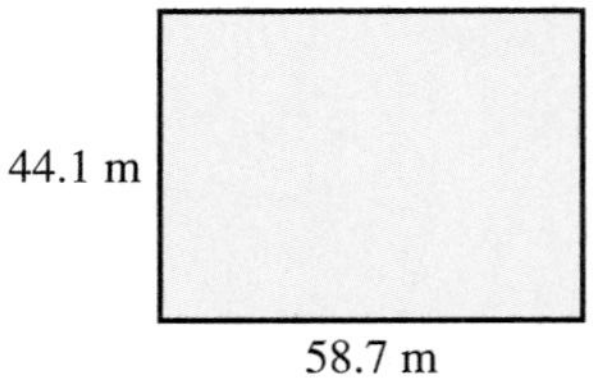

44.1 m

58.7 m

SOLUTION _____

Actual area:

$A = l \cdot w$

$A = (44.1 \text{ m})(58.7 \text{ m})$

$A = 2588.67 \text{ m}^2$

Use the formula for the area of a rectangle with the actual dimensions. The calculator value gives the area in square meters.

Estimated area:

$A = l \cdot w$

$A \approx (45 \text{ m})(60 \text{ m})$

$A \approx 2700 \text{ m}^2$

Use the estimated values in the same formula.

| Error of estimate | = | Estimated value | − | Actual value |

= 2700 m² − 2588.67 m²

= +111.33 m²

The estimated value is larger than the actual value.

| Relative error | = | Error of estimate | ÷ | Actual value |

= 111.33 m² ÷ 2588.67 m²

≈ 0.043

≈ 4.3%

First use a calculator to determine the relative error as a decimal and then convert to a percent.

Answer: The contractor's relative error was +4.3%. Many contractors prefer to be a little high on their estimate so that they do not run short on materials. ■

SELF-CHECK 1.5.4

In Example 6 use estimated dimensions of 44 m by 58 m to calculate the area of the region and then calculate the relative error of this estimate.

Two statistics that are commonly used to analyze a set of numbers are the range and the mean. The **range** of a set of numerical scores measures the number of units that separate the smallest score from the largest score and is calculated by subtracting the smallest score from the largest score. The **mean** of a set of numerical scores is an average calculated by dividing the sum of scores by the number of scores. This mean is the number most commonly referred to when the term *average* is used. Other common uses of the term *average* include the average grade of a student, the average number of miles per gallon of gasoline an automobile will achieve, and the average cost per ounce of a product. In each case, as illustrated in the following examples, division is used to obtain the desired average.

SELF-CHECK 1.5.4 ANSWER

2552 m^2 is 1.4% smaller than the actual value.

■ EXAMPLE 7   Finding the Mean and Range of Test Scores

A student made five grades of 75, 88, 94, 78, and 91 on 100-point tests. Find the mean and the range of these test scores.

SOLUTIONS

| Mean test score | = | Total points | ÷ | Number of tests |

Substitute the given values into the word equation for the mean. The numerator is the total number of points, and the denominator is the number of tests.

$$\text{Mean} = \frac{75 + 88 + 94 + 78 + 91}{5}$$

$$= \frac{426}{5}$$

$$= 85.2$$

| Range | = | Highest score | − | Lowest score |

$$= 94 - 75$$

$$= 19$$

Substitute the highest score of 94 and the lowest score of 75 into the word equation for the range.

Answer: The student's mean score is 85.2, and the range of these scores is 19. ■

A Mathematical Note

The colon has been used by the French to indicate division. Thus the use of the colon in the ratio $a : b$ indicates $a \div b$.

The ratio of a to b is the quotient $a \div b$. This ratio is sometimes written as $a : b$ and read as "the ratio of a to b." Since $a \div b = \dfrac{a}{b}$, ratios are frequently written as fractions in reduced form.

Ratio		
VERBALLY	**ALGEBRAICALLY**	**NUMERICAL EXAMPLE**
The ratio of a to b is the quotient of a divided by b.	The ratio of a to b can be denoted by either $a : b$ or $\dfrac{a}{b}$.	The ratio of 3 to 7 can be denoted by either $3 : 7$ or $\dfrac{3}{7}$.

Sometimes both the numerator and the denominator of a ratio can be expressed in terms of the same unit. In this case the common unit cancels when the ratio is reduced to its lowest terms, and the ratio is a number that is free of units.

■ EXAMPLE 8   Determining the Ratio of Two Measurements

A flagpole 8 m tall casts a 2-m shadow (see the figure to the right). Determine the ratio of the height of the flagpole to the length of its shadow.

SOLUTION

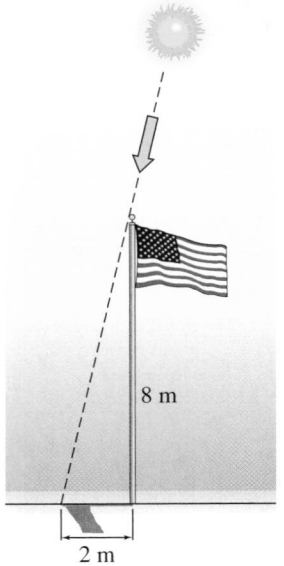

$$\frac{8 \text{ m}}{2 \text{ m}} = \frac{4}{1}$$ Divide both the numerator and the denominator of this ratio by 2 m.

$$= 4$$ The units cancel. The flagpole is taller than the shadow by a factor of 4.

Answer: The ratio of the height of the flagpole to the length of its shadow is $\dfrac{4}{1} = 4 : 1$. ■

The best price for a commodity (such as laundry detergent) that is used in various quantities over time often is determined by calculating the lowest unit price. The **unit price** of an item is the ratio of the price to the number of units.

■ EXAMPLE 9   Determining a Unit Price

A brand-name liquid laundry detergent is packaged in a 6-pound (lb) 4-ounce (oz) container that sells for $7.89 and a 4-lb container that sells for $4.80. Which size is the better buy?

SOLUTION

The size with the lower unit price is the better buy.

$$\boxed{\text{Unit price}} = \boxed{\text{Price}} \div \boxed{\text{Number of ounces}}$$

The unit price is the average price per ounce. Calculate the unit price of each box and then determine the one with the lower unit price.

SMALLER SIZE

Unit price $= \dfrac{\$4.80}{4 \text{ lb}}$

$= \dfrac{\$4.80}{64 \text{ oz}}$

$= \$0.075/\text{oz}$

LARGER SIZE

Unit price $= \dfrac{\$7.89}{6 \text{ lb } 4 \text{ oz}}$

$= \dfrac{\$7.89}{100 \text{ oz}}$

$= \$0.0789/\text{oz}$

4 lb $= 4(16)$ oz
$= 64$ oz

6 lb 4 oz $= [6(16) + 4]$ oz
$= 100$ oz

Answer: Since the smaller size has the lower price per ounce, it is the better buy.

Although many people think larger sizes are always a better buy, this is not always true.

SELF-CHECK 1.5.5 ANSWER

The ratio of the teeth on the front gear to those on the rear gear is 13 : 4.

SELF-CHECK 1.5.5

The front gear of a bicycle has 52 teeth, and the rear gear has 16 teeth. What is the ratio of the teeth on the front gear to those on the rear gear?

USING THE LANGUAGE AND SYMBOLISM OF MATHEMATICS   1.5

1. In the division $24 \div 3 = 8$, 24 is called the _____, 3 is called the _____, and 8 is called the _____.

2. The equation "The quotient of D divided by R equals T" is represented algebraically by _____.

3. The equation "The ratio of the number of men m to the number of women w is three-fifths" is represented algebraically by _____.

4. If x and y $(y \neq 0)$ have the same sign, then the quotient $\dfrac{x}{y}$ is _____.

5. If x and y $(y \neq 0)$ have unlike signs, then the quotient $\dfrac{x}{y}$ is _____.

6. To divide $\dfrac{5}{6}$ by $\dfrac{2}{3}$, we can multiply $\dfrac{5}{6}$ by $\dfrac{3}{2}$, the _____ inverse or the _____ of $\dfrac{2}{3}$.

7. The quotient $\dfrac{w}{x} \div \dfrac{y}{z} =$ _____ for _____ $\neq 0$, _____ $\neq 0$, and $z \neq$ _____.

8. The notation $a : b$ is read the _____ of a to b.

9. Division by zero is _____.

10. The _____ of a set of numbers is the largest number minus the smallest number.

11. The _____ of a set of numerical scores is an average calculated by dividing the sum of the scores by the number of scores.

12. The difference between an estimate and the actual value is the error of the estimate. The _____ error is the error of the estimate divided by the actual value.

13. The unit price of an item is the ratio of the price to the number of _____.

EXERCISES   1.5

In Exercises 1–12 calculate each quotient without using a calculator.

1. **a.** $48 \div (-6)$ **b.** $-48 \div (-6)$
   c. $-48 \div 6$ **d.** $0 \div 6$

2. **a.** $56 \div (-8)$ **b.** $-56 \div (-8)$
   c. $-56 \div 8$ **d.** $0 \div (-8)$

3. **a.** $-48 \div \left(\dfrac{1}{2}\right)$ **b.** $-48 \div 2$

   c. $-48 \div \left(-\dfrac{1}{3}\right)$ **d.** $-48 \div (-3)$

4. **a.** $48 \div \left(-\dfrac{1}{4}\right)$ **b.** $48 \div (-4)$

   c. $-48 \div \left(-\dfrac{1}{8}\right)$ **d.** $-48 \div (-8)$

5. **a.** $123 \div (0.1)$ **b.** $123 \div (-0.01)$
   c. $123 \div (-1000)$ **d.** $-123 \div (-10)$

6. **a.** $-123 \div (0.01)$ **b.** $-123 \div (100)$
   c. $123 \div (-0.001)$ **d.** $123 \div (-10,000)$

7. **a.** $\dfrac{1}{3} \div \left(-\dfrac{1}{2}\right)$ **b.** $-\dfrac{2}{3} \div \dfrac{1}{6}$

c. $-\dfrac{2}{3} \div (-6)$

d. $\dfrac{2}{3} \div \dfrac{10}{9}$

8. a. $\dfrac{1}{4} \div \left(-\dfrac{1}{5}\right)$

b. $-\dfrac{4}{5} \div \dfrac{1}{20}$

c. $-\dfrac{4}{5} \div (-20)$

d. $\dfrac{4}{5} \div \dfrac{8}{15}$

9. a. $0 \div (-7)$ **b.** $-7 \div 0$

c. $7 \div 0$ **d.** $0 \div 0$

10. a. $0 \div (-11)$ **b.** $-11 \div 0$

c. $11 \div 0$ **d.** $-0.01 \div 0$

11. a. $\dfrac{|-12|}{|3|}$

b. $\dfrac{-|12|}{|-3|}$

c. $-\sqrt{36} \div \sqrt{4}$

d. $\sqrt{100} \div \left(-\sqrt{25}\right)$

12. a. $\dfrac{|-15|}{|-5|}$

b. $\dfrac{-|15|}{|-5|}$

c. $-\sqrt{64} \div \sqrt{16}$

d. $\sqrt{100} \div \left(-\sqrt{16}\right)$

In Exercises 13 and 14 evaluate each expression for $w = 15$, $x = 5$, $y = 10$.

13. a. $\dfrac{w}{x}$ **b.** $\dfrac{-w}{x}$ **c.** $\dfrac{w}{-x}$ **d.** $\dfrac{-w}{-x}$

14. a. $\dfrac{x}{y}$ **b.** $\dfrac{-x}{y}$ **c.** $\dfrac{x}{-y}$ **d.** $\dfrac{-x}{-y}$

In Exercises 15 and 16 check both -6 and 6 to determine whether either value is a solution of these equations.

15. a. $-\dfrac{1}{3}x = x - 8$

b. $\dfrac{x}{3} = -8 - x$

16. a. $7 + x = \dfrac{-x}{6}$

b. $7 - x = \dfrac{1}{6}x$

In Exercises 17 and 18 calculate the first five terms of each sequence.

17. a. $a_n = \dfrac{120}{n}$

b. $a_n = \dfrac{-n}{2}$

18. a. $a_n = \dfrac{-120}{n}$

b. $a_n = \dfrac{600}{-n}$

Sum of Two Factors

In Exercises 19–24 complete the following table. First determine the missing factor of 36 and then compute the sum of x and y.

	x	y	$x + y$	xy
Example	1	36	37	36
19.	2			36
20.	3			36
21.	-4			36
22.	-6			36
23.	6			36
24.	12			36

25. Ratio of Diamonds   Thirteen of 52 cards in a deck of cards are diamonds. What is the ratio of diamonds to all the cards in the deck?

26. Defective Computer Chips   An inspection of 1056 experimental computer chips found 132 defective. What fractional portion of the chips were defective?

27. Tree Shadow   A tree 24 m tall casts an 8-m shadow. Determine the ratio of the height of the tree to the length of its shadow.

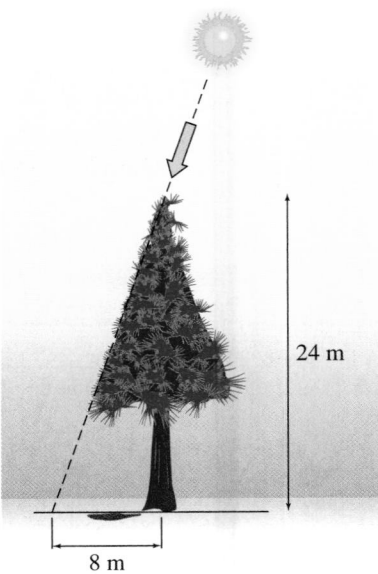

24 m

8 m

28. Shadow of a Man   A young man who is 2 m tall casts a 6-m shadow. Determine the ratio of his height to the length of his shadow.

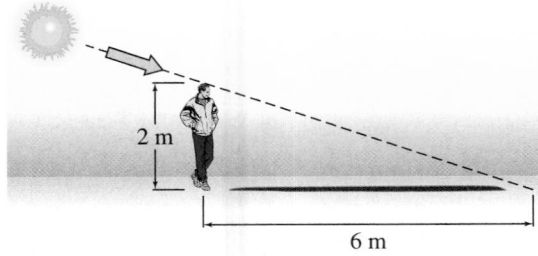

2 m

6 m

29. Banded Ducks   A wildlife study involved banding ducks and then studying those birds when they were recaptured. Of the 85 captured ducks, 17 already had been banded. What is the ratio of banded ducks to the total captured?

30. Banded Rabbits   A wildlife study involved banding rabbits' ears and then studying those banded rabbits when they were recaptured. Of the 117 captured rabbits, 26 already had been banded. What is the ratio of banded rabbits to the total captured? What is the ratio of banded rabbits to those not banded?

31. Comparison of Two Pulleys   The radius of one pulley is 24 cm and the radius of a second pulley is 15 cm. Determine the ratio of the radius of the larger pulley to that of the smaller pulley.

32. Gear Ratio   The front gear of a go-cart has 48 teeth and the rear gear has 16 teeth. What is the ratio of the teeth on the front gear to those on the rear gear?

33. Best Buy for Ketchup   One brand of ketchup is sold in two popular sizes. Which is the better buy: a 40-ounce bottle that is priced at $2.52 or a 64-ounce bottle that is priced at $3.68?

34. Best Buy for Corn Flakes   One brand of corn flakes is sold in several sizes. Which is the best buy: a 7-ounce box that is priced at $1.22, a 12-ounce box priced at $2.04, or an 18-ounce box priced at $3.15?

35. Test Average   A student scored 78, 85, 93, and 72 on four history exams. What is the student's average for these four exams?

36. Test Average   A student scored 86, 78, 82, and 80 on four algebra exams. What is the student's average for these four exams?

37. Percent of Error   The perimeter of the rectangle in the figure is estimated by using 30 cm for the width and 50 cm for the length. Determine the actual perimeter of this rectangle, the estimated perimeter, and the percent of error of this estimate.

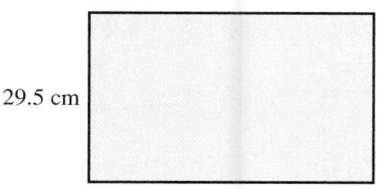

29.5 cm

48.4 cm

38. Percent of Error   Using the estimates for the width and length of the rectangle in Exercise 37, determine the actual area of this rectangle, the estimated area, and the percent error of this estimate.

39. Percent of Error   The perimeter of the isosceles triangle in the figure is estimated by using 20 cm for the length of the base and 16 cm for the length of each of the two equal sides. Determine the actual perimeter of this triangle, the estimated perimeter, and the percent of error of this estimate.

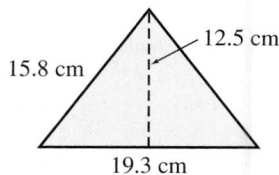

12.5 cm

15.8 cm

19.3 cm

40. Percent of Error   Using the 20-cm estimate for the base and a 12-cm estimate for the height of the triangle in Exercise 39, determine the actual area of this triangle, the estimated area, and the percent error of this estimate.

41. Range and Mean   Use a calculator to determine the range and mean of the utility bills for the months of January through June.

Month	JAN	FEB	MAR	APR	MAY	JUNE
Bill ($)	210.88	189.96	173.72	142.69	102.57	120.77

42. Range and Mean   Use a calculator to determine the range and mean of the telephone bills for the months January through June.

Month	JAN	FEB	MAR	APR	MAY	JUNE
Bill ($)	56.82	77.85	51.23	40.14	88.07	63.25

Rate of Travel

The average rate at which an object travels can be calculated by dividing the distance it travels by the time it takes to travel that distance. A car that drives 300 miles in 5 hours travels at an average rate of $\dfrac{300 \text{ miles}}{5 \text{ hours}} = 60$ mi/h. In Exercises 43–46, use the formula $R = \dfrac{D}{T}$ to compute the rate at which each object travels.

43. A plane flies 870 miles in 3 hours.

44. A hiker walks 40 miles in 2 days.

45. A snail crawls 240 cm in 10 minutes.

46. A boat sails 1200 km in 5 weeks.

Rate of Work

The rate at which a task is completed depends on the time required to complete the task. If a painter can paint 1 room in 3 hours, then his rate of work is $\dfrac{1}{3}$ room per hour. In Exercises 47–50, use the formula $R = \dfrac{W}{T}$ to compute the rate of work for each situation.

47. An assembly line produces 2400 lightbulbs in 8 hours.

48. A bricklayer lays 500 bricks in 10 hours.

49. A hose fills 1 swimming pool in 5 hours.

50. A roofer shingles 1 roof in 12 hours.

Estimation and Calculator Skills

In Exercises 51–58 mentally determine the correct sign of the quotient and then use a calculator to calculate the quotient. Approximate the quotient in Exercises 57 and 58 to the nearest thousandth. (*Hint:* See Calculator Perspective 1.5.1.)

PROBLEM	SIGN OF QUOTIENT (WITHOUT CALCULATOR)	QUOTIENT (WITH CALCULATOR)
51. $-8.2 \div 2.5$		
52. $-5.7 \div (-0.02)$		
53. $-\dfrac{12}{25} \div \left(\dfrac{-16}{15}\right)$		
54. $\dfrac{-42}{55} \div \dfrac{-14}{-66}$		

PROBLEM	SIGN OF QUOTIENT (WITHOUT CALCULATOR)	QUOTIENT (WITH CALCULATOR)
55. $(-360 \div 12) \div (-6)$		
56. $-360 \div [12 \div (-6)]$		
57. $\sqrt{37} \div \left(-\sqrt{17}\right)$		
58. $\left|-13.4\right| \div \left|4.2\right|$		

Estimation Skills

In Exercises 59–62 mentally estimate the value of each expression and then select the most appropriate answer.

59. $-28.5 \div 11.4$

    a. -17.5 **b.** 17.5   **c.** -2.5
    d. 2.5   **e.** -5.2

60. $-97.68 \div (-4.4)$

    a. -22.2 **b.** 22.2   **c.** -32.2
    d. 32.2   **e.** 52.2

61. $45.1 \div (-0.11)$

    a. -610 **b.** 610   **c.** -410
    d. 410   **e.** 31

62. $7.8 \div (-0.12)$

    a. -95 **b.** 95   **c.** -65
    d. 65   **e.** -35

In Exercises 63 and 64 mentally estimate the mean of each set of data and then select the answer that is the closest to your estimate.

63. 8.4, 9.4, 10.4, and 11.4

    a. 8.9   **b.** 9.9   **c.** 10.9   **d.** 11.9

64. 2.98, 7.05, 5.76, and 6.12

    a. 3.4775   **b.** 5.4775   **c.** 6.4775   **d.** 7.4775

65. College Tuition The following table gives the tuition costs for a student enrolling in 15 semester hours at two different colleges. Complete the last column of this table by calculating the cost per credit hour at each college.

COLLEGE	TUITION ($)	COST PER CREDIT HOUR ($)
University of Florida	1244.85	
University of South Florida	1048.05	

66. Tuition Comparison Use the table in Exercise 65 to determine the ratio of the cost of tuition at the University of Florida to the cost of tuition at the University of South Florida for a student enrolling in 15 semester hours. Interpret this result.

67. The first bar graph gives the number of items sold each quarter. The second bar graph gives the sales revenue for this item for each quarter. Calculate the price per item for each quarter.

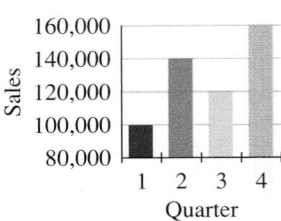
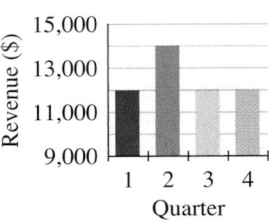

68. The first bar graph gives the number of items sold each quarter. The second bar graph gives the profit generated by the sale of this item for each quarter. Calculate the profit per item for each quarter.

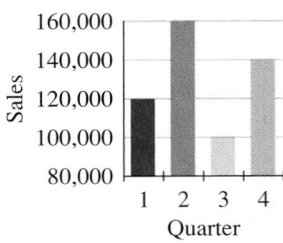
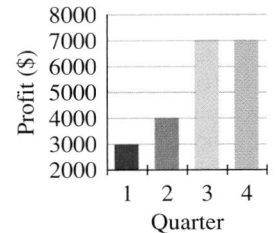

69. Volume of a Box The volume of the following box is 2552 cubic centimeters (cm^3). The width of the box is 8 cm and the length is 22 cm. Find the height of this box.

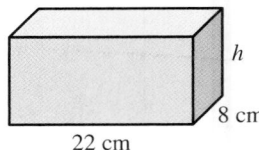

70. Volume of a Box The volume of the following box is 1512 cubic centimeters (cm^3). The length of the box is 18 cm and the height is 7 cm. Find the width of this box.

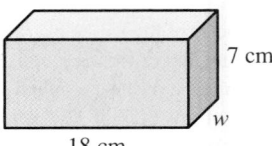

Multiple Representations

In Exercises 71–75 write each verbal statement in algebraic form.

71. a. The quotient of a divided by five.
    b. The quotient of d divided by t is seven.

72. a. The ratio of h to r.
    b. The ratio of V to R is I.

73. a. The quotient of x and y is eight.
    b. The product of x and y is eight.

74. a. Twelve divided by x is approximately equal to y.
    b. Nine divided by z is not equal to four.

75. a. Five is greater than x divided by 3.
   b. Negative two is less than or equal to eight divided by w.

In Exercises 76 and 77 fill in each blank.

76. The only real number that is neither negative nor positive is _____ .

77. The quotient of a nonzero number and its additive inverse is _____ .

Group Discussion Questions

78. Challenge Question
   a. Write four numbers with a mean of 100 and a range of zero.
   b. Write four numbers with a mean of 100 and a range of 100.

79. Challenge Question   Simplify

$$\frac{11}{12} \cdot \frac{12}{13} \cdot \frac{13}{14} \cdot \frac{14}{15} \cdot \frac{15}{16} \cdot \frac{16}{17} \cdot \frac{17}{18} \cdot \frac{18}{19} \cdot \frac{19}{20} \cdot \frac{20}{21} \cdot \frac{21}{22}$$

80. Error Analysis   In trying to evaluate the expression $\frac{2}{5} \div \frac{3}{8}$ on a TI-83 Plus calculator, a student enters the expression as shown here. What is the error and how can it be corrected?

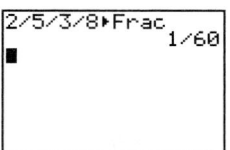

81. Discovery Question
   a. Have each person in your group mentally estimate the appropriate tip for each of the restaurant bills in the following table. Then compare how you made your estimates. Which method do you think is best?

BILL ($)	10% TIP	15% TIP	20% TIP
13.86			
25.43			
36.17			
45.82			
68.14			

   b. What does your group consider to be the appropriate amount to tip: 10, 15, or 20%?

82. Multiple Representations   Create a flowchart for dividing two real numbers. See the group discussion questions in Sections 1.2 and 1.3.

Section 1.6   Exponents and Order of Operations

Objectives:
17. Use natural number exponents.
18. Use the correct order of operations.
19. Use the distributive property.

The expression 5^2 means $5 \cdot 5$ and is read as "five squared." In the expression 5^2, 5 is called the **base** and 2 is called the **exponent.** The compact use of exponential notation to indicate repeated multiplication is defined in the following box. The expression b^n indicates that b is used as a factor n times, where the exponent n can be 1, 2, 3, 4, or any other natural number.

Exponential Notation

ALGEBRAICALLY	VERBALLY	NUMERICAL EXAMPLES
For any natural number n, $b^n = \underbrace{b \cdot b \cdot \ldots \cdot b}_{n \text{ factors of } b}$ with base b and exponent n.	For any natural number n, b^n is the product of b used as a factor n times. The expression b^n is read as "b to the nth power."	$7^4 = 7 \cdot 7 \cdot 7 \cdot 7$ $(-3)^2 = (-3)(-3)$

Some of the words and phrases used in word problems and applications to indicate exponentiation are given in Table 1.6.1.

x^2, which can be read as "x to the second power," is usually read as "x squared" because this is the area of a square with sides of length x.

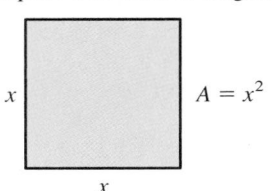

$A = x^2$

x^3, which can be read as "x to the third power," is usually read as "x cubed" because this is the volume of a cube with sides of length x.

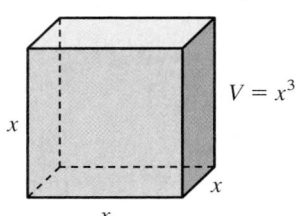

$V = x^3$

Table 1.6.1   Phrases Used to Indicate Exponentiation

KEY PHRASE	VERBAL EXAMPLE	ALGEBRAIC EXAMPLE
To a power	"3 to the 6th power"	3^6
Raised to	"y raised to the 5th power"	y^5
Squared	"4 squared"	4^2
Cubed	"x cubed"	x^3

■ EXAMPLE 1   Evaluating Exponential Expressions

Evaluate each of the following exponential expressions:

SOLUTIONS

(a) 3^4 $3^4 = 3 \cdot 3 \cdot 3 \cdot 3$
$= 81$

The base is 3. The exponent of 4 indicates that 3 is used as a factor 4 times. This is read as "3 to the 4th power equals 81."

(b) 2^5 $2^5 = 2 \cdot 2 \cdot 2 \cdot 2 \cdot 2$
$= 32$

"2 to the 5th power equals 32."

(c) 1^6 $1^6 = 1 \cdot 1 \cdot 1 \cdot 1 \cdot 1 \cdot 1$
$= 1$

"1 to the 6th power equals 1."

(d) $\left(\dfrac{2}{5}\right)^3$ $\left(\dfrac{2}{5}\right)^3 = \dfrac{2}{5} \cdot \dfrac{2}{5} \cdot \dfrac{2}{5}$
$= \dfrac{8}{125}$

The base of $\dfrac{2}{5}$ is used as a factor 3 times. This is read, "two-fifths cubed equals eight-one hundred twenty fifths."

(e) 0.12^2 $0.12^2 = (0.12) \cdot (0.12)$
$= 0.0144$

"Twelve hundredths to the second power equals one hundred forty-four ten-thousandths."

SELF-CHECK 1.6.1

Write each expression in exponential form.

1. $4 \cdot 4 \cdot 4 \cdot 4 \cdot 4$ **2.** $(-3)(-3)(-3)(-3)$
3. $x \cdot x \cdot x \cdot x \cdot x \cdot x$ **4.** $(2w)(2w)(2w)$

SELF-CHECK 1.6.1 ANSWERS

1. 4^5 **2.** $(-3)^4$
3. x^6 **4.** $(2w)^3$

An exponent is understood to refer only to the constant or variable that immediately precedes it. This can lead to misconceptions when expressions such as $(-3)^2$ and -3^2 are considered. Note the distinction between $(-3)^2$ and -3^2 in Example 2.

■ EXAMPLE 2   Evaluating Exponential Expressions

Evaluate each of the following exponential expressions:

SOLUTIONS

(a) $(-3)^2$ $(-3)^2 = (-3)(-3)$ The base is -3. The product is positive since the number of negative factors is
                   $= 9$ even (two).

(b) -3^2 $-3^2 = -(3)(3)$ The base is 3. Square 3 and then form the additive inverse of this product.
                   $= -9$ *Caution*: -3^2 is not the same as $(-3)^2$.

(c) $(-2)^4$ $(-2)^4 = (-2)(-2)(-2)(-2)$ The base is -2. The product is positive since the number of negative factors is even
                   $= 16$ (four).

(d) -2^4 $-2^4 = -(2 \cdot 2 \cdot 2 \cdot 2)$ The base is 2. Raise 2 to the 4th power and then form the additive inverse of this
                   $= -16$ product. Contrast this to part (c) where the base is -2.

(e) $(-1)^{117}$ $(-1)^{117} = -1$ The base of -1 is used as a factor an odd number of times (117), so the result is
                                           negative.

CALCULATOR PERSPECTIVE 1.6.1   **Raising Quantities to Power**

To square a value, press the x^2 button after entering the expression to be squared.
For exponents other than 2 use the $\wedge$ button followed by the desired exponent. To
evaluate $(-3)^2$ and -2^4 from Example 2 on a TI-83 Plus calculator, enter the
following keystrokes:

The expression x^6 is sometimes
read "x raised to the 6th power
and can be written as $x \wedge 6$. The
caret symbol $\boxed{\wedge}$ is used to
indicate the exponent.

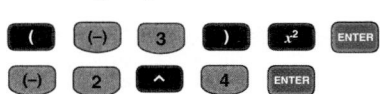

$\boxed{(}$ $\boxed{(-)}$ $\boxed{3}$ $\boxed{)}$ $\boxed{x^2}$ $\boxed{\text{ENTER}}$

$\boxed{(-)}$ $\boxed{2}$ $\boxed{\wedge}$ $\boxed{4}$ $\boxed{\text{ENTER}}$

```
(-3)²
              9
-2^4
            -16
```

SELF-CHECK 1.6.2

Evaluate each of the following exponential expressions:

1. 3^4 2. 4^3 3. $(-3)^4$ 4. -3^4

Symbols of grouping are used in mathematical expressions to separate signs, to enclose
an expression to be treated as one quantity, and to indicate the order of operations within
an expression. The common grouping symbols are the fraction bar —, which separates the
numerator from the denominator; parentheses ( ); brackets [ ]; and braces { }. For problems involving more than one operation, mathematicians have developed the following
order of operations so that each expression will have a unique interpretation and value.

Order of Operations

STEP 1 Start with the expression within the innermost pair of grouping symbols.

STEP 2 Perform all exponentiations.

STEP 3 Perform all multiplications and divisions as they appear from left to right.

STEP 4 Perform all additions and subtractions as they appear from left to right.

SELF-CHECK 1.6.2 ANSWERS

1. 81   2. 64   3. 81
4. -81

This order of operations has been adopted by the manufacturers of most calculators and incorporated into many computer languages. You should verify this hierarchy on a calculator before undertaking any important calculation.

■ EXAMPLE 3   Using the Correct Order of Operations

Evaluate each of the following expressions, using the correct order of operations.

SOLUTIONS _____

(a) $3 + 4 \cdot 7$

$$3 + 4 \cdot 7 = 3 + 28$$
$$= 31$$

Multiplication has priority over addition.

(b) $(3 + 4)(7)$

$$(3 + 4)(7) = 7(7)$$
$$= 49$$

First simplify the expression within the parentheses. Notice the distinction between part (a) and part (b) of this example.

(c) $5 \cdot 8 - 12 \div 3 + 1$

$$5 \cdot 8 - 12 \div 3 + 1 = 40 - 12 \div 3 + 1$$
$$= 40 - 4 + 1$$
$$= 36 + 1$$
$$= 37$$

Multiplication and division are performed as they appear from left to right. Then addition and subtraction are performed as they appear from left to right.

(d) $24 \div 6 \cdot 2$

$$24 \div 6 \cdot 2 = 4 \cdot 2$$
$$= 8$$

Multiplication and division are performed as they appear from left to right.

(e) $\dfrac{12 - 3(2)}{(12 - 3)(2)}$

$$\frac{12 - 3(2)}{(12 - 3)(2)} = \frac{12 - 6}{9(2)}$$
$$= \frac{6}{18}$$
$$= \frac{1}{3}$$

Simplify the numerator and the denominator separately, as the fraction bar indicates this grouping. Then reduce the fraction to lowest terms.

CALCULATOR PERSPECTIVE 1.6.2	Using Implied Parentheses

When entering expressions containing fractions, it is usually wise to include a set of parentheses around the entire numerator and the entire denominator. To evaluate $\dfrac{12 - 3(2)}{(12 - 3)(2)}$ from Example 3(e) on a TI-83 Plus calculator, enter the following keystrokes:

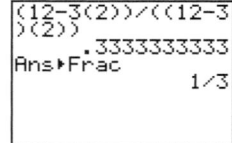

Note: Failure to use this additional set of parentheses on both the numerator and the denominator will result in an error. ■

Whenever more than one set of grouping symbols appears in an expression, you should start from the innermost pair of symbols and work toward the outermost pair of symbols.

EXAMPLE 4   Evaluating Expressions with Grouping Symbols

Evaluate each of the following expressions:

SOLUTIONS

(a) $20 + 3[(7 + 8) - (3 + 6)]$

$20 + 3[(7 + 8) - (3 + 6)]$
$= 20 + 3[15 - 9]$
$= 20 + 3[6]$
$= 20 + 18$
$= 38$

First simplify within the parentheses and then within the brackets. Multiplication has priority over addition.

(b) $149 - 2[7 + 3(11 + 2^2) - 5]$

$149 - 2[7 + 3(11 + 2^2) - 5]$
$= 149 - 2[7 + 3(11 + 4) - 5]$
$= 149 - 2[7 + 3(15) - 5]$
$= 149 - 2[7 + 45 - 5]$
$= 149 - 2[47]$
$= 149 - 94$
$= 55$

Start with the expression within the innermost pair of grouping symbols and work outward. Multiplication has priority over addition and subtraction.

SELF-CHECK 1.6.3

Evaluate each of the following expressions:

1. $18 - 4 \cdot 3$ **2.** $(18 - 4)(3)$ **3.** $\dfrac{20 - 45 \div 3}{8 + 6 \cdot 2}$ **4.** $5 \cdot 6 - 8 \div 2$

In Example 4 both brackets and parentheses were used in the same expression. This notation usually is used in printed mathematics because it is less confusing than using multiple levels of parentheses. However, $20 + 3[(7 + 8) - (3 + 6)]$ also can be denoted by $20 + 3((7 + 8) - (3 + 6))$. In fact, multiple levels of parentheses must be used with many calculators and in some computer program statements. To be interpreted correctly, these statements must have the parentheses paired properly and evaluated by starting with the innermost pair of parentheses.

CALCULATOR PERSPECTIVE 1.6.3   Using Parentheses

When entering expressions containing parentheses, be careful to keep track of pairs of left and right parentheses. To evaluate $20 + 3[(7 + 8) - (3 + 6)]$ from Example 4(a) on a TI-83 Plus calculator, enter the following keystrokes:

[2] [0] [+] [3] [(]
[(] [7] [+] [8] [)] [−]
[(] [3] [+] [6] [)] [)] [ENTER]

```
20+3((7+8)-(3+6)
)
              38
```

Note: Do not use the brackets on a calculator as they are reserved for a special purpose and are not used as general purpose grouping symbols.

SELF-CHECK 1.6.3 ANSWERS

1. 6 **2.** 42 **3.** $\dfrac{1}{4}$ **4.** 26

Order-of-operations errors can occur subtly in expressions involving exponents. One common error involves confusing the notations in parts (a) and (b) in Example 5. Note carefully the distinct meanings of each of these two expressions.

■ EXAMPLE 5   Using the Correct Order of Operations

Evaluate each of the following expressions, using the correct order of operations:

SOLUTIONS _____

(a) $2^2 + 3^2$ $2^2 + 3^2 = 4 + 9$ Exponentiation has priority over addition.
                                        $= 13$

(b) $(2 + 3)^2$ $(2 + 3)^2 = 5^2$
                                        $= 25$ First simplify within the parentheses, and then square. Notice that this is a different answer from that in part (a) of this example.

(c) $2 \cdot 5^3 - 6^2 \div 3$ $2 \cdot 5^3 - 6^2 \div 3 = 2(125) - 36 \div 3$ Exponentiation has the highest priority, followed by multiplication and division, and lastly subtraction.
                                        $= 250 - 12$
                                        $= 238$

SELF-CHECK 1.6.4

Rewrite

$$(14 - 5(13 - 11))(6 + 2(14 - 3))$$

using brackets and parentheses and then evaluate this expression. Check the result on a calculator.

SELF-CHECK 1.6.4 ANSWER

$[14 - 5(13 - 11)]$
$[6 + 2(14 - 3)] = 112$

A Mathematical Note

The French mathematician Servois (ca. 1814) introduced the terms *commutative* and *distributive*. The term *associative* is attributed to the Irish mathematician William R. Hamilton (1805–1865).

To be interpreted correctly, expressions involving different operations, such as addition and multiplication, often require grouping symbols. However, expressions involving only the operation of addition or only the operation of multiplication are usually written without grouping symbols. In these circumstances grouping symbols are not needed since all possible groupings yield the same value. For example, we can write $3 + 4 + 5$ without parentheses since both $(3 + 4) + 5$ and $3 + (4 + 5)$ are equal. Recall that it is the associative property of addition that allows terms to be regrouped in different ways. Similarly, $3 \cdot 4 \cdot 5$ can be written as either $(3 \cdot 4)(5)$ or as $(3)(4 \cdot 5)$ because of the associative property of multiplication.

There is also one property that relates multiplication and addition. By the **distributive property,** 5 times the sum of 2 and 6 equals the sum of 5 times 2 and 5 times 6.

$5(2 + 6) = 5(8)$ $5(2) + 5(6) = 10 + 30$
$\qquad\quad = 40$ $\qquad\qquad\quad = 40$

Thus $5(2 + 6) = 5(2) + 5(6)$. In general, the distributive property of multiplication over addition says $a(b + c) = ab + ac$.

Distributive Property of Multiplication over Addition

ALGEBRAICALLY	VERBALLY	NUMERICAL EXAMPLES
For all real numbers a, b, and c, $a(b + c) = ab + ac$ and $(b + c)a = ba + ca.$	Multiplication distributes over addition.	$5(2 + 6) = 5(2) + 5(6)$ $5(8) = 10 + 30$ $40 = 40$ $(2 + 6)5 = (2)5 + (6)5$ $(8)5 = 10 + 30$ $40 = 40$

Consider both $5(6 + 7)$ and $5(x + 7)$. We can use the distributive property to rewrite $5(6 + 7)$ as $5(6) + 5(7)$ and $5(x + 7)$ as $5x + 35$. Using the distributive property is not crucial in $5(6 + 7)$, but it is necessary to expand $5(x + 7)$ as $5x + 35$. We use the distributive property extensively both to expand expressions like $5(x + 7)$ to $5x + 35$ and to factor expressions like $5x + 35$ to $5(x + 7)$.

SELF-CHECK 1.6.5

Evaluate each of the following expressions:

1. $-4^2 + 5^2$ 2. $(-4)^2 + 5^2$ 3. $(-4 + 5)^2$ 4. $\sqrt{3^2 + 4^2}$

■ EXAMPLE 6   Using the Distributive Property

Use the distributive property to expand the first three expressions and to factor the last two expressions:

SOLUTIONS

(a) $9(x + 4)$ $9(x + 4) = 9(x) + 9(4)$
$= 9x + 36$ Distribute the factor of 9 to both the x term and the 4 term.

(b) $11(2y - 3)$ $11(2y - 3) = 11(2y) + 11(-3)$
$= 22y + (-33)$
$= 22y - 33$ Multiplication distributes over both addition and subtraction. We first can think of this expression as $11[2y + (-3)]$.

(c) $-(7x - 3y)$ $-(7x - 3y) = -1(7x - 3y)$
$= (-1)(7x) + (-1)(-3y)$
$= -7x + 3y$ The factor of -1 is understood. Distribute the factor of -1 to both terms.

(d) $5x - 5y$ $5x - 5y = 5(x - y)$ Five is a factor of both terms. Use the distributive property to factor out this common factor.

(e) $-4x + 8y$ $-4x + 8y = -4(x) + (-4)(-2y)$
$= -4(x - 2y)$ Factor out the common factor of -4. You can check this factorization by multiplying these two factors. ■

SELF-CHECK 1.6.5 ANSWERS
1. 9   2. 41   3. 1   4. 5

The distributive property also plays a key role in adding like terms and simplifying algebraic expressions. Using the distributive property, we can rewrite $5x + 7x$ as $(5 + 7)x$ or $12x$. The constant factor in a term is called the **numerical coefficient** of

Like terms have exactly the same
variable factors.

the term. For example, the numerical coefficient of $5x$ is 5 and the numerical coefficient of $7x$ is 7. The process of adding like terms is sometimes called **collecting like
terms** or **combining like terms.**

■ EXAMPLE 7   Adding Like Terms

Use the distributive property to add like terms.

SOLUTIONS

(a) $13x + 8x$ $13x + 8x = (13 + 8)x$
                                $= 21x$

(b) $13x - 8x$ $13x - 8x = (13 - 8)x$
                                $= 5x$

(c) $-7w + 4w$ $-7w + 4w = (-7 + 4)w$
                                $= -3w$

(d) $2a + (3b + 4a)$ $2a + (3b + 4a) = 2a + (4a + 3b)$
                                          $= (2a + 4a) + 3b$
                                          $= (2 + 4)a + 3b$
                                          $= 6a + 3b$

(e) $3\sqrt{2} + 4\sqrt{2}$ $3\sqrt{2} + 4\sqrt{2} = (3 + 4)\sqrt{2}$
                                          $= 7\sqrt{2}$

Usually we skip writing the middle step and perform this addition mentally. The distributive property justifies what we are doing.

Reorder using the commutative property of addition. Then regroup using the associative property of addition. Next add like terms using the distributive property. Again, it is common to perform this addition mentally.

Use the distributive property to factor out the common factor of $\sqrt{2}$. Then add the coefficients 3 and 4.

■ EXAMPLE 8   Estimating Expressions with Exponents

Mentally estimate $(6.07 + 3.98)^2 - (6.07^2 + 3.98^2)$.

SOLUTION

$(6.07 + 3.98)^2 - (6.07^2 + 3.98^2) \approx 10^2 - (6^2 + 4^2)$
$\approx 100 - (36 + 16)$
$\approx 100 - 52$
≈ 48

Answer: $(6.07 + 3.98)^2 - (6.07^2 + 3.98^2) \approx 48$

The actual value of 48.3172 can be obtained on a calculator.

SELF-CHECK 1.6.6

1. Expand $5(x + 3y)$.   **2.** Factor $2a - 10b$.
3. Simplify $-8y + 17y$.   **4.** Simplify $5a + (7b - 3a)$.

SELF-CHECK 1.6.6 ANSWERS

1. $5x + 15y$ **2.** $2(a - 5b)$
3. $9y$ **4.** $2a + 7b$

When using the term **quantity** in the verbal statement of an algebraic expression,
we are indicating that parentheses group the sum or difference that follows. This is significant because the expression $2 \cdot 3x + 1$ does not have the same meaning as $2(3x + 1)$.

The statement, "two times three x plus one" refers to $2 \cdot 3x + 1$, and the statement, "two times the quantity three x plus one" refers to $2(3x + 1)$.

■ **EXAMPLE 9**   Using the Language and Symbolism of Mathematics for Expressions Involving Parentheses

Give a verbal statement of each algebraic expression.

SOLUTIONS ────────────────────────────

ALGEBRAICALLY	VERBALLY
(a) $5(2x + 3)$	Five times the quantity two x plus three.
(b) $4(5x - 6)$	Four times the quantity five x minus six.
(c) $(4x + 5y)3$	The quantity four x plus five y times three.

USING THE LANGUAGE AND SYMBOLISM OF MATHEMATICS   1.6

1. _____ is read x to the 7th power.
2. y^5 is read y to the _____ _____.
3. a^8 means that a is used as a _____ 8 times.
4. In the expression w^4 the base is _____ and the exponent is _____.
5. In the expression $2 - 5 \cdot 7$ the first operation to perform is _____.
6. In the expression $(4 + 9) \cdot 8$ the first operation to perform is _____.
7. In the expression $2 + 5^2 \cdot 3$ the first operation to perform is _____.

8. In the expression $(-x)^2$ the base is _____.
9. In the expression $-x^2$ the base is _____.
10. Considering the commutative, associative, and distributive properties, the property which involves two different operations is the _____ property of _____ over _____.
11. In the expression $7v$, 7 is the _____ of v.
12. The _____ property is used to add like terms.
13. The process of adding like terms is sometimes called collecting like terms or _____ like terms.

EXERCISES   1.6

In Exercises 1 and 2 write each expression in exponential form.

1. a. $5 \cdot 5 \cdot 5 \cdot 5$ b. $(-4)(-4)(-4)$
   c. $y \cdot y \cdot y \cdot y \cdot y$ d. $(3z)(3z)(3z)(3z)(3z)(3z)$
2. a. $7 \cdot 7 \cdot 7$ b. $(-6)(-6)(-6)(-6)$
   c. $w \cdot w \cdot w \cdot w \cdot w \cdot w$ d. $(-5z)(-5z)$

In Exercises 3–44 calculate each expression without using a calculator.

3. a. 2^3 b. 3^2 c. $(-3)^2$ d. -3^2
4. a. 2^5 b. 5^2 c. $(-5)^2$ d. -5^2
5. a. 0^8 b. 1^8 c. $(-1)^8$ d. -1^8
6. a. 0^{10} b. 1^{10} c. $(-1)^{10}$ d. -1^{10}
7. a. -10^2 b. $(-10)^2$ c. $(0.1)^3$ d. $(-10)^3$
8. a. -10^3 b. $(-0.1)^2$ c. -0.1^2 d. $(-10)^4$
9. a. $\left(\dfrac{1}{2}\right)^3$ b. $\left(\dfrac{3}{5}\right)^2$ c. $\left(-\dfrac{1}{3}\right)^2$ d. $\left(\dfrac{17}{19}\right)^1$

10. a. $\left(\dfrac{1}{5}\right)^3$ b. $\left(\dfrac{4}{7}\right)^2$ c. $\left(-\dfrac{1}{8}\right)^2$ d. $\left(\dfrac{23}{27}\right)^1$

11. a. $5 + 2 \cdot 8$ b. $(5 + 2) \cdot 8$
12. a. $11 + 3 \cdot 9$ b. $(11 + 3) \cdot 9$
13. a. $17 - 3 \cdot 5$ b. $(17 - 3)(5)$
14. a. $41 - 6 \cdot 4$ b. $(41 - 6)(4)$
15. a. $(6 + 4)^2$ b. $6^2 + 4^2$
16. a. $(6 - 4)^2$ b. $6^2 - 4^2$
17. a. $25 \div 5 \cdot 5$ b. $25 \div (5 \cdot 5)$
18. a. $25 \div 5 \div 5$ b. $25 \div (5 \div 5)$
19. a. $19 - 19 \cdot 2$ b. $(19 - 19) \cdot 2$
20. a. $10 - 10 \cdot 3$ b. $(10 - 10) \cdot 3$
21. a. $4 - 7^2$ b. $(4 - 7)^2$
22. a. $1 - 9^2$ b. $(1 - 9)^2$
23. a. $3 \cdot 5 - 6 \cdot 7$ b. $-6 + 2 \cdot 4^2$
24. a. $4 \cdot 9 - 8 \cdot 11$ b. $-7 + 5 \cdot 2^4$

25. a. $15 - 3^2 - 6$ **b.** $15 - (3^2 - 6)$
26. a. $(15 - 3)^2 - 6$ **b.** $(15 - 3 - 6)^2$
27. a. $2^3 - 5^3$ **b.** $(2 - 5)^3$
28. a. $5^3 - 10^3$ **b.** $(5 - 10)^3$
29. a. $4000 - 5 \cdot 10^3$ **b.** $45 - 15 \div 5 + 10$
30. a. $36 - 24 \div 3 + 5$ **b.** $5000 - 7 \cdot 10^3$
31. a. $16 \div 4 \cdot 3 + 5$ **b.** $28 - 3[5 + 2(7 - 2)]$
32. a. $24 \div 8 \cdot 2 - 4$ **b.** $43 - 5[17 - 4(11 - 8)]$
33. a. $\dfrac{15 - 3(4)}{(15 - 3)(4)}$ **b.** $\dfrac{5 + 3(4)}{(5 + 3)(4)}$
34. a. $\dfrac{18 + 2(5)}{(18 + 2)(5)}$ **b.** $\dfrac{21 - 3(2)}{(21 - 3)(2)}$
35. a. $(3 + 4)^2 - (3^2 + 4^2)$ **b.** $(13 - 5)^2 + (13^2 - 5^2)$
36. a. $(5 + 7)^2 - (5^2 + 7^2)$ **b.** $(11 - 8)^2 + (11^2 - 8^2)$
37. a. $12 + 2[8 - 3(7 - 5)]$ **b.** $(2 + 3)^3 + (2^3 + 3^3)$
38. a. $15 + 5[19 - 4(11 - 9)]$ **b.** $(5 - 3)^3 + (5^3 - 3^3)$
39. $113 + 5[8 + 2(13 - 3^2) - 7]$
40. $217 + 6[11 - 3(50 - 7^2)]$
41. $18 + 2[(41 - 8) - (5 + 3^2)]$
42. $125 - 3[(38 - 27) - (31 - 5^2)]$
43. $\dfrac{5}{16} + \dfrac{1}{2} \cdot \dfrac{3}{8}$ **44.** $\dfrac{6}{5} - \dfrac{1}{5}\left(\dfrac{7}{8} + \dfrac{3}{8}\right)$

In Exercises 45–52 use the distributive property to expand the first expression and to factor the second expression.

Expand	Factor
45. a. $7(x + 5)$	**b.** $9x + 36$
46. a. $3(y + 9)$	**b.** $4y + 20$
47. a. $-2(3x - 7)$	**b.** $-5a - 5b$
48. a. $-4(5b - 9)$	**b.** $-7a - 21$
49. a. $(2x - 3y)(5)$	**b.** $22x - 33y$
50. a. $(4w - 5z)(6)$	**b.** $36w - 48z$
51. a. $-(b - 3)$	**b.** $-b + 5 = -1(\,?\,)$
52. a. $-(y + 5)$	**b.** $-y - 9 = -1(\,?\,)$

In Exercises 53–60 use the distributive property to add like terms.

53. a. $2x + 5x$ **b.** $2x - 5x$ **c.** $2\sqrt{3} + 5\sqrt{3}$
54. a. $3y + 8y$ **b.** $3y - 8y$ **c.** $3\sqrt{7} - 8\sqrt{7}$
55. a. $2w - 3w + 7w$ **b.** $2w - 3z + 7w$
56. a. $-4y + 5y - 9y$ **b.** $-4y + 5z - 9y$
57. a. $(2a + 3b) + (4a - 5b)$ **b.** $(2a + 3b) - (4a - 5b)$
58. a. $(3v - 4w) + (5v - 6w)$ **b.** $(3v - 4w) - (5v - 6w)$
59. $\left(2\sqrt{3} + 3\sqrt{5}\right) + \left(4\sqrt{3} - 5\sqrt{5}\right)$
60. $\left(3\sqrt{7} - 4\sqrt{11}\right) - \left(5\sqrt{7} - 6\sqrt{11}\right)$

In Exercises 61–64 check each equation for $x = -2$ and $x = 5$.

61. a. $3(x + 4) = x + 8$ **b.** $(x + 2)(x - 5) = 0$
62. a. $4(x - 4) = x - 1$ **b.** $(x + 2)(x - 5) = 1$
63. $2x - 3(x + 5) = -4x$
64. $-3x + 2(5x - 1) = 8x$

In Exercises 65 and 66 insert the correct inequality symbol, $<$ or $>$, between each pair of expressions.

65. a. 2^5 _____ 5^2 **b.** $\sqrt{9 + 16}$ _____ $\sqrt{9} + \sqrt{16}$
66. a. 2^3 _____ 3^2 **b.** $\sqrt{25 + 144}$ _____ $\sqrt{25} + \sqrt{144}$

Estimation and Calculator Skills
In Exercises 67–70 mentally estimate the value of each expression and then use a calculator to approximate each value to the nearest thousandth.

PROBLEM	MENTAL ESTIMATE	CALCULATOR APPROXIMATION
67. $(4.738 + 4.229)^2$		
68. $5.112^2 + 3.987^2$		
69. $6.129^2 - 3.178^2$		
70. $(6.129 + 3.178)^2$		

71. Volume of a Cube   The volume of a cube is given by $V = s^3$. Calculate the volume of the cube in the figure.

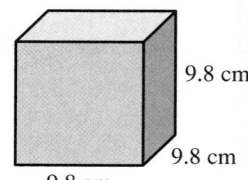

9.8 cm
9.8 cm
9.8 cm

72. Volume of a Sphere   The volume of a sphere is given by $V = \dfrac{4}{3}\pi r^3$. Calculate the volume of the sphere in the figure.

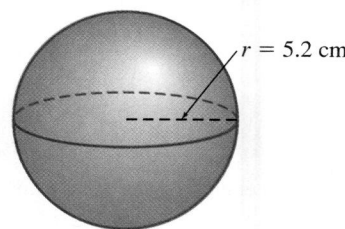

$r = 5.2$ cm

73. Heating Space   A warehouse with 500,000 cubic feet of storage space is to be used for storing crates. For purposes of planning the heating of this warehouse the designer must know the volume of air that is to be heated. Each of the 40,000 crates planned for storage in the warehouse is a cube that is 2 ft on a side. After the crates are stored, what volume of air will remain in the warehouse to be heated?

74. Surface Area of a Water Tank   *Allied Manufacturing* produces water tanks using a design with a hemisphere (half-sphere) on top of a right-circular cylinder. To plan for the painting process, the company determines the surface area to be painted on each tank. (The base of the tank is not painted.) The surface area S is given by the formula $S = 2\pi r^2 + 2\pi rh$, where r is the radius and h is the height of the cylinder. Determine the surface area in square feet of the tank shown here.

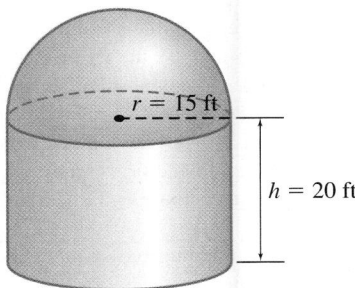

75. Use the formula $V = \dfrac{2}{3}\pi r^3 + \pi r^2 h$ to calculate the volume in cubic feet of the tank in Exercise 74.

Multiple Representations
In Exercises 76–81 write each verbal statement in algebraic form.

76. a. Seven times the quantity x minus three.
  b. Three times the quantity seven x plus two equals five.
77. a. Two times the quantity four x plus five y.
  b. Nine times the quantity six x minus eleven equals one.
78. a. y is equal to the quotient of k divided by the cube of x.
  b. h is equal to the product of k and the square of t.
79. a. c is equal to the square root of the quantity a squared plus b squared.
  b. d is equal to the square root of the quantity x squared plus 5.
80. a. The ratio of the quantity y_2 minus y_1 to the quantity x_2 minus x_1.
  b. m is the ratio of the quantity y_2 minus y_1 to the quantity x_2 minus x_1.
81. a. The sum of x squared and y squared.
  b. The square of the quantity x plus y.

Group Discussion Questions
Risks and Choices
82. Investment Comparison   Investment A yielded a dividend of $5000 at the end of the first year; and the dividend at the end of each of the next 4 years doubled. How much did this investment yield at the end of the fifth year? Investment B yielded a dividend of $25,000 at the end of the first year; and the dividend at the end of each of the next 4 years increased by $5000 per year. How much did this investment yield at the end of the fifth year? Which investment produced the greater total yield over the 5-year period? Would your answer change if the pattern continued for one more year?

83. Error Analysis   To evaluate the expression $\dfrac{16}{2+8}$ on a TI-83 Plus calculator, a student enters the expression as

shown in the figure. What is the error and how can it be corrected?

84. Calculator Discovery   Calculators and computer languages such as C++, Java, and FORTRAN require that algebraic expressions be entered on one line. Expressions with fractions must be entered using grouping symbols to indicate clearly the numerator and the denominator. Write each expression by hand as it would appear on a calculator or in a computer program. Then enter the expressions in a calculator. Compare your results with those of others in your group.

  a. $\dfrac{2(4-7)}{9+16}$ **b.** $\dfrac{3+1}{2}\cdot 8$

  c. $\dfrac{4-3}{9}+16$ **d.** $\dfrac{1+13}{2(7)}$

85. Stock Splits   If a stock splits two for one, then stockholders now own two shares for each of their original shares. The effect of this is that the split-adjusted price per share is now one-half the original cost. Thus one share costing $10 becomes two shares with a split-adjusted cost of $5 per share.
  On February 16, 1990 an original share of Cisco Systems Inc. (CSCO on the NASDAQ) cost $18. After a two for one split on 3/1/1991, one share costing $18.00 has become two shares that equally split this $18 cost; a split-adjusted price of $9 per share. Complete the following table related to Cisco's share performance.*

DATE	SPLIT RATIO	NUMBER OF SHARES	SPLIT-ADJUSTED PRICE/SHARE ($)
2/16/1990	—	1	18.00
3/1/1991	2 for 1	2	9.00
3/6/1992	2 for 1	4	
3/5/1993	2 for 1		
3/4/1994	2 for 1		
2/2/1996	2 for 1		
11/18/1997	1.5 for 1		
8/14/1998	1.5 for 1		
5/24/1999	2 for 1		
3/22/2000	2 for 1		

On March 30, 2000 the price per share of Cisco stock was $73.63. Determine the March 30, 2000 value of an $18.00 investment made February 16, 1990.

*Source: Stock Split Information under Investor Relations on the Cisco Systems Inc. website http://www.cisco.com/

KEY CONCEPTS   FOR CHAPTER 1

1. **Real Numbers:** The real numbers consist of the rational numbers and the irrational numbers.
   Rational Numbers: Numbers that can be written as a ratio of two integers are rational numbers. Expressed decimally, rational numbers are either terminating decimals or repeating decimals.
   Irrational Numbers: Real numbers, such as $\sqrt{2}$ or π, which cannot be written as a ratio of two integers, are irrational numbers. Expressed decimally, irrational numbers are infinite nonrepeating decimals.
   Subsets of the rational numbers:
   Natural Numbers: 1, 2, 3, 4, 5, 6, . . .
   Whole Numbers: 0, 1, 2, 3, 4, 5, 6, . . .
   Integers: . . . −3, −2, −1, 0, 1, 2, 3, . . .

2. **Additive Inverses:** Real numbers that are the same distance from the origin but on opposite sides of the origin are called opposites of each other, or additive inverses. If a is a real number, the opposite of a is $-a$. The opposite of $-a$ is a. The numbers -2 and 2 are opposites of each other. The sum of a number and its opposite is zero. Both $2 + (-2) = 0$ and $-2 + 2 = 0$.

3. **Additive Identity:** Zero is called the additive identity because 0 is the only real number with the property that $a + 0 = a$ and $0 + a = a$ for every real number a.

4. **Inequality and Interval Notation:**

INEQUALITY NOTATION	VERBAL MEANING	INTERVAL NOTATION
$x > a$	x is greater than a	$(a, +\infty)$
$x \geq a$	x is greater than or equal to a	$[a, +\infty)$
$x < a$	x is less than a	$(-\infty, a)$
$x \leq a$	x is less than or equal to a	$(-\infty, a]$
$a < x < b$	x is greater than a and less than b	(a, b)
$a < x \leq b$	x is greater than a and less than or equal to b	$(a, b]$
$a \leq x < b$	x is greater than or equal to a and less than b	$[a, b)$
$a \leq x \leq b$	x is greater than or equal to a and less than or equal to b	$[a, b]$

5. **Absolute Value:** The absolute value of a real number is the distance between this number and the origin on the real number line.
$$|x| = \begin{cases} x & \text{if } x \text{ is nonnegative} \\ -x & \text{if } x \text{ is negative} \end{cases}$$

6. **Square Roots:** If x is a real number and $x \neq 0$, then $\sqrt{x}$ denotes a positive number r such that $r^2 = x$. For example, $\sqrt{9} = +3$ since $3^2 = 9$.

7. **Evaluating an Algebraic Expression:** When an algebraic expression is evaluated for given values of the variables, it is often wise to use parentheses to avoid careless errors with either pencil and paper or your calculator.

8. **Addition of Real Numbers:**
   Like signs: Add the absolute values of the numbers and use the common sign of these terms.
   Unlike signs: Find the difference of the absolute values of the numbers and use the sign of the term that has the larger absolute value.

9. **Subtraction of Real Numbers:**
   Step 1   Rewrite the problem as an addition problem by changing the operation to addition and by changing the sign of the subtrahend (the number being subtracted).
   Step 2   Add using the rule for adding signed numbers.

10. **Multiplication of Real Numbers:**
    Like signs: Multiply the absolute values of the two factors and use a positive sign for the product.
    Unlike signs: Multiply the absolute values of the two factors and use a negative sign for the product.
    Zero factor: The product of 0 and any other factor is 0.

11. **Multiplicative Inverse:** The reciprocal or multiplicative inverse of any nonzero real number a can be represented by $\frac{1}{a}$. The product of a number and its multiplicative inverse is positive one.

12. **Multiplicative Identity:** One is called the multiplicative identity because 1 is the only real number with the property that $1 \cdot a = a$ and $a \cdot 1 = a$ for every real number a.

13. **Product of Negative Factors:** The product is positive if the number of negative factors is even. The product is negative if the number of negative factors is odd.

14. **Division of Real Numbers:**
    Like signs: Divide the absolute values of the two numbers and use a positive sign for the quotient.
    Unlike signs: Divide the absolute values of the two numbers and use a negative sign for the quotient.
    Zero dividend: $\frac{0}{x} = 0$ for $x \neq 0$.
    Zero divisor: $\frac{x}{0}$ is undefined for every real number x.

15. **Ratio:** The ratio a to b can be denoted by either $a : b$ or $\frac{a}{b}$.

16. **Exponential Notation:** For any natural number n,
    $$b^n = \underbrace{b \cdot b \cdot \ldots \cdot b}_{n \text{ factors of } b}$$
    with base b and exponent n.

17. **Sequence:** A sequence is an ordered set of numbers. Subscript notation ($a_1, a_2, \ldots, a_n, \ldots$, read a sub one, a sub two, and a sub n, respectively) is used often to denote the terms of a sequence.

18. **Operations with Fractions:**

Addition: $\dfrac{a}{b} + \dfrac{c}{b} = \dfrac{a+c}{b}$ for $b \neq 0$

Subtraction: $\dfrac{a}{b} - \dfrac{c}{b} = \dfrac{a-c}{b}$ for $b \neq 0$

Multiplication: $\dfrac{a}{b} \cdot \dfrac{c}{d} = \dfrac{a \cdot c}{b \cdot d}$ for $b \neq 0$ and $d \neq 0$

Division: $\dfrac{a}{b} \div \dfrac{c}{d} = \dfrac{a}{b} \cdot \dfrac{d}{c} = \dfrac{a \cdot d}{b \cdot c}$ for $b \neq 0$, $c \neq 0$, and $d \neq 0$

Reducing Fractions: $\dfrac{ac}{bc} = \dfrac{a}{b}$ for $b \neq 0$ and $c \neq 0$.

19. **Order of Operations:**
Step 1 Start with the expression within the innermost pair of grouping symbols.
Step 2 Perform all exponentiations.
Step 3 Perform all multiplications and divisions as they appear from left to right.
Step 4 Perform all additions and subtractions as they appear from left to right.

20. **Properties of Addition and Multiplication**
Commutative Properties: Properties dealing with order of operations:
- $a + b = b + a$ Addition is commutative.
- $ab = ba$ Multiplication is commutative.

Associative Properties: Properties dealing with grouping:
- $(a + b) + c = a + (b + c)$ Addition is associative.
- $(ab)(c) = a(bc)$ Multiplication is associative.

Distributive Property of Multiplication over Addition: Property describing the relationship between multiplication and addition:
- $a(b + c) = ab + ac$ Multiplication distributes over
- $(b + c)a = ba + ca$ addition.

REVIEW EXERCISES   FOR CHAPTER 1

In Exercises 1–25 perform the indicated operations.

1. **a.** $-16 + 4$ **b.** $-16 + (-4)$
   c. $-16 - 4$ **d.** $-16 - (-4)$
2. **a.** $-16(4)$ **b.** $-16(-4)$
   c. $-16 \div 4$ **d.** $-16 \div (-4)$
3. **a.** $-7 + 0$ **b.** $-7(0)$
   c. $\dfrac{0}{-7}$ **d.** $\dfrac{-7}{0}$
4. **a.** $24 + (-6)$ **b.** $24 - (-6)$
   c. $24(-6)$ **d.** $24 \div (-6)$
5. **a.** $9 + 0.01$ **b.** $9 - 0.01$
   c. $9(0.01)$ **d.** $9 \div 0.01$
6. **a.** $-4.5 + 1000$ **b.** $-4.5 - 1000$
   c. $-4.5(1000)$ **d.** $-4.5 \div 1000$
7. **a.** $-7 - 8 + 9$ **b.** $-7 - (8 + 9)$
   c. $-(7 - 8) + 9$ **d.** $-(7 - 8 + 9)$
8. **a.** $-6 + 8 - 11 - 15$ **b.** $(-6 + 8) - (11 - 15)$
   c. $-6 + (8 - 11 - 15)$ **d.** $-(6 + 8) - (11 - 15)$
9. **a.** $-36 \div 4 \cdot 3$ **b.** $-36 \div (4 \cdot 3)$
   c. $-(36 \div 4 \cdot 3)$ **d.** $-(36 \div 4) \cdot 3$
10. **a.** $(15 - 6)(9 - 11)$ **b.** $15 - 6(9 - 11)$
    c. $15 - 6 \cdot 9 - 11$ **d.** $(15 - 6)9 - 11$
11. **a.** 25% of 36 **b.** 35% of 15
    c. 10% of 12 **d.** 18% of 100
12. **a.** $(-3)^2$ **b.** -3^2
    c. $(-3)^3$ **d.** -3^3
13. **a.** $(3 + 4)^2$ **b.** $3^2 + 4^2$
    c. $(3 - 4)^2$ **d.** $3^2 - 4^2$
14. **a.** 5^2 **b.** 2^5
    c. $(-5)^2$ **d.** -2^5

15. **a.** $|3 - 11|$ **b.** $|3| - |11|$
    c. $-|3 - 11|$ **d.** $|3| + |-11|$
16. **a.** $\sqrt{64} + \sqrt{36}$ **b.** $\sqrt{64 + 36}$
    c. $\sqrt{1} + \sqrt{1} + \sqrt{1} + \sqrt{1}$ **d.** $\sqrt{1 + 1 + 1 + 1}$
17. **a.** $(-1)(-2)(-3)(-4)$
    b. $(-1)(-2)(-3)(-4)(-5)$
    c. $(0)(-1)(-2)(-3)(-4)$
    d. $(-1)(-2)(-3)(-4)(-5)(-6)$
18. **a.** 0^{37} **b.** 0^{38}
    c. $(-1)^{37}$ **d.** $(-1)^{38}$
19. **a.** $\dfrac{2}{3} + \dfrac{3}{4}$ **b.** $\dfrac{2}{3} - \dfrac{3}{4}$
    c. $\left(\dfrac{2}{3}\right)\left(\dfrac{3}{4}\right)$ **d.** $\left(\dfrac{2}{3}\right) \div \left(\dfrac{3}{4}\right)$
20. **a.** $-\dfrac{14}{15} + \dfrac{21}{25}$ **b.** $-\dfrac{14}{15} - \dfrac{21}{25}$
    c. $-\dfrac{14}{15} \cdot \dfrac{21}{25}$ **d.** $-\dfrac{14}{15} \div \dfrac{21}{25}$
21. **a.** $-48 + 12 \div 6 + 3 - 1$ **b.** $-48 - 12 \cdot 6 \div 3 + 1$
    c. $-48 + 12 - 6 \div 3 \cdot 1$ **d.** $-48 \div 12 \cdot 6 - 3 + 1$
22. $14 - 2[11 - 5(13 - 10)]$
23. $19 + 3[(40 - 7) - (4 + 3^2)]$
24. $\dfrac{-3 + 7}{8 - 10}$
25. $\dfrac{14 - 7(-5)}{-3(9) - 2(-10)}$

Estimation and Calculator Skills

In Exercises 26–33 mentally determine the sign of each expression and then use a calculator to compute the value.

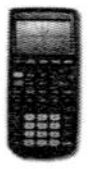

PROBLEM	SIGN	CALCULATOR VALUE
26. $-312 + (-221)$		
27. $58 - (-75)$		
28. $-41 - (-72)$		
29. $(53)(-37)(-21)$		
30. $-2^2 \cdot 4 \cdot (-8)$		
31. $(-3)^2 \cdot 5 \cdot (-7)$		
32. $-\lvert 361 \rvert - \lvert -196 \rvert$		
33. $\sqrt{196} - \sqrt{361}$		

Estimation and Calculator Skills

In Exercises 34–39 mentally estimate the value of each expression and then use a calculator to compute the exact value.

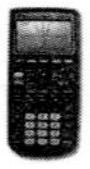

PROBLEM	MENTAL ESTIMATE	CALCULATOR VALUE
34. $8.903 + 7.146$		
35. $1049.8768 - 250.4132$		
36. $(301.06)(9.8)$		
37. $43.5969 \div 7.23$		
38. $(5.01)^3$		
39. $\sqrt{50.41}$		

In Exercises 40–45 fill in the missing entries in each column. Give the verbal meaning for each problem.

	Inequality	Graph	Interval
40.		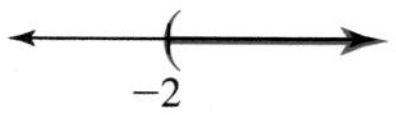	
41.			$(-\infty, 4]$
42.	$x < 3$		
43.			$(-3, 0)$
44.	$-7 \le x \le -1$		
45.		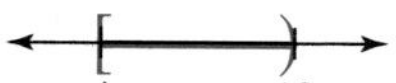	

Multiple Representations

In Exercises 46–55 write each verbal statement in algebraic form.

46. The opposite of x equals eleven.

47. The absolute value of x is less than or equal to seven.

48. The square root of twenty-six is greater than five.

49. The sum of x and five is equal to four.

50. The difference of x minus three is less than y.

51. The product of negative three and y is negative one.

52. The quotient of x and three is twelve.

53. The ratio of x to y equals three-fourths.

54. Twice the quantity x plus two equals nine.

55. Five times the quantity three x minus four equals thirteen.

In Exercises 56 and 57 calculate the first five terms of each arithmetic sequence.

56. $a_n = 10n - 3$

57. $a_n = 30 - 5n$

In Exercises 58–62 match each algebraic equation with the property it illustrates.

58. $(xy)z = x(yz)$

59. $(x + y) + z = x + (y + z)$

60. $w(xy + z) = w(yx + z)$

61. $w(xy + z) = wxy + wz$

62. $w(xy + z) = w(z + xy)$

A. Associative property of addition

B. Associative property of multiplication

C. Commutative property of addition

D. Commutative property of multiplication

E. Distributive property of multiplication over addition

In Exercises 63 and 64 evaluate the expression $\frac{y_2 - y_1}{x_2 - x_1}$ for the given values of x_1, x_2, y_1, and y_2.

63. $x_1 = 3, x_2 = -2, y_1 = 6$, and $y_2 = -4$

64. $x_1 = 5, x_2 = 4, y_1 = -7$, and $y_2 = -3$

In Exercises 65 and 66 check $x = 2$ and $x = 3$ as possible solutions of the given equation.

65. $3x + 5 = 4x + 2$

66. $3(x + 5) = 4(x + 3) + 1$

67. Determine both the perimeter and the area of the volleyball court shown here.

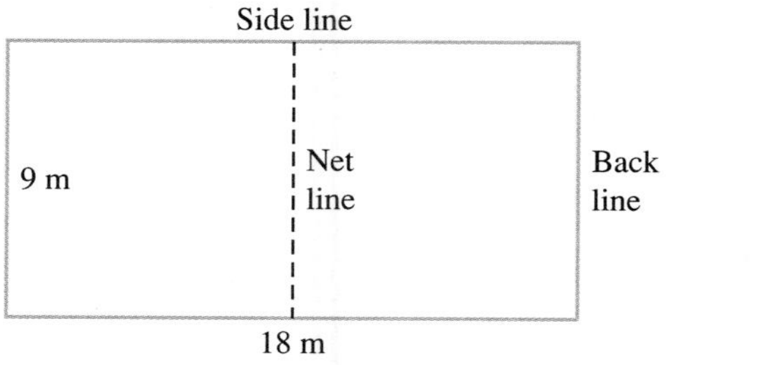

68. Determine the error and the relative error in the calculation of the perimeter if a painter measured the volleyball court in Exercise 67 as 18.1 m by 9.1 m.

69. Determine the unit cost of a CD when a box of 24 CDs costs \$56.40.

In Exercises 70–74 match each number with the description given to the right.

70. -5

71. 0

72. $\frac{3}{4}$

73. $\sqrt{4}$

74. $\sqrt{5}$

A. A whole number that is not a natural number

B. A natural number

C. An integer that is not a natural number

D. An irrational number

E. A rational number that is not an integer

75. Best Buy for a Sports Drink One brand of a sports drink is sold in two popular sizes. Which is the better buy: a 20-ounce (oz) bottle that is priced at \$0.99 or a 32-oz bottle that is priced at \$1.69?

76. **Heights of a Daughter and a Father**   A 20-year-old student is 68 in tall and her father is 72 in tall. Determine the ratio of the height of the daughter to that of her father.

77. **Computer Sale**   To make room for a new computer model the prices of all older models have been reduced by 20%. What is the sale price of a computer that normally sells for $1299?

78. **Active Ingredient**   Calculate the amount of the active ingredient in each mixture.
   a. 400 liters of a 10% insecticide mixture.
   b. 600 gallons of a 5% ethanol mixture.

79. **Rate of Work**   Use the formula $R = \dfrac{W}{T}$ to compute the rate of work for each situation.
   a. An assembly line produces 1600 golf balls in 8 hours.
   b. A pipe can fill 1 brewery vat in 4 hours.

MASTERY TEST   FOR CHAPTER 1

[1.1]   **1.** Write the additive inverse of each of the real numbers graphed as follows:
   a.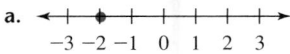
   $-3\ -2\ -1\ \ 0\ \ 1\ \ 2\ \ 3$

   b.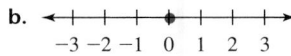
   $-3\ -2\ -1\ \ 0\ \ 1\ \ 2\ \ 3$

   c.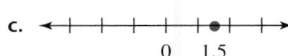
   $0\qquad 1.5$

   d.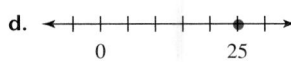
   $0\qquad\qquad 25$

[1.1]   **2.** Evaluate each of these absolute value expressions.
   a. $|23|$ **b.** $|-23|$
   c. $|23 - 23|$ **d.** $|23| + |-23|$

[1.1]   **3.** Express each of these intervals using interval notation.
   a.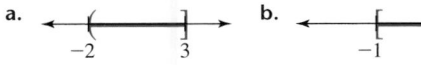
   $-2\qquad 3$
   b.
   -1

   c. $x < 4$ **d.** $5 \le x < 9$

[1.1]   **4.** Mentally estimate the value of each of the following square roots to the nearest integer and then use your calculator to approximate each value to the nearest hundredth:
   a. $\sqrt{26}$ **b.** $\sqrt{99}$ **c.** $\sqrt{4+4}$ **d.** $\sqrt{4} + \sqrt{4}$

[1.1]   **5.** Match each number with the most appropriate description.
   a. -5 **A.** A whole number that is not a natural number
   b. $-\dfrac{2}{3}$ **B.** An integer that is not a whole number
   c. 0 **C.** A rational number that is not an integer
   d. $\sqrt{9}$ **D.** A positive number that is not rational
   e. $\sqrt{11}$ **E.** A natural number

[1.2]   **6.** Calculate each sum without using a calculator.
   a. $17 + (-11)$ **b.** $-17 + (-11)$ **c.** $-17 + 11$
   d. $\dfrac{3}{8} + \dfrac{1}{8}$ **e.** $-\dfrac{3}{8} + \dfrac{1}{8}$ **f.** $-\dfrac{2}{5} + \left(-\dfrac{3}{8}\right)$
   g. $57.3 + (-57.3)$ **h.** $(-5 + 3) + [-8 + (-2)]$

[1.2]   **7. a.** The property that says $5(6 + 7) = 5(7 + 6)$ is the _____ property of _____.
   b. The property that says $5 + (6 + 7)$ $= (5 + 6) + 7$ is the _____ property of _____.
   c. Use the associative property of addition to rewrite $(x + y) + 5$
   d. Use the commutative property of addition to rewrite $5(x + y)$.

[1.2]   **8.** Evaluate expression for $x = -1$, $y = -2$, and $z = -3$.
   a. $x + y + z$ **b.** $-(x + y + z)$
   c. $-x + y + z$ **d.** $|x| + |y| + |z|$

[1.3]   **9.** Calculate each difference without using a calculator.
   a. $15 - 8$ **b.** $-15 - 8$ **c.** $-15 - (-8)$
   d. $15 - (-8)$ **e.** $\dfrac{5}{6} - \dfrac{1}{6}$ **f.** $-\dfrac{5}{6} - \dfrac{1}{6}$
   g. $\dfrac{1}{2} - \left(-\dfrac{1}{3}\right)$ **h.** $-\dfrac{3}{5} - \left(-\dfrac{5}{6}\right)$

[1.3]   **10.** Calculate the first five terms of each sequence.
   a. $a_n = n + 2$ **b.** $a_n = n - 2$
   c. $a_n = 2n + 1$ **d.** $a_n = 12 - n$

[1.3]   **11.** Check $x = 3$ to determine whether it is a solution of each equation.
   a. $x + 2 = 8 - x$ **b.** $x + 1 = x + 4$
   c. $2x + 1 = x + 4$ **d.** $x + 1 = x - 7$

[1.4]   **12.** Calculate each product without using a calculator.
   a. $5(-6)$ **b.** $-5(-6)$ **c.** $-5(6)$
   d. $-5(0)$ **e.** $\dfrac{1}{3} \cdot \dfrac{1}{5}$ **f.** $\left(-\dfrac{3}{7}\right)\left(-\dfrac{7}{9}\right)$
   g. $-\dfrac{2}{5}\left(\dfrac{15}{14}\right)$ **h.** $(-2)(-3)(-4)$

[1.4]   **13. a.** The property that says $5 \cdot (6 + 7) = (6 + 7) \cdot 5$ is the _____ property of _____.
   b. The property that says $5(6 \cdot 7) = (5 \cdot 6) \cdot 7$ is the _____ property of _____.
   c. Use the associative property of multiplication to rewrite $(5x)(y)$.
   d. Use the commutative property of multiplication to rewrite $x(a + b)$.

[1.4] **14.** Use the formula $A = l \cdot w$ to calculate the area of this rectangle.

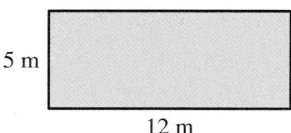

5 m

12 m

[1.5] **15.** Calculate each quotient without using a calculator.

a. $-12 \div 4$ **b.** $-12 \div (-4)$ **c.** $12 \div (-4)$

d. $0 \div 4$ **e.** $-4 \div 0$ **f.** $-\dfrac{1}{2} \div \dfrac{1}{3}$

g. $-\dfrac{5}{6} \div \left(-\dfrac{2}{3}\right)$ **h.** $\dfrac{12}{35} \div \left(-\dfrac{14}{11}\right)$

[1.5] **16. a. Defective Keyboard Trays** The computer support services at Apollo College reported that 21 of 105 keyboard trays in an open computer lab had to be repaired one year. What fractional portion of the trays was repaired?

     b. Best Buy for Salad Dressing One brand of salad dressing is sold in two popular sizes. Which is the better buy: a 16-ounce bottle that sells for $2.59 or a 12-ounce bottle that sells for $1.92?

[1.6] **17.** Calculate the value of each expression without using a calculator.

a. 5^2 **b.** 2^5 **c.** $(-5)^2$ **d.** -5^2

e. 0^7 **f.** $\left(\dfrac{3}{7}\right)^2$ **g.** 219^1 **h.** $(-1)^{219}$

[1.6] **18.** Calculate the value of each expression without using a calculator.

a. $15 - 2(8 - 5)$ **b.** $4 \cdot 7 - 2 \cdot 3 + 5$

c. $15 - 6 \div 3 + 11$ **d.** $\dfrac{14 - 3 \cdot 6}{(14 - 3) \cdot 6}$

e. $(3 + 5)^2$ **f.** $3^2 + 5^2$

g. $-8 - 3[5 - 4(6 - 9)]$

h. $\sqrt{25 - 16} - \left(\sqrt{25} - \sqrt{16}\right)$

[1.6] **19. a.** The _____ property of _____ over _____ states that $a(b + c) = ab + ac$ and $(b + c)a = ba + ca$.

b. Expand $11(3x - 4)$ **c.** Factor $15x + 20$

d. Expand $-7(2x - 5)$ **e.** Factor $16x - 24$

GROUP PROJECT   FOR CHAPTER 1

Risks and Choices: Shopping for Interest Rates on Car and House Loans

The formula for determining the monthly payment for a car or house loan is

$$P_n = \frac{AR\left(1 + \dfrac{R}{12}\right)^{12n}}{12\left(1 + \dfrac{R}{12}\right)^{12n} - 12}$$

where A represents the amount of the loan, R represents the interest rate, n represents the number of years, and P_n represents the monthly payment required to pay off the loan in n years.

Use this formula and a calculator to complete the following tables.

Consider a $10,000 car loan at several different interest rates.

INTEREST RATE (%)	LOAN AMOUNT ($)	NUMBER OF YEARS	MONTHLY PAYMENT ($)	TOTAL OF ALL PAYMENTS ($)
8.25	10,000	2		
8.50	10,000	2		
8.75	10,000	2		
9.00	10,000	2		
9.25	10,000	2		
9.50	10,000	2		

Consider a $50,000 home loan at 9.00% interest for several different time periods.

INTEREST RATE (%)	LOAN AMOUNT ($)	NUMBER OF YEARS	MONTHLY PAYMENT ($)	TOTAL OF ALL PAYMENTS ($)
9.00	50,000	10		
9.00	50,000	15		
9.00	50,000	20		
9.00	50,000	25		
9.00	50,000	30		

Use these facts to discuss the advantages and disadvantages for the borrower for each of these options.

2

LINEAR

EQUATIONS

AND PATTERNS

Table 2.1.1 is a payment schedule for a $3300 loan on a used car at 8.5% interest. The payment schedule provided by the loan company has been calculated so that the loan will be paid off at the end of 2 years. This table contains several sequences, and the calculation of the entries involves the use of linear equations.

Table 2.1.1 Two-Year Payment Schedule

PAYMENT NO.	PAYMENT DATE	STARTING BALANCE ($)	MONTHLY PAYMENT ($)	TOTAL PAID ($)	NEW BALANCE ($)
1	1/1/01	3300.00	150.00	150.00	3173.37
2	2/1/01	3173.37	150.00	300.00	3045.85
3	3/1/01	3045.85	150.00	450.00	2917.42
4	4/1/01	2917.42	150.00	600.00	2788.08
5	5/1/01	2788.08	150.00	750.00	2657.82
6	6/1/01	2657.82	150.00	900.00	2526.65
7	7/1/01	2526.65	150.00	1050.00	2394.54
8	8/1/01	2394.54	150.00	1200.00	2261.50
9	9/1/01	2261.50	150.00	1350.00	2127.51
10	10/1/01	2127.51	150.00	1500.00	1992.58
11	11/1/01	1992.58	150.00	1650.00	1856.69
12	12/1/01	1856.69	150.00	1800.00	1719.84
13	1/1/02	1719.84	150.00	1950.00	1582.01
14	2/1/02	1582.01	150.00	2100.00	1443.22
15	3/1/02	1443.22	150.00	2250.00	1303.44
16	4/1/02	1303.44	150.00	2400.00	1162.67
17	5/1/02	1162.66	150.00	2550.00	1020.90
18	6/1/02	1020.90	150.00	2700.00	878.12
19	7/1/02	878.12	150.00	2850.00	734.34
20	8/1/02	734.34	150.00	3000.00	589.54
21	9/1/02	589.54	150.00	3150.00	443.71
22	10/1/02	443.71	150.00	3300.00	296.85
23	11/1/02	296.85	150.00	3450.00	148.95
24	12/1/02	148.95	150.00	3600.00	0.00

In this chapter we study sequences with a constant rate of change and linear equations. We examine methods for solving linear equations algebraically and graphically. Some of the exercises and group projects will help you to develop the skills needed to prepare and understand your own payment schedules.

Section 2.1   The Rectangular Coordinate System and Arithmetic Sequences

Objectives:

1. Plot ordered pairs on a rectangular coordinate system.
2. Draw a scatter diagram of a set of points.
3. Identify an arithmetic sequence.

The payment schedule on the previous page contains the sequences 1, 2, 3, 4, 5 and 150, 300, 450, 600, 750. Both of these sequences have a constant change from term to term. In the first sequence the constant change is 1, and in the second sequence the constant change is 150. As a prelude to the graphing material to be covered in this section, consider the graph of the sequence of totals paid from the payment schedule on the previous page. Note the linear pattern exhibited by the points on this graph.

PAYMENT NUMBER (n)	TOTAL PAID (a_n)
First payment (1)	$a_1 = 150$
Second payment (2)	$a_2 = 150 + 150 = 300$
Third payment (3)	$a_3 = 300 + 150 = 450$
Fourth payment (4)	$a_4 = 450 + 150 = 600$
Fifth payment (5)	$a_5 = 600 + 150 = 750$

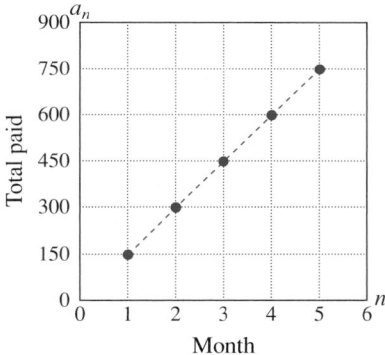

Each point visually pairs the payment number with the total paid after this payment is made. Note the linear pattern illustrated by these points.

A Mathematical Note

René Descartes (1596–1650), born to a noble French family, was known for his studies in anatomy, astronomy, chemistry, physics, and philosophy, as well as mathematics. Prior to Descartes, algebra was concerned with numbers and calculations, and geometry was concerned with figures and shapes. Descartes merged the power of these two areas into analytic geometry—his most famous discovery.

The preceding graph was shown on a **rectangular coordinate system,** also known as the **Cartesian coordinate system,** or the **coordinate plane.** On the rectangular coordinate system the horizontal number line is called the **x-axis,** the vertical number line is called the **y-axis,** and the point where they cross is called the **origin.** These axes divide the plane into four **quadrants,** labeled counterclockwise as I, II, III, and IV as illustrated in Figure 2.1.1. The points on the axes are not considered to be in any of the quadrants.

On the x-axis points to the right of the origin are positive and those to the left of the origin are negative. On the y-axis points above the origin are positive and those below are negative. Any point in the plane can be uniquely identified by specifying its horizontal and vertical location with respect to the origin. We identify a particular point by giving an ordered pair (x, y) with **coordinates** x and y. The first coordinate is called the **x-coordinate,** and the second coordinate is called the **y-coordinate.** The ordered pair $(0, 0)$ identifies the origin.

The point in Figure 2.1.1(c) is identified by the ordered pair $(5, 3)$ in quadrant I. The x-coordinate is 5 and the y-coordinate is 3. The identification of this point can be associated with the horizontal and vertical segments of the shaded rectangle—hence the name *rectangular coordinate system.*

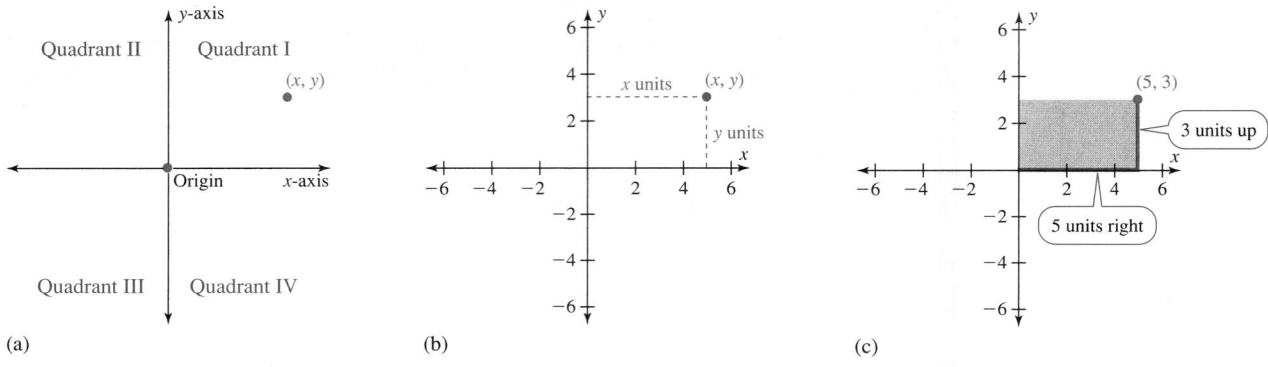

Figure 2.1.1 Cartesian coordinate system.

Parentheses are used to indicate both ordered pairs as in the point (3, 5) and intervals as in the interval (3, 5):

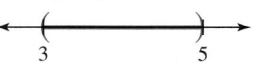

These meanings are not easily confused because the context in which they occur will make each meaning clear.

■ EXAMPLE 1   Plotting Points on a Rectangular Coordinate System

Plot $(-3, 4)$ and $(2, -5)$ on a rectangular coordinate system.

SOLUTION _____

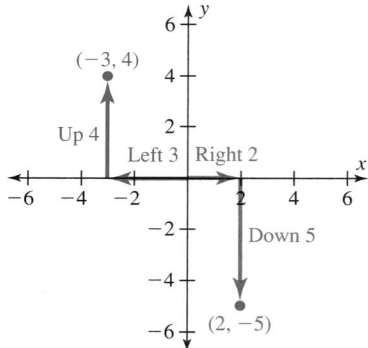

To plot $(-3, 4)$, start at the origin and move 3 units left and then move 4 units up.

To plot $(2, -5)$, start at the origin and move 2 units right and then move 5 units down.

■ EXAMPLE 2   Identifying Coordinates of Points

Identify the coordinates of each of the points A–E in the following figure and give the quadrant in which each point is located.

SOLUTIONS _____

A: $(5, 1)$; quadrant I
B: $(-2, 5)$; quadrant II
C: $(-4, -2)$; quadrant III
D: $(3, -6)$; quadrant IV
E: $(4, 0)$ is on the x-axis and thus is not in any quadrant.

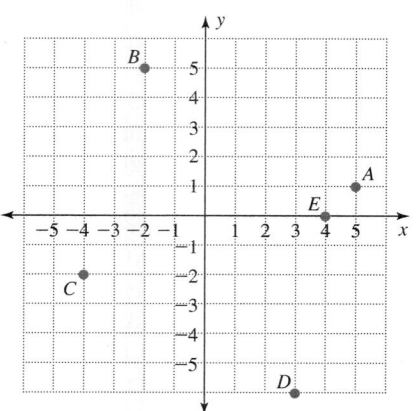

CALCULATOR PERSPECTIVE 2.1.1	Using the Integer Setting to Examine the Four Quadrants

Place a calculator in integer setting after making sure that there are no expressions entered in the ⬭ Y= screen. Then use the arrows ◖ ◗ ▲ ▼ to move to the different quadrants. Make note of the signs of the x and y values in each quadrant.

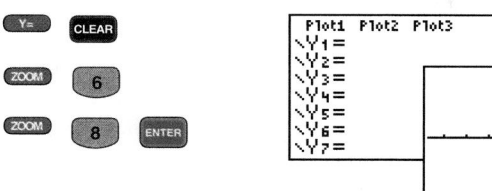

Note: What do you notice about all points (a) on the x-axis? (b) on the y-axis? (c) in each quadrant? ■

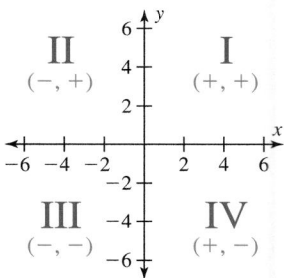

Figure 2.1.2 Quadrant sign pattern.

All the points within the same quadrant have the same sign pattern. For example, in quadrant I both coordinates are positive, which we can denote by $(+, +)$. The sign pattern for each quadrant is shown in Figure 2.1.2. Knowing the sign pattern in each quadrant is useful in trigonometry. These patterns also can be helpful in analyzing scatter diagrams in statistics.

A **scatter diagram** for a set of data points is simply a graph of these points. It allows us to examine the data for some type of pattern. For example, if the data points all lie near a line, then the pattern exhibited is called a **linear relationship.** Draw a scatter diagram for the data points given in the following table.

■ **EXAMPLE 3** Drawing a Scatter Diagram

Draw a scatter diagram for the data points given in the following table.

SOLUTION _____

x	y
-15	-12
-10	-4
-5	-1
0	6
5	8
9	15
15	18

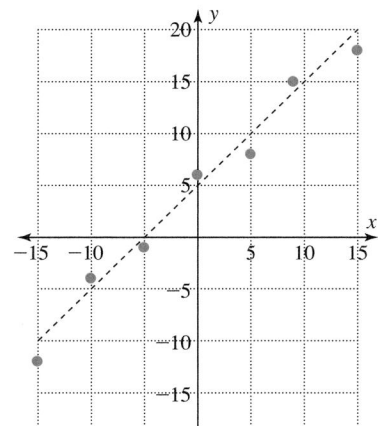

The table format is used often to give ordered pairs. The first ordered pair from this table is $(-15, -12)$.

The dashed line is placed on the graph to show that the data points all lie near this line. Therefore the relationship between x and y is approximately a linear relationship.

■

The size of the coordinates is used to determine the scale on each axis. If the coordinates differ significantly in size, then a different scale can be used on each axis. This practice is common in statistics, where the x and y variables often represent quantities measured in different units. This is illustrated in Example 4. Example 4 also illustrates that we sometimes start the input with a gap between the origin and the first value.

■ EXAMPLE 4   Drawing a Scatter Diagram

Draw a scatter diagram for the data points for the data in these financial reports from Texaco. (**http://www.texaco.com**)

SOLUTION

YEAR x	PROFIT (IN MILLIONS OF $) y
1998	580
1999	1200
2000	2500
2001	4200

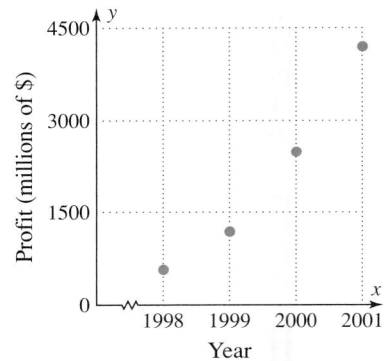

There are too few points given here to denote a specific pattern other than an upward trend in profits.

The sequence 3, 5, 7, 9, 11 also can be written as $a_1 = 3$, $a_2 = 5$, $a_3 = 7$, $a_4 = 9$, and $a_5 = 11$, or as the set of ordered pairs $\{(1, 3), (2, 5), (3, 7), (4, 9), (5, 11)\}$. Note that in ordered-pair notation the term number is the x-value and the sequence value is the y-value. A scatter diagram of these points is given in Example 5.

■ EXAMPLE 5   Drawing a Scatter Diagram of a Sequence

Draw a scatter diagram for the sequence 3, 5, 7, 9, 11.

SOLUTION

x	y
1	3
2	5
3	7
4	9
5	11

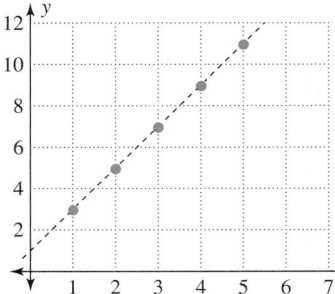

Use the table to express the first five terms as ordered pairs. Then plot these ordered pairs and note that these points all lie on a straight line.

The term number is the x-coordinate of each point. The sequence value is the y-coordinate.

The sequence 3, 5, 7, 9, 11 in Example 5 is one example of an arithmetic sequence. An **arithmetic sequence** is a sequence with a constant change from term to term. The constant change is called the **common difference** and often is denoted by *d.* The common difference d can be either positive, negative, or zero. The common difference for the sequence 3, 5, 7, 9, 11 is $+2$ since the change from one term to the next is $+2$. The graph of the terms of an arithmetic sequence will consist of individual disconnected points that lie on a straight line.

SELF-CHECK 2.1.2

Draw a scatter diagram for the sequence $-5, -2, 1, 4, 7.$

■ EXAMPLE 6   Identifying Arithmetic Sequences

Determine whether each sequence is an arithmetic sequence. If the sequence is arithmetic, determine the common difference d.
(a) $-4, -1, 2, 5, 8$
(b) $2, 4, 7, 8, 6$

SOLUTIONS

(a) NUMERICAL DIFFERENCE BETWEEN TERMS

$$-1 - (-4) = 3$$
$$2 - (-1) = 3$$
$$5 - 2 = 3$$
$$8 - 5 = 3$$

$$d = 3$$

VERBALLY

This is an arithmetic sequence because there is a common difference of 3. The points establish a linear pattern with the points rising 3 units from one term to the next.

GRAPHICALLY

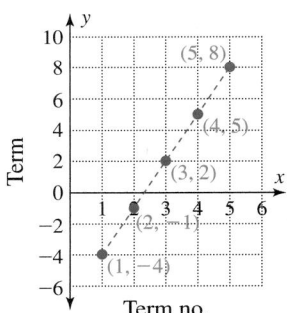

(b) NUMERICAL DIFFERENCE BETWEEN TERMS

$$4 - 2 = 2$$
$$7 - 4 = 3$$
$$2 \neq 3$$

VERBALLY

This is not an arithmetic sequence because the change from term to term is not constant. The points do not form a linear pattern. The height change between consecutive terms is not constant.

GRAPHICALLY

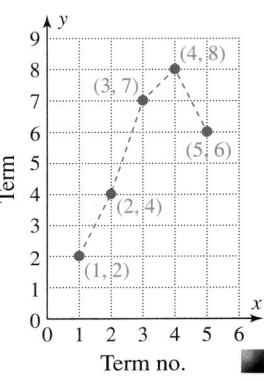

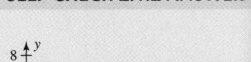
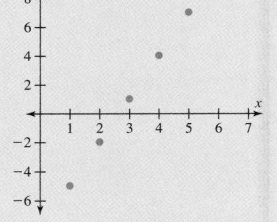

SELF-CHECK 2.1.3

Write the terms of each sequence and determine whether the sequence is arithmetic.

1.

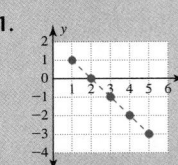

2.

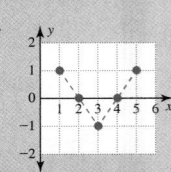

The sequence of total payments $150, $300, $450, ... in the payment schedule at the beginning of the chapter also can be represented by the algebraic equation $a_n = 150n$ where n gives the number of the term and a_n represents the value of the nth term. One of the advantages of giving an equation for a sequence is its generality. In the next example we do not need to complete the entire payment schedule in order to determine a specific monthly result. The formula allows us to compute directly any monthly result we want.

Consider a new car lease, which requires a $1200 initial payment and then a $300 payment at the first of each month for 2 years. We again let a_n represent the total paid by the nth month.

Writing out a "Word Equation" for each word problem is well worth your time. Research has shown that writing this step can produce a significant improvement in understanding how to set up word problems.

Verbally

$$\boxed{\text{Total of all payments}} = \boxed{\text{Total of monthly payments}} + \boxed{\text{Initial payment}}$$

Multiply the number of payments times the monthly amount to determine the total of the monthly payments.

Algebraically

$$a_n \quad = \quad 300n \quad + \quad 1200$$

■ **EXAMPLE 7**   Using an Equation for an Arithmetic Sequence

Use the equation $a_n = 300n + 1200$ to complete a payments table that will give a snapshot of the total of all payments after every 6-month period on a car lease with a $1200 initial payment followed by monthly payments of $300 for 2 years. Then graph this table of points.

SOLUTION _____

a_0 often denotes the initial conditions of a problem, the value at a time zero.

ALGEBRAICALLY

$a_n = 300n + 1200$
$a_0 = 300(0) + 1200 = 1200$
$a_6 = 300(6) + 1200 = 3000$
$a_{12} = 300(12) + 1200 = 4800$
$a_{18} = 300(18) + 1200 = 6600$
$a_{24} = 300(24) + 1200 = 8400$

NUMERICALLY

n	a_n
0	1200
6	3000
12	4800
18	6600
24	8400

GRAPHICALLY

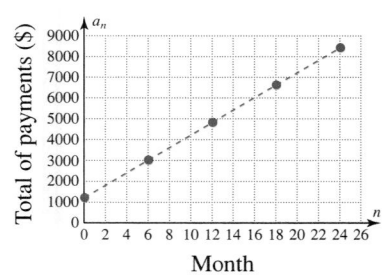

Month

SELF-CHECK 2.1.3 ANSWERS

1. 1, 0, −1, −2, −3 is an arithmetic sequence with $d = -1$.

2. 1, 0, −1, 0, 1 is not an arithmetic sequence.

VERBALLY

The cost of leasing the car is $300 more at the end of each month. The graph illustrates the linear pattern exhibited by these data points. Note that the total of the payments for leasing this car for 2 years (24 months) is $8400.

SELF-CHECK 2.1.4

Use the equation $a_n = 2n - 5$ to write the first six terms of this sequence.

The following box summarizes what we have observed regarding arithmetic sequences.

Arithmetic Sequences

Numerically: An arithmetic sequence has a constant change, d, from term to term.

Graphically: The distinct points of the graph of an arithmetic sequence all lie on a straight line. There is a constant change in height between consecutive points.

SELF-CHECK 2.1.4 ANSWER

$-3, -1, 1, 3, 5, 7$

The graph of an arithmetic sequence consists only of distinct unconnected points corresponding to the input values 1, 2, 3, 4, ... We will examine these graphs and their equations further in the next section.

USING THE LANGUAGE AND SYMBOLISM OF MATHEMATICS   2.1

1. The rectangular coordinate system also is called the _____ coordinate system.
2. On the rectangular coordinate system the horizontal axis is called the_____ axis, and the vertical axis is called the _____ axis.
3. The point where the horizontal and vertical axes cross is called the _____ and has coordinates (_____ , _____).
4. All points on the x-axis have a y-coordinate of_____ .
5. The four quadrants are numbered I, II, III, and IV in a **clockwise/counterclockwise** direction. (Select the correct choice.)

6. In quadrant _____ both coordinates are negative.
7. A _____ diagram for a set of points is a graph of these points.
8. A sequence with a constant change from term to term is called an _____ sequence.
9. In an arithmetic sequence d represents the common _____ .
10. The graph of an arithmetic sequence forms a _____ pattern.

EXERCISES 2.1

In Exercises 1 and 2 identify the coordinates of the points A, B, C, and D. Also give the quadrant in which each point is located.

1.

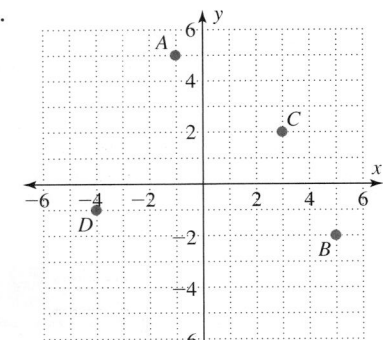

2.

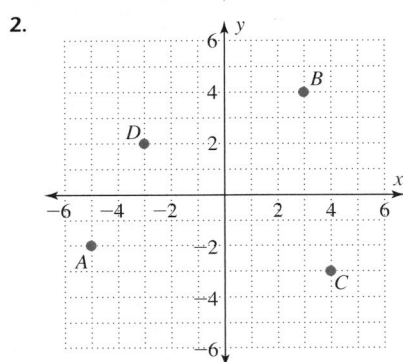

In Exercises 3 and 4 plot and label the points whose coordinates are given. Plot these points on the same coordinate system.

3. A (5, 4) B (−4, −5) C (3, −4)
    D (−6, 1) E (4, 0) F (0, −5)
4. A (2, 6) B (−3, −6) C (4, −2)
    D (−5, 3) E (−3, 0) F (0, 6)

In Exercises 5 and 6 use the sign pattern of the coordinates to determine the quadrant in which each point is located. Do not plot these points.

5. a. (−1.3, 3.7)   **b.** (−5.8, −9.3)
    c. (6.7, −2.1)   **d.** (0.4, 0.1)

6. a. (−193, −217)   **b.** $\left(\dfrac{2}{3}, -\dfrac{3}{5}\right)$
    c. (801, 913)   **d.** $\left(-\pi, \sqrt{2}\right)$

In Exercises 7 and 8 determine the axis on which each point lies. Do not plot these points.

7. a. (0, 8)   **b.** (−8, 0)
    c. $\left(-\sqrt{3}, 0\right)$ **d.** $\left(0, \sqrt{2}\right)$

8. a. $\left(\dfrac{1}{8}, 0\right)$ **b.** (0, 81)
    c. (0, −π)   **d.** $\left(-\sqrt{5}, 0\right)$

In Exercises 9 and 10 draw a scatter diagram for each set of data. Do these points lie approximately on a straight line?

9.

x	−2	−1	0	1	2	3
y	−10	−9	−4	1	6	8

10.

x	−2	−1	0	1	2	3
y	8	6	4	4	2	−2

In Exercises 11–14 graph each sequence. Do these points form a linear pattern? Is the sequence an arithmetic sequence? If the sequence is arithmetic, what is the common difference?

11. $a_1 = -4, a_2 = -2, a_3 = 0, a_4 = 2, a_5 = 4$
12. $a_1 = 5, a_2 = 3, a_3 = 1, a_4 = -1, a_5 = -3$
13. 8, 5, 2, −1, −4
14. −5, −2, 1, 4, 7

In Exercises 15–18 write the first five terms of these arithmetic sequences.

15. $a_n = 5n - 1$ **16.** $a_n = -4n + 6$
17. $a_n = 1 - 5n$ **18.** $a_n = 4n - 6$
19. Use the equation $a_n = 3n + 7$ to evaluate each expression.
    a. a_0 **b.** a_1 **c.** a_4 **d.** a_{10}
20. Use the equation $a_n = 5n - 6$ to evaluate each expression.
    a. a_0 **b.** a_2 **c.** a_{20} **d.** a_{100}
21. Use the equation $a_n = -6n + 4$ to evaluate each expression.
    a. a_0 **b.** a_1 **c.** a_4 **d.** a_{40}
22. Use the equation $a_n = -8n + 5$ to evaluate each expression.
    a. a_0 **b.** a_1 **c.** a_5 **d.** a_{50}
23. Use the equation
$$a_n = \frac{n + 1}{2}$$
to complete this table.

n	a_n
1	
2	
3	
4	
5	

24. Use the equation
$$a_n = \frac{n - 1}{2}$$
to complete this table.

n	a_n
1	
2	
3	
4	
5	

In Exercises 25 and 26 use the graph of each arithmetic sequence to write the first five terms of each sequence.

25.
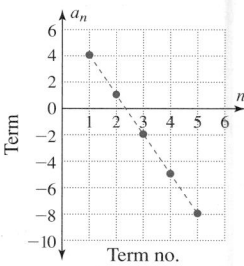

26.
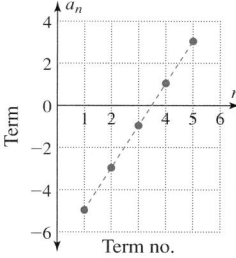

In Exercises 27–34 determine whether the sequence is arithmetic. If the sequence is arithmetic, write the common difference d.

27. $-3, 1, 5, 9, 13, \ldots$ **28.** $-11, -6, -1, 4, 9, \ldots$

29. $40, 30, 20, 10, 0$ **30.** $20, 17, 14, 11, 8, 5$

31. $3, 5, 8, 12, 17$ **32.** $2, 4, 8, 16, 32, 64$

33. a. $a_n = 2n$ **b.** $a_n = 2^n$ **34. a.** $a_n = \dfrac{n}{2}$ **b.** $a_n = \dfrac{60}{n}$

In Exercises 35–42 determine whether the sequence shown in each graph is an arithmetic sequence. For those that are arithmetic sequences find the common difference d.

35.
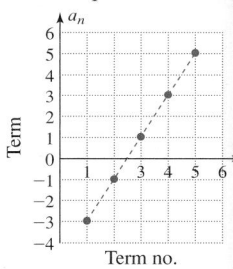

36.
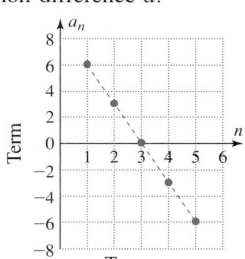

37.
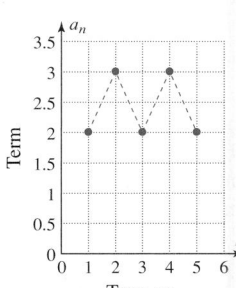

38.
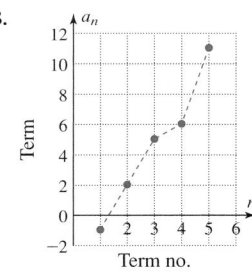

39.
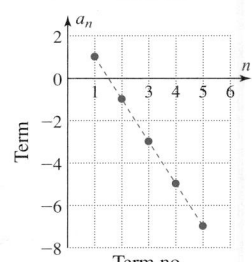

40.
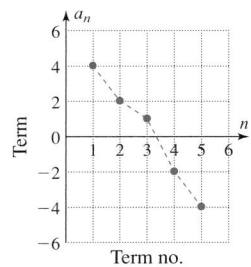

41.
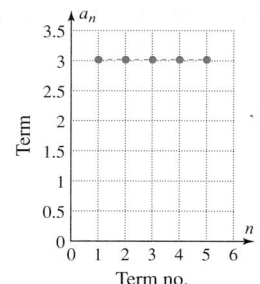

42.
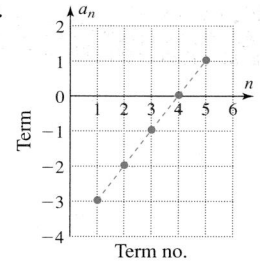

In Exercises 43–48 use the graph to complete the table. Then determine whether the sequence in the table is an arithmetic sequence.

43. Graph
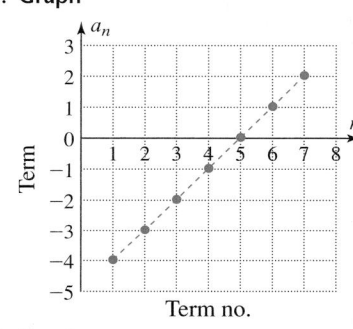

Table

n	a_n
1	
2	
3	
4	
5	
6	
7	

44. Graph
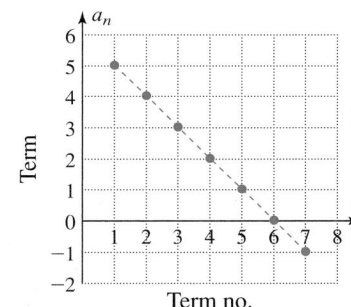

Table

n	a_n
1	
2	
3	
4	
5	
6	
7	

45. Graph
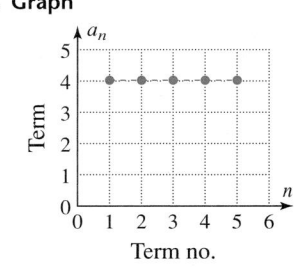

Table

n	a_n
1	
2	
3	
4	
5	

46. Graph
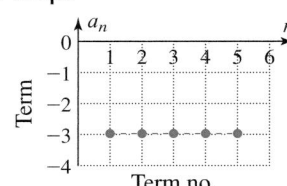

Table

n	a_n
1	
2	
3	
4	
5	

47. Graph

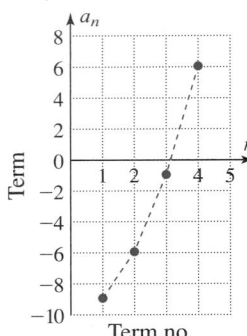

Term / Term no.

Table

n	a_n
1	
2	
3	
4	

48. Graph

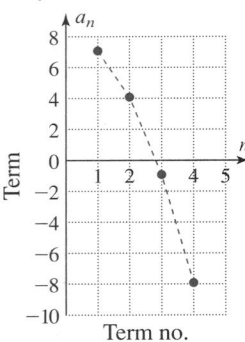

Term / Term no.

Table

n	a_n
1	
2	
3	
4	

Automobile Loan

In Exercises 49 and 50 the equation $a_n = 425n$ gives the total dollars paid on a car loan after n months. Calculate and interpret each value of a_n.

49. a. a_0 **b.** a_6 **c.** a_{12}
50. a. a_1 **b.** a_{18} **c.** a_{24}

Automobile Lease

In Exercises 51 and 52 the linear equation
$a_n = 450n + 2400$ gives the total payment in dollars after n months on the lease for a luxury sports utility vehicle. Calculate and interpret each value of a_n.

51. a. a_1 **b.** a_{18} **c.** a_{24}
52. a. a_0 **b.** a_6 **c.** a_{12}
53. A point (x, y) is in quadrant I. Complete these inequalities:
    x ___ 0, y ___ 0.
54. A point (x, y) is in quadrant II. Complete these inequalities: x ___ 0, y ___ 0.
55. A point (x, y) is in quadrant III. Complete these inequalities: x ___ 0, y ___ 0.
56. A point (x, y) is in quadrant IV. Complete these inequalities: x ___ 0, y ___ 0.
57. If $xy > 0$, then the point (x, y) is either in quadrant ___ or quadrant ___.
58. If $xy < 0$, then the point (x, y) is either in quadrant ___ or quadrant ___.
59. Consider an arithmetic sequence that has a positive common difference. Does the line through the graph of this sequence go up or down as the points move to the right?

60. Consider an arithmetic sequence that has a negative common difference. Does the line through the graph of this sequence go up or down as the points move to the right?
61. Are the even integers an arithmetic sequence? If so, what is the common difference d?
62. Are the odd integers an arithmetic sequence? If so, what is the common difference d?

Area and Perimeter

63. Determine the area and perimeter of a rectangle whose corners are located at $(0, 0)$, $(5, 0)$, $(5, 4)$, and $(0, 4)$.

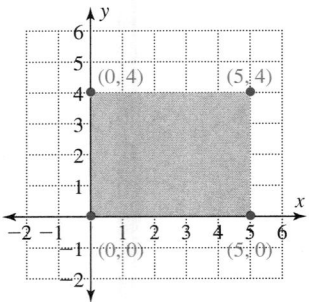

64. Determine the area and perimeter of a rectangle whose corners are located at $(0, -3)$, $(2, -3)$, $(2, 3)$, and $(0, 3)$.

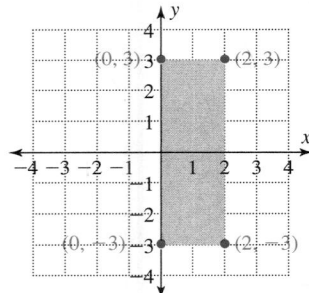

Exercises 65–67 refer to the payment schedule given at the beginning of Chapter 2.
65. Is the sequence of payment numbers 1, 2, 3, ..., 24 an arithmetic sequence? If so, what is the common difference d?
66. Is the sequence of monthly payments 150, 150, 150, ..., 150 an arithmetic sequence? If so, what is the common difference d?
67. Is the sequence of starting balances 3300.00, 3173.37, ..., 3045.85 an arithmetic sequence? If so, what is the common difference d?

Group Discussion Questions

68. Risks and Choices Suppose an employer offered you a $20,000 starting salary for your first year of employment with raises for the first 5 years determined by either option A or option B.
Option A: A $1000 raise for each of the next 5 years.
Option B: A 5% raise for each of the next 5 years.

Can you determine without any calculations which option will produce an arithmetic sequence? Explain your reasoning. Write the sequence of salaries for each option. Use the changes from term to term in each sequence to explain which option you would prefer.

69. **Writing Mathematically**   A basketball player scored 5, 10, and 15 points in her first three games. Do you think an arithmetic sequence would be a good model of her scores in the next eight games? Explain your answer.

70. **Challenge Question**   The following table is a payment schedule for a $5000 car loan at 8%. Use a calculator and the interest formula $I = P \cdot R \cdot T$ to complete the following table.

(*Hint:* One month equals $\dfrac{1}{12}$ of a year.)

Marvin Johnson in AMATYC's "Crossroads in Mathematics" says, "Writing can help mathematics students in many different ways. Students who are required to write must do considerable thinking and organizing of their thoughts before they write, thus crystallizing in their minds the concepts being studied."

MONTHLY PAYMENT NO.	STARTING BALANCE ($)	MONTHLY PAYMENT ($)	TOTAL PAID ($)	NEW BALANCE ($)
1	5000.00	435.00	435.00	4598.33
2	4598.33	435.00	870.00	4193.99
3	4193.99			
4				
5				
6				
7				
8				
9				
10				
11				
12				

Section 2.2 Function Notation and Linear Functions

Objectives:
4. Use function notation.
5. Use a linear equation to form a table of values and to graph a linear equation.
6. Write a function to model an application.

A table of values or a graph of a few points can give a valuable representation for showing the relationship between a set of *x-y* pairs. For many *x-y* pairs it may be inconvenient or impossible to put all of these points in a table. Thus it is useful to be able to express the relationship between x and y using a formula. One notation for writing a formula relating *x-y* pairs is called *function notation.*

We will use the concept of a function throughout this text. In Chapter 6 we will define a function and examine this concept in detail. For now we start by becoming familiar with function notation.

The notation $f(x)$ is referred to as **function notation** and is read as **"f of x"** or **"$f(x)$ is the output value for an input value of x."** Function notation is used to give a formula that describes a unique output that we can calculate for each input of x. We call a formula defined in this manner a function. To evaluate $f(x)$ for a specific value of x, replace x on both sides of the equation by this specific value.

Caution: $f(x)$ does not mean f times x in this context. It represents an output value for a specific input value of x.

■ EXAMPLE 1 Using Function Notation to Evaluate a Function

Evaluate each expression for $f(x) = 3x + 5$.

SOLUTIONS

(a) $f(0)$

$f(x) = 3x + 5$ Substitute each input value of x into the
$f(0) = 3(0) + 5$ formula $f(x) = 3x + 5$.
$f(0) = 0 + 5$

Answer: $f(0) = 5$.   For an input of 0 the output is 5.

(b) $f(4)$

$f(x) = 3x + 5$ Substitute 4 for x in the formula.
$f(4) = 3(4) + 5$
$f(4) = 12 + 5$

Answer: $f(4) = 17$.   For an input of 4 the output is 17.

(c) $f(-4)$

$f(x) = 3x + 5$ Substitute -4 for x in the formula.
$f(-4) = 3(-4) + 5$
$f(-4) = -12 + 5$

Answer: $f(-4) = -7$.   For an input of -4 the output is -7. ■

SELF-CHECK 2.2.1

Given $f(x) = 4x + 11$, evaluate each expression.

1. $f(0)$
2. $f(1)$
3. $f(10)$
4. $f(-10)$

SELF-CHECK ANSWERS 2.2.1

1. $f(0) = 11$
2. $f(1) = 15$
3. $f(10) = 51$
4. $f(-10) = -29$

AMATYC's "Crossroads in Mathematics" strongly recommends that functions be examined verbally, numerically, graphically, and symbolically. This book has been written to follow this guideline and to use technology to enhance your learning. Nonetheless, it is important that you understand how to produce these tables and to sketch these graphs by hand.

In the next example we will use the function $f(x) = 2x + 1$ to form a table and to sketch a graph. A major difference between the function $f(x) = 2x + 1$ and the sequence $a_n = 2n + 1$ is that x can be any real number while n is restricted to the whole numbers 0, 1, 2, 3, 4, ... Thus the graph of a sequence will consist of individual discrete points; whereas the graph of $f(x) = 2x + 1$ will be the solid line shown in the following example.

■ EXAMPLE 2   Graphing a Function

We arbitrarily selected our input values of x. Because the input values can be any real number, the graph is one continuous line.

Use the function $f(x) = 2x + 1$ to complete a table of values for $x = -2, -1, 0, 1$, and 2. Then use these input-output pairs to graph the line through these points.

SOLUTION

ALGEBRAICALLY

$f(x) = 2x + 1$
$f(-2) = 2(-2) + 1 = -3$
$f(-1) = 2(-1) + 1 = -1$
$f(0) = 2(0) + 1 = 1$
$f(1) = 2(1) + 1 = 3$
$f(2) = 2(2) + 1 = 5$

NUMERICALLY

x	$f(x)$
-2	-3
-1	-1
0	1
1	3
2	5

GRAPHICALLY

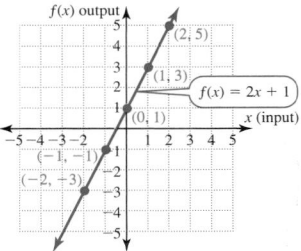

VERBALLY

The function $f(x) = 2x + 1$ defines a continuous line (no gaps) that contains the points in this table.

SELF-CHECK 2.2.2

1. Use the function $f(x) = -x + 3$ to complete this table:

x	-2	-1	0	1	2
y					

2. Graph this function.

The equation in Example 2, $f(x) = 2x + 1$, is an excellent example of a function of the form $f(x) = mx + b$. A function of this form is called a **linear function** because its graph is a straight line. We will examine linear functions in detail in Section 3.2. The linear function $f(x) = mx + b$ also can be represented by $y = mx + b$. Graphics calculators can denote several functions using y_1, y_2, y_3, and so on. In Calculator Perspective 2.2.1 we illustrate how to produce a table.

SELF-CHECK 2.2.2 ANSWERS

1.

x	-2	-1	0	1	2
y	5	4	3	2	1

2.

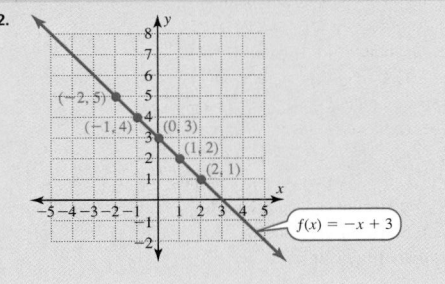

CALCULATOR	Using the Table Setup to Generate a Table
PERSPECTIVE 2.2.1	after Defining a Function

The table setup is used to fix the initial value of x and to input the increment (change) in the x-values. This setup determines the x-values in the table. The y-values are calculated based on the function in the screen. To create a table for $f(x) = 100 - x$ using input values of $x = 10, 15, 20, 25, 30, 35, 40$ on a TI-83 Plus calculator, enter the following keystrokes:

The  feature is the secondary function of the (WINDOW) key, and the TABLE feature is the secondary function of the (GRAPH) key.

The table setup (using (2nd) TBLSET) will be left to the student throughout the rest of the text.

(2nd) TBLSET (Make changes to **TblStart** and to Δ **Tbl** as shown.)

(2nd) TABLE

Note: Both the **Indpnt** and **Depend** should be set to **Auto.**

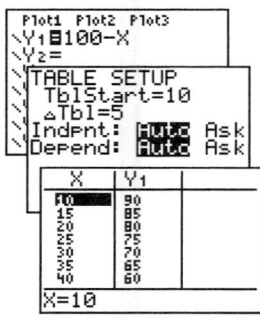

Calculator Perspective 2.2.2 illustrates how to produce a table and a graph for the equation given in Example 2.

CALCULATOR	
PERSPECTIVE 2.2.2	Generating a Table and a Graph

To generate a table and a graph for the equation $y = 2x + 1$ from Example 2 on a TI-83 Plus calculator, enter the following keystrokes:

The values shown here represent the decimal viewing window on a TI-83 Plus calculator. This window is obtained by pressing 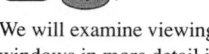 .

We will examine viewing windows in more detail in the next section.

The notation beneath the graph indicates the minimum and maximum x- and y-values used for the calculator window. $[-4.7, 4.7, 1]$ means the x-values extend from -4.7 to 4.7 with each mark on the x-axis representing 1 unit. Likewise, $[-3.1, 3.1, 1]$ means the y-values extend from -3.1 to 3.1 with each mark on the y-axis representing 1 unit.

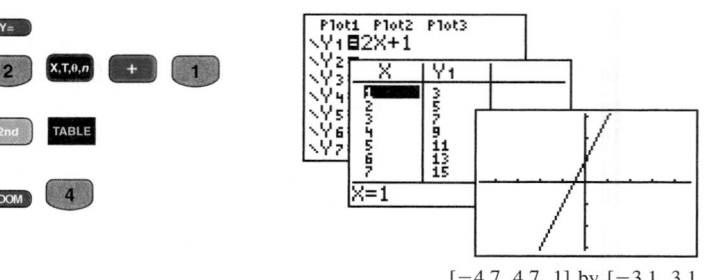

$[-4.7, 4.7, 1]$ by $[-3.1, 3.1, 1]$

Note: All the points in the table lie on this line.

■ EXAMPLE 3   Using Multiple Perspectives to Examine a Linear Function

Use a graphics calculator to examine the linear function $f(x) = -x + 2$ numerically and graphically. Then verbally describe the relationship given by this algebraic equation. Use the input values of x of $-3, -2, -1, 0, 1, 2,$ and 3 to form the table of values.

SOLUTION _____

ALGEBRAICALLY   NUMERICALLY   GRAPHICALLY

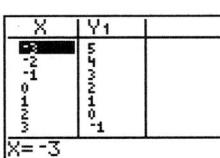   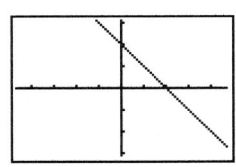

[−4.7, 4.7, 1] by [−3.1, 3.1, 1]

VERBALLY

The function defines a table of points that all lie on a straight line. Note that the line drops 1 unit for every 1-unit move to the right.

SELF-CHECK 2.2.3

Use a graphics calculator and $y = 1.75x - 2.25$ to form a table of values and to graph the line defined by this equation.

The real world has many applications of both discrete and continuous data. This book has a goal of giving you experience with both these data types.

One of the connections that you may have observed is that a linear function produces an arithmetic sequence for input values of $1, 2, 3, \ldots$. The graph of an arithmetic sequence consists of discrete points that lie on a line. The graph of a linear function $f(x) = mx + b$ consists of the entire line.

SELF-CHECK 2.2.3 ANSWER

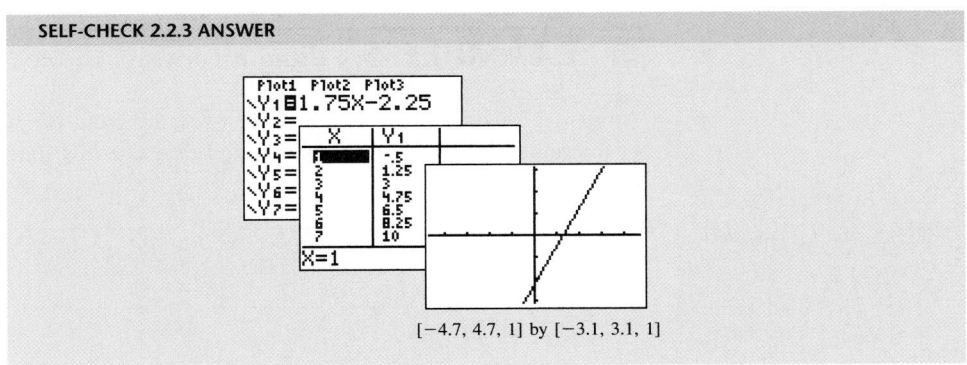

[−4.7, 4.7, 1] by [−3.1, 3.1, 1]

■ EXAMPLE 4 Comparing an Arithmetic Sequence and a Linear Function

Use the points $\{(1, -1), (2, 1), (3, 3), (4, 5), (5, 7)\}$ to graph both the arithmetic sequence $a_n = 2n - 3$ and the linear function $f(x) = 2x - 3$.

SOLUTION _____

ARITHMETIC SEQUENCE   **LINEAR FUNCTION**

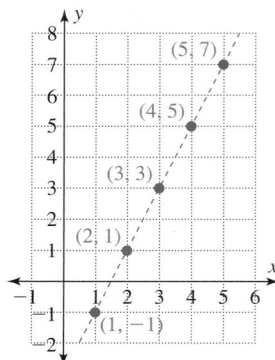 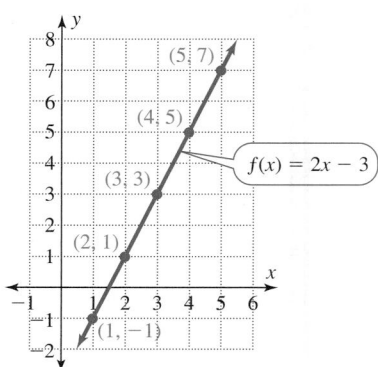

Note that the common difference of the arithmetic sequence is 2. Also note that the graph of the line goes up 2 units for each 1-unit movement to the right.

■

SELF-CHECK 2.2.4

Graph the line through the points $(0, 2)$ and $(5, 0)$.

The mathematical models presented in the examples and exercises in this section help develop skills needed for word problems that occur later in this book.

Functions are used frequently to describe the relationship between two variables. This is illustrated in the following examples and is developed further throughout this text. Creating an equation or a function to describe an application is called **mathematical modeling.**

■ EXAMPLE 5 Using a Function to Model the Length of a Board

A board 12 feet (ft) long has a piece x feet long cut off. Write a function for the length of the remaining piece in terms of x and examine this function numerically and verbally.

SELF-CHECK 2.2.4 ANSWER

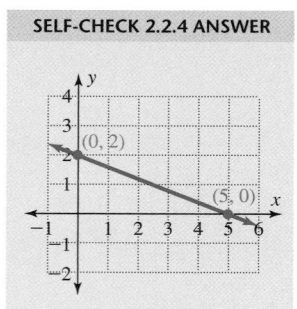

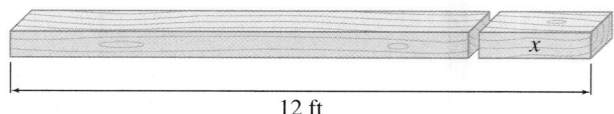

12 ft

SOLUTION _____

ALGEBRAICALLY

$f(x) = 12 - x$ If x ft are cut off, then $12 - x$ ft remain.

NUMERICALLY

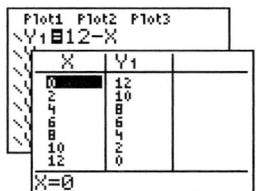

Let y_1 represent the function
$f(x) = 12 - x$. Enter $y_1 = 12 - x$.

By changing the table setup we can
examine other options.

VERBALLY

From the table we observe that if the first piece is 2 ft then the remaining piece is 10 ft. If the first piece is 4 ft, then the remaining piece is 8 ft. The input-output pairs in the table can be used to examine other possibilities. ∎

SELF-CHECK 2.2.5

The two angles shown here are complementary; their sum is 90°. Write a function for the second angle in terms of x.

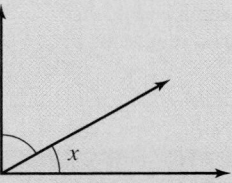

■ EXAMPLE 6   Using a Function to Model the Length of a Rectangle

A fencing company has been contracted to construct a fence on three sides of the rectangular lot shown here. The perimeter of this fenced area is 120 ft. If the width is x, write a function for the length in terms of x and examine this function numerically and verbally.

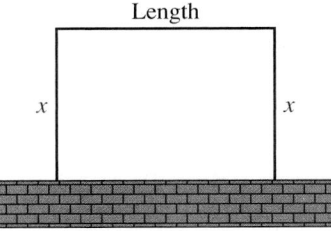

SOLUTION

SELF-CHECK 2.2.5 ANSWER
$f(x) = 90 - x$

ALGEBRAICALLY

$f(x) = 120 - 2x$ Of the 120 ft, x ft are used on each of the two sides. This leaves $120 - 2x$ ft for the third side to be fenced.

NUMERICALLY

Let y_1 represent the function
$f(x) = 120 - 2x$.

```
Plot1 Plot2 Plot3
\Y1■120-2X
   X    Y1
   0    120
   10   100
   20   80
   30   60
   40   40
   50   20
   60   0
X=0
```

VERBALLY

From the table we observe that if the width is 20 ft, then the length is 80 ft. If the width is 30 ft, then the length is 60 ft. The input-output pairs in the table can be used to examine other possibilities.

USING THE LANGUAGE AND SYMBOLISM OF MATHEMATICS   2.2

1. The notation $f(x)$ is called _____ notation.
2. The notation $f(x) = 8x - 2$ is read "_____ of _____ equals eight x minus two."
3. In the notation $f(x) = 8x - 2$, the input variable is represented by _____ and $f(x)$ represents the _____ variable.
4. The graph of $f(x) = 8x - 2$ is a _____ _____.

5. The function $f(x) = mx + b$ is called a _____ function.
6. In the notation $f(5) = 9$, the input value is _____ and the output value is _____.
7. Creating an equation or a function to describe an application is called mathematical _____.

EXERCISES   2.2

In Exercises 1–4 use the function $f(x) = 3x + 7$ to evaluate each expression.
 1. $f(0)$ 2. $f(1)$ 3. $f(-1)$ 4. $f(5)$

In Exercises 5–8 use the function $f(x) = 5x - 6$ to evaluate each expression.
 5. $f(2)$ 6. $f(-2)$ 7. $f(-10)$ 8. $f(10)$

In Exercises 9–12 use the function $f(x) = -6x + 4$ to evaluate each expression.

 9. $f(-4)$ 10. $f(4)$ 11. $f\left(\dfrac{1}{6}\right)$ 12. $f\left(-\dfrac{1}{6}\right)$

In Exercises 13–16 use the function $f(x) = -8x + 5$ to evaluate each expression.

13. $f\left(\dfrac{1}{4}\right)$ 14. $f\left(-\dfrac{1}{2}\right)$ 15. $f\left(-\dfrac{1}{8}\right)$ 16. $f(0)$

17. Use the function
$f(x) = \dfrac{x + 3}{2}$
to complete this table.

x	f(x)
-2	
-1	
0	
1	
2	

18. Use the function
$f(x) = \dfrac{x - 3}{2}$
to complete this table.

x	f(x)
-2	
-1	
0	
1	
2	

19. Use the linear function $f(x) = -2x + 4$ to complete this table and to graph this function.

x	f(x)
0	
1	
2	
3	

20. Use the linear function $f(x) = -\dfrac{1}{2}x + 1$ to complete this table and to graph this function.

x	f(x)
-2	
-1	
0	
1	
2	

21. Graph the line through $(0, 3)$ and $(2, 0)$.
22. Graph the line through $(-4, 0)$ and $(0, 2)$.

In Exercises 23–26 use the given function and a graphics calculator to complete each table.
23. $f(x) = 2x + 3$

a.
```
   X
  -3
  -2
  -1
  0
  1
  2
  3
X=-3
```

b.
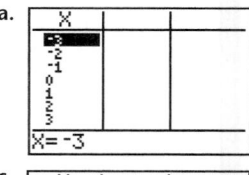
```
   X
  0
  .5
  1
  1.5
  2
  2.5
  3
X=0
```

c.
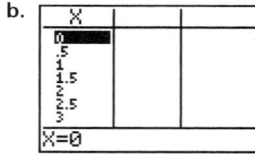
```
   X
  0
  10
  20
  30
  40
  50
  60
X=0
```

24. $f(x) = 4x - 5$

a.
b.

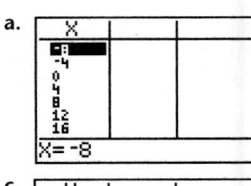

c.

25. $f(x) = \dfrac{1}{4}x - 2$

a.
b.

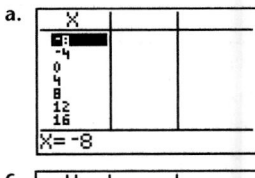

c.

26. $f(x) = -\dfrac{1}{2}x + 4$

a.
b.

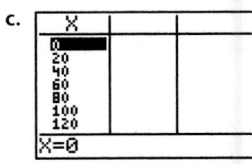

c.

27. a. Graph the first five terms of $a_n = -n + 5$.
 b. For the linear function $f(x) = -x + 5$ plot the points with x-coordinates of 1, 2, 3, 4, and 5 and then sketch the line through these points.

28. a. Graph the first five terms of $a_n = n - 4$.
 b. For the linear function $f(x) = x - 4$ plot the points with x-coordinates of 1, 2, 3, 4, and 5 and then sketch the line through these points.

29. Use the following graph to complete this table:

x	$f(x)$
-2	
-1	
0	
1	
2	

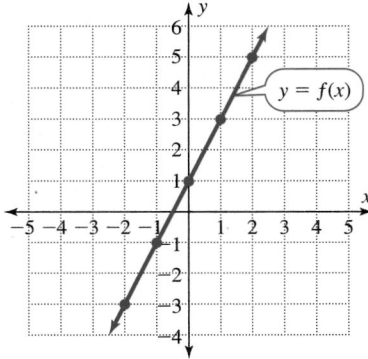

30. Use the following graph to complete this table:

x	$f(x)$
-2	
-1	
0	
1	
2	

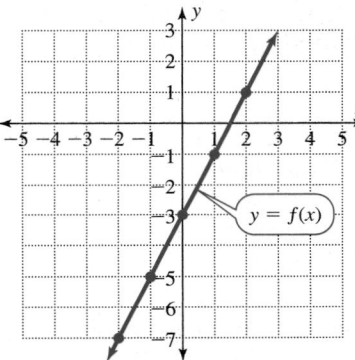

31. Use the following graph to complete this table:

x	$f(x)$
-5	
-3	
0	
2	

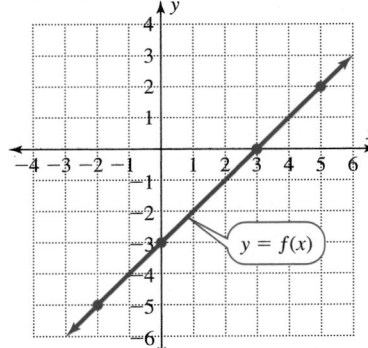

32. Use the following graph to complete this table:

x	$f(x)$
-3	
-1	
0	
3	

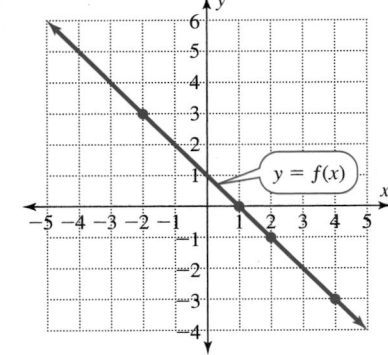

Calculator Exercises
In Exercises 33–36 use a calculator and the given equation to complete the table and to graph the equation. (*Hint:* See Calculator Perspective 2.2.1.)

33. Equation: $y = -\dfrac{x}{2}$

Numerical table:

x	−4	−2	0	2	4
y					

34. Equation: $y = \dfrac{x}{3}$

Numerical table:

x	−6	−3	0	3	6
y					

35. Equation: $y = 2.5x - 5.5$

Numerical table:

x	−1	0	1	2	3
y					

36. Equation: $y = -4.5x + 3.5$

Numerical table:

x	−1	0	1	2	3
y					

Using Functions to Create Mathematical Models
37. Modeling a Price Discount
A discount store had a holiday special with a 15% discount on the price of all sporting goods.
  a. Write a function for the amount of discount on an item with an original price of x dollars.
  b. Write a function for the new price of an item with an original price of x dollars.
  c. Complete the following table for the new price of each item whose original price is given:

x	f(x)
20	17
44	
60	
68	
90	

38. Modeling a Price Increase
A wholesaler is increasing the price of all items 2% to cover increased costs.
  a. Write a function for the amount of the increase on an item with an original price of x dollars.
  b. Write a function for the new price on an item with an original price of x dollars.
  c. Complete the following table for the new price of each item whose original price is given.

x	f(x)
20	20.40
35	
44	
80	
90	

39. Modeling the Rate of a Boat
A boat has a speed of x mi/h (miles per hour) in still water.
  a. Write a function for the rate of this boat traveling downstream in a river with a current of 5 mi/h.
  b. Write a function for the rate of this boat traveling upstream in a river with a current of 5 mi/h.

40. Modeling the Rate of an Airplane
An airplane has a speed of x mi/h in calm skies.
  a. Write a function for the rate of this airplane traveling in the same direction as a 30 mi/h wind.
  b. Write a function for the rate of this airplane traveling in the opposite direction as a 30 mi/h wind.

41. Modeling the Perimeter and Area of a Rectangle
  a. Write a function for the perimeter in centimeters (cm) of the following rectangle.
  b. Write a function for the area in square centimeters (cm^2) of the following rectangle.

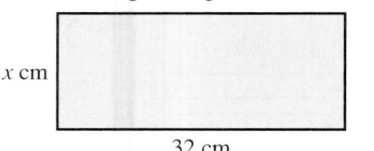

42. Modeling the Perimeter of a Triangle
Write a function for the perimeter in centimeters of the isosceles triangle shown here.

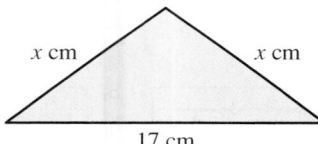

43. Modeling Supplementary Angles
The two angles shown here are supplementary; their sum is 180°. Write a function for the second angle in terms of x.

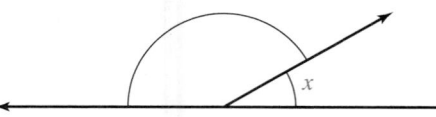

44. Modeling the Perimeter of an Equilateral Triangle
Write a function for the perimeter in centimeters of the equilateral triangle shown here.

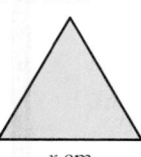

45. Modeling the Cost of Production
The overhead cost for a company is $500 per day. The cost of producing each item is $15.
  a. Write a function that gives the total cost of producing x units per day.
  b. Use this function to complete this table:

x	0	100	200	300	400	500
f(x)						

  c. Evaluate and interpret $f(250)$.

46. Modeling the Cost of Production
The overhead cost for a company is $400 per day. The cost of producing each item is $12.
 a. Write a function that gives the total cost of producing x units per day.
 b. Use this function to complete this table.

x	0	100	200	300	400	500
$f(x)$						

 c. Evaluate and interpret $f(250)$.

Group Discussion Questions

47. Discovery Question   Use a graphics calculator to graph $f(x) = \dfrac{x}{2}, f(x) = \dfrac{x}{2} + 1, f(x) = \dfrac{x}{2} + 3$, and $f(x) = \dfrac{x}{2} - 3$. What graphical relationship do you observe among these graphs?

48. Discovery Question   Use a graphics calculator to graph $f(x) = x^2$ and $f(x) = |x|$. Describe the shape of each of these graphs.

Section 2.3 Graphs of Linear Equations in Two Variables

Objectives:
 7. Check possible solutions of a linear equation.
 8. Determine the intercepts from a graph.
 9. Determine the point where two lines intersect.

The two lease options given here are summaries of two advertisements by two different automobile agencies that are both trying to lease the same automobile.

> **Lease option A:** A $2000 initial down payment followed by thirty-six monthly payments of $250.
>
> **Lease option B:** A $2500 initial down payment followed by thirty-six monthly payments of $200.

In this section we examine a method for comparing these two options. In Example 8 we determine which of these two options is the more costly. To prepare for this, we will examine linear equations in greater detail.

A solution of $y = mx + b$ is an ordered pair—not just an x-value or a y-value. The point (x, y) is a solution of the equation $y = mx + b$ if and only if (x, y) lies on the graph of this equation.

A **solution** of a linear equation of the form $y = mx + b$ is an ordered pair (x, y) that makes the equation a true statement.

■ **EXAMPLE 1** Checking Ordered Pairs in an Equation

Determine whether each ordered pair is a solution of $y = 5x - 3$.

SOLUTIONS ───────────────

(a) $(-1, -8)$

$y = 5x - 3$
$-8 \overset{?}{=} 5(-1) - 3$ Substitute -8 for y and -1 for x and then simplify.
$-8 \overset{?}{=} -5 - 3$
$-8 \overset{?}{=} -8$ checks.

Answer: $(-1, -8)$ is a solution of $y = 5x - 3$.

(b) $(1, 4)$

$y = 5x - 3$
$4 \overset{?}{=} 5(1) - 3$ Substitute 4 for y and 1 for x and then simplify.
$4 \overset{?}{=} 5 - 3$
$4 \overset{?}{=} 2$ is false.

Answer: $(1, 4)$ is not a solution of $y = 5x - 3$. ■

Both y and $f(x)$ are used to denote the output associated with an input x.

In the linear equation $y = mx + b$, x represents the input variable and y represents the output variable. Linear equations also are given frequently using function notation in the form $f(x) = mx + b$. In this notation x represents the input variable and $f(x)$ represents the output variable.

■ **EXAMPLE 2** Checking Ordered Pairs in a Function

Determine whether each ordered pair is on the graph of $f(x) = -2x + 8$.

SOLUTIONS ───────────────

(a) $(4, 0)$

$f(x) = -2x + 8$
$f(4) = -2(4) + 8$ Substitute 4 for the input value of x to determine the output value $f(4)$.
$f(4) = -8 + 8$
$f(4) = 0$

Answer: $(4, 0)$ is on the graph of $f(x) = -2x + 8$.   Because the output value is 0, $(4, 0)$ is a point on the graph of $f(x) = -2x + 8$.

(b) $(3, 5)$

$f(x) = -2x + 8$
$f(3) = -2(3) + 8$ Substitute 3 for the input value of x to determine the output value $f(3)$.
$f(3) = -6 + 8$
$f(3) = 2$

Answer: $(3, 5)$ is not on the graph of $f(x) = -2x + 8$.   Because the output value is 2, not 5, $(3, 5)$ is not a point on the graph of $f(x) = -2x + 8$. ■

A linear equation in two variables has an infinite set of ordered pairs that are solutions of the equation. Graphically all the solutions lie on a line, and all the points not on the line are not solutions.

■ EXAMPLE 3   Checking the Coordinates of a Point in an Equation

Determine whether the points A, B, and C on this graph are solutions of $y = -2x + 4$.

SOLUTIONS _____

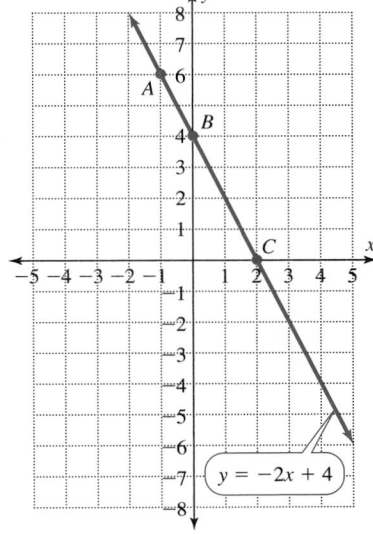

(a) Point A has coordinates $(-1, 6)$.

$y = -2x + 4$
$6 \stackrel{?}{=} -2(-1) + 4$ Substitute -1 for the input value of
$6 \stackrel{?}{=} 2 + 4$ x and 6 for the output value of y.
$6 \stackrel{?}{=} 6$ checks.

Answer: $(-1, 6)$ is a solution of $y = -2x + 4$.

(b) Point B has coordinates $(0, 4)$.

$y = -2x + 4$
$4 \stackrel{?}{=} -2(0) + 4$ $(0, 4)$ lies on the y-axis and is called the
$4 \stackrel{?}{=} 0 + 4$ y-intercept of the graph. We will
$4 \stackrel{?}{=} 4$ checks.   examine intercepts more in Section 2.6.

Answer: $(0, 4)$ is a solution of $y = -2x + 4$.

(c) Point C has coordinates $(2, 0)$.

$y = -2x + 4$
$0 \stackrel{?}{=} -2(2) + 4$ Point $(2, 0)$ lies on the x-axis and is called
$0 \stackrel{?}{=} -4 + 4$ the x-intercept of the graph. We will
$0 \stackrel{?}{=} 0$ checks.   examine intercepts more in Section 2.6.

Answer: $(2, 0)$ is a solution of $y = -2x + 4$.   ■

SELF-CHECK 2.3.1

Determine whether each ordered pair is a solution of $y = -3x + 4$.

1. $(2, -2)$ **2.** $(1, 5)$ **3.** $\left(\dfrac{4}{3}, 0\right)$ **4.** $(0, 4)$

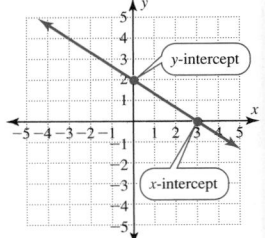

The intercepts are points with both x- and y-coordinates.

The line defined by a linear equation is completely determined by any two points on the line. To graph the line, we often select points that are easy to calculate or that have a special significance. It is also wise to plot a third point as a double check on our work. Since all points should be on the line, a third point that is not on the line is a sure indication of a computation error or an error in plotting the points. The points where a graph intersects, or crosses, the axes are called the **intercepts.** The **x-intercept** often is denoted by $(a, 0)$ and the **y-intercept** by $(0, b)$. These points frequently have special significance in applications as illustrated in Example 4.

SELF-CHECK 2.3.1 ANSWERS

1. $(2, -2)$ is a solution.

3. $\left(\dfrac{4}{3}, 0\right)$ is a solution; the x-intercept.

2. $(1, 5)$ is not a solution.

4. $(0, 4)$ is a solution; the y-intercept.

■ EXAMPLE 4 Interpreting the Intercepts of a Graph: Overhead Costs and Break-Even Values

In the graph to the right the input value of x is the number of units that a machine produces. The output y is the profit generated by the sale of these units when they are produced. Determine the intercepts and interpret the meaning of these points.

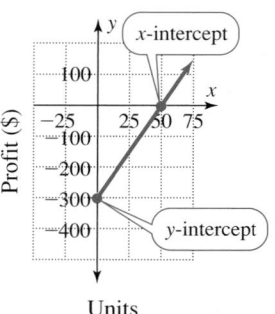

SOLUTION

The y-intercept is $(0, -300)$. If 0 units are produced, then \$300 will be lost. (Often there are overhead costs of rent, electricity, and so on even if no units are produced.)

The x-intercept is $(50, 0)$. If 50 units are produced, then \$0 will be made. (This is the break-even value for this machine; the number of units the machine must make before a profit can be realized.)

The graph of a linear function $f(x) = mx + b$ can be obtained by using any two points that fall on the line. One particularly easy point to determine is the y-intercept, which can be found by evaluating $f(0)$. Another point may be obtained by substituting any other value for x. It is wise to find a third point as a check.

■ EXAMPLE 5 Graphing a Linear Equation

Use the y-intercept and another point to graph the line defined by $f(x) = \dfrac{1}{2}x - 2$. Plot a third point as a check on your work.

SOLUTION

y-INTERCEPT

$f(x) = \dfrac{1}{2}x - 2$

$f(0) = \dfrac{1}{2}(0) - 2$ To find the y-intercept, substitute 0 for x.

$f(0) = 0 - 2$

$f(0) = -2$

$(0, -2)$ is the y-intercept.

SECOND POINT

$f(x) = \dfrac{1}{2}x - 2$

$f(4) = \dfrac{1}{2}(4) - 2$ To find a second point, find y when x is 4.

$f(4) = 2 - 2$

$f(4) = 0$

Note that $(4, 0)$ is the x-intercept.

THIRD POINT

$f(x) = \dfrac{1}{2}x - 2$

$f(6) = \dfrac{1}{2}(6) - 2$ To find a third point to double check our work, find y when x is 6.

$f(6) = 3 - 2$

$f(6) = 1$

$(6, 1)$ is also on the line.

TABLE

x	y
0	-2
4	0
6	1

GRAPH

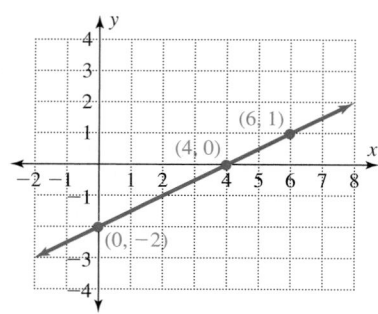

Plot these points and sketch the line through the points.

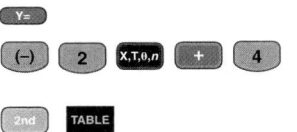

| CALCULATOR | Finding the Intercepts of a Graph Using the |
| PERSPECTIVE 2.3.1 | Table or Trace Features |

The purpose of this Calculator Perspective is to "see" the intercepts. For non-trivial intercepts we can also use the algebraic method covered in Section 2.7.

The values shown here represent the standard viewing window on a TI-83 Plus calculator. This window is obtained by pressing

 .

To find the x-intercept, look for the point with a y-coordinate of zero, and to find the y-intercept, look for the point with an x-coordinate of zero. For the linear equation $y = -2x + 4$ from Example 3 the y-intercept is highlighted in the table and the x-intercept is shown on the graph.

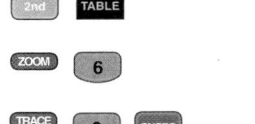

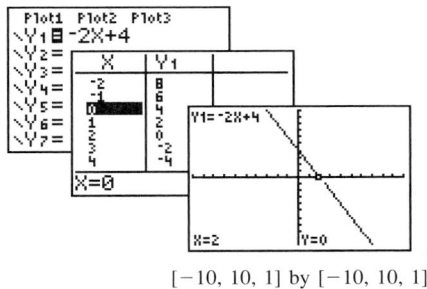

$[-10, 10, 1]$ by $[-10, 10, 1]$

Note: The x-intercept is $(2, 0)$ and the y-intercept is $(0, 4)$.

SELF-CHECK 2.3.2

1. Determine the x- and y-intercepts of this graph.

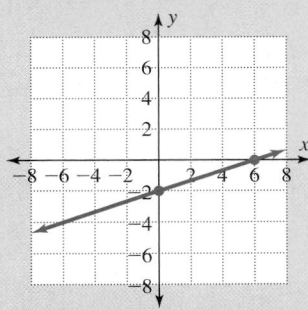

2. Determine the x- and y-intercepts of $f(x) = 5x + 10$.

Sometimes it is useful to examine two graphs simultaneously or to compare two different options. When we refer to two or more equations at the same time, we refer to this as a **system of equations. A solution of a system of linear equations in two variables** is an ordered pair that satisfies each equation in the system. If there is a unique solution to a system of linear equations, it is represented by their point of intersection, the point that is on both graphs.

Solution of a System of Two Linear Equations

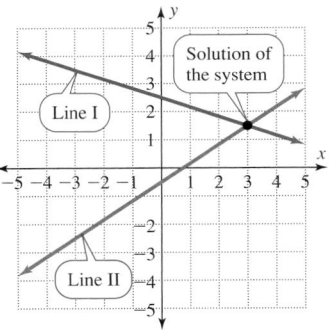

To solve a system of equations graphically, graph each equation and then estimate the coordinates of the point of intersection. Because serious errors of estimation can occur, it is wise to check an estimated solution by substituting it into each equation of the system.

■ EXAMPLE 6 Solving a Linear System Graphically

The graphs of $y = x - 1$ and $y = \dfrac{x}{3} + 1$ are shown to the right. Determine their point of intersection by inspecting the graph. Then check this point in both equations.

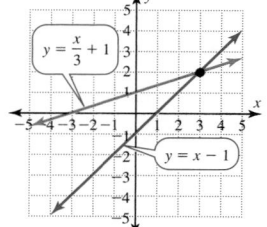

SOLUTION _____

An estimate of the solution is (3, 2).

Check:

FIRST EQUATION

$y = x - 1$

$2 \overset{?}{=} 3 - 1$

$2 \overset{?}{=} 2$ checks.

SECOND EQUATION

$y = \dfrac{x}{3} + 1$

$2 \overset{?}{=} \dfrac{3}{3} + 1$

$2 \overset{?}{=} 1 + 1$

$2 \overset{?}{=} 2$ checks.

Substitute (3, 2) into both equations to check this point.

Answer: (3, 2) is a solution of this system because it checks in both equations.

As you can verify by checking any other point, this is the only point that checks in both equations. ■

■ EXAMPLE 7 Solving a Linear System Graphically

Solve the linear system $\begin{cases} y = 2x + 5 \\ y = -2x - 3 \end{cases}$ graphically.

SOLUTION _____

First graph both lines on the same coordinate system.

$y = 2x + 5$

x	y
0	5
$-\dfrac{5}{2}$	0
-1	3

$y = -2x - 3$

x	y
0	-3
$-\dfrac{3}{2}$	0
-1	-1

Find the intercepts by letting $x = 0$ to find the y-intercept. Then let $y = 0$ to find the x-intercept. A third point was calculated by letting $x = -1$.

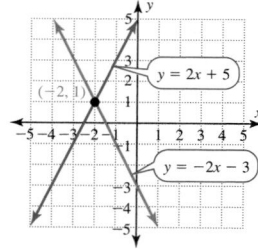

Then estimate the coordinates of the point of intersection.
The point of intersection appears to be $(-2, 1)$.

Check:

FIRST EQUATION	SECOND EQUATION	
$y = 2x + 5$	$y = -2x - 3$	
$1 \overset{?}{=} 2(-2) + 5$	$1 \overset{?}{=} -2(-2) - 3$	Substitute $(-2, 1)$ into both
$1 \overset{?}{=} -4 + 5$	$1 \overset{?}{=} 4 - 3$	equations to check this point.
$1 \overset{?}{=} 1$ checks.	$1 \overset{?}{=} 1$ checks.	

Answer: $(-2, 1)$ is a solution of this system because it checks in both equations. ■

SELF-CHECK 2.3.3

Determine whether each point is a solution of the linear system $\begin{Bmatrix} y = 3x - 5 \\ y = -x + 3 \end{Bmatrix}$.

1. $(0, -5)$ **2.** $(3, 0)$ **3.** $(2, 1)$

Calculator Perspective 2.3.2 shows how we can use a graphics calculator to assist in problems like Example 7.

CALCULATOR PERSPECTIVE 2.3.2   Approximating a Point of Intersection

To solve a system of two linear equations on a calculator, enter each equation and then use the Intersect feature on your graphics calculator to approximate the point of intersection. To solve the linear system $\begin{Bmatrix} y = 2x + 5 \\ y = -2x - 3 \end{Bmatrix}$ from Example 7 on a TI-83 Plus calculator, enter the following keystrokes:

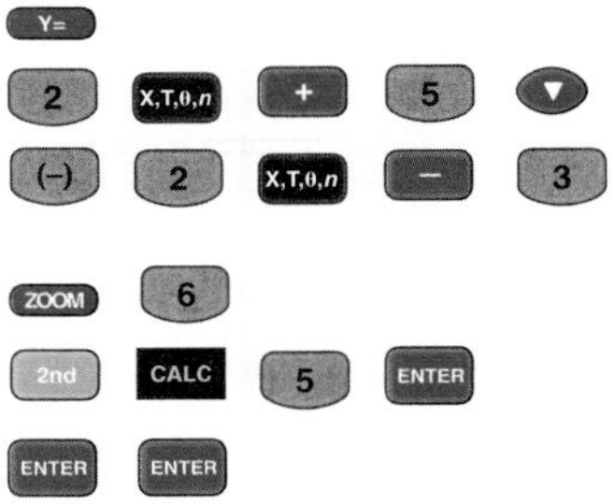

The CALC feature is the secondary function of the TRACE key.

| Y= |
2	X,T,θ,n	+	5	▼
(-)	2	X,T,θ,n	−	3
ZOOM	6
2nd	CALC	5	ENTER
ENTER	ENTER

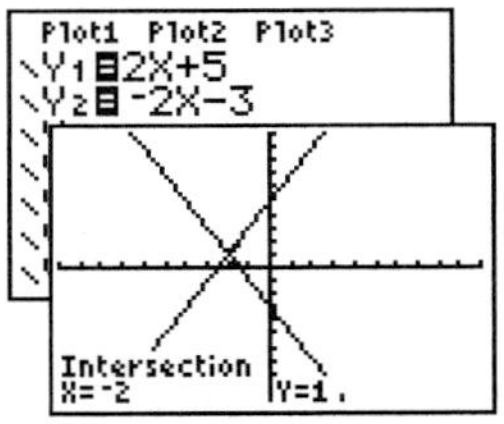

```
Plot1 Plot2 Plot3
\Y1⊟2X+5
\Y2⊟-2X-3
\
\
\
```

```
Intersection
X=-2          Y=1.
```

$[-10, 10, 1]$ by $[-10, 10, 1]$

Note: The integer setting and the table can be used sometimes to locate the coordinates of a point of intersection. ■

The important features of a graph may not always occur in the first display window you select on your calculator. Calculator Perspective 2.3.3 illustrates one way to change the display window.

CALCULATOR PERSPECTIVE 2.3.3   Changing the Window Settings

Pressing the [WINDOW] key will enable you to change the viewing window for a graph. Finding an appropriate window setting can be difficult. An example of the notation used in this text to indicate the window setting is $[-10, 10, 1]$ by $[-10, 10, 1]$. This shows that the graph will include possible x-values of -10 to 10, with a scale of 1 unit per mark on the x-axis. In this instance the y-values have the same description.

[Y=] [CLEAR]

[WINDOW]

Note that pressing [ZOOM] [6] will also set the window to these values.

[GRAPH]

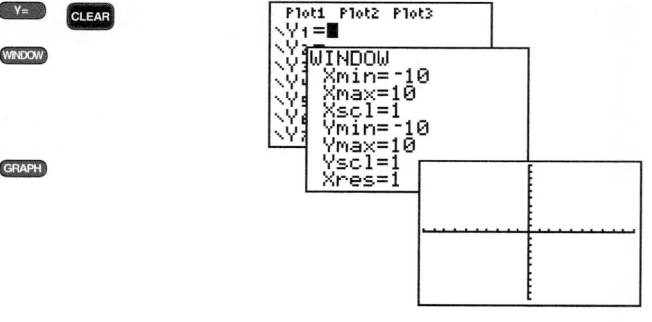

$[-10, 10, 1]$ by $[-10, 10, 1]$

To change the window settings on a TI-83 Plus calculator to $[0, 20, 5]$ by $[0, 7000, 1000]$ for Example 8, enter the following keystrokes:

[WINDOW]

[GRAPH]

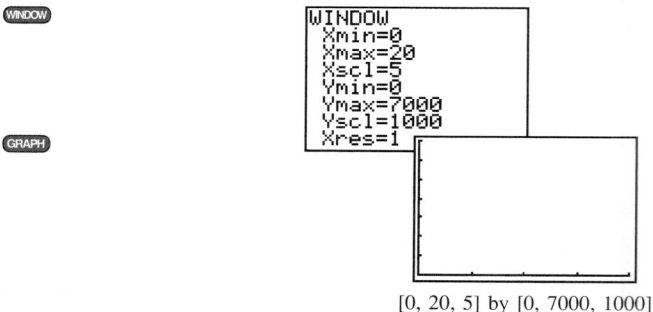

$[0, 20, 5]$ by $[0, 7000, 1000]$

Note: The values in the window setting are entered by using the numeric keys and the [ENTER] key. ∎

Comparing Car Lease Options

The two lease options given at the beginning of this section will be examined now. Assume that there are no factors other than the given costs that will affect the selection between these two options.

■ **EXAMPLE 8**   Determining Equivalent Costs

Examine these lease options numerically and graphically and determine if these costs will ever be the same.

Lease option A: A $2000 initial down payment followed by thirty-six monthly payments of $250.
Lease option B: A $2500 initial down payment followed by thirty-six monthly payments of $200.

SOLUTION

We let y represent the total cost incurred by the xth month.

VERBALLY | Total cost | = | Total of monthly payments | + | Initial payment |

The total of the monthly payments is calculated by multiplying the monthly payment times the number of months this payment has been made.

ALGEBRAICALLY

| Option A: | y | = | $250x$ | + | 2000 |
| Option B: | y | = | $200x$ | + | 2500 |

NUMERICALLY   **GRAPHICALLY**

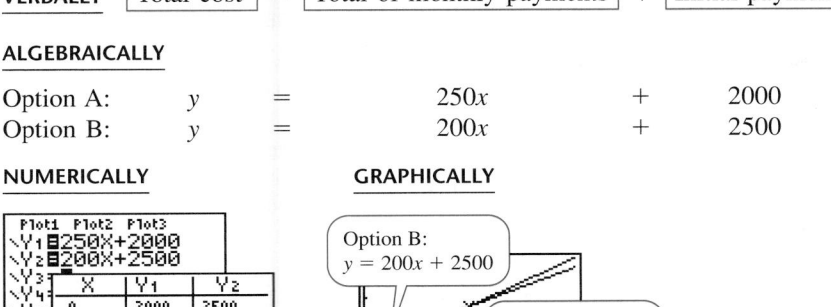

Option B:
$y = 200x + 2500$

Option A:
$y = 250x + 2000$

Intersection
X=10   Y=4500

[0, 20, 5] by [0, 7000, 1000]

A graphical approximation of the coordinates of the point of intersection is (10, 4500); this value is supported by the numerical table.

Answer: After 10 months both options will have a cost of $4500.

Note that from both the table and the graph option A starts with the lower initial cost, and the output values (the cost) are lower until the tenth month. After the tenth month the higher monthly cost for option A will produce output values (the cost) that are higher than the output values for option B.

■

USING THE LANGUAGE AND SYMBOLISM OF MATHEMATICS   2.3

1. The point $(a, 0)$ on the graph of a line is called the
_____.

2. The point $(0, b)$ on the graph of a line is called the
_____.

3. A solution of a linear equation $y = mx + b$ is an ordered pair (x, y) that makes the equation a _____ statement.

4. A linear equation of the form $y = mx + b$ has an _____ number of solutions.

5. An ordered pair that satisfies each equation in a system of linear equations in two variables is called a _____ of the system of linear equations.

6. Graphically a unique solution to a system of two linear equations is represented by their point of _____.

7. If a point is the x-intercept of a graph, then the _____ is 0.

8. If a point is the y-intercept of a graph, then the _____ is 0.

9. If a point satisfies $f(x) = mx + b$ and $x = 0$, then this point is the _____ intercept of the graph of this linear function.

10. If a point satisfies $f(x) = mx + b$ and $f(x) = 0$, then this point is the _____ intercept of the graph of this linear function.

EXERCISES 2.3

In Exercises 1–8 determine whether each point is a solution of $y = 6x - 1$.

1. $(1, 5)$ **2.** $(2, 4)$ **3.** $(0, -1)$ **4.** $\left(\frac{1}{6}, 6\right)$

5. $(2, 11)$ **6.** $(1, -1)$ **7.** $\left(\frac{1}{6}, 0\right)$ **8.** $(0, 3)$

In Exercises 9–16 determine whether each point is a solution of $y = -\frac{x}{2} + 5$.

9. $(0, 6)$ **10.** $(10, 0)$ **11.** $(-6, -8)$ **12.** $(4, 3)$
13. $(0, 5)$ **14.** $(6, 0)$ **15.** $(12, -1)$ **16.** $(-8, -9)$

In Exercises 17–20 determine whether the points A, B, C, and D on the graph to the right are solutions of the given equation.

17. $y = \frac{x}{4} + 1$

18. $y = -2x + 10$
19. $y = 3$
20. $x = -4$

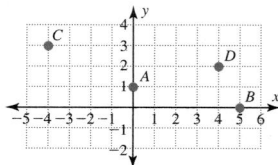

In Exercises 21–24 use the given graph to complete the following table and then to identify the x- and y-intercepts.

x	-2	0	2	
y				0

21. 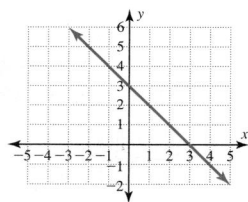 **22.**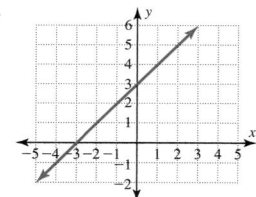

23. 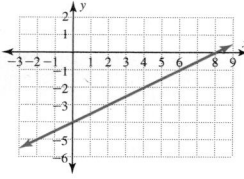 **24.**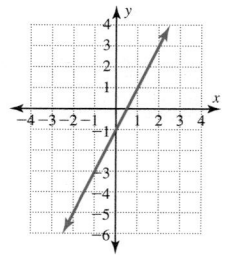

25. Graph the line with x-intercept $(3, 0)$ and y-intercept $(0, 2)$.
26. Graph the line with x-intercept $(-2, 0)$ and y-intercept $(0, -3)$.
27. Graph the line if every point on the line has an x-coordinate of -4.
28. Graph the line if every point on the line has an x-coordinate of 3.

29. Graph the line if every point on the line has a y-coordinate of 4.
30. Graph the line if every point on the line has a y-coordinate of -3.

In Exercises 31–38 plot the y-intercept of each line and one other point to graph the line. Select a third point to double check your work.
31. $y = x + 2$ **32.** $y = x - 3$
33. $y = \frac{x}{3} - 1$ **34.** $y = \frac{x}{4} + 1$
35. $f(x) = -x + 4$ **36.** $f(x) = -x - 2$
37. $f(x) = 2x + 3$ **38.** $f(x) = -2x + 1$

Calculator Exercises
In Exercises 39–42 use a calculator to complete the table of values and to graph each linear equation. Give the x- and y-intercepts for each line. (*Hint:* See Calculator Perspective 2.3.1.)

39. $y = 3x - 9$

x	y
-2	
-1	
0	
1	
2	

40. $y = -5x + 10$

x	y
-2	
-1	
0	
1	
2	

41. $f(x) = -0.5x + 1$

x	$f(x)$
-2	
-1	
0	
1	
2	

42. $f(x) = 0.9x + 1.8$

x	$f(x)$
-2	
-1	
0	
1	
2	

In Exercises 43–48 determine the point of intersection of the two lines.

43. 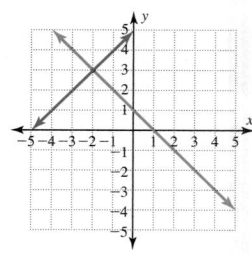 **44.**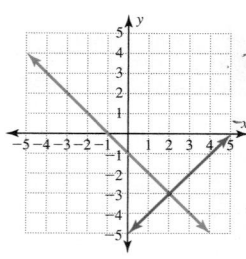

45. 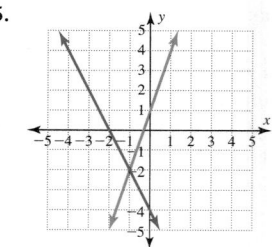 **46.**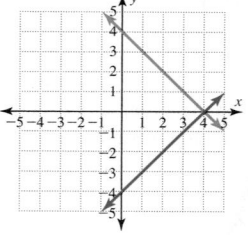

47.

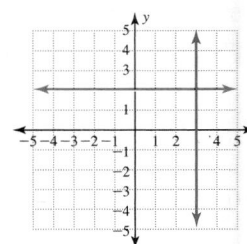

48.

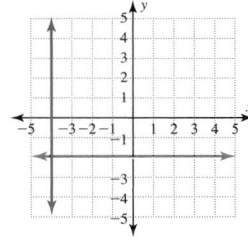

Calculator Exercises

In Exercises 49–52 solve each system of linear equations by using a calculator to graph each equation on the same coordinate system and to determine the point of intersection. Check the coordinates of this point in both the linear equations. (*Hint:* See Calculator Perspective 2.3.2.)

49. $y = 2x - 5$
$y = -2x - 1$

50. $y = x - 5$
$y = -x + 3$

51. $y = 2x - 1$
$y = \dfrac{x}{2} + 2$

52. $y = -x + 5$
$y = \dfrac{x}{3} + 1$

Interpreting the Intercepts of a Graph

In Exercises 53 and 54 determine the intercepts of each line and interpret the meaning of these points. In each graph the input value of x is the number of units of production by one assembly line at a factory, and the output value y is the profit in dollars generated by the sale of these units when they are produced.

53.

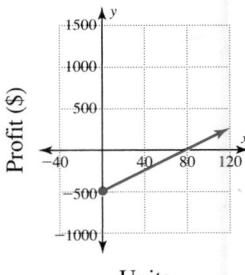

Units

54.

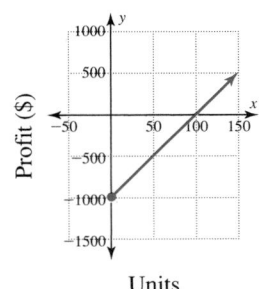

Units

A Mathematical Model for Car Payments

In Exercises 55 and 56 use a calculator and the given equation to prepare a table of the total payments for each of the months 1 through 12. In the equation x represents the number of months and $f(x)$ represents the total payment in dollars at the end of x months.

55. $f(x) = 333.33x + 457.50$

56. $f(x) = 287.67x + 393.80$

Using Models to Determine Equivalent Costs

In Exercises 57–60 determine when both options will have the same cost and also determine what this cost will be.

57. The input variable x represents the number of units produced and the output variable $f(x)$ represents the dollar cost of this production.

Option A: $f(x) = x + 7$
Option B: $f(x) = 2x + 3$

58. The input variable x represents the number of units produced and the output variable $f(x)$ represents the cost of this production.

Option A: $f(x) = 2x + 9$
Option B: $f(x) = 3x + 6$

59. The input variable x represents the number of months that an apartment has been rented. The output variable $f(x)$ represents the rental cost by the end of the xth month including an initial nonrefundable deposit. (*Hint:* Use a calculator window of [0, 5, 1] by [0, 1500, 500].)

Option A: $f(x) = 250x + 50$
Option B: $f(x) = 225x + 100$

60. The input variable x represents the number of months that an apartment has been rented. The output variable $f(x)$ represents the rental cost by the end of the xth month including an initial nonrefundable deposit. (*Hint:* Use a calculator window of [0, 4, 1] by [0, 1000, 200].)

Option A: $f(x) = 400x + 100$
Option B: $f(x) = 450x$

Group Discussion Questions

61. Discovery Question   Graph $y = x + 1$, $y = x + 5$, and $y = x + 9$ all on the same coordinate system. Compare these graphs. What is the same? What is different? What can you predict about the graph from the form $y = mx + b$?

62. Discovery Question   Graph $y = x + 1$, $y = \dfrac{x}{2} + 1$, and $y = 2x + 1$ all on the same coordinate system. Compare these graphs. What is the same? What is different? What can you predict about the graph from the form $y = mx + b$?

63. Discovery Question   Graph $y = -2$, $y = 1$, and $y = 3$ all on the same coordinate system. Compare these graphs. What is the same? What is different? What can you predict about the graph from the form $y = mx + b$?

64. Error Analysis   A student is trying to graph $y = 20$ on a TI-83 Plus calculator. However, after pressing GRAPH, the student does not see the graph on the display shown here. What is the error and how can it be corrected?

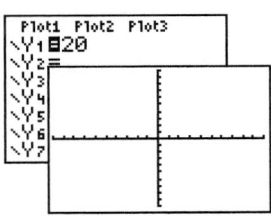

$[-10, 10, 1]$ by $[-10, 10, 1]$

Section 2.4 Solving Linear Equations in One Variable Using the Addition-Subtraction Principle

Objectives:

10. Solve linear equations in one variable using the addition-subtraction principle.
11. Use tables and graphs to solve a linear equation in one variable.

A Mathematical Note

François Viète (1540–1603) is considered by many to be the father of algebra, as we know it today. Prior to Viète, algebra was expressed rhetorically. Viète promoted the use of variables to represent unknowns, which led to the algebraic notation we now use.

One purpose of the first two sections of this chapter was to develop the concepts and skills needed to examine linear equations using both tables and graphs. We now use these skills to solve linear equations containing only one variable. We will concentrate on linear equations in one variable for the rest of this chapter and return to linear equations in two variables in Chapter 3.

This section concentrates on developing algebraic skills for solving linear equations. As already noted, $y = mx + b$, an equation with two variables, has a graph that is a straight line. Thus we refer to $y = mx + b$, which is first degree in both x and y, as a linear equation. In fact, we extend this terminology to all first-degree equations. If an equation is a **linear equation**, then each variable in the equation is first degree.

A first-degree equation in x and y means that the exponent on both x and y is 1.

Linear Equation in One Variable		
ALGEBRAICALLY	**VERBALLY**	**ALGEBRAIC EXAMPLE**
A **linear equation in one variable** x is an equation that can be written in the form $Ax = B$, where A and B are real constants and $A \neq 0$.	A linear equation in one variable is first degree in this variable.	$2x = 24$

■ EXAMPLE 1 Identifying Linear Equations

Determine which of the following choices are linear equations in one variable.

SOLUTIONS

(a) $2x = 24$ — Linear equation in one variable. — The exponent of the variable x is understood to be 1. This is a first-degree equation.

(b) $x^2 = 24$ — Not a linear equation. — The exponent on the variable x is 2.

(c) $3(y - 5) = 4y + 7$ — Linear equation in one variable. — The only variable is y, and the exponent on y is 1 on both sides of the equation. This equation can be simplified to the form $y = -22$.

(d) $7x - 2 = 5y + 1$ — Not a linear equation in one variable. — This is a linear equation, but it contains two variables, x and y.

(e) $5x - 3 + 4(x - 1)$ — Not a linear equation. — This is an algebraic expression, but it is not an equation since it has no symbol of equality.

To solve an equation whose solution is not obvious, we form simpler equivalent equations until we obtain an equation whose solution is obvious. **Equivalent equations** have the same solution set.

Think of a Balance Scale When You Solve Equations

When thinking of an equation, you may find it helpful to use the concept of a balance scale that has the left side in balance with the right side. We always must perform the same operation on both sides to preserve this balance.

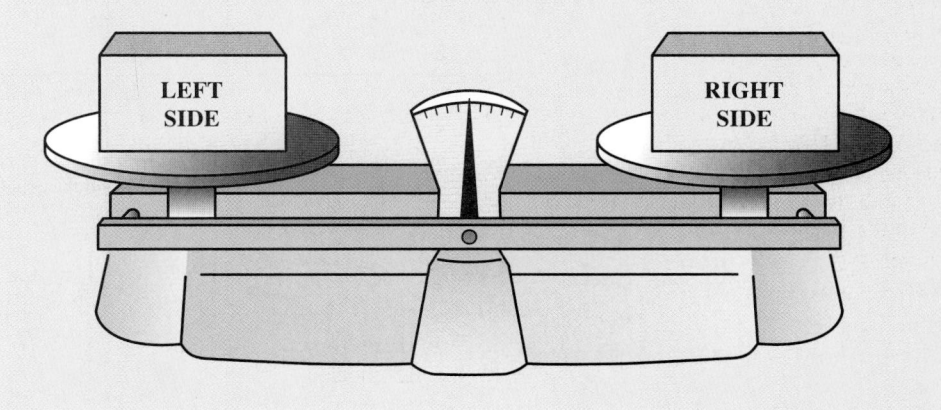

Another principle, the multiplication-division principle of equality is covered in the next section. Most students benefit by practicing on the material in this section before using the multiplication-division principle.

The general strategy for solving a linear equation, regardless of its complexity, is to isolate the variable whose value is to be determined on one side of the equation and to place all other terms on the other side. One of the principles used to accomplish this is given in the following box.

Addition-Subtraction Principle of Equality

ALGEBRAICALLY	VERBALLY	NUMERICAL EXAMPLE
If a, b, and c are real numbers, then $a = b$ is equivalent to $a + c = b + c$ and to $a - c = b - c$.	If the same number is added to or subtracted from both sides of an equation, the result is an equivalent equation.	$x + 2 = 5$ is equivalent to $x + 2 - 2 = 5 - 2$ and to $x = 3$

■ **EXAMPLE 2**   Solving a Linear Equation with One Variable Term

Solve $x - 3 = 8$.

SOLUTION _____

To remove a constant or variable from one side of an equation, we add its additive inverse to both sides of the equation.

$$x - 3 = 8$$
$$x - 3 + 3 = 8 + 3$$
$$x = 11$$

Use the addition-subtraction principle of equality to add 3 to both sides of the equation in order to isolate the variable term on the left side and the constant terms on the right side.

Check: $x - 3 = 8$

$$11 - 3 \overset{?}{=} 8$$

$$8 \overset{?}{=} 8 \text{ checks.}$$

Answer: $x = 11$.

■

■ **EXAMPLE 3** Solving a Linear Equation with Variable Terms on Both Sides

Solve $3y = 2y - 5$.

SOLUTION ――――――――――――――――――――――――――

$$3y = 2y - 5$$
$$3y - 2y = 2y - 5 - 2y$$
$$y = 2y - 2y - 5$$
$$y = -5$$

Subtract $2y$ from both sides of the equation to isolate the variable term on the left side of the equation. Then simplify both sides by combining like terms.

Check: $3y = 2y - 5$

$$3(-5) \overset{?}{=} 2(-5) - 5$$

$$-15 \overset{?}{=} -10 - 5$$

$$-15 \overset{?}{=} -15 \text{ checks.}$$

Answer: $y = -5$.

■

Example 4 contains constants and variables on both sides of the equation. We use the addition-subtraction principle of equality in order to isolate the variables on the left side of the equation and the constant terms on the right side of the equation.

■ **EXAMPLE 4** Solving a Linear Equation with Variable and Constant Terms on Both Sides

Solve $6v - 7 = 5v - 3$.

SOLUTION ――――――――――――――――――――――――――

$$6v - 7 = 5v - 3$$
$$6v - 7 - 5v = 5v - 3 - 5v$$
$$v - 7 = -3$$
$$v - 7 + 7 = -3 + 7$$
$$v = 4$$

Subtract $5v$ from both sides of the equation to isolate the variable term on the left side of the equation.

Add 7 to both sides of the equation to isolate the constant term on the right side of the equation.

Check: $6v - 7 = 5v - 3$

$$6(4) - 7 \overset{?}{=} 5(4) - 3$$

$$24 - 7 \overset{?}{=} 20 - 3$$

$$17 \overset{?}{=} 17 \text{ checks.}$$

Answer: $v = 4$.

■

Example 5 illustrates an equation that contains grouping symbols on both sides of the equation. The strategy is first to simplify the left and right sides of the equation and then to isolate the variable terms on one side of the equation and the constant terms on the other side. To solve a linear equation, try to make each step produce a simpler equation than that in the previous step.

■ **EXAMPLE 5**   Solving a Linear Equation Containing Parentheses

Solve $4(2b - 3) = 7(b - 2)$.

SOLUTION _____

$4(2b - 3) = 7(b - 2)$	Use the distributive property $a(b + c) = ab + ac$ to remove the
$4(2b) + 4(-3) = 7b + 7(-2)$	parentheses and then to simplify both sides.
$8b - 12 = 7b - 14$	
$8b - 12 - 7b = 7b - 14 - 7b$	Subtract $7b$ from both sides.
$b - 12 = -14$	
$b - 12 + 12 = -14 + 12$	Add 12 to both sides.

Answer: $b = -2$.   Does this answer check?   ■

SELF-CHECK 2.4.1

Solve the following linear equations:

1. $w + 6 = 4$
2. $8a + 9 = 7a + 16$
3. $5(3x - 1) = 2(7x + 5)$

The linear equations we have examined thus far have all had exactly one solution. These equations are examples of conditional equations. A **conditional equation** is true for some values of the variable and false for other values.

To complete our discussion we must consider two other types of equations. An equation that is true for all values of the variable is called an **identity**. An equation that is false for all values of the variable is called a **contradiction**. These two types of equations are considered in Examples 6 and 7, respectively.

■ **EXAMPLE 6**   Solving an Identity

Solve $4x + 1 - x = 2x + 1 + x$.

SOLUTION _____

$4x + 1 - x = 2x + 1 + x$	
$3x + 1 = 3x + 1$	First simplify both sides of the equation.
$3x + 1 - 3x = 3x + 1 - 3x$	Then subtract $3x$ from both sides.
$1 = 1$ is a true statement.	The last equation is an identity, so the original equation is also an identity.

Answer: Every real number is a solution since the equation is an identity.

Test a couple of values to verify that all real numbers will check.

■ **EXAMPLE 7** Solving a Contradiction

Solve $4v - 2 = 4v + 1$.

SOLUTION _____

$$4v - 2 = 4v + 1$$
$$4v - 2 - 4v = 4v + 1 - 4v \qquad \text{Subtract } 4v \text{ from both sides.}$$
$$-2 = 1 \text{ is a false statement.}$$

This last equation is a contradiction, so the original equation is also a contradiction.

Answer: Because the equation is a contradiction, there is no solution.

No matter what values you may test, none of them will check.

In Examples 6 and 7 all variables were eliminated in the process of solving the equation. If all variables are eliminated and an identity results, then every real number is a solution of the equation. If all variables are eliminated and a contradiction results, then the equation has no solution.

SELF-CHECK 2.4.2

Solve the following equations:

1. $v + 1 = 1$
2. $v + 1 = 1 + v$
3. $v + 1 = 2 + v$

SELF-CHECK 2.4.2 ANSWERS

1. $v = 0$
2. Every real number is a solution.
3. No solution.

We can use two-dimensional graphs to solve linear equations in one variable. Start by letting y_1 equal the left side of the equation and y_2 equal the right side of the equation. The x-coordinate of the intersection of the resulting lines is the solution of the original linear equation in x.

CALCULATOR PERSPECTIVE 2.4.1	Using a Table and a Graph to Solve a Linear Equation in One Variable

To solve a linear equation in x with a graphics calculator, enter each side of the equation separately under y_1 and y_2. Be sure to use x as your variable. Then use the table and/or the graph to locate the x-value for which the y-values are the same. To solve $2x - 5 = x - 7$ using a TI-83 Plus calculator, enter the following keystrokes:

Note: The display window has been set to values that allow us to display the intersection message on the screen without covering up the point of intersection.

The CALC feature is the secondary function of the TRACE key.

The original equation contains only the variable x. Thus the solution to this equation is not an ordered pair. The solution is $x = -2$. The y_1- and y_2-values are introduced to determine the x-value that makes the left side of the equation equal the right side.

The multiple representations we use to examine mathematical concepts are algebraic, graphical, numerical, and verbal representations.

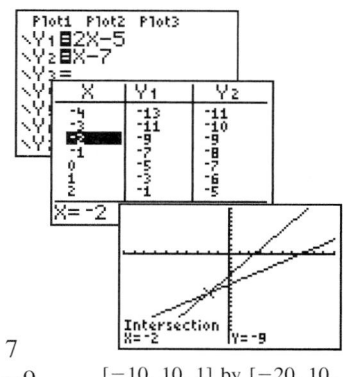

$[-10, 10, 1]$ by $[-20, 10, 1]$

Note: The solution to the equation $2x - 5 = x - 7$ is $x = -2$ since both y_1 and y_2 have the value of -9 when $x = -2$. Also -2 is the coordinate of x where the two graphs intersect.

■ **EXAMPLE 8**   Using Multiple Representations to Solve a Linear Equation

Use a graphics calculator to solve $5x - 5 = 4x - 3$.

SOLUTION _____

ALGEBRAICALLY

$$5x - 5 = 4x - 3$$
$$5x - 5 - 4x = 4x - 3 - 4x$$
$$x - 5 = -3$$
$$x - 5 + 5 = -3 + 5$$
$$x = 2$$

Check:

$$5x - 5 = 4x - 3$$
$$5(2) - 5 \overset{?}{=} 4(2) - 3$$
$$10 - 5 \overset{?}{=} 8 - 3$$
$$5 \overset{?}{=} 5 \text{ checks.}$$

GRAPHICALLY

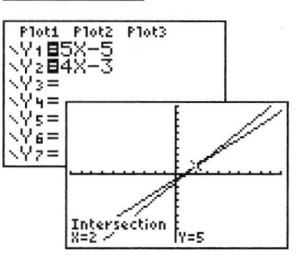

$[-10, 10, 1]$ by $[-40, 40, 4]$

NUMERICALLY

X	Y1	Y2
0	-5	-3
1	0	1
2	5	5
3	10	9
4	15	13
5	20	17
6	25	21

X=2

VERBALLY

The value of x that satisfies the equation is $x = 2$. From the table $y_1 = y_2$ when $x = 2$. From the graph the x-coordinate of the point of intersection is $x = 2$. ■

SELF-CHECK 2.4.3

Use a graphics calculator and a table to solve $4.7x - 1.5 = 3.7x + 1.5$.

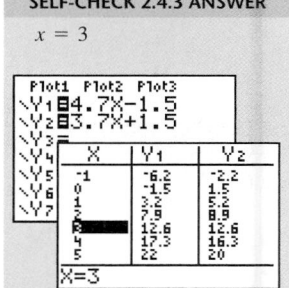

■ EXAMPLE 9   Solving a Verbally Stated Equation

If twice a number is decreased by 7, the result is the same as 4 more than the number.
Find this number.

SOLUTION

Let n = this number.

VERBALLY

| Twice a number decreased by 7 | | Is | | 4 more than the number. |

ALGEBRAICALLY

$$2n \quad - \quad 7 \quad = \quad n + 4$$

Let n represent the number and translate the **word equation** into an algebraic equation.

$$2n - 7 - n = n + 4 - n$$

Subtract n from both sides of the equation.

$$n - 7 = 4$$
$$n - 7 + 7 = 4 + 7$$

Add 7 to both sides of the equation.

$$n = 11$$

Does this value check?

Answer: The number is 11.

■

USING THE LANGUAGE AND SYMBOLISM OF MATHEMATICS   2.4

1. The equation $y = mx + b$ is a _____ equation in two variables.
2. The equation $Ax = B$ is a _____ equation in one variable.
3. If an equation is a _____ equation, then each variable is first degree.
4. An equation is first degree in x if the exponent on x is _____.
5. Equations with the same solution set are called _____ equations.

6. By the _____ principle of equality $a = b$ is equivalent to $a + c = b + c$.
7. By the addition-subtraction principle of equality, $a = b$ is equivalent to $a - c =$ _____.
8. A _____ equation is true for some values of the variable and false for other values.
9. An equation that is true for all values of the variable is called an _____.
10. An equation that is false for all values of the variable is called a _____.

EXERCISES   2.4

In Exercises 1 and 2 determine which of the following are linear equations in one variable.

1. **a.** $5x + 7$ **b.** $5x + 7 = 4x$
   c. $y = 5x + 7$ **d.** $x^2 = 25$
2. **a.** $x^2 = 4$ **b.** $4x + 5$
   c. $y = 4x + 5$ **d.** $4x + 5 = 3x$

In Exercises 3–12 solve each equation and then check your solution.

3. $x - 11 = 13$
4. $x + 11 = 13$
5. $v + 6 = 2$
6. $v - 6 = -9$
7. $2y = y - 1$
8. $3y = 2y + 4$
9. $3z + 7 = 2z - 4$
10. $5z - 1 = 4z + 3$
11. $10y + 17 = 9y + 21$
12. $8y + 17 = 7y - 21$

In Exercises 13–40 solve each equation.

13. $6x + 2 = 5x + 18$ **14.** $7x - 4 = 6x + 28$
15. $9y - 11 = 8y + 52$ **16.** $14y + 21 = 13y - 6$
17. $5v + 7 = 4v + 9$ **18.** $7v - 8 = 6v + 11$
19. $12y + 2 = 11y + 2$ **20.** $24y - 8 = 23y - 8$
21. $5m - 7 + 3m = 19 + 7m$
22. $9m - 7 = 13m - 11 - 5m$
23. $12n - 103 = 9n - 4 + 2n$
24. $17n - 18 - 7n = 9n + 18$
25. $3t + 3t + 3t = 4t + 4t$
26. $5t + 5t + 5t = 7t + 7t$
27. $10v - 8v = 9v - 8v$
28. $12v - 10v = 14v - 13v$
29. $3x - 4 + 5x = 6x + 3 + x$
30. $2x + 7 + 9x = 4x - 3 + 6x$
31. $3(v - 5) = 2(v + 5)$ **32.** $5(v + 2) = 4(v - 6)$
33. $5(w - 2) = 4(w + 3)$ **34.** $9(w + 4) = 8(w - 1)$
35. $8(m - 2) = 7(m - 3)$ **36.** $7(m + 5) = 6(m + 3)$
37. $4(n - 6) = 3(n - 5)$ **38.** $8(n - 2) = 7(n - 4)$
39. $3(5y + 2) = 2(7y - 3)$ **40.** $7(3y + 2) = 5(4y - 3)$

In Exercises 41–44 each equation is a conditional equation, an identity, or a contradiction. Identify the type of equation and then solve it.

41. a. $2x = x$ **b.** $x = x + 2$
 c. $x + 2 = x + 2$
42. a. $2x + 1 = x$ **b.** $2x = x + x$
 c. $x + 3 = 3 + x$
43. a. $v = v + 4$ **b.** $4 + v = v + 4$
 c. $4v = 3v + 4$
44. a. $x + x = 2x + 1$ **b.** $2(x + 1) = 2x + 2$
 c. $x + x = x + 2$

In Exercises 45–50 simplify the expression in the first column by adding like terms and solve the equation in the second column.

Simplify   **Solve**
45. a. $5x + 1 + 4x - 6$ **b.** $5x + 1 = 4x - 6$
46. a. $8x + 5 + (7x + 2)$ **b.** $8x + 5 = 7x + 2$
47. a. $6x - 4 - (5x + 3)$ **b.** $6x - 4 = 5x + 3$
48. a. $12x - 5 - (11x + 1)$ **b.** $12x - 5 = 11x + 1$
49. a. $3.4x - 1.7 + 2.4x + 2.3$ **b.** $3.4x - 1.7 = 2.4x + 2.3$
50. a. $2.6x - 1.9 - (1.6x + 4.1)$ **b.** $2.6x - 1.9 = 1.6x + 4.1$

Calculator Usage

In Exercises 51–54 y_1 represents the left side of the equation and y_2 represents the right side. Use the table shown from a graphics calculator to solve the equation.
51. $4.8x + 6.3 = 3.8x + 11.3$

X	Y₁	Y₂
0	6.3	11.3
1	11.1	15.1
2	15.9	18.9
3	20.7	22.7
4	25.5	26.5
5	30.3	30.3
6	35.1	34.1
X=6		

52. $12.4x - 15.8 = 11.4x - 13.8$

X	Y₁	Y₂
-2	-40.6	-36.6
-1	-28.2	-25.2
0	-15.8	-13.8
1	-3.4	-2.4
2	9	9
3	21.4	20.4
4	33.8	31.8
X=4		

53. $3.5x - 4.2 = 2.5x - 1.9$

X	Y₁	Y₂
2	2.8	3.1
2.1	3.15	3.35
2.2	3.5	3.6
2.3	3.85	3.85
2.4	4.2	4.1
2.5	4.55	4.35
2.6	4.9	4.6
X=2.6		

54. $6.3x + 7.4 = 5.3x + 5.8$

X	Y₁	Y₂
-2	-5.2	-4.8
-1.9	-4.57	-4.27
-1.8	-3.94	-3.74
-1.7	-3.31	-3.21
-1.6	-2.68	-2.68
-1.5	-2.05	-2.15
-1.4	-1.42	-1.62
X=-1.4		

In Exercises 55–60 use a graphics calculator to solve each equation by letting y_1 represent the left side of the equation and y_2 represent the right side of the equation. (*Hint:* See Calculator Perspective 2.4.1.)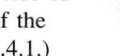

55. $2x - 2 = x - 3$ **56.** $2x + 4 = x + 1$
57. $1.8x - 4.6 = 0.8x - 2.6$ **58.** $2.5x + 2.3 = 1.5x + 0.7$
59. $\dfrac{1}{3}x - \dfrac{2}{3} = \dfrac{7}{3} - \dfrac{2}{3}x$ **60.** $\dfrac{2}{5}x - \dfrac{4}{7} = \dfrac{3}{7} - \dfrac{3}{5}x$

Estimation and Calculator Skills

In Exercises 61–64 mentally estimate the solution of each equation to the nearest integer and then use a calculator to calculate the solution.

PROBLEM	MENTAL ESTIMATE	CALCULATOR SOLUTION
61. $x - 0.918 = 0.987$		
62. $x + 39.783 = 70.098$		
63. $2x + 385.016 = x + 855.193$		
64. $5x - 1.393 = 4x + 5.416$		

Multiple Representations

In Exercises 65–70 first write an algebraic equation for each verbal statement, using the variable m to represent the number and then solve for m.

65. Seven more than three times a number equals eight less than twice the number.
66. If three is subtracted from five times a number, the result equals eight more than four times the number.

67. Twelve minus nine times a number is the same as two minus ten times the number.
68. Seven minus six times a number is the same as twelve minus seven times the number.
69. Twice the quantity three times a number minus nine is the same as five times the sum of the number and thirteen.
70. Four times the sum of a number and eleven is equal to three times the difference of the number and five.

Perimeter of a Triangle (Exercises 71 and 72)

71. The perimeter of the triangle shown in the figure is 28 centimeters (cm). Find the value of a.

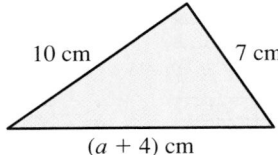

10 cm   7 cm

$(a + 4)$ cm

72. The perimeter of the triangle shown in the figure is 31 cm. Find the value of a.

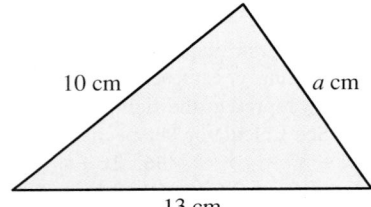

10 cm a cm

13 cm

73. **Perimeter of a Basketball Court**   The perimeter of the basketball court shown in the figure is $x + 35$ ft. Find the value of x.

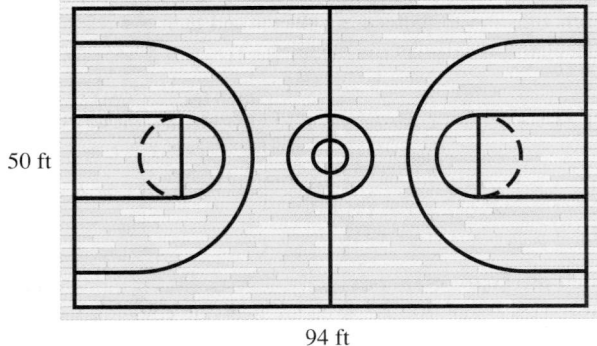

50 ft

94 ft

74. **Perimeter of a Soccer Field**   The perimeter of the soccer field shown in the figure is $x + 40$ yd. Find the value of x.

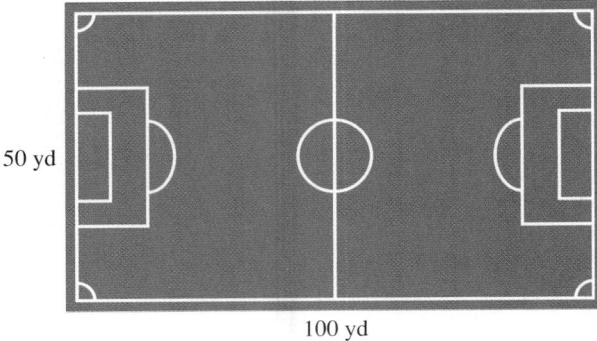

50 yd

100 yd

Group Discussion Questions

75. **Writing Mathematically**   Write a paragraph describing the addition-subtraction principle of equality, using the analogy of a balance scale.

76. **Challenge Question**
    a. Complete the equation $8x + 7 = 7x + \underline{?}$ so that the solution is $x = 4$.
    b. Complete the equation $12x - 3 = 11x + \underline{?}$ so that the solution is $x = 8$.

77. **Challenge Question**
    a. Complete the equation $3x + 5 = 2x + \underline{?}$ so that the solution is $x = 7$.
    b. Complete the equation $3x + 5 = 2x + \underline{?}$ so that the equation is a contradiction.
    c. Complete the equation $3x + 5 = 2x + \underline{?}$ so that the equation is an identity.

78. **Discovery Question**   Let y_1 represent the left side of $x^2 = 5x - 6$ and y_2 represent the right side. Then use a graphics calculator to explore this equation using the same strategy used for linear equations. Record both your observations and explanation of these observations.

Section 2.5   Solving Linear Equations in One Variable Using the Multiplication-Division Principle

Objective: **12.** Solve linear equations in one variable using the multiplication-division principle.

This section expands our ability to solve linear equations in one variable. We solve equations of the form $Ax = B$ by obtaining a coefficient of 1 for x. To obtain the coefficient of 1, we use the multiplication-division principle of equality.

Multiplication-Division Principle of Equality		
ALGEBRAICALLY	**VERBALLY**	**NUMERICAL EXAMPLE**
If a, b, and c are real numbers and $c \neq 0$, then $a = b$ is equivalent to $ac = bc$ and to $\dfrac{a}{c} = \dfrac{b}{c}$.	If both sides of an equation are multiplied or divided by the same nonzero number, the result is an equivalent equation.	$\dfrac{x}{2} = 5$ is equivalent to $2\left(\dfrac{x}{2}\right) = 2(5);$ and $3x = 12$ is equivalent to $\dfrac{3x}{3} = \dfrac{12}{3}.$

Students sometimes conclude that it is okay to perform any operation on both sides of an equation. This is incorrect. Multiplying or dividing both sides of an equation by zero is an exception. This is especially tricky if both sides of an equation are multiplied by a variable. If this variable is zero, this can cause problems such as extraneous values. We will examine this situation later in the book.

To solve $Ax = B$ with $A \neq 0$ we either divide both sides of the equation by A or equivalently multiply by $\dfrac{1}{A}$, the multiplicative inverse of A. If A is a rational fraction, then we usually multiply by the reciprocal of A. Otherwise, we usually use division.

■ **EXAMPLE 1**   Solving a Linear Equation with an Integer Coefficient

Solve $7y = 91$ and check the solution.

SOLUTION

$7y = 91$

$\dfrac{7y}{7} = \dfrac{91}{7}$ Divide both sides of the equation by 7, the coefficient of y.

$y = 13$ This produces a coefficient on y of 1; $1 \cdot y = y$.

Check: $7y = 91$

$7(13) \overset{?}{=} 91$

$91 \overset{?}{=} 91$ checks.

Answer: $y = 13$.   ■

A linear equation is not solved until the coefficient of the variable is 1. If the coefficient is -1, the equation is not yet solved.

■ **EXAMPLE 2**   Solving a Linear Equation with a Coefficient of -1

The coefficient of t is understood to be -1.

Solve $-t = 17$.

SOLUTION

$-t = 17$ The coefficient of t is -1.

$-t(-1) = 17(-1)$ Multiply both sides of the equation by -1. (The multiplicative inverse of -1 is -1.)

Answer: $t = -17$.   Does this value check?   ■

EXAMPLE 3 Estimating a Solution of an Equation

Mentally estimate the solution of $-7.15m = 42.042$.

SOLUTION

Estimate by considering the equation $-7m = 42$.

$$-7m = 42$$
$$\frac{-7m}{-7} = \frac{42}{-7}$$
$$m = -6$$

The steps shown here often can be performed mentally.

Answer: $m \approx -6$. The actual answer is -5.88.

SELF-CHECK 2.5.1

Solve the following equations:

1. $6v = -42$
2. $-v = 4$
3. $-8x = 56$

Linear equations often contain variables and constants on both sides of the equation. To solve these linear equations, first isolate the variable terms on one side of the equation by using the addition-subtraction principle. Once the equation has been simplified to the form $Ax = B$, use the multiplication-division principle to solve for the variable.

EXAMPLE 4 Solving a Linear Equation with Constant Terms on Both Sides

Solve $3x + 2 = 14$ and check the solution.

SOLUTION

$$3x + 2 = 14$$
$$3x + 2 - 2 = 14 - 2$$

Subtract 2 from both sides to isolate the constants on the right side of the equation.

$$3x = 12$$
$$\frac{3x}{3} = \frac{12}{3}$$

Divide both sides by 3 to obtain a coefficient of 1 for x.

$$x = 4$$

Check: $3x + 2 = 14$

$$3(4) + 2 \stackrel{?}{=} 14$$
$$12 + 2 \stackrel{?}{=} 14$$
$$14 \stackrel{?}{=} 14 \text{ checks.}$$

Answer: $x = 4$.

SELF-CHECK 2.5.1 ANSWERS

1. $v = -7$
2. $v = -4$
3. $x = -7$

■ EXAMPLE 5   Solving a Linear Equation with Variables on Both Sides

Solve $5v - 4 = 7v + 10$.

SOLUTION _____

$$5v - 4 = 7v + 10$$
$$5v - 4 - 7v = 7v + 10 - 7v \quad \text{Subtract } 7v \text{ from both sides to isolate the variables on the left side.}$$
$$-2v - 4 = 10$$
$$-2v - 4 + 4 = 10 + 4 \quad \text{Add 4 to both sides to isolate the constants on the right side.}$$
$$-2v = 14$$
$$\frac{-2v}{-2} = \frac{14}{-2} \quad \text{Divide both sides by } -2.$$
$$v = -7$$

Check: $5v - 4 = 7v + 10$
$$5(-7) - 4 \overset{?}{=} 7(-7) + 10$$
$$-35 - 4 \overset{?}{=} -49 + 10$$
$$-39 \overset{?}{=} -39 \text{ checks.}$$

Answer: $v = -7$.   ■

CALCULATOR PERSPECTIVE 2.5.1	Checking Solutions by Comparing the Values of Two Expressions

Store the solution to be checked under x. Then evaluate the left-side expression by typing it and pressing [ENTER]. Next evaluate the right-side expression by the same process. If the stored value of x causes both expressions to have the same value, then that stored value is a solution.

To check both 7 and -7 in the equation $5v - 4 = 7v + 10$ from Example 5 on a TI-83 Plus calculator, enter the following keystrokes, using x for a variable instead of v:

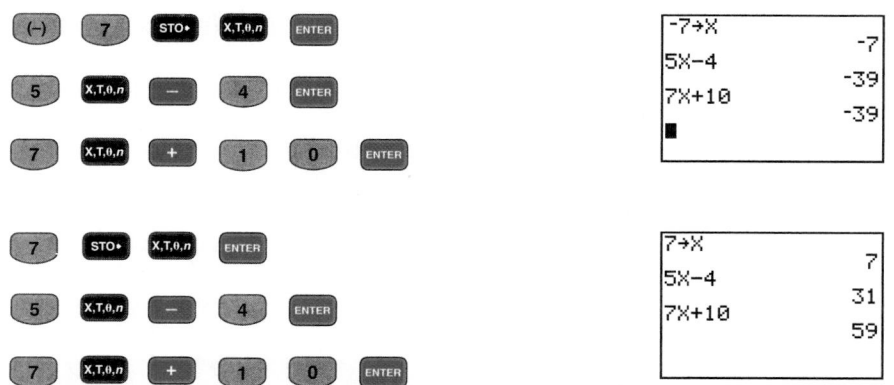

Note: Both expressions are equal for $x = -7$, so this value checks. The value $x = 7$ does not check because the expressions are not equal.   ■

The strategy used in the previous examples is described in the following box. We then use this strategy to solve equations containing parentheses and fractions.

Strategy for Solving Linear Equations

STEP 1. Simplify each side of the equation.
  a. If the equation contains fractions, simplify by multiplying both sides of the equation by the least common denominator (LCD) of all the fractions.
  b. If the equation contains grouping symbols, simplify by using the distributive property to remove the grouping symbols and then combine like terms.

STEP 2. Using the addition-subtraction principle of equality, isolate the variable terms on one side of the equation and the constant terms on the other side.

STEP 3. Using the multiplication-division principle of equality, solve the equation produced in step 2.

A good rule of thumb to use when solving a linear equation is to try to produce simpler expressions with each step.

■ EXAMPLE 6 Solving a Linear Equation Containing Parentheses

Solve $2(3x - 5) = 7(2x + 2)$.

SOLUTION

$$2(3x - 5) = 7(2x + 2)$$

Remove parentheses using the distributive property.

$$6x - 10 = 14x + 14$$

$$6x - 10 - 14x = 14x + 14 - 14x$$

$$-8x - 10 + 10 = 14 + 10$$

Isolate the variable terms on the left side and the constant terms on the right side by subtracting $14x$ from both sides and adding 10 to both sides.

$$-8x = 24$$

$$\frac{-8x}{-8} = \frac{24}{-8}$$

Solve for x by dividing both sides by -8.

$$x = -3$$

Check:
$$2(3x - 5) = 7(2x + 2)$$
$$2[3(-3) - 5] \overset{?}{=} 7[2(-3) + 2]$$
$$2(-9 - 5) \overset{?}{=} 7(-6 + 2)$$
$$2(-14) \overset{?}{=} 7(-4)$$
$$-28 \overset{?}{=} -28 \text{ checks.}$$

Answer: $x = -3$. ■

SELF-CHECK 2.5.2

Solve the following equations:

1. $5x - 3 = 27$
2. $3(v - 3) = 2(5v - 11) - 1$

SELF-CHECK 2.5.2 ANSWERS
1. $x = 6$
2. $v = 2$

■ **EXAMPLE 7**   Solving a Linear Equation Containing Parentheses

Solve $4y + 3(y - 2) = 2(y + 4) - (2y - 7)$.

SOLUTION _____

$4y + 3(y - 2) = 2(y + 4) - (2y - 7)$	Remove parentheses using the distributive property and
$4y + 3y - 6 = 2y + 8 - 2y + 7$	then combine like terms.
$7y - 6 = 15$	
$7y - 6 + 6 = 15 + 6$	Isolate the constant terms on the right side by adding 6 to
$7y = 21$	both sides.
$\dfrac{7y}{7} = \dfrac{21}{7}$	Solve for y by dividing both sides by 7.

Answer: $y = 3$.   Does this value check?   ■

If a linear equation contains fractions, then we can simplify it by converting it to an equivalent equation that does not involve fractions. To do this, multiply both sides of the equation by the LCD of all the terms.

■ **EXAMPLE 8**   Solving a Linear Equation Containing Fractions

Solve $\dfrac{z}{6} + 2 = \dfrac{z}{4}$.

SOLUTION _____

$\dfrac{z}{6} + 2 = \dfrac{z}{4}$	$6 = 2 \cdot 3$
	$4 = 2 \cdot 2$
$12\left(\dfrac{z}{6} + 2\right) = 12\left(\dfrac{z}{4}\right)$	LCD $= 2 \cdot 2 \cdot 3 = 12$
	Multiply both sides of the equation by the LCD, 12.
$12\left(\dfrac{z}{6}\right) + 12(2) = 12\left(\dfrac{z}{4}\right)$	Use the distributive property to remove the
	parentheses and then simplify.
$2z + 24 = 3z$	
$2z + 24 - 2z = 3z - 2z$	Subtract $2z$ from both sides of the equation.
$24 = z$ or	

Answer: $z = 24$.   Does this solution check?   ■

Observe in this example, and in the previous example, the execution of our basic strategy, which suggests using operations that will produce simpler expressions at each step.

◼ EXAMPLE 9 Solving a Linear Equation Containing Fractions

Solve $\dfrac{3v - 3}{6} = \dfrac{4v + 1}{15} + 2$.

SOLUTION _____

$$\frac{3v - 3}{6} = \frac{4v + 1}{15} + 2$$

$$30\left(\frac{3v - 3}{6}\right) = 30\left(\frac{4v + 1}{15} + 2\right)$$

$$\frac{30}{6}(3v - 3) = \frac{30}{15}(4v + 1) + 30(2)$$

$$5(3v - 3) = 2(4v + 1) + 30(2)$$

$$15v - 15 = 8v + 2 + 60$$

$$15v - 15 = 8v + 62$$

$$15v - 15 - 8v = 8v + 62 - 8v$$

$$7v - 15 = 62$$

$$7v - 15 + 15 = 62 + 15$$

$$7v = 77$$

$$\frac{7v}{7} = \frac{77}{7}$$

Answer: $v = 11$.

$6 = 2 \cdot 3$
$15 = 3 \cdot 5$
LCD $= 2 \cdot 3 \cdot 5 = 30$
Multiply both sides of the equation by the LCD, 30.

Use the distributive property to remove parentheses and then combine like terms.

Subtract $8v$ from both sides of the equation

Then add 15 to both sides.

Divide both sides of the equation by 7.

Does this solution check?

SELF-CHECK 2.5.3

Solve these equations.

1. $\dfrac{y}{3} - \dfrac{y}{12} = -\dfrac{1}{2}$

2. $\dfrac{4x + 2}{5} - 1 = \dfrac{7x - 3}{2}$

SELF-CHECK 2.5.3 ANSWERS

1. $y = -2$
2. $x = \dfrac{1}{3}$

USING THE LANGUAGE AND SYMBOLISM OF MATHEMATICS   2.5

1. Equations with the same solution set are called _____ equations.
2. By the _____ principle of equality, $a = b$ is equivalent to $ac = bc$ for $c \neq 0$.
3. By the multiplication-division principle of equality, $a = b$ is equivalent to $\dfrac{a}{c} =$ _____ for $c \neq 0$.
4. The equation $\dfrac{3x}{2} = 9$ is equivalent to $x =$ _____.
5. A _____ equation is true for some values of the variable and false for other values.
6. The statement "You can always multiply both sides of an equation by the same number to produce an equivalent

equation" is false because we do not obtain equivalent equations if we multiply both sides of the equation by _____.
7. The _____ property often is used to help us remove parentheses from expressions in an equation.
8. To clear an equation of fractions, we can multiply both sides of the equation by the _____ _____ _____ of all the terms in the equation.
9. The multiplicative inverse of $\dfrac{1}{4}$ is _____.
10. The multiplicative inverse of $-\dfrac{1}{4}$ is _____.

EXERCISES 2.5

In Exercises 1–6 solve each equation and then check your solution.

1. a. $7x = 42$ **b.** $42x = 7$
   c. $-x = 7$ **d.** $-x = -7$
2. a. $5y = 45$ **b.** $45y = 5$
   c. $-y = 5$ **d.** $-y = -5$
3. a. $-8t = 12$ **b.** $12t = -8$
   c. $\dfrac{t}{12} = 8$ **d.** $\dfrac{8t}{12} = 1$
4. a. $-3m = -39$ **b.** $39m = -3$
   c. $\dfrac{m}{3} = -39$ **d.** $\dfrac{3m}{39} = 1$
5. a. $-\dfrac{2}{3}m = 48$ **b.** $2(m + 1) = 18$
6. a. $\dfrac{4}{3}t = -36$ **b.** $3(m - 2) = 15$

In Exercises 7–10 solve each equation and then check your solution using your graphics calculator. (*Hint:* See Calculator Perspective 2.5.1.)

7. a. $3(2t - 1) = 7t + 1$
   b. $-3(4t - 1) = -(t - 14)$
8. a. $4(2 - 3x) = 3 - 13x$
   b. $-2(5x + 4) = -(x - 10)$
9. a. $\dfrac{y}{3} + \dfrac{2}{3} = \dfrac{y}{2} + \dfrac{3}{2}$
   b. $2 - 6(y + 1) = 4(2 - 3y) + 6$
10. a. $\dfrac{v}{4} - \dfrac{2}{5} = \dfrac{v}{10} + \dfrac{1}{2}$
   b. $13 + 3(2v - 5) = 1 - 2(3 - 6v)$

In Exercises 11–36 solve each equation.

11. $24 = -8z$ **12.** $-15 = 25z$
13. $-1 = 9m$ **14.** $1 = -7m$
15. $-47w = 0$ **16.** $0 = 31w$
17. $\dfrac{-5v}{3} = 25$ **18.** $-\dfrac{2v}{11} = 66$
19. $8x - 1 = 13x - 1$ **20.** $-17x + 5 = 17x + 5$
21. $4(3y - 5) = 5(4y + 4)$ **22.** $7(2y - 3) = 3(6y + 5)$
23. $0.12a = 13.2$ **24.** $0.07a = -3.5$
25. $2.3x + 29.3 = 1.2(4 - x)$ **26.** $5.7x + 35.7 = 1.2(x - 4)$
27. $6 - 3(2v + 2) = 4(8 - v)$
28. $2(1 - v) + 9 = 3(2v + 1)$
29. $4(x + 1) + 3(x + 2) = 7(x + 1) - 5$
30. $11(x + 3) + 4(2x - 1) = 5(3x - 2) + 13$
31. $\dfrac{w + 1}{3} = \dfrac{w - 5}{5}$ **32.** $\dfrac{24w - 67}{60} = \dfrac{3w - 8}{12}$
33. $\dfrac{x + 2}{3} - 4 = -\dfrac{x}{2}$ **34.** $\dfrac{2x}{7} = 1 - \dfrac{2x + 1}{3}$
35. $7(x - 1) - 4(2x + 3) = 2(x + 1) - 3(4 - x)$
36. $5(2x - 1) + 3(x - 3) = -4(x - 6) + 2(13 - 3x)$

In Exercises 37 and 38 each equation is a conditional equation, an identity, or a contradiction. Identify the type of each equation and then solve it.

37. a. $3(x + 1) = 3x$ **b.** $3(x + 1) = 3x + 3$
   c. $3(x + 2) = 2(x + 3)$
38. a. $4(x + 5) = 4x$ **b.** $4(x + 5) = 4x + 20$
   c. $4(x + 5) = 5(x + 4)$

In Exercises 39–42 simplify the expression in the first column and solve the equation in the second column.

Simplify	Solve
39. a. $3(2x - 4) - 5(x - 2)$	**b.** $3(2x - 4) = 5(x - 2)$
40. a. $3(2x - 4) + 5(x - 2)$	**b.** $3(2x - 4) = -5(x - 2)$
41. a. $1.5(4x - 6) + 2.5(6x - 4)$	**b.** $1.5(4x - 6) = -2.5(6x - 4)$
42. a. $1.5(4x - 6) - 2.5(6x - 4)$	**b.** $1.5(4x - 6) = 2.5(6x - 4)$

Estimation Skills (Exercises 43–46)

43. Perimeter of a Square $(P = 4s)$ The perimeter of a square is 99.837 cm. Mentally estimate the length of each side to the nearest centimeter.

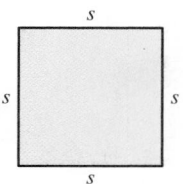

44. Perimeter of an Equilateral Triangle $(P = 3s)$ The perimeter of an equilateral triangle is 99.378 m. Mentally estimate the length of each side to the nearest meter.

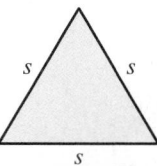

45. Hotel Room Tax $(T = 0.10C)$ A hotel tax on a bill is 10% of the room charges. Mentally estimate the pretax room charges for a room with a tax charge of $7.58.

46. Gratuity ($G = 0.20B$)   A customer added a gratuity of $13 to the restaurant charge. If the gratuity was approximately 20% of the restaurant bill, mentally estimate the bill before the gratuity was added.

Estimation and Calculator Skills

In Exercises 47–50 mentally estimate the solution of each equation to the nearest integer and then use a calculator to calculate the solution.

PROBLEM	MENTAL ESTIMATE	CALCULATOR SOLUTION
47. $-2.1x = 8.82$		
48. $4.9x = -14.945$		
49. $-0.49x = -2.009$		
50. $-0.24x = -1.896$		

Calculator Usage

In Exercises 51–54 y_1 represents the left side of the equation and y_2 represents the right side. Use the table shown from a graphics calculator to solve the equation.

51. $5.8x - 1.71 = 2.7x + 2.94$

X	Y₁	Y₂
0	-1.71	2.94
.5	1.19	4.29
1	4.09	5.64
1.5	6.99	6.99
2	9.89	8.34
2.5	12.79	9.69
	15.69	11.04

X=3

52. $7.2x + 4.2 = 4.7x + 0.95$

X	Y₁	Y₂
-1.5	-6.6	-6.1
-1.4	-5.88	-5.63
-1.3	-5.16	-5.16
-1.2	-4.44	-4.69
-1.1	-3.72	-4.22
-1	-3	-3.75
	-2.28	-3.28

X=-.9

53. $2.4(4x - 1) = 1.8(2x + 1)$

X	Y₁	Y₂
.2	-.48	2.52
.3	.48	2.88
.4	1.44	3.24
.5	2.4	3.6
.6	3.36	3.96
.7	4.32	4.32
	5.28	4.68

X=.8

54. $2.5(6x + 1) = 12x + 10.9$

X	Y₁	Y₂
2	32.5	34.9
2.2	35.5	37.3
2.4	38.5	39.7
2.6	41.5	42.1
2.8	44.5	44.5
3	47.5	46.9
	50.5	49.3

X=3.2

Multiple Representations

In Exercises 55–62 write an algebraic equation for each verbal statement and then solve for the variable.

55. Five times the sum of x and 2 is the same as seven times the quantity x minus 3.

56. Three times the sum of x and 6 equals eleven times the quantity x minus 2.

57. Twice the quantity three v minus 2 equals four times the quantity v plus 9.

58. Six times the quantity 7 minus v is the same as five times the quantity two v plus 9.

59. Twice the difference of three m minus five is four less than three times the sum of m and six.

60. Five times the sum of two m plus eleven is three more than two times the quantity m minus nine.

61. One-third the quantity two x plus five is the same as the quantity four x plus two.

62. One-half the quantity three x plus seven equals the quantity six x plus two.

63. Perimeter of a Wrestling Mat   The perimeter of one of the square wrestling mats shown in the figure is 168 ft. The length of one side of this mat is $4x + 18$ ft. Find the value of x.

64. Perimeter of a Yield Sign   The perimeter of the yield sign that follows is 60 in. Assuming that this sign is approximately an equilateral triangle, find the value of x.

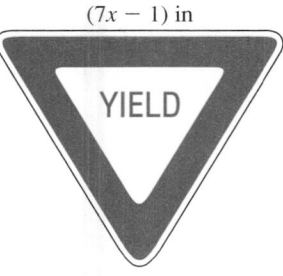

$(7x - 1)$ in

65. Perimeter of a Rectangle   The perimeter of the rectangle shown in the figure is 17 ft. Find the value of a.

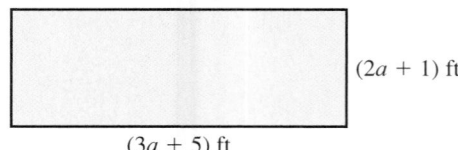

$(2a + 1)$ ft

$(3a + 5)$ ft

66. Perimeter of a Parallelogram   The perimeter of the parallelogram shown in the figure is 17 cm. Find the value of a.

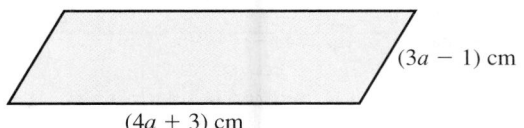

$(3a - 1)$ cm

$(4a + 3)$ cm

67. Area of a Triangle*   The area of the triangle shown in the figure is 51 square centimeters (cm²). Find the value of x.

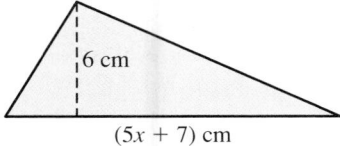

6 cm

$(5x + 7)$ cm

68. Area of a Rectangle*   The area of the rectangle shown in the figure is 102 cm². Find the value of x.

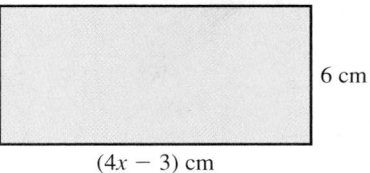

6 cm

$(4x - 3)$ cm

Group Discussion Questions

69. Writing Mathematically
Solve $5(x - 3) + 4 = 4(2x + 1) - 9$ and then write an explanation of each step of the solution. Also explain why you performed the steps in the order in which you have them listed rather than in some other order.

70. Challenge Question

a. Complete the equation $\frac{4}{5}x + 11 = \underline{?}$ so that the solution is $x = 31$.

b. Complete the equation $5(x - 3) + 1 = 2(x + 4) + \underline{?}$ so that the solution is $x = 3$.

71. Error Analysis   Examine the following argument and explain the error in the reasoning. If $x = 0$, then $2x = 0$; thus $x = 2x$. Dividing both sides of this equation by x, we conclude that $1 = 2$.

Section 2.6   Ratios, Proportions, and Direct Variation

Objectives: **13.** Use ratios and proportions to solve word problems.
14. Solve problems involving direct variation.

The ability to solve real-world problems often requires creating a mathematical model of the problem. Linear equations are used to model many problems. In this section we use ratios and proportions to form linear equations to model some applications.

A classic type of problem is determining the height of one object by comparing it to the height of a second object. We use ratios and proportions to accomplish this. A **proportion** is an equation that states two ratios are equal. For example, the proportion $\frac{a}{b} = \frac{c}{d}$ is read as "a is to b as c is to d." The four numbers a, b, c, and d are called the **terms of the proportion:** a is the first term, b is the second term, c is the third term, and d is the fourth term. The first and fourth terms are called the **extremes** and the second and third terms are called the **means.**

The word "mean" has now been used in two contexts: as an average of a set of values and as one of the middle terms of a proportion.

Proportion		
ALGEBRAICALLY	**VERBALLY**	**NUMERICAL EXAMPLE**
$\frac{a}{b} = \frac{c}{d}$ or $a:b = c:d$	This proportion is read "a is to b as c is to d." The extremes are a and d, and the means are b and c.	$\frac{2}{5} = \frac{40}{100}$ or $2:5 = 40:100$

*Formulas for the area of common geometric figures are given on the inner jacket covers of this book.

A Mathematical Note

Emmy Noether (1882–1935) was called by Albert Einstein the greatest of all the women who were creative mathematical geniuses. Emmy completed her doctoral dissertation in algebra at the University of Erlangen in 1907. She taught at the University of Göttingen in Germany from 1915 until 1933. She then moved to the United States and taught at Bryn Mawr for 2 years until her death in 1935.

Since proportions are equations, we can solve them by using the same rules and procedures that apply to all equations.

■ **EXAMPLE 1**   Solving a Proportion

Solve $\dfrac{x}{3} = \dfrac{5}{4}$.

SOLUTION

$$\frac{x}{3} = \frac{5}{4}$$

$$12\left(\frac{x}{3}\right) = 12\left(\frac{5}{4}\right) \qquad \begin{array}{l} \text{LCD} = 3 \cdot 4 = 12 \\ \text{Multiply both sides of the equation by the LCD of 12.} \end{array}$$

$$4x = 3(5)$$

$$4x = 15$$

$$\frac{4x}{4} = \frac{15}{4} \qquad \text{Divide both sides of the equation by 4.}$$

Answer: $x = \dfrac{15}{4}$.   Does this solution check?

■ **EXAMPLE 2**   Solving a Proportion

Solve $\dfrac{35}{6m} = \dfrac{5}{9}$.

SOLUTION

$$\frac{35}{6m} = \frac{5}{9} \qquad \begin{array}{l} 6m = 2 \cdot 3 \cdot m \\ 9 = 3 \cdot 3 \\ \text{LCD} = 2 \cdot 3 \cdot 3 \cdot m = 18m \\ \text{Multiply both sides by the LCD of } 18m. \end{array}$$

$$18m\left(\frac{35}{6m}\right) = 18m\left(\frac{5}{9}\right)$$

$$\frac{\overset{3}{\cancel{18m}}}{\underset{1}{\cancel{6m}}}(35) = \frac{\overset{2m}{\cancel{18m}}}{\underset{1}{\cancel{9}}}(5) \qquad \text{Then simplify both sides of the equation.}$$

$$3(35) = (2m)(5)$$

$$105 = 10m$$

$$\frac{\overset{21}{\cancel{105}}}{\underset{2}{\cancel{10}}} = \frac{10m}{10} \qquad \text{To solve for } m, \text{ divide both sides by 10.}$$

$$10.5 = m \text{ or}$$

Answer: $m = 10.5$.   Does this value check?

Ratios and proportions are so useful that many employment tests, including civil service tests, have problems that test your ability to use proportions. A typical problem involves a constant ratio under two different situations; that is,

$$\boxed{\text{Ratio for first situation}} = \boxed{\text{Ratio for second situation}}$$

This type of problem is illustrated in Example 3, which compares the ratio of distances on a first map reading to distances on a second map reading.

■ **EXAMPLE 3**   Determining Distances Indicated on a Map

2 cm on a map represent a distance of 230 km. What distance corresponds to 6.5 cm on the map?

SOLUTION _____

Let d = distance in km corresponding to 6.5 cm.

VERBALLY

$$\dfrac{\text{First map reading}}{\text{First distance}} = \dfrac{\text{Second map reading}}{\text{Second distance}}$$

First stating the strategy as a verbal equation is an excellent way to start a word problem.

ALGEBRAICALLY

A common question regarding Example 3 is to inquire if it is OK to form the ratio
$$\dfrac{\dfrac{\text{First map reading}}{\text{Second map reading}}}{\dfrac{\text{First distance}}{\text{Second distance}}}.$$
This is a good question and the answer is yes. More generally,

the equation $\dfrac{a}{b} = \dfrac{c}{d}$

is equivalent to $\dfrac{a}{c} = \dfrac{b}{d}$.

$$\dfrac{2}{230} = \dfrac{6.5}{d}$$

$$230d\left(\dfrac{2}{230}\right) = 230d\left(\dfrac{6.5}{d}\right)$$

$$2d = 230(6.5)$$

$$\dfrac{2d}{2} = \dfrac{1495}{2}$$

$$d = 747.5$$

Use the given values to translate the verbal equation into an algebraic equation.
Multiply both sides of the equation by the LCD of $230d$.

Divide both sides by 2.

Answer: 6.5 cm on the map represents 747.5 km.

Does this answer seem reasonable?

■

Describing the relationship among variables is a very important part of mathematics. Some relationships are so common that it is not surprising that we have developed multiple ways to describe the relationships. The concept of direct variation, which we examine now, is closely related to arithmetic sequences, linear equations, and ratios and proportions.

Direct Variation

If x and y are real variables and k is a real constant, then:

VERBALLY	ALGEBRAICALLY	NUMERICAL EXAMPLE		GRAPHICAL EXAMPLE
y varies directly as x with the constant of the variation k.	$y = kx$ *Example:* $y = 3x$	x	$y = 3x$	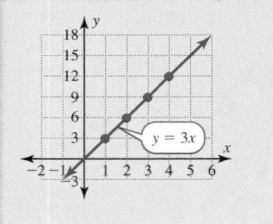
		1	3	
		2	6	
		3	9	
		4	12	
		5	15	

■ EXAMPLE 4   Translating Statements of Variation

SOLUTIONS _____

(a) Translate $C = \pi d$ into a verbal statement of variation.

C varies directly as d with the constant of variation π.

The circumference of a circle varies directly as the diameter with π being the constant of variation.

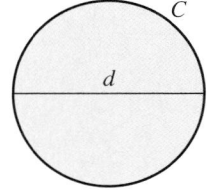

(b) Hooke's law states that the distance d a spring stretches varies directly as the mass m attached to the spring. Translate this statement of variation into an algebraic equation.

$d = km$

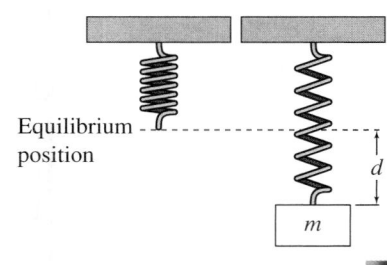

Equilibrium position

Sometimes the constant of variation is so important that this constant becomes well known or has a special symbol assigned to it. An example of this is π, which is the ratio of the circumference of a circle to its diameter. In the next example we solve for the constant of variation that represents the exchange rate between two currencies.

■ EXAMPLE 5   Determining a Currency Exchange Rate

Traveling in Australia during 2000, one could exchange 800 U.S. dollars for 1308 Australian dollars. The number of Australian dollars received varies directly as the number of U.S. dollars exchanged. The constant of variation is called the exchange rate. What was the exchange rate at this time?

SOLUTION _____

Let U = number of U.S. dollars
   A = number of Australian dollars
   k = constant of variation

VERBALLY

The number of Australian dollars varies directly as the number of U.S. dollars.

Verbally state a precise equation that is easily translated into algebraic form.

ALGEBRAICALLY

$$A = kU$$

$$\$1308 \text{ Australian} = k(\$800 \text{ U.S.})$$

$$\frac{\$1308 \text{ Australian}}{\$800 \text{ U.S.}} = k$$

$$k \approx \frac{\$1.64 \text{ Australian}}{\$1 \text{ U.S.}}$$

Translate the word equation into algebraic form using the variables identified above. Substitute in the given values for A and U.

Then solve for k, the exchange rate.

Answer: The exchange rate was $1.64 Australian for $1 United States.   ■

The next example shows how we can solve a problem involving direct variation without calculating the constant of variation. If $y = kx$, then $k = \dfrac{y}{x}$. When y varies directly as x, k is a constant whereas x and y vary. Thus if $k = \dfrac{y_1}{x_1}$ and $k = \dfrac{y_2}{x_2}$ for two sets of values, then $\dfrac{y_1}{x_1} = \dfrac{y_2}{x_2}$. This can be described by saying that x *and* y *are directly proportional* when y varies directly as x.

(x_1, y_1) represents one input-output pair and (x_2, y_2) represents a second input-output pair.

SELF-CHECK 2.6.2 ANSWER

$6\dfrac{2}{3}$ cups

SELF-CHECK 2.6.2

A recipe for old-fashioned vegetable soup suggests using 4 cups of water to prepare a serving for 6 people. How much water should be used to prepare a serving for 10?

■ **EXAMPLE 6** Using Direct Variation to Form a Proportion

If y varies directly as x, and y is 12 when x is 9, find y when x is 15.

SOLUTION

$$\frac{y_1}{x_1} = \frac{y_2}{x_2}$$
x and y are directly proportional when y varies directly as x.

$$\frac{12}{9} = \frac{y_2}{15}$$
Substitute 12 for y_1, 9 for x_1, and 15 for x_2.

$$\frac{4}{3} = \frac{y_2}{15}$$
Reduce the left side of the equation by dividing both the numerator and the denominator by 3.

$$15\left(\frac{4}{3}\right) = 15\left(\frac{y_2}{15}\right)$$
Multiply both sides by the LCD of 15.

$$5(4) = y_2$$
$$20 = y_2 \text{ or}$$
$$y_2 = 20$$
Then simplify both sides of the equation.

Check: $\dfrac{12}{9} \overset{?}{=} \dfrac{20}{15}$

$\dfrac{4}{3} \overset{?}{=} \dfrac{4}{3}$ checks.   Since $\dfrac{y_1}{x_1} = \dfrac{y_2}{x_2}$, x and y are directly proportional.

Answer: $y = 20$ when $x = 15$. ■

Ratios and proportions are the basis for many indirect measurements. In Example 7 we indirectly determine the height of a tree by measuring the length of a shadow. This same technique has been used to determine the heights of mountains on Mars.

■ **EXAMPLE 7** Modeling the Height of a Tree Using Its Shadow

At a fixed time of the day the length of a shadow varies directly as the height of the object casting the shadow. If a person 2 m tall casts a 5-m shadow, how tall is a fir tree that casts a 40-m shadow?

40 m   5 m   2 m

SOLUTION _____

Let t = height of the tree in meters.

VERBALLY

$$\frac{\text{Height of person}}{\text{Length of person's shadow}} = \frac{\text{Height of tree}}{\text{Length of tree's shadow}}$$

Verbally we form an equation based on the proportion given by this statement of direct variation. Also note that this proportion could be formed by using corresponding sides of similar right triangles.

ALGEBRAICALLY

$$\frac{2}{5} = \frac{t}{40}$$

Substitute the given values to translate the word equation into an algebraic equation.

$$40\left(\frac{2}{5}\right) = 40\left(\frac{t}{40}\right)$$

Multiply both sides of the equation by the LCD of 40.

$$8(2) = t$$
$$16 = t \text{ or}$$
$$t = 16$$

Then simplify both sides of the equation.

Does this answer seem reasonable?

Answer: The tree is 16 m tall.

The next example will relate the concepts of direct variation, linear equations, arithmetic sequences, and straight lines. Note that the sequence of y-values in the following table form an arithmetic sequence with a common difference that is equal to the constant of variation.

◼ EXAMPLE 8   Comparing Direct Variation, Linear Equations, and Arithmetic Sequences

Examine the following statement of variation verbally, algebraically, numerically, and graphically: y varies directly as x with the constant of variation 2.

SOLUTION _____

VERBALLY

y varies directly as x with the constant of variation 2.

ALGEBRAICALLY

$$y = 2x$$

NUMERICALLY

x	$y = 2x$
1	2
2	4
3	6
4	8
5	10

GRAPHICALLY

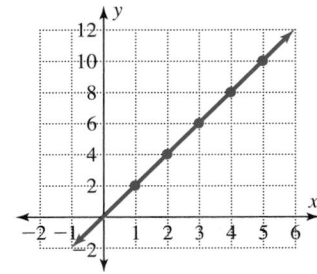

Note that the y-values form an arithmetic sequence with a common difference of 2 and the graph of these points forms a linear pattern.

USING THE LANGUAGE AND SYMBOLISM OF MATHEMATICS   2.6

1. An equation that states two ratios are equal is called a
_____ .

2. In the equation $\dfrac{a}{b} = \dfrac{c}{d}$, a, b, c, and d are called the
_____ of the proportion.

3. The means of $\dfrac{a}{b} = \dfrac{c}{d}$ are _____ and _____ .

4. The extremes of $\dfrac{a}{b} = \dfrac{c}{d}$ are _____ and
_____ .

5. If x and y are directly proportional, then y varies
_____ as x.

6. If y varies directly as x, then the equation expressing this
is a _____ equation.

7. If y varies directly as x, then inputting consecutive natural
numbers for x will produce output y-values that form an
_____ sequence.

8. If y varies directly as x, then the graph of the (x, y) ordered
pairs will form a _____ .

EXERCISES   2.6

In Exercises 1–22 solve each proportion.

1. $\dfrac{x}{20} = \dfrac{3}{5}$

2. $\dfrac{x}{16} = \dfrac{3}{2}$

3. $\dfrac{3}{5} = \dfrac{6}{x}$

4. $\dfrac{2}{7} = \dfrac{10}{x}$

5. $\dfrac{5x}{2} = \dfrac{45}{6}$

6. $\dfrac{11x}{3} = \dfrac{144}{21}$

7. $\dfrac{9}{2y} = \dfrac{27}{66}$

8. $\dfrac{6}{3y} = \dfrac{5}{6}$

9. $\dfrac{5}{6} = \dfrac{40}{3y}$

10. $\dfrac{13}{5} = \dfrac{52}{10y}$

11. $a : 3 = 4 : 6$

12. $7z : 10 = 7 : 2$

13. $5 : 11 = 15 : b$

14. $5 : 2v = 3 : 4$

15. $\dfrac{x + 1}{5} = \dfrac{8}{10}$

16. $\dfrac{x + 1}{20} = \dfrac{3}{5}$

17. $\dfrac{z + 2}{3} = \dfrac{z}{4}$

18. $\dfrac{t - 1}{5} = \dfrac{t}{2}$

19. $\dfrac{x}{3} = \dfrac{2x - 3}{5}$

20. $\dfrac{3w}{4} = \dfrac{w - 5}{1}$

21. $\dfrac{2v - 1}{5} = \dfrac{2v - 5}{1}$

22. $\dfrac{x - 1}{6} = \dfrac{2x - 4}{9}$

In Exercises 23 and 24 write an equation for each statement
of variation, using k as the constant of variation.

23. At a fixed speed the distance d that a car travels varies
directly as the time t it travels.

24. At a fixed pressure the volume V of a gas varies directly as
the absolute temperature T.

In Exercises 25 and 26 write a statement of variation for
each equation assuming k is the constant of variation.

25. $v = kw$ 26. $b = ka$

In Exercises 27–32 solve each of these problems involving
direct variation.

27. If v varies directly as w, and $v = 22$ when $w = 77$, find v
when $w = 35$.

28. If m varies directly as n, and $m = 12$ when $n = 20$, find n
when $m = 18$.

29. If p varies directly as g, and p is 0.375 when g is 1, find g
when p is 1.5.

30. If b varies directly as d, and b is 0.3125 when d is 1, find b
when d is 0.4374.

31. If v varies directly as w, and $v = 12$ when $w = 15$, find the
constant of variation k.

32. If b varies directly as a, and $b = \dfrac{2}{3}$ when $a = \dfrac{4}{5}$, find the
constant of variation k.

Multiple Representations

In Exercises 33–36 represent each direct variation algebraically, numerically, verbally, or graphically as described in each
problem. (*Hint:* See the following example.)

Algebraically

$y = -2x$

Numerically

x	y
1	−2
2	−4
3	−6
4	−8
5	−10

Graphically

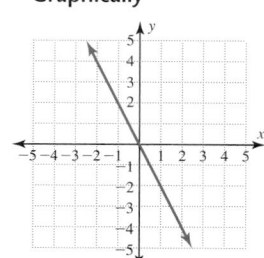

Verbally

y varies directly as x with the constant of
variation -2.

33. Represent the direct variation "y varies directly as x with the constant of variation $\frac{1}{2}$" algebraically, numerically, and graphically.

34. Represent the direct variation $y = 3x$ numerically, graphically, and verbally.

35. Represent the direct variation $y = -\frac{x}{2}$ numerically, graphically, and verbally.

36. Represent the direct variation "y varies directly as x with the constant of variation -1" algebraically, numerically, and graphically.

Modeling Using Proportions   In Exercises 37–66 use proportions to model each situation and to solve each problem.

Recipe Proportions (Exercises 37 and 38)

37. A recipe for 50 people used 3 cups of sugar. How many cups of sugar are needed for 75 people?

38. A recipe for six adults called for $\frac{3}{4}$ grams of salt. If this recipe is used to cook for four people, how much salt is required?

Similar Geometric Figures (Exercises 39 and 40)

The lengths of corresponding sides of similar geometric figures are directly proportional.

39. The triangles shown in the figure are similar. Find the length of the side labeled a.

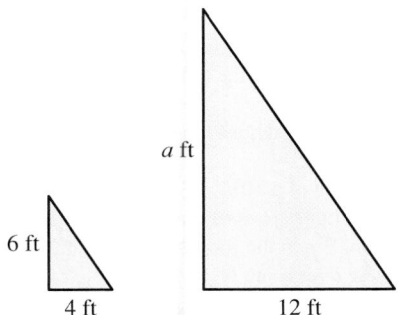

40. The two rectangles shown in the figure are similar. Find the length of the side labeled w.

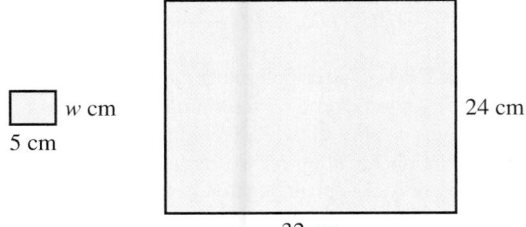

41. **Microscope Magnification**   Under a microscope an insect antenna that measures 1.5 millimeters (mm) appears to be 7.5 mm. Find the apparent length under this magnification of an object that is 2 mm long.

42. **Distance on a Map**   If 6 cm on a map represents 300 km, how much does 1 cm represent?

43. **Defective Lightbulbs**   A factory that produces lightbulbs finds that 2.5 out of every 500 bulbs are defective. In a group of 10,000 bulbs, how many would be expected to be defective?

44. **Gold Ore**   A mining company was able to recover 3 ounces of gold from 10 tons of ore. How many tons would be needed to recover 12 ounces of gold?

45. **Bricks for a Wall**   If 118 bricks are used in constructing a 2-ft section of a wall, how many bricks will be needed for a similar wall that is 24 ft long?

46. **Sand in Concrete**   If 11 cubic feet (ft^3) of sand is needed to make 32 ft^3 of concrete, how much sand is needed to make 224 ft^3 of concrete?

47. **Currency Conversion Rate**   The number of Australian dollars received varies directly as the number of U.S. dollars exchanged. In 2000, one could exchange 700 U.S. dollars for 1145 Australian dollars. Approximate to the nearest hundredth the constant of variation (the exchange rate) at that time. Obtain the current exchange rate from a website. One option is **http://www.xe.net/ucc/**.

48. **Currency Conversion Rate**   The number of Euro dollars received varies directly as the number of U.S. dollars exchanged. In 2000, one could exchange 500 U.S. dollars for 517 Euro dollars. What was the constant of variation (the exchange rate) at that time? Obtain the current exchange rate from a website. One option is **http://www.xe.net/ucc/**.

49. **Scale Drawing**   An architect's sketch of one room in a new computer lab is shown here. The dimensions of the sketch are given in centimeters. The architect used a constant scale factor of 1 cm representing 2.5 m. What will be the actual dimensions of the room?

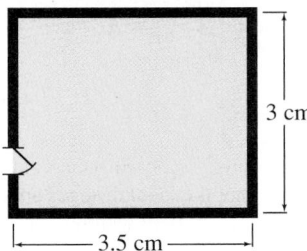

50. **Scale Drawing**   For the computer lab sketch in Exercise 49 determine the actual width of the door if the door in the sketch is 0.4 cm wide.

51. **Grass Seed**   A homeowner used 3 pounds of grass seed for 100 square feet (ft^2). How many pounds would be needed for 450 ft^2 if the amount of seed used varies directly as the area covered?

52. **Quality Control**   A quality control inspector found 4 defective computer chips in the 250 that were tested. How many defective chips would be expected in a shipment of 10,000 chips?

53. Face Cards in Blackjack   A blackjack dealer in a casino dealt 39 cards. Each of the decks from which he dealt have 12 face cards for every 52 cards. Approximately how many of these 39 cards would you expect to be face cards?

54. Maglev Train   Plans for an operational scale model of a maglev (magnetically levitating) train call for a scale of 1 to 50. The two-car train will be about 28 in long and will sell to hobbyists for $2000 to $5000. What is the length of the maglev train on which this scale model is based?

55. Width of a River   Use the dimensions shown in the following figure to determine the width w of the river. Triangles AB_1C_1 and AB_2C_2 are similar.

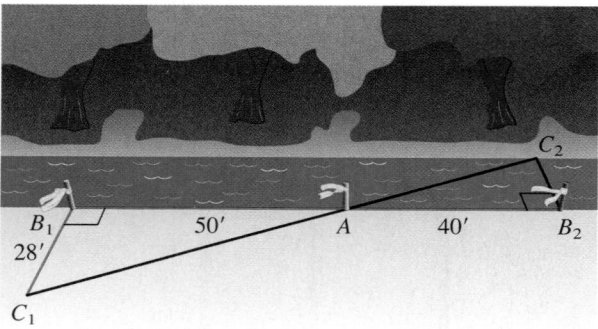

56. Dosage Instructions   A pharmaceutical company testing a new drug finds that the ideal dosages appear to be 293 milligrams (mg) for a patient weighing 45 kg, 390 mg for a patient weighing 60 kg, and 488 mg for a patient weighing 75 kg. Determine the correct dosage instructions for this drug.

57. Photograph Enlargement   A photograph is 7 cm wide and 10 cm long. What would be the width of the photograph if it were enlarged so that the length is 25 cm?

58. Photocopier Reduction   A photocopier is to be used to reduce the printed material on a page 8.5 in wide and 11 in high so that it will occupy a page 4.25 in wide. How high will be the reduced page of material?

59. Height of a Flagpole   The length of a shadow varies directly as the length of the object casting the shadow. A boy 5 ft tall casts a 6-ft shadow. How high is the flagpole beside him that casts a 21-ft shadow? (See the figure.)

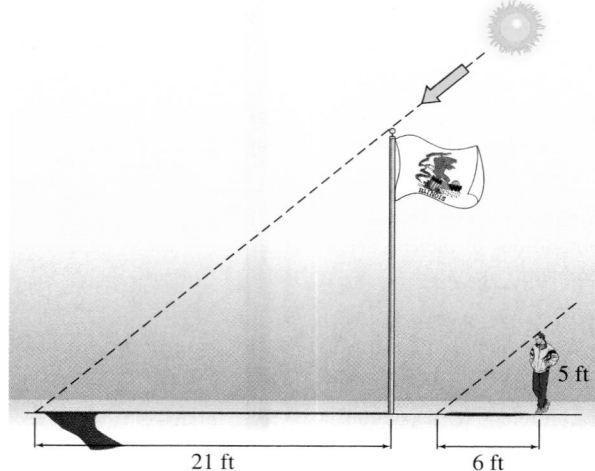

60. Height of a Building   An architect who is designing a building is concerned with the shadow that will be created by this building during a certain time of the day. If a yardstick (3 ft) casts a 10-ft shadow, what height of building would cast a 400-ft shadow?

61. Height of a Television Tower   The slope of a cable is the ratio of the rise to the run. A man wished to determine the height of a television tower. The distance (or run) from the bottom of the cable to the bottom of the tower measured 40 ft. He then measured a rise of 8 ft and found a run of 6 ft. What is the height of the tower whose tip is at the top end of the cable?

62. Scale Drawing of a House   The scale drawing of the house in the figure reveals that the roof rises 3 in over a run of 12 in. When the roof is actually built, how far will the roof rise over a run of 16 ft?

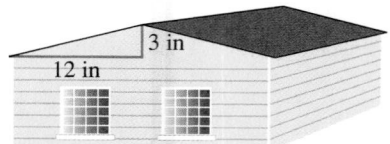

63. Baseball ERA   A baseball pitcher's ERA (Earned Run Average) is computed by determining the number of earned runs he allowed while recording 27 outs (one game). If a pitcher allowed two earned runs while recording only six outs in his first game, what would his ERA be?

64. Baseball ERA   In his first major league game a pitcher allowed five earned runs while recording nine outs. What was his ERA at the end of this game? (See Exercise 63.)

65. Doses of Medicine The child dosage of a certain medicine is $\frac{2}{3}$ of an ounce. A nurse has a bottle that contains 16 ounces of this medicine. How many children's doses of this medicine does this bottle contain?

66. Horse Bet A $2 bet on a "long shot" at the horse track collected $25.40. How much would a $5 bet on this same horse have collected?

In Exercises 67 and 68 y varies directly as x. Use the given table of values to determine the constant of variation.

67.
x	3	4	5	6	7
y	4.5	6	7.5	9	10.5

68.
x	3	4	5	6	7
y	2.4	3.2	4	4.8	5.6

In Exercises 69 and 70 y varies directly as x. Use the given graph to determine the constant of variation.

69.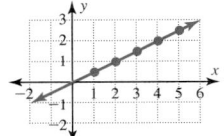

70.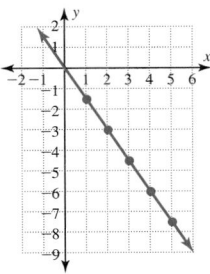

Group Discussion Questions

71. Challenge Question Complete the equation
$$\frac{2}{m-4} = \frac{4}{2m-?}$$
so that it is a proportion for all values of m.

72. Challenge Question An observer located at point A has a clear line of sight over level ground to points B and C. Describe a procedure using linear measurements that could be used to determine indirectly the distance from point B to point C.

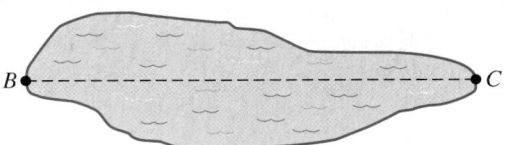

Section 2.7 Solving for a Specified Variable

Objectives:

15. Solve an equation for a specified variable.
16. Calculate the x- and y-intercepts from a linear equation and rewrite a linear equation in the form $y = mx + b$.

To enter information into calculators or computers, we often must rearrange our equations so that information fits the required format. We develop this ability in this section by using the principles already covered.

If an equation contains more than one variable, we can specify the variable that we wish to be the subject of the statement of equality. Then we can transpose the equality to write it in the desired form. This process is called **transposing an equation** or **solving for a specified variable.** To solve an equation for a specified variable, isolate this variable on one side of the equation with all other variables on the other side of the equation. In the next example we solve the linear equation $4x + 2y = 6$ for y.

A first-degree equation $Ax + By = C$ can always be rewritten in the form $y = mx + b$.

■ EXAMPLE 1   Solving a Linear Equation for *y*

Solve the linear equation $4x + 2y = 6$ for *y* and then use a graphics calculator to graph this equation.

SOLUTION

ALGEBRAICALLY

$$4x + 2y = 6$$
$$4x + 2y - 4x = 6 - 4x$$
$$2y = -4x + 6$$

Use the addition principle of equality to subtract $4x$ from both sides of the equation to isolate the *y*-term on the left side of the equation.

$$2y = 2(-2x + 3)$$
$$\frac{2y}{2} = \frac{2(-2x + 3)}{2}$$

Use the distributive property $ab + ac = a(b + c)$ to rewrite the right side of the equation.

Then use the division principle of equality to divide both sides of the equation by 2 to solve for *y*.

Answer: $y = -2x + 3$.

To enter a linear equation into a graphics calculator, first solve the equation for *y*.

GRAPHICALLY

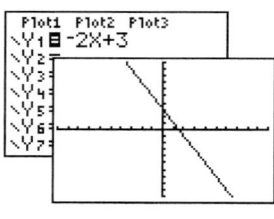

$[-10, 10, 1]$ by $[-10, 10, 1]$

Now that the linear equation is in the form $y = mx + b$, enter this equation into a graphics calculator and then graph this equation.

SELF-CHECK 2.7.1

1. Solve $x - 2y = 2$ for *y*.
2. Graph this equation on a graphics calculator.

■ EXAMPLE 2   Solving a Linear Equation for *y*

Solve the linear equation $3x - y = 3(2x - y) - 1$ for *y* and then use a graphics calculator to complete a table of values for the *x*-values: $-3, -2, -1, 0, 1, 2, 3$ and to graph the equation.

SOLUTION

ALGEBRAICALLY

$$3x - y = 3(2x - y) - 1$$
$$3x - y = 6x - 3y - 1$$
$$-y = 3x - 3y - 1$$
$$2y = 3x - 1$$

First use the distributive property to remove the parentheses from the right side.

Subtract $3x$ from both sides to move all *x*-terms to the right side.

Add $3y$ to both sides to isolate all *y*-terms on the left side.

SELF-CHECK 2.7.1 ANSWERS

1. $y = \dfrac{1}{2}x - 1$

2.

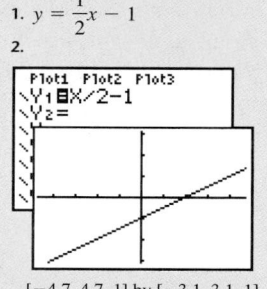

$[-4.7, 4.7, 1]$ by $[-3.1, 3.1, 1]$

The form $y = mx + b$ is used to enter linear equations into a graphics calculator. We will examine this form in more detail in Section 3.2 where we examine the meaning of both m and b.

Answer: $y = \dfrac{3}{2}x - \dfrac{1}{2}$.

Then divide both sides of the equation by 2 to solve for y.

NUMERICALLY

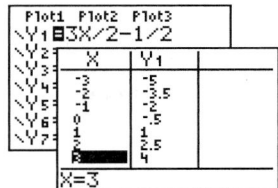

GRAPHICALLY

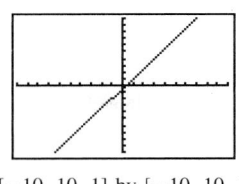

$[-10, 10, 1]$ by $[-10, 10, 1]$

Enter this equation into a graphics calculator and create the table and graph shown to the left.

The ability to solve for either x or y in an equation is a skill that we use frequently when working with linear equations. We illustrate this in the next example where we calculate the intercepts of a graph. The x- and y-intercepts are key points on the graph of an equation. These points also have key interpretations in real-world problems that are modeled by these equations. Thus it is useful to be able to calculate the coordinates of these points exactly rather than approximating them from a graph. Remember the x-intercept is of the form $(a, 0)$ with a y-coordinate of zero, and the y-intercept is of the form $(0, b)$ with an x-coordinate of zero.

■ EXAMPLE 3   Calculating the Intercepts of a Line

Calculate the x- and y-intercepts of $3x - 2y = 6$ and use these intercepts to sketch the graph of this equation.

SOLUTION _____

CALCULATION OF THE *x*-INTERCEPT

$$3x - 2y = 6$$
$$3x - 2(0) = 6 \qquad \text{To calculate the } x\text{-intercept, set } y = 0 \text{ and solve for } x.$$
$$3x - 0 = 6$$
$$3x = 6 \qquad \text{Simplify the left side and then divide both sides by 3.}$$
$$x = 2$$

Intercepts are points on the line and should be written as ordered pairs. The x-intercept is not 2, but is the point $(2, 0)$.

$(2, 0)$ is the x-intercept.

CALCULATION OF THE *y*-INTERCEPT

$$3x - 2y = 6$$
$$3(0) - 2y = 6 \qquad \text{To calculate the } y\text{-intercept, set } x = 0 \text{ and solve for } y.$$
$$0 - 2y = 6$$
$$-2y = 6 \qquad \text{Simplify the left side of the equation and then divide both sides by } -2.$$
$$y = -3$$

The y-intercept is not -3, but the point $(0, -3)$.

$(0, -3)$ is the y-intercept.

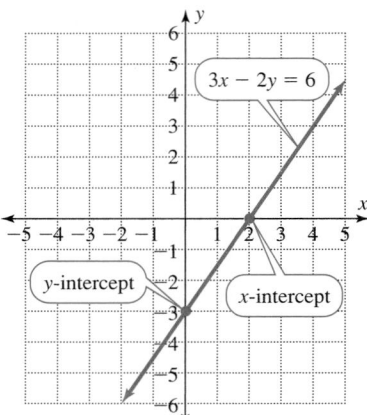

To sketch the graph of $3x - 2y = 6$, plot both the x- and y-intercepts and then draw the line through these two points.

In the next three examples we examine an equation and two common formulas and rearrange them to solve for the variable that we specify.

■ EXAMPLE 4   Solving a Linear Equation for y

Solve $3(4x - 5y + 2) = 17(2x - y) + 20$ for y.

SOLUTION _____

$3(4x - 5y + 2) = 17(2x - y) + 20$ Use the distributive property to remove the parentheses from
$12x - 15y + 6 = 34x - 17y + 20$ both sides.

$12x + 2y + 6 = 34x + 20$ Then combine like terms by adding $17y$ to both sides,
$12x + 2y = 34x + 14$ subtracting 6 from both sides, and then subtracting $12x$ from
$2y = 22x + 14$ both sides. Then divide both sides by 2.

Answer: $y = 11x + 7$.   ■

SELF-CHECK 2.7.2

Calculate the x- and y-intercepts of $2(x - y) = 4(5x + y) - 12$.

■ EXAMPLE 5   Solving the Formula for the Perimeter of an Isosceles Trapezoid for One Side

Solve $P = a + b + 2c$ for c.

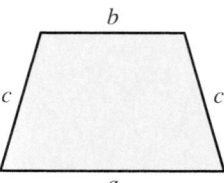

SOLUTION

$$P = a + b + 2c$$

An isosceles trapezoid (as shown) has two parallel sides and two equal nonparallel sides.

$$P - a - b = a + b + 2c - a - b$$
$$P - a - b = 2c$$

Subtract a and b from both sides in order to isolate the term containing c on one side of the equation.

$$\frac{P - a - b}{2} = \frac{2c}{2}$$

Then divide both sides of the equation by 2, the coefficient of c.

$$\frac{P - a - b}{2} = c \text{ or}$$

Answer: $c = \dfrac{P - a - b}{2}$.

In the next example there are four variables: S, n, a_1, and a_n. We solve this equation for a_1. This is an equation that we will revisit in Section 11.2.

■ **EXAMPLE 6** **Solving the Formula for the Sum of an Arithmetic Sequence for the First Term**

Solve $S = \dfrac{n}{2}(a_1 + a_n)$ for a_1.

SOLUTION

$$S = \frac{n}{2}(a_1 + a_n)$$

$$\left(\frac{2}{n}\right)S = \left(\frac{2}{n}\right)\left(\frac{n}{2}\right)(a_1 + a_n)$$

Multiply both sides of the equation by $\dfrac{2}{n}$, the reciprocal of $\dfrac{n}{2}$.

$$\frac{2S}{n} = a_1 + a_n$$

$$\frac{2S}{n} - a_n = a_1 + a_n - a_n$$

To isolate a_1 on the right side, subtract a_n (read as "a sub n") from both sides of the equation.

$$\frac{2S}{n} - a_n = a_1 \text{ or}$$

Answer: $a_1 = \dfrac{2S}{n} - a_n$.

SELF-CHECK 2.7.3

1. Solve $3(2x - 4y) = 10(2x - y) + 8$ for y.
2. Solve $P = a + b + c + 2d$ for b.
3. Solve $6ab + 3 = 3cd$ for d. Assume that $c \neq 0$.

SELF-CHECK 2.7.3 ANSWERS

1. $y = -7x - 4$
2. $b = P - a - c - 2d$
3. $d = \dfrac{2ab + 1}{c}$

The equation $I = PRT$ allows us to calculate directly the simple interest I on a principal P at an interest rate R for a time T. If we have a budgeted amount to spend on interest, then we may need to calculate the principal that we can borrow. This is illustrated in Example 7.

EXAMPLE 7   Solving the Simple Interest Formula for *P*

Solve $I = PRT$ for *P*. Then use a graphics calculator to prepare a table of principal amounts that can be borrowed when the monthly interest is $33. Use interest rates of: 7, 7.5, 8, 8.5, 9, 9.5, and 10%.

SOLUTION

If you are trying to solve for a specified variable, you may find it useful to highlight this variable. In more complicated equations this will help you to focus clearly on the variable that needs to be isolated on the left side of the equation.

$$I = PRT$$

$$\frac{I}{RT} = \frac{PRT}{RT}$$

$$\frac{I}{RT} = P \text{ or}$$

Divide both sides of the equation by *RT* to isolate the specified variable *P* on the right side of the equation.

$$P = \frac{I}{RT}$$

This is an alternate form of the equation that has been solved for *P*.

$$P = \frac{33}{R\left(\dfrac{1}{12}\right)}$$

To determine the principal that can be paid for with a monthly interest payment of $33, substitute $33 for the interest *I* and $\frac{1}{12}$ for *T* for a time of $\frac{1}{12}$ year.

$$P = \frac{33(12)}{R\left(\dfrac{1}{12}\right)(12)}$$

Then simplify by multiplying the numerator and the denominator by 12.

$$P = \frac{396}{R}$$

$$y = \frac{396}{x}$$

To use a graphics calculator, use *x* for the input variable instead of *R* and *y* for the specified output variable instead of *P*.

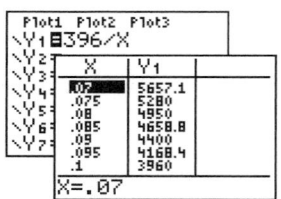

Enter the equation $y = \dfrac{396}{x}$ into a graphics calculator and then generate the table with an initial value of 0.07 for *x* with an increment of 0.005.

Answer: The amount of principal that can be borrowed for a $33 monthly interest payment varies from $5657.10 at 7% to $3960.00 at 10%. The greater the interest rate is, the less one can borrow on a fixed budget.   ■

USING THE LANGUAGE AND SYMBOLISM OF MATHEMATICS   2.7

1. The process of transposing an equation to write it in a desired form is called solving the equation for a _____ _____.

2. The point on the graph of a linear equation with a *y*-coordinate of zero is the _____.

3. The point on the graph of a linear equation with an *x*-coordinate of zero is the _____.

4. The _____ property of multiplication over addition allows us to rewrite $2(3x + y)$ as $6x + 2y$.

EXERCISES   2.7

In Exercises 1–14 solve each equation for y.

1. $2x + y = 7$

2. $5x + y = 8$

3. $3x - y = 2$

4. $4x - y = 3$

5. $-6x + 3y = -9$

6. $-30x - 5y = 20$

7. $\dfrac{x}{2} - \dfrac{y}{4} = -1$

8. $\dfrac{x}{3} + \dfrac{y}{12} = -\dfrac{1}{6}$

9. $-0.3x - 0.1y = 0.2$

10. $-0.4x + 0.8y = -0.2$

11. $5x - 2y = x - 3y + 4$

12. $-6x + 5y = 8x - 2y - 21$

13. $2(3x - y + 1) = 3(4x - y - 2)$

14. $4(2x + 3y - 5) = 3(2x + 5y + 4)$

In Exercises 15–18 calculate the x- and y-intercepts of the graphs of each of these linear equations.

15. $3x + 5y = 15$

16. $4x - 3y = 24$

17. $2(x - 3) + 4 = 3(y - 5) - 2$

18. $4(2x - 1) = 5(2y + 1) + 1$

In Exercises 19–42 solve each literal equation for the variable specified. The context in which you may encounter these formulas is indicated in parentheses. Assume that all the variables are nonzero.

19. $A = lw$;   for l (area of a rectangle)

20. $A = \dfrac{1}{2}bh$;   for h (area of a triangle)

21. $C = 2\pi r$;   for r (circumference of a circle)

22. $V = lwh$;   for w (volume of a rectangular box)

23. $V_1 T_1 = V_2 T_2$;   for V_1 (Charles's law in chemistry)

24. $P_1 V_1 = P_2 V_2$;   for P_2 (Boyle's law in chemistry)

25. $P = 2l + 2w$;   for w (perimeter of a rectangle)

26. $P = a + b + c$;   for a (perimeter of a triangle)

27. $V - E + F = 2$;   for E (Euler's theorem)

28. $F = \dfrac{9}{5}C + 32$;   for C (Fahrenheit and Celsius temperatures)

29. $A = \dfrac{1}{2}h(a + b)$;   for b (area of a trapezoid)

30. $A = \dfrac{1}{2}h(a + b)$;   for a (area of a trapezoid)

31. $C = \dfrac{5}{9}(F - 32)$;   for F (Fahrenheit and Celsius temperatures)

32. $y = mx + b$;   for x (slope-intercept form of a line)

33. $l = a + (n - 1)d$;   for a (last term of an arithmetic sequence)

34. $l = a + (n - 1)d$;   for d, $n \neq 1$ (last term of an arithmetic sequence)

35. $l = a + (n - 1)d$;   for n (last term of an arithmetic sequence)

36. $S = 2\pi r^2 + 2\pi rh$;   for h (surface area of a cylinder)

37. $S = \dfrac{a}{1 - r}$;   for a (sum of an infinite geometric sequence)

38. $S = \dfrac{a}{1 - r}$;   for r (sum of an infinite geometric sequence)

39. $y = mx + b$;   for m (slope-intercept form of a line)

40. $S = \dfrac{n}{2}(a_1 + a_n)$;   for n (sum of an arithmetic sequence)

41. $V = \dfrac{1}{3}\pi r^2 h$;   for h (volume of a cone)

42. $E = mc^2$;   for m (Einstein's formula)

Multiple Representations

In Exercises 43–46 solve each equation for y and then use a graphics calculator to complete a table of values for the x-values: $-3, -2, -1, 0, 1, 2, 3$ and to graph the equation. See the following example.

Algebraically

$3x + y - 2 = 0$

$3x + y - 2 - 3x = 0 - 3x$

$y - 2 = -3x$

$y - 2 + 2 = -3x + 2$

$y = -3x + 2$

Numerically

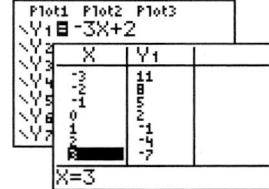

Graphically

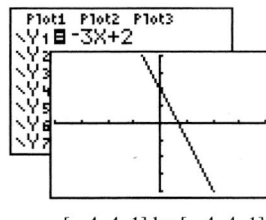

$[-4, 4, 1]$ by $[-4, 4, 1]$

43. $2x + 3y = 4x + 2y - 3$

44. $7x - 5y = 6(x - y) + 1$

45. $-5x + 4y = -2(2x - 3y) + 4$

46. $-4x + 7y = -5(x - y) + 3$

Calculator Exercises

47. Sales Tax   The sales tax T on a purchase amount P at a sales tax rate R is given by the formula $T = PR$. Solve this formula for P. Then use a calculator to prepare a table of purchase amounts that correspond to sales taxes of $50, $100, $150, $200, $250, $300, and $350 when the sales tax rate is 8%.

48. Dimensions of a Building   The area of a rectangular building is given by the formula $A = lw$. Solve this formula for w. Then use a calculator to prepare a table of widths that will produce an area of 24,000 square meters for lengths of 100, 110, 120, 130, 140, 150, and 160 m.

49. Dimensions of a Building   The perimeter of a building is given by the formula $P = 2w + 2l$. Solve this formula for w. Then use a graphics calculator to prepare a table of widths that will produce a perimeter of 400 m for lengths of: 50, 60, 70, 80, 90, 100, and 110 m.

Group Discussion Questions

50. Writing Mathematically   Write a paragraph describing why it is handier to have the formula $F = \dfrac{9}{5}C + 32$ for some applications, whereas the transposed equation $C = \dfrac{5}{9}(F - 32)$ is handier for other applications. Then complete the following table:

Celsius°	Fahrenheit°	
100°		The temperature at which water boils.
94°		Is this hot?
	98.6°	Normal body temperature.
	72°	A comfortable room temperature.
0°		Water freezes.
	0°	Zero on Fahrenheit scale.
		Both temperatures are equal.

KEY CONCEPTS   FOR CHAPTER 2

1. **Cartesian Coordinate System:** The sign pattern for the coordinates in each quadrant of the rectangular coordinate system is shown in the figure.

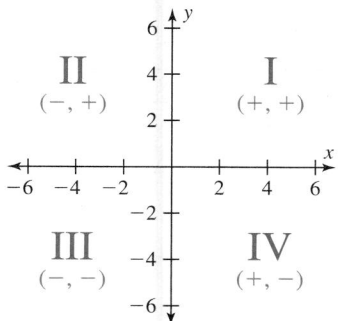

2. **Scatter Diagram:** A scatter diagram for a set of data points is a graph of these points that can be used to visually examine the data for some type of pattern.

3. **Arithmetic Sequences and Linear Equations:**
   - A sequence is an ordered set of numbers.
   - An arithmetic sequence is a sequence with a constant change from term to term.
   - The graph of the terms of an arithmetic sequence consists of discrete points that lie on a straight line.
   - The linear equation $y = mx + b$ will produce an arithmetic sequence of output values y for input values of x of $1, 2, 3, \ldots$.

4. **Function Notation:** The function notation $f(x)$ is read "f of x." $f(x)$ represents a unique output value for each input value of x.

5. **Solution of a Linear Equation:** A solution of a linear equation of the form $y = mx + b$ is an ordered pair (x, y) that makes the equation a true statement.

6. **Intercepts of a Line:**
   - The x-intercept, often denoted by $(a, 0)$, is the point where the line crosses the x-axis.
   - The y-intercept, often denoted by $(0, b)$, is the point where the line crosses the y-axis.

7. **Solution of a System of Linear Equations:**
   - A solution of a system of linear equations is an ordered pair that satisfies each equation in the system.
   - The solution of a system of linear equations is the point of intersection, the point that is on both graphs.

8. **Linear Equation in One Variable:** A linear equation in one variable x is an equation that can be written in the form $Ax = B$, where A and B are real constants and $A \neq 0$.

9. **Equivalent Equations:** Equivalent equations have the same solution set.

10. **Addition-Subtraction Principle of Equality:**
    - If the same number is added to or subtracted from both sides of an equation, the result is an equivalent equation.
    - $a = b$ is equivalent to $a + c = b + c$.
    - $a = b$ is equivalent to $a - c = b - c$.

11. **Classification of Equations:**
    Conditional Equation: A conditional equation is true for some values of the variable and false for other values.
    Contradiction: An equation that is false for all values of the variable is a contradiction.
    Identity: An equation that is true for all values of the variable is an identity.

12. **Strategy for Solving Linear Equations:**
    Step 1. Simplify each side of the equation.
    - a. If the equation contains fractions, simplify by multiplying both sides of the equation by the least common denominator (LCD) of all the fractions.
    - b. If the equation contains grouping symbols, simplify by using the distributive property to remove the grouping symbols and then combine like terms.

    Step 2. Using the addition-subtraction principle of equality, isolate the variable terms on one side of the equation and the constant terms on the other side.
    Step 3. Using the multiplication-division principle of equality, solve the equation produced in step 2.

13. **Proportions:**
    - A proportion is an equation that states two ratios are equal.
    - In the proportion $\dfrac{a}{b} = \dfrac{c}{d}$, the terms a and d are called the extremes and the terms b and c are called the means.

14. **Direct Variation:**
    - If x and y are variables and k is a constant then stating "y varies directly as x with the constant of variation k" means $y = kx$.
    - If y varies directly as x, then the graph of the (x, y) pairs will lie on a straight line.

15. **Solving for a Specified Variable:** To solve for a specified variable, transpose the equation to isolate this variable on one side of the equation with all other variables on the other side of the equation.

1. Identify the coordinates of the points A–D in the figure and give the quadrant in which each point is located.

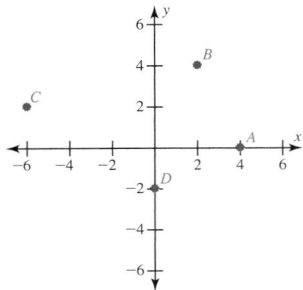

2. Draw a scatter diagram for the data points in the table to the right.

x	y
-3	-2
-2	0
-1	2
0	1
2	-1

3. For the arithmetic sequence defined by $a_n = -2n + 3$ complete this table and graph these points.

n	a_n
1	
2	
3	
4	
5	

4. For the arithmetic sequence defined by $y = 0.5x + 1.5$ complete this table and graph these points.

x	y
1	
2	
3	
4	
5	

In Exercises 5–10 determine which of these sequences are arithmetic sequences. For those that are arithmetic find the common difference d.

5. 4, 7, 10, 13, 16

6. 0, 1, 3, 6, 10

7. $a_n = -2n + 5$

8. $a_n = n^2 - 5$

9.
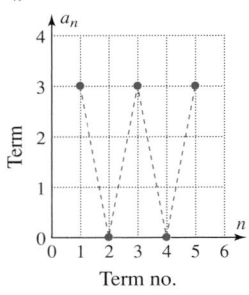

10.
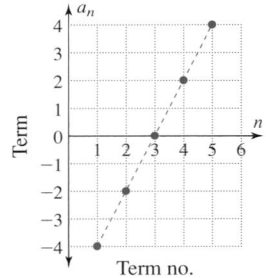

In Exercises 11–14 determine whether either $(3, -2)$ or $(-3, 2)$ is a solution of each equation.

11. $y - 3 = 4(x + 2)$ **12.** $y = x + 5$

13.
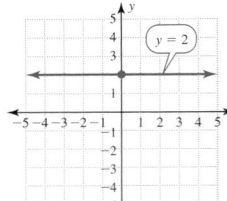

14.
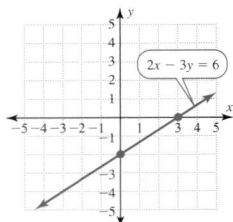

15. Determine the x- and y-intercepts of the line in Exercise 14.

16. Determine the x- and y-intercepts of $y = x + 5$.

17. Determine the x- and y-intercepts of $y - 3 = 4(x + 2)$.

In Exercises 18–21 match each exercise with the most appropriate choice.

18. $5x + 3 = 2(x + 4)$

19. $5x^2 = 3$

20. $y = 2x + 4$

21. $5x + 3 - 2(x + 4)$

A. A first-degree expression in x but not a linear equation

B. A linear equation in x

C. Not a linear equation

D. A linear equation in two variables

In Exercises 22 and 23 determine the point of intersection of the two lines.

22.
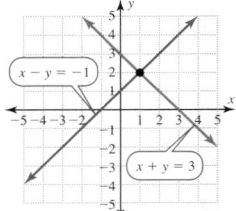

23.
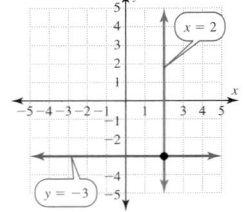

In Exercises 24 and 25 y_1 represents the left side of the equation and y_2 represents the right side of the equation. Use the table shown from a graphics calculator to solve this equation for x.

24. $3.6x + 1.2 = 1.8x + 5.7$ **25.** $8.5x - 55 = 10.5x - 103$

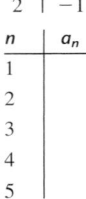

X	Y1	Y2
0	1.2	5.7
.5	3	6.6
1	4.8	7.5
1.5	6.6	8.4
2	8.4	9.3
2.5	10.2	10.2
3	12	11.1

X=0

X	Y1	Y2
20	115	107
21	123.5	117.5
22	132	128
23	140.5	138.5
24	149	149
25	157.5	159.5
26	166	170

X=20

In Exercises 26–50 solve each equation.

26. $x + 2 = 17$ **27.** $5v + 7 = 4v + 9$

28. $7m = 6m$ **29.** $-x = 3$

30. $-17m = -34$ **31.** $\dfrac{n}{3} = -12$

32. $-\dfrac{2y}{7} = 28$

33. $0.045w = -0.18$

34. $117 = 0.01t$

35. $0 = -147b$

36. $5x - 8 = 27$

37. $-8z + 7 - 2z = 5z + 4 + z$

38. $-r - r - r - r - r = r$

39. $3(2x + 7) = x + 1$

40. $(6x - 9) - (3x + 8) = 4(x - 11)$

41. $7y - 5(2 - y) = 3(2y + 1) + 5$

42. $9(2y - 3) - 13(5y - 1) = 14(3y - 1)$

43. $0.55(w - 10) = 0.80(w + 3)$

44. $\dfrac{t}{8} - 5 = \dfrac{t}{12} - 6$

45. $\dfrac{7m + 27}{6} = \dfrac{4m + 6}{5}$

46. $\dfrac{167 - 3n}{10} = 6 - \dfrac{5n - 1}{4}$

47. $1293y = 1294y + 1295$

48. $0.289x - 5 = 3.1 - 0.78(2 - x)$

49. $\dfrac{3v}{10} - \dfrac{5v}{6} = -\dfrac{8v}{15}$

50. $1.8a - 7.8a + 1.97a - 8.3a = 0$

In Exercises 51 and 52 use a graphics calculator to solve each equation by letting y_1 represent the left side of the equation and y_2 represent the right side of the equation.

51. $1.5x - 1.2 = 2.7x - 3.0$ **52.** $8.5x + 24 = 12.5x + 33.6$

In Exercises 53–56 each equation is a conditional equation, an identity, or a contradiction. Identify the type of equation, and then solve it.

53. $7(y - 3) - 4(y + 2) = 3(y - 7)$

54. $3(2v + 1) = 2(3v + 1)$

55. $4(w - 1) + (7 - 2w) = 2(w + 1) + 1$

56. $2v + 2v + 2v = 5v$

57. Use the function $f(x) = 7x - 11$ to evaluate:
   a. $f(-10)$ **b.** $f(0)$ **c.** $f(9)$ **d.** $f(\pi)$

58. Use the function $f(x) = 5(4x - 3) + 1$ to evaluate:
   a. $f(-3)$ **b.** $f(0)$ **c.** $f(6)$ **d.** $f(100)$

In Exercises 59–62 solve each equation for x.

59. $v + w = x + y$ **60.** $vw = xy$

61. $3x - 5y = 7z$ **62.** $2(x - y) + 3(2x + y) = 38$

In Exercises 63–66 simplify the expression in the first column by adding like terms, and solve the equation in the second column.

Simplify	Solve
63. a. $3x + 4 + 5x + 6$	**b.** $3x + 4 = 5x + 6$
64. a. $3x + 4 - (5x + 6)$	**b.** $3x + 4 = -5x + 6$
65. a. $-2x + 3 + 4(x - 1)$	**b.** $-2x + 3 = -4(x - 1)$
66. a. $5(x + 1) - 6(x + 2)$	**b.** $5(x + 1) = 6(x + 2)$

Estimation Skills
In Exercises 67 and 68 mentally estimate the solution of each equation to the nearest integer and then use a calculator to calculate the solution.

PROBLEM	MENTAL ESTIMATE	CALCULATOR SOLUTION
67. $0.99x - 5.1 = 4.899$		
68. $11x - 20 = 9x + 19.9$		

Multiple Representations
In Exercises 69–72 first write an algebraic equation for each verbal statement, using the variable m to represent the number, and then solve for m.

69. Four more than five times a number is forty-nine.

70. One-third the sum of a number and seven equals twelve.

71. Twice the quantity of two more than a number is the opposite of the quantity of three less than the number.

72. The sum of a number, one more than the number, and two more than the number is ninety-three.

73. Modeling the Cost of a Snow Cone Business At the end of the summer the business records for Cones-R-Us revealed that they had a fixed cost of $500 for running their business (utilities, insurance, etc.) They also calculated that each snow cone cost 25 cents for ice, juice, and a cup.
   a. Write a function that gives the total cost of selling x snow cones in a summer.
   b. Use this function to complete this table.

x	0	1000	2000	3000	4000	5000
$f(x)$						

   c. Evaluate and interpret $f(2500)$.

74. Modeling the Revenue Generated by a Snow Cone Business The snow cone business described in Exercise 73 sold snow cones for 40 cents each.
   a. Write a function that gives the total revenue generated by the selling of x snow cones this summer.
   b. Use this function to complete this table.

x	0	1000	2000	3000	4000	5000
$f(x)$						

   c. Evaluate and interpret $f(2500)$.

75. Perimeter of a Rectangle The perimeter of the rectangle shown in the figure is 48 cm. Find the value of w.

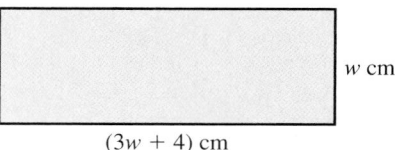

$(3w + 4)$ cm, w cm

76. Area of a Triangle The area of the triangle shown in the figure is 52 square centimeters (cm^2). If all sides of the triangle are measured in cm, find the value of b.

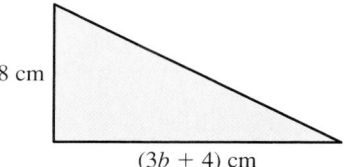

8 cm, $(3b + 4)$ cm

77. Distance on a Map   On a map of the United States 1 in represents 20 mi. If the distance on the map between Louisville, Kentucky, and Columbus, Ohio, is 10.5 in, find the distance in miles between these two cities.

78. Quality Control   A quality control inspector discovered three defective display panels in the 75 handheld computing devices that he tested. How many defective display panels would be expected in a shipment of 500 of these items?

79. Similar Polygons   The two figures shown here are similar. Thus their corresponding parts are proportional. Find the lengths of sides *a, b,* and *c.*

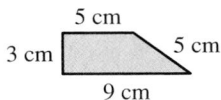

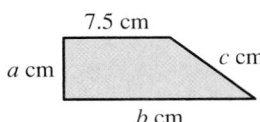

80. Modeling the Height of a Tree   If a shrub 0.64 m tall casts a 6.72-m shadow, how tall is a tree that casts a 46.2-m shadow? (See the figure.)

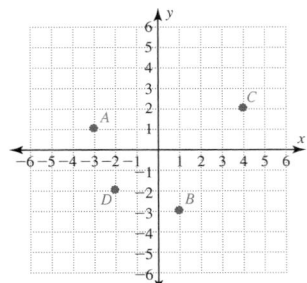

In Exercises 81 and 82 solve each problem involving direct variation.

81. If *v* varies directly as *w,* and *v* = 22 when *w* = 77, find *v* when *w* = 35.

82. If *v* varies directly as *w,* and *v* = 12 when *w* = 15, determine the constant of variation.

83. Modeling Car Payments   At the beginning of July a student purchased a used car with an initial payment of $680. She then owed $280 for the first of August and for each of the next 24 months after July. Let *y* represent the total of all payments and *x* represent the number of monthly $280 payments.

   a. Algebraically: Write an equation to express *y* in terms of *x.*

   b. Numerically: Form a table of corresponding values of *y* for *x* = 0, 1, 2, 3, 4, 5, 6.

   c. Graphically: Graph this relationship.

   d. Is the sequence of monthly payments an arithmetic sequence?

MASTERY TEST   FOR CHAPTER 2

[2.1]   **1.** Identify the coordinates of the points *A–D* in the figure and give the quadrant in which each point is located.

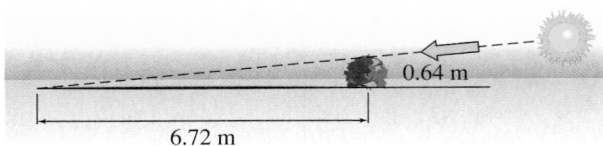

[2.1]   **2.** Draw a scatter diagram for the data points in the table to the right.

x	*y*
−2	3
0	−2
1	2
3	1
4	−3

[2.1]   **3.** Determine whether each sequence is an arithmetic sequence. If the sequence is arithmetic, write the common difference *d.*

   a. 0, 2, 4, 6, 8   **b.** 1, 2, 4, 8, 16

   c. 4, 4, 4, 4, 4   **d.** $a_n = 15 - 3n$

[2.2]   **4.** Use the function $f(x) = 11x - 7$ to evaluate each expression.

   a. $f(0)$ **b.** $f(-1)$

   c. $f(7)$ **d.** $f(10)$

[2.2]   **5.** Use each equation to complete a table with input values of 1, 2, 3, 4, and 5, and then graph these points and the line through these points.

   a. $y = x + 1$ **b.** $y = -x + 2$

   c. $y = 2x - 5$ **d.** $y = -2x + 4$

[2.2]   **6. Modeling the Total Paid on a Car Loan**   A new car loan requires an $800 down payment and a monthly payment of $375.

   a. Write a function that gives the total paid at the end of the *x*th month.

b. Use this function to complete this table.

x	0	6	12	18	24	30
$f(x)$						

c. Evaluate and interpret $f(25)$.

[2.3] **7.** Determine whether $(-2, 3)$ is a solution to each equation.

  a. $y = 2x + 7$ **b.** $y = -x - 1$
  c. $y = 3$ **d.** $x = -2$

[2.3] **8.** Determine the x- and y-intercepts of these lines.

  a.
  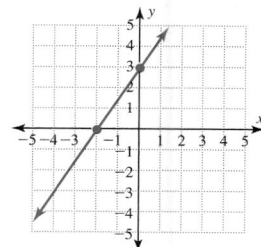

  b.
  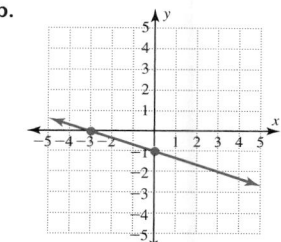

  c.
  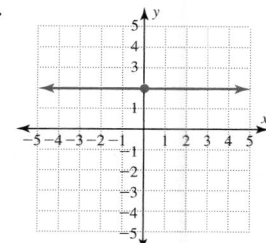

  d.
  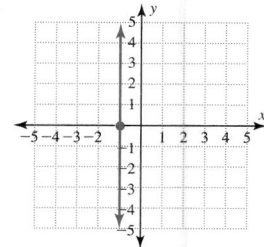

[2.3] **9.** Determine the point of intersection of the lines in each graph and then check this point in both equations.

  a.
  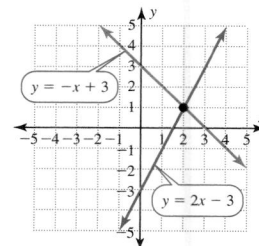

  b.
  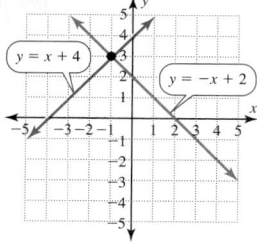

[2.4] **10.** Solve each linear equation.

  a. $x + 7 = 11$ **b.** $x + 7 = x + 11$
  c. $7x + 1 = 6x + 1$ **d.** $x + (x + 1) = 2x + 1$

[2.4] **11.** Let y_1 equal the left side of each equation and y_2 equal the right side. Then use a graphics calculator to solve each equation for x by examining the graphs of y_1 and y_2.

  a. $\dfrac{x - 4}{2} = \dfrac{-3x + 4}{2}$ **b.** $2.3x + 1.6 = 1.7x + 0.4$

[2.5] **12.** Solve each linear equation.

  a. $-x = 3x - 12$ **b.** $-\dfrac{5}{21} = \dfrac{3y}{7}$

  c. $3(2v + 4) = 4(v + 3) + 3$

  d. $\dfrac{w - 1}{6} = \dfrac{w - 3}{5}$

[2.6] **13.** Solve the proportions in parts a and b.

  a. $\dfrac{m - 1}{m - 7} = \dfrac{2}{5}$ **b.** $\dfrac{2x - 1}{7} = \dfrac{x + 1}{4}$

  c. 2 cm on a map corresponds to a distance of 75 km. What distance corresponds to a distance of 5 cm on the map?

[2.6] **14. a.** If y varies directly as x, and y is 33 when x is 22, find y when x is 8.

  b. The number of Mexican pesos varies directly as the number of U.S. dollars exchanged. In 2000 one could exchange 50 U.S. dollars for 475 pesos. What was the constant of variation (the exchange rate) at that time?

[2.7] **15.** Solve the linear equation
$2x - y = 2(3x - y - 1) + 1$ for y and then use a graphics calculator to complete this table and to graph the equation.

x	$y =$
-2	
-1	
0	
1	
2	

[2.7] **16.** Calculate the x- and y-intercepts of the graphs of these equations.

  a. $y = 2x - 6$ **b.** $y = 4$
  c. $2x - 5y = 10$ **d.** $3(2x - 1) = 4(3y - 2) - 7$

GROUP PROJECT

Risks and Choices: Comparing the Cost of Leasing Versus Purchasing an Automobile

This project requires the group to collect current data from a local automobile dealership.

Select a specific model of a new car that either can be leased or purchased. Assume that there is no trade-in for either the lease or the purchase. To facilitate a comparison, get a price for both options over a 4-year period. For each option determine the total of all initial costs (the required down payment, licenses, taxes, etc.) and the required monthly payment for the 4-year period. Use the information you have collected to complete the following items.

	INITIAL COST	MONTHLY COST
Lease Option:	_____	_____
Purchase Option:	_____	_____

Lease Option:

Write a function $f_1(x)$ for the total monthly cost of the lease option at the end of the xth month.

$f_1(x) = $ _____

Purchase Option:

Write a function $f_2(x)$ for the total monthly cost of the purchase option at the end of the xth month.

$f_2(x) = $ _____

Graph $y = f_1(x)$ and $y = f_2(x)$ on the same coordinate system and then use the graph to compare verbally the costs of the two options.

Determine the estimated value of the car at the end of the 4-year period if you purchase the car. Which option would you select? Why?

3

LINES AND SYSTEMS OF LINEAR EQUATIONS IN TWO VARIABLES

159

The ability to spot trends, to analyze trends, and to describe change are key skills in many jobs. Changes in interest rates set by the Federal Reserve affect the stock market. Changes in speed can enable a pass receiver in football to shake free of a defender. Changes in elevation on a roller coaster produce a thrilling ride. Local police often monitor the speed of commuters on their way to work. If you have money to invest, do you want to invest in a company whose profits are decreasing or one whose profits are increasing? If you are borrowing money to purchase a car, do you want to borrow at an interest rate of 8% or one of 10%?

Studying lines and the properties of lines can help us to describe many everyday events in a compact and easily understood format. This mathematical description not only can help us to understand current information, but also can help us to understand the past or the future. An important part of mathematics is analyzing trends and change. This chapter examines the slope of a line and other key skills needed to study trends and change.

Section 3.1   Slope of a Line and Applications of Slope

Objectives: 1. Determine the slope of a line.
2. Use slope to determine whether two lines are parallel, perpendicular, or neither.

Question: Which of the lines in Figure 3.1.1 is the steepest or represents the greatest change in share price per year?

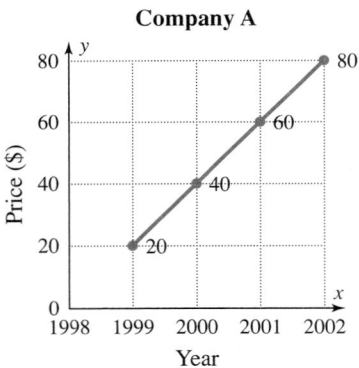
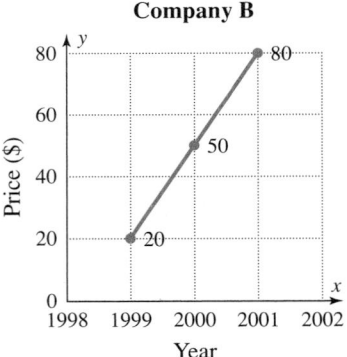
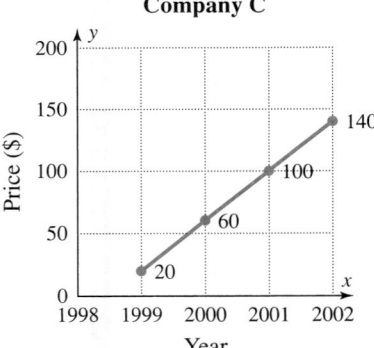

Figure 3.1.1

Answer: The graph for company B appears steeper than the graph for company A and accurately reflects that the share price of company B is increasing faster than that of company A ($30 per year for B versus $20 per year for A). Although the graph for company C does not appear as steep as the other two graphs, it is company C whose share price is increasing most rapidly. . . at $40 per year. The reason this graph does not appear as steep is that a different scale was used on the vertical axis for this graph.

This example illustrates the importance of looking beyond the picture portion of a graph and also examining the numerical information or the scale of the graph. Since our eyes can be misled when examining the steepness, we also will look at this concept algebraically.

The **slope of a line** is a measure of the steepness of the line. In Figure 3.1.2, consider the steepness of the line connecting the point (x_1, y_1) to the point (x_2, y_2). The slope of this line is defined to be the ratio of the change in y (the rise) to the change in x (the run). The slope is usually represented by the letter m.

The key concept is that slope measures the steepness of a line. This paragraph describes how the slope is calculated. (x_1, y_1) is read (x sub 1, y sub 1), and (x_2, y_2) is read (x sub 2, y sub 2). The change in y and the change in x are determined by subtraction. Change was first discussed in Section 1.3.

$$m = \frac{\text{change in } y}{\text{change in } x} = \frac{y_2 - y_1}{x_2 - x_1}$$

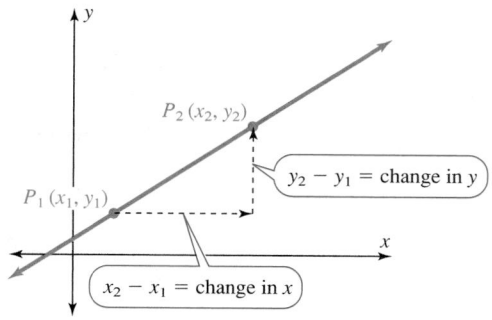

$y_2 - y_1 = $ change in y

$x_2 - x_1 = $ change in x

Figure 3.1.2

Slope of a Line Through (x_1, y_1) and (x_2, y_2)

ALGEBRAICALLY	VERBALLY	NUMERICAL EXAMPLE	GRAPHICAL EXAMPLE
$m = \dfrac{y_2 - y_1}{x_2 - x_1}$ for $x_1 \neq x_2$.	The slope of a line is the ratio of the change in y to the change in x.	The slope of the line through the points $(2, -1)$ and $(3, 1)$ is $$m = \frac{1 - (-1)}{3 - 2}$$ $$m = \frac{2}{1}$$	2-unit change in y 1-unit change in x

---

A Mathematical Note

The origin of the use of m to designate slope is unknown. In his book *Mathematical Circles Revisited*, Howard Eves says that m might have been used because slopes were first studied with respect to mountains. Some suggest that m may have been derived from the French word *monter*, which means "to mount, to climb, or to slope up."

■ EXAMPLE 1 Calculating the Slope of a Line Through Two Points

Calculate the slope of the line through the given points.

SOLUTIONS

(a) $(-5, 8)$ and $(3, -2)$

$$m = \frac{y_2 - y_1}{x_2 - x_1}$$

$$m = \frac{-2 - 8}{3 - (-5)}$$

$$m = \frac{-10}{8}$$

Substitute the given points into the formula for slope with $x_1 = -5$, $y_1 = 8$, $x_2 = 3$, and $y_2 = -2$. Note the use of parentheses in the denominator to prevent an error in sign.

Express the slope as a fraction in reduced form.

The slope of a line gives the change in y for each 1-unit change in x.

The concept of the change in y for each 1-unit change in x is important in many areas including marginal costs in economics. This concept also is related to unit pricing from Section 1.5.

$$m = -\frac{5}{4} \text{ or}$$

y decreases 5 units for each 4-unit increase in x, or y decreases 1.25 units for every 1-unit increase in x.

Answer: $m = -1.25$

(b) $(3, -2)$ and $(-5, 8)$
$$m = \frac{y_2 - y_1}{x_2 - x_1}$$

Although the order of these two points is different, these are the same points as those in part (a) and the slope is also the same. Substitute $x_1 = 3$, $y_1 = -2$, $x_2 = -5$, and $y_2 = 8$.

$$m = \frac{8 - (-2)}{-5 - 3}$$

$$m = \frac{10}{-8}$$

Answer: $m = -\frac{5}{4}$

The slope of a line is the same no matter which two points on the line are used to calculate the slope. As shown in Example 1, this slope is also the same no matter which point is taken first. This is true because the slope is the ratio of the change in y to the corresponding change in x. However, it is crucial that the x- and y-values be kept in the correct pairings. Do not match the x-coordinate of one point with the y-coordinate of another point.

Keep the x- and y-values paired in the same order.

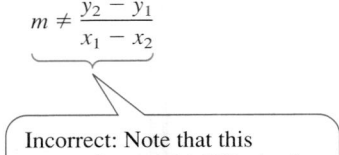

$$m = \frac{y_2 - y_1}{x_2 - x_1} \text{ or } m = \frac{y_1 - y_2}{x_1 - x_2} \qquad m \neq \frac{y_2 - y_1}{x_1 - x_2}$$

Both correct.

Incorrect: Note that this expression would differ in sign.

■ **EXAMPLE 2** Calculating the Slope of a Line Through Given Points

Calculate the slope of each line, using the points labeled on the line.

SOLUTIONS

(a)

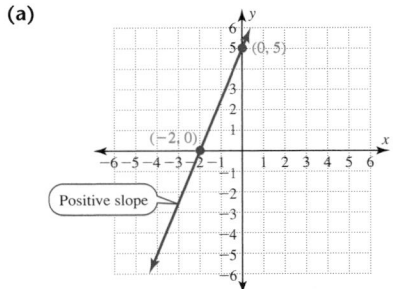

$$m = \frac{y_2 - y_1}{x_2 - x_1}$$

The line slopes upward to the right; its slope is a positive number.

$$m = \frac{5 - 0}{0 - (-2)}$$

Use $x_1 = -2$, $y_1 = 0$, $x_2 = 0$ and $y_2 = 5$.

$$m = \frac{5}{2} \text{ or}$$

y increases 5 units for each 2-unit increase in x, or y increases 2.5 units for each 1-unit increase in x.

$$m = 2.5$$

(b)

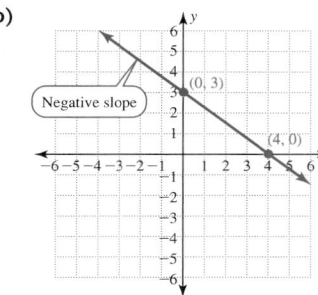

Negative slope

(0, 3)

(4, 0)

$$m = \frac{y_2 - y_1}{x_2 - x_1}$$ The line slopes downward to the right; its slope is a negative number.

$$m = \frac{0 - 3}{4 - 0}$$ Use $x_1 = 0$, $y_1 = 3$, $x_2 = 4$ and $y_2 = 0$.

$$m = -\frac{3}{4} \text{ or}$$ y decreases 3 units for each 4-unit increase in x, or y decreases by 0.75 units for each 1-unit increase in x.

Answer: $m = -0.75$

SELF-CHECK 3.1.1

1. Calculate the slope of the line through $(5, -9)$ and $(-4, -21)$.
2. Calculate the slope of the line using the points labeled on the graph.

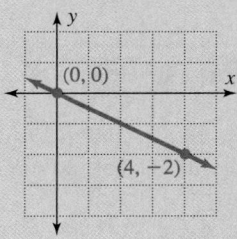

(0, 0)

(4, -2)

■ EXAMPLE 3   Comparing the Change in y to the Unit Change in x

Complete the following questions that relate the change in x, the change in y, and the slope of a line.

SOLUTIONS

(a) 2 units change in x produces 3 units change in y. Calculate the slope.

$$m = \frac{\text{Change in } y}{\text{Change in } x}$$

Answer: $m = \dfrac{3}{2}$ Substitute the given changes into the definition of slope.

(b) The slope of a line is $\dfrac{5}{4}$. How much change in y will a 1-unit increase in x produce?

$$m = \frac{5}{4}$$

$$m = \frac{1.25}{1}$$

$$m = \frac{\text{Change in } y}{\text{Change in } x}$$

Divide both the numerator and the denominator of the fraction $\dfrac{5}{4}$ by 4 to produce a change of 1 for x.

Answer: A 1-unit increase in x will produce a 1.25-unit increase in y.   The numerator, 1.25, represents the change in y for each 1-unit change in x.

(c) The slope of a line is -1.75. How much change in y will result from a 4-unit increase in x?

$$m = \frac{\text{Change in } y}{\text{Change in } x}$$

$$m = \frac{-1.75}{1}$$

The slope represents the change in y for a 1-unit change in x.

$$\frac{\text{Change in } y}{4} = \frac{-1.75}{1}$$

Solve this ratio for the change in y that results from a 4-unit change in x.

$$4\left(\frac{\text{Change in } y}{4}\right) = 4\left(\frac{-1.75}{1}\right)$$

Multiply both sides of the equation by 4.

$$\text{Change in } y = -7.$$

Answer: A 4-unit increase in x will result in a 7-unit decrease in y.

The next example illustrates how to use the relative changes in x and y to help graph a line.

■ EXAMPLE 4 Using the Slope and *y*-Intercept to Graph a Line

A line has a y-intercept of $(0, 1)$ and a slope of $\frac{5}{3}$. Use this information to determine another point on the line and to graph this line.

SOLUTION

y-intercept: $(0, 1)$

Second point: $(0 + 3, 1 + 5) = (3, 6)$

Start with the y-intercept and use the slope of $\frac{5}{3}$ to determine a second point. Do this by increasing y by 5 units for a 3-unit increase in x.

We encourage you to practice graphing lines by the method shown in Example 4. You may want to work through Self-Check 3.1.2 to be sure that you understand this method before working Exercises 27–36.

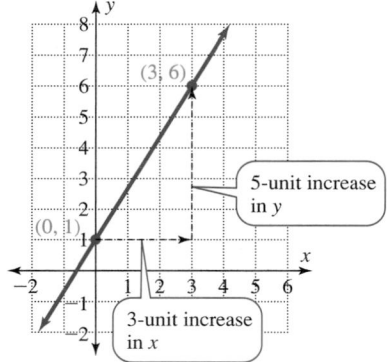

Plot both $(0, 1)$ and $(3, 6)$ and sketch the line through these points.

Check: $m = \dfrac{y_2 - y_1}{x_2 - x_1}$

$$m = \frac{6 - 1}{3 - 0}$$

$$m = \frac{5}{3} \text{ checks.}$$

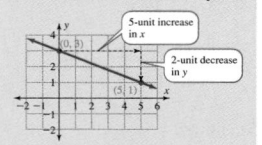

SELF-CHECK 3.1.2

The slope of a line is $-\dfrac{2}{5}$.

1. How much change in y will a 5-unit increase in x produce?
2. How much change in y will a 1-unit increase in x produce?
3. Graph a line with this slope and a y-intercept of $(0, 3)$.

All lines with positive slope go upward to the right since the *x*- and *y*-coordinates increase together. All lines with negative slope go downward to the right since the *y*-coordinate decreases as the *x*-coordinate increases. If we compare two lines graphed on the same coordinate system, the steeper line will have the slope with the larger magnitude.

Lines with Positive Slope

$m = 3$
$m = 1$
$m = \dfrac{1}{3}$

Lines with Negative Slope

$m = -4$
$m = -\dfrac{1}{4}$
$m = -1$

Example 5 will show that the slope of a horizontal line is 0 and the slope of a vertical line is undefined.

■ EXAMPLE 5   Calculating the Slopes of Horizontal and Vertical Lines

Calculate the slope of each of the following lines, using the points labeled on the line.

SOLUTIONS

(a)

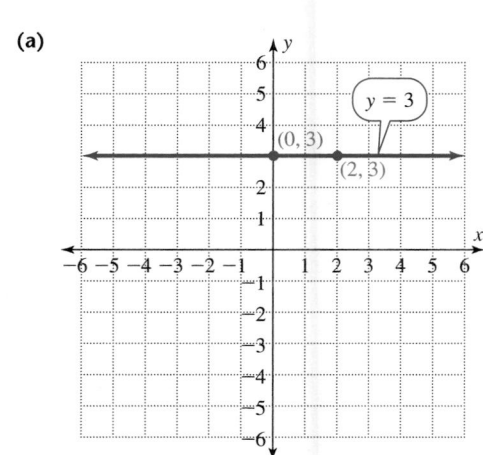

$y = 3$
$(0, 3)$
$(2, 3)$

$$m = \frac{y_2 - y_1}{x_2 - x_1}$$

$$m = \frac{3 - 3}{2 - 0}$$ Use $x_1 = 0$, $y_1 = 3$, $x_2 = 2$, and $y_2 = 3$.

$$m = \frac{0}{2}$$

Answer: $m = 0$ This line is horizontal, and its slope is 0.

(b)

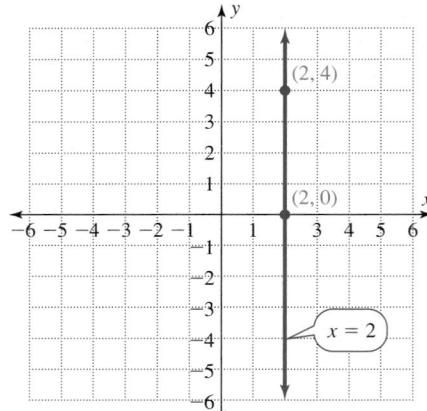

$$m = \frac{y_2 - y_1}{x_2 - x_1}$$

$$m = \frac{4 - 0}{2 - 2}$$ Use $x_1 = 2$, $y_1 = 0$, $x_2 = 2$, and $y_2 = 4$.

$$m = \frac{4}{0}$$

Answer: m is undefined.   This line is vertical, and its slope is undefined since division by 0 is undefined.

Classifying Lines by Their Slopes

NUMERICALLY	VERBALLY	GRAPHICALLY
m is positive.	The line slopes upward to the right.	
m is negative.	The line slopes downward to the right.	
m is zero.	The line is horizontal.	
m is undefined.	The line is vertical.	

The slope of a line also can be determined from its equation. One approach is to determine two points on the line and then to use these points in the definition of the slope. In the next section we will examine how to find the slope directly from an equation.

■ **EXAMPLE 6** **Calculating the Slope of a Line from Its Equation**

Determine the slope of each of the following lines:

SOLUTIONS

(a) $2x - 9y = 18$

$$2x - 9y = 18$$
$$2(0) - 9y = 18$$
$$-9y = 18$$
$$y = -2$$

$$2x - 9(0) = 18$$
$$2x = 18$$
$$x = 9$$

First determine the intercepts so that we can have two points on the line.

x	y
0	-2
9	0

$$m = \frac{y_2 - y_1}{x_2 - x_1}$$

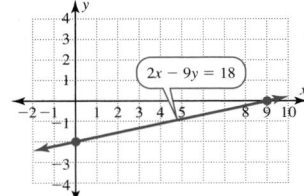

$$m = \frac{0 - (-2)}{9 - 0}$$

Substitute $(0, -2)$ for (x_1, y_1) and $(9, 0)$ for (x_2, y_2).

Answer: $m = \dfrac{2}{9}$

(b) $y = 4$

x	y
0	4
3	4

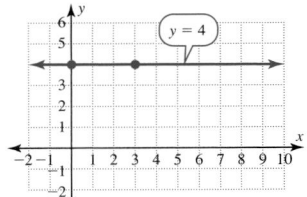

All points on this line have a y-coordinate of 4.

The equation $y = 4$ also can be written as $0x + 1y = 4$. Check the points $(0, 4)$ and $(3, 4)$ in the equation $0x + 1y = 4$.

$$m = \frac{y_2 - y_1}{x_2 - x_1}$$

$$m = \frac{4 - 4}{3 - 0}$$

$$m = \frac{0}{3}$$

Answer: $m = 0$

This is a horizontal line with a slope of 0. The y-coordinate does not change when the x-coordinate changes.

(c) $x = -3$

x	y
-3	0
-3	2

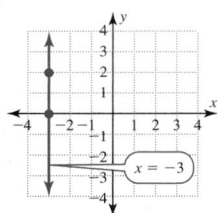

All points on this line have an x-coordinate of -3.

The equation $x = -3$ also can be written as $1x + 0y = -3$. Check the points $(-3, 0)$ and $(-3, 2)$ in the equation $1x + 0y = -3$.

$$m = \frac{y_2 - y_1}{x_2 - x_1}$$

$$m = \frac{2 - 0}{-3 - (-3)}$$

$$m = \frac{2}{0}$$

Answer: m is undefined.

This is a vertical line whose slope is undefined. The x-coordinate does not change.

SELF-CHECK 3.1.3

Calculate the slope of the lines defined by these equations.
1. $x + y = 5$
2. $y = 5x$
3. $y = 5$
4. $x = 5$

SELF-CHECK 3.1.3 ANSWERS

1. $m = -1$; this line slopes downward to the right.
2. $m = 5$; this line slopes upward to the right.
3. $m = 0$; this is a horizontal line.
4. m is undefined; this is a vertical line.

The concept of slope is used in many applications in which two quantities are changing simultaneously along a straight line. Some of these applications use the terms *rise* and *run*. Other applications involving slope may use the term *angle of elevation* or *grade*. Example 7 involves the grade of a highway.

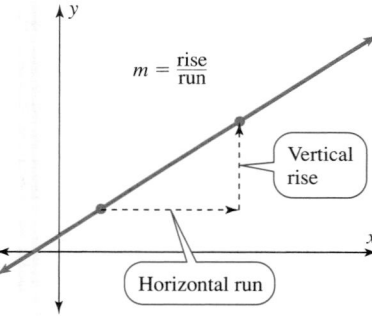

▪ EXAMPLE 7   Modeling the Slope of a Highway

For several reasons, including highway safety, an engineer has determined that the slope of a particular section of highway should not exceed a 6% grade (slope). If this maximum grade is allowed on a section of 2000 m, how much change in elevation is permitted on this section? (The change of elevation can be controlled by topping hills and filling low places.)

SOLUTION _____

Let y = change in elevation.   Sketch the problem using a convenient placement of the origin to simplify the computations (see the figure to the right).

$$m = \frac{y_2 - y_1}{x_2 - x_1}$$ Use the formula for slope.

$$\frac{6}{100} = \frac{y - 0}{2000 - 0}$$ Substitute in the changes for y and x, and write 6% as $\frac{6}{100}$.

$$\frac{6}{100} = \frac{y}{2000}$$

$$2000\left(\frac{6}{100}\right) = 2000\left(\frac{y}{2000}\right)$$ To solve this proportion for y, multiply both sides by the LCD of 2000.

$$20(6) = y$$

$$120 = y \text{ or}$$

$$y = 120$$

Answer: This section of road could change 120 m in elevation.   ▪

In Section 2.1 we noted that the points from an arithmetic sequence form a linear pattern. The next example examines the relationship between the common difference of an arithmetic sequence and the slope of the line through these points.

▪ EXAMPLE 8   Using a Sequence that Models Rent Payments

Example 8 provides an opportunity to examine the connection between the common difference of an arithmetic sequence and the slope of a line. The common difference of an arithmetic sequence gives the change in output per 1 unit of change from one term to the next in the sequence.

The formula for the total cost of a rent deposit and the monthly rent in dollars for an apartment for n months is $a_n = 400n + 250$. Determine the sequence of cost totals for each of the first 4 months, graph this sequence, and determine the slope of the line through these points.

SOLUTION _____

$$a_n = 400n + 250$$
$$a_1 = 400(1) + 250 = 650$$ Substitute 1, 2, 3, and 4 for n in
$$a_2 = 400(2) + 250 = 1050$$ the equation to calculate the total
$$a_3 = 400(3) + 250 = 1450$$ cost for the first 4 months.
$$a_4 = 400(4) + 250 = 1850$$

The dashed line through these points emphasizes that these points form an arithmetic sequence because there is a linear pattern.

The common difference: $d = 1050 - 650 = 400$

The difference between all the consecutive terms is 400.

The slope of the line: $m = \dfrac{y_2 - y_1}{x_2 - x_1}$

Use the formula for slope with the rents as the y output values and the months as the x input values.

$$m = \dfrac{1050 - 650}{2 - 1}$$

Substitute in the points $(1, 650)$ and $(2, 1050)$.

$$m = 400$$

Note the common difference of 400 is the same as the slope of 400; both denote a 400-unit increase in the output for each 1-unit increase in the input variable.

SELF-CHECK 3.1.4

In Example 8 calculate a_0 and interpret this value.

A Mathematical Note

The concept of slope and the comparison of changes is one of the most important concepts in algebra. The term "slope" is used widely in many disciplines. The website for Academic Press *Dictionary of Science and Technology* gives 53 terms using slope when a search is done on the key word "slope."
http://www.harcourt.com/ dictionary/

As this example illustrates, the common difference of an arithmetic sequence is equal to the slope of the line through the points of the sequence. The common difference of a sequence is the change from one point to the next. The slope of a line is the change in y per one unit change in x. Both the common difference of an arithmetic sequence and the slope of a line are ways of describing change. The common difference describes change from a numerical perspective, whereas slope describes change from a graphical perspective.

One use of slopes is to determine whether two lines are parallel or perpendicular. **Parallel lines** are lines that lie in the same plane but never intersect. Parallel lines have the same slope because they rise or fall at the same rate.

Perpendicular lines form a 90° angle when they intersect. (See Figure 3.1.3.) If one line slopes upward to the right, then a line perpendicular to it will slope downward to the right. Perpendicular lines have slopes that are opposite reciprocals. Opposite reciprocals, such as $\dfrac{2}{3}$ and $-\dfrac{3}{2}$, have a product of -1, as in $\left(\dfrac{2}{3}\right)\left(-\dfrac{3}{2}\right) = -1$. Thus lines that are perpendicular to each other have slopes whose product is -1. All horizontal lines are perpendicular to all vertical lines, and all vertical lines are parallel to each other.

SELF-CHECK 3.1.4 ANSWER

$a_0 = 250$; the rent deposit was $250.

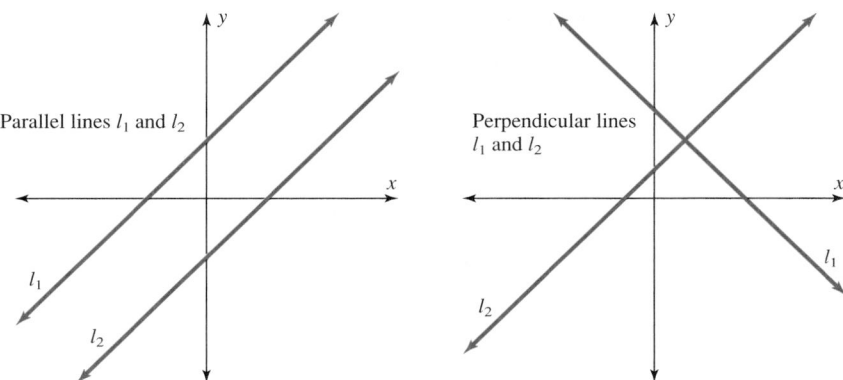

Figure 3.1.3 Parallel and Perpendicular Lines

Parallel and Perpendicular Lines

If l_1 and l_2 are distinct nonvertical* lines with slopes m_1 and m_2, respectively, then:

ALGEBRAICALLY	VERBALLY	GRAPHICALLY
$m_1 = m_2$	l_1 and l_2 are parallel because they have the same slope.	
$m_1 = -\dfrac{1}{m_2}$ or $m_1 m_2 = -1$	l_1 and l_2 are perpendicular because their slopes are negative reciprocals.	

*Also note all vertical lines are parallel to each other and all vertical lines are perpendicular to all horizontal lines.

■ EXAMPLE 9   Determining Whether Two Lines Are Parallel or Perpendicular

Determine whether the line through the first pair of points is parallel to, perpendicular to, or neither parallel nor perpendicular to the line through the second pair of points.

SOLUTIONS

(a) (1, 3) and (−1, −1)
(2, 1) and (3, 3)

$$m = \frac{y_2 - y_1}{x_2 - x_1} \qquad m = \frac{y_2 - y_1}{x_2 - x_1}$$

$$m = \frac{-1 - 3}{-1 - 1} \qquad m = \frac{3 - 1}{3 - 2}$$

$$m = \frac{-4}{-2} \qquad m = \frac{2}{1}$$

$$m = 2 \qquad m = 2$$

Calculate the slope of each line by substituting in the given points.

Since the slopes are equal, the lines are parallel.

Answer: The two lines are parallel.

(b) (3, 1) and (6, 3)
(2, −1) and (4, −4)

$$m = \frac{y_2 - y_1}{x_2 - x_1} \qquad m = \frac{y_2 - y_1}{x_2 - x_1}$$

$$m = \frac{3 - 1}{6 - 3} \qquad m = \frac{-4 - (-1)}{4 - 2}$$

$$m = \frac{2}{3} \qquad m = -\frac{3}{2}$$

Since $\left(\frac{2}{3}\right)\left(-\frac{3}{2}\right) = -1$, the two lines are perpendicular.

Answer: The two lines are perpendicular.

(c) (1, 7) and (−1, −1)
(1, 1) and (−1, 9)

$$m = \frac{y_2 - y_1}{x_2 - x_1} \qquad m = \frac{y_2 - y_1}{x_2 - x_1}$$

$$m = \frac{-1 - 7}{-1 - 1} \qquad m = \frac{9 - 1}{-1 - 1}$$

$$m = \frac{-8}{-2} \qquad m = \frac{8}{-2}$$

$$m = 4 \qquad m = -4$$

The slopes are not equal, and $4(-4) \neq -1$. Thus the lines are neither parallel nor perpendicular.

Answer: The lines are neither parallel nor perpendicular.   ■

SELF-CHECK 3.1.5

Determine whether the line through (5, 3) and (−5, −1) is parallel to, perpendicular to, or neither parallel nor perpendicular to the line through (2, −3) and (−2, 7).

SELF-CHECK 3.1.5 ANSWER

The lines are perpendicular.

In the next section we will examine parallel and perpendicular lines further and look at examples involving horizontal and vertical lines.

USING THE LANGUAGE AND SYMBOLISM OF MATHEMATICS 3.1

1. The rise between two points on a line refers to the change in the _____ variable.
2. The run between two points on a line refers to the change in the _____ variable.
3. The letter that is used to represent the slope of a line is _____.
4. The formula for the slope of a line through (x_1, y_1) and (x_2, y_2) is _____.
5. The slope of a line gives the change in y for each _____ unit change in x.
6. The slope of a horizontal line is _____.
7. The slope of a vertical line is _____.

8. If the slope of a line is positive and the x-coordinate increases, then the y-coordinate _____.
9. If the slope of a line is negative and the x-coordinate increases, then the y-coordinate _____.
10. If two lines have the same slope, the lines are _____.
11. If the slopes of two lines are negative reciprocals, the lines are _____.
12. The common difference of an arithmetic sequence is the same as the _____ of the line through the points of the sequence.
13. The slope of a highway is sometimes referred to as the _____ of the highway.

EXERCISES 3.1

In Exercises 1–6 calculate the slope of the line through the given points.

1. **a.** $(4, 3)$ and $(3, 1)$ **b.** $(-5, 2)$ and $(1, -2)$
   c. $(-1, -3)$ and $(-6, -11)$
2. **a.** $(4, 8)$ and $(2, 2)$ **b.** $(2, -7)$ and $(-10, 2)$
   c. $(-4, -3)$ and $(-10, -10)$
3. **a.** $(4, 9)$ and $(11, 9)$ **b.** $(4, 9)$ and $(4, 11)$
   c. $(0, 24)$ and $(-18, 0)$

4. **a.** $(-3, -7)$ and $(-3, 7)$ **b.** $(-3, -7)$ and $(3, -7)$
   c. $(35, 0)$ and $(0, -14)$
5. **a.** $(0, 0)$ and $(4, -6)$ **b.** $\left(\dfrac{1}{3}, \dfrac{1}{5}\right)$ and $\left(\dfrac{1}{2}, \dfrac{1}{4}\right)$
   c. $(0.37, 0.56)$ and $(0.49, 0.38)$
6. **a.** $(0, 0)$ and $(-9, 12)$ **b.** $\left(\dfrac{2}{3}, \dfrac{1}{2}\right)$ and $\left(\dfrac{3}{4}, \dfrac{4}{5}\right)$
   c. $(1.87, 2.34)$ and $(1.73, 2.41)$

In Exercises 7 and 8 calculate the slope of each line.

7. **a.** 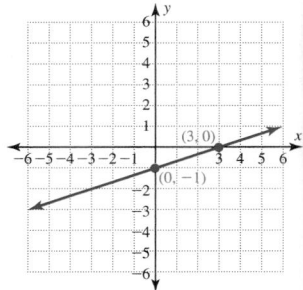   **b.** 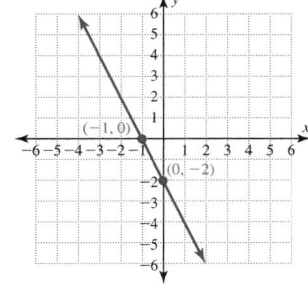   **c.** 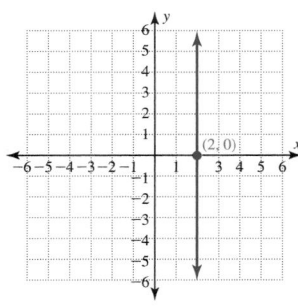   **d.**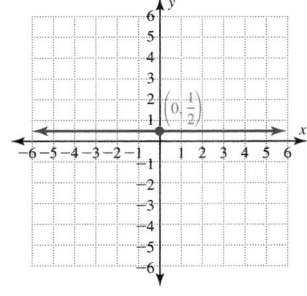

8. **a.** 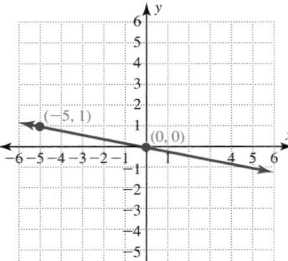   **b.** 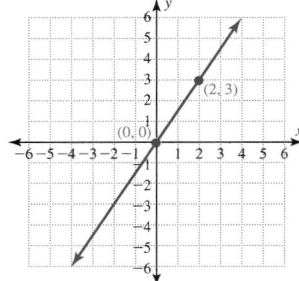   **c.** 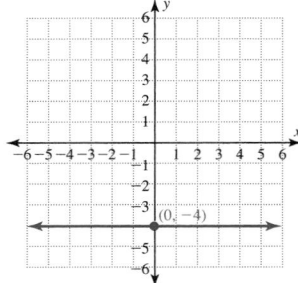   **d.**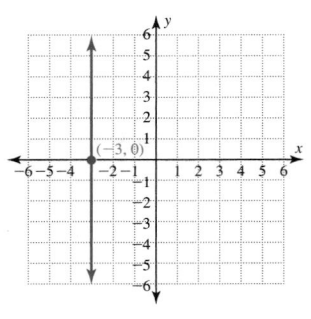

In Exercises 9–12 determine the slope of each line.

9. **a.** $3x + 5y = 15$ **b.** $-2x + 7y = 28$
   c. $5x = 8y$ **d.** $f(x) = 2x - 3$

10. **a.** $4x + 3y = 24$ **b.** $5x - 11y = 55$
    c. $7x = 4y$ **d.** $f(x) = -2x + 3$

11. **a.** $x = 11$ **b.** $y = -11$
    c. $x + y = -11$ **d.** $f(x) = 9$

12. **a.** $y = 7$ **b.** $x = -7$
    c. $x - y = -7$ **d.** $f(x) = 8$

In Exercises 13–26 complete the following table involving the change in x, the change in y, and the slope of the line defined by $y = mx + b$.

	CHANGE IN x	CHANGE IN y	SLOPE
13.	-5	8	
14.	-7	-2	
15.	3		$\dfrac{2}{3}$
16.	3		$-\dfrac{2}{3}$
17.	1		$\dfrac{2}{3}$
18.	1		$-\dfrac{2}{3}$
19.	6		$\dfrac{2}{3}$
20.	6		$-\dfrac{2}{3}$
21.		2	$\dfrac{2}{3}$
22.		2	$-\dfrac{2}{3}$
23.		6	$\dfrac{2}{3}$
24.		-6	$-\dfrac{2}{3}$
25.	6		0
26.	-6		0

In Exercises 27–30 a line has the given y-intercept and slope. Use this information to determine another point on the line and to graph the line.

27. $(0, 2)$; $m = \dfrac{3}{4}$

28. $(0, 2)$; $m = -\dfrac{3}{4}$

29. $(0, -3)$; $m = -2$

30. $(0, -3)$; $m = 2$

In Exercises 31–36 draw a line through the point $(1, 3)$ that has the given slope.

31. $m = 0$
32. $m = 1$
33. $m = 2$
34. $m = -1$
35. $m = -2$
36. m is undefined

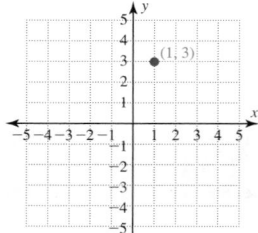

Calculator Usage
In Exercises 37–40 the graphics calculator display shows a table of x- and y-values for a linear equation $y = mx + b$. Determine the slope and the y-intercept of each line.

37.
X	Y₁	
-3	-14	
-2	-11	
-1	-8	
0	-5	
1	-2	
2	1	
3	4	
X= -3		

38.
X	Y₁	
-3	10	
-2	8	
-1	6	
0	4	
1	2	
2	0	
3	-2	
X= -3		

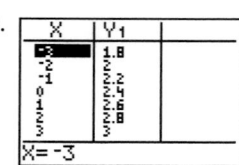

39.
X	Y₁	
-3	4.5	
-2	4	
-1	3.5	
0	3	
1	2.5	
2	2	
3	1.5	
X= -3		

40.
X	Y₁	
-3	1.8	
-2	2	
-1	2.2	
0	2.4	
1	2.6	
2	2.8	
3	3	
X= -3		

41. Using a Graphical Model for the Cost of Shirts The following graph gives the total order and shipping costs for an order of shirts from an on-line vendor. Determine the slope of this line and then interpret the meaning of this rate of change.

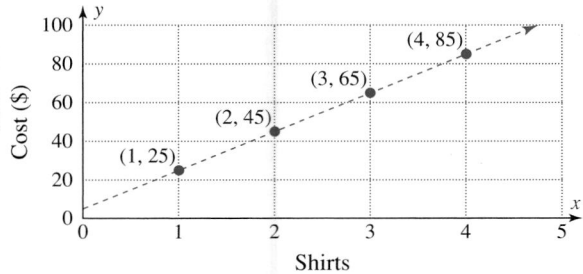

42. Using a Graphical Model for the Cost of Tuition and Fees The following graph gives the total cost of tuition and fees based on the number of semester hours that a student takes at State College. Determine the slope of this line and then interpret the meaning of this rate of change.

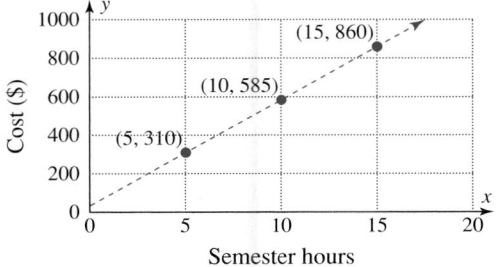

Estimation and Calculator Skills

In Exercises 43–46 mentally estimate the slope of each line and then use a calculator to approximate the slope to the nearest thousandth.

GRAPH	MENTAL ESTIMATE OF m	CALCULATOR APPROXIMATION OF m
43.		
44.		
45.		
46.		

In Exercises 47–52 m_1 is the slope of line l_1, m_2 is the slope of line l_2, and m_3 is the slope of line l_3. l_1 is parallel to l_2 and l_1 is perpendicular to l_3. Complete the following table.

	m_1	m_2	m_3
47.	$\dfrac{3}{7}$		
48.	$-\dfrac{3}{7}$		
49.	0		
50.	Undefined		
51.		$\dfrac{8}{5}$	
52.			$-\dfrac{3}{11}$

In Exercises 53–62 determine whether the line through the first pair of points is parallel to, perpendicular to, or neither parallel nor perpendicular to the line through the second pair of points. Assume that no line passes through both pairs of points.

53. $(-3, 5)$ and $(-6, 3)$
   $(3, -3)$ and $(6, -1)$
54. $(4, 4)$ and $(-4, -2)$
   $(3, 1)$ and $(-3, 9)$
55. $(5, 4)$ and $(10, 1)$
   $(6, 8)$ and $(3, 3)$
56. $(7, -1)$ and $(-7, -7)$
   $(7, 5)$ and $(-7, -1)$
57. $(0, 6)$ and $(8, 0)$
   $(-5, 0)$ and $(0, -7)$
58. $(0, -11)$ and $(9, 0)$
   $(-6, 0)$ and $(0, 6)$
59. $(0, 6)$ and $(4, 6)$
   $(0, 6)$ and $(0, 0)$
60. $(0, 0)$ and $(0, -3)$
   $(0, 0)$ and $(3, 0)$
61. $(5, 3)$ and $(5, -3)$
   $(7, 6)$ and $(7, -6)$
62. $(5, 3)$ and $(-5, 3)$
   $(7, 6)$ and $(-7, 6)$

63. Using a Sequence that Models House Payments The equation $a_n = 1250n + 5000$ gives the total payments (including the down payment) in dollars for a house for n months.
   a. Determine the sequence of total payments for the first 4 months.
   b. Graph this sequence.
   c. Determine the slope of the line through these points. Interpret the meaning of the slope in this application.
   d. Calculate a_0 and interpret the meaning of a_0.

64. Using a Sequence that Models House Payments The equation $a_n = 500n + 1500$ gives the total payments (including the down payment) in dollars for a house for n months.
   a. Determine the sequence of total payments for the first 4 months.
   b. Graph this sequence.
   c. Determine the slope of the line through these points. Interpret the meaning of the slope in this application.
   d. Calculate a_0 and interpret the meaning of a_0.

65. Slope of a Wheelchair Ramp Determine the slope of the wheelchair ramp shown in the figure.

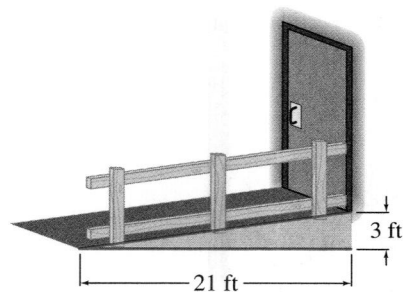

66. Slope of a Roof Determine the slope of the roof on the gable of the house whose cross section is sketched.

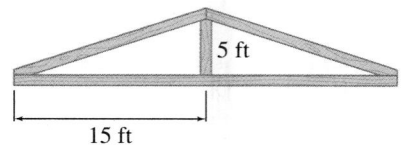

67. Modeling the Grade of a Highway For several reasons, including highway safety, an engineer has determined that the slope of a particular section of highway should not exceed a 5% grade (slope). If this maximum grade is allowed on a section of 1800 m, how much change in elevation is permitted on this section? (The change of elevation can be controlled by topping hills and filling in low places. See the following figure.)

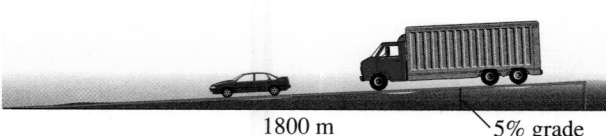

68. Modeling the Grade of a Highway How much change in elevation would be permitted on a 2400-m section of the highway in Exercise 67?

69. Modeling the Height of a Roof Brace The roof of the mountain cabin in the figure rises 14 feet (ft) over a run of 12 ft. Determine the height, *h*, of the brace placed 3 ft from the side of the house.

70. Modeling the Height of a Tree A man who is 6 ft tall is standing so that the tip of his head is exactly in the shadow line of a tree. His shadow is 4 ft long, and the shadow of the tree is 34 ft long. How tall is the tree?

71. What is the slope of the *x*-axis?

72. What is the slope of the *y*-axis?

73. A line passes through quadrants I, II, and III but not through quadrant IV. Is the slope of this line positive or negative?

74. A line passes through quadrants I, II, and IV but not through quadrant III. Is the slope of this line positive or negative?

75. A line passes through quadrants II, III, and IV but not through quadrant I. Is the slope of this line positive or negative?

76. A line passes through quadrants I, III, and IV but not through quadrant II. Is the slope of this line positive or negative?

77. A line passes through quadrants I and III but not through quadrants II and IV. Is the slope of this line positive or negative?

78. A line passes through quadrants II and IV but not through quadrants I and III. Is the slope of this line positive or negative?

79. *y* varies directly as *x* with the constant of variation 2. What is the slope of the line connecting these (x, y) data points?

80. *y* varies directly as *x* with the constant of variation -3. What is the slope of the line connecting these (x, y) data points?

Group Discussion Questions

81. Discovery Question
The equation of a line is $f(x) = mx + b$.
a. Calculate the *y*-intercept of this line.
b. Calculate the *x*-intercept of this line.
c. Use the intercepts and the formula for slope to calculate the slope of this line.
d. Use these results to determine by inspection the slope and *y*-intercept of $f(x) = 5x - 3$.

82. Discovery Question
The equation of a line is $y - y_1 = m(x - x_1)$.
a. Is (x, y_1) a point on this line? Justify your answer.
b. Determine the slope of this line and justify your answer.
c. Use these results to determine by inspection the slope and one point on the line defined by $y - 2 = 4(x - 5)$.

83. Error Analysis A student examined the graph shown on the following calculator display and concluded that the line was vertical and therefore the slope was undefined. Describe the error that the student has made.

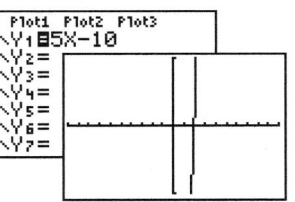

$[-10, 10, 1]$ by $[-1, 1, 1]$

Section 3.2 Special Forms of Linear Equations in Two Variables

Objectives:
3. Use the slope-intercept and point-slope forms of a linear equation.
4. Use the special forms of equations for horizontal and vertical lines.
5. Graph a line given one point and its slope.

We use a variety of graphs—bar graphs, pie graphs, graphs in the Cartesian plane, and so on—to examine mathematical concepts from a graphical perspective. Each type of graph has a context where it is more informative than other graphs. Bar graphs are good for quick comparisons of data; pie charts are good for comparing parts of a whole; and line graphs are good for showing trends.

We also use a variety of algebraic forms of linear equations. Again the context will determine which form is the most useful. Three forms of linear equations that we examine in this section are $y = mx + b$, $y - y_1 = m(x - x_1)$, and $Ax + By = C$.

W. W. Sawyer makes the point in his 1943 book *Mathematician's Delight* that the first mathematicians were practical men who built or made things. Some of our terminology can be traced to this source. For example, the word *straight* comes from Old English for "stretched," whereas the word *line* is the same as that for "linen thread." Thus a straight line is literally a stretched linen thread—as anyone who is planting potatoes or laying bricks knows.

Slope-Intercept Form

One of the most useful forms of a linear equation is the form $y = mx + b$. We have used this form frequently to enter equations into a graphics calculator. This form is also the function form for a linear equation and can be written as $f(x) = mx + b$ to stress the relationship between the x-y input-output pairs.

The form $y = mx + b$ is called the slope-intercept form because this form displays the slope m and the y-intercept $(0, b)$ directly from the equation. This form can be developed by using the slope m, the y-intercept $(0, b)$, and an arbitrary point (x, y) in the formula for the slope.

$$m = \frac{y_2 - y_1}{x_2 - x_1}$$ Start with the formula for slope and substitute in m for the slope and $(0, b)$ and (x, y) for the two points.

$$m = \frac{y - b}{x - 0}$$

$$m = \frac{y - b}{x}$$

$$mx = y - b$$ Then multiply both sides by the LCD x.

$$mx + b = y \text{ or}$$ Add b to both sides.

$$y = mx + b$$ This form of a linear equation is referred to as the slope-intercept form.

$$f(x) = mx + b$$ This is the function form for a linear function.

Slope-Intercept Form

ALGEBRAICALLY	ALGEBRAIC EXAMPLE	VERBAL EXAMPLE	GRAPHICAL EXAMPLE
$y = mx + b$ is the equation of a line with slope m and y-intercept $(0, b)$.	$y = \frac{1}{2}x + 3$	This line has slope $\frac{1}{2}$ and a y-intercept of $(0, 3)$.	

The slope-intercept form is used to enter a linear equation into a graphics calculator. We used this form in Sections 2.2 and 2.6.

■ EXAMPLE 1   Writing the Equation of a Line Given Its Slope and y-Intercept

Write the equation of a line satisfying the given conditions.

SOLUTIONS

(a) $m = \frac{2}{7}$, and $y = mx + b$ Use the slope-intercept form.

y-intercept is $(0, 4)$. $y = \frac{2}{7}x + 4$ Substitute $\frac{2}{7}$ for m and 4 for b.

Answer: $y = \frac{2}{7}x + 4$

(b) $m = -\dfrac{5}{8}$, and

y-intercept is $(0, -2)$.

$y = mx + b$ — Use the slope-intercept form.

$y = -\dfrac{5}{8}x + (-2)$ — Substitute $-\dfrac{5}{8}$ for m and -2 for b.

Answer: $y = -\dfrac{5}{8}x - 2$

(c) $m = 0$, and
y-intercept is $(0, 3)$.

$y = mx + b$ — Use the slope-intercept form.
$y = 0x + 3$ — Substitute 0 for m and 3 for b.

Answer: $y = 3$ — This is the equation of a horizontal line with slope 0. ∎

One of the advantages of the slope-intercept form of a line is that we can determine by inspection both the slope and the y-intercept. These two key pieces of information about the line are readily available. This is illustrated in Example 2.

■ EXAMPLE 2   Using the Slope-Intercept Form

Determine by inspection the slope and y-intercept of the following lines.

SOLUTIONS

(a) $f(x) = -\dfrac{5}{13}x + \dfrac{1}{3}$ $f(x) = -\dfrac{5}{13}x + \dfrac{1}{3}$ — This equation is in the slope-intercept form, $f(x) = mx + b$, with $m = -\dfrac{5}{13}$ and $b = \dfrac{1}{3}$.

Answer: $m = -\dfrac{5}{13}$, and y-intercept is $\left(0, \dfrac{1}{3}\right)$.

(b) $f(x) = -6x$

$f(x) = -6x$
$f(x) = -6x + 0$ — This equation is in the slope-intercept form, with b understood to be 0.

Answer: $m = -6$, and y-intercept is $(0, 0)$.

(c) $2x + 5y = -6$

$2x + 5y = -6$
$5y = -2x - 6$
$y = -\dfrac{2}{5}x - \dfrac{6}{5}$

To put the equation in slope-intercept form, solve for y. Subtract $2x$ from both sides and then divide both sides by 5.

Answer: $m = -\dfrac{2}{5}$, and y-intercept is $\left(0, -\dfrac{6}{5}\right)$. ∎

SELF-CHECK 3.2.1 ANSWERS

1. $y = \dfrac{3}{4}x - 1$
2. $m = \dfrac{7}{11}$, y-intercept $(0, -8)$
3. $m = -\dfrac{3}{4}$, y-intercept $(0, 3)$

SELF-CHECK 3.2.1

1. Write the equation of a line with slope $\dfrac{3}{4}$ and y-intercept $(0, -1)$.
2. Determine the slope and y-intercept of $y = \dfrac{7}{11}x - 8$.
3. Determine the slope and y-intercept of $3x + 4y = 12$.

The next example builds on Example 4 from Section 3.1. We again will use the slope of the line to help graph the line.

■ EXAMPLE 3   Graphing a Line Using Slope-Intercept Form

Graph the line defined by $f(x) = \frac{5}{3}x - 2$.

SOLUTIONS

ALGEBRAICALLY

$f(x) = \frac{5}{3}x - 2$

Slope: $m = \frac{5}{3}$ Use the slope-intercept form $f(x) = mx + b$ to determine the slope and y-intercept.

y-intercept: $(0, -2)$
Second point: $(0 + 3, -2 + 5)$ To produce a second point, start with the y-intercept and
                           $= (3, 3)$ then increase y by 5 units for a 3-unit increase in x.

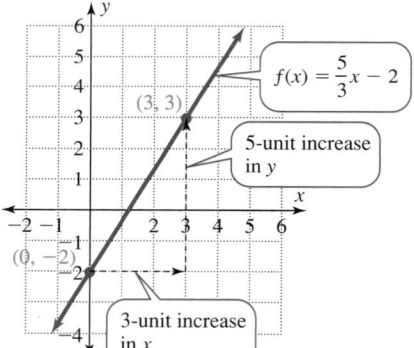

Plot both $(0, -2)$ and $(3, 3)$ and sketch the line through these points.

By graphics calculator:

All three representations (algebraic, numerical, and graphical) reveal that the y-intercept is $(0, -2)$ and the slope is $\frac{5}{3}$. Also note that we can determine from all three representations that $(3, 3)$ is a point on the graph.

NUMERICALLY

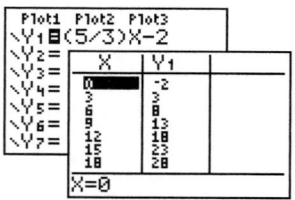

From the table a 3-unit increase in x results in a 5-unit increase in y, for a slope of $\frac{5}{3}$.

GRAPHICALLY

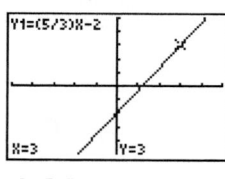

$[-5, 5, 1]$ by $[-5, 5, 1]$

On the graph observe that the y-intercept is $(0, -2)$ and a second point is $(3, 3)$.

■

Parallel and Perpendicular Lines

In Section 3.1 we noted that parallel lines have the same slope and that perpendicular lines have slopes that are opposite reciprocals. If the equations of lines are written in slope-intercept form, it is easy to identify their slopes and to determine whether the lines are parallel or perpendicular.

■ EXAMPLE 4   Determining Whether Two Lines Are Parallel or Perpendicular

Determine whether the first line is parallel to, perpendicular to, or neither parallel nor perpendicular to the second line.

SOLUTIONS ───────────────────────────────

	First Equation	*Second Equation*	
(a) $f(x) = \dfrac{2}{3}x + 7$	$f(x) = \dfrac{2}{3}x + 7$	$f(x) = -\dfrac{3}{2}x + 5$	Each slope can be determined by inspection since the equations are in slope-intercept form.
$f(x) = -\dfrac{3}{2}x + 5$	Slope: $m = \dfrac{2}{3}$	Slope: $m = -\dfrac{3}{2}$	

Answer: Since $\left(\dfrac{2}{3}\right)\left(-\dfrac{3}{2}\right) = -1$, these lines are perpendicular.

Lines are perpendicular if the product of their slopes is -1.

	First Equation	*Second Equation*	
(b) $5x - 3y = 12$	$5x - 3y = 12$	$5x = 3y + 8$	Write each equation in slope-intercept form so the slope can be determined by inspection.
$5x = 3y + 8$	$-3y = -5x + 12$	$5x - 8 = 3y$	
	$y = \dfrac{5}{3}x - 4$	$\dfrac{5}{3}x - \dfrac{8}{3} = y$	
		$y = \dfrac{5}{3}x - \dfrac{8}{3}$	
	Slope: $m = \dfrac{5}{3}$	Slope: $m = \dfrac{5}{3}$	The lines are parallel, but they are not the same line because the y-intercepts are different.

Answer: Since the slopes are equal, the lines are parallel.   ◼

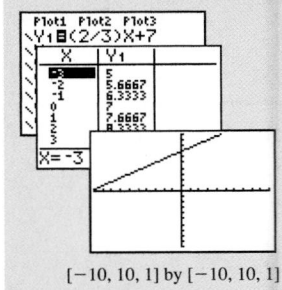

SELF-CHECK 3.2.2 ANSWER

$[-10, 10, 1]$ by $[-10, 10, 1]$

The y-coordinate of every point on a horizontal line is the same. Thus y is a constant.

SELF-CHECK 3.2.2

Use a graphics calculator to create a table using the input values $-3, -2, -1,$ 0, 1, 2, and 3 and to graph the equation in Example 4(a).

Horizontal and Vertical Lines

We examined horizontal and vertical lines in Section 3.1; now we examine their equations. Recall that all horizontal lines have a slope of 0; there is no rise between two points on a horizontal line. The equation of a horizontal line is developed next by substituting $m = 0$ and a y-intercept of $(0, b)$ into the slope-intercept form $y = mx + b$.

$$y = mx + b \qquad \text{Start with the slope-intercept form.}$$
$$y = 0x + b \qquad \text{Substitute 0 for } m.$$
$$y = b$$

The x-coordinate of every point on a vertical line is the same. Thus x has a constant value.

The equation of a horizontal line always can be written in the form $y = b$ or $f(x) = b$. The equation of a vertical line has a similar form. A vertical line with an x-intercept $(a, 0)$ must have the same x-coordinate for every point. The equation that specifies this is $x = a$.

Horizontal and Vertical Lines

ALGEBRAICALLY	NUMERICAL EXAMPLE	GRAPHICAL EXAMPLE	VERBALLY
$y = b$ is the equation of a horizontal line with a y-intercept $(0, b)$. **Example:** $y = 3$	<table><tr><td>x</td><td>y</td></tr><tr><td>-2</td><td>3</td></tr><tr><td>-1</td><td>3</td></tr><tr><td>0</td><td>3</td></tr><tr><td>1</td><td>3</td></tr><tr><td>2</td><td>3</td></tr></table>		This horizontal line has a y-intercept of $(0, 3)$ and a slope of 0.
$x = a$ is the equation of a vertical line with an x-intercept $(a, 0)$. **Example:** $x = -2$	<table><tr><td>x</td><td>y</td></tr><tr><td>-2</td><td>-2</td></tr><tr><td>-2</td><td>-1</td></tr><tr><td>-2</td><td>0</td></tr><tr><td>-2</td><td>1</td></tr><tr><td>-2</td><td>2</td></tr></table>		This vertical line has an x-intercept of $(-2, 0)$ and its slope is undefined.

SELF-CHECK 3.2.3

Determine whether the first line is parallel to, perpendicular to, or neither parallel nor perpendicular to the second line.

1. $f(x) = \dfrac{2}{5}x - 8$ $f(x) = \dfrac{5}{2}x + 8$

2. $f(x) = \dfrac{2}{5}x - 8$ $f(x) = \dfrac{2}{5}x + 8$

3. $4x - 3y = 9$ $3x + 4y = 8$

SELF-CHECK 3.2.3 ANSWERS

1. Neither parallel nor perpendicular
2. Parallel
3. Perpendicular

■ EXAMPLE 5   Writing the Equations of Horizontal and Vertical Lines

Write the equation of each of these lines.

SOLUTIONS _____

(a)

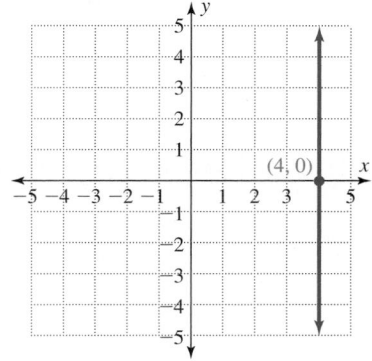

Answer: $x = 4$ This is a vertical line with an x-intercept of $(4, 0)$. Use the form $x = a$ with 4 as the value of a.

(b)

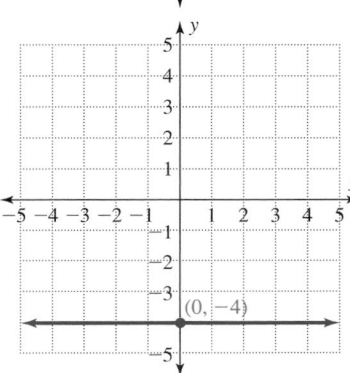

Answer: $y = -4$ This is a horizontal line with a y-intercept of $(0, -4)$. Use the form $y = b$ with -4 as the value of b.

We now examine two other forms of linear equations: the general form and the point-slope form.

General Form

A first-degree equation in both x and y means that the exponent on both these variables is 1. Sometimes the form $Ax + By + C = 0$ also is referred to as general form.

Every linear equation with variables x and y is a first-degree equation in both x and y that can be written in the form $Ax + By = C$, where A, B, and C are real number constants. The **general form** $Ax + By = C$ is used to represent linear equations in many disciplines because this form allows us to write many equations in a nice clean form free of fractions.

◼ **EXAMPLE 6**   Writing a Linear Equation in General Form

Rewrite the equation $y = \dfrac{2}{3}x + \dfrac{4}{5}$ in the general form $Ax + By = C$.

SOLUTION _____

$$y = \frac{2}{3}x + \frac{4}{5}$$

$$15y = 15\left(\frac{2}{3}x + \frac{4}{5}\right)$$
Multiply both sides of the equation by 15, the least common denominator (LCD) of all these terms.

$$15y = 15\left(\frac{2}{3}x\right) + 15\left(\frac{4}{5}\right)$$
Distribute the factor of 15 and then simplify.

The general form often is written so that the coefficient of x is positive.

$$15y = 5(2x) + 3(4)$$
$$15y = 10x + 12$$
$$-12 = 10x - 15y \text{ or}$$

Subtract $15y$ from both sides of the equation and subtract 12 from both sides of the equation.

Answer: $10x - 15y = -12$.

SELF-CHECK 3.2.4

1. Write the equation of a horizontal line through $(6, -1)$.
2. Write the equation of a vertical line through $(6, -1)$.
3. Rewrite $y = -\dfrac{1}{2}x + \dfrac{3}{7}$ in general form with integer coefficients.

Point-Slope Form

The point-slope form that is developed next is used often to write the equation of a line, given either a point on the line and its slope or two points on the line. Suppose that a line passes through the point (x_1, y_1) and has slope m. To determine the equation that relates x and y for any other point (x, y) on this line, we substitute (x, y) for (x_2, y_2) into the formula for slope to obtain

$$m = \frac{y_2 - y_1}{x_2 - x_1}$$

Start with the formula for slope. Then substitute m and the two points (x_1, y_1) and (x, y) into this formula.

$$m = \frac{y - y_1}{x - x_1}$$
$$m(x - x_1) = y - y_1 \text{ or}$$
$$y - y_1 = m(x - x_1)$$

To produce the point-slope form, multiply both sides of this equation by $x - x_1$.

Point-Slope Form $y - y_1 = m(x - x_1)$

ALGEBRAICALLY	ALGEBRAIC EXAMPLE	GRAPHICAL EXAMPLE	VERBAL EXAMPLE
$y - y_1 = m(x - x_1)$ is the equation of a line through (x_1, y_1) with slope m.	$y - 1 = \dfrac{1}{3}(x - 1)$	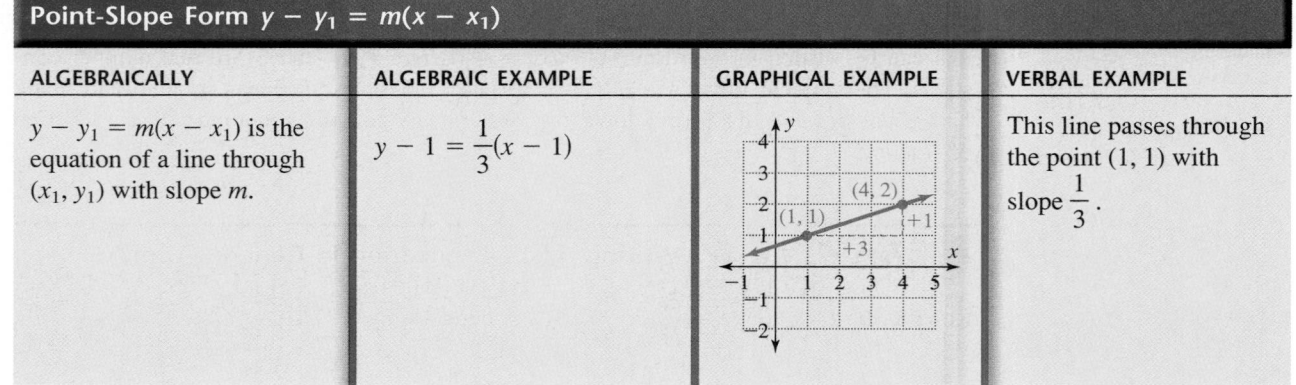	This line passes through the point $(1, 1)$ with slope $\dfrac{1}{3}$.

■ EXAMPLE 7   Using the Point-Slope Form to Write the Equation of a Line

Write in slope-intercept form the equation of a line passing through $(-5, 6)$ with slope $\frac{3}{5}$. Then use a graphics calculator to check that your equation is correct.

SOLUTION

$$y - y_1 = m(x - x_1) \qquad \text{Use the point-slope form.}$$

$$y - 6 = \frac{3}{5}(x - (-5)) \qquad \text{Substitute } (-5, 6) \text{ for } (x_1, y_1) \text{ and } \frac{3}{5} \text{ for } m.$$

$$5(y - 6) = 3(x + 5) \qquad \text{Multiply both sides by the LCD, 5.}$$
$$5y - 30 = 3x + 15$$
$$5y = 3x + 45$$

$$y = \frac{3}{5}x + 9 \qquad \text{Simplify and write the equation in the slope-intercept form, } y = mx + b.$$

Check:

NUMERICALLY

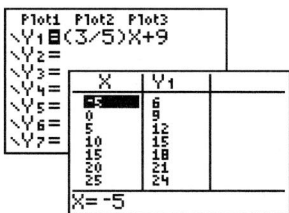

The point $(-5, 6)$ and the y-intercept $(0, 9)$ both appear in this table. Also note that a 5-unit increase in x results in a 3-unit increase in y; a slope of $\frac{3}{5}$.

GRAPHICALLY

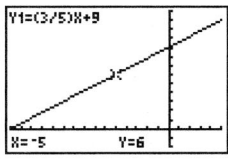

$[-15, 5, 1]$ by $[-2, 13, 1]$

This graph visually displays a slope of $\frac{3}{5}$ and passes through the point $(-5, 6)$ and the y-intercept $(0, 9)$.

SELF-CHECK 3.2.5

The equation in the point-slope form box also can be written as $y - 2 = \frac{1}{3}(x - 4)$. Show that $y - 1 = \frac{1}{3}(x - 1)$ and $y - 2 = \frac{1}{3}(x - 4)$ are equivalent by converting both equations to the general form.

SELF-CHECK 3.2.5 ANSWER

The general form for both equations is $x - 3y = -2$.

A line can be graphed if any two points on the line are known. Similarly, if we know any two points on the line, then we can write the equation of the line. This is illustrated in the next example.

■ **EXAMPLE 8** Writing the Equation of the Line Through Two Points

Write in the slope-intercept form the equation of a line through $(-6, -4)$ and $(-2, 2)$.

SOLUTION

$$m = \frac{y_2 - y_1}{x_2 - x_1} = \frac{2 - (-4)}{-2 - (-6)}$$ Step 1: Since the slope is not given, calculate it by using the two given points.

$$= \frac{6}{4} = \frac{3}{2}$$

$$y - y_1 = m(x - x_1)$$ Step 2: Use the point-slope form.

$$y - (-4) = \frac{3}{2}(x - (-6))$$ Substitute $(-6, -4)$ for (x_1, y_1) and $\frac{3}{2}$ for m.

In Example 8, you may wish to show that the same equation can be obtained by substituting in the point $(-2, 2)$ instead of the point $(-6, -4)$. It is worthwhile to note that both points were used to calculate the slope of the line. Thus either point can be used to select the specific line that has this slope.

$$y + 4 = \frac{3}{2}(x + 6)$$

$$2(y + 4) = 3(x + 6)$$ Multiply both sides by the LCD, 2.

$$2y + 8 = 3x + 18$$ Simplify and write the equation in the slope-intercept form, $y = mx + b$.

$$2y = 3x + 10$$

Answer: $y = \frac{3}{2}x + 5$ Can you use a graphics calculator to check that this is the correct equation?

■

SELF-CHECK 3.2.6

1. Determine by inspection the slope of $y + 7 = -\frac{3}{4}(x - 4)$ and one point on this line.
2. Write in the slope-intercept form the equation of a line through $(0, 4)$ and $(5, 2)$.

A line can be drawn if two points on the line are known. A line also can be drawn if one point and the slope of the line are known. This is done in Example 9 by first plotting the given point and then by using the slope to make changes in the x- and y-coordinates to produce a second point. The line then is drawn through the two points.

■ **EXAMPLE 9** Graphing a Line Through a Point with a Given Slope

Graph the line through the point $(-4, -6)$ with slope $\frac{5}{6}$.

SOLUTION

First point: $(-4, -6)$ Start with the given point $(-4, -6)$ and use the slope of $\frac{5}{6}$ to

Second point: $(-4 + 6, -6 + 5)$ determine a second point. Do this by increasing y by 5 units for

$= (2, -1)$ a 6-unit increase in x.

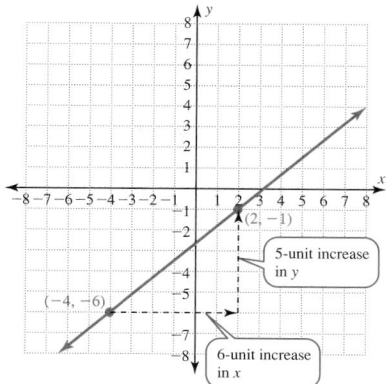

Plot both $(-4, -6)$ and $(2, -1)$ and sketch the line through these points.

SELF-CHECK 3.2.7 ANSWER

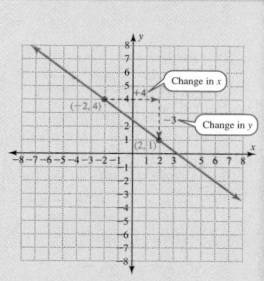

SELF-CHECK 3.2.7

Graph the line through $(-2, 4)$ with slope $-\dfrac{3}{4}$.

USING THE LANGUAGE AND SYMBOLISM OF MATHEMATICS   3.2

1. The slope-intercept form of the equation of a line is _____.

2. Parallel lines have the _____ slope.

3. Lines whose slopes are opposite reciprocals of each other are _____.

4. The line defined by $x = 5$ is a _____ line whose x-intercept is (_____, _____).

5. The line defined by $y = -5$ is a _____ line whose y-intercept is (_____, _____).

6. The general form of the equation of a line is _____.

7. When we say that a linear equation in x and y is a first-degree equation, we mean that the exponent on both x and y is _____.

8. The equation $y - y_1 = m(x - x_1)$ is the _____-_____ form of a linear equation with (x_1, y_1) representing a _____ on the line and m representing the _____ of the line.

EXERCISES   3.2

In Exercises 1–4 determine by inspection the slope and y-intercept of each line.

1. a. $f(x) = 2x + 5$ **b.** $f(x) = -\dfrac{3}{11}x - \dfrac{4}{5}$
    c. $f(x) = 6x$

2. a. $f(x) = -3x + 7$ **b.** $f(x) = \dfrac{4}{9}x - \dfrac{2}{3}$
    c. $f(x) = -6x$

3. a. $y = 2$ **b.** $x = -7$
    c. $y = -x$

4. a. $y = -2$ **b.** $x = 7$
    c. $y = x$

In Exercises 5–10 write in the slope-intercept form the equation of the line with the given slope and y-intercept.

5. $m = 4$; $(0, 7)$ **6.** $m = 3$; $(0, 5)$

7. $m = -\dfrac{2}{11}$; $(0, 5)$ **8.** $m = -\dfrac{5}{7}$; $(0, -2)$

9. $m = 0$; $(0, -6)$ **10.** $m = 0$; $(0, 4)$

In Exercises 11–16 sketch the graph of each line without using a calculator.

11. $f(x) = \dfrac{2}{3}x - 4$ **12.** $f(x) = \dfrac{3}{4}x - 5$

13. $f(x) = -\dfrac{5}{3}x + 2$ **14.** $f(x) = -\dfrac{5}{2}x + 3$

15. $f(x) = 2$ **16.** $f(x) = -3$

In Exercises 17–20 write the equation for each line in the slope-intercept form $f(x) = mx + b$.

17.   **18.**

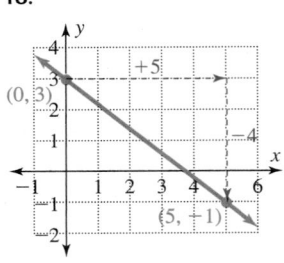

19.   **20.**

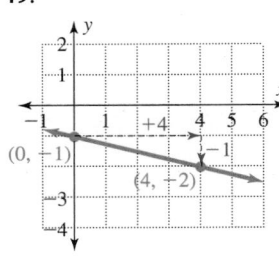 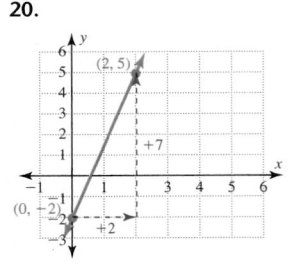

21. Using a Graphical Model for the Cost of Telephone Minutes   The following graph gives the monthly cost for a cell phone based on the number of minutes used. Use this graph to determine the linear equation $f(x) = mx + b$ of this line. Then interpret the meaning of m and b in this problem.

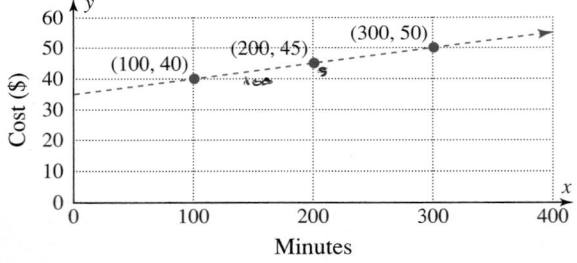

22. Using a Graphical Model for Checking Costs   The following graph gives the monthly cost for a checking account based on the number of checks written. Use this graph to determine the linear equation $f(x) = mx + b$ of this line. Then interpret the meaning of m and b in this problem.

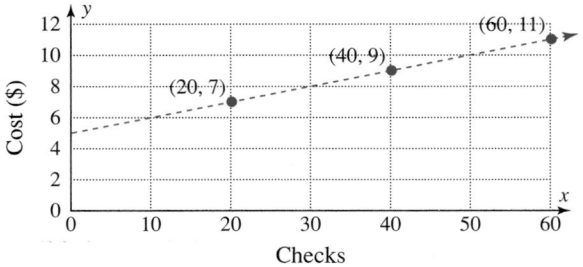

Calculator Usage

In Exercises 23–26 the graphics calculator display shows a table of values for a linear equation $f(x) = mx + b$. Use this table to determine both m and b and then write the equation of this line.

23.

X	Y₁
-3	-20
-2	-13
-1	-6
0	1
1	8
2	15
3	22

X= -3

24.

X	Y₁
-3	18
-2	13
-1	8
0	3
1	-2
2	-7
3	-12

X= -3

25.

X	Y₁
-3	9.5
-2	9
-1	8.5
0	8
1	7.5
2	7
3	6.5

X= -3

26.

X	Y₁
-3	-4.6
-2	-4.4
-1	-4.2
0	-4
1	-3.8
2	-3.6
3	-3.4

X= -3

27. Using a Table to Model the Cost of a Service Call   The following table displays the dollar cost by Reliable Heating and Air Conditioning for a service call based on the number of hours that the serviceperson is at the customer's location. Use this table to determine the linear equation $y = mx + b$ for these data points. Then interpret the meaning of m and b in this problem.

x HOURS	y COST ($)
0	75
0.5	100
1.0	125
1.5	150
2.0	175
2.5	200

28. Using a Table to Model the Salary of a Salesperson   The following table displays the total monthly salary of a salesperson at an automobile dealership. This salary consists of a base salary and a commission based on this person's sales for the month. Use this table to determine the linear equation $y = mx + b$ for these data points. Then interpret the meaning of m and b in this problem.

x SALES ($)	y SALARY ($)
0	1,200
20,000	1,400
40,000	1,600
60,000	1,800
80,000	2,000
100,000	2,200

In Exercises 29–32 write the equation of each line.

29.

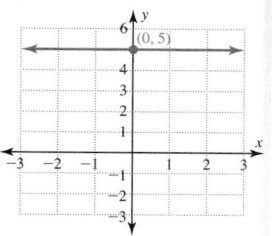

30.

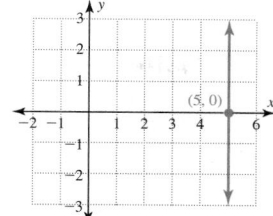

31.

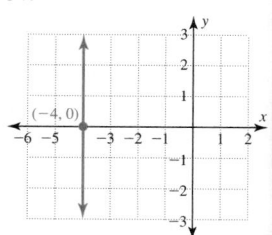

32.

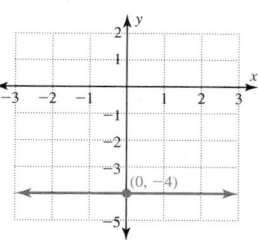

In Exercises 33–40 determine whether the first line is parallel to, perpendicular to, or neither parallel nor perpendicular to the second line.

33. $y = 5x + 11$
$y = -5x + 11$

34. $y = 5x + 11$
$y = 5x - 11$

35. $y = \dfrac{2}{3}x + 7$
$y = -\dfrac{3}{2}x - 2$

36. $y = \dfrac{3}{13}x - 4$
$y = \dfrac{13}{3}x + 4$

37. $2x + 5y = 8$
$6x + 15y - 7 = 0$

38. $3x - 4y + 8 = 0$
$4x + 3y = -15$

39. $y - 8 = 4(x + 5)$
$y + 4 = 5(x - 3)$

40. $y + 6 = 2(x - 1)$
$y - 1 = 2(x + 3)$

In Exercises 41 and 42 determine by inspection the slope of the line and one point on the line.

41. a. $y - 4 = 5(x - 3)$ **b.** $y + 5 = -7(x - 2)$

   c. $y = \dfrac{1}{2}(x + 6)$

42. a. $y - 7 = x - 2$ **b.** $y - 9 = -4(x + 11)$

   c. $y = \dfrac{2}{3}(x + 7)$

In Exercises 43–50 graph the line through the given point with the slope specified.

43. $(3, -2)$; $m = -\dfrac{3}{2}$ **44.** $(4, 2)$; $m = -\dfrac{5}{3}$

45. $(-6, 0)$; $m = 2$ **46.** $(0, -4)$; $m = \dfrac{1}{4}$

47. $(0, 0)$; $m = \dfrac{1}{4}$ **48.** $(0, 0)$; $m = -\dfrac{1}{3}$

49. $(3, 4)$; $m = 0$ **50.** $(3, 4)$;   slope is undefined

In Exercises 51–56 write in the slope-intercept form the equation of a line passing through the given point with the slope specified.

51. $(2, 3)$; $m = -4$ **52.** $(5, 2)$; $m = -2$

53. $(-1, 7)$; $m = \dfrac{2}{3}$ **54.** $(3, -4)$; $m = \dfrac{4}{5}$

55. $(0, 0)$; m is undefined   **56.** $(7, 0)$; m is undefined

In Exercises 57–66 write in the slope-intercept form the equation of a line passing through the given points.

57. $(-4, 2), (4, 4)$ **58.** $(2, 3), (-2, 1)$

59. $(0, 6), (-3, 0)$ **60.** $(5, 0), (-4, 0)$

61. $(4, 9), (4, 11)$ **62.** $(-2, 8), (-2, -7)$

63. $(-2, 8), (-5, 8)$ **64.** $(-5, -4), (3, -4)$

65. $\left(\dfrac{1}{2}, \dfrac{1}{3}\right), \left(\dfrac{1}{4}, -\dfrac{1}{3}\right)$ **66.** $\left(-\dfrac{3}{4}, \dfrac{2}{3}\right), \left(\dfrac{3}{4}, -\dfrac{2}{3}\right)$

67. Taxi Fare The following table displays the dollar cost of a taxi ride based on the number of miles traveled. Write the linear equation $y = mx + b$ for the line that contains these data points. Then interpret the meaning of m and b in this problem. (*Hint:* Although you would not pay for a ride of zero miles, the y-intercept still has a meaningful interpretation.)

x MILES	y COST ($)
1.5	4.45
4	6.45
7	8.85
12.5	13.25

68. Cost of Books The following table displays the dollar cost of a shipment of math books from a publisher to a campus bookstore. Write the linear equation $y = mx + b$ for the line that contains these data points. Then interpret the meaning of m and b in this problem. (*Hint:* Although you would not pay for a shipment of zero books, the y-intercept still has a meaningful interpretation.)

x BOOKS	y COST ($)
70	4,210
140	8,410
250	15,010
400	24,010

In Exercises 69–72 complete the slope-intercept form and the general form for each linear equation.

POINT-SLOPE FORM	SLOPE-INTERCEPT FORM	GENERAL FORM
Example: $y - 1 = 2(x - 3)$	$y = 2x - 5$	$2x - y = 5$
69. $y - 3 = 4(x + 1)$		
70. $y + 2 = -3(x - 1)$		

POINT-SLOPE FORM	SLOPE-INTERCEPT FORM	GENERAL FORM
Example: $y - 1 = 2(x - 3)$	$y = 2x - 5$	$2x - y = 5$
71. $y + 4 = -\dfrac{2}{3}(x - 5)$		
72. $y - 5 = \dfrac{7}{2}(x + 3)$		

Multiple Representations and Calculator Skills
In Exercises 73–76 use the verbal description of the line to write the equation of the line in the slope-intercept form $y = mx + b$. Use a calculator to create a table of x-y input-output values for $x = -3, -2, -1, 0, 1, 2, 3$ and then to graph this line.

73. The line has slope $-\dfrac{1}{2}$ and a y-intercept of $(0, 2)$.

74. The line has slope $\dfrac{2}{3}$ and a y-intercept of $(0, -2)$.

75. The line has slope 0 and a y-intercept of $(0, 3)$.

76. The line has slope 0 and passes through $(5, -3)$.

77. Modeling Car Payments   The down payment on a new car was $1250. The monthly payment for the car due the last day of each month is $350. Letting x represent the number of month and y equal the total amount paid after x months (including the down payment), write a linear equation in slope-intercept form that expresses the relationship between x and y. What is the significance of the slope and the y-intercept?

78. Modeling Fixed and Variable Production Costs   The monthly cost y for producing toner cartridges at a factory is $17,500 plus $5 for each of the x cartridges produced. Write a linear equation in slope-intercept form that expresses the relationship between x and y. What is the significance of the slope and the y-intercept?

79. Modeling Sales Commissions   The salary for a sales clerk is $40 per day plus a 15% commission on all sales. Write a linear equation in the slope-intercept form that gives the daily salary y in terms of the daily sales x. What is the significance of the slope and the y-intercept?

80. Modeling the Displacement of a Spring   A spring is 8 cm long. The spring stretches 2 cm for each kilogram attached. Express y the new length of the spring in terms of the mass x. Write this equation in the slope-intercept form. What is the significance of the slope and the y-intercept?

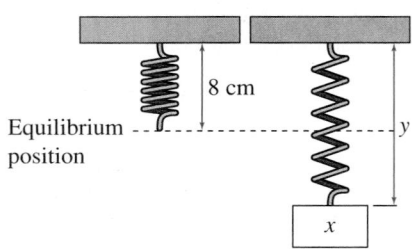

81. Modeling Texaco Dividends   In 1990 the annual dividend for Texaco Corporation was $2.25 per share. In 2000 the annual dividend per share was $1.80. Using x for the year and y as the annual dividend, write a linear equation to relate the dividend to the year. Then use this equation to estimate the dividend in 1995. Also use this equation to predict the dividend in the current year. How does your prediction compare to the actual dividend? (Source: **http://www.texaco.com**)

82. Modeling Duke Energy Dividends   In 1990 the annual dividend for Duke Energy was $2.79 per share. In 2000 the annual dividend per share was $2.20. Using x for the year and y as the annual dividend, write a linear equation to relate the dividend to the year. Then use this equation to estimate the dividend in 1995. Also use this equation to predict the dividend in the current year. How does your prediction compare to the actual dividend? (Source: **http://www.duke-energy.com/**)

Group Discussion Questions

83. Writing Mathematically   The profit y of a company over a period of time x is graphed on a rectangular coordinate system. Write a paragraph describing your interpretation when the slope is (a) negative, (b) zero, and (c) positive.

84. Challenge Question   Write in the general form the equation of the line satisfying the conditions given.
  a. Parallel to the x-axis through $(2, 3)$
  b. Parallel to the y-axis through $(2, 3)$
  c. Perpendicular to the x-axis through $(-5, 8)$
  d. Perpendicular to the y-axis through $(-5, 8)$
  e. Parallel to $y = \dfrac{3}{7}x + 7$ through $(2, -3)$
  f. Parallel to $y = \dfrac{2}{9}x - 5$ through $(-2, 3)$
  g. Perpendicular to $y - 1 = \dfrac{1}{2}(x + 3)$ through $(3, 5)$
  h. Perpendicular to $y + 2 = -\dfrac{1}{3}(x - 7)$ through $(4, -9)$

85. Challenge Question   Write in the general form the equation of the line satisfying the conditions given.
  a. y-intercept $(0, -5)$ with $m = \dfrac{1}{5}$
  b. y-intercept $(0, 5)$ with $m = -\dfrac{1}{5}$
  c. x-intercept $(3, 0)$ with $m = 7$
  d. x-intercept $(2, 0)$ with $m = -4$
  e. Through $(-7, -2)$ with $m = 0$
  f. Through $(-7, 5)$ with slope undefined

Section 3.3 Solving Systems of Linear Equations in Two Variables Graphically and Numerically

Objectives:

6. Solve a system of linear equations using graphs and tables.
7. Identify inconsistent systems and systems of dependent linear equations.

The graphical method for solving systems of linear equations, first covered in Section 2.2, is an excellent way to develop a good understanding of what the solution to a system of linear equations represents. It is easy to "see" the point of intersection of two lines, the ordered pair that satisfies the equation of each line.

Solution of a System of Two Linear Equations

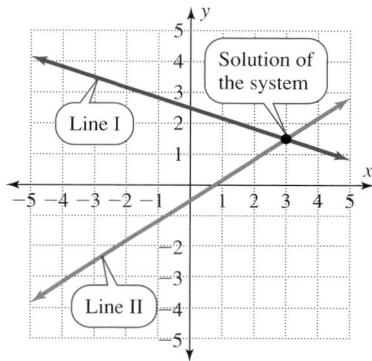

However, the graphical method sometimes can be difficult to implement, even with a graphics calculator. Answers that need several decimal places of accuracy or answers that are very small or very large can stretch the capabilities of the graphical method. On the other hand, the algebraic method, which may be less intuitive, is very accurate and straightforward to apply. This is another reason that we strongly encourage you to use multiple representations (including the examination of topics from algebraic, graphical, numerical, and verbal viewpoints) to increase your understanding of mathematics.

Before examining the algebraic methods, we review the graphical method and use it to deepen our understanding of the solution of a system of linear equations. To use your graphics calculator, you may need first to rewrite the equations in the slope-intercept form, $y = mx + b$.

■ EXAMPLE 1 Solving a Linear System Numerically and Graphically

Solve the linear system $\begin{cases} 7x - 3y = 2 \\ 9x - 5y = -2 \end{cases}$ numerically and graphically.

This is a good place to note that the solution to a linear system in two variables is an ordered pair. In Chapter 2 we used graphs to solve a single linear equation in one variable. Please note the distinction between the one-variable situation and the two-variable situation.

SOLUTION _____

First Equation
$$7x - 3y = 2$$
$$-3y = -7x + 2$$
$$y = \frac{7}{3}x - \frac{2}{3}$$

Second Equation
$$9x - 5y = -2$$
$$-5y = -9x - 2$$
$$y = \frac{9}{5}x + \frac{2}{5}$$

If you plan to use a graphics calculator, first you will need to rewrite the equations in the slope-intercept form as shown here.

NUMERICALLY

GRAPHICALLY

For help with keystrokes see the calculator perspective in Section 2.2.

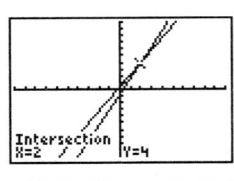

[−10, 10, 1] by [−10, 10, 1]

Both the table and the graph yield (2, 4) as the solution of this system.

The answer to a system of two linear equations should be written in ordered-pair notation. Writing this solution as (2, 4) emphasizes that there is only one point of intersection. This point has two coordinates.

Check: (2, 4) in both equations

First Equation	*Second Equation*
$7x - 3y = 2$	$9x - 5y = -2$
$7(2) - 3(4) \overset{?}{=} 2$	$9(2) - 5(4) \overset{?}{=} -2$
$14 - 12 \overset{?}{=} 2$	$18 - 20 \overset{?}{=} -2$
$2 \overset{?}{=} 2$ checks.	$-2 \overset{?}{=} -2$ checks.

Answer: (2, 4)

One Solution

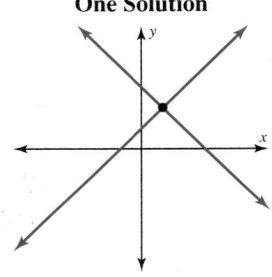

No Solution

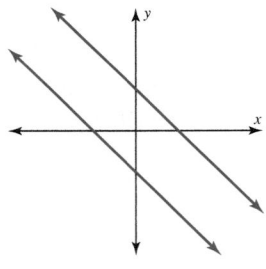

An Infinite Number of Solutions

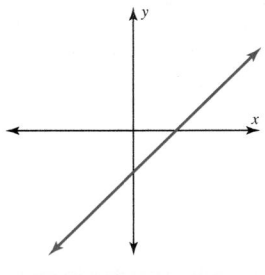

Classification of Linear Systems

Example 1 illustrates a system with exactly one simultaneous solution. There are two other possibilities that we now examine. Two lines in a plane can be related in three ways:

1. The lines are distinct and intersect at a single point. This point represents the only simultaneous solution.
2. The lines are parallel and distinct. There is no point of intersection and no simultaneous solution.
3. The lines coincide (both equations represent the same line). There are an infinite number of points on this shared line and each represents a solution. Only these points are solutions; the points not on the line are not solutions.

If a system of two equations in two variables has a solution, the system is called **consistent;** otherwise it is called **inconsistent.** If the equations have distinct graphs, the equations are called **independent;** if the graphs coincide, the equations are called **dependent.** We now present examples of an inconsistent system and a system of dependent equations.

■ EXAMPLE 2 Solving an Inconsistent System

Solve the linear system $\begin{Bmatrix} x + 2y = 2 \\ 2x + 4y = 8 \end{Bmatrix}$ algebraically, numerically, and graphically.

SOLUTION

First Equation	Second Equation
$x + 2y = 2$	$2x + 4y = 8$
$2y = -x + 2$	$4y = -2x + 8$
$y = -\dfrac{1}{2}x + 1$	$y = -\dfrac{1}{2}x + 2$

We are writing each equation in the slope-intercept form both to facilitate an algebraic comparison and to prepare the equations for graphing by either hand or using a graphics calculator.

The slope is $-\dfrac{1}{2}$ and the y-intercept is $(0, 1)$.   The slope is $-\dfrac{1}{2}$ and the y-intercept is $(0, 2)$.

Thus the two lines are parallel and distinct.

NUMERICALLY

These parallel lines are always the same distance apart. This is shown by both the table and the graph. Since these parallel lines never meet, the system has no solution and is classified as inconsistent.

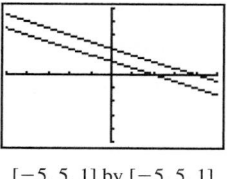

```
Plot1 Plot2 Plot3
\Y1◘(-1/2)X+1
\Y2◘(-1/2)X+2
\Y3=
\Y4=     X    Y1    Y2
\Y5=    -3    2.5   3.5
\Y6=    -2    2     3
\Y7=    -1    1.5   2.5
              1     2
              .5    1.5
              0     1
              -.5   .5
        X=-3
```

For the same x-value, the y_1 and y_2 values are always 1 unit apart.

GRAPHICALLY

$[-5, 5, 1]$ by $[-5, 5, 1]$

The lines appear parallel and are always the same distance apart.

Answer: No solution, an inconsistent system.

■ EXAMPLE 3 Solving a System of Dependent Equations

Solve the linear system $\begin{Bmatrix} 3x - 4y = 12 \\ 6x - 8y = 24 \end{Bmatrix}$ algebraically, numerically, and graphically.

SOLUTION

First Equation	Second Equation
$3x - 4y = 12$	$6x - 8y = 24$
$-4y = -3x + 12$	$-8y = -6x + 24$
$y = \dfrac{3}{4}x - 3$	$y = \dfrac{3}{4}x - 3$

Rewrite each equation in the slope-intercept form. Note that both equations have the same slope-intercept form, revealing that they are equations of the same line.

The slope is $\dfrac{3}{4}$ and the y-intercept is $(0, -3)$.   The slope is $\dfrac{3}{4}$ and the y-intercept is $(0, -3)$.

The two equations produce the same line.

NUMERICALLY

GRAPHICALLY

Although the original equations look different, the lines coincide. Thus every point satisfying one equation also will satisfy the other. The system has an infinite number of solutions and is classified as consistent with dependent equations. Only the points on this line are solutions of the linear system of equations.

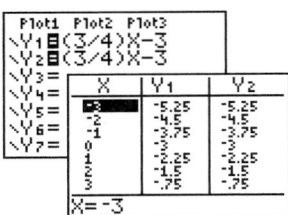

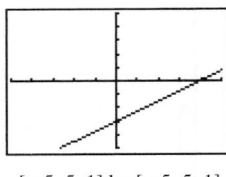

$[-5, 5, 1]$ by $[-5, 5, 1]$

For the same x-value the y_1 and y_2 values are the same.

The equations produce the same line.

Answer: There are an infinite number of solutions; the equations are dependent. ■

Now let us reflect on the first three examples. Compare the slope-intercept forms of the equations in these three classifications of linear systems. Can you predict the classification using the slope-intercept form?

Classification of Systems of Linear Equations

The linear system $\begin{cases} y = m_1x + b_1 \\ y = m_2x + b_2 \end{cases}$ is

VERBALLY	ALGEBRAICALLY	NUMERICAL EXAMPLE	GRAPHICAL EXAMPLE
1. A consistent system of independent linear equations having exactly one solution.	$m_1 \neq m_2$ *Example:* $\begin{cases} y = 2x - 1 \\ y = -3x - 6 \end{cases}$	Only one x-value has matching y-values. 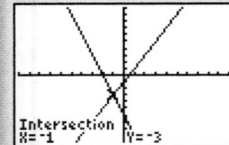	One distinct point of intersection. $[-10, 10, 1]$ by $[-10, 10, 1]$
2. An inconsistent system of linear equations having no solution.	$m_1 = m_2$ and $b_1 \neq b_2$ *Example:* $\begin{cases} y = \dfrac{1}{2}x + 1 \\ y = \dfrac{1}{2}x - 3 \end{cases}$	The difference between y_1- and y_2-values is 4 units for each x-value. 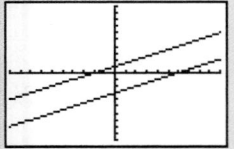	Parallel lines with no point of intersection. $[-10, 10, 1]$ by $[-10, 10, 1]$
3. A consistent system of dependent linear equations having an infinite number of solutions.	$m_1 = m_2$ and $b_1 = b_2$ *Example:* $\begin{cases} y = \dfrac{3}{2}x + \dfrac{1}{2} \\ y = \dfrac{3}{2}x + \dfrac{1}{2} \end{cases}$	The y_1- and y_2-values are equal for every value of x. 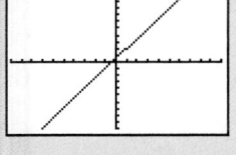	Infinite number of common points on these coincident lines. $[-10, 10, 1]$ by $[-10, 10, 1]$

We use the slope-intercept form and the information from the previous box to determine the number of solutions of each of the following systems. Note that this work does not tell us which points satisfy both equations, only how many points will be solutions.

■ EXAMPLE 4   Determining the Number of Solutions of a Linear System

Determine the number of solutions of each of the following systems:

SOLUTIONS _____

	First Equation	*Second Equation*	
(a) $\begin{cases} y = \dfrac{2}{3}x - 5 \\ y = \dfrac{3}{4}x + 7 \end{cases}$	$y = \dfrac{2}{3}x - 5$	$y = \dfrac{3}{4}x + 7$	Determine the slope from the slope-intercept form, $y = mx + b$.
	$m_1 = \dfrac{2}{3}$	$m_2 = \dfrac{3}{4}$	Because the slopes are different, the lines will intersect at exactly one point.

Answer: Because $m_1 \neq m_2$, the system has exactly one solution.

	First Equation	*Second Equation*	
(b) $\begin{cases} 3x - y = 5 \\ \quad 2y = 6x - 10 \end{cases}$	$3x - y = 5$	$2y = 6x - 10$	First write each equation in the slope-intercept form. Then use the slopes and y-intercepts to determine the number of solutions.
	$-y = -3x + 5$	$y = 3x - 5$	
	$y = 3x - 5$		
	$m_1 = 3, b_1 = -5$	$m_2 = 3, b_2 = -5$	

Answer: Because $m_1 = m_2$ and $b_1 = b_2$, the equations are dependent and the system of equations has an infinite number of solutions.

Both equations represent the same line. Only the points on this line are solutions of the system of linear equations.

	First Equation	*Second Equation*	
(c) $\begin{cases} 2x - 3y = 15 \\ 2x - 3y = -6 \end{cases}$	$2x - 3y = 15$	$2x - 3y = -6$	
	$-3y = -2x + 15$	$-3y = -2x - 6$	
	$y = \dfrac{2}{3}x - 5$	$y = \dfrac{2}{3}x + 2$	
	$m_1 = \dfrac{2}{3}, b_1 = -5$	$m_2 = \dfrac{2}{3}, b_2 = 2$	Because the slopes are equal, the lines are parallel. Because the y-intercepts are different, the lines do not coincide.

Answer: Because $m_1 = m_2$ but $b_1 \neq b_2$, the system is inconsistent and has no solution.   ■

> **SELF-CHECK 3.3.2**
>
> Determine the number of solutions to each of the following systems of linear equations:
>
> **1.** $\begin{cases} y_1 = 2x - 7 \\ y_2 = 2x + 7 \end{cases}$ **2.** $\begin{cases} y_1 = 2x - 7 \\ y_2 = 7x - 2 \end{cases}$ **3.** $\begin{cases} y_1 = 2x - 7 \\ y_2 = 2x - 7 \end{cases}$

SELF-CHECK 3.3.2 ANSWERS

1. No solution
2. Exactly one solution
3. An infinite number of solutions

The next example is included to reveal some of the weaknesses of the graphical method and an overreliance on graphics calculators. Before you read through the text explanation try to solve the system $\begin{cases} 2x - 3y = 10 \\ 3x - 5y = 0 \end{cases}$ using a graphics calculator. Please note the solution you obtain and any difficulties you encounter.

■ EXAMPLE 5   Using a Graphics Calculator to Solve a System of Linear Equations

Solve the linear system $\begin{cases} 2x - 3y = 10 \\ 3x - 5y = 0 \end{cases}$ using a graphics calculator.

SOLUTION _____

First Equation	*Second Equation*	
$2x - 3y = 10$	$3x - 5y = 0$	Write each equation in the slope-intercept form to prepare the equations for entry into a graphics calculator.
$-3y = -2x + 10$	$-5y = -3x$	
$y = \dfrac{2}{3}x - \dfrac{10}{3}$	$y = \dfrac{3}{5}x$	

First View Window

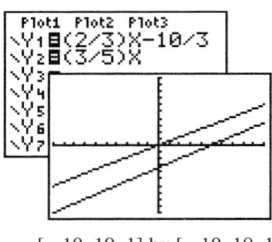

$[-10, 10, 1]$ by $[-10, 10, 1]$

Adjusted Window

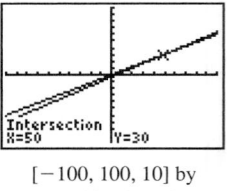

$[-100, 100, 10]$ by $[-100, 100, 10]$

You may find it easier to determine this point of intersection using the algebraic methods in the next section.

Although the lines may appear to be parallel on this window, we know the lines are not parallel because they do not have the same slope. Thus the lines must have a point of intersection that is not revealed on this window.

The new viewing rectangle contains the point of intersection which is located using the intersect feature on a graphics calculator.

The point of intersection is (50, 30).

Check:

First Equation	*Second Equation*
$2x - 3y = 10$	$3x - 5y = 0$
$2(50) - 3(30) \overset{?}{=} 10$	$3(50) - 5(30) \overset{?}{=} 0$
$100 - 90 \overset{?}{=} 10$	$150 - 150 \overset{?}{=} 0$
$10 \overset{?}{=} 10$ checks.	$0 \overset{?}{=} 0$ checks.

Answer: (50, 30)

SELF-CHECK 3.3.3

Solve the linear system $\begin{cases} 3x + 11y = 0.1 \\ 2x + 9y = 0.4 \end{cases}$ using a table on a graphics calculator.

Hint: Examine the *x*-values $-1, -.9, -.8, -.7, -.6, -.5, -.4$

SELF-CHECK 3.3.3 ANSWER

$(-0.7, 0.2)$

CALCULATOR PERSPECTIVE 3.3.1	Determining Rational Solutions to a System of Linear Equations

To solve the system of equations $\begin{cases} y = 2x - 3 \\ y = -5x + 6 \end{cases}$ on a TI-83 Plus calculator and to convert the decimal coordinates of x and y to fractions, enter the following keystrokes:

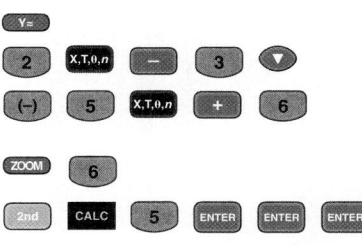

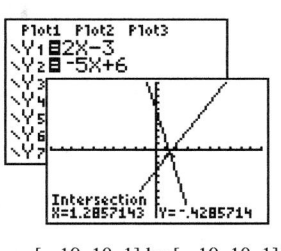

$[-10, 10, 1]$ by $[-10, 10, 1]$

The QUIT feature is the secondary function of the MODE button.

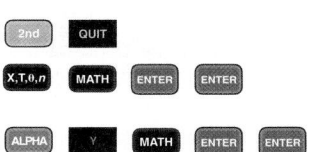

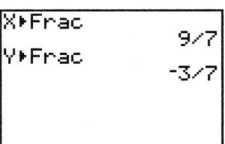

SELF-CHECK 3.3.4

Solve the linear system $\begin{cases} 5x - y = 6 \\ 3x + 2y = 4 \end{cases}$ using a graphics calculator.

One of the prominent uses of systems of linear equations is to solve problems involving two quantities that are both changing. Often changing one variable (the input variable) will control the change in the other variable (the output variable). We now present two examples of this. The first example involves supplementary angles and the second involves the daily cost of using two different pasta machines.

SELF-CHECK 3.3.4 ANSWER

The point of intersection is $\left(\dfrac{16}{13}, \dfrac{2}{13}\right)$.

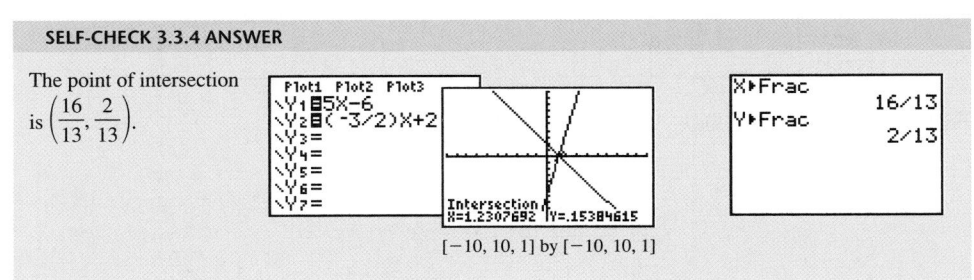

$[-10, 10, 1]$ by $[-10, 10, 1]$

■ EXAMPLE 6   Modeling the Angles in a Metal Part

A metal fabrication shop must assemble a part so that the angles shown in the figure are supplementary. To function properly the specifications for the part require that the larger angle is twice the smaller angle. Determine the number of degrees in each angle.

SOLUTION

Let x = the number of degrees in the smaller angle

Let y = the number of degrees in the larger angle

Select a variable to represent each unknown.

VERBALLY

(1) The angles are supplementary.

(2) The larger angle is twice the smaller angle.

ALGEBRAICALLY

$$x + y = 180$$
$$\text{or } y = 180 - x$$

Angles are supplementary if the total of their measures is 180°.

$$y = 2x$$

Specifications require that the larger angle is twice the smaller angle.

NUMERICALLY

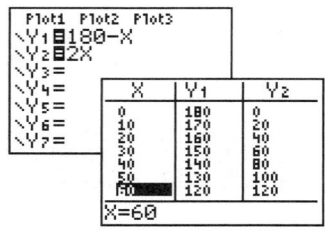

```
Plot1 Plot2 Plot3
\Y1 ■ 180-X
\Y2 ■ 2X
\Y3=
\Y4=
\Y5=
\Y6=
\Y7=
```

X	Y1	Y2
0	180	0
10	170	20
20	160	40
30	150	60
40	140	80
50	130	100
60	120	120

X=60

```
TABLE SETUP
 TblStart=0
 △Tbl=10
Indpnt: Auto Ask
Depend: Auto Ask
```

Enter both equations into a graphics calculator and examine a table of values to determine the solution of the system. Note the table setup that was used for this table.

Answer: The smaller angle is 60° and the larger angle is 120°.

Does this answer check?   ■

You may find it easier to use algebraic methods than to select an appropriate table of values for Example 6. We will revisit this problem in Section 3.4.

SELF-CHECK 3.3.5 ANSWER

The smaller angle is 30° and the larger angle is 60°.

SELF-CHECK 3.3.5

Rework Example 6 with the same conditions except that the angles must be complementary (sum is 90°).

■ EXAMPLE 7   Modeling Business Options with Fixed and Variable Costs

A small restaurant can make a new specialty pasta using two different machines. For the first machine there is a fixed daily cost of $46, plus a variable cost of $5 for each order of pasta produced. The second machine has a fixed daily cost of $22 and a variable cost of $8 for each order produced. Using x to represent the number of orders of pasta produced and $C(x)$ to represent the total daily cost of producing these orders, determine the number of orders for which the daily costs will be the same on the two machines.

SOLUTION

Let x = the number of orders of pasta produced

Let $C(x)$ = the total daily cost of producing x orders of pasta

VERBALLY

Machine 1: Total daily cost equals variable cost plus fixed cost.
Machine 2: Total daily cost equals variable cost plus fixed cost.

ALGEBRAICALLY

$C(x) = 5x + 46$ The function notation $C(x)$ stresses that for the input x, the number of pasta orders,
$C(x) = 8x + 22$ we can calculate the output $C(x)$, the daily cost.

NUMERICALLY

Plot1	Plot2	Plot3
\Y1■5X+46		
\Y2■8X+22		
\Y3=		

X	Y1	Y2
3	61	46
4	66	54
5	71	62
6	76	70
7	81	78
8	86	86
9	91	94

X=8

GRAPHICALLY

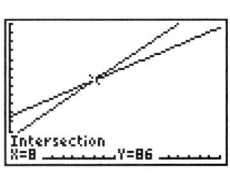

Intersection
X=8   Y=86

[0, 20, 1] by [0, 150, 10]

A hint for selecting an appropriate viewing window is to use the table to estimate the point of intersection and then to set the viewing window centered around this point. It may still be necessary to adjust the scale on either axis.

Write the answers to word problems as full sentences. This helps to reinforce that you understand the question and to increase your ability to use mathematics in the workforce.

Answer: The total daily costs are the same on the two machines when eight orders are produced. ■

USING THE LANGUAGE AND SYMBOLISM OF MATHEMATICS   3.3

1. An ordered pair that satisfies each equation in a system of linear equations is called a _____ of the system of linear equations.

2. Graphically the solution to a system of two linear equations is represented by their point of _____.

3. An _____ system of two linear equations has no solution. Graphically these lines are _____ with no point in common.

4. A consistent system of two independent linear equations has exactly _____ solution. Graphically these lines _____ at exactly _____ point.

5. A consistent system of two dependent linear equations has an _____ _____ of solutions. Graphically these lines _____.

6. Use the linear system $\begin{cases} y = m_1x + b_1 \\ y = m_2x + b_2 \end{cases}$ to select the correct choice that describes the relationship between the two lines given:

   a. $m_1 = m_2$ and $b_1 = b_2$ The lines intersect at **zero/one/ infinitely many** point(s).

   b. $m_1 = m_2$ and $b_1 \neq b_2$ The lines **are/are not** parallel. The lines intersect at **zero/one/ infinitely many** point(s).

   c. $m_1 \neq m_2$ The lines **are/are not** parallel. The lines intersect at **zero/one/ infinitely many** point(s).

EXERCISES 3.3

In Exercises 1–4 determine the point of intersection, the ordered pair that is a solution of the system of linear equations.

1.

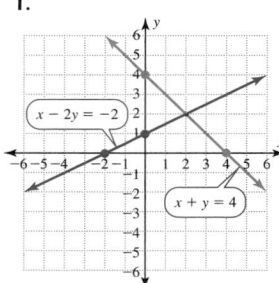

2.

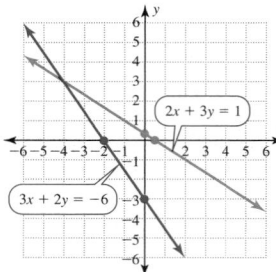

3.

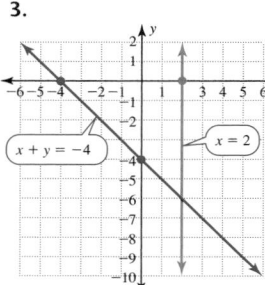

4.

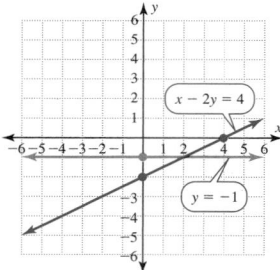

In Exercises 5–12 determine whether the given ordered pair is a solution of the system of linear equations.

5. $(-1, -2)$
$7x - 3y + 1 = 0$
$5x - 4y - 3 = 0$

6. $(-4, 2)$
$4x + 5y + 6 = 0$
$6x + 5y + 14 = 0$

7. $\left(\dfrac{1}{2}, \dfrac{1}{3}\right)$
$4x - 3y = 1$
$6x - 6y = 1$

8. $\left(\dfrac{1}{4}, -\dfrac{2}{3}\right)$
$8x + 3y = 0$
$4x - 3y = 3$

9. $(0.1, -0.2)$
$4x - 3y = 1$
$2x + y = 0$

10. $(-0.3, 0.5)$
$5x + 3y = 0$
$6x - y = -2.3$

11. $(0, 6)$
$2x - y = -6$
$3x + y = 3$

12. $(9, 0)$
$3x - 8y = 27$
$2x + 9y = 18$

13. Using a Graphical Model of Rental Car Costs The following graph compares the cost of a one-day rental for a luxury car at two different rental car companies. The cost is based on the number of miles driven. Give the solution to the corresponding system of equations. Then interpret the meaning of the x- and y-coordinates of this solution.

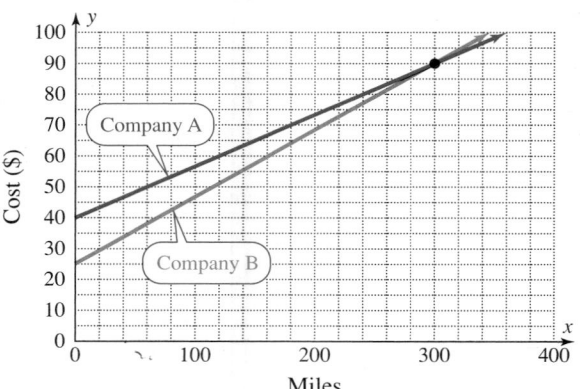

14. Using a Graphical Model of Digital Phone Plans The following graph compares the monthly cost of two different digital phone plans based on the number of minutes of use. Give the solution to the corresponding system of equations. Then interpret the meaning of the x- and y-coordinates of this solution.

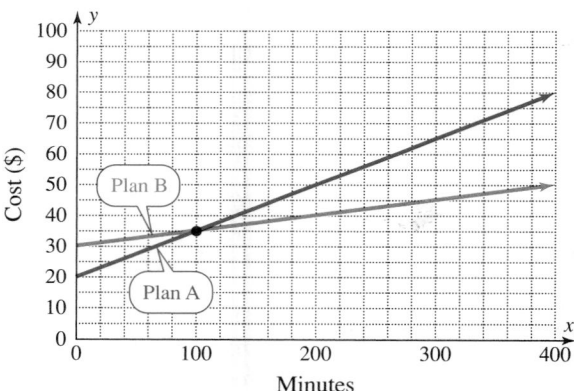

In Exercises 15–30 graphically solve each system of linear equations.

15. $y = x - 6$
$y = -x + 6$

16. $y = x - 1$
$y = -x + 3$

17. $y = -\dfrac{x}{3} + 2$
$y = -\dfrac{x}{2} + 1$

18. $y = -\dfrac{x}{2} + 2$
$y = \dfrac{x}{2}$

19. $x + 3 = 0$
$y - 2 = 0$

20. $y = 2$
$x - 5 = 0$

21. $y = -2x + 10$
$x = 6$

22. $y = 2x - 6$
$y = -4$

23. $y = 2x - 7$
$y = -\dfrac{3}{2}x + 7$

24. $y = -3x + 7$
$y = -\dfrac{2}{5}x + \dfrac{9}{5}$

25. $2x - 3y - 6 = 0$
$4x - 3y + 3 = 0$

26. $5x - 4y + 16 = 0$
$3x - 2y - 6 = 0$

27. $y = 3x$
$y = 3x - 6$

28. $y = 2x + 1$
$y = 2x - 1$

29. $x - 2y - 2 = 0$
$2x - 4y - 4 = 0$

30. $2x - 3y - 3 = 0$
$6x - 9y - 9 = 0$

In Exercises 31–40 use the slope-intercept form of each line to determine the number of solutions of the system and then to classify each system as a consistent system of independent equations, an inconsistent system, or a consistent system of dependent equations.

31. $y = \dfrac{3}{7}x - 17$

$y = -\dfrac{7}{8}x + 8$

32. $y = \dfrac{5}{9}x + 4$

$y = \dfrac{9}{5}x - 4$

33. $y = \dfrac{3}{8}x + 11$

$y = \dfrac{3}{8}x - 5$

34. $y = -\dfrac{9}{5}x + 7$

$y = -\dfrac{9}{5}x + 4$

35. $6x + 2y = 4$
$9x = 6 - 3y$

36. $8x = 12 - 20y$
$6x + 15y = 9$

37. $2x + 3y = 3$
$4x - 3y = -3$

38. $9x + 6y = -12$
$12x + 8y = -16$

39. $\dfrac{x}{6} + \dfrac{y}{4} = \dfrac{1}{3}$

$\dfrac{x}{3} + \dfrac{y}{2} = \dfrac{1}{4}$

40. $\dfrac{x}{4} - \dfrac{y}{5} = 1$

$\dfrac{x}{6} + \dfrac{y}{7} = 1$

Calculator Usage
In Exercises 41–44 solve each linear system using a graphics calculator. See Calculator Perspective 3.3.1.

41. $14x - 7y = -5$
$7x + 21y = 29$

42. $13x + 7y = 8$
$13x - 14y = -7$

43. $6x - 11y = 7$
$3x + 22y = -1$

44. $6x + 12y = -7$
$6x - 8y = 5$

Calculator Usage
In Exercises 45–48 use the graphics calculator display to solve each system of linear equations.

45.

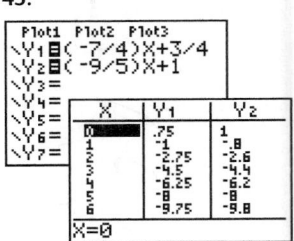

46.

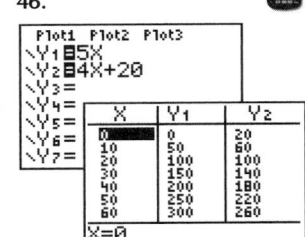

47.

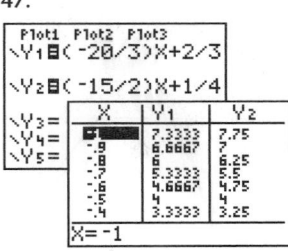

48.

49. Using a Table that Models Flight Distances Two planes are leaving Chicago's O'Hare Airport. A small single-engine plane that travels 150 mi/h leaves at noon. Two hours later a larger commercial jet that travels 450 mi/h takes off and flies in the same direction. The following tables display the distance of each plane from O'Hare Airport, where x is the number of hours past 2 P.M. Give the solution to the corresponding system of equations. Then interpret the meaning of the x- and y-coordinates of this solution.

SMALL PLANE		LARGE JET	
x HOURS	y DISTANCE	x HOURS	y DISTANCE
0	300	0	0
0.5	375	0.5	225
1.0	450	1.0	450
1.5	525	1.5	675
2.0	600	2.0	900
2.5	675	2.5	1125

50. Using a Table that Models Bus Fares When planning a student trip, the student council has to choose between two busing services. Service A charges $4.50 per person, whereas Service B charges a fee of $200 plus $0.50 per person. The tables display the charges for each service based on the number of students on the trip. Give the solution to the corresponding system of equations. Then interpret the meaning of the *x*- and *y*-coordinates of this solution.

SERVICE A

x STUDENTS	y COST ($)
10	45
20	90
30	135
40	180
50	225
60	270

SERVICE B

x STUDENTS	y COST ($)
10	205
20	210
30	215
40	220
50	225
60	230

Calculator Usage

In Exercises 51–54 the graphics calculator display shows a table for a system of two linear equations, $y_1 = m_1x + b_1$ and $y_2 = m_2x + b_2$. Classify each system as a consistent system of independent equations, an inconsistent system, or a consistent system of dependent equations.

51.

X	Y₁	Y₂
1	-1	1
1.2	-.3	.7
1.4	.4	.4
1.6	1.1	.1
1.8	1.8	-.2
2	2.5	-.5
	3.2	-.8

X=2.2

52.

X	Y₁	Y₂
1	-1	-1
1.2	-.3	-.3
1.4	.4	.4
1.6	1.1	1.1
1.8	1.8	1.8
2	2.5	2.5
	3.2	3.2

X=2.2

53.

X	Y₁	Y₂
1	-1	1
1.2	-.3	1.7
1.4	.4	2.4
1.6	1.1	3.1
1.8	1.8	3.8
2	2.5	4.5
	3.2	5.2

X=2.2

54.

X	Y₁	Y₂
-10	28	-28
-5	18	-18
0	8	-8
5	-2	2
10	-12	12
15	-22	22
20	-32	32

X=20

Multiple Representations

In Exercises 55 and 56 complete the numerical and graphical descriptions for each problem and then write a full sentence that answers the question.

55. Verbally

One number is seven more than twice another number. The sum of these numbers is thirteen. Find these numbers.

Algebraically

$y = 2x + 7$
$x + y = 13$

Numerically

x	$y_1 = 2x + 7$	$y_2 = 13 - x$
−3	1	16
−2		
−1		
0		
1		
2		
3	13	10

Graphically

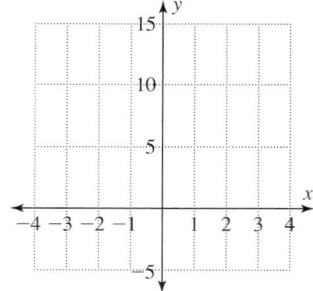

56. Verbally

One number is five less than three times another number. The sum of these numbers is eleven. Find these numbers.

Algebraically

$y = 3x - 5$
$x + y = 11$

Numerically

x	$y_1 = 3x - 5$	$y_2 = 11 - x$
−1	−8	12
0		
1		
2		
3		
4		
5	10	6

Graphically

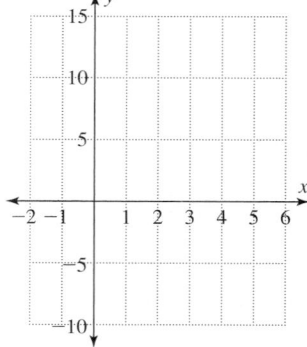

57. Modeling the Angles in Two Parts   The two angles shown in Exercise 58 represent parts that should be assembled so that the angles are supplementary and the larger angle is 5 more than three times the smaller angle. Determine the number of degrees in each angle.

58. Modeling the Angles in Two Parts   The two angles shown here represent parts that should be assembled so that the angles are supplementary and the larger angle is 6 less than twice the smaller angle. Determine the number of degrees in each angle.

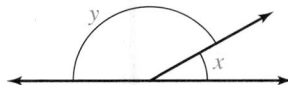

59. Modeling Fixed and Variable Costs   One machine has a fixed daily cost of $75 and a variable cost of $3 per item produced, whereas a second machine has a fixed daily cost of $60 and a variable cost of $4.50 per item produced. Using y to represent the total daily cost of producing these items, determine the number of items x for which the total daily costs will be the same.

60. Modeling Fixed and Variable Costs   Rework Exercise 59 assuming that machine 1 has a fixed cost of $20 and a variable cost of $1.25 per item and machine 2 has a fixed cost of $15 and a variable cost of $1.75 per item.

61. Write a system of two linear equations with a solution of (3, 5) so that one line is horizontal and the other is vertical.

62. Write a system of two linear equations with a solution of $(-4, -2)$ so that one line is horizontal and the other is vertical.

63. Write in the slope-intercept form a system of two linear equations with a solution of (0, 3) so that one line has a slope of $\dfrac{1}{2}$ and the other line has a slope of -4.

64. Write in the slope-intercept form a system of two linear equations with a solution of (0, -2) so that one line has a slope of $\dfrac{1}{3}$ and the other line has a slope of -1.

In Exercises 65 and 66 determine the solution of the system of equations and write the equation of both lines in the general form.

65.

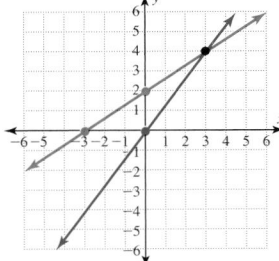

66.

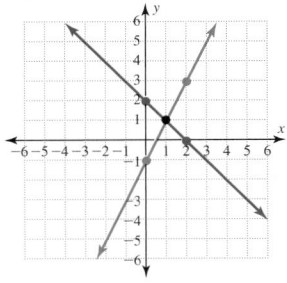

Group Discussion Questions

67. Discovery Question
  a. Consider a system of two linear equations of the form $Ax + By = C$. Can the system have zero solutions? If so, sketch an example.
  b. Consider a system of two linear equations of the form $Ax + By = C$. Can the system have exactly one solution? If so, sketch an example.
  c. Consider a system of two linear equations of the form $Ax + By = C$. Can the system have exactly two solutions? If so, sketch an example.
  d. Consider a system of two linear equations of the form $Ax + By = C$. Can the system have exactly three solutions? If so, sketch an example.
  e. Consider a system of two linear equations of the form $Ax + By = C$. Can the system have more than three solutions? If so, sketch an example.

68. Discovery Question
  a. Consider a system of three linear equations of the form $Ax + By = C$. Can the system have zero solutions? If so, sketch an example.
  b. Consider a system of three linear equations of the form $Ax + By = C$. Can the system have exactly one solution? If so, sketch an example.
  c. Consider a system of three linear equations of the form $Ax + By = C$. Can the system have exactly two solutions? If so, sketch an example.
  d. Consider a system of three linear equations of the form $Ax + By = C$. Can the system have exactly three solutions? If so, sketch an example.
  e. Consider a system of three linear equations of the form $Ax + By = C$. Can the system have more than three solutions? If so, sketch an example.

69. Error Analysis   A student graphed the following system and concluded that the lines were parallel and the system was inconsistent with no solution. Describe the error that the student has made and then solve this system of equations.

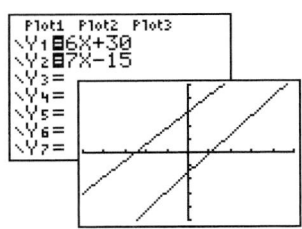

$[-10, 10, 2]$ by $[-50, 50, 10]$

Section 3.4 Solving Systems of Linear Equations in Two Variables by the Substitution Method

Objective:

8. Solve a system of linear equations by the substitution method.

A Mathematical Note

Grace Hopper (1906–1992) was one of the foremost pioneers in the field of computer programming. She was one of the developers of the COBOL programming language. She retired from the U.S. Navy in 1986 as a rear admiral 43 years after joining as a lieutenant. In 1977 the author James Hall had the pleasure of hearing her humorously describe what she called the first computer bug—a moth that in 1945 flew into a relay of the early Mark II computer and caused it to malfunction. For a photo of this moth see **http://ei.cs.vt.edu/~history/ Bug.GIF.**

Although the graphical method is an excellent means of visualizing the solution of a system of equations, it has some limitations. Even with a computer or graphics calculator, this method can be time-consuming. This method also can produce some error because of limitations in estimating the point of intersection. Thus algebraic methods that are quicker and yield exact solutions often are preferred. The first algebraic method presented in this chapter is based on the **substitution principle,** which states that a quantity may be substituted for its equal. The substitution method is particularly appropriate when it is easy to solve one equation for either x or y.

Substitution Method

STEP 1. Solve one of the equations for one variable in terms of the other variable.

STEP 2. Substitute the expression obtained in step 1 into the other equation (eliminating one of the variables), and solve the resulting equation.

STEP 3. Substitute the value obtained in step 2 into the equation obtained in step 1 (back-substitution) to find the value of the other variable.

The ordered pair obtained in steps 2 and 3 is the solution.

■ **EXAMPLE 1** Solving a Linear System by the Substitution Method

Solve $\begin{cases} 3x + 4y = 8 \\ x + 2y = 10 \end{cases}$ by the substitution method.

SOLUTION

The substitution method, which is shown here for systems of linear equations, also can be used to solve some nonlinear systems.

The addition method for solving linear systems is covered in the next section.

1	$x + 2y = 10$	Solve the second equation for x. (Note that this is relatively easy because x has a coefficient of 1 in this equation.)
	$x = 10 - 2y$	
2	$3x + 4y = 8$	
	$3(10 - 2y) + 4y = 8$	Substitute for x in the first equation. (Note that x has been eliminated.) Then solve this equation for y.
	$30 - 6y + 4y = 8$	
	$30 - 2y = 8$	
	$-2y = 8 - 30$	
	$-2y = -22$	
	$y = 11$	This is the y-coordinate of the solution.
3	$x = 10 - 2y$	
	$x = 10 - 2(11)$	Back-substitute 11 for y into the equation that was solved for x in step 1.
	$x = 10 - 22$	
	$x = -12$	This is the x-coordinate of the solution.

The answer to a consistent system of independent linear equations should be written in ordered pair notation. This notation emphasizes that the system has only one solution and that the x- and y-coordinates must be used in the correct order.

Check:

First Equation	*Second Equation*
$3x + 4y = 8$	$x + 2y = 10$
$3(-12) + 4(11) \stackrel{?}{=} 8$	$-12 + 2(11) \stackrel{?}{=} 10$
$-36 + 44 \stackrel{?}{=} 8$	$-12 + 22 \stackrel{?}{=} 10$
$8 \stackrel{?}{=} 8$ checks.	$10 \stackrel{?}{=} 10$ checks.

Answer: $(-12, 11)$

The next example revisits Example 6 from the previous section. Solving even a relatively simple system of linear equations using a graphics calculator can take some time as we decide what input values will generate a useful table or graph. Fortunately, the substitution method provides an algebraic solution that is quick and exact. This is illustrated in Example 2. (Compare this to Example 6 in Section 3.3.)

■ EXAMPLE 2   Determining the Number of Degrees in Two Supplementary Angles

A metal fabrication shop must assemble a part so that the angles shown in the figure are supplementary. To function properly the specifications for the part require that the larger angle is twice the smaller angle. Determine the number of degrees in each angle.

SOLUTION ⎯⎯⎯⎯⎯⎯⎯⎯⎯⎯⎯⎯⎯⎯⎯⎯⎯⎯⎯⎯⎯⎯⎯⎯⎯⎯⎯⎯⎯⎯

Let $x =$ the number of degrees in the smaller angle
Let $y =$ the number of degrees in the larger angle

Select a variable to represent each unknown.

VERBALLY	ALGEBRAICALLY	
(1) The angles are supplementary.	$x + y = 180$	Angles are supplementary if the total of their measures is 180°.
(2) The larger angle is twice the smaller angle.	$y = 2x$	
	$x + 2x = 180$	Solve the system using the substitution method. Substitute $2x$ for y in the first equation and then solve for x.
	$3x = 180$	
	$x = 60$	
	$y = 2x$	
	$y = 2(60)$	Back-substitute 60 for x in the second equation.
	$y = 120$	

Answer: The smaller angle is 60° and the larger angle is 120°.

Does this answer check? ■

SELF-CHECK 3.4.1

1. Rework Example 2 with the same conditions, except that the angles must be complementary (sum is 90°).

2. Solve $\begin{cases} 3x + y = 6 \\ 2x + y = 2 \end{cases}$ by the substitution method.

A system of equations that contains fractional coefficients may be easier to solve if we first convert the coefficients to integers by multiplying through by the LCD. For example, multiplying both sides of $\frac{x}{4} + \frac{y}{3} = 1$ by 12 yields $3x + 4y = 12$, which has integer coefficients.

■ **EXAMPLE 3**   Solving a Linear System with Fractional Coefficients

Solve $\begin{cases} \dfrac{x}{2} + \dfrac{y}{6} = \dfrac{2}{3} \\ \dfrac{x}{4} - \dfrac{y}{5} = \dfrac{7}{4} \end{cases}$.

SOLUTION ───────────────────────────────────

First Equation $\quad$ *Second Equation*	We start by producing a simpler system of equations with integer coefficients.

First Equation

$$\frac{x}{2} + \frac{y}{6} = \frac{2}{3}$$

$$6\left(\frac{x}{2} + \frac{y}{6}\right) = 6\left(\frac{2}{3}\right)$$

$$6\left(\frac{x}{2}\right) + 6\left(\frac{y}{6}\right) = 6\left(\frac{2}{3}\right)$$

(1) $\quad 3x + y = 4$

Second Equation

$$\frac{x}{4} - \frac{y}{5} = \frac{7}{4}$$

$$20\left(\frac{x}{4} - \frac{y}{5}\right) = 20\left(\frac{7}{4}\right)$$

$$20\left(\frac{x}{4}\right) - 20\left(\frac{y}{5}\right) = 20\left(\frac{7}{4}\right)$$

(2) $\quad 5x - 4y = 35$

Multiply both sides of the first equation by the LCD of 6 and both sides of the second equation by the LCD of 20.

(1) $\begin{cases} 3x + y = 4 \\ 5x - 4y = 35 \end{cases}$
(2)

This system with integer coefficients has the same solution as the original system of equations.

(1) $\qquad 3x + y = 4$
$\qquad\qquad\quad y = 4 - 3x$

Solve Equation **(1)** for y since y has a coefficient of 1.

(2) $\qquad 5x - 4y = 35$
$\qquad 5x - 4(4 - 3x) = 35$
$\qquad 5x - 16 + 12x = 35$
$\qquad\quad 17x - 16 = 35$
$\qquad\qquad\quad 17x = 51$
$\qquad\qquad\qquad x = 3$

Substitute for y in Equation **(2)**. Then solve this equation for x.

This is the x-coordinate of the solution.

(3) $\qquad\qquad y = 4 - 3x$
$\qquad\qquad y = 4 - 3(3)$
$\qquad\qquad y = 4 - 9$
$\qquad\qquad y = -5$

Back-substitute 3 for x in the equation that was solved for y in step 1.

This is the y-coordinate of the solution.

Answer: $(3, -5)$

Does this solution check in the original equations? ■

SELF-CHECK 3.4.2

Solve $\begin{cases} \dfrac{x}{2} + \dfrac{y}{3} = -4 \\ \dfrac{x}{6} - \dfrac{y}{6} = -3 \end{cases}$

Examples 4 and 5 illustrate what will happen when the substitution method is used to solve an inconsistent system of linear equations or a consistent system of dependent linear equations.

■ EXAMPLE 4   Solving an Inconsistent System

Solve $\left\{ \begin{array}{r} y = 3x - 6 \\ 6x - 2y = 9 \end{array} \right\}$ by the substitution method.

SOLUTION _____

$$y = 3x - 6$$ The first equation is already solved for y.
$$6x - 2y = 9$$
$$6x - 2(3x - 6) = 9$$ Substitute for y in the second equation.
$$6x - 6x + 12 = 9$$ Note that the simplification of this equation eliminates all the variables and produces a contradiction.

$$12 = 9 \text{ (a contradiction)}$$ The conditions given by the two equations are contradictory. Thus this system has no solution; it is an inconsistent system.

Answer: No solution.   ◼

■ EXAMPLE 5   Solving a Consistent System of Dependent Equations

Solve $\left\{ \begin{array}{r} 2x = 4y + 6 \\ 3x - 6y = 9 \end{array} \right\}$ by the substitution method.

SOLUTION _____

$$2x = 4y + 6$$ Solve the first equation for x.
$$x = 2y + 3$$
$$3x - 6y = 9$$
$$3(2y + 3) - 6y = 9$$ Then substitute this value for x into the second equation.
$$6y + 9 - 6y = 9$$ Note that the simplification of this equation eliminates all variables and produces an identity.

$$9 = 9 \text{ (an identity)}$$ This identity means that the system contains dependent equations and has an infinite number of solutions.

$(3, 0), \left(0, -\dfrac{3}{2} \right)$ and $(-1, -2)$ are three of the points that lie on the common line and satisfy both equations. $(0, 0), (1, 1),$ and $(2, 3)$ are three points that do not lie on the common line and do not satisfy either equation.

Answer: An infinite number of solutions.   ◼

SELF-CHECK 3.4.3 ANSWER

An infinite number of solutions; three of these solutions are $(2.5, 0), (0, -2),$ and $(1.25, -1)$.

SELF-CHECK   3.4.3

Solve $\left\{ \begin{array}{r} 5y = 4x - 10 \\ 12x - 15y = 30 \end{array} \right\}$.

The following box summarizes what happens when each of the three types of linear systems is solved by the substitution method.

Algebraic Solution of the Three Types of Linear Systems

1. **Consistent system of independent equations:** The solution process will produce unique *x*- and *y*-values.
2. **Inconsistent system:** The solution process will produce a contradiction.
3. **Consistent system of dependent equations:** The solution process will produce an identity.

The next example compares the algebraic, numerical, and graphical solutions of a consistent system of dependent equations.

■ EXAMPLE 6   Using Multiple Perspectives to Examine a System of Dependent Equations

Solve $\begin{cases} y = 2x - 3 \\ 6x - 3y = 9 \end{cases}$ algebraically, numerically, and graphically. Then describe the answer verbally.

SOLUTION _____

ALGEBRAICALLY

Substitution Method

(1) $\qquad\qquad y = 2x - 3$
(2) $\qquad\quad 6x - 3y = 9$
$\qquad 6x - 3(2x - 3) = 9$
$\qquad\quad 6x - 6x + 9 = 9$
$\qquad\qquad\qquad 9 = 9$

This identity signals a system of dependent equations.

NUMERICALLY

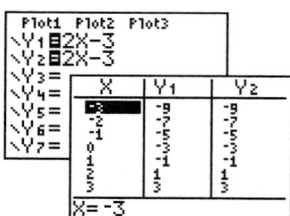

Writing Both Equations in Slope-Intercept Form

(1) $\qquad\quad y = 2x - 3$
(2) $\quad 6x - 3y = 9$
$\qquad\quad -3y = -6x + 9$
$\qquad\qquad y = 2x - 3$

Equation (2) simplified to exactly the same form as Equation (1). Thus this is a system of dependent equations.

GRAPHICALLY

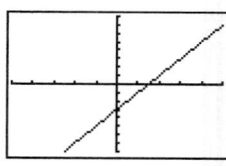

$[-5, 5, 1]$ by $[-8, 8, 1]$

VERBALLY

Answer: There is an infinite number of solutions that satisfy both equations. Because both equations are just different forms of the same equation, they produce exactly the same output values, the same table of values, and the same graph.   ■

The mean of a set of test scores is one type of average for these scores. The range is the difference between the high and low scores. The next example involves both the mean and the range of two test scores.

■ EXAMPLE 7   Determining Test Scores with a Given Mean and Range

A student has two test scores in an algebra class. The mean of these test scores is 82 and their range is 12. Use this information to determine both test scores.

SOLUTION _____

Let x = the higher test score

Let y = the lower test score

VERBALLY		**ALGEBRAICALLY**
(1)	The mean of the two test scores is 82.	$\dfrac{x + y}{2} = 82$
(2)	The range of the two test scores is 12.	$x - y = 12$

1	$x - y = 12$	Solve the second equation for x.
	$x = y + 12$	
2	$\dfrac{x + y}{2} = 82$	Simplify the first equation by multiplying both sides by the LCD of 2.
	$x + y = 164$	
	$(y + 12) + y = 164$	Then substitute for x in Equation **(2)**.
	$2y + 12 = 164$	Simplify this equation and solve for y.
	$2y = 152$	
	$y = 76$	This is the lower of the two test scores.
3	$x = y + 12$	
	$x = 76 + 12$	Back-substitute 76 for y into the equation that was solved for x in step 1.
	$x = 88$	This is the higher of the two test scores.

The answer to a word problem should be written as a full sentence. The question in Example 7 did not mention either x or y, so the answer should not use these symbols that were introduced in the solution process.

Answer: The two scores are 88 and 76.   ■

SELF-CHECK 3.4.4

Rework Example 7 assuming a mean of 82 and a range of 20.

SELF-CHECK 3.4.4 ANSWER

The two scores are 92 and 72.

The next example is solved by using the substitution method. The solution process includes the use of fractions, which many students would prefer to avoid. This is one reason that we will examine the addition method for solving this system in the next section.

■ EXAMPLE 8 Solving a Linear System by the Substitution Method

Solve $\begin{cases} 3x - 2y = 5 \\ 9x + 4y = -10 \end{cases}$ by the substitution method.

SOLUTION ──────────────────────────────

⬚1 $3x - 2y = 5$

$-2y = -3x + 5$

$y = \dfrac{3}{2}x - \dfrac{5}{2}$

The coefficient of y is the smallest in magnitude, so we elect to solve for y.

⬚2 $9x + 4y = -10$

$9x + 4\left(\dfrac{3}{2}x - \dfrac{5}{2}\right) = -10$

$9x + 6x - 10 = -10$

$15x = 0$

$x = \dfrac{0}{15}$

$x = 0$

Substitute this expression for y in the second equation. Then solve this equation for x.

This is the x-coordinate of the solution.

⬚3 $y = \dfrac{3}{2}x - \dfrac{5}{2}$

$y = \dfrac{3}{2}(0) - \dfrac{5}{2}$

$y = 0 - \dfrac{5}{2}$

$y = -\dfrac{5}{2}$

Back-substitute 0 for x in the equation that was solved for y in step 1.

This is the y-coordinate of the solution.

Answer: $\left(0, -\dfrac{5}{2}\right)$

Does this answer check? ■

USING THE LANGUAGE AND SYMBOLISM OF MATHEMATICS 3.4

1. The substitution principle states that a quantity may be substituted for its _____.
2. The solution for a system of two linear equations in two variables should be written in _____ _____ notation.
3. A contradiction is an equation that is always _____.
4. An identity is an equation that is always _____.
5. The solution of a consistent system of independent linear equations by the substitution method will produce **a conditional equation/a contradiction/an identity.** (Select the correct choice.)
6. The solution of a consistent system of dependent linear equations by the substitution method will produce **a conditional equation/a contradiction/an identity.** (Select the correct choice.)
7. The solution of an inconsistent system of independent linear equations by the substitution method will produce **a conditional equation/a contradiction/an identity.** (Select the correct choice.)
8. Using the substitution method produced the equations $x = 3$ and $y = 4$. Thus the system of two linear equations is a _____ system of _____ linear equations with one solution that is the ordered pair _____.
9. Using the substitution method produced the equation $3 = 4$. Thus the system of two linear equations is an _____ system of _____ linear equations with _____ solution.
10. Using the substitution method produced the equation $3 = 3$. Thus the system of two linear equations is a _____ system of _____ linear equations with an _____ number of solutions.

EXERCISES   3.4

In Exercises 1–42 solve each system of linear equations by the substitution method.

1. $y = 2x - 3$
$x + y = 9$

2. $y = 3x - 4$
$x + 2y = 6$

3. $y = 5x - 10$
$2x + 3y = 4$

4. $y = 9 - 3x$
$x + 2y = 8$

5. $x = 6 - 5y$
$2x + 9y = 4$

6. $x = 4y - 9$
$2x + 3y = 15$

7. $3x - y = 1$
$x + 2y = 2$

8. $x - 3y = -14$
$2x + y = 7$

9. $2x - y - 6 = 0$
$x - y - 6 = 0$

10. $2x + y - 11 = 0$
$x + 2y - 13 = 0$

11. $2x + 3y = -7$
$x + 4y = -6$

12. $3x + 4y = -2$
$x - 4y = -6$

13. $2x - 5y = 9$
$x - 3 = 0$

14. $4x - 7y = -1$
$x + 2 = 0$

15. $3x - 2y = 1$
$y + 7 = 0$

16. $5x - 4y = 3$
$y - 4 = 0$

17. $y = -2x$
$5x - y = -7$

18. $y = 6x$
$8x - 5y = -11$

19. $3x - 2y = 0$
$11x - 9y = 5$

20. $5x + 4y = 0$
$6x + 5y = 2$

21. $x + y = 21$
$x - y = 3$

22. $2x + y = 3$
$x + 2y = 9$

23. $2x - 7y = 42$
$x = 0$

24. $5x - 9y = 35$
$y = 0$

25. $5x - 2y = 11$
$3x + 3y = 15$

26. $3x + 4y + 2 = 0$
$5x - 20y + 30 = 0$

27. $5x - 4y + 12 = 0$
$2x - 3y + 2 = 0$

28. $3x + 4y + 5 = 0$
$-2x + 5y - 11 = 0$

29. $y = x + 3$
$\dfrac{x}{2} - \dfrac{y}{5} = 3$

30. $y = x + 4$
$\dfrac{x}{4} - \dfrac{y}{2} = 0$

31. $x = 2y + 4$
$3x - 6y = 12$

32. $y = 4 - 2x$
$6x + 3y = 12$

33. $y = 5x + 3$
$10x - 2y = 6$

34. $x = 5y - 3$
$4x - 20y = -1$

35. $4x + 7y = 0$
$7x - 4y = 0$

36. $9x - 5y = 0$
$5x + 9y = 0$

37. $x = 2y - 5$

$\dfrac{x}{6} + \dfrac{y}{8} = 1$

38. $\dfrac{x}{6} + \dfrac{y}{12} = \dfrac{1}{6}$

$\dfrac{x}{8} - \dfrac{y}{4} = -2$

39. $\dfrac{x}{2} + \dfrac{y}{5} = \dfrac{4}{5}$

$\dfrac{x}{6} - \dfrac{y}{2} = \dfrac{5}{6}$

40. $\dfrac{x}{8} - \dfrac{y}{5} = \dfrac{1}{10}$

$y = \dfrac{5}{8}x - \dfrac{1}{4}$

41. $\dfrac{x}{4} - \dfrac{y}{2} = \dfrac{7}{24}$

$\dfrac{x}{3} + \dfrac{y}{2} = 0$

42. $\dfrac{x}{2} - \dfrac{y}{3} = -\dfrac{7}{12}$

$\dfrac{x}{8} + \dfrac{y}{9} = 0$

In Exercises 43–46 select the system of linear equations (choices A, B, C, or D) that represents the word problem. Then solve this system by the substitution method and answer the problem using a full sentence.

Verbally

43. Find two numbers whose sum is 120 and whose difference is 30.

44. Find two numbers whose sum is 30 and the first number is 120 more than the second number.

45. Find two numbers whose sum is 30 and whose difference is 30.

46. Find two numbers whose sum is 20 and whose difference is 30.

Algebraically

A. $x + y = 20$
$x - y = 30$

B. $x + y = 120$
$x - y = 30$

C. $x + y = 30$
$x - y = 30$

D. $x + y = 30$
$x - y = 120$

Multiple Representations

In Exercises 47 and 48 complete the numerical and graphical descriptions for each problem and then write a full sentence that answers the question.

47. Verbally

The sum of two numbers is 8 and their difference is 2. Find these numbers.

Algebraically

$x + y = 8$
$x - y = 2$

Numerically

x	$y_1 = 8 - x$	$y_2 = x - 2$
2	6	0
3		
4		
5		
6		
7		
8	0	6

Graphically

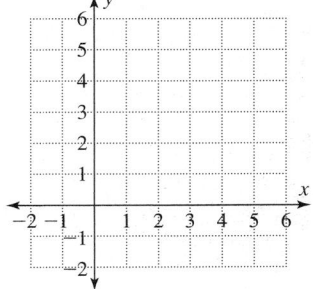

48.

Verbally	Algebraically	Numerically			Graphically

Verbally	Algebraically
The sum of two numbers is 2 and their difference is 6. Find these numbers.	$x + y = 2$ $x - y = 6$

Numerically

x	$y_1 = 2 - x$	$y_2 = x - 6$
0	2	-6
1		
2		
3		
4		
5		
6	-4	0

Graphically

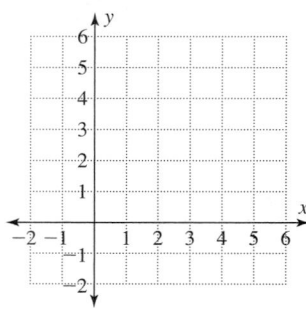

49. Modeling the Angles in Two Parts To function properly, two parts must be assembled so that the two angles are complementary and one angle is 18° larger than the other angle. Determine the number of degrees in each angle. (See Exercise 50.)

50. Modeling the Angles in Two Parts To function properly, two parts must be assembled so that the two angles are complementary as shown in the figure and one angle is 12° more than twice the other angle. Determine the number of degrees in each angle.

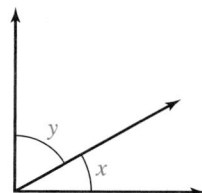

Mean and Range

In Exercises 51–54 write a system of linear equations using the variables x and y and use this system of equations to solve the problem.

51. A student has two test scores in a psychology class. The mean of these scores is 76 and their range is 28. Use this information to determine both test scores.

52. A student has two test scores in a history class. The mean of these scores is 90 and their range is 12. Use this information to determine both test scores.

53. Tiger Woods had identical scores for three of the first four rounds at a PGA golf tournament. After shooting a tournament record in the final round his mean score was 66 and the range of his scores was 8. Use this information to determine his score in each of the four rounds.

54. An amateur playing in a pro-am golf tournament in Scottsdale, Arizona, had identical scores for three of the first four rounds to enter the final day of play in first place. However, her score on the final day was her highest of the four rounds, and she finished fifth for the tournament. Her mean score was 73 and the range of her scores was 8. Use this information to determine her score in each of the four rounds.

55. Modeling the Costs of Two Checking Plans First Federal has two personal checking account plans. The first plan costs $10 per month and 20 cents per check. The second plan costs $20 per month and 10 cents per check. Using x to represent the number of checks and y to represent the monthly cost, write a system of linear equations representing the cost of each plan. Use this system of equations to determine the conditions for which these two plans will have the same monthly cost. What is this cost?

56. Modeling the Tuition at Two Colleges The college tuition at one state university is $95 per credit hour plus a student activity fee of $200. The tuition at their arch rival across the state is $100 per credit hour plus a student activity fee of $125. Using x to represent the number of credit hours and y to represent the total cost of tuition and fees, write a system of equations representing the cost of a semester of tuition and fees at each university. Use this system of equations to determine the conditions for which these costs will be the same. What is this cost?

57. Using a Graphical Model for Taxi Fares The following graph compares the cost of a taxi ride for two different taxicab companies. The cost is based on the number of miles driven. Use the y-intercept and an additional point to determine the equation of each line. Then solve the system of equations and interpret the meaning of the x- and y-coordinates of this solution.

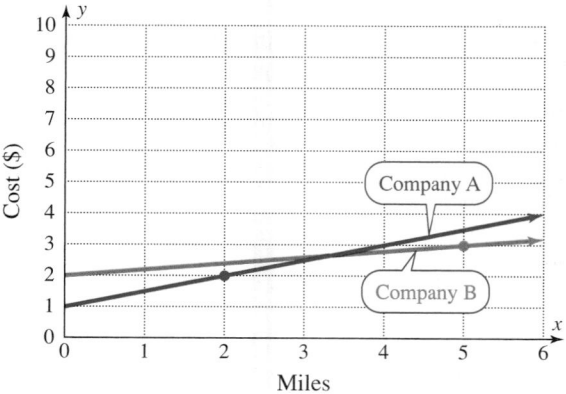

58. Using a Graphical Model for Population Growth Plainsville and Springfield were both founded in 1900. The graph compares the populations of each city based on the number of years since 1900. Use the y-intercept and an additional point to determine the equation of each line. Then solve the system of equations and interpret the meaning of the *x*- and *y*-coordinates of this solution.

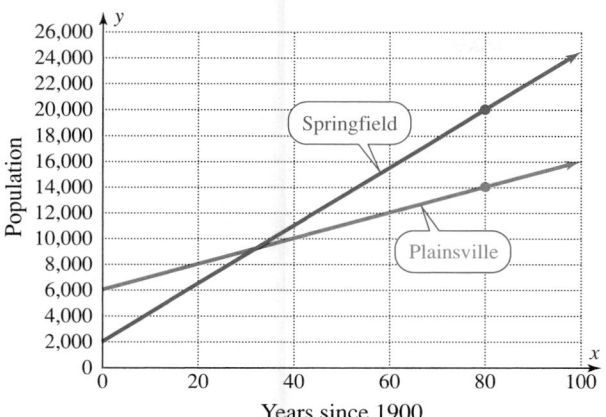

Years since 1900

59. Using a Table that Models Investment Growth   The following tables display the values of two different investments made in 1975. Investment A was an initial deposit of $1500 and has earned a simple interest rate of 12%. Investment B was an initial deposit of $2000 and has earned a simple interest rate of 7%. Use the *y*-intercept and an additional point to determine the equation of each line where *x* represents the number of years since 1975. Then solve the system of equations and interpret the meaning of the *x*- and *y*-coordinates of this solution.

INVESTMENT A

x YEAR	y VALUE
0	1500
5	2400
10	3300
15	4200
20	5100
25	6000

INVESTMENT B

x YEAR	y VALUE
0	2000
5	2700
10	3400
15	4100
20	4800
25	5500

60. Using a Table that Models Sales Growth   The following tables display the sales in millions of dollars of two companies founded in 1980. Use the *y*-intercept and an additional point to determine the equation of each line where *x* represents the number of years since 1980. Then solve the system of equations and interpret the meaning of the *x*- and *y*-coordinates of this solution.

COMPANY A

x YEAR	y SALES ($) (in millions)
0	5
5	22.5
10	40
15	57.5
20	75

COMPANY B

x YEAR	y SALES ($) (in millions)
0	17
5	27
10	37
15	47
20	57

Inconsistent Systems

In Exercises 61 and 62 determine the value of *B* so that each system of equations will be an inconsistent system.

61. $4x - 8y = 12$
$3x + By = 10$

62. $4x + 2y = 10$
$10x + By = 20$

Systems of Dependent Equations

In Exercises 63 and 64 determine the value of *C* so that each system of equations will have dependent equations.

63. $6x - 15y = 9$
$4x - 10y = C$

64. $5x - 15y = 10$
$4x - 12y = C$

Calculator Usage

In Exercises 65–68 use a calculator to assist you in solving each system of linear equations.

65. $y = 1.05x - 2.1325$
$7.14x - 8.37y = 10.1835$

66. $y = 4.55x - 10.696$
$2.08x - 1.17y = 1.0972$

67. $x = 4.91y - 5.899$
$2.1x - 9.9y = -12.84$

68. $2.409x + 2.409y = 4.818$
$3.56x - 3.56y = 14.24$

Group Discussion Questions

69. Discovery Question

Given the system of equations $\begin{cases} x - 5y = -1 \\ 3x - y = 11 \end{cases}$:

a. Solve this system by solving the first system for *x* and then substituting for *x* in the second equation.
b. Solve this system by solving the second system for *y* and then substituting for *y* in the first equation.
c. Compare these solutions and generalize about implementing the substitution method for other systems of linear equations.

70. Discovery Question   Given the system $\begin{cases} x + 4y = -5 \\ 2x + 7y = 10 \end{cases}$, have part of your group solve this system using a graphical method and have the rest of the group solve this system using the substitution method. Time each group to determine which method seems more efficient for this problem. Did both groups get the same answer?

71. Challenge Question   Solve each of these systems for (x, y) in terms of *a* and *b* for $a \neq 0$.

a. $x - ay = b$
$x + ay = 2b$

b. $3ax + y = b$
$2ax - y = 4b$

Section 3.5 Solving Systems of Linear Equations in Two Variables by the Addition Method

Objective: 9. Solve a system of linear equations by the addition method.

As noted in Section 3.4 the substitution method is well suited to systems that contain at least one variable with a coefficient of 1 or -1. For other systems it may be easier to use the addition method, which is described in the following box. This method is based on the **addition-subtraction principle of equality,** which states that equal values can be added to or subtracted from both sides of an equation to produce an equivalent equation. This method is also called the **elimination method** because the strategy is to eliminate a variable in one of the equations.

Addition Method

STEP 1. Write both equations in the general form $Ax + By = C$.

STEP 2. If necessary, multiply each equation by a constant so that the equations have one variable for which the coefficients are additive inverses.

STEP 3. Add the new equations to eliminate a variable and then solve the resulting equation.

STEP 4. Substitute this value into one of the original equations (back-substitution), and solve for the other variable.

The ordered pair obtained in steps 3 and 4 is the solution that should check in both equations.

■ EXAMPLE 1   Solving a Linear System by the Addition Method

Solve $\left\{ \begin{array}{l} 2x + y = 10 \\ 3x - y = 5 \end{array} \right\}$ by the addition method.

SOLUTION

Since $3x - y = 5$, $3x - y$ can be added to one side and 5 can be added to the other side of a balanced scale and maintain the balance.

$$
\begin{array}{ll}
2x + y = 10 & \text{The equations are already in the proper form, and the} \\
\underline{3x - y = 5} & \text{coefficients of } y \text{ are additive inverses. Add these} \\
5x = 15 & \text{equations to eliminate } y. \text{ Solve this equation for } x. \\
x = 3 & \text{This is the } x\text{-coordinate of the solution.}
\end{array}
$$

$$
\begin{array}{ll}
2x + y = 10 & \\
2(3) + y = 10 & \text{Back-substitute 3 for } x \text{ in the first equation. Then} \\
6 + y = 10 & \text{solve for } y. \\
y = 4 & \text{This is the } y\text{-coordinate of the solution.}
\end{array}
$$

$2x + y$ $=$ 10

$(2x + y) + (3x - y) =$ $10 + 5$

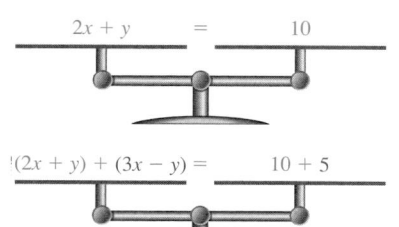

Check:

First Equation	Second Equation
$2x + y = 10$	$3x - y = 5$
$2(3) + 4 \overset{?}{=} 10$	$3(3) - 4 \overset{?}{=} 5$
$6 + 4 \overset{?}{=} 10$	$9 - 4 \overset{?}{=} 5$
$10 \overset{?}{=} 10$ checks.	$5 \overset{?}{=} 5$ checks.

Please remember to write the solution using ordered-pair notation.

Answer: (3, 4)   ◼

The next example revisits Example 8 from the previous section. When you compare these two solutions, you likely will prefer the addition method shown next.

If you wish to compare the substitution method to the addition method, you may want to examine Example 2 side-by-side with Example 8 from Section 3.4.

◼ **EXAMPLE 2**   Solving a Linear System by the Addition Method

Solve $\begin{cases} 3x - 2y = 5 \\ 9x + 4y = -10 \end{cases}$.

SOLUTION _____

$3x - 2y = 5$	$6x - 4y = 10$	Multiply both sides of the first equation by 2 to obtain
$9x + 4y = -10$	$\underline{9x + 4y = -10}$	coefficients of y that are additive inverses. Add these
		equations to eliminate y.
	$15x \qquad = 0$	Divide both sides of the equation by 15 to solve for x.
	$x = 0$	This is the x-coordinate of the solution.

$3x - 2y = 5$
$3(0) - 2y = 5$ Back-substitute 0 for x in the first equation of the original
$-2y = 5$ system. Then solve for y.

$y = -\dfrac{5}{2}$ This is the y-coordinate of the solution.

Answer: $\left(0, -\dfrac{5}{2}\right)$ This is the same solution we obtained in Example 8 in
                                                                        Section 3.4.   ◼

In Example 3 both equations are multiplied by a constant so that the coefficients of y will be additive inverses. The choice to eliminate y is arbitrary since it would be just as easy to eliminate x.

◼ **EXAMPLE 3**   Solving a Linear System by the Addition Method

Solve $\begin{cases} 5x + 3y = -6 \\ 3x + 2y = -5 \end{cases}$ by the addition method.

SOLUTION _____

$5x + 3y = -6$	$10x + 6y = -12$	Multiply both sides by 2.
$3x + 2y = -5$	$\underline{-9x - 6y = 15}$	Multiply both sides by -3.
	$x \qquad = 3$	Add these equations to eliminate y.

$$5x + 3y = -6$$
$$5(3) + 3y = -6 \qquad \text{Back-substitute 3 for } x \text{ in the first equation of the original system.}$$
$$15 + 3y = -6$$
$$3y = -21$$
$$y = -7 \qquad \text{Then solve for } y.$$

Answer: $(3, -7)$ Does this solution check?

Although any common multiple of the coefficients of y can be used to eliminate the y variable, we encourage you to use the least common multiple of the coefficients.

In Example 4 the decision to eliminate y means that we need to produce new coefficients of y that are multiples of both of the original coefficients of y. Since 12 is the least common multiple of both 3 and 4, we use this to guide the steps taken in this example.

■ EXAMPLE 4   Solving a Linear System by the Addition Method

Solve $\begin{cases} 7x = 3y + 8 \\ 4y = -5(x + 5) \end{cases}$ by the addition method.

SOLUTION

$$7x = 3y + 8 \qquad 7x - 3y = 8 \qquad 28x - 12y = 32 \qquad \text{First write both equations in the form } Ax + By = C.$$
$$4y = -5(x + 5) \qquad 5x + 4y = -25 \qquad \underline{15x + 12y = -75} \qquad \text{To produce the coefficients of } -12 \text{ and 12 for } y, \text{ multiply the}$$
$$43x \qquad\quad = -43 \qquad \text{first equation by 4 and the second equation by 3. Add these}$$
$$x = -1 \qquad \text{equations to eliminate } y.$$
$$\text{Solve for } x.$$

$$4y = -5(x + 5)$$
$$4y = -5(-1 + 5) \qquad\qquad\qquad\qquad\qquad \text{Back-substitute } -1 \text{ for } x \text{ in the second equation of the original}$$
$$4y = -5(4) \qquad\qquad\qquad\qquad\qquad\qquad\quad \text{system.}$$
$$4y = -20$$
$$y = -5 \qquad\qquad\qquad\qquad\qquad\qquad\qquad \text{Then solve for } y.$$

Answer: $(-1, -5)$ Does this solution check?

The next example involves fractional coefficients. Thus the first step we take is to multiply through by the LCD to produce equivalent equations with integer coefficients.

SELF-CHECK 3.5.1

1. Solve $\begin{cases} 4x + 3y = 7 \\ -2x - 3y = -11 \end{cases}$.

2. Solve $\begin{cases} 2x - 7y = 1 \\ x + 5y = 9 \end{cases}$.

3. Solve $\begin{cases} 8x - 3y = 5 \\ -5x + 7y = 43 \end{cases}$.

SELF-CHECK 3.5.1 ANSWERS

1. $(-2, 5)$ 2. $(4, 1)$
3. $(4, 9)$

◼ EXAMPLE 5   Solving a Linear System with Fractional Coefficients

Solve $\begin{cases} \dfrac{x}{6} + \dfrac{y}{9} = 2 \\ \dfrac{x}{8} - \dfrac{y}{3} = 9 \end{cases}$ by the addition method.

SOLUTION

(1) $\dfrac{x}{6} + \dfrac{y}{9} = 2$ $18\left(\dfrac{x}{6} + \dfrac{y}{9}\right) = 18(2)$ Multiply both sides by 18, the LCD. Distribute the factor of 18 and simplify this equation.

(2) $\dfrac{x}{8} - \dfrac{y}{3} = 9$ $-24\left(\dfrac{x}{8} - \dfrac{y}{3}\right) = -24(9)$ Multiply both sides by -24, the opposite of the LCD. Distribute the factor of -24 and simplify this equation.

(1) $3x + 2y = 36$ Add these equations to eliminate x.
(2) $\underline{-3x + 8y = -216}$
                 $10y = -180$
                   $y = -18$ Solve for y.

           $3x + 2y = 36$
         $3x + 2(-18) = 36$ Back-substitute -18 for y in the first equation
           $3x - 36 = 36$ with integer coefficients.
               $3x = 72$
                 $x = 24$ Then solve for x.

Answer: $(24, -18)$ Does this answer check in the original system?

◼

Remember that the algebraic solution of an inconsistent system of linear equations will produce a contradiction. The algebraic solution of a consistent system of dependent linear equations will produce an identity. In Section 3.4 we examined these two possibilities using the substitution method. Now we will see what happens when the addition method is used to solve an inconsistent system of linear equations or a consistent system of dependent linear equations.

◼ EXAMPLE 6   Solving an Inconsistent System

Solve $\begin{cases} 5x + 10y = 11 \\ x + 2y = 3 \end{cases}$ by the addition method.

SOLUTION

$5x + 10y = 11$ $5x + 10y = 11$ Multiply both sides of the second
$x + 2y = 3$ $\underline{-5x - 10y = -15}$ equation by -5. Adding these
                 $0 = -4$, a contradiction   equations eliminates both variables and produces a contradiction.

Answer: No solution.   Since the result is a contradiction, the original system is inconsistent and has no solution.

◼

■ **EXAMPLE 7** Solving a Consistent System of Dependent Equations

Solve $\begin{cases} 6x - 15y = 3 \\ -4x + 10y = -2 \end{cases}$ by the addition method.

SOLUTION

$\begin{aligned} 6x - 15y &= 3 \\ -4x + 10y &= -2 \end{aligned}$ $\begin{aligned} 12x - 30y &= 6 \\ -12x + 30y &= -6 \\ \hline 0 &= 0, \text{ an identity} \end{aligned}$ Multiply both sides of the first equation by 2 and then multiply both sides of the second equation by 3.

Adding these equations eliminates both variables and produces an identity.

Answer: An infinite number of solutions.

Since the result is an identity, the original equations are dependent, and the system has an infinite number of solutions. ■

SELF-CHECK 3.5.2

1. Solve $\begin{cases} \dfrac{x}{2} - \dfrac{y}{8} = 3 \\ \dfrac{x}{4} + \dfrac{y}{2} = -3 \end{cases}$.

2. Solve $\begin{cases} 4x - 8y = 20 \\ 3x - 6y = 15 \end{cases}$.

The next problem involves two unknown numbers. By using x and y to represent these numbers, we then can form a system of two linear equations to describe this problem.

■ **EXAMPLE 8** Solving a Numeric Word Problem

Find two numbers whose sum is 60 and whose difference is 14.

SOLUTION

Let $x =$ the larger number

Let $y =$ the smaller number

VERBALLY	ALGEBRAICALLY	
(1) "two numbers whose sum is 60"	$\begin{aligned} x + y &= 60 \\ x - y &= 14 \\ \hline 2x &= 74 \end{aligned}$	Add the two equations to eliminate y.
(2) "two numbers whose difference is 14"	$x = 37$	Divide both sides of the equation by 2 to solve for x.

$\begin{aligned} x + y &= 60 \\ 37 + y &= 60 \\ y &= 23 \end{aligned}$

Back-substitute 37 for x in the first equation and solve for y.

Answer: The numbers are 37 and 23.

Write the answer as a full sentence. Does this answer check? ■

USING THE LANGUAGE AND SYMBOLISM OF MATHEMATICS   3.5

1. The addition-subtraction principle of equality states that _____ values can be added to or subtracted from both sides of an equation to produce an equivalent equation.
2. The addition method for solving a system of linear equations is also sometimes referred to as the _____ method because the strategy is to _____ one of the variables when the two equations are added.
3. In order to eliminate the y-variable when adding two linear equations, we first must make the coefficients of y in the two equations _____ of each other.

4. Using the addition method to solve a system of two linear equations produced the equation $0 = 5$. Thus the system of equations is an _____ system with _____ solution.
5. Using the addition method to solve a system of two linear equations produced the equation $0 = 0$. Thus the system is a _____ system of _____ equations with an _____ number of solutions.

EXERCISES   3.5

In Exercises 1–30 solve each system of linear equations by the addition method.

1. $x + 2y = 6$
   $-x + 3y = 4$

2. $5x - y = 10$
   $2x + y = 4$

3. $5x + 2y = -26$
   $3x - 2y = -38$

4. $-5x + 3y = 27$
   $5x + 2y = -7$

5. $6x + 2y = -1$
   $12x - y = 3$

6. $5x - y = -1$
   $15x + 2y = 7$

7. $x + 2y = 1$
   $3x + 4y = 0$

8. $x + 6y = 0$
   $3x + 8y = 5$

9. $2x + 3y = -9$
   $-4x + 5y = -37$

10. $3x + 7y = 25$
    $-6x + 3y = 18$

11. $5x - 3y = 5$
    $4x + 6y = 46$

12. $2x + 15y = 3$
    $3x - 5y = -1$

13. $2x + 5y = -3$
    $3x + 8y = -5$

14. $2x - 3y = 4$
    $11x - 5y = -1$

15. $2x + 3y = 15$
    $5x + 4y = -1$

16. $4x + 3y = 11$
    $5x + 2y = -9$

17. $2x - 13y = 38$
    $5x + 27y = 95$

18. $23x - 2y = -30$
    $47x + 3y = 45$

19. $2x = 11y$
    $5x = 19y$

20. $3x = 17y$
    $4x = 9y$

21. $2x + 6 = 0$
    $3x + 2y = 1$

22. $2y + 14 = 0$
    $2x + 3y = -13$

23. $\dfrac{x}{2} + \dfrac{y}{3} = 5$

    $\dfrac{x}{3} - \dfrac{y}{2} = -1$

24. $\dfrac{x}{8} + \dfrac{y}{8} = 0$

    $\dfrac{x}{2} - \dfrac{y}{4} = -3$

25. $6x - 8y = 10$
    $-15x + 20y = -20$

26. $2x - 6y = 8$
    $-3x + 9y = -12$

27. $6x = 9 - 3y$
    $4y = 12 - 8x$

28. $3x = 6 - 12y$
    $7 - 4x = 16y$

29. $\dfrac{x}{2} - \dfrac{y}{2} = \dfrac{3}{4}$

    $\dfrac{x}{2} + \dfrac{y}{4} = \dfrac{5}{8}$

30. $\dfrac{x}{3} + \dfrac{y}{3} = 0$

    $10x + 5y = -1$

In Exercises 31–40 solve each system of linear equations by either the substitution method or the addition method.

31. $y = 2x - 1$
    $7x - 4y = -1$

32. $x = 3y - 5$
    $-5x + 6y = -11$

33. $3x + 7y = 20$
    $5x - 7y = -4$

34. $4x + 3y = 26$
    $5x - 2y = -25$

35. $y = -x$
    $5x - 7y = 6$

36. $y = \dfrac{5}{3}x$
    $x + y = 2$

37. $\dfrac{x}{4} - \dfrac{y}{3} = \dfrac{5}{12}$

    $\dfrac{x}{2} - \dfrac{2y}{3} = 1$

38. $\dfrac{x}{4} - \dfrac{y}{6} = 6$

    $\dfrac{x}{6} + \dfrac{y}{3} = -4$

39. $2x + 3y = 4x + 5y - 6$
    $x - 5y = 7x + 2y - 21$

40. $3x + 4(2y - 1) = x + 8y - 2$
    $5x + 7(y + 5) = 4(x + y + 5) + 1$

In Exercises 41–44 select the system of linear equations (choices A, B, C, or D) that represents the word problem. Then solve this system by the addition method and answer the problem using a full sentence.

Verbally

41. The sum of three times one number and a second number is 1. The difference of the first number minus the second number is 7. Find these numbers.

42. The sum of twice one number plus a second number is 14. The first number plus three times the second number is 22. Find these numbers.

43. The sum of five times one number plus twice a second number is 2. The sum of twice the first number plus five times the second number is 26. Find the numbers.

44. The sum of twice one number plus a second number is 14. The difference of the first number minus the second number is 4. Find the numbers.

In Exercises 45–48 write a system of linear equations using the variables x and y and use this system to solve the problem.

45. Find two numbers whose sum is 88 and whose difference is 28.

46. Find two numbers whose sum is 133 and whose difference is 11.

47. Find two numbers whose sum is 102 if one number is twice the other number.

48. Find two numbers whose sum is 213 if one number is twice the other number.

Mean and Range

In Exercises 49 and 50 write a system of linear equations using the variables x and y and use this system of equations to solve the problem.

49. Basketball Scoring A basketball player had four games with identical scores and a fifth game that was his high for the whole season. His mean score for the five games was 25 and the range was 15. Use this information to determine his score in each of the five games.

50. Factory Defects A quality control worker in a factory measured 10 wheel rims. The diameters of the first nine were all equal, but the tenth had a defect and was too small. The mean of these measurements was 24.9 cm and the range was 1 cm. Use this information to determine each measurement.

51. Modeling College Enrollment Patterns In 1985, 38% of all graduates at Mason High School attended a four-year college and 18% of all graduates attended a community college. In 2001, 30% of all graduates attended a four-year college and 42% of all graduates attended a community college. Use the two points (1985, 38) and (2001, 30) to write an equation for the

Algebraically

A. $2x + y = 14$
$x - y = 4$

B. $2x + y = 14$
$x + 3y = 22$

C. $3x + y = 1$
$x - y = 7$

D. $5x + 2y = 2$
$2x + 5y = 26$

line showing the trend in the percent of graduates attending a four-year college. Also use the two points (1985, 18) and (2001, 42) to write an equation for the line showing the trend in the percent of graduates attending a community college. Then solve this system of equations and interpret the meaning of the x- and y-coordinates of this solution.

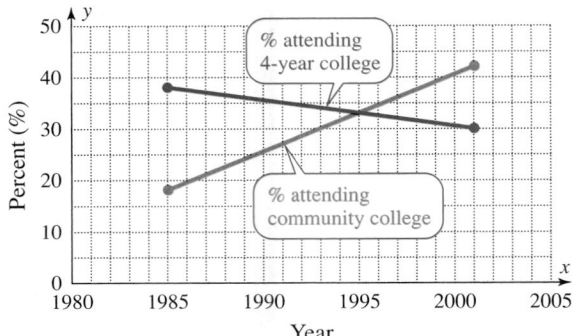

52. Modeling Generation of Electricity from the Wind In 1990, only 3% of all electricity needed in one community was generated by wind and 39% was generated by coal. By 2000, 20% of the electricity needed was generated by wind and 16% was generated by coal. Use the two points (1990, 3) and (2000, 20) to write an equation for the line showing the trend in the percent of electricity generated by wind. Also use the two points (1990, 39) and (2000, 16) to write an equation for the line showing the trend in the percent of electricity generated by coal. Then solve this system of equations and interpret the meaning of the x- and y-coordinates of this solution.

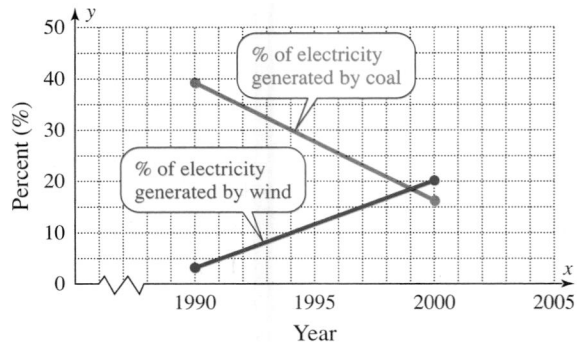

Calculator Usage

In Exercises 53–56 use a calculator to assist you in solving each system of linear equations.

53. $75x - 45y = -150$
$25x + 15y = 550$

54. $22x + 19y = 29$
$33x + 14y = -29$

55. $8.93x - 7.21y = -0.363$
$6.04x + 7.21y = 66.231$

56. $2.14x - 3.15y = -4.131$
$3.07x + 4.21y = 0.184$

Group Discussion Questions

57. Writing Mathematically

a. Two linear equations produce the same line when they are graphed. Write a paragraph describing what would happen if you used the addition method to solve this system.

b. Two linear equations produce distinct parallel lines when they are graphed. Write a paragraph describing what would happen if you used the addition method to solve this system.

58. Writing Mathematically   Write a numeric word problem for each system of linear equations.

a. $3x - 4y = 5$
$4x + y = 13$

b. $6x + 2y = -16$
$5x + y = -18$

59. Challenge Question   Solve this system for (x, y) in terms of a, b, and c. Assume a, b, and c are all nonzero.
$ax + by = c$
$2ax - by = 2c$

Section 3.6 More Applications of Linear Systems

Objective: **10.** Use systems of linear equations to solve word problems.

It is likely that almost all the mathematics you will encounter outside the classroom will be stated in words. Thus to profit from your algebra, you must be able to work with word problems. The problems in this section are carefully selected to develop gradually and to broaden your problem-solving skills. Some problems are selected for the same purpose as drill exercises on a piano or in basketball—to hone the skills needed for a real performance that will come later.

A basic strategy for solving word problems is outlined in the following box. Until you become quite proficient in solving word problems, we suggest that you use the steps in this box as a check list. Learning how to break down a seemingly complicated problem into a series of manageable steps is an important lifetime skill that extends well beyond word problems. When you start the first step, read actively, not passively. Circle key words, underline key phrases, make sketches, or jot down key facts as you read.

A Mathematical Note

The problem-solving strategy given here and used throughout this book is not new or unique. George Poyla (1888–1985) is well known for teaching problem-solving techniques. His best-selling book *How to Solve It* includes four steps to problem solving:

Step 1 Understand the problem.
Step 2 Devise a plan.
Step 3 Carry out the plan.
Step 4 Check back.

Compare these steps to those given in the box to the right.

Strategy for Solving Word Problems

STEP 1. Read the problem carefully to determine what you are being asked to find.

STEP 2. Select a variable to represent each unknown quantity. Specify precisely what each variable represents.

STEP 3. If necessary, translate the problem into word equations. Then translate the word equations into a system of algebraic equations.

STEP 4. Solve the equation or the system of equations and answer the question asked by the problem.

STEP 5. Check the reasonableness of your answer.

We illustrate this strategy on a numeric word problem in Example 1. We have examined problems similar to this one in previous sections. In this section you will practice and extend the skills you have already used.

■ EXAMPLE 1   Solving a Numeric Word Problem with Phone Bills

The bill for a cellular phone in February was $10 more than twice the January bill. The total that was required to pay both these bills was $175. What was the bill for each month?

SOLUTION _____

Let x = the dollar amount of the January bill

Let y = the dollar amount of the February bill

Select a variable to represent each unknown.

VERBALLY

(1) The February bill is $10 more than twice the January bill.

(2) The total bill was $175:

ALGEBRAICALLY

$$y = 2x + 10$$

$$x + y = 175$$

Form word equations and then translate them into algebraic equations.

NUMERICALLY

```
Plot1  Plot2  Plot3
\Y1◼2X+10
\Y2◻175-X
\Y3
\Y4       X  | Y1  | Y2
\Y5    25 | 60  | 150
\Y6    30 | 70  | 145
\Y7    35 | 80  | 140
        40 | 90  | 135
        45 | 100 | 130
        50 | 110 | 125
        55 | 120 | 120
       X=55
```

$$x + (2x + 10) = 175$$
$$3x + 10 = 175$$
$$3x = 165$$
$$x = 55$$

$$y = 2x + 10$$
$$y = 2(55) + 10$$
$$y = 110 + 10$$
$$y = 120$$

Solve the system using the substitution method by substituting $2x + 10$ for y or examine the system on a graphics calculator.

Back-substitute 55 for x in the first equation.

Answer: The January bill was $55 and the February bill was $120.

Does this answer check?

■

SELF-CHECK 3.6.1

The sum of two numbers is 15.25. The larger number is 14 more than four times the smaller number. Find these numbers.

SELF-CHECK 3.6.1 ANSWER

The two numbers are 0.25 and 15.

If a word problem involves any shape or design, it is wise to make a sketch of this shape. Observing the visual relationship of the variables can assist us in writing equations for the problem. The next example involves two supplementary angles.

▮ EXAMPLE 2   Determining the Number of Degrees in Two Supplementary Angles

A metal fabrication shop must assemble a part so that the angles shown in the figure are supplementary. To function properly the specifications for the part require that the larger angle is 40° more than the smaller angle. Determine the number of degrees in each angle.

SOLUTION

Let x = the number of degrees in the smaller angle

Let y = the number of degrees in the larger angle

Select a variable to represent each unknown.

VERBALLY	ALGEBRAICALLY	
(1) The angles are supplementary.	$x + y = 180$	Angles are supplementary if the total of their measures is 180°.
(2) The larger angle is 40° more than the smaller angle.	$y = x + 40$	Specifications require that the larger angle is 40° more than the smaller angle.

NUMERICALLY

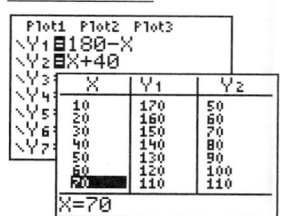

$$x + (x + 40) = 180$$
$$2x + 40 = 180$$
$$2x = 140$$
$$x = 70$$

Solve the system using the substitution method by substituting $x + 40$ for y in the first equation.

Then solve this equation for x.

$$y = x + 40$$
$$y = 70 + 40$$
$$y = 110$$

Back-substitute 70 for x in the second equation.

Answer: The smaller angle is 70° and the larger angle is 110°.

Does this answer check?   ▮

SELF-CHECK 3.6.2

Rework Example 2 with the same conditions except that the angles must be complementary (sum is 90°).

As you solve problems throughout this book try to recognize how problems share the general principles of mathematics. Students who recognize the general principles can profit from their mathematics more than those who memorize individual problems.

Another tool that is used often to form equations is the **mixture principle.** The mixture principle is a great example of how one general mathematical principle can be used to form equations for seemingly unrelated problems. One of the main goals of this book is to help you see these connections. Mathematics is difficult for those who try to memorize it bit by bit; it is much easier when you understand how to use its general principles.

Mixture Principle for Two Ingredients
Amount in first + Amount in second = Amount in mixture

Applications of the Mixture Principle
1. Amount of product A + Amount of product B = Total amount of mixture
2. Variable cost + Fixed cost = Total cost
3. Interest on bonds + Interest on CD's = Total interest
4. Distance by first plane + Distance by second plane = Total distance
5. Antifreeze in first solution + Antifreeze in second solution = Total amount of antifreeze

SELF-CHECK 3.6.2 ANSWER

The smaller angle is 25° and the larger angle is 65°.

■ **EXAMPLE 3** Determining the Costs for a Pizza Business

A family pizza business opened a new store. The daily cost for making pizzas includes a fixed daily cost and a variable cost per pizza. The total cost for making 200 pizzas on Monday was $750. The total cost for making 250 pizzas on Tuesday was $900. Determine the fixed daily cost and variable cost per pizza for this business.

SOLUTION

Let x = the variable cost per pizza

Let y = the fixed dollar cost per day

Select a variable to represent each unknown.

VERBALLY

(1) **Monday:** Variable cost + Fixed cost = Total cost
(2) **Tuesday:** Variable cost + Fixed cost = Total cost

ALGEBRAICALLY

$$200x + y = 750$$
$$250x + y = 900$$

Note that each of these word equations is based on the mixture principle. The variable cost for each day is the number of pizzas times the cost per pizza.

NUMERICALLY

```
Plot1 Plot2 Plot3
\Y1=750-200X
\Y2=900-250X
\Y3=
\Y4=     X    Y1    Y2
\Y5=     0    750   900
\Y6=    .5    650   775
\Y7=     1    550   650
        1.5   450   525
         2    350   400
        2.5   250   275
         3    150   150
X=3
```

(1) $\quad -200x - y = -750$
(2) $\quad \underline{250x + y = 900}$

$$50x = 150$$
$$x = 3$$

Multiply each side of Equation (**1**) by -1 and then use the addition method to eliminate y.
Solve this equation for x.

(2) $\quad 250x + y = 900$
$\quad\quad 250(3) + y = 900$
$\quad\quad\quad 750 + y = 900$
$\quad\quad\quad\quad\quad\quad y = 150$

Back-substitute 3 for x in Equation (**2**).

Answer: The fixed cost per day is $150 and $3 is the variable cost per pizza. ■

SELF-CHECK 3.6.3

Rework Example 3 assuming that the total costs were $800 for Monday and $975 for Tuesday.

Another general principle you may have observed is the rate principle: $A = RB$; Amount = Rate × Base. Applications using this principle include problems involving interest rates, rates of travel, or rates of work.

Example 4 uses the mixture principle to form both equations. The total principal is composed of two separate investments and the total interest is the combined interest from the two investments.

■ EXAMPLE 4   Calculating the Amount of Two Investments

A student saving for college was given \$6000 by her grandparents. She invested part of this money in a savings account that earned interest at the rate of 7% per year. The rest was invested in a bond that paid interest at the rate of 8.5% per year. If the combined interest at the end of one year was \$495, how much was invested at each rate?

SOLUTION

Let x = principal invested in the savings account

Let y = principal invested in the bond

Select a variable to represent each unknown.

VERBALLY

ALGEBRAICALLY

(1) $\dfrac{\text{Principal in the}}{\text{savings account}} + \dfrac{\text{Principal in}}{\text{the bond}} = \dfrac{\text{Total}}{\text{principal}}$

$$x + y = 6000$$

(2) $\dfrac{\text{Interest on the}}{\text{savings account}} + \dfrac{\text{Interest on}}{\text{the bond}} = \dfrac{\text{Total}}{\text{interest}}$

$$0.07x + 0.085y = 495$$

Both word equations are based on the mixture principle. Using $I = PRT$ and one year for time, the interest on the savings account is $I = (x)(0.07)(1) = 0.07x$ and the interest on the bond is $I = (y)(0.085)(1) = 0.085y$.

NUMERICALLY

(1) $\quad -0.07x - 0.07y = -420$
(2) $\quad \underline{0.07x + 0.085y = 495}$
$\qquad\qquad\qquad 0.015y = 75$

Multiply each side of Equation (1) by -0.07 and then use the addition method to solve this system.

```
Plot1 Plot2 Plot3
\Y1■6000-X
\Y2■(495-0.07X)/
0.085
\Y3=
\Y4=
\Y5=
\Y6=
```

X	Y₁	Y₂
0	6000	5823.5
500	5500	5411.8
1000	5000	5000
1500	4500	4588.2
2000	4000	4176.5
2500	3500	3764.7
3000	3000	3352.9

X=1000

$$y = \frac{75}{0.015}$$

Divide both sides by 0.015 to solve for y.

$$y = 5000$$

(2) $\qquad x + y = 6000$
$\qquad x + 5000 = 6000$
$\qquad\qquad\quad x = 1000$

Back-substitute 5000 for y in the first equation and solve for x.

Answer: She invested \$1000 in the savings account and \$5000 in the bond.

Check: $1000 + 5000 = 6000$ checks.

$0.07(1000) + 0.085(5000) = 495$ checks.   ■

SELF-CHECK 3.6.4

Rework Example 4 assuming that the combined interest at the end of 1 year was \$472.50.

SELF-CHECK 3.6.4 ANSWER

She invested \$2500 in the savings account and \$3500 in the bond.

Example 5 uses the mixture principle to calculate the total distance as the sum of the distances traveled by two different planes.

■ EXAMPLE 5   Determining the Speed Flown by Two Airplanes

Two airplanes depart from an airport simultaneously, one flying 100 km/h faster than the other one. These planes travel in opposite directions, and after 1.5 hours they are 1275 km apart. Determine the speed of each plane. (*Hint: D = RT*; distance equals rate times the time.)

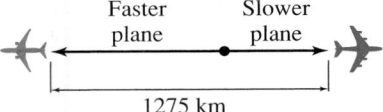

SOLUTION _____

Let r_1 = rate of the slower plane in km/h

Let r_2 = rate of the faster plane in km/h

Select a variable to represent each unknown.

VERBALLY

(1) $\dfrac{\text{Rate of the}}{\text{faster plane}} = \dfrac{\text{Rate of the}}{\text{slower plane}} + 100 \text{ km/h}$

(2) $\dfrac{\text{Distance by}}{\text{slower plane}} + \dfrac{\text{Distance by}}{\text{faster plane}} = \dfrac{\text{Total distance}}{\text{between planes}}$

ALGEBRAICALLY

$$r_2 = r_1 + 100$$

$$1.5r_1 + 1.5r_2 = 1275$$

Note the second word equation is based on the mixture principle. Using $D = RT$ and a time of 1.5 hours, the distance is $1.5r_1$ for the slower plane and $1.5r_2$ for the faster plane.

NUMERICALLY

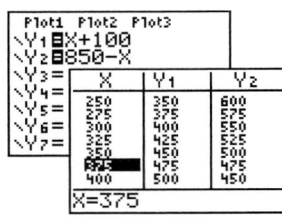

On a calculator use *x* to represent r_1 and *y* to represent r_2.

(1) $r_1 - r_2 = -100$
(2) $\underline{r_1 + r_2 = 850}$

$$2r_1 = 750$$
$$r_1 = 375$$

(1) $r_2 = r_1 + 100$
$$r_2 = 375 + 100$$
$$r_2 = 475$$

Rearrange the terms in the first equation and simplify the second equation by dividing both sides of the equation by 1.5. Then use the addition method to solve this system.

Back-substitute 375 for r_1 in the first equation.

Answer: The planes traveled at 375 and 475 km/h.

Check: $375 + 100 = 475$ checks.

$1.5(375) + 1.5(475) = 1275$ checks.   ■

SELF-CHECK 3.6.5

Rework Example 5 assuming that the planes are 1900 km apart after 2 hours.

When two liquids of different concentrations are mixed, the total volume is found by adding the volumes of the liquids that are mixed. The total of the chemicals found in the mixture is calculated by adding the chemicals found in each of the individual solutions. This is illustrated in Example 6.

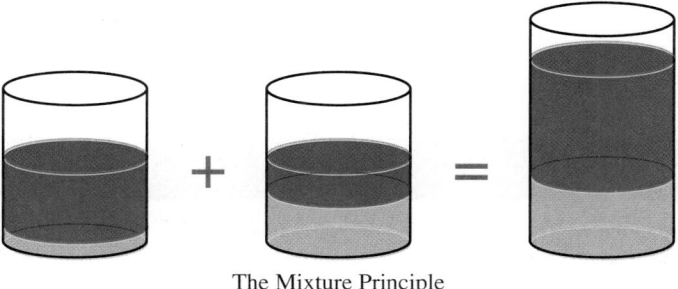

The Mixture Principle

■ EXAMPLE 6   Calculating the Amount of Two Solutions to Mix

A water treatment plant uses batch deliveries of two levels of chlorine concentrations to mix in order to create the level of concentration needed on any given day. Determine how many liters of a 3% chlorine solution should be mixed with a 5% chlorine solution to produce 500 liters of a 4.4% chlorine solution.

SOLUTION

Let x = the number of liters of the 3% chlorine solution   Select a variable to represent each unknown.

Let y = the number of liters of the 5% chlorine solution

VERBALLY

(1) Volume of first solution + Volume of second solution = Total volume of mixture

(2) Volume of chlorine in first solution + Volume of chlorine in second solution = Total chlorine of mixture

NUMERICALLY

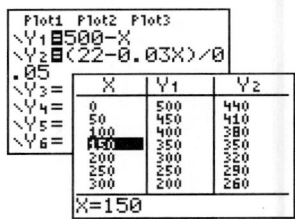

To use a graphics calculator, rewrite Equation (1) as $y = 500 - x$ and rewrite Equation (2) as $y = \dfrac{22 - 0.03x}{0.05}$.

ALGEBRAICALLY

$$x + y = 500$$

$$0.03x + 0.05y = 0.044(500)$$

(1) $-0.03x - 0.03y = -15$
(2) $\underline{\quad 0.03x + 0.05y = 22 \quad}$
$$0.02y = 7$$

$$\frac{0.02y}{0.02} = \frac{7}{0.02}$$
$$y = 350$$

(1) $x + y = 500$
$x + 350 = 500$
$x = 150$

The volume of chlorine in x liters of 3% solution is $0.03x$. The volume of chlorine in y liters of the 5% solution is $0.05y$. The volume of chlorine in 500 liters of 4.4% solution is $0.044(500)$.

Multiply both sides of the first equation by -0.03 and then use the addition method to solve this system.

Divide both sides of this equation by 0.02.

Back-substitute 350 for y in the first equation and solve for x.

Answer: Use 150 liters of the 3% solution and 350 liters of the 5% solution to produce 500 liters of a 4.4% chlorine solution. ■

SELF-CHECK 3.6.6

Rework Example 6 assuming that the mixture produced 500 liters of a 4.5% chlorine solution.

USING THE LANGUAGE AND SYMBOLISM OF MATHEMATICS   3.6

1. The first step in the word problem strategy given in this book is to read the problem carefully to determine what you are being asked to _____ .

2. The second step in the word problem strategy given in this book is to select a _____ to represent each unknown quantity.

3. The third step in the word problem strategy given in this book is to model the problem verbally with word equations and then to translate these word equations into _____ equations.

4. The fourth step in the word problem strategy given in this book is to _____ the equation or system of equations and to answer the question asked by the problem.

5. The fifth step in the word problem strategy given in this book is to check your answer to make sure the answer is _____ .

6. The _____ principle states that the amount obtained by combining two parts is equal to the amount obtained from the first part plus the amount obtained from the second part.

7. In the formula $D = RT$, D represents distance, R represents _____ , and T represents _____ .

8. In the formula $I = PRT$, I represents interest, P represents _____ , R represents _____ , and T represents _____ .

EXERCISES   3.6

In Exercises 1–42, solve each problem by using a system of equations and the word problem strategy developed in this section.

1. **Numeric Word Problem** Find two numbers whose sum is 100 if one number is 16 more than twice the smaller number.

2. **Numeric Word Problem** Find two numbers whose sum is 100 if one number is 16 more than three times the smaller number.

3. **Dimensions of a Rectangular Poster** The perimeter of a rectangular poster is 204 cm. Find the dimensions of the poster if the length is 6 cm more than the width.

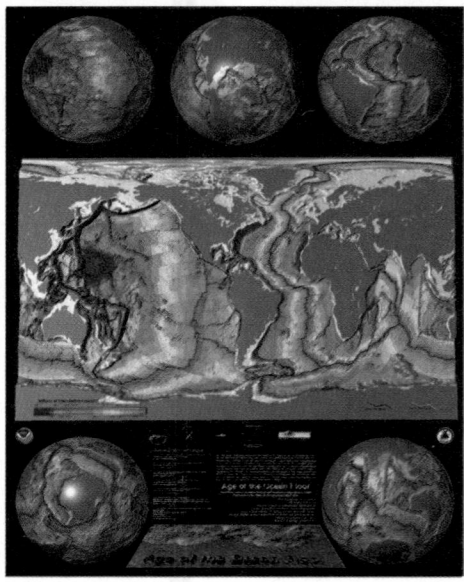

4. **Dimensions of a Rectangular Pane of Glass** The perimeter of a rectangular pane of glass is 68 cm. Find the dimensions of the pane of glass if the length is 2 cm less than three times the width.

5. **Dimensions of an Isosceles Triangle** An isosceles triangle has two equal sides. The triangle shown here has a perimeter of 72 cm. The base is 12 cm longer than the two equal sides. Find the dimensions of this triangle.

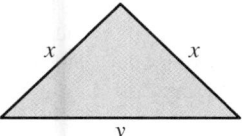

6. **Angles of an Isosceles Triangle** The sum of the interior angles of a triangle is 180°. An isosceles triangle has two equal angles. Find the angles of the isosceles triangle shown here if angle y is 33° larger than angle x.

7. **Credit Card Codes** If you randomly make up a credit card number it is unlikely to work because there are codes used with each legitimate card. A simple code rejected for membership cards at a golf club was to have the sum of two numbers on the card be eleven and the difference of these numbers be three. This was foolish because the two numbers always would be the same. What are the numbers that would satisfy both these conditions?

8. **Credit Card Codes** One credit card company decided that two numbers on its card would have a sum of twelve. A second company used the same two numbers on its card but required that the difference of these numbers be two. Are there any numbers that work for both companies? If so, what are these numbers?

9. **Coded Messages** A simple method for sending coded messages is to convert all letters to numbers, send the numbers, and then the receiver converts the numbers back to letters. The sender sometimes requires an initiation code before authorizing the release of a particular message. The two-number initiation code for one receiver must have the sum of five times the first number plus three times the second number equal eighty-four. The two-number initiation code for a second receiver must have four times the first number minus the second number equal twenty-three. Are there any numbers that will work for both these receivers? If so, what are these numbers?

10. **Coded Messages** To receive a message, a spy must send a pair of numbers that satisfy the conditions of the two independent supervisors on duty that day. Supervisor A must receive the code and confirm that the sum of seven times the first number plus four times the second number is two hundred ninety-four. Supervisor B must receive the code and confirm that six times the first number minus the second number is ninety-seven. What two numbers should be sent when supervisors A and B are on duty?

11. **Pollution Gauge Reading** The readings from two different pollution gauges are recorded at the end of each shift in a factory. These readings are compared to see which reading is higher and totaled to get an overview of factorywide pollution. On one reading the first gauge totaled 7.5 units more than twice the reading on the other gauge. If the total of these readings was 42 units, what was the reading on each gauge?

12. **Mean and Range** A company has six employees. Five of these employees make exactly the same amount each month, which is less than their supervisor makes. The mean (average) hourly salary of all employees is $26/h and the range of these salaries is $12/h. Determine the hourly salary of each employee.

13. **Cable TV Bills** A cable TV bill consists of a fixed monthly charge and a variable charge, which depends on the number of pay-per-view movies ordered. In January the $44 bill included four pay-per-view movies. In February the $54.50 bill included seven pay-per-view movies. Determine the fixed monthly charge and the charge for each pay-per-view movie.

14. **Rental Car Costs** The cost of a rental car includes a fixed daily cost and a variable cost based on the miles driven. For a customer driving 180 mi the cost for one day will be $71. For a customer driving 150 mi the cost for one day will be $65. Determine the fixed daily cost and the cost per mile for this car.

15. **Costs for Union Employees** A union contract requires at least one union steward for every twenty union employees or fraction thereof. Regular union employees earn $120 per day and union stewards earn $144 per day. A company has a need for 50 total employees and has a daily payroll budget of $6072. How many of each type of employee will meet their needs and consume all their budget?

16. **Equipment Costs** A bulldozer can move 25 tons of material per hour and an end loader can move 18 tons per hour. The bulldozer costs $75/h to operate and the end loader costs $50/h to operate. During one hour a construction company has moved 204 tons at a cost of $600. How many bulldozers and how many end loaders were operating during this hour?

17. **Number of Basketball Tickets**   The total receipts for a basketball game are $1400 for 788 tickets sold. Adults paid $2.50 for admission and students paid $1.25. The ticket takers counted only the number of tickets not the type of tickets. Determine the number of each type of ticket sold.

18. **Loads of Topsoil**   Topsoil sells for $55 a truckload, and fill dirt sells for $40 a truckload. A landscape architect estimates that 20 truckloads will be needed for a certain job. The estimated cost is $920. How many loads of topsoil and how many loads of fill dirt are planned for this job?

19. **Complementary Angles**   The angles shown in Exercise 20 are complementary and one angle is 28° larger than the other angle. Determine the number of degrees in each angle.

20. **Complementary Angles**   The angles shown in the figure are complementary and one angle is 15° more than twice the other angle. Determine the number of degrees in each angle.

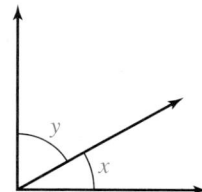

21. **Supplementary Angles**   The two angles shown in Exercise 22 represent parts that should be assembled so that the angles are supplementary and the larger angle is 5° more than four times the smaller angle. Determine the number of degrees in each angle.

22. **Supplementary Angles**   The two angles shown here represent parts that should be assembled so that the angles are supplementary and the larger angle is 9° less than twice the smaller angle. Determine the number of degrees in each angle.

23. **Order for Left- and Right-Handed Desks**   When ordering new desks for classrooms, the business manager for a college uses the fact that there are nine times as many right-handed people as left-handed people. How many desks of each type must be ordered if 600 new desks are needed?

24. **Length of a Board**   A board 12 ft long is to be cut into two pieces so that one piece is 3 ft longer than the other piece. Determine the length of each piece.

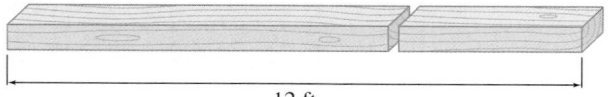

12 ft

25. **Fixed and Variable Costs**   A family business makes custom rocking chairs, which they market as retirement gifts. One month the total fixed and variable costs for producing 250 rockers was $32,250. The next month the total fixed and variable costs for producing 220 rockers was $28,500. Determine the fixed cost and the variable cost per rocking chair.

26. **Fixed and Variable Costs**   A music producer has structured the business costs as a fixed cost plus a variable cost per CD shipped. To produce and ship 15,000 CDs, the cost is $75,000. To produce and ship 25,000 CDs, the cost is $95,000. Determine the fixed cost and the variable cost per CD.

27. **Interest on Two Investments**   Part of a $12,000 investment earned interest at a rate of 7%, and the rest earned interest at a rate of 9%. The combined interest earned at the end of one year was $890. How much was invested at each rate?

28. **Interest on Two Investments**   A broker made two separate investments on behalf of a client. The first investment earned 8% and the second earned 5% for a total gain of $600. If these investments had earned 4% and 7% respectively, then the gain would have been $480. How much was actually invested at each rate?

29. **Interest on Two Investments**   An investment of $10,000 earned a net income of $345 in one year. Part of the investment was in bonds and earned income at a rate of 8%. The rest of the investment was in stocks and lost money at the rate of 5%. How much was invested in bonds and how much was invested in stocks?

30. **Interest on Two Investments**   A broker made two separate investments on behalf of a client. The first investment earned 12% and the second investment lost 5% for a total gain of $1040. If the amounts in these two investments had been switched, then the result would have been a gain of $360. Determine how much was actually invested at each rate.

31. **Rates of Two Trains**   Two trains depart simultaneously from a station, traveling in opposite directions. One averages 10 km/h more than the other. After $\frac{1}{2}$ hour they are 89 km apart. Determine the speed of each train.

32. **Rates of Two Airplanes**   A jet plane and a tanker that are 525 mi apart head toward each other so that the jet can refuel. The jet flies 200 mi/h faster than the tanker. Determine the speed of each aircraft if they meet in 45 minutes. (*Hint:* Use consistent units of measurement.)

33. **Rates of Two Trains**   Two trains depart from a station, traveling in opposite directions. The train that departs at 6:30 A.M. travels 15 km/h faster than the train that departs at 6 A.M. At 7 A.M. they are 135 km apart. Determine the rate of each train.

34. **Rates of Two Buses**   Two buses that are 90 mi apart travel toward each other on an interstate highway. The slower bus, which departs at 9:30 A.M. travels 5 mi/h slower than the bus that departs at 9:45 A.M. At 10:30 A.M. the buses pass each other. Determine the rate of each bus.

$$24 \text{ km} = (B+C)(1\text{hr})$$
$$24 \text{ km} = (B-C)(4\text{hrs})$$

35. Rate of a River Current   When a boat travels downstream with a river current the rate of the boat in still water and the rate of the current are added. When a boat travels upstream, the rate of the current is subtracted from the rate of the boat.

A paddlewheel riverboat takes 1 hour to go 24 km downstream and another 4 hours to return upstream. Determine the rate of the boat and the rate of the current.

36. Rate of a River Current   A small boat can go 40 km downstream in an hour but only 10 km upstream in an hour. Determine the rate of the boat and the rate of the current. (*Hint:* See Exercise 35.)

37. Mixture of Disinfectant   A hospital needs 80 liters of a 12% solution of disinfectant. How many liters of a 33% solution and a 5% solution should be mixed to obtain this 12% solution?

38. Mixture of Medicine   The dosage of a medicine ordered by a doctor is 40 milliliters (ml) of a 16% solution. A nurse has available both a 20% solution and a 4% solution of this medicine. How many milliliters of each could be mixed to prepare this 40-ml dosage?

39. Gold Alloy   A goldsmith has 80 grams of an alloy that is 50% pure gold. How many grams of an alloy that is 80% pure gold must be combined with the 80 grams of alloy that is 50% pure gold to form an alloy that will be 72% pure gold?

40. Saline Mixture   A 30% salt solution is prepared by mixing a 20% salt solution and a 45% salt solution. How many liters of each must be used to produce 60 liters of the 30% salt solution?

41. Mixture of Fruit Drinks   A fruit drink concentrate is 15% water. How many liters of pure water should be added to 12 liters of concentrate in order to produce a mixture that is 50% water?

42. Insecticide Mixture   A nurseryman is preparing an insecticide by mixing a 95% solution with water (0% solution). How much solution and how much water are needed to fill a 500-gallon tank with 3.8% solution?

Group Discussion Questions

43. Writing Mathematically   Write a word problem that can be solved using the system of equations

$$\begin{cases} x + y = 500 \\ 5x + 8y = 3475 \end{cases} \text{ and that models:}$$

 a. A numeric problem (see Exercise 1)
 b. A problem involving ticket sales (see Exercise 17)
 c. A distance problem (*Hint:* Be careful on your description of how the rates are related.)

44. Writing Mathematically   Write a word problem that can be solved using the system of equations

$$\begin{cases} x + y = 500 \\ 0.05x + 0.08y = 34.75 \end{cases} \text{ and that models:}$$

 a. A combined investment problem (see Exercise 27)
 b. A mixture of two liquids at different concentrations (see Exercise 37)

45. Discovery Question
 a. Compare the two systems of equations in Exercises 43 and 44. Are these systems equivalent; that is, can you convert one system to the other? If so, does this mean that this one system can represent all five of the word problems created in Exercises 43 and 44?
 b. Discuss the use of one general mathematical principle, such as the mixture principle, to solve many seemingly unrelated problems.
 c. Although pure numeric-based word problems have limited direct application, they are widely used in textbooks. Discuss a rationale of why this is helpful for students.

KEY CONCEPTS   FOR CHAPTER 3

1. **Slope of a Line:**
   - The slope m of a line through (x_1, y_1) and (x_2, y_2) with
     $x_1 \neq x_2$ is $m = \dfrac{\text{change in } y}{\text{change in } x} = \dfrac{y_2 - y_1}{x_2 - x_1} = \dfrac{\text{rise}}{\text{run}}$.

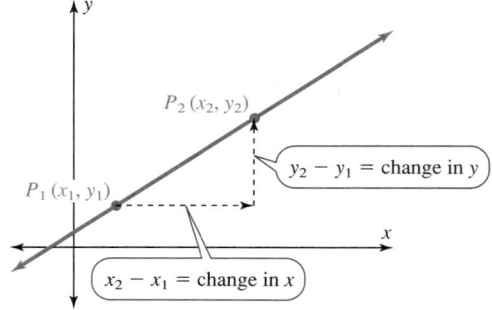

   - The slope of a line gives the change in y for each 1-unit change in x.
   - A line with positive slope goes upward to the right.
   - A line with negative slope goes downward to the right.
   - The slope of a horizontal line is 0.
   - The slope of a vertical line is undefined.
   - The slopes of parallel lines are the same.
   - The slopes of perpendicular lines are opposite reciprocals.
   - The product of the slopes of perpendicular lines is -1.
   - The slope of $y = mx + b$ is m.
   - The slope of $y - y_1 = m(x - x_1)$ is m.
   - The common difference of an arithmetic sequence is the same as the slope of the line through the points formed by this sequence.
   - The grade of a highway refers to the slope of the highway.

2. **Forms of Linear Equations:**   A linear equation in x and y is first degree in both x and y.
   - Slope-intercept form: $y = mx + b$ or $f(x) = mx + b$ with slope m and y-intercept $(0, b)$,
   - Vertical line: $x = a$ for a real constant a.
   - Horizontal line: $y = b$ for a real constant b.
   - General form: $Ax + By = C$.
   - Point-slope form: $y - y_1 = m(x - x_1)$ through (x_1, y_1) with slope m.

3. **Solution of a System of Linear Equations in x and y:**
   - A solution is an ordered pair that satisfies each equation in the system.
   - A solution of a system of linear equations is represented graphically by a point that is on both lines in the system.
   - In a table of values, the x-coordinate of a solution produces two y-values that are the same.

4. **Terminology Describing Systems of Linear Equations:**
   - Consistent system: The system of equations has at least one solution.
   - Inconsistent system: The system of equations has no solution.
   - Independent equations: The equations cannot be reduced and rewritten as the same equation.
   - Dependent equations: Both the equations can be written in exactly the same form.

5. **Classifications of Systems of Linear Equations:**
   - Exactly one solution: A consistent system of independent equations.
   - No solution: An inconsistent system of independent equations.
   - An infinite number of solutions: A consistent system of dependent equations.

6. **Using the Slope-Intercept Form to Classify Systems of Linear Equations:**
   For the linear system $\begin{cases} y = m_1x + b_1 \\ y = m_2x + b_2 \end{cases}$ with:
   - $m_1 \neq m_2$
     The system is consistent with independent equations and exactly one solution.
   - $m_1 = m_2$ and $b_1 \neq b_2$
     The system is inconsistent with independent equations and no solution.
   - $m_1 = m_2$ and $b_1 = b_2$
     The system is consistent with dependent equations and an infinite number of solutions.

7. **Methods for Solving Systems of Linear Equations:**
   - Graphical method; the point(s) on both lines
   - Numerical method; for the same x-value the two y-values are equal.
   - Algebraically by the substitution method
   - Algebraically by the addition method

8. **Algebraic Solution of the Three Types of Linear Systems:**
   - **A consistent system of independent equations:** The solution process will produce unique x- and y-coordinates.
   - **An inconsistent system:** The solution process will produce a contradiction.
   - **A consistent system of dependent equations:** The solution process will produce an identity.

9. **Solving a System of Linear Equations Containing Fractions:**
   - It may be easier to start by multiplying both sides of the equation by the LCD of all the fractions in the equation.

10. **Substitution Principle:**
    - A quantity may be substituted for its equal.

11. **Addition-Subtraction Principle of Equality:**
    - Equal values can be added to or subtracted from both sides of an equation to produce an equivalent equation.

12. **Mixture Principle:**
    - $\begin{pmatrix} \text{Amount} \\ \text{in first} \end{pmatrix} + \begin{pmatrix} \text{Amount} \\ \text{in second} \end{pmatrix} = \begin{pmatrix} \text{Amount} \\ \text{in mixture} \end{pmatrix}$

13. Strategy for Solving Word Problems:
Step 1. Read the problem carefully to determine what you are being asked to find.
Step 2. Select a variable to represent each unknown quantity. Specify precisely what each variable represents.
Step 3. If necessary, translate the problem into word equations. Then translate the word equations into a system of algebraic equations.

Step 4. Solve the equation or the system of equations and answer the question asked by the problem.
Step 5. Check the reasonableness of your answer.

REVIEW EXERCISES   FOR CHAPTER 3

In Exercises 1–10 calculate the slope of each line.

1. The line through (1, −3) and (4, 3).

2. The line through (1, 4) and (6, 1).

3. $f(x) = \frac{4}{7}x + 3$

4. $y - 2 = -(x - 8)$

5. $y = -7$

6.

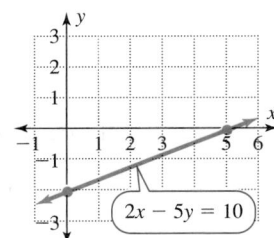

$2x - 5y = 10$

7.

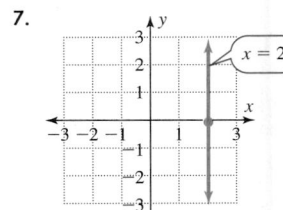

$x = 2$

8. The line that contains the points in this table.

x	y
−3	4
−2	4
−1	4
0	4
1	4
2	4
3	4

9. The line that contains the points in this table.

x	y
4	−3
4	−2
4	−1
4	0
4	1
4	2
4	3

10. The line that contains the points in this table.

X	Y₁
-3	-.5
-2	0
-1	.5
0	1
1	1.5
2	2
3	2.5

X= -3

11. Complete the following table involving the change in x, the change in y, and the slope of the line $y = mx + b$.

	CHANGE IN x	CHANGE IN y	SLOPE
a.	5		$\frac{4}{5}$
b.		4	$\frac{4}{5}$
c.	−5		$\frac{4}{5}$
d.	1		$\frac{4}{5}$
e.	−1		$\frac{4}{5}$

12. Using a Sequence that Models House Payments The equation $a_n = 800n + 3000$ gives the total amount paid in dollars (including the down payment) for a house for n months.
a. Determine the sequence of total amounts paid at the end of each of the first 4 months.
b. Graph this sequence.
c. Determine the slope of the line through these points. Interpret the meaning of the slope in this application.
d. Calculate a_0 and interpret the meaning of a_0.

13. Slope of a Wheelchair Ramp Determine the slope of the wheelchair ramp shown in the figure.

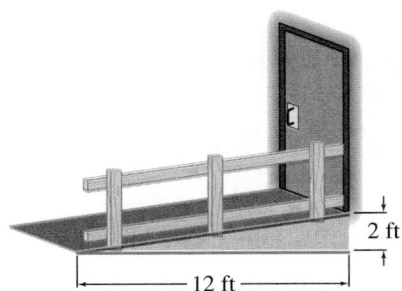

2 ft

12 ft

14. Determine whether the first line is parallel to, perpendicular to, or neither parallel nor perpendicular to the second line.
a. l_1 passes through (4, −2) and (6, 2).
l_2 passes through (5, 3) and (9, 11).
b. l_1 defined by $y = 2$
l_2 defined by $x = -2$
c. l_1 defined by $y = 5x - 4$
l_2 defined by $y - 4 = -5(x + 1)$

In Exercises 15–21 write the equation of each line in general form with integer coefficients.

15. $y - 4 = 2(x - 3)$ **16.** $y = \dfrac{2}{3}x - \dfrac{1}{5}$

17. A line with a y-intercept $(0, 6)$ and a slope of $-\dfrac{1}{2}$.

18. A line with an x-intercept $(3, 0)$ and a y-intercept $(0, -2)$.
19. A horizontal line through $(4, 9)$.
20. A vertical line through $(4, 9)$.
21. A line containing these points.

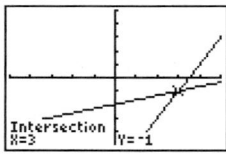

In Exercises 22–34 determine the solution for each system of linear equations.

22.   **23.**

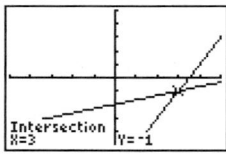   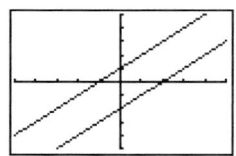

24.   **25.**

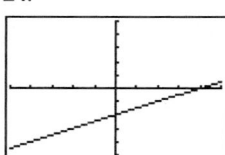

26.   **27.**

28. $2x - y = -1$ **29.** $3x + 2y = 5$
         $3x - 2y = -7$ $5x - 4y = 8$

30. $\dfrac{x}{2} + \dfrac{y}{3} = 1$ **31.** $1.2x + 2.3y = 168$
                                                   $2.5x - 3.1y = -123.5$
         $\dfrac{x}{4} + \dfrac{y}{5} = 8$

32. $4x - 7y = 3$ **33.** $7x + 10y = 4$
         $4x - 7y = 8$ $14x - 5y = -7$
34. $1.1x - 3.3y = 4.4$
         $0.8x - 2.4y = 3.2$

In Exercises 35 and 36 mentally estimate the slope of each line and then use a calculator to calculate the slope.

35.   **36.**

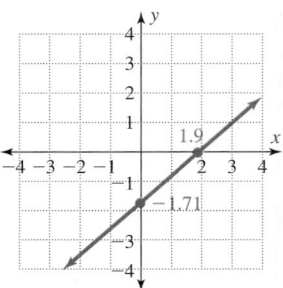   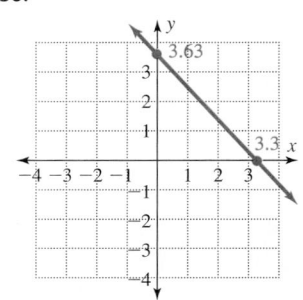

In Exercises 37–39 match each word problem with the corresponding system of equations. Then use these equations to solve the problem.

Verbally	**Algebraically**

37. The sum of two numbers is 9. Twice the smaller number plus the larger number is 12. Find these numbers.

38. The difference of two numbers is 9. The larger number plus three times the smaller number is 13. Find these numbers.

39. The sum of two numbers is 13 and their difference is 9. Find these numbers.

A. $x - y = 9$
        $x + 3y = 13$

B. $x + y = 13$
        $x - y = 9$

C. $x + y = 9$
        $2x + y = 12$

40. Fixed and Variable Printing Costs   A mathematics department is preparing a color supplement for their beginning algebra class. Two options are presented to the department chair. Option 1 is to pay 35 cents per page to a local printer. Option 2 is to buy a color printer for $2200 and to print the supplement in the department for a variable cost of 13 cents per page. What total page count would produce the same cost for both options?

41. Supplementary Angles   The angles labeled x and y in the figure are supplementary. (Their sum is 180°.) The larger angle is 42° more than twice the smaller angle. Determine the number of degrees in each angle.

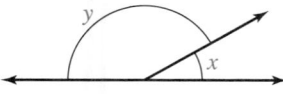

42. Income on Two Investments   An investment of $8000 earned a net gain of $220 in one year. Part of the investment was in bonds and earned income at the rate of 7%. The rest of the investment was in stocks and lost money at the rate of 10%. How much was invested in bonds?

43. Dimensions of an Isosceles Triangle   The perimeter of an isosceles triangle (two equal sides) is 37 cm. The base is 5 cm shorter than each of the other two sides. Find the length of the base. (See the figure.)

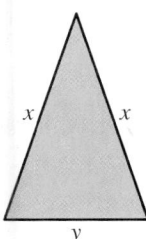

44. Rates of Two Trains Two trains depart simultaneously from a station, traveling in opposite directions. One averages 8 km/h more than the other, and after $\frac{1}{2}$ hour they are 100 km apart. Determine the speed of each train.

45. Mixture of Chemicals Petrochemicals worth $6 per liter are mixed with other chemicals worth $9 per liter to produce 100 liters of chemicals worth $7.65 per liter. How many liters of each are used?

46. Modeling the Costs of Two Music Clubs Two popular music clubs have an introductory offer. Club A charges a $6 initiation fee plus $0.95 per CD. Club B does not charge any initiation fee but charges $1.95 per CD. The following graph compares the cost of membership in each club based on the number of CDs purchased.
  a. Give the equation in the slope-intercept form for the line representing the costs for club A.
  b. Interpret the meaning of the slope and y-intercept of the line for club A.
  c. Give the equation in the slope-intercept form for the line representing the costs for club B.
  d. Interpret the meaning of the slope and the y-intercept of the line for club B.
  e. Determine the exact solution to the corresponding system of equations.

f. Interpret the meaning of the x- and y-coordinates of this solution.

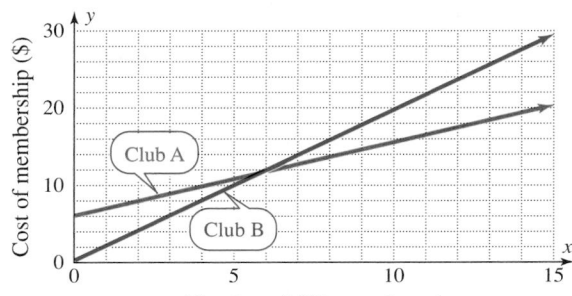

Number of CDs purchased

47. Modeling Automotive Service Charges The following tables display the charges by two automotive repair shops based on the number of hours required for a repair.
  a. Give the equation in slope-intercept form for the line representing the costs for shop A.
  b. Interpret the meaning of the slope and the y-intercept of the line for shop A.
  c. Give the equation in slope-intercept form for the line representing the costs for shop B.
  d. Interpret the meaning of the slope and the y-intercept of the line for shop B.
  e. Give the solution to the corresponding system of equations.
  f. Interpret the meaning of the x- and y-coordinates of this solution.

SHOP A

x HOURS	y COST ($)
1	50
2	80
3	110
4	140

SHOP B

x HOURS	y COST ($)
1	56
2	82
3	108
4	134

MASTERY TEST FOR CHAPTER 3

[3.1] **1.** Calculate the slope of the line through the given points.
  a. $(3, -1)$ and $(5, 3)$ **b.** $(3, -1)$ and $(1, 5)$
  c. $(3, -1)$ and $(5, -1)$ **d.** $(3, -1)$ and $(3, 5)$

[3.1] **2.** Determine whether the line defined by the first equation is parallel to, perpendicular to, or neither parallel nor perpendicular to the line defined by the second equation.

  a. $y = \frac{1}{2}x - 3$

  $y = -2x + 3$

  b. $y = \frac{1}{2}x - 3$

  $y = \frac{1}{2}x + 3$

  c. $y = 3$

  $x = 4$

  d. $y = 2x + 3$

  $y = -2x + 3$

[3.2] **3.** Write in the slope-intercept form the equation of a line satisfying the given conditions.
  a. through $(1, 4)$ with slope -2
  b. through $(-1, 3)$ with slope 4
  c. y-intercept $(0, 5)$ and slope $\frac{2}{3}$
  d. y-intercept $(0, -2)$ and slope $-\frac{5}{3}$

[3.2] **4. a.** Write the equation of a horizontal line through $(-4, 3)$.

b. Write the equation of a vertical line through $(-4, 3)$.

c. Write the equation of a line through $(-4, 3)$ and parallel to $x = 5$.

d. Write the equation of a line through $(-4, 3)$ and perpendicular to $x = 5$.

[3.2] **5.** Graph a line satisfying the given conditions.

a. through $(3, 1)$ with slope -1

b. through $(0, 0)$ with slope $\dfrac{1}{3}$

c. y-intercept $(0, 2)$ and slope $-\dfrac{3}{4}$

d. y-intercept $(0, -4)$ and slope $\dfrac{5}{3}$

[3.3] **6.** Graphically determine the simultaneous solution for each system of linear equations.

a.

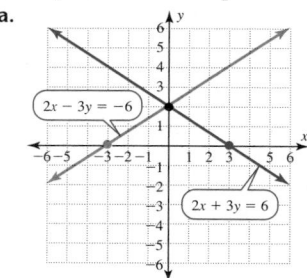

b. $f(x) = 2x + 1$
$f(x) = x + 3$

Use a numerical table to determine the simultaneous solution for each system of linear equations.

c.

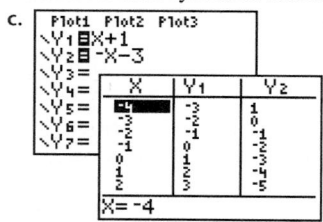

d. $y = 2x - 19$
$y = 3x - 31$

[3.3] **7.** Select the choice A, B, or C that best describes each system of linear equations.

a. $f(x) = 2x - 3$
$f(x) = 3x + 7$

b. $f(x) = 2x - 3$
$f(x) = 2x + 7$

c. $f(x) = 2x - 3$
$f(x) = 2x - 3$

A. A consistent system of dependent equations

B. A consistent system of independent equations

C. An inconsistent system of independent equations

[3.4] **8.** Solve each system of linear equations by the substitution method.

a. $y = -4x - 1$
$2x + y = 3$

b. $y = 4x - 1$
$2x + 3y = -2$

c. $x + 3y = 1$
$2x - 3y = 5$

d. $x - 2y = 5$
$5x + 10y = 12$

[3.5] **9.** Solve each system of linear equations by the addition method.

a. $3x - 7y = -13$
$2x + 7y = 3$

b. $3x + 19y = 16$
$6x - 11y = -17$

c. $3x - 2y = 6$
$-6x + 4y = -12$

d. $y = \dfrac{3}{5}x + 2$
$9x - 15y = 20$

[3.6] **10.** Use systems of linear equations to solve each word problem.

a. Complementary Angles   Two angles are complementary. If the larger angle is $12°$ more than five times the smaller angle, determine the number of degrees in each angle.

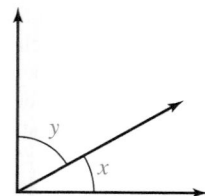

b. Value of Two Investments   An investment of $10,000 earned a net income of $625 in one year. Part of the investment was in bonds and earned income at a rate of 7%. The rest of the investment was in a savings account that earned income at a rate of 4%. How much was invested in bonds and how much in the savings account?

c. Mixture of Two Medicines   The dosage of medicine ordered by a doctor is 25 milliliters (mL) of a 50% solution. A nurse has available a 30% solution and an 80% solution of this medicine. How many milliliters of each should be mixed to produce this 25-mL dosage?

d. Rates of Two Planes   A jet plane and a tanker are 350 mi apart. They head toward each other so that the jet can refuel. The jet flies 250 mi/h faster than the tanker. Determine the speed of each aircraft if they meet in 30 minutes.

GROUP PROJECT   FOR CHAPTER 3

OSHA Standards for Stair-Step Rise and Run

This project requires a tape measure, a ruler, or a yardstick.

The OSHA standards for stairs for single-family residences vary from the standards for other buildings. The standards for single-family residences set a maximum on the rise of 8 in and a minimum on the run of 9 in.

OSHA* Standard 1910.24
Definitions:

> **Rise:** *The vertical distance from the top of a tread to the top of the next higher tread.*
> **Run:** *The horizontal distance from the leading edge of a tread to the leading edge of an adjacent tread.*

1. There are ten steps to the first landing in the stairs of the home of the author, James Hall. Each of these steps has a rise of 8 in and a run of $9\frac{3}{4}$ in.
   a. Do these stairs meet the stated standard for single-family residences?
   b. What is the total rise in feet and inches of these ten steps?
   c. What is the total run in feet and inches of these ten steps?
   d. Calculate the slope of the line from the leading edge of one step to the leading edge of the next step to the nearest hundredth.
2. a. Using the OSHA definitions, measure the rise and run of a set of stairs on your campus.
   b. How many steps are on this set of stairs?
   c. What is the total rise of these steps?
   d. What is the total run of these steps?
   e. Calculate the slope of the line from the leading edge of one step to the leading edge of the next step.
   f. Which stairs are more steep, those given for the Hall house or those in your school?

*Occupational Safety and Health Act

4

LINEAR

INEQUALITIES

AND SYSTEMS

OF LINEAR

INEQUALITIES

Restrictions and limitations on quantities can be expressed by a variety of words in signs, messages, and written contracts. Algebraically, we can express these restrictions and limitations by inequalities. These inequalities enable us to use the conciseness and precision of algebra and then to apply the visual power of graphs to understand better these relationships.

Section 4.1   Solving Linear Inequalities Using the Addition-Subtraction Principle

Objectives:

1. Check a possible solution of an inequality.
2. Use linear equations with two variables to solve a linear inequality with one variable.
3. Solve linear inequalities using the addition-subtraction principle.

A Mathematical Note

The symbols $>$ and $<$ for greater than or less than are due to Thomas Harriot (1560–1621). These symbols were not immediately accepted, as many mathematicians preferred the symbols ☐ and ☐.

This section presents linear inequalities and three methods for solving these inequalities—a graphical method, a numerical method, and the addition-subtraction principle. Although the steps used to solve linear inequalities are very similar to the steps used to solve linear equations, there are some important distinctions. We build on the similarities and carefully point out these distinctions.

Linear Inequalities

VERBALLY	ALGEBRAICALLY	ALGEBRAIC EXAMPLES	GRAPHICALLY
A linear inequality in one variable is an inequality that is the first degree in that variable.	For real constants A and B, with $A \neq 0$. $Ax > B$	$x > 2$	
	$Ax \geq B$	$x \geq 2$	
	$Ax < B$	$x < 2$	
	$Ax \leq B$	$x \leq 2$	

Remember the convention used is a parenthesis indicates an endpoint is not included in the interval. A bracket indicates an endpoint is included in the interval.

A conditional inequality contains a variable and is true for some, but not all, real values of the variable. A value that makes an inequality a true statement is a solution of the inequality. To solve an inequality is to find all the solutions of the inequality. Since inequalities often have an infinite interval of solutions, it is common to represent these solutions with a graph or by using interval notation (see Section 1.1).

■ **EXAMPLE 1**   Checking Possible Solutions of an Inequality

Determine whether the given values are solutions of $6x - 2 < 5x - 4$.

SOLUTIONS

(a) $x = -3$

$$6x - 2 < 5x - 4$$
$$6(-3) - 2 \not< 5(-3) - 4$$
$$-18 - 2 \not< -15 - 4$$
$$-20 \not< -19 \text{ is true}$$

Substitute the given value for x in the inequality and determine if this makes the inequality a true statement.

Answer: -3 is a solution.

(b) $x = -2$

$$6x - 2 < 5x - 4$$
$$6(-2) - 2 \not< 5(-2) - 4$$
$$-12 - 2 \not< -10 - 4$$
$$-14 \not< -14 \text{ is false}$$

If the inequality were $6x - 2 \le 5x - 4$, then $x = -2$ would be a solution.

Answer: -2 is not a solution.   ■

SELF-CHECK 4.1.1

Determine whether either -4 or 4 is a solution of the inequality in Example 1.

Some of the more common phrases used to indicate inequalities are given in the following box.

Phrases Used to Indicate Inequalities

PHRASE	VERBAL MEANING	INEQUALITY NOTATION	INTERVAL NOTATION	GRAPHICAL NOTATION
■ "x is at least a" ■ "x is a minimum of a"	x is greater than or equal to a	$x \ge a$	$[a, +\infty)$	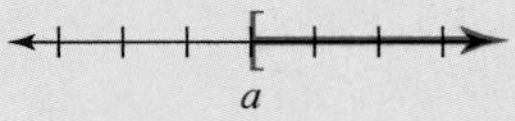
■ "x is at most a" ■ "x is a maximum of a" ■ "x never exceeds a"	x is less than or equal to a	$x \le a$	$(-\infty, a]$	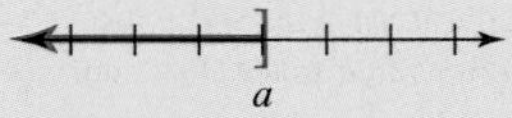
■ "x exceeds a"	x is greater than a	$x > a$	$(a, +\infty)$	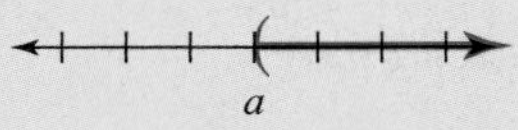

SELF-CHECK 4.1.1 ANSWER

-4 is a solution; 4 is not a solution.

■ **EXAMPLE 2**   Translating Verbal Statements into Algebraic Inequalities

Write each of these statements in both inequality notation and interval notation, and then graph the solution of this inequality.

SOLUTIONS

	INEQUALITY	INTERVAL	GRAPH
(a) x is at least 4	$x \geq 4$	$[4, +\infty)$	
(b) x is at most 3	$x \leq 3$	$(-\infty, 3]$	
(c) x exceeds 2	$x > 2$	$(2, +\infty)$	
(d) x cannot exceed 2	$x \leq 2$	$(-\infty, 2]$	
(e) x is a minimum of 5	$x \geq 5$	$[5, +\infty)$	

Remember that correct usage of interval notation requires the smaller value on the left and the larger value on the right. The notation (2, 7] is correct, whereas [7, 2) is incorrect.

The solution to a linear inequality in one variable can be determined graphically by letting y_1 represent the left side of the inequality and y_2 represent the right side of the inequality. In Example 3 we will solve $2x - 5 \leq 1$ by letting y_1 represent $2x - 5$ and y_2 represent 1 and then using a graph to determine where $y_1 \leq y_2$.

■ **EXAMPLE 3**   Using a Graph to Solve a Linear Inequality

Solve $2x - 5 \leq 1$ by graphing $y_1 = 2x - 5$ and $y_2 = 1$.

SOLUTION

First graph both $y_1 = 2x - 5$ and $y_2 = 1$. Then determine the point of intersection of these two lines; in this case the point $(3, 1)$. Observe where $y_1 \leq y_2$. This is where the graph of y_1 is below y_2 or intersects y_2, in this case for $x \leq 3$.

The key is to determine where the graph of y_1 is below the graph of y_2. Then write the x-values that make this true.

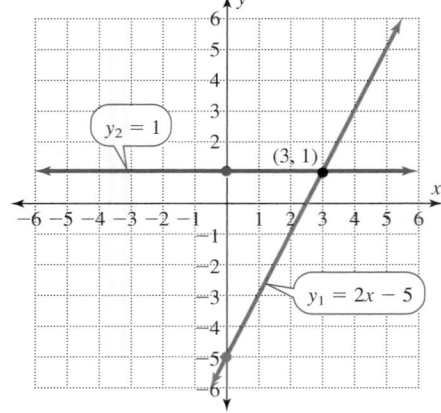

Note that the solution consists of the interval of values ←———┤——+—→ . This is the portion of the x-axis for which y_1 is below y_2; $y_1 \leq y_2$. The endpoint of this interval is the x-value where y_1 and y_2 are equal.

Answer: $(-\infty, 3]$

CALCULATOR PERSPECTIVE 4.1.1	Using a Table and a Graph to Solve a Linear Inequality

To solve the inequality $2x + 1 > -x + 4$ on a TI-83 Plus calculator, enter the following keystrokes:

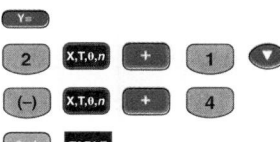

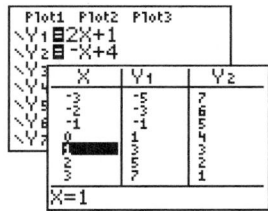

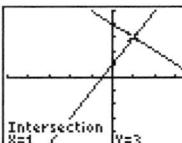

[−5, 5, 1] by [−5, 5, 1]

Note that y_1 has a positive slope and y_2 has a negative slope. This can help to keep the two lines properly identified.

Note: The solution to the inequality is the set of x-values for which $y_1 > y_2$. This occurs when x is greater than 1. Both the table and the graph support this result.

Answer: $(1, +\infty)$

■ EXAMPLE 4 Using a Calculator to Solve a Linear Inequality

Use a graphics calculator to solve $x - 2 > -x - 6$ both numerically and graphically.

SOLUTION _____

Let $y_1 = x - 2$
Let $y_2 = -x - 6$

Let y_1 represent the left side of the inequality and y_2 represent the right side of the inequality.

NUMERICALLY

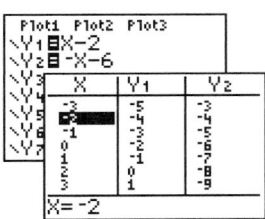

GRAPHICALLY

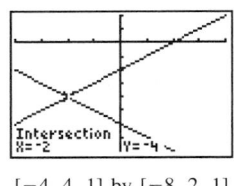

[−4, 4, 1] by [−8, 2, 1]

Then select an appropriate table and window for your calculator.

In the table $y_1 > y_2$ for $x > -2$. Also the graph of y_1 is above the graph of y_2 for $x > -2$.

Answer: $(-2, +\infty)$ whose graph is

-5 -4 -3 -2 -1 0 1 ■

SELF-CHECK 4.1.2

Use this table to solve the inequality for the x-values for which $y_1 < y_2$.

X	Y₁	Y₂
1	7	8
1.1	7.3	8.1
1.2	7.6	8.2
1.3	7.9	8.3
1.4	8.2	8.4
1.5	8.5	8.5
1.6	8.8	8.6

X=1

The procedure for solving a linear inequality algebraically is very similar to the procedure we have already used to solve linear equations. The basic idea is to isolate the variable terms on one side of the inequality and the constant terms on the other side. Inequalities that have the same solution are called **equivalent inequalities.** We use the addition-subtraction principle to produce equivalent inequalities that are simpler than the original inequality.

Addition-Subtraction Principle for Inequalities

VERBALLY	ALGEBRAICALLY*	NUMERICAL EXAMPLE
If the same number is added to or subtracted from both sides of an inequality, the result is an equivalent inequality.	If a, b, and c are real numbers, then $a < b$ is equivalent to $a + c < b + c$ and to $a - c < b - c$.	$x - 2 < 5$ is equivalent to $x - 2 + 2 < 5 + 2$ and to $x < 7$.

*Similar statements can be made for the inequalities $\leq$, $>$, and $\geq$.

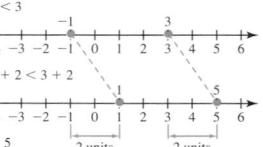

$-1 < 3$

$-1 + 2 < 3 + 2$

$1 < 5$

2 units   2 units

Figure 4.1.1

The logic behind the addition-subtraction principle can be explained by examining the two graphs shown in Figure 4.1.1. Adding 2 to both sides of an inequality shifts both points to the right 2 units. Thus their position relative to each other is preserved. Likewise, subtracting 3 from both sides of an inequality shifts both points to the left 3 units. Thus their position relative to each other also is preserved.

SELF-CHECK 4.1.2 ANSWER

$(-\infty, 1.5)$ whose graph is

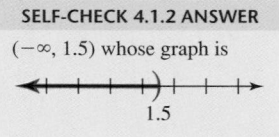

1.5

■ **EXAMPLE 5** Solving a Linear Inequality with One Variable Term

Solve $y + 3 \geq 7$ and graph the solution set.

SOLUTION _____

It is wise to check a value on both sides of 4 to help ensure that the correct interval is selected.

$$y + 3 \geq 7$$
$$y + 3 - 3 \geq 7 - 3 \quad \text{To isolate the variable on the left side, subtract 3 from both sides of the}$$
$$y \geq 4 \quad \text{inequality. This inequality is equivalent to the original inequality.}$$

Answer: $[4, +\infty)$ whose graph is

1 2 3 4 5 6 7

■ **EXAMPLE 6**   Solving a Linear Inequality with Variables on Both Sides

Solve $6v - 2 < 5v - 3$ and graph the solution set.

SOLUTION _____

$$6v - 2 < 5v - 3$$
$$6v - 5v - 2 < 5v - 5v - 3 \qquad \text{Subtract } 5v \text{ from both sides of the inequality.}$$
$$v - 2 < -3$$
$$v - 2 + 2 < -3 + 2 \qquad \text{Add 2 to both sides of the inequality.}$$
$$v < -1$$

Answer: $(-\infty, -1)$ whose graph is

The strategy for solving linear inequalities is much the same as the one used to solve linear equations. At each step try to produce simpler expressions than on the previous step. In Example 7 we start by using the distributive property to remove parentheses.

■ **EXAMPLE 7**   Solving a Linear Inequality Containing Parentheses

Solve $5(x + 2) \geq 3(2x + 3) - 1$ and graph the solution set.

SOLUTION _____

$$5(x + 2) \geq 3(2x + 3) - 1$$
$$5x + 10 \geq 6x + 9 - 1 \qquad \text{Use the distributive property to remove the parentheses and then}$$
$$5x + 10 \geq 6x + 8 \qquad \text{combine like terms.}$$
$$5x + 10 - 5x \geq 6x + 8 - 5x \qquad \text{Subtract } 5x \text{ from both sides of the inequality.}$$
$$10 \geq x + 8$$
$$10 - 8 \geq x + 8 - 8 \qquad \text{Subtract 8 from both sides of the inequality.}$$
$$2 \geq x \text{ or} \qquad \text{The statements } a \geq b \text{ and } b \leq a \text{ have the same meaning.}$$
$$x \leq 2$$

Answer: $(-\infty, 2]$ whose graph is

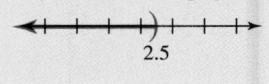
SELF-CHECK 4.1.3

Solve and graph the solution set of these inequalities.

1. $2x + 4 > x$ **2.** $3(x - 1.5) - 2 < 2(x - 2)$

It is common to express the inequality so that x is the subject of the sentence that gives the solution.

In Example 7 note that we solved the problem so that the coefficient of x would be 1. The solution $2 \geq x$ was then rewritten in the equivalent form $x \leq 2$, which is more common since x is the subject of the sentence "x is less than or equal to 2." All the exercises in this section can be solved by using the addition-subtraction principle to produce inequalities whose variables have a coefficient of 1. We will work with other coefficients, including negative ones, in Section 4.2.

■ EXAMPLE 8   Using Multiple Perspectives to Solve a Linear Inequality

Solve $\dfrac{x}{2} - 1 \geq 1 - \dfrac{x}{2}$ algebraically, graphically, and numerically and describe the solution verbally.

SOLUTION

ALGEBRAICALLY

$$\frac{x}{2} - 1 \geq 1 - \frac{x}{2}$$

$$\frac{x}{2} - 1 + \frac{x}{2} \geq 1 - \frac{x}{2} + \frac{x}{2}$$

$$x - 1 \geq 1$$

$$x - 1 + 1 \geq 1 + 1$$

$$x \geq 2$$

GRAPHICALLY

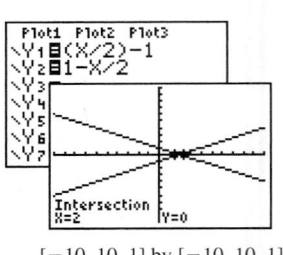

$[-10, 10, 1]$ by $[-10, 10, 1]$

NUMERICALLY

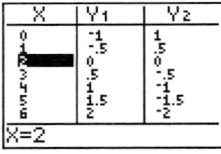

VERBALLY

All values of x greater than or equal to 2 satisfy this inequality. From the graph, y_1 is on or above y_2 for $x \geq 2$. In the table $y_1 \geq y_2$ for $x \geq 2$.

Answer: $[2, +\infty)$ whose graph is

```
<----+---+---+---[---+---+---+--->
    -1   0   1   2   3   4   5
```

■ EXAMPLE 9   Solving a Verbally Stated Inequality

If twice one number is increased by 11, the result is at least five more than the original number. What numbers satisfy this statement?

SOLUTION

Let $n =$ the original number.

Twice a number increased by 11	Is greater than or equal to	5 more than the number.
$2n + 11$	$\geq$	$n + 5$

Let n represent the number, and translate this inequality into an algebraic inequality.

$$2n + 11 - n \geq n + 5 - n$$

Subtract n from both sides of the inequality.

$$n + 11 \geq 5$$

$$n + 11 - 11 \geq 5 - 11$$

Subtract 11 from both sides of the inequality.

$$n \geq -6$$

Answer: The original number is greater than or equal to -6.

It might be wise to check both -7 and 1; -7 does not check and 1 does check.

USING THE LANGUAGE AND SYMBOLISM OF MATHEMATICS   4.1

1. An inequality that is first degree in x is called a _____ inequality in x.
2. An inequality is first degree in x if the exponent on x is _____.
3. Inequalities with the same solution are called _____ inequalities.
4. A _____ inequality contains a variable and is true for some, but not all, real values of the variable.
5. A value that makes an inequality a true statement is a _____ of the inequality.
6. By the _____ - _____ principle, $a < b$ is equivalent to $a + c < b + c$.

7. By the _____ - _____ principle, $a < b$ is equivalent to $a - c < b - c$.
8. The statement "x is at least five" is represented by the inequality x _____ 5.
9. The statement "x is at most five" is represented by the inequality x _____ 5.
10. The statement "x never exceeds five" is represented by the inequality x _____ 5.
11. The statement "x is a maximum of five" is represented by the inequality x _____ 5.
12. The statement "x is a minimum of five" is represented by the inequality x _____ 5.

EXERCISES   4.1

In Exercises 1 and 2 determine which of the following are linear inequalities in one variable.

1. **a.** $6x - 3$ **b.** $6x - 3 \le 4x$
   c. $y = 6x - 3$ **d.** $x^2 \ge 36$
2. **a.** $4x + 1$ **b.** $x^2 > 4$
   c. $4x + 1 > 9$ **d.** $y \le 4x + 1$

In Exercises 3 and 4 express each interval using interval notation.

3. **a.** $x \le -4$ **b.** $x > 5$
   c. (number line: arrow from left through parenthesis at 5, marks 0 1 2 3 4 5 6)   **d.** x is at least -1
4. **a.** $x \ge 1.5$ **b.** $x < -3$
   c. (number line: arrow from parenthesis at -2 to right, marks -5 -4 -3 -2 -1 0 1)   **d.** x is at most 5

5. Determine if $x = 3$ is a solution of each inequality.
   a. $x < 3$ **b.** $x \le 3$
   c. $x > 3$ **d.** $x \ge 3$
6. Determine if $x = 2$ is a solution of each inequality.
   a. $x \le 2$ **b.** $x < 2$
   c. $x \ge 2$ **d.** $x > 2$
7. Determine if $x = 3$ is a solution of each inequality.
   a. $x > -3$ **b.** $x > 5$
   c. $3x - 5 \ge 2(x - 1)$ **d.** $5x - 3 \le 2x + 3$
8. Determine if $x = 2$ is a solution of each inequality.
   a. $x > 3$ **b.** $x \ge -1$
   c. $3x - 5 \ge 2(x - 1)$ **d.** $5x - 3 \le 2x + 3$

In Exercises 9–12 use the graph to determine the x-values that satisfy each equation and inequality.

9. **a.** $y_1 = y_2$
   b. $y_1 < y_2$
   c. $y_1 > y_2$
10. **a.** $y_1 = y_2$
    b. $y_1 < y_2$
    c. $y_1 > y_2$

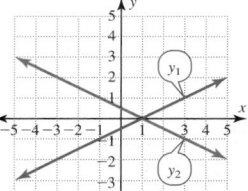
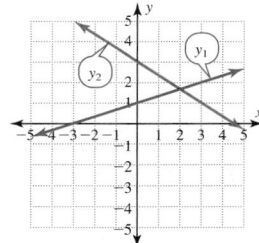

11. **a.** $y_1 = y_2$
    b. $y_1 \le y_2$
    c. $y_1 \ge y_2$
12. **a.** $y_1 = y_2$
    b. $y_1 \le y_2$
    c. $y_1 \ge y_2$

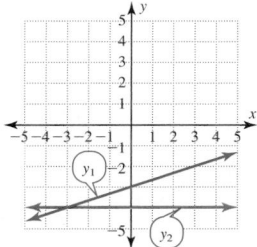
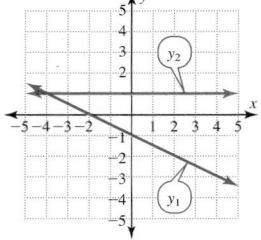

13. **Using a Graph that Models Overtime Spending Limits**
    Robinson Construction Company has budgeted a spending limit of $15,000 per year on overtime for their foremen. The graph of $y_1 = 15,000$ displays the spending limit and the graph of $y_2 = f(x)$ displays the cost for overtime based on the number of overtime hours worked by the foremen. Use the following graph to solve:
    a. $y_1 = y_2$
    b. $y_1 < y_2$

c. $y_1 > y_2$

d. Interpret the meaning of each of these solutions.

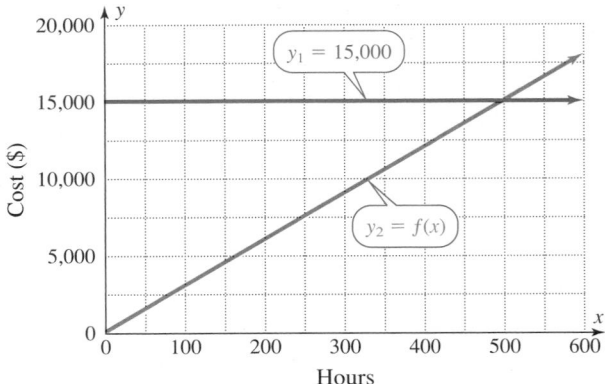

b. $y_1 < y_2$

c. $y_1 > y_2$

d. Interpret the meaning of each of these solutions.

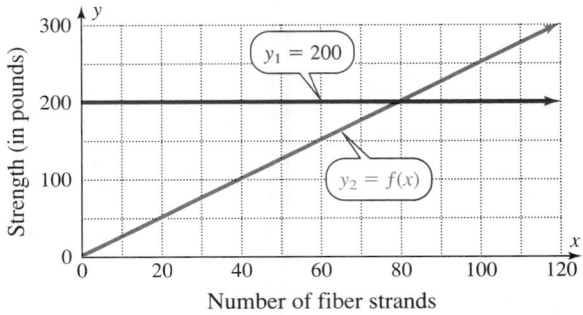

In Exercises 15–18 use the table of values to determine the x-values that satisfy each equation and inequality.

15. a. $y_1 = y_2$

  b. $y_1 \leq y_2$

  c. $y_1 \geq y_2$

16. a. $y_1 = y_2$

  b. $y_1 \leq y_2$

  c. $y_1 \geq y_2$

X	Y₁	Y₂
0	2	12
1	7	15
2	12	18
3	17	21
4	22	24
5	27	27
6	32	30
X=0		

X	Y₁	Y₂
0	-5	-9
1	-1	-3
2	3	3
3	7	9
4	11	15
5	15	21
6	19	27
X=0		

17. a. $y_1 = y_2$

  b. $y_1 < y_2$

  c. $y_1 > y_2$

18. a. $y_1 = y_2$

  b. $y_1 < y_2$

  c. $y_1 > y_2$

X	Y₁	Y₂
0	-1	2
.1	-.4	1.6
.2	.2	1.2
.3	.8	.8
.4	1.4	.4
.5	2	0
.6	2.6	-.4
X=0		

X	Y₁	Y₂
1	-3	0
1.2	-2.4	-.4
1.4	-1.8	-.8
1.6	-1.2	-1.2
1.8	-.6	-1.6
2	0	-2
2.2	.6	-2.4
X=1		

19. Using Tables that Model Supply Costs The following two tables give the costs, including delivery, for ordering copier paper from two different office supply firms. Use these tables to solve:

  a. $y_1 = y_2$

  b. $y_1 < y_2$

  c. $y_1 > y_2$

  d. Interpret the meaning of each of these solutions.

14. Using a Graph that Models Strength Limits for a Cable The contract for a composite fiber cable ordered for a sailboat from a cable manufacturer specifies that each cable must support a minimum load of 200 pounds. The graph of $y_1 = 200$ displays the strength limit and the graph of $y_2 = f(x)$ displays the strength of a cable based on the number of separate fiber strands braided into the cable. Use the following graph to solve:

  a. $y_1 = y_2$

ACE OFFICE SUPPLY		MAX OFFICE SUPPLY	
x Boxes of Paper	y_1 Cost ($)	x Boxes of Paper	y_2 Cost ($)
4	112	4	136
8	224	8	224
12	336	12	312
16	448	16	400
20	560	20	488
24	672	24	576

20. Using Tables that Model Distances Traveled The following two tables give the distances traveled from

Boston by a bus and by a train. In each table the time represents the number of hours after 8 A.M. eastern time. Use these tables to solve:

a. $y_1 = y_2$
b. $y_1 < y_2$
c. $y_1 > y_2$
d. Interpret the meaning of each of these solutions.

BUS

x Hours After 8:00 A.M.	y_1 Distance from Boston
1.0	50
1.5	75
2.0	100
2.5	125
3.0	150
3.5	175
4.0	200

TRAIN

x Hours After 8:00 A.M.	y_2 Distance from Boston
1.0	0
1.5	35
2.0	70
2.5	105
3.0	140
3.5	175
4.0	210

In Exercises 21–24 algebraically solve each inequality and show each step of your solution process.

21. $x + 7 > 11$
22. $x - 7 < -11$
23. $2x + 3 \le x + 4$
24. $5x - 3 \ge 4x - 9$

In Exercises 25–28 solve each inequality algebraically, graphically, and numerically and describe the solution verbally. (*Hint:* See Example 8.)

25. $2x - 1 \ge x + 2$
26. $0.5x + 2 > -0.5x - 2$
27. $-x + 3 < -2x + 5$
28. $3(x - 1) \le 2(x - 2)$

In Exercises 29–44 solve each inequality algebraically.

29. $x - 7 \le -3$
30. $x - 3 \le -7$
31. $5x - 8 < 4x - 8$
32. $11x + 12 < 10x + 12$
33. $7 \ge x$
34. $-4 \ge x$
35. $x - 4 > 2x - 3$
36. $4x + 9 > 5x + 3$
37. $5 - 9x < 6 - 10x$
38. $12 - 11x < 10 - 12x$
39. $5(x - 2) \le 4(x - 2)$
40. $7(x - 3) \ge 6(x - 4)$
41. $5(2 - x) > 4(3 - x)$
42. $2(3 - 2x) > 3(2 - x)$
43. $9(x - 5) \ge 4(2x - 2) + 3$
44. $11(x + 4) \ge 5(2x + 3) - 7$

Estimation and Calculator Skills
In Exercises 45–48 mentally estimate the solution of each inequality and then use a calculator to calculate the solution.

PROBLEM	MENTAL ESTIMATE	CALCULATOR SOLUTION
45. $x - 0.727 > 0.385$		
46. $x + 29.783 \le 70.098$		
47. $2x + 25.87 < x + 3.94$		
48. $5x - 1.61 \ge 4x + 5.42$		

In Exercises 49–60 solve each inequality by the method of your choice.

49. $5(w - 2) + 14 > 4(w + 3)$
50. $7(x - 3) + 16 \ge 6(x - 2)$

51. $3(5n - 2) \le 7(2n - 1)$
52. $13(2n - 1) < 5(5n - 3)$
53. $4(6m + 7) - 2m \ge 7(3m + 1) + 1$
54. $5(2m + 8) + 3m > 3(4m + 5) - 4$
55. $-3(2 - m) < -4(3 - m) + 4.5$
56. $-11(2 - w) - 10.5 \le -12(3 - w)$
57. $3y - 4(5 + 2y) > y - 6(3 + y)$
58. $19 - 3(4 - y) \ge 11 - 4(1 - y)$
59. $\frac{1}{2}x - \frac{1}{3} \ge \frac{5}{3} - \frac{1}{2}x$
60. $\frac{2}{3}x - \frac{4}{7} < \frac{3}{7} - \frac{1}{3}x$

Multiple Representations
In Exercises 61–66 first write an algebraic inequality for each verbal statement and then solve this inequality.
61. Five minus two times x is less than or equal to seven minus three times x.
62. Six plus four times x is greater than or equal to three times x plus five.
63. Twice w plus three results in a minimum of eleven more than w.
64. Five times w never exceeds three more than four times w.
65. Three times the quantity m plus five exceeds twelve more than twice m.
66. Six times the quantity m minus eleven is a maximum of ninety-six more than five times m.

67. Total Employees Because of cutbacks in the state budget, a driver's examination office is limited to employing at most 12 people. The office currently has one supervisor, one receptionist, and two clerks. The rest of the employees are driver's license examiners. How many driver's license examiners can this office have?

68. Total Grade Points To earn a grade of B, a student needs her exam points to total at least 320 points. Her first three test grades are 71, 79, and 78. What score, s, must she earn on the fourth test to obtain a B?

69. Perimeter of a Rectangular Pen A farmer has 84 ft of woven wire fence available to enclose three sides of a rectangular pen shown here. What are the possible values for the length x?

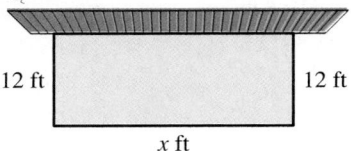

12 ft   12 ft

x ft

70. Perimeter of a Triangular Pen A buried pet security system is placed in the triangular corner of a yard. The pet owner can afford only 54 m of the buried wire. What are the possible values for the length of x?

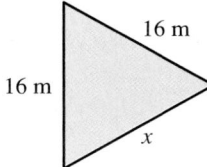

16 m

16 m

x

71. Modeling Costs and Revenue The daily cost of producing x units of computer mice includes a fixed cost of $450 per day and a variable cost of $4 per unit. The income produced by selling x units is $5 per unit. Letting y_1 represent the income and y_2 represent the cost, graph y_1 and y_2. Determine the values of x for which $y_1 < y_2$, the loss interval for this company. Also, determine the values of x for which $y_1 > y_2$, the profit interval for this company.

72. Modeling Costs and Revenue The daily cost of producing x units of cellular phones includes a fixed cost of $600 per day and a variable cost of $14 per unit. The income produced by selling x units is $15 per unit. Letting y_1 represent the income and y_2 represent the cost, graph y_1 and y_2. Determine the values of x for which $y_1 < y_2$, the loss interval for this company. Also determine the values of x for which $y_1 > y_2$, the profit interval for this company.

73. Dimensions of a Triangle The sum of the lengths of any two sides of a triangle is greater than the length of the

third side. Write three different inequalities expressing this concept for the triangle given below.

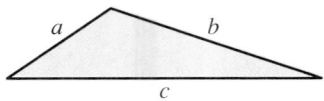

Group Discussion Questions

74. Writing Mathematically Write an algebraic expression for each verbal expression. Then explain the different meanings of these expressions.
  a. 3 more than x
  b. 3 is more than x
  c. 2 less than x
  d. 2 is less than x

75. Challenge Question You have solved an inequality and tested one value from your solution. This value checked, yet your solution is incorrect. Give two distinct examples that illustrate how this is possible.

Section 4.2 Solving Linear Inequalities Using the Multiplication-Division Principle

Objective: 4. Solve linear inequalities using the multiplication-division principle.

This section expands our ability to solve linear inequalities in one variable. We use the addition-subtraction principle to isolate the variable terms on one side of an inequality and the constant terms on the other side. Then we use the multiplication-division principle to obtain a coefficient of 1 for the variable.

Multiplication-Division Principle for Inequalities

VERBALLY	ALGEBRAICALLY*	NUMERICAL EXAMPLES
Order Preserving: If both sides of an inequality are multiplied or divided by a *positive number,* the result is an inequality that has the same solution as the original inequality.	If a, b, and c are real numbers and c is positive, then $a > b$ is equivalent to $ac > bc$.	$\dfrac{x}{2} > 3$ is equivalent to $2\left(\dfrac{x}{2}\right) > 2(3)$ and to $x > 6$
Order Reversing: If both sides of an inequality are multiplied or divided by a *negative number and the order of inequality is reversed,* the result is an inequality that has the same solution as the original inequality.	If a, b, and c are real numbers and c is negative, then $a > b$ is equivalent to $ac < bc$.	$-\dfrac{x}{3} > 5$ is equivalent to $(-3)\left(-\dfrac{x}{3}\right) < (-3)(5)$ and to $x < -15$

*Similar statements can be made for division and for the inequalities $<$, $\leq$, and $\geq$.

The logic behind the multiplication-division principle can be understood by examining the two graphs shown in Figures 4.2.1 and 4.2.2.

An Order-Preserving Multiplication

$$1 < 3$$
$$1(2) < 3(2)$$
$$2 < 6$$

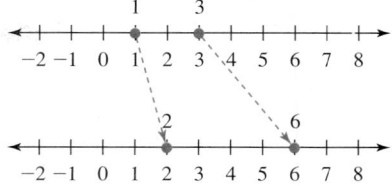

Figure 4.2.1 Multiplication-Division Principle with Positive Numbers

Multiplying both sides of the inequality by $+2$ appears to stretch out the distance between the points, whereas their position relative to each other is preserved. Likewise, dividing by $+2$ would preserve the order relation although the distance between the points would appear to shrink.

One way to illustrate visually the order-reversing case is with a meterstick. Label the meterstick with $-$, 0, and $+$ and points a and b with a to the left of b. Multiplying by a negative 1 changes the $-$ side to $+$, and the $+$ side to $-$. Illustrate this by rotating the meterstick $180°$ about the 0 point. This reverses the order in which the points a and b now occur on the meterstick.

An Order-Reversing Multiplication

$$-1 < 3$$
$$-1(-2) > 3(-2)$$
$$2 > -6$$

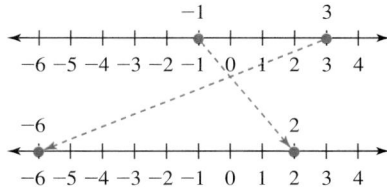

Figure 4.2.2 Multiplication-Division Principle with Negative Numbers

Multiplying both sides of the inequality by -2 changes the sign of each side of the inequality and thus reverses the order of their position relative to each other. Likewise, dividing by -2 would reverse the order relation.

■ EXAMPLE 1   Solving a Linear Inequality with a Positive Coefficient

Solve $3x \leq 6$ and graph the solution set.

SOLUTION _____

$$3x \leq 6$$
$$\frac{3x}{3} \leq \frac{6}{3}$$ Dividing both sides of the inequality by $+3$ preserves the order.
$$x \leq 2$$

Answer: $(-\infty, 2]$ whose graph is   ■

To solve an inequality for the variable w, we must isolate w on one side of the inequality, not $-w$. In the next example we solve for w by multiplying both sides of the inequality by -1. Note that this step reverses the order of the inequality.

■ **EXAMPLE 2** Solving a Linear Inequality with a Coefficient of -1

Solve $-w < 15$ and graph the solution set.

SOLUTION _____

$$-w < 15$$
$$(-1)(-w) > (-1)(15)$$ Multiplying both sides of the inequality by -1 reverses the order of the inequality.
$$w > -15$$

Answer: $(-15, +\infty)$ whose graph is

-15

SELF-CHECK 4.2.1

Solve these inequalities and graph their solutions.

1. $6w \geq -12$

2. $-\dfrac{1}{2}y < 2$

■ **EXAMPLE 3** Solving a Linear Inequality with Variables on Both Sides

Solve $5m - 7 < 3m - 13$.

SOLUTION _____

$$5m - 7 < 3m - 13$$
$$5m - 7 - 3m < 3m - 13 - 3m$$ Subtracting $3m$ preserves the order.
$$2m - 7 < -13$$
$$2m - 7 + 7 < -13 + 7$$ Adding 7 preserves the order.
$$2m < -6$$
$$\frac{2m}{2} < \frac{-6}{2}$$ Dividing by $+2$ preserves the order.
$$m < -3$$

Answer: $(-\infty, -3)$ whose graph is

-6 -5 -4 -3 -2 -1 0

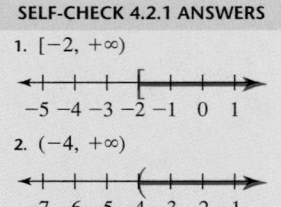
With practice some of the steps in the following example can be combined. This is illustrated by the alternative solution.

■ **EXAMPLE 4**   Solving a Linear Inequality with Variables
on Both Sides

Solve $6x - 2 \geq 8x + 12$.

SOLUTION ___

$$6x - 2 \geq 8x + 12$$
$$6x - 2 - 8x \geq 8x + 12 - 8x \quad \text{Subtracting } 8x \text{ serves the order.}$$
$$-2x - 2 \geq 12$$
$$-2x - 2 + 2 \geq 12 + 2 \quad \text{Adding 2 preserves the order.}$$
$$-2x \geq 14$$
$$\frac{-2x}{-2} \leq \frac{14}{-2} \quad \text{Dividing by } -2 \text{ reverses the order.}$$
$$x \leq -7$$

ALTERNATIVE SOLUTION _____

A shortened version of the solution process follows.

$$6x - 2 \geq 8x + 12$$
$$-2x \geq 14 \quad \text{Subtracting } 8x \text{ and adding 2 preserves the order.}$$
$$x \leq -7 \quad \text{Dividing by } -2 \text{ reverses the order.}$$

Answer: $(-\infty, -7]$ whose graph is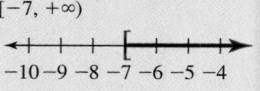
$-10\ -9\ -8\ -7\ -6\ -5\ -4$

■

SELF-CHECK 4.2.2

Solve $4y - 20 \leq 9y + 15$ and graph the solution.

■ **EXAMPLE 5**   Solving a Linear Inequality Containing
Parentheses

Solve $6(3 - 4n) < 5(2 - 3n) - 28$.

SOLUTION _____

$$6(3 - 4n) < 5(2 - 3n) - 28 \quad \text{Use the distributive property to remove the parentheses and then}$$
$$18 - 24n < 10 - 15n - 28 \quad \text{combine like terms.}$$
$$18 - 24n < -15n - 18$$
$$-24n < -15n - 36 \quad \text{Subtracting 18 preserves the order.}$$
$$-9n < -36 \quad \text{Adding } 15n \text{ preserves the order.}$$
$$n > 4 \quad \text{Dividing by } -9 \text{ reverses the order.}$$

Answer: $(4, +\infty)$ whose graph is
$1\ 2\ 3\ 4\ 5\ 6\ 7$

■

SELF-CHECK 4.2.2 ANSWER

$[-7, +\infty)$

$-10\ -9\ -8\ -7\ -6\ -5\ -4$

Linear inequalities with fractions, like linear equations with fractions, can be simplified by converting the inequality to an equivalent inequality that does not involve fractions. This can be accomplished by multiplying both sides of the inequality by the LCD (least common denominator) of all the terms.

■ EXAMPLE 6   Solving a Linear Inequality Containing Fractions

Solve $\dfrac{x}{15} \le \dfrac{x}{10} - \dfrac{5}{6}$.

SOLUTION —

$$\frac{x}{15} \le \frac{x}{10} - \frac{5}{6}$$

$$30\left(\frac{x}{15}\right) \le 30\left(\frac{x}{10}\right) - 30\left(\frac{5}{6}\right)$$

$$2x \le 3x - 25$$

$$-x \le -25$$

$$(-1)(-x) \ge (-1)(-25)$$

$$x \ge 25$$

$15 = 3 \cdot 5$
$10 = 2 \cdot 5$
$6 = 2 \cdot 3$
LCD $= 2 \cdot 3 \cdot 5 = 30$
Multiplying by 30, the LCD, preserves the order.

Subtracting $3x$ from both sides of the inequality preserves the order.

Multiplying by -1 reverses the order.

Answer: $[25, +\infty)$ whose graph is

22 23 24 25 26 27 28

SELF-CHECK 4.2.3

Solve $-\dfrac{14}{15} > -\dfrac{2}{3}x$ and graph the solution.

SELF-CHECK 4.2.3 ANSWER

$\left(\dfrac{7}{5}, +\infty\right)$

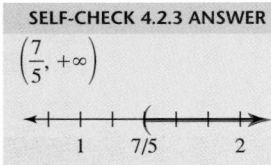

1   7/5   2

The ability to solve inequalities graphically and numerically is even more important when the inequalities become more complicated. Note that in Example 7 all perspectives yield the same result.

■ EXAMPLE 7   Using Multiple Perspectives to Solve a Linear Inequality

Solve $\dfrac{x}{2} + \dfrac{1}{6} \le \dfrac{x}{6} + \dfrac{1}{2}$ algebraically, graphically, and numerically and give the answer using interval notation.

SOLUTION _____

ALGEBRAICALLY

$$\frac{x}{2} + \frac{1}{6} \le \frac{x}{6} + \frac{1}{2}$$

$$6\left(\frac{x}{2}\right) + 6\left(\frac{1}{6}\right) \le 6\left(\frac{x}{6}\right) + 6\left(\frac{1}{2}\right)$$

$$3x + 1 \le x + 3$$

$$3x \le x + 2$$

$$2x \le 2$$

$$x \le 1$$

GRAPHICALLY

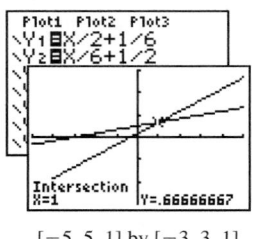

$[-5, 5, 1]$ by $[-3, 3, 1]$

NUMERICALLY

X	Y₁	Y₂
-3	⁻1.333	0
-2	⁻.8333	.16667
-1	⁻.3333	.33333
0	.16667	.5
1	.66667	.66667
2	1.1667	.83333
3	1.6667	1

X=1

VERBALLY

All values of x less than or equal to 1 satisfy this inequality. From the graph, y_1 is below or on y_2 for $x \le 1$. From the table y_1 is less than or equal to y_2 for $x \le 1$.

Answer: $(-\infty, 1]$ whose graph is

-2 -1 0 1 2 3 4

USING THE LANGUAGE AND SYMBOLISM OF MATHEMATICS 4.2

1. By the _____ - _____ principle for inequalities $x > y$ is equivalent to $2x > 2y$.
2. By the _____ - _____ principle for inequalities $x > y$ is equivalent to $-2x < -2y$.
3. If a, x, and y are real numbers and $a > 0$, then $x > y$ is equivalent to ax _____ ay.

4. If a, x, and y are real numbers and $a < 0$, then $x > y$ is equivalent to ax _____ ay.
5. To clear an inequality of fractions, we can multiply both sides of the inequality by the _____ _____ _____ of all the terms in the inequality.
6. The _____ property often is used to help us remove parentheses from expressions in an inequality.

EXERCISES 4.2

In Exercises 1 and 2 fill in each blank with the correct inequality symbol.

1. a. If $x > y$, then $x + 2$ _____ $y + 2$.
   b. If $x > y$, then $x - 2$ _____ $y - 2$.
   c. If $x > y$, then $\dfrac{x}{2}$ _____ $\dfrac{y}{2}$.
   d. If $x > y$, then $-2x$ _____ $-2y$.
2. a. If $x < y$, then $2x$ _____ $2y$.
   b. If $x < y$, then $\dfrac{x}{-2}$ _____ $\dfrac{y}{-2}$.
   c. If $x < y$, then y _____ x.
   d. If $x < y$, then $-x$ _____ $-y$.

In Exercises 3–6 use the graph to determine the x-values that satisfy each equation and inequality.

3. a. $y_1 = y_2$
   b. $y_1 < y_2$
   c. $y_1 > y_2$

4. a. $y_1 = y_2$
   b. $y_1 < y_2$
   c. $y_1 > y_2$

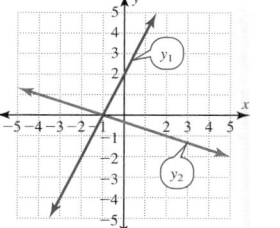
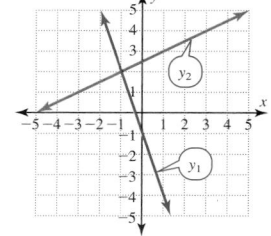

5. a. $y_1 = y_2$
   b. $y_1 \leq y_2$
   c. $y_1 \geq y_2$

6. a. $y_1 = y_2$
   b. $y_1 \leq y_2$
   c. $y_1 \geq y_2$

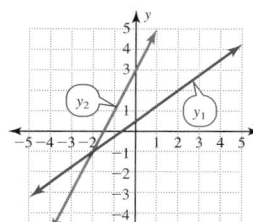
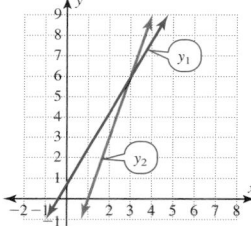

7. **Using a Graph that Models Rental Costs** The costs for renting a rug-shampooing machine from two different rental companies are given by the following graphs. The graph of $y_1 = f_1(x)$ gives the cost by Dependable Rental Company based on the number of hours of use. The graph of $y_2 = f_2(x)$ gives the cost by Anytime Rental Company based on the number of hours of use. Use these graphs to solve:
   a. $y_1 = y_2$
   b. $y_1 < y_2$
   c. $y_1 > y_2$
   d. Interpret the meaning of each of these solutions.

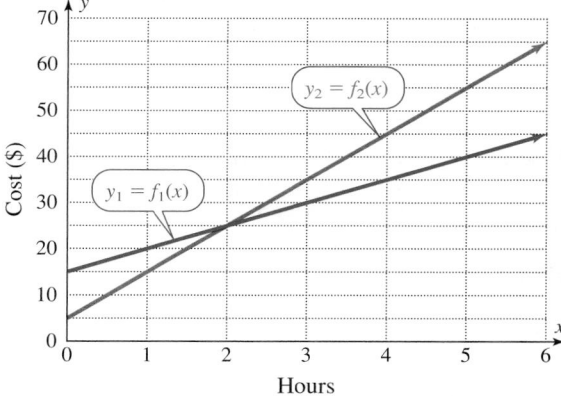

8. **Using a Graph that Models the Strength of Two Cables** A cable manufacturer makes cables that are wound around two different cores and braided from different fibers. Part of the strength comes from the core and the rest of the strength from the fiber strands. The graph of $y_1 = f_1(x)$ gives the strength of the first cable based on the number of separate fiber strands braided into this cable. The graph of $y_2 = f_2(x)$ gives the strength of the second cable based on the number of separate fiber strands braided into this cable. Use these graphs to solve:

a. $y_1 = y_2$
b. $y_1 < y_2$
c. $y_1 > y_2$
d. Interpret the meaning of each of these solutions.

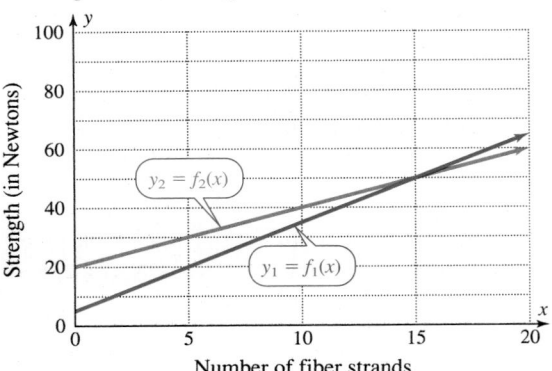

Number of fiber strands

In Exercises 9–12 use the table of values to determine the x-values that satisfy each equation or inequality.

9. a. $y_1 = y_2$
   b. $y_1 < y_2$
   c. $y_1 > y_2$

10. a. $y_1 = y_2$
   b. $y_1 < y_2$
   c. $y_1 > y_2$

X	Y₁	Y₂
0	-3	3
1	2	6
2	7	9
3	12	12
4	17	15
5	22	18
6	27	21

X=0

X	Y₁	Y₂
-3	-11	-14
-2	-7	-7
-1	-3	0
0	1	7
1	5	14
2	9	21
3	13	28

X=-3

11. a. $y_1 = y_2$
   b. $y_1 \le y_2$
   c. $y_1 \ge y_2$

12. a. $y_1 = y_2$
   b. $y_1 \le y_2$
   c. $y_1 \ge y_2$

X	Y₁	Y₂
-20	-67	-87
-15	-52	-72
-10	-37	-47
-5	-22	-22
0	-7	3
5	8	28
10	23	53

X=-20

X	Y₁	Y₂
-2	7	-23
-1	5	-19
0	3	-15
1	1	-11
2	-1	-7
3	-3	-3
4	-5	1

X=-2

13. Using Tables that Model Sweatshirt Costs The following two tables give the costs for ordering college logo sweatshirts from two sporting goods stores. In each case there is a setup charge for designing the logo as well as a cost per sweatshirt. Use these tables to solve:

a. $y_1 = y_2$
b. $y_1 < y_2$
c. $y_1 > y_2$
d. Interpret the meaning of each of these solutions.

MEL'S SPORTS

x Number of Sweatshirts	y₁ Cost ($)
10	310
20	560
30	810
40	1060
50	1310
60	1560

MICHAEL'S SPORTING GOODS

x Number of Sweatshirts	y₂ Cost ($)
10	320
20	565
30	810
40	1055
50	1300
60	1545

14. Using Tables that Model Bottle Production The following two tables give the number of glass bottles produced after a number of minutes by two different machines. For each machine there is an initial start-up time before any bottles will come out of either machine. Use these tables to solve:

a. $y_1 = y_2$
b. $y_1 < y_2$
c. $y_1 > y_2$
d. Interpret the meaning of each of these solutions.

MACHINE 1

x Minutes After Start-up	y₁ Bottles Produced
4	40
6	80
8	120
10	160
12	200
14	240
16	280

MACHINE 2

x Minutes After Start-up	y₂ Bottles Produced
4	20
6	70
8	120
10	170
12	220
14	270
16	320

In Exercises 15–18 algebraically solve each inequality and show each step of your solution process.

15. $-2x < 4$
16. $-3x \ge -12$
17. $-3x + 4 \ge 7$
18. $-5x - 7 \le 32$

Multiple Representations

In Exercises 19–22 solve each inequality algebraically, graphically, and numerically, and describe the solution verbally. (*Hint:* See Example 7.)

19. $2x < 5x + 6$
20. $3x + 1 < -x + 5$
21. $3(x - 5) \le 1 - x$
22. $2(x - 1) \ge 4 - x$

In Exercises 23–56 solve each inequality algebraically.

23. $2x \ge 4$
24. $4x < 2$
25. $\dfrac{1}{2}v < -4$
26. $\dfrac{1}{3}v > -\dfrac{2}{3}$
27. $-\dfrac{3}{4}w < -12$
28. $-\dfrac{4}{3}w \le -12$
29. $-x \ge 0$
30. $-x \le 5$
31. $1.1m < 2.53$
32. $2.1m \le 1.68$
33. $-\dfrac{n}{4} > \dfrac{1}{2}$
34. $-\dfrac{n}{6} \ge \dfrac{2}{3}$
35. $-8z \ge 40$
36. $-9z < -36$
37. $-\dfrac{t}{3} \ge 0$
38. $-\dfrac{2t}{5} > 0$
39. $-10 > -5x$
40. $-21 > -7x$
41. $5t \le 21 - 2t$
42. $8t < 3t + 35$
43. $9x + 7 > 5x - 13$
44. $23x - 14 > 18x + 56$
45. $11y - 8 > 14y + 7$
46. $8y + 19 \ge 2y - 11$
47. $5 - 9y \le 19 - 2y$
48. $13 - 7y \le 1 - y$
49. $2(3t - 4) > 2(t - 2)$
50. $7(2t - 4) \ge 3(3t + 2)$
51. $4(3 - x) \ge 7(2 - x)$
52. $5(1 - 2x) < 9(3 - x)$
53. $3(2y - 1) \le 5(3y + 4) + 4$

54. $6(3y + 2) \le 7(2y - 3) + 1$
55. $-2(3w + 7) < 3(5 - w) + 2$
56. $-5(4w - 2) > 3(3w + 5) - 5$

In Exercises 57–60 solve each inequality by the method of your choice.

57. $-3(5 - 2x) > -2(5x + 1) + 3$
58. $-4(2 - 3x) \ge -7(3 - x) - 2$
59. $6(7x - 8) - 4(3x + 2) \le 8(5x + 3)$
60. $-5y - 7(2y - 3) < 3y - 2(5 - y) - 5$

Estimation and Calculator Skills

In Exercises 61–64 mentally estimate the solution of each inequality and then use a calculator to calculate the solution.

PROBLEM	MENTAL ESTIMATE	CALCULATOR SOLUTION
61. $5.02x \le 35.642$		
62. $-4.95x \ge 31.185$		
63. $-0.47x > -4.183$		
64. $-0.99x < 118.8$		

Multiple Representations

In Exercises 65–68 first write an algebraic inequality for each verbal statement, and then solve this inequality.

65. Twice the quantity x plus five is less than six.
66. Three times the quantity x minus four is greater than fifteen.
67. Four times the quantity y minus seven is greater than or equal to six times the quantity y plus three.
68. Five times the quantity y plus nine is less than or equal to eight times the quantity y minus three.
69. **Temperature Limit**   The temperature in a computer room must be at least 41° Fahrenheit. Find the allowable temperature in degrees Celsius. The relationship between Fahrenheit and Celsius temperature is $F = \dfrac{9}{5}C + 32$.
70. **Budgetary Limit**   A company has set a limit of at most $450 to be spent on its monthly electric bill. If electricity costs $0.12 per kilowatt, find the number of kilowatts the company can use each month.

71. **Interest Payment**   The most a family can budget each month for an interest payment is $450. Their interest payment for one month will be approximately 0.75% of the amount the family can borrow. How much can the family borrow?
72. **Taxi Fare**   The minimum charge for a taxi ride is $2.40, plus an additional $0.15 per tenth of a mile. How far can you travel if you have at most $8.40 to spend on taxi fare?
73. **Modeling Costs and Revenue**   The cost of producing an order of bricks includes a fixed cost of $250 and a variable cost of $0.50 per brick. All bricks are custom ordered and the charge for manufacturing these bricks includes a setup fee of $150 plus a charge of $1 per brick. Let y_1 represent the income received from an order for x bricks and y_2 represent the cost of producing x bricks. Determine the values of x for which $y_1 < y_2$, the loss interval for this order. Also determine the values of x for which $y_1 > y_2$, the profit interval for this order.
74. **Modeling Costs and Revenue**   The cost of producing x flooring tiles includes a fixed cost of $400 and a variable cost of $1.25 per tile. All tiles are custom ordered and the charge for manufacturing these tiles includes a setup fee of $300 plus a charge of $2 per tile. Let y_1 represent the income received from an order for x tiles and y_2 represent the cost of producing x tiles. Determine the values of x for which $y_1 < y_2$, the loss interval for this order. Also determine the values of x for which $y_1 > y_2$, the profit interval for this order.

Group Discussion Questions

75. **Error Analysis**   Examine the following argument and explain the error in the reasoning.

$x < 0$
$x + x < x$ Add x to both sides.
$2x < x$
$2 < 1$ Divide both sides by x.

76. **Challenge Question**   If x and y are real numbers and $x > y$, is there a real number a so that $ax > ay$ is false and $ax < ay$ is also false? If this is possible, list all values of a that make this possible.
77. **Challenge Question**   Solve each inequality for x.
    a. $2x + 3y < 5y - 2x$ b. $8x + 5y > 4x - 3y$
    c. $2(x + 3y) \le -3(x + 3y)$ d. $5(x - 2y) > 2(7x - 5y)$

Section 4.3   Solving Compound Inequalities

Objectives: 5. Identify an inequality that is a contradiction or an unconditional inequality.
6. Solve compound inequalities involving intersection and union.

Unconditional Inequalities and Contradictions

Most of the inequalities that we consider are conditional inequalities. **Conditional inequalities** contain a variable and are true for some, but not all, real values of the variable. An inequality that is always true is called an **unconditional inequality,** whereas an inequality that is always false is called a **contradiction.**

◾ EXAMPLE 1   Classifying Inequalities

Identify each inequality as a conditional inequality, an unconditional inequality, or a contradiction.

SOLUTIONS

(a) $x < x + 1$ This inequality is true for every value of x and therefore is an unconditional inequality.

Every real number is less than one more than that number. Thus the solution set is the set of all real numbers, $\mathbb{R}$. The graph of the solution set is the entire real number line.

(b) $y < y - 1$ This inequality is false for every value of y and therefore is a contradiction.

No real number is less than the number one unit less than itself. The solution set is the empty set, a set with no elements. There are no solutions to plot on the number line.

(c) $z > 4$ This inequality is a conditional inequality since it is true for real numbers greater than 4 but is false if 4 or any value less than 4 is substituted for z.

The graph of the solution set is:

1 2 3 4 5 6 7

If simplifying an inequality results in an inequality with only constants and no variable, then the inequality is either a contradiction or an unconditional inequality. This is illustrated in the next example.

◾ EXAMPLE 2   Solving an Inequality When the Variable Is Eliminated

Solve each inequality.

SOLUTIONS

(a) $3x + 4 < 3x + 1$

$$3x + 4 < 3x + 1$$
$$3x + 4 - 3x < 3x + 1 - 3x$$
$$4 < 1 \text{ is a false statement.}$$

Subtracting $3x$ preserves the order. Because the last inequality is a contradiction, the original inequality is also a contradiction.

Answer: No solution. Since the inequality is a contradiction, there are no points to graph.

(b) $4v + 3 \geq 4v + 1$

$$4v + 3 \geq 4v + 1$$
$$4v + 3 - 4v \geq 4v + 1 - 4v$$
$$3 \geq 1 \text{ is a true statement.}$$

Subtracting $4v$ preserves the order. Since the last inequality is an unconditional inequality, the original inequality is also an unconditional inequality.

Answer: All real numbers. Because the inequality is an unconditional inequality, the graph of the solution set is the set of all real numbers, $\mathbb{R}$.

SELF-CHECK 4.3.1 ANSWERS

1. No solution.
2. $[6, +\infty)$
3. All real numbers, $\mathbb{R}$.

SELF-CHECK 4.3.1

Solve each inequality.

1. $2x + 5 \leq 2x - 1$
2. $2x + 5 \leq 3x - 1$
3. $2x + 5 \geq 2x - 1$

CALCULATOR PERSPECTIVE 4.3.1	Using a Table and a Graph to Solve Contradictions and Unconditional Inequalities

To solve the inequality $3x + 4 < 3x + 1$ from Example 2(a) on a TI-83 Plus calculator, enter the following keystrokes:

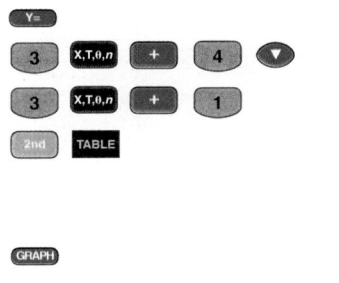 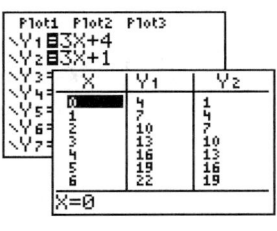

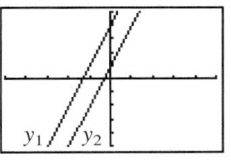

Note that y_1 has a y-intercept of $(0, 4)$ and y_2 has a y-intercept of $(0, 1)$. Thus the graph of y_1 is above the graph of y_2.

$[-5, 5, 1]$ by $[-5, 5, 1]$

Note: From the table we see that for every value of x, y_1 is larger than y_2 by 3. The graph of y_1 is also above the graph of y_2 for each x-value. There are no values of x for which $y_1 < y_2$. The inequality $3x + 4 < 3x + 1$ is a contradiction.

To solve the inequality $4v + 3 \geq 4v + 1$ from Example 2(b) on a TI-83 Plus calculator, enter the following keystrokes:

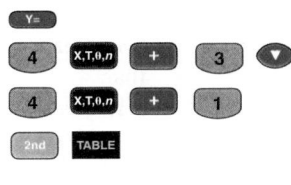

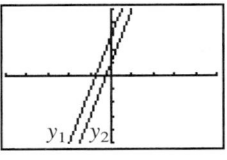

Note that y_1 has a y-intercept of $(0, 3)$ and y_2 has a y-intercept of $(0, 1)$. Thus the graph of y_1 is above the graph of y_2.

$[-5, 5, 1]$ by $[-5, 5, 1]$

Note: From the table we see that for every value of x, y_1 is larger than y_2 by 2. The graph of y_1 is also above the graph of y_2 for each x-value. For each value of x, $y_1 \geq y_2$. The inequality $4v + 3 \geq 4v + 1$ is an unconditional inequality.

Compound Inequalities

A compound inequality is formed when we place more than one restriction on a variable—for example, on the golf hole as shown in the following Figure, a golfer must hit a drive more than 153 yd but less than 195 yd to clear the water and to land safely on this island green.

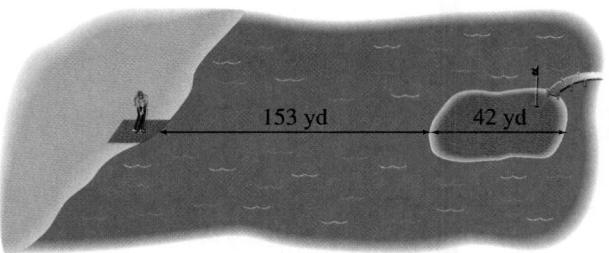

You can read a compound inequality with x as the subject of the sentence as in "x is greater than a and less than b."

The compound inequality $153 < x < 195$ is read "x is greater than 153 and less than 195." The compound inequality $a < x < b$ is equivalent to $x > a$ and $x < b$.

■ EXAMPLE 3   Interpreting Compound Inequalities

Write each of these compound inequalities using interval notation, graph the inequality, and give a verbal statement for the inequality.

SOLUTIONS

	INTERVAL NOTATION	GRAPH	VERBALLY
(a) $-1 \le x < 4$	$[-1, 4)$		x is greater than or equal to -1 and less than 4.
(b) $0 < x \le 5$	$(0, 5]$		x is greater than 0 and is less than or equal to 5.

SELF-CHECK 4.3.2

1. Write $w > -3$ and $w \le 4$ as a single compound inequality.
2. Graph $-2 \le x < 1$.
3. Write the inequality corresponding to the interval $[-5, 1]$.

SELF-CHECK 4.3.2 ANSWERS
1. $-3 < w \le 4$
2. (graph from -3 to 3)
3. $-5 \le x \le 1$

Compound inequalities can be solved by considering each of the component inequalities using either the graphical, numerical, or algebraic methods.

■ EXAMPLE 4   Solving a Compound Inequality

Solve $-3 < 2x - 1 \leq 5$ both numerically and graphically.

SOLUTION _____

Focus on what the table and graph are telling you. For the table look at the y-values from -3 to 5 and then give the x-values that produce these y-values. On the graph highlight the portion of the graph between -3 and 5 and then give the x-values that produce this portion of the graph.

NUMERICALLY

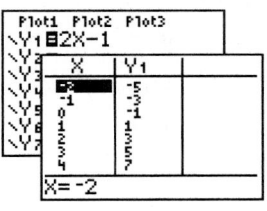

GRAPHICALLY

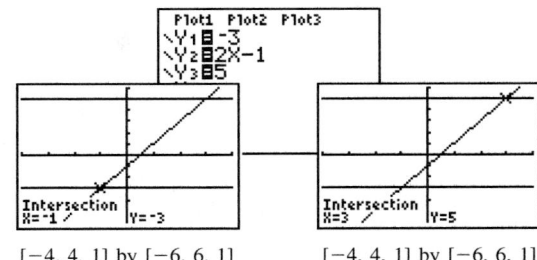

$[-4, 4, 1]$ by $[-6, 6, 1]$ $[-4, 4, 1]$ by $[-6, 6, 1]$

In the table $-3 < y_2 \leq 5$ for the x-values: $-1 < x \leq 3$. In the graph y_2 is between $y_1 = -3$ and $y_3 = 5$ for the x-values: $-1 < x \leq 3$.

Answer: $(-1, 3]$ whose graph is

```
←+—(—+—+—+—]—+→
 -2 -1  0  1  2  3  4
```

■

SELF-CHECK 4.3.3

Use this table to solve the inequality for the x-values for which $-5 \leq y_1 < 4$.

X	Y₁	
-3	-8	
-2	-5	
-1	-2	
0	1	
1	4	
2	7	
3	10	

X= -3

An algebraic procedure that can be used to solve many compound linear inequalities is to perform the same operation on each of the component inequalities. This procedure is illustrated in the following examples.

■ EXAMPLE 5   Solving a Compound Inequality

Solve each inequality.

SOLUTIONS _____

(a) $-4 < w - 2 \leq 1$

$$-4 < w - 2 \leq 1$$
$$-4 + 2 < w - 2 + 2 \leq 1 + 2 \qquad \text{Adding 2 preserves the order.}$$
$$-2 < w \leq 3$$

Answer: $(-2, 3]$ whose graph is

```
←+—(—+—+—+—+—]—+→
 -3 -2 -1  0  1  2  3
```

SELF-CHECK 4.3.3 ANSWER

$[-2, 1)$ whose graph is

```
←+—[—+—+—)—+—+→
 -3 -2 -1  0  1  2  3
```

Write your final answer with the smaller number in the left member of an inequality and the larger value in the right member. This is the same order these numbers occur on the number line.

(b) $-4 < -2x \leq 2$

$$-4 < -2x \leq 2$$

$$\frac{-4}{-2} > \frac{-2x}{-2} \geq \frac{2}{-2} \qquad \text{Dividing by } -2 \text{ reverses the order.}$$

$$2 > x \geq -1 \qquad \text{The inequalities } b > x > a \text{ and } a < x < b \text{ have}$$

$$-1 \leq x < 2 \qquad \text{the same meaning. Both of these inequalities state that } x \text{ is between } a \text{ and } b.$$

Answer: $[-1, 2)$ whose graph is

■ **EXAMPLE 6** Solving a Compound Inequality with Two Middle Terms

Solve $-5 \leq 3v + 4 \leq 1$.

SOLUTION _____

$$-5 \leq 3v + 4 \leq 1$$

$$-5 - 4 \leq 3v + 4 - 4 \leq 1 - 4 \qquad \text{First subtract 4 from each member of the inequality to isolate the}$$

$$-9 \leq 3v \leq -3 \qquad \text{variable in the middle term. Subtracting preserves the order.}$$

$$\frac{-9}{3} \leq \frac{3v}{3} \leq \frac{-3}{3} \qquad \text{Dividing by } +3 \text{ preserves the order.}$$

$$-3 \leq v \leq -1$$

Answer: $[-3, -1]$ whose graph is

SELF-CHECK 4.3.4

Solve these inequalities:
1. $0 \leq -z < 2$
2. $-1 < 2m + 1 \leq 3$

The intersection of two highways consists of the road common to both highways. The intersection of two sets consists of the elements common to both sets.

For some compound inequalities the variable cannot be isolated between the inequalities as in the previous example. In such cases the compound inequality must be split into two simple inequalities that can be solved individually. The final solution is then formed by the intersection of these individual sets. The **intersection** of two sets is indicated often by the connecting word *and* because the intersection of set A and set B consists of the elements that are in set A *and* in set B.

SELF-CHECK 4.3.4 ANSWERS

1. $(-2, 0]$ whose graph is

2. $(-1, 1]$ whose graph is

Intersection of Two Sets

ALGEBRAIC NOTATION	VERBALLY	NUMERICAL EXAMPLE	GRAPHICAL EXAMPLE
$A \cap B$	The intersection of A and B is the set that contains the elements in *both A and B*.	$(-3, 4) \cap [0, 6]$ $= [0, 4)$	The intersection of and is

Remember the convention used is a parenthesis indicates an endpoint is not included in the interval. A bracket indicates an endpoint is included in the interval.

■ **EXAMPLE 7**   Solving an Inequality Whose Solution Is the Intersection of Two Sets

Solve $5v - 8 \le 2v + 4 < 4v - 2$.

SOLUTION _____

This compound inequality is equivalent to:

$5v - 8 \le 2v + 4$ and $2v + 4 < 4v - 2$ Solve each of the inequalities individually.
$3v - 8 \le 4$ $\qquad\qquad -2v + 4 < -2$
$\quad 3v \le 12$ $\qquad\qquad\quad -2v < -6$
$\quad\;\; v \le 4$ $\qquad\qquad\qquad\;\; v > 3$

The intersection of $v \le 4$ and $v > 3$ is $3 < v \le 4$.

Answer: $(3, 4]$ whose graph is   ■

The United States is a union of states, the set containing all these states.

In many problems we are asked to form the *union* of the solution sets of two inequalities. The union of two sets is indicated often by the connecting word *or* since the union of set A and set B consists of the elements in set A *or* in set B.

Union of Two Sets

ALGEBRAIC NOTATION	VERBALLY	NUMERICAL EXAMPLE	GRAPHICAL EXAMPLE
$A \cup B$	The union of A and B is the set that contains the elements in either A or B or both.	$(-3, 4) \cup [0, 6]$ $= (-3, 6]$	The union of and is

Note: Elements are listed only once in the union, even if they occur in both sets.

The word "or" can be misunderstood if you think of this word only in the exclusive sense (as in "black *or* white") rather than in the inclusive sense used in the definition of the union of two sets.

■ EXAMPLE 8 Solving an Inequality Whose Solution Is the Union of Two Sets

Solve $2(w - 5) \geq 3w - 8$ or $3(w + 2) > w + 8$.

SOLUTION _____

$2(w - 5) \geq 3w - 8$	or	$3(w + 2) > w + 8$
$2w - 10 \geq 3w - 8$		$3w + 6 > w + 8$
$2w \geq 3w + 2$		$3w > w + 2$
$-w \geq 2$		$2w > 2$
$w \leq -2$		$w > 1$

Solve each of these inequalities individually. Then graph all the points that are in one set or the other set.

Answer: $(-\infty, -2] \cup (1, +\infty)$ whose graph is

$$\xleftarrow{\hspace{0.5cm}} \overset{\quad -3\; -2\; -1\;\; 0\;\;\; 1\;\;\; 2\;\;\; 3}{\rule{5cm}{0.4pt}} \xrightarrow{\hspace{0.5cm}}$$

The union symbol, $\cup$, is used to indicate that the numbers can be in either the first interval or the second interval. ■

SELF-CHECK 4.3.5 ANSWERS

1. $(4, 6)$ whose graph is

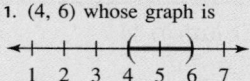

2. $[2, 7]$ whose graph is

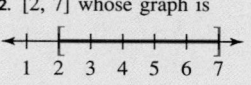

SELF-CHECK 4.3.5

Determine these sets.

1. $[2, 6) \cap (4, 7]$
2. $[2, 6) \cup (4, 7]$

USING THE LANGUAGE AND SYMBOLISM OF MATHEMATICS   4.3

1. A _____ inequality is true for some values of the variable in the inequality but false for other values.
2. An inequality that is always true is an _____ inequality.
3. An inequality that is always false is a _____ .
4. The _____ inequality $a \leq x \leq b$ is equivalent to $x \geq a$ _____ $x \leq b$.

5. The _____ of sets A and B is the set of elements that are in both A and B.
6. The _____ of sets A and B is the set of elements that are either in set A or in set B.

EXERCISES   4.3

In Exercises 1–3 match each inequality with the choice which best describes this inequality.
1. $x > x + 5$
2. $x > 5$
3. $x < x + 5$

    A. A conditional inequality
    B. An unconditional inequality
    C. A contradiction

In Exercises 4–12 write an inequality for each statement.
4. x is greater than or equal to 1 and less than 3.
5. x is greater than -5 and less than or equal to 1.
6. x is greater than or equal to -4 and less than or equal to -2.
7. x is greater than 0 and less than 6.
8. x is at least 30 and at most 45.
9. x is at least 150 and at most 450.
10. x is less than 3 or greater than 6.

11. x is less than or equal to 1 or x is greater than or equal to 5.
12. x is less than or equal to -2 or x is greater than 4.
13. **Modeling the Length of a Golf Shot** Write a compound inequality that expresses the length of the tee shot that a golfer must make to clear the water and to land safely on the green of the golf hole shown here.

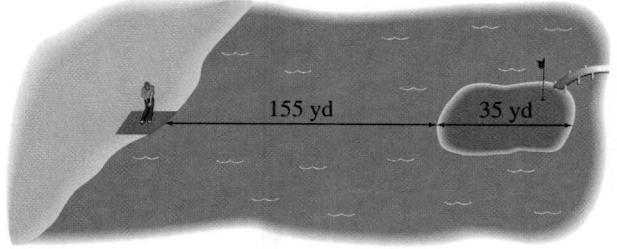

155 yd   35 yd

14. Modeling the Length of a Golf Shot A golfer must hit a shot at least 170 yd to clear a lake in front of a green. The depth of the green is only 20 yd. Write a compound inequality that expresses the length of the tee shot that a golfer must make to clear the water and to land safely on the green of this golf hole. (*Hint:* See the figure in Exercise 13.)

15. Modeling the Length of a Golf Shot A golfer is faced with a decision on the 15th fairway of the Lake of the Woods golf course. A safe shot of at most 150 yd will land short of a lake surrounding the green. A more aggressive shot must carry at least 220 yd and at most 270 yd to clear the lake and to stay on the green. Write a union of two intervals that gives the lengths of shots which will not land in the lake.

16. Modeling the Length of a Golf Shot A golfer is faced with a decision on the 3rd fairway of the Springfield golf course. A safe shot of at most 100 yd will land short of a lake surrounding the green. A more aggressive shot must carry at least 195 yd and at most 220 yd to clear the lake and to stay on the green. Write a union of two intervals that gives the lengths of shots which will not land in the lake.

In Exercises 17–22 determine $A \cap B$ and $A \cup B$ for the given intervals A and B.

17. $A = (-4, 7], B = [5, 11)$
18. $A = (-3, 6), B = [2, 9]$
19. $A = (-4, +\infty), B = [5, 11)$
20. $A = (-3, +\infty), B = [2, 9]$
21. $A = (-\infty, -2], B = [-4, +\infty)$
22. $A = (-\infty, 5), B = (1, +\infty)$

In Exercises 23 and 24 complete the following table:

INEQUALITY NOTATION	INTERVAL NOTATION	GRAPH	VERBALLY
23. $1 \le x < 6$			
24. $-4 < x \le 2$			

In Exercises 25 and 26 write each inequality expression as a compound inequality.

25. a. $x > -2$ and $x \le 5$ **b.** $x \ge 0$ and $x \le 4$
26. a. $x \ge -1$ and $x < \pi$ **b.** $x > -\pi$ and $x < 11$

In Exercises 27–30 use the graph to solve each compound inequality.

27. $2 \le x + 3 < 5$

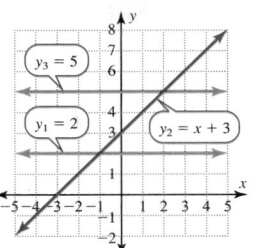

28. $-4 \le x - 2 \le 2$

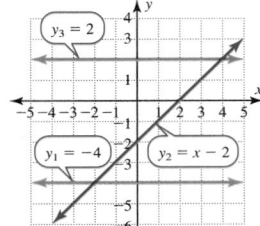

29. $x - 2 < 2x - 1 < x + 3$

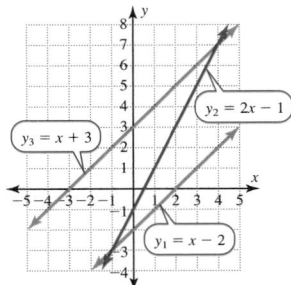

30. $-\dfrac{x}{2} - 3 < \dfrac{x}{2} + 2 < -\dfrac{x}{2} + 4$

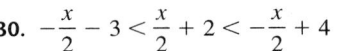

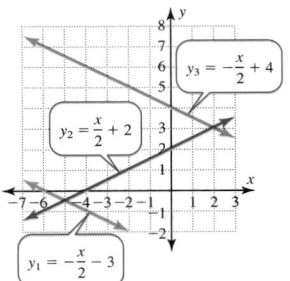

31. Using a Graph that Models the Strength of a Belt The drive belt for a wood chipper needs a minimum strength of 200 pounds. However, there is also a required maximum strength of 400 pounds as the equipment is designed for the belt to shear to prevent major damage to other components if the machine encounters a foreign object such as a steel bar. The graphs of $y_1 = 200$ and $y_2 = 400$ display the strength limits and the graph of $y_3 = f(x)$ displays the strength of the belt based on the number of separate fiber strands embedded into the belt. Use the following graphs to solve $200 \le y_3 \le 400$ and to interpret the meaning of this solution.

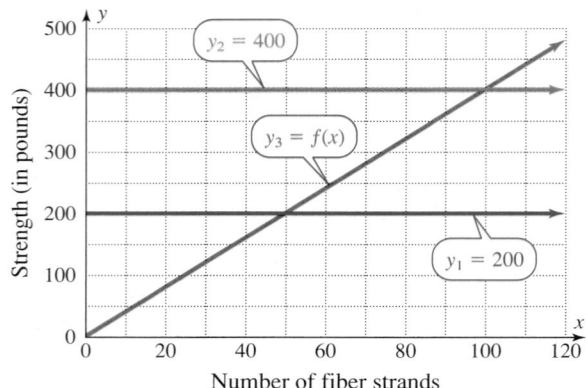

Number of fiber strands

32. Using a Graph that Models Company Bonuses   A local automobile dealership calculated a holiday bonus for each salesperson based on the number of automobiles that the person sold during the year. The bonus records show that each salesperson received at least \$400 and that the most anyone received was \$1500. The graphs of $y_1 = 400$ and $y_2 = 1500$ display the minimum and maximum bonus amounts and the graph of $y_3 = f(x)$ displays the bonus paid based on the number of cars sold. Use the following graph to solve $400 \le y_3 \le 1500$ and to interpret the meaning of this solution.

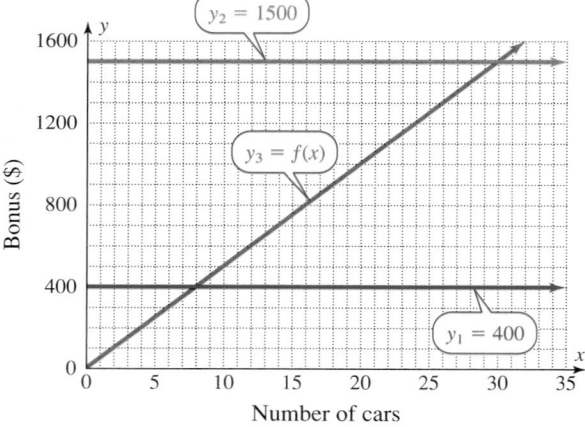

Number of cars

In Exercises 33–36 use the table to solve each inequality.

33. $-2 < 4 - x < 3$

34. $0 < 3 - x < 4$

35. $-3 < 2x + 1 \le 7$

36. $0 \le \dfrac{x}{2} + 1 \le 2$

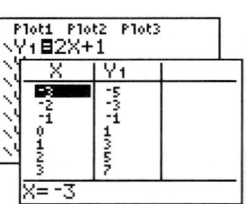

In Exercises 37–42 each inequality is a conditional inequality, an unconditional inequality, or a contradiction. Identify the type of each inequality and solve it.

37. $2x + 3 < 2(x + 1) + 1$ **38.** $2(x + 3) \ge 3(x - 2)$

39. $3x + 2 \le 3(x + 1)$ **40.** $2(x + 3) < 2(x - 2)$

41. $4x + 1 \le 3(x + 1)$ **42.** $2(x + 3) < 2(x + 4)$

Multiple Representations

In Exercises 43–46 solve each inequality algebraically, graphically, and numerically and describe the solution verbally.

43. $0 \le x + 5 < 6$ **44.** $-1 < x - 3 \le 1$

45. $-7 \le x - 8 \le -6$ **46.** $6 < x + 7 < 9$

In Exercises 47–68 solve each inequality.

47. $-30 < 5x < -20$ **48.** $21 \le 7x \le 42$

49. $-4 < -4x \le 8$ **50.** $-15 \le -5x < 5$

51. $-14 \le -\dfrac{7v}{2} \le 0$ **52.** $-15 \le -\dfrac{5v}{3} \le 0$

53. $3 \le 2w + 3 < 11$ **54.** $1 < 3w - 2 \le 13$

55. $5 < 4x - 3 \le 29$ **56.** $9 \le 7x + 2 \le 37$

57. $-4 < 6 - 5t < 11$ **58.** $-1 \le 5 - 6t < 17$

59. $-2 \le \dfrac{m}{2} - 3 \le -1$ **60.** $-3 < \dfrac{m}{2} - 2 < -1$

61. $-2 < \dfrac{m - 3}{2} \le -1$ **62.** $-3 \le \dfrac{m - 2}{3} < -1$

63. $\dfrac{x - 3}{10} \ge \dfrac{x - 2}{3}$ **64.** $\dfrac{x - 3}{3} < \dfrac{x + 1}{9}$

65. $x - 1 < 2x < x + 2$ **66.** $2x - 1 \le 3x \le 2x + 5$

67. $3x - 4 \le 4x + 1 \le 3x + 2$

68. $x - 1 \le 3x + 1 \le x + 5$

Estimation and Calculator Skills

In Exercises 69–72 mentally estimate the solution of each inequality and then use a calculator to calculate the solution.

PROBLEM	MENTAL ESTIMATE	CALCULATOR SOLUTION
69. $5.719 < 3.01x < 9.331$		
70. $2.01 \le x + 0.99 \le 5.08$		
71. $-4.179 \le -1.99x < 5.771$		
72. $7.81 < -7.1x \le 13.49$		

In Exercises 73–78 solve each inequality.

73. $7x - 1 < 5x + 7 < 6x + 3$

74. $x - 1 < 2$ and $2 > x + 1$

75. $5(x + 4) < 28$ and $3(x - 4) > -11$

76. $x - 1 < \dfrac{2x + 1}{3}$ or $\dfrac{4x - 1}{3} \geq x - 1$

77. $2x - 1 < 3x + 1$ or $5x - 6 < 3x - 12$

78. $7(4x - 5) - 16 < 5(8x - 3)$ and
$11 - 4(2x - 3) \geq -3(2 - 7x)$

In Exercises 79–82 translate each problem into a linear inequality and then solve this inequality.

79. Temperature Limits   The warranty for a computer specifies that the temperature of its environment must be maintained between 41 and 95° F. Find the temperatures that are within this range, expressed in degrees Celsius. Solve $41 < \dfrac{9}{5}C + 32 < 95$ to find the acceptable range of Celsius temperatures.

80. Temperature Range   The range of acceptable Celsius temperatures for a piece of equipment is from 15 to 30°. Solve $15 \leq \dfrac{5}{9}(F - 32) \leq 30$ to find the acceptable range of Fahrenheit temperatures.

81. Perimeter of a Triangle   Two sides of a triangle must be 8 m and 12 m. The perimeter must be at least 24 m and at most 39 m. Determine the length that can be used for the third side.

82. Perimeter of a Rectangle   The width of a rectangle must be 9 m. If the perimeter must be between 44 and 64 m, determine the length that can be used for the rectangle.

Group Discussion Questions

83. Writing Mathematically   If possible, list all values of x for which $0.9 < x < 1.0$. If this is not possible, explain why.

84. Challenge Question   Can you determine, without direct calculations, which is larger: $25(\pi - 2)$ or $25(4 - \pi)$?

Section 4.4 Solving Absolute Value Equations and Inequalities

Objectives:

7. Solve absolute value equations and inequalities algebraically.

8. Solve absolute value equations and inequalities using tables and graphs.

The absolute value of x, denoted by $|x|$, is the distance on the number line between 0 and x. Thus the absolute value inequality $|x| < 4$ represents the interval of points for which the distance from the origin is less than 4 units. Likewise, the absolute value inequality $|x| > 4$ represents the two intervals containing points for which the distance from the origin is more than 4 units. The following box illustrates these concepts.

Absolute Value Expressions

For any real number x and any nonnegative real number d:

ALGEBRAICALLY	VERBALLY	ALGEBRAIC EXAMPLE	GRAPHICAL EXAMPLE
$|x| = d$	The distance from 0 to x is d units.	$|x| = 4$ if $x = -4$ or $x = 4$	4 units left of 0 4 units right of 0 (number line at -4, 0, 4)
$|x| < d$	The distance from 0 to x is less than d units.	$|x| < 4$ if $-4 < x < 4$	Points less than 4 units from 0 (number line at -4, 0, 4)
$|x| > d$	The distance from 0 to x is more than d units.	$|x| > 4$ if $x < -4$ or $x > 4$	Points more than 4 units left of 0 Points more than 4 units right of 0 (number line at -4, 0, 4)

SELF-CHECK 4.4.1

Use interval notation to represent the real numbers that are solutions of these inequalities.

1. $|x| < 5$
2. $|x| \geq 5$

The distance between two real numbers on the number line can be determined by subtraction.

Questions: Can you determine the distance between these real numbers on the number line?

A: The distance from 2 to 5.
B: The distance from -5 to -2.
C: The distance from -2 to -5.

Graphically   Absolute Value of Difference

Answers: **A:**

3 units

$|5 - 2| = |3| = 3$

The distance between two points is always positive. The distance between two points is the same in either direction.

B:

3 units

$|-2 - (-5)| = |3| = 3$

C:

3 units

$|-5 - (-2)| = |-3| = 3$

If a is larger than b, then the distance from a to b is given by the difference $a - b$. To indicate that the distance between a and b is always nonnegative, we can denote this distance by $|a - b|$. In particular, $|x - 0|$, or $|x|$, can be interpreted as the distance between x and the origin. Likewise, $|a + b| = |a - (-b)|$ equals the distance between a and $-b$. This concept of distance is examined in the next two examples.

■ **EXAMPLE 1**   **Representing Distance Using Absolute Value Notation**

Use absolute value notation to represent the distance between each pair of points.

SOLUTIONS

(a) -5 and 1

$|1 - (-5)| = |6| = 6$ or
$|-5 - 1| = |-6| = 6$

6 units

(b) 1 and b,
with $b > 1$

$|b - 1|$

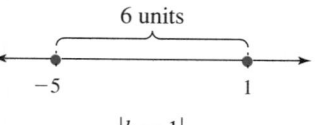

$|b - 1|$

Because $|1 - b| = |b - 1|, |1 - b|$
also represents this distance.

(c) c and 1, $|c - 1|$
with $c < 1$

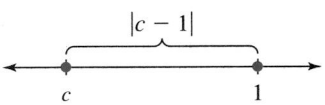

$|1 - c|$ also can be used to represent this distance.

(d) v and w $|v - w|$

SELF-CHECK 4.4.2

Use absolute value to represent the distance between these real numbers.

1. 8 and x
2. x and y
3. v and $-w$

■ EXAMPLE 2   Using Distance to Interpret an Absolute Value Equation

Using the geometric concept of distance, interpret (a) $|x - 3| = 4$, (b) $|x - 3| < 4$, and (c) $|x - 3| > 4$.

SOLUTIONS _____

(a) $|x - 3| = 4$ indicates the distance between x and 3 is 4 units. This means that x is either 4 units to the left of 3 or 4 units to the right of 3.

ALGEBRAICALLY   **GRAPHICALLY**

Left 4 Units *or* *Right 4 Units*
$x - 3 = -4$ or $x - 3 = +4$
$x = -1$ or $x = 7$

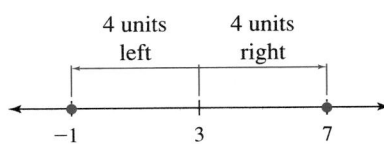

Answer: $x = -1$ or $x = 7$ Do these values check?

(b) $|x - 3| < 4$ indicates the distance between x and 3 is less than 4 units. This means that x is less than 4 units to the left of 3 and less than 4 units to the right of 3.

ALGEBRAICALLY   **GRAPHICALLY**

Left Less   *and* *Right Less*
than 4 Units   *than 4 Units*
$-4 < x - 3 < 4$
$-1 < x < 7$

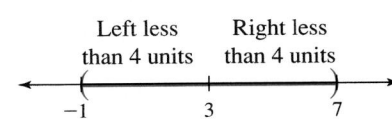

Answer: $(-1, 7)$ Can you check this interval?

(c) $|x - 3| > 4$ indicates the distance between x and 3 is more than 4 units. This means that x is either more than 4 units to the left of 3 or more than 4 units to the right of 3.

<table>
<tr><td>**ALGEBRAICALLY**</td><td>**GRAPHICALLY**</td></tr>
</table>

Left More *or* *Right More*	Left more Right more
than 4 Units *than 4 Units*	than 4 units than 4 units
$x - 3 < -4$ or $x - 3 > 4$	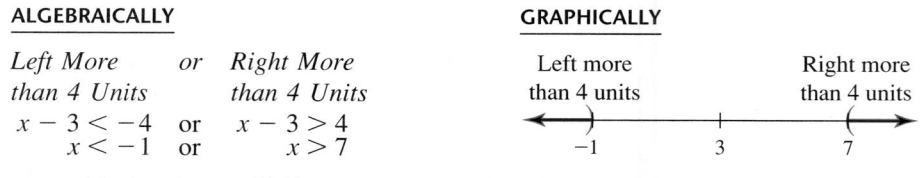
$x < -1$ or $x > 7$	

Answer: $(-\infty, -1) \cup (7, +\infty)$

Can you check the union of these two intervals?

The results that we have observed in Example 2 can be summarized in the following box.

Solving Absolute Value Equations and Inequalities*

For any real numbers x and a and positive real number d:

Note that $|x - a| > d$ corresponds to the union of two separate intervals. This requires the word "or" to describe $x - a < -d$ or $x - a > d$. Again the visual aid of a graph and a description in terms of distance should help you to use the correct notation.

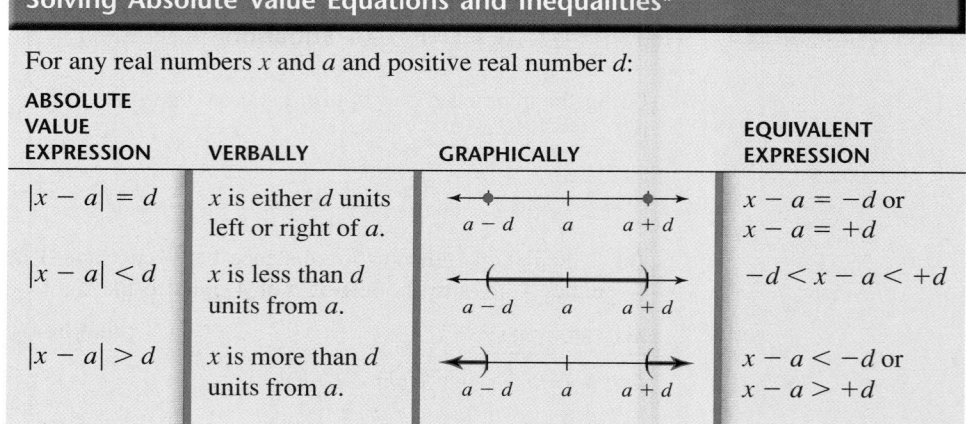

ABSOLUTE VALUE EXPRESSION	VERBALLY	GRAPHICALLY	EQUIVALENT EXPRESSION
$|x - a| = d$	x is either d units left or right of a.		$x - a = -d$ or $x - a = +d$
$|x - a| < d$	x is less than d units from a.		$-d < x - a < +d$
$|x - a| > d$	x is more than d units from a.		$x - a < -d$ or $x - a > +d$

*Similar statements also can be made about the order relations less than or equal to ($\leq$) and greater than or equal to ($\geq$). Expressions with d negative are examined in the group exercises at the end of this section.

The results in this box can be applied to solve more complicated expressions within absolute value symbols as illustrated in Example 3.

■ EXAMPLE 3   Solving an Absolute Value Equation

Solve $|2x - 3| = 31$.

SOLUTION

Left 31 Units *or* *Right 31 Units*	
$2x - 3 = -31$ or $2x - 3 = +31$	To obtain all solutions for this equation, we must
$2x = -28$ $2x = 34$	consider both of these equations.
$x = -14$ $x = 17$	

Answer: $x = -14$ or $x = 17$

Solve $|4x - 6| = 10$.

We first used this strategy in Section 2.3 to solve linear equations in one variable.

We can use two-dimensional graphs to solve absolute value equations and inequalities in one variable. In the following calculator perspective we start by using y_1 to represent the absolute value expression.

CALCULATOR PERSPECTIVE 4.4.1	Using a Graph and a Table to Solve an Absolute Value Equation or Inequality

To solve $|x - 3| = 4$, $|x - 3| < 4$, and $|x - 3| > 4$ from Example 2 using a TI-83 Plus calculator, enter the following keystrokes:

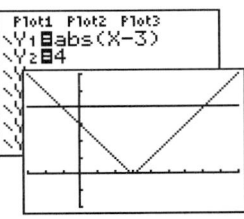

$[-3, 9, 1]$ by $[-2, 6, 1]$

Press the down arrow key ▼ repeatedly to obtain the next screen.

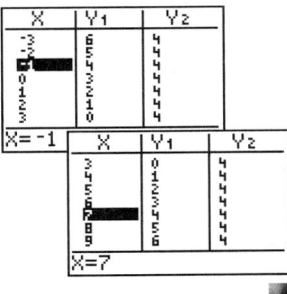

Note: Both in the table and on the graph $|x - 3| = 4$ for $x = -1$ or $x = 7$. Also $|x - 3| < 4$ for the x-values between -1 and 7, the interval $(-1, 7)$; and $|x - 3| > 4$ for $x < -1$ or for $x > 7$, the set $(-\infty, -1) \cup (7, +\infty)$.

■ **EXAMPLE 4**   Using Multiple Perspectives to Solve an Absolute Value Inequality

Solve $|4x - 6| < 10$ algebraically, numerically, and graphically.

SOLUTION

ALGEBRAICALLY

Left Less   and   Right Less
than 10   than 10
Units   Units
$$-10 < 4x - 6 < 10$$
$$-4 < 4x < 16$$
$$-1 < x < 4$$

NUMERICALLY

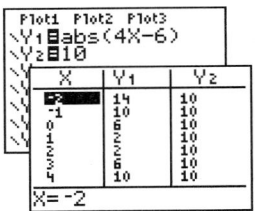

GRAPHICALLY

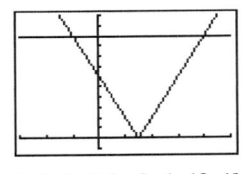

$[-3, 5, 1]$ by $[-1, 12, 1]$

VERBALLY

From the table $y_1 < y_2$ for $-1 < x < 4$ and from the graph y_1 is below y_2 for these same x-values.

Answer: $(-1, 4)$ ◾

◼ **EXAMPLE 5** Using Multiple Perspectives to Solve an Absolute Value Inequality

Solve $\left| \dfrac{x + 2}{2} \right| > 1$.

SOLUTION

ALGEBRAICALLY

Left More	*or*	*Right More*
than 1 Unit		*than 1 Unit*

$$\dfrac{x + 2}{2} < -1 \quad \text{or} \quad \dfrac{x + 2}{2} > 1$$

$$x + 2 < -2 \quad \text{or} \quad x + 2 > 2$$

$$x < -4 \quad \text{or} \quad x > 0$$

NUMERICALLY

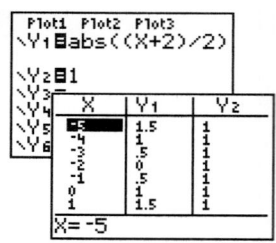

GRAPHICALLY

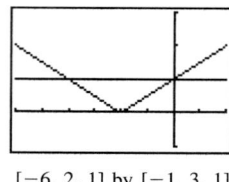

$[-6, 2, 1]$ by $[-1, 3, 1]$

VERBALLY

From the table $y_1 > y_2$ for $x < -4$ or for $x > 0$ and from the graph y_1 is above y_2 for these same x-values.

Answer: $(-\infty, -4) \cup (0, +\infty)$ ◾

SELF-CHECK 4.4.4

Use the tables and graphs in Examples 4 and 5 to solve:

1. $|4x - 6| \geq 10$

2. $\left| \dfrac{x + 2}{2} \right| \leq 1$.

$|a| = |b|$ is equivalent to $a = b$ or $a = -b$.

If $|a| = |b|$, then a and b are equal in magnitude but their signs either can agree or disagree. Thus $|a| = |b|$ implies that either $a = b$ or $a = -b$. This result is used to solve the next example.

■ EXAMPLE 6   Using Multiple Perspectives to Solve an Equation Involving Two Absolute Value Expressions

Solve $|3x - 5| = |5x - 7|$.

SOLUTION

ALGEBRAICALLY

$3x - 5 = 5x - 7$ or $3x - 5 = -(5x - 7)$
$\quad -2x = -2$ $\qquad\qquad\;\; 3x - 5 = -5x + 7$
$\qquad x = 1$ $\qquad\qquad\qquad\; 8x = 12$
$\qquad\qquad\qquad\qquad\qquad\qquad x = \dfrac{3}{2}$

NUMERICALLY

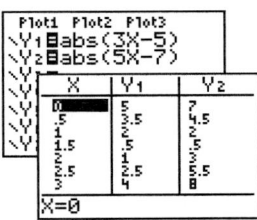
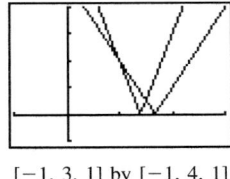

GRAPHICALLY

$[-1, 3, 1]$ by $[-1, 4, 1]$

VERBALLY

From the table $y_1 = y_2$ for $x = 1$ and $x = 1.5$ and the graphs of y_1 and y_2 intersect at $x = 1$ and $x = 1.5$.

Answer: $x = 1$ or $x = \dfrac{3}{2}$

Some of the earlier examples in this section start with an absolute value inequality and then we produce the interval of values that satisfy this inequality. The next example goes in the other direction. We start with an interval of values and then produce an absolute value inequality that represents this interval.

■ EXAMPLE 7   Representing an Interval Using Absolute Value Notation

Write an absolute value inequality to represent the interval $(-3, 7)$.

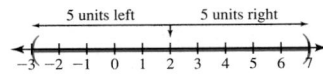

SOLUTION

The distance from one end of this interval to the other is $|7 - (-3)| = |10| = 10$. Thus the midpoint is 5 units to the right of -3 and 5 units to the left of 7. This midpoint is 2 since $-3 + 5 = 2$ and $7 - 5 = 2$.

$(-3, 7)$ is the interval of points less than 5 units from 2.

In absolute value notation, $|x - 2| < 5$.

Answer: $|x - 2| < 5$ represents the interval $(-3, 7)$.

SELF-CHECK 4.4.5 ANSWERS

1. $x = -1$ or $x = \dfrac{1}{3}$

2. $|x - 2| \geq 5$

SELF-CHECK 4.4.5

1. Solve $|x - 1| = |2x|$.
2. Use Example 7 to write an absolute value inequality to represent $(-\infty, -3] \cup [7, +\infty)$.

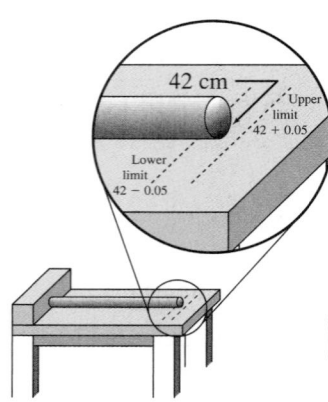

In industrial applications there is generally an allowance for a small variation, or lee-way, between the standard size for a part or component and the actual size. This accept-able variation is called the tolerance. For a 42 cm steel rod, for example, the tolerance may be ±0.05 cm. A worker on an assembly line might merely lay the rod on a table that is marked to indicate the upper and lower limits of tolerance, as shown in the fig-ure. However, an engineer doing calculations would need to describe this tolerance algebraically. The expression $|r - 42| \leq 0.05$ is an algebraic statement that the length of the rod, r, and the desired length of 42 cm can differ by at most 0.05 cm.

■ **EXAMPLE 8**   **Representing a Tolerance Interval Using Absolute Value Notation**

The amount of a drug that is placed in an IV bag for a patient should be 20 milliliters, with a tolerance of ±0.5 mL. Express this interval as a compound linear inequality and as an absolute value inequality using d to represent the volume of the drug.

SOLUTION ──

ALGEBRAICALLY

$-0.5 \leq d - 20 \leq 0.5$
$|d - 20| \leq 0.5$

GRAPHICALLY

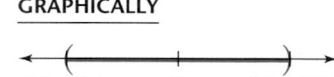

20 − 0.5   20   20 + 0.5

VERBALLY

The drug can be from 0.5 mL below 20 mL to 0.5 mL above 20 mL. ■

USING THE LANGUAGE AND SYMBOLISM OF MATHEMATICS   4.4

1. The absolute value of x is the _____ on the number line between 0 and x.
2. Fill in these blanks with either =, <, or >.
   a. The distance from 0 to x is less than d units. $|x|$ ____ d
   b. The distance from 0 to x is d units. $|x|$ ____ d
   c. The distance from 0 to x is greater than d units. $|x|$ ____ d

3. Write an absolute value equation that indicates x is d units left or right of a. _____
4. The acceptable variation between the standard size for a part and the actual size of the part is called the _____.

EXERCISES   4.4

In Exercises 1–6 write an absolute value equation or inequality to represent each set of points.

1. a.
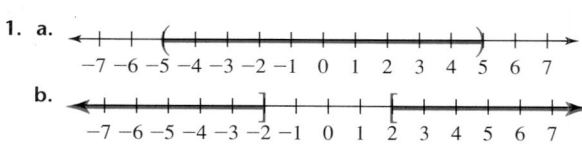

b.
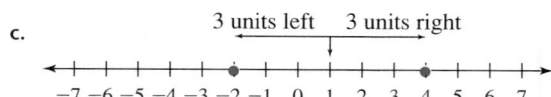

c.   3 units left   3 units right

−7 −6 −5 −4 −3 −2 −1 0 1 2 3 4 5 6 7

d. The points between −4 and 4.

2. a.
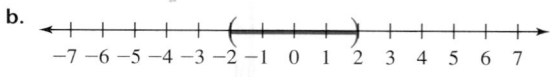
−7 −6 −5 −4 −3 −2 −1 0 1 2 3 4 5 6 7

b.
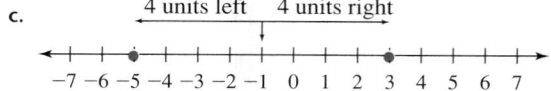
−7 −6 −5 −4 −3 −2 −1 0 1 2 3 4 5 6 7

c.   4 units left   4 units right

−7 −6 −5 −4 −3 −2 −1 0 1 2 3 4 5 6 7

d. The numbers that are at least −3 and at most 3.

3. a.

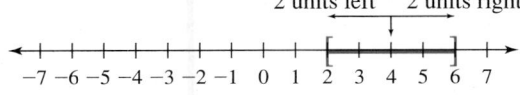

b.

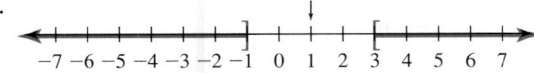

c.

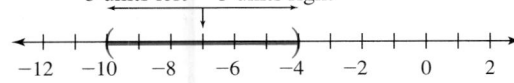

d. The numbers that are at least -5 and at most 5.

4. a.

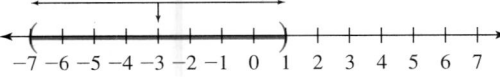

b.

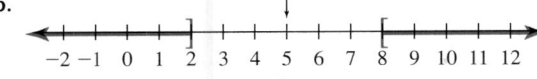

c.

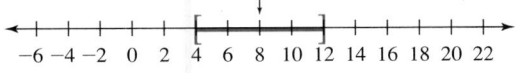

d. The points between -8 and 8.

5. a. $[-3, 3]$ **b.** $(-\infty, -3) \cup (3, +\infty)$
   c. $(-7, 7)$
6. a. $(-9, 9)$ **b.** $(-\infty, -9] \cup [9, +\infty)$
   c. $[-\pi, \pi]$

In Exercises 7 and 8 solve each equation.
7. a. $|x| = 6$ **b.** $|x| = 0$
   c. $|x - 2| = 5$ **d.** $|x + 2| = 5$
8. a. $|x| = 13$ **b.** $|x - 7| = 8$
   c. $|2x| = 0$ **d.** $|x + 7| = 8$

In Exercises 9–12 use the graph to solve each inequality.
9. a. $|x - 2| = 4$
   b. $|x - 2| < 4$
   c. $|x - 2| > 4$

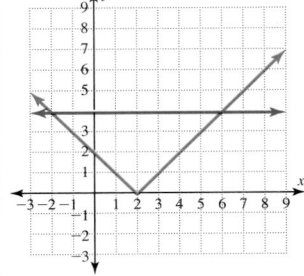

10. a. $|x + 3| = 2$
   b. $|x + 3| \le 2$
   c. $|x + 3| > 2$

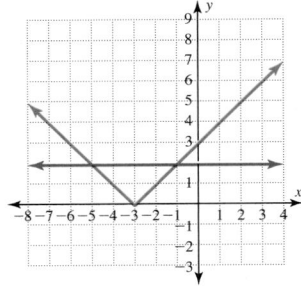

11. a. $\left|\dfrac{x - 4}{2}\right| = 1$

   b. $\left|\dfrac{x - 4}{2}\right| < 1$

   c. $\left|\dfrac{x - 4}{2}\right| \ge 1$

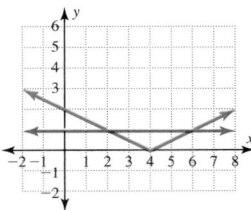

12. a. $|3x - 6| = 3$
   b. $|3x - 6| \le 3$
   c. $|3x - 6| \ge 3$

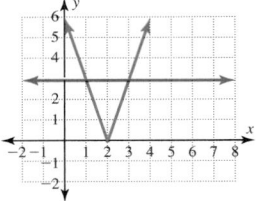

In Exercises 13–16 use the table of values to determine the x-values that satisfy each equation and inequality.

13. a. $|x + 2| = 2$ **14. a.** $|2x + 1| = 3$
   b. $|x + 2| \le 2$ **b.** $|2x + 1| < 3$
   c. $|x + 2| \ge 2$ **c.** $|2x + 1| > 3$

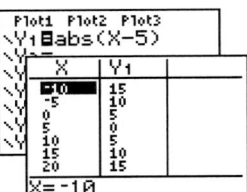

15. a. $|x - 5| = 10$ **16. a.** $|x + 25| = 50$
   b. $|x - 5| < 10$ **b.** $|x + 25| \le 50$
   c. $|x - 5| \ge 10$ **c.** $|x + 25| > 50$

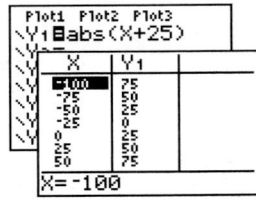

In Exercises 17–20 solve each inequality algebraically, graphically, and numerically, and describe the solution verbally.

17. $|x + 1| < 1$ **18.** $|x + 2| < 1$
19. $|2x - 1| < 3$ **20.** $|2x + 3| < 5$

In Exercises 21–40 solve equation and inequality.
21. $|3x + 4| \ge 2$ **22.** $|3x - 5| \ge 1$
23. $\left|\dfrac{-7x}{2}\right| \le 14$ **24.** $\left|\dfrac{-5y}{11}\right| \le 55$
25. $|2x + 3| - 4 < 1$ **26.** $|2x - 5| - 2 < 3$
27. $|2(x + 3) - 4| < 1$ **28.** $|2(x - 5) - 2| < 3$
29. $|2x + 1| + 4 > 11$ **30.** $|2x - 1| + 6 > 9$

31. $\left|\dfrac{x-1}{7}\right| \geq 14$ **32.** $\left|\dfrac{2x-1}{3}\right| \geq 6$

33. $|2(2x-1)-(x-3)| < 11$

34. $|3(2x+1)-4(3x-2)| < 5$

35. $|2x| = |x-1|$ **36.** $|3x| = |4-x|$

37. $|x-1| = \left|\dfrac{x+1}{2}\right|$ **38.** $|x+3| = \left|\dfrac{x-1}{2}\right|$

39. $|2x+3| = |3x-1|$ **40.** $|4x+2| = |5x-1|$

Estimation and Calculator Skills

In Exercises 41–44 mentally estimate the solution of each equation to the nearest integer and then use a calculator to calculate the solution.

PROBLEM	MENTAL ESTIMATE	CALCULATOR SOLUTION
41. $|2.01x| = 9.849$		
42. $|3.9x| = 39.39$		
43. $|x+7.93| = 15.03$		
44. $|x-9.01| = 29.13$		

In Exercises 45–50 write an absolute value inequality to represent these intervals.

45. $(-12, 6)$ **46.** $(-3, 11)$

47. $[-4, 26]$ **48.** $[12, 32]$

49. $(-\infty, -2) \cup (6, +\infty)$ **50.** $(-\infty, -6] \cup [4, +\infty)$

Using Absolute Value Inequalities to Model Tolerance Intervals

In Exercises 51–56 express the tolerance interval as an absolute value inequality, and determine the lower and upper limits of the interval.

51. Tolerance of a Piston Rod The length of a piston rod in an automobile engine is specified by contract to the manufacturer to be 15 cm with a tolerance of ± 0.001 cm.

52. Tolerance of a False Tooth A dental clinic has received an order for a dental implant that is 1.3 cm long with a tolerance of ± 0.05 cm.

53. Tolerance of a Soft Drink Bottle The specifications for a bottle-filling machine in a soft drink plant call for the machine to dispense 16 ounces into a bottle with a tolerance of ± 0.25 ounce.

54. Tolerance of a Fuel Tank A military contract for an aircraft fuel tank specifies a capacity of 420 liters with a tolerance of ± 1 liter.

55. Tolerance of a Temperature Control A temperature control at a pharmaceutical manufacturing plant is set to keep the temperature at $25°$ C with a tolerance of $\pm 2°$.

56. Tolerance of a Voltage Control The electrical supply to an important internet provider is monitored to keep the line feed at 120 volts with a tolerance of ± 1 volt.

Group Discussion Questions

57. Special Cases Involving Absolute Value Solve these inequalities.
  a. $|x| = -1$ **b.** $|x| < -1$ **c.** $|x| > -1$
  Then generalize these results to give the solution of each of these inequalities for real numbers x and a and any negative real number d.
  a. $|x-a| = d$ **b.** $|x-a| < d$ **c.** $|x-a| > d$

58. Error Analysis A classmate in your algebra class says that you can take the absolute value of any expression by just removing the signs of each term. For example, the student replaced $|-x|$ by just x in one equation. The same student also replaced $|x-2|$ by $x+2$. Give examples to show that this logic is flawed.

Section 4.5 Graphing Systems of Linear Inequalities in Two Variables

Objectives:

9. Graph a linear inequality in two variables.

10. Graph a system of linear inequalities.

A Mathematical Note

Winifred Edgerton Merrill (1862–1951) was the first American woman to receive a Ph.D. in mathematics. She studied mathematical astronomy and was later instrumental in the foundation of Barnard College for Women.

In this section we examine linear inequalities in two variables and systems of these inequalities. This topic is a natural extension of our work with systems of linear equations in Chapter 3 and our work in this chapter with inequalities. In Exercise 62 we examine how inequalities can be used to describe possible options for businesses.

A **solution to a linear inequality** in two variables is an ordered pair of values that, when substituted into the inequality, makes a true statement.

■ **EXAMPLE 1**   Checking Possible Solutions of an Inequality

Determine whether each ordered pair is a solution of $2x + 3y \leq 6$.

SOLUTIONS

(a) $(5, 0)$

$2x + 3y \leq 6$

$2(5) + 3(0) \overset{?}{\leq} 6$ Substitute the given coordinates into the inequality and determine whether a true statement results.

$10 \overset{?}{\leq} 6$ is false.

Answer: $(5, 0)$ is not a solution.

(b) $(0, 2)$

$2x + 3y \leq 6$

$2(0) + 3(2) \overset{?}{\leq} 6$

$6 \overset{?}{\leq} 6$ The inequality $\leq$, less than or equal to, includes the equality, so the point $(0, 2)$ is a solution.

Answer: $(0, 2)$ is a solution.

(c) $(0, 0)$

$2x + 3y \leq 6$

$2(0) + 3(0) \overset{?}{\leq} 6$

$0 \overset{?}{\leq} 6$

Answer: $(0, 0)$ is a solution.

SELF-CHECK 4.5.1

Determine whether these points are solutions of $5x + y > 10$.

1. $(2, 1)$

2. $(2, 0)$

3. $(0, 0)$

The line $Ax + By = C$ separates the plane into two regions called half-planes. (See the following figure.) One half-plane will satisfy $Ax + By < C$, and the other half-plane will satisfy $Ax + By > C$. If the inequality is $\leq$ or $\geq$, the line will be part of the solution. If the inequality is $<$ or $>$, the line will be the boundary separating the half-planes but it will not be part of the solution.

Standard convention is to use a solid line for $\leq$ or $\geq$ when the line is part of the solution and to use a dashed line for $<$ or $>$ when the line is not part of the solution.

SELF-CHECK 4.5.1 ANSWERS

1. Solution

2. Not a solution

3. Not a solution

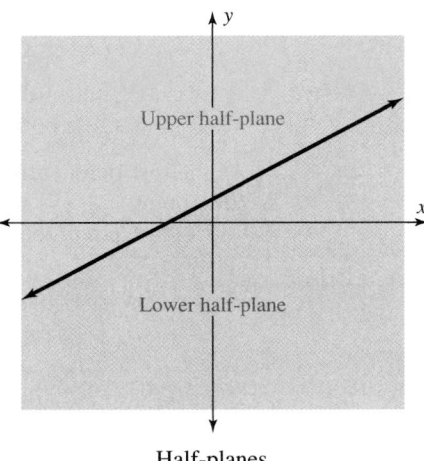

Half-planes

To graph a linear inequality, we first graph the line that forms the boundary for the half-planes. Then a test point can be used to determine which half-plane satisfies the inequality.

■ EXAMPLE 2   Checking Possible Solutions of an Inequality

Determine which of the points $A–F$ shown in the graph below satisfy the inequality $y > \dfrac{x}{2} - 1$.

SOLUTION _____

Equality: The dashed line represents all ordered pairs whose y-coordinate equals $\dfrac{x}{2} - 1$.

Upper half-plane: The upper half-plane represents all ordered pairs where the y coordinate is greater than (above) $\dfrac{x}{2} - 1$.

Thus A, B, and C represent solutions of $y > \dfrac{x}{2} - 1$.

Lower half-plane: The lower half-plane represents all ordered pairs where the y coordinate is less than (below) $\dfrac{x}{2} - 1$.

Thus D, E, and F are *not* solutions of $y > \dfrac{x}{2} - 1$.

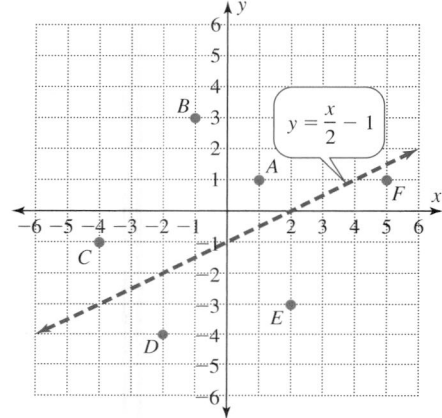

You also can check these points by substituting their coordinates into the given inequality.

■

How many solutions does the inequality $y > \dfrac{x}{2} - 1$ have? It is important to realize that all the points in the upper half-plane for this problem are solutions of the inequality. You could never list all of these solutions. Thus we represent these points algebraically or graphically.

Graphing a Linear Inequality

STEP 1. Graph the equality $Ax + By = C$ using
  a. A solid line for $\leq$ or $\geq$
  b. A dashed line for $<$ or $>$

STEP 2. Choose an arbitrary test point not on the line; $(0, 0)$ is often convenient. Substitute this test point into the inequality.

STEP 3. a. If the test point satisfies the inequality, shade the half-plane containing this point.
  b. If the test point does not satisfy the inequality, shade the other half-plane.

■ EXAMPLE 3   Graphing a Linear Inequality

Graph the solution of $3x - 5y < -15$.

SOLUTION

1. Draw a dashed line for $3x - 5y = -15$ because the equality is not part of the solution. The line passes through the intercepts $(-5, 0)$ and $(0, 3)$.

2. Test the origin:

$$3x - 5y < -15$$
$$3(0) - 5(0) \stackrel{?}{<} -15$$
$$0 \stackrel{?}{<} -15 \text{ is false.}$$

3. Shade the half-plane that does *not* include the test point $(0, 0)$.

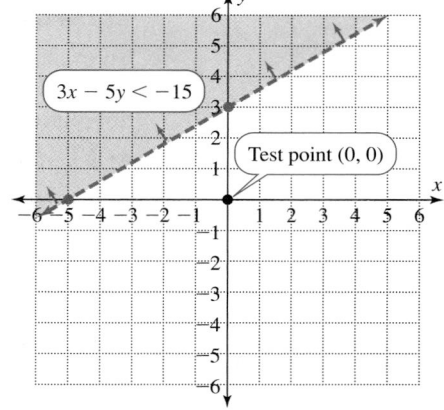

■ EXAMPLE 4   Graphing a Linear Inequality

Graph the solution of $5x \geq 3y$.

SOLUTION

1. Draw a solid line for $5x = 3y$ since the original statement includes the equality. The line passes through $(0, 0)$ and $(3, 5)$.

2. Test the point $(1, 0)$. [Do not test $(0, 0)$ since the origin lies on the line.]

$$5x \geq 3y$$
$$5(1) \stackrel{?}{\geq} 3(0)$$
$$5 \stackrel{?}{\geq} 0 \text{ is true.}$$

3. Shade the half-plane that includes the test point $(1, 0)$.

Observe that we used $(0, 0)$ as a test point in Example 3 but not in Example 4. We always select a test point that is not on the line separating the two half-planes.

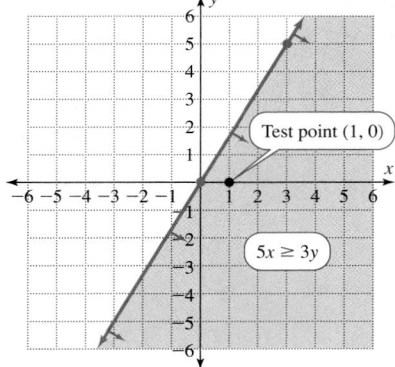

SELF-CHECK 4.5.2

Graph $2x + y \leq 10$.

CALCULATOR PERSPECTIVE 4.5.1 | **Graphing a Linear Inequality**

To graph a linear inequality on a TI-83 Plus calculator, we must first solve the inequality for y. Then enter the right side of the inequality next to y_1 and select the appropriate shading. To graph the solution to $3x - 5y < -15$ from Example 3 on a TI-83 Plus calculator, first rewrite the inequality as $y > \frac{3}{5}x + 3$ and then enter the following keystrokes:

Note that on a TI-83 Plus calculator the symbol ◥ denotes upper half-plane should be shaded and the symbol ◢ denotes the lower half-plane should be shaded.

Use the arrow keys to move the cursor to the left of y_1 and press ENTER until the marker indicates the graph will be shaded above the line.

ZOOM 6

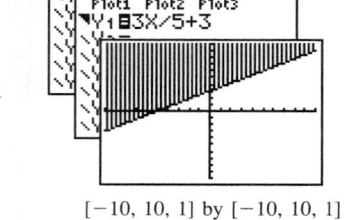

$[-10, 10, 1]$ by $[-10, 10, 1]$

Note: For $\geq$ and $>$ the graph will be shaded above the line, and for $\leq$ and $<$ the graph should be shaded below the line. The calculator does not distinguish between $\geq$ or $>$, nor does it distinguish between $\leq$ or $<$. ■

SELF-CHECK 4.5.2 ANSWER

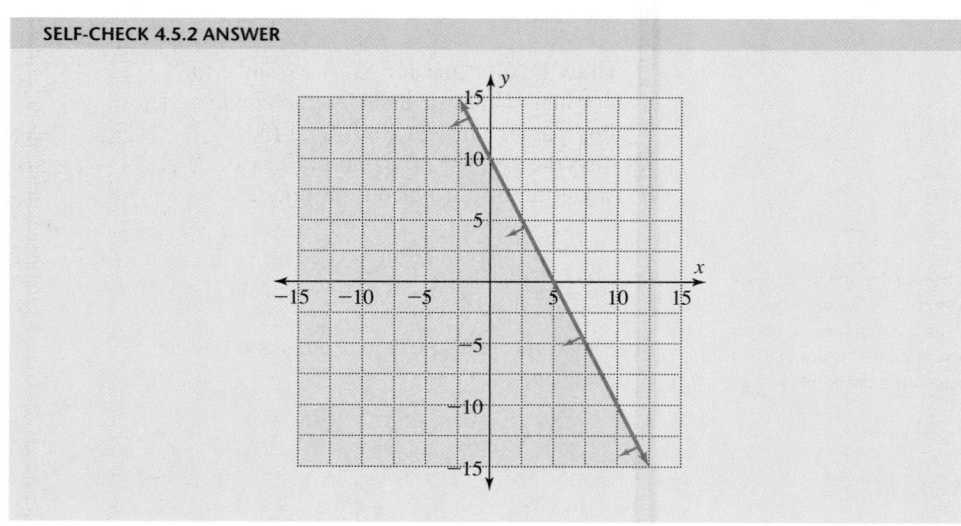

■ **EXAMPLE 5**   Using a Graphics Calculator to Solve Inequalities

Use a graphics calculator to solve **(a)** $2x + 3y \le 6$ and **(b)** $2x + 3y \ge 6$.

SOLUTIONS

(a) $2x + 3y \le 6$

$$3y \le -2x + 6$$

$$y \le -\frac{2}{3}x + 2$$

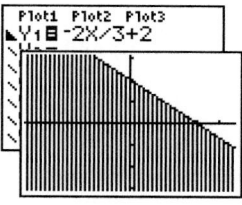

$[-4.7, 4.7, 1]$ by $[-3.1, 3.1, 1]$

(b) $2x + 3y \ge 6$

$$3y \ge -2x + 6$$

$$y \ge -\frac{2}{3}x + 2$$

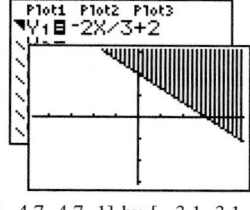

$[-4.7, 4.7, 1]$ by $[-3.1, 3.1, 1]$

To graph these inequalities on a calculator, first solve for y, then enter the equation into y and select the appropriate shading for each inequality. The solution of the first inequality is the lower half-plane and the solution of the second inequality is the upper half-plane. ■

SELF-CHECK 4.5.3

Use a graphics calculator to solve $4x - 5y \le 20$.

To graph a system of linear inequalities, we first graph each inequality on the same coordinate system. The solution of the system is the intersection of these two individual regions. To clarify which points satisfy each inequality, we use slanted lines to indicate the solutions of the individual inequalities. The solution of the system, the cross-hatched region where these lines intersect, then will be shaded for greater emphasis.

■ **EXAMPLE 6**   Graphing a System of Linear Inequalities

Graph the solution of $\left\{\begin{array}{l} x \ge -3 \\ x \le 2 \end{array}\right\}$.

SOLUTION

Inspection reveals that both $x = -3$ and $x = 2$ represent vertical lines. The region containing $(0, 0)$ satisfies both inequalities. Thus the solution set is the violet strip between these solid lines.

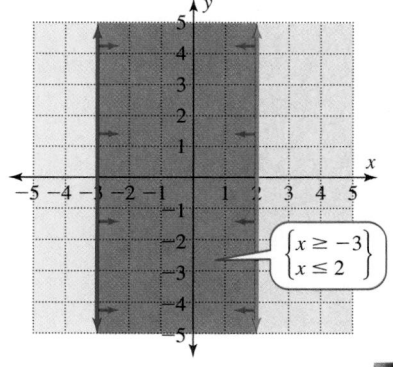

$\left\{\begin{array}{l} x \ge -3 \\ x \le 2 \end{array}\right\}$

■

SELF-CHECK 4.5.3 ANSWER

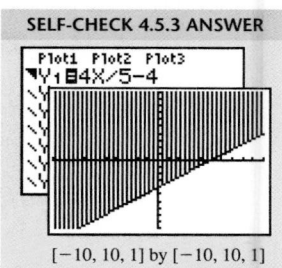

$[-10, 10, 1]$ by $[-10, 10, 1]$

The solution to the next example proceeds step by step. First one inequality is graphed and then the second inequality is graphed. Finally the intersection of these regions is shaded as the solution for the system.

EXAMPLE 7   Graphing a System of Linear Inequalities

Graph the solution of $\begin{cases} x + y \geq 2 \\ 3x - 2y < 6 \end{cases}$.

SOLUTION ───────────────────────────────────────

(a) ① To graph $x + y \geq 2$, draw a solid line for
    $x + y = 2$. Use the intercepts $(2, 0)$ and $(0, 2)$
    to draw the line.
    ② Test the point $(0, 0)$ in $x + y \geq 2$
    $$0 + 0 \not\geq 2 \text{ is false.}$$
    ③ Shade the half-plane that does *not* contain $(0, 0)$.
    (Shown here in light red.)

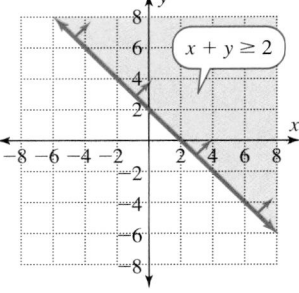

(b) ① To graph $3x - 2y < 6$, draw a dashed line for
    $3x - 2y = 6$. Use the intercepts $(2, 0)$ and
    $(0, -3)$ to draw the line.
    ② Test the point $(0, 0)$ in $3x - 2y < 6$
    $$3(0) - 2(0) \not< 6$$
    $$0 \not< 6 \text{ is true.}$$
    ③ Shade the half-plane that contains $(0, 0)$.
    (Shown here in light blue.)

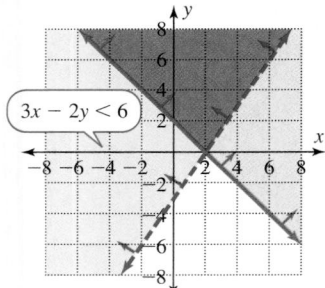

(c) The intersection of these two regions is the set of
    all the points satisfying this system of
    inequalities. (Shown here in violet.)

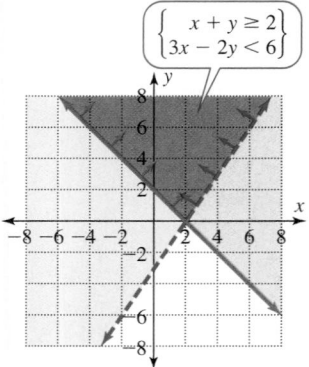

■   **EXAMPLE 8**   Using a Graphics Calculator to Solve a System of Inequalities

Use a graphics calculator to graph the solution of $\begin{cases} x + y \geq 2 \\ 3x - 2y < 6 \end{cases}$.
This is the same system as in Example 7.

SOLUTION _____

$x + y \geq 2$ $3x - 2y < 6$ To graph these inequalities on a calculator, first
$\quad y \geq 2 - x$ $\quad -2y < -3x + 6$ solve each inequality for y.

$\qquad\qquad\qquad\qquad y > \dfrac{3}{2}x - 3$

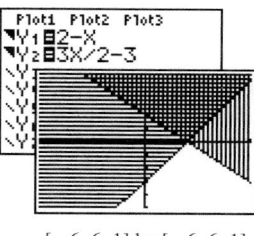

$[-6, 6, 1]$ by $[-6, 6, 1]$

Graphics calculators are powerful tools, but we must interpret properly the information they display. For inequalities be careful to determine from the given inequalities whether the boundary lines on the display should be included in the solution set.

Answer: The solution set is represented by the cross-hatched area. To interpret this solution properly, note that the line representing $y = \dfrac{3}{2}x - 3$ is not part of the solution.

■

SELF-CHECK   4.5.4

1. Graph the solution of $\begin{cases} y < 4 \\ y > -3 \end{cases}$.

2. Use a graphics calculator to graph the solution of $\begin{cases} x + y \leq 3 \\ 2x - 3y \leq 6 \end{cases}$.

SELF-CHECK 4.5.4 ANSWERS

1.
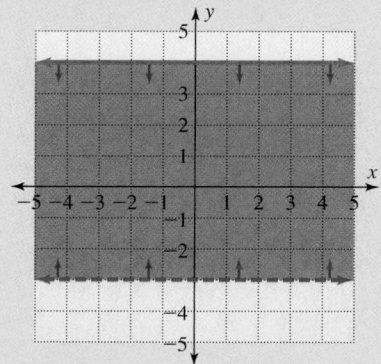

2.
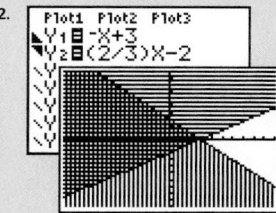

$[-10, 10, 1]$ by $[-10, 10, 1]$

The next example contains a system of three inequalities. We solve this system using pencil-and-paper methods. Also, you may want to try this example using the graphics calculator.

■ **EXAMPLE 9** Graphing a System of Linear Inequalities

Graph the solution of $\begin{cases} y \leq 4 \\ x - y \leq 2 \\ y \geq -x \end{cases}$.

SOLUTION _____

First graph each line on the same coordinate system. Then test the point $(1, 1)$ in each inequality.

$y \leq 4$ $x - y \leq 2$ $y \geq -x$

$1 \overset{?}{\leq} 4$ is true. $1 - 1 \overset{?}{\leq} 2$ $1 \overset{?}{\geq} -1$ is true.

                    $0 \overset{?}{\leq} 2$ is true.

Use arrows to indicate each individual region formed by these lines and then use shading to indicate the intersection of these regions. This triangular region is the solution of the system.

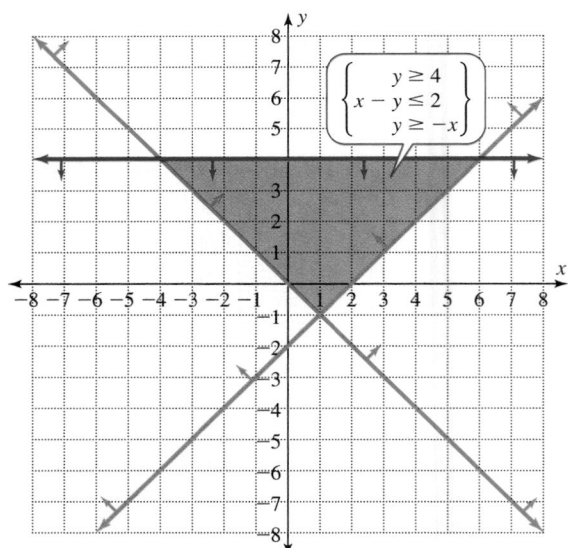

USING THE LANGUAGE AND SYMBOLISM OF MATHEMATICS   4.5

1. A solution to a linear inequality in two variables is an
   _____ _____ of values that, when
   substituted into the inequality, makes a _____
   statement.
2. The line $Ax + By = C$ separates the plane into two regions
   called _____ - _____ .
3. For the linear inequalities $\leq$ and $\geq$ we graph the boundary
   equation using a _____ line.
4. For the linear inequalities $<$ and $>$ we graph the boundary
   equations using a _____ line.

5. If a test point satisfies an inequality, shade the half-plane
   that _____ this point.
6. If a test point does not satisfy an inequality, shade the half-
   plane that does _____ _____ this point.
7. If ◤ $y_1 = 3x - 7$ is used to graph an inequality on a
   graphics calculator, then the calculator will shade the
   _____ half-plane.
8. If ◣ $y_1 = 3x - 7$ is used to graph an inequality on a
   graphics calculator, then the calculator will shade the
   _____ half-plane.

EXERCISES   4.5

In Exercises 1–4 determine whether the given point is a
solution of the inequality.

1. Check $(0, 0)$ in these inequalities.
   a. $2x + 3y < 1$ b. $2x + 3y \leq 1$
   c. $2x + 3y > 1$ d. $2x + 3y \geq 1$
2. Check $(1, 2)$ in these inequalities.
   a. $2x + 3y < 8$ b. $2x + 3y \leq 8$
   c. $2x + 3y > 8$ d. $2x + 3y \geq 8$
3. Check $(2, -3)$ in these inequalities.
   a. $3x - y < 9$ b. $3x - y \leq 9$
   c. $3x - y > 9$ d. $3x - y \geq 9$
4. Check $(-4, 1)$ in these inequalities.
   a. $x + 5y < 2$ b. $x + 5y \leq 2$
   c. $x + 5y > 2$ d. $x + 5y \geq 2$
5. Determine whether the points A–D are solutions of the
   inequality graphed here.

   $A\ (0, 0)$
   $B\ (4, -2)$
   $C\ (0, 2)$
   $D\ (-5, 0)$

   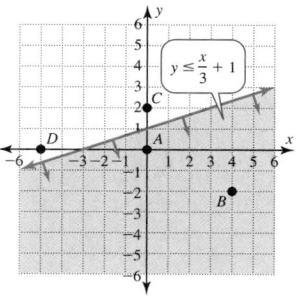

6. Determine whether the points A–D are solutions of the
   inequality graphed here.

   $A\ (4, 2)$
   $B\ (0, 4)$
   $C\ (-2, 0)$
   $D\ (-4, -4)$

   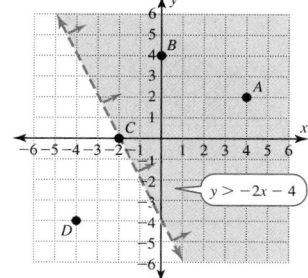

7. Determine whether the points A–D are solutions of the
   inequality graphed here.

   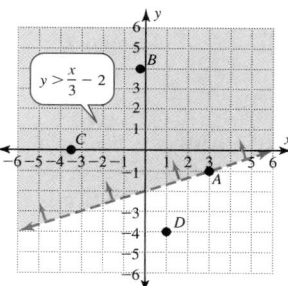

8. Determine whether the points A–D are solutions of the
   inequality graphed here.

   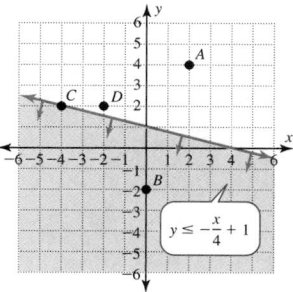

9. Which of the following inequalities corresponds to the
   graph in the figure?
   a. $x - 3y \leq 3$
   b. $x - 3y < 3$
   c. $x - 3y \geq 3$
   d. $x - 3y > 3$

   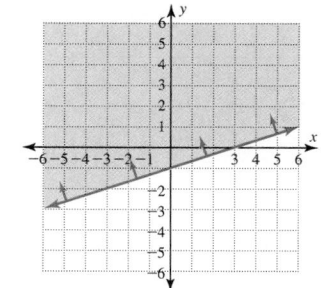

10. Which of the following inequalities corresponds to the graph in the figure?
   a. $x + 2y \leq 4$
   b. $x + 2y < 4$
   c. $x + 2y \geq 4$
   d. $x + 2y > 4$

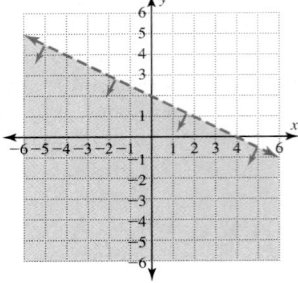

11. Determine which of the points $A-D$ satisfy the system of inequalities $\begin{cases} -x + 2y > 4 \\ 2x - y \geq -2 \end{cases}$.

   A $(4, 5)$ B $(0, 2)$
   C $(2, 6)$ D $(-3, -2)$

12. Determine which of the points $A-D$ satisfy the system of inequalities $\begin{cases} x - y \geq -4 \\ x + y < 4 \end{cases}$.

   A $(0, 4)$ B $(-4, 0)$
   C $(0, -4)$ D $(0, 0)$

In Exercises 13–16 shade the regions on the graphs of $y_1 = m_1x + b_1$ and $y_2 = m_2x + b_2$ which satisfy each system of inequalities. In Exercises 15 and 16, $m_1 = m_2$.

a. $\begin{cases} y_1 \geq m_1x + b_1 \\ y_2 \geq m_2x + b_2 \end{cases}$ b. $\begin{cases} y_1 \geq m_1x + b_1 \\ y_2 \leq m_2x + b_2 \end{cases}$

c. $\begin{cases} y_1 \leq m_1x + b_1 \\ y_2 \leq m_2x + b_2 \end{cases}$ d. $\begin{cases} y_1 \leq m_1x + b_1 \\ y_2 \geq m_2x + b_2 \end{cases}$

13.

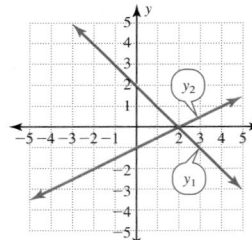

14.

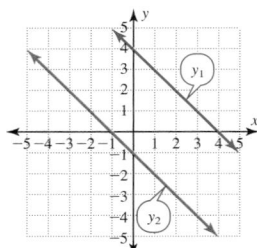

15.

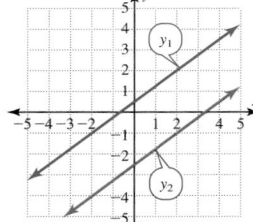

16.

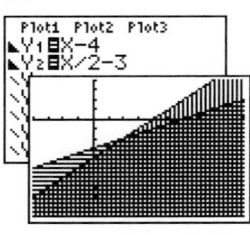

In Exercises 17–24 graph each linear inequality.
17. $x - y \geq 5$ 18. $2x + y \leq 6$
19. $3x - 2y - 12 < 0$ 20. $-5x + 2y + 10 < 0$
21. $x > 3y$ 22. $5x < -y$

23. $\dfrac{1}{2}x + \dfrac{1}{3}y \leq 1$ 24. $\dfrac{1}{4}x - \dfrac{1}{5}y \geq 1$

In Exercises 25–46 graph the solution of each system of linear inequalities.

25. $x - y \geq 4$ 26. $x + 2y \leq 2$
    $2x - y < 6$ $2x - y \geq 4$
27. $2x - 5y - 10 < 0$ 28. $3x + 2y - 12 > 0$
    $2x - y - 7 \geq 0$ $2x + 5y - 10 < 0$

29. $\dfrac{x}{2} - \dfrac{y}{2} < 1$ 30. $\dfrac{x}{3} - \dfrac{y}{3} \geq -1$

    $\dfrac{x}{2} + \dfrac{y}{2} > -1$ $\dfrac{x}{4} + \dfrac{y}{4} \geq -1$

31. $x \geq 1$ 32. $x > -5$
    $x < 4$ $x \leq -2$
33. $y < -2$ 34. $y > 3$
    $y \geq -6$ $y < 4$
35. $x > -2$ 36. $y < 5$
    $x \leq 3$ $y > 3$
37. $y \geq -2$ 38. $x \leq -4$
    $y \leq -4$ $x \geq -1$
39. $x \geq 0$ 40. $x \geq 0$
         $y \geq 0$ $y \geq 0$
    $2x + 3y < 6$ $5x + 2y < 10$
41. $x + y > 0$ 42. $2x + 3y \leq 0$
    $x - y < 0$ $3x - 2y \geq 0$
    $y < 4$ $y \geq -4$
43. $x \geq -2$ 44. $x > 1$
    $x \leq 2$ $x < 3$
    $y \geq -3$ $y > -5$
    $y \leq 3$ $y < -2$
45. $3x + 3y - 15 \leq 0$ 46. $x - y - 2 \leq 0$
    $6x + 2y - 18 \leq 0$ $2x + 2y - 8 \leq 0$
         $x \geq 0$ $x \geq 0$
         $y \geq 0$ $y \geq 0$

In Exercises 47 and 48 write the system of inequalities which is graphed on the given graphics calculator display. (Assume that the inequalities are either $\leq$ or $\geq$ and are not $<$ or $>$.)

47.

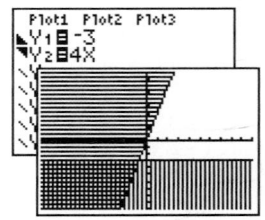

$[-4, 10, 1]$ by $[-10, 4, 1]$

48.

$[-10, 10, 1]$ by $[-10, 10, 1]$

Multiple Representations

In Exercises 49–52 write an algebraic inequality for each verbal statement.

49. The x-coordinate is at least two more than the y-coordinate.

50. The y-coordinate is at least three more than the x-coordinate.

51. The y-coordinate is at most four more than the x-coordinate.

52. The x-coordinate is at most one more than the y-coordinate.

In Exercises 53–56 write a system of algebraic equations for these verbal statements.

53. Both the x- and y-coordinates are positive and the sum of the two coordinates does not exceed 10.

54. Both the x- and y-coordinates are positive and the sum of the two coordinates does not exceed 8.

55. Both the x- and y-coordinates are nonnegative and the sum of x and twice y is at most 5.

56. Both the x- and y-coordinates are nonnegative and the sum of twice x and three times y is at most 6.

In Exercises 57–60 write a system of inequalities that represents each situation and then graph this system of inequalities.

57. **Length of Two Pieces of Rope** A store sold two pieces of rope from a spool containing 150 m. Lengths x and y must both be positive and their total is at most 150 m.

58. **Theater Tickets** A theater can seat at most 500 customers. Neither the number of adult tickets x nor the number of child tickets y can be negative and their total is at most 500.

59. **Production Units** A company makes a profit of $40 for each of the x stereos it ships and $50 for each of the y televisions it ships. The number of units of each item shipped is nonnegative and the profit per day for the factory has never exceeded $4400.

60. **Factory Production** A printer makes a profit of $2 for each of the x books produced and $0.25 for each of the y magazines produced. The number produced of each is nonnegative and the profit per day for the printer has never exceeded $15,000.

Group Discussion Questions

61. **Writing Mathematically** Write two different word problems that describe a situation that can be modeled by this system of inequalities.

$$\begin{cases} x \geq 0 \\ y \geq 0 \\ x + y \leq 100 \end{cases}$$

62. **Production Choices** The following system of inequalities models the restrictions on the number x of wooden table chairs and the number y of wooden rocking chairs made at a furniture factory each week. These restrictions are caused by limitations on the factory's equipment and their labor supply.

$$\begin{cases} x \geq 0 \\ y \geq 0 \\ x + y \leq 50 \\ 3x + y \leq 90 \end{cases}$$

a. Determine each corner point of the region formed by graphing this system of inequalities.

b. The factory makes $20 for each table chair and $25 for each rocking chair. Write an expression for the profit involving x and y.

c. Evaluate the profit expression from part **b** at each of the corner points from part **a**. Which of these points produces the greatest profit for the factory?

KEY CONCEPTS   FOR CHAPTER 4

1. Solution of an Inequality
- A solution of an inequality in one variable is a value of the variable that makes the inequality a true statement.
- A solution of an inequality in two variables is an ordered pair that makes the inequality a true statement.

2. Types of Inequalities
- Conditional Inequality: A conditional inequality contains a variable and is true for some, but not all, real values of the variable(s).
- Unconditional Inequality: An inequality that is always true.
- Contradiction: An inequality that is always false.
- Equivalent Inequalities: Inequalities that have the same solution are called equivalent inequalities.
- Linear Inequality: An inequality that is the first degree in each variable.

3. Principles Used to Solve Inequalities*
- **Addition-Subtraction Principle:**
  If a, b, and c are real numbers, then $a < b$ is equivalent to $a + c < b + c$ and to $a - c < b - c$.
- **Multiplication-Division Principle:**
  If a, b, and c are real numbers and $c > 0$, then $a < b$ is equivalent to $ac < bc$. If $c < 0$, then $a < b$ is equivalent to $ac > bc$.†

4. Intersection and Union of Two Sets
Intersection: $A \cap B$ is the set of points in both A **and** B.
Union: $A \cup B$ is the set of points in either set A **or** B (or both).

5. Compound Inequality
$a \leq x \leq b$ is equivalent to $x \geq a$ and $x \leq b$.

6. Absolute Value Equations and Inequalities
For any real numbers x and a and positive real number d:
- $|x - a| = d$ is equivalent to $x - a = -d$ or $x - a = d$
- $|x - a| < d$ is equivalent to $-d < x - a < d$
- $|x - a| > d$ is equivalent to $x - a < -d$ or $x - a > d$
- $|x - a| = -d$ is a contradiction and has no solution
- $|x - a| < -d$ is a contradiction and has no solution
- $|x - a| > -d$ is an unconditional inequality and the solution set is the set of all real numbers

7. Half-Planes
The line $Ax + By = C$ separates the plane into two regions called half-planes. One half-plane satisfies $Ax + By < C$ and the other half-plane satisfies $Ax + By > C$.

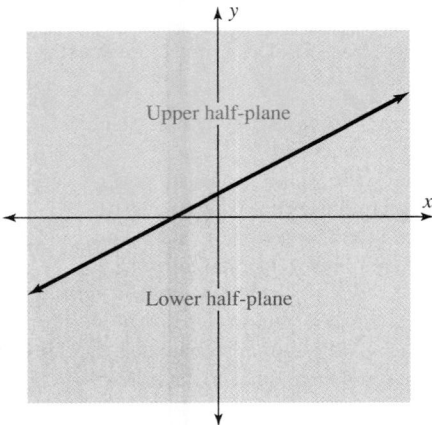

8. Graphing a Linear Inequality
Step 1. Graph the equality $Ax + By = C$ using
a. A solid line for $\leq$ or $\geq$.
b. A dashed line for $<$ or $>$.
Step 2. Choose an arbitrary test point not on the line; $(0, 0)$ is often convenient. Substitute this test point into the inequality.
Step 3.
a. If the test point satisfies the inequality, shade the half-plane containing this point.
b. If the test point does not satisfy the inequality, shade the other half-plane.

9. Graphing a System of Linear Inequalities
Graph each inequality in the system on the same coordinate system. Use arrows to indicate each individual region formed by these lines. Then use shading to indicate the intersection of these regions. The solution of the system is represented by this intersection.

REVIEW EXERCISES   FOR CHAPTER 4

1. Determine if $x = -4$ is a solution of each of these inequalities.
a. $2x > -8$
b. $x + 3 \leq -1$
c. $5x + 3 < 2x - 3$
d. $5(x - 2) \geq 3(2x - 4) + 7$

2. Determine which of the points A–D satisfy the inequality $y < \dfrac{x}{3} - 2$.

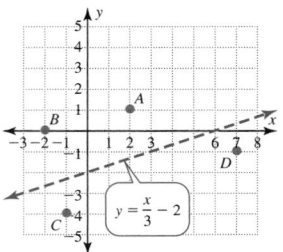

*Similar statements can be made for the inequalities $\leq$, $>$ and $\geq$.
†Similar statements can be made for division.

3. Determine if the point $(-2, 5)$ is a solution of each of these inequalities.
a. $y \le 2x - 12$ **b.** $y \ge 3x - 12$
c. $3x + 5y < 15$ **d.** $-2x + 4y > 22$

4. Use the graphs of $y_1 = 0.2x + 0.4$ and $y_2 = 1$ to solve each equation and inequality.
a. $0.2x + 0.4 = 1$
b. $0.2x + 0.4 > 1$
c. $0.2x + 0.4 < 1$

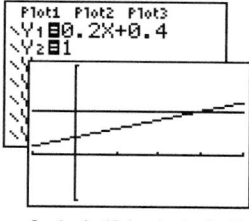

$[-1, 4, 1]$ by $[-1, 2, 1]$

5. Use the tables for $y_1 = 5(x + 2)$ and $y_2 = 2x + 7$ to solve each equation and inequality.
a. $5(x + 2) = 2x + 7$
b. $5(x + 2) < 2x + 7$
c. $5(x + 2) > 2x + 7$

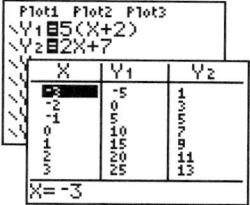

6. Modeling Electricity Production from Wind and Coal
In 1990, only 3% of all electricity needed in one community was generated by wind and 43% was generated by coal. By 2000, 30% of the electricity needed was generated by wind and 20% was generated by coal. The graph of y_1 represents the percent of electricity generated by wind. The graph of y_2 represents the percent of electricity generated by coal. Use the graphs that follow to solve:
a. $y_1 = y_2$
b. $y_1 < y_2$
c. $y_1 > y_2$
d. Interpret the meaning of each of these solutions.

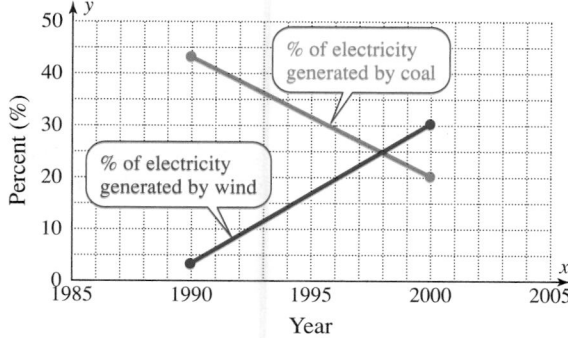

7. Using Tables that Model Repair Shop Charges The following tables display the charges by two TV repair shops based on the number of hours required for a repair. Use these tables to solve:

a. $y_1 = y_2 T$
b. $y_1 < y_2$
c. $y_1 > y_2$
d. Interpret the meaning of each of these solutions.

SHOP A		SHOP B	
x Hours	y_1 Cost ($)	x Hours	y_2 Cost ($)
1	50	1	55
2	85	2	85
3	120	3	115
4	155	4	145

8. Multiple Perspectives Complete the following table.

VERBALLY	INEQUALITY NOTATION	INTERVAL NOTATION	GRAPH
a. x exceeds 3			
b.	$x \le 4$		
c.		$[2, +\infty)$	
d.			(graph)

9. Multiple Perspectives Complete the following table.

VERBALLY	INEQUALITY NOTATION	INTERVAL NOTATION	GRAPH
a. x is greater than 2 and less than 6			
b.	$-3 < x \le 4$		
c.		$[-2, 0)$	
d.			(graph)

10. Use the following table for $y_1 = \dfrac{x - 1}{2}$ to solve $-1 \le \dfrac{x - 1}{2} < 1$.

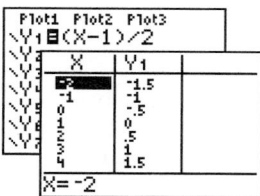

11. Use the given graph to solve $-1 \le \dfrac{2 - x}{3} \le 1$.

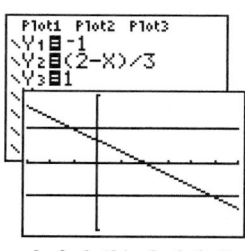

$[-3, 6, 1]$ by $[-2, 2, 1]$

12. Multiple Representations   Write an algebraic inequality for each verbal statement.
  a. $x + 7$ is at most 11.
  b. Twice the quantity x minus 1 exceeds 13.
  c. Three x minus 5 is at least 7.
  d. Four x plus 9 never exceeds 21.

Multiple Perspectives
In Exercises 13–18 solve each inequality algebraically, graphically, and numerically and describe the solution verbally.

13. $x + 3 > 1$

14. $\dfrac{x}{2} - 1 \le 1$

15. $\dfrac{x}{2} + 4 \ge 1 - \dfrac{x}{2}$

16. $2x - 1 < x + 2$

17. $2(x + 2) \le 3(x + 1)$

18. $|x - 2| < 2$

In Exercises 19–42 solve each inequality.

19. $x + 7 < 5$

20. $x - 11 > -10$

21. $-10 \ge 2x$

22. $-3x \le 12$

23. $3x + 7 > 5x + 13$

24. $7 \ge 7 - 9y$

25. $-\dfrac{x}{2} < \dfrac{1}{4}$

26. $\dfrac{3x}{7} + \dfrac{4}{5} > \dfrac{3x}{5} + \dfrac{2}{7}$

27. $7y + 14 \ge 2(3y + 8)$

28. $-5(2y - 2) \le 7 - 11y$

29. $2(3t - 4) > 3(t - 6) + 1$

30. $2(11t - 3) < 5(3t + 2) - 20$

31. $5(x - 4) + 6 < (x + 4) - 30$

32. $7(x - 3) + 6 \ge 12(x + 4) + 2$

33. $4(y - 1) - 7(y + 1) \le -3(y + 2) - 5$

34. $1 - 4(y - 2) \ge 3(y - 7) - 5(4 - y)$

35. $\dfrac{v}{2} - \dfrac{3v + 4}{4} > \dfrac{v + 40}{4}$

36. $\dfrac{v}{8} + 6 > \dfrac{v}{12} + 5$

37. $9 < x + 7 \le 13$

38. $0 \le \dfrac{4}{5}x < 20$

39. $42 \le \dfrac{-3m}{7} < 60$

40. $-5 \le 9 - x \le 5$

41. $-7 \le 5x + 3 \le 13$

42. $-10 < 5 - 3x < 8$

In Exercises 43–46 solve each absolute value equation or inequality.

43. $|x - 3| = 4$

44. $|2x - 5| = 3$

45. $|5x - 2| > 8$

46. $|x - 5| < 3$

In Exercises 47–50 each inequality is a conditional inequality, an unconditional inequality, or a contradiction. Identify the type of each inequality and solve it.

47. $3v + 3v > 5v$

48. $3v + 3v \le 5v$

49. $3(v + 1) < 3v + 4$

50. $5(3v - 1) > 3(5v - 1)$

In Exercises 51–54 determine $A \cap B$ and $A \cup B$ for the given intervals A and B.

51. $A = (1, 4), B = (2, 7)$

52. $A = [2, 5), B = [3, 6]$

53. $A = (-\infty, 2], B = [-3, +\infty)$

54. $A = (-\infty, 5), B = [-2, 6)$

In Exercises 55 and 56 write each inequality expression as a compound inequality.

55. $x > -3$ and $x \le 4$

56. $x \ge 0$ and $x < \pi$

In Exercises 57 and 58 solve each inequality.

57. $x - 2 < 3$ and $3 > x + 2$

58. $x - 2 < 4x + 2$ or $6x - 6 < 4x - 12$

In Exercises 59–64 graph the solution of each system of inequalities.

59. $x \ge -1$
     $x \le 3$

60. $y \le 4$
     $y \ge 1$

61. $2x - 3y < 6$
     $2x + 5y > 10$

62. $2x + y < 4$
     $2x + y > 1$

63. $3x + 4y \le 12$
     $3x - 4y \le 12$

64. $2x + y \le 4$
     $x \ge 0$
     $y \ge 0$

In Exercises 65–68 write an absolute value inequality to represent each interval.

65. $(-5, 5)$

66. $[-4, 4]$

67. $(-2, 10)$

68. $[-4, 20]$

Multiple Representations
In Exercises 69–71 write an algebraic inequality for each verbal statement, then solve this inequality and answer the question.

69. Four times the quantity x plus three is greater than eight. Solve for x.

70. Three times the quantity x minus seventeen is less than or equal to twice the quantity x plus eleven. Solve for x.

71. Three x minus two is greater than or equal to seven and is less than nineteen. Solve for x.

In Exercises 71–77 solve each word problem.

72. Basketball Average   A basketball player scored 17 points, 27 points, and 18 points in the first three games of the season. How many points will he have to score in the fourth game to average at least 20 points for the first four games?

73. Average Salary   When she was hired, an employee was guaranteed that she would average at least $500 per week. The first three weeks she made $560, $450, and $480. How much money must she make the fourth week to meet the guarantee?

74. Perimeter of a Rectangle   The width of a rectangle must be exactly 12 cm. If the perimeter must be between 44 and 64 cm, determine the length that can be used for the rectangle.

12 cm

L

75. Modeling the Length of a Golf Shot   Write a compound inequality that expresses the length of the tee shot that a golfer must make to clear the water and to land safely on the green of the golf hole shown here.

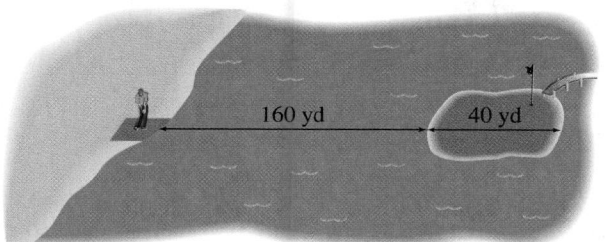

160 yd   40 yd

76. Tolerance of Rocket Fuel   To adjust the flight of a Mars probe, a rocket is allowed to burn 7.2 grams of fuel with a tolerance of 0.25 grams. Express this tolerance interval as an absolute value inequality and determine the lower and upper limits of the interval.

77. Modeling Costs and Revenue   The cost of printing x advertising posters includes a fixed cost of $75 and a variable cost of $0.50 per poster. All posters are custom ordered and the charge for printing these posters includes a setup fee of $50 plus a charge of $1 per poster. Let y_1 represent the income received from an order for x posters and y_2 represent the cost of printing x posters. Determine the values of x for which $y_1 < y_2$, the loss interval for this order. Also determine the values of x for which $y_1 \geq y_2$, the profit interval for this order.

In Exercises 78–80 mentally estimate the solution of each inequality and then use a calculator to calculate the solution.

PROBLEM	MENTAL ESTIMATE	CALCULATOR SOLUTION
78. $x - 2.53 < 7.52$		
79. $-2.53x \geq 5.0853$		
80. $10.89 < 9.9x \leq 29.601$		

MASTERY TEST   FOR CHAPTER 4

[4.1]   **1.** Determine whether $x = 10$ is a solution of each inequality.
   a. $4x - 1 \geq x + 29$ **b.** $2(x + 15) > -2(x + 5)$
   c. $5(x + 2) < 2(x + 20)$ **d.** $5(x - 2) \leq 3(x - 10)$

[4.1]   **2.** In parts **a** and **b** use the graph to solve each inequality.

   a. $\dfrac{x}{3} - 2 < -\dfrac{1}{2}(x - 1)$

   b. $\dfrac{x}{3} - 2 \geq -\dfrac{1}{2}(x - 1)$

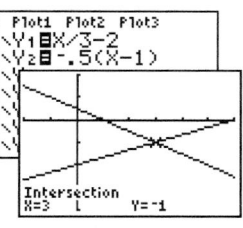

$[-2, 6, 1]$ by $[-4, 2, 1]$

In parts **c** and **d** use the table to solve each inequality.
   c. $4(x + 1) \leq 2(x + 6)$
   d. $4(x + 1) > 2(x + 6)$

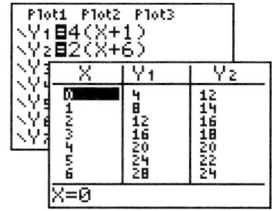

[4.1]   **3.** Algebraically solve each of these inequalities.
   a. $x + 12 \leq 17$ **b.** $x - 9 < -11$
   c. $4x + 3 > 3x - 8$ **d.** $-2x - 5 \geq -3x - 1$

[4.2]   **4.** Algebraically solve each of these inequalities.
   a. $-11x \leq 165$
   b. $\dfrac{3x}{7} > 105$
   c. $8(x + 3) < 4(x + 4) - (10 - 2x)$
   d. $-5 \leq \dfrac{4x - 9}{3}$

[4.3]   **5.** Identify each of these inequalities as either a conditional inequality, an unconditional inequality, or a contradiction, and then solve the inequality.
   a. $x + x < x$
   b. $2(3x + 5) \geq 3(2x - 6)$
   c. $3(8x + 1) > 4(6x + 5)$
   d. $2x + 3 \leq 3x + 3$

[4.3]   **6.** Solve each of these inequalities.
   a. $-30 < -6x \leq 48$
   b. $-30 \leq x - 6 < 48$
   c. $-1 \leq 4x + 3 \leq 19$
   d. $2x + 1 \leq -7$ or $3x - 2 \geq 7$

[4.4]   **7.** Algebraically solve each equation and inequality.
   a. $|x| = 8$
   b. $|2x + 5| = 49$
   c. $|2x - 3| < 19$
   d. $|x - 5| > 4$

[4.4] **8.** In parts **a** and **b** use the graph to solve each inequality.

    a. $|2x - 3| < 3$

    b. $|2x - 3| > 3$

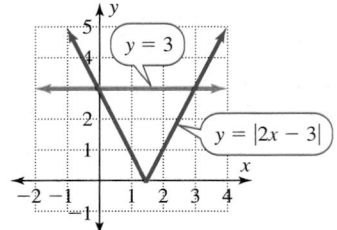

In parts **c** and **d** use the table to solve each inequality.

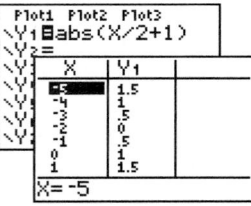

    c. $\left|\dfrac{x}{2} + 1\right| \le 1$

    d. $\left|\dfrac{x}{2} + 1\right| \ge 1$

[4.5] **9.** Graph each of these linear inequalities.

    a. $2x - 3y < 6$ **b.** $y \ge \dfrac{x}{2}$

    c. $x \le -2$ **d.** $y > -2$

[4.5] **10.** Graph the solution to each system of inequalities.

    a. $\begin{cases} x \ge 2 \\ x \le 4 \end{cases}$ **b.** $\begin{cases} y > -2 \\ y < 3 \end{cases}$

    c. $\begin{cases} x + 2y \ge 4 \\ -2x + y < 4 \end{cases}$ **d.** $\begin{cases} x \ge 0 \\ y \ge 0 \\ 2x + y \le 4 \end{cases}$

GROUP PROJECT   FOR CHAPTER 4

Risks and Choices: Comparing the Costs for Two Checking Accounts

This project requires the group to collect information about the cost of a checking account from two local banks.

1. Two options for checking accounts presented by a local bank follows.

Option A: Checking is free with a $3000 minimum daily account balance.
Option B: The checking account has a $10 per month charge plus a charge of $0.05 per check.

    a. An indirect charge for option A is that an individual using this option is giving up the use of $3000 per month to obtain this "free" service. One use that could be made of this $3000 is to invest it in a savings account with this same bank. This bank currently has a savings account that pays interest monthly at the rate of 5% per year. How much interest could $3000 earn in this savings account each month?

    b. Write a linear function that gives the monthly cost of option B in terms of the number of checks written.

    c. Write a linear function that gives the monthly cost of option A in terms of the number of checks written. (Assume that the cost is the value of the interest that is being lost as calculated in part **a.**)

    d. Write a system of linear equations that can be solved to determine when these two options produce the same monthly cost.

    e. Graph the system of equations from part **d.**

    f. Use the equations from part **d** to prepare a table of values for the costs of both options for various numbers of checks written (e.g., 10, 20, 30, . . . 80).

    g. Determine the number of checks that will produce the same monthly cost for both options. Show how this value can be determined algebraically, graphically, and numerically.

    h. Write an inequality that describes when option A is less expensive than option B. Then solve this inequality. Show this solution on the graph from part **e.**

    i. Write an inequality that describes when option B is less expensive than option A. Then solve this inequality. Show this solution on the graph from part **e.**

2. Determine two different checking account options from either one bank or two competing banks in your area. Then compare the costs of these two options using the steps from part **1.**

5

EXPONENTS AND OPERATIONS WITH POLYNOMIALS

The graphics calculator display shown here was used to calculate the number of transistors required to prepare an order of 125,000 computer game station controllers. Each controller has 990,000 transistors.

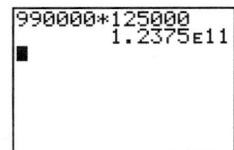

This calculator display gives the product in scientific notation. This chapter examines scientific notation, properties of exponents, and the basic operations with polynomials. Polynomials, the simplest of algebraic expressions, play a fundamental role in the study of algebra and are useful in modeling many problems in business and in the sciences.

Section 5.1   Product and Power Rules for Exponents

Objectives: 1. Use the product rule for exponents.
2. Use the power rule for exponents.

Question: To determine the value of each \$1 invested at a 7% annual rate of return for a period of 10 years we can use 1.07 as a factor 10 times (1.07 is 100% plus 7%). Represent this computation on paper and determine the value at the end of the 10-year period with a calculator.

Answer: We can represent this product in either the expanded form or in exponential notation.

EXPANDED FORM	EXPONENTIAL NOTATION
(1.07) (1.07) (1.07) (1.07) (1.07) (1.07) (1.07) (1.07) (1.07) (1.07)	$(1.07)^{10}$

The following calculator display confirms the fact that both results are identical. In either case the value of \$1 nearly doubles in 10 years, producing a new value of about \$1.97.

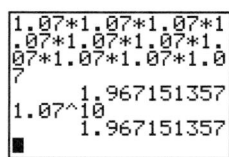

Try this yourself. Which of these two algebraic representations do you prefer?

One of the themes of this book is that algebra presents the language and symbolism of mathematics—a language and symbolism invented by humankind to make things easier to represent and examine. Obviously a student of algebra must study this notation in order to use it effectively—but please keep in mind that without algebra mathematical concepts will be very cumbersome and lengthy to describe. Exponential notation, first examined in Section 1.6, is one example of a compact algebraic notation.

In this section we examine the product and power rules, which enable us to quickly multiply exponential expressions with the same base. Before examining these rules, we need to practice interpreting exponential notation.

The most common error made in evaluating exponential expressions is using the wrong base. If there are no symbols of grouping, only the constant or variable immediately to the left of the exponent is the base. Recall from Section 1.6 that $(-3)^2 = (-3)(-3) = 9$, whereas $-3^2 = -(3 \cdot 3) = -9$.

■ **EXAMPLE 1**   Identifying the Base of an Exponential Expression

Write each of the following exponential expressions in expanded form.

SOLUTIONS

The subtle but significant differences in the expressions in each part of Example 1 are very important for you to understand. To understand the properties of exponents, note the base of the exponent in each part of this example. You also can practice by completing the following self-check.

(a) x^4	$x \cdot x \cdot x \cdot x$	The base is x and the exponent is 4.
(b) $-x^4$	$-x \cdot x \cdot x \cdot x$	The base is x, not $-x$ and the exponent is 4.
(c) $(-x)^4$	$(-x)(-x)(-x)(-x)$	The base is $-x$ and the exponent is 4.
(d) $(xy)^3$	$(xy)(xy)(xy)$	The base is xy and the exponent is 3.
(e) xy^3	$x \cdot y \cdot y \cdot y$	The base of the exponent 3 is y, not xy.

■ **EXAMPLE 2**   Identifying Bases and Exponents

Write each of the following expressions in exponential form.

SOLUTIONS

(a) $x \cdot x \cdot y \cdot y$	$x^2 y^2$
(b) $x \cdot x + y \cdot y$	$x^2 + y^2$
(c) $(x + y)(x + y)$	$(x + y)^2$
(d) $-3 \cdot a \cdot a \cdot a \cdot a$	$-3a^4$
(e) $(-3a)(-3a)(-3a)(-3a)$	$(-3a)^4$

SELF-CHECK 5.1.1

Write each of these exponential expressions in expanded form.

1. $-5a^3$ 2. $(-5a)^3$
3. $a^3 b^2$ 4. -7^2

Product Rule

The properties of exponents allow us to simplify many computations involving exponential expressions. These properties follow directly from the definition of natural number exponents. We start by examining the logic behind the product rule for exponents. The product $x^3 \cdot x^4$ is written here in expanded form.

$$x^3 x^4 = \underbrace{\boxed{x \cdot x \cdot x}}_{\substack{\text{Three factors} \\ \text{of } x}} \cdot \underbrace{\boxed{x \cdot x \cdot x \cdot x}}_{\substack{\text{Four factors} \\ \text{of } x}} = \underbrace{\boxed{x \cdot x \cdot x \cdot x \cdot x \cdot x \cdot x}}_{\substack{\text{Seven factors} \\ \text{of } x}} = x^7$$

Thus $x^3 \cdot x^4 = x^7$. The fact that these expressions are equivalent means that they have the same value for all values of x. This is confirmed by comparing the values of y_1 and y_2 in the following table.

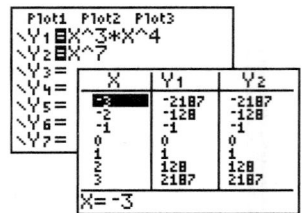

In general the product $x^m x^n = x^{m+n}$ has the base x used as a factor a total of $m + n$ times.

$$x^m x^n = \underbrace{\boxed{x \cdot x \cdot \ldots \cdot x}}_{\substack{m \text{ factors} \\ \text{of } x}} \cdot \underbrace{\boxed{x \cdot x \cdot \ldots \cdot x}}_{\substack{n \text{ factors} \\ \text{of } x}} = \underbrace{\boxed{x \cdot x \cdot \ldots \cdot x}}_{\substack{m + n \text{ factors} \\ \text{of } x}} = x^{m+n}$$

Product Rule for Exponents

ALGEBRAICALLY	VERBALLY	ALGEBRAIC EXAMPLE
For any real number x and natural numbers m and n, $x^m \cdot x^n = x^{m+n}$.	To multiply two factors with the same base, use the common base and add the exponents.	$x^3 \cdot x^4 = x^7$

Although it is useful to memorize "add the exponents" for multiplying two factors, it is also important to remember that the bases must be the same.

■ EXAMPLE 3   Using the Product Rule

Use the product rule to simplify each expression.

SOLUTIONS

(a) $z^3 z^{10}$

$z^3 z^{10} = z^{3+10}$
$\qquad = z^{13}$

Add the exponents with the common base z.

(b) $v^2 v^4 v^5$

$v^2 v^4 v^5 = v^{2+4+5}$
$\qquad = v^{11}$

Add the exponents with the common base v.
The product rule can be extended to any number of factors that have the same base.

(c) $-x^7 x^8$

$-x^7 x^8 = -x^{7+8}$
$\qquad = -x^{15}$

$-x^7 x^8 = -(x^7 x^8)$ The base for each exponent is x, not $-x$.

(d) $(3y^4)(5y^5)$

$(3y^4)(5y^5) = (3)(5)y^{4+5}$
$\qquad\qquad = 15y^9$

Add the exponents with the common base y.

(e) $y^3 x^2$

$y^3 x^2 = x^2 y^3$

Since these factors do not have the same base, this expression cannot be simplified. However, factors are usually written in alphabetical order. ■

SELF-CHECK 5.1.2

Simplify each expression.

1. $x^2 \cdot x^3$ 2. $y^{15}y^{16}$
3. m^2n^4 4. $-z^2z^9$

Power Rule

The next important property of exponents that we examine is the power rule for exponents. We first examine the logic for raising a power to a power. The expression $(x^2)^4$ is expanded next.

$$(x^2)^4 = \boxed{\overset{\text{Four factors of } x^2}{x^2 \cdot x^2 \cdot x^2 \cdot x^2}}$$

$$= \boxed{\overset{\substack{\text{Four groups, each with two} \\ \text{factors of } x}}{(x \cdot x)(x \cdot x)(x \cdot x)(x \cdot x)}}$$

$$= \boxed{\overset{2 \cdot 4 \text{ factors of } x}{x \cdot x \cdot x \cdot x \cdot x \cdot x \cdot x \cdot x}}$$

$$= x^8$$

Thus $(x^2)^4 = x^8$.

The fact that $(x^2)^4$ is equivalent to x^8 is confirmed by comparing the values of y_1 and y_2 in the following table.

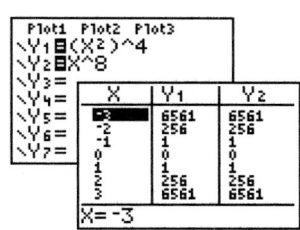

In general $(x^m)^n = \boxed{\overset{n \text{ factors of } x^m}{x^m \cdot x^m \cdot \ldots \cdot x^m}}$

$$= \boxed{\overset{\substack{n \text{ groups, each with } m \text{ factors of } x}}{(x \cdot x \cdot \ldots \cdot x)(x \cdot x \cdot \ldots \cdot x) \cdot \ldots \cdot (x \cdot x \cdot \ldots \cdot x)}}$$

$$= \boxed{\overset{m \cdot n \text{ factors of } x}{x \cdot x \cdot x \cdot \ldots \cdot x \cdot x \cdot x}}$$

$$= x^{mn}$$

SELF-CHECK 5.1.2 ANSWERS

1. x^5 2. y^{31}
3. m^2n^4 4. $-z^{11}$

Power Rule for Exponents

ALGEBRAICALLY	VERBALLY	ALGEBRAIC EXAMPLE
For any real number x, and natural numbers m and n, $(x^m)^n = x^{mn}$	To raise a power to a power, multiply the exponents.	$(x^2)^4 = x^8$

■ EXAMPLE 4 Using the Power Rule

Simplify each expression.

SOLUTIONS

(a) $(b^3)^4$ 　　$(b^3)^4 = b^{3\cdot4}$ 　　To raise a power to a power, multiply the exponents.
　　　　　　　　$= b^{12}$

(b) $-(x^3)^2$ 　　$-(x^3)^2 = -x^{3\cdot2}$ 　　Multiply the exponents. The base is x, not $-x$.
　　　　　　　　$= -x^6$

SELF-CHECK 5.1.3

Simplify each expression.
1. $(x^2)^5$ 　　　　2. $(x^5)^2$
3. $(-x^5)^2$ 　　　4. $(-x^2)^5$
5. $-(x^5)^2$

The power rule can be applied to products and quotients as developed in the following paragraphs.

A Product to a Power

The expression $(xy)^3$ is expanded here.

$$(xy)^3 = \boxed{(xy)(xy)(xy)}$$ Three factors of xy

$$= \boxed{(x \cdot x \cdot x)} \cdot \boxed{(y \cdot y \cdot y)}$$ Three factors of x, Three factors of y

$$= x^3y^3$$

Rearrange the factors, using the associative and commutative properties of multiplication, which were discussed in Chapter 1.

Thus $(xy)^3 = x^3y^3$.

In general

$$(xy)^m = \boxed{(xy)(xy) \cdot \ldots \cdot (xy)}$$ m factors of xy

$$= \boxed{(x \cdot x \cdot \ldots \cdot x)} \cdot \boxed{(y \cdot y \cdot \ldots \cdot y)}$$ m factors of x, m factors of y

$$= x^my^m$$

■ EXAMPLE 5   Simplifying a Product to a Power

Simplify each expression.

SOLUTIONS

(a) $(ab)^5$

$(ab)^5 = a^5 b^5$

For a product to a power $(xy)^m = x^m y^m$.

(b) $(a^2 b)^5$

$(a^2 b)^5 = (a^2)^5 b^5$

$= a^{10} b^5$

For a product to a power $(xy)^m = x^m y^m$.
Use the power rule $(x^m)^n = x^{m \cdot n}$.

(c) $(2t)^3$

$(2t)^3 = 2^3 t^3$

$= 8t^3$

For a product to a power $(xy)^m = x^m y^m$.

(d) $(-3t)^4$

$(-3t)^4 = (-3)^4 t^4$

$= 81t^4$

Notice the subtle distinctions among parts (d), (e), and (f). In part (d) the base is $-3t$.

(e) $-(3t)^4$

$-(3t)^4 = -(3^4 \cdot t^4)$

$= -(81t^4)$

$= -81t^4$

In part (e) the base is $3t$.

(f) $-3t^4$

$-3t^4 = -3t^4$

In part (f) the exponent applies only to the base of t; thus this expression is already in simplified form. ■

A Quotient to a Power

We now examine a quotient to a power starting with the expansion of $\left(\dfrac{x}{y}\right)^3$.

Three factors of $\dfrac{x}{y}$

$$\left(\frac{x}{y}\right)^3 = \boxed{\left(\frac{x}{y}\right)\left(\frac{x}{y}\right)\left(\frac{x}{y}\right)}$$

Three factors of x

$$= \boxed{\frac{x \cdot x \cdot x}{y \cdot y \cdot y}}$$

Three factors of y

Multiply, using the procedure for multiplying fractions.

$$= \frac{x^3}{y^3}$$

Thus $\left(\dfrac{x}{y}\right)^3 = \dfrac{x^3}{y^3}$.

m factors of $\dfrac{x}{y}$

In general $\left(\dfrac{x}{y}\right)^m = \boxed{\left(\dfrac{x}{y}\right)\left(\dfrac{x}{y}\right) \cdots \left(\dfrac{x}{y}\right)}$

m factors of x

$$= \boxed{\frac{x \cdot x \cdot \ldots \cdot x}{y \cdot y \cdot \ldots \cdot y}}$$

m factors of y

$$= \frac{x^m}{y^m}$$

Both the product to a power and the quotient to a power rules are summarized in the following box.

Raising Products and Quotients to a Power

ALGEBRAICALLY	VERBALLY	ALGEBRAIC EXAMPLE
For any real numbers x and y and any natural number m, $$(xy)^m = x^m y^m$$ $$\left(\frac{x}{y}\right)^m = \frac{x^m}{y^m} \quad \text{for } y \neq 0.$$	To raise a product to a power, raise each factor to this power. To raise a quotient to a power, raise both the numerator and the denominator to this power.	$$(xy)^3 = x^3 y^3$$ $$\left(\frac{x}{y}\right)^3 = \frac{x^3}{y^3}$$

■ EXAMPLE 6   Simplifying a Quotient to a Power

Simplify each expression assuming each variable is not zero.

SOLUTIONS _____

(a) $\left(\dfrac{a}{b}\right)^5$
$$\left(\frac{a}{b}\right)^5 = \frac{a^5}{b^5}$$
Use the quotient to a power rule, $\left(\dfrac{x}{y}\right)^m = \dfrac{x^m}{y^m}$.

(b) $\left(\dfrac{v^2}{w}\right)^3$
$$\left(\frac{v^2}{w}\right)^3 = \frac{(v^2)^3}{w^3}$$
Use the quotient to a power rule, $\left(\dfrac{x}{y}\right)^m = \dfrac{x^m}{y^m}$.
$$= \frac{v^6}{w^3}$$
Use the power rule, $(x^m)^n = x^{mn}$.

(c) $\left(\dfrac{3}{v}\right)^4$
$$\left(\frac{3}{v}\right)^4 = \frac{3^4}{v^4}$$
Use the quotient to a power rule, $\left(\dfrac{x}{y}\right)^m = \dfrac{x^m}{y^m}$.
$$= \frac{81}{v^4}$$

(d) $\left(-\dfrac{5}{y}\right)^2$
$$\left(-\frac{5}{y}\right)^2 = \left(\frac{-5}{y}\right)^2$$
The fraction $-\dfrac{a}{b}$ equals $\dfrac{-a}{b}$.
$$= \frac{(-5)^2}{y^2}$$
Use the quotient to a power rule, $\left(\dfrac{x}{y}\right)^m = \dfrac{x^m}{y^m}$.
$$= \frac{25}{y^2}$$

(e) $-\left(\dfrac{5}{y}\right)^2$
$$-\left(\frac{5}{y}\right)^2 = -\left(\frac{5^2}{y^2}\right)$$
Use the quotient to a power rule, $\left(\dfrac{x}{y}\right)^m = \dfrac{x^m}{y^m}$.
$$= -\frac{25}{y^2}$$
The exponent applies only to the base $\dfrac{5}{y}$, not to $-\dfrac{5}{y}$. ■

Example 7 reviews some problems that were covered when the order of operations was discussed in Section 1.6. Although these problems are not difficult, they easily can be misinterpreted. It is important to pay careful attention to the small changes in notation that dictate the correct order of operations.

■ EXAMPLE 7   Evaluating Algebraic Expressions

Evaluate each expression for $x = -3$ and $y = -5$.

SOLUTIONS

(a) $(-x)^2$

$$\begin{aligned}(-x)^2 &= [-(-3)]^2 \\ &= (3)^2 \\ &= 9\end{aligned}$$

Note that in parts **(a)** and **(b)** the difference in notation leads to different answers. In part **(a)** the base for the exponent is $-x$, but in part **(b)** the base is just x.

(b) $-x^2$

$$\begin{aligned}-x^2 &= -(x^2) \\ &= -[(-3)^2] \\ &= -(9) \\ &= -9\end{aligned}$$

(c) $x^2 + y^2$

$$\begin{aligned}x^2 + y^2 &= (-3)^2 + (-5)^2 \\ &= 9 + 25 \\ &= 34\end{aligned}$$

In parts **(c)** and **(d)** the different notations indicate distinct orders of operation. This produces different answers.

(d) $(x + y)^2$

$$\begin{aligned}(x + y)^2 &= [-3 + (-5)]^2 \\ &= (-8)^2 \\ &= 64\end{aligned}$$

■

SELF-CHECK 5.1.4

Simplify each expression.

1. $(xy)^5$

2. $(x^3 y)^5$

3. $\left(\dfrac{x}{y}\right)^5$

4. $\left(\dfrac{x}{y^4}\right)^5$

Evaluate each expression for $x = -3$ and $y = -5$.

5. xy^2

6. $(xy)^2$

SELF-CHECK 5.1.4 ANSWERS

1. $x^5 y^5$ 2. $x^{15} y^5$

3. $\dfrac{x^5}{y^5}$ 4. $\dfrac{x^5}{y^{20}}$

5. -75 6. 225

In Section 1.6 we used the distributive property to add like terms. In Example 8 we review this topic and compare it to multiplying factors using the product rule.

◼ EXAMPLE 8 Comparing Addition and Multiplication

Add like terms and simplify the products.

SOLUTIONS _____

		Sum	*Product*
(a) Add $x + x$	Multiply $x \cdot x$	$x + x = 2x$	$x \cdot x = x^2$
(b) Add $v^3 + v^3$	Multiply $v^3 \cdot v^3$	$v^3 + v^3 = 2v^3$	$v^3 \cdot v^3 = v^6$
(c) Add $3x^2 + 5x^2$	Multiply $(3x^2)(5x^2)$	$3x^2 + 5x^2 = 8x^2$	$(3x^2)(5x^2) = (3)(5)(x^2)(x^2)$
			$= 15x^4$
(d) Add $x^2 + x^4$	Multiply $x^2 \cdot x^4$	$x^2 + x^4$ cannot be simplified further as these terms are unlike.	$x^2 \cdot x^4 = x^6$

USING THE LANGUAGE AND SYMBOLISM OF MATHEMATICS   5.1

1. In the exponential expression x^n, x is the _____ and n is the _____.
2. In the product $x \cdot x$ both x and x are called _____.
3. In the sum $x + x$ both x and x are called _____.
4. In the exponential expression $-x^4$ the base is _____.
5. In the exponential expression $(-x)^4$ the base is _____.
6. In the exponential expression $(xy)^4$ the base is _____.
7. In the exponential expression xy^4 the base of the exponent 4 is _____.
8. In the expression $x^5 = x \cdot x \cdot x \cdot x \cdot x$, x^5 is referred to as the _____ form and $x \cdot x \cdot x \cdot x \cdot x$ is referred to as the _____ form.

9. The product rule for exponents states that $x^m \cdot x^n =$ _____ for any real number x and natural numbers m and n.
10. The power rule for exponents states that $(x^m)^n =$ _____ for any real number x and natural numbers m and n.
11. For any real numbers x and y and any natural number m, $(xy)^m =$ _____.
12. For any real numbers x and y, $y \neq 0$, and any natural number m, $\left(\dfrac{x}{y}\right)^m =$ _____.

EXERCISES   5.1

Multiple Representations
In Exercises 1–4 write each expression in exponential form.

1. **a.** $x \cdot x \cdot x \cdot x \cdot x$
   b. $x \cdot y \cdot y \cdot y \cdot y$
   c. $(x \cdot y)(x \cdot y)(x \cdot y)$
2. **a.** $-3 \cdot a \cdot a \cdot a \cdot a$
   b. $(-3a)(-3a)(-3a)(-3a)$
   c. $-(3a)(3a)(3a)(3a)$
3. **a.** $(a + b)(a + b)(a + b)$
   b. $a \cdot a \cdot a + b \cdot b \cdot b$
   c. $a + b \cdot b \cdot b$
4. **a.** $a \cdot a \cdot a + b$
   b. $(a + b)(a + b)$
   c. $(ab)(ab)(ab)(ab)$

Multiple Representations
In Exercises 5–10 write each exponential expression in expanded form.

5. **a.** m^6 **b.** $(-m)^6$ **c.** $-m^6$
6. **a.** mn^6 **b.** $(mn)^6$ **c.** $6m^2$
7. **a.** $(m + n)^2$ **b.** $m^2 + n^2$ **c.** $m + n^2$
8. **a.** $m^2 + n^3$ **b.** m^2n^3 **c.** $(m + n)^3$
9. **a.** $\dfrac{a^2}{b^3}$ **b.** $\left(\dfrac{a}{b}\right)^4$ **c.** $\dfrac{a}{b^4}$
10. **a.** $\dfrac{x^3}{y^2}$ **b.** $\left(\dfrac{x}{y}\right)^3$ **c.** $\dfrac{x^3}{y}$

In Exercises 11–16 simplify each expression.

11. **a.** 6^2 **b.** $(-6)^2$
    c. -6^2 **d.** -1^4
12. **a.** 4^2 **b.** $(-4)^2$
    c. -4^2 **d.** $(-1)^5$
13. **a.** -1^{10} **b.** $(-1)^{10}$
    c. $(-1)^{11}$ **d.** 10^3
14. **a.** -1^{12} **b.** $(-1)^{12}$
    c. $(-1)^{13}$ **d.** 10^4
15. **a.** $5^2 + 12^2$ **b.** $(5 + 12)^2$
16. **a.** $11^2 - 8^2$ **b.** $(11 - 8)^2$

In Exercises 17–20 use a calculator to complete each table. Then state the equation that is confirmed by each table.

17.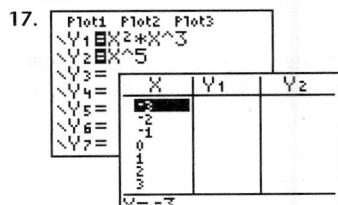

18.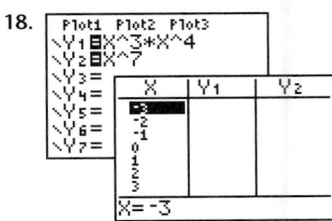

19.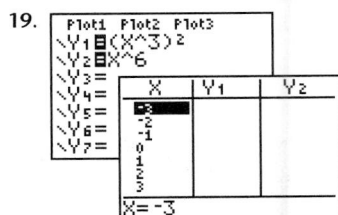

20.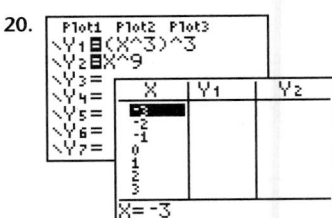

In Exercises 21–42 simplify each expression.

21. **a.** $x^9 x^{11}$ **b.** $-x \cdot x^2 \cdot x^3$
22. **a.** $y^7 y^{23}$ **b.** $-y \cdot y^3 \cdot y^5$
23. **a.** $(2m^3)(3m^2)$ **b.** $(-5y^2)(4y^3)$
24. **a.** $(-7v^4)(-6v^5)$ **b.** $(-3b)(-4b)(6b^2)$
25. **a.** $(y^2)^7$ **b.** $(y^4)^5$

26. **a.** $(x^{10})^9$ **b.** $(x^{11})^{20}$
27. **a.** $(xy)^3$ **b.** $(2n)^5$
28. **a.** $(xy)^4$ **b.** $(2m)^6$
29. **a.** $(-n^2)^6$ **b.** $-(n^2)^6$
30. **a.** $(-3m)^4$ **b.** $-(3m)^4$
31. **a.** $(5abc)^2$ **b.** $(-3bcd)^2$
32. **a.** $(-5xyz)^2$ **b.** $(2xyz)^3$
33. **a.** $(xy^2)^3$ **b.** $(x^2y^3)^4$
34. **a.** $(m^2n)^4$ **b.** $(m^2n^3)^5$
35. **a.** $\left(\dfrac{3}{4}\right)^2$ **b.** $\left(\dfrac{-3}{10}\right)^4$
36. **a.** $\left(\dfrac{4}{5}\right)^2$ **b.** $\left(\dfrac{-2}{5}\right)^3$
37. **a.** $\left(\dfrac{4}{w}\right)^3$; $w \neq 0$ **b.** $\left(\dfrac{5}{v}\right)^2$; $v \neq 0$
38. **a.** $-\left(\dfrac{3}{y}\right)^2$; $y \neq 0$ **b.** $\left(-\dfrac{4}{z}\right)^4$; $z \neq 0$
39. **a.** $\left(\dfrac{x}{y}\right)^{11}$; $y \neq 0$ **b.** $\left(\dfrac{v}{w}\right)^9$; $w \neq 0$
40. **a.** $\left(\dfrac{m}{n}\right)^7$; $n \neq 0$ **b.** $\left(\dfrac{-x}{y}\right)^8$; $y \neq 0$
41. **a.** $\left(\dfrac{2x}{3y}\right)^2$; $y \neq 0$ **b.** $\left(\dfrac{3v}{10w}\right)^3$; $w \neq 0$
42. **a.** $\left(\dfrac{6m}{5n}\right)^2$; $n \neq 0$ **b.** $\left(\dfrac{5a}{4b}\right)^4$; $b \neq 0$

In Exercises 43–54 evaluate each expression for $x = 2$ and $y = 3$.

43. **a.** $-x^2$ **b.** $(-x)^2$
44. **a.** $-y^2$ **b.** $(-y)^2$
45. **a.** $(xy)^2$ **b.** xy^2
46. **a.** $(xy)^3$ **b.** xy^3
47. **a.** $x^3 + y^3$ **b.** $(x + y)^3$
48. **a.** $x^2 + y^2$ **b.** $(x + y)^2$
49. **a.** $5x^2y^2$ **b.** $(5xy)^2$
50. **a.** $(3xy)^2$ **b.** $3x^2y^2$
51. **a.** $x^3 - y^3$ **b.** $(x - y)^3$
52. **a.** x^y **b.** y^x
53. **a.** $6y^x$ **b.** $5x^y$
54. **a.** $-\left(\dfrac{x}{y}\right)^2$ **b.** $\left(-\dfrac{x}{y}\right)^2$

In Exercises 55–58 add the like terms and simplify the product.

Add	Multiply
55. **a.** $v + v + v$	**b.** $v \cdot v \cdot v$
56. **a.** $w^2 + w^2$	**b.** $w^2 \cdot w^2$
57. **a.** $4m^3 + 6m^3$	**b.** $(4m^3)(6m^3)$
58. **a.** $2x + 3x + 10x$	**b.** $(2x)(3x)(10x)$

Estimation and Calculator Skills

In Exercises 59–64 mentally estimate the value of each expression and then use a calculator to calculate the value.

PROBLEM	MENTAL ESTIMATE	CALCULATOR VALUE
59. -6.98^2		
60. $(-9.9)^4$		
61. $(1.01 + 1.01)^2$		
62. $1.01^2 + 1.01^2$		
63. $7.99^2 - 5.9^2$		
64. $(7.99 - 5.9)^2$		

Compound Interest

In Exercises 65 and 66 use the formula $A = P(1 + r)^t$ to determine the value of a principal P, which is compounded for t years at an interest rate r.

65. What is the value of a principal P of $1 compounded for 10 years at a rate of 8%?

66. What is the value of a principal P of $1 compounded for 12 years at a rate of 7%?

Modeling the Generation of Electricity by a Windmill

Use the following information to work Exercises 67–70.

There are several environmental reasons to promote more generation of electricity by windmills. There are also several concerns with this generation method: noise, visual pollution, reliability, efficiency, cost, and the variability of the wind from hour to hour. To make sound business decisions, a company needs the facts related to all these concerns. One manufacturer of windmills has prepared the following table that gives the number of kilowatt hours generated from a given speed of wind. The output of the windmill actually varies directly as the cube of the speed of the wind. Thus the windmill is rather inefficient at low wind speeds and very efficient at higher wind speeds that are within its safe operating range. The function the company used to create the following table is $P(x) = 0.01x^3$.

x WIND SPEED (MI/H)	$P(x)$ ELECTRICITY (KW)
0	0
5	1.25
10	10
15	33.75
20	80
25	156.25
30	270

67. **a.** Evaluate and interpret $P(20)$.
  b. Evaluate and interpret $P(27)$.

68. What value of x will produce $P(x) = 270$? Interpret this result.

69. Compare the results of a company installing one of these windmills in an area with an average wind speed of 10 mi/h with a company that installs the same type of windmill in an area with an average wind speed of 20 mi/h. How many times greater will be the power generated by the higher 20-mi/h wind?

70. Suppose that the company made an error and the function it should have used to produce this table is $P(x) = 0.01x^2$. Use this formula to produce a new table of values for the same input values of x.

In Exercises 71–78 simplify each expression.

71. $(x^2y)^3$
72. $(5x^2y^3)^2$
73. $(-2xy^2)^4$
74. $(x^2y^3)(x^4y^5)$
75. $(5x)^2(2x)^3$
76. $(2x)^2(3x)^3$
77. $\left(\dfrac{3x}{2y^2}\right)^3$; $y \neq 0$
78. $\left(\dfrac{-2x^3}{3y}\right)^4$; $y \neq 0$

Group Discussion Questions

79. **Challenge Question** Use the product and power rules for exponents to simplify each expression assuming m is a natural number.

  a. $x^m x^2$
  b. $(x^m)^2$
  c. $x^{m+1}x^{m+2}$
  d. $(x^{m+1})^2$

80. **Challenge Question** Use the following figure to determine the area of each region:

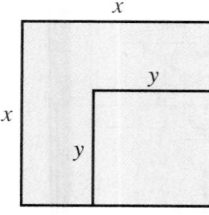

  a. The area of the larger square.
  b. The area of the smaller square.
  c. The area inside the larger square and outside the smaller square.

81. Challenge Question

a. Each side of one square is double that of another square. Compare their areas.

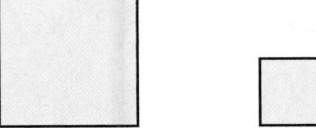

b. Each side of one cube is double that of another cube. Compare their volumes.

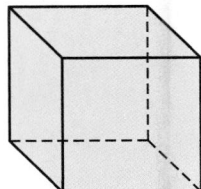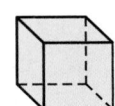

82. Challenge Question   One person becomes infected with a new strain of bacteria. Over the course of the next 24 hours the victim will infect 4 more persons. On the following day the original victim and the new victims will each infect 4 more people. Assume that the infection continues in this manner for 8 days. Determine the total number of people infected each of the 8 days in this table. Can you determine a formula for the number infected on the nth day?

DAY (n)	POPULATION INFECTED (P)
0	1
1	$1 + 4 = 5$
2	
3	
4	
5	
6	
7	
8	

Section 5.2   Quotient Rule and Zero Exponents

Objectives:

3. Use the quotient rule for exponents.
4. Simplify expressions with zero exponents.
5. Combine the properties of exponents to simplify expressions.

Quotient Rule

Another property of exponents that allows us to simplify some computations involving exponential expressions is the quotient rule. The logic behind the quotient rule is now examined. The quotient of $\dfrac{x^5}{x^3}$ is written next in the expanded form.

For $x \neq 0$ the quotient
$$\frac{x^5}{x^3} = \frac{\overbrace{x \cdot x \cdot x \cdot x \cdot x}^{\text{Five factors of } x}}{\underbrace{x \cdot x \cdot x}_{\text{Three factors of } x}}$$

The fraction is reduced by dividing both the numerator and the denominator by three factors of x.

$$= \underbrace{\boxed{x \cdot x}}_{\text{Two factors of } x}$$

$$= x^2$$

Thus $\dfrac{x^5}{x^3} = x^2$. The fact that these two expressions are equivalent for $x \neq 0$ is confirmed by comparing the values of y_1 and y_2 in the following table:

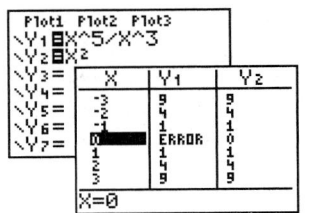

Note that $\dfrac{x^5}{x^3}$ is undefined for $x = 0$;

for all other values $\dfrac{x^5}{x^3} = x^2$.

In general, if m and n are natural numbers and $m > n$, then the quotient $\frac{x^m}{x^n}$ can be reduced by dividing out each of the n common factors of x. This will leave $m - n$ factors of x in the numerator; that is $\frac{x^m}{x^n} = x^{m-n}$ for $x \neq 0$ and $m > n$.

$\frac{x^m}{x^n}$ is undefined if $x = 0$.

Quotient Rule for Exponents

ALGEBRAICALLY	VERBALLY	ALGEBRAIC EXAMPLE
For any nonzero real number x and natural numbers m and n with $m > n, \frac{x^m}{x^n} = x^{m-n}$	To divide two expressions with the same base, use the common base and subtract the exponents.	$\frac{x^5}{x^3} = x^2$

EXAMPLE 1 Using the Quotient Rule

Use the quotient rule to simplify each expression. Assume that the denominators are not zero.

SOLUTIONS

(a) $\frac{z^{10}}{z^3}$

$$\frac{z^{10}}{z^3} = z^{10-3} = z^7$$

Subtract the exponents with the same base.

(b) $\frac{m^4n^7}{m^2n^6}$

$$\frac{m^4n^7}{m^2n^6} = m^{4-2}n^{7-6} = m^2n^1 = m^2n$$

Subtract the exponents on the common bases, m and n.

(c) $-\frac{7^8}{7^6}$

$$-\frac{7^8}{7^6} = -(7^{8-6}) = -(7^2) = -49$$

Note how much easier it is to apply the quotient rule than it is to evaluate 7^8 and 7^6 and then divide.

(d) $\frac{x^5}{y^3}$

$$\frac{x^5}{y^3} = \frac{x^5}{y^3}$$

Because the numerator and the denominator do not have the same base, this expression cannot be simplified. ■

SELF-CHECK 5.2.1

Simplify each expression.

1. $\frac{m^{14}}{m^9}$
2. $\frac{x^5y^5}{x^3y^2}$
3. $\frac{5^8}{5^6}$
4. $\frac{111^{113}}{111^{112}}$

SELF-CHECK 5.2.1 ANSWERS

1. m^5
2. x^2y^3
3. 25
4. 111

The expression $\dfrac{x^m}{x^n}$ also can be simplified when $m < n$. The reasoning is similar to that when $m > n$. If $x \neq 0$ and m and n are natural numbers, then $\dfrac{x^m}{x^n} = \dfrac{1}{x^{n-m}}$ for $m < n$. The example $\dfrac{x^4}{x^7}$ is examined next.

Negative exponents are covered in the next section. The properties of exponents are so important that many exercises are given on each individual property. These properties will be revisited extensively throughout the book.

$$\frac{x^4}{x^7} = \frac{\overset{1}{\cancel{x} \cdot \cancel{x} \cdot \cancel{x} \cdot \cancel{x}}}{\underset{1}{\cancel{x} \cdot \cancel{x} \cdot \cancel{x} \cdot \cancel{x}} \cdot x \cdot x \cdot x} = \frac{1}{x \cdot x \cdot x}$$

$$= \frac{1}{x^3}$$

Thus $\dfrac{x^4}{x^7} = \dfrac{1}{x^3}$. The fact that these two expressions are equivalent for $x \neq 0$ is confirmed by comparing the values of y_1 and y_2 in the following table:

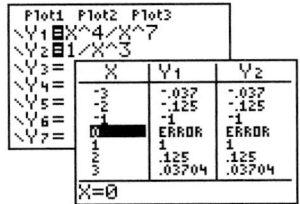

Both expressions are undefined for $x = 0$.

■ EXAMPLE 2   Simplifying Exponential Expressions

Simplify each of the following expressions.

SOLUTIONS

(a) $\dfrac{x^3}{x^5}$

$$\frac{x^3}{x^5} = \frac{1}{x^{5-3}}$$
$$= \frac{1}{x^2}$$

Reduce the fraction by dividing both the numerator and the denominator by the three common factors of x. This is accomplished by subtracting the smaller exponent from the larger exponent.

(b) $\dfrac{x^4 y^4}{x^6 y}$

$$\frac{x^4 y^4}{x^6 y} = \frac{y^{4-1}}{x^{6-4}}$$
$$= \frac{y^3}{x^2}$$

For each base subtract the smaller exponent from the larger exponent.

(c) $\dfrac{x^8}{y^9}$

$$\frac{x^8}{y^9} = \frac{x^8}{y^9}$$

The bases are not the same, so this expression cannot be simplified.

(d) $\dfrac{11^8}{11^{10}}$

$$\frac{11^8}{11^{10}} = \frac{1}{11^{10-8}}$$
$$= \frac{1}{11^2}$$
$$= \frac{1}{121}$$

Note that it is easier first to apply the properties of exponents than to start by evaluating 11^8.

Zero Exponent

We start our discussion of the meaning of zero exponents by noting that $\dfrac{x^m}{x^m} = 1$ for $x \neq 0$. If we try to subtract exponents, we obtain $\dfrac{x^m}{x^m} = x^{m-m} = x^0$. If this is true for $x \neq 0$, then $x^0 = \dfrac{x^m}{x^m} = 1$. This suggests that we define $x^0 = 1$ for $x \neq 0$. This is further confirmed by examining y_1 in the following table.

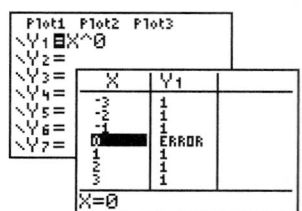

x^0 is undefined for $x = 0$.

Definition of Zero Exponent

ALGEBRAICALLY	VERBALLY	ALGEBRAIC EXAMPLE
For any nonzero real number x, $x^0 = 1$. 0^0 is undefined.	Any nonzero real number raised to the 0 power is 1.	$17^0 = 1$

0^0 is undefined.

■ EXAMPLE 3   Using Zero Exponents

Simplify each expression assuming that all the bases are nonzero.

Example 3 illustrates that we must be very careful to identify correctly the base when we use exponential expressions.

SOLUTIONS

(a) 3^0	$3^0 = 1$	The base is 3.
(b) $(-3)^0$	$(-3)^0 = 1$	The base is -3.
(c) -3^0	$-3^0 = -(3)^0 = -1$	The base is 3 (not -3).
(d) $3x^0$	$3x^0 = 3(1) = 3$	The base is x.
(e) $(3x)^0$	$(3x)^0 = 1$	The base is $3x$.
(f) $(x + y)^0$	$(x + y)^0 = 1$	The base is $x + y$.
(g) $x^0 + y^0$	$x^0 + y^0 = 1 + 1 = 2$	The different results in parts (f) and (g) are a result of the different order of operations.

The rules for exponents are summarized in the following box. In Section 5.3 these rules will be applied to all integral exponents: positive, zero, and negative.

Summary of the Properties of Exponents

The whole number possibilities for m and n include 0, 1, 2, 3, 4, ...

For any nonzero real numbers x and y and whole number exponents m and n,

Product rule: $x^m \cdot x^n = x^{m+n}$

Power rules: $(x^m)^n = x^{mn}$

$(xy)^m = x^m y^m$

$\left(\dfrac{x}{y}\right)^m = \dfrac{x^m}{y^m}$

Quotient rule: $\dfrac{x^m}{x^n} = x^{m-n}$;   for $m > n$

$\dfrac{x^m}{x^n} = \dfrac{1}{x^{n-m}}$;   for $m < n$

Zero exponent: $x^0 = 1$; 0^0 is undefined

Problems that involve several of these properties sometimes can be simplified correctly by using more than one sequence of steps. Study the examples in the text and those given by your instructor. Try to use a sequence of steps that will minimize your effort as well as produce a correct result.

■ EXAMPLE 4   Simplifying Exponential Expressions

Simplify each expression assuming that $x \neq 0$ and $y \neq 0$.

SOLUTIONS

(a) $\left(\dfrac{x^{12}}{x^4}\right)^2$

$\left(\dfrac{x^{12}}{x^4}\right)^2 = (x^{12-4})^2$ Use the quotient rule: Subtract the exponents.

$= (x^8)^2$ Use the power rule: Multiply the exponents.

$= x^{16}$

(b) $[(2a^7)(5a^4)]^3$

$[(2a^7)(5a^4)]^3 = [(2)(5)a^{7+4}]^3$ Use the product rule: Add the exponents.

$= (10a^{11})^3$ Use the power rule: Multiply the exponents.

$= 10^3 a^{(3)(11)}$

$= 1000a^{33}$

(c) $\dfrac{12x^7y^7}{18x^5y^8}$ $\qquad$ $\dfrac{12x^7y^7}{18x^5y^8} = \dfrac{2x^{7-5}}{3y^{8-7}}$ $\qquad$ Divide both the numerator and the denominator by 6. Also use the quotient rule and subtract the exponents on the common bases.

$$= \dfrac{2x^2}{3y}$$

SELF-CHECK 5.2.4

Simplify each expression for $x \neq 0$.

1. $\left(\dfrac{x^7}{x^5}\right)^3$ $\qquad\qquad$ **2.** $[(2x^3)(3x^2)]^2$

The ability to estimate calculations is useful in many contexts. One important usage is to check calculator results to see if they are reasonable. Many calculator keystroke errors, especially misplacing a decimal point or keying in an incorrect operation, can produce significant errors. On the other hand, good mental approximations should produce relatively small errors. (Relative error was first covered in Section 1.5.) Thus mental estimations can help you to spot your own calculator errors.

▗ EXAMPLE 5 Estimating a Quotient and Calculating the Relative Error

Estimate $\dfrac{5.01^6}{4.98^4}$ and then determine the relative error of this estimate.

SOLUTION

Estimated Value $\qquad\qquad\qquad\qquad\qquad\qquad$ *Calculator Value*

$5.01 \approx 5$ $\qquad$ $\dfrac{5^6}{5^4} = 5^2$

$4.98 \approx 5$ $\qquad\qquad = 25$

Mental Estimate: $\dfrac{5.01^6}{4.98^4} \approx 25$

```
5.01^6/4.98^4
        25.71040892
```

Calculator Approximation:

$$\dfrac{5.01^6}{4.98^4} \approx 25.71$$

Error of estimate	=	Estimated value	−	Actual value
	$\approx$	25	−	25.71
	$\approx$	−0.71		

The estimated value is smaller than the actual value so the error of the estimate is negative.

Relative error	=	Error of estimate	÷	Actual value
	$\approx$	(−0.71)	÷	25.71
	$\approx$	−0.0276		
	$\approx$	−2.8%		

This relative error, whose magnitude is nearly 3% is likely to be much smaller than an error produced by an order of operations error or an incorrect calculator keystroke. For error checking this mental estimate is very useful.

Answer: The estimate of 25 has a relative error of about −2.8% compared to the actual value of approximately 25.71.

The next example compares the subtraction of like terms to the division of expressions with a common base. Please note the distinction between these operations.

■ EXAMPLE 6   Comparing Subtraction and Division

Subtract like terms and simplify the quotients. Assume that $x \neq 0$.

SOLUTIONS

	Difference	*Quotient*
(a) Subtract: $10x^2 - 2x^2$	$10x^2 - 2x^2 = 8x^2$	$\dfrac{10x^2}{2x^2} = 5$
Divide: $\dfrac{10x^2}{2x^2}$		
(b) Subtract: $8x^2 - 4x$	$8x^2 - 4x$ cannot be simplified further since these terms are unlike.	$\dfrac{8x^2}{4x} = 2x$
Divide: $\dfrac{8x^2}{4x}$		

The compound interest formula $A = P(1 + r)^t$ can be used to find the value A of an initial investment of a principal P at an annual interest rate r compounded for t years. In the next example we will use this formula and a calculator to compute the value of an investment over a 6-year period. Note that the expression $P(1 + r)^t$ involves an exponent that is a variable. Example 7 illustrates one way to evaluate this expression for different values of t. We will examine expressions with variable exponents further in Chapter 10.

■ EXAMPLE 7   Calculating the Value of an Investment

Use the formula $A = P(1 + r)^t$ to find the value of a $7500 investment compounded at 7% at the end of each year over a 6-year period.

SOLUTION

$A = P(1 + r)^t$
$A = 7500(1 + 0.07)^t$
$A = 7500(1.07)^t$

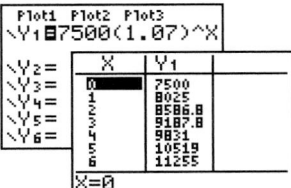

Using the compound interest formula, substitute $7500 for the principal P and 0.07 for the interest rate of 7%.

Enter this formula into a calculator using x to represent the number of years, t from the formula and y_1 to represent A.

Note for a zero exponent $(1.07)^0 = 1$ and the value of $P = 7500. This is the value at $t = 0$, the initial time of the investment.

The last row shows the value of the investment after 6 years is approximately $11,255. You can obtain additional significant digits by moving the cursor over this value and obtaining 11255.4776389.

SELF-CHECK 5.2.5

Rework Example 7 assuming that the annual interest rate is 8%.

SELF-CHECK 5.2.5 ANSWER

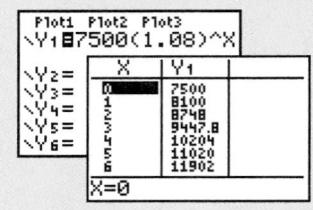

USING THE LANGUAGE AND SYMBOLISM OF MATHEMATICS   5.2

1. Two numbers that are added are called _____ or _____ .

2. Two numbers that are multiplied are called _____ .

3. The answer to a division of two numbers is called their _____ .

4. The quotient rule for exponents states that $\dfrac{x^m}{x^n} =$ _____ for any real number x, $x \neq$ _____ , and for natural numbers m and n with $m > n$. If $m < n$, then we can write $\dfrac{x^m}{x^n} =$ _____ . To divide two expressions with the same base, use the common base and _____ the exponents.

5. The value of x^0 is _____ for $x \neq$ _____ .

6. 0^0 is _____ .

EXERCISES   5.2

In Exercises 1–14 simplify each expression.

1. a. $0 - 5$ **b.** $0 \cdot 5$ **c.** 5^0
   d. 0^5 **e.** $5 \div 0$

2. a. $0 + 6$ **b.** $6 \cdot 0$ **c.** 6^0
   d. 0^6 **e.** $6 \div 0$

3. a. $\dfrac{10^8}{10^5}$ **b.** $\dfrac{9^9}{9^7}$ **c.** $\dfrac{7^4}{7^6}$
   d. $\dfrac{8^8}{8^9}$ **e.** $\dfrac{11^6}{11^6}$

4. a. $\dfrac{5^2}{5^5}$ **b.** $\dfrac{6^7}{6^9}$ **c.** $\dfrac{7^7}{7^5}$
   d. $\dfrac{4^8}{4^5}$ **e.** $\dfrac{3^{11}}{3^{11}}$

5. a. 10^0 **b.** -10^0 **c.** $(-10)^0$
   d. $(10 + 8)^0$ **e.** $10^0 + 8^0$

6. a. 7^0 **b.** -7^0 **c.** $(-7)^0$
   d. $(10 - 7)^0$ **e.** $10^0 - 7^0$

7. a. $2^0 + 8^0$ **b.** $(2 + 8)^0$ **c.** $2^0 - 8^0$
   d. $(2 - 8)^0$ **e.** $(8 - 2)^0$

8. a. $4^0 + 6^0$ **b.** $(4 + 6)^0$ **c.** $4^0 - 6^0$
   d. $(4 - 6)^0$ **e.** $(6 - 4)^0$

9. a. 3^0 **b.** 0^3 **c.** 1^3
   d. 3^1 **e.** -3^0

10. a. 5^0 **b.** 0^5 **c.** 1^5
   d. 5^1 **e.** $(-5)^0$

11. a. x^0;   for $x \neq 0$ **b.** x^0;   for $x = 0$

12. a. $5y^0$;   for $y \neq 0$ **b.** $(5y)^0$;   for $y = 0$

13. a. $-3x^0$;   for $x \neq 0$ **b.** $(-3x)^0$;   for $x = 0$

14. a. $-y^0$;   for $y \neq 0$ **b.** $(-y)^0$;   for $y = 0$

In Exercises 15–18 use a calculator to complete each table. Then state the equation that is confirmed by each table.

15.
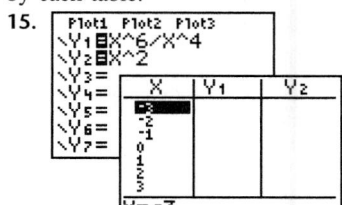

16.
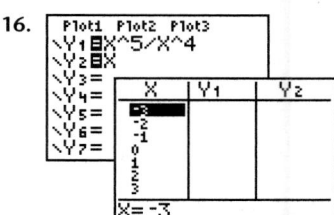

17.
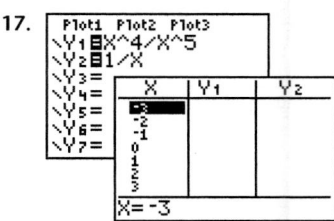

18.
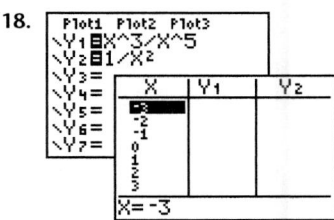

In Exercises 19–50 simplify each expression. Assume that all bases are nonzero.

19. $\dfrac{x^{15}}{x^5}$ **20.** $\dfrac{y^{18}}{y^6}$ **21.** $\dfrac{m^{42}}{m^{18}}$

22. $\dfrac{m^{83}}{m^{72}}$ **23.** $\dfrac{y^{48}}{y^{48}}$ **24.** $\dfrac{v^{113}}{v^{113}}$

25. $\dfrac{-x^{11}}{x^7}$ **26.** $\dfrac{-y^{23}}{y^{17}}$ **27.** $\dfrac{t^8}{t^{12}}$

28. $\dfrac{w^5}{w^{11}}$ **29.** $\dfrac{v^{10}}{v^{15}}$ **30.** $\dfrac{n^{15}}{n^{18}}$

31. $-\dfrac{a^6}{a^9}$ **32.** $\dfrac{-b^6}{b^8}$ **33.** $\dfrac{a^3}{b^4}$

34. $\dfrac{a^4}{b^3}$ **35.** $\dfrac{14m^5}{21m^2}$ **36.** $\dfrac{35m^8}{42m^3}$

37. $\dfrac{36v^5}{66v^8}$ **38.** $\dfrac{20v^6}{12v^2}$ **39.** $\dfrac{15a^2b^3}{20a^2b^2}$

40. $\dfrac{33a^5b^5}{22a^3b^8}$ **41.** $\dfrac{8m^7n^7}{6m^9n^2}$ **42.** $\dfrac{6x^6y^6}{3x^3y^{10}}$

43. $-\dfrac{27v^4w^9}{45v^6w^6}$ **44.** $-\dfrac{35a^9b^7}{28a^8b^9}$ **45.** $\left(\dfrac{x^5}{x^2}\right)^2$

46. $\left(\dfrac{a^7}{a^4}\right)^2$ **47.** $\left(\dfrac{v^8}{v^5}\right)^4$ **48.** $\left(\dfrac{w^7}{w^3}\right)^5$

49. $\left(\dfrac{m^3}{m^9}\right)^2$ **50.** $\left(\dfrac{b^2}{b^6}\right)^3$

In Exercises 51–54 subtract like terms and simplify the quotients. Assume that the variables are nonzero.

Subtract	Divide
51. a. $38x - 2x$	**b.** $\dfrac{38x}{2x}$
52. a. $x^3 - x^3$	**b.** $\dfrac{x^3}{x^3}$
53. a. $3x^2 - 3x$	**b.** $\dfrac{3x^2}{3x}$
54. a. $8x^3 - 2x^2$	**b.** $\dfrac{8x^3}{2x^2}$

Estimation and Calculator Skills
In Exercises 55–58 mentally estimate the value of each expression and then use a calculator to approximate each value to the nearest hundredth.

PROBLEM	MENTAL ESTIMATE	CALCULATOR VALUE
55. $\dfrac{8.014}{7.99^2}$		
56. $\dfrac{1.99^2}{2.01^4}$		
57. $(1.99)^3(5.02)^2$		
58. $[(1.99)(5.02)]^4$		

In Exercises 59–80 simplify each expression. Assume that the variables are nonzero.

59. a. $(4x)^0$ **b.** $4x^0$ **c.** $(4x - 3y)^0$
d. $(4x)^0 - (3y)^0$ **e.** $4x^0 - 3y^0$

60. a. $(-5y)^0$ **b.** $-5y^0$ **c.** $(-5y + 2z)^0$
d. $(-5y)^0 + (2z)^0$ **e.** $-5y^0 + 2z^0$

61. a. $\dfrac{2^{45}}{2^{43}}$ **b.** $\dfrac{2^{43}}{2^{45}}$

c. $\dfrac{2^{43}}{2^{43}}$ **d.** $-\dfrac{2^{45}}{2^{45}}$

62. a. $\dfrac{5^{87}}{5^{84}}$ **b.** $\dfrac{5^{87}}{5^{88}}$

c. $\dfrac{5^{87}}{5^{87}}$ **d.** $-\dfrac{5^{88}}{5^{88}}$

63. $(5a + 3b)^0 + (5a)^0 + (3b)^0 + 5a^0 + 3b^0$
64. $(7v - 8w)^0 + (7v)^0 + (-8w)^0 + 7v^0 - 8w^0$

65. $[(3x^2)(2x^3)]^2$

66. $[(2b^4)(5b^3)]^3$

67. $[(5a^4)(4a^5)]^3$

68. $[(1.1x^5)(10x^7)]^2$

69. $\left(\dfrac{10x^7}{5x^4}\right)^5$

70. $\left(\dfrac{21v^9}{7v^4}\right)^3$

71. $[(6m^4)(2m^5)]^2$

72. $[(5m^7)(3m^5)]^2$

73. $(5x^3)^2(2x^5)^3$

74. $(10v^4)^3(6v^5)^2$

75. $\left(\dfrac{12x^5}{6x^3}\right)\left(\dfrac{15x^7}{5x^3}\right)$

76. $\left(\dfrac{24a^9}{8a^3}\right)\left(\dfrac{42a^7}{14a^6}\right)$

77. $\dfrac{36a^7b^8}{12a^3b^3}$

78. $\dfrac{64m^9n^9}{16mn^7}$

79. $\dfrac{(4x^2y)^3}{(8xy^2)^2}$

80. $\dfrac{(2x^3y^2)^5}{(4x^2y^3)^3}$

Compound Interest

In Exercises 81 and 82 use the given table to determine the value of a \$5000 investment for $x = 5$ years. In Exercise 81 the interest rate is 8.5% and in Exercise 82 the interest rate is 6%.

81.

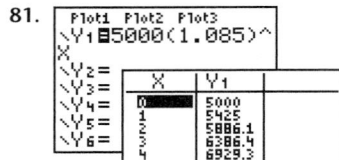

82.

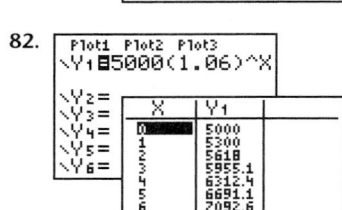

Compound Interest

In Exercises 83 and 84 use the formula $A = P(1 + r)^t$ to determine the value of each investment for $t = 5$

years. A represents the value after t years of an original amount P invested at an interest rate r.

83. \$4000 at 5%

84. \$4000 at 8%

Group Discussion Questions

85. Challenge Question Use the properties of exponents to simplify each expression assuming that m is a natural number and x is a nonzero real number.

a. $\dfrac{x^{m+3}}{x^{m+1}}$

b. $\dfrac{x^{3m}}{x^m}$

c. $\left(\dfrac{x^{2m+1}}{x^m}\right)^3$

d. $[(x^{m+1})(x^{m+2})]^4$

e. $\dfrac{(-1)^{m+2}}{(-1)^m}$

86. Challenge Question Insert parentheses in the expression $3 + 4 \cdot 5^2$ to create an order of operations that produces a result of

a. 175 **b.** 403 **c.** 529 **d.** 1225

87. Discovery Question Use a calculator to complete the following table for x^{-1}, then discuss what you think this notation represents.

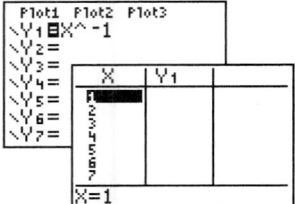

88. Discovery Question

a. Evaluate 6^0, 5^0, 4^0, 3^0, 2^0, and, 1^0. If you are defining 0^0 based on these observations, what value would you give for 0^0?

b. Evaluate 0^6, 0^5, 0^4, 0^3, 0^2, and, 0^1. If you are defining 0^0 based on these observations, what value would you give for 0^0?

c. Use a calculator to evaluate 0^0. What result did you get?

d. 0^0 is undefined. Based on your observations in parts a and b explain why you think 0^0 is undefined.

Section 5.3 Negative Exponents and Scientific Notation

Objectives: 6. Simplify expressions with negative exponents.

7. Use scientific notation.

Negative Exponents

To develop a definition of negative exponents, consider two approaches to applying the quotient rule to $\dfrac{x^4}{x^5}$ for $x \neq 0$.

Option 1

$$\frac{x^4}{x^5} = \frac{1}{x^{5-4}}$$

$$= \frac{1}{x}$$

Option 2

$$\frac{x^4}{x^5} = \frac{x^{4-5}}{1}$$

$$= x^{-1}$$

In option 1 we use the approach from Section 5.2 to subtract the smaller exponent from the larger exponent. In option 2 we extend the quotient rule to examine what will happen when we subtract the larger exponent from the smaller exponent.

For these two options to obtain the same result we must define $x^{-1} = \frac{1}{x}$ for $x \neq 0$.

This is further confirmed by examining y_1 and y_2 in the following table:

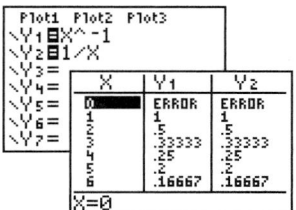

x^{-1} and $\frac{1}{x}$ are both undefined for $x = 0$.

Similarly, $\frac{x^3}{x^5} = \frac{1}{x^2}$ or $\frac{x^3}{x^5} = x^{-2}$. Thus we define $x^{-2} = \frac{1}{x^2}$ for $x \neq 0$. This suggests the following general definition of negative exponents in the next box.

A Mathematical Note

John Wallis, in *Arithmetica Infinitorum* (1655), was the first writer to explain the use of zero and negative exponents.

Definition of Negative Exponents

ALGEBRAICALLY	VERBALLY	ALGEBRAIC EXAMPLE
For any nonzero real number x and natural number n, $x^{-n} = \frac{1}{x^n}$.	A nonzero base with a negative exponent can be rewritten by reciprocating the base and using the corresponding positive exponent.	$x^{-3} = \frac{1}{x^3}$

■ EXAMPLE 1   Simplifying Expressions with Negative Exponents

Simplify each of the following expressions:

SOLUTIONS

(a) 3^{-4}

$$3^{-4} = \frac{1}{3^4}$$ Reciprocate the base. Note that the answer is positive.

$$= \frac{1}{81}$$

(b) $(-3)^{-4}$

$$(-3)^{-4} = \frac{1}{(-3)^4} = \frac{1}{81}$$

The base is -3. Reciprocate the base.

(c) -3^{-4}

$$-3^{-4} = -(3^{-4}) = -\frac{1}{3^4} = -\frac{1}{81}$$

The base is 3 (not -3). Reciprocate the base. The negative coefficient makes the answer negative.

(d) $2^{-1} + 3^{-1}$

$$2^{-1} + 3^{-1} = \frac{1}{2} + \frac{1}{3} = \frac{3}{6} + \frac{2}{6} = \frac{5}{6}$$

Simplify each term and then add these fractions.

(e) $(2 + 3)^{-1}$

$$(2 + 3)^{-1} = (5)^{-1} = \frac{1}{5}$$

Simplify inside the parentheses first; then reciprocate the base.

SELF-CHECK 5.3.1

Simplify each expression.

1. 2^{-3} **2.** -2^3 **3.** $3^{-1} + 4^{-1}$ **4.** $(3 + 4)^{-1}$

A special case of the definition of negative exponents is useful for working with fractions. The fact that $\left(\frac{x}{y}\right)^{-1} = \frac{y}{x}$ for $x \neq 0$ and $y \neq 0$ is shown next.

$$\left(\frac{x}{y}\right)^{-1} = \frac{1}{\frac{x}{y}} = 1 \div \frac{x}{y} = \frac{1}{1} \cdot \frac{y}{x} = \frac{y}{x}$$

The general case is given in the following box.

Fraction to a Negative Power

ALGEBRAICALLY	VERBALLY	NUMERICAL EXAMPLE
For any nonzero real numbers x and y and natural number n, $\left(\frac{x}{y}\right)^{-n} = \left(\frac{y}{x}\right)^{n}$.	A nonzero fraction to a negative exponent can be rewritten by reciprocating the fraction and using the corresponding positive exponent.	$\left(\frac{3}{5}\right)^{-2} = \left(\frac{5}{3}\right)^{2} = \frac{25}{9}$

SELF-CHECK 5.3.1 ANSWERS

1. $\frac{1}{8}$ **2.** -8 **3.** $\frac{7}{12}$ **4.** $\frac{1}{7}$

■ **EXAMPLE 2**   Simplifying Expressions with Negative Exponents

Simplify the following expressions for $x \neq 0$ and $y \neq 0$.

SOLUTIONS

(a) $\left(\dfrac{2}{3}\right)^{-1}$ $\left(\dfrac{2}{3}\right)^{-1} = \dfrac{3}{2}$ Reciprocate the base.

We encourage you to rewrite $\left(\dfrac{5}{3}\right)^{-2}$ as $\left(\dfrac{3}{5}\right)^{2}$ rather than as $\dfrac{5^{-2}}{3^{-2}}$. Both are correct, but you should find the first simplification easier to use.

(b) $\left(\dfrac{5}{3}\right)^{-2}$ $\left(\dfrac{5}{3}\right)^{-2} = \left(\dfrac{3}{5}\right)^{2}$ Reciprocate the base and then apply the power rule,

$= \dfrac{3^2}{5^2}$ $\left(\dfrac{x}{y}\right)^{m} = \dfrac{x^m}{y^m}$.

$= \dfrac{9}{25}$

(c) $\left(\dfrac{x}{y}\right)^{-3}$ $\left(\dfrac{x}{y}\right)^{-3} = \left(\dfrac{y}{x}\right)^{3}$ Reciprocate the base and then apply the power rule.

$= \dfrac{y^3}{x^3}$

(d) $\dfrac{x^{-3}}{y}$ $\dfrac{x^{-3}}{y} = \dfrac{x^{-3}}{1} \cdot \dfrac{1}{y}$ The base for the exponent -3 is just x. Reciprocate this base but not the y.

$= \left(\dfrac{1}{x^3}\right)\left(\dfrac{1}{y}\right)$

$= \dfrac{1}{x^3 y}$ ■

Rules for Negative Exponents

All the expressions with negative exponents can be written in terms of positive exponents. Thus it can be shown that all the properties of exponents that have been stated for only whole number exponents are in fact valid for all integral exponents—positive integer exponents, zero exponents, and negative integer exponents.

The product, power, and quotient rules for exponents are restated in the following box in algebraic form—a form that also works for negative exponents.

Properties of Integer Exponents

For any nonzero real numbers x and y and integers m and n,

Product rule: $x^m \cdot x^n = x^{m+n}$

Power rule: $(x^m)^n = x^{mn}$

Quotient rule: $\dfrac{x^m}{x^n} = x^{m-n}$

■ EXAMPLE 3   Using the Properties of Exponents

Simplify each of the following expressions. Assume that $x \neq 0$.

SOLUTIONS

Use the properties of exponents whenever possible rather than apply the definitions. The properties of exponents are shortcuts which can save you time. When more than one property of exponents is involved, there may be several ways to simplify the expression. As you practice, try to gain efficiency by studying the examples given by your instructor and presented in this book.

(a) $\dfrac{x^4}{x^{-7}}$

$\dfrac{x^4}{x^{-7}} = x^{4-(-7)}$

$= x^{11}$

Use the quotient rule. Subtract the smaller exponent from the larger exponent to obtain an expression with a positive exponent.

(b) $(x^{-4})^{-2}$

$(x^{-4})^{-2} = x^{(-4)(-2)}$

$= x^8$

Use the power rule. Multiply the exponents.

(c) $(2a^{-2}b^4)^{-3}$

$(2a^{-2}b^4)^{-3} = 2^{-3}(a^{-2})^{-3}(b^4)^{-3}$

$= 2^{-3}a^6 b^{-12}$

$= \dfrac{a^6}{2^3 b^{12}}$

$= \dfrac{a^6}{8b^{12}}$

Product to a power rule.

Use the power rule: Multiply the exponents.

Express in terms of positive exponents by reciprocating the bases.

(d) $\left(\dfrac{12x^{-2}y^4}{15x^5 y^{-6}}\right)^{-2}$

$\left(\dfrac{12x^{-2}y^4}{15x^5 y^{-6}}\right)^{-2} = \left(\dfrac{4y^{10}}{5x^7}\right)^{-2}$

$= \left(\dfrac{5x^7}{4y^{10}}\right)^2$

$= \dfrac{5^2(x^7)^2}{4^2(y^{10})^2}$

$= \dfrac{25x^{14}}{16y^{20}}$

First simplify the expression inside the parentheses, observing carefully the order of operations. Use the quotient rule on each base by subtracting the smaller exponent from the larger exponent. Next reciprocate the base to remove the negative exponent.

Raise each factor to the second power.

■

SELF-CHECK 5.3.2

Simplify each expression.

SELF-CHECK 5.3.2 ANSWERS

1. $\dfrac{3}{10}$ 2. $\dfrac{49}{4}$ 3. $\dfrac{1}{100}$

4. $\dfrac{16}{25}$ 5. $\dfrac{1}{x^3}$ 6. $\dfrac{x^3}{y^7}$

7. $\dfrac{25x^{10}}{4y^{12}}$

1. $\left(\dfrac{10}{3}\right)^{-1}$ 2. $\left(\dfrac{2}{7}\right)^{-2}$ 3. $\dfrac{5^{-2}}{4}$ 4. $\left(\dfrac{5}{4}\right)^{-2}$

Simplify each expression to a form involving only positive exponents. Assume that $x \neq 0$ and $y \neq 0$.

5. $x^4 x^{-7}$ 6. $\dfrac{x^2 y^{-3}}{x^{-1}y^4}$ 7. $\left(\dfrac{14x^{-3}y^2}{35x^2 y^{-4}}\right)^{-2}$

Be careful to distinguish between negative exponents and negative coefficients.

One of the most common errors made is to confuse the meaning of negative coefficients with the meaning of negative exponents. Negative coefficients designate negative numbers. Negative exponents designate reciprocals, *not* negative numbers.

■ EXAMPLE 4   Evaluating Algebraic Expressions

Evaluate each expression for $x = 3$ and $y = 5$.

SOLUTIONS

(a) $-x^4$

$$-x^4 = -(3^4)$$
$$= -81$$

Compare parts (a) and (b) to note the distinction between a negative coefficient and a negative exponent.

(b) x^{-4}

$$x^{-4} = 3^{-4}$$
$$= \frac{1}{3^4}$$
$$= \frac{1}{81}$$

(c) $(x + y)^{-1}$

$$(x + y)^{-1} = (3 + 5)^{-1}$$
$$= 8^{-1}$$
$$= \frac{1}{8}$$

Simplify inside the parentheses and then reciprocate the base of 8.

(d) $x^{-1} + y^{-1}$

$$x^{-1} + y^{-1} = 3^{-1} + 5^{-1}$$
$$= \frac{1}{3} + \frac{1}{5}$$
$$= \frac{5}{15} + \frac{3}{15}$$
$$= \frac{8}{15}$$

Simplify each term and then add these fractions using the LCD of 15.

(e) $-x + y^{-1}$

$$-x + y^{-1} = -3 + 5^{-1}$$
$$= -3 + \frac{1}{5}$$
$$= -3 + 0.2$$
$$= -2.8$$

■

SELF-CHECK 5.3.3

Evaluate each expression for $x = 2$ and $y = 4$.

1. $-x^5$ 2. x^{-5} 3. $x^{-2} + y^{-2}$
4. $-(x + y)^2$ 5. $(x + y)^{-2}$

Scientific Notation

One of the most useful applications of exponents is scientific notation. Scientific notation is used extensively in the sciences to represent the numbers of either macroscopic or microscopic proportions. For example, the mean distance between Earth and the planet Uranus is 2,870,000,000,000 m, whereas the width of one line etched on an IBM computer chip is 0.000000032 m.

Our decimal system of representing numbers is based on powers of 10. Scientific notation uses exponents to represent powers of 10 and thus provides an alternative format for representing numbers. Each of the powers of 10 from 10,000 to 0.001 is written here in exponential notation:

POWERS OF 10

$$10,000 = 10^4$$
$$1000 = 10^3$$
$$100 = 10^2$$
$$10 = 10^1$$
$$1 = 10^0$$
$$0.1 = 10^{-1}$$
$$0.01 = 10^{-2}$$
$$0.001 = 10^{-3}$$

A Mathematical Note

Extremely small numbers and extremely large numbers are difficult to comprehend. For example, try to grasp the value of each of the following amounts: a million dollars, a billion dollars, and a trillion dollars. Spent at the rate of a dollar a second, a million dollars would last almost 12 days; a billion dollars would last almost 32 years; and a trillion dollars would last almost 32,000 years.

As these numbers illustrate, the exponent on 10 determines the position of the decimal point. **Scientific notation** uses this fact to express any decimal number as a product of a number between 1 and 10 (or -1 and -10 if the number is negative) and an appropriate power of 10. Table 5.3.1 shows some representative numbers written in scientific notation.

Table 5.3.1

NUMBER	SCIENTIFIC NOTATION
$12,345 = 1.2345 \times 10,000$	1.2345×10^4
$123.45 = 1.2345 \times 100$	1.2345×10^2
$12.345 = 1.2345 \times 10$	1.2345×10^1
$0.12345 = 1.2345 \times (0.1)$	1.2345×10^{-1}
$0.0012345 = 1.2345 \times (0.001)$	1.2345×10^{-3}
$-1234.5 = -1.2345 \times 1000$	-1.2345×10^3

If a number is written in scientific notation, then it can be rewritten in standard decimal notation by multiplying by the power of 10. This multiplication is easily accomplished by shifting the decimal point the number of places indicated by the exponent on the 10. Positive exponents indicate larger magnitudes and shift the decimal point to the right. Negative exponents indicate smaller magnitudes and shift the decimal point to the left.

■ EXAMPLE 5   Writing Numbers in Standard Decimal Notation

Write each of the following numbers in standard decimal notation.

SOLUTIONS

(a) 5.789×10^2 $5.789 \times 10^2 = 578.9$

Two places right

Shift the decimal point two places to the right.

(b) 8.6×10^4 $8.6 \times 10^4 = 86,000$

Four places right

Fill in zeros so that the decimal point can be moved four places to the right.

(c) 4.61×10^{-4} $4.61 \times 10^{-4} = 0.000461$

Four places left

Fill in zeros so that the decimal point can be moved four places to the left.

(d) -1.4×10^7 $-1.4 \times 10^7 = -14,000,000$

Seven places right

Shift the decimal point seven places to the right.

The steps for writing a number in scientific notation are given in the following box. Remember that positive exponents indicate larger magnitudes and negative exponents indicate smaller magnitudes.

Writing a Number in Scientific Notation

VERBALLY	NUMERICAL EXAMPLE
1. Move the decimal point immediately to the right of the first nonzero digit of the number.	
2. Multiply by a power of 10 determined by counting the number of places the decimal point has been moved.	
a. The exponent on 10 is 0 or positive if the magnitude of the original number is 1 or greater.	$3.456 = 3.456 \times 10^0$ $345.6 = 3.456 \times 10^2$
b. The exponent on 10 is negative if the magnitude of the original number is less than 1.	$0.03456 = 3.456 \times 10^{-2}$

Numbers with larger magnitude have positive exponents on 10. Numbers whose magnitude is between 0 and 1 have negative exponents on 10.

■ EXAMPLE 6   Writing Numbers in Scientific Notation

Write each of the following numbers in scientific notation.

SOLUTIONS

(a) 9876

$9876. = 9.876 \times 10^3$

Three places left

The decimal point is moved three places to position it behind the first nonzero digit, 9. The exponent 3 is positive since the original number is greater than 1.

(b) 325.49

$325.49 = 3.2549 \times 10^2$

Two places left

The exponent on 10 is positive 2 since the decimal point was moved two places and the original number is greater than 1.

(c) 0.0009876

$0.0009876 = 9.876 \times 10^{-4}$

Four places right

The exponent on 10 is -4 since the decimal point was moved four places and the original number is less than 1.   ■

Calculators often use a special syntax to represent scientific notation. Many calculators use E to precede the power of 10 that would be used to represent a number in scientific notation. This is illustrated in the following table:

STANDARD DECIMAL NOTATION	SCIENTIFIC NOTATION	CALCULATOR SYNTAX
12,345	1.2345×10^4	$1.2345E4$
1.2345	1.2345×10^0	$1.2345E0$
0.00567	5.67×10^{-3}	$5.67E - 3$

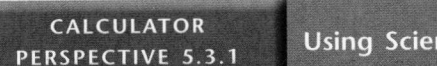

CALCULATOR PERSPECTIVE 5.3.1   Using Scientific Notation

When a value becomes sufficiently large or small, many graphics calculators will automatically display values in scientific notation. To evaluate the products 99×125 and $990,000 \times 125,000$ on a TI-83 Plus calculator, enter the following keystrokes:

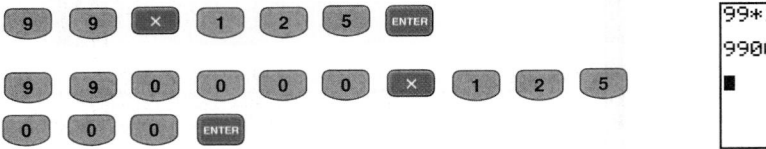

Note: Scientific notation is not needed to display 12,375 but is automatically used by the calculator to display the product of 123,750,000,000.

The [EE] feature is the secondary function of the [,] key.

The [EE] feature can be used to enter values in scientific notation. To evaluate the product $(1.5 \times 10^{12})(3.2 \times 10^{-8})$ on a TI-83 Plus calculator, enter the following keystrokes:

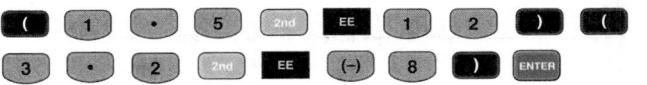

Note: Although the values entered were given in scientific notation, the magnitude of the result did not require scientific notation on the display. The result is shown in standard decimal notation.

To set a TI-83 Plus calculator to display all results in scientific notation and to evaluate the product 99×125, enter the following keystrokes:

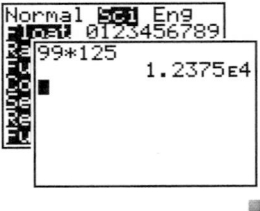

Note: In **Normal** mode, as shown previously, this result would be in the standard decimal form. To display all the results in scientific notation, set the calculator to **Sci** mode.

Scientific notation also can be used to perform both pencil-and-paper calculations or quick estimates of otherwise lengthy calculations.

■ **EXAMPLE 7**   Using Scientific Notation to Estimate a Product

Use scientific notation to perform a pencil-and-paper estimate of $(149,736)(0.00399)$.

SOLUTION

Estimated Factors	*Estimated Product*	*Calculator Product*
$149,736 \approx 150,000$	$(1.5 \times 10^5)(4.0 \times 10^{-3})$	149736*.003999
$\quad = 1.5 \times 10^5$	$= (1.5)(4)(10^5)(10^{-3})$	$\qquad$ 598.794264
$0.003999 \approx 0.004$	$= 6 \times 10^2$	■
$\quad = 4.0 \times 10^{-3}$	$= 600$	

Answer: The calculator product of 598.794264 seems reasonable based on our estimate of 600.   ■

SELF-CHECK 5.3.5

Write each of these numbers in scientific notation.

1. 80,000 $\qquad\qquad$ **2.** 0.008

3. 0.723 $\qquad\qquad$ **4.** 72.3

5. Use scientific notation to estimate $(4,997,000)(0.03002)$.

Optional Material on Accuracy and Precision

A word of warning is appropriate for those using calculators to perform calculations. Calculators will typically display from 8 to 12 digits, and it is tempting always to copy all of these digits when giving answers. However, it is not appropriate to copy digits that are not significant. Roughly speaking, the answer to a problem cannot be assumed to be more accurate than the measurements that produced that answer.

The **accuracy** of a number refers to the number of **significant digits.** For numbers written in scientific notation all written digits are considered significant. For numbers obtained by measurement the digits considered reasonably trustworthy are those that are significant.

The **precision** of a measurement refers to the smallest unit used in the measuring device and thus to the position of the last significant digit. The accuracy and precision of some sample measurements are given in Table 5.3.2.

Table 5.3.2   Accuracy and Precision

NUMBER (m)	SCIENTIFIC NOTATION (m)	ACCURACY	PRECISION
0.045	4.5×10^{-2}	2 significant digits	Thousandths of a meter
0.0450	4.50×10^{-2}	3 significant digits	Ten-thousandths of a meter
45,000	4.5×10^4	2 significant digits	Thousands of meters
123.45	1.2345×10^2	5 significant digits	Hundredths of a meter
0.05	$5. \times 10^{-2}$	1 significant digit	Hundredths of a meter

There are many different rounding rules used for specific tasks and professions. All of these rules are based on trying to report an answer that is reasonably accurate. Most of these rules are similar to the standardized rules given in the following box.

Rounding Rules

- For calculations involving multiplication, division, or exponentiation, round the answer to the same number of significant digits as the measurement having the least number of significant digits.
- For calculations involving addition or subtraction round the answer to the same precision as the least precise measurement.

■ EXAMPLE 8   Calculating Red Blood Cell Production

The average human produces about 2×10^{12} new red blood cells each day. How many new red blood cells are produced each hour?

SOLUTION

Let n = the number of new red blood cells produced in an hour.

$$\boxed{\begin{array}{c}\text{Number produced}\\\text{per hour}\end{array}} = \boxed{\begin{array}{c}\text{Number produced}\\\text{per day}\end{array}} \div \boxed{24 \text{ hours per day}}$$

First describe the strategy in a **word equation.**

$$n = \frac{2 \times 10^{12}}{24}$$

```
(2E12)/24
    8.333333333E10
■
```

Then substitute in the given values and use a calculator to approximate n.

$$n \approx 8.333\,333\,333 \times 10^{10}$$
$$n \approx 80,000,000,000$$

It is not reasonable to report the answer to the nearest unit when the original estimate was rounded to the nearest trillion units. The answer also has been rounded to one significant digit.

Answer: The body produces about 80,000,000,000 new red blood cells each hour. ■

USING THE LANGUAGE AND SYMBOLISM OF MATHEMATICS   5.3

1. For any real number x, $x \neq$ _____ , and natural number n, $x^{-n} =$ _____ .
2. For any nonzero real numbers x and y and integers m and n,
   a. $x^m \cdot x^n =$ _____
   b. $(x^m)^n =$ _____
   c. $\dfrac{x^m}{x^n} =$ _____
3. Scientific notation is used to express a decimal number as a product of a number between 1 and 10 (or -1 and -10 if the number is negative) and an appropriate power of _____ .
4. In scientific notation the exponent on _____ determines the position of the decimal point.
5. The accuracy of a number is determined by the number of _____ digits in the number.
6. In scientific notation _____ written digits are considered significant.
7. The _____ of a measurement refers to the smallest unit used in the measuring device and thus to the position of the last significant digit.

EXERCISES 5.3

In Exercises 1–4 use a calculator to complete each table. Then state the equation that is confirmed by each table.

1.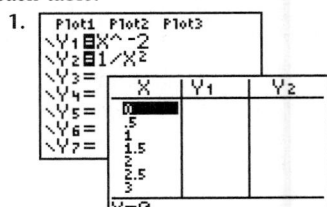

2.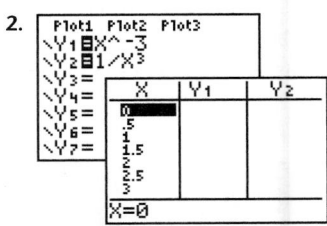

3.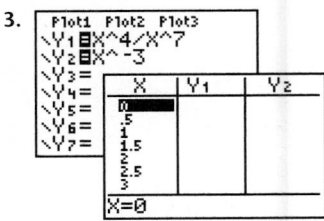

4.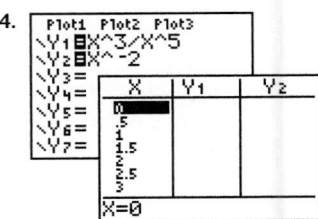

In Exercises 5–20 simplify each expression by writing it in a form that is free of negative exponents. Assume that all the bases are nonzero real numbers.

5. a. 3^{-2} b. 2^{-5}
   c. -2^5 d. -5^2

6. a. 4^{-3} b. 3^{-4}
   c. -4^3 d. -3^4

7. a. $\left(\dfrac{6}{5}\right)^{-1}$ b. $\left(\dfrac{6}{5}\right)^{-2}$

   c. $\left(\dfrac{6}{5}\right)^{1}$ d. $\left(\dfrac{6}{5}\right)^{0}$

8. a. $\left(\dfrac{3}{4}\right)^{-1}$ b. $\left(\dfrac{3}{4}\right)^{-2}$

   c. $\left(\dfrac{3}{4}\right)^{1}$ d. $\left(\dfrac{3}{4}\right)^{0}$

9. a. 10^{-2} b. 10^{-3}
   c. 10^{-4} d. -10^{-2}

10. a. 2^{-4} b. -2^{-4}
    c. $(-4)^2$ d. -4^2

11. a. $(2+5)^{-1}$ b. $2^{-1}+5^{-1}$
    c. $-2+5^{-1}$ d. $(2+5)^0$

12. a. $(5+10)^{-1}$ b. $5^{-1}+10^{-1}$
    c. $-5+10^{-1}$ d. $(5+10)^0$

13. a. $\left(\dfrac{1}{2}+\dfrac{1}{5}\right)^{-1}$ b. $\left(\dfrac{1}{2}\right)^{-1}+\left(\dfrac{1}{5}\right)^{-1}$

    c. $\left(\dfrac{1}{2}+\dfrac{1}{5}\right)^{-2}$ d. $\left(\dfrac{1}{2}\right)^{0}+\left(\dfrac{1}{5}\right)^{0}$

14. a. $\left(\dfrac{1}{3}+\dfrac{1}{4}\right)^{-1}$ b. $\left(\dfrac{1}{3}\right)^{-1}+\left(\dfrac{1}{4}\right)^{-1}$

    c. $\left(\dfrac{1}{3}+\dfrac{1}{4}\right)^{-2}$ d. $\left(\dfrac{1}{3}+\dfrac{1}{4}\right)^{0}$

15. a. $\left(\dfrac{x}{y}\right)^{-1}$ b. $\left(\dfrac{x}{y}\right)^{-2}$

    c. $\left(\dfrac{x}{y}\right)^{0}$ d. $\dfrac{x^{-2}}{y}$

16. a. $\left(\dfrac{v}{w}\right)^{-1}$ b. $\left(\dfrac{v}{w}\right)^{-2}$

    c. $\left(\dfrac{v}{w}\right)^{0}$ d. $\dfrac{v}{w^{-2}}$

17. a. $(3x)^{-2}$ b. $3x^{-2}$ c. $(-3x)^2$

18. a. $(2x)^{-3}$ b. $2x^{-3}$ c. $(-2x)^3$

19. a. $(m+n)^{-1}$ b. $m^{-1}+n^{-1}$ c. $m+n^{-1}$

20. a. $(2x+y)^{-1}$ b. $2x+y^{-1}$ c. $2x^{-1}+y^{-1}$

In Exercises 21–24 write each number in standard decimal notation.

21. a. 4.58×10^4 b. 4.58×10^{-4}
    c. 4.58×10^6 d. 4.58×10^{-6}

22. a. 1.7×10^1 b. 1.7×10^{-1}
    c. 1.7×10^5 d. 1.7×10^{-5}

23. a. 8.1×10^3 b. 8.1×10^{-3}
    c. -8.1×10^3 d. -8.1×10^{-7}

24. a. 3.6×10^7 b. 3.6×10^{-7}
    c. -3.6×10^7 d. -3.6×10^{-7}

In Exercises 25–28 write each number in scientific notation.

25. a. 9700   **b.** 97,000,000
    c. 0.97   **d.** 0.00097
26. a. 18,900   **b.** 189
    c. 0.0000189   **d.** 0.00189
27. a. 35,000,000,000   **b.** 0.0000000035
    c. −3500   **d.** −0.035
28. a. 470,000,000,000 **b.** 0.000000047
    c. −47,000   **d.** −0.0047

Applications of Scientific Notation

In Exercises 29–32 write the numbers in each statement in standard decimal notation.

29. Distance to Neptune   The mean distance from our sun to the planet Neptune, which was examined by the *Voyager 2* spacecraft in 1989, is 4.493×10^9 km.

30. Operations by a Computer   A 1990 advertisement for one computer claimed that the computer could perform 4.5×10^8 floating-point operations per second.

31. Computer Switching Speed   The switching speeds of some of the most sophisticated computers are measured in picoseconds. A picosecond is one-trillionth of a second, or 1.0×10^{-12} second.

32. Wavelength of Red Light   The wavelength of red light is 7000 angstroms, which is 7.0×10^{-7} m.

In Exercises 33–36 write the numbers in each statement in scientific notation.

33. Temperature of the Sun   Temperatures inside the sun are estimated to be 14,000,000 °C.

34. Speed of Light   The speed of light is approximately 299,790,000 meters per second.

35. Tidal Friction   The friction between the ocean and the ocean floor due to tides is causing the earth's rotation to slow down by about 0.00000002 second per day.

36. Speed of an Electrical Signal   An electrical signal will travel a kilometer in about 0.00000334 second.

37. Moore's Law   Gordon Moore gave a talk in 1965, four years after the first integrated circuit, which predicted that the number of transistors per integrated circuit would double every 18 months. Remarkably, computer memory and processing speed have followed this doubling prediction for 35 plus years, yielding fantastic advances in computers. This prediction has become known as Moore's law. Use the following table to answer each part of this question.

YEAR	TYPICAL MEMORY IN BYTES	PROCESSING SPEED IN OPERATIONS PER SECOND
1980	6.4×10^4	6×10^6
1985	2.5×10^5	1.6×10^7
1990	8.0×10^7	3.3×10^7
1995	5.0×10^8	2.0×10^8
2000	2.0×10^{10}	1.5×10^9

(*Source:* http://www.intel.com/research/silicon/mooreslaw.htm)

a. Write in standard decimal form the number of operations per second of a typical computer in 1980.

b. How many times faster was a computer manufactured in 2000 than a computer manufactured in 1980?

c. How many times as much memory did a computer manufactured in 2000 have than a computer manufactured in 1980?

38. Modeling Time with a Cesium Clock The NIST (National Institute of Standards & Technology) F-1 Cesium fountain clock built in 1999 is accurate to within 1 second in 20 million years. Accuracy of this level is important for many scientific processes such as Global Positioning Systems on airplanes and the design of microelectronic devices. *(Source: http://www.nist.gov)* How many seconds does the clock vary from the actual time in:

a. a century   **b.** a year   **c.** a day

Estimation and Calculator Skills

In Exercises 39–44 use scientific notation and pencil and paper to estimate the value of each expression. Then use a calculator to approximate the value of each expression.

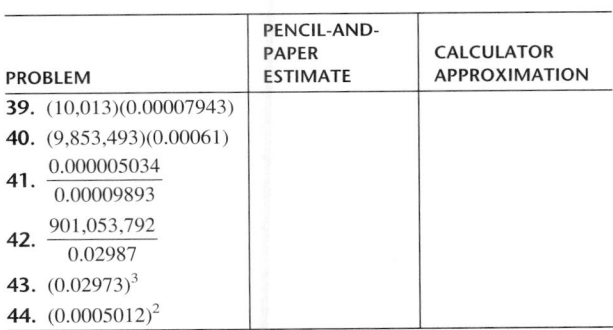

PROBLEM	PENCIL-AND-PAPER ESTIMATE	CALCULATOR APPROXIMATION
39. $(10,013)(0.00007943)$		
40. $(9,853,493)(0.00061)$		
41. $\dfrac{0.000005034}{0.00009893}$		
42. $\dfrac{901,053,792}{0.02987}$		
43. $(0.02973)^3$		
44. $(0.0005012)^2$		

In Exercises 45 and 46 write each result on the calculator screen in standard decimal notation.

45.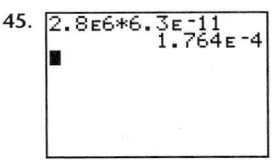

46.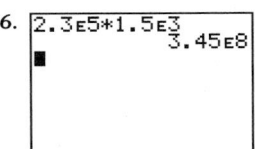

Calculator Skills

In Exercises 47 and 48 use a calculator to approximate each expression accurate to three significant digits. Write the answer in standard decimal notation.

47. a. $(4.32 \times 10^{15})(8.49 \times 10^{-17})$ **b.** $(7.16 \times 10^3)^2$

48. a. $\dfrac{9.71 \times 10^{13}}{6.53 \times 10^{11}}$ **b.** $[(6.7)(4.98 \times 10^2)]^2$

In Exercises 49–56 simplify each expression, writing each in a form that is free of negative exponents. Assume that all variables are nonzero.

49. a. $v^{-3}v^{12}$ **b.** $\dfrac{v^{12}}{v^{-3}}$ **c.** $\dfrac{v^{-12}}{v^3}$

50. a. $w^{21}w^{-7}$ **b.** $\dfrac{w^{21}}{w^{-7}}$ **c.** $\dfrac{w^{-21}}{w^7}$

51. a. $x^3x^0x^{-7}$ **b.** $(x^3x^{-7})^0$ **c.** $(3x^{-7})^2$
52. a. y^0yy^{-9} **b.** $(y^9y^{-1})^0$ **c.** $(2x^{-9})^3$

53. a. $\dfrac{6x^{-4}}{2x^{-3}}$ **b.** $(2x^{-3})(6x^4)$ **c.** $(2x^{-3})^{-4}$

54. a. $\dfrac{15x^7}{5x^{-2}}$ **b.** $(15x^7)(5x^{-2})$ **c.** $(5x^{-2})^{-3}$

55. a. $\left(\dfrac{3x}{5y}\right)^2$ **b.** $\left(\dfrac{3x}{5y}\right)^{-2}$ **c.** $\dfrac{a^3b^{-2}}{a^{-5}b^4}$

56. a. $\left(\dfrac{5v}{7w}\right)^2$ **b.** $\left(\dfrac{5v}{7w}\right)^{-2}$ **c.** $\dfrac{v^{-7}w^8}{v^9w^{-6}}$

In Exercises 57–66 simplify each expression, assuming that all variables are nonzero.

57. $[(5x^3)(-4x^{-2})]^{-1}$ **58.** $[(6x^{-4})(5x^7)]^{-1}$
59. $(2x^{-3}y^4)^3(3x^4y^{-2})^2$ **60.** $(5x^3y^{-4})(2x^{-5}y^3)^4$

61. $\left(\dfrac{12x^3}{6x^{-2}}\right)^2$ **62.** $\left(\dfrac{24x^5}{8x^{-3}}\right)^3$

63. $\left(\dfrac{m^3n^{-7}}{m^7n^{-11}}\right)^{-3}$ **64.** $\left(\dfrac{v^{-6}w^4}{v^{-9}w^{-7}}\right)^{-5}$

65. $\dfrac{(2x^{-1}y^2)(4x^2y^{-3})^{-2}}{(12x^{-2}y^{-2})^{-1}}$ **66.** $\left[\dfrac{(5x^{-3}y^4)^{-2}(6x^2y^{-5})}{15x^2y^{-4}}\right]^{-2}$

In Exercises 67–70 use the given values of x and y to evaluate each expression.

	$x^2 + y^2$	$(x + y)^2$	$x^{-1} + y^{-1}$	$(x + y)^{-1}$
67. $x = 2,$ $y = 3$				
68. $x = 3,$ $y = 4$				
69. $x = -3,$ $y = -4$				
70. $x = -2,$ $y = -4$				

Group Discussion Questions

71. Challenge Question   Use the properties of exponents to simplify each expression assuming that m is a natural number and x is a nonzero real number.

a. $\dfrac{x^{m+3}}{x^{m+1}}$ **b.** $\dfrac{x^{3m}}{x^m}$ **c.** $\left(\dfrac{x^{2m+1}}{x^m}\right)^3$

d. $[(x^{m+1})(x^{m+2})]^4$ **e.** $\dfrac{(-1)^{m+2}}{(-1)^m}$

72. Risks and Choices

a. If you live to be a billion seconds old, how many years will you have lived?

b. If you live to be 80 years old, how many seconds will you have lived?

c. If smoking cigarettes decreases one individual's lifetime from 80 to 65 years, how many seconds has this individual lost of his or her potential lifetime?

d. In part **c** assume that this person smoked $20 \times 365 \times 50$ cigarettes. How many seconds of lifetime has each cigarette cost this person?

73. Multiple Representations   Engineering notation is similar to scientific notation. Engineering notation expresses a decimal number as a product of a number between 1 and 1000 (or -1 and -1000 if the number is negative) and a power of 10, which is a multiple of 3 as in 12.4×10^3 or 453.7×10^6. Set a calculator first to the scientific notation (Sci), then to engineering notation (Eng) and complete this table.

STANDARD DECIMAL NOTATION	SCIENTIFIC NOTATION	ENGINEERING NOTATION
a. 45		
b. 45678		
c. 456789		
d. 4567890		
e. 0.045		
f. 0.00045		

74. Discovery Question

a. If x is negative and m is an integer, can x^m be positive? If so, give the values of m for which x^m is positive. Can x^m be negative? If so, give the values of m for which x^m is negative.

b. Can x^{-2} represent a negative value? If so, give the values of x for which this is true.

c. Can x^{-3} represent a negative value? If so, give the values of x for which this is true.

Section 5.4   Adding and Subtracting Polynomials

Objectives: **8.** Use the terminology associated with polynomials.

9. Add and subtract polynomials.

This section examines important algebraic expressions called polynomials. We already have used polynomials such as $2x + 3$, $\dfrac{5}{9}(F - 32)$, $2\pi r$, and πr^2 throughout the previous sections of this book. We now extend our earlier work on operations with polynomials and define some key terminology.

Monomials and Polynomials

Recall that factors are constants or variables that are multiplied together to form a product. A single number or an indicated product of factors is called a **term.** The terms in an algebraic expression are separated from each other by plus or minus symbols.

Factors are multiplied; terms are added or subtracted.

The expression $7x^2 + 9xy - 3y^2$ has three terms: $7x^2$, $9xy$, and $-3y^2$. The second term, $9xy$, has factors 9, x, and y.

A **monomial** is a real number, a variable, or a product of real numbers and variables. Because a variable can be used repeatedly as a factor, whole number exponents can occur in monomials. A **polynomial** is a monomial or a sum of monomials.

Monomials and Polynomials		
	VERBALLY	**ALGEBRAIC EXAMPLES**
Monomials:	A monomial is a real number, a variable, or a product of real numbers and variables with whole number exponents.	-5, π, x, A, $5x$, $7xy$, and πr^2 are monomials.
Polynomials:	A polynomial is a monomial or a sum of monomials.	-5 and $7xy$ are polynomials. $7x^2 + 9xy - 3y^2$ is a polynomial.

The variables in a polynomial can have exponents of 0, 1, 2, 3, ... but they cannot have negative exponents or fractional exponents.

A monomial in the variable x has the form ax^n, where a is some constant coefficient, x is a variable, and n is a whole number. Because the exponent n cannot be negative, a monomial cannot have a variable in the denominator.

■ EXAMPLE 1 Identifying Monomials

Determine which of the following expressions are monomials.

SOLUTIONS

(a) $-7x^4$ Monomial   The coefficient of x^4 is -7.

(b) $\frac{4}{7}x^3y^2$ Monomial   The numerical coefficient is the constant $\frac{4}{7}$. The exponent on x is $+3$ and on y the exponent is $+2$.

(c) $\frac{4x^3}{7y^2}$ or $\frac{4}{7}x^3y^{-2}$ Not a monomial   This expression contains a variable in the denominator; in the optional form the exponent on y is not a whole number.

(d) 5 Monomial   All constants are monomials. 5 also can be written as $5x^0$.

(e) $5x^{\frac{1}{2}}$ Not a monomial $5x^{\frac{1}{2}}$ (or $5\sqrt{x}$)* has an exponent that is not a whole number. ■

Although the prefix "poly" means many, polynomials can have a single term.

Although the word *coefficient* is not usually applied to a constant term, it can be. For example, the coefficient of 5 is 5. (Note that $5 = 5x^0$, and the coefficient of $5x^0$ is 5.)

Polynomials containing one, two, and three terms are called **monomials, binomials,** and **trinomials,** respectively. Example 2 illustrates these classifications.

■ EXAMPLE 2 Identifying Polynomials

Determine whether each expression is a polynomial. Classify each polynomial according to the number of terms it contains.

SOLUTIONS

(a) $5x^2 + 3x - 7$ Trinomial   The three terms are $5x^2$, $3x$, and -7.
(b) $x^5y - 4x^3y^2$ Binomial   The two terms are x^5y and $-4x^3y^2$.

*The relationship between exponential and radical notation will be examined in Chapter 8.

(c) $\dfrac{x+5}{x-5}$ Not a polynomial   A polynomial cannot contain a variable in the denominator.

(d) $\dfrac{7xy}{9}$ Monomial   This polynomial has only one term with a coefficient of $\dfrac{7}{9}$.

(e) 18.93 Monomial   All constants are monomials.

(f) $3x^4 - 9x^3 + 7x^2 + 8x - 1$ Polynomial with five terms   Polynomials with more than three terms are not assigned special names.

SELF-CHECK 5.4.1

Determine whether each expression is a polynomial. Classify each polynomial according to the number of terms it contains.

1. $7x + 3$ **2.** $\dfrac{7x}{3}$ **3.** $\dfrac{3}{7x}$ **4.** $7x^3 + x - 9$

One place the degree of a polynomial plays a key role is in determining the number of possible solutions to an *nth* degree polynomial equation. For example, a first-degree linear equation has one solution and a second-degree equation has two solutions.

The **degree of a monomial** is the sum of the exponents for all the variables in this term. A nonzero constant is understood to have degree zero ($4 = 4x^0$ with exponent 0), but no degree is assigned to the monomial 0.

■ EXAMPLE 3   Determining the Degree of a Monomial

Determine the coefficient and the degree of each monomial.

SOLUTIONS

	Coefficient	Degree	
(a) $-5x^3$	-5	3	The numerical coefficient is -5, not 5.
(b) $5x^3y^7$	5	10	The sum of the exponents is $3 + 7 = 10$.
(c) 5	5	0	$5 = 5x^0$ with exponent 0.
(d) $-x$	-1	1	$-x = -1x^1$ with coefficient of -1 and exponent of 1, usually not written.
(e) mn^3	1	4	The coefficient is understood to be 1. The sum of the exponents is $1 + 3 = 4$.

The **degree of a polynomial** is the same as the degree of the term with the highest degree. To find this highest degree, examine each term individually—do *not* sum the degrees of the terms.

■ EXAMPLE 4   Determining the Degree of a Polynomial

Determine the degree of each of the following polynomials.

SOLUTIONS

(a) $5x^3 + 7x^2$	3	The degrees of the individual terms are 3 and 2.
(b) $-11x^2 + 7x + 8$	2	The degrees of the individual terms are 2, 1, and 0.
(c) $4x^6y - 3x^3y^5$	8	The degrees of the individual terms are 7 and 8.
(d) $a^3 + 5a^2b - 3ab^2 + b^3$	3	Each of these terms is of degree 3.

A polynomial in x is in **descending order** if the exponents on x decrease from left to right. A polynomial in x is in **ascending order** if the exponents on x increase from left to right. For example, $3x^2 - 7x^3 + 4 - 9x$ can be written in descending order as $-7x^3 + 3x^2 - 9x + 4$ or in ascending order as $4 - 9x + 3x^2 - 7x^3$. Polynomials are easier to compare and work with if they are written in standard form. A polynomial is in **standard form** if (1) the variables in each term are written in alphabetical order and (2) the terms are arranged in descending powers of the first variable. The **leading term** is the first term of a polynomial in standard form.

In Section 5.6 it is definitely best to write polynomials in standard form before performing long division of polynomials.

■ EXAMPLE 5   Writing Polynomials in Standard Form

Write each of the following polynomials in standard form.

SOLUTIONS

(a) $8y^3z^2x$ $8xy^3z^2$ Write the factors in alphabetical order.

(b) $2x^3 + 7 + x^4 - 5x^2$ $x^4 + 2x^3 - 5x^2 + 7$ Arrange the terms in descending order. The leading term is x^4.

(c) $y^2 + 4yx + x^2$ $x^2 + 4xy + y^2$ Write each term in alphabetical order and then arrange the terms in decreasing powers of x.

A polynomial whose only variable is x is called a **polynomial in x.** A polynomial function in x can be represented by the function notation $f(x)$. For example, $f(x) = 5x^2 - 13$ represents a polynomial function. Sometimes we let letters other than f represent functions. For example, we can let $P(x) = 2x^2 - 4x + 11$. To evaluate this polynomial for $x = 3$, we then substitute 3 for x in the polynomial.
$P(3) = 2(3)^2 - 4(3) + 11 = 17$

SELF-CHECK 5.4.2

Determine the coefficient and the degree of each monomial.

1. $-3x^4$ 2. xyz 3. π

Write each polynomial in standard form and give the degree of each.

4. $-9 + 3x^2 + 8x$
5. $-9y^2zx^5$
6. $-4v + 9v^5 + 3v^2 - v^3 + 1$
7. Given $P(x) = -x^2 + 14x - 33$, use a calculator to evaluate $P(7)$.

SELF-CHECK 5.4.2 ANSWERS

1. -3; 4   2. 1; 3   3. π; 0
4. $3x^2 + 8x - 9$;   degree 2
5. $-9x^5y^2z$;   degree 8
6. $9v^5 - v^3 + 3v^2$
    $-4v + 1$;   degree 5

7.
```
7→X
                    7
-X²+14X-33
                   16
```

■ EXAMPLE 6   Evaluating a Profit Polynomial

The profit in dollars made by selling x units is given by the polynomial $P(x) = -x^2 + 14x - 33$. Evaluate and interpret each expression.

SOLUTIONS

(a) $P(0)$ $P(0) = -(0)^2 + 14(0) - 33$ Substitute 0 for x in $P(x)$.
                $= -33$
         Answer: Selling 0 units results in a loss of $33.

(b) $P(3)$ $\qquad$ $P(3) = -(3)^2 + 14(3) - 33$
$$= -9 + 42 - 33$$
$$= 0$$

Answer: The seller breaks even if 3 units are sold.

(c) $P(10)$ $\qquad$ $P(10) = -(10)^2 + 14(10) - 33$
$$= -100 + 140 - 33$$
$$= 7$$

Answer: The seller makes $7 if 10 units are sold. ∎

Adding and Subtracting Polynomials

The distributive property $ax + bx = (a + b)x$ justifies adding the coefficients of like terms. In Section 1.6 we added $13x + 8x$ as $(13 + 8)x = 21x$.

We first used the distributive property in Section 1.6 to add like terms and to remove parentheses from a group of terms. This is the same process used to add and subtract polynomials.

■ **EXAMPLE 7** Adding and Subtracting Polynomials

(a) Add $5x^2 - 7x + 9$ and $3x^2 + 6x - 8$ and then
(b) Subtract $5x^2 - 7x + 9$ from $3x^2 + 6x - 8$.

SOLUTIONS

(a) $(5x^2 - 7x + 9) + (3x^2 + 6x - 8)$ — Use a plus symbol to indicate the addition.
$= 5x^2 - 7x + 9 + 3x^2 + 6x - 8$ — Remove the parentheses.
$= 5x^2 + 3x^2 - 7x + 6x + 9 - 8$ — Use the commutative and associative properties of addition to reorder the terms.

$= (5 + 3)x^2 + (-7 + 6)x + (9 - 8)$ — Then use the distributive property to combine like terms.

$= 8x^2 - x + 1$ — The answer is written in standard form.

(b) $(3x^2 + 6x - 8) - (5x^2 - 7x + 9)$ — Use a minus symbol to indicate the subtraction.
$= 3x^2 + 6x - 8 - 5x^2 + 7x - 9$ — Remove the parentheses using the distributive property.

$= 3x^2 - 5x^2 + 6x + 7x - 8 - 9$ — Reorder the terms.
$= (3 - 5)x^2 + (6 + 7)x + (-8 - 9)$ — Then use the distributive property to combine like terms.

$= -2x^2 + 13x - 17$ — The answer is written in standard form. ∎

SELF-CHECK 5.4.3

1. Add $5x - 9$ and $8x + 7$.
2. Subtract $2x^2 - 4$ from $5x^2 + 7x$.

If two polynomials are equal, then their graphs will be identical and a table of values will show identical values for the two polynomials. We examine tables of values in the next example, but we wait until Chapter 6 to examine the graphs of polynomials.

■ EXAMPLE 8   Using Tables to Compare Two Polynomials

Add $2(3x + 5)$ to $4(2x - 3)$ and check the sum by using a table of values.

SOLUTION

ALGEBRAICALLY

$2(3x + 5) + 4(2x - 3)$ Indicate the addition and then use the distributive property to remove the
$= 6x + 10 + 8x - 12$ parentheses.
$= 6x + 8x + 10 - 12$ Reorder the terms.
$= (6 + 8)x - 2$ Then use the distributive property to combine like terms.
$= 14x - 2$

GRAPHICALLY

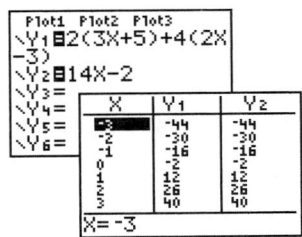

Let y_1 represent the indicated sum of the two terms and y_2 represent the calculated sum. Since the values of y_1 and y_2 in the table are identical, this indicates that $2(3x + 5) + 4(2x - 3)$ is equal to $14x - 2$.

Answer: $2(3x + 5) + 4(2x - 3) = 14x - 2$.

USING THE LANGUAGE AND SYMBOLISM OF MATHEMATICS   5.4

1. _____ are constants or variables that are multiplied together to form a product.
2. Terms in an algebraic expression are separated from each other by _____ or _____ symbols.
3. The whole numbers are _____.
4. A _____ is a real number, a variable, or a product of real numbers and variables with whole number exponents.
5. A _____ is a monomial or a sum of monomials.
6. In the monomial ax^n, a is a _____ of x^n.
7. a. A polynomial containing exactly one term is a

   _____.
   b. A polynomial containing exactly two terms is a

   _____.
   c. A polynomial containing exactly three terms is a

   _____.

8. The degree of a monomial is the _____ of the exponents for all the variables in the monomial.
9. The degree of a polynomial is the same as the degree of the term with the _____ degree.
10. a. The degree of a nonzero constant is _____.
    b. No degree is assigned to the constant _____.
11. A polynomial is in standard form if (1) the variables in each term are written in _____ order and (2) the terms are arranged in _____ powers of the first variable.
12. a. The coefficient of x is _____.
    b. The coefficient of $-x$ is _____.
    c. The degree of $-x$ is _____.

EXERCISES 5.4

In Exercises 1–6 determine whether each expression is a monomial. If the expression is a monomial, give its coefficient and its degree.

1. **a.** -12 **b.** $-12x$ **c.** $-x^{12}$ **d.** x^{-12}

2. **a.** 3 **b.** $3x$ **c.** x^3 **d.** x^{-3}

3. **a.** $\dfrac{2}{5}x^4$ **b.** $\dfrac{5x^4}{2}$ **c.** $-\dfrac{5x^4}{2}$ **d.** $\dfrac{2}{5x^4}$

4. **a.** $\sqrt{2}$ **b.** $\sqrt{2}x$ **c.** $2\sqrt{x}$ **d.** $\dfrac{x}{\pi}$

5. **a.** $4xy^2z^3$ **b.** $-5x^2y^2$ **c.** $\dfrac{5x^2}{y^2}$ **d.** $\dfrac{x^3y^4}{6}$

6. **a.** $9x^3yz^4$ **b.** $\dfrac{a^2b^2c^2}{9}$ **c.** $\dfrac{a^2b^2}{c^2}$ **d.** $a^{-3}bc^4$

In Exercises 7–10 determine whether each expression is a polynomial. Classify each polynomial according to the number of terms it contains and give its degree.

7. **a.** $3x^2 - 7x$ **b.** $x^3 - 8x^2 + 9$
   c. $-9x^4$ **d.** -9

8. **a.** $\dfrac{-4x^5}{7}$ **b.** $x^4 - 5$

   c. $x^2 - 5x + 7$ **d.** 0

9. **a.** $11xy$ **b.** $11x + y$

   c. $\dfrac{11x}{y}$ **d.** $x^3 + y^2 - 11$

10. **a.** $4x^2 - 3xy + y^2$ **b.** $-3xy$

    c. $-3x + y$ **d.** $\dfrac{3x}{y}$

In Exercises 11 and 12 write each polynomial in standard form.

11. **a.** $8yx^2z^3$ **b.** $-9 + x^2 - 5x$
    c. $-2x^2 + 7 - 4x^3 + 9x$

12. **a.** $-11c^3a^4b$ **b.** $-7x + 8 - 3x^2$
    c. $13x + 7x^2 - 9x^3 - 11$

In Exercises 13 and 14 determine if the table indicates that the polynomials defined by y_1 and y_2 are equal. The polynomials y_1 and y_2 are either first, second, or third degree.

13. **a.**

X	Y₁	Y₂
-3	-16	-16
-2	-13	-13
-1	-10	-10
0	-7	-7
1	-4	-4
2	-1	-1
3	2	2

X = -3

b.

X	Y₁	Y₂
-3	10	-30
-2	4	-12
-1	0	0
0	-2	6
1	-2	6
2	0	0
3	4	-12

X = -3

14. **a.**

X	Y₁	Y₂
-3	6	6
-2	0	0
-1	-4	-4
0	-6	-6
1	-6	-6
2	-4	-4
3	0	0

X = -3

b.

X	Y₁	Y₂
-3	-24	48
-2	-6	12
-1	0	0
0	0	0
1	6	-12
2	24	-48

X = -3

In Exercises 15–20 determine each sum and difference.

15. **a.** $7x + 5x + 9x$
    b. $8x^2 - 4x^2 + 3x^2$

16. **a.** $8y + 3y + 12y$
    b. $7y^2 - 5y^2 - 6y^2$

17. **a.** $(3x^2 - 7x) + (4x^2 - 5x)$
    b. $(3x^2 - 7x) - (4x^2 - 5x)$

18. **a.** $(-5x + 3) + (8x - 11)$
    b. $(-5x + 3) - (8x - 11)$

19. **a.** $(5x^2 - 7x + 9) + (4x^2 + 6x - 3)$
    b. $(5x^2 - 7x + 9) - (4x^2 + 6x - 3)$

20. **a.** $(7x^2 + 6x - 13) + (3x^2 - 9x - 4)$
    b. $(7x^2 + 6x - 13) - (3x^2 - 9x - 4)$

In Exercises 21–52 determine each sum or difference and write the result in standard form.

21. $(6x + 5) + (8x - 7)$
22. $(-5x + 3) + (4x - 11)$
23. $(-7v + 3) - (2v - 11)$
24. $(-4v - 13) - (-7v + 11)$
25. $(5a^2 - 7a + 9) + (4a^2 + 6a - 3)$
26. $(7a^2 + 6a - 13) + (3a^2 - 9a - 4)$
27. $(8x^2 - 3x - 4) - (2x^2 - 2x - 9)$
28. $(11x^2 + 7x - 1) - (3x^2 - 12x + 8)$
29. $(-w^2 + 4) - (-4w^2 + 7w - 6)$
30. $(17w - 13) - (3w^2 - 4w + 6)$
31. $(-2m^4 + 3m^3 + m^2 - 1) + (m^4 - m^2 - 2m + 5)$
32. $(4m^3 + 7m^2 - 9m - 1) + (2m^3 - m^2 + 7)$
33. $(-3n^4 + 7n^3 + n - 8) - (n^4 - 2n^3 - n^2 + 4)$
34. $(-2n^4 - n^2 + 3n - 5) - (5n^4 + 7n^3 - 2n + 5)$
35. $(-7y^4 + 3y^6 - y + 4y^5 - 11 + 2y^2) -$
    $(3y^5 - 5y + 7y^6 - 9y^2 + 7 - y^4)$
36. $(8y^3 + 3 - 5y^2 + y^6 - 4y^4 + 9y) -$
    $(5 + 3y^2 - 2y^6 + y - 2y^4 - y^5)$
37. $2(x^2 - 3x - 5) - 7(x^2 + 2x - 1)$
38. $8(x^2 + 5x - 2) - 5(2x^2 - 3x - 4)$
39. $3(2a^2 - 5a + 1) - 2(3a^2 - a - 5)$
40. $(7a^2 - a + 3) - 5(5a^2 + a - 2)$
41. $(5x + 7) + (4x - 8) + (3x - 11)$
42. $(6x - 4) + (2x - 3) + (5x + 9)$
43. $(13x - 8) - (7x - 9)$
44. $(17x + 5) - (14x - 7)$
45. $(5x^2 + 7xy - 9y^2) + (6x^2 - 3xy + y^2)$
46. $(11x^2 - 9xy - 4y^2) + (8x^2 + xy - 5y^2)$
47. $(-7x^2 + 6xy + 8y^2) - (13x^2 - xy - 3y^2)$
48. $(4x^2 - 5xy + 12y^2) - (7x^2 - 8xy - 11y^2)$
49. $(5x^3 - 7x + 9 + 5x^2) + (2x^2 + 13 + x^3 - x) -$
    $(4x - 3x^3 + x^2 - 8)$
50. $(6x - 11x^3 + 7 - x^2) - (14 - 2x^2 - x^3 - 5x) +$
    $(9x - 12 - x^3 + 8x^2)$
51. $(x^4 - 2x^3y + x^2y^2 + xy^3 - 3y^4) -$
    $(2x^4 + x^3y - 5x^2y^2 - 7y^4) - (7x^4 + 3x^2y^2 + 2y^4)$
52. $(3x^4 - 9x^3y + 7x^2y^2 - y^4) - (7x^4 + 4x^2y^2 - 11y^4) -$
    $(-11x^4 - 9x^3y + y^4)$

In Exercises 53 and 54 evaluate each expression for
$P(x) = 5x^2 + 3x - 2$.
53. a. $P(0)$ **b.** $P(2)$ **c.** $P(10)$ **d.** $P(8)$
54. a. $P(1)$ **b.** $P(5)$ **c.** $P(20)$ **d.** $P(100)$

Profit Polynomial
In Exercises 55–60 evaluate and interpret each expression.
The profit in dollars made by selling x units is given by
$P(x) = -x^2 + 22x - 40$.
55. $P(0)$ **56.** $P(1)$ **57.** $P(2)$
58. $P(10)$ **59.** $P(20)$ **60.** $P(22)$

In Exercises 61 and 62 use a calculator to evaluate
the given polynomial for each x-value.

61.
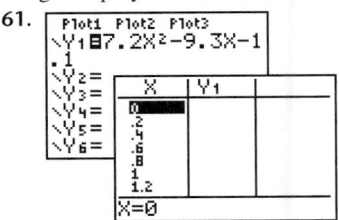

62.
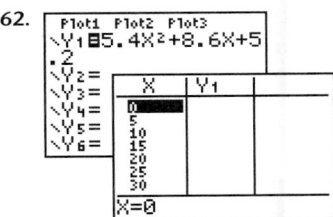

In Exercises 63–66 calculate each sum. Then check
your result by creating a table of values from $x = -3$
to 3. Use y_1 to represent the given problem and y_2 to
represent the calculated sum.
63. $(3.49x^2 + 7.81x) + (5.53x^2 - 3.79x)$
64. $(5.76x^2 + 4.3x - 1.9) + (2.04x^2 + 5.7x - 2.1)$
65. $(6.11x^2 - 5.98x - 3.51) + (5.07x^2 + 4.01x - 2.47)$
66. $(4.5x - 9.1) + (3.7x + 6.7) + (6.3x - 4.6)$

Modeling Perimeter and Area Using Polynomials
In Exercises 67–72, use the following figure to write a
polynomial describing each of the quantities indicated.

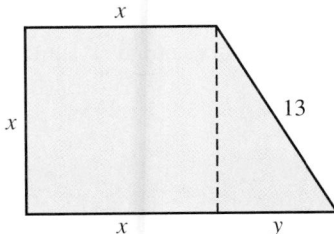

67. The perimeter of the entire figure.
68. The perimeter of the square in the figure.
69. The perimeter of the triangle in the figure.
70. The area of the square region.

71. The area of the triangular region.
72. The area of the entire region.

Multiple Representations
In Exercises 73–80 write a polynomial for each verbal
expression.
73. Two times x cubed minus five x plus eleven.
74. Nine times y squared plus seven y minus nine.
75. Three times the sixth power of w plus five times the fourth power of w.
76. Eight times the fifth power of v minus v cubed plus eleven v.
77. x squared minus y squared.
78. a cubed plus b cubed.
79. Seven more than twice w.
80. Eight less than the opposite of four y.

Modeling the World Production of Chlorofluorocarbons
The 1987 Montreal Protocol on Substances that Deplete the
Ozone Layer restricted the use of CFCs, which damage the
ozone layer above the Earth. The following table gives the
world production of CFCs for selected years after 1950 as
measured in thousands of tons.

x Years After 1950	$C(x)$ Thousands of Tons of CFCs
0	60
10	100
20	700
30	900
40	1250
50	800

Source: http://sedac.ciesin.org/sitemap.html

81. This data can be modeled by the function
$C(x) = -0.05x^3 + 3.0x^2 - 13.0x + 60$. Use this function
to approximate the tons of CFC produced in 1965.
82. Use the function in Exercise 81 to predict the tons of CFC
that will be produced in 2005.

Group Discussion Questions
83. Challenge Question   Give an example that satisfies the
given conditions.
  a. A fifth-degree monomial in x with a coefficient of 2.
  b. A monomial of degree 0.
  c. Two first-degree monomials in x whose sum is a constant.
  d. Two first-degree monomials in x whose difference is a constant.
  e. Two fourth-degree binomials in x whose sum is a third-degree monomial.
84. Error Analysis   A classmate subtracted $3x^2 - x - 8$
from $7x^2 + 10x + 9$ by writing the expression
$7x^2 + 10x + 9 - 3x^2 - x - 8$ to obtain $4x^2 + 9x + 1$.
Explain the *error* in this work and make a suggestion as to
how the classmate could organize the work to avoid this
error.

Section 5.5 Multiplying Polynomials

Objective: **10.** Multiply polynomials.

The product of two monomials was considered when the product rule for exponents was discussed. Recall that the product rule states that $x^m x^n = x^{m+n}$.

■ EXAMPLE 1   Multiplying Monomials

Simplify $(7x^3 y^2)(-8x^4 y^6)$.

SOLUTION ──────────────────────────

Note that the coefficients are multiplied and the exponents are added.

$$
\begin{aligned}
(7x^3 y^2)(-8x^4 y^6) &= (7)(-8)(x^3 x^4)(y^2 y^6) \\
&= -56x^{3+4} y^{2+6} \\
&= -56x^7 y^8
\end{aligned}
$$

Regroup and reorder the factors using the associative and commutative properties of multiplication.
Using the product rule for exponents, add the exponents on the common bases. ■

The distributive property $a(b + c) = ab + ac$ plays a key role in the multiplication of polynomials.

The distributive property is used to find the product of a monomial times a polynomial. This use of the distributive property is illustrated in Example 2 by the product of a monomial times a binomial.

■ EXAMPLE 2   Multiplying a Monomial Times a Polynomial

Multiply these polynomials: **(a)** $5x^3(7x - 9)$;   **(b)** $(4x^2 - 5xy + 11y^2)(2xy)$

SOLUTIONS ──────────────────────────

(a) $5x^3(7x - 9) = (5x^3)(7x) + (5x^3)(-9)$
$\qquad\qquad\qquad = (5)(7)(x^3 x) + (5)(-9)x^3$
$\qquad\qquad\qquad = 35x^4 - 45x^3$

Distribute the multiplication of $5x^3$ to each term of the binomial. Then simplify each term, using the product rule for exponents.

(b) $(4x^2 - 5xy + 11y^2)(2xy)$
$\qquad = (4x^2)(2xy) + (-5xy)(2xy) + (11y^2)(2xy)$
$\qquad = 8x^3 y - 10x^2 y^2 + 22xy^3$

Distribute the multiplication of $2xy$ to each term of the trinomial. Then simplify, using the product rule for exponents. ■

The procedure for multiplying a polynomial by a monomial is summarized in the following box.

Multiplying a Monomial Times a Polynomial

To multiply a monomial times a polynomial, use the distributive property to multiply the monomial times each term of the polynomial.

SELF-CHECK 5.5.1

Multiply the polynomial factors.

1. $(-9a^2bc)(-4ab^2c^3)$
2. $9x^3(8x - 5)$
3. $-3x^2(7x^2 - 2xy - 3y^2)$
4. $(7x^2 - 2xy - 3y^2)(4y)$

Multiplying two polynomials with more than one term each requires the repeated use of the distributive property. An illustration is provided in Example 3.

■ EXAMPLE 3   Multiplying Two Binomials

Find the product of $(7x + 3)(2x + 5)$.

SOLUTION

$$(7x + 3)(2x + 5) = (7x)(2x + 5) + (3)(2x + 5)$$
$$= (7x)(2x) + (7x)(5) + (3)(2x) + (3)(5)$$
$$= 14x^2 + 35x + 6x + 15$$
$$= 14x^2 + 41x + 15$$

First distribute the multiplication of $2x + 5$ to each term of $7x + 3$. Then distribute $7x$ times each factor of $2x + 5$, and also distribute 3 times each factor of $2x + 5$. Note that each term of $7x + 3$ is multiplied times each term of $2x + 5$.

Finally, simplify by combining like terms.   ■

Fortunately all the steps shown in Example 3 to justify the multiplication of $(7x + 3)(2x + 5)$ can be shortened by following the procedure in the following box. With practice some of the steps shown in Example 3 can be performed mentally.

A Mathematical Note

Plato (c. 427–348 B.C.) was a pupil of Socrates. In 388 B.C. he formed a school called the Academy in Athens. Inscribed above the gate to this academy was the motto: "Let no man ignorant of geometry enter." Many of our algebraic formulas are based on early geometric discoveries.

Multiplying Two Polynomials

To multiply one polynomial times another, multiply each term of the first polynomial times each term of the second polynomial and then combine like terms.

We have examined the multiplication of polynomials using algebraic properties and now give a geometric viewpoint.

Multiplying Binomials: A Geometric Viewpoint

A geometric viewpoint of the multiplication of two binomials is shown next. The area of the rectangle is considered two different ways. First calculate the area of the whole rectangle and then compute the area by adding the areas of the four parts. Since the area

---

SELF-CHECK 5.5.1 ANSWERS

1. $36a^3b^3c^4$
2. $72x^4 - 45x^3$
3. $-21x^4 + 6x^3y + 9x^2y^2$
4. $28x^2y - 8xy^2 - 12y^3$

Area of Whole Rectangle

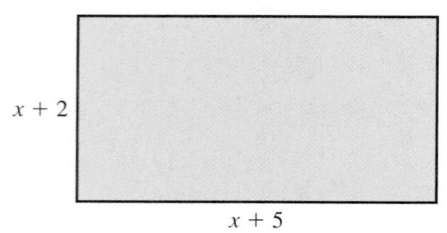

Area of Four Parts

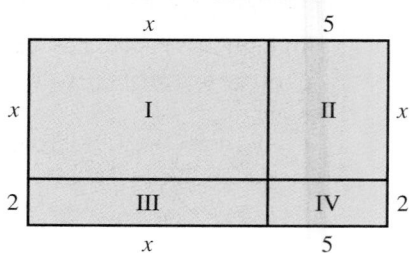

is the same either way, these two polynomials must be equal.

Area = length × width   Total area = area I + area II + area III + area IV
$A = (x + 5)(x + 2)$ $A = x^2 + 5x + 2x + 2(5)$
                                                             $A = x^2 + 7x + 10$

Thus $(x + 5)(x + 2) = x^2 + 7x + 10$.

■ EXAMPLE 4   Multiplying a Binomial Times a Trinomial

Multiply $3x - 2$ times $4x^2 - 5x + 6$.

SOLUTION

$(3x - 2)(4x^2 - 5x + 6)$ The distributive property justifies forming the
$= (3x)(4x^2 - 5x + 6) + (-2)(4x^2 - 5x + 6)$ product of each term of $4x^2 - 5x + 6$ times each
$= (3x)(4x^2) + (3x)(-5x) + (3x)(6) + (-2)(4x^2) + (-2)(-5x) + (-2)(6)$ term of $3x - 2$.
$= 12x^3 - 15x^2 + 18x - 8x^2 + 10x - 12$ Simplify by combining like terms.   ■
$= 12x^3 - 23x^2 + 28x - 12$

It is generally easier to keep our steps organized and to obtain accurate results if we
write polynomials in standard form. This is especially wise when multiplying polynomials.

■ EXAMPLE 5   Multiplying a Trinomial Times a Trinomial

Multiply $4 + x^2 - 2x$ times $x + 1 + x^2$.

SOLUTION

$(4 + x^2 - 2x)(x + 1 + x^2)$ First write each factor in standard form.
$= (x^2 - 2x + 4)(x^2 + x + 1)$
$= (x^2)(x^2 + x + 1) + (-2x)(x^2 + x + 1) + (4)(x^2 + x + 1)$ Distribute the factor $x^2 + x + 1$ times each term of
$= (x^2)(x^2) + (x^2)(x) + (x^2)(1) + (-2x)(x^2) + (-2x)(x) + (-2x)(1)$ $x^2 - 2x + 4$.
$\quad + (4)(x^2) + (4)(x) + (4)(1)$ Again use the distributive property to perform each
$= x^4 + x^3 + x^2 - 2x^3 - 2x^2 - 2x + 4x^2 + 4x + 4$ multiplication.
$= x^4 + x^3 - 2x^3 + x^2 - 2x^2 + 4x^2 - 2x + 4x + 4$ Then use the associative and commutative properties of
$= x^4 - x^3 + 3x^2 + 2x + 4$ addition to reorder the terms and then add like terms.
                                                                                                                    ■

The next example illustrates the multiplication of three binomial factors. We start by multiplying the first two factors and then multiply this product times the third factor.

■ EXAMPLE 6   Finding the Product of Three Binomials

Multiply $(x - 3)(x + 2)(x - 5)$.

SOLUTION

$(x - 3)(x + 2)(x - 5)$
$= [(x)(x + 2) + (-3)(x + 2)](x - 5)$
$= [x^2 + 2x - 3x - 6](x - 5)$
$= (x^2 - x - 6)(x - 5)$
$= (x^2)(x - 5) + (-x)(x - 5) + (-6)(x - 5)$
$= x^3 - 5x^2 - x^2 + 5x - 6x + 30$
$= x^3 - 6x^2 - x + 30$

First find the product of $(x - 3)(x + 2)$ by distributing $x + 2$ to each term of $x - 3$. Then distribute $x - 5$ to each term of $x^2 - x - 6$. ■

Because of the associative property of multiplication, we also could determine the product in Example 6 by multiplying $(x + 2)(x - 5)$ and then multiplying $(x - 3)$ times that result.

SELF-CHECK 5.5.2 ANSWERS
1. $4x^2 - 9y^2$
2. $x^3 - 8$
3. $v^4 + v^3 - 5v^2 + 8v - 2$

The number of units that can be sold for some products depends on the price of the item. For example, lowering the price of new cars will usually generate more sales. Lowering interest rates (the cost of borrowing money) is a standard tool of the Federal Reserve to promote more borrowing in the economy. The revenue generated by selling a number of units is found by multiplying the price per unit times the number of units sold. Thus some companies can exert control over their revenue by controlling their prices. Example 7 expresses the revenue for automobile sales as a function of the price of the car.

■ EXAMPLE 7   Modeling the Revenue for Automobile Sales

For a price of x dollars per car (x from $20,000 to $30,000) a car manufacturer estimates that it can sell $10,000 - 0.2x$ cars. Write a revenue polynomial $R(x)$ based on this estimate and evaluate $R(20,000)$ and $R(30,000)$. Then use a graphics calculator to check these values.

SOLUTIONS

Let x = price per car in dollars
Let $10,000 - 0.2x$ = the number of cars that will sell at a price x

| Revenue | = | Number of items sold | · | Price per item | Write a word equation which is the basis of the algebraic equation. |

$$R(x) = (10{,}000 - 0.2x)(x)$$ This is the revenue polynomial.

$$R(20{,}000) = [10{,}000 - 0.2(20{,}000)](20{,}000)$$ Evaluate $R(x)$ when the price per car is $20,000.
$$= [10{,}000 - 4{,}000](20{,}000)$$
$$= (6{,}000)(20{,}000)$$
$$= 120{,}000{,}000$$ The revenue produced by selling cars at $20,000 each is estimated to be $120,000,000.

$$R(30{,}000) = [10{,}000 - 0.2(30{,}000)](30{,}000)$$ Evaluate $R(x)$ when the price per car is $30,000.
$$= (10{,}000 - 6{,}000)(30{,}000)$$
$$= (4{,}000)(30{,}000)$$
$$= 120{,}000{,}000$$ The revenue produced by selling cars at $30,000 each also is estimated to be $120,000,000.

```
Plot1 Plot2 Plot3
\Y1◻(10000-0.2X)
(X)
\Y2=
\Y3=          X      Y1
\Y4=       20000  1.2E8
\Y5=       22000  1.23E8
\Y6=       24000  1.25E8
           26000  1.25E8
           28000  1.23E8
           30000  1.2E8
           32000  1.15E8
         X=20000
```

We can examine many values in a table generated by a graphics calculator. From Section 5.3, 1.2E8 is a form of scientific notation representing $1.2 \times 10^8 = 120{,}000{,}000$.

Analysis: Although the revenue generated by a price of $20,000 and a price of $30,000 are both $120,000,000, the revenue is not constant for all input values. Note that at a price of $26,000 the revenue generated is $125,000,000. ◼

SELF-CHECK 5.5.3

Use the revenue polynomial in Example 6 to evaluate and interpret $R(27{,}000)$.

SELF-CHECK 5.5.3 ANSWER

$R(27{,}000) = \$124{,}200{,}000$. At a price of $27,000 per car the manufacturer estimates the revenue from sales of this car to be $124,200,000.

Equal polynomials will have identical output values for all input values. In Example 8 we use tables to compare the output values of two polynomials as a check that these polynomials are equal.

◼ EXAMPLE 8　Using Tables to Compare Two Polynomials

Multiply $(2x + 3)(2x - 3)$ and check this product by using a table of values.

SOLUTION

ALGEBRAICALLY

$$(2x + 3)(2x - 3) = (2x)(2x - 3) + (3)(2x - 3)$$ Distribute the factor of $2x - 3$ to each term of $2x + 3$.
$$= (2x)(2x) + (2x)(-3) + (3)(2x) + (3)(-3)$$ Again use the distributive property to perform each multiplication.
$$= 4x^2 - 6x + 6x - 9$$
$$= 4x^2 + 0 - 9$$ Then add like terms.
$$= 4x^2 - 9$$

NUMERICALLY

```
Plot1 Plot2 Plot3
\Y1=(2X+3)(2X-3)
\Y2=4X2-9
\Y3=
\Y4=
\Y5=
\Y6=
```

X	Y1	Y2
-3	27	27
-2	7	7
-1	-5	-5
0	-9	-9
1	-5	-5
2	7	7
3	27	27

X=-3

Let y_1 represent the indicated product of the two factors and y_2 represent the calculated product. Because the values of y_1 and y_2 in the table are identical, this indicates that the polynomials $(2x + 3)(2x - 3)$ and $4x^2 - 9$ are equal.

Answer: $(2x + 3)(2x - 3) = 4x^2 - 9$ ◼

Working a problem by one method and then using another method to check your work is a very useful test strategy. To check the product of two polynomials, you can use tables as illustrated in Example 8.

The distributive property is one of the most important properties of algebra; a property that is used throughout this book. In Section 1.6 we used this property to expand the expression in Example 6(a), $9(x + 4) = 9x + 36$. We also used this property to factor the expression in Example 6(c), $5x - 5y = 5(x - y)$. To factor a polynomial, we usually start by factoring out the greatest common factor. The **greatest common factor (GCF)** of the terms of a polynomial is the common factor that contains the largest possible coefficient and the largest possible exponent on each variable factor. For example, the GCF of $12x^4y^5$ and $15x^2y^6$ is $3x^2y^5$.

■ **EXAMPLE 9**   Using the Distributive Property to Factor Out a Greatest Common Factor

Use the distributive property to expand the expressions in parts **(a)**, **(c)**, and **(e)** and to factor the expressions in parts **(b)**, **(d)**, and **(f)**.

SOLUTIONS

(a) Expand $3x^2(5x - 4y)$.

$3x^2(5x - 4y) = (3x^2)(5x) + (3x^2)(-4y)$
$= 15x^3 - 12x^2y$

Distribute the factor of $3x^2$ to each term of the trinomial. Then simplify using the product rule for exponents.

(b) Factor $15x^3 - 12x^2y$.

$15x^3 - 12x^2y = (3x^2)(5x) + (3x^2)(-4y)$
$= 3x^2(5x - 4y)$

$3x^2$ is the greatest common factor of each term. Use the distributive property to factor out this common factor.

(c) Expand $8xy(2x + 3y)$.

$8xy(2x + 3y)$
$= (8xy)(2x) + (8xy)(3y)$
$= 16x^2y + 24xy^2$

Distribute the factor of $8xy$ to each term of the binomial. Then simplify using the product rule for exponents.

(d) Factor $16x^2y + 24xy^2$.

$16x^2y + 24xy^2$
$= (8xy)(2x) + (8xy)(3y)$
$= 8xy(2x + 3y)$

$8xy$ is the greatest common factor of each term. Use the distributive property to factor out this common factor.

(e) Expand $2x(5x - 4) + 3(5x - 4)$.

$2x(5x - 4) + 3(5x - 4)$
$= (2x)(5x) + (2x)(-4) + (3)(5x) + (3)(-4)$
$= 10x^2 - 8x + 15x - 12$
$= 10x^2 + 7x - 12$

Distribute the factor of $2x$ to each term of $5x - 4$ and distribute the factor 3 to each term of $5x - 4$. Then simplify using the product rule for exponents. Lastly, add like terms.

(f) Factor $2x(5x - 4) + 3(5x - 4)$.

$2x(5x - 4) + 3(5x - 4)$
$= (2x + 3)(5x - 4)$

Use the distributive property to factor out the common factor of $5x - 4$.   ◼

SELF-CHECK 5.5.4

Use the distributive property to factor these polynomials.

1. $5x^2 - 10x$
2. $6x^2 - 3x + 33$
3. $7x(2x - 5) - 3(2x - 5)$

USING THE LANGUAGE AND SYMBOLISM OF MATHEMATICS   5.5

1. **a.** The degree of $6x^2$ is _____.
   b. The degree of $4x^3$ is _____.
   c. The degree of the product $6x^2(4x^3) = 24x^5$ is _____.

2. The _____ property is the property used to rewrite $3(7x^2 + xy + y^2)$ as $3(7x^2) + 3(xy) + 3(y^2)$.

3. The _____ property is the property used to rewrite $5x(4x^2) + 5x(9x) + 5x(-7)$ as $5x(4x^2 + 9x - 7)$.

4. The _____ property is the property used to rewrite $5x(4x - y) + y(4x - y)$ as $(5x + y)(4x - y)$.

5. **a.** The polynomial $4x - 5$ has _____ terms.
   b. The polynomial $3x + 2$ has _____ terms.
   c. The product $(4x - 5)(3x + 2) = 12x^2 - 7x - 10$ has _____ terms.

6. **a.** The binomial $4x - 5$ is of degree _____.
   b. The binomial $3x + 2$ is of degree _____.
   c. The product of $4x - 5$ and $3x + 2$ is the trinomial $12x^2 - 7x - 10$ of degree _____.

7. If two polynomials with x as the input variable are equal, then the two polynomials will have _____ output values for all input values of x.

8. The polynomial $P(x) = x^2 - x + 3$ is read "P _____ _____ equals x squared minus x plus 3."

EXERCISES   5.5

In Exercises 1–42 multiply the polynomial factors.

1. $2x(-5x)$
2. $-3y(8y)$
3. $(-9v^2)(-7v^3)$
4. $(11w^3)(-w^5)$
5. $(a^3b^4)(4a^6b)$
6. $(-11x^2y^3)(-8xy^6)$
7. $4x^2(6x - 5)$
8. $7x^3(9x + 4)$
9. $-4v^2(8v + 3)$
10. $-9v^4(2v - 11)$
11. $2x(7x^2 - 9x - 4)$
12. $3x(8x^2 + 4x - 5)$
13. $-11x^2(4x^2 - 5x - 8)$
14. $-14x^2(2x^2 - 3x + 5)$
15. $(2x^2 - xy - y^2)(5xy)$
16. $(3x^2 + xy - y^2)(2xy)$
17. $3a^2b(2a^2 + 4ab - b^2)$
18. $5v^2w^2(8v^2 - vw + 3w^2)$
19. $(y + 2)(y + 3)$
20. $(y - 4)(y - 6)$
21. $(2v + 5)(3v - 4)$
22. $(2v - 3)(5v + 2)$
23. $(6x + 5)(9x + 3)$
24. $(7x + 11)(4x + 6)$
25. $(4m - 1)(6m - 7)$
26. $(9m - 4)(7m - 5)$
27. $(x + y)(2x - y)$
28. $(x - y)(3x + y)$
29. $(2x + 4y)(5x - 7y)$
30. $(2x - 5y)(3x + 4y)$
31. $(x + 2)(x^2 - 3x - 4)$
32. $(2x - 1)(x^2 + x - 3)$
33. $(2x - 4)(3x^2 + 2x - 5)$
34. $(3x + 2)(5x^2 - 4x + 6)$
35. $(2m - 5)(m^2 + 3m + 2)$
36. $(3m + 4)(m^2 - 2m - 7)$
37. $(x + y)(x^2 - xy + y^2)$
38. $(x - y)(x^2 + xy + y^2)$
39. $5x(x + 3)(x - 2)$
40. $7x(x - 4)(x + 2)$
41. $(2x - 1)(x + 3)(x + 4)$
42. $(3x + 2)(x - 1)(x + 5)$

Modeling the Area of a Region Using Polynomials

In Exercises 43–48 write a polynomial for the area of each figure. See the formulas on the inside of the back cover.

43.

$(4x - 1)$ cm

$(2x + 3)$ cm

44.

$(4x + 1)$ cm

$(5x - 2)$ cm

45.

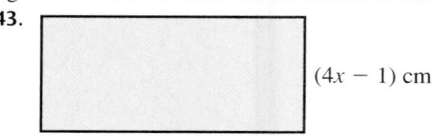

$(x - 1)$ cm

$(2x + 4)$ cm

46.

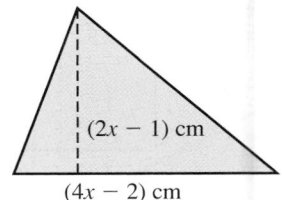

$(2x - 1)$ cm

$(4x - 2)$ cm

47.

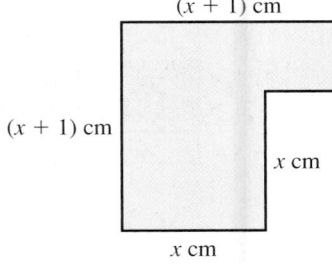

$(x + 1)$ cm

$(x + 1)$ cm

x cm

x cm

48.

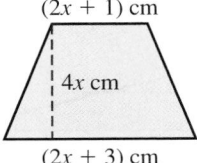

$(2x + 1)$ cm

$4x$ cm

$(2x + 3)$ cm

In Exercises 49 and 50 write a polynomial for an area of the shaded region. Assume that the curved portions are semicircles of the same size.

49.

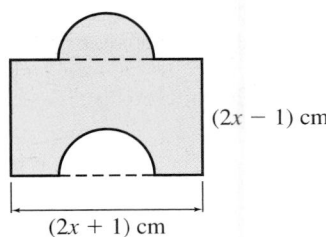

$(2x - 1)$ cm

$(2x + 1)$ cm

50.

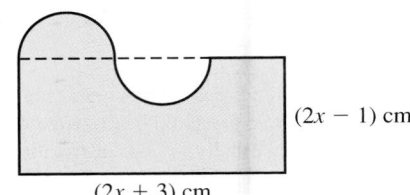

$(2x - 1)$ cm

$(2x + 3)$ cm

In Exercises 51 and 52 use a graphics calculator to fill in the table of values for $y_1 = P_1(x)$ and $y_2 = P_2(x)$. Do these results indicate that the polynomials $P_1(x)$ and $P_2(x)$ are equal?

51.

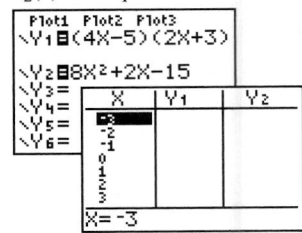

52.

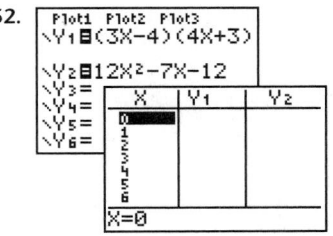

In Exercises 53–62 use the distributive property to expand the expression in the first column and to factor the GCF out of the polynomial in the second column.

	Expand		Factor
53.	**a.** $4x(9x + 3)$		**b.** $36x^2 + 12x$
54.	**a.** $5v(8v - 7)$		**b.** $40v^2 - 35v$
55.	**a.** $3x^3y(2x + 7y)$		**b.** $6x^4y + 21x^3y^2$
56.	**a.** $8xy^2(3x^2 - 2y)$		**b.** $24x^3y^2 - 16xy^3$
57.	**a.** $2mn(m^2 - mn + 5n^2)$		**b.** $2m^3n - 2m^2n^2 + 10mn^3$
58.	**a.** $3ab(a^2 + 2ab - 6b^2)$		**b.** $3a^3b + 6a^2b^2 - 18ab^3$
59.	**a.** $x(x - 3) + 2(x - 3)$		**b.** $x(x - 3) + 2(x - 3)$
60.	**a.** $x(x + 4) - 5(x + 4)$		**b.** $x(x + 4) - 5(x + 4)$
61.	**a.** $3x(5x - 2) - 4(5x - 2)$		**b.** $3x(5x - 2) - 4(5x - 2)$
62.	**a.** $(2x + 7)(5x) +$ $(2x + 7)(3)$		**b.** $(2x + 7)(5x) +$ $(2x + 7)(3)$

Modeling the Revenue of a Business

In Exercises 63–66 write a revenue polynomial $R(x)$ for each problem. The price per unit is x and the number of units that will sell at that price is given by the polynomial in the second column. Then evaluate $R(50)$ for each problem.

	PRICE	NUMBER OF UNITS THAT WILL SELL
63.	x	$280 - 2x$
64.	x	$700 - 5x$
65.	x	$1000 - 0.5x$
66.	x	$5000 - 0.4x$

In Exercises 67–74 multiply the polynomial factors.

67. $(5 - 2x^2)(3x^2 + 4)$ **68.** $(2 - x)(x - x^2 + 3)$

69. $(x + 4 - x^2)(5 - x + x^2)$ **70.** $(3 + 2x - x^2)(4x - x^2 + 1)$

71. $(v + 5)^2$ **72.** $(v - 6)^2$

73. $(x + 2)^3$ **74.** $(x - 5)^3$

Multiple Representations

In Exercises 75–78 write each verbal statement in algebraic form, and then perform the indicated operation.

75. Find the product of $-5xy$ and $4x - y$.

76. Find the product of $7xy$ and $3x + 2y$.

77. Square the quantity $x + y$.

78. Square the quantity $x - y$.

Modeling the Volume of a Solid Using Polynomials

In Exercises 79 and 80 write a polynomial for the volume of each solid.

79.

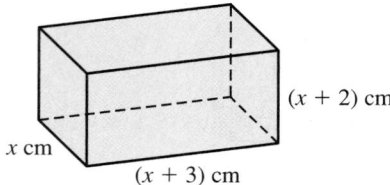

$(x + 2)$ cm

x cm

$(x + 3)$ cm

80.

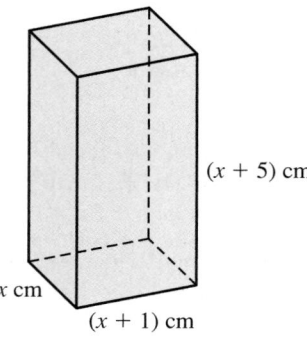

$(x + 5)$ cm

x cm

$(x + 1)$ cm

In Exercises 81–84 use the distributive property to complete the blanks and to write the original polynomial as a product of binomial factors.

81. $9x^2 + 12x + 6x + 8 = (9x^2 + 12x) + (6x + 8)$
$$= 3x(\underline{\hspace{2cm}}) + 2(\underline{\hspace{2cm}})$$
$$= (3x + 2)(\underline{\hspace{2cm}})$$

82. $20x^2 + 8x - 5x - 2 = (20x^2 + 8x) + (-5x - 2)$
$$= (\underline{\hspace{1.5cm}})(4x) + (\underline{\hspace{1.5cm}})(-1)$$
$$= (\underline{\hspace{1.5cm}})(4x - 1)$$

83. $10x^2 + 2xy + 5xy + y^2 = (10x^2 + 2xy) + (5xy + y^2)$
$$= (5x + y)(\underline{\hspace{0.5cm}}) + (5x + y)(\underline{\hspace{0.5cm}})$$
$$= (5x + y)(\underline{\hspace{1.5cm}})$$

84. $3x^2 - 2xy + 21xy - 14y^2 = (3x^2 - 2xy) + (21xy - 14y^2)$
$$= (\underline{\hspace{1cm}})(x) + (\underline{\hspace{1cm}})(7y)$$
$$= (\underline{\hspace{1.5cm}})(x + 7y)$$

Group Discussion Questions

85. Challenge Question   For each rectangular region first write a polynomial that represents the area of the whole rectangle. Then write a polynomial that represents the sum of the areas of all the parts. Compare these polynomials. Are they equal?

Area of Whole Rectangle   Area of Component Parts

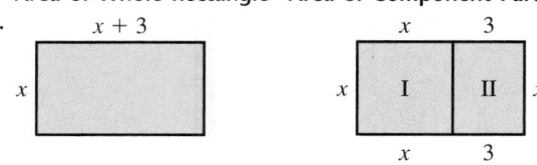

a. $x + 3$ x 3

x x | I | II | x

      x 3

b. $x + 3$ x 3

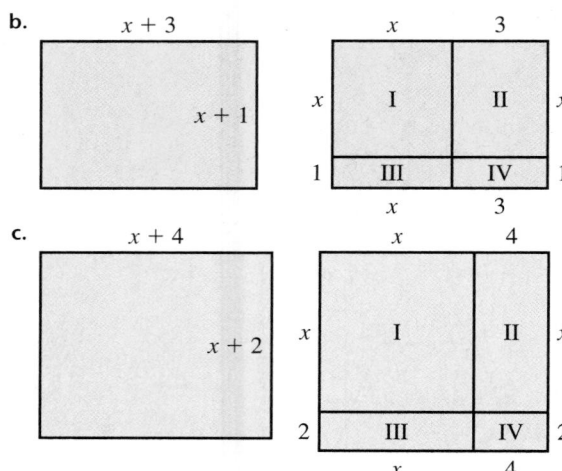

$x + 1$ x | I | II | x
       1 | III | IV | 1
          x 3

c. $x + 4$ x 4

$x + 2$ x | I | II | x
       2 | III | IV | 2
          x 4

86. Challenge Question
  a. Can you give two binomials whose sum is a monomial?
  b. Can you give two binomials whose sum is a binomial?
  c. Can you give two binomials whose sum is a trinomial?
  d. Can you give two binomials whose product is a trinomial?
  e. Can you give two binomials whose product is a binomial?

87. Discovery Question
  a. Complete each product.
    i. $(x + 1)(x - 1) = $ _____
    ii. $(x + 2)(x - 2) = $ _____
    iii. $(x + 3)(x - 3) = $ _____
    iv. $(x + 4)(x - 4) = $ _____
    v. $(x + 5)(x - 5) = $ _____
  b. Make a general conjecture based on this work. State this conjecture in algebraic form and be prepared to describe this conjecture verbally to your instructor.
  c. Test this conjecture on $(x + 12)(x - 12)$.
  d. Test this conjecture on $(2v + 3)(2v - 3)$.

88. Challenge Question   Expand the product of the factors $(x - a)(x - b)(x - c) \cdot \ldots \cdot (x - y)(x - z)$.

Section 5.6   Using Division to Factor Polynomials

Objectives:

11. Divide polynomials.
12. Use division to factor a polynomial.

A Mathematical Note

George Boole (1815–1864), an English mathematician, created a whole new area of algebra now called Boolean algebra. Boolean algebra plays a key role in the logic and design of computer circuitry.

This section examines the division of polynomials. We build on our familiarity with the division of real numbers and monomials. Recall that $\frac{15}{3} = 5$ because 15 equals the product of the factors 3 and 5. Thus we also use division to express some polynomials as a product of factors. This helps us to prepare for more work with factoring in later sections, including Chapter 7.

We start by reviewing the quotient of two monomials which was first considered in Section 5.2. Recall that the quotient rule states that $\frac{x^m}{x^n} = x^{m-n}$ for $x \neq 0$. We assume that throughout this section the variables are restricted to values that avoid division by 0.

In this example we must have $x \neq 0$ and $y \neq 0$ to avoid division by 0.

■ **EXAMPLE 1**   Dividing a Monomial by a Monomial

Simplify $\frac{8x^2y^5}{4xy^3}$ and then check your answer by multiplying the divisor and the quotient.

SOLUTION _____

Note that the coefficients are divided and the exponents are subtracted.

$$\frac{8x^2y^5}{4xy^3} = \frac{2x^{2-1}y^{5-3}}{1}$$ Divide both the numerator and the denominator by 4. Using the quotient rule for exponents, subtract the exponents on the common bases.

$$= 2xy^2$$ This is the quotient.

Check: $(4xy^3)(2xy^2) = 8x^2y^5$ The answer checks because $8x^2y^5$ is expressed as the product of two factors, the divisor and the quotient. ■

The quotient of a polynomial divided by a monomial often is written in fractional form. Since $\frac{a+b}{c} = \frac{a}{c} + \frac{b}{c}$ $(c \neq 0)$, each term of the polynomial in the numerator is divided by the monomial in the denominator. Then each fraction is written in the reduced form.

Dividing a Polynomial by a Monomial

To divide a polynomial by a monomial, divide each term of the polynomial by the monomial.

■ **EXAMPLE 2**   Factoring Out the GCF of a Binomial

Factor $15x^3 + 12x^2$ by dividing $15x^3 + 12x^2$ by the greatest common factor of both terms, $3x^2$. Then multiply these factors to check your factorization.

SOLUTION _____

In this example we must have $x \neq 0$ to avoid division by 0.

$$\frac{15x^3 + 12x^2}{3x^2} = \frac{15x^3}{3x^2} + \frac{12x^2}{3x^2}$$ Divide each term of the numerator by $3x^2$ and then simplify.

$$= 5x + 4$$ This quotient is the other factor.

$$15x^3 + 12x^2 = 3x^2(5x + 4)$$ This is the factorization of $15x^3 + 12x^2$.

Check:

$$3x^2(5x + 4) = (3x^2)(5x) + (3x^2)(4)$$ $15x^3 + 12x^2$ is the product of the factors $3x^2$ and $5x + 4$.

$$= 15x^3 + 12x^2$$ ■

Some divisions do not have a zero remainder. In fact, the quotient $-6y^2 + 9y - \dfrac{4}{y}$ in Example 3 is not even a polynomial $\left(-\dfrac{4}{y}\text{ has a variable in the denominator}\right)$. Nonetheless, the answers to all division problems can be checked by multiplication, as illustrated in Example 3.

■ **EXAMPLE 3**   Dividing a Trinomial by a Monomial

Find $(20y + 30y^4 - 45y^3) \div (-5y^2)$ and check your result.

SOLUTION _____

In this example we must have $y \neq 0$ to avoid division by 0.

$$\frac{30y^4 - 45y^3 + 20y}{-5y^2} = \frac{30y^4}{-5y^2} - \frac{45y^3}{-5y^2} + \frac{20y}{-5y^2}$$ First write the division in fractional form, expressing the numerator in descending order.

$$= -6y^2 + 9y - \frac{4}{y}$$ Divide each term of the numerator by $-5y^2$ and then simplify.

Check:

$$(-5y^2)\left(-6y^2 + 9y - \frac{4}{y}\right)$$ The answer to a division problem can be checked by multiplication.

$$= (-5y^2)(-6y^2) + (-5y^2)(9y) + (-5y^2)\left(-\frac{4}{y}\right)$$ Distribute the multiplication of $-5y^2$ to each term and then simplify.

$$= 30y^4 - 45y^3 + 20y$$ ■

The procedure for dividing one polynomial by another polynomial with two or more terms is very similar to the long division procedure used to divide integers. We refer to this as long division of polynomials. To use this procedure, be sure first to write both polynomials in standard form.

SELF-CHECK 5.6.1

1. Divide $21x^3y^2$ by $3x^2y^2$ and then write $21x^3y^2$ as a product of $3x^2y^2$ and another factor.

2. Simplify $\dfrac{10x^3y + 4x^2y^2}{2x^2y}$ and then write $10x^3y + 4x^2y^2$ as a product of $2x^2y$ and this quotient.

3. Reduce $\dfrac{18x^3y^2}{45xy^5}$ to its lowest terms.

Long Division of Polynomials

STEP 1. Write the polynomials in long division format, expressing each in standard form.

STEP 2. Divide the first term of the divisor into the first term of the dividend. The result is the first term of the quotient.

STEP 3. Multiply the first term of the quotient times every term in the divisor, and write this product under the dividend, aligning like terms.

STEP 4. Subtract this product from the dividend and bring down the next term.

STEP 5. Use the result of step 4 as a new dividend and repeat steps 2–4 until either the remainder is 0 or the degree of the remainder is less than the degree of the divisor.

To illustrate the close relationship between the long division of integers and the long division of polynomials, we compare these procedures side by side. Before examining this comparison, note that if 10 is substituted for x in $(6x^2 + 7x + 2) \div (2x + 1)$, we obtain $[6(10)^2 + 7(10) + 2] \div [2(10) + 1]$, or $(600 + 70 + 2) \div (20 + 1) = 672 \div 21$. Thus for this value of x both problems denote the same thing.

Long Division of Integers
Problem: Divide 672 by 21.
Step 1. Write the division in the long division format.

$21\overline{)672}$

Step 2. Divide to obtain the first term in the quotient.

$\begin{array}{r} 3 \\ 21\overline{)672} \end{array}$

Long Division of Polynomials
Problem: Divide $6x^2 + 7x + 2$ by $2x + 1$.

$2x + 1\overline{)6x^2 + 7x + 2}$

$\begin{array}{r} 3x \\ 2x + 1\overline{)6x^2 + 7x + 2} \end{array}$ $6x^2$ divided by $2x$ is $3x$.

SELF-CHECK 5.6.1 ANSWERS

1. $7x$; $21x^3y^2 = (3x^2y^2)(7x)$
2. $5x + 2y$; $10x^3y + 4x^2y^2 = 2x^2y(5x + 2y)$
3. $\dfrac{2x^2}{5y^3}$

Step 3. Multiply the first term in the quotient times the divisor.

$$\begin{array}{r} 3 \\ 21\overline{)672} \\ 63 \end{array}$$

$$\begin{array}{r} 3x \\ 2x+1\overline{)6x^2+7x+2} \\ 6x^2+3x \end{array}$$

Note the alignment of similar terms.

Step 4. Subtract this product from the dividend and bring down the next term.

$$\begin{array}{r} 3 \\ 21\overline{)672} \\ 63 \\ \hline 42 \end{array}$$

$$\begin{array}{r} 3x \\ 2x+1\overline{)6x^2+7x+2} \\ 6x^2+3x \\ \hline 4x+2 \end{array}$$

Step 5. Repeat steps 2–4 to obtain the next term in the quotient.

$$\begin{array}{r} 32 \\ 21\overline{)672} \\ 63 \\ \hline 42 \\ 42 \\ \hline 0 \end{array}$$

$$\begin{array}{r} 3x+2 \\ 2x+1\overline{)6x^2+7x+2} \\ 6x^2+3x \\ \hline 4x+2 \\ 4x+2 \\ \hline 0 \end{array}$$

Because the remainder is 0, the division is finished.

Answer: $672 \div 21 = 32$ \qquad $(6x^2+7x+2) \div (2x+1) = 3x+2$

We can use the fact that $12 \div 4 = 3$ to write $12 = 4 \cdot 3$. In Example 4 we use the same reasoning to write $2x^2 + x - 15$ in factored form.

■ EXAMPLE 4 Using Long Division to Factor a Trinomial

In this example we must have $x \neq -3$ so that the divisor $x + 3$ is not 0.

Use long division to complete this factorization: $2x^2 + x - 15 = (x + 3)(\quad)$.

SOLUTION ————————————

① $x + 3\overline{)2x^2 + x - 15}$

Set up the format for long division, writing both polynomials in standard form.

② $\begin{array}{r} 2x \\ x+3\overline{)2x^2+x-15} \end{array}$

Divide $2x^2$ (the first term of the dividend) by x (the first term of the divisor) to obtain $2x$ (the first term of the quotient). Align similar terms.

③ $\begin{array}{r} 2x \\ x+3\overline{)2x^2+x-15} \\ 2x^2+6x \end{array}$

Multiply $2x$ times every term in the divisor, aligning under similar terms in the dividend.

④ $\begin{array}{r} 2x \\ x+3\overline{)2x^2+x-15} \\ 2x^2+6x \\ \hline -5x-15 \end{array}$

Subtract $2x^2 + 6x$ from the dividend by changing the sign of each term and adding.

$$\begin{array}{r} 2x - 5 \\ x + 3 \overline{)2x^2 + x - 15} \\ \underline{2x^2 + 6x} \\ -5x - 15 \\ \underline{-5x - 15} \\ 0 \end{array}$$

5   Divide $-5x$ (the first term in the last row) by x (the first term of the divisor) to obtain -5 (the next term in the quotient). Then multiply each term of the divisor by -5, aligning under similar terms in the dividend. Subtract to obtain the remainder of 0.

$$\boxed{(2x^2 + x - 15)} \div \boxed{(x + 3)} = \boxed{2x - 5}$$

Dividend   Divisor   Quotient

The quotient $2x - 5$ is the other factor of $2x^2 + x - 15$.

Answer: $2x^2 + x - 15 = (x + 3)(2x - 5)$

■ EXAMPLE 5   Using Division to Factor a Polynomial

In this example we must have $v \neq \dfrac{2}{3}$ so that the divisor $3v - 2$ is not 0.

Divide $(7v + 6v^3 + 2 - 19v^2)$ by $3v - 2$ and write the dividend, $(7v + 6v^3 + 2 - 19v^2)$, in factored form.

SOLUTION

$$\begin{array}{r} 2v^2 \\ 3v - 2 \overline{)6v^3 - 19v^2 + 7v + 2} \\ \underline{6v^3 - 4v^2} \\ -15v^2 + 7v \end{array}$$

Write both polynomials in standard form. Divide to obtain the first term in the quotient, $\dfrac{6v^3}{3v} = 2v^2$. Multiply $2v^2$ times $3v - 2$ to obtain $6v^3 - 4v^2$ and then subtract.

$$\begin{array}{r} 2v^2 - 5v \\ 3v - 2 \overline{)6v^3 - 19v^2 + 7v + 2} \\ \underline{6v^3 - 4v^2} \\ -15v^2 + 7v \\ \underline{-15v^2 + 10v} \\ -3v + 2 \end{array}$$

Divide to obtain the second term in the quotient, $-\dfrac{15v^2}{3v} = -5v.$

Multiply $-5v$ times $3v - 2$ to obtain $-15v^2 + 10v$ and then subtract.

$$\begin{array}{r} 2v^2 - 5v - 1 \\ 3v - 2 \overline{)6v^3 - 19v^2 + 7v + 2} \\ \underline{6v^3 - 4v^2} \\ -15v^2 + 7v \\ \underline{-15v^2 + 10v} \\ -3v + 2 \\ \underline{-3v + 2} \\ 0 \end{array}$$

Divide to obtain the third term in the quotient, $\dfrac{-3v}{3v} = -1.$

Multiply -1 times $3v - 2$ to obtain $-3v + 2$ and then subtract.

$$\dfrac{6v^3 - 19v^2 + 7v + 2}{3v - 2} = 2v^2 - 5v - 1$$

This quotient is the other factor of $6v^3 - 19v^2 + 7v + 2$.

The answer to a division problem can be checked by multiplication.

Answer: $6v^3 - 19v^2 + 7v + 2 = (3v - 2)(2v^2 - 5v - 1)$

If we divide 22 by 5, we get 4 with a remainder of 2. Thus

$$\frac{22}{5} = 4 + \frac{2}{5}.$$

Because 5 does not divide 22 evenly, we say that 5 is not a factor of 22. In Example 6 we divide two polynomials that produce a nonzero remainder.

■ EXAMPLE 6   Dividing a Polynomial with a Nonzero Remainder

Divide $x^2 + 2x - 13$ by $x + 5$.

SOLUTION ─────────────────────────────────

$$
\begin{array}{r}
x - 3 \\
x + 5 \overline{)x^2 + 2x - 13} \\
\underline{x^2 + 5x} \\
-3x - 13 \\
\underline{-3x - 15} \\
2
\end{array}
$$

Divide x^2 by x to obtain x.
Multiply x by $x + 5$ to obtain $x^2 + 5x$ and then subtract.
Divide $-3x$ by x to obtain -3.
Multiply -3 by $x + 5$ to obtain $-3x - 15$ and then subtract to obtain the remainder of 2.

Answer: $\dfrac{x^2 + 2x - 13}{x + 5} = x - 3 + \dfrac{2}{x + 5}$

The quotient is expressed as $x - 3$ plus the remainder, 2, divided by $x + 5$.

Check: $(x + 5)\left(x - 3 + \dfrac{2}{x + 5}\right)$

To check, we multiply the divisor $x + 5$ by the quotient $x - 3 + \dfrac{2}{x + 5}$.

$$= (x + 5)(x - 3) + (\cancel{x + 5})\left(\frac{2}{\cancel{x + 5}}\right)$$
$$= (x)(x - 3) + (5)(x - 3) + 2$$
$$= x^2 - 3x + 5x - 15 + 2$$
$$= x^2 + 2x - 13$$

■

The first step in the long division procedure is to write the polynomials in standard form. Some polynomials have missing terms. For example, the polynomial $x^3 - 5x - 14$ is missing the x^2 term. This polynomial can be rewritten as $x^3 + 0x^2 - 5x - 14$. This format can help to prevent careless errors in the long division procedure.

■ EXAMPLE 7   Dividing a Polynomial with Missing Terms

Divide $x^3 - 5x - 14$ by $x - 3$.

SOLUTION _____

In this example we must have $x \neq 3$ so that the divisor $x - 3$ is not 0.

$$
\begin{array}{r}
x^2 + 3x + 4 \\
x - 3 \overline{)x^3 + 0x^2 - 5x - 14} \\
\underline{x^3 - 3x^2} \\
3x^2 - 5x \\
\underline{3x^2 - 9x} \\
4x - 14 \\
\underline{4x - 12} \\
-2
\end{array}
$$

Use 0 as the coefficient of the missing x^2 term.
Divide x^3 by x to obtain x^2.
Multiply x^2 times $x - 3$ to obtain $x^3 - 3x^2$ and then subtract.
Divide $3x^2$ by x to obtain $3x$. Multiply $3x$ times $x - 3$ to obtain $3x^2 - 9x$ and then subtract.
Divide $4x$ by x to obtain 4. Multiply 4 times $x - 3$ to obtain $4x - 12$.
Subtract to obtain the remainder of -2.

Answer: $\dfrac{x^3 - 5x - 14}{x - 3} =$

Because the remainder is not 0, $x - 3$ is not a factor of $x^3 - 5x - 14$.

$$x^2 + 3x + 4 - \dfrac{2}{x - 3}$$

In the next example we examine a problem with a second-degree trinomial as a divisor.

■ EXAMPLE 8   Dividing by a Second-Degree Trinomial

Divide $x^4 - x^3 + 6x^2 - 7x + 15$ by $x^2 - 2x + 3$ and write the dividend, $x^4 - x^3 + 6x^2 - 7x + 15$, in factored form.

SOLUTION _____

$$
\begin{array}{r}
x^2 + x + 5 \\
x^2 - 2x + 3 \overline{)x^4 - x^3 + 6x^2 - 7x + 15} \\
\underline{x^4 - 2x^3 + 3x^2} \\
x^3 + 3x^2 - 7x \\
\underline{x^3 - 2x^2 + 3x} \\
5x^2 - 10x + 15 \\
\underline{5x^2 - 10x + 15} \\
0
\end{array}
$$

Divide x^4 by x^2 to obtain the first term of x^2 in the quotient.
Multiply x^2 times $x^2 - 2x + 3$ to obtain $x^4 - 2x^3 + 3x^2$ and then subtract.
Divide x^3 by x^2 to obtain x.
Multiply x times $x^2 - 2x + 3$ to obtain $x^3 - 2x^2 + 3x$ and then subtract.
Divide $5x^2$ by x^2 to obtain 5.
Multiply 5 times $x^2 - 2x + 3$ to obtain $5x^2 - 10x + 15$ and then subtract to produce the remainder of 0.

Answer: $x^4 - x^3 + 6x^2 - 7x + 15 = (x^2 - 2x + 3)(x^2 + x + 5)$

SELF-CHECK 5.6.2 ANSWERS

1. $12x^2 + x - 35 = (3x - 5)(4x + 7)$
2. $4x^2 - 9 = (2x + 3)(2x - 3)$

SELF-CHECK 5.6.2

Use long division to complete each factorization.

1. $(12x^2 + x - 35) = (3x - 5)(\quad)$
2. $4x^2 - 9 = (2x + 3)(\quad)$

Equal polynomials have identical graphs and identical tables of values. In Example 9 we use tables and graphs to check our division.

■ EXAMPLE 9   Using Tables and Graphs to Compare Two Polynomials

Simplify $\dfrac{8x^2 - 10x - 3}{4x + 1}$ and check this quotient by using tables and graphs.

SOLUTION _____

ALGEBRAICALLY

In this example we must have $x \neq -\dfrac{1}{4}$ so that the divisor is not 0.

$$
\begin{array}{r}
2x - 3 \\
4x + 1 \overline{)8x^2 - 10x - 3} \\
\underline{8x^2 + 2x} \\
-12x - 3 \\
\underline{-12x - 3} \\
0
\end{array}
$$

$$\frac{8x^2 - 10x - 3}{4x + 1} = 2x - 3$$

Write both polynomials in the long division format. Divide $8x^2$ by $4x$ to obtain $2x$, the first term in the quotient. Multiply $2x$ times $4x + 1$ to obtain $8x^2 + 2x$ and then subtract. Divide $-12x$ by $4x$ to obtain -3, the second term in the quotient. Multiply -3 times $4x + 1$ to obtain $-12x - 3$ and then subtract. A remainder of 0 means the division is exact.

NUMERICALLY

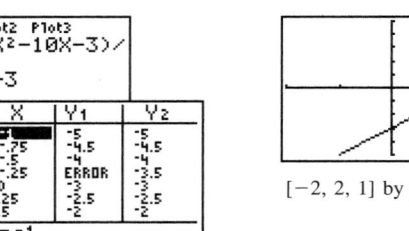

GRAPHICALLY

$[-2, 2, 1]$ by $[-5, 5, 1]$

Let y_1 represent the fraction and y_2 represent the quotient. Both the table of values and the graphs are identical except at $x = -0.25$ which is the value that causes division by 0.

Answer: $\dfrac{8x^2 - 10x - 3}{4x + 1} = 2x - 3$ for $x \neq -\dfrac{1}{4}$. For $x = -\dfrac{1}{4}$ the fraction is undefined because of division by 0.   ■

SELF-CHECK 5.6.3

Simplify $\dfrac{5x^2 + 9x - 2}{5x - 1}$ and check this quotient by using tables and graphs.

SELF-CHECK 5.6.3 ANSWERS

For $x \neq \dfrac{1}{5}$, $\dfrac{5x^2 + 9x - 2}{5x - 1}$

$= x + 2$. This answer is supported by the table and graphs shown here.

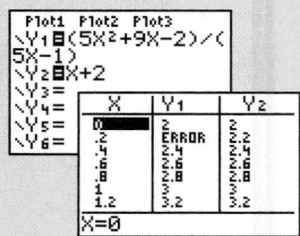

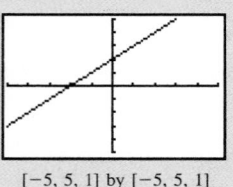

$[-5, 5, 1]$ by $[-5, 5, 1]$

USING THE LANGUAGE AND SYMBOLISM OF MATHEMATICS   5.6

1. In the division $\frac{15}{3} = 5$ the dividend is 15, the divisor is 3, and the _____ is 5.

2. In the division $\frac{x^2 - x - 6}{x - 3} = x + 2$ the dividend is $x^2 - x - 6$, the divisor is _____, and the _____ is $x + 2$.

3. The division $\frac{35}{5} = 7$ can be checked by writing $5 \cdot 7 = 35$. In the expression $5 \cdot 7 = 35$, 5 and 7 are factors and 35 is their _____.

4. The division $\frac{x^2 - x - 6}{x - 3} = x + 2$ can be checked by writing $(x - 3)(x + 2) = x^2 - x - 6$. In the expression $(x - 3)(x + 2) = x^2 - x - 6$ the factors are _____ and _____ and the product is _____.

5. The first step in the long division procedure for dividing polynomials is to write both the divisor and the dividend in _____ form.

6. When using the long division procedure for dividing polynomials, we continue the steps until either the remainder is _____ or the degree of the remainder is less than the degree of the _____.

7. The polynomial $3x^4 + 5x^2 - 7x + 2$ is said to have a missing x^3 term. If this polynomial is rewritten with an x^3 term, then the coefficient of x^3 will be _____.

EXERCISES   5.6

All exercises are assumed to restrict the variables to values that avoid division by 0.

In Exercises 1–14 find each quotient.

1. $\frac{18x^3}{6x}$

2. $\frac{24x^5}{3x^2}$

3. $\frac{35a^3b^2}{-7ab}$

4. $\frac{48a^4b^4}{-8ab^2}$

5. $\frac{15a^2 - 20a}{5a}$

6. $\frac{24a^5 - 16a^3}{4a}$

7. $\frac{9a^3 - 15a^2 + 6a}{3a^2}$

8. $\frac{24a^5 - 6a^4 - 12a^2}{6a^3}$

9. $\frac{16m^4 - 8m^3 + 10m^2 + 6m}{2m}$

10. $\frac{24m^4 - 18m^3 + 36m^2 - 6m}{3m}$

11. $\frac{45x^5y^2 - 54x^4y^5 + 99x^2y^4}{9x^2y^2}$

12. $\frac{-77x^7y^2 + 55x^5y^4 - 33x^3y^6}{11x^3y^2}$

13. $\frac{54v^2 - 36v^4 + 12v - 18v^3}{-6v}$

14. $\frac{15v^3 - 6v + 9v^4 + 27v^2}{-3v}$

In Exercises 15–28 find the quotient and write the dividend in the factored form. (*Hint:* See Example 8.)

15. $(35v^3w - 7v^2w - 28vw^2) \div 7vw$

16. $(54m^3n - 18m^2n + 27mn^2) \div 9mn$

17. $(100x^{20} - 50x^{12} + 30x^9) \div 10x^9$

18. $(66x^{21} - 48x^{14} - 18x^7) \div 6x^7$

19. $x + 2 \overline{)x^2 + 9x + 14}$

20. $x + 5 \overline{)x^2 + 8x + 15}$

21. $\frac{v^2 + 2v - 24}{v - 4}$

22. $\frac{v^2 - 2v - 35}{v - 7}$

23. $\frac{6w^2 + w - 12}{2w + 3}$

24. $\frac{12w^2 + 14w - 10}{3w + 5}$

25. $\frac{20m^2 - 43m + 14}{5m - 2}$

26. $\frac{28m^2 - 27m + 5}{4m - 1}$

27. $\frac{63y^2 - 130y + 63}{7y - 9}$

28. $\frac{32y^2 + 36y - 35}{8y - 5}$

In Exercises 29 and 30 perform the indicated division. (*Hint:* See Example 7.)

29. $\frac{4m^3 - 9m^2 + 10m + 7}{m - 3}$

30. $\frac{6m^3 - 19m^2 - 23m + 15}{m - 4}$

In Exercises 31–50 use division to complete each factorization.

31. $-96b^3 - 84b^2 = -12b(\ )$

32. $-81b^4 + 63b^3 = -9b(\ )$

33. $63v^4 - 21v^3 = 3v^2(\ )$

34. $40v^3 - 35v^2 = 5v^2(\ )$

35. $21b^3 - 14b^2 + 63b = 7b(\ )$

36. $72b^5 - 63b^4 - 9b^2 = 9b^2(\ )$

37. $36m^3n^3 - 40m^2n^4 = 4m^2n^3(\ )$

38. $-30m^5n^5 + 75m^4n^6 = -5m^4n^5(\ )$

39. $30x^2 - 45x^3 + 10x + 35x^4 = 5x(\ )$

40. $14x^2 - 35x^3 + 28x^4 - 42x = 7x(\ )$

41. $x^2 - 12x + 35 = (x - 5)(\ )$

42. $x^2 - 16x + 60 = (x - 6)(\ )$

43. $6a^2 - 11a - 10 = (2a - 5)(\quad)$
44. $15a^2 - 19a - 10 = (5a + 2)(\quad)$
45. $3v^3 - 4v^2 - 8 = (\quad)(v - 2)$
46. $4v^3 - 6v^2 - v - 51 = (\quad)(v - 3)$
47. $x^3 - 8 = (x - 2)(\quad)$
48. $x^3 + 125 = (x + 5)(\quad)$
49. $20y^3 + 16y^2 - 3y - 54 = (5y - 6)(\quad)$
50. $18y^3 - 63y^2 + 106y - 55 = (6y - 5)(\quad)$

51. Area of a Rectangle The area of the rectangle shown here is $4x^2 + 4x - 3$ cm². Find the width of this rectangle.

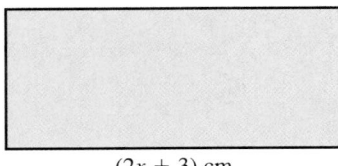

$(2x + 3)$ cm

52. Area of a Rectangle The area of the rectangle shown here is $9x^2 - 1$ cm². Find the width of this rectangle.

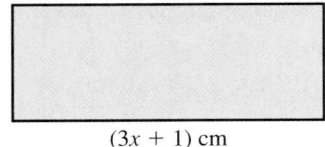

$(3x + 1)$ cm

53. Area of a Triangle The area of the triangle shown here is $12x^2 - 17x - 7$ cm². Find the altitude of this triangle as shown by the dashed line.

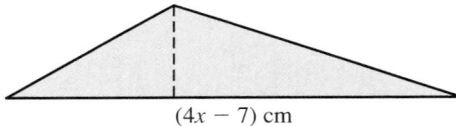

$(4x - 7)$ cm

54. Area of a Triangle The area of the triangle shown here is $5x^2 + 13x + 6$ cm². Find the altitude of this triangle as shown by the dashed line.

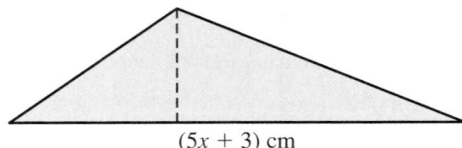

$(5x + 3)$ cm

55. Volume of a Box The volume of the box shown here is $8x^3 + 14x^2 + 3x$ cm³. Find the height of this box.

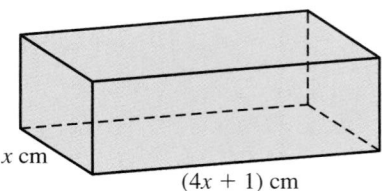

x cm

$(4x + 1)$ cm

56. Volume of a Box The volume of the box shown here is $18x^3 + 33x^2 + 5x$ cm³. Find the height of this box.

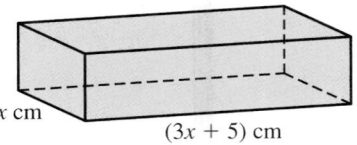

x cm

$(3x + 5)$ cm

In Exercises 57 and 58 use a graphics calculator to fill in the table of values and to graph $y_1 = P_1(x)$ and $y_2 = P_2(x)$ using the indicated window. Do these results indicate that $P_1(x) = P_2(x)$?

57.
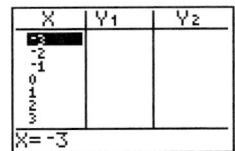

Table:

X	Y₁	Y₂
0		
.5		
1		
1.5		
2		
2.5		
3		

X=0

Graph: Use $[-2, 6, 1]$ by $[-2, 6, 1]$.

58.

Ploti Plot2 Plot3
\Y1⬛(X^3-8)/(X²+
2X+4)
\Y2⬛X-2
\Y3=
\Y4=
\Y5=
\Y6=

Table:

X	Y₁	Y₂
-3		
-2		
-1		
0		
1		
2		
3		

X=-3

Graph: Use $[-4.7, 4.7, 1]$ by $[-3.1, 3.1, 1]$.

59. Entries in a Computer Array A two-dimensional array in a computer program has $30k^2 + 39k + 12$ entries. If there are $6k + 3$ rows in the array, how many columns are in the array?

60. Entries in a Computer Array A two-dimensional array in a computer program has $4k^2 + 81k + 20$ entries. If there are $k + 20$ columns in the array, how many rows are in the array?

61. What polynomial, when divided by $x + 3$, yields a quotient of $x^2 - 3x - 5$?

62. What polynomial, when divided by $x - 7$, yields a quotient of $2x^2 + x - 6$?

63. What polynomial, when divided into $6x^2 - 2x - 20$, produces a quotient of $2x - 4$?

64. What polynomial, when divided into $20x^2 + 7x - 6$, produces a quotient of $5x - 2$?

65. a. If a fourth-degree trinomial in x is added to a second-degree binomial in x, the degree of the sum is

_____ .

b. If a fourth-degree trinomial in x is multiplied by a second-degree binomial in x, the degree of the product is _____ .

c. If a fourth-degree trinomial in x is divided by a second-degree binomial in x, the degree of the quotient is

_____ .

66. a. If a third-degree trinomial in x is added to a second-degree binomial in x, the degree of the sum is

_____ .

b. If a third-degree trinomial in x is multiplied by a second-degree binomial in x, the degree of the product is _____ .

c. If a third-degree trinomial in x is divided by a second-degree binomial in x, the degree of the quotient is

_____ .

67. When a polynomial is divided by $2x + 3$, the result is $x + 2 + \dfrac{4}{2x + 3}$. Determine this polynomial.

68. When a polynomial is divided by $3x - 2$, the result is $2x + 1 + \dfrac{2}{3x - 2}$. Determine this polynomial.

Modeling the Average Cost per Unit

In Exercises 69–72 determine $A(t)$, the average cost per unit for the units produced by an assembly line in t hours. Use the formula $A(t) = C(t) \div N(t)$, where $C(t)$ is the total cost of operating an assembly line for t hours and $N(t)$ is the number of units produced in t hours.

69. $C(t) = 3t^2 - 5t - 2;\quad N(t) = 3t + 1$

70. $C(t) = t^3 + 27;\quad N(t) = t + 3$

71. $C(t) = t^3 + 64;\quad N(t) = t + 4$

72. $C(t) = t^4 + 4t^2 - 5;\quad N(t) = t^2 + 5$

Group Discussion Questions

73. Challenge Question

a. Can you give two binomials whose quotient is a monomial?

b. Can you give two binomials whose quotient is a binomial?

c. Can you give two binomials whose quotient is a trinomial?

74. Discovery Question

a. Complete each product.

i. $(x + 1)^2 =$ _____

ii. $(x + 2)^2 =$ _____

iii. $(x + 3)^2 =$ _____

iv. $(2x + 5)^2 =$ _____

v. $(3x + 1)^2 =$ _____

b. Make a general conjecture based on this work. State this conjecture in algebraic form using $(x + y)^2$. Be prepared to describe this conjecture verbally to your instructor.

c. Test this conjecture on $(5x + 3)^2$.

d. Use the following figure to explain your conjecture geometrically.

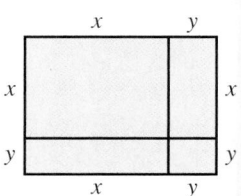

75. Discovery Question

a. Complete each of these factorizations.

i. $x^2 - 1 = (x + 1)(\ \ )$

ii. $x^2 - 4 = (x + 2)(\ \ )$

iii. $x^2 - 9 = (\ \ )(x - 3)$

iv. $x^2 - 16 = (\ \ )(x - 4)$

v. $x^2 - 25 = (\ \ )(\ \ )$

b. Make a general conjecture based on this work. State this conjecture in algebraic form using $x^2 - y^2$. Be prepared to describe this conjecture verbally to your instructor.

c. Test this conjecture by factoring $x^2 - 169$.

d. Use the following figure to explain your conjecture geometrically.

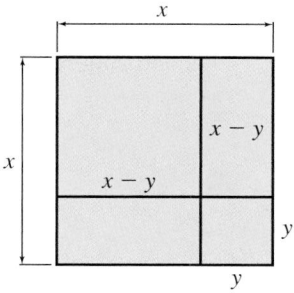

76. Discovery Question

a. Complete each of these factorizations.

i. $x^2 + 2x + 1 = (x + 1)(\qquad)$

ii. $x^2 + 4x + 4 = (x + 2)(\qquad)$

iii. $x^2 + 6x + 9 = (x + 3)(\qquad)$

iv. $x^2 + 8x + 16 = (x + 4)(\qquad)$

v. $x^2 + 10x + 25 = (\ \ )(\qquad)$

b. Make a general conjecture based on this work. State this conjecture in algebraic form using $x^2 + 2xy + y^2$. Be prepared to describe this conjecture verbally to your instructor.

c. Test this conjecture by factoring $x^2 + 16x + 64$.

d. Compare this result to the conjecture you developed in Exercise 74.

Section 5.7   Special Products and Factors

Objectives:
13. Multiply by inspection a sum times a difference and the square of a binomial.
14. Factor by inspection a difference of two perfect squares and perfect square trinomials.

In this section we present a very useful topic on multiplying and factoring some polynomial patterns that occur frequently. The recognition of patterns either algebraically, numerically, or graphically is an important part of mathematics. Once patterns are recognized, we often can take advantage of these patterns to shorten our work or to gain insights on the problems we are examining.

A thorough treatment of factoring is presented in Chapter 7. However, it is important to note some factoring was done in Example 6 of Section 1.6 when the distributive property was introduced. Factoring also was examined in Section 5.5 when the multiplication of polynomials was introduced. The special forms introduced in this section to facilitate multiplication are just as useful for factoring. The key point is that multiplying and factoring are two different uses of the same set of algebraic information.

We will examine factoring in greater depth in Chapters 6 and 7. There we will examine numerical and graphical perspectives on factoring and also will factor polynomials that do not exhibit the nice patterns shown in this section.

Factoring is the process of rewriting an algebraic expression as a product of its factors. Factoring is the reverse process of multiplying.

Multiplying

$(x + 1)(x - 1) = x^2 - 1$

Factoring

$x^2 - 1 = (x + 1)(x - 1)$

The factors of a polynomial break down the polynomial into component parts which literally reveal the "genetic code" of the polynomial and thus play a key role in much of our further work with polynomials.

A Difference of Two Squares

The products and factors shown next are problems repeated from the previous two sections. By writing them together, we want to observe the pattern shown and then to take advantage of this pattern.

Multiplying

$(x + 2)(x - 2) = x^2 - 4$
$(x + 3)(x - 3) = x^2 - 9$
$(x + 4)(x - 4) = x^2 - 16$
$(x + 5)(x - 5) = x^2 - 25$
$(x + y)(x - y) = x^2 - y^2$

Factoring

$x^2 - 4 = (x + 2)(x - 2)$
$x^2 - 9 = (x + 3)(x - 3)$
$x^2 - 16 = (x + 4)(x - 4)$
$x^2 - 25 = (x + 5)(x - 5)$
$x^2 - y^2 = (x + y)(x - y)$

One advantage of recognizing these special factors is that we can perform the multiplication by inspection without applying the distributive property step by step. Note that this product is a binomial with no middle term.

Difference of Two Squares		
ALGEBRAICALLY	**VERBALLY**	**ALGEBRAIC EXAMPLE**
$(x + y)(x - y) = x^2 - y^2$	The product of a sum times a difference is a difference of two squares.	$(x + 11)(x - 11) = x^2 - 121$
$x^2 - y^2 = (x + y)(x - y)$	The difference of two squares factors as a sum times a difference.	$x^2 - 49 = (x + 7)(x - 7)$

This relationship is illustrated geometrically by the areas of the figures shown below.

Factoring: A Geometric Viewpoint

The difference between the areas of two squares with sides of x and y is $x^2 - y^2$. This is illustrated by the L-shaped region at the end of the first row of figures. This region consists of two rectangular regions, both of width $x - y$. One has a length x and the other has a length y. These two rectangles can be combined to form a rectangle with a width $x - y$ and a length $x + y$. The area of this rectangle is $(x - y)(x + y)$.

Thus $x^2 - y^2 = (x + y)(x - y)$.

■ EXAMPLE 1   Multiplying a Sum Times a Difference

Multiply these binomial factors.

SOLUTIONS _____

(a) $(x + 9)(x - 9)$

$$(x + 9)(x - 9) = (x)^2 - (9)^2$$
$$= x^2 - 81$$

The product is the difference of the squares of x and 9.

(b) $(3v - 8)(3v + 8)$

$$(3v - 8)(3v + 8) = (3v)^2 - 8^2$$
$$= 9v^2 - 64$$

Since multiplication is commutative, this product equals $(3v + 8)(3v - 8)$. The product is the difference of the squares.

(c) $(5x + 7y)(5x - 7y)$

$$(5x + 7y)(5x - 7y) = (5x)^2 - (7y)^2$$
$$= 25x^2 - 49y^2$$

It is not necessary for you to write the extra step shown in these examples. Try to think this step and write only the final product. ■

SELF-CHECK 5.7.1

Multiply these binomial factors.

1. $(w + 12)(w - 12)$
2. $(8v - 1)(8v + 1)$
3. $(2a + 3b)(2a - 3b)$

SELF-CHECK 5.7.1 ANSWERS

1. $w^2 - 144$
2. $64v^2 - 1$
3. $4a^2 - 9b^2$

It is very easy to factor polynomials which are the difference of two perfect squares. The key point is to be sure you are using the correct form before applying this pattern.

■ EXAMPLE 2   Identifying the Difference of Two Squares

Determine which of the following polynomials is the difference of two squares.

SOLUTIONS

(a) $9x^2 - 4y^2$

This binomial fits the form of the difference of two squares as $9x^2 - 4y^2 = (3x)^2 - (2y)^2$.

Note that $9x^2 = (3x)^2$ and $4y^2 = (2y)^2$.

(b) $100x^2 - 1$

This binomial fits the form of the difference of two squares as $100x^2 - 1 = (10x)^2 - (1)^2$.

Note that $100x^2 = (10x)^2$ and $1 = (1)^2$.

(c) $4x^2 + 9$

This binomial is not the difference of two squares.

This binomial is the sum of two perfect squares, not their difference.

(d) $x^2 - 7$

This binomial is not the difference of two perfect squares.

7 is not a perfect square integer. However, we could factor this using irrational factors as $x^2 - 7 = x^2 - (\sqrt{7})^2 = (x + \sqrt{7})(x - \sqrt{7})$.

(e) $x^2 - 2xy - y^2$

This is a trinomial, not a binomial, and is not a difference of two perfect squares.

The difference of two squares is always a binomial, never a trinomial. ■

The fact that $(x + y)(x - y)$ is equal to $x^2 - y^2$ was used to multiply the binomials in Example 1. Now we reverse this process to factor the binomials in Example 3.

■ EXAMPLE 3   Factoring the Difference of Two Squares

Show that each binomial fits the form of the difference of two squares and then factor it.

SOLUTIONS

(a) $25v^2 - 1$

$$25v^2 - 1 = (5v)^2 - (1)^2$$
$$= (5v + 1)(5v - 1)$$

First write each binomial in the form $x^2 - y^2$ and then factor as a sum times a difference.

(b) $81z^2 - 25$

$$81z^2 - 25 = (9z)^2 - (5)^2$$
$$= (9z + 5)(9z - 5)$$

(c) $9v^2 - 49w^2$

$$9v^2 - 49w^2 = (3v)^2 - (7w)^2$$
$$= (3v + 7w)(3v - 7w)$$

(d) $y^4 - 9$

$$y^4 - 9 = (y^2)^2 - (3)^2$$
$$= (y^2 + 3)(y^2 - 3)$$ ■

SELF-CHECK 5.7.2

Factor each of these binomials.

1. $a^2 - 36$
2. $49m^2 - 100n^2$
3. $9x^4 - 25$

Perfect Square Trinomials

The products and factors listed here are similar to those we have examined in the previous two sections. By examining them as organized below we want to observe the patterns shown and then to take advantage of these patterns.

Multiplying	**Factoring**
$(x + 2)^2 = x^2 + 4x + 4$	$x^2 + 4x + 4 = (x + 2)^2$
$(x + 3)^2 = x^2 + 6x + 9$	$x^2 + 6x + 9 = (x + 3)^2$
$(x + 4)^2 = x^2 + 8x + 16$	$x^2 + 8x + 16 = (x + 4)^2$
$(x + 5)^2 = x^2 + 10x + 25$	$x^2 + 10x + 25 = (x + 5)^2$
$(x + y)^2 = x^2 + 2xy + y^2$	$x^2 + 2xy + y^2 = (x + y)^2$
$(x - 2)^2 = x^2 - 4x + 4$	$x^2 - 4x + 4 = (x - 2)^2$
$(x - 5)^2 = x^2 - 10x + 25$	$x^2 - 10x + 25 = (x - 5)^2$
$(x - y)^2 = x^2 - 2xy + y^2$	$x^2 - 2xy + y^2 = (x - y)^2$

The preceding patterns are summarized in the following box.

Perfect Square Trinomials

ALGEBRAICALLY	VERBALLY	ALGEBRAIC EXAMPLE
Square of a Sum $(x + y)^2 = x^2 + 2xy + y^2$ and $x^2 + 2xy + y^2 = (x + y)^2$	A trinomial that is the square of a binomial has: 1. a first term that is a square of the first term of the binomial, 2. a middle term that is twice the product of the two terms of the binomial, 3. a last term that is a square of the last term of the binomial.	$(x + 1)^2 = x^2 + 2x + 1$ $x^2 + 12x + 36 = (x + 6)^2$
Square of a Difference $(x - y)^2 = x^2 - 2xy + y^2$ and $x^2 - 2xy + y^2 = (x - y)^2$		$(x - 1)^2 = x^2 - 2x + 1$ $x^2 - 12x + 36 = (x - 6)^2$

SELF-CHECK 5.7.2 ANSWERS

1. $(a + 6)(a - 6)$
2. $(7m + 10n)(7m - 10n)$
3. $(3x^2 + 5)(3x^2 - 5)$

The formula for the square of a sum is illustrated geometrically by the areas of the figures shown here.

Area of a Whole Square   **Area of Four Parts**

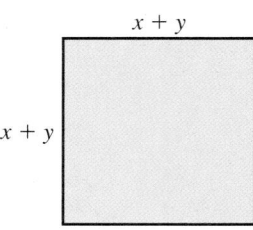

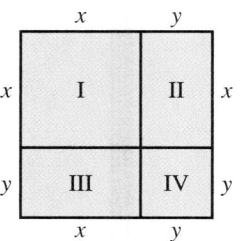

Area = $(x + y)(x + y)$ Total area = area I + area II + area III + area IV

$A = (x + y)^2$ $A = x^2 + xy + xy + y^2$

$A = x^2 + 2xy + y^2$

Thus $(x + y)^2 = x^2 + 2xy + y^2$.

■ EXAMPLE 4 Squaring Binomials

Square these binomials.

SOLUTIONS

(a) $(x + 7)^2$ $(x + 7)^2 = x^2 + 2(7)(x) + (7)^2$

$= x^2 + 14x + 49$

The first term is the square of x, the second term is twice the product of 7 and x. The third term is the square of 7.

(b) $(v - 4)^2$ $(v - 4)^2 = v^2 + 2(-4)(v) + (-4)^2$

$= v^2 - 8v + 16$

The first term is the square of v. The second term is twice the product of -4 and v. The third term is the square of -4.

(c) $(2x + 3y)^2$ $(2x + 3y)^2 = (2x)^2 + 2(2x)(3y) + (3y)^2$

$= 4x^2 + 12xy + 9y^2$

The first term is the square of $2x$. The second term is twice the product of $2x$ and $3y$. The third term is the square of $3y$.

SELF-CHECK 5.7.3

Square these binomials.

1. $(x - 3)^2$
2. $(7m + n)^2$
3. $(3a - 4b)^2$

SELF-CHECK 5.7.3 ANSWERS

1. $x^2 - 6x + 9$
2. $49m^2 + 14mn + n^2$
3. $9a^2 - 24ab + 16b^2$

Before factoring some perfect square trinomials, we will first practice identifying these special forms.

EXAMPLE 5 Identifying Perfect Square Trinomials

Determine which of the following polynomials are perfect square trinomials.

SOLUTIONS

(a) $5x^2 - 10x + 1$

This trinomial is not a perfect square.

The coefficient of the first term, 5, is not a perfect square.

(b) $9x^2 + 6x + 1$

This trinomial fits the form of a perfect square, as illustrated here:

$$9x^2 + 6x + 1 = (3x)^2 + 2(3x)(1) + 1^2$$

The first and last terms are perfect squares since $9x^2 = (3x)^2$ and $1 = (1)^2$. Also note that the middle term, $6x$, equals $2(3x)(1)$.

(c) $4x^2 - 12x + 3$

This trinomial is not a perfect square.

The last term, 3, is not a perfect square.

(d) $4x^2 - 20xy + 25y^2$

This trinomial fits the form of a perfect square, as illustrated here:

$$4x^2 - 20xy + 25y^2 = (2x)^2 - 2(2x)(5y) + (5y)^2$$

The first and last terms are perfect squares because $4x^2 = (2x)^2$ and $25y^2 = (5y)^2$. Also note that the middle term, $-20xy$, equals $-2(2x)(5y)$.

(e) $36x^2 + 25y^2$

This is a binomial, not a perfect square trinomial.

The square of a sum or a difference is always a trinomial, never a binomial.

(f) $x^2 - 6xy - 9y^2$

This trinomial is not a perfect square.

Because the last term has a negative coefficient, it cannot be a perfect square.

(g) $4x^2 + 3x + 9$

This trinomial is not a perfect square.

Although both the first and the last terms are perfect squares, the middle term does not fit the correct form. To fit the correct form, the middle term would have to be $2(2x)(3) = 12x$.

The fact that $(x + y)^2$ equals $x^2 + 2xy + y^2$ and that $(x - y)^2$ equals $x^2 - 2xy + y^2$ was used to square the binomials in Example 4. Now we reverse this process to factor the trinomials in Example 6.

EXAMPLE 6 Factoring Perfect Square Trinomials

Show that each trinomial fits the form of a perfect square trinomial and then factor it.

SOLUTIONS

(a) $x^2 + 16x + 64$

$$x^2 + 16x + 64 = (x)^2 + 2(8)(x) + (8)^2$$
$$= (x + 8)^2$$

First write this trinomial in the form $x^2 + 2xy + y^2$.

(b) $25v^2 - 60vw + 36w^2$

$$25v^2 - 60vw + 36w^2 = (5v)^2 - 2(5v)(6w) + (-6w)^2$$
$$= (5v - 6w)^2$$

First write this trinomial in the form $x^2 - 2xy + y^2$.

SELF-CHECK 5.7.4

Factor each of these trinomials.

1. $v^2 - 14v + 49$
2. $a^2 + 12ab + 36b^2$
3. $4m^2 + 12mn + 9n^2$

SELF-CHECK 5.7.4 ANSWERS

1. $(v - 7)^2$
2. $(a + 6b)^2$
3. $(2m + 3n)^2$

USING THE LANGUAGE AND SYMBOLISM OF MATHEMATICS   5.7

1. A binomial is a polynomial with _____ terms.
2. A trinomial is a polynomial with _____ terms.
3. Using the operation of _____, we can rewrite $(x + 2)(x - 2)$ as $x^2 - 4$.
4. Using the process of _____, we can rewrite $x^2 - 4$ as $(x + 2)(x - 2)$.

5. Match each of these polynomials with the description which best fits each polynomial,
   a. $a^2 - b^2$ A. The square of a sum
   b. $(a - b)^2$ B. The sum of two squares
   c. $(a + b)^2$ C. The difference of two squares
   d. $a^2 + b^2$ D. The square of a difference

EXERCISES   5.7

In Exercises 1–30 use the special product forms to multiply each of these polynomials by inspection.

1. $(7a + 1)(7a - 1)$
2. $(v + 9)(v - 9)$
3. $(z - 10)(z + 10)$
4. $(w - 13)(w + 13)$
5. $(2w - 3)(2w + 3)$
6. $(3w + 5)(3w - 5)$
7. $(9a + 4b)(9a - 4b)$
8. $(7a + 3b)(7a - 3b)$
9. $(4x - 11y)(4x + 11y)$
10. $(5x - 12y)(5x + 12y)$
11. $(x^2 - 2)(x^2 + 2)$
12. $(x^2 + 3)(x^2 - 3)$
13. $(4m + 1)^2$
14. $(7m + 1)^2$
15. $(n - 9)^2$
16. $(n - 10)^2$
17. $(3t + 2)^2$
18. $(8t + 3)^2$
19. $(5v - 8)^2$
20. $(7v - 3)^2$
21. $(4x + 5y)^2$
22. $(7x + 2y)^2$
23. $(6a - 11b)^2$
24. $(8a - 7b)^2$
25. $(a + bc)^2$
26. $(a - bc)^2$
27. $(x^2 - 3)^2$
28. $(x^2 + 5)^2$
29. $(x^2 + y)^2$
30. $(x^2 - 3y)^2$

In Exercises 31–40 factor by inspection each binomial which is the difference of two squares.

31. $49x^2 - 1$
32. $64x^2 - 1$
33. $a^2 - 100$
34. $b^2 - 144$
35. $81x^2 - 25$
36. $36x^2 - 49$
37. $4a^2 - 9b^2$
38. $25v^2 - 49w^2$
39. $x^4 - 4$
40. $x^4 - 9$

In Exercises 41–54 factor by inspection each trinomial which is a perfect square trinomial.

41. $9x^2 + 6x + 1$
42. $9x^2 - 6x + 1$
43. $25y^2 - 10y + 1$
44. $25y^2 + 10y + 1$
45. $m^2 - 26m + 169$
46. $m^2 + 26m + 169$
47. $4w^2 + 12w + 9$
48. $4w^2 - 12w + 9$
49. $9x^2 + 30xy + 25y^2$
50. $9x^2 - 30xy + 25y^2$
51. $x^4 - 12x^2 + 36$
52. $x^4 + 14x^2 + 49$
53. $121m^2 + 88mn + 16n^2$
54. $169m^2 - 78mn + 9n^2$

In Exercises 55 and 56 use a graphics calculator to fill in the table of values for $y_1 = P_1(x)$ and $y_2 = P_2(x)$. Do these results indicate that the polynomials $P_1(x)$ and $P_2(x)$ are equal?

55.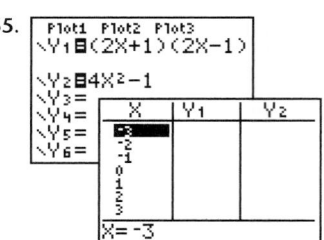

56.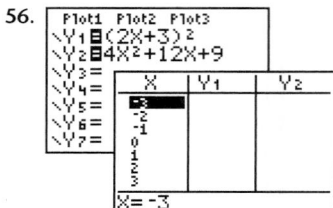

In Exercises 57–60 fill in the blanks in each factorization. First factor out the GCF. Then complete the factorization.

57. $45x^2 - 20 = 5(\quad) + 5(\quad)$
$= 5(\quad)$
$= 5(3x + 2)(\quad)$

58. $4x^2 - 40x + 100 = 4(\quad) + 4(\quad) + 4(\quad)$
$= 4(\quad)$
$= 4(\quad)^2$

59. $98x^3 - 18x = 2x(\quad) + 2x(\quad)$
$= 2x(\quad)$
$= 2x(\quad)(7x - 3)$

60. $3x^3y + 12x^2y^2 + 12xy^3 = 3xy(\quad) + 3xy(\quad) + 3xy(\quad)$
$= 3xy(\quad)$
$= 3xy(\quad)^2$

In Exercises 61 and 62 fill in the blanks in each factorization.

61. $(x - 1)^2 - 4 = [(\quad) + 2][(\quad) - 2]$
$\qquad = (x + 1)(\quad)$

62. $(2x + 1)^2 - 4 = [(2x + 1) + (\quad)][(\quad) - (2)]$
$\qquad = (2x + 3)(\quad)$

In Exercises 63–70 simplify each expression.

63. $(v + 7)^2 - (v^2 - 7^2)$
64. $(w - 5)^2 - (w^2 - 5^2)$
65. $(2x + y)^2 - [(2x)^2 + y^2]$
66. $(x + y)^2 - (x - y)^2$
67. $(v + w)(v - w) - (v - w)^2$
68. $(3v - w)(3v + w) - (3v - w)^2$
69. $(a - b)^2 - (b - a)^2$
70. $(ab + 6)(ab - 6)$

Multiple Representations

In Exercises 71–76 write each verbal statement in algebraic form and then perform the indicated operation.

71. Square the quantity $3x + 7y$.
72. Square the quantity $8x - 9y$.
73. Subtract the square of $x - 6$ from the square of $x + 8$.
74. Subtract the square of $x + 7$ from the square of $x - 9$.
75. Multiply $a^2 + 1$ times the product of $a + 1$ and $a - 1$.
76. Multiply $a^2 + 4$ times the product of $a - 2$ and $a + 2$.

Group Discussion Questions

77. Mental Arithmetic Sometimes impressive feats of mental arithmetic have a very simple algebraic basis.

　　a. Use the fact that $(103)(97) = (100 + 3)(100 - 3)$ to compute this product mentally.
　　b. Use the fact that $(96)(104) = (100 - 4)(100 + 4)$ to compute this product mentally.
　　c. Use the fact that $(99)(99) = (100 - 1)^2$ to compute this product mentally.
　　d. Use the fact that $(101)(101) = (100 + 1)^2$ to compute this product mentally.
　　e. Make up two problems of your own which you can compute mentally.

78. Discovery Question Substitute $-y$ for y in the special factoring formula $(x + y)^2 = x^2 + 2xy + y^2$ and simplify the result. What do you observe?

79. Discovery Question

　　a. Factor each of these polynomials and then graph $y = P(x)$ using a viewing window of $[-6, 6, 1]$ by $[-20, 10, 5]$. What relationship do you observe between the factors of $P(x)$ and attributes of the graph?

Factor $P(x)$	Graph $y = P(x)$
i.　$x^2 - 1$	$y = x^2 - 1$
ii.　$x^2 - 4$	$y = x^2 - 4$
iii.　$x^2 - 9$	$y = x^2 - 9$
iv.　$x^2 - 16$	$y = x^2 - 16$

　　b. Given the following graph of $y = P(x)$, can you predict the factors of $P(x)$? Test your prediction with a graphics calculator.

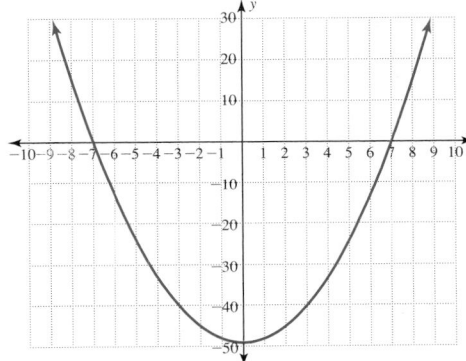

80. Error Analysis A student divided $10x^2 + 132x - 270$ by $5x - 9$ and obtained $2x + 3$. The student then checked this work on a graphics calculator by comparing the following graphs of y_1 and y_2. Seeing only one line on the display the student concluded this meant that $y_1 = y_2$. Describe the calculator error that the student has made and then work this division problem correctly.

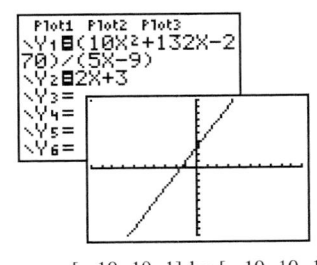

$[-10, 10, 1]$ by $[-10, 10, 1]$

1. **Exponential Notation:**
   For any real base x and natural number n,
   a. $x^n = x \cdot x \cdot \ldots \cdot x$ (n factors of x).
   b. $x^0 = 1$;   for $x \neq 0$; 0^0 is undefined.
   c. $x^{-n} = \dfrac{1}{x^n}$;   for $x \neq 0$.

2. **Base of an Exponent:** If there are no symbols of grouping, only the constant or variable immediately to the left of an exponent is the base. Thus $-x^2$ and $(-x)^2$ have distinct meanings.

3. **Summary of the Properties of Exponents:** For any nonzero real numbers x and y and integral exponents m and n,
   Product rule: $x^m \cdot x^n = x^{m+n}$
   Power rule: $(x^m)^n = x^{mn}$
   Product to a Power: $(xy)^m = x^m y^m$
   Quotient to a Power: $\left(\dfrac{x}{y}\right)^m = \dfrac{x^m}{y^m}$
   Quotient rule: $\dfrac{x^m}{x^n} = x^{m-n}$
   Negative power: $\left(\dfrac{x}{y}\right)^{-n} = \left(\dfrac{y}{x}\right)^n$

4. **Scientific Notation:**
   a. A number is in scientific notation when it is expressed as the product of a number between 1 and 10 (or -1 and -10 if negative) and an appropriate power of 10.
   b. On many calculators the ⌷EE⌷ feature is used to enter the power of 10.
   c. If a number such as 6.89E5 appears on a calculator display, this represents 6.89×10^5 in scientific notation.

5. **To Write a Number in Scientific Notation:**
   a. Move the decimal point immediately to the right of the first nonzero digit of the number.
   b. Multiply by a power of 10 determined by counting the number of places the decimal point has been moved.
   i. The exponent on 10 is 0 or positive if the magnitude of the original number is 1 or greater.
   ii. The exponent on 10 is negative if the magnitude of the original number is less than 1.

6. **Monomial:**
   A monomial is a real number, a variable, or a product of real numbers and variables.

7. **Polynomial:**
   A polynomial is a monomial or a sum of monomials.

8. **Classification of Polynomials:**
   a. Monomials contain one term.
   b. Binomials contain two terms.
   c. Trinomials contain three terms.

9. **Degree of a Monomial:**
   The degree of a monomial is the sum of the exponents on all the variables in the term.

10. **Degree of a Polynomial:**
    The degree of a polynomial is the same as the degree of the term of the highest degree.

11. **A Polynomial Is in Standard Form if:**
    a. The variables in each term are written in alphabetical order.
    b. The terms are arranged in descending powers of the first variable.

12. **Adding and Subtracting Polynomials:**
    To add or subtract polynomials, combine like terms.

13. **Product of Polynomials:**
    a. To multiply a monomial times a polynomial, use the distributive property to multiply the monomial times each term of the polynomial.
    b. To multiply two polynomials, use the distributive property to multiply each term of the first polynomial times each term of the second polynomial, then combine like terms.

14. **Quotient of Polynomials:**
    a. To divide a polynomial by a monomial, divide each term of the polynomial by the monomial.
    b. To divide a polynomial by a polynomial of more than one term, use long division of polynomials.

15. **Factors of a Polynomial:**
    If one factor of a polynomial is known, we can divide the polynomial by this factor to obtain a second factor.

16. **Special Products and Factors:**
    a. $(x + y)(x - y) = x^2 - y^2$ The product of a sum times a difference is the difference of their squares.
    b. $x^2 - y^2 = (x + y)(x - y)$ The difference of two squares factors as a sum times a difference.
    c. $(x + y)^2 = x^2 + 2xy + y^2$ The square of a sum.
    d. $x^2 + 2xy + y^2 = (x + y)^2$ A perfect square trinomial.
    e. $(x - y)^2 = x^2 - 2xy + y^2$ The square of a difference.
    f. $x^2 - 2xy + y^2 = (x - y)^2$ A perfect square trinomial.

REVIEW EXERCISES   FOR CHAPTER 5

All exercises are assumed to restrict the variables to values that avoid division by 0 and avoid 0^0.

1. Write each expression in exponential form.
   a. $xyyy$
   b. $(xy)(xy)(xy)$
   c. $xx + yy$
   d. $(x + y)(x + y)$

2. Write each exponential expression in expanded form.
   a. $x^2 + y^2$
   b. $(x + y)^2$
   c. $-x^4$
   d. $(-x)^4$

In Exercises 3–12 simplify each expression.

3. a. 3^2 b. 2^3
   c. -3^2 d. 3^{-2}
4. a. 3^0 b. 0^3
   c. $3^0 + 4^0$ d. $(3 + 4)^0$
5. a. $\left(\dfrac{1}{2}\right)^{-1} + \left(\dfrac{1}{3}\right)^{-1}$ b. $\left(\dfrac{1}{2} + \dfrac{1}{3}\right)^{-1}$
   c. $\left(\dfrac{2}{3}\right)^{-1}$ d. $\dfrac{2^{-1}}{3}$
6. a. -1^6 b. $(-1)^6$
   c. 6^{-1} d. -6^{-1}
7. a. $x^3 + x^3$ b. $x^3 x^3$
   c. $\dfrac{x^3}{x^3}$ d. $x^3 - x^3$
8. a. 10^2 b. $(-10)^2$
   c. 10^{-2} d. -10^{-2}
9. a. $5^0 + 6^0 + 7^0$ b. $(5 + 6)^0 + 7^0$
   c. $(5 + 6 + 7)^0$
10. a. $2x^0 + 3y^0 + 4z^0$ b. $(2x + 3y + 4z)^0$
    c. $(2x)^0 + (3y)^0 + (4z)^0$
11. a. $x^2 x^5$ b. $\dfrac{x^5}{x^2}$
    c. $(x^2)^5$
12. a. $(4x^3y^2)(12x^5y^4)$ b. $\dfrac{12x^5y^4}{4x^3y^2}$
    c. $(4x^3y^2)^2$

In Exercises 13–18 evaluate each expression for $x = 3$ and $y = 4$.

13. a. $-x^2$ b. $(-x)^2$
14. a. xy^2 b. $(xy)^2$
15. a. $x^2 + y^2$ b. $(x + y)^2$
16. a. $x^{-1} + y^{-1}$ b. $(x + y)^{-1}$
17. a. $x^0 + y^0$ b. $(x + y)^0$
18. a. xy^{-2} b. $(xy)^{-2}$

In Exercises 19–32 simplify each expression.

19. $(5x^2y^3z^4)(6xy^4z^7)$ 20. $(5x^2y^3z^4)^2$
21. $\dfrac{24a^2b^4c^5}{40ab^7c^3}$ 22. $\left(\dfrac{12m^3}{4n^2}\right)^{-2}$

23. $(3x^2)(4x^3)(5x^{11})$ 24. $\dfrac{(3x^2)^4}{9x^5}$
25. $(-2x^5)^3(-5x^3)^2$ 26. $(7y^8)(8y^{-6})$
27. $\dfrac{16w^{-4}}{8w^{-6}}$ 28. $(5m^{-6})^{-4}$
29. $(6x)^{-3}\left(\dfrac{x}{6}\right)^{-4}$ 30. $\left(\dfrac{x^{188}}{x^{-439}}\right)^0$
31. $\dfrac{(2a^3b^3)(8a^2b^5)}{(4ab^2)^2}$ 32. $(x^2y^{-3}z^{-4})^{-2}(x^4y^{-1}z^{-2})^{-1}$

33. **Strength of Satellite Signals** The strength of the signals from *Voyager 2* to Earth on its 1989 flyby of Neptune were only 1.0×10^{-13} of a watt. Write this strength in standard decimal notation.

34. **Speed of Satellite Signals** The speed of the signals sent by *Voyager 2* was about the speed of light, 2.9979×10^8 meters per second. Write this speed in standard decimal notation.

35. Express the number shown on the following calculator display in standard decimal notation.

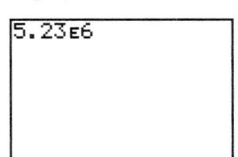

```
5.23E6
```

36. Express 9,578,420,000 in scientific notation.
37. Express 0.00437 in scientific notation.
38. **Estimation Skills** Use scientific notation to mentally select the best approximation of $(40,010)(0.0000000001989)$.
    a. 8.0×10^{-8} b. 8.0×10^{-6} c. 8.0×10^{-4}
    d. 6.0×10^{-5} e. 6.0×10^5

In Exercises 39–42 classify each polynomial according to the number of terms it contains, and give its degree.

39. π
40. $3x^6 - 17x^2$
41. $-9x^3 + 7x^2 + 8x - 11$
42. $x^5y - 7x^4y^2 + 23x^3y^3$
43. Write $11x^4 - 3x^2 + 9x^3 + 7x^5 + 4 - 8x$ in standard form.
44. Write a fifth-degree monomial in x with a coefficient of negative seven.
45. Write a second-degree binomial whose leading coefficient is one and with a constant term of negative three.

In Exercises 46–64 perform the indicated operations.

46. $(7x^2 - 9x + 13) + (4x^2 + 6x - 11)$
47. $(9x^3 - 5x^2 - 7) - (4x^3 + 8x - 11)$
48. $(3x^4 - 8x^3 + 7x^2 + 9x - 4) + (2x^4 + 6x^2 + 9)$
49. $7x^5 + 9x^3 + 6x - 3 - (4x^5 - 3x^4 - 7x^3 - x^2 - x + 8)$
50. $(x^2 - 8x + 7) - (2x^2 + 7x + 11) + (3x^2 + 4x - 8)$
51. $5x^2(7x^3 - 9x^2 + 3x + 1)$

52. $(5v + 1)(7v - 1)$

53. $(5y - 7)^2$

54. $(9y + 5)^2$

55. $(3a + 5b)(3a - 5b)$

56. $(3m + 5)(2m^2 - 6m + 7)$

57. $\dfrac{-36a^3b^7}{9a^4b^4}$

58. $\dfrac{15m^4n^4 - 21m^4n^5 - 3m^3n^6}{3m^2n^3}$

59. $\dfrac{v^2 - 6v + 8}{v - 4}$

60. $\dfrac{21w^2 - 40w - 21}{3w - 7}$

61. $\dfrac{(a + b)^7}{(a + b)^6}$

62. $(3x + 4y)^2 - (3x + 4y)(3x - 4y)$

63. $(2x + 3y)^2 - (2x - 3y)^2$

64. $\dfrac{6x^2 + x - 2}{2x - 1} - \dfrac{8x^2 + 18x - 35}{4x - 5}$

65. The answer to $\dfrac{x^3 + 64}{x + 4}$ is $x^2 - 4x + 16$. Check this answer.

66. The answer to $\dfrac{x^2 + 4x - 5}{x - 2}$ is $x + 6 + \dfrac{7}{x - 2}$. Check this answer.

In Exercises 67–72 use the distributive property to expand the first expression and to factor the second expression.

Expand	Factor
67. a. $3x(x - 4)$	**b.** $3x^2 - 12x$
68. a. $5x(2x + 3)$	**b.** $10x^2 + 15x$
69. a. $2mn(m^2 - 3mn - 5n^2)$	
b. $2m^3n - 6m^2n^2 - 10mn^3$	
70. a. $2xy(x^2 - 3xy + y^2)$	**b.** $2x^3y - 6x^2y^2 + 2xy^3$
71. a. $2x(x + 4y) - 3y(x + 4y)$	
b. $2x(x + 4y) - 3y(x + 4y)$	
72. a. $(3x + 2)(7x) + (3x + 2)(9)$	
b. $(3x + 2)(7x) + (3x + 2)(9)$	

In Exercises 73–76 use the special products and factors covered in Section 5.7 to expand the first expression and to factor the second expression.

Expand	Factor
73. a. $(5x - 8)(5x + 8)$	**b.** $49x^2 - 121$
74. a. $(2x + 13y)(2x - 13y)$	**b.** $9x^2 - 16y^2$
75. a. $(6x - 7)^2$	**b.** $9x^2 - 60x + 100$
76. a. $(2x + 9y)^2$	**b.** $25x^2 + 40xy + 16y^2$

In Exercises 77–82 use division to complete each factorization.

77. $6x^3 + 15x^2 - 12x = 3x(\quad)$

78. $2a^2b + 6ab^2 - 8ab = 2ab(\quad)$

79. $x^2 + 2x - 35 = (x + 7)(\quad)$

80. $x^2 + 2x - 35 = (\quad)(x - 5)$

81. $x^2 - 121 = (x + 11)(\quad)$

82. $x^3 - 8 = (\quad)(x^2 + 2x + 4)$

83. Area of a Rectangle Determine the area of the following rectangle.

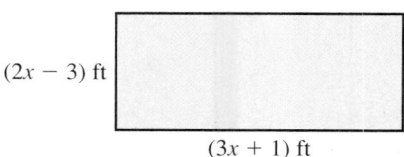

$(2x - 3)$ ft

$(3x + 1)$ ft

84. Height of a Box The volume of the box shown here is $2x^3 + 2x^2$ cm^3. Determine the height of this box.

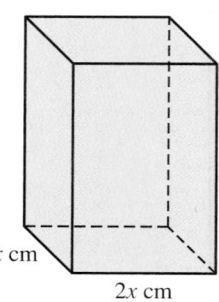

x cm

$2x$ cm

85. Modeling Revenue The price per unit of an item is x dollars and the number of units that will sell at this price is estimated to be $350 - 4x$.

a. Write $R(x)$, a revenue polynomial for this item.

b. Evaluate and interpret $R(0)$.

c. Evaluate and interpret $R(40)$.

d. Evaluate and interpret $R(80)$.

86. Modeling the Value of an Investment Use the formula $A = P(1 + r)^t$ to find the value of a $10,000 investment compounded at 7.5% for eight years.

87. Examine the following four tables of values. Which of these tables represent equal polynomials?

a.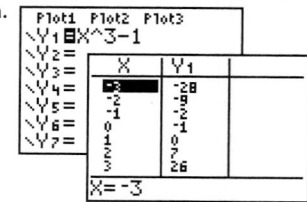

b.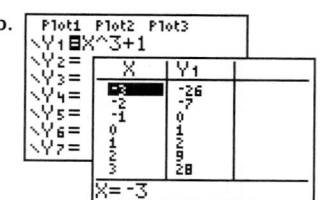

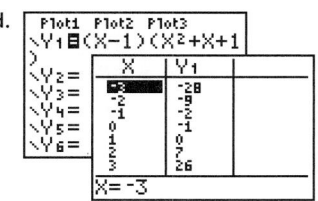

c.
```
Plot1 Plot2 Plot3
\Y1▄(X+1)(X²-X+1)
>
\Y2=        X   | Y1
\Y3=       -3   | -26
\Y4=       -2   | -7
\Y5=       -1   | 0
\Y6=        0   | 1
            1   | 2
            2   | 9
            3   | 28
          X=-3
```

d.
```
Plot1 Plot2 Plot3
\Y1▄(X-1)(X²+X+1)
>
\Y2=        X   | Y1
\Y3=       -3   | -28
\Y4=       -2   | -9
\Y5=       -1   | -2
\Y6=        0   | -1
            1   | 0
            2   | 7
            3   | 26
          X=-3
```

MASTERY TEST FOR CHAPTER 5

All exercises are assumed to restrict the variables to values that avoid division by 0 and avoid 0^0.

[5.1] **1.** Simplify each expression.
  a. $x^5 x^8$
  b. $(5y^4)(8y^3)$
  c. $-v^4 v^5 v^6$
  d. $(mn^2)(m^3 n)$

[5.1] **2.** Simplify each expression.
  a. $(w^5)^8$
  b. $(2y^4)^5$
  c. $(-v^2 w^3)^4$
  d. $\left(\dfrac{-3m}{2n}\right)^3$

[5.2] **3.** Simplify each expression.
  a. $\dfrac{z^8}{z^2}$
  b. $\dfrac{4x^9}{2x^3}$
  c. $-\dfrac{v^{23}}{v^{23}}$
  d. $\dfrac{-6a^4 b^6}{3a^2 b^2}$

[5.2] **4.** Simplify each expression.
  a. 1^0
  b. $\left(\dfrac{2}{5}\right)^0$
  c. $(3x + 5y)^0$
  d. $3x^0 + 5y^0$

[5.2] **5.** Simplify each expression.
  a. $(x^3 y^5)(x^2 y^6)$
  b. $\left(\dfrac{x^2 y^3}{xy}\right)^3$
  c. $\dfrac{(6x^2)^2}{(3x^3)^3}$
  d. $\left[(2x^2)(3x^4)\right]^2$

[5.3] **6.** Simplify each expression.
  a. 3^{-1}
  b. 7^{-2}
  c. $\left(\dfrac{2}{3}\right)^{-2}$
  d. $\left(\dfrac{1}{3}\right)^{-1} + \left(\dfrac{1}{6}\right)^{-1}$

[5.3] **7. a.** Write these numbers in standard decimal notation: 3.57×10^5; 7.35×10^{-5}
  b. Write these numbers in scientific notation: 0.000509; 93,050,000

[5.4] **8.** Classify each of these polynomials according to the number of terms it contains, and give its degree.
  a. $-5y^3 - 13y$
  b. 273
  c. $2x^2 - 7x + 1$
  d. $17x^5 - 4x^3 + 9x + 8$

[5.4] **9.** Simplify each expression.
  a. $(4x - 9y) + (3x + 8y)$
  b. $(2x^2 - 3x + 7) - (5x^2 - 9x - 11)$
  c. $(5x^4 - 9x^3 + 7x^2 + 13) + (4x^4 + 12x^2 + 9x - 5)$
  d. $(3x^3 - 7x + 1) - (12x^2 + 9x - 5)$

[5.5] **10.** Find each of the indicated products.
  a. $(5x^2 y^3)(11x^4 y^7)$
  b. $-3x^2(5x^3 - 2x^2 + 7x - 9)$
  c. $(x + 5)(x^2 + 3x + 1)$
  d. $(x - 3y)(x^2 + xy + y^2)$

[5.6] **11.** Find each of the indicated quotients.
  a. $\dfrac{15a^4 b^2}{5a^2 b^2}$
  b. $\dfrac{36m^4 n^3 - 48m^2 n^5}{12m^2 n^2}$
  c. $x - 2\overline{)x^2 + 9x - 22}$
  d. $\dfrac{6x^3 + 10x^2 - 32}{3x - 4}$

[5.6] **12.** Use division to complete each factorization.
  a. $x^2 + 4x - 77 = (x + 11)(\quad)$
  b. $20x^2 + 13x - 15 = (\quad)(5x - 3)$
  c. $x^3 - 64 = (x - 4)(\quad)$
  d. $3x^3 + 4x^2 + 5x - 6 = (x^2 + 2x + 3)(\quad)$

[5.7] **13.** Multiply these factors by inspection.
  a. $(5x + 11)(5x - 11)$
  b. $(x - 9y)(x + 9y)$
  c. $(2x + 7)^2$
  d. $(3x - 10y)^2$

[5.7] **14.** Factor these polynomials.
  a. $81m^2 - 16$
  b. $25x^2 - 9y^2$
  c. $49x^2 + 42x + 9$
  d. $16x^2 - 88xy + 121y^2$

GROUP PROJECT   FOR CHAPTER 5

A Polynomial Model for the Construction of a Box

Supplies Needed for Each Group

- Heavy weight paper (120 pound weight is suggested) for constructing a box
- Metric rulers with cm marked for measuring and to use as an aid in folding the paper
- Scissors for cutting out the corners of the paper
- Masking tape to use to tape up the sides of the box at the corners
- Rice to place into the box and to be measured for volume
- Graduated cylinders marked in milliliters (1 milliliter = 1 cubic centimeter)
- Funnels to be used to pour rice into the graduated cylinders

LAB
Constructing a Box and Measuring the Volume of Rice that the Box Contains

1. Measure the width of your sheet of paper to the nearest tenth of a centimeter. _____
2. Measure the length of your sheet of paper to the nearest tenth of a centimeter. _____
3. Assign each group different sized squares to cut out (e.g., 4 cm, 4.5 cm, 5 cm, etc.). Mark off equal squares to be cut from each corner and record the width of each square to the nearest tenth of a centimeter. _____
4. Carefully cut the squares from each corner of your sheet of paper as illustrated.
5. Using a ruler to maintain a straight fold, fold the paper along lines that you have drawn between the corners.
6. Using the masking tape, tape up the corners as shown in the illustration. Take care to form vertical sides.

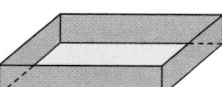

7. Using graduated cylinders, measure the rice in milliliters (1 milliliter = 1 cubic centimeter) and carefully pour the rice into your box until it is full (level the top of the rice with your ruler as needed). Record the volume of rice your box contains in cubic centimeters. _____
8. Calculate the volume of the box using the formula $V = l \cdot w \cdot h$ and the measurements of the length, width, and height, recorded earlier. _____
9. Compare the volume of rice measured to the calculated volume of the box. What is the relative error of the rice measured to that predicted by the formula? _____ Is your relative error acceptable? Be prepared to defend your answer or to explain any excessive error.
10. Create a table of values using the results from the box of each group. Use x to represent the width of the square cut from each corner of the sheet of paper. Use y to represent the volume of rice contained in each box. (Discard any values with excessive error.)

Creating a Mathematical Model for the Volume of a Box

1. Use x to represent the height of each box. Write an interval to express the complete domain of possible values of x for the sheet of paper used by your group. _____
2. Write a function $W(x)$ that gives the width of the box as a function of the height x. _____
3. Write a function $L(x)$ that gives the length of the box as a function of the height x. _____
4. Write a function $V(x)$ that gives the volume of the box as a function of the height x. _____
5. Use the volume function to create a table of values over the domain of x-values listed earlier.
6. Use the volume function to create a graph over values of x that includes the domain of values listed earlier.
7. Do all values of x produce the same volume? If so, why? If not, is there any value that seems to produce more volume than other values of x?

The purpose of this diagnostic review is to help you gauge your mastery of Beginning Algebra, material that is needed for the Intermediate Algebra portion of this book in Chapters 6–11. This review is intended to give you a realistic assessment of your areas of strength and weakness. Some of these questions require you to interpret graphs or calculator screens. A few questions may cover questions that are not discussed at your school. You may wish to ask your instructor for questions that your school stresses. There are examples of all these questions in the exercises in this book.

   The answer to each of these questions follows this diagnostic review. Each answer is keyed to an example in this book. You can refer to these examples to find explanations and additional exercises for practice.

Arithmetic Review

In Exercises 1–12 calculate the value of each expression without using a calculator.

1. a. $12 + (-4)$ **b.** $12 - (-4)$
   c. $12(-4)$ **d.** $12 \div (-4)$

2. a. $-12 + (-6)$ **b.** $-12 - (-6)$
   c. $-12(-6)$ **d.** $-12 \div (-6)$

3. a. $-12 + 0$ **b.** $-12 - 0$
   c. $-12(0)$ **d.** $-12 \div 0$

4. a. $\dfrac{4}{5} + \dfrac{3}{10}$ **b.** $\dfrac{4}{5} - \dfrac{3}{10}$
   c. $\dfrac{4}{5}\left(\dfrac{3}{10}\right)$ **d.** $\dfrac{4}{5} \div \dfrac{3}{10}$

5. a. 2^3 **b.** 3^2
   c. $(-1)^6$ **d.** 6^{-1}

6. a. $\left(\dfrac{2}{3}\right)^2$ **b.** $\left(\dfrac{2}{3}\right)^{-1}$
   c. $\left(\dfrac{2}{3}\right)^{-2}$ **d.** $\left(\dfrac{2}{3}\right)^0$

7. a. 0^5 **b.** 5^0
   c. $\dfrac{0}{5}$ **d.** $\dfrac{5}{0}$

8. a. $|27|$ **b.** $|-27|$
   c. $-|27|$ **d.** $|27 - 27|$

9. a. $\sqrt{49}$ **b.** $\sqrt{\dfrac{4}{9}}$
   c. $\sqrt{9} + \sqrt{16}$ **d.** $\sqrt{9 + 16}$

10. a. 10^2 **b.** 10^4
   c. 10^0 **d.** 10^{-1}

11. a. 1.23×10^2 **b.** 1.23×10^4
   c. 1.23×10^{-1} **d.** 1.23×10^{-2}

12. a. $-0.9 + 0.03$ **b.** $-0.9 - 0.03$
   c. $-0.9(0.03)$ **d.** $-0.9 \div 0.03$

Order of Operations

In Exercises 13–19 calculate the value of each expression without using a calculator.

13. a. $-13 + 14 - 7 + 8$ **b.** $-13 + 14 - (7 + 8)$
   c. $-(13 + 14) - (7 + 8)$ **d.** $-(13 + 14 - 7) + 8$

14. a. $6 - 4(11 - 8)$ **b.** $(6 - 4)11 - 8$
   c. $(6 - 4)(11 - 8)$ **d.** $6 - 4(11) - 8$

15. a. $\dfrac{2 + 3}{6 + 9}$ **b.** $2 + \dfrac{3}{6 + 9}$
   c. $\dfrac{2}{6} + \dfrac{3}{9}$ **d.** $2 + \dfrac{3}{6} + 9$

16. a. $-5^2 + 3^2 + 4^2$ **b.** $(-5)^2 + 3^2 + 4^2$
   c. $(-5 + 3 + 4)^2$ **d.** $-(5 + 3 + 4)^2$

17. a. $-5^0 + 3^0 + 4^0$ **b.** $(-5)^0 + 3^0 + 4^0$
   c. $(-5 + 3 + 4)^0$ **d.** $-(5 + 3 + 4)^0$

18. a. $\left(\dfrac{1}{2}\right)^{-1} + \left(\dfrac{1}{3}\right)^{-1}$ **b.** $\left(\dfrac{1}{2} + \dfrac{1}{3}\right)^{-1}$
   c. $\left(\dfrac{1}{2 + 3}\right)^{-1}$ **d.** $\left(\dfrac{1 + 1}{2 + 3}\right)^{-1}$

19. a. $4 - 6 \div 2 + 5^2$ **b.** $(4 - 6) \div 2 + 5^2$
   c. $4 - (6 \div 2 + 5)^2$ **d.** $(4 - 6 \div 2 + 5)^2$

Evaluating Algebraic Expressions

In Exercises 20–23 evaluate each expression for $x = -4$, $y = -9$, and $z = -16$ without using a calculator.

20. a. $x + y$ **b.** $x - y$
   c. xy **d.** $\dfrac{z}{x}$

21. a. $x + y - z$ **b.** $x - (y + z)$
   c. $-(x + y + z)$ **d.** $|x| + |y| + |z|$

22. a. $2x - y$ **b.** $x^2 - y$
   c. $x^2 - y^2$ **d.** $(x - y)^2$

23. a. $-2y + z$ **b.** $-2y - z$
   c. $x^2 - 3x - 5$ **d.** $\dfrac{y + z}{x + 1}$

Using the Properties of Exponents

In Exercises 24–26 simplify each expression. Assume all variables are restricted to values that avoid division by zero.

24. a. $(3x + 7y)^0$ **b.** $(3x)^0 + (7y)^0$
   c. $3(x + 7y)^0$ **d.** $3x^0 + 7y^0$

25. a. $x^3 + x^3$ **b.** $x^3 x^8$
   c. $\dfrac{x^8}{x^3}$ **d.** $(x^8)^3$

26. a. $7x^2y - 5x^2y$ **b.** $(8x^2y^3)(4xy)$
   c. $\dfrac{8x^2y^3}{4xy}$ **d.** $(8x^2y^3)^2$

27. The result of each of the following products is shown on the calculator display. Write each product in standard decimal form.

a. 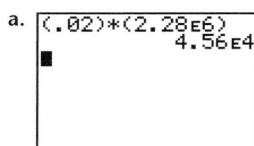   **b.**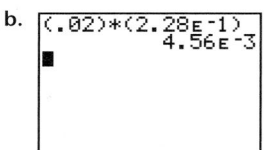

Operations with Polynomials

In Exercises 28–32 simplify each expression.

28. a. $(3x + 7y) + (2x - 4y)$ **b.** $(3x + 7y) - (2x - 4y)$
   c. $(3x + 7y) - 2(x - 4y)$ **d.** $(3x + 7y)(2x - 4y)$

29. a. $(4x^2 - 5x + 3) + (3x^2 + 4x - 9)$
   b. $(4x^2 - 5x + 3) - (3x^2 + 4x - 9)$

30. a. $(x + 4)(x^2 - 2x - 3)$ **b.** $(4x + 7)(4x - 7)$

31. a. $(3x - 5)^2$ **b.** $(x - 3)(x + 2)(x - 4)$

32. a. $\dfrac{6x^3 + 4x^2}{2x}$ **b.** $\dfrac{6x^2 + 7x - 5}{3x + 5}$

In Exercises 33 and 34 use division to complete each factorization.

33. a. $3ax^2 - 6ax + 9a = 3a(\quad)$
   b. $x^2 - x - 12 = (x + 3)(\quad)$

34. a. $2x^2 - 19x + 24 = (2x - 3)(\quad)$
   b. $x^3 - 27 = (x - 3)(\quad)$

In Exercises 35 and 36 factor each special form.

35. a. $x^2 - 100$ **b.** $4x^2 - 9y^2$

36. a. $25x^2 + 40x + 16$ **b.** $4x^2 - 28xy + 49y^2$

37. Write a first-degree monomial in x with a coefficient of 2.

38. Write a third-degree binomial in x whose leading coefficient is 5 and with a constant term of 11.

Properties and Subsets of the Real Numbers

39. Identify all the numbers from the set
$$\left\{-7, -4.73, -\pi, -\sqrt{4}, -\frac{3}{7}, 0, \sqrt{7}, 17\right\} \text{ that are:}$$

   a. Rational numbers   **b.** Irrational numbers
   c. Integers   **d.** Natural numbers

40. Name the property that justifies each statement.
   a. $5(x + z) = 5x + 5z$ **b.** $5(x + z) = 5(z + x)$
   c. $5(x + z) = (x + z)(5)$ **d.** $5 + (x + z) = (5 + x) + z$

41. a. The additive identity is _____.
   b. The additive inverse of 7 is _____.

42. a. The multiplicative identity is _____.
   b. The multiplicative inverse of 7 is _____.

Interval Notation

43. Write the interval notation for each of these inequalities.
   a. $2 \le x < 4$ **b.** $-2 < x \le 5$
   c. $-1 \le x \le 0$ **d.** $x \ge 3$

44. Write the interval notation for each of these intervals.

   a.
$$\text{---} \overset{-4\ -3\ -2\ -1\ \ 0\ \ 1\ \ 2}{\longleftarrow(\!+\!+\!+\!+\!+\!+\!)\longrightarrow}$$

   b.
$$\overset{-1\ \ 0\ \ 1\ \ 2\ \ 3\ \ 4\ \ 5}{\longleftarrow\!+\!+\!+\!)\!+\!+\!+\!\longrightarrow}$$

   c. $[5, 11) \cup (6, 13]$ **d.** $[5, 11) \cap (6, 13]$

Sequences

45. Calculate the first five terms of each sequence starting with $n = 1$.
   a. $a_n = n + 3$ **b.** $a_n = 3n$
   c. $a_n = n^2 - 1$ **d.** $a_n = 8 - n$

46. Determine whether each sequence is an arithmetic sequence. If the sequence is arithmetic, write the common difference.
   a. 0, 3, 6, 9, 12   **b.** 8, 6, 4, 2, 0
   c. 1, 3, 9, 27, 81   **d.** 3, 3, 3, 3, 3

Function Notation

47. Evaluate each expression given $f(x) = 5x - 3$.
   a. $f(0)$ **b.** $f(1)$
   c. $f(-1)$ **d.** $f(10)$

48. Profit Polynomial   The profit polynomial $P(x) = 200x - 2x^2$ gives the profit in dollars for the sale of x units. Evaluate and interpret these expressions.
   a. $P(0)$ **b.** $P(10)$
   c. $P(25)$ **d.** $P(50)$

Function Models

49. A 16-ft board has two pieces of length x cut off. Write a function for the length of the remaining piece in terms of x.

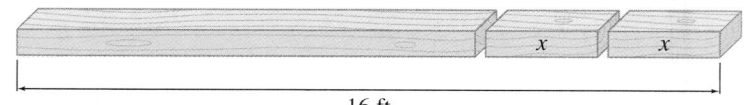

16 ft

50. For a price of x dollars per item a manufacturer estimates that it can sell $5000 - 0.01x$ units. Write a revenue polynomial $R(x)$ based on this estimate and evaluate and interpret $R(1000)$.

Solving Equations

51. Check $x = 5$ to determine whether it is a solution of each equation.
   a. $2x - 4 = x + 1$ **b.** $2(x - 4) = x - 7$
   c. $|x - 3| = |x + 3|$ **d.** $(x - 5)(x + 3) = 0$

Solve each of these linear equations.

52. a. $2x + 4 = x + 1$ **b.** $2(x - 4) = x - 7$
    c. $2(x - 5) = 2x + 9$ **d.** $2(x - 5) = 2x - 10$
53. a. $-x = 23$ **b.** $3(x - 4) = x - 2$

    c. $5(x - 4) = 2(x - 1)$ **d.** $\dfrac{x}{2} = \dfrac{x}{3} - 2$

54. Solve each equation for y.
    a. $4x - 2y = 6$ **b.** $4(x - 2y) = 3(2x - y) + 5$
55. Use the following table to solve
        $1.15x + 2.5 = 3.65x - 1.5$.

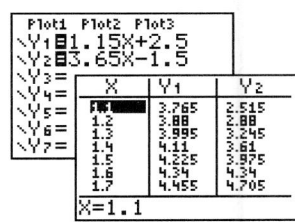

56. Use the following graph to solve
        $1.5x + 5.75 = -0.5x - 0.25$.

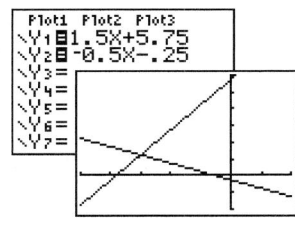

$[-5, 2, 1]$ by $[-2, 6, 1]$

In Exercises 57 and 58 simplify each expression in part **a** and solve each equation in part **b**.

	Simplify	*Solve*
57. a.	$2(x - 3) + 4(x - 3)$	**b.** $2(x - 3) + 4(x - 3) = 0$
58. a.	$5(x + 4) + 3(x + 4)$	**b.** $5(x + 4) + 3(x + 4) = 0$

59. Determine whether the ordered pair $(-1, 2)$ is a solution to each linear equation.
    a. $3x + y = -1$ **b.** $y = -2x + 4$
    c. $x = -1$ **d.** $y = 2$

Graphs of Linear Equations

60. Determine the x- and y-intercepts of each line.

a.

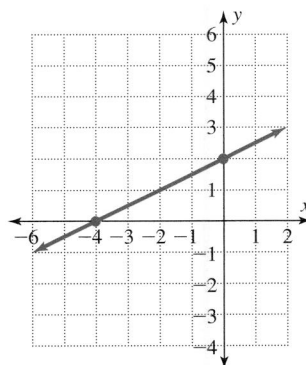

b.

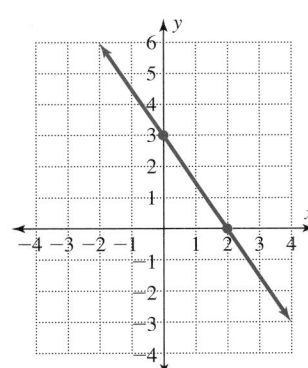

c. $y = 5x + 3$ **d.** $3x - 4y = 12$

61. Determine the point of intersection of these lines.

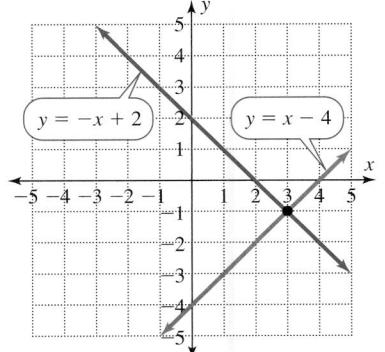

$y = -x + 2$ $y = x - 4$

62. The following table gives values for two linear equations with $y_1 = 5x + 3$ and $y_2 = 4x + 2$. Determine the point that satisfies both equations.

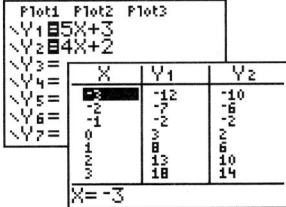

63. Graph each of these lines.

    a. The line through $(2, 4)$ and $(-1, -3)$

    b. The line with slope $m = \dfrac{3}{4}$ and with a y-intercept point $(0, 4)$

    c. The line with slope $m = -\dfrac{3}{2}$ and through $(1, 2)$

    d. The line defined by $3x - y = 6$

64. Match each equation with the description that best describes the graph of this linear equation.

    a. $y = \dfrac{1}{2}x + 5$

    b. $y = -2x + 3$

    c. $y = -2$

    d. $x = -3$

    A. A horizontal line

    B. A vertical line

    C. A line parallel to $y = -2x + 5$

    D. A line perpendicular to $y = -2x + 5$

65. Determine the slope of each line.

    a. The line through $(2, 3)$ and $(4, 1)$

    b. $y = \dfrac{3}{5}x - 4$

    c. $y = 4$

    d. $x = 4$

66. Write the equation of a line with a slope $m = 3$ and a y-intercept $(0, -4)$.

67. Write the equation of a line with a slope $m = -2$ through $(2, 6)$.

68. Write the equation of a line through $(1, 1)$ and $(2, 3)$.

69. Write the equation of a horizontal line through $(2, 3)$.

70. Write the equation of a vertical line through $(2, 3)$.

71. Write the equation of a line through $(0, 4)$ and parallel to $y = -3x + 7$.

Systems of Linear Equations

72. Select the choice that best describes each system of linear equations.

    a. $y = 4x - 5$
        $y = 5x + 4$

    b. $y = 4x - 5$
        $y = 4x + 5$

    c. $y = 4x - 5$
        $y = 4x - 5$

    A. A consistent system of independent equations

    B. A consistent system of dependent equations

    C. An inconsistent system of independent equations

73. Solve this system of linear equations by the substitution method.

$$y = 2x + 3$$
$$3x + 2y = 20$$

74. Solve this system of linear equations by the addition method.

$$3x - 5y = 7$$
$$3x + 5y = -13$$

75. Solve this system of linear equations by the addition method.

$$-3x + 5y = 8$$
$$6x + 7y = 1$$

Applications of Equations

76. Direct Variation If y varies directly as x and y is 8 when x is $\dfrac{1}{4}$, find y when x is 2.

77. Complementary Angles Two angles are complementary. If the larger angle is 6° more than 6 times the smaller angle, determine the number of degrees in each angle.

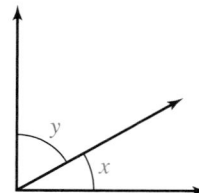

78. Amount of Two Investments A high school student invested $2000 of summer income to save for college. A portion was deposited in a savings account that earned only 4%. The rest was invested in a local electrical utility and earned 8%. The total income for one year was $136. How much did the student put in the savings account?

79. Rates of Two Airplanes A jet plane and a tanker that are 450 mi apart head toward each other so that the jet can refuel. The jet flies 100 mi/h faster than the tanker. Determine the speed of each aircraft if they meet in 45 minutes. (*Hint:* Use consistent units of measurement.)

80. Mixture of Medicine The dosage of a medicine ordered by a doctor is 40 ml of a 22% solution. A nurse has available a 10% solution and a 25% solution of this medicine. How many ml of each solution could be mixed to prepare this 40-ml dosage?

81. Height of a Box The volume of the box shown here is $6x^3 - 4x^2$ cm³. Determine the height of this box.

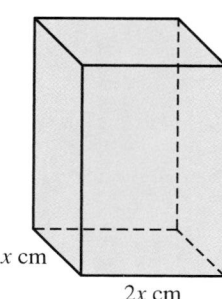

x cm

$2x$ cm

82. Perimeter and Area

　a. Calculate the perimeter of the rectangle shown to the right.

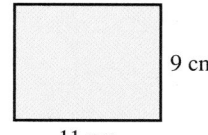

9 cm

11 cm

　b. Calculate the area of the rectangle shown to the right.

83. Tree Shadow

　A man 2 m tall casts a 6-m shadow. Nearby is a tree that simultaneously casts a 39-m shadow. How tall is the tree?

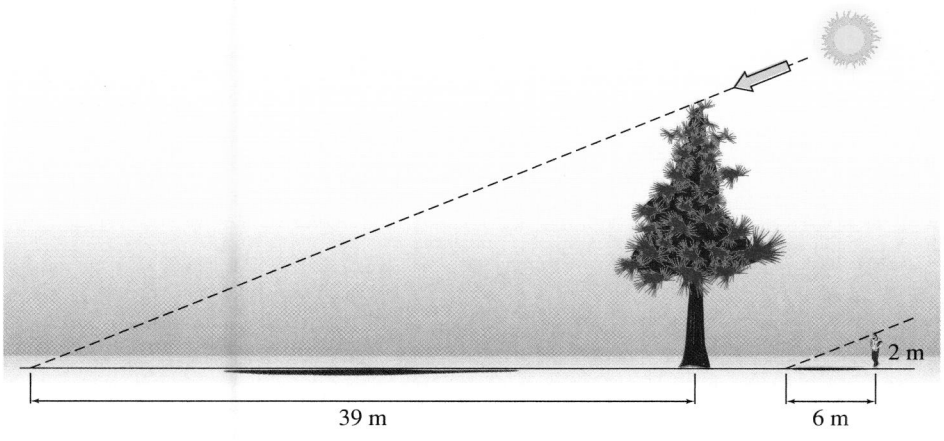

39 m　　　　6 m　　2 m

Linear Inequalities

84. Use this table to solve each inequality.

　a. $4x - 1 \geq 3x + 1$

　b. $4x - 1 \leq 3x + 1$

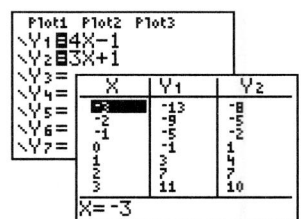

85. Use the following graph to solve each inequality.

　a. $2x + 1 \geq -x - 2$

　b. $2x + 1 \leq -x - 2$

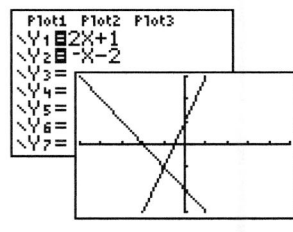

$[-5, 5, 1]$ by $[-3, 3, 1]$

86. Algebraically solve:

$$2(x + 3) < 4(x - 1) - 3(x + 2)$$

87. Identify each of these inequalities as either a conditional inequality, an unconditional inequality, or a contradiction. Then solve each inequality.

　a. $x + x < 0$

　b. $x + x < 2x$

　c. $x + x \leq 2x$

88. Solve each equation and inequality.

　a. $|2x - 1| = 7$

　b. $|2x + 1| < 5$

　c. $|x - 3| \geq 5$

89. Solve $2 < 3x - 4 \leq 5$.

90. Solve $2x + 1 \leq -5$ or $3x - 2 \geq 4$.

91. Graph the solution of $3x - 4y \leq 12$.

92. Graph the solution of this system of inequalities.

$$x + 2y \geq 2$$
$$-x + 2y < 4$$

93. Graph the solution of this system of inequalities.

$$x \geq 0$$
$$y \geq 0$$
$$3x + y \leq 3$$

Question	Answer	Reference Example	Question	Answer	Reference Example
1.a	8	[1.2–4]	12.a	−0.87	[1.2–4]
1.b	16	[1.3–2]	12.b	−0.93	[1.3–4]
1.c	−48	[1.4–2]	12.c	−0.027	[1.4–2]
1.d	−3	[1.5–1]	12.d	−30	[1.5–1]
2.a	−18	[1.2–2]	13.a	2	[1.6–3]
2.b	−6	[1.3–2]	13.b	−14	[1.6–3]
2.c	72	[1.4–2]	13.c	−42	[1.6–3]
2.d	2	[1.5–1]	13.d	−12	[1.6–3]
3.a	−12	[1.1–2]	14.a	−6	[1.6–4]
3.b	−12	[1.1–2]	14.b	14	[1.6–4]
3.c	0	[1.4–2]	14.c	6	[1.6–4]
3.d	Undefined	[1.5–3]	14.d	−46	[1.6–4]
4.a	$\dfrac{11}{10}$	[1.2–2]	15.a	$\dfrac{1}{3}$	[1.6–3]
4.b	$\dfrac{1}{2}$	[1.3–4]	15.b	$\dfrac{11}{5}$	[1.6–3]
4.c	$\dfrac{6}{25}$	[1.4–2]	15.c	$\dfrac{2}{3}$	[1.6–3]
4.d	$\dfrac{8}{3}$	[1.5–2]	15.d	$\dfrac{23}{2}$	[1.6–3]
5.a	8	[1.6–1]	16.a	0	[1.6–5]
5.b	9	[1.6–1]	16.b	50	[1.6–5]
5.c	1	[1.6–2]	16.c	4	[1.6–5]
5.d	$\dfrac{1}{6}$	[5.3–1]	16.d	−144	[1.6–5]
			17.a	1	[5.2–3]
6.a	$\dfrac{4}{9}$	[1.6–1]	17.b	3	[5.2–3]
			17.c	1	[5.2–3]
6.b	$\dfrac{3}{2}$	[5.3–2]	17.d	−1	[5.2–3]
			18.a	5	[5.3–1]
6.c	$\dfrac{9}{4}$	[5.3–2]	18.b	$\dfrac{6}{5}$	[5.3–1]
6.d	1	[5.2–3]	18.c	5	[5.3–1]
7.a	0	[1.6–1]	18.d	$\dfrac{5}{2}$	[5.3–1]
7.b	1	[5.2–3]			
7.c	0	[1.5–3]	19.a	26	[1.6–3]
7.d	Undefined	[1.5–3]	19.b	24	[1.6–3]
8.a	27	[1.1–3]	19.c	−60	[1.6–3]
8.b	27	[1.1–3]	19.d	36	[1.6–3]
8.c	−27	[1.1–3]	20.a	−13	[1.2–8]
8.d	0	[1.1–3]	20.b	5	[1.3–7]
9.a	7	[1.1–7]	20.c	36	[1.4–4]
			20.d	4	[1.5–4]
9.b	$\dfrac{2}{3}$	[1.1–9]	21.a	3	[1.3–7]
9.c	7	[1.1–7]	21.b	21	[1.3–7]
9.d	5	[1.1–7]	21.c	29	[1.3–7]
10.a	100	[1.6–1]	21.d	29	[1.3–7]
10.b	10,000	[1.6–1]	22.a	1	[5.1–7]
10.c	1	[5.2–3]	22.b	25	[5.1–7]
10.d	0.1	[5.3–2]	22.c	−65	[5.1–7]
11.a	123	[5.3–5]	22.d	25	[5.1–7]
11.b	12,300	[5.3–5]	23.a	2	[5.1–7]
11.c	0.123	[5.3–5]	23.b	34	[5.1–7]
11.d	0.0123	[5.3–5]	23.c	23	[5.1–7]

Question	Answer	Reference Example	Question	Answer	Reference Example
23.d	$\dfrac{25}{3}$	[5.1–7]	41.a	0	[1.1–1]
			41.b	-7	[1.1–1]
24.a	1	[5.2–3]	42.a	1	[1.4–5]
24.b	2	[5.2–3]	42.b	$\dfrac{1}{7}$	[1.4–5]
24.c	3	[5.2–3]			
24.d	10	[5.2–3]	43.a	[2, 4)	[1.1–5]
25.a	$2x^3$	[5.1–8]	43.b	$(-2, 5]$	[1.1–5]
25.b	x^{11}	[5.1–3]	43.c	$[-1, 0]$	[1.1–5]
25.c	x^5	[5.2–1]	43.d	$[3, \infty)$	[1.1–5]
25.d	x^{24}	[5.1–5]	44.a	$(-3, 1]$	[1.1–5]
26.a	$2x^2y$	[5.4–7]	44.b	$(-\infty, 2)$	[1.1–5]
26.b	$32x^3y^4$	[5.1–3]	44.c	$[5, 13]$	[4.3–8]
26.c	$2xy^2$	[5.2–4]	44.d	(6, 11)	[4.3–7]
26.d	$64x^4y^6$	[5.1–5]	45.a	4, 5, 6, 7, 8	[1.3–8]
27.a	45,600	[5.3–7]	45.b	3, 6, 9, 12, 15	[1.4–8]
27.b	0.00456	[5.3–7]	45.c	0, 3, 8, 15, 24	[1.4–8]
28.a	$5x + 3y$	[5.4–7]	45.d	7, 6, 5, 4, 3	[1.3–8]
28.b	$x + 11y$	[5.4–7]	46.a	Arithmetic, $d = 3$	[2.1–6]
28.c	$x + 15y$	[5.4–7]	46.b	Arithmetic, $d = -2$	[2.1–6]
28.d	$6x^2 + 2xy - 28y^2$	[5.5–3]	46.c	Not arithmetic	[2.1–6]
29.a	$7x^2 - x - 6$	[5.4–7]	46.d	Arithmetic, $d = 0$	[2.1–6]
29.b	$x^2 - 9x + 12$	[5.4–7]	47.a	-3	[2.2–1]
30.a	$x^3 + 2x^2 - 11x - 12$	[5.5–4]	47.b	2	[2.2–1]
30.b	$16x^2 - 49$	[5.7–1]	47.c	-8	[2.2–1]
31.a	$9x^2 - 30x + 25$	[5.7–4]	47.d	47	[2.2–1]
31.b	$x^3 - 5x^2 - 2x + 24$	[5.7–5]	48.a	0, selling 0 units produces a profit of $0	[5.4–6]
32.a	$3x^2 + 2x$	[5.6–2]	48.b	1800, selling 10 units produces a profit of $1800	[5.4–6]
32.b	$2x - 1$	[5.6–4]			
33.a	$x^2 - 2x + 3$	[5.6–3]	48.c	3750, selling 25 units produces a profit of $3750	[5.4–6]
33.b	$x - 4$	[5.6–3]			
34.a	$x - 8$	[5.6–3]	48.d	5000, selling 50 units produces a profit of $5000	[5.4–6]
34.b	$x^2 + 3x + 9$	[5.6–6]			
35.a	$(x + 10)(x - 10)$	[5.7–3]	49	$f(x) = 16 - 2x$	[2.2–5]
35.b	$(2x + 3y)(2x - 3y)$	[5.7–3]	50	$R(x) = 5000x - 0.01x^2$	[5.5–7]
36.a	$(5x + 4)^2$	[5.7–6]		$R(1000) = 4,990,000$	
36.b	$(2x - 7y)^2$	[5.7–6]		At a price of $1000 per item the company estimates a revenue of $4,990,000.	
37	$2x$	[5.4–1]			
38	$5x^3 + 11$	[5.4–4]			
39.a	$-7, -4.73, -\sqrt{4},$ $-\dfrac{3}{7}, 0, 17$	[1.1–8]	51.a	Solution	[1.3–9]
			51.b	Not a solution	[1.3–9]
			51.c	Not a solution	[1.3–9]
39.b	$-\pi, \sqrt{7}$	[1.1–8]	51.d	Solution	[1.3–9]
39.c	$-7, -\sqrt{4}, 0, 17$	[1.1–8]	52.a	$x = -3$	[2.4–4]
39.d	17	[1.1–8]	52.b	$x = 1$	[2.4–5]
40.a	Distributive property of multiplication over addition	[1.6–6]	52.c	No solution	[2.4–7]
			52.d	Every real number is a solution	[2.4–6]
40.b	Commutative property of addition	[1.2–6]	53.a	$x = -23$	[2.5–2]
			53.b	$x = 5$	[2.5–4]
40.c	Commutative property of multiplication	[1.4–1]	53.c	$x = 6$	[2.5–6]
			53.d	$x = -12$	[2.5–8]
40.d	Associative property of addition	[1.2–6]	54.a	$y = 2x - 3$	[2.7–1]

continued

Question	Answer	Reference Example
54.b	$y = -\dfrac{2}{5}x - 1$	[2.7–4]
55.	$x = 1.6$	[2.4–8]
56.	$x = -3$	[2.4–8]
57.a	$6x - 18$	[1.6–7]
57.b	$x = 3$	[2.5–6]
58.a	$8x + 32$	[1.6–7]
58.b	$x = -4$	[2.5–6]
59.a	Solution	[2.3–1]
59.b	Not a solution	[2.3–1]
59.c	Solution	[2.3–1]
59.d	Solution	[2.3–1]
60.a	$(-4, 0), (0, 2)$	[2.3–4]
60.b	$(2, 0), (0, 3)$	[2.3–4]
60.c	$\left(-\dfrac{3}{5}, 0\right), (0, 3)$	[2.7–3]
60.d	$(4, 0), (0, -3)$	[2.7–3]
61.	$(3, -1)$	[3.3–1]
62.	$(-1, -2)$	[3.3–1]
63.a	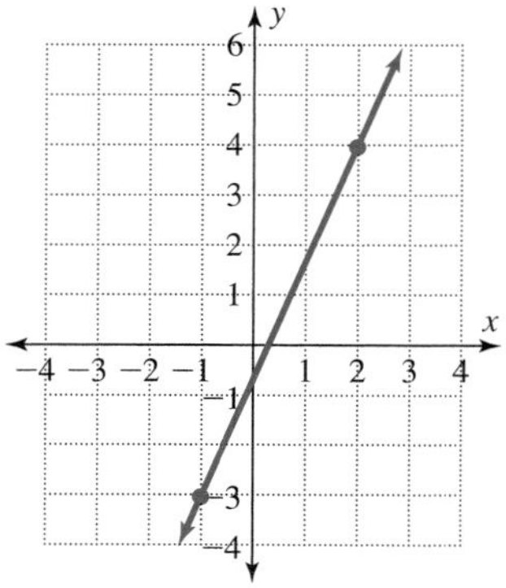	[2.3–5]

63.b   [3.1–4]

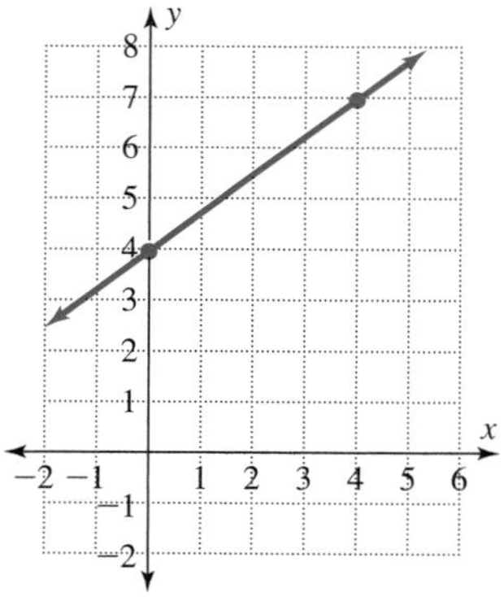

Question	Answer	Reference Example
63.c	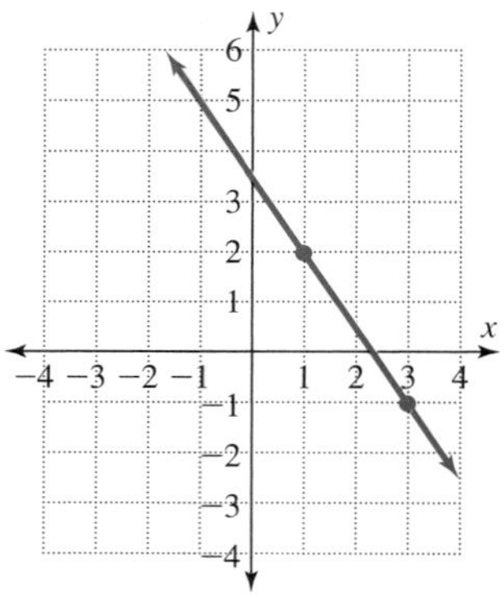	[3.2–9]

63.d   [3.2–3]

Question	Answer	Reference Example
64.a	D	[3.2–4]
64.b	C	[3.2–4]
64.c	A	[3.2–5]
64.d	B	[3.2–5]
65.a	-1	[3.1–1]
65.b	$\dfrac{3}{5}$	[3.2–2]
65.c	0	[3.1–6]
65.d	Undefined	[3.1–6]
66.	$y = 3x - 4$	[3.2–1]
67.	$y = -2x + 10$	[3.2–7]
68.	$y = 2x - 1$	[3.2–8]
69.	$y = 3$	[3.2–1]
70.	$x = 2$	[3.2–5]
71.	$y = -3x + 4$	[3.2–1]
72.a	A	[3.3–4]
72.b	C	[3.3–4]
72.c	B	[3.3–4]
73.	$(2, 7)$	[3.4–1]
74.	$(-1, -2)$	[3.5–1]
75.	$(-1, 1)$	[3.5–2]
76.	$y = 64$	[2.6–6]
77.	$12°, 78°$	[3.6–2]

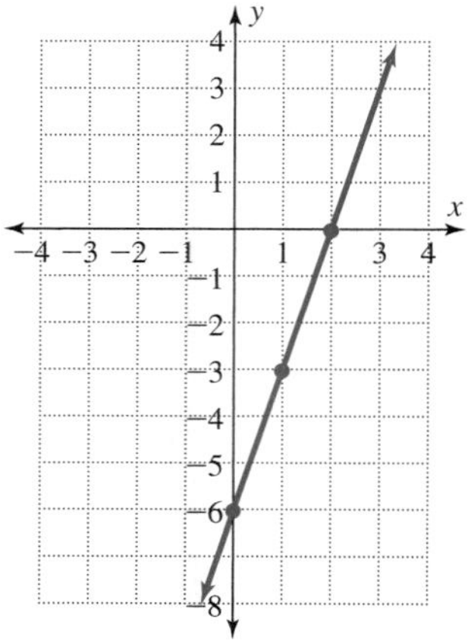

Question	Answer	Reference Example
78.	$600 was invested in the savings account.	[3.6–4]
79.	The tanker flew 250 mi/h and the jet flew 350 mi/h.	[3.6–5]
80.	Use 8 ml of 10% solution and 32 ml of 25% solution.	[3.6–6]
81.	$(3x - 2)$ cm	[5.6–2]
82.a	40 cm	[1.2–6]
82.b	99 cm^2	[1.4–1]
83.	13 m	[2.6–7]
84.a	$[2, \infty)$	[4.1–4]
84.b	$(-\infty, 2]$	[4.1–4]
85.a	$[-1, \infty)$	[4.1–4]
85.b	$(-\infty, -1]$	[4.1–4]
86.	$(-\infty, -16)$	[4.1–7]
87.a	Conditional inequality, $(-\infty, 0)$	[4.3–1]
87.b	Contradiction, no solution	[4.3–1]
87.c	Unconditional inequality, $\mathbb{R}$	[4.3–1]
88.a	$x = -3, x = 4$	[4.4–3]
88.b	$(-3, 2)$	[4.4–4]
88.c	$(-\infty, -2] \cup [8, \infty)$	[4.4–5]
89.	$(2, 3]$	[4.3–6]
90.	$(-\infty, -3] \cup [2, \infty)$	[4.3–8]
91.		[4.5–3]

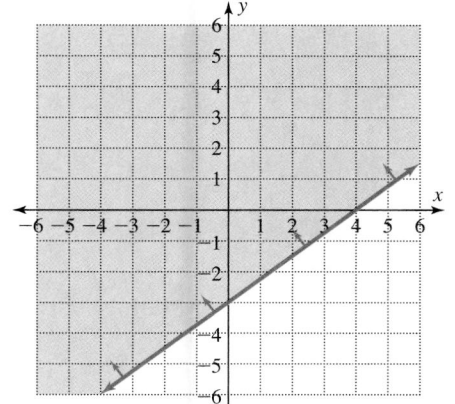

Question	Answer	Reference Example
92.		[4.5–8]

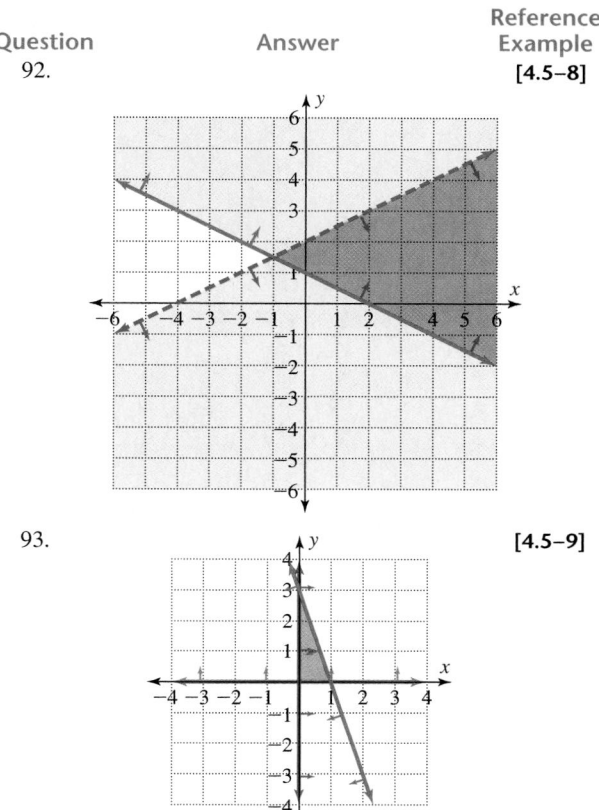

93.		[4.5–9]

6

USING

COMMON

ALGEBRAIC

FUNCTIONS

379

Many mathematicians view the concept of a function as the single most important concept in mathematics.* Due to the importance of this function concept, this chapter examines functions from many different perspectives, including algebraic, numerical, and graphical.

Section 6.1   Functions and Representations of Functions

Objectives:
1. Determine whether a relation is a function.
2. Determine the domain and range of a function.
3. Use function notation.

A Mathematical Note

The word *function* was introduced by the German mathematician Gottfried Wilhelm Freiherr von Leibniz (c. 1682). Leibniz is credited as a developer of calculus, together with the English mathematician Sir Isaac Newton (c. 1680). Leibniz contributed both original terminology and notation, as well as his results in the areas of algebra and calculus.

Function notation was first introduced in Section 2.2 when we examined linear functions written in the form $f(x) = mx + b$. In this section we define a function, reexamine function notation, and examine functions from multiple perspectives:

- Mapping representation of a function.
- Ordered-pair representation of a function.
- Table representation of a function.
- Graphical representation of a function.
- Function notation representation of a function.

Definition of a Function

In mathematics we use the word *function* to designate a correspondence that has very specific properties. A function must match each value of the input variable with exactly one value of an output variable.

Function

A **function** is a correspondence that matches each input value with exactly one value of the output variable. The set of all possible input values is called the **domain** of the function or the set of **independent** values. The set of all output values is called the **range** of the function or the set of **dependent** values.

When you first begin studying the function concept, you may want to think about three different interrelated parts of a function:

1. The domain of the function, the set of all possible input values.
2. The range of the function, the set of all possible output values.
3. The correspondence (rule or formula) which pairs each domain element with exactly one range element.

A key point is to understand the importance of the words "exactly one." Each input value is paired with one and only one output value.

Mapping Representation of a Function

Functions can be denoted by a variety of notations. In the next example we use mapping notation. The mapping or arrow notation clarifies the pairing that a function creates between the domain elements and the range elements.

The Concept of Function: Aspects of Epistemology and Pedagogy, by Ed Dubinsky and Gvershon Harel, MAA Notes Volume 25, 1992.

■ EXAMPLE 1   Identifying Functions Using Mapping Notation

Classify each correspondence as either a function or as a correspondence that is not a function. For each function identify the domain and range.

SOLUTION _____

(a) | D | R |
     |---|---|
     | 3 | → 15 |
     | 9 | → 39 |
     | 13 | → 55 |

This correspondence is a function.
Domain $D = \{3, 9, 13\}$
Range $R = \{15, 39, 55\}$

Each domain element is paired with exactly one output value in the range.

Compare parts (b) and (c) of Example 1. It is important to understand the distinction between these examples.

(b) D   R
     16 → 4
        ↘ −4
     0 → 0

This correspondence is not a function.

The element 16 is not paired with exactly one element in the range; it is paired with both 4 and −4.

(c) D   R
     −4 → 16
     4 ↗
     0 → 0

This correspondence is a function.
Domain $D = \{-4, 0, 4\}$
Range $R = \{0, 16\}$

−4 is paired only with 16; 4 is paired only with 16; and 0 is paired only with 0. Thus each element in D is paired with exactly one element in R. ■

SELF-CHECK 6.1.1

Classify each correspondence as either a function or a correspondence that is not a function. Give the domain and range of each function.

1. D   R
    0 → 0
    3 → 3
    −3 ↗

2. D   R
    0 → 0
    3 → 3
       ↘ −3

The mapping notation helps to emphasize that a function can be viewed as a process that takes input values and produces a unique output.

Ordered-Pair Representation of a Function

The arrows in the mapping notation for functions have the advantage of clearly stressing the active nature of a function in pairing input values with output values. Conceptually this is a great notation. Practically, mapping notation has some limitations. If there are many domain values it can take a long time to draw all the arrows for the function. Arrows are also more awkward to type than other notations. The ordered-pair notation shown in Example 2 can convey the same information more concisely. The first coordinate of an ordered pair is the input value, the second coordinate is the output value, and the parentheses establish how these elements are paired.

A **relation** is defined as any set of ordered pairs. Thus some relations are functions and some are not. To be a function, a relation must pair each input value with exactly one output value.

SELF-CHECK 6.1.1 ANSWERS

1. Function with
   $D = \{-3, 0, 3\}$
   and $R = \{0, 3\}$
2. Not a function

■ **EXAMPLE 2** Comparing Mapping Notation and Ordered-Pair Notation

(a) Convert this function given in mapping notation to ordered-pair notation.

SOLUTION _____

Mapping Notation   *Ordered-Pair Notation*
D   R

$-2 \to -1$ $(-2, -1)$ Both notations indicate a function whose domain
$-1 \to 3$ $(-1, 3)$ is $D = \{-2, -1, 0, 1, 2\}$ and whose range is R
$0 \to 5$ $(0, 5)$ $= \{-1, 3, 5, 7, 9\}$. The two notations also pair
$1 \to 7$ $(1, 7)$ the elements exactly the same way.
$2 \to 9$ $(2, 9)$

This function can be written as the set of ordered pairs:

$$\{(-2, -1), (-1, 3), (0, 5), (1, 7), (2, 9)\}.$$

(b) Convert this function given as a set of ordered pairs to mapping notation.

SOLUTION _____

Ordered-Pair Notation   *Mapping Notation*
$\{(-5, 5), (5, 5), (-7, 7), (7, 7)\}$ D   R

$-5 \to 5$ Both notations indicate a function
$5 \nearrow$ whose domain is $D = \{-5, 5, -7,$
                     $7\}$ and whose range is $R = \{5, 7\}$.
$-7 \to 7$
$7 \nearrow$ ■

SELF-CHECK 6.1.2

Classify each relation as either a function or a correspondence that is not a function. Give the domain and range of each function.

1. $\{(1, 6), (3, 8), (\pi, 10), (9, 12)\}$
2. $\{(1, 6), (1, 8), (\pi, 10), (9, 12)\}$

Graphical Representation of a Function

The Cartesian coordinate system provides a pictorial means of presenting the relationship between two variables. Since each point in a plane can be uniquely identified by an ordered pair (x, y), graphs can be used to represent mathematical relations. The x-coordinate of the ordered pair (x, y) is called the **independent variable,** and the y-coordinate is called the **dependent variable.**

■ EXAMPLE 3   Graphing Relations

Graph each of these relations and determine whether the relation is a function.

SOLUTIONS

(a) $\{(2, -2), (2, -1), (2, 0), (2, 1), (2, 2)\}$

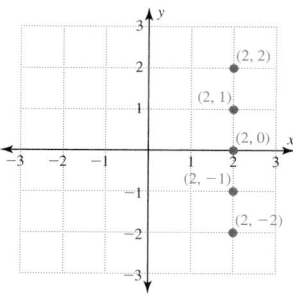

Answer: Not a function

This relation is not a function. The input of 2 has more than one output paired with it.

(b)

D		R
-2	$\rightarrow$	-3
-1	$\rightarrow$	-1
0	$\rightarrow$	1
1	$\rightarrow$	3
2	$\rightarrow$	5

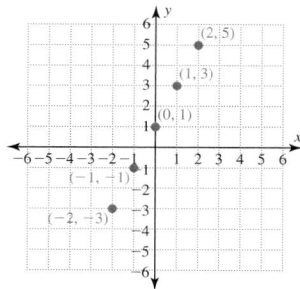

Answer: A function

This relation is a function with domain $D = \{-2, -1, 0, 1, 2\}$ and range $R = \{-3, -1, 1, 3, 5\}$.

(c)

x	y
-2	2
-1	2
0	2
1	2
2	2

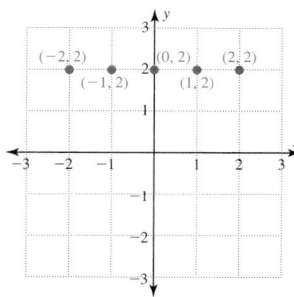

Answer: A function

This relation is a function with domain $D = \{-2, -1, 0, 1, 2\}$ and range $R = \{2\}$. Each input value has exactly one output paired with it. Each input is paired with 2.

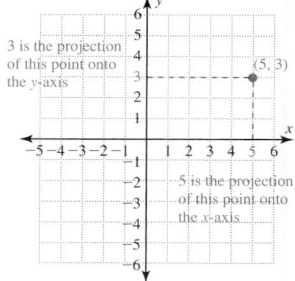

If a function is defined by a graph, then the ordered pairs can be determined by examining the points that form the graph. The domain can be found by projecting these points onto the *x*-axis, and the range can be found by projecting these points onto the *y*-axis. This is illustrated by the point (5, 3) shown to the left.

Domain and Range from the Graph of a Function

Domain: The domain of a function is the projection of its graph onto the *x*-axis.
Range: The range of a function is the projection of its graph onto the *y*-axis.

■ **EXAMPLE 4**   Determining the Domain and Range from the Graph of a Function

Write the domain and the range of the functions defined by these graphs.

SOLUTIONS _____

(a)

Domain: Projection onto x-*axis*

Range: Projection onto y-*axis*

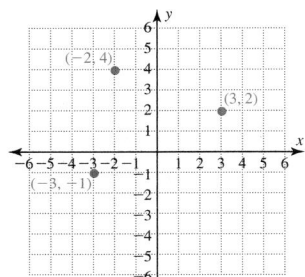

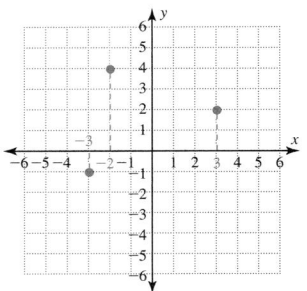

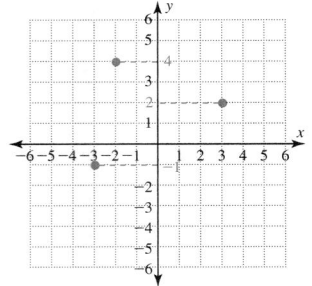

Domain = {−3, −2, 3}

This function consists of the
ordered pairs
{(−3, −1), (−2, 4), (3, 2)}.

Range = {−1, 2, 4}

(b)

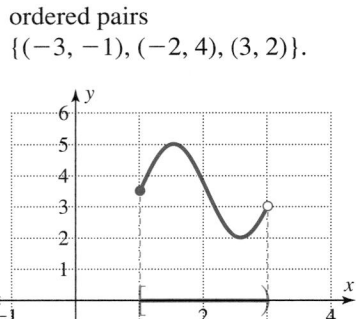

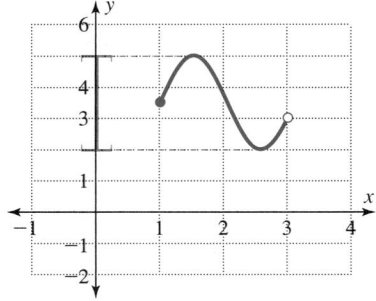

Domain = [1, 3)

The domain includes the
endpoint 1 but does not
include the endpoint 3.

Range = [2, 5]

(c)

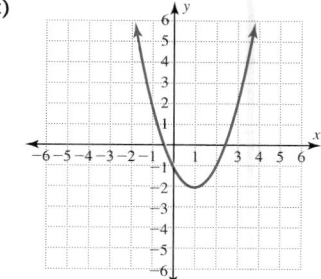

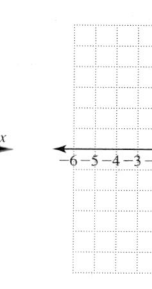

Domain = ℝ, the set of all real numbers.

Range = [−2, +∞)

SELF-CHECK 6.1.3

Write the domain and the range of the functions defined by these graphs.

1.
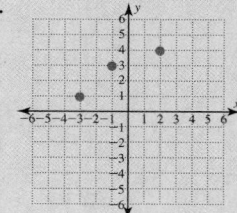

2.
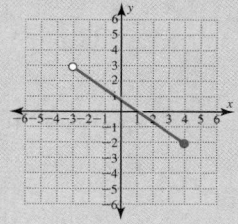

The Vertical Line Test

An easy way to determine if a graph represents a function is to imagine a vertical line sweeping across the graph. A vertical line hitting a graph in exactly one point pairs that input value of x with exactly one output value of y. A vertical line hitting a graph in two points pairs that input value of x with two output values of y—thus the graph does not represent a function.

A Graph of a Function

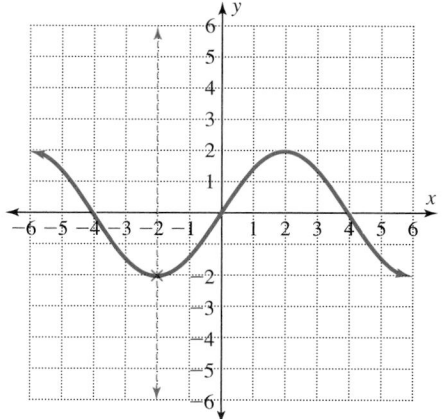

A Graph That Is Not a Function

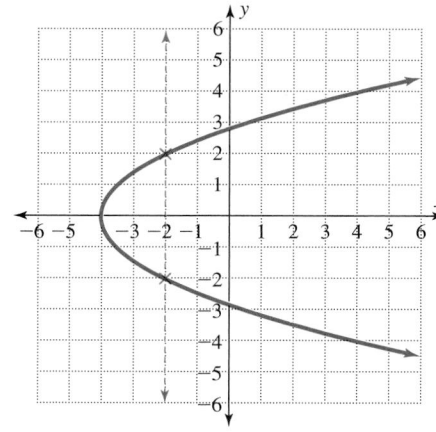

The Vertical Line Test

A graph represents a function if it is impossible to have any vertical line intersect the graph at more than one point.

■ EXAMPLE 5 Using the Vertical Line Test

Use the vertical line test to determine if the graphs of these relations represent functions.

SOLUTIONS

(a)
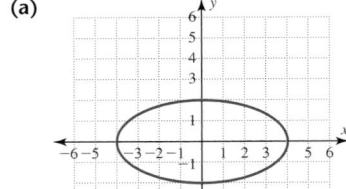
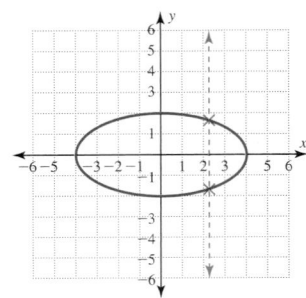

Answer: Not a function

It is possible to draw a vertical line, as the one shown here, that intersects the graph in two points.

(b)
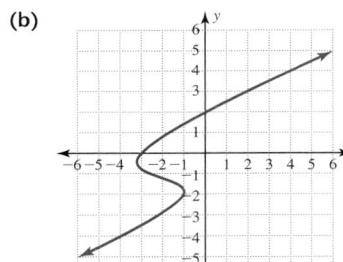
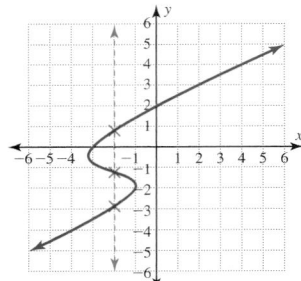

Answer: Not a function

The vertical line shown here intersects the graph at three points.

(c)
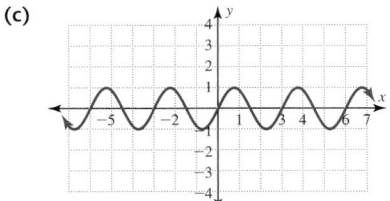
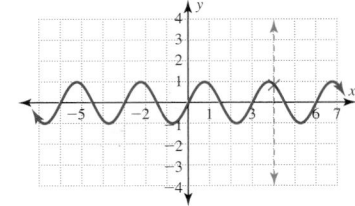

Answer: A function

Every vertical line will intersect this graph in exactly one point. This function pairs each input value with exactly one output value.

SELF-CHECK 6.1.4

Use the vertical line test to determine whether each relation is a function.

1.
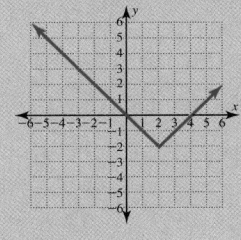

2.
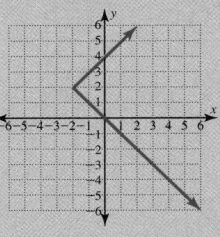

Function Notation Representation of a Function

For a function with many domain elements, it is not practical to list all possible ordered pairs. Thus we often examine the relationship between the domain elements and the range elements to find a pattern or formula that describes how each element in the domain is paired with an element in the range. Function notation (first covered in Section 2.2) can be used to state these formulas.

Function Notation

The expression $f(x)$ is read as "f of x" or "the value of f at x." The variable x represents an input value from the domain, and $f(x)$ is the corresponding output value.

It is common to use the letter f to represent the function although any letter can be used. In Example 6 we use function notation to represent a linear function. We first examined linear functions in Section 2.2.

■ EXAMPLE 6   Evaluating Function Notation

Evaluate $f(x) = 2x - 4$ for each input value and then graph this linear function.

SOLUTIONS

(a) $x = 0$ $f(x) = 2x - 4$
$f(0) = 2(0) - 4$ Substitute each input value of x into the formula $f(x) = 2x - 4$.
$f(0) = 0 - 4$
$f(0) = -4$ $(0, -4)$ is a point on the graph.

(b) $x = 2$ $f(x) = 2x - 4$
$f(2) = 2(2) - 4$
$f(2) = 4 - 4$
$f(2) = 0$ $(2, 0)$ is a point on the graph.

(c) $x = 3$ $f(x) = 2x - 4$
$f(3) = 2(3) - 4$
$f(3) = 6 - 4$
$f(3) = 2$ $(3, 2)$ is a point on the graph.

SELF-CHECK 6.1.4 ANSWERS

1. A function
2. A relation but not a function

The graph of $f(x) = 2x - 4$ is one continuous line, not just the three distinct points calculated here. These points allow us to represent the graph of all the points on this line.

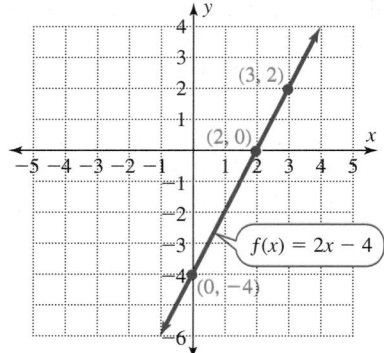

The equation $f(x) = 2x - 4$ is in the slope-intercept form ($y = mx + b$) of a line with slope $m = 2$ and y-intercept $(0, -4)$.

The letter x represents the input value. The output value is represented either by y or by $f(x)$.

In the next example the formula which produces this graph is given by $y = f(x)$. For each point (x, y) on the graph x represents an input value and y represents $f(x)$; $f(x)$ is the output of the function f for the input x.

■ **EXAMPLE 7** Using a Graph to Evaluate Input and Output Values

Use the given graph to:

(a) Evaluate $f(-3)$.
(b) Evaluate $f(2)$.
(c) Determine the value(s) of x for which $f(x) = -3$.
(d) Determine the value(s) of x for which $f(x) = 2$.

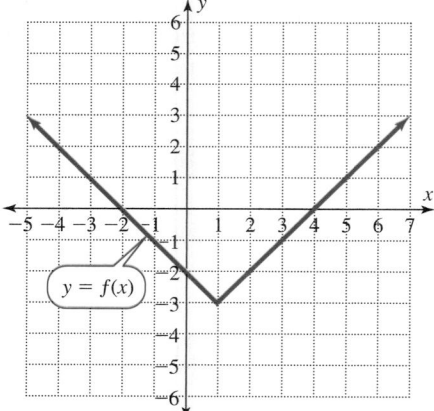

SOLUTIONS

(a) $f(-3) = 1$
(b) $f(2) = -2$
(c) $x = 1$
(d) $x = -4$ and $x = 6$

For each x-coordinate the y-coordinate represents $f(x)$. $(-3, 1)$ is a point on the graph, so $f(-3) = 1$. $(2, -2)$ is a point on the graph, so $f(2) = -2$. The y-coordinate is -3, where the x-coordinate is 1; $f(1) = -3$. There are two x-coordinates that have a y-coordinate of 2. $f(-4) = 2$ and $f(6) = 2$. ■

The formula that defines a function can be used to produce a graph of the function and then this graph can be used to determine the domain and the range of the function. This can be facilitated by the use of a graphics calculator.

■ EXAMPLE 8 Using a Graphics Calculator to Determine the Domain and the Range of a Function

Use a graphics calculator to determine the domain and the range of $f(x) = x^2 - 2$.

SOLUTION

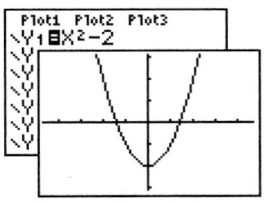

$[-4.7, 4.7, 1]$ by $[-3.1, 3.1, 1]$

Use $y = x^2 - 2$ to represent the equation. (Again either y or $f(x)$ can represent the output variable.)

The graph is continuing to spread to both the left and the right so the projection of the graph onto the x-axis will include the whole x-axis. This means the domain is the set of all real numbers.

The range, the projection of the graph onto the y-axis, is the interval $[-2, +\infty)$. This notation indicates that -2 is included in the range.

Answers: Domain $= \mathbb{R}$
Range $= [-2, +\infty)$

SELF-CHECK 6.1.6

Use the graph in Example 7 to answer these questions.

1. Evaluate $f(-2)$.
2. Evaluate $f(0)$.
3. Evaluate $f(1)$.
4. Determine the domain of this function.
5. Determine the range of this function.

Multiple Representations of a Function

As we noted at the beginning of this section, there are many ways to represent a function. In the next example we compare the representations given earlier in this section.

■ EXAMPLE 9 Using Multiple Representations for a Function

An investment in a bond tripled in value over a twenty-five-year period. Four individuals invested, respectively, $50, $100, $200, and $500. Using these four values as the input values, represent this function using function notation, mapping notation, a table, ordered pairs, and a graph.

SOLUTION _____

Verbal Representation
Triple each input value to obtain the output value.

Function Notation
$f(x) = 3x$ for each x in the domain
$D = \{50, 100, 200, 500\}$

Mapping Notation

x		$f(x)$
50	→	150
100	→	300
200	→	600
500	→	1500

This example contains only selected discrete points. Thus the graph contains only these discrete points. The dashed line is not part of the graph but does help to show the linear pattern.

Table

x	$f(x)$
50	150
100	300
200	600
500	1500

Ordered-Pair Notation
$\{(50, 150), (100, 300), (200, 600), (500, 1500)\}$

Graph

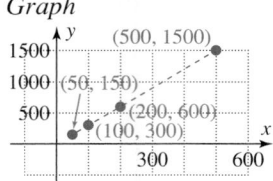

SELF-CHECK 6.1.7 ANSWERS

1.

x	$f(x)$
-2	9
0	5
1	3
3	-1

2. $\{(-2, 9), (0, 5), (1, 3), (3, -1)\}$

SELF-CHECK 6.1.7

Using the domain $D = \{-2, 0, 1, 3\}$ and $f(x) = -2x + 5$
1. Give a table to represent this function.
2. Represent this function as a set of ordered pairs.

USING THE LANGUAGE AND SYMBOLISM OF MATHEMATICS   6.1

1. A function is a correspondence that matches each input value with exactly _____ value of the _____ variable.
2. The set of all input values of a function is called the _____ of the function or the set of _____ values.
3. The set of all output values of a function is called the _____ of the function or set of _____ values.
4. In ordered-pair notation the domain of a function is represented by the set of _____ coordinates.
5. In ordered-pair notation the range of a function is represented by the set of _____ coordinates.

6. If a function is defined by a graph, the domain of the function is represented by the projection of the graph onto the _____ axis.
7. If a function is defined by a graph, the range of the function is represented by the projection of the graph onto the _____ axis.
8. The _____ line test can be used to inspect visually a graph to determine if it represents a function.
9. The notation $f(x)$ is called _____ notation.
10. The notation $f(x) = 3x^2$ is read "_____ of _____ equals three x squared." In this notation the input variable is represented by _____ and $f(x)$ represents the _____ variable.

EXERCISES   6.1

In Exercises 1–4 determine whether each relation is a function. For each function identify the domain and the range.

1.
a.
D		R
1	→	-1
1	→	1
2	→	2
3	→	3

b.
D		R
-1	→	1
1	→	1
2	→	2
-3	→	3

c.
D		R
3	→	-1
2	→	1
1	→	3
0	→	0

2.
a.
D		R
3	→	-5
3	→	0
3	→	4

b.
D		R
-5	→	3
0	→	3
4	→	3

c.
D		R
4	→	2
1	→	1
0	→	0
-1	→	0

3. a. $\{(1, 1), (-1, 1), (2, 0)\}$

b. $\{(1, 1), (1, -1), (0, 2)\}$

c.

x	y
-3	-1
-2	0
-1	1
0	2
1	3

4. a. $\{(2, 3), (2, -3), (0, 13), (13, 0)\}$

b. $\{(3, 2), (-3, 2), (0, 13), (13, 0)\}$

c.

x	y
-2	-8
-1	-1
0	0
1	1
2	8

In Exercises 5–10 use the vertical line test to determine whether each graph represents a function.

5. a.
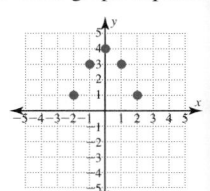

b.
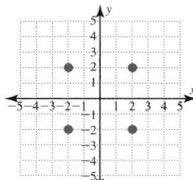

c.
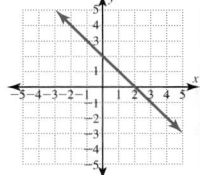

6. a.
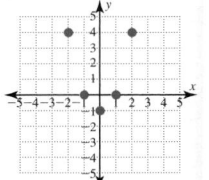

b.
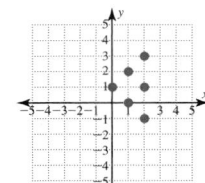

c.
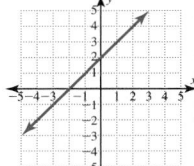

7. a.
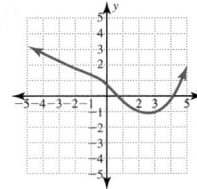

b.
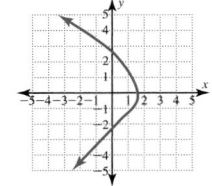

c.
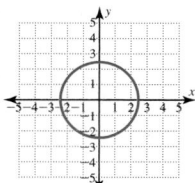

8. a.
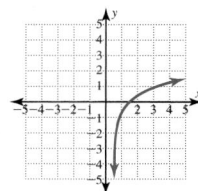

b.
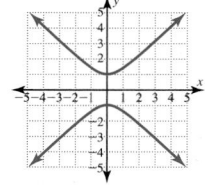

c.
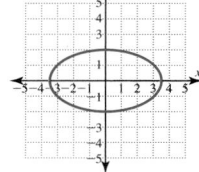

9. a.
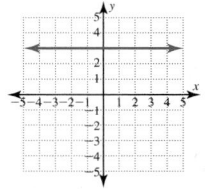

b.
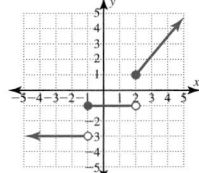

c.
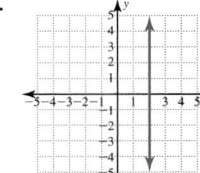

10. a.
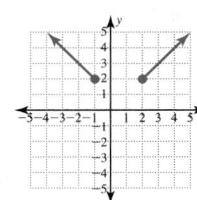

b.
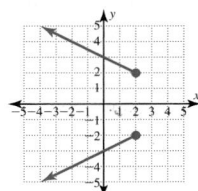

c.
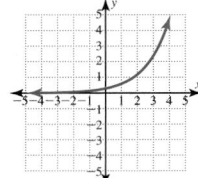

Given $f(x) = 2x - 7$, evaluate each expression in Exercises 11 and 12.

11. a. $f(-10)$ **b.** $f(0)$ **c.** $f(5)$

12. a. $f(-4)$ **b.** $f(-1)$ **c.** $f(10)$

Given $f(x) = \sqrt{x}$, evaluate each expression in Exercises 13 and 14.

13. a. $f(0)$ **b.** $f(9)$ **c.** $f\left(\dfrac{1}{4}\right)$

14. a. $f(16)$ **b.** $f(100)$ **c.** $f\left(\dfrac{9}{25}\right)$

15. Use the following table to answer each question.

x	y
−1	1
2	3
5	5
8	7

    a. Is 3 an input value or an output value?
    b. Is −1 an input value or an output value?
    c. If 2 is the input value, what is the output value?
    d. If the output value is 7, what is the input value?

16. Use the following set of ordered pairs to answer each question.
    $\{(0, 4), (3, 5), (5, 1), (2, 0)\}$
    a. Is 3 an input value or an output value?
    b. Is 4 an input value or an output value?
    c. If 5 is the input value, what is the output value?
    d. If the output value is 5, what is the input value?

17. Use this graph to answer each question.

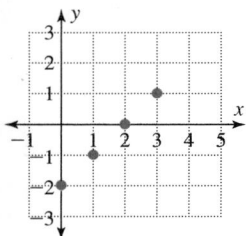

    a. Is 3 an input value or an output value?
    b. Is −1 an input value or an output value?
    c. If 2 is the input value, what is the output value?
    d. If the output value is 1, what is the input value?

18. Use this graph to answer each question.

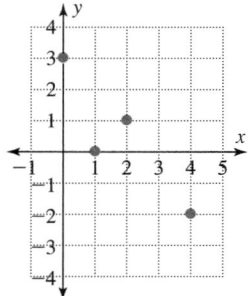

    a. Is −2 an input value or an output value?
    b. Is 4 an input value or an output value?
    c. If 2 is the input value, what is the output value?
    d. If the output value is 0, what is the input value?

In Exercises 19 and 20 use the given table to evaluate each value of $f(x)$.

19. a. $f(-3)$ **b.** $f(1)$ **c.** $f(2)$
20. a. $f(-2)$ **b.** $f(0)$ **c.** $f(3)$

x	f(x)
−3	8
−2	3
−1	0
0	−1
1	−2
2	−5
3	−10

In Exercises 21 and 22 use the given table to determine the value of x that produces the given value of $f(x)$.

21. a. $f(x) = -1$ **b.** $f(x) = 0$ **c.** $f(x) = 8$
22. a. $f(x) = 3$ **b.** $f(x) = -2$ **c.** $f(x) = -10$

In Exercises 23 and 24 use the given graph to evaluate each value of $f(x)$.

23. a. $f(-2)$ **b.** $f(0)$
    c. $f(3)$
24. a. $f(-1)$ **b.** $f(1)$
    c. $f(2)$

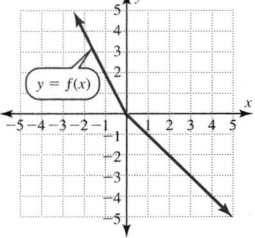

In Exercises 25 and 26 use the given graph to determine the value of x that produces the given value of $f(x)$.

25. a. $f(x) = 4$ **b.** $f(x) = 0$
    c. $f(x) = -2$
26. a. $f(x) = 2$ **b.** $f(x) = -4$
    c. $f(x) = -3$

In Exercises 27–30 use the given graph to determine the domain and the range of each function.

27. a.
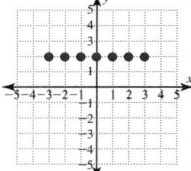

    b.
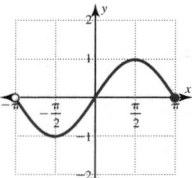

    c.
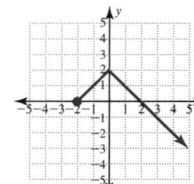

28. a.
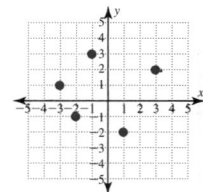

    b.
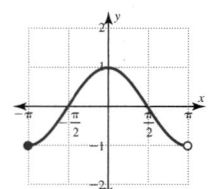

    c.
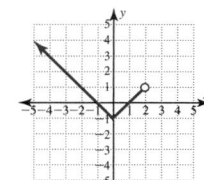

29. a.

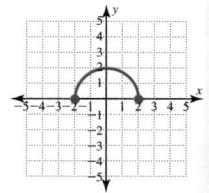

b.

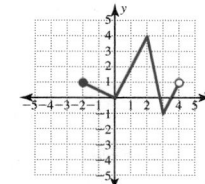

c.

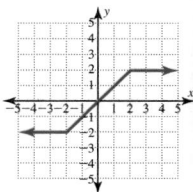

30. a.

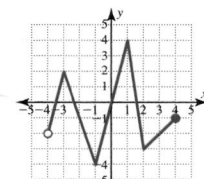

b.

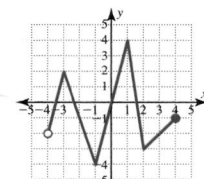

c.

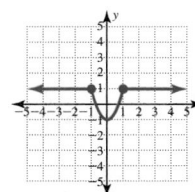

Multiple Representations

In Exercises 31–38 represent each function multiple ways as described in each problem.

31. Use the function defined by the following table to complete each part of this exercise.

x	−5	−3	−2	0	1	4
y	4	2	0	−2	−3	−4

    a. Express this function using mapping notation.
    b. Express this function using ordered-pair notation.
    c. Graph this function.

32. Use the function defined by the following table to complete each part of this exercise.

x	−5	−3	−2	0	1	4
y	3	3	3	3	3	3

    a. Express this function using mapping notation.
    b. Express this function using ordered-pair notation.
    c. Graph this function.

33. Use the function defined by the following graph to complete each part of this exercise.

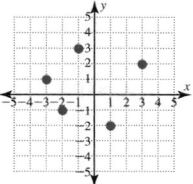

    a. Express this function using mapping notation.
    b. Express this function using ordered-pair notation.
    c. Express this function using a table format.

34. Use the function defined by this graph to complete each part of this exercise.

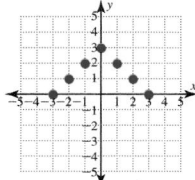

    a. Express this function using mapping notation.
    b. Express this function using ordered-pair notation.
    c. Express this function using a table format.

35. Use the function defined by the given mapping notation to complete each part of this exercise.

D		R
−1	→	1
1	→	3
2	→	−1
4	→	−2

    a. Express this function using a table format.
    b. Express this function using ordered-pair notation.
    c. Graph this function.

36. Use the function defined by the given mapping notation to complete each part of this exercise.

D		R
−4	→	−4
−2	→	−2
0	→	0
1	→	1
3	→	3

    a. Express this function using a table format.
    b. Express this function using ordered-pair notation.
    c. Graph this function.

37. Use the function $\{(-5, 4), (-3, 4), (1, 4), (2, 4), (3, 4)\}$ to complete each part of this exercise.
    a. Express this function using a table format.
    b. Express this function using mapping notation.
    c. Graph this function.

38. Use the function $\{(-4, -2), (-2, 4), (4, -3), (3, 1)\}$ to complete each part of this exercise.
   a. Express this function using a table format.
   b. Express this function using mapping notation.
   c. Graph this function.

39. Test Item Analysis An instructor analyzing the results from an Intermediate Algebra quiz prepared the following histogram. Use the function defined by this histogram to complete each part of this exercise.

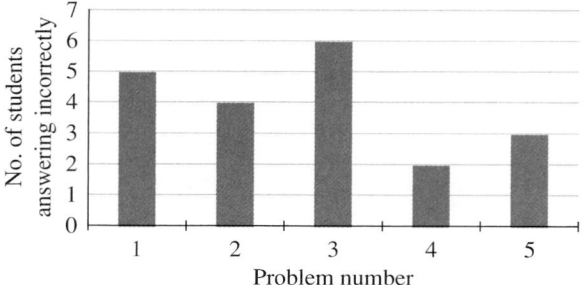

   a. If the input x represents the problem number, the output y represents _____.
   b. How many students missed Problem 5?
   c. Which problem was missed by 5 students?
   d. Which problem was answered incorrectly by the most students?

40. Test Item Analysis An instructor analyzing the results from the same Intermediate Algebra quiz from Exercise 39 prepared the following histogram. Use the function defined by the given histogram to complete each part of this exercise.

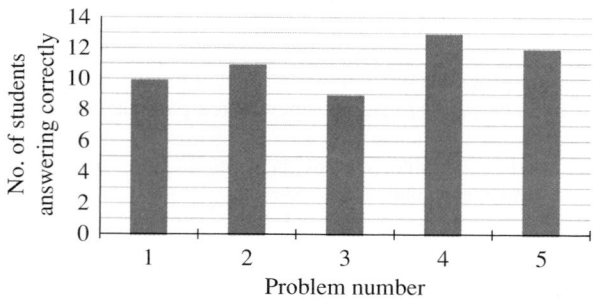

   a. If the input x represents the problem number, the output y represents _____.
   b. How many students answered Problem 5 correctly?
   c. Which problem was correctly answered by 11 students?
   d. How many students took this quiz? (*Hint:* See both Exercises 39 and 40.)

41. Boyle's Law Boyle's law gives the relationship between the pressure and the volume of a gas at a constant temperature. The following histogram shows the pressure and volume of a gas. Use the function defined by this histogram to complete each part of this exercise.

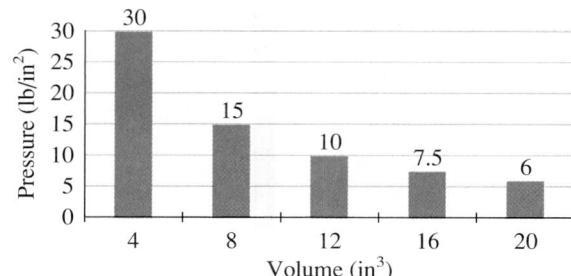

   a. If the input x represents the volume of gas, the output y represents _____.
   b. What is the volume of gas if the pressure is 7.5 pounds per square inch?
   c. What is the pressure of a gas that has a volume of 8 cubic inches?

42. World Series Scores The New York Yankees defeated the New York Mets in five games to win the 2000 World Series. The following histogram shows the number of runs scored by the Yankees in each of the five games. Use the function defined by this histogram to complete each part of this exercise.

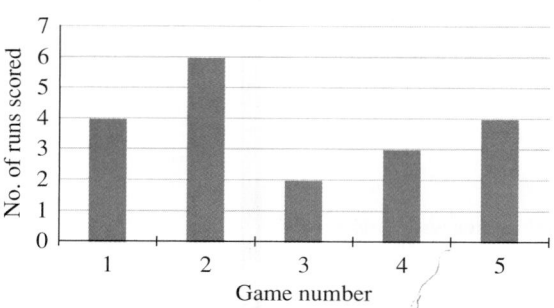

   a. If the input x represents the game number, the output y represents _____.
   b. How many runs did the Yankees score in game 4?
   c. In which game(s) did the Yankees score four runs?

43. Percent of U.S. Adults Who Smoke The following table shows the percent of U.S. adults who smoke.

YEAR x	PERCENT OF U.S. ADULTS WHO SMOKE y
1965	41.9
1974	37.0
1985	29.9
1990	25.3
1995	24.6

Source: Centers for Disease Control.

   a. Graph these points.
   b. What percent of U.S. adults smoked in 1985?
   c. Write a sentence describing the percent of U.S. adults who smoked in 1990.

44. Women's Earnings as Percent of Men's   The following table shows women's annual earnings as a percent of men's annual earnings.

YEAR x	ANNUAL EARNINGS FOR WOMEN AS A PERCENT OF MEN'S y
1979	59.7
1984	63.7
1989	68.7
1994	72.0
1999	72.2

Source: U.S. Department of Labor.

a. Graph these points.
b. What were the earnings for women as a percent of men's earnings in 1989?
c. Write a sentence to describe earnings for women in 1999.

45. Tuition and Fees at 2-Year Colleges in the United States   The cost of tuition and fees at 2-year colleges in the United States can be approximated by the function $C(x) = 200x + 1000$, where x represents the number of years since 1970.
Source: U.S. Department of Education.
a. Use this function to calculate the tuition for years 1970, 1980, 1990, and 2000. Then graph these four points.
b. Estimate the cost of tuition and fees in 2010.

46. Leasing a Car   The total cost to lease a car for x months is given by the function $C(x) = 300x + 1500$.
a. Use this function to calculate the total cost to lease a car for each of the first 6 months. Then graph these six points.
b. Give the cost to lease the car for 24 months.

In Exercises 47–52 use a graphics calculator to graph each function and to determine the domain and range of each function.

47. $f(x) = \sqrt{x}$ **48.** $f(x) = -\sqrt{x}$
49. $f(x) = -x^2 + 1$ **50.** $f(x) = (x - 1)^2$
51. $f(x) = \sqrt{-x}$ **52.** $f(x) = \sqrt{2 - x}$

In Exercises 53–56 match each graph with its domain and range, letters A–D.

53. 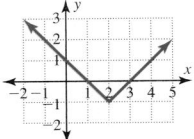   **54.**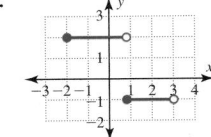

55. 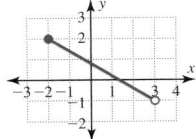   **56.**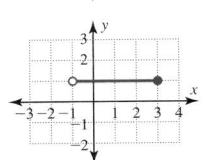

A. $D = (-1, 3], R = \{1\}$ B. $D = [-2, 3), R = (-1, 2]$
C. $D = [-2, 3), R = \{-1, 2\}$ D. $D = \mathbb{R}$ (all real numbers), $R = [-1, \infty)$

In Exercises 57–68 graph a function with the given domain D and range R. (For some exercises there may be several correct graphs.)

57. $D = \{2\}, R = \{3\}$ **58.** $D = \{-3\}, R = \{2\}$
59. $D = (-3, 2], R = \{-2\}$ **60.** $D = [-4, 4), R = \{3\}$
61. $D = (-3, 3), R = \{-1, 2\}$
62. $D = [-2, 2], R = \{-4, 4\}$
63. $D = [-4, 3), R = (-3, 4]$
64. $D = (-3, 4], R = [-2, 2)$
65. $D = \mathbb{R}$ (all real numbers), $R = [-2, +\infty)$ **66.** $D = \mathbb{R}$ (all real numbers), $R = (-\infty, 2]$
67. $D = (-1, 1), R = \mathbb{R}$ **68.** $D = (0, 2), R = \mathbb{R}$
69. a. Draw a semicircle that represents a function.
    b. Draw a semicircle that does not represent a function.
70. Can a circle ever be a function? Explain your answer.

Group Discussion Questions

71. Calculator Discovery   Each of these graphs is produced by an equation of the form $y = |x| + c$ for a specific value of c. Write the equation for each graph and use a graphics calculator to test each function.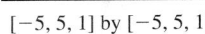

a. 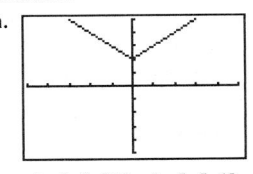   b.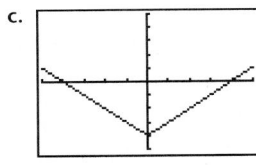

$[-5, 5, 1]$ by $[-5, 5, 1]$ $[-5, 5, 1]$ by $[-5, 5, 1]$

c. 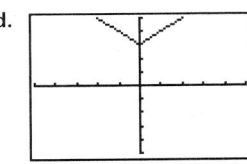   d.

$[-5, 5, 1]$ by $[-5, 5, 1]$ $[-5, 5, 1]$ by $[-5, 5, 1]$

72. Calculator Discovery   Use each set of data to write a function $y = f(x)$ that will produce the given table of values. Use a graphics calculator to test each function.

a.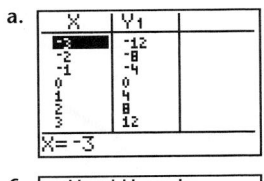

X	Y1
-3	-12
-2	-8
-1	-4
0	0
1	4
2	8
3	12

X=-3

b.

X	Y1
-3	9
-2	4
-1	1
0	0
1	1
2	4
3	9

X=-3

c.

X	Y1
-3	-11
-2	-7
-1	-3
0	1
1	5
2	9
3	13

X=-3

d.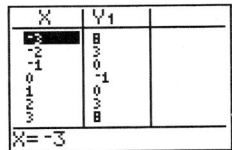

X	Y1
-3	8
-2	3
-1	0
0	-1
1	0
2	3
3	8

X=-3

Section 6.2 Linear and Absolute Value Functions

Objectives:

4. Graph a linear function.
5. Write a linear function to model given information.
6. Graph an absolute value function.

Each common function in algebra has an equation of a standard form and a graph with a characteristic shape. (See inside the jackets of this book.)

Each of the functions in this section and the next has an equation of a standard form and a graph with a characteristic shape. An important goal for these sections is for you to become familiar with some of the most basic functions that we are most likely to encounter in later applications. To profit from this knowledge outside the classroom, we need to recognize the shape of a graph when we see an equation. It is also important to be able to write the equation for the graph of a given set of data. In this section we review the linear function introduced in Section 2.2 and examined in Chapter 3. Then we introduce the absolute value function.

First-Degree Functions—Straight Lines

A first-degree equation in x and y means that the exponent on both the x and y terms is one.

Every linear function has a graph that is a straight line and the equation can be written in the form $y = mx + b$.

A linear function is so named because its graph is a straight line. Every straight line in the x-y plane can be defined by an equation that is first degree in both x and y. The slope-intercept form, $y = mx + b$, was discussed in Section 3.2. This form is the one that we use to enter a linear equation into a graphics calculator. This form also reveals by inspection two of the most important features of a line—its slope m and its y-intercept $(0, b)$. The slope-intercept form also can be written in function notation as $f(x) = mx + b$. For these reasons the slope-intercept form is the most utilized form in this book.

We start the examination of linear functions by reviewing the slope of a line (see Section 3.1). The **slope of a line** is a measure of the steepness of the line. The slope of the line connecting the point (x_1, y_1) to the point (x_2, y_2) is defined to be the ratio of the change in y (the rise) to the change in x (the run). Slope is usually represented by the letter m.

Slope of a Line Through (x_1, y_1) and (x_2, y_2)

The slope of a line gives the change in y for each 1-unit change in x.

ALGEBRAICALLY	VERBALLY	NUMERICAL EXAMPLE	GRAPHICAL EXAMPLE
$m = \dfrac{y_2 - y_1}{x_2 - x_1}$ for $x_1 \neq x_2$.	The slope of a line is the ratio of the change in y to the change in x.	The slope of the line through the points $(2, -1)$ and $(3, 1)$ is $m = \dfrac{1 - (-1)}{3 - 2}$ $m = \dfrac{2}{1}$	

■ **EXAMPLE 1**   Calculating the Slope of a Line Through Given Points

Calculate the slope of each line using the points labeled on the line.

SOLUTIONS

(a)

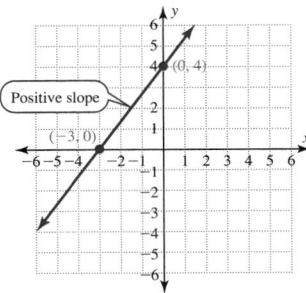

$$m = \frac{y_2 - y_1}{x_2 - x_1}$$

The line slopes upward to the right; its slope is a positive number.

$$m = \frac{4 - 0}{0 - (-3)}$$

Use $x_1 = -3$, $y_1 = 0$, $x_2 = 0$, and $y_2 = 4$.

$$m = \frac{4}{3}$$

y increases 4 units for each 3-unit increase in x.

(b)

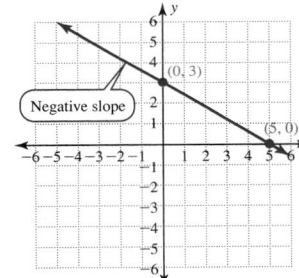

$$m = \frac{y_2 - y_1}{x_2 - x_1}$$

The line slopes downward to the right; its slope is a negative number.

$$m = \frac{0 - 3}{5 - 0}$$

Use $x_1 = 0$, $y_1 = 3$, $x_2 = 5$, and $y_2 = 0$.

$$m = -\frac{3}{5}$$

y decreases 3 units for each 5-unit increase in x.

(c)

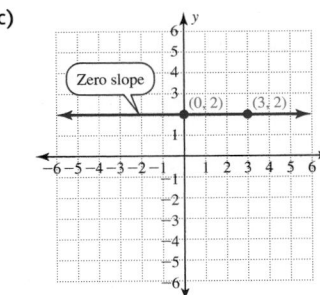

$$m = \frac{y_2 - y_1}{x_2 - x_1}$$

$$m = \frac{2 - 2}{3 - 0}$$

Use $x_1 = 0$, $y_1 = 2$, $x_2 = 3$, and $y_2 = 2$.

$$m = \frac{0}{3}$$

$$m = 0$$

This line is horizontal and its slope is 0.

(d)

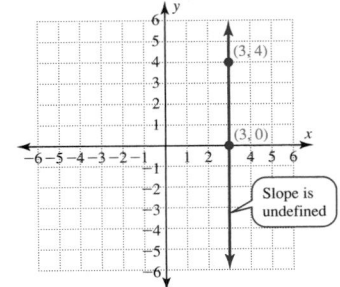

$$m = \frac{y_2 - y_1}{x_2 - x_1}$$

$$m = \frac{4 - 0}{3 - 3}$$

Use $x_1 = 3$, $y_1 = 0$, $x_2 = 3$, and $y_2 = 4$.

$$m = \frac{4}{0}$$

m is undefined.

This line is vertical and its slope is undefined since division by 0 is undefined.

A vertical line is not a function of x; it fails the vertical line test for a function.

■

The next example illustrates how to determine the slope of a line directly from its defining equation. This method is faster than first finding two points on the line and then using the definition of a slope to calculate the slope of the line.

■ EXAMPLE 2   Calculating the Slope of a Line from Its Equation

Determine the slope of each of the following lines.

SOLUTIONS

(a) $3x - 6y = 18$

$$3x - 6y = 18$$
$$-6y = -3x + 18$$
$$y = \frac{1}{2}x - 3$$

$$m = \frac{1}{2}$$

First rewrite the equation in slope-intercept form $y = mx + b$. Then identify m to determine the slope of the line.

The slope of this line is $\frac{1}{2}$.

(b) $y = 5$

$$m = 0$$

This is the equation of a horizontal line; thus the slope must be 0.

$$y = 0x + 5$$

The equation also can be written in the form $y = mx + b$, which reveals that the slope is 0.

(c) $x = -2$

m is undefined

This is a vertical line whose slope is undefined. The equation of a vertical line cannot be written in the form $y = mx + b$ with a coefficient of 1 for y. A vertical line is not a function of x; it fails the vertical line test for a function. ■

SELF-CHECK 6.2.1

1. If a line is horizontal, its slope is _____.
2. If a line is vertical, its slope is _____.
3. If a line slopes upward to the right, its slope is _____.
4. If a line slopes downward to the right, its slope is _____.
5. Calculate the slope of the line using the points labeled here.

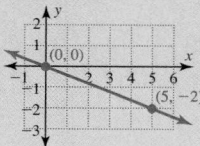

Calculate the slope of the lines defined by these equations.

6. $x + y = 7$
7. $y = 7x$
8. $y = 7$
9. $x = 7$

SELF-CHECK 6.2.1 ANSWERS

1. 0
2. Undefined
3. Positive
4. Negative
5. $m = -\frac{2}{5}$; y decreases 2 units for each 5-unit increase in x.
6. $m = -1$; this line slopes downward to the right.
7. $m = 7$; this line slopes upward to the right.
8. $m = 0$; this is a horizontal line.
9. m is undefined; this is a vertical line.

In the next example we review (see Section 3.2) how to use the slope and the y-intercept to graph a line.

■ EXAMPLE 3 Graphing a Line Using the Slope-Intercept Form

Graph the line defined by $y = \dfrac{5}{4}x + 2$ by using the y-intercept and the slope of the line.

SOLUTION _____

$$y = \frac{5}{4}x + 2$$

y-intercept: $(0, 2)$

Slope: $m = \dfrac{5}{4}$

Second point: $(0 + 4, 2 + 5)$
$= (4, 7)$

Use the slope-intercept form $y = mx + b$ to identify the y-intercept $(0, 2)$ and the slope $\dfrac{5}{4}$.

Start with the y-intercept and use the slope of $\dfrac{5}{4}$ to determine a second point. Do this by increasing y by 5 units for a 4-unit increase in x.

Plot both $(0, 2)$ and $(4, 7)$ and sketch the line through these points.

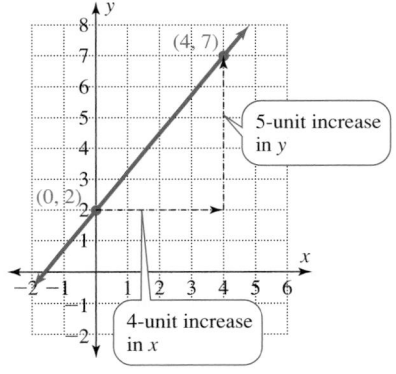

Check: $m = \dfrac{y_2 - y_1}{x_2 - x_1}$

$m = \dfrac{7 - 2}{4 - 0}$

$m = \dfrac{5}{4}$ checks.

SELF-CHECK 6.2.2

Graph the line defined by $y = -\dfrac{2}{5}x + 4$.

SELF-CHECK 6.2.2 ANSWER

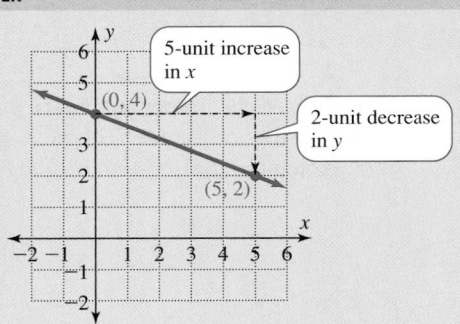

Other forms of linear equations are listed in the following table. In Example 4 we convert an equation given in a general form to a slope-intercept form so that we can enter the equation into a graphics calculator.

Forms of Linear Equations

NAME OF FORM	ALGEBRAIC FORM	KEY FEATURES
Slope-intercept form	$y = mx + b$	m is the slope of the line and $(0, b)$ is the y-intercept of the line.
Point-slope form	$y - y_1 = m(x - x_1)$	m is the slope of the line and (x_1, y_1) is a point on the line.
General form	$Ax + By = C$	This form can be used to write a "clean" form that is free of fractions.
Horizontal line	$y = b$	This is a horizontal line with a y-intercept $(0, b)$.
Vertical line	$x = a$	This is a vertical line with an x-intercept $(a, 0)$.

■ **EXAMPLE 4**   **Using a Graphics Calculator to Graph a Linear Function**

Use a graphics calculator to graph $5x + 3y = 15$.

SOLUTION _____

$5x + 3y = 15$ This is a linear equation in the general form $Ax + By = C$.

$\quad\quad 3y = -5x + 15$ First we rewrite the equation in the slope-intercept form.

$\quad\quad\ y = -\dfrac{5}{3}x + 5$

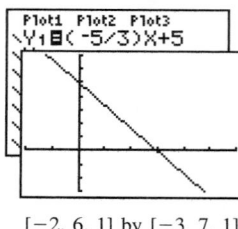

[−2, 6, 1] by [−3, 7, 1]

Then we use the slope-intercept form to enter the equation into a graphics calculator.

This line has a slope of $-\dfrac{5}{3}$ and a y-intercept of $(0, 5)$.

Note that a line with a negative slope goes downward to the right.

■

One of the important goals of this book is to enable you to write functions that model real applications. The next example uses data on engine performance taken from the September 2001 *Popular Science* magazine.

■ **EXAMPLE 5**   Writing a Linear Equation to Model Engine Performance

The horsepower (hp) of a 100-hp engine falls by about 3 hp for every 1000 ft above sea level that you travel.

(a) Write a linear equation to describe the horsepower of an automobile engine as a function of elevation.

(b) Use this function to calculate the horsepower of an engine at the top of the road to Pikes Peak, elevation 14,000 ft.

(c) Use this function to prepare a horsepower performance table for this engine for elevations from sea level to 6000 ft in increments of 1000 ft.

Turbocharging the air pressure back to that at sea level can compensate for this loss of power. Aircraft engines often have turbochargers for this purpose.

SOLUTIONS

This application is equivalent mathematically to writing the equation of a line whose slope is given as $m = -\dfrac{3}{1000}$ and whose y-intercept is (0, 100).

(a) Let x = the elevation of the engine in feet
Let y = the horsepower of the engine

y-intercept: (0, 100)

Slope: $m = \dfrac{-3}{1000}$

$m = -0.003$

Equation: $P(x) = -0.003x + 100$
$y = -0.003x + 100$

We are using elevation as the input variable and horsepower as the output variable.

Sea level is at an elevation of 0 ft. The horsepower at that level is 100 hp.
The change in horsepower is a decrease of 3 hp for each 1000 ft increase in elevation.

Write the equation in the form $P(x) = mx + b$ or $y = mx + b$ to represent the horsepower at x ft. Use the y-intercept of (0, 100) and the slope of -0.003.

(b) $P(x) = -0.003x + 100$
$P(14,000) = -0.003(14,000) + 100$
$P(14,000) = 58$
The engine has 58 hp at 14,000 ft.

To determine the horsepower at 14,000 ft, substitute 14,000 into the function
$P(x) = -0.003x + 100$.

The horsepower of this engine has decreased to 58 hp by the time it has climbed to 14,000 ft.

(c)

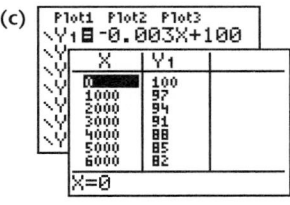

Engineers and architects often take advantage of the properties of geometry and algebra to prepare their plans and to create safe and functional projects. In Example 6 we will use information about parallel and perpendicular lines. This information is reviewed below (see Section 3.1).

If l_1 and l_2 are two distinct nonvertical lines with slopes m_1 and m_2, respectively, then:

Parallel lines have the same slope. Perpendicular lines have slopes that are opposite reciprocals.

■ Parallel lines: $m_1 = m_2$

■ Perpendicular lines: $m_1 = -\dfrac{1}{m_2}$

■ EXAMPLE 6   Modeling the Path of a Communications Cable

A communications corporation is planning to install a new fiber optic cable parallel to the cable shown in this coordinate system. Use this coordinate system to write the following equations.

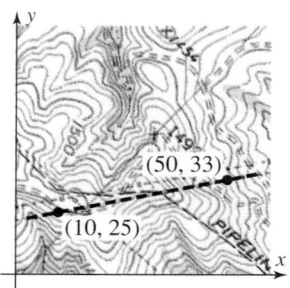

(a) Write an equation for the path of the current cable that passes through the points (10, 25) and (50, 33).
(b) Write an equation for the path of the planned cable. It will pass through the point (25, 30) and will be parallel to the existing cable.
(c) Future expansion may require yet another cable with a path perpendicular to the given cable and passing through the point (25,30). Write the equation for the path of this cable.

SOLUTIONS

(a) $m = \dfrac{y_2 - y_1}{x_2 - x_1} = \dfrac{33 - 25}{50 - 10}$

      Step 1: Since the slope is not given, first use the given points to calculate the slope.

$= \dfrac{8}{40}$

$= \dfrac{1}{5}$

$y - y_1 = m(x - x_1)$

$y - 25 = \dfrac{1}{5}(x - 10)$

Step 2: Using the point-slope form, substitute (10, 25) for the point (x_1, y_1) and $\dfrac{1}{5}$ for the slope m.

$y - 25 = \dfrac{1}{5}x - 2$

Step 3: Solve this equation for y to obtain the slope-intercept form of the equation.

$y = \dfrac{1}{5}x + 23$

This is the equation of the path of the current cable.

(b) $y - y_1 = m(x - x_1)$

Step 1: The slope of the path parallel to the given path must also be $\dfrac{1}{5}$.

$y - 30 = \dfrac{1}{5}(x - 25)$

Step 2: Using the point-slope form, substitute (25, 30) for the point (x_1, y_1) and $\dfrac{1}{5}$ for the slope m.

$y - 30 = \dfrac{1}{5}x - 5$

$y = \dfrac{1}{5}x + 25$

Step 3: Solve this equation for y to obtain the slope-intercept form of the equation. This is the equation of the path of the new cable that is planned.

(c) $y - y_1 = m(x - x_1)$
$y - 30 = -5(x - 25)$

Step 1: The slope of the path perpendicular to the given path must be -5, the opposite reciprocal of $\dfrac{1}{5}$.

$y - 30 = -5x + 125$

Step 2: Using the point-slope form, substitute (25, 30) for the point (x_1, y_1) and -5 for the slope m.

$y = -5x + 155$

Step 3: Solve this equation for y to obtain the slope-intercept form of the equation. This is the equation of the path of the cable that may result from future expansion. ■

Absolute Value Functions—V-Shaped Graphs

An absolute value function is named for the algebraic equation which defines the function. Every absolute value function has a characteristic **V**-shape. This **V**-shape is illustrated by the graphs of $f(x) = |x|$ and $f(x) = -|x|$ shown in Example 7. For the **V**-shape opening upward the lowest point of the curve is called the **vertex.** The vertex of the **V**-shape opening downward is the highest point on the curve.

The graph of the absolute value function $y = |ax + b| + c$ has a V-shape.

◼ **EXAMPLE 7** Graphing an Absolute Value Function

Create a table of values for $f(x) = |x|$ and $f(x) = -|x|$ and then graph these functions. Determine the domain and the range of each function.

SOLUTIONS

ALGEBRAICALLY	NUMERICALLY	GRAPHICALLY	VERBALLY

(a) $f(x) = |x|$

x	$f(x) = |x|$
-3	3
-2	2
-1	1
0	0
1	1
2	2
3	3

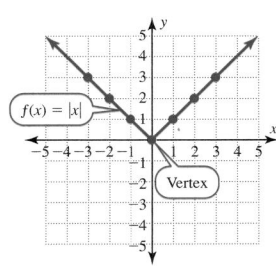

This **V**-shaped graph opens upward with the vertex of the **V**-shape at $(0, 0)$. The domain, the projection of this graph onto the x-axis, is $D = \mathbb{R}$. The range, the projection of this graph onto the y-axis, is $R = [0, +\infty)$.

(b) $f(x) = -|x|$

x	$f(x) = -|x|$
-3	-3
-2	-2
-1	-1
0	0
1	-1
2	-2
3	-3

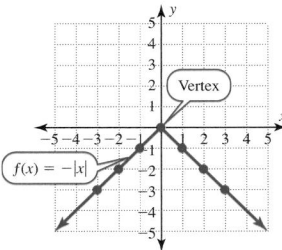

This **V**-shaped graph opens downward with the vertex of the **V**-shape at $(0, 0)$. The domain, the projection of this graph onto the x-axis, is $D = \mathbb{R}$. The range, the projection of this graph onto the y-axis, is $R = (-\infty, 0]$. Recall from Section 1.1 that in interval notation the smaller value always is written first. $(-\infty, 0]$ is the correct notation whereas $(0, -\infty]$ is incorrect. ◼

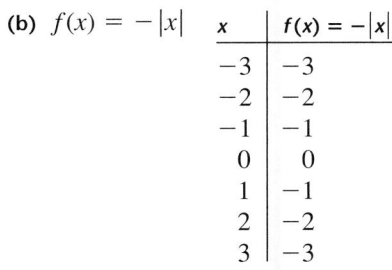

The following calculator perspective is used to produce the characteristic V-shape of an absolute value function. Notice that the absolute value function is denoted algebraically by $|x|$ and on a graphics calculator by abs(x).

CALCULATOR PERSPECTIVE 6.2.1	Graphing an Absolute Value Function

To graph the absolute value function $f(x) = |x - 3|$ on a TI-83 Plus calculator, enter the following keystrokes:

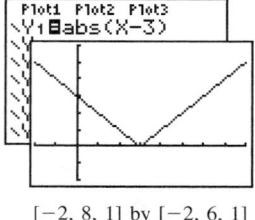

$[-2, 8, 1]$ by $[-2, 6, 1]$

Note: This absolute value function is V-shaped with the vertex of $(3, 0)$.

■ EXAMPLE 8 Determining the Domain and the Range of an Absolute Value Function

Use a graphics calculator to graph $f(x) = |x - 2| - 3$ and then use this graph to determine the domain, the range, and the vertex of this function.

SOLUTION

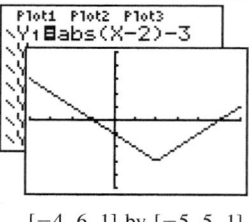

$[-4, 6, 1]$ by $[-5, 5, 1]$

The domain is the projection of the graph onto the x-axis; in this case the projection is the entire x-axis. The range is the projection of the graph onto the y-axis; in this case the projection is the interval $[-3, \infty)$. The vertex, the lowest point on this graph, is $(2, -3)$.

Answers: $D = \mathbb{R}$
$R = [-3, \infty)$
Vertex: $(2, -3)$

SELF-CHECK 6.2.4

Given $f(x) = |x + 1| - 2$:

1. Evaluate $f(3)$.
2. Use a graphics calculator to graph $y = f(x)$.
3. Determine the vertex of this graph.
4. Determine the domain and the range of this function.

USING THE LANGUAGE AND SYMBOLISM OF MATHEMATICS   6.2

1. A function of the form $f(x) = mx + b$ has a graph whose shape is a _____.
2. The form $f(x) = mx + b$ is called the _____-_____ form of a linear equation with m representing the _____ and $(0, b)$ representing the _____.
3. A linear equation of the form $y - y_1 = m(x - x_1)$ is called the _____-_____ form.
4. A linear equation of the form $Ax + By = C$ is called the _____ form.
5. The formula for the slope of a line through (x_1, y_1) and (x_2, y_2) is _____ for $x_1 \neq x_2$.
6. Parallel lines have the _____ slope.
7. Lines whose slopes are _____ _____ of each other are perpendicular.
8. The line defined by $x = 7$ is a _____ line whose x-intercept is _____.
9. The line defined by $y = -7$ is a _____ line whose y-intercept is _____.
10. The slope of a _____ line is 0.
11. The slope of a _____ line is undefined.
12. A function of the form $f(x) = |ax + b| + c$ has a graph that is _____ shaped.
13. The lowest point on the graph of $f(x) = |ax + b| + c$ is called the _____.

EXERCISES   6.2

In Exercises 1–8 calculate the slope of the line through the given points.

1. **a.** $(5, 6)$ and $(3, 10)$ **b.** $(-5, 2)$ and $(-2, 5)$
2. **a.** $(1, 5)$ and $(2, 9)$ **b.** $(1, 5)$ and $(-1, 7)$
3. **a.** $(4, 8)$ and $(7, 6)$ **b.** $(2, -7)$ and $(4, -2)$
4. **a.** $(4, 9)$ and $(1, 2)$ **b.** $(4, 9)$ and $(0, 12)$
5. **a.** $(4, 8)$ and $(7, 8)$ **b.** $(4, 8)$ and $(4, -8)$
6. **a.** $(-2, 3)$ and $(-2, -1)$ **b.** $(2, -5)$ and $(-1, -5)$
7. **a.** $\left(\dfrac{1}{3}, \dfrac{1}{5}\right)$ and $\left(-\dfrac{2}{3}, \dfrac{6}{5}\right)$
   b. $(-1.22, 2.13)$ and $(1.13, -2.57)$
8. **a.** $\left(\dfrac{5}{4}, \dfrac{7}{3}\right)$ and $\left(-\dfrac{3}{4}, -\dfrac{2}{3}\right)$
   b. $(1.87, 2.34)$ and $(1.49, 1.2)$

In Exercises 9–12 calculate the slope of each line.

9. **a.** 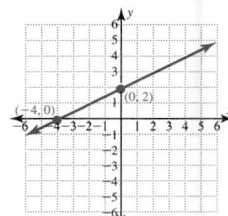 **b.**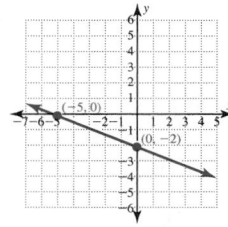

10. **a.** 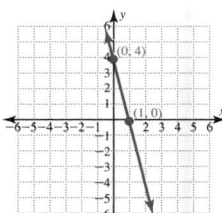 **b.**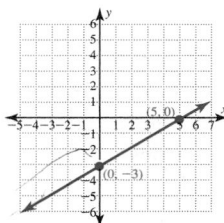

11. **a.** 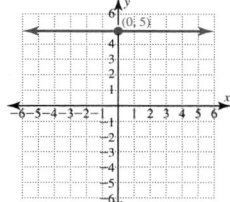 **b.**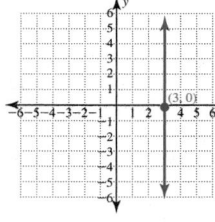

12. **a.** 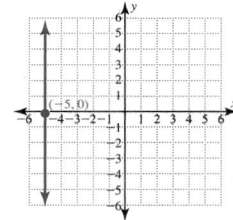 **b.**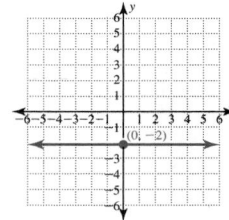

In Exercises 13–22 determine the slope of each line.

13. **a.** $y = \dfrac{3}{5}x - 4$ **b.** $y = -\dfrac{4}{3}x + 9$
14. **a.** $y = 0.56x + 1.4$ **b.** $y = -2.3x + 4.5$
15. **a.** $y - 2 = 3(x + 4)$ **b.** $y + 2 = -3(x - 4)$
16. **a.** $y + 3 = 4(x - 1)$ **b.** $y - 1 = -5(x + 8)$
17. **a.** $3x + 4y = 12$ **b.** $-2x + 5y = 10$
18. **a.** $5x - 7y = 35$ **b.** $3x + 5y = 45$
19. **a.** $3x = 8y$ **b.** $-2x = 5y$
20. **a.** $-5x = 9y$ **b.** $3x = 7y$
21. **a.** $x = 8$ **b.** $y = 8$
22. **a.** $x = -8$ **b.** $y = -8$

23. Given $f(x) = \dfrac{1}{4}x - 5$, evaluate each expression.
    a. $f(-12)$ **b.** $f(-2)$ **c.** $f(0)$ **d.** $f(8)$

24. Given $f(x) = -\dfrac{2}{5}x + 3$, evaluate each expression.
    a. $f(-5)$ **b.** $f(0)$ **c.** $f(2)$ **d.** $f(10)$

In Exercises 25–28 complete the given table of values and sketch the graph of each function.

x	−3	−2	−1	0	1	2	3
y							

25. $f(x) = 2x - 5$ **26.** $f(x) = -3x + 1$

27. $f(x) = -x + 3$ **28.** $f(x) = 3x + 7$

In Exercises 29–32 use the slope and y-intercept to sketch the graph of each linear function.

29. $f(x) = \frac{2}{3}x - 2$ **30.** $f(x) = -\frac{2}{3}x + 2$

31. $f(x) = -\frac{4}{3}x + 3$ **32.** $f(x) = \frac{4}{3}x - 3$

In Exercises 33–40 sketch the graph of each line.

33. $y = 2$ **34.** $y = -2$

35. $x = -3$ **36.** $x = 3$

37. $y - 2 = \frac{1}{2}(x + 1)$ **38.** $y + 2 = \frac{1}{3}(x - 1)$

39. $2x - 3y = 6$ **40.** $2x + 3y = -6$

In Exercises 41–56 write a linear equation that satisfies the given conditions. Use slope-intercept form where this is appropriate.

41. slope of 3 and y-intercept $(0, -5)$

42. slope of 4 and y-intercept $(0, -3)$

43. slope of $-\frac{2}{7}$ and y-intercept $(0, 1)$

44. slope of $-\frac{3}{8}$ and y-intercept $(0, 2)$

45. through $(-2, 3)$ with slope $\frac{3}{7}$

46. through $(4, -7)$ with slope $\frac{3}{7}$

47. through $(1, -3)$ and $(-1, 1)$

48. through $(-3, 2)$ and $(-1, -6)$

49. a horizontal line through $(2, 3)$

50. a horizontal line through $(4, 1)$

51. a vertical line through $(2, 3)$

52. a vertical line through $(4, 1)$

53. through $(1, 2)$ and parallel to $y = 3x - 5$

54. through $(1, 2)$ and parallel to $y = -2x + 7$

55. through $(1, 2)$ and perpendicular to $y = 3x - 5$

56. through $(1, 2)$ and perpendicular to $y = -2x + 7$

57. Given $f(x) = |3x - 2|$, evaluate each expression.
  a. $f(-10)$ **b.** $f(-1)$ **c.** $f(0)$ **d.** $f(5)$

58. Given $f(x) = |2x + 3|$, evaluate each expression.
  a. $f(-3)$ **b.** $f(-1)$ **c.** $f(10)$ **d.** $f(3)$

In Exercises 59–62 complete the given table of values and sketch the graph of each function. Determine the domain, the range, and the vertex of each graph.

x	−3	−2	−1	0	1	2	3
y							

59. $f(x) = |x + 2|$ **60.** $f(x) = |x| + 2$

61. $f(x) = |x| - 3$ **62.** $f(x) = |x - 4|$

In Exercises 63–66 use a graphics calculator to assist you in matching each function with its graph A–D.

63. $f(x) = |x + 1| - 2$ **64.** $f(x) = |x - 2| + 3$

65. $f(x) = |x - 1| + 2$ **66.** $f(x) = |x + 2| - 3$

A. 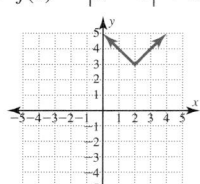   **B.**

C. 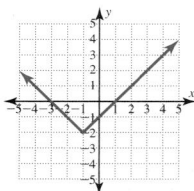   **D.**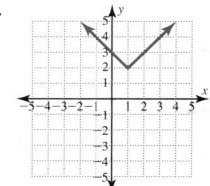

In Exercises 67–70 use a graphics calculator to graph each function and then determine the domain and the range of the function.

67. $f(x) = 2x + 3$ **68.** $f(x) = -2x - 3$

69. $f(x) = 3$ **70.** $f(x) = -3$

In Exercises 71–74 use a graphics calculator to graph each function and then determine the domain, the range, and the vertex of the function.

71. $f(x) = 2|x - 3|$ **72.** $f(x) = \frac{1}{2}|x + 3|$

73. $f(x) = -|x - 1| + 2$ **74.** $f(x) = -|x + 2| - 1$

75. Modeling Windchill Temperature   Temperature and wind combine to cause heat loss from body surfaces. With a 10 mi/h wind, the function $T(x) = 1.2x - 21.6$ can be used to approximate the windchill temperature, where x is the actual temperature and $T(x)$ is the temperature that your body is sensing due to the wind.
Source: National Weather Service
http://www.nws.noaa.gov/
Evaluate and interpret each of these expressions.
  a. $T(0)$ **b.** $T(5)$ **c.** $T(20)$

76. Modeling Smoking Trends   The percent of U.S. adults who smoke can be approximated by the function $P(n) = -0.61n + 42.04$, where n is the number of years past 1965.
Source: Centers for Disease Control **http://www.cdc.gov/**
Evaluate and interpret each of these expressions.
  a. $P(0)$ **b.** $P(15)$ **c.** $P(35)$

77. Modeling the Rolling of a Pair of Dice   When rolling a pair of dice, the probability that the sum of the numbers on these dice is x is given by $P(x) = \dfrac{6 - |7 - x|}{36}$, where x can be any integer from 2 to 12. Evaluate and interpret each of these expressions.
  a. $P(2)$ **b.** $P(7)$ **c.** $P(12)$

78. Modeling Electrical Voltage   The voltage in an electrical circuit designed by a beginning electronics class is a function of time. This function is given by $V(t) = -120|2x - 1| + 120$, where t is the time in seconds and $V(t)$ is the voltage. Evaluate and interpret each of these expressions:

   a. $V(0)$ **b.** $V(0.5)$ **c.** $V(1.0)$

79. Modeling Depreciation of Carpentry Tools   The value of a $500 set of tools depreciates by $50 per year.

   a. Write a function that gives the value of these tools after x years.

   b. Use this function to calculate the value of these tools after 7 years.

   c. Use this function to prepare a table of values for this function from year 0 to year 6.

80. Modeling the Length of Hair   The length of human hair grows by approximately 1.3 cm per month. A young girl has a ponytail 20 cm long and wants a ponytail much longer.

   a. Write a function that gives the length of her ponytail after x months.

   b. Use this function to determine how many months it will take her to have a ponytail that is 50 cm long.

   c. Use this function to prepare a table of values for this function from 0 to 3 years in increments of 6 months.

81. Using a Table to Model the Cost of a Collect Call   The following table displays the dollar cost of a collect telephone call, based on the length of the call in minutes. Use this table to determine the linear equation $f(x) = mx + b$ for these data points. Then interpret the meaning of m and b in this problem.

x MINUTES	f(x) COST ($)
3	3.80
7	6.20
10	8.00
16	11.60

82. Using a Table to Model Printing Costs   The next table displays the total dollar cost of a science manual, based on the number of manuals printed for the order. Use this table to determine the linear equation $f(x) = mx + b$ for these data points. Then interpret the meaning of m and b in this problem.

x MANUALS	f(x) COST ($)
25	325
50	525
75	725
100	925

83. Modeling the Path of a Pipeline   A pipeline is planned to pass through towns with coordinates $(0, 30)$ and $(70, 20)$ on the grid shown here.

   a. Write a linear equation of the path of this pipeline.

   b. Will this pipeline pass through the town whose coordinates are $(35, 25)$?

   c. Write a linear equation of the path of a connecting pipeline that would pass through $(35, 25)$, which is perpendicular to the path of the pipeline in part **a.**

84. Modeling the Path of a Railroad   A new highway is planned to be parallel to an existing railroad. The following grid shows that the railroad passes through the points $(12, 16)$ and $(34, 49)$.

   a. Write a linear equation of the path of this railroad.

   b. Write a linear equation of the path of this highway if it passes through the origin on this grid.

   c. Will the highway pass through the town whose coordinates are $(40, 60)$?

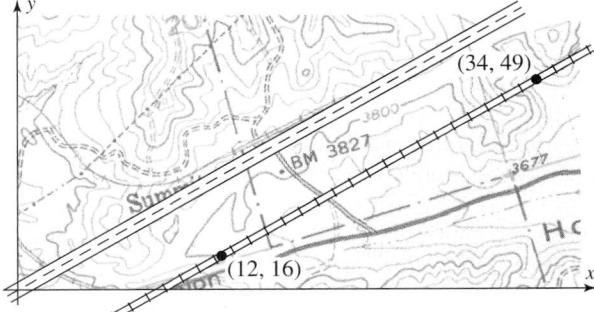

Group Discussion Questions

85. Calculator Discovery Use a graphics calculator to graph each function: $f(x) = |x|, f(x) = |x - 1|$, $f(x) = |x - 2|, f(x) = |x - 3|$, and $f(x) = |x - 4|$. Describe the relation between the graphs of $f(x) = |x|$ and $f(x) = |x - c|$.

86. Error Analysis A student produced the following calculator window for the function shown and concluded that the function is a linear function and the range of the function is the set of all real numbers. Determine the correct analysis and the source of the error in this answer.

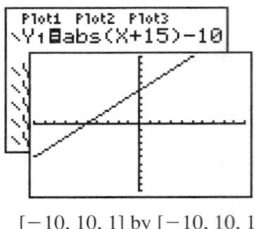

$[-10, 10, 1]$ by $[-10, 10, 1]$

Section 6.3 Quadratic, Square Root, Cubic, and Cube Root Functions

Objectives:

7. Graph quadratic, square root, cubic, and cube root functions.
8. Determine the domain of a function from the defining equation.

This section introduces four new shapes. These shapes are defined by quadratic, square root, cubic, and cube root functions. As we noted in the previous section, each common function in algebra has an equation of a standard form and a graph with a characteristic shape. We start with the parabolic shape that is defined by a quadratic function.

Quadratic Functions—Parabolas

A function of the form $f(x) = ax^2 + bx + c$ is called a **quadratic function.** The term ax^2 is called the **quadratic** or the **second-degree term.** The shape of a quadratic function is called a **parabola.** This parabolic shape is illustrated by the graphs of $f(x) = x^2$ and $f(x) = -x^2$ shown in Example 1. The **vertex** of a parabola opening upward is the lowest point on the parabola. The vertex of a parabola opening downward is the highest point on the parabola. We examine these functions in greater depth later in this book.

A Mathematical Note

Ian Stewart writes in *Nature's Numbers* (a 1995 book by Basic Books) that his "book ought to have been called *Nature's Numbers and Shapes.*" Even though shapes always can be reduced to numbers, "it is often better to think of shapes that make use of our powerful visual capabilities." The parabola is one of the most common shapes in nature. The stream of water shown in this photo and the path of a thrown ball both exhibit a parabolic shape.

■ EXAMPLE 1   Graphing a Quadratic Function

Create a table of values for $f(x) = x^2$ and $f(x) = -x^2$ and then graph each function. Determine the domain and the range of each function.

SOLUTIONS

ALGEBRAICALLY **NUMERICALLY** **GRAPHICALLY** **VERBALLY**

(a) $f(x) = x^2$

x	$f(x) = x^2$
−3	9
−2	4
−1	1
0	0
1	1
2	4
3	9

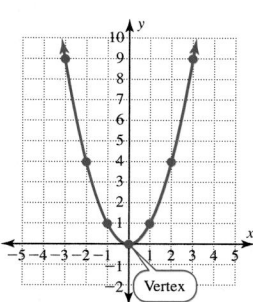

This parabola opens upward with a vertex of (0, 0). This function is defined for all real numbers. The domain $D = \mathbb{R}$. The range, the projection of the graph onto the y-axis, is $R = [0, +\infty)$.

(b) $f(x) = -x^2$

x	$f(x) = -x^2$
−3	−9
−2	−4
−1	−1
0	0
1	−1
2	−4
3	−9

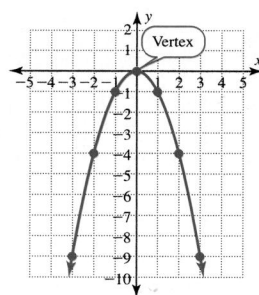

This parabola opens downward with a vertex of (0, 0). This function is defined for all real numbers. The domain $D = \mathbb{R}$. The range, the projection of the graph onto the y-axis, is $R = (-\infty, 0]$.

It is important to be able to predict the shape of a graph from the defining equation. Knowing the shape of a graph can help us to determine an appropriate viewing window to use on a graphics calculator. Without this knowledge, we may select an inappropriate window that fails to show key features of the graph. In Example 2 we know the graph will be a parabola, so we select a window that will show the vertex of the parabola.

■ **EXAMPLE 2** Using a Graphics Calculator to Graph a Parabola

The vertex of a parabola is a key point on the parabola. In this example we are approximating the coordinates of the vertex by examining the graph. In later sections of this chapter we will be able to determine the vertex of a parabola by examining its defining equation.

Use a graphics calculator to graph $f(x) = x^2 - 4$. Then determine the domain, the range, and the vertex of this function.

SOLUTION

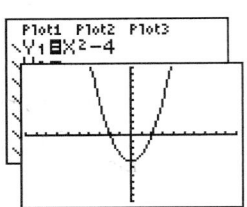

[−10, 10, 1] by [−10, 10, 1]

This parabola opens upward with a vertex of $(0, -4)$. The function $f(x) = x^2 - 4$ is defined for all values of x, so the domain $D = \mathbb{R}$. The range, the projection of the graph onto the y-axis, is $R = [-4, +\infty)$.

Answers: $D = \mathbb{R}$
$R = [-4, +\infty)$

Vertex: $(0, -4)$

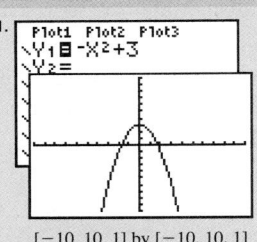
SELF-CHECK 6.3.1

1. Use a graphics calculator to graph $f(x) = -x^2 + 3$.
2. Determine the vertex of this graph.
3. Determine the domain and range of this function.

Square Root Functions

A square root function is named for the square root symbol that defines the function. The basic shape of a square root function is illustrated by the graphs of $f(x) = \sqrt{x}$ and $f(x) = -\sqrt{x}$ shown in Example 3. Note that the domain of each function is restricted to prevent negative numbers under the square root symbol.

■ **EXAMPLE 3** Graphing a Square Root Function

Create a table of values for $f(x) = \sqrt{x}$ and $f(x) = -\sqrt{x}$ and then graph these functions.

SOLUTIONS

ALGEBRAICALLY

(a) $f(x) = \sqrt{x}$

NUMERICALLY

x	$f(x) = \sqrt{x}$
0	0
1	1
4	2
1	1
4	2
9	3

GRAPHICALLY

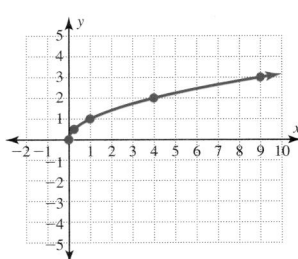

VERBALLY

This square root function opens to the right in quadrant I. For $x \geq 0$ we also have $\sqrt{x} \geq 0$. Thus the domain $D = [0, +\infty)$ and the range $R = [0, +\infty)$.

(b) $f(x) = -\sqrt{x}$

x	$f(x) = -\sqrt{x}$
0	0
$\frac{1}{4}$	$-\frac{1}{2}$
1	-1
4	-2
9	-3

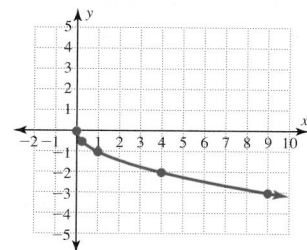

This square root function opens to the right in quadrant IV. For $x \geq 0$ we also have $-\sqrt{x} \leq 0$. Thus the domain $D = [0, +\infty)$ and the range $R = (-\infty, 0]$.

Later in this section we will examine how to determine the domain of a function by algebraic means. However, in the next example we will determine the domain of $f(x) = \sqrt{2x - 6}$ by examining the graph of this function.

■ **EXAMPLE 4**   Using a Graphics Calculator to Graph a Square Root Function

Use a graphics calculator to graph $f(x) = \sqrt{2x - 6}$. Then determine the domain and the range of this function.

SOLUTION _____

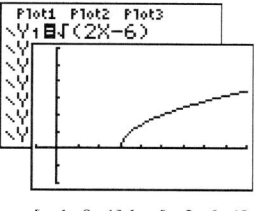

The projection of this graph onto the x-axis is the interval $[3, +\infty)$. Thus the domain $D = [3, +\infty)$. The range, the projection of the graph onto the y-axis, is $R = [0, +\infty)$.

$[-1, 9, 1]$ by $[-2, 6, 1]$

Answers: $D = [3, +\infty)$
$R = [0, +\infty)$ ■

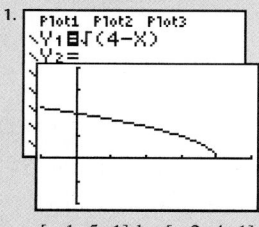
SELF-CHECK 6.3.2

1. Use a graphics calculator to graph $f(x) = \sqrt{4 - x}$.
2. Determine the domain and the range of this function.

Cubic Functions

We now extend what we have observed about first-degree functions and second-degree functions. First-degree functions produce graphs that are straight lines and second-degree functions produce graphs that are parabolas. We now examine the shape of third-degree functions. A third-degree function also is called a cubic function. **Cubic functions** have the form $f(x) = ax^3 + bx^2 + cx + d$. The basic shape of the cubic functions $f(x) = x^3$ and $f(x) = -x^3$ is illustrated in Example 5.

■ **EXAMPLE 5** Graphing a Cubic Function

Create a table of values for $f(x) = x^3$ and $f(x) = -x^3$ and then graph each function.
Determine the domain and the range of each function.

SOLUTIONS

ALGEBRAICALLY	NUMERICALLY	GRAPHICALLY	VERBALLY

(a) $f(x) = x^3$

x	$f(x) = x^3$
-3	-27
-2	-8
-1	-1
0	0
1	1
2	8
3	27

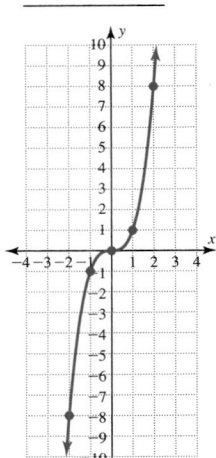

This cubic function is defined for all real numbers. The domain $D = \mathbb{R}$. The range, the projection of the graph onto the y-axis, is $R = \mathbb{R}$.

(b) $f(x) = -x^3$

x	$f(x) = -x^3$
-3	27
-2	8
-1	1
0	0
1	-1
2	-8
3	-27

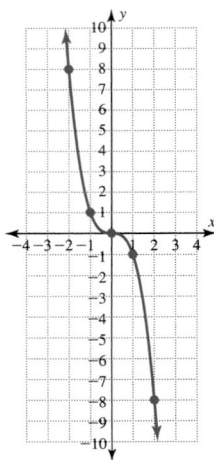

This cubic function is defined for all real numbers. The domain $D = \mathbb{R}$. The range, the projection of the graph onto the y-axis, is $R = \mathbb{R}$.

SELF-CHECK 6.3.3 ANSWERS

1. x-intercepts:
   $(-2, 0)$ and $(3, 0)$
   y-intercept: $(0, -6)$
2. x-intercepts:
   $(1, 0)$ and $(7, 0)$
   y-intercept: $(0, -2)$

SELF-CHECK 6.3.3

Use a graphics calculator to graph each function and to determine the x- and y-intercepts of each graph.

1. $f(x) = x^2 - x - 6$
2. $f(x) = -2|x - 4| + 6$

The graph of a cubic function can have as many as three x-intercepts. The graph in the next example contains three x-intercepts.

■ **EXAMPLE 6**   Using a Graph to Determine the Intercepts of a Cubic Function

Use a graphics calculator to graph $f(x) = x^3 + 4x^2 - 4x - 16$ and then determine the x- and y-intercepts of this graph.

SOLUTION

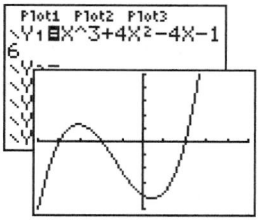

The graph crosses the x-axis at $(-4, 0)$, $(-2, 0)$, and $(2, 0)$. The graph crosses the y-axis at $(0, -16)$.

$[-5, 5, 1]$ by $[-20, 20, 4]$

Answers: x-intercepts: $(-4, 0)$, $(-2, 0)$, and $(2, 0)$
                  y-intercept: $(0, -16)$

■

Cube Root Functions

> The domain of $f(x) = \sqrt{x}$ is $D = [0, \infty)$ and the domain of $f(x) = \sqrt[3]{x}$ is $D = \mathbb{R}$.

A cube root function is named for the cube root symbol that defines the function. The basic shape of a cube root function is illustrated by the graphs of $f(x) = \sqrt[3]{x}$ and $f(x) = -\sqrt[3]{x}$ shown in Example 6. An important difference between the graph of a cube root function and the graph of a square root function is that cube root functions are defined for all real numbers while square root functions are restricted to values that make the expression under the square root symbol nonnegative.

■ **EXAMPLE 7**   Graphing a Cube Root Function

Create a table of values for $f(x) = \sqrt[3]{x}$ and $f(x) = -\sqrt[3]{x}$ and then graph each function. Determine the domain and the range of each function.

SOLUTIONS

ALGEBRAICALLY	NUMERICALLY		GRAPHICALLY	VERBALLY
(a) $f(x) = \sqrt[3]{x}$	x	$f(x) = \sqrt[3]{x}$	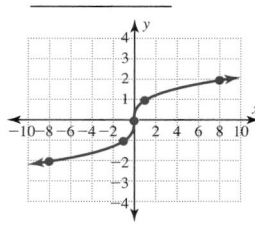	This cube root function is defined for all real numbers. The domain $D = \mathbb{R}$. The range, the projection of the graph onto the y-axis, is $R = \mathbb{R}$.
	-27	-3		
	-8	-2		
	-1	-1		
	0	0		
	1	1		
	8	2		
	27	3		

ALGEBRAICALLY	NUMERICALLY	GRAPHICALLY	VERBALLY

(b) $f(x) = -\sqrt[3]{x}$

x	$f(x) = \sqrt[3]{x}$
-27	3
-8	2
-1	1
0	0
1	-1
8	-2
27	-3

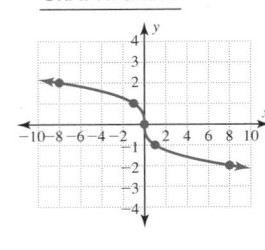

This cube root function is defined for all real numbers. The domain $D = \mathbb{R}$. The range, the projection of the graph onto the y-axis, is $R = \mathbb{R}$.

CALCULATOR PERSPECTIVE 6.3.1   Graphing $y = \sqrt[3]{x}$

To graph the function $f(x) = \sqrt[3]{x}$ from Example 6(a) on a TI-83 Plus calculator, enter the following keystrokes:

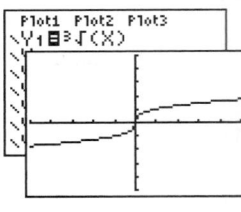

$[-5, 5, 1]$ by $[-5, 5, 1]$

Note: The cube root feature is found next to option 4 under the MATH menu.

 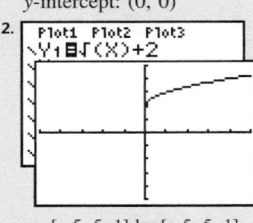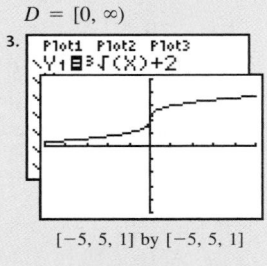
SELF-CHECK 6.3.4

1. Use a graphics calculator to graph $f(x) = x^3 + 2x^2 - 3x$ and to determine the x- and y-intercepts.

Use a graphics calculator to graph each function and to determine its domain.

2. $f(x) = \sqrt{x} + 2$ 3. $f(x) = \sqrt[3]{x} + 2$

Each type of function has its own characteristic shape. The following table summarizes the shapes we have examined in this section and the previous section. The major purpose of these sections is to establish the connection between an equation and the shape of the graph. We will examine these functions in more detail later.

NAME OF FUNCTION	ALGEBRAIC EXAMPLE	GRAPHIC EXAMPLE
Linear function	$f(x) = x$	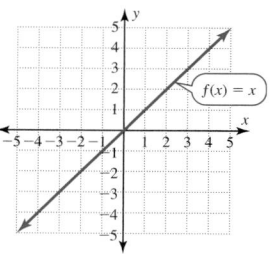
Absolute value function	$f(x) = \lvert x \rvert$	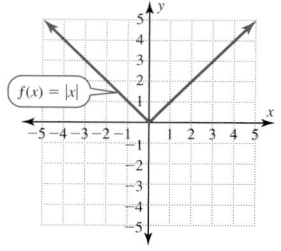
Quadratic function	$f(x) = x^2$	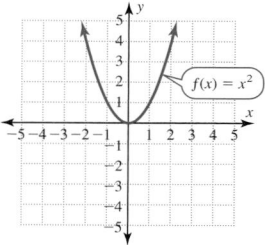
Square root function	$f(x) = \sqrt{x}$	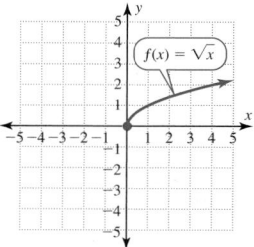
Cubic function	$f(x) = x^3$	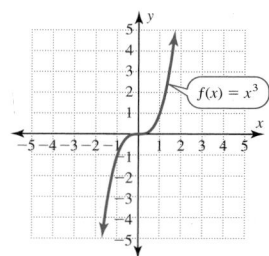
Cube root function	$f(x) = \sqrt[3]{x}$	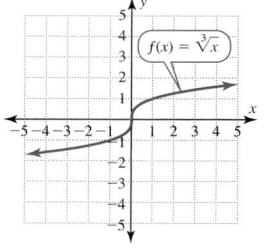

When a function is defined by an equation $y = f(x)$, the accepted convention is that the domain is the set of all real numbers for which $f(x)$ is also a real number. A function restricted to real numbers in the range is called a **real-valued function.** This rules out two types of input values from the domain as described in the following box.

Domain of $y = f(x)$

The domain of a real-valued function $y = f(x)$ is the set of all real numbers for which $f(x)$ is also a real number. This means the domain excludes all values that

1. cause division by 0 or
2. cause a negative number under a square root symbol.

Using this algebraic approach, we can determine the domain of each of the functions given in this section. We urge you to compare the domain that is determined algebraically to the domain that can be determined from the graph of the function. Remember the domain of a graph is its projection onto the x-axis. In the previous box, we stated that the domain of a real-valued function is the set of all real numbers except those that cause division by 0 or that cause a negative number under a square root symbol. The function $f(x) = \dfrac{1}{x}$ is not defined for $x = 0$. Thus the domain of this function is all real numbers except 0. We will wait until Chapter 9 to examine functions of this type.

We will wait until Chapter 9 to examine rational functions like $f(x) = \dfrac{1}{x}$. For these functions we must avoid values that cause division by 0.

■ EXAMPLE 8   Determining the Domain of a Function

Determine the domain of each function.

SOLUTIONS

(a) $f(x) = 5x + 30$ $D = \mathbb{R}$

The equation $f(x) = 5x + 30$ is defined for every real input value of x. Note that the equation defines a linear function so the domain is all real numbers.

(b) $f(x) = (x - 5)^2$ $D = \mathbb{R}$

The equation $f(x) = (x - 5)^2$ is defined for every real input value of x. The equation defines a quadratic function whose graph is a parabola. Thus the domain is all real numbers.

(c) $f(x) = |32 - 8x|$ $D = \mathbb{R}$

The equation $f(x) = |32 - 8x|$ is defined for every real input value of x. The function is an absolute value function whose graph is V-shaped. Thus the domain is all real numbers.

(d) $f(x) = \sqrt{5x + 30}$
$$5x + 30 \geq 0$$
$$5x \geq -30$$
$$x \geq -6$$
$$D = [-6, \infty)$$

The expression under the square root must be nonnegative. Subtracting 30 from both sides of the inequality preserves the order of the inequalities. Dividing both sides of the inequality by 5 preserves the order of the inequality.

■ **EXAMPLE 9**   Using Multiple Perspectives to Examine the Domain of a Square Root Function

Determine the domain of $f(x) = \sqrt{6 - 3x}$.

SOLUTION

ALGEBRAICALLY	NUMERICALLY	GRAPHICALLY	VERBALLY

ALGEBRAICALLY

$$6 - 3x \geq 0$$
$$-3x \geq -6$$
$$x \leq 2$$

NUMERICALLY

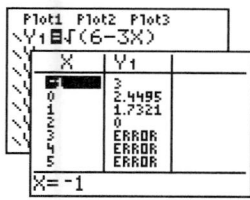

GRAPHICALLY

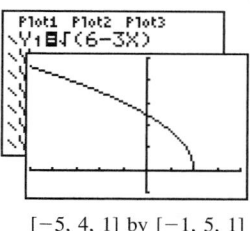

$[-5, 4, 1]$ by $[-1, 5, 1]$

VERBALLY

The expression under the square root symbol must be nonnegative. The domain is the set of all x-values less than or equal to 2. Values greater than 2 produce an error message on the calculator. There are no points on the graph for values of x greater than 2.

Answer: $D = (-\infty, 2]$ ■

SELF-CHECK 6.3.5 ANSWERS

1. $D = \mathbb{R}$
2. $D = \mathbb{R}$
3. $[4, \infty)$
4. $D = \mathbb{R}$

SELF-CHECK 6.3.5

Determine the domain of each function.

1. $f(x) = 6 - 3x$
2. $f(x) = (x + 3)^2$
3. $f(x) = \sqrt{2x - 8}$
4. $f(x) = \sqrt[3]{2x - 8}$

USING THE LANGUAGE AND SYMBOLISM OF MATHEMATICS   6.3

1. A function of the form $f(x) = ax^2 + bx + c$ has a graph whose shape is a _____.
2. A function of the form $f(x) = ax^2 + bx + c$ is called a _____ function.
3. The highest point on the graph of $f(x) = -x^2$ is called the _____ of the parabola.
4. A function of the form $f(x) = ax^3 + bx^2 + cx + d$ is called a _____ function.

5. A real-valued function $y = f(x)$ has real values of x in the domain and is restricted to _____ numbers in the range.
6. The domain of the real-valued function $y = f(x)$ must exclude values that would cause division by _____ or values that would cause a _____ number under a square root symbol.

EXERCISES   6.3

1. Given $f(x) = (x - 1)^2 + 1$, evaluate each expression.
   a. $f(-4)$ b. $f(-2)$ c. $f(0)$ d. $f(4)$
2. Given $f(x) = (x + 2)^2 - 1$, evaluate each expression.
   a. $f(-10)$ b. $f(-2)$ c. $f(0)$ d. $f(8)$
3. Given $f(x) = \sqrt{4 - x}$, evaluate each expression.
   a. $f(-21)$ b. $f(-5)$ c. $f(0)$ d. $f(4)$
4. Given $f(x) = \sqrt{8 - 2x}$, evaluate each expression.

   a. $f(-4)$ b. $f\left(-\dfrac{1}{2}\right)$ c. $f(2)$ d. $f(3.5)$

5. Given $f(x) = x^3 - 4$, evaluate each expression.
   a. $f(-10)$ b. $f(-1)$ c. $f(0)$ d. $f(5)$
6. Given $f(x) = x^3 + 1$, evaluate each expression.
   a. $f(-3)$ b. $f(-1)$ c. $f(10)$ d. $f(3)$
7. Given $f(x) = \sqrt[3]{x + 3}$, evaluate each expression.
   a. $f(5)$ b. $f(-2)$ c. $f(-11)$ d. $f(24)$
8. Given $f(x) = \sqrt[3]{3x - 2}$, evaluate each expression.

   a. $f(1)$ b. $f\left(\dfrac{1}{3}\right)$ c. $f(22)$ d. $f(-41)$

In Exercises 9–12 complete the given table of values and sketch the graph of each function. Determine the domain, the range, and the vertex of each graph. Use a graphics calculator only as a check on your work.

x	−3	−2	−1	0	1	2	3
y							

9. $f(x) = x^2 - 2$ **10.** $f(x) = x^2 + 2$
11. $f(x) = -(x + 1)^2$ **12.** $f(x) = -(x - 1)^2$

In Exercises 13–16 complete the given table of values and sketch the graph of each function. Determine the domain and the range of each function. Use a graphics calculator only as a check on your work.

x	−3	−2	−1	0	1	2	3
y							

13. $f(x) = (x + 2)^3$ **14.** $f(x) = (x - 2)^3$
15. $f(x) = -x^3 + 2$ **16.** $f(x) = -x^3 - 3$

In Exercises 17 and 18 complete the given table of values and sketch the graph of each function. Determine the domain and the range of each function. Use a graphics calculator only as a check on your work.

x	0	1	4	9	16	25
y						

17. $f(x) = \sqrt{x} - 2$ **18.** $f(x) = -\sqrt{x} + 3$

In Exercises 19 and 20 complete the given table of values and sketch the graph of each function. Determine the domain and the range of each function. Use a graphics calculator only as a check on your work.

x	−8	−1	$-\frac{1}{8}$	0	$\frac{1}{8}$	1	8
y							

19. $f(x) = \sqrt[3]{2 - x}$ **20.** $f(x) = \sqrt[3]{x} - 2$

In Exercises 21–28 match each function with its graph. (*Hint:* Work each problem by recognizing the shape of the graph from the defining equation. Use a graphics calculator only as a check on your work.)

21. $f(x) = 2x + 1$ **22.** $f(x) = -2x + 1$
23. $f(x) = \sqrt{2 - x}$ **24.** $f(x) = \sqrt{x - 2}$
25. $f(x) = |x - 2| + 1$ **26.** $f(x) = -x^2 + 1$
27. $f(x) = x^3 + 2$ **28.** $f(x) = \sqrt[3]{x} + 2$

A.

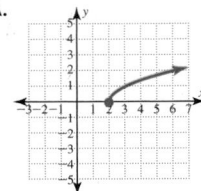

B.

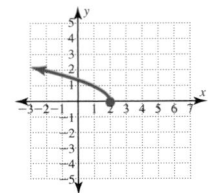

C.

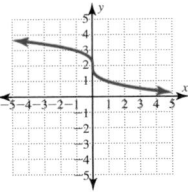

D.

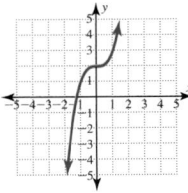

E.

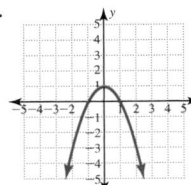

F.

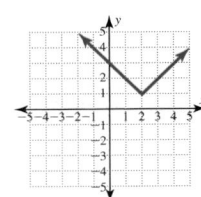

G.

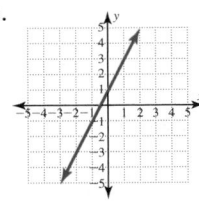

H.

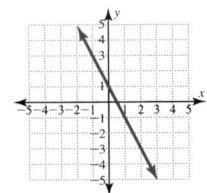

In Exercises 29–38 match each description of a function with its graph.
29. A linear function whose graph has a positive slope
30. A linear function whose graph has a negative slope
31. An absolute value function whose graph opens upward
32. An absolute value function whose graph opens downward
33. A quadratic function whose graph opens upward
34. A quadratic function whose graph opens downward
35. A square root function
36. A cube root function
37. A cubic function with one x-intercept
38. A cubic function with three x-intercepts

A.

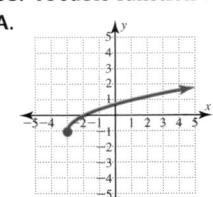

B.

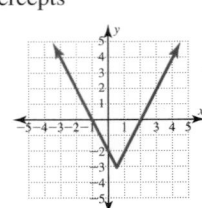

C.

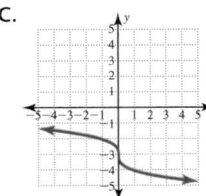

D.

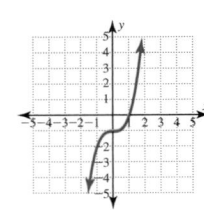

E.

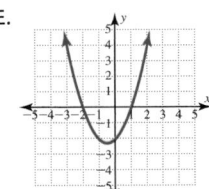

F.

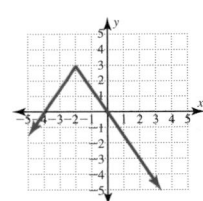

G.

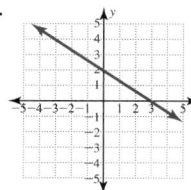

H.

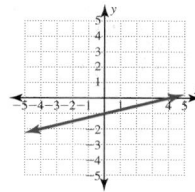

I.

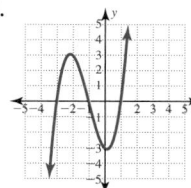

J.

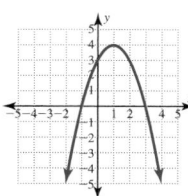

In Exercises 39–52 use a graphics calculator to graph each function and then determine the domain and the range of each function.

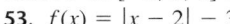

39. $f(x) = 5$
40. $f(x) = -5$
41. $f(x) = x^2 - 5$
42. $f(x) = 4 - x^2$
43. $f(x) = (x + 2)^2 + 1$
44. $f(x) = (x - 1)^2 - 2$
45. $f(x) = -|x + 2| + 1$
46. $f(x) = |x - 3| + 2$
47. $f(x) = -\sqrt{x + 3}$
48. $f(x) = \sqrt{x - 3}$
49. $f(x) = x^3 + 4$
50. $f(x) = -(x + 4)^3$
51. $f(x) = \sqrt[3]{x} - 3$
52. $f(x) = \sqrt[3]{x - 3}$

In Exercises 53–60 use a graphics calculator to graph each function and then determine the x- and y-intercepts of each graph. (*Hint:* Use a viewing window of $[-5, 5, 1]$ by $[-5, 5, 1]$.)

53. $f(x) = |x - 2| - 3$
54. $f(x) = -2|x + 1| + 4$
55. $f(x) = -x^2 + 2x + 3$
56. $f(x) = x^2 - 2x - 3$
57. $f(x) = x^3 - x$
58. $f(x) = -x^3 + 4x$
59. $f(x) = -\frac{1}{2}x^3 + x^2 + \frac{5}{2}x - 3$
60. $f(x) = \frac{1}{2}x^3 + 2x^2 - \frac{1}{2}x - 2$

61. **Modeling the Height of a Parachutist** The height of a parachutist t seconds after jumping from a plane from 5000 ft is given by $h(t) = -16t^2 + 5000$. Evaluate and interpret each of these expressions.
 a. $h(0)$ b. $h(5)$ c. $h(10)$

62. **Modeling the Height of a Baseball** The height in feet of a baseball t seconds after being thrown by a centerfielder toward home plate can be approximated by $h(t) = -16t^2 + 25x + 7$. Evaluate and interpret each of these expressions.
 a. $h(0)$ b. $h(0.78)$ c. $h(1.8)$

63. **Modeling the Cost to Run a Power Line** The cost in dollars to run a power line from a power station on one side of a river to a factory on the other side of the river and x miles downstream is given by $C(x) = 1200\sqrt{x^2 + 1}$. Evaluate and interpret each of these expressions.
 a. $C(0)$ b. $C(2)$ c. $C(5)$

64. **Modeling the Period of a Pendulum** The time in seconds for one complete period of a pendulum of length x cm is approximated by the function $T(x) = 0.2\sqrt{x}$. Evaluate and interpret each of these expressions.
 a. $T(100)$ b. $T(64)$ c. $T(25)$

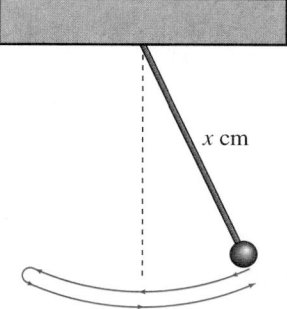

x cm

65. **Modeling the Volume of a Box** The volume in cubic centimeters of the following box of height x centimeters is given by the function $V(x) = x(20 - 2x)(10 - 2x)$. Evaluate and interpret each of these expressions.
 a. $V(1)$ b. $V(3)$ c. $V(4)$

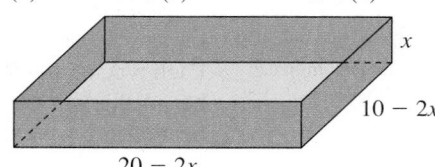

x

$10 - 2x$

$20 - 2x$

66. **Modeling the Volume of a Sphere** The volume in cubic centimeters of a sphere of radius r centimeters is given by the function $V(r) = \frac{4}{3}\pi r^3$. Evaluate and interpret each of these expressions.
 a. $V(1)$ b. $V(5)$ c. $V(10)$

In Exercises 67–78 determine the domain of each function by inspecting the defining equation. (*Hint:* See Example 8. Use a graphics calculator only as a check on your work.)

67. $f(x) = |2x - 3|$
68. $f(x) = 2x - 3$
69. $f(x) = x^2 - 4$
70. $f(x) = \sqrt{x - 4}$
71. $f(x) = \sqrt{x + 4}$
72. $f(x) = \sqrt[3]{x - 1}$
73. $f(x) = x^3 - 2$
74. $f(x) = x^2 + 4$
75. $f(x) = 4x$
76. $f(x) = |4x - 1|$
77. $f(x) = \sqrt[3]{2 - x}$
78. $f(x) = \sqrt{3x + 6}$

Group Discussion Questions

79. **Calculator Discovery** Use a graphics calculator to graph each function: $f(x) = x^2, f(x) = x^2 + 1$, $f(x) = x^2 + 2, f(x) = x^2 + 3$, and $f(x) = x^2 + 4$. Describe the relation between the graphs of $f(x) = x^2$ and $f(x) = x^2 + c$.

80. **Calculator Discovery** Use a graphics calculator to graph each function: $f(x) = x^2, f(x) = (x + 1)^2$, $f(x) = (x + 2)^2, f(x) = (x + 3)^2$, and $f(x) = (x + 4)^2$. Describe the relation between the graphs of $f(x) = x^2$ and $f(x) = (x + c)^2$.

81. Error Analysis A student produced the following calculator window for the function shown and concluded that the function was a linear function and the range of the function is the set of all real numbers. Determine the correct analysis and the source of the error in this answer.

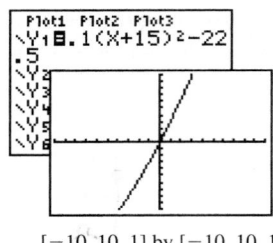

$[-10, 10, 1]$ by $[-10, 10, 1]$

82. Error Analysis A student graphing $f(x) = \sqrt[3]{x} + 2$ obtained the following display. Explain how you can tell by inspection that an error has been made.

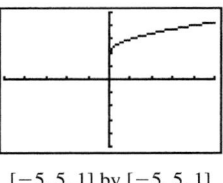

$[-5, 5, 1]$ by $[-5, 5, 1]$

Section 6.4 (Optional) Horizontal and Vertical Translations of Functions

Objectives:

9. Predict the vertical shift in the graph of a function or the shift in the y-values in a table of values for a function.
10. Predict the horizontal shift in the graph of a function or the shift in the x-values in a table of values for a function.

Each function from a family of functions shares the same characteristic shape.

Each type of algebraic equation $y = f(x)$ generates a specific family of graphs. Each member of this family shares a characteristic shape with all other members of this family. In Section 6.2 we examined some of the basic shapes. Now we use one basic graph from each family to generate other graphs in this family. The first way we will do this is by shifting graphs up and down. Later in this section we will shift graphs left and right.

Vertical Shifts

A Mathematical Note

Funds donated to the American Mathematical Society in memory of Ruth Satter, a researcher for Bell Labs, were used to establish a prize for outstanding contributions in mathematics research by a woman. This prize has been awarded every 2 years, starting in 1991.

We first examine shifting graphs up and down. We then start by observing the patterns illustrated by the following functions. Observing patterns is an important part of mathematics. By recognizing patterns, we can gain insights that deepen our understanding and shorten our work. The functions compared next are $y_1 = |x|$ and $y_2 = |x| + 2$.

NUMERICALLY

x	$y_1 = |x|$	$y_2 = |x| + 2$
-3	3	5
-2	2	4
-1	1	3
0	0	2
1	1	3
2	2	4
3	3	5

GRAPHICALLY

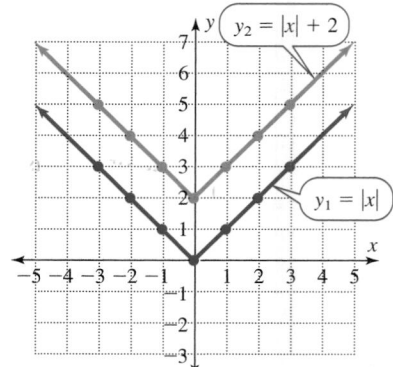

VERBALLY

Each y_2-value is 2 more than the corresponding y_1-value.

The graph of y_2 can be obtained by shifting the graph of y_1 up 2 units.

The graph of $y_2 = |x| + 2$ can be obtained by vertically shifting or translating the graph $y_1 = |x|$ up 2 units. **Vertical translations** or **vertical shifts** are described in the following box.

Vertical Shifts of $y = f(x)$

For a function $y_1 = f(x)$ and a positive real number c:

ALGEBRAICALLY	GRAPHICALLY	NUMERICALLY
$y_2 = f(x) + c$	To obtain the graph of $y_2 = f(x) + c$, shift the graph of $y_1 = f(x)$ up c units.	For the same x-value, $y_2 = y_1 + c$.
$y_3 = f(x) - c$	To obtain the graph of $y_3 = f(x) - c$, shift the graph of $y_1 = f(x)$ down c units.	For the same x-value $y_3 = y_1 - c$.

■ **EXAMPLE 1**   Examining a Vertical Shift of a Quadratic Function

As noted in Section 6.3, each quadratic function in the form $f(x) = ax^2 + bx + c$ graphs as a parabola. The vertex of $y_1 = x^2$ is $(0, 0)$. The vertex of $y_2 = x^2 - 4$ is $(0, -4)$.

Compare the functions $y_1 = x^2$ and $y_2 = x^2 - 4$ numerically, graphically, and verbally.

SOLUTIONS

NUMERICALLY

x	$y_1 = x^2$	$y_2 = x^2 - 4$
-3	9	5
-2	4	0
-1	1	-3
0	0	-4
1	1	-3
2	4	0
3	9	5

GRAPHICALLY

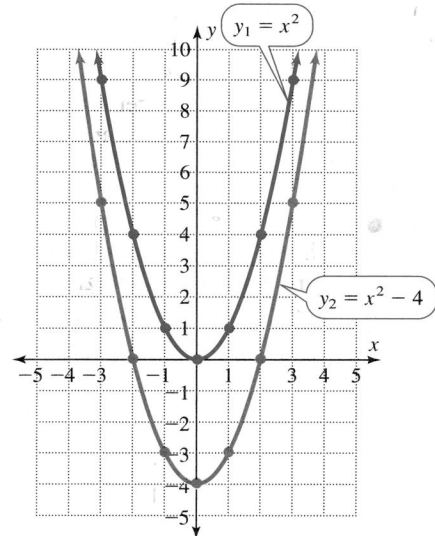

VERBALLY

For each value of x, y_2 is 4 units less than y_1.

The parabola defined by $y_2 = x^2 - 4$ can be obtained by shifting the parabola defined by $y_1 = x^2$ down 4 units. The range of y_1 is $[0, +\infty)$. The range of y_2 is $[-4, +\infty)$.

SELF-CHECK 6.4.1 ANSWERS

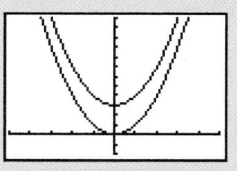

$[-5, 5, 1]$ by $[-2, 12, 1]$

For each value of x, y_2 is 3 units more than y_1. The parabola defined by $f(x) = x^2 + 3$ can be obtained by shifting the parabola defined by $f(x) = x^2$ up 3 units.

SELF-CHECK 6.4.1

Use a graphics calculator to compare $f(x) = x^2$ and $f(x) = x^2 + 3$, both numerically and graphically.

| CALCULATOR PERSPECTIVE 6.4.1 | Performing a Vertical Translation of a Graph |

To perform the vertical translation of $y_1 = x^2$ from Example 1 on a TI-83 Plus calculator, enter the following keystrokes:

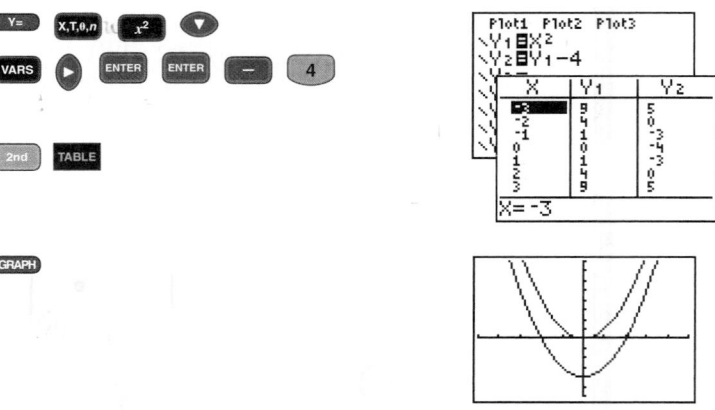

$[-5, 5, 1]$ by $[-6, 8, 1]$

Note: The table illustrates that y_2 is 4 units less than y_1 at each x-value. The graphs show that the graph of y_2 can be obtained by shifting the graph of y_1 down 4 units. ■

The next two examples illustrate how to recognize functions that are vertical shifts of each other. In Example 2 we examine two graphs and in Example 3 we examine two tables of values.

■ **EXAMPLE 2**   Identifying a Vertical Shift

Use the graphs of $y_1 = f_1(x)$ and $y_2 = f_2(x)$ to write an equation for y_2 in terms of $f_1(x)$.

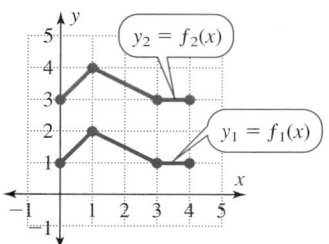

SOLUTION _____

The graph of $y_2 = f_2(x)$ can be obtained by shifting the graph of $y_1 = f_1(x)$ up 2 units. Thus $y_2 = f_1(x) + 2$.
Key points shifted are:
$(0, 1)$ is shifted up 2 units to $(0, 3)$
$(1, 2)$ is shifted up 2 units to $(1, 4)$
$(3, 1)$ is shifted up 2 units to $(3, 3)$
$(4, 1)$ is shifted up 2 units to $(4, 3)$

Answer: $y_2 = f_1(x) + 2$ ■

■ EXAMPLE 3   Identifying a Vertical Shift

Use the table of values for $y_1 = f_1(x)$ and $y_2 = f_2(x)$ to write an equation for y_2 in terms of $f_1(x)$.

x	$y_1 = f_1(x)$	$y_2 = f_2(x)$
-3	18	13
-2	6	1
-1	1	-4
0	0	-5
1	0	-5
2	-2	-7
3	-9	-14

SOLUTION _____

For each value of x, y_2 is 5 units less than y_1. Thus $y_2 = f_1(x) - 5$.

Answer: $y_2 = f_1(x) - 5$

■

SELF-CHECK 6.4.2

1. Use the graph of $y_1 = f(x)$ to graph $y_2 = f(x) + 2$.

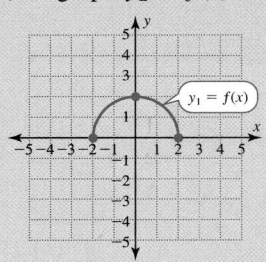

2. Use the table for $y_1 = f(x)$ to complete the table for $y_2 = f(x) + 1$.

x	y_1	y_2
-2	-8	
-1	-1	
0	0	
1	1	
2	8	

SELF-CHECK 6.4.2 ANSWERS

1.

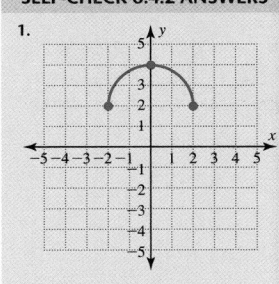

2.

x	y_1	y_2
-2	-8	-7
-1	-1	0
0	0	1
1	1	2
2	8	9

Horizontal Shifts

We have observed that vertical shifts affect the y-values of the points on a graph. Now we examine horizontal shifts that affect the x-values of the points on a graph. We start our inspection of horizontal shifts by examining the functions $y_1 = |x|$ and $y_2 = |x + 2|$.

NUMERICALLY

x	$y_1 = |x|$
-3	3
-2	2
-1	1
0	0
1	1
2	2
3	3

x	$y_2 = |x + 2|$
-5	3
-4	2
-3	1
-2	0
-1	1
0	2
1	3

GRAPHICALLY

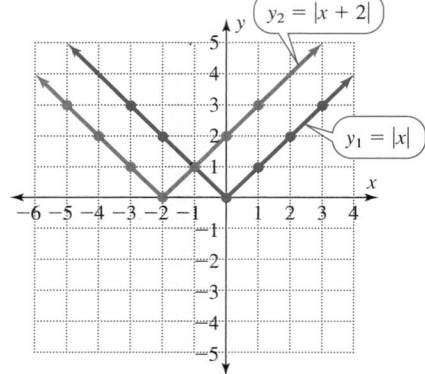

VERBALLY

For the same y-value in y_1 and y_2 the x-value is 2 units less for y_2 than for y_1.

The graph of y_2 can be obtained by shifting the graph of y_1 to the left 2 units.

Horizontal Shifts of $y = f(x)$

Caution: Adding a positive c shifts the graph to the *left*. Reconsider $y = |x + 2|$. Note that output values of $y = |x + 2|$ occur 2 units sooner for x than they do for $y = |x|$. Two units sooner means 2 units to the left on the number line.

For a function $y_1 = f(x)$ and a positive real number c:

ALGEBRAICALLY	GRAPHICALLY	NUMERICALLY
$y_2 = f(x + c)$	To obtain the graph of $y_2 = f(x + c)$, shift the graph of $y_1 = f(x)$ to the left c units.	For the same y-value in y_1 and y_2 the x-value is c units less for y_2 than for y_1.
$y_3 = f(x - c)$	To obtain the graph of $y_3 = f(x - c)$, shift the graph of $y_1 = f(x)$ to the right c units.	For the same y-value in y_1 and y_2 the x-value is c units more for y_2 than for y_1.

■ **EXAMPLE 4**   Examining a Horizontal Shift of a Quadratic Function

Compare the functions $f_1(x) = x^2$ and $f_2(x) = (x - 2)^2$ numerically, graphically, and verbally.

SOLUTIONS

NUMERICALLY

x	$f_1(x) = x^2$	x	$f_2(x) = (x - 2)^2$
-3	9	-1	9
-2	4	0	4
-1	1	1	1
0	0	2	0
1	1	3	1
2	4	4	4
3	9	5	9

GRAPHICALLY

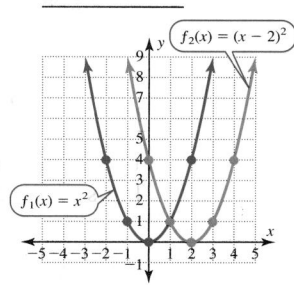

Both graphs are parabolas. The vertex of $f_1(x) = x^2$ is $(0, 0)$. The vertex of $f_2(x) = (x - 2)^2$ is $(2, 0)$.

VERBALLY

For the same x-value in f_1 and f_2, the x-value is 2 units more for f_2 than for f_1.

The graph of $f_2(x) = (x - 2)^2$ can be obtained by shifting the parabola defined by $f_1(x) = x^2$ to the right 2 units. The domain for both functions is $\mathbb{R}$, the set of all real numbers. ■

SELF-CHECK 6.4.3

Use a graphics calculator to compare the graph of $f(x) = x^2$ to the graph of $f(x) = (x + 1)^2$.

The next example examines two graphs which are horizontal translations or shifts of each other.

SELF-CHECK 6.4.3 ANSWER

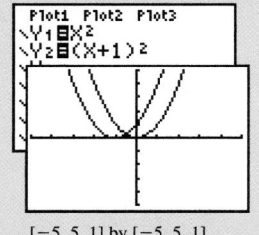

$[-5, 5, 1]$ by $[-5, 5, 1]$

The parabola $f(x) = (x + 1)^2$ can be obtained by shifting the parabola defined by $f(x) = x^2$ to the left 1 unit.

■ **EXAMPLE 5**   Identifying a Horizontal Shift

Use the graphs of $y_1 = f_1(x)$ and $y_2 = f_2(x)$ to write an equation for y_2 in terms of $f_1(x)$.

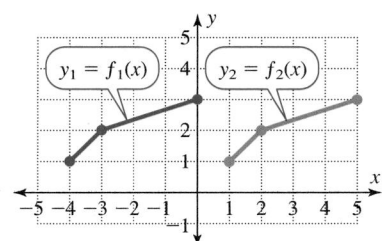

SOLUTION

Note that the graph of $y_2 = f_2(x)$ can be obtained by shifting the graph of $y_1 = f_1(x)$ to the right 5 units. Thus $y_2 = f_1(x - 5)$.

Key points shifted are:
$(-4, 1)$ is shifted right 5 units to $(1, 1)$
$(-3, 2)$ is shifted right 5 units to $(2, 2)$
$(0, 3)$ is shifted right 5 units to $(5, 3)$

Answer: $y_2 = f_1(x - 5)$ ■

Example 6 is used to examine horizontal shifts when two functions are defined by a table of values.

■ **EXAMPLE 6** Using Translations to Report Sales Data

When working with data for which the input values are years, analysts often make use of a horizontal translation to work with smaller input values but with equivalent results. The following tables represent the sales figures for a company founded in 1997. Use the information in the tables to write an equation for $y_2 = f_2(x)$ as a translation of $y_1 = f_1(x)$.

x YEAR	y_1 SALES (\$)
1997	100,000
1998	120,000
1999	150,000
2000	200,000
2001	270,000
2002	350,000

x YEAR	y_2 SALES (\$)
0	100,000
1	120,000
2	150,000
3	200,000
4	270,000
5	350,000

SOLUTION

Note that 1997 corresponds to year 0, 1998 to year 1, and so on. We can obtain a table of values for y_2 based on the table for y_1 by shifting the input x-values to the left 1997 units. Thus $y_2 = f_1(x + 1997)$.

Answer: $y_2 = f_1(x + 1997)$ ■

SELF-CHECK 6.4.4

The following two tables show the number of students who completed a graphics design program that started in 1980 at a midwestern community college.

x YEAR	y_1 STUDENTS
1985	2
1990	6
1995	14
2000	20

x YEAR	y_2 STUDENTS
5	2
10	6
15	14
20	20

Use the information in the tables to write an equation for $y_2 = f_2(x)$ as a translation of $y_1 = f_1(x)$.

SELF-CHECK 6.4.4 ANSWER

$y_2 = f_1(x + 1980)$

The next example gives some additional practice recognizing horizontal and vertical shifts from the equation defining the function.

■ **EXAMPLE 7**   Identifying Horizontal and Vertical Shifts

Describe how to shift the graph of $y = f(x)$ to obtain the graph of each function.

_____ SOLUTIONS _____

(a) $y = f(x) + 8$ Shift the graph of $y = f(x)$ up 8 units.
(b) $y = f(x) - 8$ Shift the graph of $y = f(x)$ down 8 units.
(c) $y = f(x + 8)$ Shift the graph of $y = f(x)$ left 8 units.
(d) $y = f(x - 8)$ Shift the graph of $y = f(x)$ right 8 units.   ■

SELF-CHECK 6.4.5

1. Use the graph of $y_1 = f(x)$ to graph $y_2 = f(x - 3)$.

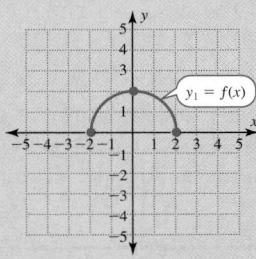

2. Use the table for $y_1 = f(x)$ to complete the table for $y_2 = f(x - 5)$.

x	$f(x)$	x	$f(x - 5)$
-3	18		18
-2	6		6
-1	1		1
0	0		0
1	0		0
2	-2		-2
3	9		9

SELF-CHECK 6.4.5 ANSWERS

1.

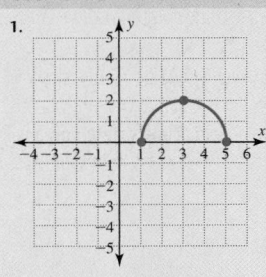

2.

x	$f(x - 5)$
2	18
3	6
4	1
5	0
6	0
7	-2
8	9

Combining Horizontal and Vertical Shifts

The next example illustrates how to combine horizontal and vertical shifts to produce a new graph.

■ **EXAMPLE 8**   Combining Horizontal and Vertical Shifts

Use the graphs of $y_1 = f_1(x)$ and $y_2 = f_2(x)$ to write an equation for y_2 in terms of $f_1(x)$. Give the domain and range of each function.

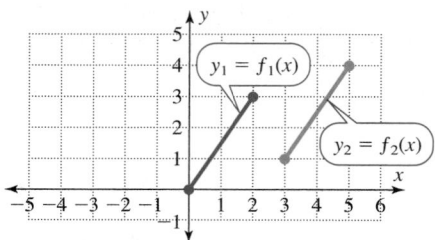

SOLUTION

The graph of $y_2 = f_2(x)$ can be obtained by shifting the graph of $y_1 = f_1(x)$ to the right 3 units and up 1 unit. Note how this shift affects the domain and range.

Key points shifted are:

$(0, 0)$ is shifted right 3 units and up 1 unit to $(3, 1)$

$(2, 3)$ is shifted right 3 units and up 1 unit to $(5, 4)$

Answer: $y_2 = f(x - 3) + 1$

Domain of y_1: $D = [0, 2]$
Range of y_1: $R = [0, 3]$
Domain of y_2: $D = [3, 5]$
Range of y_2: $R = [1, 4]$ ■

The parabola $f(x) = x^2$ has a vertex of $(0, 0)$. This parabola can be shifted horizontally and vertically to produce many other parabolas. The vertex is a key point on each of these parabolas. The next example gives us practice in identifying the vertex of a parabola from the defining equation.

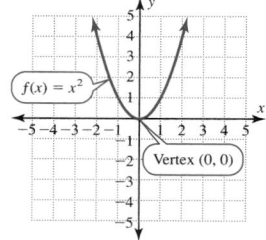

■ **EXAMPLE 9**   **Determining the Vertex of a Parabola**

Determine the vertex of each parabola using the fact that the vertex of $f(x) = x^2$ is $(0, 0)$.

SOLUTIONS

(a) $f(x) = x^2 + 7$ Vertex of $(0, 7)$ This is a vertical shift up 7 units of every point on $f(x) = x^2$.

(b) $f(x) = (x + 7)^2$ Vertex of $(-7, 0)$ This is a horizontal shift left of 7 units of every point on $f(x) = x^2$.

In Example 9(c) you can think of the shifts following the same order as the order of operations in the expression. First we subtract inside the parentheses, producing a shift 7 units to the right. Later we add 9, producing a shift 9 units up.

(c) $f(x) = (x - 7)^2 + 9$ Vertex of $(7, 9)$ This is a horizontal shift 7 units right and a vertical shift 9 units up of every point on $f(x) = x^2$.

(d) $f(x) = (x + 8)^2 - 6$ Vertex of $(-8, -6)$ This is a horizontal shift 8 units left and a vertical shift 6 units down of every point on $f(x) = x^2$. ■

SELF-CHECK 6.4.6 ANSWERS

1. $y = \sqrt{x + 10}$
2. $y = \sqrt{x} - 10$
3. $y = \sqrt{x - 5} - 6$
4. $y = \sqrt{x + 7} + 8$
5a. Vertex of $(0, 7)$
5b. Vertex of $(-7, 0)$
5c. Vertex of $(7, 9)$
5d. Vertex of $(-8, -6)$

SELF-CHECK 6.4.6

Write an equation to shift the graph of $y = \sqrt{x}$ as described in each problem.

1. A shift left 10 units.
2. A shift down 10 units.
3. A shift right 5 units and down 6 units.
4. A shift left 7 units and up 8 units.
5. Use a graphics calculator to graph the parabolas in Example 9 in order to check the vertex of each parabola.

USING THE LANGUAGE AND SYMBOLISM OF MATHEMATICS   6.4

1. Each member of a family of functions has a graph with the same characteristic _____.
2. A shift of a graph up or down is called a _____ shift or a _____ translation.
3. A shift of a graph left or right is called a _____ shift or a _____ translation.
4. For a function $y = f(x)$ and a positive real number c:
   a. $y = f(x) + c$ will produce a _____ translation c units _____.
   b. $y = f(x) - c$ will produce a _____ translation c units _____.
   c. $y = f(x + c)$ will produce a _____ translation c units _____.
   d. $y = f(x - c)$ will produce a _____ translation c units _____.

5. For a function $y = f(x)$ and positive real numbers h and k:
   a. $y = f(x + h) + k$ will produce a _____ translation h units _____ and a _____ translation k units _____.
   b. $y = f(x - h) - k$ will produce a _____ translation h units _____ and a _____ translation k units _____.
6. The shape of the graph of $f(x) = x^2$ is called a _____ and the lowest point on this graph is called its _____.

EXERCISES   6.4

Exercises 1–4 give functions which are translations of $f(x) = |x|$. Match each graph to the correct function. All graphs are displayed with the window $[-10, 10, 1]$ by $[-10, 10, 1]$.

1. $f(x) = |x| + 3$
2. $f(x) = |x + 3|$
3. $f(x) = |x| - 3$
4. $f(x) = |x - 3|$

A.

B.
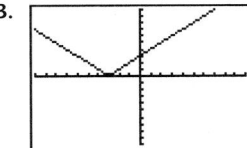

C.
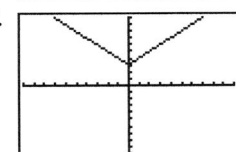
D.
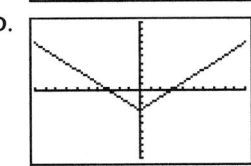

Exercises 5–8 give functions which are translations of $f(x) = x^2$. Match each graph to the correct function. All graphs are displayed with the window $[-10, 10, 1]$ by $[-10, 10, 1]$.

5. $f(x) = x^2 - 5$
6. $f(x) = (x - 5)^2$
7. $f(x) = (x + 5)^2$
8. $f(x) = x^2 + 5$

A.
B.

C.
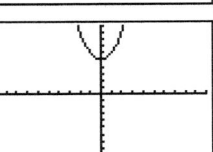
D.
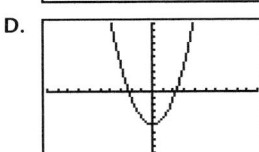

Exercises 9–12 give functions which are translations of $f(x) = \sqrt{x}$. Match each graph to the correct function. All graphs are displayed with the window $[-10, 10, 1]$ by $[-10, 10, 1]$.

9. $f(x) = \sqrt{x - 4}$
10. $f(x) = \sqrt{x + 4}$
11. $f(x) = \sqrt{x} + 4$
12. $f(x) = \sqrt{x} - 4$

A.
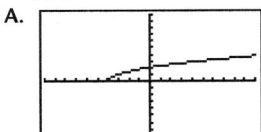
B.
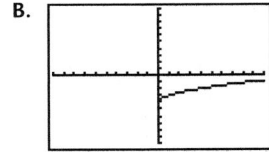

C.
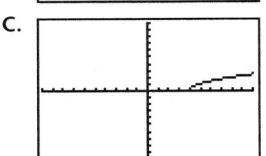
D.
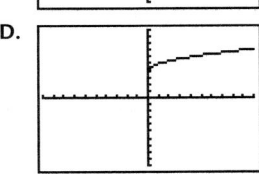

Exercises 13–16 give functions which are translations of $f(x) = |x|$. Match each graph to the correct function. All graphs are displayed with the window $[-10, 10, 1]$ by $[-10, 10, 1]$.

13. $f(x) = |x - 1| + 2$
14. $f(x) = |x + 1| - 2$
15. $f(x) = |x + 1| + 2$
16. $f(x) = |x - 1| - 2$

A.
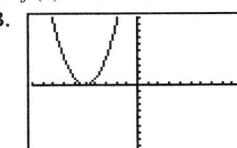
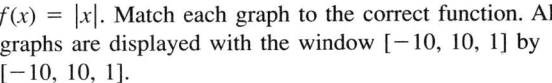
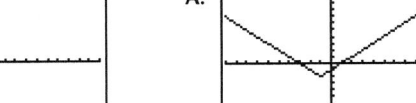
B.
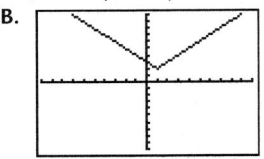

C.
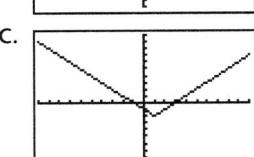
D.
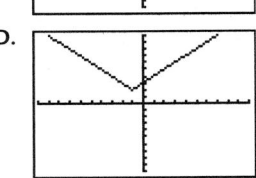

In Exercises 17–22 use the given graph of $y = f(x)$ and horizontal and vertical shifts to graph each function. (*Hint:* First translate the three points labeled on the graph.)

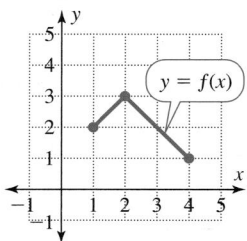

17. $y = f(x) - 3$ **18.** $y = f(x) + 2$
19. $y = f(x + 4)$ **20.** $y = f(x - 1)$
21. $y = f(x - 1) + 2$ **22.** $y = f(x + 1) - 2$

In Exercises 23 and 24 each graph is a translation of the graph of $y = \dfrac{x}{2}$. Write the equation of each graph. Identify the y-intercept of each graph.

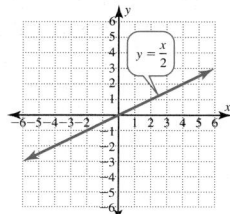

23. 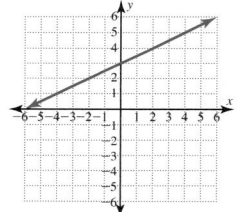   **24.**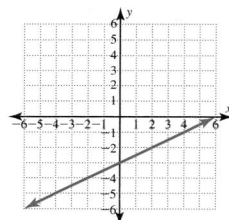

In Exercises 25 and 26 use the given table of values for $y = f(x)$ to complete each table.

x	y
-3	0
-2	8
-1	6
0	0
1	-4
2	0

25.

x	y = f(x) + 10
-3	
-2	
-1	
0	
1	
2	

26.

x	y = f(x) - 10
-3	
-2	
-1	
0	
1	
2	

In Exercises 27 and 28 use the given tables to write an equation for y_2 and y_3.

x	$y_1 = f(x)$	x	y_2	x	y_3
-3	4	-3	-3	-3	11
-2	3	-2	-4	-2	10
-1	8	-1	1	-1	15
0	7	0	0	0	14
1	0	1	-7	1	7
2	-1	2	-8	2	6

27. Write an equation for y_2 in terms of $f(x)$.
28. Write an equation for y_3 in terms of $f(x)$.
29. Use the given table for $f(x) = 2x - 3$ to complete the table for y_2.

x	$y_1 = f(x)$	x	$y_2 = f(x - 4)$
0	-3	4	-3
1	-1		-1
2	1		1
3	3		3
4	5		5
5	7	9	7

30. Use the given table for $f(x) = -|x|$ to complete the table for y_2.

x	$y_1 = f(x)$	x	$y_2 = f(x + 1)$
-3	-3	-4	-3
-2	-2		-2
-1	-1		-1
0	0		0
1	-1		-1
2	-2		-2
3	-3	2	-3

In Exercises 31 and 32 use the given tables to write an equation for y_2 and y_3.

x	$y_1 = f(x)$	x	y_2	x	y_3
-3	3	0	3	-4	3
-2	2	1	2	-3	2
-1	1	2	1	-2	1
0	0	3	0	-1	0
1	1	4	1	0	1
2	2	5	2	1	2
3	3	6	3	2	3

31. Write an equation for y_2 in terms of $f(x)$.
32. Write an equation for y_3 in terms of $f(x)$.

In Exercises 33–38 determine the vertex of each parabola using the fact that the vertex of $f(x) = x^2$ is at $(0, 0)$. Use a graphics calculator to check your answers.

33. $f(x) = (x - 6)^2$

34. $f(x) = (x + 6)^2$

35. $f(x) = x^2 + 11$

36. $f(x) = x^2 - 11$

37. $f(x) = (x + 5)^2 - 8$

38. $f(x) = (x - 8)^2 + 5$

In Exercises 39–44 determine the vertex of the graph of each absolute value function using the fact that the vertex of $f(x) = |x|$ is at $(0, 0)$. Use a graphics calculator to check your answers.

39. $f(x) = |x| - 13$

40. $f(x) = |x| + 14$

41. $f(x) = |x + 15|$

42. $f(x) = |x - 10|$

43. $f(x) = |x - 7| - 6$

44. $f(x) = |x + 9| + 8$

Exercises 45–50 describe a translation of the graph $y = f(x)$. Match each description to the correct function.

45. A translation eleven units left

46. A translation eleven units right

47. A translation eleven units down

48. A translation eleven units up

49. A translation eleven units right and eleven units up

50. A translation eleven units left and eleven units down

A. $y = f(x - 11) + 11$

B. $y = f(x + 11) - 11$

C. $y = f(x - 11)$

D. $y = f(x) - 11$

E. $y = f(x + 11)$

F. $y = f(x) + 11$

In Exercises 51–56 determine the domain and range of each function f_2 given that the domain of a function f_1 is $D = [0, 5)$ and the range is $R = [2, 4)$.

51. The graph of f_2 is obtained from the graph of f_1 by shifting it 3 units right.

52. The graph of f_2 is obtained from the graph of f_1 by shifting it 5 units left.

53. The graph of f_2 is obtained from the graph of f_1 by shifting it 7 units down.

54. The graph of f_2 is obtained from the graph of f_1 by shifting it 8 units up.

55. The graph of f_2 is obtained from the graph of f_1 by shifting it 4 units left and 5 units up.

56. The graph of f_2 is obtained from the graph of f_1 by shifting it 6 units right and 4 units down.

In Exercises 57–64 match each function with its graph. Use the shape of the graph of each function and your knowledge of translations to make your choices. All graphs are displayed with the window $[-10, 10, 1]$ by $[-10, 10, 1]$.

57. $f(x) = x - 6$

58. $f(x) = x + 4$

59. $f(x) = |x + 5|$

60. $f(x) = |x| - 3$

61. $f(x) = \sqrt{x} + 3$

62. $f(x) = (x + 4)^3$

63. $f(x) = \sqrt[3]{x} - 5$

64. $f(x) = (x - 3)^2$

A.

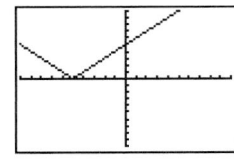

B.

C.

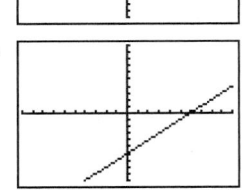

D.

E.

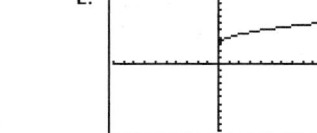

F.

G.

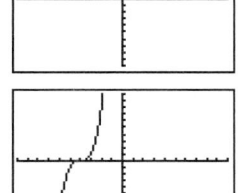

H.

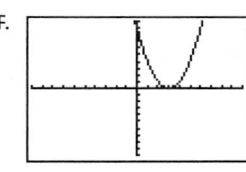

65. Write the first five terms of each sequence. (*Hint:* See Example 8 in Section 3.1.)

a. $a_n = 2n$

b. $a_n = 2n + 3$

c. $a_n = 2(n + 3)$

d. $a_n = -2n$

66. Write the first five terms of each sequence. (*Hint:* See Example 8 in Section 3.1.)

a. $a_n = n^2$

b. $a_n = n^2 - 9$

c. $a_n = (n - 9)^2$

d. $a_n = (n - 1)^2 + 3$

67. Shifting the Temperature Scale   Many properties of nature are related to the temperature of the object being studied. The Celsius temperature scale is the same incrementally as the Kelvin temperature scale except that it is 273° higher (water freezes 0° Celsius and 273° Kelvin). The equation $V(t) = 4t$ gives the volume in milliliters of a gas as a function of the temperature in degrees Celsius. Write an equation for the volume as a function of the temperature in degrees Kelvin and complete the following table.

t CELSIUS	$V(t) = 4t$	t KELVIN	$V = ?$
200	800		800
150	600		600
100	400		400
50	200		200
0	0	273	0

68. Shifting a Time Reference   A company has a function which approximates its revenue in terms of the year those revenues were generated. The company was founded in January 1995. Thus 1995 is taken as year 0 for the company, 1996 as year 1, and so on. A formula for its revenue per year in millions of dollars is given by $R(t) = t^2$, where t is the number of years the company has been in business. Convert this formula to give the revenue in terms of the calendar year.

t YEAR	$R(t) = t^2$ REVENUE
0	0
1	1
2	4
3	9
4	16
5	25

69. Shifting a Pricing Function
The price in dollars of five different types of networking cables is given in the table to the right. Due to increased overhead cost, the price of each item is increased by $1. Form a table that gives the new price of each item.

x ITEM NO.	P(x) PRICE ($)
1	24.95
2	33.79
3	12.98
4	26.78
5	19.95

70. Shifting the Distance for Golf Clubs Thanks to a new technology in making golf clubs, many golfers have been able to add 5 yd to the distance they can hit with each club. Form a table that gives the new distance for each new club.

IRON	DISTANCE FOR OLD CLUBS
2	200
3	190
4	180
5	165
6	150

Group Discussion Questions

71. Calculator Discovery Use a graphics calculator to graph each pair of functions on the same graphics screen. Then describe the relationship between each pair of functions.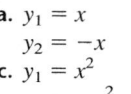

a. $y_1 = x$
$y_2 = -x$

b. $y_1 = |x|$
$y_2 = -|x|$

c. $y_1 = x^2$
$y_2 = -x^2$

d. $y_1 = \sqrt{x}$
$y_2 = -\sqrt{x}$

72. Calculator Discovery Use a graphics calculator to graph each function: $f(x) = |x|, f(x) = 2|x|, f(x) = 3|x|,$ and $f(x) = 4|x|$. Can you describe the relationship between the graph of $f(x) = |x|$ and $f(x) = c|x|$ for any positive constant c?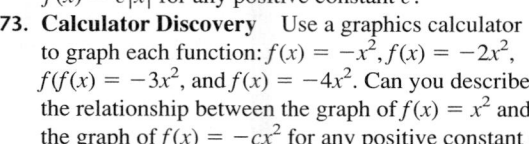

73. Calculator Discovery Use a graphics calculator to graph each function: $f(x) = -x^2, f(x) = -2x^2, f(f(x)) = -3x^2,$ and $f(x) = -4x^2$. Can you describe the relationship between the graph of $f(x) = x^2$ and the graph of $f(x) = -cx^2$ for any positive constant c?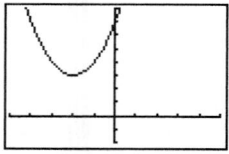

74. Error Analysis A student graphing $f(x) = |x + 2| + 3$ obtained the following display. Explain how you can tell by inspection that an error has been made.

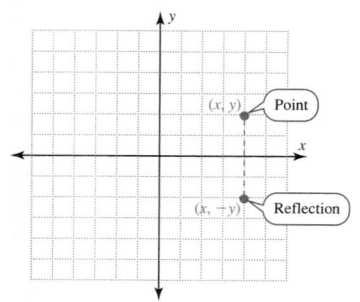

$[-5, 5, 1]$ by $[-2, 8, 1]$

Section 6.5 (Optional) Stretching, Shrinking, and Reflecting Graphs of Functions

Objectives: 11. Recognize the reflection of a graph.
12. Use stretching and shrinking factors to graph functions.

A key concept of algebra is that there are some basic families of graphs. In Sections 6.2 and 6.3 we learned how to create and recognize a fundamental representative from each family. In Section 6.4 we learned how to shift a graph to create an exact copy of this graph at another location in the plane. We now learn how to modify a given graph to create a stretching or a shrinking of this shape. We start by examining the reflection of a graph.

Reflecting a Graph Across the x-Axis

The reflection of a point (x, y) across the x-axis is a mirror image of the point on the opposite side of the x-axis. The reflection of the point (x, y) about the x-axis is the point $(x, -y)$.

We now examine the reflection of an entire graph by examining the functions $y_1 = |x|$ and $y_2 = -|x|$.

NUMERICALLY

x	$y_1 = |x|$	$y_2 = -|x|$
-3	3	-3
-2	2	-2
-1	1	-1
0	0	0
1	1	-1
2	2	-2
3	3	-3

GRAPHICALLY

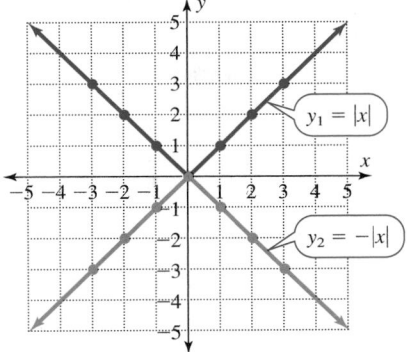

VERBALLY

Each y_2-value is the opposite of the corresponding y_1-value.

The graph of y_2 can be obtained by reflecting across the x-axis each point of the graph of y_1.

A reflection of a graph is simply a reflection of each point of the graph. The following box describes a reflection of a graph across the x-axis.

Reflection of $y = f(x)$ Across the x-Axis		
GRAPHICALLY	**NUMERICALLY**	**ALGEBRAICALLY**
To obtain the graph of $y = -f(x)$, reflect the graph of $y = f(x)$ across the x-axis.	For each value of x, y_2 is the additive inverse of y_1. That is, $y_2 = -y_1$.	Original function: $y_1 = f(x)$ Reflection: $\qquad y_2 = -f(x)$

■ EXAMPLE 1   Examining a Reflection of a Quadratic Function

Compare the functions $y_1 = x^2$ and $y_2 = -x^2$ numerically, graphically, and verbally.

SOLUTION

NUMERICALLY

x	$y_1 = x^2$	$y_2 = -x^2$
-3	9	-9
-2	4	-4
-1	1	-1
0	0	0
1	1	-1
2	4	-4
3	9	-9

GRAPHICALLY

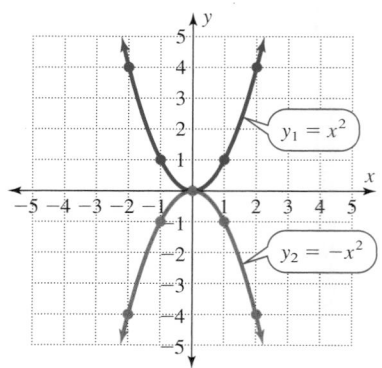

VERBALLY

For each value of x, $y_2 = -y_1$.

The parabola defined by $y_2 = -x^2$ can be obtained by reflecting the graph of $y_1 = x^2$ across the x-axis.

CALCULATOR PERSPECTIVE 6.5.1	Reflecting a Graph Across the x-Axis

To create the table and the graph of $y_2 = -y_1$ from Example 1 on a TI-83 Plus calculator, enter the following keystrokes:

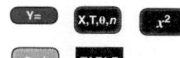

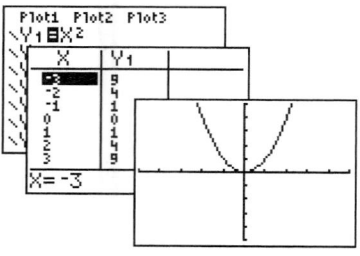

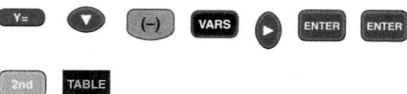

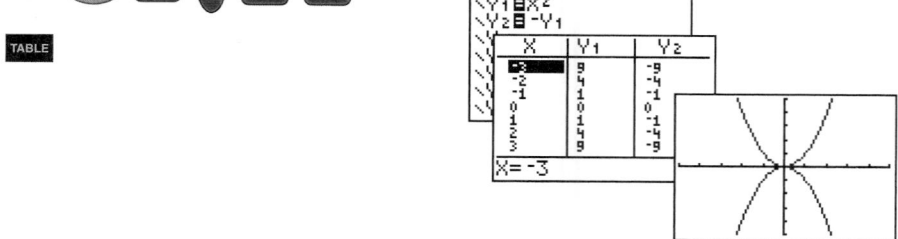

The next example illustrates how to form the reflection of a graph when no formula is given for $y = f(x)$.

■ EXAMPLE 2   Forming the Reflection of a Graph

Use the graph of $y_1 = f(x)$ to graph $y_2 = -f(x)$.

SOLUTION _____

Example 2 clearly illustrates the reflection idea when we visually compare these two graphs. According to David Bock, Graphics Research Programmer for the National Center for Supercomputing Applications, "This concept is used extensively in three-dimensional computer graphics and animation to create and position symmetrical objects and models."

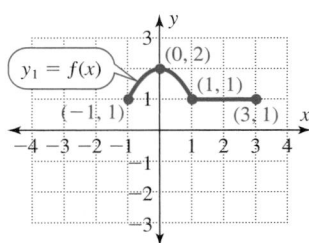

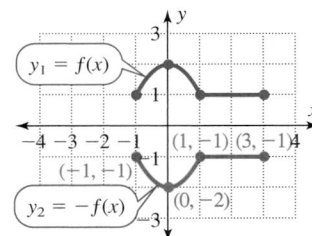

Start by reflecting the key points labeled on the graph of $y_1 = f(x)$. Then use the shape of $y_1 = f(x)$ to sketch its reflection across the x-axis. ■

Example 3 is used to examine reflections when two functions are defined by a table of values.

■ EXAMPLE 3   Identifying a Reflection Across the x-Axis

Use the table of values for $y_1 = x^3 - x^2$ and y_2 to write an equation for y_2.

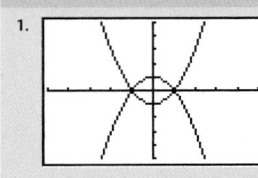
x	$y_1 = x^3 - x^2$	y_2
-3	-36	36
-2	-12	12
-1	-2	2
0	0	0
1	0	0
2	4	-4
3	18	-18

SOLUTION _____

For each value of x, y_2 is the additive inverse of y_1.

Answer: $y_2 = -y_1$ or $y_2 = -x^3 + x^2$. ■

SELF-CHECK 6.5.2

1. Use the graph of $y_1 = f(x)$ to graph $y_2 = -f(x)$.

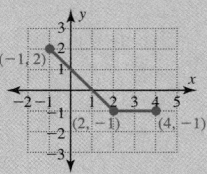

2. Use the table for $y_1 = f(x)$ to complete the table for $y_2 = -f(x)$.

x	y_1	y_2
-2	7	
-1	-4	
0	3	
1	9	
2	-6	

Stretching and Shrinking Graphs

We now examine how to change the size of a function by applying a stretching or shrinking factor. These stretching or shrinking factors affect the vertical scale of the graph. They do not affect the horizontal scale and they are not rigid translations. We will start our inspection of scaling factors by examining the functions $y_1 = |x|$ and $y_2 = 3|x|$.

NUMERICALLY

x	$y_1 = |x|$	$y_2 = 3|x|$
-3	3	9
-2	2	6
-1	1	3
0	0	0
1	1	3
2	2	6
3	3	9

GRAPHICALLY

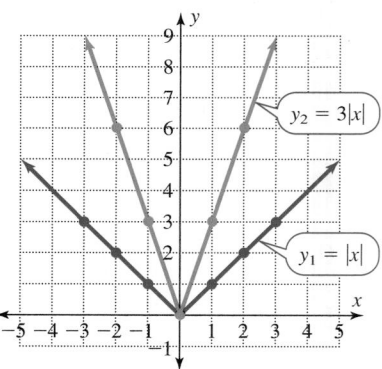

VERBALLY

Each value of y_2 is 3 times the corresponding y_1 value.

The graph of y_2 is the same basic V-shape as y_1 but rises 3 times as rapidly.

The scaling factor c in $y = cf(x)$ affects the vertical scale of the graph—not the horizontal scale.

Comparing the graphs of $y_2 = 3|x|$ to $y_1 = |x|$, we call 3 a **stretching factor** because this factor vertically stretches the graph of $y_1 = |x|$. The following box describes **vertical scaling factors** that can either stretch or shrink a graph. Then in Example 4 we examine a scaling factor of $\frac{1}{4}$ that vertically shrinks the graph.

Vertical Stretching and Shrinking Factors of $y = f(x)$

For a function $y_1 = f(x)$ and a positive real number c:

ALGEBRAICALLY	GRAPHICALLY	NUMERICALLY
If $c > 1$: Original function: $y_1 = f(x)$ Scaled function: $y_2 = cf(x)$	To obtain the graph of $y_2 = cy_1$, vertically stretch the graph of $y_1 = f(x)$ by a factor of c.	For each value of x, $y_2 = cy_1$.
If $0 < c < 1$: Original function: $y_1 = f(x)$ Scaled function: $y_2 = cf(x)$	To obtain the graph of $y_2 = cy_1$ vertically shrink the graph of $y_1 = f(x)$ by a factor of c.	For each value of x, $y_2 = cy_1$.

■ EXAMPLE 4   Examining a Shrinking of a Quadratic Function

Compare the functions $y_1 = x^2$ and $y_2 = \dfrac{1}{4}x^2$ numerically, graphically, and verbally.

SOLUTION

NUMERICALLY

x	$y_1 = x^2$	$y_2 = \dfrac{1}{4}x^2$
-3	9	$\dfrac{9}{4}$
-2	4	1
-1	1	$\dfrac{1}{4}$
0	0	0
1	1	$\dfrac{1}{4}$
2	4	1
3	9	$\dfrac{9}{4}$

Note that points on the x-axis are not moved by scaling factors.

GRAPHICALLY

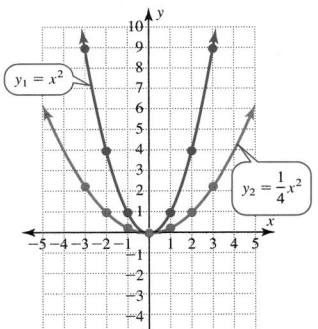

Both graphs are parabolas opening upward. At each x-value the height of y_2 is $\dfrac{1}{4}$ the height of y_1. The vertex of each parabola is $(0, 0)$.

VERBALLY

For each value of x,

$y_2 = \dfrac{1}{4}y_1$.

Note: $\dfrac{1}{4}(0) = 0$.

The parabola defined by $y_2 = \dfrac{1}{4}x^2$ can be obtained by vertically shrinking the parabola defined by $y_1 = x^2$ by a factor of $\dfrac{1}{4}$. Note: $(0, 0)$ is the vertex of both parabolas. ■

■ **EXAMPLE 5** Examining a Stretching of a Square Root Function

Use a graphics calculator to compare the functions $y_1 = \sqrt{x}$ and $y_2 = 2\sqrt{x}$ numerically, graphically, and verbally.

SOLUTION _____

NUMERICALLY

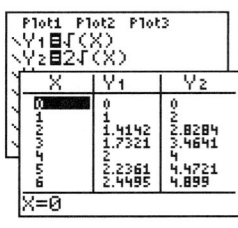

GRAPHICALLY

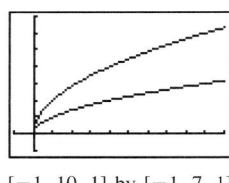

$[-1, 10, 1]$ by $[-1, 7, 1]$

Both graphs have the characteristic shape of the square root function. For each the domain $D = [0, +\infty)$ and the range $R = [0, +\infty)$.

VERBALLY

For each value of x, $y_2 = 2y_1$.

The graph of $y_2 = 2\sqrt{x}$ can be obtained by vertically stretching by a factor of 2 the graph of $y_1 = \sqrt{x}$.

■

SELF-CHECK 6.5.3 ANSWERS

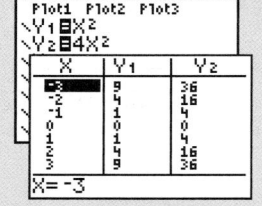

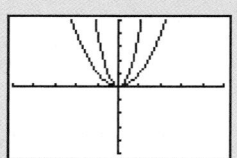

For each value of x, $y_2 = 4y_1$. The parabola defined by $y_2 = 4x^2$ can be obtained by vertically stretching by a factor of 4 the parabola defined by $y_1 = x^2$.

SELF-CHECK 6.5.3

Use a graphics calculator to compare $y_1 = x^2$ and $y_2 = 4x^2$ both numerically and graphically.

The next example uses a table of values to identify a stretching factor.

EXAMPLE 6 Identifying a Stretching Factor

Use the table of values for $y_1 = f_1(x)$ and $y_2 = f_2(x)$ to write an equation for y_2 in terms of $f_1(x)$.

x	$y_1 = f_1(x)$	$y_2 = f_2(x)$
-2	5	15
-1	4	12
0	2	6
1	2	6
2	4	12

SOLUTION _____

For each value of x, $y_2 = 3y_1$. Thus $y_2 = 3f_1(x)$.

Answer: $y_2 = 3f_1(x)$

■

SELF-CHECK 6.5.4

1. Use the graph of $y_1 = f(x)$ to graph $y_2 = 2f(x)$. *Hint*: Start by stretching the labeled points.

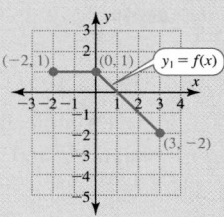

2. Use the table for $y_1 = f(x)$ to compute the table for $y_2 = 5f(x)$.

x	y_1	y_2
-2	4	
-1	3	
0	1	
1	-2	
2	-5	

Every graph in the family of absolute value functions can be obtained from the V-shaped graph of $y = |x|$. To obtain other members of this family, we can use horizontal and vertical translations, stretching and shrinking factors, and reflections. In Example 7 a function that both shrinks and reflects $y = |x|$ is examined.

■ **EXAMPLE 7**   Combining a Shrinking Factor and a Reflection

Use the graph of $y_1 = |x|$ to graph $y_2 = -\dfrac{1}{2}|x|$.

SOLUTION _____

NUMERICALLY

x	$y_1 = |x|$	$y_2 = -\dfrac{1}{2}|x|$
-4	4	-2
-2	2	-1
0	0	0
2	2	-1
4	4	-2

GRAPHICALLY

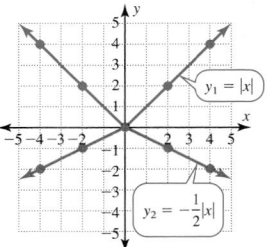

Both graphs have the characteristic V-shape. The vertex of both V-shaped graphs is (0, 0), but they open in opposite directions.

VERBALLY

Every y_2 value can be obtained by multiplying the corresponding y_1 value by $-\dfrac{1}{2}$. $y_2 = -\dfrac{1}{2}y_1$.

The graph of $y_2 = -\dfrac{1}{2}|x|$ can be obtained by vertically shrinking the graph of $y_1 = |x|$ by a factor of $\dfrac{1}{2}$ and then reflecting this graph across the x-axis.

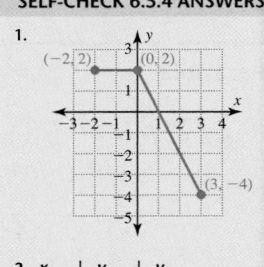

SELF-CHECK 6.5.5

1. Use a graphics calculator to graph $y_1 = x^2$ and $y_2 = -2x^2$ on the same calculator screen.
2. Compare the graphs of y_1 and y_2.

The next example illustrates how to analyze a graph that involves horizontal and vertical translations and a reflection of a parabola.

■ **EXAMPLE 8** Combining Translations and a Reflection

Use the graph of $y_1 = x^2$ to write an equation for y_2 in terms of $f_1(x)$. Give the domain and the range of each function.

You can think of the shifts and reflections following the same order as the order of operations in the expression. First we subtract 2 inside the parentheses to produce a shift 2 units to the right. Next we multiply by -1 to produce a reflection across the x-axis. Finally we add 1, to produce a shift 1 unit up.

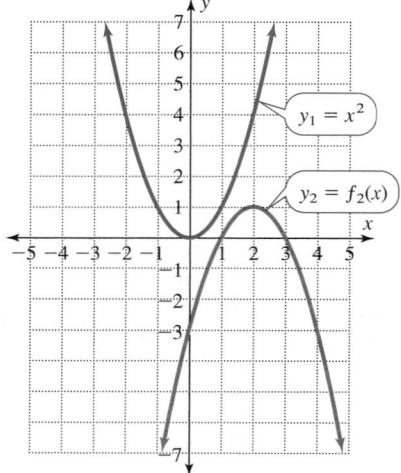

SOLUTION _____

The vertex of f_1 is $(0, 0)$ and the vertex of f_2 is $(2, 1)$. The parabola defined by $y_2 = f_2(x)$ is the same size as the parabola defined by $y_1 = x^2$. There is no stretching or shrinking in this example. The graph of $y_2 = f_2(x)$ can be obtained from the graph of the parabola defined by $y_1 = x^2$ by first shifting this graph to the right 2 units, then reflecting this graph across the x-axis, and finally shifting this graph up 1 unit.

Answer: $y_2 = -(x - 2)^2 + 1$

For y_1 the domain $D = \mathbb{R}$ and the range $R = [0, +\infty)$.
For y_2 the domain $D = \mathbb{R}$ and the range $R = (-\infty, 1]$. ■

The parabola $f(x) = x^2$ has a vertex of $(0, 0)$. The next example provides practice identifying the vertex of a parabola from the defining equation.

■ **EXAMPLE 9** Determining the Vertex of a Parabola

Determine the vertex of each parabola using the fact that the vertex of $f(x) = x^2$ is at $(0, 0)$.

SELF-CHECK 6.5.5 ANSWERS

1.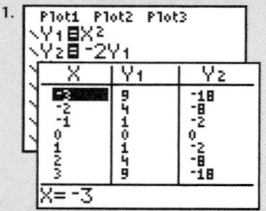

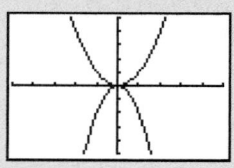

[−5, 5, 1] by [−5, 5, 1]

2. The graph of $y_2 = -2x^2$ can be obtained by vertically stretching the parabola defined by $y_1 = x^2$ by a factor of 2 and then reflecting this graph across the x-axis.

SOLUTIONS

(a) $f(x) = 7x^2$ Vertex of $(0, 0)$ The stretching factor of 7 does not move the vertex of $f(x) = x^2$.

Points on the x-axis are not moved by scaling factors.

(b) $f(x) = (x + 9)^2$ Vertex of $(-9, 0)$ This is a horizontal shift 9 units left of every point on $f(x) = x^2$. The vertex is shifted from $(0, 0)$ to $(-9, 0)$.

You may wish to use a graphics calculator to visualize the results in Example 9.

(c) $f(x) = -(x - 5)^2 + 8$ Vertex of $(5, 8)$ The vertex $(0, 0)$ is shifted to $(5, 0)$ by the horizontal shift. The reflection leaves this vertex at $(5, 0)$. The vertical translation shifts the vertex to $(5, 8)$.

USING THE LANGUAGE AND SYMBOLISM OF MATHEMATICS   6.5

1. Vertical and horizontal shifts of a graph create another graph whose shape is the _____ as the original graph and whose size is the _____ as the original graph.
2. The reflection of the point (x, y) across the x-axis is the point _____.
3. If $y = f(x)$ and $c > 1$, then we call c a _____ factor in the function $y = cf(x)$.

4. If $y = f(x)$ and $0 < c < 1$, then we call c a _____ factor in the function $y = cf(x)$.
5. If $y = f(x)$ and $c = -1$, then we call the graph of $y = cf(x)$ a _____ of the graph of $y = f(x)$ across the _____ -axis.
6. If we examine the shapes of $y = f(x)$ and $y = 2f(x)$, then we will observe that the two graphs have the _____ basic shape but that $y = 2f(x)$ is obtained by vertically _____ $y = f(x)$ by a factor of _____.

EXERCISES   6.5

Exercises 1–8 give some basic functions and reflections of these functions. Match each graph to the correct function. All graphs are displayed with the window $[-5, 5, 1]$ by $[-5, 5, 1]$.

1. $f(x) = |x|$ **2.** $f(x) = -|x|$ **3.** $f(x) = x$ **4.** $f(x) = -x$

5. $f(x) = \sqrt{x}$ **6.** $f(x) = -\sqrt{x}$ **7.** $f(x) = x^3$ **8.** $f(x) = -x^3$

A. 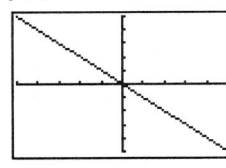   **B.** 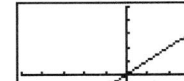   **C.** 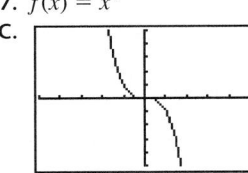   **D.**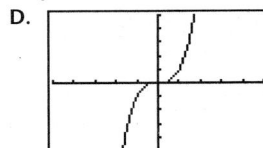

E. 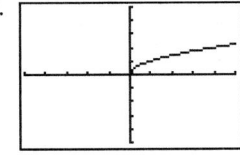   **F.** 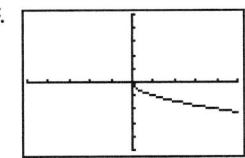   **G.** 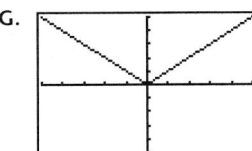   **H.**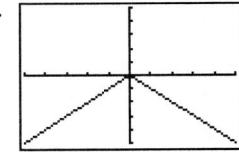

Exercises 9–12 give functions that are obtained by either stretching or shrinking the graph of $f(x) = x^2$. Match each graph to the correct function. All graphs are displayed with the window $[-5, 5, 1]$ by $[-5, 5, 1]$.

9. $f(x) = 2x^2$ **10.** $f(x) = 5x^2$ **11.** $f(x) = \frac{1}{2}x^2$ **12.** $f(x) = \frac{1}{5}x^2$

A.   **B.** 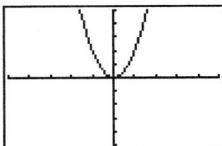   **C.** 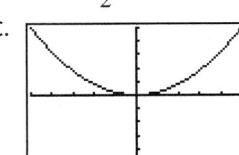   **D.**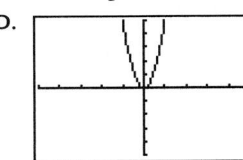

Exercises 13–16 give functions that are obtained by reflecting across the x-axis, stretching, or shrinking the graph of $f(x) = x$. Match each graph to the correct function. All graphs are displayed with the window $[-5, 5, 1]$ by $[-5, 5, 1]$.

13. $f(x) = 5x$

14. $f(x) = \frac{1}{5}x$

15. $f(x) = -2x$

16. $f(x) = -\frac{1}{2}x$

A.
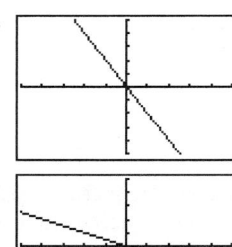

B.
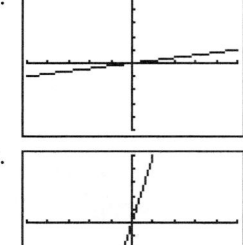

C.
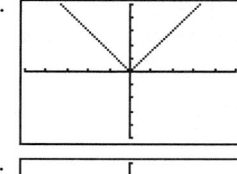

D.
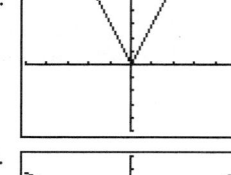

Exercises 17–20 give functions that are obtained by either stretching or shrinking the graph of $f(x) = |x|$. Match each graph to the correct function. All graphs are displayed with the window $[-5, 5, 1]$ by $[-5, 5, 1]$.

17. $f(x) = \frac{3}{4}|x|$

18. $f(x) = 3|x|$

19. $f(x) = \frac{3}{2}|x|$

20. $f(x) = \frac{1}{4}|x|$

A.
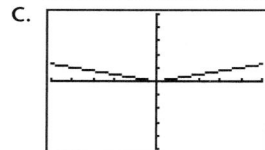

B.

C.

D.
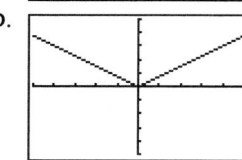

Exercises 21–24 give functions that are obtained by either stretching or shrinking the graph of $y = \sqrt{x}$. Match each graph to the correct function. All graphs are displayed with the window $[-1, 6, 1]$ by $[-1, 6, 1]$.

21. $f(x) = 4\sqrt{x}$

22. $f(x) = \frac{1}{4}\sqrt{x}$

23. $f(x) = \frac{3}{4}\sqrt{x}$

24. $f(x) = \frac{9}{4}\sqrt{x}$

A.
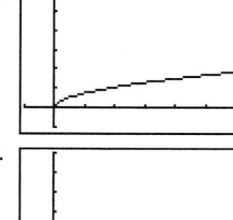

B.

C.

D.
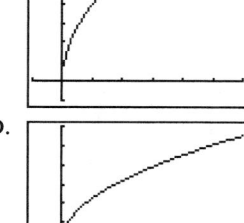

In Exercises 25–30 use the given graph of $y = f(x)$ to graph each function. (*Hint:* First graph the three key points on the new graph.)

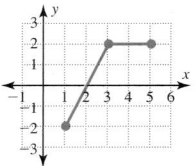

25. $y = -f(x)$

26. $y = \frac{1}{2}f(x)$

27. $y = 2f(x)$

28. $y = -3f(x)$

29. $y = f(x) + 2$

30. $y = f(x + 2)$

In Exercises 31–34 use the given table of values for $y = f(x)$ to complete each table.

x	f(x)
-2	18
-1	9
0	3
1	1
2	3

31.

x	$y = -f(x)$
-2	
-1	
0	
1	
2	

32.

x	$y = \frac{1}{3}f(x)$
-2	
-1	
0	
1	
2	

33.

x	$y = 2f(x)$
-2	
-1	
0	
1	
2	

34.

x	$y = -3f(x)$
-2	
-1	
0	
1	
2	

In Exercises 35–42 use the given table to write an equation for y_2 in terms of $f(x)$.

x	$y_1 = f(x)$
0	8
1	4
2	6
3	12
4	20
5	24

35.

x	y_2
0	4
1	2
2	3
3	6
4	10
5	12

36.

x	y_2
0	-8
1	-4
2	-6
3	-12
4	-20
5	-24

37.

x	y_2
0	-2
1	-1
2	-1.5
3	-3
4	-5
5	-6

38.

x	y_2
0	80
1	40
2	60
3	120
4	200
5	240

39.

x	y_2
0	0
1	−4
2	−2
3	4
4	12
5	16

40.

x	y_2
0	13
1	9
2	11
3	17
4	25
5	29

41.

x	y_2
−3	8
−2	4
−1	6
0	12
1	20
2	24

42.

x	y_2
2	8
3	4
4	6
5	12
6	20
7	24

In Exercises 43–50 determine the vertex of each parabola using the fact that the vertex of $f(x) = x^2$ is at (0, 0). Use a graphics calculator to check your answer.

43. $f(x) = 8x^2$

44. $f(x) = \dfrac{1}{7}x^2$

45. $f(x) = 8x^2 - 7$

46. $f(x) = \dfrac{1}{7}(x - 7)^2$

47. $f(x) = (x - 8)^2 - 7$

48. $f(x) = (x + 6)^2 + 9$

49. $f(x) = -(x - 8)^2 + 7$

50. $f(x) = -(x + 6)^2 - 9$

51. Translate or reflect the graph of $f(x) = x^2$ as described by each equation.
  a. Graph $f(x) = (x - 3)^2$ and describe how to obtain this graph from $f(x) = x^2$.
  b. Graph $f(x) = -(x - 3)^2$ and describe how to obtain this graph from $f(x) = (x - 3)^2$.
  c. Graph $f(x) = -(x - 3)^2 + 2$ and describe how to obtain this graph from $f(x) = -(x - 3)^2$.

52. Translate or reflect the graph of $f(x) = |x|$ as described by each equation.
  a. Graph $f(x) = |x + 2|$ and describe how to obtain this graph from $f(x) = |x|$.
  b. Graph $f(x) = -|x + 2|$ and describe how to obtain this graph from $f(x) = |x + 2|$.
  c. Graph $f(x) = -|x + 2| - 3$ and describe how to obtain this graph from $f(x) = -|x + 2|$.

Exercises 53–58 describe a translation, a reflection, a stretching, or a shrinking of $y = f(x)$. Match each description to the correct function.

53. A vertical stretching of $y = f(x)$ by a factor of 7

54. A vertical shrinking of $y = f(x)$ by a factor of $\dfrac{1}{7}$

55. A reflection of $y = f(x)$ across the x-axis

56. A horizontal shift of $y = f(x)$ left 7 units

57. A horizontal shift of $y = f(x)$ right 7 units

58. A vertical shift of $y = f(x)$ up 7 units

A. $y = f(x + 7)$
B. $y = 7f(x)$
C. $y = \dfrac{1}{7}f(x)$
D. $y = -f(x)$
E. $y = f(x) + 7$
F. $y = f(x - 7)$

In Exercises 59–64 determine the range of each function given that the range of $y = f(x)$ is [2, 6).

59. $y = f(x) + 2$

60. $y = f(x) - 2$

61. $y = 2f(x)$

62. $y = \dfrac{1}{2}f(x)$

63. $y = -f(x)$

64. $y = -3f(x)$

65. Write the first five terms of each sequence. (*Hint*: See Example 8 in Section 3.1.)
  a. $a_n = n^3$
  b. $a_n = -n^3$
  c. $a_n = 2n^3$
  d. $a_n = n^2 - 2$

66. Write the first five terms of each sequence. (*Hint*: See Example 8 in section 3.1.)
  a. $a_n = |n|$
  b. $a_n = |n - 5|$
  c. $a_n = -|n - 5|$
  d. $a_n = 2|n - 5|$

67. Modeling a Sequence of Retirement Bonuses As part of a bonus plan a company gives each secretary a number of shares in the company at retirement. The number of shares given equals the number of years the secretary has worked.
  a. Write a formula for A_n, the sequence of the number of shares that would be given for a retirement after n years.
  b. Write a formula for V_n, the value of the shares that would be given for a retirement after n years if the value of each share is $50.

68. Comparing the Distance Traveled by Two Airplanes The formula $D = RT$ can be used to determine the distance D flown by an airplane flying at a rate R for time T. Complete the table if the rate of the second plane is double that of the first plane.

PLANE 1		PLANE 2	
T Hours	D Distance	T Hours	D Distance
1	200	1	
2	400	2	
3	600	3	
4	800	4	

69. Comparing the Growth of Two Investments The formula $I = PRT$ can be used to determine the interest earned on an investment of P dollars for T years at simple interest rate R. A second investment of the same amount is invested at a rate that is three-fourths that of the first investment. Complete the table for the second investment.

INVESTMENT I		INVESTMENT II	
T Years	I Interest	T Years	I Interest
5	4,000	5	
10	8,000	10	
15	12,000	15	
20	16,000	20	

70. Comparing the Production of Two Assembly Lines
The following graph shows the number of lightbulbs produced by the workers at an assembly line during a very busy day shift. The night shift, which has fewer workers than the day shift, produces one-third as many bulbs as the day shift. Sketch the graph of the number of lightbulbs produced by the workers on the night shift.

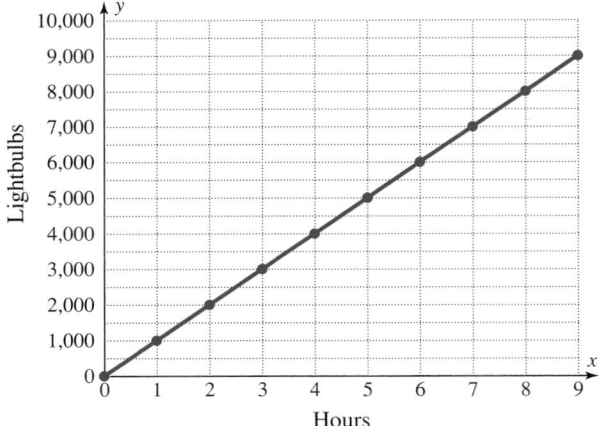

Group Discussion Questions
71. Discovery Question
a. Graph all the points in this table.
b. Sketch a parabola containing all these points.
c. Use your knowledge of translations, reflections, stretchings, and shrinkings to write the equation of this parabola.
d. Check your equation using a graphics calculator.

x	y
-2	11
-1	4
0	-1
1	-4
2	-5
3	-4
4	-1

72. Discovery Question
a. Graph all the points in this table.
b. Sketch a parabola containing all these points.
c. Use your knowledge of translations, reflections, stretchings, and shrinkings to write the equation of this parabola.
d. Check your equation using a graphics calculator.

x	y
-3	0.5
-2	-2
-1	-3.5
0	-4
1	-3.5
2	-2
3	0.5

73. Error Analysis A student graphing $y = 0.5|x|$ obtained the following display. Explain how you can tell by inspection that an error has been made.

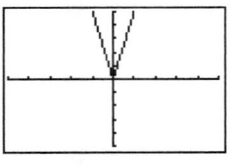

$[-5, 5, 1]$ by $[-5, 5, 1]$

74. Error Analysis A student graphing $y = -3\sqrt{x}$ obtained the following display. Explain how you can tell by inspection that an error has been made.

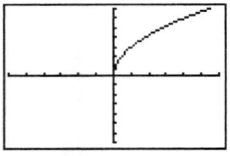

$[-6, 6, 1]$ by $[-7, 7, 1]$

Section 6.6 (Optional) Curve Fitting—Selecting a Formula that Best Fits a Set of Data

Objectives: 13. Select a linear function that best fits a set of data.
14. Select a quadratic function that best fits a set of data.

In this chapter we have examined some of the most basic functions that you are likely to encounter in the real world of applications. We have noted that each type of function produces a graph with a characteristic shape. By using a basic representative from each family of graphs, we then can use translations, reflections, stretching, and shrinking to produce any member of this family of graphs. If we have a formula, then we can produce a graph or calculate outputs for any input value.

What we wish to do now is much more sophisticated. We want to take a table of values and select what type of curve fits that data. Then we want to select the one curve from that family of curves that best fits the data. The topic of selecting a curve that best fits a set of data points is called **curve fitting.** Actually this section will only examine fitting straight lines and parabolas to data. Other types of curve fitting are covered in other mathematics courses.

The next example suggests a visual way to begin our curve-fitting process. We start by drawing a scatter diagram for the given data points and then examining this graph for a visual pattern.

■ **EXAMPLE 1** **Selecting the Type of Graph that Best Fits a Set of Data**

Scatter diagrams were first discussed in Section 2.1. A scatter diagram for a set of data points is simply a graph of these points.

Draw a scatter diagram for each set of data points and then select the shape of graph which describes the visual patterns formed by these points.

SOLUTIONS

(a)

x	y
−3	−9
−2	−7
−1	−5
0	−3
1	−1
2	1
3	3

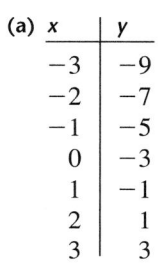

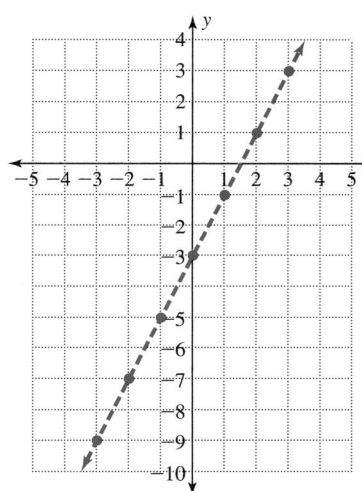

First note the visual pattern formed by the points is linear. Then determine the key information in order to write the equation in the slope-intercept form, $y = mx + b$.

In this case each 1-unit change in x yields 2-unit change in y. Thus the slope $m = \dfrac{2}{1}$. The y-intercept is $(0, -3)$. The equation of this specific line is $y = 2x - 3$. Do all the given points check in this equation?

Shape: A line
Equation: $y = 2x - 3$

(b)

x	y
−3	4
−2	−1
−1	−4
0	−5
1	−4
2	−1
3	4

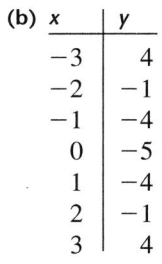

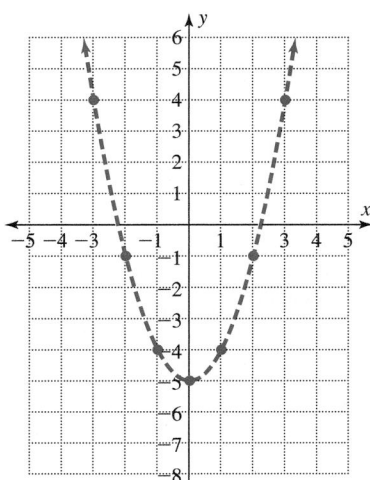

First note the visual pattern formed by the points is parabolic. Then determine the key information needed to write the quadratic equation.

Note the vertex is $(0, -5)$ suggesting that we consider shifting the function $y = x^2$ down 5 units by forming the function $y = x^2 - 5$. Do all the given points check in this equation?

Shape: Parabola
Equation: $y = x^2 - 5$ ■

SELF-CHECK 6.6.1

Draw a scatter diagram for each set of data points and then select the shape of graph that describes the visual pattern formed by these points.

1.

x	y
−6	5
−4	4
−2	3
0	2
2	1
4	0
6	−1

2.

x	y
−3	−3
−2	2
−1	5
0	6
1	5
2	2
3	−3

SELF-CHECK 6.6.1 ANSWERS

1.

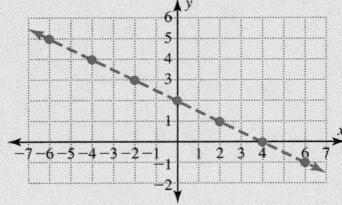

Shape: A line
Equation: $y = -0.5x + 2$

2.

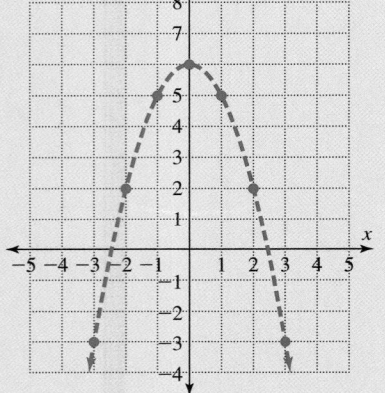

Shape: A parabola
Equation: $y = -x^2 + 6$

In Example 1(a) we used the slope-intercept form $y = mx + b$ to write the equation of the line containing these points. We will now revisit this problem by using the point-slope form, $y - y_1 = m(x - x_1)$ and two points to write this equation. Both of these approaches are fine when the data points all lie on the same line.

■ EXAMPLE 2 Writing the Equation of the Line Through Two Points

Write in slope-intercept form the equation of a line through $(-3, -9)$ and $(-2, -7)$, the first two points from Example 1(a).

SOLUTION

$$m = \frac{y_2 - y_1}{x_2 - x_1}$$

To use the point-slope form, first calculate the slope by substituting the two given points into the formula for the slope.

$$m = \frac{-7 - (-9)}{-2 - (-3)}$$

$$m = \frac{2}{1}$$

$$m = 2$$

$$y - y_1 = m(x - x_1)$$
$$y - (-9) = 2(x - (-3))$$
$$y + 9 = 2(x + 3)$$
$$y + 9 = 2x + 6$$

Then substitute $(-3, -9)$ for (x_1, y_1) and 2 for m into the point-slope form.

Simplify and write the equation in the slope-intercept form $y = mx + b$. Note that this is the same equation produced in Example 1(a).

Answer: $y = 2x - 3$

SELF-CHECK 6.6.2

Use the equation $y = 2x - 3$ from Example 2 to generate a table of values to verify that this function produces all the points in the table in Example 1(a).

The data that we obtain from real-life problems is usually not as simple as the nice neat integers given in Example 1. With messier data the thought process is the same, but it is very useful to have a calculator to handle the details. The following two calculator perspectives illustrate how to use a graphics calculator to draw a scatter diagram for a set of data points.

CALCULATOR PERSPECTIVE 6.6.1   Using Lists to Store a Set of Data Points

To enter the data points from Example 1(a) on a TI-83 Plus calculator, enter the following keystrokes:

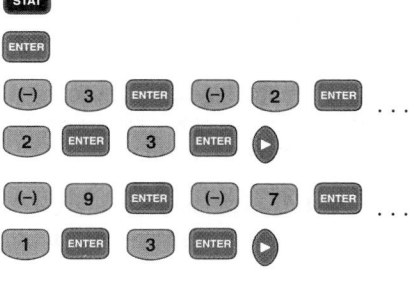

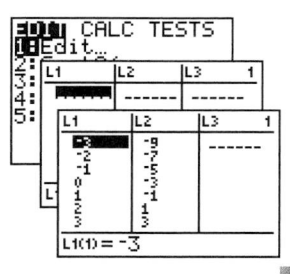

SELF-CHECK 6.6.2 ANSWER

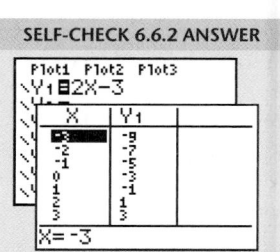

Once data points are entered into a graphics calculator, we then can use the calculator to draw the scatter diagram. This is illustrated by the next calculator perspective.

CALCULATOR **PERSPECTIVE 6.6.2**	**Drawing a Scatter Diagram**	

To draw the scatter diagram for the data points from Example 1(a) on a TI-83 Plus calculator, enter the following keystrokes:

The STAT PLOT feature is the secondary function of the Y= key.

 STAT PLOT (Check to see that the menu is set up as shown for **Plot1.**)

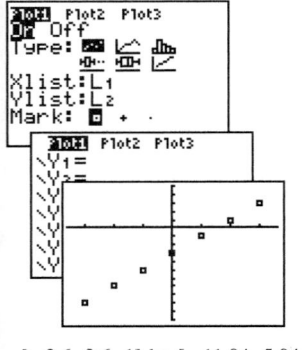

Y= (Check to see that the Y= screen is blank.)

ZOOM   9

Note: Pressing ZOOM   9 will have the calculator automatically select an appropriate window to display the points.

$[-3.6, 3.6, 1]$ by $[-11.04, 5.04, 1]$

■ EXAMPLE 3   Using a Graphics Calculator to Draw a Scatter Diagram

Use a graphics calculator to draw a scatter diagram for this set of data points as repeated from Example 1(b).

x	y
-3	4
-2	-1
-1	-4
0	-5
1	-4
2	-1
3	4

SOLUTION _____

(a) First enter the data points using STAT-EDIT feature.

(b) Then use the STAT-PLOT feature to draw a scatter diagram of the data points listed in parts (a) and (b).

Note that this calculator display corresponds to the parabolic pattern shown in Example 1(b).

$[-3.6, 3.6, 1]$ by $[-6.53, 5.53, 1]$

The next example is used to examine a scatter diagram which illustrates the relationship between the speed in miles per hour of an automobile and its rate of fuel consumption measured in miles per gallon. At higher speeds there is more wind resistance and less engine efficiency. Thus greater speeds generally mean less fuel efficiency as illustrated by the data in Example 4.

■ EXAMPLE 4   Using a Graphics Calculator to Draw a Scatter Diagram

Use a graphics calculator to draw a scatter diagram for the data points in the table. Then describe the visual pattern formed by these points.

x (mi/h)	y (mi/gal)
20	23
25	22
30	21
35	20
40	19
45	18
50	17

SOLUTION _____

(a) First enter the data points using the STAT-EDIT feature.

L1	L2	L3 1
20	23	------
25	22	
30	21	
35	20	
40	19	
45	18	
50	17	

L1(1)=20

(b) Then use the STAT-PLOT feature to draw a scatter diagram of the data points listed in part (a).

These points appear to form a linear pattern with a negative slope. The meaning of the negative slope is that fuel efficiency is declining as speed increases.

[17, 53, 1] by [15.98, 24.02, 1]

A Mathematical Note

Sir Francis Galton introduced the concept of linear regression in 1877. He was studying data which showed that tall parents have children whose heights tend to "regress" or revert back to the mean height of the population.

SELF-CHECK 6.6.3

1. Use a graphics calculator to draw a scatter diagram for these data points.

x	−4	−3	−2	−1	0	1	2	3	4
y	.5	0	−.5	−1	−1.5	−1	−.5	0	.5

2. What type of pattern is formed by these points?

SELF-CHECK 6.6.3 ANSWERS

1.

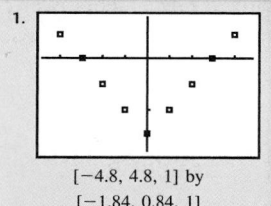

[−4.8, 4.8, 1] by
[−1.84, 0.84, 1]

2. These points form the characteristic V-shape of an absolute value function.

We have shown how to store data points in a graphics calculator and have examined how to use a calculator to draw a scatter diagram of these points. We will add the last piece to our curve-fitting skills. We examine how to select a line or parabola that best fits a set of data points. The topic of fitting an equation to a set of data points is known in mathematics as **regression analysis.** We limit our discussion of the details of this topic and take advantage of the capabilities of a graphics calculator to perform this work for us.

Before reading the next calculator perspective, enter the data from Example 4 in a graphics calculator and form the scatter diagram as shown in this example.

CALCULATOR PERSPECTIVE 6.6.3	Calculating and Graphing the Line of Best Fit

To calculate the line of best fit for the data in Example 4 on a TI-83 Plus calculator and then to graph this line, enter the following keystrokes:

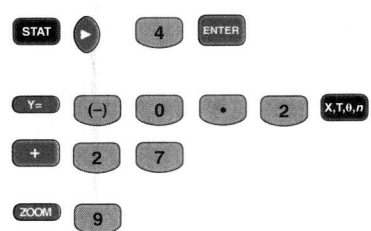

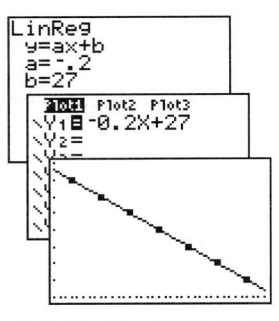

[17, 53, 1] by [15.98, 24.02, 1]

Note that the equation of the line of best fit also can be inserted into the screen from the memory of the calculator by pressing .

Note: The line of best fit and the data points are both displayed on the graph. The keystrokes for calculating the parabola of best fit are identical to those for calculating the line of best fit except that we select option **5** under the STAT **CALC.**

The criteria for deciding which curve best fits a set of data is covered in most introductory statistics courses. We will use calculators to help us focus on the key concepts here—rather than on the theoretical background.

Real Data for Real People   The data in Example 4 is phony in the sense that the authors made it up exactly to fit a line—it was manipulated to meet an ideal model. Textbooks are full of idealized data. It is easier to teach concepts piece by piece with nice clean data. Students do not get lost in the messy details and can concentrate on the concepts. However, students often sense that problems contain phony data—even though there are very good reasons for these problems. It is important to realize that we can use the same procedures with real data—it just may be a little messier. The next examples address working with real data. Real data rarely lie perfectly on an idealized curve. Thus we try to select the curve that best fits the data, even when the curve will miss some of the data points.

SCATTER DIAGRAMS OF REAL DATA

DATA FORMING A LINEAR PATTERN	DATA FORMING A PARABOLIC PATTERN

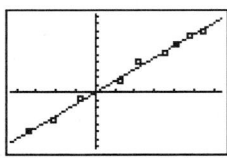

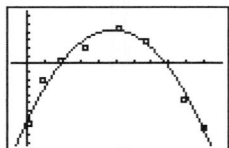

These data points do not lie exactly on a line or on a parabola, but respectively they do exhibit a strong linear and a strong parabolic pattern. In the next two examples we will use a graphics calculator to determine the curve which best fits these two sets of data.

■ EXAMPLE 5   Using a Graphics Calculator to Determine a Line of Best Fit

Enter these data points into a graphics calculator, draw the scatter diagram, and calculate the line of best fit. Use this linear equation to approximate y when $x = 2$.

x	y
−3.4	−5.0
−2.1	−3.6
−.8	−0.8
1.3	1.5
2.2	3.9
3.5	5.1
4.1	6.2
4.8	7.5
5.5	8.0

SOLUTION

(a) Enter the data points using the STAT-EDIT feature.

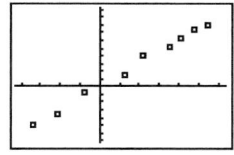

Be sure to enter all the data points since these points will fill more than one display screen.

(b) Use the STAT-PLOT feature to draw a scatter diagram of the given data points.

Use the visual pattern to select the most appropriate family of curves. In this case the visual pattern is linear.

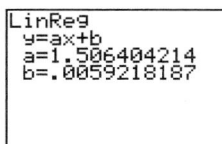

[−4.29, 6.39, 1] by [−7.21, 10.21, 1]

Since these points appear to form a linear pattern we will try to find a linear function which best fits this data.

(c) Use the STAT-CALC feature to calculate the line of best fit.

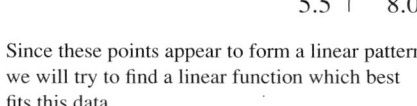

(d) Enter the line of best fit on the ◼(Y=) screen and then graph this line. Round each coefficient to the nearest thousandth.

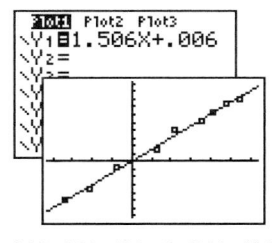

[−4.29, 6.39, 1] by [−7.21, 10.21, 1]

This line of best fit does *not* contain each data point, but it fits these points better than any other line.

(e) $y = 1.506x + 0.006$
     $y \approx 1.506(2) + 0.006$
     $y \approx 3.018$
     $y \approx 3.0$

To approximate y for $x = 2$, substitute 2 into the equation for the line of best fit.

Since the original data was accurate only to the tenths place, the estimate is rounded to the nearest tenth.

Answer: The line $y = 1.506x + 0.006$ is approximately the line of best fit. For $x = 2$, $y \approx 3.0$.

The equation of the line of best fit is written in $y = mx + b$ form with m and b rounded to the nearest thousandth. ◼

SELF-CHECK 6.6.4

1. Use a graphics calculator to determine the line of best fit for these data points.

x	-3	-2	-1	0	1	2	3
y	-6	-3	-2	2	2	6	6

Write your answer in slope-intercept form with m and b rounded to the nearest thousandth.

2. Use this linear equation to approximate y when $x = -1.3$.

The data points given in Example 6 form a parabolic pattern. Therefore a graphics calculator is used to determine the quadratic function of best fit. The steps are nearly identical to those used in Example 5. The only difference is the curve selected is a parabola rather than a line.

■ EXAMPLE 6   Using a Graphics Calculator to Determine a Parabola of Best Fit

Enter these points into a graphics calculator, draw the scatter diagram, and calculate the parabola of the best fit. Use this quadratic equation to approximate y when $x = 6$.

x	y
0.1	-7.7
0.9	-2
1.9	0.4
3.3	2.1
5.2	4.5
6.7	2.9
8.9	-4.6
10	-8.2

SOLUTION

(a) Enter the data points using the STAT-EDIT feature.

L1	L2	L3	1
.1	-7.7	------	
.9	-2		
1.9	.4		
3.3	2.1		
5.2	4.5		
6.7	2.9		
8.9	-4.6		

L1(1)=.1

Not all data points can be displayed on one screen.

(b) Use the STAT-PLOT feature to draw a scatter diagram of the given data points. Use the visual pattern to select the most appropriate family of curves. In this case the visual pattern is parabolic.

$[-0.89, 10.99, 1]$ by $[-10.359, 6.659, 1]$

Since the points appear to form a parabola opening downward, we will try to find a quadratic function which best fits this data.

(c) Use the STAT-CALC feature to calculate the parabola of best fit.

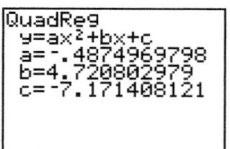

QuadReg
y=ax²+bx+c
a=-.4874969798
b=4.720802979
c=-7.171408121

This parabola of best fit does *not* contain each data point, but it does fit these points better than any other parabola.

Reminder: To calculate the parabola of best fit, select option **5** under the STAT **CALC** menu. (See Calculator Perspective 6.6.3.)

(d) Enter the equation of the parabola of best fit on the ⬛Y=⬛ screen and then graph this parabola. Round each coefficient to the nearest thousandth.

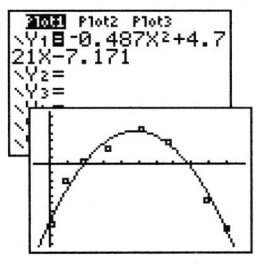

[−0.89, 10.99, 1] by [−10.359, 6.659, 1]

(e) $y = -0.487x^2 + 4.721x - 7.171$
$y \approx -0.487(6)^2 + 4.721(6) - 7.171$
$y \approx 3.623$
$y \approx 3.6$

To approximate y for $x = 6$, substitute 6 into the equation of the parabola of best fit. Since the original data was accurate only to the tenths place, the estimate is rounded to the nearest tenth.

Answer: $y = -0.487x^2 + 4.721x - 7.171$ is approximately the parabola of the best fit. For $x = 6$, $y \approx 3.6$.

The equation of the parabola of best fit is written in $y = ax^2 + bx + c$ form with each coefficient rounded to the nearest thousandth.

The material in this section has presented a procedure for taking data points and producing a function which best fits these points. The procedure involves two main steps:

1. Use a scatter diagram and the visual pattern formed by data points to select the most appropriate family of curves.
2. If the pattern is linear or parabolic, use a calculator to select the best linear or quadratic function. In later courses you may examine curve fitting for other shapes.

SELF-CHECK 6.6.5

1. Use a graphics calculator to determine the parabola of best fit for these data points.

x	0	1	2	3	4	5	6
y	−4	0	4	4	2	0	−4

Write your answer in the form $y = ax^2 + bx + c$ with each coefficient rounded to the nearest thousandth.
2. Use this quadratic equation to approximate y when $x = 1.5$.

SELF-CHECK 6.6.5 ANSWERS

1. $y = -0.881x^2 +$
   $5.214x - 3.905$
2. For $x = 1.5$, $y \approx 2$

USING THE LANGUAGE AND SYMBOLISM OF MATHEMATICS   6.6

1. A _____ diagram for a set of data points is formed by graphing these points on a coordinate system.
2. If the pattern formed by the points of a scatter diagram is approximately a _____, then we will select an equation of the form $y = mx + b$.
3. If the pattern formed by the points of a scatter diagram is approximately a _____, then we will select an equation of the form $y = ax^2 + bx + c$.
4. The topic of selecting a curve that best fits a set of data points is called _____ fitting.
5. Selecting an equation that best fits a set of data points is called _____ analysis.
6. In the _____-_____ form $y = mx + b$, m represents the _____ of a line with y-intercept _____.

EXERCISES   6.6

Exercises 1–6 give a family of functions. Select the scatter diagram that best matches the corresponding description. All graphs are displayed with the window [0, 10, 1] by [0, 10, 1].

1. A line with positive slope
2. A line with negative slope
3. A parabola opening upward
4. A parabola opening downward
5. An absolute value function
6. A square root function

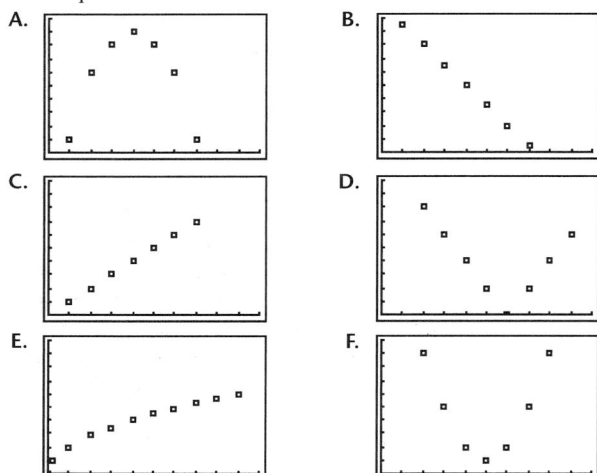

A.   B.

C.   D.

E.   F.

In Exercises 7–10 draw a scatter diagram for each set of data points and then select the shape that best describes the pattern formed by these points.

7.
x	−3	−2	−1	0	1	2	3
y	4	4	3	3	2	2	1

A. line   B. parabola

8.
x	−3	−2	−1	0	1	2	3
y	−6	−2	1	2	1	−3	−8

A. line   B. parabola

9.
x	−3	−2	−1	0	1	2	3
y	10	4	0	−2	−2	0	4

A. line   B. parabola

10.
x	−3	−2	−1	0	1	2	3
y	16	8	0	6	−8	−8	−10

A. line   B. parabola

In Exercises 11–14 write in slope-intercept form the equation of a line that satisfies the given conditions.

11. The line with slope $m = \dfrac{4}{7}$ and y-intercept $(0, 3)$

12. The line with slope $m = -\dfrac{3}{2}$ and y-intercept $(0, 5)$

13. The line that goes down 3 units for every one unit increase in x and passes through $(0, -6)$

14. The line that goes up 2 units for every one unit increase in x and passes through $(0, -4)$

In Exercises 15–18 use the point-slope form $y - y_1 = m(x - x_1)$ and the given information to write the equation of the line. Write each answer in slope-intercept form.

15. A line through $(2, -4)$ with slope $-\dfrac{2}{3}$

16. A line through $(-3, 5)$ with slope $\dfrac{4}{5}$

17. A line through $(2, -5)$ and $(-1, 1)$
18. A line through $(-1, 4)$ and $(2, 2)$

19. **Modeling the Interest on a Savings Account**   The following data points all lie on a line. Write the equation of this line.

x AMOUNT IN SAVINGS ACCOUNT	y DOLLAR VALUE OF YEARLY INTEREST
500	20
750	30
1000	40
1250	50
1500	60
1750	70
2000	80

20. **Modeling the Stretch by a Steel Spring**   The following data points all lie on a line. Write the equation of this line.

x MASS (KG) ON THE SPRING	y STRETCH (CM) BY THE SPRING
5	0.4
10	0.8
15	1.2
20	1.6
25	2.0
30	2.4
35	2.8

In Exercises 21–24 use a graphics calculator to draw a scatter diagram for the data points and to determine the line of best fit. Express your answer in the form $y = mx + b$ with the m and b rounded to the nearest hundredth.

21.

x	y
1	3
2	5
3	8
4	9
5	11
6	12

Use this linear equation to estimate y when x is 3.5.

22.

x	y
10	18
15	16
19	12
26	9
29	6
34	2

Use this linear equation to estimate y when x is 22.

23.

x	y
−10	5
−7	1
−5	−3
−2	−7
0	−10
3	−14

Use this linear equation to estimate y when x is 1.

24.

x	y
1	4
4	7
9	8
12	11
15	15
19	19

Use this linear equation to estimate y when x is 10.

In Exercises 25–28 use a graphics calculator to draw a scatter diagram for the data points and to determine the parabola of best fit. Express your answer in the form $y = ax^2 + bx + c$ with a, b, and c rounded to the nearest hundredth.

25.

x	y
0	11
1	2
2	−4
3	−5
4	−3
5	1

Use this quadratic equation to estimate y when x is 2.5.

26.

x	y
2	5.3
3	3.1
5	1
6	1.4
7	2.9
8	5.5

Use this quadratic equation to estimate y when x is 4.

27.

x	y
−5	−18
−4	−12
−3	−9
−2	−8
−1	−9
0	−13

Use this quadratic equation to estimate y when x is 1.

28.

x	y
−5	−5.1
−3	−2.8
−1	−1.9
1	−2.6
3	−5.2
5	−9.1

Use this quadratic equation to estimate y when x is 6.

In Exercises 29–32 use a graphics calculator to draw a scatter diagram for the data points. Then determine either the line of best fit or the parabola of best fit depending on which is the more appropriate.

29.

x	y
1	5
2	2
3	1
4	0
5	2
6	3

Use this equation to estimate y when x is 4.5.

30.

x	y
1	5
2	3
3	0
4	−2
5	−5
6	−7

Use this equation to estimate y when x is 2.5.

31.

x	y
5	16
10	17
20	19
30	20
50	22
70	25

Use this equation to estimate y when x is 25.

32.

x	y
5	−148
10	−92
20	−18
30	7
50	−90
70	−385

Use this equation to estimate y when x is 40.

33. Modeling the Effect of Smoking on Birth Weights

This table compares the number of cigarettes smoked per day by eight expectant mothers to the birth weight of their children.

a. Draw a scatter diagram for these data points and determine the line of best fit.

b. Use this linear equation to estimate the birth weight of a child whose mother smoked 18 cigarettes per day.

x NO. OF CIGARETTES	y BIRTH WEIGHT (kg)
5	3.20
10	3.15
15	3.10
20	3.06
25	3.09
30	3.05
35	3.01
40	3.00

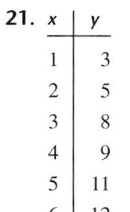

34. Modeling Body Height Based on Femur Length

Forensic scientists can estimate the height of a person based on the length of different bones from our body. This table of values compares the length of the femur (the thigh bone) to the height of the man with this femur.

a. Draw a scatter diagram for these data points and determine the line of best fit.

b. Use this linear equation to estimate the height of a man whose femur length is 44.7 cm.

x LENGTH OF FEMUR (cm)	y HEIGHT (cm)
43.0	164
43.5	166
44.0	166
44.5	168
45.0	170
45.5	169
46.0	172
46.5	173
47.0	175
47.5	175

35. Modeling Test Grades Based on Homework Grades

An algebra teacher graded all the homework problems for eight random students in her class and compared the homework grade to their second hour exam grade.

a. Draw a scatter diagram for these data points and determine the line of best fit.

b. Use this linear equation to estimate the exam grade of a student who earned a homework grade of 20.

x HOMEWORK GRADE	y EXAM GRADE
22	74
30	95
25	82
28	54
15	50
19	62
32	92
26	80

36. Modeling Profit Based on Sales   A new company recorded its sales units versus its profit (or loss) for eight consecutive quarters.

a. Draw a scatter diagram for this data and determine the parabola of best fit.

b. Use this equation to estimate the profit in dollars generated by the sale of 7000 units in a quarter. (*Hint:* Be careful with units.)

x UNITS SOLD (IN HUNDREDS)	y PROFIT (IN THOUSANDS OF DOLLARS)
0	−30
25	50
40	70
50	70
60	75
75	40
80	30
100	−60

37. Modeling Distance of a Fall Based on Time   An experiment is run in a physics course to compare the distance an object falls to the time of the fall.

a. Draw a scatter diagram for this data and determine the parabola of best fit.

b. Use this quadratic equation to estimate the distance an object would fall in 4.5 secs.

x TIME (IN SECONDS)	y DISTANCE (IN FEET)
0	0
1	15.9
2	62.5
3	146.0
4	257.1
5	400.9
6	580.2
7	783.6

38. Modeling the Load Strength of a Beam   The following experimental data was collected which relates the maximum load in kilograms that a beam can support to its depth in centimeters.

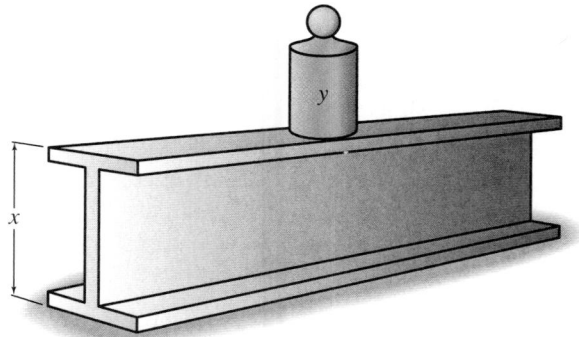

a. Draw a scatter diagram for these data points and determine the parabola of best fit.

b. Use this quadratic equation to estimate the maximum load that the beam can support if its depth is 15 cm.

x DEPTH (in cm)	y LOAD (in kg)
10	4,900
12	7,200
14	9,900
16	12,500
18	16,000
20	20,000
22	24,000
24	28,500

Group Discussion Questions

39. Local Linearity   Over a small segment of a curve almost any curve will appear linear.

  a. Determine the line of best fit for these data points:

x	0	1	2
y	4	3	0

  b. Determine the parabola of best fit for these data points:

x	−3	−2	−1	0	1	2	3
y	−5	0	3	4	3	0	−5

  c. Which curve do you think best fits the points in part **a**, a line or a parabola?

  d. Discuss your choice in part **c** with respect to "local linearity."

Risks and Choices   For engineers and scientists, choosing an appropriate model for a set of data is extremely important. In Exercises 40 and 41 we compare a parabola of best fit to a line of best fit. Points that lie along one branch of a parabola may appear to be approximately linear, but a parabola will still be a better fit for these points.

40.

x	−7	−6	−5	−4	−3	−2	−1	0	1
y	4	3	2	1	1	0.5	0.4	0.5	0.8

For the data points shown in the table use a graphics calculator to:

  a. Draw a scatter diagram of these points.
  b. Determine the line of best fit in the form $y = mx + b$ with m and b rounded to the nearest thousandth.
  c. Graph this line of best fit on the scatter diagram.
  d. Determine the parabola of best fit in the form $y = ax^2 + bx + c$ with the coefficients rounded to the nearest thousandth.
  e. Graph this parabola of best fit on the scatter diagram.
  f. Explain why these graphs indicate that the parabola fits the data points better than the line.
  g. For $x = -2.5$ use the line of best fit to calculate y.
  h. For $x = -2.5$ use the parabola of best fit to calculate y.
  i. Compare the points calculated in parts **g** and **h**. Which of these points is the closest to the given data points?

41.

x	−2	−1	0	1	2	3	4	5	7
y	0.5	0.4	0.5	0.8	1	2	3	4	7

For the data points shown in the table use a graphics calculator to:

  a. Draw a scatter diagram of these points.
  b. Determine the line of best fit in the form $y = mx + b$ with m and b rounded to the nearest thousandth.
  c. Graph this line of best fit on the scatter diagram.
  d. Determine the parabola of best fit in the form $y = ax^2 + bx + c$ with the coefficients rounded to the nearest thousandth.
  e. Graph this parabola of best fit on the scatter diagram.
  f. Explain why these graphs indicate that the parabola fits the data points better than the line.
  g. For $x = 2.5$ use the line of best fit to calculate y.
  h. For $x = 2.5$ use the parabola of best fit to calculate y.
  i. Compare the points calculated in parts **g** and **h**. Which of these points is the closest to the given data points?

Section 6.7   Linear Factors of a Polynomial, Zeros of a Polynomial Function, and *x*-Intercepts of a Graph

Objectives: **15.** Factor the GCF out of a polynomial.
  16. Use zeros of a function to factor a polynomial.
  17. Use the *x*-intercepts of the graph of a polynomial function to factor a polynomial.

The material in this section can serve as a foundation for Chapter 7. In Chapter 7 we will examine other algebraic methods for factoring polynomials.

The polynomials that we factored in Section 1.6 and in Chapter 5 were all factored by either using the distributive property or observing algebraic patterns. In this section we examine how to factor a wider variety of polynomials. We examine factoring from several viewpoints: algebraic, numerical, and graphical.

  We start by using the skills from Section 1.6 and Chapter 5 to factor the greatest common factor (GCF) out of each term of a polynomial. Loosely speaking, the GCF of a polynomial is the "largest" factor that is common to each term. To be more precise, the **GCF of a polynomial** is the common factor that contains:

1. the largest possible numerical coefficient and
2. the largest possible exponent on each variable factor.

The best way to start factoring most polynomials is to determine the GCF and then to factor out this GCF. For many problems you will be able to determine the GCF by inspection. In the next example we show the GCF as a factor of each term and then use the distributive property to factor the polynomial.

▪ EXAMPLE 1 Factoring Out the Greatest Common Factor

Factor out the GCF of each polynomial.

SOLUTIONS

(a) $20x^2y - 30xy^2$

$20x^2y - 30xy^2 = 10xy(2x) + 10xy(-3y)$
$= 10xy(2x - 3y)$

The GCF is $10xy$. Use the distributive property $ab + ac = a(b + c)$ to factor out the GCF of this binomial.

(b) $12a^3b - 18a^2b^2 + 30ab^3$

$12a^3b - 18a^2b^2 + 30ab^3$
$= 6ab(2a^2) + 6ab(-3ab) + 6ab(5b^2)$
$= 6ab(2a^2 - 3ab + 5b^2)$

The GCF is $6ab$. Use the distributive property to factor out the GCF of this trinomial.

(c) $(x + 2y)(b) + (x + 2y)(3c)$

$(x + 2y)(b) + (x + 2y)(3c)$
$= (x + 2y)(b + 3c)$

The GCF is $x + 2y$. Use the distributive property to factor out the GCF of this binomial. ▪

SELF-CHECK 6.7.1

Factor out the GCF of each polynomial.
1. $10x^3 - 15x^2$
2. $6x^3 - 9x^2 + 12x$
3. $(x - y)(2x) + (x - y)(3y)$

Equal polynomials have identical graphs and identical tables of values. In Example 2 we use tables and graphs to compare a binomial to its factored form. This table and graph also reveal some information that we can use to factor other polynomials.

SELF-CHECK 6.7.1 ANSWERS
1. $5x^2(2x - 3)$
2. $3x(2x^2 - 3x + 4)$
3. $(x - y)(2x + 3y)$

▪ EXAMPLE 2 Using Tables and Graphs to Compare Two Polynomials

Factor $(x + 2)(x) - (x + 2)(3)$ and check this factorization by using tables and graphs.

SOLUTION _____

$$(x + 2)(x) - (x + 2)(3) = (x + 2)(x - 3)$$

Check:

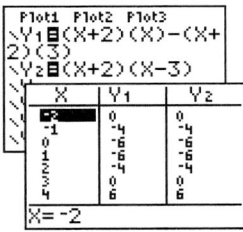 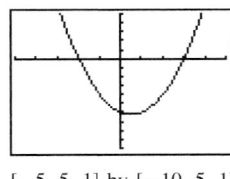

$[-5, 5, 1]$ by $[-10, 5, 1]$

The GCF is $x + 2$. Use the distributive property to factor out the GCF of this binomial. Let y_1 represent the original form and y_2 represent the factored form $(x + 2)(x - 3)$. Both the table of values and the graphs are identical for all values of x.

Answer: $(x + 2)(x) - (x + 2)(3) = (x + 2)(x - 3)$

An examination of the table and graph in Example 2 reveals that both y_1 and y_2 are 0 for $x = -2$ and $x = 3$. Graphically, this means the x-intercepts are $(-2, 0)$ and $(3, 0)$. For a function f a value of the input variable x that makes the output $f(x)$ equal to zero is called a **zero of the function.** For the function $f(x) = (x + 2)(x - 3)$ in Example 2, both -2 and 3 are zeros of this function because $f(-2) = 0$ and $f(3) = 0$. A summary of this information for the function $f(x) = (x + 2)(x - 3)$ is given here.

FACTORS OF $f(x) = (x + 2)(x - 3)$	ZEROS OF $f(x) = (x + 2)(x - 3)$	INTERCEPTS OF THE GRAPH OF $f(x) = (x + 2)(x - 3)$
$x + 2$ is a factor of $f(x)$	-2 is a zero of $f(x)$	$(-2, 0)$ is an x-intercept of the graph of $y = f(x)$
$x - 3$ is a factor of $f(x)$	3 is a zero of $f(x)$	$(3, 0)$ is an x-intercept of the graph of $y = f(x)$

A generalization of this relationship is given in the following table.

Equivalent Statements About Linear Factors of a Polynomial

For a real constant c and a real polynomial $P(x)$, the following statements are equivalent.

ALGEBRAICALLY	NUMERICALLY	GRAPHICALLY
$x - c$ is a factor of $P(x)$	$P(c) = 0$, that is, c is a zero of $P(x)$	$(c, 0)$ is an x-intercept of the graph of $y = P(x)$

If $P(x)$ is a real polynomial, then c is zero of $P(x)$ if and only if $x - c$ is a factor of $P(x)$.

In Example 3 we use the zeros of a polynomial to factor this polynomial.

■ **EXAMPLE 3** Using Zeros to Factor a Polynomial

Use a graphics calculator to find the zeros of $x^2 - 7x + 12$. Use these zeros to factor $x^2 - 7x + 12$.

SOLUTION

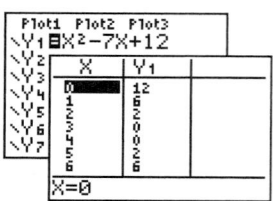

Enter $x^2 - 7x + 12$ as y_1 and create a table of values for y_1 until you determine the two zeros of y_1.
From the table, y_1 is zero for $x = 3$ and $x = 4$.

The zeros of $x^2 - 7x + 12$ are $x = 3$ and $x = 4$. Thus the factors of $x^2 - 7x + 12$ are $x - 3$ and $x - 4$.

Answer: $x^2 - 7x + 12 = (x - 3)(x - 4)$

You can check this factorization by multiplying $x - 3$ times $x - 4$.

A real polynomial $P(x)$ (see Section 6.3) has real number outputs for each real number input for x.

A polynomial that has real numbers as the coefficients for each term is a real polynomial. Each real polynomial function of degree n has exactly n zeros. What can vary from polynomial to polynomial is how many of these zeros are real numbers. The zeros that are not real numbers will be complex numbers, which we will examine in Chapter 8. In Example 4 we examine a third-degree polynomial. Note that this polynomial has three zeros.

SELF-CHECK 6.7.2

1. Use a graphics calculator to determine the zeros of $x^2 + x - 2$.
2. Then factor this trinomial.

■ **EXAMPLE 4** Using Zeros to Factor a Polynomial

SELF-CHECK 6.7.2 ANSWERS

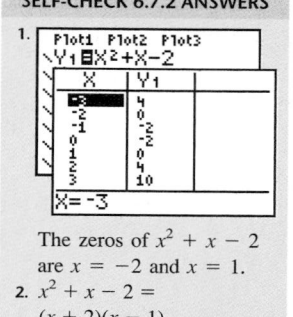

The zeros of $x^2 + x - 2$ are $x = -2$ and $x = 1$.
2. $x^2 + x - 2 = (x + 2)(x - 1)$

Use a graphics calculator to find the zeros of $f(x) = x^3 - 5x^2 - x + 5$. Then factor this polynomial.

SOLUTION

Enter $x^3 - 5x^2 - x + 5$ as y_1 and create a table of values for y_1 until you determine the three zeros of y_1.
From the table y_1 is zero for $x = -1$, $x = 1$, and $x = 5$.

The zeros of $x^3 - 5x^2 - x + 5$ are $x = -1$, $x = 1$, and $x = 5$. Thus the factors are $x + 1$, $x - 1$, and $x - 5$.

Answer: $x^3 - 5x^2 - x + 5 = (x + 1)(x - 1)(x - 5)$ You can check this factorization by multiplying these three factors.

SELF-CHECK 6.7.3

1. Use a graphics calculator to determine the zeros of $x^3 + 3x^2 - 4x$.
2. Then factor this polynomial.

Before reading the next calculator perspective, enter $y = x^3 - 5x^2 - x + 5$ from Example 4 and select the standard viewing window.

CALCULATOR PERSPECTIVE 6.7.1   Calculating the Zeros of a Function

Many graphics calculators have the ability to calculate a zero of a function if an appropriate viewing window is selected to display the zero. To calculate the zero $x = -1$ from Example 4 on a TI-83 Plus calculator, enter the following keystrokes:

[2nd] [CALC] [2]

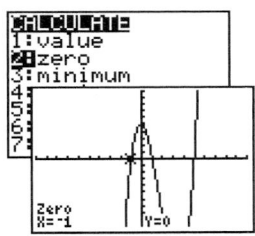

◄ (to move the marker to the left of the zero) [ENTER]

► (to move the marker to the right of the zero) [ENTER]

◄ (to move the marker close to a guess of the zero) [ENTER]

1.

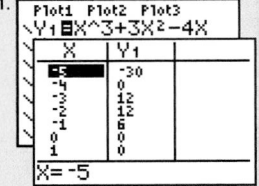

The zeros of $x^3 + 3x^2 - 4x$ are $x = -4$, $x = 0$, and $x = 1$.

2. $x^3 + 3x^2 - 4x = x(x + 4)(x - 1)$

The method used to factor the polynomials in Examples 3 and 4 was first to find the zeros of the polynomial. This was done by using a graphics calculator to form a table of values. This process can become time-consuming when it is not obvious what table of values is appropriate. Sometimes it can be faster to graph the polynomial function and to use the *x*-intercepts of the graph to factor the polynomial. This is illustrated in Example 5.

■ EXAMPLE 5   Using *x*-Intercepts to Factor a Polynomial

Use a graphics calculator to graph $y = x^2 - 3x - 18$ and to determine the *x*-intercepts of this graph. Then factor $x^2 - 3x - 18$.

SOLUTION

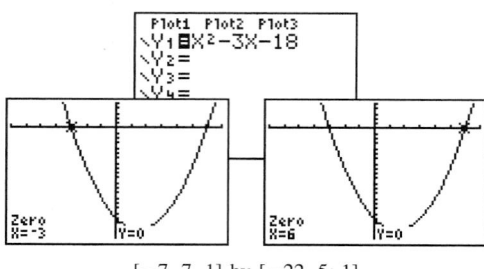

[−7, 7, 1] by [−22, 5, 1]

Enter $x^2 − 3x − 18$ as y_1 and create a graph of $y_1 = x^2 − 3x − 18$.

Use a graphics calculator to calculate the zeros.

The x-intercepts of $y = x^2 − 3x − 18$ are $(−3, 0)$ and $(6, 0)$.
Thus the factors of $x^2 − 3x − 18$ are $x + 3$ and $x − 6$.

Answer: $x^2 − 3x − 18 = (x + 3)(x − 6)$

You can check this factorization by multiplying $x + 3$ times $x − 6$. ∎

The next example is a fourth-degree polynomial whose graph has four real x-intercepts. The polynomial has four corresponding linear factors.

■ **EXAMPLE 6**   Using x-Intercepts to Factor a Polynomial

Use a graphics calculator to graph $y = x^4 − 20x^2 + 64$ and to determine the x-intercepts of this graph. Then factor $x^4 − 20x^2 + 64$.

SOLUTION

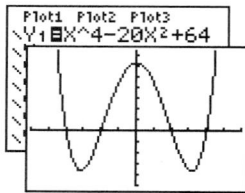

[−6, 6, 1] by [−50, 75, 10]

Enter $y = x^4 − 20x^2 + 64$ as y_1 and create a graph of $y_1 = x^4 − 20x^2 + 64$.

Use a graphics calculator to calculate the zeros, which are $x = −4$, $x = −2$, $x = 2$, and $x = 4$. These zeros correspond to the x-intercepts of the graph.

The x-intercepts of $y = x^4 − 20x^2 + 64$ are $(−4, 0)$, $(−2, 0)$, $(2, 0)$, and $(4, 0)$.

Answer: $x^4 − 20x^2 + 64 = (x + 4)(x + 2)(x − 2)(x − 4)$

You can check this factorization by multiplying these factors. ∎

SELF-CHECK 6.7.4 ANSWERS

1.
```
Plot1  Plot2  Plot3
\Y1▤X²+X−20
```

[−7, 7, 1] by [−30, 10, 5]

The x-intercepts are $(−5, 0)$ and $(4, 0)$.

2. $x^2 + x − 20 = (x + 5)(x − 4)$

SELF-CHECK 6.7.4

1. Use a graphics calculator to graph $y = x^2 + x − 20$ and to determine the x-intercepts of this graph.
2. Then factor $x^2 + x − 20$.

Observe the pattern illustrated by the factorizations given below. We can use this pattern to factor polynomials that contain more than one variable.

FACTORIZATIONS OF POLYNOMIALS IN *x*	FACTORIZATIONS OF POLYNOMIALS IN *x* AND *y*
$x^2 - 4 = (x + 2)(x - 2)$	$x^2 - 4y^2 = (x + 2y)(x - 2y)$
$x^2 - 3x - 18 = (x + 3)(x - 6)$	$x^2 - 3xy - 18y^2 = (x + 3y)(x - 6y)$
$x^2 - 7x + 12 = (x - 3)(x - 4)$	$x^2 - 7xy + 12y^2 = (x - 3y)(x - 4y)$

To factor $x^2 - xy - 56y^2$ in the next example, we first factor $x^2 - x - 56$.

■ EXAMPLE 7   Factoring a Polynomial with Two Variables

Use a graphics calculator to assist you in factoring $x^2 - xy - 56y^2$.

SOLUTION _____

Use a graphics calculator to factor $x^2 - x - 56$ then use this pattern to factor $x^2 - xy - 56y^2$.

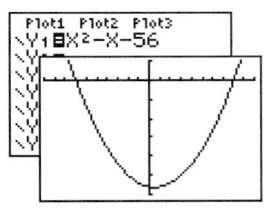

[−10, 10, 1] by [−60, 10, 10]

Enter $x^2 - x - 56$ as y_1 and create a graph of $y_1 = x^2 - x - 56$.

Use a graphics calculator to determine the *x*-intercepts $(-7, 0)$ and $(8, 0)$.

The *x*-intercepts of $y_1 = x^2 - x - 56$ are $(-7, 0)$ and $(8, 0)$.
Thus $x^2 - x - 56 = (x + 7)(x - 8)$.

Answer: $x^2 - xy - 56y^2 = (x + 7y)(x - 8y)$

You can check this factorization by multiplying these factors.
Then use this factorization of $x^2 - x - 56$ to factor $x^2 - xy - 56y^2$. ■

SELF-CHECK 6.7.5

Use a graphics calculator to assist you in factoring each polynomial.

1. $x^2 + 4x - 96$
2. $x^2 + 6xy - 55y^2$

USING THE LANGUAGE AND SYMBOLISM OF MATHEMATICS   6.7

1. The GCF of a polynomial is the _____ _____ factor of the polynomial.
2. The GCF of a polynomial is a common factor of each term of the polynomial that has the largest possible numerical _____ and the largest possible _____ on each variable factor.
3. When we factor out a common factor from each term of a polynomial we are using the _____ property.
4. An *x*-intercept of a graph is a point on the graph whose *y*-coordinate is _____.
5. If $f(x)$ is a function and $f(c) = 0$, then c is called a _____ of the function.

6. If $P(x)$ is a polynomial and $P(5) = 0$ then _____ is a factor of $P(x)$ and _____ is an x-intercept of the graph of $y = P(x)$.

7. If $x + 4$ is a factor of the polynomial $P(x)$, then _____ is a zero of the function $y = P(x)$, and _____ is an x-intercept of the graph of $y = P(x)$.

8. If $(8, 0)$ is an x-intercept of the graph of the polynomial function $y = P(x)$, then $P(8) = $ _____ , and _____ is a factor of $P(x)$.

9. Each coefficient of a real polynomial $P(x)$ is a _____ number.

10. Each real polynomial function of degree n has exactly _____ zeros. Some (perhaps all) of these zeros can be real numbers. If some of the zeros are not real numbers then they will be complex numbers. (We will examine complex numbers in Chapter 8.)

EXERCISES 6.7

In Exercises 1–10 complete each factorization by factoring out the GCF.

1. $12x^3 - 18x^2 = 6x^2(\quad)$
2. $15x^4 + 20x^3 = 5x^3(\quad)$
3. $10a^2b + 6ab^2 = 2ab(\quad)$
4. $21a^3b - 15a^2b^2 = 3a^2b(\quad)$
5. $10x^3 - 15x^2y + 20xy^2 = 5x(\quad)$
6. $12x^3 + 16x^2y + 20xy^2 = 4x(\quad)$
7. $2x(x - 2y) + y(x - 2y) = (\quad)(x - 2y)$
8. $5x(x + 3y) + 4y(x + 3y) = (\quad)(x + 3y)$
9. $7x(2x - 5) - 9(2x - 5) = (\quad)(2x - 5)$
10. $(3x + 4)(5v) - (3x + 4)(4w) = (3x + 4)(\quad)$

In Exercises 11–20 factor out the GCF of each polynomial.

11. $14x^2 + 77x$
12. $22x^3 - 33x^2$
13. $8x^2y - 20xy^2$
14. $18x^2y^2 - 30xy^3$
15. $6x^3 - 8x^2 + 10x$
16. $6x^3 - 9x^2 - 21x$
17. $(5x - 2)(x) + (5x - 2)(3)$
18. $(3x + 4)(7x) - (3x + 4)(5)$
19. $(x + 2y)(6x) - (x + 2y)(7y)$
20. $(5x - 2y)(3x) + (5x - 2y)(4y)$

In Exercises 21–26 use the given table to factor the polynomial entered into the calculator as y_1.

21.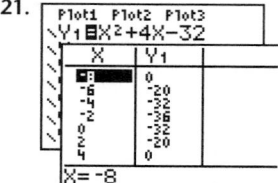

22.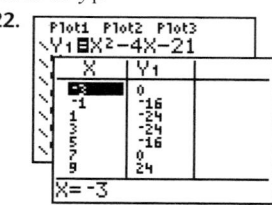

23.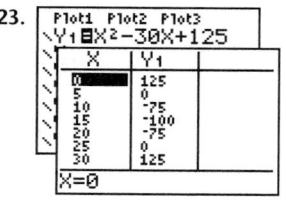

24.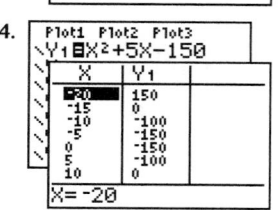

25.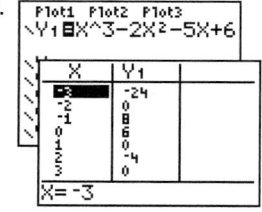

26.

In Exercises 27–32 use the graph of the polynomial function $y = P(x)$ to factor the polynomial $P(x)$.

27. $P(x) = x^2 - x - 30$

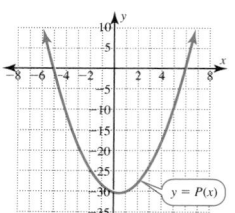

28. $P(x) = x^2 - 6x - 7$

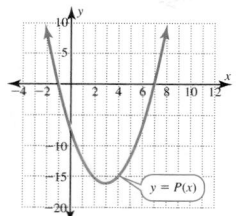

29. $P(x) = x^2 - 7x + 10$

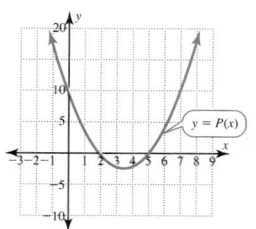

30. $P(x) = x^2 - 5x - 24$

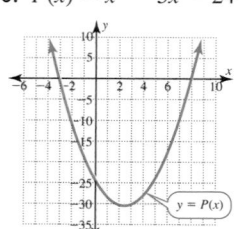

31. $P(x) = x^3 - x^2 - 6x$

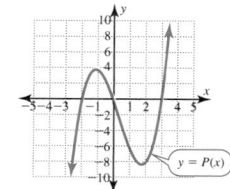

32. $P(x) = x^3 - x^2 - 12x$

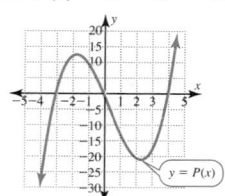

In Exercises 33–38 use the factored form of each polynomial $P(x)$ to list the zeros of $P(x)$ and the x-intercepts of the graph of $y = P(x)$.

33. $P(x) = (x + 5)(x - 9)$

34. $P(x) = (x + 7)(x - 11)$

35. $P(x) = (x + 3)(x) - (x + 3)(17)$

36. $P(x) = (x + 4)(x) - (x + 4)(15)$

37. $P(x) = (x + 3)(x + 1)(x - 8)$

38. $P(x) = (x + 5)(x - 6)(x - 12)$

In Exercises 39–44 use the given polynomial and a graphics calculator to complete each table. Use these results to factor the polynomial.

39. $x^2 + 6x - 27$

x	-12	-9	-6	-3	0	3	6
y							

40. $x^2 - 6x + 5$

x	0	1	2	3	4	5	6
y							

41. $x^2 - 35x + 300$

x	0	5	10	15	20	25	30
y							

42. $x^2 + 26x + 168$

x	-20	-18	-16	-14	-12	-10	-8
y							

43. $x^3 - 10x^2 + 24x$

x	0	1	2	3	4	5	6
y							

44. $x^3 - 35x^2 + 350x - 1000$

x	0	5	10	15	20	25	30
y							

In Exercises 45–52 use the given polynomial $P(x)$ and a graphics calculator to graph $y = P(x)$. Use this graph to factor $P(x)$.

45. $P(x) = x^2 - 11x + 28$ **46.** $P(x) = x^2 + 13x + 40$

47. $P(x) = x^2 + 20x - 125$ **48.** $P(x) = x^2 - 2x - 48$

49. $P(x) = x^3 + 3x^2 - 49x + 45$

50. $P(x) = x^3 + 8x^2 - 4x - 32$

51. $P(x) = x^4 - 5x^2 + 4$ **52.** $P(x) = x^4 - 10x^2 + 9$

In Exercises 53–56 complete the following table for each polynomial.

POLYNOMIAL $P(x)$	FACTORED FORM OF $P(x)$	ZEROS OF $P(x)$	x-INTERCEPTS OF THE GRAPH OF $y = P(x)$
53. $x^2 + 2x - 63$	$(x - 7)(x + 9)$		
54. $x^2 - 17x + 66$		6 and 11	
55. $x^2 - 2x - 80$			$(-8, 0), (10, 0)$
56. $x^3 - x^2 - 30x$		$-5, 0,$ and 6	

57. Factor $x^2 - xy - 72y^2$ given that $x^2 - x - 72 = (x + 8)(x - 9)$.

58. Factor $x^2 + 20xy + 96y^2$ given that $x^2 + 20x + 96 = (x + 12)(x + 8)$.

59. Factor $x^2 - 24xy + 143y^2$ given that $x^2 - 24x + 143 = (x - 11)(x - 13)$.

60. Factor $x^2 - 25xy - 150y^2$ given that $x^2 - 25x - 150 = (x + 5)(x - 30)$.

Group Discussion Questions

61. Discovery Question Determine the zeros for each of these polynomials.

a. $(2x + 1)(x - 4)$

b. $(2x - 1)(x + 3)$

c. $(4x - 1)(x + 2)$

d. $(4x + 1)(x - 1)$

Use your observations from these problems to write the factors of a polynomial with these zeros.

e. $\dfrac{1}{3}$ and 5

f. $-\dfrac{1}{3}$ and -5

g. $\dfrac{1}{6}$ and -2

h. $\dfrac{2}{3}$ and $\dfrac{1}{3}$

Explain what you have observed to your instructor.

62. Discovery Question

a. Graph $y = x^2 - x - 42$ and use this graph to factor $x^2 - x - 42$.

b. Graph $y = -x^2 + x + 42$ and use this graph to factor $-x^2 + x + 42$.

c. Graph $y = x^2 + 5x - 24$ and use this graph to factor $x^2 + 5x - 24$.

d. Graph $y = -x^2 - 5x + 24$ and use this graph to factor $-x^2 - 5x + 24$.

e. What is the relationship between the graph of $y = P(x)$ and the graph of $y = -P(x)$?

f. What is the relationship between the zeros of $P(x)$ and $-P(x)$?

g. What is the relationship between the factored form of $P(x)$ and $-P(x)$?

KEY CONCEPTS FOR CHAPTER 6

1. **Functions**
   - A function is a correspondence that matches each input value with exactly one value of the output variable.
   - The input variable also is called the independent variable.
   - The output variable also is called the dependent variable.
   - The domain of a function is the set of all input values.
   - The range of a function is the set of all output values.

2. **Notations for Representing Functions**
   - Mapping notation
   - Ordered-pair notation
   - Table of values
   - Graphs
   - Function notation
   - Verbally

3. **Domain and Range from the Graph of a Function**
   - The domain of a function is the projection of its graph onto the x-axis.
   - The range of a function is the projection of its graph onto the y-axis.

4. **Vertical Line Test**
   - A graph represents a function if it is impossible to have any vertical line intersect the graph at more than one point.

5. **Function Notation**
   - The notation $f(x)$ is read "f of x" or "the value of f at x."
   - In the function $f(x)$, x represents an input value and $f(x)$ represents the corresponding output value.

6. **Domain of a Real-Valued Function**
   - A function that produces only real output values for real input values is called a real-valued function.
   - The domain of a real-valued function $f(x)$ is the set of all real numbers for which $f(x)$ is also a real number. This means the domain excludes all values that:
     i. cause division by zero or
     ii. cause a negative number under a square root symbol.

7. **Slope of a Line**
   The slope of a line through (x_1, y_1) and (x_2, y_2) with $x_1 \neq x_2$ is $m = \dfrac{y_2 - y_1}{x_2 - x_1}$.

8. **Basic Functions Examined in This Chapter**
   Each basic function from algebra has an equation of a standard form and a characteristic shape.
   - Linear functions, the shape of the graph is a straight line
     i. Slope-intercept form, $y = mx + b$
     ii. Point-slope form, $y - y_1 = m(x - x_1)$
     iii. General form, $Ax + By = C$

- Absolute value functions, the shape of the graph is a V-shape.
- Quadratic functions, the shape of the graph is a parabola
- Square root functions
- Cubic functions
- Cube root functions

9. **Vertical and Horizontal Translations of $y = f(x)$**
   If c is a positive real number,
   - $y = f(x) + c$ shifts the graph of $y = f(x)$ up c units.
   - $y = f(x) - c$ shifts the graph of $y = f(x)$ down c units.
   - $y = f(x + c)$ shifts the graph of $y = f(x)$ left c units.
   - $y = f(x - c)$ shifts the graph of $y = f(x)$ right c units.

10. **Stretching, Shrinking, and Reflecting $y = f(x)$**
    - $y = -f(x)$ reflects the graph of $y = f(x)$ across the x-axis.
    - $y = cf(x)$ stretches the graph of $y = f(x)$ vertically by a factor of c for $c > 1$.
    - $y = cf(x)$ shrinks the graph of $y = f(x)$ vertically by a factor of c for $0 < c < 1$.

11. **Vertex of a Parabola**
    - The vertex of a parabola opening upward is the lowest point on the parabola.
    - The vertex of a parabola opening downward is the highest point on the parabola.

12. **Curve Fitting**
    - A scatter diagram for a set of data points is a graph of these points.
    - Selecting a curve that best fits a set of data is called curve fitting.
    - On a TI-83 Plus graphics calculator use linear regression, **LinReg**, to calculate a line of best fit for a set of data.
    - On a TI-83 Plus graphics calculator use quadratic regression, **QuadReg**, to calculate the parabola of best fit for a set of data.

13. **GCF of a Polynomial**
    - The greatest common factor of a polynomial is the factor of each term of the polynomial that contains:
      i. the largest possible numerical coefficient and
      ii. the largest possible exponent on each variable factor.

14. **Equivalent Statements About Linear Factors of a Polynomial**
    For a real constant c and a real polynomial $P(x)$, the following statements are equivalent:
    - $x - c$ is a factor of $P(x)$.
    - $P(c) = 0$, c is a zero of $P(x)$.
    - $(c, 0)$ is an x-intercept of the graph of $y = P(x)$.

REVIEW EXERCISES   FOR CHAPTER 6

In Exercises 1 and 2 determine whether each relation is a function.

1. a.

x	y
9	1
8	2
6	4
3	5

   b.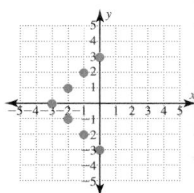

   c. $\{(1, 3), (1, 8), (1, 9)\}$ **d.** $\{(1, \pi), (2, 4\pi), (3, 9\pi)\}$

2. a.   **b.**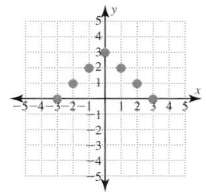

   c. 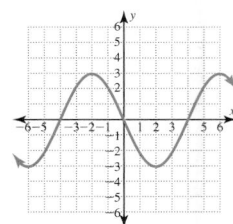   **d.**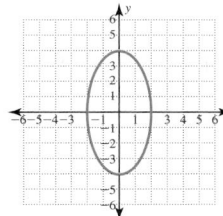

3. Determine the domain and range of each function.

   a. $\{(5, 8), (6, 8), (7, 8)\}$ **b.**

   c.

x	−3	−2	−1	0	1
y	11	9	4	1	6

   d. $f(x) = \sqrt{x - 2} + 3$

4. Evaluate each expression for $f(x) = (x + 2)^2 + 3$

   a. $f(-2)$ **b.** $f(0)$
   c. $f(1)$ **d.** $f(8)$

5. Evaluate each expression given the table of values for $f(x)$.

   a. $f(-2)$ **b.** $f(0)$
   c. $f(2)$ **d.** $f(3)$

x	y
−3	5
−2	4
−1	3
0	2
1	6
2	8
3	10

6. Evaluate each expression given the graph of $y = f(x)$.

   a. $f(-4)$ **b.** $f(-2)$
   c. $f(0)$ **d.** $f(3)$

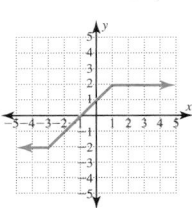

7. Use the function defined by the table to the right to complete each part of this exercise.

   a. Express this function using mapping notation.
   b. Express this function using ordered-pair notation.
   c. Graph this function by making a scatter diagram of this data.
   d. Write an equation for the line containing these points.

x	y
−10	4
−5	3
0	2
5	1
10	0
15	−1
20	−2

8. Use the function graphed in the scatter diagram to the right to complete each part of this exercise.

   a. Express this function using mapping notation.
   b. Express this function using a table format.
   c. Express this function using ordered-pair notation.
   d. Write an equation for the line containing these points.

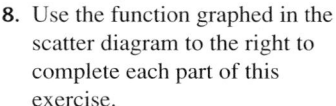

9. Use the slope and y-intercept to sketch the graph of each function.

   a. $f(x) = 2x - 3$ **b.** $f(x) = -3$
   c. $2x - y = 3$ **d.** $3x + 2y = 0$

10. Determine the domain of each function.

   a. $f(x) = x - 5$ **b.** $f(x) = x^2 - 5$
   c. $f(x) = \sqrt{x - 5}$ **d.** $f(x) = \sqrt[3]{x - 5}$

11. Match each function with its graph. All graphs are displayed with the window $[-10, 10, 1]$ by $[-10, 10, 1]$.

   a. $f(x) = x^2 - 4$ **b.** $f(x) = |x - 4|$
   c. $f(x) = \sqrt{x - 4}$ **d.** $f(x) = x - 4$

   A. 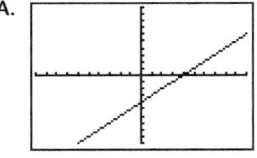   **B.**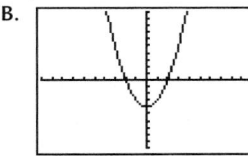

   C.   **D.**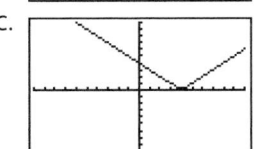

12. Use a graphics calculator to graph each function and then determine the domain and range of each function.

   a. $f(x) = -x^2 + 3$ **b.** $f(x) = \sqrt{4 - x}$
   c. $f(x) = x^3 - 4$ **d.** $f(x) = 1$

13. Height of a Falling Object   A workman dropped a wrench from the edge of a skyscraper to the sidewalk below. The height of the wrench after t secs is given by $h(t) = -16t^2 + 1296$. Evaluate and interpret each of these expressions.

   a. $h(0)$ **b.** $h(2)$
   c. $h(5)$ **d.** $h(9)$

14. Match each function with its graph. All graphs are displayed with the window $[-6, 6, 1]$ by $[-6, 16, 1]$.
  a. $f(x) = x^2$ **b.** $f(x) = x^2 - 5$
  c. $f(x) = x^2 + 5$

A. 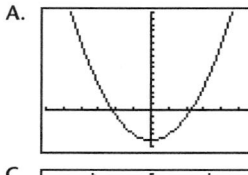   B.

C.

15. Match each function with its graph. All graphs are displayed with the window $[-5, 10, 1]$ by $[-5, 5, 1]$.
  a. $f(x) = \sqrt{x}$ **b.** $f(x) = \sqrt{x - 4}$
  c. $f(x) = \sqrt{x + 4}$

A. 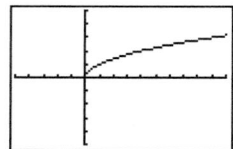   B.

C.

16. Match each function with its graph. All graphs are displayed with the window $[-5, 5, 1]$ by $[-5, 5, 1]$
  a. $f(x) = |x - 1| + 2$ **b.** $f(x) = |x - 2| + 1$
  c. $f(x) = |x + 2| - 1$

A. 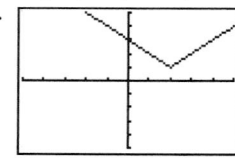   B.

C.

17. Use the given graph of $y = f(x)$ to graph each function.
  a. $y = f(x - 3)$
  b. $y = f(x) + 4$
  c. $y = f(x - 3) + 4$

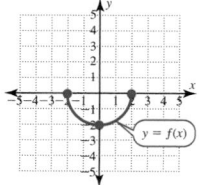

18. Determine the vertex of each parabola using the fact that the vertex of $f(x) = x^2$ is at $(0, 0)$.
  a. $f(x) = x^2 + 7$ **b.** $f(x) = (x - 9)^2$
  c. $f(x) = (x + 11)^2 - 12$

19. Use the given table of values to complete each table.

x	y = f(x)
-2	5
-1	9
0	3
1	-4
2	0

a.

x	y = f(x) + 5
-2	
-1	
0	
1	
2	

b.

x	y = f(x + 5)
	5
	9
	3
	-4
	0

20. The graph of each of these functions is a translation of the graph of $y = f(x)$. Match each function to the correct description.
  a. $y = f(x) - 17$ **A.** A translation 17 units right
  b. $y = f(x - 17)$ **B.** A translation 17 units up
  c. $y = f(x) + 17$ **C.** A translation 17 units left
  d. $y = f(x + 17)$ **D.** A translation 17 units down

21. Use the given table and match each function with its table.

x	y = f(x)
0	5
1	-3
2	0
3	2
4	-6

  a. $y = f(x) + 2$ **b.** $y = f(x + 2)$
  c. $y = f(x - 1) + 3$ **d.** $y = f(x + 3) - 1$

A.

x	y
-3	4
-2	-4
-1	-1
0	1
1	-7

B.

x	y
-2	5
-1	-3
0	0
1	2
2	-6

C.

x	y
1	8
2	0
3	3
4	5
5	-3

D.

x	y
0	7
1	-1
2	2
3	4
4	-4

22. Use the graph of $y = f(x)$ shown to the right and match each function with its graph.

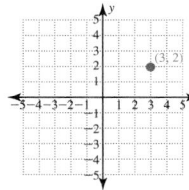

(3, 2)

a. $y = -f(x)$ **b.** $y = 2f(x)$

c. $y = -2f(x)$ **d.** $y = \frac{1}{2}f(x)$

A.

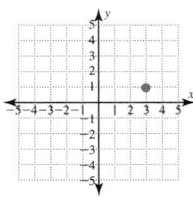

B.

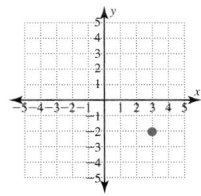

C.

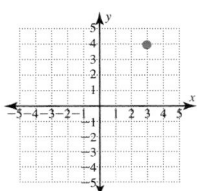

D.

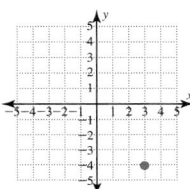

23. Use the graph of $y = f(x)$ shown to the right and match each function with its graph.

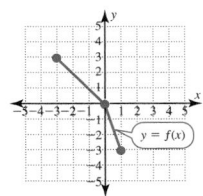

$y = f(x)$

a. $y = -f(x)$ **b.** $y = -\frac{1}{3}f(x)$

c. $y = \frac{1}{3}f(x)$ **d.** $y = 2f(x)$

A.

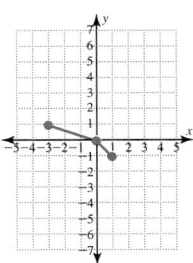

B.

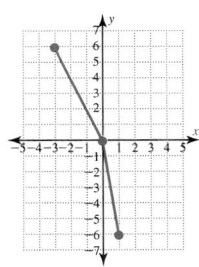

C.

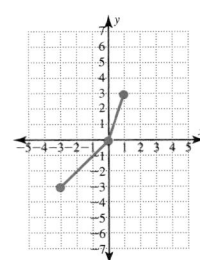

D.

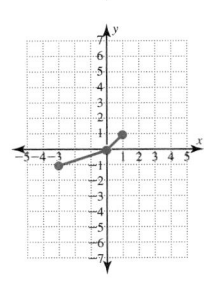

24. Use the given table of values for $y = f(x)$ to complete each table.

x	y = f(x)
0	0
1	8
2	4
3	12
4	24

a.

x	y = -f(x)
0	
1	
2	
3	
4	

b.

x	$y = \frac{1}{4}f(x)$
0	
1	
2	
3	
4	

c.

x	y = 4f(x)
0	
1	
2	
3	
4	

d.

x	y = -3f(x)
0	
1	
2	
3	
4	

25. Match each function below with the description which compares its graph to the graph of $y = f(x)$.

a. $y = 2f(x)$

b. $y = -f(x)$

c. $y = f(x + 2)$

d. $y = f(x) + 2$

A. A reflection of $y = f(x)$ across the x-axis

B. A horizontal shift of $y = f(x)$ 2 units left

C. A vertical shift of $y = f(x)$ 2 units up

D. A vertical stretching of $y = f(x)$ by a factor 2

26. Write the first five terms of each sequence.

a. $a_n = n^2$ **b.** $a_n = n^2 + 1$

c. $a_n = 2n^2$ **d.** $a_n = -n^2$

27. Determine the vertex of each absolute value function using the fact that the vertex of $f(x) = |x|$ is at (0, 0).

a. $f(x) = 4|x|$ **b.** $f(x) = |x + 4|$

c. $f(x) = |x| + 4$ **d.** $f(x) = -4|x|$

28. Write in slope-intercept form the equation of a line with slope $m = 3$ and y-intercept (0, 5).

29. Write in slope-intercept form the equation of a line through (6, 9) with slope $m = \frac{2}{3}$.

30. Write in slope-intercept form the equation of a line through (2, 5) and (4, −1).

31. A line passes through the point (0, −4) and goes up 3 units for every 2 unit increase in x. Write the equation of this line.

32. Write the equation of a vertical line through (2, −7).

33. Write the equation of a line through the origin and parallel to $y = 5x - 9$.

In Exercises 34 and 35 use a graphics calculator to draw a scatter diagram for the data points and to determine the line of best fit. Express your answer in the form $y = mx + b$ with m and b rounded to the nearest hundredth.

34.

x	y
-2.5	-2.5
-2	-1
-1.5	0
-1	1.5
$-.5$	2.5
0	4
$.5$	5

35.

x	y
-1.4	9
$-.9$	7.4
$-.4$	5.8
$.1$	4
$.6$	2.5
1.1	1
1.6	$-.5$

In Exercises 36 and 37 use a graphics calculator to draw a scatter diagram for the data points and to determine the parabola of best fit. Express your answer in the form $y = ax^2 + bx + c$ with a, b, and c rounded to the nearest hundredth.

36.

x	y
0	4.5
1	8
2	14
3	22
4	32
5	45
6	60

37.

x	y
0	8
1	7
2	3
3	-2
4	-10
5	-20
6	-32

38. Modeling the Stretch by a Steel Spring The following data points were taken from an experiment in which a mass was hung from a spring. The distance the spring stretched was measured for each mass. Use a graphics calculator to write the equation for the line of best fit for this data. Then use this linear equation to estimate the distance the spring would stretch for a mass of 12 kg.

x MASS (kg) ON THE SPRING	y STRETCH (cm) BY THE SPRING
5	0.3
10	0.9
15	1.4
20	1.6
25	2.2
30	2.5
35	2.7

39. Modeling Length of a Pendulum The following experimental data was collected which relates the length of a pendulum to its period. Use a graphics calculator to draw a scatter diagram for this data and determine the parabola of best fit. Then use this quadratic equation to estimate the length of a pendulum needed for a clock to produce a period of 2 secs.

x TIME (IN SECS)	y LENGTH (IN CM)
0.8	15
1.1	30
1.4	45
1.5	60
1.7	75
1.9	90
2.2	105
2.3	120

40. $(2x^2 + 3x) + 5x = 2x^2 + (3x + 5x)$ because of the _____ property of addition.

41. $2x^2 + 3x = x(2x + 3)$ because of the _____ property of multiplication over addition.

42. $(5x + 7)(4x) + (5x + 7)(11) = (5x + 7)(4x + 11)$ because of the _____ property of multiplication over addition.

In Exercises 43–48 complete each factorization by factoring out the GCF.

43. $25x^4 - 30\,x^3 = 5x^3(\quad)$
44. $15a^2b + 12ab^2 = 3ab(\quad)$
45. $6x^3 - 10x^2 + 14x = 2x(\quad)$
46. $9x(2x - 3) + 7(2x - 3) = (\quad)(2x - 3)$
47. $(5x + 9)(4x) - (5x + 9)(11) = (5x + 9)(\quad)$
48. $(8x - y)(6x) + (8x - y)(5y) = (8x - y)(\quad)$

In Exercises 49–54 factor out the GCF from each polynomial.

49. $22x^2 + 55x$ **50.** $7x^2y - 14xy^2$
51. $9x^2 - 12x + 27$ **52.** $6a^3b + 10a^2b^2 + 22ab^3$
53. $(2x - 9)(3x) + (2x - 9)(17)$
54. $(4x)(7x - y) - (5y)(7x - y)$

In Exercises 55–58 use the given table to factor the polynomial entered into the calculator as y_1.

55.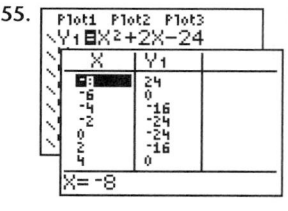

56.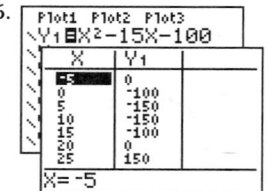

57.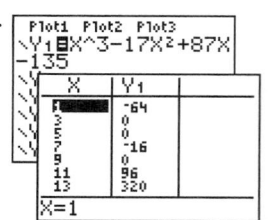

58.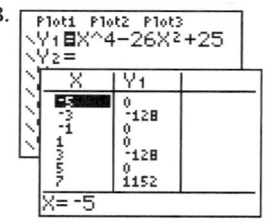

In Exercises 59–61 use the graph of the polynomial function $y = P(x)$ to factor the polynomial $P(x)$.

59. $P(x) = x^2 + 6x + 5$

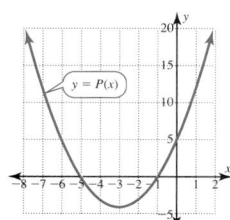

60. $P(x) = x^2 + 3x - 18$

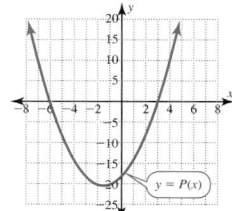
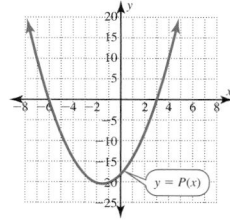

In Exercises 62 and 63 use the given polynomial and a graphics calculator to complete each table. Use these results to factor the polynomial.

62. $x^2 - 15x + 36$

x	0	3	6	9	12	15	18
y							

63. $x^2 - 16x + 48$

x	0	4	8	12	16	20	24
y							

61. $P(x) = x^3 + 3x^2 - 10x$

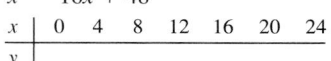

In Exercises 64–66 use the given polynomial and a graphics calculator to graph $y = P(x)$. Use this graph to factor $P(x)$.

64. $P(x) = x^2 + x - 56$
65. $P(x) = x^3 - 16x$
66. $P(x) = x^4 + 5x^3 - 7x^2 - 29x + 30$

In Exercises 67–70 complete the following table for each polynomial.

POLYNOMIAL $P(x)$	FACTORED FORM	ZEROS OF $P(x)$	x-INTERCEPTS OF THE GRAPH OF $y = P(x)$
67. $x^2 + 7x - 44$	$(x + 11)(x - 4)$		
68. $x^2 - 8x - 65$		-5 and 13	
69. $x^2 - 3x - 108$			$(-9, 0), (12, 0)$
70. $x^3 - 7x^2 - 9x + 63$		$-3, 3, 7$	

71. Factor $x^2 - 2xy - 143y^2$ given that $x^2 - 2x - 143 = (x + 11)(x - 13)$.
72. Factor $x^2 - 8xy - 84y^2$ given that $x^2 - 8x - 84 = (x + 6)(x - 14)$.

MASTERY TEST FOR CHAPTER 6

[6.1] **1.** Determine whether each relation is a function.

a.

D	R
-3	9
3	
-4	16
4	

b.

x	y
9	-3
9	3
16	-4
16	4

c.
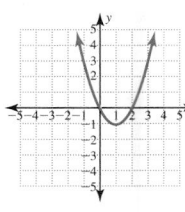

d.
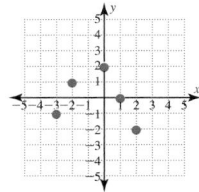

e.
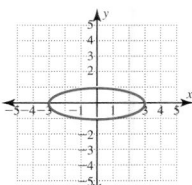

f. $\{(6, 6), (5, 6), (3, 6)\}$

[6.1] **2.** Determine the domain and the range of each of these functions.

a. $\{(1, 8), (2, 4), (7, 11), (8, 13)\}$ **b.** $f(x) = 7$
c. $f(x) = \sqrt{x - 3}$

d.

x	y
-3	0
-2	1
-1	2
0	7

e.
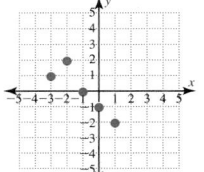

f.
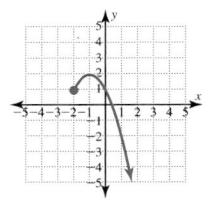

[6.1] **3.** Evaluate each of the following expressions for $f(x) = 3x^2 + x - 12$.

a. $f(0)$ **b.** $f(2)$
c. $f(-2)$ **d.** $-f(2)$

[6.2] **4.** Graph each linear function.

a. $y = \dfrac{3}{4}x - 5$ **b.** $y = -\dfrac{5}{3}x + 2$

c. $y = 4$ **d.** $2x + 3y = -6$

[6.2] **5. Using a Table to Model the Cost of a Collect Call**
The following table displays the dollar cost of a collect telephone call based on the length of the call in minutes. Use this table to determine the linear equation $y = mx + b$ for these data points. Then interpret the meaning of m and b in this problem.

x MINUTES	y COST ($)
1	1.65
5	3.25
7	4.05
12	6.05

[6.2] **6.** Complete the given table of values and sketch the graph of each function. Determine the domain, the range, and the vertex of each graph.

x	−3	−2	−1	0	1	2	3
y							

a. $f(x) = |x - 2| + 3$
b. $f(x) = |x + 1| - 5$

[6.3] **7.** Match each function with its graph. All graphs are displayed with the window $[-2, 8, 1]$ by $[-5, 5, 1]$.

a. $f(x) = (x - 3)^3 + 1$ **b.** $f(x) = -\dfrac{x}{3} + 1$

c. $f(x) = \sqrt{3 - x}$ **d.** $f(x) = \sqrt[3]{x - 3}$

e. $f(x) = |x - 3| + 1$ **f.** $f(x) = (x - 3)^2 - 1$

A.
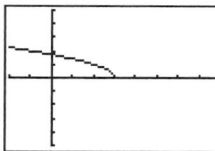

B.

C.
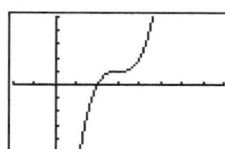

D.
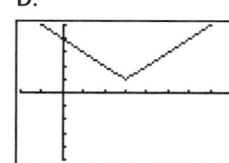

E.
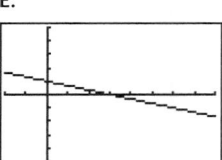

F.

[6.3] **8.** Determine the domain of each function.

a. $f(x) = 6$ **b.** $f(x) = 6x$

c. $f(x) = 6x^2$ **d.** $f(x) = \sqrt{x - 6}$

e. $f(x) = \sqrt[3]{x - 6}$ **f.** $f(x) = |x + 6|$

[6.4] **9.** For parts **a** and **b**, match each function with its graph. All graphs are displayed with the window $[-5, 5, 1]$ by $[-3, 7, 1]$.

a. $f(x) = |x| - 2$ **b.** $f(x) = |x| + 2$

A.
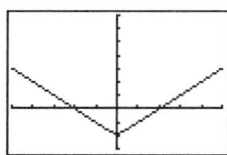

B.
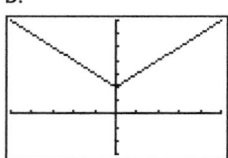

For parts **c** and **d,** use the given table to complete each table.

x	f(x)
0	−5
1	−4
2	3
3	22
4	59

c.

x	f(x) − 2
0	
1	
2	
3	
4	

d.

x	f(x) + 2
0	
1	
2	
3	
4	

[6.4] **10.** For parts **a** and **b**, match each function with its graph. All graphs are displayed with the window $[-7, 7, 1]$ by $[-3, 7, 1]$.

a. $f(x) = (x - 3)^2$ **b.** $f(x) = (x + 3)^2$

A.
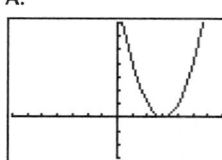

B.
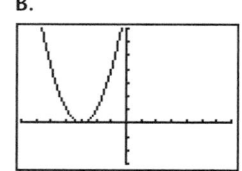

For parts **c** and **d,** use the given table to complete each table.

x	f(x)
0	−5
1	−4
2	3
3	22
4	59

c.

x	f(x − 3)
	−5
	−4
	3
	22
	59

d.

x	f(x + 3)
	−5
	−4
	3
	22
	59

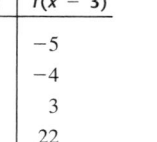
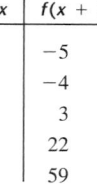

[6.5] **11.** Determine the reflection across the *x*-axis of each graph.

a.

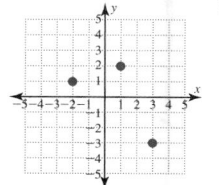

b.

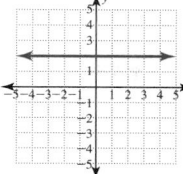

c.

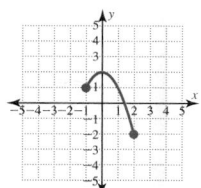

d.

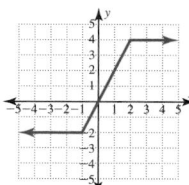

[6.5] **12.** Use the graph of $y = f(x)$ shown to the right and match each function with its graph.

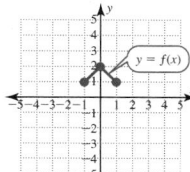

$y = f(x)$

a. $y = 2f(x)$

b. $-2f(x)$

c. $y = \frac{1}{2}f(x)$

d. $y = -\frac{1}{2}f(x)$

A.

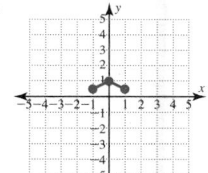

B.

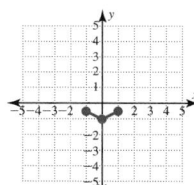

C.

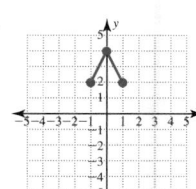

D.

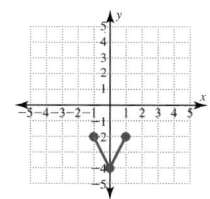

[6.6] **13.** Use a graphics calculator to draw a scatter diagram for these data points and to determine the line of best fit.

x	y
0	−50
5	15
10	80
15	120
20	200
25	250
30	300

[6.6] **14.** Use a graphics calculator to draw a scatter diagram for these data points and to determine the parabola of best fit.

x	y
−3	10
−2	8
−1	6
0	5
1	5
2	8
3	10

[6.7] **15.** Factor out the GCF from each polynomial.

a. $18x^3 - 99x^2$

b. $4x^3y^2 - 6x^2y^3 + 10xy^4$

c. $(7x - 2)(3x) + (7x - 2)(4)$

d. $(2x + 3y)(x) - (2x + 3y)(5y)$

[6.7] **16.** Use the table of values to factor the polynomial entered into the calculator as y_1.

a.

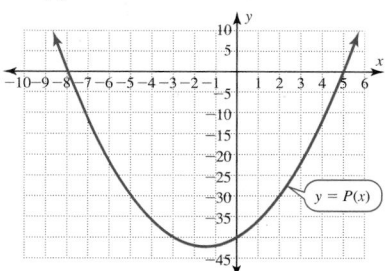

b.

[6.7] **17.** Use the graph of $y = P(x)$ to factor $P(x)$.

a. $P(x) = x^2 + 3x - 40$

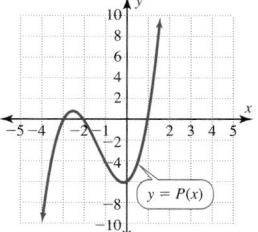

$y = P(x)$

b. $P(x) = x^3 + 4x^2 + x - 6$

$y = P(x)$

GROUP PROJECT   FOR CHAPTER 6

An Algebraic Model for Real Data

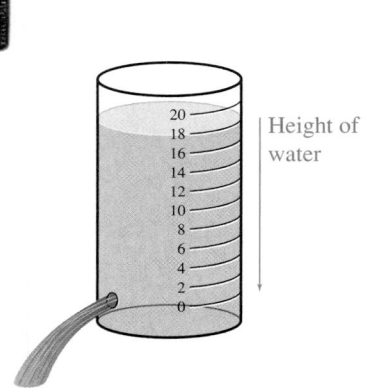

Supplies needed for the lab:

- A clear cylindrical container such as a 2-liter soft drink container with a hole drilled near the bottom of the straight-walled portion and with the height labeled in cm on an attached paper strip or on the plastic with a permanent marker. Label the numerical scale from top to bottom, so that 0 is on the hole. *Suggestion:* Experiment with the hole size—approximately one-fourth of an inch—until the total drain time is less than 3 minutes.
- Water and a bucket to drain the water into.
- A watch that displays seconds.
- A graphics calculator with a linear and a quadratic regression feature.

Lab: Recording the Time to Drain a Cylindrical Container

Place water into a cylindrical container up to the fill line of the cylindrical portion of this container while keeping the drain hole closed. Set your time to zero seconds and record this time and the initial height of the water. Then open the drain hole and remove the cap from the top of the container. As the water drains from the container, record the time for each height in the table. (You may decide to repeat this experiment by recording the height every 10 or more secs.)

1. Record your data in a table similar to the one shown to the right.
2. Use this data to create a scatter diagram using an appropriate scale for each axis.

x TIME (SEC)	y HEIGHT (CM)
0	20
	18
	16
	14
	12
	10
	8
	6
	4
	2
	0

Creating an Algebraic Model for This Data

1. **a.** Which of these models, a linear model or a quadratic model, do you think is the better model for this experiment?
   b. What information in the table or in the graph supports your choice in part (**a**)? Why do you believe this model is better?
   c. Without referring to any of the data that you collected, describe any observations that you made during the course of the experiment that you think support your choice in part (**a**).
2. **a.** Using your graphics calculator, calculate either the line of the best fit or the parabola of the best fit (depending on which is the better choice). (*Hint:* Watch out for scientific notation.)
   b. Graph this curve on the scatter diagram.
3. Letting $y = H(x)$ represent the curve of best fit from **2(a):**
   a. Evaluate and interpret $H(50)$.
   b. Evaluate and interpret $H(70)$.
   c. Determine the value of x for which $H(x) = 11$. Interpret the meaning of this value.

7

ANOTHER LOOK AT FACTORING POLYNOMIALS

The ability to factor polynomials allows us to examine these polynomials piece by piece. Complicated problems can be broken down into simpler components. We have already examined the connection among the factors of a polynomial, the zeros of a polynomial function, and the x-intercepts of the graph of a polynomial function. This chapter builds on these connections and the work done on factoring in earlier sections. Then we examine some additional algebraic factoring methods that enable us to factor polynomials with more than one variable and help to prepare us with the skills we need to work with rational expressions. (See Chapter 9.)

Developing the skills to factor some polynomials algebraically will deepen your mathematical ability. Although some calculators now can perform most factorizations shown in this chapter, the ability to perform mental factorizations will give you greater insight and speed your work in other mathematical topics. This chapter will continue to develop a hands-on understanding of algebraic factoring and the relevance that factoring has to other topics that we have examined numerically and graphically.

Section 7.1   Factoring Trinomials of the Form $x^2 + bxy + cy^2$

Objectives:

1. Factor trinomials of the form $x^2 + bx + c$.
2. Factor trinomials of the form $x^2 + bxy + cy^2$.

A Mathematical Note

Prime numbers play a key role in cryptography. In the frequently used RSA algorithm, prime numbers are used for both encryption to input a message and decryption to decode this message. The security of this procedure comes from the computational difficulty of factoring extremely large numbers. To be secure, the prime numbers must have at least 100 digits.

There are different methods for factoring trinomials presented in different textbooks. Thus we urge you to follow the advice of your instructor. The method shown in this book stresses the connection of factoring to the distributive property.

In Chapter 6 we factored some trinomials of the form $x^2 + bx + c$ by using tables to display the zeros of the polynomial or by using graphs to display the x-intercepts of the graph of $y = x^2 + bx + c$. We now present an algebraic method for factoring trinomials of the form $x^2 + bx + c$ or the form $x^2 + bxy + cy^2$. The procedure given in this section enables you either to factor a trinomial or to determine that it is prime. A polynomial is **prime over the integers** if its only factorization must involve either 1 or -1 as one of the factors.

The trinomial $ax^2 + bx + c$ is referred to as a **quadratic trinomial**. The ax^2 term is called the **second-degree term,** or the **quadratic term;** bx is the **first-degree term** or the **linear term;** and c is the **constant term.** For example, $6x^2 - 7x - 5$ is a quadratic trinomial with $a = 6$, $b = -7$, and $c = -5$. The quadratic term is $6x^2$, the linear term is $-7x$, and the constant term is -5.

Different methods for factoring trinomials are presented in different textbooks. Thus it is likely that your class will have students who have studied different methods. We urge you to follow the advice of your instructor. Work toward understanding the method you select as you practice this method to gain speed and proficiency. As more factoring is done with calculators, it is this understanding of factoring and its connections with other topics that has lasting importance.

Because of the importance of connecting concepts, we present a method for factoring trinomials that builds on our earlier work using the distributive property to factor out the GCF (greatest common factor) of a polynomial.

To develop the logic we will use to factor trinomials of the form $x^2 + bx + c$, we will first examine the following three products:

$$(x + 24)(x + 1) = x(x + 1) + 24(x + 1)$$
$$= x^2 + x + 24x + 24$$
$$= x^2 + 25x + 24$$

$$(x + 2)(x + 12) = x(x + 12) + 2(x + 12)$$
$$= x^2 + 12x + 2x + 24$$
$$= x^2 + 14x + 24$$

$$(x + 3)(x + 8) = x(x + 8) + 3(x + 8)$$
$$= x^2 + 8x + 3x + 24$$
$$= x^2 + 11x + 24$$

Note in each of these three cases that the coefficients of the two middle terms have a product of 24. In the first case $1 \cdot 24 = 24$, in the second case $12 \cdot 2 = 24$, and in the third case $8 \cdot 3 = 24$. In general $(x + c_1)(x + c_2) = x^2 + (c_1 + c_2)x + c_1c_2$. Thus a factorization of $x^2 + bx + c$ must have $c_1c_2 = c$ and $c_1 + c_2 = b$. The method of factoring $x^2 + bx + c$ that we now illustrate will examine the factors of c in search of a pair of factors whose sum is b.

Factorable Trinomials

ALGEBRAICALLY	VERBALLY	EXAMPLE
$x^2 + bx + c = (x + c_1)(x + c_2)$ where $c_1c_2 = c$ and $c_1 + c_2 = b$.	A trinomial $x^2 + bx + c$ is factorable into a pair of binomial factors with integer coefficients if and only if there are two integers whose product is c and whose sum is b. Otherwise, the trinomial is prime over the integers.	$x^2 + 14x + 24 = (x + 2)(x + 12)$ where $2 \cdot 12 = 24$ and $2 + 12 = 14$.

Factoring polynomials reverses the steps used in multiplication. Both use the distributive property.

Thus to factor $x^2 + 10x + 24$ in Example 1, we can examine all possible pairs of factors of 24 in search of a pair of factors whose sum is 10. Note that the steps used in this example to factor the trinomial into two binomials are identical to the steps used to multiply the two binomials—but in the reverse order.

■ EXAMPLE 1   Factoring a Quadratic Trinomial

Factor $x^2 + 10x + 24$.

SOLUTION

Examine the factors of 24 for a pair of factors whose sum is 10.

FACTORS OF 24		SUM OF FACTORS
1	24	25
2	12	14
3	8	11
4	6	10

The only factors of 24 with a sum of 10 are 4 and 6. Thus we rewrite the middle term of $10x$ as $6x + 4x$.

$$x^2 + 10x + 24 = x^2 + 6x + 4x + 24$$
$$= x(x + 6) + 4(x + 6)$$
$$= (x + 4)(x + 6)$$

After rewriting the linear term $10x$ as $6x + 4x$, use the distributive property to factor out the GCF of x from the first two terms and the GCF of 4 out from the last two terms.
Then use the distributive property to factor out the GCF of $x + 6$ from both terms.
Does this factorization check?

Answer: $x^2 + 10x + 24 = (x + 4)(x + 6)$

Because the second step of this factoring procedure involves grouping together pairs of terms we will refer to this procedure as factoring by grouping. We will examine factoring by grouping in a wider context in Section 7.4.

The procedure used in Example 1 is summarized in the following box. A key step in this procedure is to use the sign pattern of the terms to determine the sign pattern for each possible pair of factors of the constant term.

Factoring Trinomials of the Form $x^2 + bx + c$ by Grouping

PROCEDURE

STEP 1. Be sure you have factored out the GCF if it is not 1.

STEP 2. Find a pair of factors of c whose sum is b. If there is not a pair of factors whose sum is b, the trinomial is prime over the integers.
- If the constant c is positive, the factors of c must have the same sign. These factors will share the same sign as the linear coefficient, b.
- If the constant c is negative, the factors of c must be opposite in sign. The sign of b will determine the sign of the factor with the larger absolute value.

STEP 3. Rewrite the linear term of $x^2 + bx + c$ so that b is the sum of the pair of factors from Step 2.

STEP 4. Factor the polynomial from Step 3 by grouping the terms and factoring out the GCF from each pair of terms.

EXAMPLE

Factor $x^2 + 5x - 24$.

FACTORS OF 24		SUM OF FACTORS
-1	24	23
-2	12	10
-3	8	5
-4	6	2

$$x^2 + 5x - 24 = x^2 + 8x - 3x - 24$$
$$= x(x + 8) - 3(x + 8)$$
$$= (x - 3)(x + 8)$$

◼ EXAMPLE 2 Factoring a Quadratic Trinomial

Factor $x^2 + 7x - 60$.

SOLUTION

Examine the factors of -60 for a pair of factors whose sum is 7.

FACTORS OF -60		SUM OF FACTORS
-1	60	59
-2	30	28
-3	20	17
-4	15	11
-5	12	7
-6	10	4

The constant term, -60, is negative so the factors of -60 are of opposite sign. Because 7, the coefficient of the linear term, is positive, the positive factor of -60 must be larger in absolute value than the negative factor.

$$x^2 + 7x - 60 = x^2 - 5x + 12x - 60$$
$$= x(x - 5) + 12(x - 5)$$
$$= (x + 12)(x - 5)$$

Use this table to rewrite the linear term $7x$ as $-5x + 12x$. Factor the GCF of x from the first two terms and factor the GCF of 12 from the last two terms. Then use the distributive property again to factor $x - 5$ out of both terms.

Answer: $x^2 + 7x - 60 = (x + 12)(x - 5)$

Does this factorization check? ◼

Example 3 provides further practice factoring a trinomial of the form $x^2 + bx + c$. In this example both b and c are negative.

■ EXAMPLE 3   Factoring a Quadratic Trinomial

Factor $x^2 - 4x - 45$.

SOLUTION

Examine the factors of -45 for a pair of factors whose sum is -4.

FACTORS OF -45		SUM OF FACTORS
1	-45	-44
3	-15	-12
5	-9	-4

The constant term, -45, is negative so the factors of -45 are of opposite sign. Because -4, the coefficient of the linear term, is negative, the negative factor of -45 must be larger in absolute value than the positive factor.

$$x^2 - 4x - 45 = x^2 + 5x - 9x - 45$$
$$= x(x + 5) - 9(x + 5)$$
$$= (x - 9)(x + 5)$$

Answer: $x^2 - 4x - 45 = (x - 9)(x + 5)$

Rewrite the linear term, $-4x$, as $5x - 9x$. Factor the GCF of x from the first two terms and the GCF of -9 from the last two terms. Then use the distributive property again to factor out $x + 5$ from both terms.

Does this factorization work?   ■

Each step of the factoring procedure is shown in Example 3. If you can mentally select the correct pair of factors without using all these steps, that is great. However, the harder the problem the greater the value of this procedure. This procedure does not involve any guessing; it is a systematic procedure that always works.

SELF-CHECK 7.1.1

Factor each trinomial.

1. $v^2 + 8v + 12$ 2. $m^2 - 7m + 12$ 3. $n^2 - 11n - 12$

SELF-CHECK 7.1.1 ANSWERS

1. $(v + 6)(v + 2)$
2. $(m - 3)(m - 4)$
3. $(n + 1)(n - 12)$

Recall from Chapter 6 that we also can examine the polynomial from Example 3 both numerically and graphically.

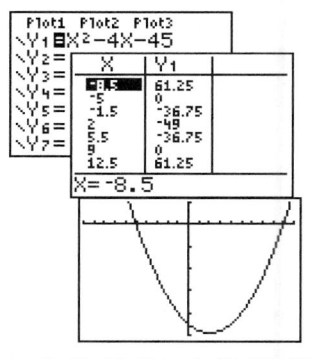

[−10, 10, 1] by [−50, 10, 10]

POLYNOMIAL $P(x)$	FACTORED FORM	ZEROS OF $P(x)$	x-INTERCEPTS OF THE GRAPH OF $y = P(x)$
$x^2 - 4x - 45$	$(x - 9)(x + 5)$	9 and -5	$(9, 0)$ and $(-5, 0)$

One thing that can be difficult to do by just examining a table of values or the graph of a polynomial function is to determine if the polynomial is prime over the integers. This can be accomplished routinely by the algebraic method used in Example 2 and Example 3. If the list of all possible factors does not produce the correct middle term, then the polynomial must be prime. This is illustrated in Example 4.

■ **EXAMPLE 4** Identifying a Prime Trinomial

Factor $v^2 + 9v + 12$.

SOLUTION _____

Examine the factors of 12 for a pair of factors whose sum is 9.

FACTORS OF 12		SUM OF FACTORS
1	12	13
2	6	8
3	4	7

Both factors must be positive because both 12 and 9 are positive.

A distinct advantage of this factoring procedure is that it can identify polynomials that are prime. Once all possible pairs of factors of c are listed, either the correct pair of factors is listed or the polynomial must be prime.

None of the pairs of factors of 12 has a sum of 9. Because we have systematically examined and eliminated all integral factors of 12, we know that this polynomial is prime over the integers.

Answer: $v^2 + 9v + 12$ is prime over the integers. ■

As we noted in Section 6.7 we also can use the procedure for factoring $x^2 + bx + c$ to factor trinomials of the form $x^2 + bxy + cy^2$. This is illustrated next and in the following examples.

FACTORIZATIONS OF POLYNOMIALS IN x	FACTORIZATIONS OF POLYNOMIALS IN x and y
$x^2 + 10x + 24 = (x + 6)(x + 4)$	$x^2 + 10xy + 24y^2 = (x + 6y)(x + 4y)$
$x^2 + 7x - 60 = (x + 12)(x - 5)$	$x^2 + 7xy - 60y^2 = (x + 12y)(x - 5y)$
$v^2 + 9v + 12$ is prime.	$v^2 + 9vw + 12w^2$ is prime.

■ **EXAMPLE 5** Factoring a Trinomial with Two Variables

Factor $x^2 + 19xy + 48y^2$.

SOLUTION _____

Examine the factors of 48 for a pair of factors whose sum is 19.

FACTORS OF 48		SUM OF FACTORS
1	48	49
2	24	26
3	16	19

The coefficient of the last term, 48, is positive, so the factors of 48 have the same sign. Because 19, the coefficient of the middle term, is positive, both factors of 49 must be positive. In this case we stop our list as soon as we discover a pair of factors whose sum is 19.

Although we could list other factors, there is no need. Once the correct pair of factors is found we can stop the list and use these factors.

$$x^2 + 19xy + 48y^2 = x^2 + 3xy + 16xy + 48y^2$$
$$= x(x + 3y) + 16y(x + 3y)$$
$$= (x + 16y)(x + 3y)$$

Rewrite the middle term, 19xy, as 3xy + 16xy. Then complete the factorization.

Answer: $x^2 + 19xy + 48y^2 = (x + 16y)(x + 3y)$

Does this factorization check? ■

■ EXAMPLE 6 Factoring a Trinomial with Two Variables

Factor $m^2 - 4mn - 32n^2$.

SOLUTION

Examine the factors of -32 for a pair of factors whose sum is -4.

FACTORS OF -32		SUM OF FACTORS
1	-32	-31
2	-16	-14
4	-8	-4

The last coefficient, -32, is negative so the factors of -32 are of opposite sign. Because -4, the coefficient of the middle term, is negative, the negative factor of -32 must be larger in absolute value than the positive factor.

$$\begin{aligned} m^2 - 4mn - 32n^2 &= m^2 + 4mn - 8mn - 32n^2 \\ &= m(m + 4n) - 8n(m + 4n) \\ &= (m - 8n)(m + 4n) \end{aligned}$$

Rewrite the middle term, $-4mn$, as $4mn - 8mn$. Then complete the factorization.

Answer: $m^2 - 4mn - 32n^2 = (m - 8n)(m + 4n)$

Does this factorization check? ■

SELF-CHECK 7.1.2

Factor each trinomial.

1. $x^2 + 18x + 48$ **2.** $a^2 - 2ab - 48b^2$

SELF-CHECK 7.1.2 ANSWERS
1. $x^2 + 18x + 48$ is prime
2. $(a + 6b)(a - 8b)$

If each term of a trinomial has a GCF other than 1, then the first step in factoring this trinomial is to factor out the GCF.

■ EXAMPLE 7 Completely Factoring a Trinomial

Factor $2x^3 + 6x^2 - 80x$.

SOLUTION

$$2x^3 + 6x^2 - 80x = 2x(x^2 + 3x - 40)$$

First factor out $2x$, the GCF of the trinomial. Then factor the trinomial $x^2 + 3x - 40$.

Examine the factors of -40 for a pair of factors whose sum is 3.

FACTORS OF -40		SUM OF FACTORS
-1	40	39
-2	20	18
-4	10	6
-5	8	3

The last term, -40, is negative so the factors of -40 are of opposite sign. Because 3, the coefficient of the middle term, is positive, the positive factor of -40 must be larger in absolute value than the negative factor.

$$\begin{aligned} 2x^3 + 6x^2 - 80x &= 2x(x^2 + 3x - 40) \\ &= 2x(x^2 - 5x + 8x - 40) \\ &= 2x[x(x - 5) + 8(x - 5)] \\ &= 2x(x + 8)(x - 5) \end{aligned}$$

Rewrite the linear term, $3x$, as $-5x + 8x$. Then complete the factorization.

Answer: $2x^3 + 6x^2 - 80x = 2x(x + 8)(x - 5)$

Does this factorization check? ■

■ EXAMPLE 8   Completely Factoring a Trinomial

Factor $3x^4 + 15x^3y - 42x^2y^2$.

SOLUTION

$3x^4 + 15x^3y - 42x^2y^2 = 3x^2(x^2 + 5xy - 14y^2)$ First factor out $3x^2$, the GCF of the trinomial. Then factor $x^2 + 5xy - 14y^2$.

Examine the factors of -14 for a pair of factors whose sum is 5.

FACTORS OF -14		SUM OF FACTORS
-1	14	13
-2	7	5

The coefficient of the last term, -14, is negative so the factors of -14 are of opposite sign. Because 5, the coefficient of the middle term, is positive, the positive factor of -14 must be larger in absolute value than the negative factor.

$$3x^4 + 15x^3y - 42x^2y^2 = 3x^2(x^2 + 5xy - 14y^2)$$
$$= 3x^2(x^2 - 2xy + 7xy - 14y^2)$$
$$= 3x^2[x(x - 2y) + 7y(x - 2y)]$$
$$= 3x^2(x + 7y)(x - 2y)$$

Rewrite the linear term, $5xy$, as $-2xy + 7xy$. Then complete the factorization.

Answer: $3x^4 + 15x^3y - 42x^2y^2 = 3x^2(x + 7y)(x - 2y)$ Does this factorization check?

SELF-CHECK 7.1.3

Completely factor $5x^3 - 75x^2 + 270x$.

The examples in this section show each step of the factoring procedure. As you practice, you are likely to see patterns and be able to do some of the steps mentally. It would be wise to check with your instructor to confirm how much work you should show on your paper and to be sure that the shortcuts you have noted are indeed correct.

USING THE LANGUAGE AND SYMBOLISM OF MATHEMATICS   7.1

1. A polynomial is _____ over the integers if its only factorization must involve either 1 or -1 as one of the factors.
2. A trinomial of the form $ax^2 + bx + c$ is called a _____ trinomial.
3. In the trinomial $ax^2 + bx + c$:
   a. The second-degree term, ax^2, is called the _____ term.
   b. The first-degree term, bx, is called the _____ term.
   c. The term, c, is called the _____ term.
4. The factoring method presented in this text utilizes the _____ property to factor out the _____ of a polynomial.

5. A trinomial $x^2 + bx + c$ is factorable over the integers if there are two integers whose _____ is c and whose _____ is b.
6. When attempting to factor a quadratic trinomial $x^2 + bx + c$, we consider the factors of c. If the constant c is positive, the factors of c must have the _____ sign. These factors will share the same sign as the linear coefficient, b. If the constant c is negative, the factors of c must be _____ in sign.

EXERCISES 7.1

In Exercises 1–8 fill in the blanks to complete the factorization of each trinomial.

1. $x^2 - 12x + 35 = x^2 - 5x - 7x + 35$
$$= x(\quad) - 7(\quad)$$
$$= (x - 7)(\quad)$$

2. $x^2 - 6x - 72 = x^2 + 6x - 12x - 72$
$$= x(\quad) - 12(\quad)$$
$$= (x - 12)(\quad)$$

3. $x^2 + 2x - 80 = x^2 - 8x + \underline{\hspace{2cm}} - 80$
$$= x(\quad) + 10(\quad)$$
$$= (x + 10)(\quad)$$

4. $x^2 + 12x + 32 = x^2 + 4x + \underline{\hspace{2cm}} + 32$
$$= x(\quad) + 8(\quad)$$
$$= (x + 8)(\quad)$$

5. $5x^3 + 25x^2 + 20x = (\quad)(x^2 + 5x + 4)$
$$= (\quad)(x^2 + x + \underline{\hspace{1.5cm}} + 4)$$
$$= (\quad)[x(\quad) + 4(\quad)]$$
$$= (\quad)(x + 4)(\quad)$$

6. $3x^4 - 12x^3 - 36x^2 = (\quad)(x^2 - 4x - 12)$
$$= (\quad)(x^2 - 6x + \underline{\hspace{1.5cm}} - 12)$$
$$= (\quad)[x(\quad) + 2(\quad)]$$
$$= (\quad)(x + 2)(\quad)$$

7. $x^2 - 5xy - 36y^2 = x^2 + 4xy - 9xy - 36y^2$
$$= x(\quad) - 9y(\quad)$$
$$= (x - 9y)(\quad)$$

8. $x^2 - 12xy + 35y^2 = x^2 - 7xy - 5xy + 35y^2$
$$= x(\quad) - 5y(\quad)$$
$$= (x - 5y)(\quad)$$

In Exercises 9–14 use the given table to factor each polynomial. (*Hint:* Some of these trinomials are prime.)

9. **a.** $x^2 + 7x + 12$ **b.** $x^2 + 8x + 12$
   c. $x^2 + 10x + 12$ **d.** $x^2 + 13x + 12$

FACTORS OF 12		SUM OF FACTORS
1	12	13
2	6	8
3	4	7

10. **a.** $x^2 + x - 12$ **b.** $x^2 + 2x - 12$
    c. $x^2 + 4x - 12$ **d.** $x^2 + 11x - 12$

FACTORS OF -12		SUM OF FACTORS
-1	12	11
-2	6	4
-3	4	1

11. **a.** $x^2 - 3x - 28$ **b.** $x^2 - 12x - 28$
    c. $x^2 - 15x - 28$ **d.** $x^2 - 27x - 28$

FACTORS OF -28		SUM OF FACTORS
1	-28	-27
2	-14	-12
4	-7	-3

12. **a.** $x^2 - 10x + 28$ **b.** $x^2 - 11x + 28$
    c. $x^2 - 16x + 28$ **d.** $x^2 - 29x + 28$

FACTORS OF 28		SUM OF FACTORS
-1	-28	-29
-2	-14	-16
-4	-7	-11

13. **a.** $x^2 - 36$ **b.** $x^2 + 9x - 36$
    c. $x^2 + 16x - 36$ **d.** $x^2 + 18x - 36$

FACTORS OF -36		SUM OF FACTORS
-1	36	35
-2	18	16
-3	12	9
-4	9	5
-6	6	0

14. **a.** $x^2 + 36$ **b.** $x^2 + 12x + 36$
    c. $x^2 + 13x + 36$ **d.** $x^2 + 20x + 36$

FACTORS OF 36		SUM OF FACTORS
1	36	37
2	18	20
3	12	15
4	9	13
6	6	12

In Exercises 15–18 each trinomial is of the form $x^2 + bx + c$ and each pair of trinomials has the same value of c. Make a table of the factors of c and use this table to factor each trinomial.

15. **a.** $x^2 + 11x + 10$ **b.** $x^2 + 7x + 10$
16. **a.** $x - 18x + 80$ **b.** $x^2 - 24x + 80$
17. **a.** $x^2 + 5x - 24$ **b.** $x^2 + 2x - 24$
18. **a.** $x^2 - 6x - 16$ **b.** $x^2 - 15x - 16$

In Exercises 19–40 factor each trinomial.

19. $a^2 - 10a - 11$ 20. $b^2 + 6b + 5$
21. $y^2 - 18y + 17$ 22. $x^2 + 3x - 4$
23. $x^2 - 8x + 7$ 24. $x^2 + 14x + 13$
25. $y^2 - 2y - 99$ 26. $y^2 + 30y - 99$
27. $p^2 - 29p + 100$ 28. $p^2 + 20p + 100$
29. $t^2 + 14t + 48$ 30. $t^2 - 26t + 48$

31. $n^2 + 5n - 36$

32. $n^2 - 9n - 36$

33. $x^2 + 5xy + 6y^2$

34. $x^2 - 7xy + 6y^2$

35. $x^2 + xy - 6y^2$

36. $x^2 - 3xy - 10y^2$

37. $x^2 - 10xy + 25y^2$

38. $x^2 - 14xy + 49y^2$

39. $a^2 - 16ab - 36b^2$

40. $a^2 + 5ab - 36b^2$

In Exercises 41–50 some of the polynomials are factorable over the integers and some are prime over the integers. Factor those that are factorable and label the others as prime.

41. $x^2 + 33x + 90$

42. $x^2 - 13x - 90$

43. $x^2 - 22x + 90$

44. $x^2 + 10x - 90$

45. $m^2 - 21mn + 98n^2$

46. $m^2 + 20mn + 98n^2$

47. $m^2 - 7mn - 9n^2$

48. $m^2 - 12mn - 98n^2$

49. $v^2 + 25$

50. $v^2 - 25$

In Exercises 51–60 first factor out the GCF and then complete the factorization.

51. $5x^3 + 15x^2 + 10x$

52. $7y^3 - 14y^2 - 21y$

53. $4ax^3 - 20ax^2 + 24ax$

54. $7bx^3 - 56bx^2 - 63bx$

55. $10az^2 + 290az + 1000a$

56. $4xy^2 + 8xy - 192x$

57. $2abx^2 + 4abx - 96ab$

58. $33v^2w - 77vw - 66w$

59. $x^2(a + b) - 6x(a + b) + 8(a + b)$

60. $(2a - 3)y^2 + 9(2a - 3)y - 11(2a - 3)$

In Exercises 61–66 first factor out -1 from each polynomial and then complete the factorization.

61. $-x^2 + 2x + 35$

62. $-x^2 + 34x + 35$

63. $-m^2 - 3m + 18$

64. $-m^2 + m + 12$

65. $-x^2 + 12xy + 45y^2$

66. $-x^2 - xy + 20y^2$

In Exercises 67 and 68 use the given table to factor the polynomial entered into the calculator as y_1.

67.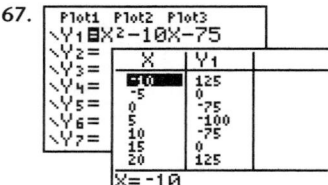

68.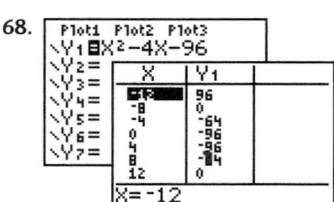

In Exercises 69 and 70 use the graph of the polynomial function $y = P(x)$ to factor the polynomial $P(x)$.

69. $P(x) = x^2 - 64$

70. $P(x) = x^2 - 49$

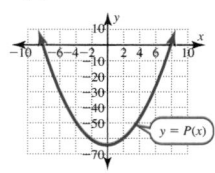

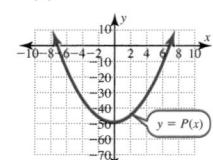

In Exercises 71–74 complete the following table for each polynomial.

POLYNOMIAL $P(x)$	FACTORED FORM	ZEROS OF $P(x)$	x-INTERCEPTS OF THE GRAPH OF $y = P(x)$
71. $x^2 + 2x - 15$			
72. $x^2 - 17x + 60$			
73. $x^2 - 3x - 54$			
74. $x^2 - x - 72$			

Group Discussion Questions

75. Discovery Question Multiply these factors and determine the zeros of each polynomial.

a. $2\left(x + \dfrac{1}{2}\right)(3)\left(x - \dfrac{1}{3}\right)$

b. $4\left(x + \dfrac{1}{4}\right)(3)\left(x - \dfrac{2}{3}\right)$

c. $4\left(x - \dfrac{3}{4}\right)(5)\left(x + \dfrac{2}{5}\right)$

Use your observations from these problems to write the factors of a polynomial with these zeros.

d. $\dfrac{2}{3}$ and $\dfrac{1}{5}$

e. $-\dfrac{1}{2}$ and $-\dfrac{3}{5}$

f. $\dfrac{2}{5}$ and $-\dfrac{1}{7}$

Explain what you have observed to your instructor.

76. Calculator Discovery To factor $x^2 + 21x - 72$, we need to find a pair of factors of -72 whose sum is 21. Let x and y_1 represent this pair of factors.

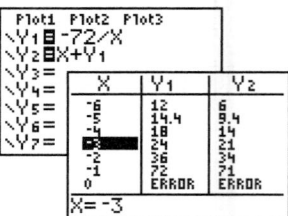

Thus $xy_1 = -72$ or $y_1 = \dfrac{-72}{x}$. Let $y_2 = x + y_1$ and use a table to find where y_2 is 21.

We see that $y_2 = 21$ when $x = -3$ and $y_1 = 24$. So,

$$x^2 + 21x - 72 = x^2 - 3x + 24x - 72$$
$$= x(x - 3) + 24(x - 3)$$
$$= (x + 24)(x - 3)$$

Try this method to factor $x^2 + 10x - 96$.

Section 7.2   Factoring Trinomials of the Form $ax^2 + bxy + cy^2$

Objectives: **3.** Factor trinomials of the form $ax^2 + bx + c$.
4. Factor trinomials of the form $ax^2 + bxy + cy^2$.

This section builds on our work in Section 7.1 and extends the method of factoring trinomials by grouping. We now factor trinomials in which the leading coefficient is not 1. To develop the logic to factor trinomials of the form $ax^2 + bx + c$, we start by examining the pattern shown in the following three products.

$$(2x + 5)(3x + 1) = 2x(3x + 1) + 5(3x + 1) \qquad (6x + 1)(x + 5) = 6x(x + 5) + 1(x + 5)$$
$$= 6x^2 + 2x + 15x + 5 \qquad\qquad\qquad = 6x^2 + 30x + x + 5$$
$$= 6x^2 + 17x + 5 \qquad\qquad\qquad\quad = 6x^2 + 31x + 5$$

$$(6x + 5)(x + 1) = 6x(x + 1) + 5(x + 1)$$
$$= 6x^2 + 6x + 5x + 5$$
$$= 6x^2 + 11x + 5$$

Note that in each of these three cases the coefficients of the two middle terms have a product of 30. In the first case $2 \cdot 15 = 30$, in the second case $30 \cdot 1 = 30$, and in the third case $6 \cdot 5 = 30$. In each case the two middle terms consist of a pair of factors of ac; $ac = 30$ in each of these three examples.

Thus to factor $6x^2 + 13x + 5$ in Example 1, we can examine all possible pairs of factors of 30 in search of a pair of factors whose sum is 13. Note that the steps used in this example to factor the trinomial into two binomials are identical to the steps used to multiply the two binomials—but in the reverse order.

■ EXAMPLE 1   Factoring a Quadratic Trinomial

Factor $6x^2 + 13x + 5$ after first rewriting this polynomial as $6x^2 + 3x + 10x + 5$.

SOLUTION _____

Examine the factors of 30 for a pair of factors whose sum is 13.

FACTORS OF 30		SUM OF FACTORS
1	30	31
2	15	17
3	10	13
5	6	11

The only factors of 30 with a sum of 13 are 3 and 10.

$$6x^2 + 13x + 5 = 6x^2 + 3x + 10x + 5$$
$$= 3x(2x + 1) + 5(2x + 1)$$
$$= (3x + 5)(2x + 1)$$

Thus we rewrite the middle term of $13x$ as $3x + 10x$. Use the distributive property to factor out the GCF of $3x$ from the first two terms and the GCF of 5 from the last two terms.
Then use the distributive property to factor out the GCF of $2x + 1$ from both terms.

Answer: $6x^2 + 13x + 5 = (3x + 5)(2x + 1)$

Does this factorization check?   ■

The method used to work Example 1 can be generalized to apply to all trinomials of the form $ax^2 + bx + c$. This method is summarized in the following box.

Factoring Trinomials of the Form $ax^2 + bx + c$ with $a > 0$ by Grouping*

PROCEDURE

STEP 1. Be sure you have factored out the GCF if it is not 1.

STEP 2. Find a pair of factors of ac whose sum is b. If there is not a pair of factors whose sum is b, the trinomial is prime over the integers.

- If the constant c is positive, the factors of ac must have the same sign. These factors will share the same sign as the linear coefficient, b.
- If the constant c is negative, the factors of ac must be opposite in sign. The sign of b will determine the sign of the factor with the larger absolute value.

STEP 3. Rewrite the linear term of $ax^2 + bx + c$ so that b is the sum of the pair of factors from Step 2.

STEP 4. Factor the polynomial from Step 3 by grouping the terms and factoring out the GCF from each pair of terms.

EXAMPLE

Factor $6x^2 - x - 12$.

FACTORS OF -72		SUM OF FACTORS
1	-72	-71
2	-36	-34
3	-24	-21
4	-18	-14
6	-12	-6
8	-9	-1

$$6x^2 - x - 12 = 6x^2 + 8x - 9x - 12$$
$$= 2x(3x + 4) - 3(3x + 4)$$
$$= (2x - 3)(3x + 4)$$

*If $a < 0$, first factor out -1.

■ EXAMPLE 2   Factoring a Quadratic Trinomial

Factor $3x^2 + 16x + 5$.

SOLUTION _____

The product of $3 \cdot 5 = 15$. Examine the factors of 15 for a pair of factors whose sum is 16.

FACTORS OF 15		SUM OF FACTORS
1	15	16
3	5	8

From the form $ax^2 + bx + c$ this trinomial has $a = 3$, $b = 16$, and $c = 5$. Therefore, $ac = (3)(5) = 15$.

$$3x^2 + 16x + 5 = 3x^2 + x + 15x + 5$$
$$= x(3x + 1) + 5(3x + 1)$$
$$= (x + 5)(3x + 1)$$

Use this table to rewrite the linear term, $16x$, as $x + 15x$. Then complete the factorization.

Answer: $3x^2 + 16x + 5 = (x + 5)(3x + 1)$

Does this factorization check?   ■

The steps shown to factor the trinomial in Example 2 are exactly the same steps used to multiply these binomials—only the order of the steps is reversed.

Example 3 provides further practice factoring a trinomial of the form $ax^2 + bx + c$. In this example b is negative and ac is positive, so both factors of ac must be negative.

■ EXAMPLE 3   Factoring a Quadratic Trinomial

Factor $15x^2 - 41x + 14$.

SOLUTION

The product of $15 \cdot 14 = 210$. Examine the factors of 210 for a pair of factors whose sum is -41.

FACTORS OF 210		SUM OF FACTORS
-1	-210	-211
-2	-105	-107
-3	-70	-73
-5	-42	-47
-6	-35	-41
-7	-30	-37
-10	-21	-31
-14	-15	-29

The last term, 14, is positive so the factors of 210 have the same sign. Because the linear coefficient, -41, is negative, both factors of 210 must be negative.

Obviously when you determine the correct factors, you can stop your list. The whole list is shown here so that you will have it for reference.

$$15x^2 - 41x + 14 = 15x^2 - 6x - 35x + 14$$
$$= 3x(5x - 2) - 7(5x - 2)$$
$$= (3x - 7)(5x - 2)$$

Rewrite the linear term, $-41x$, as $-6x - 35x$. Then complete the factorization.

Answer: $15x^2 - 41x + 14 = (3x - 7)(5x - 2)$

Does this factorization check?   ■

SELF-CHECK 7.2.1

Factor each trinomial.

1. $5x^2 + 12x + 7$ 2. $15x^2 - 37x + 14$ 3. $15x^2 - 107x + 14$

SELF-CHECK 7.2.1 ANSWERS

1. $(5x + 7)(x + 1)$
2. $(15x - 7)(x - 2)$
3. $(15x - 2)(x - 7)$

In Example 4 the trinomial $ax^2 + bx + c$ has the product of a and c negative. Thus the factors of ac must have opposite signs.

■ EXAMPLE 4   Factoring a Quadratic Trinomial

Factor $10x^2 + x - 2$.

SOLUTION

Examine the factors of $(10)(-2) = -20$ for a pair of factors whose sum is 1.

FACTORS OF -20		SUM OF FACTORS
-1	20	19
-2	10	8
-4	5	1

The constant term, -2, is negative, so the factors of -20 are of opposite sign. Because the linear coefficient, 1, is positive, the positive factor of -20 must be larger in absolute value than the negative factor.

$$10x^2 + x - 2 = 10x^2 - 4x + 5x - 2$$
$$= 2x(5x - 2) + 1(5x - 2)$$
$$= (2x + 1)(5x - 2)$$

Rewrite the linear term, x, as $-4x + 5x$. Then complete the factorization.

Answer: $10x^2 + x - 2 = (2x + 1)(5x - 2)$

Does this factorization check?   ■

A distinct advantage of the factoring procedure demonstrated here is that it can determine whether a trinomial of the form $ax^2 + bx + c$ is prime. Once all possible factors of ac are listed, either the correct pair of factors is listed or the polynomial must be prime. In Example 5 the complete list of possible factors does not produce the correct middle term so the polynomial is prime over the integers.

■ EXAMPLE 5   A Prime Trinomial

Factor $4x^2 + 15x + 6$.

SOLUTION _____

Examine the factors of $(4)(6) = 24$ for a pair of factors whose sum is 15.

FACTORS OF 24		SUM OF FACTORS
1	24	25
2	12	14
3	8	11
4	6	10

None of the factor pairs of 24 has a sum of 15. Because we have systematically examined and eliminated all integral factor pairs of 24, we know that this polynomial is prime over the integers.

Answer: $4x^2 + 15x + 6$ is prime over the integers.   ◼

SELF-CHECK 7.2.2

Factor each trinomial.

1. $4x^2 + 9x + 5$ **2.** $4x^2 + 10x + 5$

SELF-CHECK 7.2.2 ANSWERS

1. $(4x + 5)(x + 1)$
2. $4x^2 + 10x + 5$ is prime.

In Example 6 we first factor out a factor of -1 and then complete the factorization of the polynomial. This is a customary step when the leading coefficient is negative.

■ EXAMPLE 6   Factoring a Trinomial with a Lead Coefficient of −1

Factor $-6a^2 + 23a - 21$.

SOLUTION _____

$$-6a^2 + 23a - 21 = -1(6a^2 - 23a + 21)$$

First factor out -1 so that the leading coefficient of the trinomial will be positive.

Examine the factors of $6(21) = 126$ for a pair of factors whose sum is -23.

FACTORS OF 126		SUM OF FACTORS
-1	-126	-127
-2	-63	-65
-3	-42	-45
-6	-21	-27
-7	-18	-25
-9	-14	-23

The coefficient of the last term, 21, is positive. Therefore, the factors of 126 have the same sign. Because the coefficient of the middle term, -23, is negative, both factors of 126 must be negative.

$$
\begin{aligned}
-6a^2 + 23a - 21 &= -1(6a^2 - 23a + 21) \\
&= -1(6a^2 - 9a - 14a + 21) \\
&= -1[3a(2a - 3) - 7(2a - 3)] \\
&= -1(3a - 7)(2a - 3)
\end{aligned}
$$

Rewrite the linear term, $-23a$, as $-9a - 14a$. Then complete the factorization.

Answer: $-6a^2 + 23a - 21 = -1(3a - 7)(2a - 3)$

Does this factorization check?   ◼

In Example 7 we use the method of factoring a trinomial by grouping to factor a polynomial with two variables.

■ EXAMPLE 7 Factoring a Trinomial with Two Variables

Factor $6v^2 - vw - 15w^2$.

SOLUTION

Examine the factors of $6(-15) = -90$ for a pair of factors whose sum is -1.

FACTORS OF −90		SUM OF FACTORS
1	−90	−89
2	−45	−43
3	−30	−27
5	−18	−13
6	−15	−9
9	−10	−1

The coefficient of the last term, -15, is negative so the factors of are of opposite sign. Because the coefficient of vw, -1, is negative, the negative factor of -90 must be larger in absolute value than the positive factor.

$$6v^2 - vw - 15w^2 = 6v^2 + 9vw - 10vw - 15w^2$$
$$= 3v(2v + 3w) - 5w(2v + 3w)$$
$$= (3v - 5w)(2v + 3w)$$

Rewrite $-vw$ as $9vw - 10vw$. Then complete the factorization.

Answer: $6v^2 - vw - 15w^2 = (3v - 5w)(2v + 3w)$

Does this factorization check? ■

If each term of a trinomial has a GCF other than 1, then the first step in factoring this trinomial is to factor out the GCF.

SELF-CHECK 7.2.3 ANSWERS

1. $(2a - 21b)(3a - b)$
2. $(v - 3w)(6v + 5w)$
3. $3(2a - 7b)(a - b)$
4. $7a(3w + 4z)(4w - 3z)$

SELF-CHECK 7.2.3

Completely factor each trinomial.

1. $6a^2 - 65ab + 21b^2$
2. $6v^2 - 13vw - 15w^2$
3. $6a^2 - 27ab + 21b^2$
4. $84aw^2 + 49awz - 84az^2$

■ EXAMPLE 8 Completely Factoring a Trinomial

Factor $20ax^2 + 38ax + 12a$.

SOLUTION

$$20ax^2 + 38ax + 12a = 2a(10x^2 + 19x + 6)$$

First factor out $2a$, the GCF of the trinomial. Then factor the trinomial $10x^2 + 19x + 6$.

Examine the factors of $(10)(6) = 60$ for a pair of factors whose sum is 19.

FACTORS OF 60		SUM OF FACTORS
1	60	61
2	30	32
3	20	23
4	15	19

In this table we stopped listing the factors once we determined a pair of factors whose sum is 19.

$$20ax^2 + 38ax + 12a = 2a(10x^2 + 19x + 6)$$
$$= 2a(10x^2 + 4x + 15x + 6)$$
$$= 2a[2x(5x + 2) + 3(5x + 2)]$$
$$= 2a(2x + 3)(5x + 2)$$

Rewrite the linear term, $19x$, as $4x + 15x$. Then complete the factorization.

Answer: $20ax^2 + 38ax + 12a = 2a(2x + 3)(5x + 2)$

Does this factorization check? ■

The systematic factoring procedure shown in this section always works and does not involve any guessing or intuition. With practice, you may be able to use your intuition and to mentally skip some of these steps. You also may want to use your calculator to produce a table of possible factors. This is outlined in Exercise 61 of this section.

USING THE LANGUAGE AND SYMBOLISM OF MATHEMATICS   7.2

1. The first step in factoring $ax^2 + bx + c$ is to check for a _____ other than 1 or -1.

2. When factoring the trinomial $ax^2 + bx + c$ by grouping, we use a pair of factors of ac whose sum is _____.

3. When factoring the trinomial $ax^2 + bx + c$, the pair of factors of ac are of the _____ sign when $ac > 0$.

4. When factoring the trinomial $ax^2 + bx + c$, the pair of factors of ac are of _____ sign when $ac < 0$.

EXERCISES   7.2

In Exercises 1–8 fill in the blanks to complete the factorization of each trinomial.

1. $3a^2 + 17a + 10 = 3a^2 + 15a + 2a + 10$
$= 3a(\quad) + 2(\quad)$
$= (3a + 2)(\quad)$

2. $3a^2 - 11a + 10 = 3a^2 - 6a - 5a + 10$
$= 3a(\quad) - 5(\quad)$
$= (3a - 5)(\quad)$

3. $3m^2 - 10m - 8 = 3m^2 - 12m + \underline{\quad} - 8$
$= 3m(\quad) + 2(\quad)$
$= (3m + 2)(\quad)$

4. $3m^2 + 23m - 8 = 3m^2 - m + \underline{\quad} - 8$
$= m(\quad) + 8(\quad)$
$= (m + 8)(\quad)$

5. $48am^2 - 16am - 84a = (\quad)(12m^2 - 4m - 21)$
$= (\quad)(12m^2 - 18m + \underline{\ } - 21)$
$= (\quad)[6m(\quad) + 7(\quad)]$
$= (\quad)(6m + 7)(\quad)$

6. $42am^3 + 7am^2 - 245am = (\quad)(6m^2 + m - 35)$
$= (\quad)(6m^2 - 14m + \underline{\ } - 35)$
$= (\quad)[2m(\quad) + 5(\quad)]$
$= (\quad)(2m + 5)(\quad)$

7. $15x^2 - 37xy + 20y^2 = 15x^2 - 12xy - 25xy + 20y^2$
$= 3x(\quad) - 5y(\quad)$
$= (3x - 5y)(\quad)$

8. $20x^2 + 37xy + 15y^2 = 20x^2 + 12xy + 25xy + 15y^2$
$= 4x(\quad) + 5y(\quad)$
$= (4x + 5y)(\quad)$

In Exercises 9–14 use the given table to factor each polynomial. (*Hint:* Some of these trinomials are prime.)

9. **a.** $3m^2 + 10m + 8$ **b.** $3m^2 + 11m + 8$
c. $3m^2 + 14m + 8$ **d.** $3m^2 + 25m + 8$

FACTORS OF 24		SUM OF FACTORS
1	24	25
2	12	14
3	8	11
4	6	10

10. **a.** $8v^2 - 14v + 3$ **b.** $6v^2 - 11v + 4$
c. $12v^2 - 25v + 2$ **d.** $6v^2 - 25v + 4$

FACTORS OF 24		SUM OF FACTORS
-1	-24	-25
-2	-12	-14
-3	-8	-11
-4	-6	-10

11. **a.** $4n^2 + 39n - 10$ **b.** $2n^2 + 3n - 20$
c. $5n^2 + 6n - 8$ **d.** $40n^2 + 8n - 1$

FACTORS OF -40		SUM OF FACTORS
-1	40	39
-2	20	18
-4	10	6
-5	8	3

12. **a.** $4n^2 - 3n - 10$ **b.** $4n^2 - 39n - 10$
c. $5n^2 - 18n - 8$ **d.** $5n^2 - 12n - 8$

FACTORS OF -40		SUM OF FACTORS
1	-40	-39
2	-20	-18
4	-10	-6
5	-8	-3

13. **a.** $3x^2 - 25xy + 12y^2$ **b.** $4x^2 - 15xy + 9y^2$
c. $2x^2 - 13xy + 18y^2$ **d.** $6x^2 - 37xy + 6y^2$

FACTORS OF 36		SUM OF FACTORS
-1	-36	-37
-2	-18	-20
-3	-12	-15
-4	-9	-13
-6	-6	-12

14. a. $3x^2 + 16xy - 12y^2$ **b.** $18x^2 + 9xy - 2y^2$
  c. $6x^2 + 11xy - 6y^2$ **d.** $9x^2 + 35xy - 4y^2$

FACTORS OF −36		SUM OF FACTORS
−1	36	35
−2	18	16
−3	12	9
−4	9	5
−6	6	0

In Exercises 15–18 each trinomial is of the form $ax^2 + bx + c$ and each trinomial has the same value of ac. Make a table of the factors of ac and use this table to factor each trinomial.

15. a. $3x^2 - 8x + 5$ **b.** $5x^2 - 16x + 3$
  c. $15x^2 - 16x + 1$ **d.** $3x^2 - 10x + 5$
16. a. $7v^2 + 34v - 5$ **b.** $35v^2 + 34v - 1$
  c. $5v^2 + 2v - 7$ **d.** $5v^2 + 9v - 7$
17. a. $10y^2 - 69y - 7$ **b.** $14y^2 - 69y - 5$
  c. $2y^2 - 33y - 35$ **d.** $70y^2 - 9y - 1$
18. a. $6v^2 + 25v + 25$ **b.** $150v^2 + 53v + 1$
  c. $6v^2 + 35v + 25$ **d.** $75v^2 + 77v + 2$

In Exercises 19–40 factor each polynomial.
19. $2x^2 + 15x - 7$ **20.** $11m^2 + 100m + 9$
21. $4w^2 - 9w + 5$ **22.** $9w^2 + 9w - 10$
23. $9z^2 - 9z - 10$ **24.** $3z^2 - 35z + 50$
25. $10b^2 + 29b + 10$ **26.** $15b^2 - 22b + 8$
27. $12m^2 - mn - 35n^2$ **28.** $12m^2 + 13mn - 35n^2$
29. $18y^2 + 55y - 28$ **30.** $18y^2 + 3y - 28$
31. $55y^2 - 29y - 12$ **32.** $24v^2 - 43v + 18$
33. $20x^2 + 37xy + 15y^2$ **34.** $15x^2 - 37xy + 20y^2$
35. $14x^2 - 39xy + 10y^2$ **36.** $15x^2 + 61xy + 22y^2$
37. $33m^2 + 13mn - 6n^2$ **38.** $35m^2 + 4mn - 4n^2$
39. $6v^2 + 13vw + 6w^2$ **40.** $10v^2 - 29vw + 10w^2$

In Exercises 41–46 first factor out −1 from each polynomial and then complete the factorization.
41. $-12x^2 + 20x - 3$ **42.** $-12x^2 - 20x - 3$
43. $-35y^2 + 4y + 4$ **44.** $-35y^2 + 24y - 4$

45. $-12z^2 - 47z - 40$ **46.** $-12z^2 + 17z + 40$

In Exercises 47–56 first factor out the GCF and then completely factor the polynomial.
47. $-38ax^2 + 32ax + 6a$ **48.** $-51ax^2 - 99ax + 6a$
49. $60x^5 - 145x^4 + 75x^3$ **50.** $48x^4 + 112x^3 + 60x^2$
51. $15m^3n - 6m^2n^2 - 21mn^3$
52. $44m^3n - 8m^2n^2 - 52mn^3$
53. $210av^3 + 77av^2w - 210avw^2$
54. $350av^2w + 240avw^2 - 350aw^3$
55. $35x^2(a + b) - 29x(a + b) + 6(a + b)$
56. $21x^2(a - b) + 41x(a - b) + 10(a - b)$

Group Discussion Questions
57. Discovery Question Write each of the trinomials in the form $ax^2 + 0x + c$ and then use the method of factoring a trinomial by grouping to factor each polynomial.
  a. $x^2 - 25$
  b. $x^2 - 64$
  c. $4x^2 - 1$
  d. $4x^2 - 25$
Use your observations from these problems to write the factors of these polynomials by inspection.
  e. $x^2 - 49$
  f. $9x^2 - 1$
  g. $9x^2 - 16$
Explain what you have observed to your instructor.
58. Discovery Question
  a. Use $2x^2 + 3x + 8x + 12$ to factor $2x^2 + 11x + 12$.
  b. Use $2x^2 + 8x + 3x + 12$ to factor $2x^2 + 11x + 12$ and then compare the results to part **a**.
  c. Use $-2x^2 + 3x - 4x + 6$ to factor $-2x^2 - x + 6$.
  d. Use $-2x^2 - 4x + 3x + 6$ to factor $-2x^2 - x + 6$ and then compare the results to part **c**.
  e. What conclusion can you reach from this work? Explain your conclusion to your instructor.

59. Discovery Question Complete the following table for each polynomial.

POLYNOMIAL $P(x)$	FACTORED FORM	ZEROS OF $P(x)$	x-INTERCEPTS OF THE GRAPH OF $y = P(x)$
Example $18x^2 - 27x + 10$	$(6x - 5)(3x - 2)$	$\frac{5}{6}$ and $\frac{2}{3}$	$\left(\frac{5}{6}, 0\right), \left(\frac{2}{3}, 0\right)$
a. $28x^2 + 13x - 6$		$-\frac{3}{4}$ and $\frac{2}{7}$	
b. $9x^2 + 39x + 22$			$\left(-\frac{11}{3}, 0\right), \left(-\frac{2}{3}, 0\right)$
c. $39x^2 - 19x + 2$			$\left(\frac{2}{13}, 0\right), \left(\frac{1}{3}, 0\right)$
d. $40x^2 + x - 6$		$-\frac{2}{5}$ and $\frac{3}{8}$	
e. $20x^2 + 13x - 21$	$(5x + 7)(4x - 3)$		

60. Challenge Question Can you write a trinomial $ax^2 + bx + c$ with a, b, and c all odd, that is not prime? If you can, give your example. If you cannot, explain why you found this task impossible.

61. Calculator Discovery To factor $18x^2 + 39x + 20$, we need to find a pair of factors of $18 \cdot 20 = 360$ whose sum is 39. Let x and y_1 represent this pair of factors. Thus $xy_1 = 360$ or $y_1 = \dfrac{360}{x}$. Let $y_2 = x + y_1$ and use a table to find where y_2 is 39.

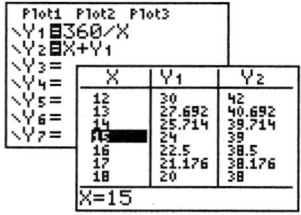

We see that $y_2 = 39$ when $x = 15$ and $y_1 = 24$. So

$$18x^2 + 39x + 20 = 18x^2 + 15x + 24x + 20$$
$$= 3x(6x + 5) + 4(6x + 5)$$
$$= (3x + 4)(6x + 5)$$

Try this method to factor $12x^2 - 53x + 56$.

Section 7.3 Factoring Special Forms

Objectives:

5. Factor perfect square trinomials.
6. Factor the difference of two squares.
7. Factor the sum or difference of two cubes.

The connection of factoring to other topics in mathematics is one of the primary reasons we study factoring. The special forms examined in this section occur in many different contexts throughout mathematics.

Certain forms of polynomials occur so frequently that you can save considerable time and effort by memorizing these forms. You can continue to use the grouping method to factor quadratic trinomials, but you can gain insight and efficiency by recognizing these patterns.

Perfect Square Trinomials

We start by reexamining the special forms that were first introduced in Section 5.7.

A Mathematical Note

Pierre de Fermat's last theorem is one of the most famous theorems in mathematics. It states that there do not exist positive integers x, y, z, and n such that $x^n + y^n = z^n$ for $n > 2$. Two of the features that make this theorem so interesting are (1) it is easily understood—it even resembles the Pythagorean theorem $x^2 + y^2 = z^2$ and thus shows the Pythagorean theorem does not generalize to a higher-degree equation, (2) yet it resisted the best attempts of mathematicians to prove until 1993 when Andrew Wiles, an American mathematician, proved this result.

Perfect Square Trinomials		
ALGEBRAICALLY	**VERBALLY**	**ALGEBRAIC EXAMPLE**
Square of a Sum $x^2 + 2xy + y^2 = (x + y)^2$ **Square of a Difference** $x^2 - 2xy + y^2 = (x - y)^2$	A trinomial that is the square of a binomial has: 1. a first term that is a square of the first term of the binomial. 2. a middle term that is twice the product of the two terms of the binomial. 3. a last term that is a square of the last term of the binomial.	$x^2 + 12x + 36 = (x + 6)^2$ $x^2 - 12x + 36 = (x - 6)^2$

■ EXAMPLE 1 Factoring Perfect Square Trinomials

Determine which of these polynomials are perfect square trinomials and factor those that fit this form.

SOLUTIONS _____

(a) $25v^2 + 10v + 1$

$$\begin{aligned}
25v^2 + 10v + 1 &= (5v)^2 + 10v + (1)^2 \\
&= (5v)^2 + 2(5v)(1) + (1)^2 \\
&= (5v + 1)^2
\end{aligned}$$

Write the first and last terms as perfect squares. Then check to see if the middle term fits the form $x^2 + 2xy + y^2$. Use the form $x^2 + 2xy + y^2 = (x + y)^2$ to write the factored form.

(b) $m^2 - 12mn + 36n^2$

$$\begin{aligned}
m^2 - 12mn + 36n^2 &= m^2 - 12mn + (6n)^2 \\
&= m^2 - 2(m)(6n) + (6n)^2 \\
&= (m - 6n)^2
\end{aligned}$$

This trinomial fits the form $x^2 - 2xy + y^2 = (x - y)^2$ with $x = m$ and $y = 6n$.

(c) $x^2 - 10x - 25$

This trinomial is not a perfect square.

Because the last term has a negative coefficient it cannot be a perfect square of an integer.

(d) $x^2 + 8x + 10$

This trinomial is not a perfect square.

The last term, 10, is not a perfect square of an integer. ■

SELF-CHECK 7.3.1

Factor each of these trinomials.
1. $w^2 - 18w + 81$
2. $121w^2 + 22w + 1$
3. $25v^2 + 60vw + 36w^2$

Difference of Two Squares

Another special form that is encountered frequently in subsequent mathematics courses is the difference of two squares. This form also was examined in Section 5.7.

Difference of Two Squares

ALGEBRAICALLY	VERBALLY	ALGEBRAIC EXAMPLE
$x^2 - y^2 = (x + y)(x - y)$	The difference of the squares of two terms factors as the sum of these terms times their difference.	$x^2 - 49 = (x + 7)(x - 7)$

SELF-CHECK 7.3.1 ANSWERS
1. $(w - 9)^2$
2. $(11w + 1)^2$
3. $(5v + 6w)^2$

■ **EXAMPLE 2** Factoring the Difference of Two Squares

Determine which of these polynomials are the difference of two squares and factor those that fit this form.

SOLUTIONS _____

(a) $100a^2 - 9$

$$100a^2 - 9 = (10a)^2 - (3)^2$$
$$= (10a + 3)(10a - 3)$$

First write this binomial in the form $x^2 - y^2$ and then factor as a sum times a difference.

(b) $9v^2 - 16w^2$

$$9v^2 - 16w^2 = (3v)^2 - (4w)^2$$
$$= (3v + 4w)(3v - 4w)$$

(c) $9v^2 + w^2$

This binomial is not the difference of two squares.

This binomial is the sum of two perfect squares, not their difference.

(d) $x^2 - 5$

This binomial is not the difference of two squares.

5 is not a perfect square integer. However, we could factor this using irrational factors as $x^2 - 5 = x^2 - (\sqrt{5})^2 = (x + \sqrt{5})(x - \sqrt{5})$.

■

SELF-CHECK 7.3.2

Factor each of these binomials.

1. $36y^2 - 25$
2. $100a^2 - 121b^2$
3. $16a^2b^2 - 1$

Sum of Two Squares

The sum of two squares is another special form that can be advantageous to recognize. Consider the binomial $x^2 + y^2$ in the trinomial form $x^2 + 0xy + y^2$. An attempt to factor this by the grouping method cannot produce a middle term of 0. The factors of ac must be of the same sign. Thus their sum cannot be 0. Therefore, the sum of two squares is prime over the integers.

Sum of Two Squares		
ALGEBRAICALLY	**VERBALLY**	**ALGEBRAIC EXAMPLE**
$x^2 + y^2$ is prime.*	The sum of two squares is prime.	$16x^2 + 9$ is prime.

* For binomials of degree greater than 2 this statement is not always true.

■ **EXAMPLE 3**   Factoring Special Forms

Factor each of these polynomials.

SOLUTIONS

Each one of these polynomials could be factored by the method used in the previous section. (We can consider $4a^2 - 25b^2$ as $4a^2 + 0ab - 25b^2$.) Note how much more efficient it is to use the special forms to quickly select the correct factors.

(a) $4a^2 - 20ab + 25b^2$

$\begin{aligned} 4a^2 &- 20ab + 25b^2 \\ &= (2a)^2 - 2(2a)(5b) + (5b)^2 \\ &= (2a - 5b)^2 \end{aligned}$

First write this trinomial in the form $x^2 - 2xy + y^2$ and then factor this perfect square trinomial.

(b) $4a^2 - 25b^2$

$\begin{aligned} 4a^2 &- 25b^2 \\ &= (2a)^2 - (5b)^2 \\ &= (2a + 5b)(2a - 5b) \end{aligned}$

First write this binomial in the form $x^2 - y^2$ and then factor this difference of two squares.

(c) $4a^2 + 25b^2$

$4a^2 + 25b^2$ is prime over the integers.

This binomial can be written as $(2a)^2 + (5b)^2$. As the sum of two squares it must be prime over the integers. ■

SELF-CHECK 7.3.3

Factor each of these polynomials.

1. $169v^2 - 100w^2$ 2. $169v^2 + 100w^2$ 3. $169v^2 + 260vw + 100w^2$

Sum or Difference of Two Cubes

We now examine the special forms $x^3 + y^3$ and $x^3 - y^3$, which are known as the sum and the difference of two cubes. We start by examining $x^3 - 8$. Since an x-intercept of the graph of $y = x^3 - 8$ corresponds to a factor of $x^3 - 8$, we begin by examining this graph.

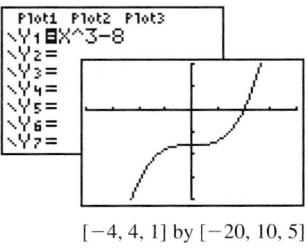

$[-4, 4, 1]$ by $[-20, 10, 5]$

From this graph we note that $(2, 0)$ is an x-intercept of the graph of this third-degree polynomial function. Thus $x - 2$ is a factor of $x^3 - 8$. We now use long division to obtain the other factor of $x^3 - 8$. (Long division was covered in Section 5.6.)

$$\begin{array}{r} x^2 + 2x + 4 \\ x - 2 \overline{)\,x^3 + 0x^2 + 0x - 8} \\ \underline{x^3 - 2x^2} \\ 2x^2 + 0x \\ \underline{2x^2 - 4x} \\ 4x - 8 \\ \underline{4x - 8} \\ 0 \end{array}$$

Use 0 as the coefficient of the missing x^2 and x terms. Then complete the long division.

Thus $x^3 - 8 = (x - 2)(x^2 + 2x + 4)$. We now try to factor $x^2 + 2x + 4$.

Examine the factors of 4 for a pair of factors whose sum is 2.

FACTORS OF 4		SUM OF FACTORS
1	4	5
2	2	4

None of the factor pairs of 4 have a sum of 2. Therefore $x^2 + 2x + 4$ is prime over the integers. The complete factorization of $x^3 - 8$ is

$$x^3 - 8 = (x - 2)(x^2 + 2x + 4).$$

Next we list some other products and factors that fit the form of the sum or difference of two cubes. You may wish to multiply these factors to verify these products.

PRODUCTS	FACTORS
$(x + 1)(x^2 - x + 1) = x^3 + 1$	$x^3 + 1 = (x + 1)(x^2 - x + 1)$
$(x - 1)(x^2 + x + 1) = x^3 - 1$	$x^3 - 1 = (x - 1)(x^2 + x + 1)$
$(x + 2)(x^2 - 2x + 4) = x^3 + 8$	$x^3 + 8 = (x + 2)(x^2 - 2x + 4)$
$(x - 2)(x^2 + 2x + 4) = x^3 - 8$	$x^3 - 8 = (x - 2)(x^2 + 2x + 4)$

This pattern exhibited by these polynomials is summarized in the following box.

Sum of Difference of Two Cubes

ALGEBRAICALLY	VERBALLY	ALGEBRAIC EXAMPLE
Sum of Two Cubes $x^3 + y^3 = (x + y)(x^2 - xy + y^2)$ **Difference of Two Cubes** $x^3 - y^3 = (x - y)(x^2 + xy + y^2)$	Write the binomial factor as the sum (or difference) of the two cube roots. Then use the binomial factor to obtain each term of the trinomial: 1. The square of the first term of the binomial is the first term of the trinomial factor. 2. The opposite of the product of the two terms of the binomial is the second term of the trinomial factor. 3. The square of the last term of the binomial is the third term of the trinomial factor.	$x^3 + 8 = (x + 2)(x^2 - 2x + 4)$ $x^3 - 8 = (x - 2)(x^2 + 2x + 4)$

It is helpful to be able to recognize the perfect squares 1, 4, 9, 16, 25, and so on, when you are examining a binomial to determine whether it is a sum or a difference of two squares. Likewise, it is helpful to be able to recognize the perfect cubes 1, 8, 27, 64, 125, and so on, when you are examining a binomial to determine whether it is a sum or a difference of two cubes.

■ **EXAMPLE 4**   Identifying the Sum or Difference of Two Cubes

Examine each binomial to determine whether it is a sum or a difference of two cubes or neither.

SOLUTIONS

(a) $x^3 + 64$ — This binomial is the sum of two cubes. — In the form $x^3 + y^3$, $x^3 + 64 = x^3 + 4^3$.

(b) $x^3 + 16$ — This binomial is not the sum of two cubes. — The last term, 16, is not a perfect cube.

(c) $27x^3 - 1$ — This binomial is the difference of two cubes. — $27x^3 - 1 = (3x)^3 - 1^3$

(d) $8x^3 - 125$ — This binomial is the difference of two cubes. — $8x^3 - 125 = (2x)^3 - 5^3$

■

■ **EXAMPLE 5**   Factoring the Sum of Two Cubes

Factor $x^3 + 64$.

SOLUTION

$$x^3 + 64 = x^3 + 4^3$$
$$= (x + 4)[(\ \ )^2 - (\ \ )(\ \ ) + (\ \ )^2]$$
$$= (x + 4)(x^2 - (x)(4) + 4^2)$$
$$= (x + 4)(x^2 - 4x + 16)$$

First express this binomial in the form $x^3 + y^3$. Write the binomial factor and then use this factor and the form $x^3 + y^3 = (x + y)(x^2 - xy + y^2)$ to complete the factorization.
The trinomial $x^2 - xy + y^2$ is prime over the integers.

Square of the first term, x.
Opposite of the product of the two terms.
Square of the last term, 4.

■

■ **EXAMPLE 6**   Factoring the Difference of Two Cubes

Factor $27x^3 - 1$.

SOLUTION

$$27x^3 - 1 = (3x)^3 - (1)^3$$
$$= (3x - 1)[(\ \ )^2 + (\ \ )(\ \ ) + (\ \ )^2]$$
$$= (3x - 1)[(3x)^2 + (3x)(1) + (1)^2]$$
$$= (3x - 1)(9x^2 + 3x + 1)$$

First express this binomial in the form $x^3 - y^3$. Write the binomial factor and then use this factor and the form $x^3 - y^3 = (x - y)(x^2 + xy + y^2)$ to complete the factorization.
The trinomial $9x^2 + 3x + 1$ is prime over the integers.

Square of the first term, $3x$.
Opposite of the product of the two terms.
Square of the last term, 1.

■

The next example contains two variables and fits the form of the difference of two cubes.

■ EXAMPLE 7 Factoring the Difference of Two Cubes

Factor $125a^3 - 64b^3$.

SOLUTION _____

$$125a^3 - 64b^3 = (5a)^3 - (4b)^3$$
$$= (5a - 4b)[(\quad)^2 + (\quad)(\quad) + (\quad)^2]$$
$$= (5a - 4b)[(5a)^2 + (5a)(4b) + (4b)^2]$$
$$= (5a - 4b)(25a^2 + 20ab + 16b^2)$$

First express this binomial in the form $x^3 - y^3$. Write the binomial factor and then use this factor and the form $x^3 - y^3 = (x - y)(x^2 + xy + y^2)$ to complete the factorization.

Square of the first term, $5a$.
Opposite of the product of the two terms.
Square of the last term, $4b$.

SELF-CHECK 7.3.4

Factor these binomials.

1. $x^3 + 125$ **2.** $64x^3 - 1$ **3.** $8x^3 - 125$ **4.** Factor $1000x^3 + 27y^3$.

Some polynomials with a degree greater than 2 also can be factored as special forms. When factoring polynomials of a higher degree, be sure to continue your work until the polynomial is factored completely. A polynomial is factored completely if each factor other than a constant factor is prime over the integers.

■ EXAMPLE 8 Factoring a Higher-Degree Binomial

Factor $16y^4 - 1$.

SELF-CHECK 7.3.4 ANSWERS
1. $(x + 5)(x^2 - 5x + 25)$
2. $(4x - 1)(16x^2 + 4x + 1)$
3. $(2x - 5)(4x^2 + 10x + 25)$
4. $(10x + 3y) \cdot$
$(100x^2 - 30xy + 9y^2)$

SOLUTION _____

$$16y^4 - 1 = (4y^2)^2 - 1^2$$
$$= (4y^2 + 1)(4y^2 - 1)$$
$$= (4y^2 + 1)[(2y)^2 - 1^2]$$
$$= (4y^2 + 1)(2y + 1)(2y - 1)$$

First write this binomial in the form $x^2 - y^2$ and then factor as a sum times a difference.
$4y^2 + 1$ is the sum of two squares and is prime.
$4y^2 - 1$ is the difference of two squares and factors as a sum times a difference.

USING THE LANGUAGE AND SYMBOLISM OF MATHEMATICS   7.3

1. The special form $x^2 + 2xy + y^2$, which factors as $(x + y)^2$, is called a _____ _____ _____ .
2. The special form $x^2 - y^2$, which factors as $(x + y)(x - y)$, is called a _____ _____ _____ _____ .
3. The special form $x^2 + y^2$, which is prime, is called a _____ _____ _____ _____ .
4. The special form $x^3 - y^3$, which factors as $(x - y)(x^2 + xy + y^2)$, is called a _____ _____ _____ _____ .

EXERCISES   7.3

In Exercises 1–4 fill in the blanks to complete the factorization of each perfect square trinomial.

1. $a^2 + 6a + 9 = a^2 + 2(\quad)(a) + (\quad)^2$
   $= (a + \quad)^2$
2. $a^2 - 10a + 25 = a^2 - 2(\quad)(a) + (\quad)^2$
   $= (a - \quad)^2$
3. $4w^2 - 12w + 9 = (\quad)^2 - 2(\quad)(3) + 3^2$
   $= (\quad - 3)^2$
4. $9w^2 + 30w + 25 = (\quad)^2 + 2(\quad)(5) + 5^2$
   $= (\quad + 5)^2$

In Exercises 5–18 completely factor each trinomial, using the forms for perfect square trinomials. If necessary, first factor out the GCF.

5. $m^2 - 2m + 1$
6. $m^2 + 2m + 1$
7. $36v^2 + 12v + 1$
8. $100v^2 - 20v + 1$
9. $25x^2 - 20xy + 4y^2$
10. $81x^2 + 180xy + 100y^2$
11. $121m^2 + 88mn + 16n^2$
12. $169m^2 - 78mn + 9n^2$
13. $-x^2 + 16xy - 64y^2$
14. $-25x^2 + 60xy - 36y^2$
15. $9 + 16x^2 - 24x$
16. $121 + x^2 - 22x$
17. $2x^3 + 16x^2 + 32x$
18. $3x^3 + 42x^2 + 147x$

In Exercises 19–32 completely factor each binomial, using the form for the difference of two squares. If necessary, first factor out the GCF.

19. $w^2 - 49$
20. $w^2 - 36$
21. $9v^2 - 1$
22. $16v^2 - 1$
23. $81m^2 - 25$
24. $36m^2 - 49$
25. $4a^2 - 9b^2$
26. $100a^2 - 49b^2$
27. $16v^2 - 121w^2$
28. $25v^2 - 144w^2$
29. $36 - m^2$
30. $81 - m^2$
31. $20x^2 - 45y^2$
32. $75x^2 - 48y^2$

In Exercises 33–44 some of the polynomials are factorable over the integers and some are prime over the integers. Factor those that are factorable and label the others as prime. If necessary, first factor out the GCF.

33. $25x^2 - 64y^2$
34. $16x^2 - 49y^2$
35. $25x^2 + 64y^2$
36. $16x^2 + 49y^2$
37. $16m^2 + 64n^2$
38. $81m^2 + 9n^2$
39. $100x^2 - 81y^2$
40. $100x^2 + 180xy + 81y^2$
41. $100x^2 - 180xy + 81y^2$
42. $100x^2 + 81y^2$
43. $100av^2 - 36aw^2$
44. $81bv^2 - 225bw^2$

In Exercises 45–52 completely factor each binomial, using the form for the sum or difference of two cubes.

45. $x^3 + 27$
46. $x^3 - 27$
47. $m^3 - 125$
48. $m^3 + 125$
49. $64a^3 - b^3$
50. $64a^3 + b^3$
51. $125x^3 + 8y^3$
52. $64a^3 - 125b^3$

In Exercises 53–74 completely factor each polynomial.

53. $x_1^2 - x_2^2$ (the area between two squares)

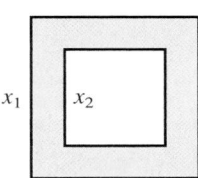

54. $\pi r_1^2 - \pi r_2^2$ (the area between two concentric circles)

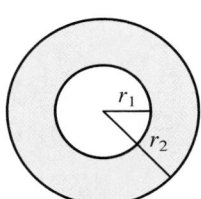

55. $\pi r_1^2 h - \pi r_2^2 h$ (the volume of steel in a steel pipe)

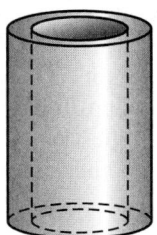

56. $4x^2 + 12x + 9$ (the area of a square)

57. $x^3 + 1000$ (total volume of two boxes)
58. $-16t^2 + 96t - 144$ (the vertical distance of a projectile)
59. $x^4 - 9$ **60.** $x^6 - 36$
61. $25y^4 - 1$ **62.** $x^4 - 16$
63. $81y^4 - 1$ **64.** $m^2n^2 - 25$
65. $(a + 2b)^2 - (2a + b)^2$
66. $(a - 5b)x^2 - (a - 5b)y^2$
67. $(a + b)x^2 + 6(a + b)x + 9(a + b)$
68. $(a - b)x^2 - 10(a - b)x + 25(a - b)$
69. $(x^2 - y^2)a^2 - (x^2 - y^2)b^2$
70. $(x^2 + 20xy + 100y^2)a^2 - 16(x^2 + 20xy + 100y^2)$

71. $z^3 - 1000$ **72.** $8ax^3 - 216ay^3$
73. $343v^3 + w^3$ **74.** $32a^3 + 4$

Calculator Usage

In Exercises 75–78 use your calculator to factor out the GCF and then complete each factorization.

75. $7.46x^2 - 89.52x + 268.56 = 7.46(\quad) = 7.46(\quad)^2$
76. $5.83x^2 + 34.98xy + 52.47y^2 = 5.83(\quad) = 5.83(\quad)^2$
77. $3.64v^2 - 91 = 3.64(\quad) = 3.64(\quad)(\quad)$
78. $-9.21a^2 + 147.36 = -9.21(\quad) = -9.21(\quad)(\quad)$

Group Discussion Questions

79. Discovery Question Complete the following table.

POLYNOMIAL $P(x)$	FACTORED FORM OF $P(x)$	ZEROS OF $P(x)$	x-INTERCEPTS OF THE GRAPH OF $y = P(x)$	SOLUTIONS OF $P(x) = 0$
a. $x^2 - 4x + 3$	$(x - 1)(x - 3)$	1 and 3	$(1, 0)$ and $(3, 0)$	$x = 1$ and $x = 3$
b. $x^2 - 4x - 5$				
c. $2x^2 - 7x - 15$				
d. $4x^2 - 4x - 3$				

What conclusion can you reach from this work regarding the solutions of $P(x) = 0$? Explain your conclusion to your instructor.

80. Challenge Question The binomial $a^6 + b^6$ can be written as the sum of two squares: $(a^3)^2 + (b^3)^2$. Yet $a^6 + b^6$ is not prime. Factor $a^6 + b^6$.

Section 7.4 Factoring by Grouping and a General Strategy for Factoring Polynomials

Objectives:

8. Factor polynomials by the method of grouping.
9. Determine the most appropriate technique for factoring a polynomial.

The factoring techniques developed in the previous sections are used primarily with binomials and trinomials. For polynomials with four or more terms a technique often used is called **factoring by grouping.** Instead of examining the entire polynomial at one time we first group together terms that we already know how to factor. Thus factoring by grouping relies heavily on the ability to spot common factors and special forms. This grouping method is an extension of the method used to factor trinomials.

■ **EXAMPLE 1** **Factoring by Grouping**

Factor $3ac - bc + 6a - 2b$.

SOLUTION ⎯⎯⎯⎯⎯⎯

$$3ac - bc + 6a - 2b = (3ac - bc) + (6a - 2b)$$
$$= c(3a - b) + 2(3a - b)$$
$$= (c + 2)(3a - b)$$

There is no common factor other than 1 for all four terms. Thus we group together the first two terms, which have a GCF of c. Also group together the last two terms, which have a GCF of 2. Then factor out $3a - b$ from both terms. ■

To determine an advantageous grouping for a polynomial, examine the polynomial for groups that are special forms. This is illustrated in Example 2 where the first two terms form the difference of two squares.

■ EXAMPLE 2   Factoring by Grouping

Factor $x^2 - y^2 + x - y$.

SOLUTION _____

$$x^2 - y^2 + x - y = (x^2 - y^2) + (x - y)$$
$$= (x - y)(x + y) + (x - y)(1)$$
$$= (x - y)(x + y + 1)$$

Note that the first two terms form the difference of two squares. Group them together, and factor $x^2 - y^2$. Then factor out $x - y$ from both terms.

Caution: This term of 1 must be included to account for the term $(x - y)(1)$ in the previous step.

The terms of a polynomial can be grouped several ways. It is possible that one grouping will lead to a factorization, whereas other groupings will prove useless. For example, it is tempting to group $z^2 - a^2 + 4ab - 4b^2$ as $(z^2 - a^2) + (4ab - 4b^2)$, which would result in $(z + a)(z - a) + 4b(a - b)$. However, this grouping, which has a GCF of 1, is not useful. A more useful grouping of this polynomial is shown in the next example.

■ EXAMPLE 3   Factoring by Grouping

To select an appropriate grouping start by looking for special forms or other groups that you already know how to factor.

Factor $z^2 - a^2 + 4ab - 4b^2$.

SOLUTION _____

$$z^2 - a^2 + 4ab - 4b^2 = z^2 - (a^2 - 4ab + 4b^2)$$
$$= z^2 - (a - 2b)^2$$
$$= [z + (a - 2b)][z - (a - 2b)]$$
$$= (z + a - 2b)(z - a + 2b)$$

Group the last three terms together, noting the sign change of each term within the parentheses. Then factor the perfect square trinomial within the parentheses. This polynomial fits the form of the difference of two squares. Factor this special form as a sum times a difference and then simplify.

Sometimes it is advantageous to reorder the terms of a polynomial to facilitate a useful grouping. This is illustrated in Example 4.

■ EXAMPLE 4 Factoring by Grouping

Factor $y^3 + x + x^3 + y$.

SOLUTION

$$
\begin{aligned}
y^3 + x + x^3 + y &= (x^3 + y^3) + (x + y) \\
&= (x + y)(x^2 - xy + y^2) + (x + y)(1) \\
&= (x + y)[(x^2 - xy + y^2) + 1] \\
&= (x + y)(x^2 - xy + y^2 + 1)
\end{aligned}
$$

Reorder the terms and group together the sum of the two perfect cubes. Then factor this special form. Then factor out the GCF, $x + y$. Be sure to include the last term of 1. ■

SELF-CHECK 7.4.1

Factor each polynomial.

1. $rt - st + rv - sv$ **2.** $x^2 - 10x + 25 - 36y^2$

It is important to practice factoring a variety of polynomials so that you can quickly select the appropriate technique. The exercises at the end of this section contain a mix of problems intended to promote the development of a general factoring strategy. The following box outlines the methods of factoring covered earlier in this chapter.

A key step in determining how to factor a polynomial is noting the number of terms in the polynomial.

Strategy for Factoring a Polynomial over the Integers

After factoring out the GCF, proceed as follows:

Binomials: Factor special forms:

$x^2 - y^2 = (x + y)(x - y)$	Difference of two squares
$x^3 - y^3 = (x - y)(x^2 + xy + y^2)$	Difference of two cubes
$x^3 + y^3 = (x + y)(x^2 - xy + y^2)$	Sum of two cubes
$x^2 + y^2$ is prime	The sum of two squares is prime if x^2 and y^2 are only second-degree terms and have no common factor other than 1.

Trinomials: Factor the forms that are perfect squares:

$x^2 + 2xy + y^2 = (x + y)^2$	Perfect square trinomial; A square of a sum
$x^2 - 2xy + y^2 = (x - y)^2$	Perfect square trinomial; A square of a difference

Otherwise, factor the trinomial by the grouping method.

Polynomials of Four or More Terms: Factor by grouping

■ **EXAMPLE 5**   Completely Factoring a Binomial

Factor $5x^3y - 5xy^3$.

SOLUTION _____

$$
\begin{aligned}
5x^3y - 5xy^3 &= 5xy(x^2 - y^2) && \text{Factor out the GCF, } 5xy. \\
&= 5xy(x - y)(x + y) && \text{Noting that } x^2 - y^2 \text{ is the difference of two perfect squares,} \\
& && \text{factor this special form.}
\end{aligned}
$$

■ **EXAMPLE 6**   Completely Factoring a Binomial

Factor $6a^4b^2 - 6ab^5$.

SOLUTION _____

$$
\begin{aligned}
6a^4b^2 - 6ab^5 &= 6ab^2(a^3 - b^3) && \text{Factor out the GCF, } 6ab^2. \\
&= 6ab^2(a - b)(a^2 + ab + b^2) && \text{Next factor the difference of two perfect cubes.}
\end{aligned}
$$

SELF-CHECK 7.4.2

Completely factor each polynomial.
1. $12ax^3 - 75axy^2$ 　　　　　　　　　　2. $11x^4y + 88xy^4$

The next example contains four terms. After factoring out the GCF, use the grouping method to factor the polynomial.

■ **EXAMPLE 7**   Completely Factoring a Polynomial

Factor $3x^3y^2 - 3xy^4 - 3x^2y^2 + 3xy^3$.

SOLUTION _____

$$
\begin{aligned}
& 3x^3y^2 - 3xy^4 - 3x^2y^2 + 3xy^3 \\
&= 3xy^2(x^2 - y^2 - x + y) && \text{Factor out the GCF, } 3xy^2. \\
&= 3xy^2[(x^2 - y^2) - 1(x - y)] && \text{Group together the first two terms that fit the form of a} \\
& && \text{difference of two squares. Also factor out } -1 \text{ from the last two} \\
& && \text{terms } -x + y \text{ to rewrite them as } -1(x - y). \\
&= 3xy^2[(x - y)(x + y) - 1(x - y)] && \text{Factor the difference of two squares as a sum times a} \\
&= 3xy^2(x - y)(x + y - 1) && \text{difference. Then factor out the common factor, } x - y. \text{ Be sure} \\
& && \text{to include the last term of } -1 \text{ in the last factor.}
\end{aligned}
$$

The substitution shown in Example 8 is not necessary and may be skipped. However, the substitution is shown to emphasize that this polynomial can be considered a trinomial of the form $a^2 - 8ab + 15b^2$. In order to apply the special forms and the methods of factoring presented here, it is important to recognize the variety of expressions that fit these forms.

■ EXAMPLE 8   Using Substitution to Factor a Polynomial

Factor $(x + 3y)^2 - 8b(x + 3y) + 15b^2$.

SOLUTION

$(x + 3y)^2 - 8b(x + 3y) + 15b^2$ is of the form $a^2 - 8ab + 15b^2$.

Substitute a for the quantity $x + 3y$.

To factor $a^2 - 8ab + 15b^2$, examine the factors of 15 for a pair of factors whose sum is -8.

FACTORS OF 15		SUM OF FACTORS
-1	-15	-16
-3	-5	-8

Then factor $a^2 - 8ab + 15b^2$ by the grouping method.

$$a^2 - 8ab + 15b^2 = a^2 - 3ab - 5ab + 15b^2$$
$$= a(a - 3b) - 5b(a - 3b)$$
$$= (a - 5b)(a - 3b)$$

$$(x + 3y)^2 - 8b(x + 3y) + 15b^2 = [(x + 3y) - 5b][(x + 3y) - 3b]$$
$$= (x + 3y - 5b)(x + 3y - 3b)$$

Substitute $x + 3y$ back for a and then simplify each factor.

Answer: $(x + 3y)^2 - 8b(x + 3y) + 15b^2 = (x + 3y - 5b)(x + 3y - 3b)$ ■

SELF-CHECK 7.4.3

Completely factor each polynomial.
1. $7m^3 - 35m^2 - 42m$ 2. $(a + 3b)^2 - 6(a + 3b) + 8$

A polynomial is prime over the integers if its only factorization must involve 1 or -1 as one of the factors. You should continue all factorizations until each factor, other than a monomial factor, is prime. The polynomials presented in the examples and exercises in this section either are prime or can be factored by the strategy given in this section. Other methods of factoring are usually considered in college algebra courses.

SELF-CHECK 7.4.3 ANSWERS
1. $7m(m + 1)(m - 6)$
2. $(a + 3b - 2)(a + 3b - 4)$

USING THE LANGUAGE AND SYMBOLISM OF MATHEMATICS   7.4

1. The method often used to factor polynomials with four or more terms is called factoring by _____.
2. GCF stands for _____ _____ _____.

3. Match each polynomial with its factored form.
   a. $x^2 + 2xy + y^2$, a perfect square trinomial
   b. $x^2 - 2xy + y^2$, a perfect square trinomial
   c. $x^2 - y^2$, a difference of two squares
   d. $x^3 + y^3$, a sum of two cubes
   e. $x^3 - y^3$, a difference of two cubes

   A. $(x + y)(x - y)$
   B. $(x - y)(x^2 + xy + y^2)$
   C. $(x + y)(x^2 - xy + y^2)$
   D. $(x + y)^2$
   E. $(x - y)^2$

EXERCISES 7.4

In Exercises 1–4 fill in the blanks to complete the factorization of each polynomial by the method of grouping.

1. $x^2 - xy + 5x - 5y = x(\quad) + 5(\quad)$
$= (x + 5)(\quad)$

2. $xy + xz - 2y - 2z = x(\quad) - 2(\quad)$
$= (x - 2)(\quad)$

3. $x^2 - y^2 + 2y - 1 = x^2 - (\quad)$
$= x^2 - (\quad)^2$
$= [x + (\quad)][x - (\quad)]$
$= (x + y - 1)(\quad)$

4. $x^2 - 5xy - 6y^2 + x - 6y = (x^2 - 5xy - 6y^2) + (x - 6y)$
$= (x - 6y)(\quad) + (x - 6y)(1)$
$= (x - 6y)(\quad)$

In Exercises 5–26 factor each polynomial completely using the technique of grouping.

5. $ac + bc + ad + bd$
6. $xy + xz + 2y + 2z$
7. $3a - 6b + 5ac - 10bc$
8. $6a^2 + 3ab + 2a + b$
9. $ab + bc - ad - cd$
10. $ac + bc + a + b$
11. $v^2 - vw - 7v + 7w$
12. $mn - 7m - n + 7$
13. $4a^2 + 12a + 9 - 16b^2$
14. $16x^2 - a^2 - 2a - 1$
15. $3mn + 15m - kn - 5k$
16. $az^2 + bz^2 + aw^2 + bw^2$
17. $x^3 - y^3 - x + y$
18. $x^3 + y^3 + x^2 - y^2$
19. $a^2 + 2a + 1 + ab + b$
20. $x^3 + y^3 - 3x - 3y$
21. $az^3 + bz^3 + aw^2 + bw^2$
22. $s^3 + 11s^2 + s + 11$
23. $9b^2 - 24b + 16 - a^2$
24. $ax - ay - az + bx - by - bz$
25. $ay^2 + 2ay - y + a - 1$
26. $x^2 - 14x + 49 - 16y^2$

In Exercises 27–74 completely factor each polynomial using the strategy outlined in this section.

27. $64y^2 - 9z^2$
28. $25a^2 - 144b^2$
29. $16x^2 + 49y^2$
30. $3ax^2 + 33ax + 72a$
31. $12x^2 - 27x + 15$
32. $121m^2 + 49n^2$
33. $49a^2 - 28a + 4$
34. $25a^2 - 10a + 1$
35. $x(a - b) + y(a - b)$
36. $a(x - y) - b(x - y)$
37. $10w^2 - 6w - 21$
38. $3s^2 + 3s + 3t - 3t^2$
39. $25v^2 - vw + 36w^2$
40. $bw^3 - b$
41. $4x^{10} + 12x^5y^3 + 9y^6$
42. $9s^2 - 63$
43. $12x^3y - 12xy^3$
44. $25y^2 - 30yz + 9z^2$
45. $cx + cy + dx + dy$
46. $5a^2bc - 5b^3c$
47. $ax^2 + ax + bxy + by$
48. $20x^3y - 245xy^3$
49. $x^6 + 4x^3y + 4y^2$
50. $3ax^2 + 3ay^2$
51. $4bx^3 - 32b$
52. $100s^4 + 120s^3t + 36s^2t^2$
53. $5x^2 - 55$
54. $(4x^2 - 12xy + 9y^2) + (72ay - 48ax) - 25a^2$

55. $(25x^2 - 10xy + y^2) + (10xz - 2yz) - 24z^2$
56. $63a^3b - 175ab$
57. $49b^2 + 126bc + 81c^2$
58. $5ab^3 - 5a$
59. $-71ax^4 + 71a$
60. $-12x^2 + 12xy - 3y^2$
61. $8x^6 - y^3$
62. $x^3 - y^3 + x - y$
63. $x^2 + 2xy + y^2 - 16z^2$
64. $9x^2 - 6x + 1 - 25y^2$
65. $2x^2 + 2x + 2y - 2y^2$
66. $12ax^2 - 10axy - 12ay^2$
67. $8ax^2 - 648ay^4$
68. $7s^5t - 7st^5$
69. $-6ax^3 + 24ax$
70. $3ax^2 - 3ay^2 + 6ay - 3a$
71. $x^3 + 4x^2y + 4xy^2$
72. $200x^2 + 2$
73. $(a - 2b)x^2 - 2(a - 2b)x - 24(a - 2b)$
74. $36(a - b)x^2 - 6(a - b)xy - 20(a - b)y^2$

In Exercises 75–77 use a graphics calculator to graph $y = P(x)$, to determine the x-intercepts of this graph, and to complete this table.

POLYNOMIAL $P(x)$	x-INTERCEPTS OF THE GRAPH OF $y = P(x)$	ZEROS OF $P(x)$	FACTORED FORM
75. $x^3 - 3x^2 - 10x + 24$			
76. $(x^3 - 4x) + (5x^2 - 20)$			
77. $4x^3 - 9x$			

Group Discussion Questions

78. **Challenge Question**　Factor each polynomial completely assuming that m and n are natural numbers.
a. $81y^{2n} - 16$
b. $x^{2m} - y^{2n}$
c. $4x^{2m} + 20x^my^n + 25y^{2n}$
d. $10x^{2m} + 4x^my^n - 6y^{2n}$

79. **Challenge Question**　Use the distributive property to expand the first expression and to factor the second expression.

EXPAND	FACTOR
a. $x^{-4}(x^2 - 3)$	$x^{-2} - 3x^{-4} = x^{-4}(\quad)$
b. $2x^{-5}(x^2 - 2)$	$2x^{-3} - 4x^{-5} = 2x^{-5}(\quad)$
c. $x^{-4}(2x - 1)(x - 3)$	$2x^{-2} - 7x^{-3} + 3x^{-4}$
	$= x^{-4}(\quad)$
	$= x^{-4}(\quad)(\quad)$

80. **Challenge Question**　Factor $2025x^4 + 2700x^2y + 900y^2$ first by factoring this perfect square trinomial as the square of a sum. Then start over and factor the GCF from this trinomial. Complete each of these factorizations and compare these results, stating which factorization you find easier.

Section 7.5 Solving Equations by Factoring

Objectives:

10. Use factoring to solve selected second- and third-degree equations.
11. Use the x-intercepts of the graph of a quadratic function to solve the corresponding quadratic equation and inequalities.

A Mathematical Note

The importance of quadratic equations has been noted by many ancient civilizations. A surviving document from Egypt (c. 2000 B.C.) contains quadratic equations. The Hindu mathematician Brahmagupta (c. 628) included a version of the quadratic formula for solving quadratic equations in his works.

A quadratic equation in x is a second-degree equation that can be written in the standard form $ax^2 + bx + c = 0$, where a, b, and c represent real constants and $a \neq 0$. (If we allowed a to equal zero, the equation would not be quadratic; instead it would be the linear equation $bx + c = 0$.) In the quadratic equation $ax^2 + bx + c = 0$, ax^2, is called the **quadratic term,** bx is called the **linear term,** and c is called the **constant term.**

Quadratic Equation		
ALGEBRAICALLY	**VERBALLY**	**ALGEBRAIC EXAMPLE**
If a, b, and c are real constants and $a \neq 0$, then $ax^2 + bx + c = 0$ is the standard form of a quadratic equation in x.	A quadratic equation in x is a second-degree equation in x.	$3x^2 - 4x + 1 = 0$ is a quadratic equation with: Quadratic term: $3x^2$ Linear term: $-4x$ Constant term: 1

■ EXAMPLE 1 Identifying Quadratic Equations

Determine which of these equations are quadratic equations in one variable. Write each quadratic equation in standard form.

SOLUTIONS

(a) $5y - 6 = -y^2$

This equation is a quadratic equation in y and can be written in standard form as $y^2 + 5y - 6 = 0$.

In this standard form $a = 1$, $b = 5$, and $c = -6$. It is customary to write the equation so that a is positive.

(b) $5w = 12$

This equation is a linear equation in w; it is not a quadratic equation.

Quadratic equations are second degree. Linear equations are first degree.

(c) $5v^2 = 3v - 4$

This equation is a quadratic equation in v and can be written in standard form as $5v^2 - 3v + 4 = 0$.

The quadratic term is $5v^2$, the linear term is $-3v$, and the constant term is 4.

(d) $x^2 + 7x = x^3 - 5$

This equation is a third-degree equation; it is not a quadratic equation.

(e) $x^2 + 3x + 7$

This is not an equation, just a polynomial expression.

(f) $x^2 = 6x$

This equation is a quadratic equation and can be written in standard form as $x^2 - 6x = 0$.

In standard form, $a = 1$, $b = -6$, and c is understood to be 0.

Caution: Some quadratic equations cannot be solved by factoring over the integers.

The easiest method for solving some quadratic equations is to use factoring and the zero-factor principle. Because some quadratic polynomials are not factorable over the integers, some quadratic equations cannot be solved by factoring over the integers. We will cover methods for solving those quadratic equations in Chapter 8.

The word "or" is used in the inclusive sense to mean $a = 0$ or $b = 0$ (or both a and b are 0).

Zero-Factor Principle

ALGEBRAICALLY	VERBALLY	ALGEBRAIC EXAMPLE
For real numbers a and b: If $a = 0$ or $b = 0$, then $ab = 0$. If $ab = 0$ then $a = 0$ or $b = 0$.	The product of zero and any other factor is zero. If the product of two factors is zero, at least one of the factors must be zero.	$0 \cdot a = 0$ If $(x - 3)(x + 4) = 0$, then either $x - 3 = 0$ or $x + 4 = 0$.

■ EXAMPLE 2   Solving a Quadratic Equation in Factored Form

Solve $(v - 2)(v + 3) = 0$.

SOLUTION

$(v - 2)(v + 3) = 0$ By the zero-factor principle, this is equivalent to stating
$v - 2 = 0$ or $v + 3 = 0$ that either the first factor or the second factor is zero.
$v = 2$ $v = -3$ Notice that the solution set $\{-3, 2\}$ contains two numbers.

Answer: $v = 2, v = -3$ Do these values check? ■

The strategy used to solve quadratic equations by factoring is described in the following box and illustrated in Example 3.

Solving Quadratic Equations by Factoring

VERBALLY	ALGEBRAIC EXAMPLE
1. Write the equation in standard form, with the right side 0.	$x^2 + x = 6$ $x^2 + x - 6 = 0$
2. Factor the left side of the equation.	$(x - 2)(x + 3) = 0$
3. Set each factor equal to 0.	$x - 2 = 0$ or $x + 3 = 0$
4. Solve the resulting first-degree equations.	$x = 2$ $x = -3$

If a quadratic equation contains fractions, then we can simplify it by converting it to an equivalent equation that does not involve fractions. To do this, multiply both sides of the equation by the LCD of all the terms.

■ **EXAMPLE 3** Solving a Quadratic Equation Containing Fractions

Solve $\dfrac{1}{8}x^2 - \dfrac{1}{4}x = 1$.

A common theme in algebra is to take a given problem and convert it to an equivalent but easier problem. Multiplying both sides of the equation by the LCD produces an equivalent equation that does not contain fractions.

SOLUTION

$$\frac{1}{8}x^2 - \frac{1}{4}x = 1$$

$$\frac{1}{8}x^2 - \frac{1}{4}x - 1 = 0 \qquad \text{First write the equation in standard form.}$$

$$8\left(\frac{1}{8}x^2 - \frac{1}{4}x - 1\right) = 8(0) \qquad \text{Multiply both sides by the LCD, 8, to obtain integer coefficients.}$$

$$8\left(\frac{1}{8}x^2\right) + 8\left(-\frac{1}{4}x\right) + 8(-1) = 0 \qquad \text{Distribute the factor of 8 to each term on the left side of the equation.}$$

$$x^2 - 2x - 8 = 0 \qquad \text{Next factor the left side of the equation.}$$

$$x^2 + 2x - 4x - 8 = 0$$

$$x(x + 2) - 4(x + 2) = 0$$

$$(x - 4)(x + 2) = 0$$

FACTORS OF −8		SUM OF FACTORS
1	−8	−7
2	−4	−2

$$x - 4 = 0 \quad \text{or} \quad x + 2 = 0 \qquad \text{Set each factor equal to zero.}$$

$$x = 4 \qquad\qquad x = -2 \qquad \text{Solve each of these linear equations.}$$

Answer: $x = 4$, $x = -2$ Do these values check?

SELF-CHECK 7.5.1

Solve each quadratic equation.

1. $(2x - 9)(x + 5) = 0$ **2.** $6x^2 - 37x + 22 = 0$ **3.** $x^2 - \dfrac{5}{2}x - \dfrac{3}{2} = 0$

Caution: The zero-factor principle works only if the product is zero.

It is important that one side of a quadratic equation is zero before you try to apply the zero-factor principle. Failure to observe this is the source of many student errors.

■ **EXAMPLE 4** Solving a Quadratic Equation Not in Standard Form

Solve $(6x + 1)(x - 2) = 8$.

SOLUTION

$$(6x + 1)(x - 2) = 8 \qquad \text{The first step is to multiply the factors on the left side of the equation and write the equation in standard form with the right side 0. Then factor the left side of the equation.}$$

$$6x^2 - 11x - 2 = 8$$

$$6x^2 - 11x - 10 = 0$$

$$6x^2 + 4x - 15x - 10 = 0$$

$$2x(3x + 2) - 5(3x + 2) = 0$$

$$(2x - 5)(3x + 2) = 0$$

FACTORS OF −60		SUM OF FACTORS
1	−60	−59
2	−30	−28
3	−20	−17
4	−15	−11

$$2x - 5 = 0 \quad \text{or} \quad 3x + 2 = 0 \qquad \text{Set each factor equal to 0. Then solve each of the linear equations for } x.$$

$$2x = 5 \qquad\qquad 3x = -2$$

$$x = \frac{5}{2} \qquad\qquad x = -\frac{2}{3}$$

Answer: $x = \dfrac{5}{2}$, $x = -\dfrac{2}{3}$ Do these solutions check?

SELF-CHECK 7.5.1 ANSWERS

1. $x = \dfrac{9}{2}, x = -5$

2. $x = \dfrac{2}{3}, x = \dfrac{11}{2}$

3. $x = -\dfrac{1}{2}, x = 3$

If we can factor a quadratic equation in standard form, then we can find its solution. If we reverse this procedure, we can create a quadratic equation whose solutions are given. This procedure is used routinely by teachers and authors to create equations with known answers and a known level of difficulty. This procedure is also understood by engineers who wish to design a product with known behavior characteristics. Engineers start with the behaviors they want and create equations that model products that exhibit the desired behaviors.

Engineers can create products with desired behavior characteristics by creating equations to model the products they are designing.

■ EXAMPLE 5   Constructing a Quadratic Equation Whose Solutions Are Given

Construct a quadratic equation in x with solutions of $\dfrac{2}{3}$.

SOLUTION _____

$x = -1$ or	$x = \dfrac{2}{3}$	The solutions are -1 and $\dfrac{2}{3}$.
$x + 1 = 0$	$3x = 2$	Rewrite each equation so that the right side is 0.
$x + 1 = 0$	$3x - 2 = 0$	
	$(x + 1)(3x - 2) = 0$	These factors equal 0, so their product is 0.
	$3x^2 + x - 2 = 0$	

Answer: $3x^2 + x - 2 = 0$ is a quadratic equation with solutions of -1 and $\dfrac{2}{3}$. ■

SELF-CHECK 7.5.2

1. Solve $x^2 - 7x + 12 = 2$.
2. Write a quadratic equation with solutions of $x = 4$ and $x = \dfrac{3}{7}$.
3. Write a cubic equation with solutions $x = -3$, $x = 4$, and $x = 10$.

From Example 5 we also can make the following observations about $P(x) = 3x^2 + x - 2$.

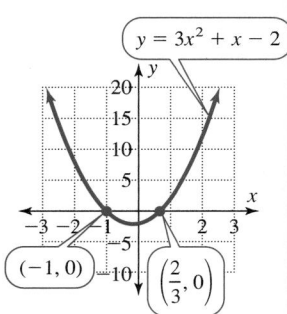

FACTORS OF $P(x) = 3x^2 + x - 2$	SOLUTIONS OF $P(x) = 0$	ZEROS OF $P(x)$	x-INTERCEPTS OF THE GRAPH OF $y = P(x)$
$x + 1$ is a factor of $P(x)$	$x = -1$ is a solution of $P(x) = 0$	-1 is a zero of $P(x)$	$(-1, 0)$ is an x-intercept
$3x - 2$ is a factor of $P(x)$	$x = \dfrac{2}{3}$ is a solution of $P(x) = 0$	$\dfrac{2}{3}$ is a zero of $P(x)$	$\left(\dfrac{2}{3}, 0\right)$ is an x-intercept

A generalization of this relationship is given in the following box.

Equivalent Statements about Linear Factors of a Polynomial

For a real constant c and a real polynomial $P(x)$ the following statements are equivalent.

ALGEBRAICALLY		NUMERICALLY	GRAPHICALLY
$x - c$ is a factor of $P(x)$.	$x = c$ is a solution of $P(x) = 0$.	$P(c) = 0$; that is, c is a zero of $P(x)$.	$(c, 0)$ is an x-intercept of the graph of $y = P(x)$.

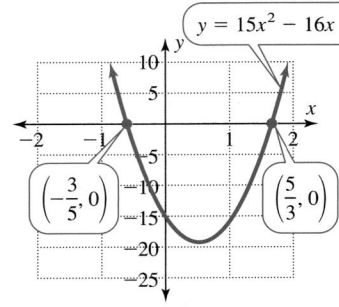

Examine the graph of $y = 15x^2 - 16x - 15$ shown to the left. From the graph the x-intercepts are $\left(-\dfrac{3}{5}, 0\right)$ and $\left(\dfrac{5}{3}, 0\right)$. Thus the solutions of the corresponding equation are $x = -\dfrac{3}{5}$ and $x = \dfrac{5}{3}$. Using these solutions, we can work backward to form the equation.

$$x = -\frac{3}{5} \qquad\qquad x = \frac{5}{3}$$
$$5x = -3 \qquad\qquad 3x = 5$$
$$5x + 3 = 0 \qquad\qquad 3x - 5 = 0$$
$$(5x + 3)(3x - 5) = 0$$
$$15x^2 - 16x - 15 = 0$$

The value of this observation is that if we can determine the x-intercepts, then we can use the x-intercepts to produce the factored form of the polynomial. The following calculator perspective shows how we can use a calculator to help us determine the x-intercepts of the graph and the zeros of the function.

CALCULATOR
PERSPECTIVE 7.5.1

Using the Graph of a Quadratic Function to Determine Its Zeros

To find the zeros of the trinomial $5x^2 + 13x - 6$ on a TI-83 Plus calculator, enter the following keystrokes:

[Y=] [5] [X,T,θ,n] [x²] [+] [1] [3] [X,T,θ,n] [−] [6]

[GRAPH] [2nd] [CALC] [2] (Then follow the procedure to calculate the zero.)

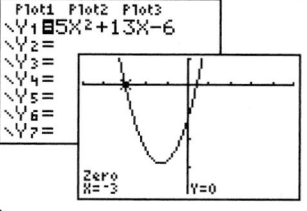

$[-5, 5, 1]$ by $[-20, 5, 5]$

-3 is a zero of the polynomial; so $x = -3$ is a solution of $5x^2 + 13x - 6 = 0$.

[GRAPH] [2nd] [CALC] [2] (Then follow the procedure to calculate the other zero.)

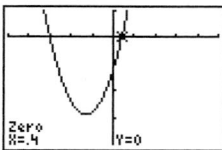

[2nd] [QUIT] [X,T,θ,n] [MATH] [ENTER] [ENTER]

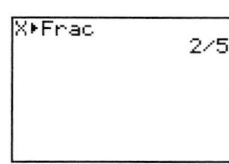

$\frac{2}{5}$ is a zero of the polynomial; so $x = \frac{2}{5}$ is a solution of $5x^2 + 13x - 6 = 0$.

Note: Once we know the zeros of the polynomial function, we also can factor the polynomial or solve the corresponding polynomial equation. ■

SELF-CHECK 7.5.3

1. Use a graphics calculator to graph $y = 21x^2 - 2x - 3$.
2. Find the x-intercepts of this graph.
3. Use these x-intercepts to factor $21x^2 - 2x - 3$.

The zero-factor principle also can be extended to solve equations of a higher degree. This is illustrated in the next example, where we compare the algebraic, numerical, and graphical methods for solving a cubic equation.

SELF-CHECK 7.5.3 ANSWERS

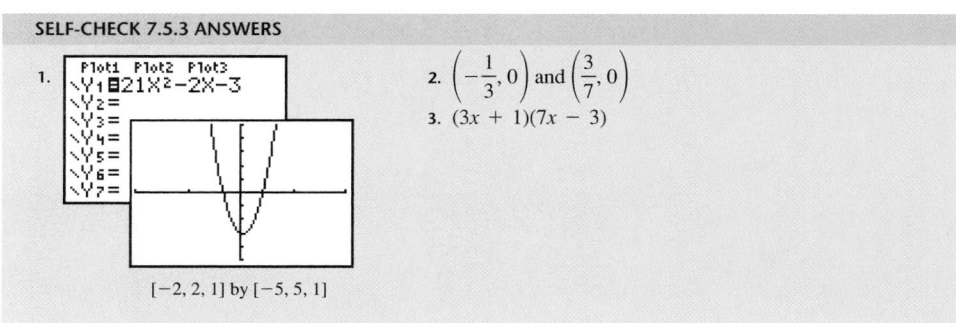

1.
2. $\left(-\frac{1}{3}, 0\right)$ and $\left(\frac{3}{7}, 0\right)$
3. $(3x + 1)(7x - 3)$

$[-2, 2, 1]$ by $[-5, 5, 1]$

■ **EXAMPLE 6** Using Multiple Perspectives to Solve a Cubic Equation

Solve $x^3 - 4x^2 - 4x + 16 = 0$ algebraically, numerically, and graphically.

SOLUTION

ALGEBRAICALLY

$$x^3 - 4x^2 - 4x + 16 = 0$$
$$x^2(x - 4) - 4(x - 4) = 0$$
$$(x^2 - 4)(x - 4) = 0$$
$$(x + 2)(x - 2)(x - 4) = 0$$

$$x + 2 = 0, \quad x - 2 = 0 \quad \text{or} \quad x - 4 = 0$$
$$x = -2, \quad x = 2 \quad \text{or} \quad x = 4$$

First use the grouping method to factor the left side of the equation.

Then note that $x^2 - 4$ is a difference of two squares and factors as a sum times a difference.

Set each factor equal to 0 and solve the resulting linear equations.

NUMERICALLY

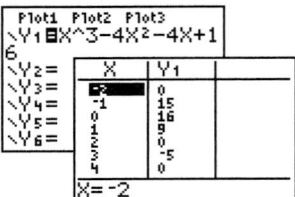

Enter the cubic polynomial into a graphics calculator as y_1. Then use the TABLE feature to find the zeros of this polynomial function.
Note that this function is 0 for $x = -2$, $x = 2$, and $x = 4$.

The zeros of $y = x^3 - 4x^2 - 4x + 16$ are -2, 2, and 4.

GRAPHICALLY

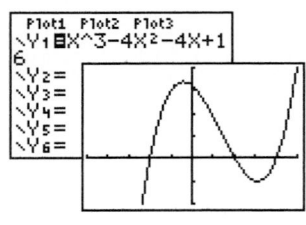

$[-5, 5, 1]$ by $[-10, 20, 5]$

Use a graphics calculator to graph $y_1 = x^3 - 4x^2 - 4x + 16$. Then examine this graph to determine its x-intercepts. This graph has x-intercepts of $(-2, 0)$, $(2, 0)$, and $(4, 0)$.

The x-intercepts of the graph of $y = x^3 - 4x^2 - 4x + 16$ are $(-2, 0)$, $(2, 0)$, and $(4, 0)$.

Answer: The solutions of $x^3 - 4x^2 - 4x + 16 = 0$ are $x = -2$, $x = 2$, and $x = 4$.

Algebraically, we used the factors of $x^3 - 4x^2 - 4x + 16$ to solve $x^3 - 4x^2 - 4x + 16 = 0$. Numerically, we found the zeros of $y = x^3 - 4x^2 - 4x + 16$ to solve $x^3 - 4x^2 - 4x + 16 = 0$. Graphically, we determined the x-intercepts of the graph of $y = x^3 - 4x^2 - 4x + 16$ to solve $x^3 - 4x^2 - 4x + 16 = 0$. ■

SELF-CHECK 7.5.4 ANSWER

$x = -5, x = -1, x = 1$

SELF-CHECK 7.5.4

Solve $x^3 + 5x^2 - x - 5 = 0$.

A quadratic equation is a second-degree equation that can be written in the form $ax^2 + bx + c = 0$. The second-degree equations that we have examined so far have had two distinct real solutions. We now examine some other possibilities starting with an example that has a double real root. If a quadratic polynomial is a perfect square trinomial, then both factors will be the same and the solutions of the quadratic equation will be the same. In this case we call the solution a **double solution** or a **solution of multiplicity two**.

■ EXAMPLE 7   Solving a Quadratic Equation with a Double Solution

Solve $x^2 = 4x - 4$.

SOLUTION _____

ALGEBRAICALLY

$$x^2 = 4x - 4$$
$$x^2 - 4x + 4 = 0$$
$$(x - 2)^2 = 0$$
$$x - 2 = 0 \quad \text{or} \quad x - 2 = 0$$
$$x = 2 \qquad\qquad x = 2$$

GRAPHICALLY

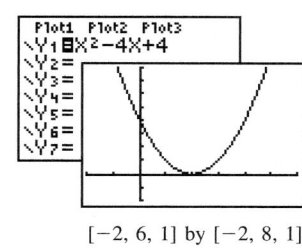

[−2, 6, 1] by [−2, 8, 1]

Write the equation in standard form.

The equation $y_1 = x^2 - 4x + 4$ defines a quadratic function whose graph is a parabola.

The form $y = (x - 2)^2$ indicates a parabola whose vertex is shifted horizontally to the point (2, 0). This vertex is the only x-intercept of this parabola.

Answer: $x = 2$ is a double solution.   ■

SELF-CHECK 7.5.5

Solve $9x^2 + 25 - 30x = 0$.

SELF-CHECK 7.5.5 ANSWER

$x = \dfrac{5}{3}$ is a double solution.

Quadratic equations always have two solutions (if we count double solutions as two solutions). Three possibilities exist for these solutions. These possibilities are given in the following box. This box summarizes what we have examined in this chapter and provides some insight into the material in the next chapter. In the box we assume that $a > 0$ in the equation $ax^2 + bx + c = 0$. If $a < 0$, we could multiply both sides of the equation by -1 to convert the equation to one with $a > 0$.

The Nature of the Solutions of a Quadratic Equation

There are three possibilities for the solutions of $ax^2 + bx + c = 0$. (In the following graphs we assume that $a > 0$.)

SOLUTIONS OF $ax^2 + bx + c = 0$	THE PARABOLA $y = ax^2 + bx + c$	GRAPHICAL EXAMPLE
1. Two distinct real solutions	Two x-intercepts	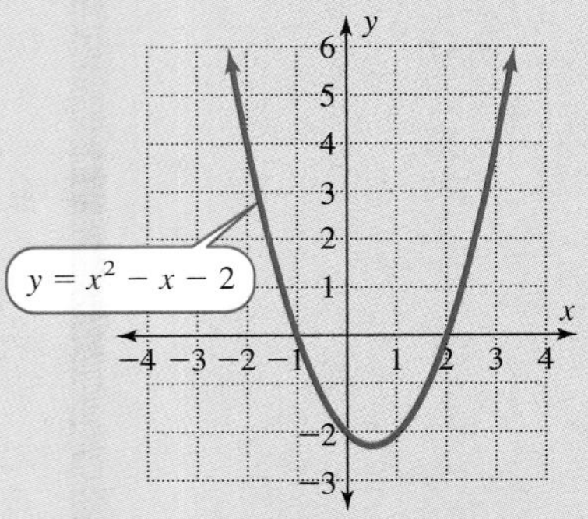 $y = x^2 - x - 2$
2. A double real solution	One x-intercept with the vertex on the x-axis	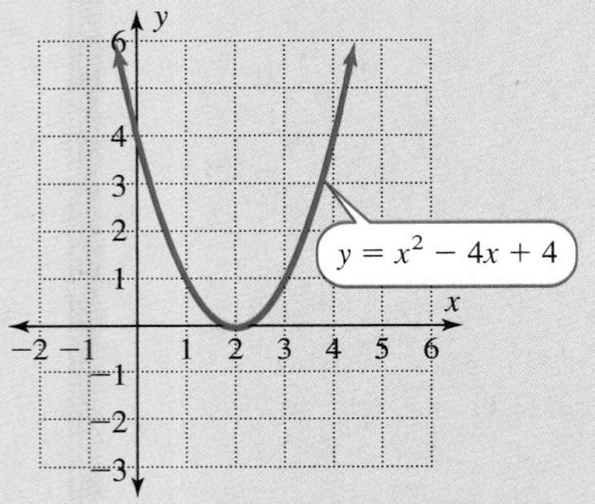 $y = x^2 - 4x + 4$
3. Neither solution is real, both solutions are complex numbers with imaginary parts.[*]	No x-intercepts	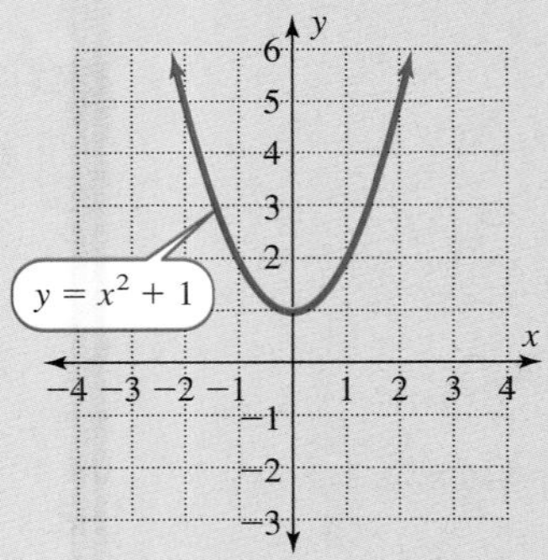 $y = x^2 + 1$

*Complex numbers are covered in Chapter 8.

In Chapter 8 we examine other algebraic methods for solving quadratic equations. The factoring method used in this chapter has been used to find real solutions that are rational numbers. In Chapter 8 we will examine quadratic equations with irrational solutions and quadratic equations with imaginary solutions.

We now use the graph of a quadratic polynomial function $P(x) = ax^2 + bx + c$ to solve the inequalities $P(x) < 0$ and $P(x) > 0$. We use the information in the following table to solve the inequalities in Example 8.

Solutions of Equations and Inequalities

If $P(x)$ is a real polynomial, then:

VERBALLY	ALGEBRAICALLY	GRAPHICALLY
c is a solution of $P(x) = 0$	$P(c) = 0$	$(c, 0)$ is an x-intercept of the graph of $y = P(x)$.
c is a solution of $P(x) < 0$	$P(c) < 0$	At $x = c$ the graph is below the x-axis.
c is a solution of $P(x) > 0$	$P(c) > 0$	At $x = c$ the graph is above the x-axis.

■ EXAMPLE 8 Using x-Intercepts to Solve Quadratic Equations and Inequalities

Use a graphics calculator to find the zeros of $x^2 + 2x - 35$ and to solve

(a) $x^2 + 2x - 35 = 0$
(b) $x^2 + 2x - 35 \leq 0$
(c) $x^2 + 2x - 35 \geq 0$

SOLUTION

Once the quadratic equation is solved, the graph of the corresponding function will reveal the solutions of the quadratic inequalities.

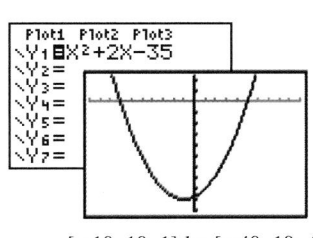

$[-10, 10, 1]$ by $[-40, 10, 4]$

Answers: (a) $x = -7$, $x = 5$

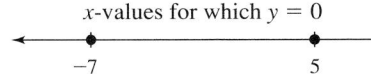

x-values for which $y = 0$

Enter $x^2 + 2x - 35$ as y_1 and graph this function. Determine the x-intercepts of $y_1 = x^2 + 2x - 35$.

The x-intercepts are $(-7, 0)$ and $(5, 0)$.

The solutions of $x^2 + 2x - 35 = 0$ are revealed by the x-intercepts.

When the graph is below the x-axis, $P(x) < 0$. When the graph is above the x-axis, $P(x) > 0$.

(b) $[-7, 5]$

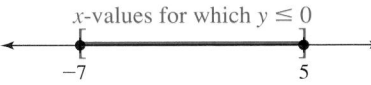

x-values for which $y \leq 0$

The graph of $y_1 = x^2 + 2x - 35$ is on or below the x-axis for the x-values given by $[-7, 5]$.

(c) $(-\infty, -7] \cup [5, \infty)$

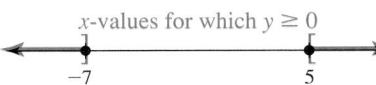

x-values for which $y \geq 0$

The graph of $y_1 = x^2 + 2x - 35$ is on or above the x-axis for the x-values given by $(-\infty, -7] \cup [5, \infty)$.

SELF-CHECK 7.5.6

1. Use a graphics calculator to graph $y = 2x^2 - 11x - 6$.
2. Determine the x-intercepts of this graph.
3. Determine the zeros of $2x^2 - 11x - 6$.
4. Solve $2x^2 - 11x - 6 = 0$.
5. Factor $2x^2 - 11x - 6$.
6. Solve $2x^2 - 11x - 6 \geq 0$.
7. Solve $2x^2 - 11x - 6 < 0$.

SELF-CHECK 7.5.6 ANSWERS

1.

```
Plot1 Plot2 Plot3
\Y1◘2X²-11X-6
\Y2=
\Y3=
\Y4=
\Y5=
\Y6=
\Y7=
```

$[-2, 8, 1]$ by $[-25, 10, 5]$

2. $\left(-\dfrac{1}{2}, 0\right), (6, 0)$

3. $-\dfrac{1}{2}$ and 6

4. $x = -\dfrac{1}{2}$ and $x = 6$

5. $(2x + 1)(x - 6)$

6. $\left(-\infty, -\dfrac{1}{2}\right] \cup [6, \infty)$

7. $\left(-\dfrac{1}{2}, 6\right)$

We now examine an application that yields a quadratic equation.

■ EXAMPLE 9 Determining the Dimensions of a Rectangle

Forty meters of rope are used to enclose a rectangular area of 91 m². Find the dimensions of this rectangular area.

```
┌─────────────────────────┐
│                         │
w│        91 m²           │
│                         │
└─────────────────────────┘
```

SOLUTION ────────────────────────────────

Let w = the width of the rectangle in meters
$20 - w$ = the length of the rectangle in meters

The dimensions are the width and length. Because the perimeter is 40 m, $2w + 2l = 40$ and $l = 20 - w$.

VERBALLY

Area equals 91 m².
(width)(length) = 91

The area of a rectangle is given by $A = w \cdot l$.

ALGEBRAICALLY

$$w(20 - w) = 91$$
$$20w - w^2 = 91$$
$$0 = w^2 - 20w + 91$$
$$w^2 - 20w + 91 = 0$$
$$(w - 7)(w - 13) = 0$$

$$w - 7 = 0 \quad \text{or} \quad w - 13 = 0$$
$$w = 7 \quad \text{or} \quad w = 13$$
$$20 - w = 13 \quad \text{or} \quad 20 - w = 7$$

Answer: The dimensions of this rectangular region are 7 m by 13 m.

Substitute the expressions identified for the width and length.

Write this quadratic equation in standard form. Solve this quadratic equation by factoring.

Either the width is 7 m and the length is 13 m or the width is 13 m and the length is 7 m. Do these dimensions check for both the perimeter and the area?

USING THE LANGUAGE AND SYMBOLISM OF MATHEMATICS   7.5

1. A second-degree equation in x that can be written as $ax^2 + bx + c = 0$ is called a _____ _____.
2. In the equation $ax^2 + bx + c = 0$, ax^2 is called the _____ term, bx is called the _____ term, and c is called the _____ term.
3. By the zero-factor principle, if $(x - 3)(x + 6) = 0$, then _____ $= 0$ or _____ $= 0$.
4. To clear an equation of fractions, multiply each term by the _____ of all the terms.
5. All quadratic equations have _____ solutions. If both solutions are equal, we call the solution a _____ solution or a solution of _____ two.

6. If the parabola $y = ax^2 + bx + c$ has its vertex on the x-axis, then the equation $ax^2 + bx + c = 0$ has a _____ real solution.
7. If $x = c$ is a real solution of $P(x) = 0$, then _____ is an x-intercept of the graph of $y = P(x)$.
8. If $x = c$ is a real solution of $P(x) = 0$, then _____ is a factor of the polynomial $P(x)$.
9. If the graph of $y = P(x)$ crosses the x-axis at $x = c$, then $P(c)$ _____ 0.
10. If the graph of $y = P(x)$ is above the x-axis at $x = c$, then $P(c)$ _____ 0.
11. If the graph of $y = P(x)$ is below the x-axis at $x = c$, then $P(c)$ _____ 0.

EXERCISES   7.5

Exercises 1 and 2 match each algebraic description with the most appropriate description.

1. a. $2x - 3 = 0$
   b. $(2x - 3)^2 = 0$
   c. $(2x - 3)(x + 4) = 0$
   d. $(2x - 3)(x + 4)$

   A. A quadratic equation with two distinct real solutions
   B. A quadratic polynomial, but not an equation
   C. A quadratic equation with a double real solution
   D. A linear equation with one real solution

2. a. $x^2 - 25 = 0$
   b. $x^2 - 10x + 25 = 0$
   c. $x^2 - 10x + 25$
   d. $x^3 - 25x = 0$

   A. A quadratic equation with a double real solution
   B. A quadratic equation with two distinct real solutions
   C. A cubic equation with three solutions
   D. A quadratic polynomial, but not an equation

In Exercises 3 and 4 write each quadratic equation in the standard form $ax^2 + bx + c = 0$ and identify a, b, and c.

3. a. $2x^2 - 7x + 3 = 0$ b. $8x^2 = 3x$
   c. $7x^2 = -5$ d. $(x - 1)(2x + 1) = 3$
4. a. $3x^2 = 5x - 2$ b. $3x^2 = 17$
   c. $2x^2 = 17x$ d. $(3x + 1)(x - 2) = 3$

In Exercises 5–50 solve each equation.
5. $(m - 8)(m + 17) = 0$
6. $(m + 15)(m - 3) = 0$
7. $(2n - 5)(3n + 1) = 0$
8. $(5n + 2)(n - 6) = 0$
9. $(z - 1)(z + 2)(2z - 7) = 0$
10. $z(z + 6)(6z + 1) = 0$
11. $v^2 - 121 = 0$
12. $v^2 - 169 = 0$
13. $x^2 + 3x + 2 = 0$
14. $x^2 + 5x + 4 = 0$
15. $y^2 - 3y = 18$
16. $t^2 + 4t = 12$
17. $3v^2 = -v$
18. $5v^2 = v$
19. $2w^2 = 7w + 15$
20. $3w^2 = 17w + 6$
21. $6x^2 + 19x + 10 = 0$
22. $10x^2 - 17x + 3 = 0$
23. $x^2 = 10x - 25$
24. $x^2 = 14x - 49$
25. $9m^2 = 42m - 49$
26. $25m^2 = 90m - 81$
27. $70z^2 = 5z + 15$
28. $8v^2 + 24v = 0$

29. $9x^2 = 25$
30. $4x^2 = 9$
31. $r(r + 3) = 10$
32. $(r + 6)(r - 1) = -10$
33. $\dfrac{m^2}{18} - \dfrac{m}{6} - 1 = 0$
34. $\dfrac{m^2}{20} - \dfrac{m}{4} + \dfrac{1}{5} = 0$
35. $\dfrac{x^2}{12} - \dfrac{x}{3} - 1 = 0$
36. $\dfrac{x^2}{18} + \dfrac{x}{6} - 1 = 0$
37. $(v - 12)(v + 1) = -40$
38. $(v + 5)(v + 3) = 5v + 25$
39. $(2w - 3)^2 = 25$
40. $(3w + 2)^2 = 16$
41. $(3x - 8)(x + 1) = (x + 1)(x - 3)$
42. $(4x + 5)(x + 5) = 45x$
43. $(3x + 2)(3x - 4) = -3(4x + 3)$
44. $(t + 1)(t - 1)(3t + 1) = 0$
45. $v(v^2 - 5v - 24) = 0$
46. $v(v^2 + 14v + 24) = 0$
47. $w(6w^2 + 5w - 6) = 0$
48. $w(5w^2 + w - 4) = 0$
49. $14y^3 = 3y - 19y^2$
50. $10y^3 = 29y^2 - 10y$

In Exercises 51 and 52 construct a quadratic equation in x with the given solutions.
51. a. -3 and 3 **b.** $\dfrac{3}{7}$ and -2
c. 0 and 6 **d.** A double solution of -2
52. a. A double solution of 3 **b.** 0 and -4
c. $-\dfrac{7}{5}$ and $\dfrac{2}{3}$ **d.** $-\dfrac{2}{5}$ and $\dfrac{5}{2}$

In Exercises 53 and 54 construct a cubic (third-degree) equation in x with the given solutions.
53. a. $0, 1,$ and 2 **b.** $-1, 0,$ and 1
54. a. $2, -3,$ and 4 **b.** $-5, 6,$ and 10

Exercises 55–58 have two parts. In part **(a)** solve the equation, and in part **(b)** perform the indicated operations and simplify the result.

SOLVE	SIMPLIFY
55. a. $(5m - 3)(m - 2) = 0$	**b.** $(5m - 3)(m - 2)$
56. a. $(4x - 3)(5x + 2) = 0$	**b.** $(4x - 3)(5x + 2)$
57. a. $4(x - 3)(5x + 2) = 0$	**b.** $4(x - 3)(5x + 2)$
58. a. $x(x - 3)(5x + 2) = 0$	**b.** $x(x - 3)(5x + 2)$

In Exercises 59 and 60 use the given graph to factor each polynomial.
59. $P(x) = 12x^2 + 11x - 15$ **60.** $P(x) = 15x^2 - 11x - 12$

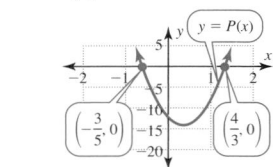

In Exercises 61–64 use the graph of $y = P(x)$ to estimate the solutions of:
a. $P(x) = 0$
b. $P(x) < 0$
c. $P(x) > 0$

61. $P(x) = x^2 + x - 6$
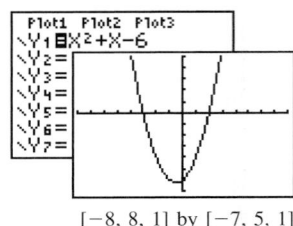
$[-8, 8, 1]$ by $[-7, 5, 1]$

62. $P(x) = 4x^3 - 4x^2 - 5x + 3$
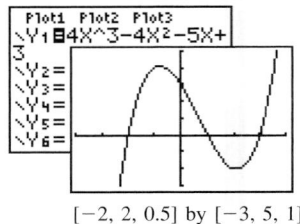
$[-2, 2, 0.5]$ by $[-3, 5, 1]$

63. $P(x) = 4x^2 - 20x + 25$
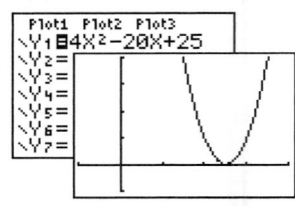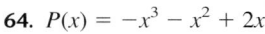
$[-1, 4, 1]$ by $[-1, 4, 1]$

64. $P(x) = -x^3 - x^2 + 2x$
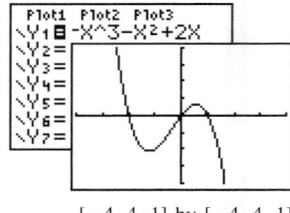
$[-4, 4, 1]$ by $[-4, 4, 1]$

In Exercises 65 and 66 use the given table to factor each polynomial.
65. $P(x) = 50x^2 - 45x + 7$
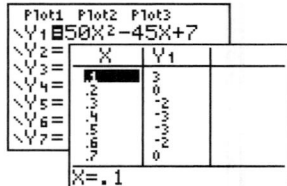

66. $P(x) = 25x^2 + 10x - 3$

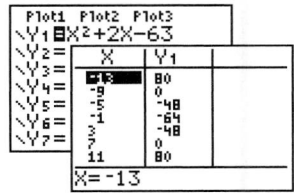

In Exercises 67–70 use the given table to solve each equation and inequality.

67. a. $x^2 + 2x - 63 = 0$
b. $x^2 + 2x - 63 < 0$
c. $x^2 + 2x - 63 > 0$

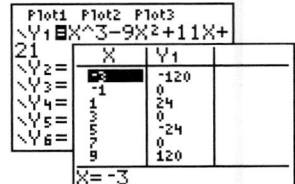

68. a. $8x^2 - 2x - 3 = 0$
b. $8x^2 - 2x - 3 \leq 0$
c. $8x^2 - 2x - 3 \geq 0$

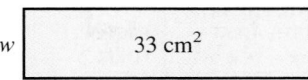

69. a. $x^3 - 9x^2 + 23x - 15 = 0$
b. $x^3 - 9x^2 + 23x - 15 < 0$
c. $x^3 - 9x^2 + 23x - 15 > 0$

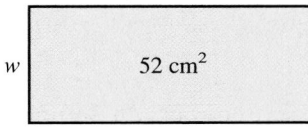

70. a. $x^3 - 9x^2 + 11x + 21 = 0$
b. $x^3 - 9x^2 + 11x + 21 \leq 0$
c. $x^3 - 9x^2 + 11x + 21 \geq 0$

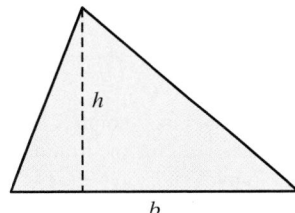

71. The length of the following rectangle is 2 cm more than three times the width. Find the dimensions of this rectangle if the area is 33 cm².

w [33 cm²]

72. The length of the following rectangle is 5 cm more than twice the width. Find the dimensions of this rectangle if the area is 52 cm².

w [52 cm²]

73. The base of the following triangle is 2 m longer than the height. Find the base if the area of this triangle is 24 m².

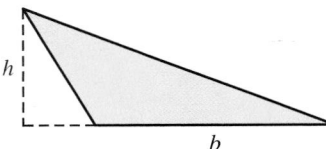

74. The base of the following triangle is 10 cm longer than the height. Determine the height of the triangle if its area is 48 cm².

75. A metal sheet 60 cm wide is used to form a trough by bending up each side as illustrated in the figure. Determine the height of each side if the cross-sectional area is 450 cm².

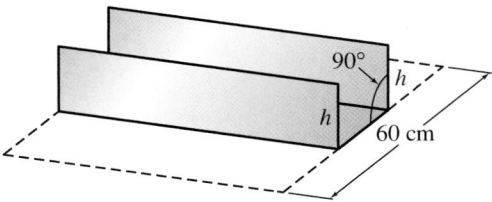

76. A rectangular pad of concrete is outlined by a 52-m stripe of paint. Find the dimensions of this pad if the surface area is 168 m².

In Exercises 77–80 use a graphics calculator to graph $y = P(x)$, to determine the x-intercepts of this graph, and to complete this table.

POLYNOMIAL $P(x)$	x-INTERCEPTS OF THE GRAPH OF $y = P(x)$	FACTORED FORM	ZEROS OF $P(x)$	SOLUTIONS OF $P(x) = 0$
77. $x^2 - 4x - 21$				
78. $2x^2 - 15x - 8$				
79. $x^2 - 12x + 36$				
80. $x^3 + 2x^2 - 48x$				

In Exercises 81–84 use a graphics calculator to graph $y = P(x)$ and to complete this table.

POLYNOMIAL $P(x)$	SOLUTIONS OF $P(x) = 0$	SOLUTIONS OF $P(x) < 0$	SOLUTIONS OF $P(x) \geq 0$
81. $(x - 2)(x + 7)$			
82. $-2(x + 4)(x - 2)$			
83. $-3x^2 - 19x + 14$			
84. $x^2 + 7x - 8$			

Group Discussion Questions

85. Discovery Question Each of these polynomials is prime over the integers but can be factored over the irrational numbers. Factor these polynomials and solve each equation.

$P(x)$	FACTORS OF $P(x)$	SOLUTIONS OF $P(x) = 0$
Example $x^2 - 3$	$(x + \sqrt{3})(x - \sqrt{3})$	$x = -\sqrt{3}, x = \sqrt{3}$
a. $x^2 - 5$	$(x + \sqrt{5})(x - \sqrt{5})$	
b. $x^2 - 7$		$x = -\sqrt{7}, x = \sqrt{7}$
c. $x^2 - 11$		

86. Challenge Question Solve each equation for x.
  a. $(x + y)(x - 3y) = 0$
  b. $x^2 - 5xy - 6y^2 = 0$
  c. $6x^2 - xy = 2y^2$
  d. $x^2 + 2xy = 15y^2$

KEY CONCEPTS   FOR CHAPTER 7

1. **Quadratic Trinomial:**
   $ax^2 + bx + c$ is a quadratic trinomial; ax^2 is called the second-degree term or the quadratic term, bx is the first-degree term or the linear term, and c is the constant term.

2. **Prime Polynomial:**
   A polynomial is prime over the integers if its only factorization with integer coefficients must involve 1 or -1 as one of the factors.

3. **Factorable Trinomials:**
   A trinomial $ax^2 + bx + c$ is factorable into a pair of binomial factors with integer coefficients if and only if there are two integers whose product is ac and whose sum is b. Otherwise, the trinomial is prime over the integers.

4. **GCF of a Polynomial:**
   The greatest common factor of a polynomial is the factor of each term that contains:
   i. The largest possible numerical coefficient and
   ii. The largest possible exponent on each variable factor.

5. **Factoring $ax^2 + bx + c$ by Grouping:**
   Step 1. Be sure you have factored out the GCF if it is not 1.
   Step 2. Find two factors of ac whose sum is b.
   Step 3. Rewrite the linear term of $ax^2 + bx + c$ so that b is the sum of the factors from Step 2.
   Step 4. Factor the polynomial from Step 3 by grouping the terms and factoring the GCF out of each pair of terms.

6. **Strategy for Factoring a Polynomial over the Integers:**
   After factoring out the GCF, proceed as follows:
   a. For binomials, factor special forms:
   $x^2 - y^2 = (x + y)(x - y)$ Difference of two squares
   $x^3 - y^3 = (x - y)(x^2 + xy + y^2)$ Difference of two cubes
   $x^3 + y^3 = (x + y)(x^2 - xy + y^2)$ Sum of two cubes
   ($x^2 + y^2$, the sum of two squares, is prime.)
   b. For trinomials, factor forms that are perfect squares:
   $x^2 + 2xy + y^2 = (x + y)^2$ Square of a sum
   $x^2 - 2xy + y^2 = (x - y)^2$ Square of a difference
   Otherwise, factor by the grouping method.

   c. For polynomials of four or more terms, factor by grouping.

7. **Quadratic Equation:**
   If a, b, and c are real constants and $a \neq 0$, then $ax^2 + bx + c = 0$ is the standard form of a quadratic equation in x.

8. **Zero-Factor Principle:**
   If a and b are real algebraic expressions, then $ab = 0$, if and only if $a = 0$ or $b = 0$.

9. **Solving Quadratic Equations by Factoring:**
   Step 1. Write the equation in standard form, with the right side 0.
   Step 2. Factor the left side of the equation.
   Step 3. Set each factor equal to 0.
   Step 4. Solve the resulting first-degree equations.

10. **Equivalent Statements about Linear Factors of a Polynomial:**
    For a real constant c and a real polynomial $P(x)$ the following statements are equivalent:
    ▪ $x - c$ is a factor of $P(x)$.
    ▪ $x = c$ is a solution to $P(x) = 0$.
    ▪ $P(c) = 0$; that is, c is a zero of $P(x)$.
    ▪ $(c, 0)$ is an x-intercept of the graph of $y = P(x)$.

11. **Solution of Multiplicity Two:**
    If both solutions of a quadratic equation are the same, we call the solution a double solution or a solution of multiplicity two.

12. **Nature of the Solutions of a Quadratic Equation:**
    ▪ If the parabola $y = ax^2 + bx + c$ has two x-intercepts, $ax^2 + bx + c = 0$ has two distinct real solutions.
    ▪ If the parabola $y = ax^2 + bx + c$ has one x-intercept with the vertex on the x-axis, $ax^2 + bx + c = 0$ has a double real solution.
    ▪ If the parabola $y = ax^2 + bx + c$ has no x-intercepts, $ax^2 + bx + c = 0$ has no real solution. Both solutions are imaginary. (This case will be covered in Chapter 8.)

REVIEW EXERCISES FOR CHAPTER 7

In Exercises 1–4 use the given table to factor each polynomial or to indicate that the polynomial is prime.

1. a. $x^2 - 48x - 100$ **b.** $x^2 - 15x - 100$
    c. $x^2 - 100$ **d.** $x^2 - 20x - 100$

FACTORS OF -100		SUM OF FACTORS
1	-100	-99
2	-50	-48
4	-25	-21
5	-20	-15
10	-10	0

2. a. $x^2 + 22x - 48$ **b.** $x^2 + 13x - 48$
    c. $x^2 + 2x - 48$ **d.** $x^2 + x - 48$

FACTORS OF -48		SUM OF FACTORS
-1	48	47
-2	24	22
-3	16	13
-4	12	8
-6	8	2

3. a. $100x^2 + 52x + 1$ **b.** $4x^2 + 52x + 25$
    c. $5x^2 + 29xy + 20y^2$ **d.** $10x^2 + 29xy + 10y^2$

FACTORS OF 100		SUM OF FACTORS
1	100	101
2	50	52
4	25	29
5	20	25
10	10	20

4. a. $48x^2 - 19x + 1$ **b.** $3x^2 - 26x + 16$
    c. $2x^2 - 49xy + 24y^2$ **d.** $16x^2 - 16xy + 3y^2$

FACTORS OF 48		SUM OF FACTORS
-1	-48	-49
-2	-24	-26
-3	-16	-19
-4	-12	-16
-6	-8	-14

In Exercises 5–42 factor each polynomial completely over the integers.

5. $4x - 36$ **6.** $2x^2 - 10x$
7. $12ax^2 - 24ax$ **8.** $x^2 - 4$
9. $4x^2 - 1$ **10.** $x^2 + 4x + 4$
11. $7x^2 - 28$ **12.** $m^2 - 64$
13. $x^2 - 11x + 18$ **14.** $m^2 + 10m + 21$
15. $ax + bx + ay + by$ **16.** $ax + 2ay - 3x - 6y$
17. $x^2 - xy - 42y^2$ **18.** $v^2 + 2v + 1 - w^2$
19. $9y^2 + 30y + 25$ **20.** $x^2 - 16$
21. $8x^3 - y^3$ **22.** $2x^4 - 16x$

23. $11x^2 - 44y^2$ **24.** $11x^2 - 11y^2 + 33x + 33y$
25. $4ax^2 - 9ay^2$
26. $5a(2x - 3y) + 3b(3y - 2x)$
27. $4x^2 + 2xy - 30y^2$ **28.** $6x^2 - 7xy - 20y^2$
29. $x^3y - 5x^2y + 4xy$ **30.** $6x^2 + 61xy + 10y^2$
31. $4x^2 + 12xy + 9y^2$ **32.** $x^2(a + b) - 25(a + b)$
33. $a^2(x + 5y) + 2ab(x + 5y) + b^2(x + 5y)$
34. $a^2(2x - 3y) - b^2(2x - 3y)$
35. $v^2 + 9w^2$ **36.** $20x^2 + 50x - 30$
37. $v^4 - v^3 - v + 1$ **38.** $v^4 - 1$
39. $64x^2 + 1$ **40.** $64x^2 - 1$
41. $64x^3 + 1$ **42.** $64x^3 - 1$

In Exercises 43–56 solve each equation.

43. $(x - 5)(x + 1) = 0$ **44.** $(2x - 3)(3x + 2) = 0$
45. $x(x - 7)(7x - 2) = 0$ **46.** $v^2 - 4v - 21 = 0$
47. $10y^2 + 13y - 3 = 0$ **48.** $6w^2 = 11w + 21$

49. $\dfrac{v^2}{30} = \dfrac{v}{15} + \dfrac{1}{2}$ **50.** $(x + 6)(x - 2) = 9$

51. $x(x^2 - 36) = 0$ **52.** $2x(x^2 - 10x + 25) = 0$
53. $(3w + 2)(w + 1) = (2w + 3)(w - 2)$

54. $\dfrac{x^2}{5} - \dfrac{9x}{10} + 1 = 0$

55. $\dfrac{z^2}{12} = \dfrac{z + 3}{3}$ **56.** $\dfrac{x^2}{9} + \dfrac{9}{4} = -x$

In Exercises 57–62 construct a quadratic equation in x with the solutions specified.

57. $-7, 7$ **58.** $2, -11$ **59.** $\dfrac{3}{5}, \dfrac{-1}{2}$ **60.** $0, 8$

61. A double root of -2 **62.** A double root of $\dfrac{3}{4}$

63. Construct a cubic equation in x with solutions of 0, 5, and 7.

64. Construct a cubic equation in x with a double solution of 2 and a solution of 3.

65. a. The difference in the distance traveled in one revolution by two bicycle tires of radius r_1 and r_2 is $2\pi r_1 - 2\pi r_2$. Factor this polynomial.

b. The lengths of two booms on circular irrigation systems are L_1 and L_2. The difference in the areas covered by these irrigation systems is $\pi L_1^2 - \pi L_2^2$. Factor this polynomial.

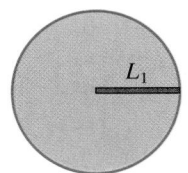 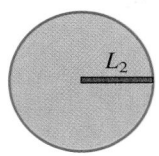

66. The difference in the volumes of two spherical oxygen tanks is $\frac{4}{3}\pi r_1^3 - \frac{4}{3}\pi r_2^3$. Factor this polynomial.

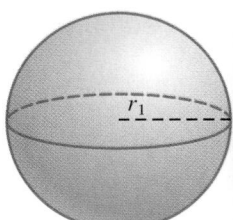 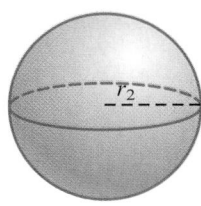

In Exercises 67 and 68 use the given table to solve each equation.

67. $x^2 + 6x - 216 = 0$

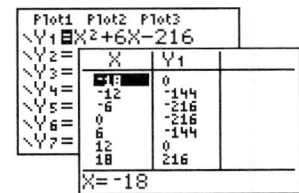

68. $4x^2 + 4x - 35 = 0$

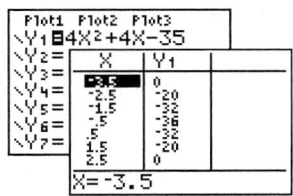

In Exercises 69 and 70 use the graph of $y = P(x)$ to solve:
   a. $P(x) = 0$ **b.** $P(x) < 0$ and   **c.** $P(x) > 0$

69. $P(x) = x^2 - 5x - 500$

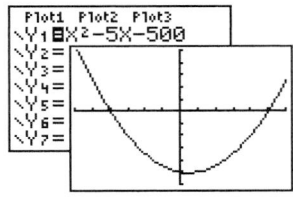

$[-30, 30, 5]$ by $[-600, 500, 100]$

70. $P(x) = x^3 - 4x^2 + x + 6$

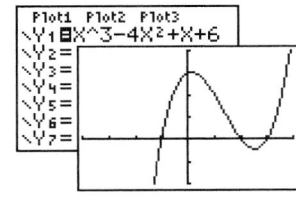

$[-4, 4, 1]$ by $[-4, 8, 2]$

In Exercises 71–74 use a graphics calculator to graph $y = P(x)$ and then use this graph to assist you in completing this table.

POLYNOMIAL $P(x)$	FACTORED FORM	ZEROS OF $P(x)$	SOLUTIONS OF $P(x) = 0$	x-INTERCEPTS OF THE GRAPH OF $y = P(x)$
71. $x^2 + 4x - 77$				
72. $35x^2 + 13x - 12$				
73. $x^2 - 20x - 525$				
74. $16x^2 - 72x + 81$				

75. The base of the triangle shown in the figure to the right is 3 cm longer than the height. Determine the height of the triangle if its area is 14 cm^2.

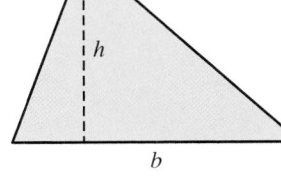

76. A metal sheet 60 cm wide is used to form a trough by bending up each side as illustrated in the figure to the right. Determine the height of each side if the cross-sectional area is 400 cm^2.

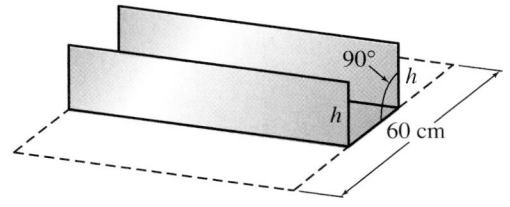

[7.1] **1.** Completely factor each trinomial.
 a. $w^2 - 4w - 45$ **b.** $w^2 + 14w + 45$
 c. $v^2 - 10v + 24$ **d.** $v^2 + 5v - 36$

[7.1] **2.** Completely factor each trinomial.
 a. $x^2 - xy - 12y^2$ **b.** $x^2 - 13xy + 12y^2$
 c. $a^2 + 30ab + 144b^2$ **d.** $a^2 + 2ab - 48b^2$

[7.2] **3.** Completely factor each trinomial.
 a. $6x^2 - 17x - 14$ **b.** $6x^2 - 31x + 14$
 c. $6x^2 + 85x + 14$ **d.** $10x^2 + 17x - 6$

[7.2] **4.** Completely factor each trinomial.
 a. $12x^2 + 25xy + 12y^2$ **b.** $12x^2 - 7xy - 12y^2$
 c. $12x^2 - 145xy + 12y^2$ **d.** $40x^2 + 7xy - 3y^2$

[7.3] **5.** Factor each perfect square trinomial.
 a. $x^2 + 14xy + 49y^2$ **b.** $x^2 - 16xy + 64y^2$
 c. $9x^2 + 60xy + 100y^2$ **d.** $25x^2 - 110xy + 121y^2$

[7.3] **6.** Factor each difference of two squares.
 a. $x^2 - 4y^2$ **b.** $400a^2 - b^2$
 c. $16v^2 - 49w^2$ **d.** $36x^2 - 25y^2$

[7.3] **7.** Factor each sum or difference of two cubes.
 a. $64v^3 - 1$ **b.** $v^3 + 125$
 c. $8x^3 + 125y^3$ **d.** $27a^3 - 1000b^3$

[7.4] **8.** Use grouping to factor each polynomial.
 a. $2ax + 3a + 2bx + 3b$
 b. $14ax - 6bx - 35ay + 15by$
 c. $a^2 - 4b^2 + a - 2b$
 d. $x^2 + 10xy + 25y^2 - 4$

[7.4] **9.** Completely factor each polynomial.
 a. $5x^2 - 245$
 b. $19x^3 - 19$
 c. $2ax^2 + 20ax + 50a$
 d. $-5ax^2 + 20axy - 15ay^2$

[7.5] **10.** Solve each equation.
 a. $(2x + 1)(x - 3) = 0$
 b. $x^2 - 2x - 99 = 0$
 c. $(x + 3)(x - 2)(4x - 9) = 0$
 d. $(3x + 1)(2x + 7) = 4(x + 1)$

[7.5] **11.** Use the x-intercepts of the graph of $y = P(x)$ to complete this table.

POLYNOMIAL $P(x)$	x-INTERCEPTS OF THE GRAPH OF $y = P(x)$	FACTORED FORM	ZEROS OF $P(x)$	SOLUTIONS OF $P(x) = 0$
a. $x^2 - 2x - 35$	$(-5, 0), (7, 0)$			
b. $16x^2 + 2x - 3$	$\left(-\frac{1}{2}, 0\right), \left(\frac{3}{8}, 0\right)$			
c. $16x^2 - 40x + 25$	$\left(\frac{5}{4}, 0\right)$			
d. $x^3 - 7x^2 + 6x$	$(0, 0), (1, 0), (6, 0)$			

GROUP PROJECT   FOR CHAPTER 7

Risks and Choices: Creating a Mathematical Model and Using This Model to Find an Optimal Solution

A fencing contractor in Sacramento, California, was contracted to install 150 ft of chain-link fencing to surround a
playground area next to an existing grade school building.
This is illustrated in the figure to the right. For security reasons the specifications indicate that the only entrances to
this area will come through the school. The length of the
wall along the school where no fencing is needed is 80 ft. The question that we explore is, What
dimensions should the contractor use to enclose the maximum possible rectangular area inside this
fence?

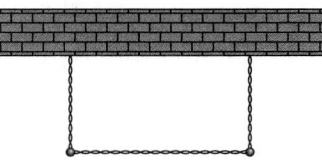

Exploration: Examining the Relationship Between the Width of the Playground and the Area of the Playground

Creating Numerical, Graphical, and Algebraic Models for This Problem

1. Complete the given table to form a numerical
   model for this problem.
2. **a.** Using the first two columns of this table,
   create a scatter diagram for this data using
   an appropriate scale for each axis. (See the
   Student Study Guide for a sample.) This
   will produce a graphical model for the
   length as a function of the width.

   b. Now write an equation for $L(W)$, which
   gives the length of the rectangular
   playground area as a function of the
   width. This will produce an algebraic
   model for the length as a function of the
   width. (*Suggestion:* Use your algebraic
   model to double-check your table and
   your graph.)

   c. What is the domain of acceptable input
   values for W in the function $L(W)$? (*Hint:*
   There is only so much fencing.)

W WIDTH (FT)	L LENGTH (FT)	A AREA (SQ FT)
10	130	1300
15		
20		
25		
30		
35		
40		
45		
50		
55		
60		
65		
70		

   d. What is the range of possible output values of $L(W)$ for all the input values listed in part
   2(c)?
3. **a.** Using the first and third columns of this table, plot this data on a scatter diagram using
   an appropriate scale for each axis. (See the Student Study Guide for a sample.) This
   produces a graphical model for the area as a function of the width.

   b. Now write an equation for $A(W)$, which gives the area of the rectangular playground as a
   function of the width. This will produce an algebraic model for the area as a function of
   the width. (*Suggestion:* Use your algebraic model to double-check your table and your
   graph.)

   c. What is the domain of acceptable input values for W in the function $A(W)$?

   d. What is the range of possible output values of $A(W)$ for all the input values listed in part
   3(c)?

Using the Models to Answer Questions about This Problem

1. **a.** Evaluate and interpret $L(32)$.
   b. Evaluate and interpret $A(32)$.
2. **a.** Evaluate and interpret $L(0)$.
   b. Evaluate and interpret $A(0)$.
3. **a.** Evaluate and interpret $L(75)$.
   b. Evaluate and interpret $A(75)$.
4. Determine the exact value of W for which $L(W) = 105$.
5. Determine the exact value of W for which $A(W) = 2772$.
6. Determine the width and the length of the fence that will enclose the maximum possible playground area. What is the maximum possible area?
7. Explain how you determined the dimensions that yield the maximum area. Which of the three types of mathematical models (numerical, graphical, or algebraic) did you use to answer this question? Why did you use this model?

8

RADICAL EXPRESSIONS, COMPLEX NUMBERS, AND QUADRATIC EQUATIONS

527

One classic engineering problem that involves fractional exponents is the relationship between the volume of a box beam and the strength of the beam. Bridges built to scale based upon a scaled-down model may not be a good idea. If a box beam in a bridge has its length, width, and height tripled, its cross-sectional area will be 9 times as great, and its volume will be 27 times as great.

Note that the volume increased 3 times as much as the area. The problem with this is that most measures of the strength of a box beam vary directly as the cross-sectional area. However, part of the stress on the beam comes from its own weight. Thus the stress a beam puts on itself increases at a greater rate than its strength. Later in this chapter we examine an application that gives the strength of a beam as directly proportional to $v^{2/3}$ where the exponent on the volume v is the fraction two-thirds.

In this chapter we examine fractional exponents, the relationship of fractional exponents to radicals, and the definition of imaginary and complex numbers. Then we examine quadratic equations and inequalities and some applications of these equations.

Section 8.1   Rational Exponents and Radicals

Objectives: 1. Interpret and use rational exponents.
2. Interpret and use radical notation.

Fractional Exponents

To develop a definition of fractional exponents we start by examining $x^{1/2}$. We want $x^{1/2}$ to obey the same rules for exponents as those given in Section 5.3 for integer exponents. Thus we want to be able to use the product rule to simplify $x^{1/2} \cdot x^{1/2}$.

$$x^{1/2} \cdot x^{1/2} = x^{1/2 + 1/2} \qquad \text{Use the product rule } x^m \cdot x^n = x^{m+n} \text{ to}$$
$$= x^1 \qquad\qquad \text{add the exponents.}$$

From Section 1.1, $\sqrt{x}$ denotes the principal square root of x. If $x > 0$, the principal square root is the positive square root of x. We now define $x^{1/2}$ to be $\sqrt{x}$.

A number which when multiplied by itself yields the product x is known as the square root of x. Thus it is reasonable to define $x^{1/2}$ as $\sqrt{x}$, the principal square root of x.

The following box examines the definition of $x^{1/2}$ algebraically, numerically, and graphically.

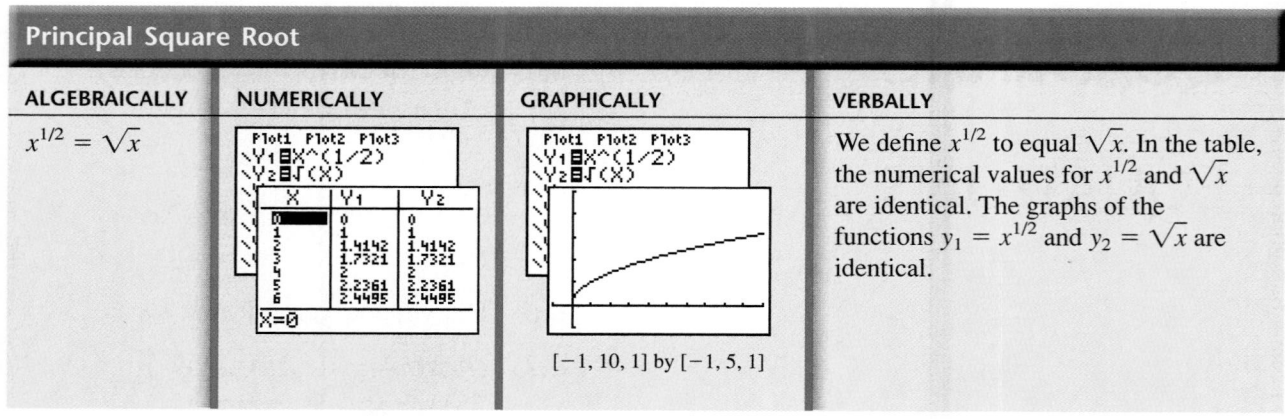

Principal Square Root

ALGEBRAICALLY	NUMERICALLY	GRAPHICALLY	VERBALLY
$x^{1/2} = \sqrt{x}$			We define $x^{1/2}$ to equal $\sqrt{x}$. In the table, the numerical values for $x^{1/2}$ and $\sqrt{x}$ are identical. The graphs of the functions $y_1 = x^{1/2}$ and $y_2 = \sqrt{x}$ are identical.

$[-1, 10, 1]$ by $[-1, 5, 1]$

We now generalize the meaning of fractional exponents in order to define $x^{1/n}$ for any natural number n and fraction $\frac{1}{n}$. We want to be able to use the power rule for exponents with $x^{1/n}$ as shown here.

$$\left(x^{1/n}\right)^n = x^{(1/n \cdot n/1)}$$
$$= x^1$$

Use the power rule $(x^m)^n = x^{mn}$ to multiply the exponents.

Because $x^{1/n}$ is used as a factor n times in the expression $\left(x^{1/n}\right)^n$, it is reasonable to define $x^{1/n}$ as $\sqrt[n]{x}$, the **principal nth root of x.** In the radical notation, $\sqrt[n]{x}$, x is called the **radicand,** $\sqrt{}$ is the radical symbol, and n is called the **index** or the **order of the radical.** Both the fractional exponent form $x^{1/n}$ and the radical notation form $\sqrt[n]{x}$ are available on most calculators.

We define $x^{1/n}$ to equal $\sqrt[n]{x}$, the principal nth root of x. If no index is written, as in $\sqrt{x}$, this notation is understood to mean the square root of x.

9 has two square roots: 3 and -3. Only 3 equals the principal square root of 9, denoted by $\sqrt{9}$ or by $9^{1/2}$.

Principal nth Root

The principal nth root of the real number x is denoted by either $x^{1/n}$ or $\sqrt[n]{x}$.

EXAMPLES

VERBALLY		RADICAL NOTATION	EXPONENTIAL NOTATION
For $x > 0$:	The principal nth root is positive for all natural numbers n.	$\sqrt{9} = 3$ $\sqrt[3]{8} = 2$	$9^{1/2} = 3$ $8^{1/3} = 2$
For $x = 0$:	The principal nth root of 0 is 0.	$\sqrt{0} = 0$ $\sqrt[3]{0} = 0$	$0^{1/2} = 0$ $0^{1/3} = 0$
For $x < 0$:	If n is odd, the principal nth root is negative.	$\sqrt[3]{-8} = -2$ $\sqrt[5]{-1} = -1$	$(-8)^{1/3} = -2$ $(-1)^{1/5} = -1$
	If n is even, there is no real nth root.* (The nth roots will be imaginary.)	$\sqrt{-1}$ is not a real number.	$(-1)^{1/2}$ is not a real number.

*The nth roots will be imaginary numbers which are covered in Section 8.4.

■ EXAMPLE 1   Representing and Evaluating Principal nth Roots

Represent each of these principal nth roots using both radical notation and exponential notation and evaluate each expression.

SOLUTIONS

	Radical Notation	*Exponential Notation*	
(a) The principal square root of 25	$\sqrt{25} = 5$	$25^{1/2} = 5$	Check: $5 \cdot 5 = 25$
(b) The principal cube root of 27	$\sqrt[3]{27} = 3$	$27^{1/3} = 3$	Check: $3 \cdot 3 \cdot 3 = 27$
(c) The principal cube root of -27	$\sqrt[3]{-27} = -3$	$(-27)^{1/3} = -3$	Check: $(-3)(-3)(-3) = -27$
(d) The principal fourth root of 16	$\sqrt[4]{16} = 2$	$16^{1/4} = 2$	Check: $(2)(2)(2)(2) = 16$
(e) The principal fifth root of 1	$\sqrt[5]{1} = 1$	$1^{1/5} = 1$	Check: $(1)(1)(1)(1)(1) = 1$
(f) The principal square root of -25	$\sqrt{-25}$ is not a real number.	$(-25)^{1/2}$ is not a real number.	
(g) The opposite of the principal square root of 25	$-\sqrt{25} = -5$	$-(25)^{1/2} = -5$	Because $\sqrt{25} = 5, -\sqrt{25} = -5$

SELF-CHECK 8.1.1

Write each of these expressions in radical notation and evaluate each expression.

1. $64^{1/2}$ **2.** $64^{1/3}$ **3.** $64^{1/6}$ **4.** $-64^{1/2}$ **5.** $(-64)^{1/2}$ **6.** $(-64)^{1/3}$

To extend the definition of exponents to include any rational number $\frac{m}{n}$, we start with $x^{1/n} = \sqrt[n]{x}$ and then use the power rule for exponents.

$$x^{1/n} = \sqrt[n]{x}$$
$$\left(x^{1/n}\right)^m = \left(\sqrt[n]{x}\right)^m \quad \text{Raise both sides of the equation to the } m\text{th power.}$$
$$x^{m/n} = \left(\sqrt[n]{x}\right)^m \quad \text{Then use the power rule for exponents to multiply the exponents.}$$

In keeping with our earlier work with negative exponents, we also interpret $x^{-m/n}$ as $\frac{1}{x^{m/n}}$. These new definitions are summarized in the following box.

Rational Exponents

For a real number x and natural numbers m and n:

EXAMPLES

ALGEBRAICALLY	RADICAL NOTATION	EXPONENTIAL NOTATION
If $x^{1/n}$ is a real number,* $x^{m/n} = \left(x^{1/n}\right)^m = \left(\sqrt[n]{x}\right)^m$	$(-8)^{2/3} = \left(\sqrt[3]{-8}\right)^2$ $= (-2)^2$ $= 4$	$(-8)^{2/3} = \left[(-8)^{1/3}\right]^2$ $= (-2)^2$ $= 4$
or $x^{m/n} = \left(x^m\right)^{1/n} = \sqrt[n]{x^m}$	$(y^3)^{2/7} = \sqrt[7]{(y^3)^2}$ $= \sqrt[7]{y^6}$	$(y^3)^{2/7} = y^{(3/1)(2/7)}$ $= y^{6/7}$
$x^{-m/n} = \frac{1}{x^{m/n}} = \frac{1}{\sqrt[n]{x^m}}; x \neq 0$	$16^{-3/4} = \frac{1}{\left(\sqrt[4]{16}\right)^3}$ $= \frac{1}{2^3}$ $= \frac{1}{8}$	$16^{-3/4} = \frac{1}{16^{3/4}}$ $= \frac{1}{\left(16^{1/4}\right)^3}$ $= \frac{1}{2^3}$ $= \frac{1}{8}$

*If $x < 0$ and n is even, then $x^{1/n}$ is not a real number.

SELF-CHECK 8.1.1 ANSWERS

1. $\sqrt{64} = 8$
2. $\sqrt[3]{64} = 4$
3. $\sqrt[6]{64} = 2$
4. $-\sqrt{64} = -8$
5. $\sqrt{-64}$ is not a real number.
6. $\sqrt[3]{-64} = -4$

■ EXAMPLE 2 Evaluating Expressions with Fractional Exponents

Represent each of these expressions in radical notation and exponential notation and evaluate each expression.

SOLUTIONS

		Radical Notation	*Exponential Notation*
(a)	$27^{2/3}$	$\left(\sqrt[3]{27}\right)^2 = 3^2 = 9$	$27^{2/3} = \left(27^{1/3}\right)^2 = 3^2 = 9$
(b)	$32^{3/5}$	$\left(\sqrt[5]{32}\right)^3 = 2^3 = 8$	$32^{3/5} = \left(32^{1/5}\right)^3 = 2^3 = 8$
(c)	$9^{-1/2}$	$\dfrac{1}{\sqrt{9}} = \dfrac{1}{3}$	$9^{-1/2} = \dfrac{1}{9^{1/2}} = \dfrac{1}{3}$
(d)	$27^{-2/3}$	$\dfrac{1}{\left(\sqrt[3]{27}\right)^2} = \dfrac{1}{3^2} = \dfrac{1}{9}$	$27^{-2/3} = \dfrac{1}{27^{2/3}} = \dfrac{1}{\left(27^{1/3}\right)^2} = \dfrac{1}{3^2} = \dfrac{1}{9}$

A Mathematical Note

The radical sign, $\sqrt{}$, is composed of two parts: $\sqrt{}$ and $\overline{}$. The symbol $\sqrt{}$ comes from the letter r, the first letter of the Latin word *radix,* which means root. Thus the $\sqrt{}$ indicates that a root is to be taken of the quantity underneath the bar (also called the vinculum). On many calculators we can access the symbol $\sqrt{}$ but not the bar. To indicate the square root of a quantity, we must use parentheses instead of the bar. For example, $\sqrt{2x+1}$ can be represented on calculators by $\sqrt{}\ (2x+1)$.

SELF-CHECK 8.1.2

Write each of these expressions in radical notation and evaluate each expression.

1. $64^{3/2}$ 2. $64^{2/3}$ 3. $(-64)^{2/3}$ 4. $64^{-2/3}$

The expressions in Example 2 were carefully selected so that we could illustrate the notation but still evaluate the expressions using pencil and paper. In Calculator Perspective 8.1.1 and in Example 3 we use a calculator to approximate each expression.

CALCULATOR PERSPECTIVE 8.1.1	Evaluating Expressions with Rational Exponents and Radicals

To evaluate $27^{2/3}$, $32^{3/5}$, and $9^{-1/2}$ from Example 2 on a TI-83 Plus calculator, enter the following keystrokes:

       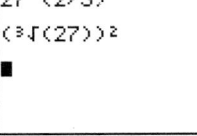

    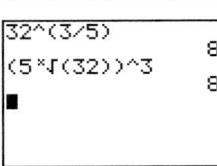

1. $64^{3/2} = \left(\sqrt{64}\right)^3$ $= 8^3 = 512$

2. $64^{2/3} = \left(\sqrt[3]{64}\right)^2$ $= 4^2 = 16$

3. $(-64)^{2/3} = \left(\sqrt[3]{-64}\right)^2$ $= (-4)^2 = 16$

4. $64^{-2/3} = \dfrac{1}{\left(\sqrt[3]{64}\right)^2}$ $= \dfrac{1}{4^2} = \dfrac{1}{16}$

   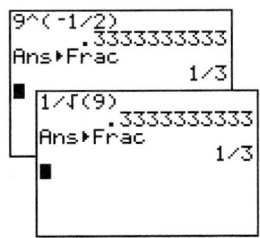

■ EXAMPLE 3   Approximating Radical Expressions with a Calculator

Use a graphics calculator to approximate each expression to the nearest thousandth.

(a) $\sqrt[3]{40}$ (b) $\sqrt[4]{50}$ (c) $\sqrt[5]{-45}$

SOLUTIONS

The notation $4\sqrt[x]{\ }(50)$ on a TI-83 Plus calculator denotes the principal fourth root of 50. Many students prefer to enter expressions like this in the exponential form illustrated by 50^(1/4).

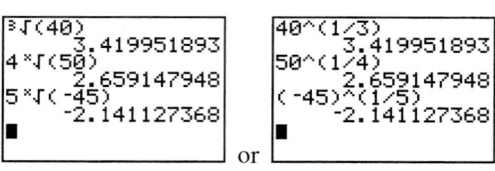

or

Answers: (a) $\sqrt[3]{40} \approx 3.420$

(b) $\sqrt[4]{50} \approx 2.659$

(c) $\sqrt[5]{-45} \approx -2.141$

SELF-CHECK 8.1.3

Use a graphics calculator to approximate each expression to the nearest hundredth.

1. $\sqrt[7]{28}$ 2. $\sqrt[3]{-28}$ 3. $\sqrt[4]{28}$ 4. $\sqrt[4]{-28}$

We have defined fractional exponents in a way that all the properties of integral exponents now apply to fractional exponents. These properties are summarized next.

One reason it is useful to examine radical expressions in exponential form is that we can apply the properties of exponents to simplify these expressions.

Properties of Exponents

Let x, y, m, and n be real numbers, with $x \neq 0$ and $y \neq 0$.

Product rule: $x^m \cdot x^n = x^{m+n}$

Power rule: $(x^m)^n = x^{mn}$

Product to a power: $(xy)^m = x^m y^m$

Quotient to a power: $\left(\dfrac{x}{y}\right)^m = \dfrac{x^m}{y^m}$

Quotient rule: $\dfrac{x^m}{x^n} = x^{m-n}$

Negative power: $\left(\dfrac{x}{y}\right)^{-n} = \left(\dfrac{y}{x}\right)^n$

SELF-CHECK 8.1.3 ANSWERS

1. $\sqrt[7]{28} \approx 1.61$
2. $\sqrt[3]{-28} \approx -1.61$
3. $\sqrt[4]{28} \approx 2.30$
4. $\sqrt[4]{-28}$ is not a real number

To avoid the special care that must be exercised when the base is negative and we are evaluating an even number root, we restrict the variables to positive values in Examples 4 to 8. We present examples with negative bases in Section 8.4.

■ EXAMPLE 4   Using the Properties of Exponents

Simplify each of the following expressions. Express all answers in terms of positive exponents. Assume that z represents a positive real number.

SOLUTIONS

(a) $9^{3/8} \cdot 9^{1/8}$

$$9^{3/8} \cdot 9^{1/8} = 9^{(3/8)+(1/8)}$$
$$= 9^{1/2}$$
$$= 3$$

Use the product rule to add the exponents:
$\dfrac{3}{8} + \dfrac{1}{8} = \dfrac{4}{8} = \dfrac{1}{2}$
The principal square root of 9 is 3.

(b) $8^{1/5} \cdot 4^{1/5}$

$$8^{1/5} \cdot 4^{1/5} = (8 \cdot 4)^{1/5}$$
$$= 32^{1/5}$$
$$= 2$$

Product to a power: $x^m y^m = (xy)^m$.
The principal fifth root of 32 is 2.

(c) $\dfrac{625^{7/8}}{625^{1/8}}$

$$\dfrac{625^{7/8}}{625^{1/8}} = 625^{(7/8)-(1/8)}$$
$$= 625^{3/4}$$
$$= \left(625^{1/4}\right)^3$$
$$= 5^3$$
$$= 125$$

Use the quotient rule to subtract the exponents:
$\dfrac{7}{8} - \dfrac{1}{8} = \dfrac{6}{8} = \dfrac{3}{4}$

The principal fourth root of 625 is 5.

(d) $\left(z^{-2/3}\right)^{-3}$

$$\left(z^{-2/3}\right)^{-3} = z^{(-2/3)(-3)}$$
$$= z^2$$

Use the power rule to multiply the exponents:
$\left(-\dfrac{2}{3}\right)(-3) = 2$

■

■ EXAMPLE 5   Using the Properties of Exponents

Simplify each expression. Express each answer in terms of positive exponents. Assume that the variables represent only positive real numbers.

SOLUTIONS

(a) $\left(a^{-8}b^{12}\right)^{-1/4}$

$$\left(a^{-8}b^{12}\right)^{-1/4} = \left(a^{-8}\right)^{-1/4}\left(b^{12}\right)^{-1/4}$$
$$= a^2 b^{-3}$$
$$= \dfrac{a^2}{b^3}$$

Use the product to a power rule for exponents $(xy)^m = x^m y^m$ to take the $-\dfrac{1}{4}$ power of each factor.
Use the power rule to multiply the exponents:
$-8\left(-\dfrac{1}{4}\right) = 2$ and $12\left(-\dfrac{1}{4}\right) = -3.$
Rewrite the expression without negative exponents: $b^{-3} = \dfrac{1}{b^3}$.

(b) $\left(\dfrac{16x^4}{25}\right)^{-3/2}$

$$\left(\dfrac{16x^4}{25}\right)^{-3/2} = \left[\left(\dfrac{16x^4}{25}\right)^{1/2}\right]^{-3} \qquad x^{m/n} = \left(x^{1/n}\right)^m$$

$$= \left[\dfrac{16^{1/2}x^{4/2}}{25^{1/2}}\right]^{-3} \qquad \text{Take the } \dfrac{1}{2} \text{ power of each factor in the}$$
numerator and the denominator and then
simplify. Note that $16^{1/2} = 4$ and $25^{1/2} = 5$.

$$= \left[\dfrac{4x^2}{5}\right]^{-3}$$

$$= \left[\dfrac{5}{4x^2}\right]^3 \qquad \text{Rewrite the expression without negative exponents.}$$

$$= \dfrac{5^3}{4^3(x^2)^3} \qquad \text{Raise each factor to the third power. To raise a power to a power, multiply exponents.}$$

$$= \dfrac{125}{64x^6}$$

SELF-CHECK 8.1.4

Simplify each expression. Assume that x is a positive real number.

1. $\left(8^{-4/5}\right)^{-5/2}$ 2. $25^{5/16} \cdot 25^{3/16}$ 3. $25^{1/3} \cdot 5^{1/3}$ 4. $\left(\dfrac{8x^6}{125}\right)^{2/3}$

Example 6 uses the distributive property to simplify an expression involving both fractional and negative exponents.

■ **EXAMPLE 6**   Multiplying Using the Distributive Property

Simplify $2x^{2/3}\left(3x^{1/3} - 5x^{-2/3}\right)$. Express the answer in terms of positive exponents. Assume that all variables represent positive real numbers.

SOLUTION ───────────────

$$2x^{2/3}\left(3x^{1/3} - 5x^{-2/3}\right) = \left(2x^{2/3}\right)\left(3x^{1/3}\right) - \left(2x^{2/3}\right)\left(5x^{-2/3}\right) \quad \text{First use the distributive property.}$$
$$= 6x^{3/3} - 10x^0 \qquad \text{Then simplify each term, adding the exponents of the factors with the same base. Note that } x^0 = 1.$$
$$= 6x - 10$$

SELF-CHECK 8.1.4 ANSWERS
1. 64
2. 5
3. 5
4. $\dfrac{4x^4}{25}$

The product of exponential expressions with two or more terms can be found using some of the skills developed for multiplying polynomials. In particular, we use special forms, such as the product of a sum times a difference:

$$(x + y)(x - y) = x^2 - y^2$$

■ EXAMPLE 7   Multiplying a Sum by a Difference

Simplify $(a^{1/2} + b^{1/2})(a^{1/2} - b^{1/2})$. Assume that all variables represent positive real numbers.

SOLUTION _____

$$(a^{1/2} + b^{1/2})(a^{1/2} - b^{1/2}) = (a^{1/2})^2 - (b^{1/2})^2$$
$$= a - b$$

By inspection, observe that this expression fits the special form $(x + y)(x - y) = x^2 - y^2$, with $x = a^{1/2}$ and $y = b^{1/2}$.
To raise a power to a power, multiply exponents. ■

Example 8 uses the special factors for the sum of two cubes:
$(x + y)(x^2 - xy + y^2) = x^3 + y^3$.

■ EXAMPLE 8   Multiplying the Factors of the Sum of Two Cubes

Simplify $(a^{1/3} + b^{1/3})(a^{2/3} - a^{1/3}b^{1/3} + b^{2/3})$. Assume that all variables represent positive real numbers.

SOLUTION _____

$$(a^{1/3} + b^{1/3})(a^{2/3} - a^{1/3}b^{1/3} + b^{2/3}) = (a^{1/3})^3 + (b^{1/3})^3$$
$$= a + b$$

By inspection, observe that this expression fits the special form $(x + y)(x^2 - xy + y^2) = x^3 + y^3$, with $x = a^{1/3}$ and $y = b^{1/3}$.
To raise a power to a power, multiply exponents. ■

SELF-CHECK 8.1.5

Simplify each expression.

1. $4x^{5/8}(3x^{11/8} - 5x^{3/8} - 7x^{-5/8})$

2. $(a^{1/3} - b^{1/3})(a^{2/3} + a^{1/3}b^{1/3} + b^{2/3})$

We now compare $\sqrt[3]{-1}$ and $\sqrt{-1}$. $\sqrt[3]{-1} = -1$ because $(-1)(-1)(-1) = -1$. However, $\sqrt{-1}$ is not defined as a real number because there is no real number x such that $x^2 = -1$. This is true because if $x \geq 0$, then $x^2 \geq 0$ and if $x < 0$, then we still have $x^2 > 0$. Note the message displayed on the calculator window shown next for $\sqrt{-1}$.

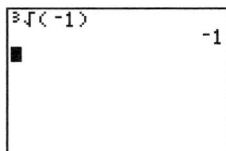
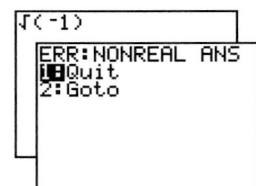

This example illustrates an important distinction regarding radicals of even and odd order. Even roots of negative numbers are not defined as real numbers while odd roots of negative numbers are defined as negative real numbers. This can be seen graphically by examining the graphs of radical functions.

■ EXAMPLE 9   Using a Graphics Calculator to Determine the Domain of Radical Functions

Use a graphics calculator to graph $y = \sqrt{x}$ and $y = \sqrt[3]{x}$ and to determine the domain of each function.

SOLUTION

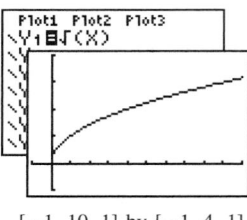

$[-1, 10, 1]$ by $[-1, 4, 1]$

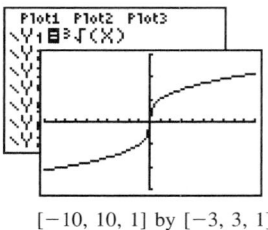

$[-10, 10, 1]$ by $[-3, 3, 1]$

From Section 6.1 the domain of a function is the projection of its graph into the x-axis. The projection of the graph of $y = \sqrt{x}$ onto the x-axis is $[0, +\infty)$. The projection of the graph of $y = \sqrt[3]{x}$ onto the x-axis is $\mathbb{R}$.

Answer: The domain of $y = \sqrt{x}$ is $[0, +\infty)$. The domain of $y = \sqrt[3]{x}$ is $\mathbb{R}$ (the set of all real numbers).   ■

Not only is the domain of $f(x) = \sqrt{x}$ the set $[0, +\infty)$, the domain of $f(x) = \sqrt[n]{x}$ is $[0, +\infty)$ for any even natural number n. We examine even roots of negative numbers further in Section 8.4. Also the domain of $f(x) = \sqrt[n]{x}$ is $\mathbb{R}$ for any odd natural number.

USING THE LANGUAGE AND SYMBOLISM OF MATHEMATICS   8.1

1. In the radical notation $\sqrt[n]{x}$, x is called the _____, $\sqrt{}$ is called the _____ symbol, and n is called the _____ or the _____ of the radical.
2. In radical notation, the expression $x^{1/5}$ can be written as _____.

3. In exponential notation, the expression $\sqrt[4]{x^3}$ can be written as _____.
4. The principal nth root of x is denoted by either _____ or _____.
5. The principal nth root of x is not a real number if n is _____ and x is _____.

EXERCISES   8.1

In Exercises 1 and 2 represent each expression using both radical notation and exponential notation. Assume that x and y are positive real numbers.
   1. a. The principal cube root of w
      b. The principal fourth root of x
      c. The principal seventh root of v
   2. a. The principal sixth root of x
      b. The principal ninth root of v
      c. The principal twelfth root of y

In Exercises 3 and 4 represent each expression using both radical notation and exponential notation and evaluate each expression.

   3. a. The principal square root of 4
      b. The principal cube root of 1000
      c. The principal fifth root of 32
   4. a. The principal square root of 100
      b. The principal fourth root of 81
      c. The principal sixth root of 1

In Exercises 5–8 represent each expression using exponential notation and evaluate each expression without using a calculator. Use a calculator only to check your answer.
   5. a. $\sqrt{16}$ b. $\sqrt[4]{16}$ c. $\sqrt[5]{0}$
   6. a. $\sqrt{49}$ b. $\sqrt[3]{125}$ c. $\sqrt[7]{1}$

7. a. $\sqrt[3]{-8}$ **b.** $\sqrt[5]{\dfrac{-1}{32}}$ **c.** $\sqrt{0.01}$

8. a. $\sqrt[3]{-0.001}$ **b.** $\sqrt{\dfrac{4}{9}}$ **c.** $\sqrt[5]{-32}$

In Exercises 9–26 represent each expression using radical notation and evaluate each expression without using a calculator. Use a calculator only to check your answer.

9. a. $36^{1/2}$ **b.** $36^{-1/2}$ **c.** $-36^{1/2}$
10. a. $100^{1/2}$ **b.** $100^{-1/2}$ **c.** $-100^{1/2}$
11. a. $27^{1/3}$ **b.** $(-27)^{1/3}$ **c.** $27^{-1/3}$
12. a. $16^{1/4}$ **b.** $16^{-1/4}$ **c.** $-16^{1/4}$
13. a. $0.09^{1/2}$ **b.** $-0.09^{1/2}$ **c.** $400^{-1/2}$
14. a. $0.49^{1/2}$ **b.** $-0.49^{1/2}$ **c.** $900^{-1/2}$
15. a. $0.008^{1/3}$ **b.** $(-0.008)^{1/3}$ **c.** $64^{-1/3}$
16. a. $0.027^{1/3}$ **b.** $(-0.027)^{1/3}$ **c.** $125^{-1/3}$
17. a. $16^{1/4}$ **b.** $16^{-1/4}$ **c.** $-16^{1/2}$
18. a. $81^{1/4}$ **b.** $81^{-1/4}$ **c.** $-81^{1/2}$
19. a. $\left(\dfrac{8}{125}\right)^{2/3}$ **b.** $\left(\dfrac{9}{4}\right)^{3/2}$ **c.** $\left(\dfrac{8}{125}\right)^{-2/3}$
20. a. $\left(\dfrac{16}{81}\right)^{3/4}$ **b.** $\left(\dfrac{49}{25}\right)^{3/2}$ **c.** $\left(\dfrac{16}{81}\right)^{-3/4}$
21. a. $25^{1/2} + 144^{1/2}$ **b.** $(25 + 144)^{1/2}$ **c.** $(25 - 16)^{1/2}$
22. a. $9^{1/2} + 16^{1/2}$ **b.** $(9 + 16)^{1/2}$ **c.** $25^{1/2} - 16^{1/2}$
23. a. $8^{1/3} + 1^{1/3}$ **b.** $(26 + 1)^{1/3}$ **c.** $(1001 - 1)^{1/3}$
24. a. $27^{1/3} - 1^{1/3}$ **b.** $(65 - 1)^{1/3}$ **c.** $(7 + 1)^{1/3}$
25. a. $0.000001^{1/2}$ **b.** $0.000001^{1/3}$ **c.** $0.000001^{1/6}$
26. a. $0.0001^{1/2}$ **b.** $0.0001^{1/4}$ **c.** $(-0.00001)^{1/5}$

In Exercises 27–30 use a graphics calculator to approximate each expression to the nearest thousandth.

27. a. $\sqrt{35}$ **b.** $\sqrt[3]{35}$ **c.** $\sqrt[6]{35}$
28. a. $\sqrt{52}$ **b.** $\sqrt[3]{52}$ **c.** $\sqrt[5]{52}$
29. a. $\sqrt[3]{-21}$ **b.** $\sqrt[3]{-21}$ **c.** $-\sqrt{21}$
30. a. $\sqrt[3]{-15}$ **b.** $\sqrt[3]{-15}$ **c.** $-\sqrt{15}$

In Exercises 31 and 32 determine which expression does not represent a real number.

31. a. $-\sqrt{17}$ **b.** $\sqrt{-17}$ **c.** $\sqrt[3]{-17}$
32. a. $\sqrt{-23}$ **b.** $\sqrt[3]{-23}$ **c.** $-\sqrt[3]{23}$

In Exercises 33–56 simplify each expression. Express all answers in terms of positive exponents. Assume that all variables represent positive real numbers.

33. $5^{1/2} \cdot 5^{3/2}$ **34.** $7^{5/3} \cdot 7^{1/3}$ **35.** $\left(8^{5/3}\right)^{2/5}$

36. $\left(32^{7/10}\right)^{2/7}$ **37.** $\dfrac{11^{4/3}}{11^{1/3}}$ **38.** $\dfrac{9^{5/4}}{9^{3/4}}$

39. $\left(27^{1/12} \cdot 27^{-5/12}\right)^{-2}$ **40.** $\left(4^{1/5} \cdot 4^{2/5}\right)^{5/2}$
41. $x^{1/3} \cdot x^{1/2}$ **42.** $y^{1/4} \cdot y^{1/5}$
43. $\dfrac{x^{1/2}}{x^{1/3}}$ **44.** $\dfrac{y^{1/4}}{y^{1/5}}$
45. $\left(z^{3/4}\right)^{2/7}$ **46.** $\left(z^{5/12}\right)^{4/15}$
47. $\dfrac{w^{-2/3}}{w^{-5/3}}$ **48.** $\dfrac{w^{3/4}}{w^{-3/8}}$

49. $\left(v^{-10}w^{-15}\right)^{-1/5}$ **50.** $\left(v^{-26}w^{-39}\right)^{-1/13}$
51. $\left(16v^{-2/5}\right)^{3/2}$ **52.** $\left(25v^{-4/9}\right)^{3/2}$
53. $\left(\dfrac{16n^{2/3}}{81n^{-2/3}}\right)^{-3/4}$ **54.** $\left(\dfrac{27n^{3/5}}{n^{-3/5}}\right)^{-2/3}$
55. $\dfrac{\left(27x^2y\right)^{1/2}\left(3xy\right)^{1/2}}{5x^{1/2}y^2}$ **56.** $\dfrac{\left(25x^2y^3z\right)^{1/3}\left(5xy^2z^2\right)^{1/3}}{3x^2y^{-1/3}z^{-2}}$

In Exercises 57–74 perform the indicated multiplications and express the answers in terms of positive exponents. Assume that all variables represent positive real numbers.

57. $x^{3/5}\left(x^{2/5} - x^{-3/5}\right)$ **58.** $x^{2/3}\left(x^{4/3} + x^{-2/3}\right)$
59. $y^{-7/4}\left(2y^{11/4} - 3y^{7/4}\right)$ **60.** $y^{-9/7}\left(5y^{23/7} - 6y^{16/7}\right)$
61. $3w^{5/11}\left(2w^{17/11} - 5w^{6/11} - 9w^{-5/11}\right)$
62. $4w^{-5/3}\left(3w^{11/3} - 7w^{8/3} + 2w^{5/3}\right)$
63. $\left(a^{1/2} + 3\right)\left(a^{1/2} - 3\right)$
64. $\left(2a^{1/2} - 3b^{1/2}\right)\left(2a^{1/2} + 3b^{1/2}\right)$
65. $\left(b^{3/5} - c^{5/3}\right)\left(b^{3/5} + c^{5/3}\right)$
66. $\left(b^{3/5} + c^{5/3}\right)^2$
67. $\left(b^{3/5} - c^{5/3}\right)^2$
68. $\left(x^{1/2} - x^{-1/2}\right)^2$
69. $\left(x^{-1/2} + x^{1/2}\right)^2$
70. $\left(x^{2/3} + x\right)\left(x^{2/3} - x\right)$
71. $\left(y^{1/3} + 2\right)\left(y^{2/3} - 2y^{1/3} + 4\right)$
72. $\left(3y^{1/3} - 5\right)\left(9y^{2/3} + 15y^{1/3} + 25\right)$
73. $\left[\left(3^{1/3} + 5^{1/3}\right)\left(3^{2/3} - 15^{1/3} + 5^{2/3}\right)\right]^{5/3}$
74. $\left[\left(14^{1/3} - 5^{1/3}\right)\left(14^{2/3} + 70^{1/3} + 5^{2/3}\right)\right]^{3/2}$

In Exercises 75 and 76 match each function with its graph. (*Hint:* This can be done by inspection by examining the domain of each function.)

75. a. $f(x) = \sqrt[4]{x}$ **b.** $f(x) = \sqrt[5]{x}$

A.
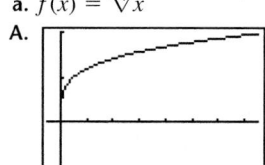
$[-1, 15, 2]$ by $[-1, 2, 1]$

B.
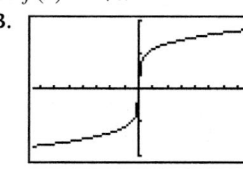
$[-15, 15, 2]$ by $[-2, 2, 1]$

76. a. $f(x) = \sqrt[6]{x}$ **b.** $f(x) = \sqrt[7]{x}$

A.
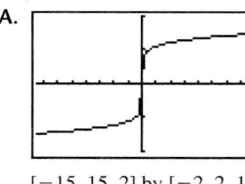
$[-15, 15, 2]$ by $[-2, 2, 1]$

B.
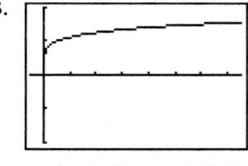
$[-1, 15, 2]$ by $[-2, 2, 1]$

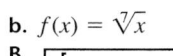

Group Discussion Questions

77. Challenge Question Use the properties of exponents to simplify each expression assuming m is a natural number and that x is a positive real number.

a. $x^{m/3}x^{m/2}$　　　　**b.** $\dfrac{16x^{m/2}}{2x^{m/3}}$

c. $\left(x^{m/3}y^{m/2}\right)^6$

78. Challenge Question Use the distributive property to expand the first expression and to factor the second expression.

EXPAND	FACTOR
a. $x^{1/2}(x^2 + 3x - 4)$	$x^{5/2} + 3x^{3/2} - 4x^{1/2} = x^{1/2}(\quad)$
	$= x^{1/2}(\quad)(\quad)$
b. $x^{-1/2}(x^2 - 25)$	$x^{3/2} - 25x^{-1/2} = x^{-1/2}(\quad)$
	$= x^{-1/2}(\quad)(\quad)$

79. Challenge Question The following results were obtained on a TI-83 Plus calculator. Explain these results. (*Hint:* Write each expression in radical notation.)

a. $(-70) \wedge .2 \approx -2.34$

b. $(-70) \wedge .3$ is undefined

c. $(-70) \wedge .4 \approx 5.47$

d. $(-70) \wedge .5$ is undefined

e. $(-70) \wedge .6 \approx -12.80$

80. Challenge Question

a. Give an example of a real number for which $\left(x^2\right)^{1/2} \neq x$.

b. Give an example of a real number for which $x^{1/2}$ is defined but $x^{-1/2}$ is undefined.

81. Discovery Question

a. Compare the graphs of $y_1 = \sqrt{x^2}$ and $y_2 = |x|$.

b. Compare the graphs of $y_1 = \sqrt[3]{x^3}$ and $y_2 = x$.

Section 8.2 Addition and Simplification of Radical Expressions

Objectives:　**3.** Add and subtract radical expressions.
　　　　　　　　　4. Simplify radical expressions.

Like Radicals

We add like radicals in exactly the same way that we add like terms of a polynomial. Like radicals must have the same index and the same radicand; only the coefficients of the like terms can differ. In exponential form, like radicals have the same base and the same exponents.

■ EXAMPLE 1　Classifying Radicals as Like or Unlike

Classify these radicals as like or unlike.

SOLUTIONS _____

(a) $-3\sqrt{2}, 5\sqrt{2},$ and $\dfrac{\sqrt{2}}{3}$　　Like radicals　　The radicand of each of these square roots is 2. In exponential form, each radical is of the form $2^{1/2}$ with a base of 2 and an exponent of $\dfrac{1}{2}$.

(b) $-\sqrt{xy^2}$ and $4\sqrt{xy^2}$　　Like radicals　　The radicand of each of these square roots is xy^2. The coefficients of the radicals are -1 and 4.

(c) $7\sqrt{5}$ and $7\sqrt[3]{5}$　　Unlike radicals　　These radicals are unlike because they do not have the same index. One is a square root, and the other is a cube root.

(d) $11\sqrt{7}$ and $11\sqrt{8}$　　Unlike radicals　　These square roots are unlike because the radicands 7 and 8 are different.

(e) $\sqrt[3]{xy^2}$ and $\sqrt[3]{x^2y}$　　Unlike radicals　　These radicals are unlike because the radicands are not equal ($xy^2 \neq x^2y$). ■

SELF-CHECK 8.2.1

Classify each pair of radicals as either like or unlike.

1. $-\dfrac{\sqrt{7}}{2}$ and $\dfrac{\sqrt{7}}{3}$ 2. $\sqrt[3]{7}$ and $-\sqrt{7}$ 3. $\sqrt[3]{2}$ and $\sqrt[3]{3}$

Addition and Subtraction of Radicals

To add like radicals we add their coefficients.

Using the distributive property, we can rewrite $3\sqrt{2} + 4\sqrt{2}$ as $(3 + 4)\sqrt{2}$, which equals $7\sqrt{2}$. Note that we can add like radicals by adding their coefficients, just as we add like terms of a polynomial. This similarity is illustrated by the examples in the following box.

Comparison of the Addition of Polynomials and the Addition of Radical Terms

ADDITION OF POLYNOMIALS	ADDITION OF RADICALS
$3x + 4x = 7x$	$3\sqrt{2} + 4\sqrt{2} = 7\sqrt{2}$
$9x + 5y - 6x = 3x + 5y$	$9\sqrt{7} + 5\sqrt{11} - 6\sqrt{7} = 3\sqrt{7} + 5\sqrt{11}$

■ EXAMPLE 2   Adding and Subtracting Radicals

Perform the following additions and subtractions.

SOLUTIONS

(a) $\sqrt{3} + 5\sqrt{3}$

$$\sqrt{3} + 5\sqrt{3} = (1 + 5)\sqrt{3}$$
$$= 6\sqrt{3}$$

Use the distributive property to add the coefficients of these like terms.

(b) $\sqrt[3]{7} - 5\sqrt[3]{7} + 12\sqrt[3]{7}$

$$\sqrt[3]{7} - 5\sqrt[3]{7} + 12\sqrt[3]{7} = (1 - 5 + 12)\sqrt[3]{7}$$
$$= 8\sqrt[3]{7}$$

Use the distributive property to combine like terms.

(c) $4\sqrt{5} + 7\sqrt[3]{5}$

$4\sqrt{5} + 7\sqrt[3]{5}$

This expression cannot be simplified since the terms are unlike.

(d) $3\sqrt[3]{x^2y} - 4\sqrt[3]{x^2y}$

$$3\sqrt[3]{x^2y} - 4\sqrt[3]{x^2y} = (3 - 4)\sqrt[3]{x^2y}$$
$$= -\sqrt[3]{x^2y}$$

Subtract the coefficients of these like terms.

(e) $\left(5\sqrt{11} + 6\sqrt[3]{11}\right) - \left(2\sqrt[3]{11} - 7\sqrt{11}\right)$

$$\left(5\sqrt{11} + 6\sqrt[3]{11}\right) - \left(2\sqrt[3]{11} - 7\sqrt{11}\right)$$
$$= 5\sqrt{11} + 6\sqrt[3]{11} - 2\sqrt[3]{11} + 7\sqrt{11}$$
$$= (5 + 7)\sqrt{11} + (6 - 2)\sqrt[3]{11}$$
$$= 12\sqrt{11} + 4\sqrt[3]{11}$$

First remove the parentheses and then group like terms together. Then add the coefficients of the like terms.   ■

SELF-CHECK 8.2.1 ANSWERS

1. Like
2. Unlike
3. Unlike

> **SELF-CHECK 8.2.2**
>
> Perform the indicated operations.
>
> **1.** $4\sqrt{6} - 9\sqrt{6}$ **2.** $8\sqrt[3]{xy} - 4\sqrt[3]{xy} + \sqrt[3]{xy}$

Simplifying Radical Expressions

One way to simplify a radical expression is to use the properties in the following box to make the radicand as small as possible. *Warning:* If $x^{1/n}$ or $\sqrt[n]{x}$ is not a real number, then the properties in this box are not true.

Properties of Radicals

If $\sqrt[n]{x}$ and $\sqrt[n]{y}$ are both real numbers*, then

RADICAL NOTATION	VERBALLY	RADICAL EXAMPLE
$\sqrt[n]{xy} = \sqrt[n]{x}\sqrt[n]{y}$	The nth root of a product equals the product of the nth roots.	$\sqrt{50} = \sqrt{25}\sqrt{2}$
$\sqrt[n]{\dfrac{x}{y}} = \dfrac{\sqrt[n]{x}}{\sqrt[n]{y}}$ for $y \neq 0$	The nth root of a quotient equals the quotient of the nth roots.	$\sqrt{\dfrac{4}{9}} = \dfrac{\sqrt{4}}{\sqrt{9}}$

*If either $\sqrt[n]{x}$ or $\sqrt[n]{y}$ is not a real number, these properties do not hold true.

The properties of radicals in the box will be used to simplify the expressions in Example 3. The key to using these properties is to recognize perfect nth powers. When working with square roots, look for perfect square factors, such as 1, 4, 9, 16, and 25. Similarly, when working with cube roots, look for perfect cube factors, such as 1, 8, 27, 64, and 125.

■ EXAMPLE 3   Simplifying Radicals

Simplify each of these radical expressions.

SOLUTIONS

In Example 3(a), $5\sqrt{2}$ is considered a simpler form than $\sqrt{50}$ because it has a smaller radicand.

(a) $\sqrt{50}$ $\begin{aligned} \sqrt{50} &= \sqrt{25 \cdot 2} \\ &= \sqrt{25}\sqrt{2} \\ &= 5\sqrt{2} \end{aligned}$ 25 is a perfect square factor of 50. $\sqrt{xy} = \sqrt{x}\sqrt{y}$

Caution: In Example 3(c), do not make the classic error of writing the answer incorrectly as $2\sqrt{5}$. Leaving off the index of 3 on the radical symbol completely changes the value of this expression.

(b) $\sqrt{45}$ $\begin{aligned} \sqrt{45} &= \sqrt{9 \cdot 5} \\ &= \sqrt{9}\sqrt{5} \\ &= 3\sqrt{5} \end{aligned}$ 9 is a perfect square factor of 45. $\sqrt{xy} = \sqrt{x}\sqrt{y}$

(c) $\sqrt[3]{40}$ $\begin{aligned} \sqrt[3]{40} &= \sqrt[3]{8 \cdot 5} \\ &= \sqrt[3]{8}\sqrt[3]{5} \\ &= 2\sqrt[3]{5} \end{aligned}$ 8 is a perfect cube factor of 40. $\sqrt[3]{xy} = \sqrt[3]{x}\sqrt[3]{y}$

A second way to simplify a radical expression is to use the property for the radical of a quotient to remove radicals from the denominator. Sometimes this property can also be used to remove fractions from the radicand.

■ EXAMPLE 4   Simplifying Radicals

Simplify each of these radical expressions.

SOLUTIONS

(a) $\dfrac{\sqrt{75}}{\sqrt{3}}$ $\dfrac{\sqrt{75}}{\sqrt{3}} = \sqrt{\dfrac{75}{3}}$ $\dfrac{\sqrt{x}}{\sqrt{y}} = \sqrt{\dfrac{x}{y}}$

$= \sqrt{25}$ Reduce the fraction in the radicand and then take the square root.

$= 5$

(b) $\sqrt{\dfrac{4}{9}}$ $\sqrt{\dfrac{4}{9}} = \dfrac{\sqrt{4}}{\sqrt{9}}$ $\sqrt{\dfrac{x}{y}} = \dfrac{\sqrt{x}}{\sqrt{y}}$

$= \dfrac{2}{3}$

In Example 4(c), $\dfrac{\sqrt[3]{25}}{2}$ is considered a simpler form than $\sqrt[3]{\dfrac{25}{8}}$ because it does not contain a radical in the denominator.

(c) $\sqrt[3]{\dfrac{25}{8}}$ $\sqrt[3]{\dfrac{25}{8}} = \dfrac{\sqrt[3]{25}}{\sqrt[3]{8}}$ $\sqrt[3]{\dfrac{x}{y}} = \dfrac{\sqrt[3]{x}}{\sqrt[3]{y}}$

$= \dfrac{\sqrt[3]{25}}{2}$ The numerator $\sqrt[3]{25}$ cannot be reduced further since it has no perfect cube factor other than 1. ∎

In Example 3(a), the expressions $\sqrt{50}$ *and* $5\sqrt{2}$ are exactly equal. Any calculator approximation of these irrational numbers will not be exact but can be used to check to see if the expressions appear to be equal. We illustrate this in Self-Check 8.2.3.

SELF-CHECK 8.2.3

1. Use a calculator to compare the values $\sqrt{50}$ and $5\sqrt{2}$ from Example 3(a).

Simplify each of these radical expressions.

2. $\sqrt{27}$ 3. $\sqrt[3]{16}$ 4. $\sqrt{\dfrac{36}{169}}$ 5. $\dfrac{\sqrt[3]{16}}{\sqrt[3]{2}}$

SELF-CHECK 8.2.3 ANSWERS

1. $\sqrt{50} = 5\sqrt{2}$

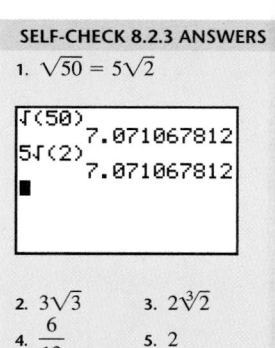

2. $3\sqrt{3}$ 3. $2\sqrt[3]{2}$

4. $\dfrac{6}{13}$ 5. 2

One reason for writing radicals always in simplified form is to identify like terms. Radicals that appear to be unlike can sometimes be simplified so that the terms are like terms. Remember to look for perfect square factors when working with square roots and for perfect cube factors when working with cube roots.

■ EXAMPLE 5 Simplifying Radicals and Then Combining Like Radicals

Simplify these radicals first, and then combine like radicals.

SOLUTIONS

(a) $7\sqrt{20} - 2\sqrt{45}$

$$7\sqrt{20} - 2\sqrt{45} = 7\sqrt{4 \cdot 5} - 2\sqrt{9 \cdot 5}$$
$$= 7\sqrt{4}\sqrt{5} - 2\sqrt{9}\sqrt{5}$$
$$= 7(2\sqrt{5}) - 2(3\sqrt{5})$$
$$= 14\sqrt{5} - 6\sqrt{5}$$
$$= 8\sqrt{5}$$

Note that 4 is a perfect square factor of 20 and that 9 is a perfect square factor of 45. Simplify using the property that $\sqrt{xy} = \sqrt{x}\sqrt{y}$.

Then combine like terms.

(b) $11\sqrt[3]{16} - 2\sqrt[3]{54}$

$$11\sqrt[3]{16} - 2\sqrt[3]{54} = 11\sqrt[3]{8 \cdot 2} - 2\sqrt[3]{27 \cdot 2}$$
$$= 11\sqrt[3]{8}\sqrt[3]{2} - 2\sqrt[3]{27}\sqrt[3]{2}$$
$$= 11(2\sqrt[3]{2}) - 2(3\sqrt[3]{2})$$
$$= 22\sqrt[3]{2} - 6\sqrt[3]{2}$$
$$= 16\sqrt[3]{2}$$

Note that 8 is a perfect cube factor of 16 and that 27 is a perfect cube factor of 54.
Simplify these radicals.
Then combine like terms.

(c) $\sqrt{18} + \sqrt{12}$

$$\sqrt{18} + \sqrt{12} = \sqrt{9 \cdot 2} + \sqrt{4 \cdot 3}$$
$$= \sqrt{9}\sqrt{2} + \sqrt{4}\sqrt{3}$$
$$= 3\sqrt{2} + 2\sqrt{3}$$

9 is a perfect square factor of 18, and 4 is a perfect square factor of 12.

These radicals are unlike, so the expression cannot be simplified further. ■

SELF-CHECK 8.2.4

Simplify each expression.

1. $2\sqrt{50} - 3\sqrt{200}$

2. $\sqrt[3]{24} - \sqrt[3]{3000}$

We must exercise care when taking even roots of variables because $\sqrt{-1}$, $\sqrt[4]{-1}$, and other even roots of negative values are not defined as real numbers. As we illustrate next, some of the properties that hold when $\sqrt[n]{x}$ is a real number will fail when $\sqrt[n]{x}$ is not a real number. Here is a situation where $\sqrt[n]{x^n} \neq x$:

$$\sqrt{(-5)^2} = \sqrt{25} = 5$$

Thus

$$\sqrt{(-5)^2} \neq -5$$

The definition of $|x|$ given in Section 1.1 is

$$|x| = \begin{cases} x & \text{if } x \text{ is nonnegative} \\ -x & \text{if } x \text{ is negative} \end{cases}$$

If $x < 0$ and n is an even number, then $\sqrt[n]{x^n} \neq x$. Since x^2 is positive both when x is positive and when x is negative, the principal square root of x^2, $\sqrt{x^2}$, will be positive. We can use absolute value notation to denote correctly the result in all cases:

$$\sqrt{x^2} = |x|$$

To handle the general nth root of both positive and negative values of x we can define $\sqrt[n]{x^n} = |x|$ when n is even. This is highlighted in the following box.

SELF-CHECK 8.2.4 ANSWERS

1. $-20\sqrt{2}$
2. $-8\sqrt[3]{3}$

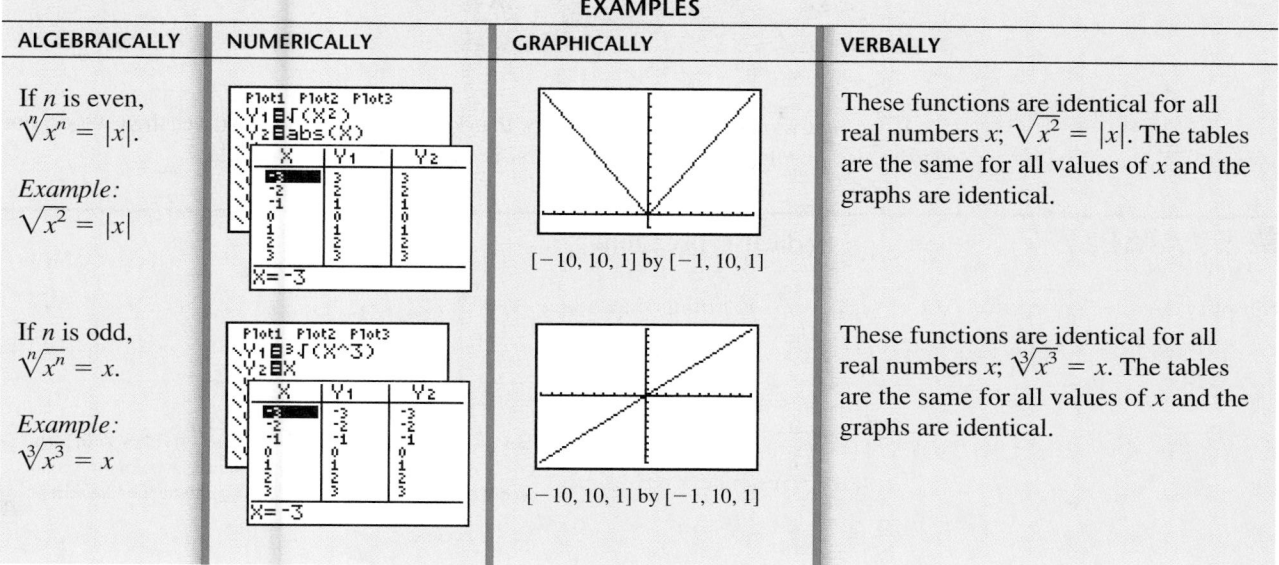

$$\sqrt[n]{x^n}$$

For any real number x and natural number n,

EXAMPLES

ALGEBRAICALLY	NUMERICALLY	GRAPHICALLY	VERBALLY
If n is even, $\sqrt[n]{x^n} = \|x\|$. *Example:* $\sqrt{x^2} = \|x\|$		$[-10, 10, 1]$ by $[-1, 10, 1]$	These functions are identical for all real numbers x; $\sqrt{x^2} = \|x\|$. The tables are the same for all values of x and the graphs are identical.
If n is odd, $\sqrt[n]{x^n} = x$. *Example:* $\sqrt[3]{x^3} = x$		$[-10, 10, 1]$ by $[-1, 10, 1]$	These functions are identical for all real numbers x; $\sqrt[3]{x^3} = x$. The tables are the same for all values of x and the graphs are identical.

There are two correct ways to handle $\sqrt[n]{x^n}$. The first option is the simpler one: we merely avoid the difficulty by restricting x to positive values. However, since odd roots pose no difficulty, the second option is to allow x to be negative and to use absolute value notation for the restricted case $\sqrt[n]{x^n}$ when n is even.

■ EXAMPLE 6   Evaluating Radical Expressions

Simplify each of these radical expressions. Use absolute value notation wherever necessary.

SOLUTIONS

(a) $\sqrt{(-7)^2}$

$\sqrt{(-7)^2} = |-7|$
$= 7$

$\sqrt[n]{x^n} = |x|$ if n is even: $\sqrt{(-7)^2} = \sqrt{49} = 7$

(b) $\sqrt[3]{(-2)^3}$

$\sqrt[3]{(-2)^3} = -2$

$\sqrt[n]{x^n} = x$ if n is odd: $\sqrt[3]{(-2)^3} = \sqrt[3]{-8} = -2$

(c) $\sqrt{64x^6}$ for $x > 0$

$\sqrt{64x^6} = \sqrt{(8x^3)^2}$
$= 8x^3$

First write $64x^6$ as the perfect square of $8x^3$. Absolute value notation is not needed because $x > 0$. If we know that $x > 0$, then $|8x^3| = 8x^3$.

(d) $\sqrt{64x^6}$ if x is any real number

$\sqrt{64x^6} = |8x^3|$
$= 8|x|^3$

First write $64x^6$ as the perfect square of $8x^3$. Because $\sqrt[n]{x^n} = |x|$ if n is even, we must use absolute value notation in this problem.

(e) $\sqrt[3]{64x^6}$

$\sqrt[3]{64x^6} = \sqrt[3]{(4x^2)^3}$
$= 4x^2$

First write $64x^6$ as the perfect cube of $4x^2$. $\sqrt[n]{x^n} = x$ if n is odd.

If we restrict the variables to values that keep the radicand positive, then we do not need to use absolute value notation. This is the situation in Example 7.

■ EXAMPLE 7   Simplifying Radical Expressions

Simplify $\sqrt{x^2 + 2xy + y^2} + \sqrt{x^2 - 2xy + y^2}$, assuming that $x > y > 0$.

SOLUTION _____

$$\sqrt{x^2 + 2xy + y^2} + \sqrt{x^2 - 2xy + y^2} = \sqrt{(x + y)^2} + \sqrt{(x - y)^2}$$
$$= (x + y) + (x - y)$$
$$= 2x$$

Express the radicands as perfect squares. If $x > y > 0$, then $x + y > 0$ and $x - y > 0$; thus absolute value notation is not needed for these square roots. Then combine the like terms. ■

Carpenters, cooks, and many other people who make things often perform mental computations either to check their calculations or to make quick decisions about the task in front of them. Example 8 involves estimating the length of a side of a square.

■ EXAMPLE 8   Estimating the Dimensions of a Square

Estimate the length of each side of a square piece of plywood needed to cover a damaged square window with an area of 85 cm^2. Then use a calculator to approximate this length.

SOLUTION _____

$A = s^2$ Formula for the area of a square.
$s = \sqrt{A}$ Formula for the length of a side of a square.

ESTIMATED ANSWER CALCULATOR APPROXIMATION

$$85 \approx 81$$
$$\sqrt{85} \approx \sqrt{81} = 9$$
$$s \approx 9 \text{ cm}$$

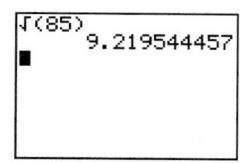

Answer: The calculator approximation of 9.219544457 seems reasonable based upon the mental estimate of 9. The length of each side of the square piece of plywood is approximately 9.2 cm.

Because $81 < 85$, the estimate of 9 is less than the actual value of $\sqrt{85}$.

USING THE LANGUAGE AND SYMBOLISM OF MATHEMATICS 8.2

1. Like radicals must have the same _____ and the same _____.

2. The statement that "the nth root of a product equals the product of the nth roots" is represented algebraically by $\sqrt[n]{xy} =$ _____.

3. The statement that "the nth root of a quotient equals the quotient of the nth roots" is represented algebraically by $\sqrt[n]{\dfrac{x}{y}} =$ _____.

4. If n is even, $\sqrt[n]{x^n} =$ _____.

5. If n is odd, $\sqrt[n]{x^n} =$ _____.

EXERCISES 8.2

In Exercises 1 and 2 determine which radical is a like radical to the given radical.

1. $7\sqrt{5}$
    a. $2\sqrt[3]{5}$ **b.** $3\sqrt{7}$
    c. $7\sqrt{3}$ **d.** $-2\sqrt{5}$

2. $4\sqrt[3]{7}$
    a. $4\sqrt{7}$ **b.** $11\sqrt[3]{7}$
    c. $5\sqrt[3]{4}$ **d.** $-5\sqrt[3]{4}$

In Exercises 3–20 find the sum or difference. Assume that all variables represent positive real numbers so that absolute value notation is not necessary.

3. $\sqrt{25} + \sqrt{49}$ **4.** $\sqrt{36} + \sqrt{64}$

5. $2\sqrt{9} - 3\sqrt{16}$ **6.** $5\sqrt{100} - 3\sqrt{4}$

7. $19\sqrt{2} + 11\sqrt{2}$ **8.** $7\sqrt{3} - 8\sqrt{3}$

9. $9\sqrt{7} - 13\sqrt{7}$ **10.** $6\sqrt{5} + 8\sqrt{5}$

11. $4\sqrt[3]{6} + 2\sqrt[3]{6} - 11\sqrt[3]{6}$

12. $5\sqrt[3]{11} - 2\sqrt[3]{11} + 12\sqrt[3]{11}$

13. $5\sqrt[4]{13} + 6\sqrt[4]{13} - \sqrt[4]{13}$

14. $\sqrt[5]{9} - 8\sqrt[5]{9} - 7\sqrt[5]{9}$

15. $7\sqrt{5} - 5\sqrt{7} - \sqrt{5} + \sqrt{7}$

16. $7\sqrt{2} - \sqrt{3} + 3\sqrt{2} + 2\sqrt{3}$

17. $6\sqrt{7x} - 9\sqrt{7x}$

18. $4\sqrt{11y} - 7\sqrt{11y}$

19. $7\sqrt[4]{17w} - 8\sqrt[4]{17w} + \sqrt[4]{17w}$

20. $6\sqrt[5]{5w} - 11\sqrt[5]{5w} + 5\sqrt[5]{5w}$

In Exercises 21–28 simplify each radical expression.

21. $\sqrt{75}$ **22.** $\sqrt{28}$ **23.** $\sqrt{63}$ **24.** $\sqrt{72}$

25. $\sqrt[3]{24}$ **26.** $\sqrt[3]{40}$ **27.** $\sqrt[4]{48}$ **28.** $\sqrt[5]{96}$

In Exercises 29–50 find the sum or difference. Assume that all variables represent positive real numbers so that absolute value notation is not necessary.

29. $\sqrt{28} + \sqrt{63}$ **30.** $\sqrt{12} - \sqrt{27}$

31. $\sqrt{75} - \sqrt{48}$ **32.** $\sqrt{44} + \sqrt{99}$

33. $3\sqrt{50v} - 7\sqrt{32v}$ **34.** $6\sqrt{45v} - 7\sqrt{320v}$

35. $5\sqrt{28w} - 4\sqrt{63w}$ **36.** $3\sqrt{27b} - \sqrt{75b}$

37. $\sqrt[3]{24} - \sqrt[3]{375}$ **38.** $\sqrt[3]{54} - \sqrt[3]{128}$

39. $20\sqrt[3]{-81t^3} - 7\sqrt[3]{24t^3}$ **40.** $11\sqrt[3]{-16t^3} - 5\sqrt[3]{54t^3}$

41. $9\sqrt[3]{40z^2} - 2\sqrt[3]{5000z^2}$

42. $7\sqrt{1210w^3} - 5\sqrt{1440w^3}$

43. $\sqrt{\dfrac{5}{4}} + \sqrt{\dfrac{5}{9}}$ **44.** $\sqrt{\dfrac{7}{9}} - \sqrt{\dfrac{7}{25}}$

45. $2\sqrt{\dfrac{11x}{25}} - \sqrt{\dfrac{11x}{49}}$ **46.** $3\sqrt{\dfrac{13x}{49}} - 5\sqrt{\dfrac{13x}{4}}$

47. $7\sqrt{0.98} + 2\sqrt{0.75} - \sqrt{0.12} + 5\sqrt{0.72}$

48. $4\sqrt{0.20} + 3\sqrt{0.90} - 7\sqrt{0.80} - \sqrt{1.60}$

49. $\sqrt{x^2 + 2xy + y^2} - \sqrt{x^2} - \sqrt{y^2}$

50. $\sqrt{x^2 + y^2} - \sqrt{x^2 - 2xy + y^2}$ for $x > y > 0$

In Exercises 51–58 simplify each expression. Use absolute value notation only when it is needed.

	FOR $x \geq 0$	FOR ANY REAL NUMBER x
51. $\sqrt{25x^2}$	**a.**	**b.**
52. $\sqrt{36x^2}$	**a.**	**b.**
53. $\sqrt[3]{8x^3}$	**a.**	**b.**
54. $\sqrt[3]{-125x^3}$	**a.**	**b.**
55. $\sqrt{1,000,000x^6}$	**a.**	**b.**
56. $\sqrt[3]{1,000,000x^6}$	**a.**	**b.**
57. $\sqrt[5]{x^{30}}$	**a.**	**b.**
58. $\sqrt[6]{x^{30}}$	**a.**	**b.**

In Exercises 59–62 match each function with its graph. All graphs are displayed with the window $[-5, 5, 1]$ by $[-5, 5, 1]$.

59. $f(x) = \sqrt{(x + 2)^2}$ **60.** $f(x) = |x| + 2$
61. $f(x) = \sqrt[3]{(x - 3)^3}$ **62.** $f(x) = x + 3$

A. 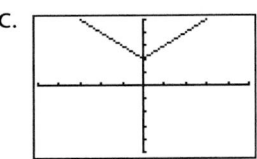 **B.**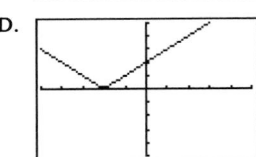

C. **D.**

Calculator Exercises

In Exercises 63–68 complete the following table by:
  a. Estimating each radical expression to the nearest integer.
  b. Determining whether this integer estimate is less than or greater than the actual value.
  c. Using a calculator to approximate each expression to the nearest thousandth.

	INTEGER ESTIMATE	INEQUALITY	APPROXI- MATION
Example $\sqrt[4]{79}$	3	$3 > \sqrt[4]{79}$	2.981
63. $\sqrt[3]{63}$			
64. $\sqrt[4]{82}$			
65. $\dfrac{1 + \sqrt{9.05}}{2}$			
66. $\dfrac{1 - \sqrt{9.05}}{2}$			
67. $\dfrac{-4 - \sqrt{35.97}}{2}$			
68. $\dfrac{-4 + \sqrt{35.97}}{2}$			

69. The best mental estimate of the length of each side of a square of area 26 cm^2 (see the figure) is _____.
  a. 3 cm **b.** 4 cm **c.** 5 cm
  d. 12 cm **e.** 13 cm

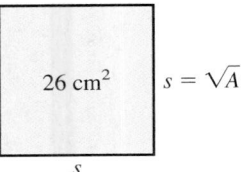

70. The best mental estimate of the length of each side of a cubical box whose volume is 26 cm^3 (see the figure) is
_____.
  a. 3 cm **b.** 4 cm **c.** 5 cm
  d. 12 cm **e.** 13 cm

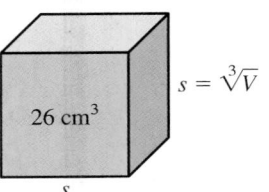

71. Determine to the nearest tenth of a centimeter the length c of the hypotenuse of the right triangle shown in the figure. The length c can be calculated by $c = \sqrt{a^2 + b^2}$.

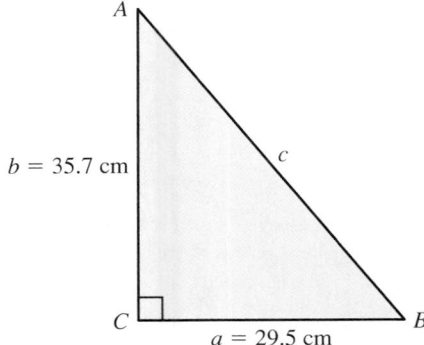

72. Length of a Brace The length d of a diagonal brace from the lower corner of a rectangular storage box to the opposing upper corner is given by
$$L = \sqrt{w^2 + l^2 + h^2}$$
where w, l, and h are, respectively, the width, length, and height of the box. Determine the length of a diagonal of a box with a width of 5.2 m, length of 9.4 m, and height of 6.5 m.

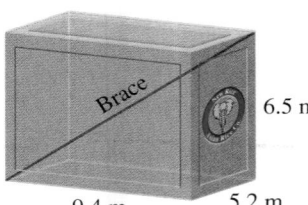

Group Discussion Questions

73. Discovery Question
a. Use a graphics calculator to compare tables of values for $f(x) = \sqrt[4]{x^2}$ and $f(x) = \sqrt{x}$.
b. Use a graphics calculator to compare tables of values for $f(x) = \sqrt[9]{x^3}$ and $f(x) = \sqrt[3]{x}$.
c. Write each of the expressions in parts (a) and (b) in exponential form and then use the properties of exponents to explain the relationship between each pair of expressions.
d. Use your observations in part (c) to simplify
$$f(x) = \sqrt[15]{x^3 y^6 z^9}$$
e. Explain to your teacher how you reduced the order of the radical in part (d).

74. Discovery Question
a. Compare the calculator values of $\sqrt{\sqrt{256}}$ and $\sqrt[4]{256}$. If these expressions are equal, explain why this is so.
b. Is $\sqrt{\sqrt{\sqrt{256}}}$ equal to either $\sqrt[6]{256}$ or $\sqrt[8]{256}$? Explain your choice.
c. What is another way of writing $\sqrt{\sqrt{\sqrt{\sqrt{x}}}}$? Explain your reasoning to your teacher.

75. Challenge Question
a. Using a calculator, evaluate $\sqrt{10}$. Then take the square root of this result, then the square root of this result, and so on, until you have taken 10 square roots.
b. Using a calculator, evaluate $\sqrt{0.1}$. Then take the square root of this result, then the square root of this result, and so on, until you have taken 10 square roots.
c. Make a conjecture based on your observations in parts (a) and (b).

Section 8.3 Multiplication and Division of Radical Expressions

Objectives:
5. Multiply radical expressions.
6. Divide and simplify radical expressions.

To multiply and divide some radical expressions, we use the properties:

$$\sqrt[n]{x}\sqrt[n]{y} = \sqrt[n]{xy} \quad \text{for } x \geq 0 \text{ and } y \geq 0$$

$$\frac{\sqrt[n]{x}}{\sqrt[n]{y}} = \sqrt[n]{\frac{x}{y}} \quad \text{for } x \geq 0 \text{ and } y > 0$$

The restriction that $y \neq 0$ is needed in the expression $\frac{\sqrt[n]{x}}{\sqrt[n]{y}} = \sqrt[n]{\frac{x}{y}}$ to prevent division by 0.

When multiplying or dividing radicals using these properties, it is important to remember two things:

1. The radicals must be of the same order before these properties can be applied.
2. Do not apply these properties if one or both of the radicands are negative and n is even.

■ EXAMPLE 1 Multiplying Radicals

Perform each indicated multiplication and then simplify the product. Assume $x \geq 0$ and $y \geq 0$.

SOLUTIONS

(a) $\sqrt{6}\sqrt{15}$
$$\begin{aligned}\sqrt{6}\sqrt{15} &= \sqrt{90} \\ &= \sqrt{9(10)} \\ &= \sqrt{9}\sqrt{10} \\ &= 3\sqrt{10}\end{aligned}$$

$\sqrt{x}\sqrt{y} = \sqrt{xy}$

9 is a perfect square factor of 90.
$\sqrt{xy} = \sqrt{x}\sqrt{y}$

(b) $\sqrt{9x}\sqrt{x}$
$$\begin{aligned}\sqrt{9x}\sqrt{x} &= \sqrt{9x^2} \\ &= 3x\end{aligned}$$

$\sqrt{x}\sqrt{y} = \sqrt{xy}$

Absolute value notation is not needed for the answer since $x \geq 0$.

(c) $\left(\sqrt[3]{20y^2}\right)\left(\sqrt[3]{50y^2}\right)$ $\left(\sqrt[3]{20y^2}\right)\left(\sqrt[3]{50y^2}\right) = \sqrt[3]{1000y^4}$

$\sqrt[3]{x}\sqrt[3]{y} = \sqrt[3]{xy}$

$1000y^3$ is a perfect cube factor of $1000y^4$. Because this root is odd, this simplification also would work if $y < 0$.

$$= \sqrt[3]{(1000y^3)(y)}$$
$$= \sqrt[3]{1000y^3}\sqrt[3]{y}$$
$$= 10y\sqrt[3]{y}$$

The multiplication of some radical expressions is similar to the multiplication of polynomials. Watch for special forms that can be multiplied by inspection and use the distributive property when it is appropriate.

■ EXAMPLE 2   Multiplying Radicals with More Than One Term

Perform each indicated multiplication and then simplify the product. Assume that the variables represent positive real numbers so that absolute value notation is not necessary.

SOLUTIONS

(a) $2\sqrt{3}(3\sqrt{3} - 5)$

$2\sqrt{3}(3\sqrt{3} - 5) = (2\sqrt{3})(3\sqrt{3}) - (2\sqrt{3})(5)$
$\qquad = 6(\sqrt{3})^2 - 10\sqrt{3}$
$\qquad = 6(3) - 10\sqrt{3}$
$\qquad = 18 - 10\sqrt{3}$

Distribute the multiplication of $2\sqrt{3}$ and then simplify.

(b) $(4\sqrt{11} - 2)(3\sqrt{11} + 7)$

$(4\sqrt{11} - 2)(3\sqrt{11} + 7)$
$= 4\sqrt{11}(3\sqrt{11}) + 4\sqrt{11}(7) - 2(3\sqrt{11}) - 2(7)$
$= 12(\sqrt{11})^2 + 28\sqrt{11} - 6\sqrt{11} - 14$
$= 12(11) + 22\sqrt{11} - 14$
$= 132 + 22\sqrt{11} - 14$
$= 118 + 22\sqrt{11}$

Use the distributive property to multiply each term in the first factor times each term in the second factor.

Simplify and then add the like terms.

(c) $(\sqrt{2a} + \sqrt{3b})(\sqrt{2a} - \sqrt{3b})$

$(\sqrt{2a} + \sqrt{3b})(\sqrt{2a} - \sqrt{3b}) = (\sqrt{2a})^2 - (\sqrt{3b})^2$
$\qquad = 2a - 3b$

This is a special product of the form $(x + y)(x - y) = x^2 - y^2$ with $x = \sqrt{2a}$ and $y = \sqrt{3b}$.

(d) $(\sqrt[3]{a} - 1)(\sqrt[3]{a^2} + \sqrt[3]{a} + 1)$

$(\sqrt[3]{a} - 1)(\sqrt[3]{a^2} + \sqrt[3]{a} + 1) = (\sqrt[3]{a})^3 - (1)^3$
$\qquad = a - 1$

This is a special product of the form
$(x - y)(x^2 + xy + y^2) = x^3 - y^3$
with $x = \sqrt[3]{a}$ and $y = 1$. ■

SELF-CHECK 8.3.1

Simplify each product. Assume $x \ge 0$.

1. $\sqrt{6x}\sqrt{10x}$
2. $\sqrt[4]{8x^3}\sqrt[4]{2x}$
3. $(2\sqrt{3} - \sqrt{5})^2$
4. $(5\sqrt{x} + 4)(5\sqrt{x} - 4)$

The multiplication of conjugate radicals produces a product that does not contain a radical expression.

Conjugate Radicals

The special products play an important role in simplifying many algebraic expressions. Once again we use the product $(x + y)(x - y) = x^2 - y^2$. The radical expressions $\sqrt{x} + \sqrt{y}$ and $\sqrt{x} - \sqrt{y}$ are called **conjugates** of each other. Knowing that $(\sqrt{x} + \sqrt{y})(\sqrt{x} - \sqrt{y}) = x - y$ is very useful because the product is an expression free of radicals.

RADICAL EXPRESSIONS	CONJUGATE RADICAL EXPRESSIONS
$\sqrt{2} + \sqrt{3}$	$\sqrt{2} - \sqrt{3}$
$2\sqrt{a} - 5\sqrt{b}$	$2\sqrt{a} + 5\sqrt{b}$
$-3 + 4\sqrt{11}$	$-3 - 4\sqrt{11}$
$1 + \sqrt{3x}$	$1 - \sqrt{3x}$

■ EXAMPLE 3   Multiplying Conjugates

Find the product of $\sqrt{7} + \sqrt{5}$ and its conjugate.

SOLUTION

$$(\sqrt{7} + \sqrt{5})(\sqrt{7} - \sqrt{5}) = (\sqrt{7})^2 - (\sqrt{5})^2$$
$$= 7 - 5$$
$$= 2$$

The conjugate of $\sqrt{7} + \sqrt{5}$ is $\sqrt{7} - \sqrt{5}$. Note that both $\sqrt{7}$ and $\sqrt{5}$ are irrational numbers, but the product of these conjugates is the natural number 2. ■

SELF-CHECK 8.3.2

1. Write the conjugate of $\sqrt{2a} - 5$.
2. Write the conjugate of $7 - \sqrt{y}$.
3. Multiply $-5 + \sqrt{3}$ times its conjugate.
4. Check the product in part 3 on your calculator.

Dividing Radicals

In Section 8.2 we simplified the radical expression $\sqrt{\dfrac{4}{9}}$ as follows:

$$\sqrt{\frac{4}{9}} = \frac{\sqrt{4}}{\sqrt{9}} = \frac{2}{3}$$

Rationalizing the denominator of an expression produces an expression whose denominator does not contain a radical expression.

However, we do not always obtain a perfect square when we divide. Sometimes we wish to remove the radical from the denominator when obtaining something other than a perfect square in the denominator. The process of doing this is called **rationalizing the denominator.** Because $\sqrt{x^2} = x$ for $x \geq 0$, we can multiply a square root by itself to obtain a perfect square and thus remove the radical in the denominator. This is illustrated in Example 4.

SELF-CHECK 8.3.2 ANSWERS

1. $\sqrt{2a} + 5$
2. $7 + \sqrt{y}$
3. $(-5 + \sqrt{3})(-5 - \sqrt{3})$
   $= 22$

4.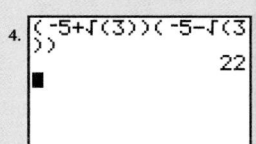

▮ EXAMPLE 4 Dividing Radicals Involving Square Roots

Perform the indicated divisions and express the quotients in rationalized form. Assume $a \geq 0$ and $b > 0$.

SOLUTIONS

(a) $\sqrt{7} \div \sqrt{11}$

$\sqrt{7} \div \sqrt{11} = \dfrac{\sqrt{7}}{\sqrt{11}}$

Write the quotient in fractional form.

$= \dfrac{\sqrt{7}}{\sqrt{11}} \cdot \dfrac{\sqrt{11}}{\sqrt{11}}$

$= \dfrac{\sqrt{77}}{11}$

Multiply both the numerator and the denominator by $\sqrt{11}$ to make the radicand in the denominator a perfect square. Multiplying by the multiplicative identity 1 in the form $\dfrac{\sqrt{11}}{\sqrt{11}}$ does not change the value of this fraction. Note $\sqrt{11}\sqrt{11} = 11$ and simplify the denominator. The denominator is now rationalized because it is the rational number 11 instead of the irrational number $\sqrt{11}$.

(b) $\dfrac{\sqrt{2a}}{\sqrt{3b}}$

$\dfrac{\sqrt{2a}}{\sqrt{3b}} = \dfrac{\sqrt{2a}}{\sqrt{3b}} \cdot \dfrac{\sqrt{3b}}{\sqrt{3b}}$

$= \dfrac{\sqrt{6ab}}{3b}$

Multiply both the numerator and the denominator by $\sqrt{3b}$ to make the radicand in the denominator a perfect square. Multiplying by the multiplicative identity 1 in the form $\dfrac{\sqrt{3b}}{\sqrt{3b}}$ does not change the value of this fraction. ▮

SELF-CHECK 8.3.3

Use a calculator to compare, from Example 4, the values of $\dfrac{\sqrt{7}}{\sqrt{11}}$ and $\dfrac{\sqrt{77}}{11}$.

For a radicand to be a perfect cube, every factor in the radicand must be a perfect cube. This is illustrated in Example 5 where we rationalize an expression involving cube roots.

▮ EXAMPLE 5 Dividing Radicals Involving Cube Roots

Divide $\dfrac{\sqrt[3]{v}}{\sqrt[3]{wx^2}}$ and express the quotient in rationalized form. Assume that $w \neq 0$ and $x \neq 0$.

SELF-CHECK 8.3.3 ANSWER

$\dfrac{\sqrt{7}}{\sqrt{11}} = \dfrac{\sqrt{77}}{11}$

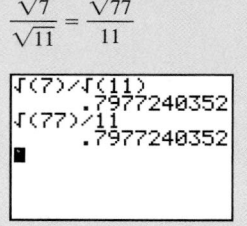

SOLUTION

$\dfrac{\sqrt[3]{v}}{\sqrt[3]{wx^2}} = \dfrac{\sqrt[3]{v}}{\sqrt[3]{wx^2}} \cdot \dfrac{\sqrt[3]{w^2x}}{\sqrt[3]{w^2x}}$

Multiply both the numerator and the denominator by the factors needed to produce the perfect cube w^3x^3.

$= \dfrac{\sqrt[3]{vw^2x}}{\sqrt[3]{w^3x^3}}$

$= \dfrac{\sqrt[3]{vw^2x}}{wx}$

Simplify the denominator. This form is rationalized because the denominator is written free of radicals. ▮

We now examine division by radical expressions with two terms. The key to simplifying these expressions is to use conjugates, because the product of the conjugates $\sqrt{x} + \sqrt{y}$ and $\sqrt{x} - \sqrt{y}$ contains no radical. For $x \geq 0$ and $y \geq 0$,
$$(\sqrt{x} + \sqrt{y})(\sqrt{x} - \sqrt{y}) = x - y.$$

■ EXAMPLE 6 Rationalizing a Denominator with Two Terms

Simplify $\dfrac{6}{\sqrt{7} - \sqrt{5}}$ by rationalizing the denominator.

SOLUTION

$$\frac{6}{\sqrt{7} - \sqrt{5}} = \frac{6}{\sqrt{7} - \sqrt{5}} \cdot \frac{\sqrt{7} + \sqrt{5}}{\sqrt{7} + \sqrt{5}}$$

Multiply the numerator and the denominator by $\sqrt{7} + \sqrt{5}$, the conjugate of the denominator.

$$= \frac{6(\sqrt{7} + \sqrt{5})}{7 - 5}$$

Multiplying by the multiplicative identity 1 in the form $\frac{\sqrt{7} + \sqrt{5}}{\sqrt{7} + \sqrt{5}}$ does not change the value of this fraction.

$$= \frac{6(\sqrt{7} + \sqrt{5})}{2}$$

$(\sqrt{7} - \sqrt{5})(\sqrt{7} + \sqrt{5}) = 7 - 5$

$$= 3(\sqrt{7} + \sqrt{5})$$
$$= 3\sqrt{7} + 3\sqrt{5}$$

Divide the numerator and the denominator by their common factor, 2.

SELF-CHECK 8.3.4 ANSWERS

1. $\sqrt[3]{\dfrac{2}{25}} = \dfrac{\sqrt[3]{10}}{5}$

```
³√(2/25)
         .430886938
³√(10)/5
         .430886938
■
```

2. $2\sqrt{6}$
3. $2\sqrt[3]{36}$
4. $6\sqrt{6} + 12$

SELF-CHECK 8.3.4

1. Determine an exact rationalized form for $\sqrt[3]{\dfrac{2}{25}}$ and then check this value on your calculator.

Simplify each of these expressions.

2. $\dfrac{12}{\sqrt{6}}$

3. $\dfrac{12}{\sqrt[3]{6}}$

4. $\dfrac{12}{\sqrt{6} - 2}$

Example 7 involves a denominator with variables in the radicands. The restriction that $x \neq y$ is made to prevent division by 0.

■ **EXAMPLE 7** Rationalizing a Denominator with Variables

Simplify $\dfrac{\sqrt{x}}{\sqrt{x} + \sqrt{y}}$ by rationalizing the denominator. Assume that x and y represent positive real numbers and that $x \neq y$.

SOLUTION

$$\dfrac{\sqrt{x}}{\sqrt{x} + \sqrt{y}} = \dfrac{\sqrt{x}}{\sqrt{x} + \sqrt{y}} \cdot \dfrac{\sqrt{x} - \sqrt{y}}{\sqrt{x} - \sqrt{y}}$$

Multiply both the numerator and the denominator by $\sqrt{x} - \sqrt{y}$, the conjugate of the denominator.

$$= \dfrac{\sqrt{x}(\sqrt{x} - \sqrt{y})}{(\sqrt{x} + \sqrt{y})(\sqrt{x} - \sqrt{y})}$$

$$= \dfrac{x - \sqrt{xy}}{x - y}$$

Distribute the factor of $\sqrt{x}$ in the numerator and multiply the conjugates in the denominator. ■

One context in which we frequently encounter radical expressions is as solutions of quadratic equations. Example 8 checks a radical expression to determine whether it is a solution of the given quadratic equation.

■ **EXAMPLE 8** Checking a Solution of a Quadratic Equation

Determine whether or not $1 + \sqrt{3}$ is a solution of $x^2 - 2x - 2 = 0$.

SOLUTION

$$x^2 - 2x - 2 = 0$$

Substitute the given value of x into the equation to determine

$$(1 + \sqrt{3})^2 - 2(1 + \sqrt{3}) - 2 \overset{?}{=} 0$$

if it checks.

$$1 + 2\sqrt{3} + 3 - 2 - 2\sqrt{3} - 2 \overset{?}{=} 0$$

$$0 \overset{?}{=} 0$$

This is a true statement.

Answer: Because $1 + \sqrt{3}$ checks, it is a solution of the equation. ■

SELF-CHECK 8.3.5

Determine whether or not the conjugate of $1 + \sqrt{3}$ is a solution of $x^2 - 2x - 2 = 0$.

USING THE LANGUAGE AND SYMBOLISM OF MATHEMATICS 8.3

1. Two properties sometimes used to multiply and divide radicals are
   a. $\sqrt[n]{x}\sqrt[n]{y} =$ _____ for $x \geq 0$ and $y \geq 0$.
   b. $\dfrac{\sqrt[n]{x}}{\sqrt[n]{y}} =$ _____ for $x \geq 0$ and $y > 0$.

2. Before the properties mentioned in part **1** above can be applied, the radicals must be of the _____ _____.

3. The properties mentioned in part **1** cannot be applied if one or both of the radicands is negative and n is _____.

4. The radical expressions $\sqrt{x} + \sqrt{y}$ and $\sqrt{x} - \sqrt{y}$ are called _____ of each other.

5. The process of removing the radical from the denominator of an expression is called _____ the denominator.

EXERCISES   8.3

In Exercises 1–4 perform each indicated multiplication and simplify the product. Then use a calculator to evaluate both the original expression and your answer as a check on your answer.

1. $\sqrt{2}\sqrt{6}$

2. $\sqrt{3}\sqrt{6}$

3. $\sqrt{2}(5\sqrt{2}-1)$

4. $\sqrt{7}(2\sqrt{7}+1)$

In Exercises 5–40 perform the indicated multiplication and simplify the product. Assume that the variables represent nonnegative real numbers so that absolute value notation is not necessary.

5. $(2\sqrt{3})(4\sqrt{5})$

6. $(3\sqrt{7})(4\sqrt{2})$

7. $(7\sqrt{15})(4\sqrt{21})$

8. $(4\sqrt{10})(9\sqrt{15})$

9. $\sqrt{3}(\sqrt{2}+\sqrt{5})$

10. $\sqrt{5}(\sqrt{3}+\sqrt{7})$

11. $3\sqrt{5}(2\sqrt{15}-7\sqrt{35})$

12. $2\sqrt{7}(5\sqrt{14}-\sqrt{21})$

13. $\sqrt{2}(5\sqrt{6})(8\sqrt{3})$

14. $-\sqrt{5}(3\sqrt{15})(\sqrt{3})$

15. $\sqrt{8w}\sqrt{2w}$

16. $\sqrt{6v}\sqrt{21v}$

17. $(2\sqrt{6z})(4\sqrt{3z})$

18. $(5\sqrt{14z})(11\sqrt{7z})$

19. $3\sqrt{x}(2\sqrt{x}-5)$

20. $4\sqrt{x}(3\sqrt{x}-11)$

21. $(\sqrt{3x}-\sqrt{y})(\sqrt{3x}+4\sqrt{y})$

22. $(\sqrt{2x}+\sqrt{y})(\sqrt{2x}-5\sqrt{y})$

23. $(2\sqrt{3}-\sqrt{2})^2$

24. $(3\sqrt{2}-\sqrt{3})^2$

25. $(\sqrt{a}+5\sqrt{3b})^2$

26. $(\sqrt{3a}+3\sqrt{5b})^2$

27. $(\sqrt{v-2}+3)(\sqrt{v-2}-3)$

28. $(\sqrt{2v+3}-5)(\sqrt{2v+3}+5)$

29. $\sqrt[3]{7}\sqrt[3]{49}$

30. $\sqrt[3]{36}\sqrt[3]{6}$

31. $\sqrt[3]{9v}\sqrt[3]{-3v^2}$

32. $\sqrt[3]{-4v}\sqrt[3]{2v^2}$

33. $-\sqrt[3]{4}(2\sqrt[3]{2}+\sqrt[3]{5})$

34. $-\sqrt[3]{6}(3\sqrt[3]{4}-2\sqrt[3]{9})$

35. $(\sqrt[3]{xy^2})^2$

36. $(\sqrt[3]{4ab^2})^2$

37. $4(3\sqrt{7}-2\sqrt{5}-3)$

38. $-2(4\sqrt{5}-3\sqrt{11}+4)$

39. $-\sqrt{2}(3\sqrt{2}-5\sqrt{6}-7)$

40. $3(5\sqrt{2}-\sqrt{6}+8)$

In Exercises 41–48 multiply each radical expression by its conjugate and simplify the result. Assume that the variables represent nonnegative real numbers so that absolute value notation is unnecessary.

41. $\sqrt{5}-\sqrt{2}$

42. $7+\sqrt{6}$

43. $5+\sqrt{11}$

44. $3-\sqrt{11}$

45. $\sqrt{x}+\sqrt{3y}$

46. $\sqrt{2x}-\sqrt{y}$

47. $\sqrt{2v+1}+\sqrt{3v-1}$

48. $\sqrt{5v+2}-\sqrt{4v-3}$

In Exercises 49–56 simplify each indicated division. Rationalize the denominator only if this step is necessary. Then use a calculator to evaluate both the original expression and your answer as a check on your answer.

49. $\sqrt{\dfrac{3}{4}}$

50. $\sqrt{\dfrac{7}{36}}$

51. $\dfrac{2}{\sqrt{6}}$

52. $\dfrac{5}{\sqrt{15}}$

53. $\dfrac{\sqrt{5}}{\sqrt{8}}$

54. $\dfrac{\sqrt{11}}{\sqrt{13}}$

55. $\dfrac{\sqrt{10}}{\sqrt{2}}$

56. $\dfrac{\sqrt{33}}{\sqrt{3}}$

In Exercises 57–78 perform each indicated division by rationalizing the denominator and then simplifying. Assume that all variables represent positive real numbers.

57. $18\div\sqrt{6}$

58. $25\div\sqrt{5}$

59. $\dfrac{26}{\sqrt{10}}$

60. $\dfrac{9}{\sqrt{15}}$

61. $\dfrac{15}{\sqrt{3x}}$

62. $\dfrac{14}{\sqrt{7y}}$

63. $\dfrac{3}{1+\sqrt{7}}$

64. $\dfrac{12}{3-\sqrt{5}}$

65. $\dfrac{36}{5-\sqrt{13}}$

66. $\dfrac{2}{1+\sqrt{3}}$

67. $\dfrac{-15}{\sqrt{7}-\sqrt{2}}$

68. $\dfrac{-6}{\sqrt{5}-\sqrt{3}}$

69. $\dfrac{\sqrt{a}}{\sqrt{a}-\sqrt{b}}$

70. $\dfrac{\sqrt{b}}{\sqrt{a}+\sqrt{b}}$

71. $\dfrac{12}{\sqrt[3]{3}}$

72. $\dfrac{4}{\sqrt[3]{2}}$

73. $\sqrt[3]{\dfrac{3}{4}}$

74. $\sqrt[3]{\dfrac{2}{25}}$

75. $\sqrt[3]{\dfrac{v}{9w^2}}$

76. $\sqrt[3]{\dfrac{2v^2}{3w}}$

77. $\dfrac{5}{\sqrt{3x}+\sqrt{2y}}$

78. $\dfrac{7}{\sqrt{5x}-\sqrt{3y}}$

79. Determine whether $1+\sqrt{2}$ is a solution of $x^2-2x-1=0$.

80. Determine whether $1-\sqrt{2}$ is a solution of $x^2-2x-1=0$.

In Exercises 81 and 82 perform the indicated operations and simplify the result.

81. a. $(2+\sqrt[3]{2})(4-2\sqrt[3]{2}+\sqrt[3]{4})$

   b. $(\sqrt[3]{x}+\sqrt[3]{y})(\sqrt[3]{x^2}-\sqrt[3]{xy}+\sqrt[3]{y^2})$

82. a. $(\sqrt[3]{2}-5)(\sqrt[3]{4}+5\sqrt[3]{2}+25)$

   b. $(\sqrt[3]{a}-\sqrt[3]{b})(\sqrt[3]{a^2}+\sqrt[3]{ab}+\sqrt[3]{b^2})$

Calculator Exercises

In Exercises 83–86 complete the following table by:

a. Estimating each radical expression to the nearest integer.

b. Determining whether this integer estimate is less than or greater than the actual value.

c. Using a calculator to approximate each expression to the nearest thousandth.

	INTEGER ESTIMATE	INEQUALITY	APPROXI-MATION
Example $\dfrac{6}{\sqrt{4.1}}$	3	$3 > \dfrac{6}{\sqrt{4.1}}$	2.963
83. $\dfrac{100}{\sqrt{24}}$			
84. $\dfrac{144}{\sqrt{65}}$			
85. $\dfrac{50}{\sqrt{37} - \sqrt{0.98}}$			
86. $\dfrac{10}{\sqrt{50} - \sqrt{24}}$			

Group Discussion Questions

87. Discovery Question   Approximations of π   Use a calculator to determine which of these approximations is closest to π.

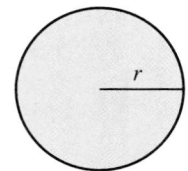

a. $\dfrac{22}{7}$ b. $\dfrac{355}{113}$ c. $\sqrt{\sqrt{\dfrac{2143}{22}}}$

88. Challenge Question

a. A circle and a square have the same area. Determine the ratio of the perimeter of the circle to the perimeter of the square.

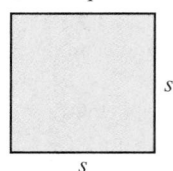

b. A circle and a square have the same perimeter. Determine the ratio of the area of the circle to the area of the square.

89. Challenge Question

a. Determine whether or not $x_1 = \dfrac{-b + \sqrt{b^2 - 4ac}}{2a}$ is a solution of the quadratic equation $ax^2 + bx + c = 0$.

b. x_2 is the conjugate of x_1. Write x_2 and determine whether it is a solution of $ax^2 + bx + c = 0$.

Section 8.4 Complex Numbers

Objectives:

7. Express complex numbers in standard form.
8. Add, subtract, multiply, and divide complex numbers.

Complex Numbers

Every real number is either negative, 0, or positive. Because the square of a negative number is positive, the square of 0 is 0, and the square of a positive number is positive, there is no real number x for which $x^2 = -1$. The desire to solve equations of the form $x^2 = -1$ led mathematicians to define the number i so that $i^2 = -1$. We use i to represent $\sqrt{-1}$, and $-i$ to represent $-\sqrt{-1}$. When these numbers were first developed in the seventeenth century, they were called imaginary numbers, because mathematicians were not familiar with them and did not know a concrete application for them. One of the first concrete applications of imaginary numbers was developed in 1892, when Charles P. Steinmetz used them in his theory of alternating currents. We often use imaginary numbers in the computation of problems whose final answers are real numbers, just as we use fractions in the computation of problems whose answers are natural numbers.

The square root of any negative number is imaginary and can be expressed in terms of the imaginary unit i. For $x > 0$, we define $\sqrt{-x}$ to be $i\sqrt{x}$.

The Imaginary Number i

$\sqrt{-1}$ has two square roots: the principal square root is i and the other square root is $-i$.

$i = \sqrt{-1}$, so $i^2 = -1$.

$-i = -\sqrt{-1}$, so $(-i)^2 = i^2 = -1$.

For any positive real number x, $\sqrt{-x} = i\sqrt{x}$.

■ **EXAMPLE 1** Writing Imaginary Numbers Using the i Notation

Write each imaginary number in terms of i.

SOLUTIONS

We usually write $5i$, not $i5$, just as we usually write $7x$, not $x7$.

We usually write $i\sqrt{3}$ instead of $\sqrt{3}\,i$ so we do not accidentally interpret this expression as $\sqrt{3i}$.

(a) $\sqrt{-25}$ $\sqrt{-25} = i\sqrt{25}$ $\sqrt{-x} = i\sqrt{x}$
$$= i(5)$$
$$= 5i$$

(b) $\sqrt{-3}$ $\sqrt{-3} = i\sqrt{3}$ ■

Imaginary numbers and complex numbers really do exist. To the layperson who does not need to use irrational numbers like $\sqrt{2}$ or an imaginary number i these numbers may seem mysterious. One goal of this chapter is to remove some of this mystery.

Using the real numbers, the imaginary number i, and the operations of addition, subtraction, multiplication, and division, we obtain numbers that can be written in the form $a + bi$, where a and b are real numbers.

Any number that can be written in the **standard form** $a + bi$ is called a **complex number.** If $b = 0$, then $a + bi$ is just the real number a. If $b \neq 0$, then $a + bi$ is called **imaginary.** If $a = 0$ and $b \neq 0$, then bi is called **pure imaginary.** Thus the complex numbers include both the real numbers and the pure imaginary numbers.

Complex Numbers

If a and b are real numbers and $i = \sqrt{-1}$, then:

ALGEBRAIC FORM	NUMERICAL EXAMPLE
$a + bi$ is a complex number with a **real term** a and an **imaginary term** bi.	$5 + 6i$ has a real term 5 and an imaginary term $6i$.

Real numbers

Complex numbers

Imaginary numbers

Figure 8.4.1

Every real number is a complex number, and every imaginary number is a complex number. The real numbers and the imaginary numbers are distinct subsets of the complex numbers with no elements in common. The relationships among the real, complex, and imaginary numbers are shown in Fig. 8.4.1 and illustrated by Table 8.4.1.

A common misunderstanding is that the imaginary numbers contain only the pure imaginary numbers. As these examples of complex numbers illustrate, $3 - 4i$ is an imaginary number.*

Every number in Table 8.4.1 is a complex number.

Table 8.4.1 Classifying Complex Numbers

COMPLEX NUMBER	STANDARD FORM	REAL TERM	COEFFICIENT OF THE IMAGINARY TERM	CLASSIFICATION
6	$6 + 0i$	6	0	Real
$-7i$	$0 - 7i$	0	-7	Pure imaginary
$3 - 4i$	$3 - 4i$	3	-4	Imaginary
0	$0 + 0i$	0	0	Real
$-\sqrt{25}$	$-5 + 0i$	-5	0	Real
$\sqrt{-25}$	$0 + 5i$	0	5	Pure imaginary

CALCULATOR PERSPECTIVE 8.4.1	Switching Between Real Mode and Complex Mode

Many graphics calculators have the ability to work with complex numbers. The TI-83 Plus calculator has a **Real** mode and an $a + bi$ (complex) mode. Note that when the calculator is set in **Real** mode, attempting to evaluate $\sqrt{-25}$ from Example 1(a) generates an error because $\sqrt{-25}$ is not a real number.

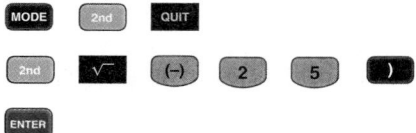

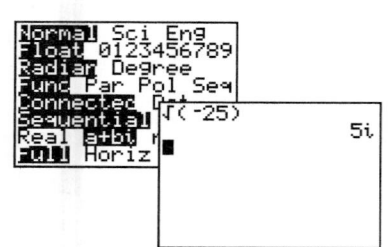

To switch the calculator to $a + bi$ (complex) mode, enter the following keystrokes:

[ENTER] [MODE]

(Use the arrow keys and the [ENTER] key to change modes as shown.) [2nd] [QUIT]

[2nd] [√] [(-)] [2] [5] [)] [ENTER]

When the calculator is set in $a + bi$ mode, the calculator correctly evaluates $\sqrt{-25}$ as $5i$.

Two complex numbers are equal if and only if both real terms are equal and both imaginary terms are equal. That is, $a + bi = c + di$ if and only if $a = c$ and $b = d$.

*Mathematics Dictionary. Fourth Edition, Multilingual Edition by Robert James, Glenn James, 1976, Van Nostrand Reinhold Company, Inc.

■ EXAMPLE 2   Equality of Complex Numbers

Determine a and b such that the complex numbers are equal.

SOLUTIONS

(a) $a + bi = 6 + 11i$ $a = 6, b = 11$ Equate the real parts and the
(b) $-5 + bi = a + 9i$ $a = -5, b = 9$ imaginary parts.
(c) $(a + 3) - 8i = 4 + 2bi$ $a + 3 = 4$ and $2b = -8$
                               $a = 1$ $b = -4$ ■

SELF-CHECK 8.4.1

Determine a and b such that the following complex numbers are equal.

1. $3 + bi = a - 17i$ 2. $-\sqrt{36} + \sqrt{-49} = a + bi$

Addition of Complex Numbers

The arithmetic of complex numbers is very similar to the arithmetic of binomials.

Since complex numbers consist of two terms, a real term and an imaginary term, the arithmetic of complex numbers is very similar to the arithmetic of binomials. The similarity is illustrated below using the operation of addition.

ADDITION OF BINOMIALS

$$(2x + 3y) + (4x + 7y) = (2x + 4x) + (3y + 7y)$$
$$= 6x + 10y$$

ADDITION OF COMPLEX NUMBERS

$$(2 + 3i) + (4 + 7i) = (2 + 4) + (3i + 7i)$$
$$= 6 + 10i$$

SELF-CHECK 8.4.1 ANSWERS
1. $a = 3, b = -17$
2. $a = -6, b = 7$

Add complex numbers by adding the real terms and the imaginary terms separately. That is,

$$(a + bi) + (c + di) = (a + c) + (b + d)i.$$

Subtraction is performed similarly.

■ EXAMPLE 3   Adding and Subtracting Complex Numbers

Perform the indicated operations.

SOLUTIONS

(a) $(2 + 5i) + (8 + 4i)$

$(2 + 5i) + (8 + 4i) = 2 + 5i + 8 + 4i$ Group like terms together and then add
$\qquad\qquad\qquad = (2 + 8) + (5 + 4)i$ like terms. Add the real terms and then
$\qquad\qquad\qquad = 10 + 9i$ add the imaginary terms.

(b) $6 + (11 - 4i)$

$6 + (11 - 4i) = (6 + 11) + (0 - 4)i$ 6 also can be written in the form
$\qquad\qquad = 17 - 4i$ $6 + 0i$.

(c) $(8 - 5i) - 6i$

$(8 - 5i) - 6i = (8 + 0) + (-5 - 6)i$ $-6i$ also can be written in the form
$\qquad\qquad = 8 - 11i$ $0 - 6i$.

(d) $(-7 + 6i) - (4 + 10i)$

$(-7 + 6i) - (4 + 10i) = -7 + 6i - 4 - 10i$ Use the distributive property to remove
$\qquad\qquad\qquad\quad = (-7 - 4) + (6 - 10)i$ the second pair of parentheses. Then
$\qquad\qquad\qquad\quad = -11 - 4i$ add like terms.   ■

Since $i^2 = -1$, higher powers of i always can be simplified to i, -1, $-i$, or 1. The first four powers of i are keys to simplifying higher powers to standard form and therefore should be memorized.

FIRST FOUR POWERS OF *i*

$i^1 = i$
$i^2 = i \cdot i = -1$
$i^3 = i^2 \cdot i = (-1)i = -i$
$i^4 = i^2 \cdot i^2 = (-1)(-1) = 1$

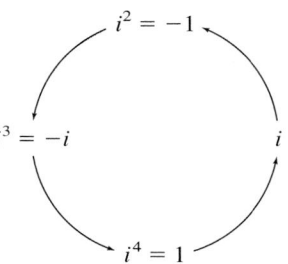

The powers of i repeat in cycles of four: i^4, i^8, i^{12}, and so on; all are equal to 1. We use this fact to simplify i^n, where n is any integer exponent, to either i, -1, $-i$, or 1.

■ EXAMPLE 4   Powers of *i*

Simplify each power of i.

SOLUTIONS

(a) i^7

$i^7 = i^4 \cdot i^3$
$\quad = (1)(-i)$
$\quad = -i$

First extract the largest multiple of 4 from each exponent. Then simplify, replacing i^4, i^8, i^{12}, and so on, by 1.

(b) i^{13}

$i^{13} = i^{12}i^1$
$\quad = (1)(i)$
$\quad = i$

(c) i^{406}

$i^{406} = i^{404} \cdot i^2$
$\quad = (i^4)^{101} \cdot (-1)$
$\quad = (1)^{101}(-1)$
$\quad = (1)(-1)$
$\quad = -1$

$$\begin{array}{r} 101 \\ 4\overline{)406} \\ \underline{404} \\ 2 \end{array}$$

■

SELF-CHECK 8.4.2

Simplify each of these expressions.

1. $(21 + 4i) - (9 - 7i)$ **2.** i^{33}

When working with complex numbers, it is important to remember the restriction on the formula given in Section 8.2.

$$\left(\sqrt[n]{x}\right)^m = \sqrt[n]{x^m} \quad \text{if and only if } \sqrt[n]{x} \text{ is a real number.}$$

Specifically we noted that $\left(\sqrt{-5}\right)^2 \neq \sqrt{(-5)^2}$.

$\left(\sqrt{-5}\right)^2 = \left(i\sqrt{5}\right)^2$ whereas $\sqrt{(-5)^2} = \sqrt{25}$
$\qquad\quad = 5i^2$ $\qquad\qquad\qquad\qquad = 5$
$\qquad\quad = -5$

It is also crucial to remember that $\sqrt{x}\sqrt{y} \neq \sqrt{xy}$ for negative values of x and y. For example:

$$\sqrt{-2}\sqrt{-3} = \left(i\sqrt{2}\right)\left(i\sqrt{3}\right) \quad \text{whereas} \quad \sqrt{(-2)(-3)} = \sqrt{+6}$$
$$= i^2\sqrt{6} \qquad\qquad\qquad\qquad = \sqrt{6}$$
$$= -\sqrt{6}$$

Before performing operations with complex numbers, it is best to write each complex number in the standard $a + bi$ form. Then all answers should be simplified to standard form.

Note that, in the preceding equation, we wrote $i\sqrt{2}$ rather than $\sqrt{2}i$. The latter form could easily be confused with $\sqrt{2i}$, which has i under the radical symbol. To make it clear that the factor i is not in the radicand, it is best to put the i in front of the radical.

To avoid the potential errors noted above, we recommend that you write every complex number in standard form before proceeding with any operations. In Example 5 we multiply complex numbers as if they were binomials with a real term and an imaginary term. We use the distributive property extensively to form these products.

■ EXAMPLE 5 Multiplying Complex Numbers

Calculate each product.

SOLUTIONS

(a) $3(4 - 5i)$

$3(4 - 5i) = 3(4) - 3(5i)$ Distribute the factor 3.
$\qquad\qquad = 12 - 15i$

(b) $\sqrt{-2}\left(\sqrt{3} - \sqrt{-2}\right)$

$\sqrt{-2}\left(\sqrt{3} - \sqrt{-2}\right) = i\sqrt{2}\left(\sqrt{3} - i\sqrt{2}\right)$ First write each factor in the standard $a + bi$ form.
$\qquad\qquad = i\sqrt{2}\left(\sqrt{3}\right) - i\sqrt{2}\left(i\sqrt{2}\right)$ Then distribute the factor $i\sqrt{2}$.
$\qquad\qquad = i\sqrt{6} - 2i^2$ Simplify each term, noting that $\sqrt{2}\sqrt{2} = 2$.
$\qquad\qquad = i\sqrt{6} - 2(-1)$ Replace i^2 by -1.
$\qquad\qquad = 2 + i\sqrt{6}$ Reorder the terms to write the answer in the standard form.

(c) $(5 - 6i)(5 + 6i)$

$(5 - 6i)(5 + 6i) = (5)^2 - (6i)^2$ Note that this expression is of the form $(x - y)(x + y) = x^2 - y^2$.
$\qquad\qquad = 25 - 36i^2$
$\qquad\qquad = 25 - 36(-1)$ Replace i^2 by -1.
$\qquad\qquad = 61$

(d) $(2 - 7i)(5 + 3i)$

$(2 - 7i)(5 + 3i) = 2(5 + 3i) - 7i(5 + 3i)$ Distribute the factor of $5 + 3i$. Then distribute the factors of 2 and of $-7i$.
$\qquad\qquad = 10 + 6i - 35i - 21i^2$
$\qquad\qquad = 10 - 29i - 21(-1)$ Combine like terms and replace i^2 by -1.
$\qquad\qquad = 10 - 29i + 21$
$\qquad\qquad = 31 - 29i$

(e) $(a + bi)(a - bi)$

$(a + bi)(a - bi) = a^2 - (bi)^2$ Note that this expression is of the form $(x + y)(x - y) = x^2 - y^2$.
$\qquad\qquad = a^2 - b^2i^2$
$\qquad\qquad = a^2 + b^2$ Replace i^2 by -1.

SELF-CHECK 8.4.3 ANSWER

$36 - 8i$

SELF-CHECK 8.4.3

Calculate the product $(5 - 3i)(6 + 2i)$.

A Mathematical Note

The term *conjugates* for $a + bi$ and $a - bi$ was suggested by Augustin-Louis Cauchy (1789–1857). Cauchy was a prolific writer in mathematics, perhaps second only to Euler.

Complex Conjugates

Since $i = \sqrt{-1}$, any expression containing i can be treated like a radical expression. In keeping with our earlier definition of conjugates, we define the **conjugate** of $a + bi$ to be $a - bi$.

COMPLEX NUMBER	COMPLEX NUMBER IN STANDARD FORM	COMPLEX CONJUGATE IN STANDARD FORM	COMPLEX CONJUGATE
$5 - 6i$	$5 - 6i$	$5 + 6i$	$5 + 6i$
$\sqrt{2} + i$	$\sqrt{2} + i$	$\sqrt{2} - i$	$\sqrt{2} - i$
$7i$	$0 + 7i$	$0 - 7i$	$-7i$
8	$8 + 0i$	$8 - 0i$	8

The product of a complex number and its conjugate is always a real number.

The product of a complex number and its conjugate uses the special product of a sum times a difference to yield the difference of two squares. This product is always a real number, as we illustrated in Example 5(e). We now look at some specific examples of the product of conjugates.

■ EXAMPLE 6 Multiplying Complex Conjugates

Multiply each complex number by its conjugate.

SOLUTIONS

(a) $3 + 4i$

$$(3 + 4i)(3 - 4i) = 9 - 16i^2$$
$$= 9 + 16$$
$$= 25$$

The conjugate of $3 + 4i$ is $3 - 4i$. Multiply using the fact that this expression is of the form $(x + y)(x - y) = x^2 - y^2$.

(b) $5 - 2i$

$$(5 - 2i)(5 + 2i) = 25 - 4i^2$$
$$= 25 + 4$$
$$= 29$$

The conjugate of $5 - 2i$ is $5 + 2i$.

(c) $7i$

$$(7i)(-7i) = -49i^2$$
$$= 49$$

The conjugate of $7i$ is $-7i$.

(d) 8

$$8 \cdot 8 = 64$$

The conjugate of 8 is 8. ■

Dividing Complex Numbers

The fact that the product of a complex number and its conjugate is always a real number plays a key role in the division of complex numbers as outlined in the following box.

Division of Complex Numbers

VERBALLY	NUMERICAL EXAMPLE
Step 1 Write the division problem as a fraction.	$20 \div (1 + 3i) = \dfrac{20}{1 + 3i}$
Step 2 Multiply both the numerator and the denominator by the conjugate of the denominator.	$= \dfrac{20}{1 + 3i} \cdot \dfrac{1 - 3i}{1 - 3i}$ $= \dfrac{20(1 - 3i)}{1 + 9}$
Step 3 Simplify the result, and express it in standard $a + bi$ form.	$= 2(1 - 3i)$ $= 2 - 6i$

■ EXAMPLE 7   Dividing Complex Numbers

Simplify $(8 - i) \div (1 - 2i)$.

SOLUTION

$$(8 - i) \div (1 - 2i) = \frac{8 - i}{1 - 2i}$$

Write the division problem as a fraction.

$$= \frac{8 - i}{1 - 2i} \cdot \frac{1 + 2i}{1 + 2i}$$

Multiply the numerator and the denominator by $1 + 2i$, the conjugate of the denominator.

$$= \frac{8 + 15i - 2i^2}{1 - 4i^2}$$

Multiply the numerators and multiply the conjugates in the denominator.

$$= \frac{10 + 15i}{5}$$

Simplify, replacing i^2 by -1.

$$= \frac{10}{5} + \frac{15}{5}i$$

Write the result in standard $a + bi$ form.

$$= 2 + 3i$$ ■

SELF-CHECK 8.4.4

1. Multiply $11 + 10i$ times its conjugate.
2. Calculate the quotient $\dfrac{3 + 2i}{2 - 3i}$.

CALCULATOR PERSPECTIVE 8.4.2	Operations with Complex Numbers

The $\boxed{i}$ feature is the secondary function of the $\boxed{\cdot}$ key.

Use Calculator Perspective 8.4.1 to switch the mode to $a + bi$. To evaluate the expressions with complex numbers from examples in this section on a TI-83 Plus calculator, enter the following keystrokes:

Example 3(a): $(2 + 5i) + (8 + 4i)$

$\boxed{(}$ $\boxed{2}$ $\boxed{+}$ $\boxed{5}$ $\boxed{2nd}$ $\boxed{i}$ $\boxed{)}$ $\boxed{+}$

$\boxed{(}$ $\boxed{8}$ $\boxed{+}$ $\boxed{4}$ $\boxed{2nd}$ $\boxed{i}$ $\boxed{)}$ $\boxed{ENTER}$

```
(2+5i)+(8+4i)
            10+9i
■
```

Example 5(d): $(2 - 7i)(5 + 3i)$

$\boxed{(}$ $\boxed{2}$ $\boxed{-}$ $\boxed{7}$ $\boxed{2nd}$ $\boxed{i}$ $\boxed{)}$

$\boxed{(}$ $\boxed{5}$ $\boxed{+}$ $\boxed{3}$ $\boxed{2nd}$ $\boxed{i}$ $\boxed{)}$ $\boxed{ENTER}$

```
(2-7i)(5+3i)
           31-29i
■
```

Example 7: $(8 - i) \div (1 - 2i)$

$\boxed{(}$ $\boxed{8}$ $\boxed{-}$ $\boxed{2nd}$ $\boxed{i}$ $\boxed{)}$ $\boxed{\div}$

$\boxed{(}$ $\boxed{1}$ $\boxed{-}$ $\boxed{2}$ $\boxed{2nd}$ $\boxed{i}$ $\boxed{)}$ $\boxed{ENTER}$

```
(8-i)/(1-2i)
            2+3i
■
```

Note: It is sometimes useful to press $\boxed{MATH}$ $\boxed{ENTER}$ $\boxed{ENTER}$ to convert a complex number in decimal form to fractional form. ■

Calculators are powerful tools that allow us to leverage our mathematical power. It is important for us to know how to use our calculator and to know some of its limitations. This is illustrated in Example 8.

■ EXAMPLE 8   Evaluating Powers of *i* and Interpreting Calculator Results

The TI-83 Plus calculator screen shown in the figure displays an evaluation of i^{27}. Evaluate i^{27} by hand and interpret this calculator screen.

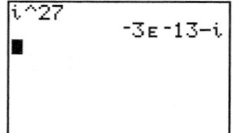

SOLUTION

$$i^{27} = i^{24} \cdot i^3$$
$$= (1)(-i)$$
$$= -i$$

First extract the largest multiple of 4 from the exponent of 27.
Replace i^{24} by 1 and i^3 by $-i$.

The calculator display in $a + bi$ form is:
$$\left(-3.0 \times 10^{-13}\right) - i \approx 0 - i$$

Answer: $i^{27} = -i$

The calculator display gives the result in $a + bi$ form with the real part a given in scientific notation.
$-3.0 \times 10^{-13} = -0.0000000000003 \approx 0$
Therefore the calculator has introduced a small error because the real part is exactly 0; $i^{27} = 0 - i$. ■

One of the uses you are most likely to have for complex numbers is as solutions of quadratic equations. Example 9 illustrates checking a complex solution of a quadratic equation.

■ EXAMPLE 9   Checking a Complex Number Solution of a Quadratic Equation

Verify that $1 + 2i$ is a solution of $x^2 - 2x + 5 = 0$.

SOLUTION

$$x^2 - 2x + 5 = 0$$
$$\left(1 + 2i\right)^2 - 2\left(1 + 2i\right) + 5 \stackrel{?}{=} 0 \quad \text{Substitute } 1 + 2i \text{ for } x.$$
$$\left(1 + 4i + 4i^2\right) - 2 - 4i + 5 \stackrel{?}{=} 0 \quad \text{Simplify and then add like terms.}$$
$$1 + 4i - 4 - 2 - 4i + 5 \stackrel{?}{=} 0$$
$$0 \stackrel{?}{=} 0$$

Answer: This value checks, so it is a solution of the equation. ■

SELF-CHECK   8.4.5

Use a calculator to check the solution in Example 9.

USING THE LANGUAGE AND SYMBOLISM OF MATHEMATICS 8.4

1. The imaginary number i is used to represent _____ .
2. The complex number $a + bi$ has a real term, _____ , and an imaginary term, _____ .
3. A complex number is **always/sometimes/never** a real number. (Select the correct choice.)

4. An imaginary number is **always/sometimes/never** a complex number. (Select the correct choice.)
5. The standard form of a complex number is _____ .
6. Powers of i repeat in cycles of _____ .
7. The complex _____ of $3 + 5i$ is $3 - 5i$.

EXERCISES 8.4

In Exercises 1–10 simplify each expression and write the result in the standard $a + bi$ form.

1. **a.** $-\sqrt{36}$ **b.** $\sqrt{-36}$
   c. $-\sqrt{-36}$ **d.** $\sqrt{36}$
2. **a.** $-\sqrt{49}$ **b.** $\sqrt{-49}$
   c. $-\sqrt{-49}$ **d.** $\sqrt{49}$
3. **a.** $\sqrt{-9} + \sqrt{16}$ **b.** $\sqrt{9} + \sqrt{-16}$
   c. $-\sqrt{9} - \sqrt{-16}$ **d.** $\sqrt{-9} - \sqrt{16}$
4. **a.** $\sqrt{4} + \sqrt{-25}$ **b.** $\sqrt{-4} - \sqrt{25}$
   c. $-\sqrt{4} - \sqrt{-25}$ **d.** $-\sqrt{-4} + \sqrt{-25}$
5. **a.** $\sqrt{-9 - 16}$ **b.** $-\sqrt{-9 - 16}$
   c. $\sqrt{-6}\sqrt{-6}$ **d.** $\sqrt{(-6)(-6)}$
6. **a.** $\sqrt{-25 - 144}$ **b.** $\sqrt{-25} + \sqrt{-144}$
   c. $\sqrt{-7}\sqrt{-7}$ **d.** $\sqrt{(-7)(-7)}$
7. **a.** $\sqrt{-4}\sqrt{-25}$ **b.** $\sqrt{(-4)(-25)}$
   c. $-\sqrt{4}\sqrt{-25}$ **d.** $(-\sqrt{4})(-\sqrt{25})$
8. **a.** $\sqrt{-9}\sqrt{-100}$ **b.** $\sqrt{(-9)(-100)}$
   c. $-\sqrt{9}\sqrt{-100}$ **d.** $(-\sqrt{9})(-\sqrt{100})$
9. **a.** $\sqrt{-\dfrac{25}{9}}$ **b.** $\dfrac{\sqrt{-25}}{\sqrt{9}}$
   c. $\dfrac{\sqrt{25}}{\sqrt{-9}}$ **d.** $\sqrt{\dfrac{-25}{-9}}$
10. **a.** $\sqrt{\dfrac{-36}{-49}}$ **b.** $\dfrac{\sqrt{36}}{\sqrt{-49}}$
    c. $\sqrt{\dfrac{36}{49}}$ **d.** $\dfrac{\sqrt{-36}}{\sqrt{49}}$

In Exercises 11 and 12 identify the real values of a and b that will make each statement true.

11. **a.** $a + bi = 18 - 5i$ **b.** $a - 2i = -6 + bi$
    c. $5a - 9i = 20 + (b + 2)i$ **d.** $3a - bi = 72$
12. **a.** $a + bi = -6 + 17i$ **b.** $-3 + bi = a + 11i$
    c. $14a - 3bi = (5a - 1) + 2i$ **d.** $3a - bi = 72i$

In Exercises 13–42 perform the indicated operations and express the result in the standard $a + bi$ form.

13. $(1 + 2i) + (8 - 3i)$ 14. $(6 - 7i) + (13 - 4i)$
15. $(5 + 3i) - (2 + 2i)$ 16. $(7 + i) - (5 - 2i)$
17. $6(10 + 4i)$ 18. $-9(3 - 12i)$
19. $(3 + 3i) + \dfrac{1}{2}(8 - 6i)$ 20. $(3 + 3i) - \dfrac{1}{3}(6 - 9i)$
21. $(2i)(3i)$ 22. $(5i)(7i)$

23. $2i(3 - 5i)$ 24. $-3i(4 - 2i)$
25. $i(3 + 2i) + 2(5 - 3i)$ 26. $4(3 - 2i) - i(5 + 3i)$
27. $(2 - 7i)(2 + 7i)$ 28. $(6 + 9i)(6 - 9i)$
29. $(5 - 2i)(4 + 7i)$ 30. $(6 + i)(3 - 5i)$
31. $(5 + i)^2$ 32. $(6 - i)^2$
33. $(4 - 3i)^2$ 34. $(3 + 5i)^2$
35. $(4 + 7i)(7 - 4i)$ 36. $(11 - 3i)(2 + 5i)$
37. $\sqrt{-3}(\sqrt{2} + \sqrt{-3})$
38. $(\sqrt{2} - \sqrt{-5})(\sqrt{2} - 3\sqrt{-5})$
39. $2\sqrt{-75} + \sqrt{-27}$
40. $3\sqrt{-8} - 5\sqrt{-98}$
41. $\sqrt{4} + \sqrt{-9} - \sqrt{9} - \sqrt{-25}$
42. $\sqrt{64} + \sqrt{-36} - \sqrt{9} - \sqrt{-1}$

In Exercises 43–46 multiply each complex number by its conjugate.

43. $2 + 5i$ 44. $3 - 8i$ 45. $13i$ 46. 13

In Exercises 47–52 perform the indicated operations and express the result in standard $a + bi$ form.

47. $\dfrac{4}{1 + i}$ 48. $\dfrac{6}{1 - i}$
49. $\dfrac{4 - i}{4 + i}$ 50. $\dfrac{5 + i}{5 - i}$
51. $85 \div (7 - 6i)$ 52. $185 \div (11 + 8i)$

In Exercises 53–56 simplify each power of i.
53. i^9 54. i^{11} 55. i^{58} 56. i^{81}
57. Given $a = 1$, $b = 2$, and $c = 5$, evaluate both
$$\frac{-b - \sqrt{b^2 - 4ac}}{2a} \quad \text{and} \quad \frac{-b + \sqrt{b^2 - 4ac}}{2a}.$$
58. Given $a = 2$, $b = -3$, and $c = 2$, evaluate both
$$\frac{-b - \sqrt{b^2 - 4ac}}{2a} \quad \text{and} \quad \frac{-b + \sqrt{b^2 - 4ac}}{2a}.$$

Calculator Usage
In Exercises 59–66 use a calculator to evaluate each expression.
59. $3(4 - 5i) - 7(6 + 2i)$ 60. $\dfrac{4 - 5i}{6 + 7i}$
61. $(1 - 2i)^3$ 62. $(1 - i)^4$
63. $\left(-\dfrac{1}{2} + \dfrac{\sqrt{3}}{2}i\right)^3$ 64. $\left(-\dfrac{1}{2} - \dfrac{\sqrt{3}}{2}i\right)^3$
65. $\dfrac{58}{\sqrt{4} - \sqrt{-25}}$ 66. $\sqrt[3]{-64} - \sqrt{-64}$

In Exercises 67–70 interpret each calculator display.

67.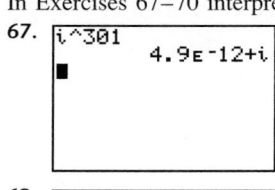

68.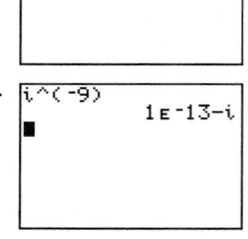

69.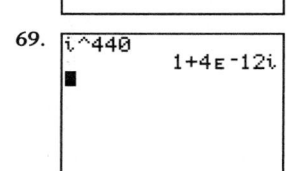

70.

In Exercises 71–74 determine whether the given complex number is a solution of the quadratic equation.

71. $x^2 - 6x + 13 = 0$; $3 - 2i$

72. $x^2 - 6x + 13 = 0$; $3 + 2i$

73. $x^2 - 4x + 13 = 0$; $2 - 3i$

74. $x^2 - 4x + 13 = 0$; $2 + 3i$

75. Given $z = 3 - i$:
   a. What is the conjugate of z?
   b. What is the additive inverse of z?
   c. What is the multiplicative inverse of z?

76. a. Is $\sqrt{3}$ a real number?
   b. Is $\sqrt{3}$ an imaginary number?
   c. Is $\sqrt{3}$ a complex number?
   d. Is $\sqrt{-3}$ a real number?
   e. Is $\sqrt{-3}$ an imaginary number?
   f. Is $\sqrt{-3}$ a complex number?

77. a. What can you say about a complex number that is equal to its conjugate?
   b. Give an example of two imaginary numbers whose sum is a real number.
   c. Give an example of two imaginary numbers whose product is a real number.
   d. Name a complex number that cannot be used as a divisor.

Group Discussion Questions

78. Discovery Question   Complete the following diagram from i to i^{12} to illustrate that powers of i repeat in cycles of four.

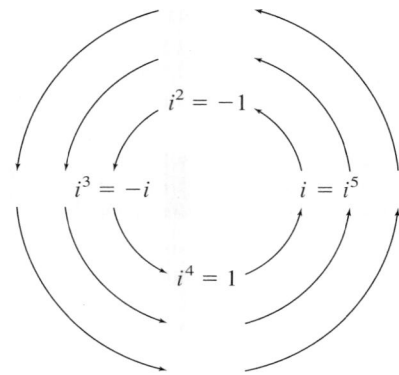

79. Challenge Question   Use a calculator to answer these questions.
   a. Is 2 a solution of $x^3 - 8 = 0$?
   b. Is $-1 + i\sqrt{3}$ a solution of $x^3 - 8 = 0$?
   c. Is $-1 - i\sqrt{3}$ a solution of $x^3 - 8 = 0$?
   d. How many cube roots does 8 have? What are these cube roots, and which is the principal cube root of 8?

80. Challenge Question
   Simplify $i + i^2 + i^3 + i^4 + \ldots + i^{99} + i^{100}$.

Section 8.5   Quadratic Equations and Inequalities

Objectives: 9. Solve a quadratic equation using the quadratic formula.
   10. Use the discriminant to determine the nature of the solutions of a quadratic equation.
   11. Solve a quadratic inequality.

We already have covered solving quadratic equations by factoring. We also have used graphs and tables to solve quadratic equations and inequalities. The new material presented in this section is the use of extraction of roots and completing the square to develop the quadratic formula. This formula can be used to solve any quadratic equation.

This section extends the material on solving quadratic equations that was presented in Section 7.5. In Section 7.5 we solved quadratic equations of the form $ax^2 + bx + c = 0$ by factoring and noted the relationship between the real solutions of a quadratic equation and the x-intercepts of the graph of the corresponding quadratic function. We also noted how we could use the graph of a parabola to solve quadratic inequalities.

   In this section we examine quadratic equations whose solutions are complex numbers; some of these complex solutions are real numbers and some are imaginary numbers. We also examine the graphical significance of these solutions. A major goal of this section is to develop algebraic methods for solving all quadratic equations—including those with imaginary solutions.

Extraction of Roots

The notation $x = \pm\sqrt{k}$ is read "x equals plus or minus the (principal) square root of k."

We start by examining quadratic equations of the form $x^2 = k$. The solutions of this equation are values whose square is k—that is, the two square roots of k. The solutions of $x^2 = k$ are $x = \sqrt{k}$ and $x = -\sqrt{k}$. To denote both possible square roots of k, we write $x = \pm\sqrt{k}$. The notation $\pm$ is read "plus or minus."

QUADRATIC EQUATION	SOLUTIONS	NATURE OF SOLUTIONS
$x^2 = 9$	$x = \pm\sqrt{9}$ $x = 3$ or $x = -3$	Two real solutions that are integers
$x^2 = 2$	$x = \pm\sqrt{2}$ $x = \sqrt{2}$ or $x = -\sqrt{2}$	Two real solutions that are irrational numbers
$x^2 = -4$	$x = \pm\sqrt{-4}$ $x = 2i$ or $x = -2i$	Two imaginary solutions that are complex conjugates
$x^2 = -3$	$x = \pm\sqrt{-3}$ $x = i\sqrt{3}$ or $x = -i\sqrt{3}$	Two imaginary solutions that are complex conjugates

Solving $x^2 = k$ as $x = \pm\sqrt{k}$ is called solving by **extraction of roots.** In Example 1 we use this method to solve $(x - 1)^2 = 3$.

■ EXAMPLE 1   Using Multiple Perspectives to Solve a Quadratic Equation

Solve $(x - 1)^2 = 3$ algebraically and graphically.

SOLUTION

ALGEBRAICALLY

$(x - 1)^2 = 3$ ⟶ Solve by extracting both square roots of 3.

$\quad x - 1 = \pm\sqrt{3}$

$\qquad\quad x = 1 \pm \sqrt{3}$ ⟶ Solve for x by adding 1 to both sides of the equation.

$x = 1 - \sqrt{3}$ or $x = 1 + \sqrt{3}$

Be careful to include the "$\pm$" symbol. A quadratic equation has two solutions. One will come from the "plus" choice and one from the "minus" choice.

GRAPHICALLY

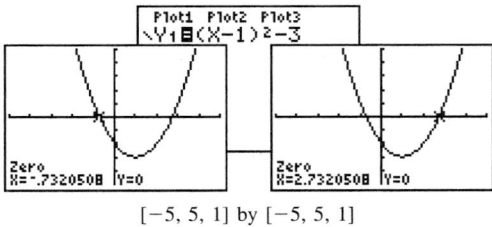

$[-5, 5, 1]$ by $[-5, 5, 1]$

The x-intercepts of $y_1 = (x - 1)^2 - 3$ are $\left(1 - \sqrt{3}, 0\right)$ and $\left(1 + \sqrt{3}, 0\right)$.

Solving $(x - 1)^2 = 3$ is equivalent to solving $(x - 1)^2 - 3 = 0$. Use a calculator to determine the x-intercepts of the parabola defined by $y_1 = (x - 1)^2 - 3$. One x-intercept is $\left(1 - \sqrt{3}, 0\right)$ and the other is $\left(1 + \sqrt{3}, 0\right)$. You can obtain decimal approximations of $1 - \sqrt{3} \approx -0.7320508$ and $1 + \sqrt{3} \approx 2.7320508$ by using the calculator to find the zeros of this quadratic function. You can also test the exact values $1 - \sqrt{3}$ and $1 + \sqrt{3}$ by using the trace feature to evaluate the function for these input values.

Answer: The exact solutions of $(x - 1)^2 = 3$ are $x = 1 - \sqrt{3}$ and $x = 1 + \sqrt{3}$. ■

A thorough examination of Example 1 reveals several things. Each real solution of $ax^2 + bx + c = 0$, including irrational solutions, corresponds to an x-intercept of the parabola defined by $y = ax^2 + bx + c$ and a zero of the function $f(x) = ax^2 + bx + c$. Also note that the algebraic method can produce exact answers rather than just approximations. As you work through the steps in Example 1, you may also discover that often it can be faster to solve a quadratic equation using pencil and paper by an algebraic method than it is to use a graphics calculator to approximate the solutions numerically or graphically.

SELF-CHECK 8.5.1

1. Solve $(x + 2)^2 = 5$ by extraction of roots.
2. Determine the x-intercepts of the parabola $y = (x + 2)^2 - 5$.
3. Determine the zeros of $f(x) = (x + 2)^2 - 5$.

A Mathematical Note

Many methods that we now use quickly and efficiently in algebraic form have their roots in ancient geometric methods. In particular, the method of completing the square was used by both Greek and Arab mathematicians in the geometric form shown in the Geometric Viewpoint. Al-Khowârizmi (c. 825) was a noted Arab mathematician, astronomer, and author who illustrated this method in his writings.

Completing the Square: A Geometric Viewpoint

The left side of a quadratic equation always can be written as a perfect square. The process of writing the left side of the equation in this form is called **completing the square.** The term "completing the square" has a very natural geometric basis that is illustrated in the following geometric viewpoint. Here we examine $x^2 + 6x$. This expression is the left side of the equation that we examine in Example 2(a).

COMPLETING THE SQUARE FOR $x^2 + 6x$

ALGEBRAICALLY	GEOMETRICALLY
1. Start with: $x^2 + 6x$	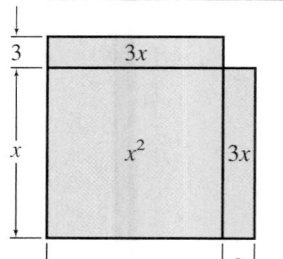
2. Add 9: $x^2 + 6x + 9$	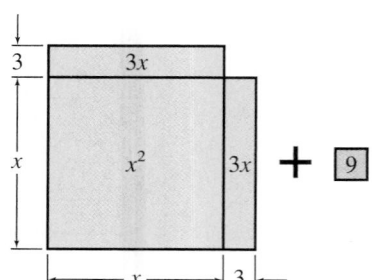

---

3. Write as a perfect square:
$$x^2 + 6x + 9 = (x + 3)^2$$

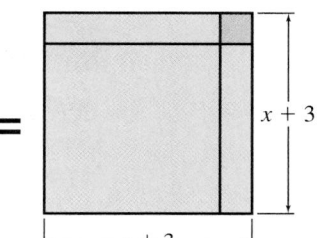

We are using completing the square to put the left side of a quadratic equation in a form that allows us to use the method of extraction of roots. Then we use the method of completing the square to develop the quadratic formula.

Completing the Square

Solving a quadratic equation by factoring over the integers is a great method if the coefficients are small integers and the solutions are nice rational numbers. For messier coefficients and solutions that are irrational real numbers or imaginary numbers it is nice to have a formula that we always can use. The work that follows presents the logic for the development of the quadratic formula.

■ EXAMPLE 2   Completing the Square

Rewrite each equation so that the left side of the equation is a perfect square.

SOLUTIONS

(a) $x^2 + 6x = -7$

$$x^2 + 6x = -7$$
$$x^2 + 6x + 9 = -7 + 9$$
$$(x + 3)^2 = 2$$

In each case the constant we add to both sides of the equation is the square of one-half the coefficient of x.

$$\left(\frac{6}{2}\right)^2 = 3^2 = 9$$

(b) $x^2 - 8x = -11$

$$x^2 - 8x = -11$$
$$x^2 - 8x + 16 = -11 + 16$$
$$(x - 4)^2 = 5$$

$$\left(\frac{-8}{2}\right)^2 = (-4)^2 = 16$$

(c) $x^2 + 3x = 1$

$$x^2 + 3x = 1$$
$$x^2 + 3x + \frac{9}{4} = 1 + \frac{9}{4}$$
$$\left(x + \frac{3}{2}\right)^2 = \frac{13}{4}$$

$$\left(\frac{3}{2}\right)^2 = \frac{9}{4}$$

SELF-CHECK 8.5.2

Rewrite each equation so that the left side is written as a perfect square.

1. $x^2 + 2x = 4$
2. $x^2 - 10x = -17$
3. $x^2 + 4x = -5$
4. $x^2 + 5x = -5$

SELF-CHECK 8.5.2 ANSWERS

1. $(x + 1)^2 = 5$
2. $(x - 5)^2 = 8$
3. $(x + 2)^2 = -1$
4. $\left(x + \frac{5}{2}\right)^2 = \frac{5}{4}$

The process of completing the square can be used to solve any quadratic equation. The key steps in this process are shown in the following box.

Solving Quadratic Equations by Completing the Square	
VERBALLY	**NUMERICAL EXAMPLE**
Step 1 Write the equation with the constant term on the right side.	$x^2 + 3x - 4 = 0$ $x^2 + 3x = 4$
Step 2 Divide both sides of the equation by the coefficient of x^2 to obtain a coefficient of 1 for x^2.	$x^2 + 3x + \left(\dfrac{3}{2}\right)^2 = 4 + \left(\dfrac{3}{2}\right)^2$
Step 3 Take one-half of the coefficient of x, square this number, and add the result to both sides of the equation.	$\left(x + \dfrac{3}{2}\right)^2 = \dfrac{16}{4} + \dfrac{9}{4}$
Step 4 Write the left side of the equation as a perfect square.	$\left(x + \dfrac{3}{2}\right)^2 = \dfrac{25}{4}$ $x + \dfrac{3}{2} = \pm\dfrac{5}{2}$
Step 5 Solve this equation by extraction of roots.	$x = -\dfrac{3}{2} \pm \dfrac{5}{2}$ $x = -\dfrac{3}{2} - \dfrac{5}{2}$ or $x = -\dfrac{3}{2} + \dfrac{5}{2}$ $x = -4$ or $x = 1$

The steps outlined in this box are used to solve Example 3 and then to create a formula that can be used to solve any quadratic equation.

■ EXAMPLE 3 Using Completing the Square to Determine Two Irrational Solutions

Solve $4x^2 - 20x + 7 = 0$ by completing the square.

SOLUTION _____

$4x^2 - 20x + 7 = 0$

$4x^2 - 20x = -7$ Shift the constant to the right side of the equation.

$x^2 - 5x = -\dfrac{7}{4}$ Divide both sides of the equation by the coefficient of x^2, which is 4.

$x^2 - 5x + \left(-\dfrac{5}{2}\right)^2 = -\dfrac{7}{4} + \left(-\dfrac{5}{2}\right)^2$ Take one-half the coefficient of x: $\dfrac{1}{2}(-5) = -\dfrac{5}{2}$. Square this number: $\left(-\dfrac{5}{2}\right)^2$. Then add this result to both sides of the equation.

$\left(x - \dfrac{5}{2}\right)^2 = -\dfrac{7}{4} + \dfrac{25}{4}$ Write the left side as a perfect square.

$\left(x - \dfrac{5}{2}\right)^2 = \dfrac{9}{2}$

$x - \dfrac{5}{2} = \pm\sqrt{\dfrac{9}{2}}$ Extract the roots.

$x = \dfrac{5}{2} \pm \dfrac{3\sqrt{2}}{2}$ Add $\dfrac{5}{2}$ to both sides of the equation. Simplify, noting that $\sqrt{\dfrac{9}{2}} = \dfrac{3}{\sqrt{2}} = \dfrac{3\sqrt{2}}{2}$.

Answer: $x = \dfrac{5 - 3\sqrt{2}}{2}$ or $x = \dfrac{5 + 3\sqrt{2}}{2}$

SELF-CHECK 8.5.3

Replace the question mark in each step of the solution. $5x^2 - 3x - 2 = 0$.

1. $5x^2 - 3x = ?$

2. $x^2 + ?x = \dfrac{2}{5}$

3. $x^2 - \dfrac{3}{5}x + \dfrac{9}{100} = \dfrac{2}{5} + ?$

4. $\left(x - \dfrac{3}{10}\right)^2 = ?$

5. $x - \dfrac{3}{10} = \pm?$

$x = \dfrac{3}{10} \pm \dfrac{7}{10}$

$x = \dfrac{3}{10} - \dfrac{7}{10}$ or

$x = \dfrac{3}{10} + \dfrac{7}{10}$

$x = -\dfrac{4}{10}$ or $x = \dfrac{10}{10}$

6. $x = -\dfrac{2}{5}$ or $x = ?$

The Quadratic Formula

We could continue to solve each individual quadratic equation by completing the square. Rather than repeating these steps for each problem, we can solve the general quadratic equation $ax^2 + bx + c = 0$ by completing the square. Then we can use this general solution as a formula that can be applied to any quadratic equation.

$$ax^2 + bx + c = 0$$

Start with a quadratic equation in standard form with a, b, and c real numbers and $a > 0$.

$$ax^2 + bx = -c$$

Shift the constant to the right side of the equation.

$$x^2 + \frac{b}{a}x = -\frac{c}{a}$$

Divide both sides by the coefficient of x^2, a. To divide by a, we assume $a \neq 0$.

$$x^2 + \frac{b}{a}x + \left(\frac{b}{2a}\right)^2 = -\frac{c}{a} + \left(\frac{b}{2a}\right)^2$$

Add the square of one-half the coefficient of x to both sides of the equation.

$$\left(x + \frac{b}{2a}\right)^2 = -\frac{4ac}{4a^2} + \frac{b^2}{4a^2}$$

Write the left side as a perfect square and the right side in terms of a common denominator.

$$\left(x + \frac{b}{2a}\right)^2 = \frac{b^2 - 4ac}{4a^2}$$

Simplify by combining the terms on the right side of the equation.

$$x + \frac{b}{2a} = \pm\sqrt{\frac{b^2 - 4ac}{4a^2}}$$

Extract the roots.

$$x = -\frac{b}{2a} \pm \frac{\sqrt{b^2 - 4ac}}{2a}$$

Simplify the radical and subtract $\dfrac{b}{2a}$ from both sides of the equation.

$$x = \frac{-b \pm \sqrt{b^2 - 4ac}}{2a}$$

This is the quadratic formula.

$$x = \frac{-b + \sqrt{b^2 - 4ac}}{2a} \quad \text{or} \quad x = \frac{-b - \sqrt{b^2 - 4ac}}{2a}$$

SELF-CHECK 8.5.3 ANSWERS

1. 2   2. $\dfrac{-3}{5}$ 3. $\dfrac{9}{100}$

4. $\dfrac{49}{100}$ 5. $\dfrac{7}{10}$ 6. 1

The quadratic formula is given in the following box. You should memorize this formula. Note that the symbol "$\pm$" is used to express the two solutions concisely.

Quadratic Formula

The solutions of the quadratic equation $ax^2 + bx + c = 0$ with real coefficients a, b, and c, with $a \neq 0$, are

with $x = \dfrac{-b \pm \sqrt{b^2 - 4ac}}{2a}$

If $a = 0$, the equation $ax^2 + bx + c = 0$ becomes the linear equation $bx + c = 0$.

Before using this formula, be sure to write the quadratic equation you are trying to solve in standard form; otherwise, it is easy to make an error in the sign of a, b, or c.

■ **EXAMPLE 4**   **Using the Quadratic Formula to Determine Two Rational Solutions**

Use the quadratic formula to solve $x^2 - 5x - 6 = 0$ and to determine the x-intercepts of the graph of $y = x^2 - 5x - 6$.

SOLUTION _____

ALGEBRAICALLY

$x^2 - 5x - 6 = 0$

$x = \dfrac{-(-5) \pm \sqrt{(-5)^2 - 4(1)(-6)}}{2(1)}$

Substitute $a = 1$, $b = -5$, and $c = -6$ into $x = \dfrac{-b \pm \sqrt{b^2 - 4ac}}{2a}$.

$x = \dfrac{5 \pm \sqrt{25 + 24}}{2}$

Simplify and find both solutions.

$x = \dfrac{5 \pm \sqrt{49}}{2}$

$x = \dfrac{5 \pm 7}{2}$

$x = \dfrac{5 - 7}{2}$ or $x = \dfrac{5 + 7}{2}$

$x = -1$ $x = 6$

GRAPHICALLY

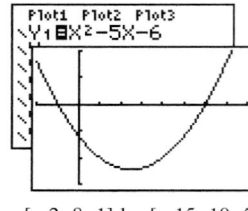

[−2, 8, 1] by [−15, 10, 5]

The x-intercepts of the parabola defined by $y = x^2 - 5x - 6$ correspond to the solutions of $x^2 - 5x - 6 = 0$. A calculator also can be used to determine that the x-intercepts are $(-1, 0)$ and $(6, 0)$.

These x-intercepts confirm that the solutions of $x^2 - 5x - 6 = 0$ are $x = -1$ and $x = 6$.

Answer: The solutions are $x = -1$ and $x = 6$.
The x-intercepts of the graph are $(-1, 0)$ and $(6, 0)$. ■

Note that we also could have solved the equation in Example 4 by factoring it as $(x + 1)(x - 6) = 0$. The primary advantage of the quadratic formula is that it can be used to solve problems that cannot be solved by factoring over the integers. This is illustrated by Example 5.

Before using the quadratic formula, be sure to write the equation in the standard form $ax^2 + bx + c = 0$.

■ EXAMPLE 5   Using the Quadratic Formula to Determine Two Irrational Solutions

Use the quadratic formula to solve $4x^2 - 4x = 1$ and then determine the x-intercepts of $y = 4x^2 - 4x - 1$.

SOLUTION

ALGEBRAICALLY

$4x^2 - 4x = 1$

$4x^2 - 4x - 1 = 0$ First write the equation in standard form.

$x = \dfrac{-(-4) \pm \sqrt{(-4)^2 - 4(4)(-1)}}{2(4)}$ Substitute $a = 4$, $b = -4$, and $c = -1$ into the quadratic formula.

$x = \dfrac{4 \pm \sqrt{16 + 16}}{8}$

$x = \dfrac{4 \pm \sqrt{32}}{8}$

$x = \dfrac{4 \pm 4\sqrt{2}}{8}$ Note that $\sqrt{32} = \sqrt{16} \cdot \sqrt{2} = 4\sqrt{2}$.

$x = \dfrac{4(1 \pm \sqrt{2})}{8}$

$x = \dfrac{1 \pm \sqrt{2}}{2}$ Simplify and write both solutions separately.

$x = \dfrac{1 - \sqrt{2}}{2}$ or $x = \dfrac{1 + \sqrt{2}}{2}$ These are the exact solutions and then we give calculator approximations of these solutions.

$x \approx -0.2071$ $x \approx 1.2071$

GRAPHICALLY

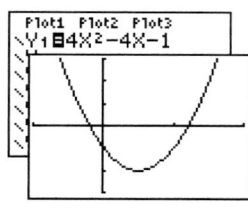

Plot1 Plot2 Plot3
\Y1■4X²-4X-1

$[-1, 2, 1]$ by $[-3, 3, 1]$

A calculator either can be used to confirm that the x-intercepts are exactly $\left(\dfrac{1 - \sqrt{2}}{2}, 0\right)$ and $\left(\dfrac{1 + \sqrt{2}}{2}, 0\right)$ or can be used to determine approximations of these exact values. These x-intercepts confirm that the solutions of $4x^2 - 4x = 1$ are $x = \dfrac{1 - \sqrt{2}}{2}$ and $x = \dfrac{1 + \sqrt{2}}{2}$.

Answer: The solutions are $x = \dfrac{1 - \sqrt{2}}{2}$ and $x = \dfrac{1 + \sqrt{2}}{2}$.

The x-intercepts are $\left(\dfrac{1 - \sqrt{2}}{2}, 0\right)$ and $\left(\dfrac{1 + \sqrt{2}}{2}, 0\right)$.

SELF-CHECK 8.5.4

Use the quadratic formula to solve $x^2 - 2x = 5$.

Example 6 gives a quadratic equation with imaginary solutions. It is not possible to obtain these solutions by factoring over the integers or by graphing the corresponding parabola on the real coordinate plane. This quadratic equation illustrates the importance of the quadratic formula as a method for solving any quadratic equation.

■ **EXAMPLE 6**   **Using the Quadratic Formula to Determine Two Imaginary Solutions**

Use the quadratic formula to solve $3x^2 = 4x - 2$ and then determine the x-intercepts of $y = 3x^2 - 4x + 2$.

SOLUTION _____

ALGEBRAICALLY

$3x^2 = 4x - 2$

$3x^2 - 4x + 2 = 0$ First write the equation in standard form.

$x = \dfrac{-(-4) \pm \sqrt{(-4)^2 - 4(3)(2)}}{2(3)}$ Substitute $a = 3$, $b = -4$, and $c = 2$ into the quadratic formula.

$x = \dfrac{4 \pm \sqrt{16 - 24}}{6}$

$x = \dfrac{4 \pm \sqrt{-8}}{6}$

$x = \dfrac{4 \pm 2i\sqrt{2}}{6}$ Note that $\sqrt{-8} = \sqrt{-4}\sqrt{2} = 2i\sqrt{2}$.

$x = \dfrac{2(2 \pm i\sqrt{2})}{6}$ Use the distributive property to factor 2 from each term of the numerator.

$x = \dfrac{2 \pm i\sqrt{2}}{3}$ Reduce this fraction by dividing both the numerator and the denominator by 2.

Because of the "±" symbol in the quadratic formula, imaginary solutions will always appear as complex conjugates.

$x = \dfrac{2}{3} - \dfrac{\sqrt{2}}{3}i$ or $x = \dfrac{2}{3} + \dfrac{\sqrt{2}}{3}i$ These complex solutions are imaginary numbers. Note that they are complex conjugates.

GRAPHICALLY

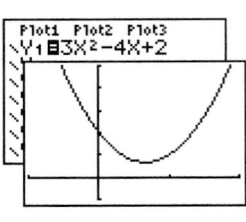

[-1, 2, 1] by [-1, 5, 1]

The parabola defined by $y = 3x^2 - 4x + 2$ has no x-intercepts. Because there are no real x-intercepts the solutions of $3x^2 = 4x - 2$ must be imaginary. You can check these solutions algebraically by substituting them into the original equation.

SELF-CHECK 8.5.4 ANSWER

$x = 1 - \sqrt{6}$ or
$x = 1 + \sqrt{6}$

Answer: The solutions are $x = \dfrac{2}{3} - \dfrac{\sqrt{2}}{3}i$ and $x = \dfrac{2}{3} + \dfrac{\sqrt{2}}{3}i$.

There are no x-intercepts for this graph. ■

James Joseph Sylvester (1814–1897) was born in England as James Joseph. He changed his last name to Sylvester when he moved to the United States. At Johns Hopkins University he led efforts to establish graduate work in mathematics in the United States. He also founded the *American Journal of Mathematics*. Among his lasting contributions to mathematics are the many new terms he introduced, including the term *discriminant*.

The Discriminant

Every quadratic equation has either two distinct roots or a double root. These roots may be either real numbers or imaginary numbers. The nature of the roots can be determined by examining the radicand, $b^2 - 4ac$, of the quadratic formula

$$x = \frac{-b \pm \sqrt{b^2 - 4ac}}{2a}.$$

Because $b^2 - 4ac$ can be used to discriminate between real solutions and imaginary solutions, it is called the **discriminant.**

The Nature of the Solutions of a Quadratic Equation

There are three possibilities for the solutions of $ax^2 + bx + c = 0$.

VALUE OF THE DISCRIMINANT	SOLUTIONS OF $ax^2 + bx + c = 0$	THE PARABOLA $y = ax^2 + bx + c$	GRAPHICAL EXAMPLE
1. $b^2 - 4ac > 0$	Two distinct real solutions	Two x-intercepts	
2. $b^2 - 4ac = 0$	A double real solution	One x-intercept with the vertex on the x-axis	
3. $b^2 - 4ac < 0$	Neither solution is real; both solutions are complex numbers with imaginary parts. These solutions will be complex conjugates.	No x-intercepts	

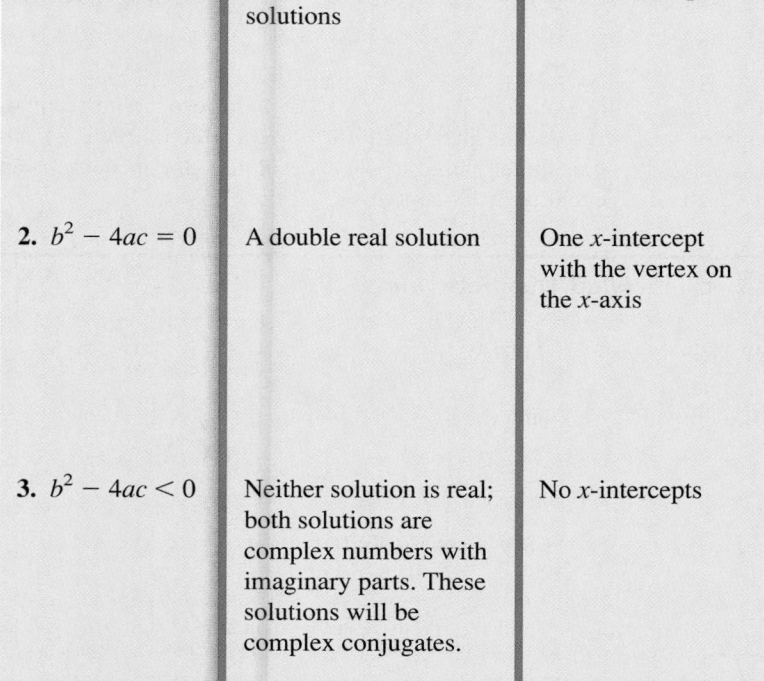
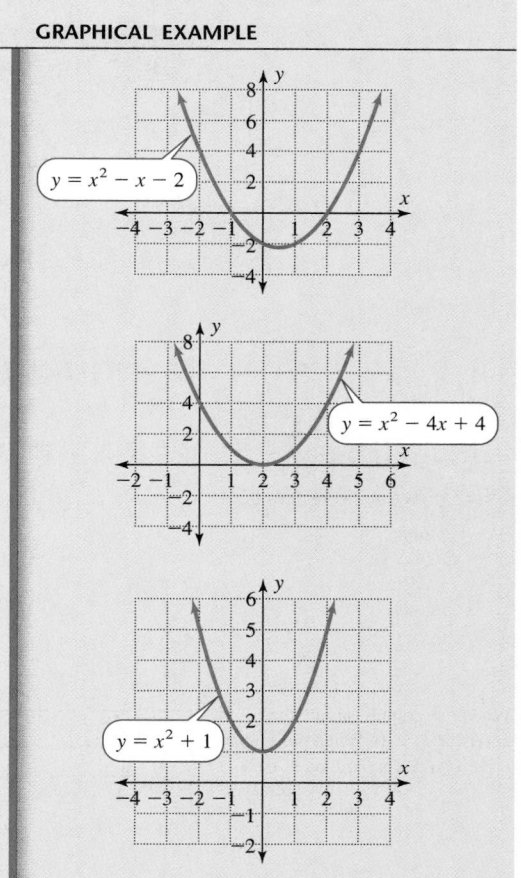

■ EXAMPLE 7 Determining the Nature of the Solutions of a Quadratic Equation

Use the discriminant to determine the nature of the solutions of each quadratic equation.

SOLUTIONS

(a) $x^2 - 6x + 8 = 0$

$b^2 - 4ac = (-6)^2 - 4(1)(8)$

$= 36 - 32$

$= 4$

Substitute $a = 1$, $b = -6$, and $c = 8$ into the discriminant. The discriminant is greater than 0, so there are two real solutions.

There are two distinct real solutions.

(b) $x^2 - 6x + 9 = 0$

$b^2 - 4ac = (-6)^2 - 4(1)(9)$

$= 36 - 36$

$= 0$

Substitute $a = 1$, $b = -6$, and $c = 9$ into the discriminant.

The discriminant is 0, so there is a double real solution.

There is a double real solution.

(c) $x^2 - 6x + 10 = 0$

$b^2 - 4ac = (-6)^2 - 4(1)(10)$

$= 36 - 40$

$= -4$

Substitute $a = 1$, $b = -6$, and $c = 10$ into the discriminant.

The discriminant is negative, so the two solutions are complex numbers with imaginary parts.

The solutions are complex conjugates.

SELF-CHECK 8.5.5

Use the discriminant to determine the nature of the solutions of each equation.

1. $4x^2 = 8x - 4$
2. $4x^2 = 8x - 5$
3. $x^2 + x = 1$

Solving Quadratic Inequalities

Once we have solved a quadratic equation it is easy to solve a corresponding inequality. We use the graph of the corresponding parabola to determine which interval(s) satisfy the inequality. This is the same strategy used to solve absolute value inequalities in Section 4.4 and the quadratic inequality in Example 5 of Section 7.5.

■ EXAMPLE 8 Solving Quadratic Inequalities

Solve each quadratic equation and inequality.

(a) $x^2 = 7$, $x^2 < 7$, and $x^2 > 7$

(b) $x^2 + x + 1 = 0$, $x^2 + x + 1 < 0$, and $x^2 + x + 1 > 0$

SELF-CHECK 8.5.5 ANSWERS

1. A double real solution
2. Complex conjugates
3. Two distinct real solutions

SOLUTION

When the graph is above the x-axis, $P(x) > 0$. When the graph is below the x-axis, $P(x) < 0$.

(a) $x^2 = 7$

$x = \pm\sqrt{7}$

First solve the equation algebraically. This equation is solved by extraction of roots.

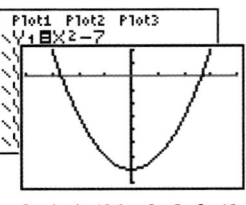

[−4, 4, 1] by [−8, 2, 1]

First write the equation $x^2 = 7$ in the standard form $x^2 - 7 = 0$. Then graph the function defined by $y = x^2 - 7$. From the work already done we know that the x-intercepts are at $\left(-\sqrt{7}, 0\right)$ and $\left(\sqrt{7}, 0\right)$. Note that this parabola is below the x-axis in the interval $\left(-\sqrt{7}, \sqrt{7}\right)$ and above the x-axis in the interval $\left(-\infty, -\sqrt{7}\right) \cup \left(\sqrt{7}, +\infty\right)$.

Answers:

(a) The solutions of $x^2 = 7$ are $x = -\sqrt{7}$ or $x = \sqrt{7}$.

x-values for which $y = 0$

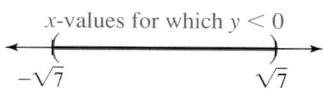

The solutions of $x^2 - 7 = 0$ are revealed by the *x*-intercepts.

The solution of $x^2 < 7$ is $\left(-\sqrt{7}, \sqrt{7}\right)$.

x-values for which $y < 0$

The graph of $y_1 = x^2 - 7$ is below the *x*-axis for the *x*-values given by $\left(-\sqrt{7}, \sqrt{7}\right)$.

The solution of $x^2 > 7$ is $\left(-\infty, -\sqrt{7}\right) \cup \left(\sqrt{7}, +\infty\right)$.

x-values for which $y > 0$

The graph of $y_1 = x^2 - 7$ is above the *x*-axis for the *x*-values given by $\left(-\infty, -\sqrt{7}\right) \cup \left(\sqrt{7}, +\infty\right)$.

SOLUTION

(b) $x^2 + x + 1 = 0$

$$x = \frac{-1 \pm \sqrt{1^2 - 4(1)(1)}}{2(1)}$$

$$x = \frac{-1 \pm \sqrt{1 - 4}}{2}$$

$$x = \frac{-1 \pm i\sqrt{3}}{2}$$

First solve the equation algebraically. Substitute into the quadratic formula using $a = 1$, $b = 1$, and $c = 1$.

Both solutions are imaginary. There are no real solutions.

The parabola defined by $y = x^2 + x + 1$ is always above the *x*-axis. There are no real *x*-values for which the graph is below the *x*-axis.

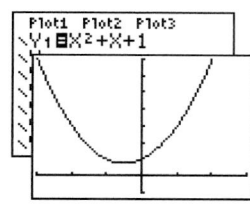

$[-3, 3, 1]$ by $[-1, 7, 1]$

Answers:

(b) The solutions of $x^2 + x + 1 = 0$ are $x = \dfrac{-1 - i\sqrt{3}}{2}$ and $x = \dfrac{-1 + i\sqrt{3}}{2}$.

The inequality $x^2 + x + 1 < 0$ has no real solutions.
The solution of $x^2 + x + 1 > 0$ is $\mathbb{R}$.

SELF-CHECK 8.5.6

1. Solve $x^2 - 3x - 3 = 0$.
2. Solve $x^2 - 3x - 3 < 0$.
3. Solve $x^2 - 3x - 3 > 0$.

SELF-CHECK 8.5.6 ANSWERS

1. $x = \dfrac{3 - \sqrt{21}}{2}, x = \dfrac{3 + \sqrt{21}}{2}$

2. $\left(\dfrac{3 - \sqrt{21}}{2}, \dfrac{3 + \sqrt{21}}{2}\right)$

3. $\left(-\infty, \dfrac{3 - \sqrt{21}}{2}\right) \cup \left(\dfrac{3 + \sqrt{21}}{2}, \infty\right)$

If we extend factoring beyond the integers to include irrational numbers or imaginary numbers, then we can solve even more equations by factoring. For example:

QUADRATIC EQUATION	FACTORS	SOLUTIONS
$x^2 - 2 = 0$	$\left(x + \sqrt{2}\right)\left(x - \sqrt{2}\right) = 0$	$x = -\sqrt{2}, x = \sqrt{2}$
$x^2 + 1 = 0$	$(x + i)(x - i) = 0$	$x = -i, x = i$

Each solution of a quadratic equation corresponds to a linear factor over the complex numbers.

This information is not intended to encourage you to solve even more quadratic equations by factoring. The intent is to help you see the relationship between linear factors and complex solutions. Each solution of a quadratic equation corresponds to a linear factor over the complex numbers.

Your instructor can use this information to create equations designed to have certain solutions. Also electrical engineers (complex numbers have concrete meanings in electrical circuits) can create equations that help them to design products with desired characteristics. Example 9 illustrates how to create an equation with given complex solutions. This topic is an extension of Example 7 from Section 7.5.

■ EXAMPLE 9   Constructing a Quadratic Equation with Given Complex Solutions

Construct a quadratic equation whose solutions are the given complex conjugates.

SOLUTIONS

(a) $x = -3i, x = 3i$

$$x = -3i \quad \text{or} \quad x = 3i$$
$$x + 3i = 0 \qquad x - 3i = 0$$
$$(x + 3i)(x - 3i) = 0$$
$$x^2 - 9i^2 = 0$$
$$x^2 - 9(-1) = 0$$
$$x^2 + 9 = 0$$

The solutions are $-3i$ and $3i$.
Rewrite each linear equation so that the right side of the equation is 0. If the factors of a quadratic equation are 0, then their product is 0.

This quadratic equation has the given solutions.

(b) $x = 3 + i, x = 3 - i$

$$x = 3 + i \quad \text{or} \quad x = 3 - i$$
$$x - (3 + i) = 0 \qquad x - (3 - i) = 0$$
$$(x - 3 - i)(x - 3 + i) = 0$$
$$[(x - 3) - i][(x - 3) + i] = 0$$
$$(x - 3)^2 - (i)^2 = 0$$
$$(x^2 - 6x + 9) + 1 = 0$$
$$x^2 - 6x + 10 = 0$$

The solutions are $3 + i$ and $3 - i$.
Rewrite each equation so that the right side is 0.
If the factors of a quadratic equation are 0, then their product is 0. Noting the special form, multiply by inspection.

This quadratic equation has the given solutions. ■

SELF-CHECK 8.5.7

Write a quadratic equation in x that has the given roots.

1. $5 - \sqrt{2}$ and $5 + \sqrt{2}$
2. $2 + 3i$ and $2 - 3i$

SELF-CHECK 8.5.7 ANSWERS

1. $x^2 - 10x + 23 = 0$
2. $x^2 - 4x + 13 = 0$

USING THE LANGUAGE AND SYMBOLISM OF MATHEMATICS 8.5

1. The symbol "$\pm$" is read _____ _____ _____.

2. Solving $x^2 = k$ as $x = \pm\sqrt{k}$ is called solving by _____ _____ _____.

3. Each real solution of $ax^2 + bx + c = 0$, including irrational solutions, corresponds to an _____-_____ of the parabola defined by $y = ax^2 + bx + c$ and a _____ of the function $f(x) = ax^2 + bx + c$.

4. The act of rewriting an expression or equation so that it contains a perfect square trinomial is called _____ _____ _____.

5. One method used to solve the equation $ax^2 + bx + c = 0$ is to use the formula $x = \dfrac{-b \pm \sqrt{b^2 - 4ac}}{2a}$, which is called the _____ _____.

6. In the quadratic formula the radicand, $b^2 - 4ac$, is called the _____.

7. When solving the quadratic equation $ax^2 + bx + c = 0$, if $b^2 - 4ac > 0$, then the equation will have _____ real solutions and the graph of $y = ax^2 + bx + c$ will have _____ x-intercepts.

8. When solving the quadratic equation $ax^2 + bx + c = 0$, if $b^2 - 4ac = 0$, then the equation will have a real solution of multiplicity _____ and the graph of $y = ax^2 + bx + c$ will be _____ to the x-axis.

9. When solving the quadratic equation $ax^2 + bx + c = 0$, if $b^2 - 4ac < 0$, then the equation will have _____ real solutions and the graph of $y = ax^2 + bx + c$ will have _____ x-intercepts.

EXERCISES 8.5

In Exercises 1–4 match the graph of each quadratic function with the corresponding quadratic equation. All graphs are displayed with the window $[-10, 10, 1]$ by $[-10, 10, 1]$.

1. A quadratic equation whose solutions are complex conjugates: $x = 3 - i\sqrt{5}, x = 3 + i\sqrt{5}$

2. A quadratic equation with two irrational solutions: $x = 1 - \sqrt{5}, x = 1 + \sqrt{5}$

3. A quadratic equation with a double real solution of $x = 3$

4. A quadratic equation with two rational solutions: $x = -3$, $x = 3$

A. 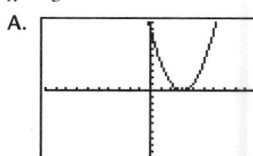   B.

C. 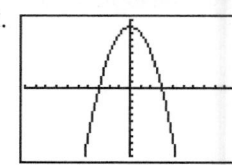   D.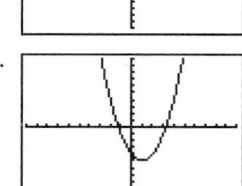

In Exercises 5–14 solve each quadratic equation by extraction of roots.

5. $v^2 = 81$
6. $v^2 = 169$
7. $z^2 = 18$
8. $z^2 = 50$
9. $(x - 1)^2 = 4$
10. $(x + 3)^2 = 25$
11. $z^2 = -16$
12. $z^2 = -36$
13. $(2x + 1)^2 = -9$
14. $(2x - 1)^2 = -25$

In Exercises 15–28 solve each quadratic equation using the quadratic formula.

15. $8t^2 = 2t + 1$
16. $2t^2 + 5t = -3$
17. $w^2 - 8 = 0$
18. $5w^2 - 4 = 0$
19. $-5m^2 = -6m$
20. $-6m^2 = 5m$
21. $-2w^2 + 6w = 5$
22. $w^2 = 4w - 5$
23. $x(x - 6) = -13$
24. $3v(v + 1) = -2 - v$
25. $9v^2 + 12v + 4 = 0$
26. $25v^2 - 30v + 9 = 0$
27. $-5w^2 + 2w + 1 = 0$
28. $-3w^2 - 7w - 3 = 0$

In Exercises 29–34 use the given parabola to solve each equation and inequality.

29. a. $x^2 - 4x - 5 = 0$
    b. $x^2 - 4x - 5 < 0$
    c. $x^2 - 4x - 5 > 0$

30. a. $-x^2 - 2x + 8 = 0$
    b. $-x^2 - 2x + 8 < 0$
    c. $-x^2 - 2x + 8 > 0$

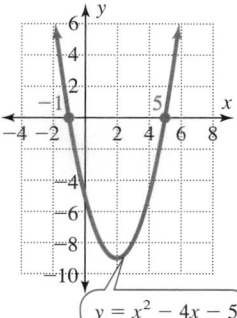

$y = x^2 - 4x - 5$

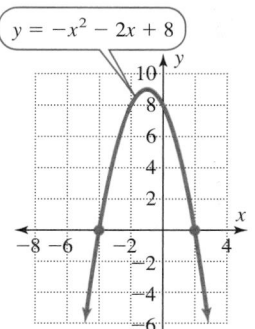

$y = -x^2 - 2x + 8$

31. a. $-x^2 + 2x + 6 = 0$
 b. $-x^2 + 2x + 6 < 0$
 c. $-x^2 + 2x + 6 > 0$

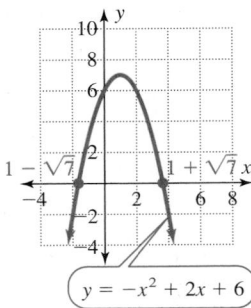

$y = -x^2 + 2x + 6$

32. a. $4x^2 - 11 = 0$
 b. $4x^2 - 11 < 0$
 c. $4x^2 - 11 > 0$

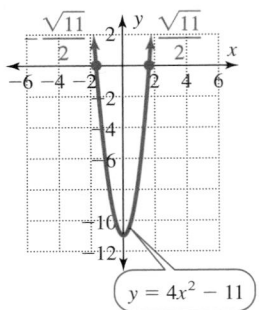

$y = 4x^2 - 11$

33. a. $x^2 + 2x + 5 < 0$
 b. $x^2 + 2x + 5 > 0$

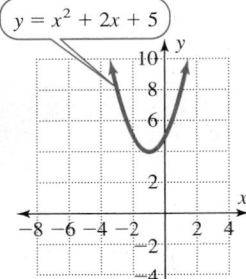

$y = x^2 + 2x + 5$

34. a. $-2x^2 + 2x - 3 < 0$
 b. $-2x^2 + 2x - 3 > 0$

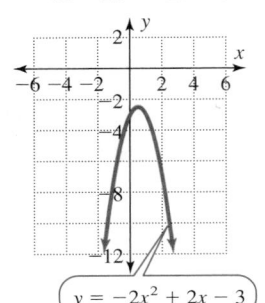

$y = -2x^2 + 2x - 3$

In Exercises 35–40 solve each equation and then use a graphics calculator to solve each inequality.

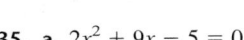

35. a. $2x^2 + 9x - 5 = 0$
 b. $2x^2 + 9x - 5 < 0$
 c. $2x^2 + 9x - 5 > 0$
36. a. $2x^2 - 11x - 6 = 0$
 b. $2x^2 - 11x - 6 < 0$
 c. $2x^2 - 11x - 6 > 0$
37. a. $x^2 - 7 = 0$
 b. $x^2 - 7 < 0$
 c. $x^2 - 7 > 0$
38. a. $x^2 - 11 = 0$
 b. $x^2 - 11 < 0$
 c. $x^2 - 11 > 0$
39. a. $x^2 + 7 = 0$
 b. $x^2 + 7 < 0$
 c. $x^2 + 7 > 0$
40. a. $x^2 + 11 = 0$
 b. $x^2 + 11 < 0$
 c. $x^2 + 11 > 0$

In Exercises 41–46 rewrite each equation so that the left side of the equation is expressed as a perfect square.

41. $x^2 + 10x = -22$
42. $x^2 - 10x = -17$
43. $x^2 - 14x = -48$
44. $x^2 + 14x = -24$
45. $x^2 - 18x = 9$
46. $x^2 + 20x = 1$

In Exercises 47 and 48 replace the question mark in each step. The solution of the given equation is outlined using the method of completing the square.

47. $x^2 + 2x - 4 = 0$
 a. $x^2 + 2x = ?$
 b. $x^2 + 2x + ? = 5$
 c. $(x + ?)^2 = 5$
 d. $x + ? = \pm\sqrt{5}$
 e. $x = ?$

48. $x^2 + 6x - 5 = 0$
 a. $x^2 + 6x = ?$
 b. $x^2 + 6x + ? = 14$
 c. $(x + ?)^2 = 14$
 d. $x + ? = \pm\sqrt{14}$
 e. $x = ?$

In Exercises 49–56 solve each quadratic equation by completing the square.

49. $z^2 - 4z = 0$
50. $z^2 - 8z = 0$
51. $x^2 + 4x - 5 = 0$
52. $x^2 + 2x = 4$
53. $z^2 + 2z + 2 = 0$
54. $z^2 + 6z + 10 = 0$
55. $x^2 - \dfrac{3}{2}x = \dfrac{7}{16}$
56. $x^2 - \dfrac{2}{3}x = \dfrac{8}{9}$

In Exercises 57–59 match the graph of each quadratic function with the corresponding quadratic equation. All graphs are displayed with the window $[-5, 5, 1]$ by $[-5, 5, 1]$.

57. A quadratic equation whose discriminant is 0.
58. A quadratic equation whose discriminant is negative.
59. A quadratic equation whose discriminant is positive.

A. Graph of $y = x^2 - 5x + 4$ **B.** Graph of $y = x^2 + 3$

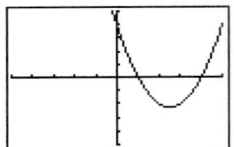   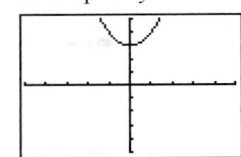

C.   Graph of $y = x^2 + 4x + 4$

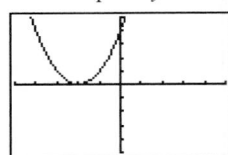

In Exercises 60–62 calculate the discriminant of each quadratic equation and determine the nature of the solutions of this equation.

60. $x^2 - 6x + 9 = 0$ **61.** $x^2 - 6 = 0$ **62.** $x^2 + 6 = 0$

In Exercises 63–66 calculate the exact solutions of $P(x) = 0$ and then complete the table.

POLYNOMIAL $P(x)$	SOLUTIONS OF $P(x) = 0$	x-INTERCEPTS OF THE GRAPH OF $y = P(x)$	FACTORED FORM OF $P(x)$	ZEROS OF $P(x)$
63. $x^2 + 2x - 99$				
64. $6x^2 + x - 2$				
65. $x^2 - 13$				
66. $x^2 + 5x + 2$				

In Exercises 67–74 construct a quadratic equation in x that has the given solutions.

67. 5 and $-\dfrac{1}{2}$

68. -6 and $\dfrac{2}{3}$

69. $-\sqrt{7}$ and $\sqrt{7}$

70. $-\sqrt{11}$ and $\sqrt{11}$

71. $-4i$ and $4i$

72. $-7i$ and $7i$

73. $2 - \sqrt{3}$ and $2 + \sqrt{3}$

74. $3 - \sqrt{2}$ and $3 + \sqrt{2}$

75. One number is 6 more than another number. Find these numbers if their product is 8.

76. The sum of a number and its reciprocal is 4. Find this number.

77. The rectangle shown in the figure is 4 cm longer than it is wide. Find the dimensions if the area is 8 cm².

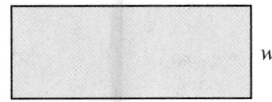

78. The perimeter of the square shown in the figure is numerically 2 more than the area. Find the length in meters of one side of the square.

In Exercises 79 and 80 solve each cubic equation.

79. $x(x^2 - 5x + 4) = 0$

80. $-y^3 = 25y$

Group Discussion Questions

81. Discovery Question   Solve the equation $x^2 - 36 = 0$ by:
   a. Factoring
   b. Extraction of roots
   c. The quadratic formula
   d. Graphing $y = x^2 - 36$ with a graphics calculator
   e. Which method did you prefer? Explain why you prefer this method.

82. Discovery Question   Complete the square of $x^2 + 10x$ both algebraically and geometrically.

ALGEBRAICALLY	GEOMETRICALLY
$x^2 + 10x$	

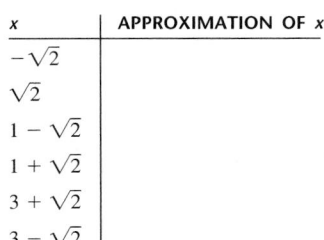

83. Challenge Question   Solve each equation for x.
   a. $x^2 - 3y^2 = 0$
   b. $x^2 + 3y^2 = 0$
   c. $x^2 + 5xy + 3y^2 = 0$
   d. $x^2 + xy + 3y^2 = 0$

84. Discovery Question
   a. Use a calculator to complete the approximations in this table accurate to the nearest ten-thousandth.

x	APPROXIMATION OF x
$-\sqrt{2}$	
$\sqrt{2}$	
$1 - \sqrt{2}$	
$1 + \sqrt{2}$	
$3 + \sqrt{2}$	
$3 - \sqrt{2}$	

   b. Use a graphics calculator to approximate the solutions of $x^2 - 6x + 7 = 0$ to the nearest ten thousandth. Then use these approximations and the table in part (a) to predict the exact solutions of this equation. Do these solutions check?

Section 8.6 Equations with Radicals

Objectives:
   12. Solve equations involving radical expressions.
   13. Use the Pythagorean theorem.
   14. Calculate the distance between two points.

Equations that contain variables in a radicand are called **radical equations.** Since these equations occur frequently in various disciplines, it is important to know how to solve them. In this section we examine radical equations that result in either linear equations

A Mathematical Note

Niels Abel, 1802–1829, was a Norwegian mathematician who was raised in extreme poverty and died of tuberculosis at the age of 26. Nonetheless, he established several new mathematical concepts. Among his accomplishments was his proof, using radicals, that the general fifth-degree polynomial equation is impossible to solve exactly.

or quadratic equations. An example of such a radical equation is $\sqrt{x} = 3$. It is easy to verify that $x = 9$ is a solution of this equation because $\sqrt{9} = 3$. The key to solving radical equations is to raise both sides of the equation to the same power.

Power Theorem

For any real numbers x and y and natural number n:

ALGEBRAICALLY	VERBALLY	EXAMPLE
If $x = y$, then $x^n = y^n$.	If two expressions are equal, then their nth powers are equal.	If $\sqrt{x} = 3$, then $(\sqrt{x})^2 = (3)^2$ $x = 9$ $(-3)^2 = 3^2$ but $-3 \neq 3$

Caution: The equations $x = y$ and $x^n = y^n$ are not always equivalent. The equation $x^n = y^n$ can have a solution that is not a solution of $x = y$.

Because $x^n = y^n$ can be true when $x \neq y$, we must check all possible solutions in the original equation. Any value that occurs as a solution of the last equation of a solution process but is not a solution of the original equation is called an **extraneous value**.

Solving an equation with radicals is a two-stage process. Stage I is to solve for possible solutions. Stage II is to check each of these possibilities to eliminate any extraneous values.

Solving Radical Equations Containing a Single Radical

VERBALLY	ALGEBRAIC EXAMPLE
Step 1 Isolate a radical term on one side of the equation.	Solve: $\sqrt{x-1} + 2 = 6$. $\sqrt{x-1} = 4$
Step 2 Raise both sides to the nth power.	$(\sqrt{x-1})^2 = 4^2$
Step 3 Solve the resulting equation.	$x - 1 = 16$ $x = 17$
Step 4 Check the possible solutions in the original equation to determine whether they are really solutions or are extraneous.	Check: $\sqrt{17-1} + 2 \overset{?}{=} 6$ $\sqrt{16} + 2 \overset{?}{=} 6$ $4 + 2 \overset{?}{=} 6$ $6 \overset{?}{=} 6$ is true Answer: $x = 17$

■ EXAMPLE 1 Using Multiple Perspectives to Solve a Radical Equation

Solve $\sqrt{7 - x} = 4$ algebraically, numerically, and graphically.

SOLUTION

ALGEBRAICALLY

$$\sqrt{7 - x} = 4$$
$$\left(\sqrt{7 - x}\right)^2 = 4^2$$ Square both sides of the equation.
$$7 - x = 16$$ Simplify and then solve the equation for x.
$$-x = 9$$
$$x = -9$$

Check: $\sqrt{7 - (-9)} \stackrel{?}{=} 4$ Substitute -9 back into the original equation to check this value.
$$\sqrt{16} \stackrel{?}{=} 4$$
$$4 \stackrel{?}{=} 4 \text{ checks}$$
$$x = -9 \text{ checks.}$$

NUMERICALLY

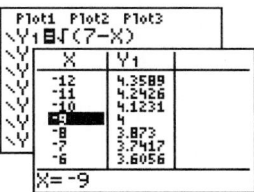

Enter $\sqrt{7 - x}$ for y_1 and examine the table for a value of x that results in a value of 4 for y_1.
This function equals 4 for an x-value of -9.
What this table does not answer is whether there are other solutions. The algebraic method shows that there is only one solution.

For $x = -9$, $y_1 = 4$.

GRAPHICALLY

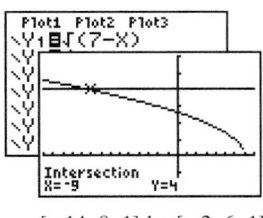

$[-14, 8, 1]$ by $[-2, 6, 1]$

Use a graphics calculator to graph $y_1 = \sqrt{7 - x}$ and $y_2 = 4$. Then determine the x-value of the point of intersection of these two graphs.
These graphs intersect for an x-value of -9.

The x-coordinate of the point of intersection is $x = -9$.

Answer: $x = -9$

Example 2 illustrates a problem that yields an extraneous value.

SELF-CHECK 8.6.1

Solve $\sqrt{t - 3} = 5$.

■ **EXAMPLE 2** Solving a Radical Equation with No Solution

Solve $\sqrt{2x + 3} = -5$ algebraically and graphically.

SOLUTION

ALGEBRAICALLY

$$\sqrt{2x + 3} = -5$$
$$\left(\sqrt{2x + 3}\right)^2 = (-5)^2$$
$$2x + 3 = 25$$
$$2x = 22$$
$$x = 11$$

This equation cannot have a solution since a principal root is always nonnegative. However, we will continue the solution process in order to see what happens.
Square both sides of the equation.
Simplify, and then solve for x.

The step of checking a possible solution is not a luxury; it is a necessity to avoid extraneous values when you solve a radical equation.

Check: $\sqrt{2(11) + 3} \stackrel{?}{=} -5$
$$\sqrt{25} \stackrel{?}{=} -5$$
$$5 \stackrel{?}{=} -5 \text{ does not check}$$
$x = 11$ does not check.

Substitute 11 back into the original equation to check this value. The principal square root of 25 is $+5$, not -5. Thus 11 is an extraneous value in the original equation.

GRAPHICALLY

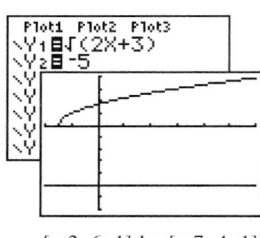

[−2, 6, 1] by [−7, 4, 1]

Use a graphics calculator to graph $y_1 = \sqrt{2x + 3}$ and $y_2 = -5$. Note that these two graphs do not intersect. Thus there is no solution for $\sqrt{2x + 3} = -5$.

The graphs of the two functions do not intersect.

Answer: No solution

Example 3 involves a cube root. The basic procedure for solving this equation is the same as that used for square roots.

■ **EXAMPLE 3** Solving an Equation with One Radical Term

Solve $\sqrt[3]{a - 1} = 4$.

SOLUTION

$$\sqrt[3]{a - 1} = 4$$
$$\left(\sqrt[3]{a - 1}\right)^3 = 4^3$$
$$a - 1 = 64$$
$$a = 65$$

Cube both sides of the equation.
Simplify and then solve the linear equation.

Check: $\sqrt[3]{65 - 1} \stackrel{?}{=} 4$
$$\sqrt[3]{64} \stackrel{?}{=} 4 \text{ checks}$$

Answer: $a = 65$

The solution of the next radical equation produces a quadratic equation. Although the quadratic equation has two solutions, only one of these values checks in the original equation. The other value is extraneous.

■ EXAMPLE 4   Solving an Equation with One Radical Term

Solve $x = \sqrt{x + 6}$.

SOLUTION

$$x = \sqrt{x + 6}$$
$$x^2 = x + 6$$ Square both sides of the equation.
$$x^2 - x - 6 = 0$$ Write the quadratic equation in standard form.
$$(x - 3)(x + 2) = 0$$ Factor the left side of the equation.
$$x - 3 = 0 \quad \text{or} \quad x + 2 = 0$$ Set each factor equal to 0.
$$x = 3 \qquad\qquad x = -2$$ Solve for x.

Check:

$x = 3$: $3 \overset{?}{=} \sqrt{3 + 6}$ $x = -2$: $-2 \overset{?}{=} \sqrt{-2 + 6}$
$\qquad\quad 3 \overset{?}{=} \sqrt{9}$ $\qquad\qquad\quad -2 \overset{?}{=} \sqrt{4}$
$\qquad\quad 3 \overset{?}{=} 3$ is true $\qquad\quad -2 \overset{?}{=} 2$ is false -2 is an extraneous value.

Answer: $x = 3$ ■

SELF-CHECK 8.6.2

Solve each equation.

1. $\sqrt{1 - p} = -1$ 2. $\sqrt[3]{x - 5} = -2$ 3. $x = \sqrt{3x + 4}$

Whenever a radical equation contains another term on the same side of the equation as the radical, we begin by isolating the radical term on one side of the equation. Then we simplify both sides by squaring if the radicals are square roots, cubing if the radicals are cube roots, and so forth.

■ EXAMPLE 5   Solving an Equation with One Radical Term

Solve $\sqrt{2w - 3} + 9 = w$.

SOLUTION

$$\sqrt{2w - 3} + 9 = w$$
$$\sqrt{2w - 3} = w - 9$$ First isolate the radical term on the left side of the equation by subtracting 9 from both sides.
$$2w - 3 = w^2 - 18w + 81$$ Square both sides of the equation. (Be careful not to omit the middle term when you square the binomial on the right side.)
$$w^2 - 20w + 84 = 0$$ Write the quadratic equation in standard form.
$$(w - 14)(w - 6) = 0$$ Factor and solve for w.
$$w - 14 = 0 \quad \text{or} \quad w - 6 = 0$$
$$w = 14 \qquad\qquad w = 6$$

SELF-CHECK 8.6.2 ANSWERS

1. No solution
2. $x = -3$
3. $x = 4$

Check: $w = 14$: $\sqrt{2(14) - 3} + 9 \stackrel{?}{=} 14$

$$\sqrt{25} + 9 \stackrel{?}{=} 14$$

$$5 + 9 \stackrel{?}{=} 14$$

$$14 \stackrel{?}{=} 14 \text{ is true}$$

$w = 6$: $\sqrt{2(6) - 3} + 9 \stackrel{?}{=} 6$

$$\sqrt{9} + 9 \stackrel{?}{=} 6$$

$$3 + 9 \stackrel{?}{=} 6$$

$$12 \stackrel{?}{=} 6 \text{ is false} \qquad \text{6 is an extraneous value.}$$

Answer: $w = 14$

A Mathematical Note

The Greek mathematician Pythagoras (c. 500 B.C.) taught orally and required secrecy of his initiates. Thus records of the society formed by Pythagoras are anecdotal. His society produced a theory of numbers that was part science and part numerology. It assigned numbers to many abstract concepts—for example, 1 for reason, 2 for opinion, 3 for harmony, and 4 for justice. The star pentagon was the secret symbol of the Pythagoreans.

SELF-CHECK 8.6.3

Solve $\sqrt{4x + 15} - 2 = x$.

One of the theorems most often used in mathematics is the Pythagorean theorem, which states an important relationship among the sides of a right triangle. A **right triangle** is a triangle containing a 90° angle; that is, two sides of the triangle are perpendicular. In a right triangle the two shorter sides are called the **legs** and the longer side is called the **hypotenuse**. The lengths of the three sides can be denoted, respectively, by a, b, and c. (See the figure.)

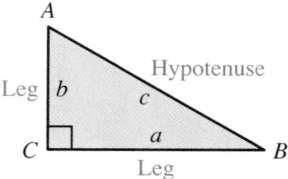

Pythagorean Theorem

If triangle ABC is a right triangle, then:

ALGEBRAICALLY	GEOMETRIC EXAMPLE	VERBALLY
$a^2 + b^2 = c^2$	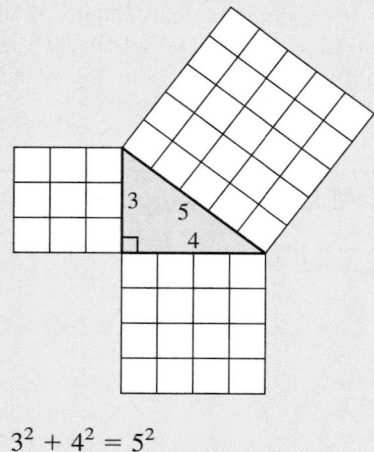	The sum of the areas of the squares formed on the legs of a right triangle is equal to the area of the square formed on the hypotenuse.
(The converse of this theorem is also true. If $a^2 + b^2 = c^2$, then triangle ABC is a right triangle.)	$3^2 + 4^2 = 5^2$ $9 + 16 = 25$	

SELF-CHECK 8.6.3 ANSWER

$x = \sqrt{11}$

■ **EXAMPLE 6** Using the Pythagorean Theorem to Solve an Application

An actress has a spotlight located 8 ft above her head.

(a) Express the distance d between the top of her head and the spotlight as a function of x, the distance in feet she has moved along the stage floor.

(b) If the limit of the light's effective range from her head is 17 ft, how far can she move from underneath this light?

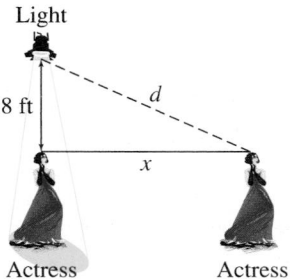

SOLUTION

(a) $d^2 = x^2 + 8^2$ Use the Pythagorean theorem and the information from the right triangle in the sketch.

$d = \sqrt{x^2 + 64}$ Because d must be positive, we take only the principal square root when we solve for d. This is an equation giving d as a function of x.

(b) $17 = \sqrt{x^2 + 64}$ Substitute 17 for d to determine how far the actress can move from the light. To solve for x, square both sides of this radical equation.

$17^2 = x^2 + 64$

$289 = x^2 + 64$

$225 = x^2$ Solve for x by extraction of roots—only the positive value has meaning in this distance application.

$x^2 = 225$

$x = \sqrt{225}$

$x = 15$ She can move 15 ft in any direction.

Answer: The function that gives the distance from the light to her head is $d = \sqrt{x^2 + 64}$. She can move up to 15 ft and stay within the effective range of the light. ■

SELF-CHECK 8.6.4

1. Two legs of a right triangle are 5 and 12 cm. Determine the length of the hypotenuse.
2. The hypotenuse of a right triangle is 25 cm and one leg is 7 cm. How long is the other leg?
3. In Example 6, express the length x as a function of the distance d.

An important formula that can be developed using the Pythagorean theorem is the formula for the distance between two points in a plane. We develop this formula by first considering the horizontal and vertical changes between two points. Then we compute the direct distance between them using the Pythagorean theorem.

DISTANCE BETWEEN TWO POINTS

SPECIAL CASE	GENERAL CASE

Calculate the distance between $P(2, 2)$ and $Q(5, 6)$.

Calculate the distance between $P(x_1, y_1)$ and $Q(x_2, y_2)$.

Step 1 Find the horizontal change from P to Q.

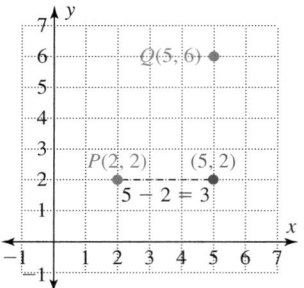   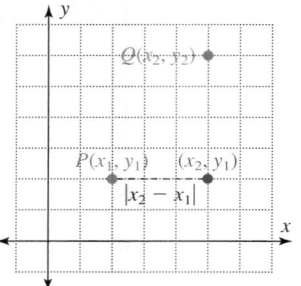

Horizontal distance $= 5 - 2 = 3$

Horizontal distance $= |x_2 - x_1|$

Step 2 Find the vertical change from P to Q.

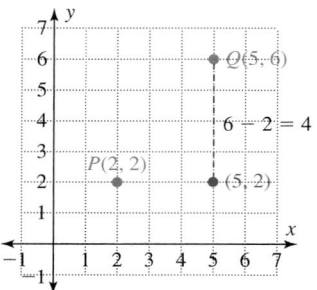   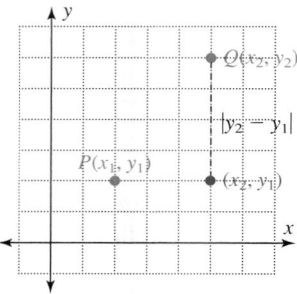

Vertical distance $= 6 - 2 = 4$

Vertical distance $= |y_2 - y_1|$

Step 3 Use the Pythagorean theorem to find the length of hypotenuse PQ.

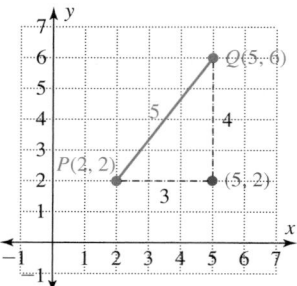   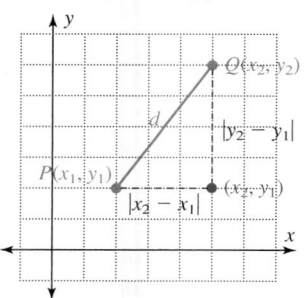

$$d = \sqrt{3^2 + 4^2}$$
$$= \sqrt{9 + 16}$$
$$= \sqrt{25}$$
$$= 5$$

$$d = \sqrt{|x_2 - x_1|^2 + |y_2 - y_1|^2}$$
$$d = \sqrt{(x_2 - x_1)^2 + (y_2 - y_1)^2}$$

Absolute value notation is not needed in the distance formula because the squares in this formula must be nonnegative. The formula is applicable in all cases, even if P and Q are on the same horizontal or vertical line.

We refer to the distance *between* (x_1, y_1) and (x_2, y_2) rather than the distance *from* (x_1, y_1) to (x_2, y_2) because the distance between these points is the same in either direction. In the formula, note that
$(x_2 - x_1)^2 = (x_1 - x_2)^2$ and
$(y_2 - y_1)^2 = (y_1 - y_2)^2$.

Distance Formula

The distance d between (x_1, y_1) and (x_2, y_2) is given by $d = \sqrt{(x_2 - x_1)^2 + (y_2 - y_1)^2}$.

■ EXAMPLE 7   Calculating the Distance Between Two Points

Calculate the distance between $(-3, 1)$ and $(5, -1)$.

SOLUTION

$d = \sqrt{(x_2 - x_1)^2 + (y_2 - y_1)^2}$

$d = \sqrt{[5 - (-3)]^2 + (-1 - 1)^2}$ Substitute the given values into the distance formula.

$d = \sqrt{8^2 + (-2)^2}$

$d = \sqrt{64 + 4}$

$d = \sqrt{68}$

$d = \sqrt{4}\sqrt{17}$

Answer: $d = 2\sqrt{17}$

This exact distance is illustrated in the figure. A calculator approximation of this distance is $2\sqrt{17} \approx 8.25$.

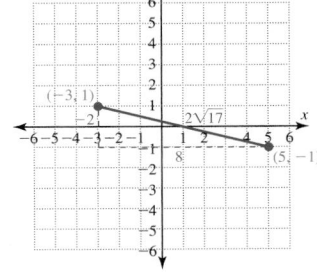

SELF-CHECK 8.6.5

Calculate the distance between $(2, -3)$ and $(-13, 5)$.

SELF-CHECK 8.6.5 ANSWER

17

Engineering, architecture, and construction often involve placing new structures in a specific location relative to other buildings and utility lines. Example 8 is a skill builder for problems of this type.

■ EXAMPLE 8   Applying the Distance Formula

Find all points with an x-coordinate of 6 that are 5 units from $(3, 3)$.

SOLUTION

Identify the desired point(s) by $(6, y)$.

$d = \sqrt{(x_2 - x_1)^2 + (y_2 - y_1)^2}$ All points with an x-coordinate of 6 can be written as $(6, y)$.

$5 = \sqrt{(6 - 3)^2 + (y - 3)^2}$ Substitute the given values into the distance formula.

$5 = \sqrt{9 + y^2 - 6y + 9}$ Then simplify the radicand.

$5 = \sqrt{y^2 - 6y + 18}$ Solve the radical equation for y.

$25 = y^2 - 6y + 18$ Square both sides of the equation.

$0 = y^2 - 6y - 7$ Subtract 25 from both sides of the equation.

$(y + 1)(y - 7) = 0$ Factor the trinomial.

$y + 1 = 0$ or $y - 7 = 0$ Set each factor equal to 0.

$\quad y = -1 \qquad\qquad y = 7$ Both values check.

Answer: The points $(6, -1)$ and $(6, 7)$ are both 5 units from $(3, 3)$.

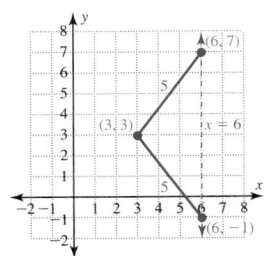

USING THE LANGUAGE AND SYMBOLISM OF MATHEMATICS   8.6

1. A radical equation is an equation that contains variables in a _____ .
2. The power theorem states that if $x = y$, then _____ , provided that x and y are real numbers and n is a natural number.
3. The reason that we must check all possible solutions to a radical equation in the original equation is that when we raise both sides to the nth power, we can introduce _____ values.
4. In a right triangle, the two shorter sides are called the _____ and the longer side is called the _____ .
5. The Pythagorean theorem states that for a right triangle with sides of lengths a, b, and c, _____ .
6. The distance between points (x_1, y_1) and (x_2, y_2) is given by $d =$ _____ .

EXERCISES   8.6

In Exercises 1–4 use the given graphs to solve each radical equation.

1. $\sqrt{x - 4} = 1$

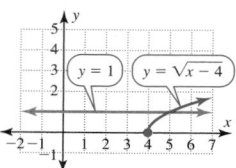

2. $\sqrt{5x - 1} = 3$

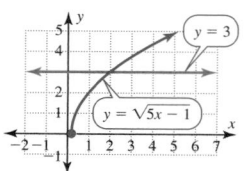

3. $\sqrt[3]{3x - 5} + 2 = 0$

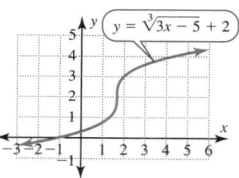

4. $\sqrt[3]{3x - 4} - 2 = 0$

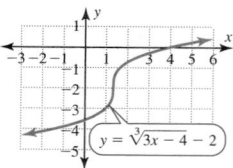

In Exercises 5–8 use the given table to solve each radical equation.

5. $\sqrt{2x - 11} = 3$

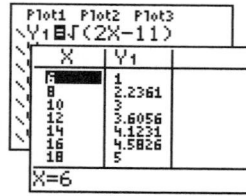

6. $\sqrt{5x + 19} = 7$

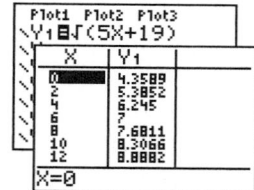

7. $\sqrt{4x + 1} = x - 1$

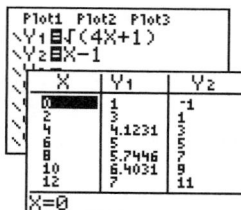

8. $\sqrt{2x + 1} = x - 1$

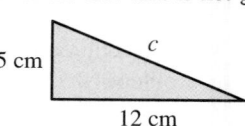

In Exercises 9–30 solve each equation.

9. $\sqrt{t - 4} = 3$
10. $\sqrt{t - 2} = 12$
11. $\sqrt{c + 7} + 23 = 43$
12. $\sqrt{c + 5} - 3 = 6$
13. $\sqrt{3x - 21} + 7 = 2$
14. $\sqrt{25 + 2x} + 11 = 5$
15. $\sqrt[3]{2w + 1} = -2$
16. $\sqrt[3]{5w + 2} = -3$
17. $\sqrt[4]{6v - 2} = 2$
18. $\sqrt[4]{7v - 3} = 3$
19. $\sqrt[5]{w^2 - 4w} = 2$
20. $\sqrt[5]{2w^2 + w - 16} = -1$
21. $\sqrt{y^2 - y + 13} - y = 1$
22. $\sqrt{y^2 + y + 11} - y = 1$
23. $\sqrt{7t + 2} = 2t$
24. $\sqrt{-9t - 2} = 2t$
25. $\sqrt{w^2 - 2w + 1} = 2w$
26. $\sqrt{2w + 1} = w + 1$
27. $\sqrt{2x + 1} + 5 = 2x$
28. $\sqrt{7x - 3} + 1 = 3x$
29. $\sqrt{6u + 7} = \sqrt{11u + 7}$
30. $\sqrt{14u - 12} = \sqrt{10u - 8}$

In Exercises 31–36 use the Pythagorean theorem to find the length of the side that is not given.

31.

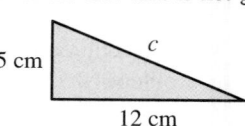

32.

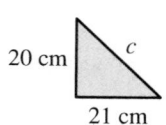

33.

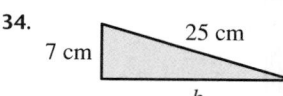

34.

35.

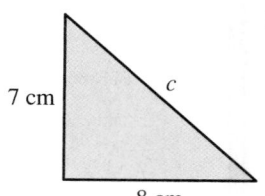

36.

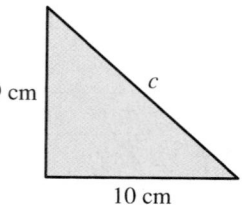

In Exercises 37–44 use the distance formula to calculate the distance between each pair of points.

37. $(-2, 6)$ and $(3, 6)$
38. $(-3, 2)$ and $(1, -1)$

39. $(2, -7)$ and $(-6, 8)$

40. $(0, -20)$ and $(-9, 20)$

41. $\left(-\dfrac{1}{2}, \dfrac{2}{3}\right)$ and $\left(\dfrac{1}{2}, -\dfrac{1}{3}\right)$

42. $\left(\dfrac{4}{5}, \dfrac{3}{2}\right)$ and $\left(-\dfrac{1}{5}, \dfrac{1}{2}\right)$

43. $(0, 0)$ and $\left(-\sqrt{2}, \sqrt{7}\right)$

44. $\left(\sqrt{11}, -\sqrt{14}\right)$ and $(0, 0)$

In Exercises 45–48 use the distance formula to determine the perimeter of the triangle formed by these points. Then use the Pythagorean theorem to determine if the triangle formed is a right triangle.

45. $(-3, -1)$, $(4, -4)$, and $(-1, 1)$

46. $(-6, -1)$, $(2, -1)$, and $(0, 1)$

47. $(-5, -1)$, $(2, -2)$, and $(4, 4)$

48. $(3, -3)$, $(-2, -1)$, and $(5, 2)$

49. Measuring a Buried Pipe   A pipe-cleaning firm contracted to clean a pipe buried in a lake. Access points to the pipe are at points A and B on the edge of the lake, as shown in the figure. The contractor placed a stake as a reference point and then measured the coordinates in meters from this reference point to A and B. The coordinates of A and B are, respectively, $(3.0, 5.2)$ and $(37.8, 29.6)$. Approximate to the nearest tenth of a meter the distance between A and B.

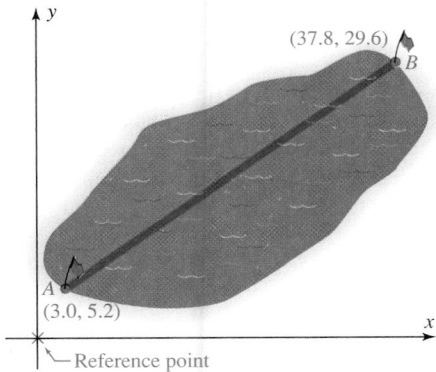

50. Measuring an Engine Block   A machinist is measuring a metal block that is part of an automobile engine. To find the distance from the center of hole A to the center of hole B, the machinist determines the coordinates shown on the drawing with respect to a reference point at the lower left corner of the metal block. Calculate the distance from the center of A to the center of B. The dimensions are given in cm.

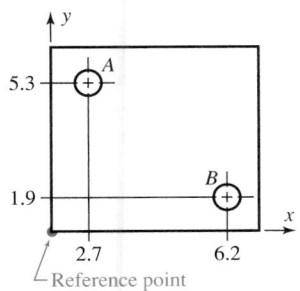

51. Find all points with an x-coordinate of 5 that are 10 units from the point $(13, 2)$.

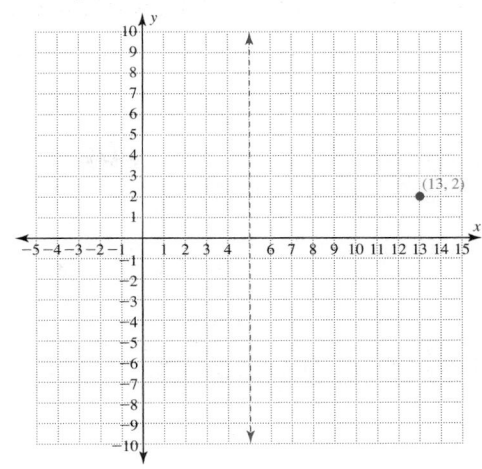

52. Find all points with a y-coordinate of 2 that are 13 units from the point $(-2, 7)$.

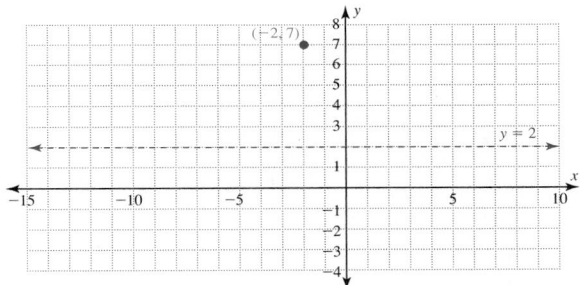

In Exercises 53–56 solve each problem for x.

53. The square root of the quantity x plus 5 is equal to 3.

54. The cube root of the quantity x plus 2 is equal to -2.

55. Four times the square root of x equals 3 times the quantity x minus 5.

56. If the square root of the quantity x plus 4 is added to 2, the result equals x.

In Exercises 57–64 solve each equation.

57. $\sqrt{2x + 6} = x$

58. $\sqrt{x + 5} = x$

59. $\sqrt{2x^2 + 3x - 1} = x$

60. $\sqrt{2x^2 + 6x + 3} - 2 = x$

61. $\sqrt[3]{x^3 - 6x^2 + 12x} = x$

62. $\sqrt[3]{x^3 + 9x^2 + 27x} = x$

63. $\sqrt{4w^2 + 12w + 6} = 2w$

64. $\sqrt{9w^2 - 12w + 8} = 3w$

Production of Solar Cells

In Exercises 65–68 use the fact that the dollar cost C of producing n solar cells per shift is given by the formula $C = 18\sqrt[3]{n^2} + 450$.

65. Find the overhead cost for one shift. That is, find the cost of producing 0 solar cells.

66. Find the cost of producing 27 solar cells.
67. Find the number of solar cells produced when the cost is $738.
68. Find the number of solar cells produced when the cost is $1098.

Strength of a Box Beam

In Exercises 69–72 use the fact that the maximum load of a square box beam is related to its volume by the function $S(V) = 750\sqrt[3]{V^2}$. In this formula V is given in cubic centimeters and S is given in newtons.

69. Evaluate and interpret $S(0)$.
70. Evaluate and interpret $S(1000)$.
71. If $S(V) = 750$, calculate and interpret V.
72. If $S(V) = 300,000$, calculate and interpret V.
73. **Height of a Wall**   A 17-ft ladder is leaning against a chimney.

a. Express the distance from the base of the chimney to the top of the ladder as a function of x, the distance from the base of the ladder to the base of the chimney.
b. If the bottom of the ladder is 8 ft from the base of the chimney, how far is it from the bottom of the chimney to the top of the ladder?

74. **Length of a Guy Wire**   A television tower has a guy wire attached 40 ft above the base of the tower.
a. Express the length of the guy wire as a function of x, the distance from the base of the tower to the anchor point of the guy wire.
b. If the anchor point of the guy wire is 42 ft from the base of the tower, how long is the guy wire?

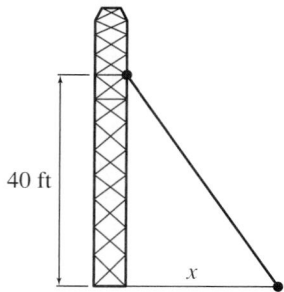

75. **Positioning Effective Lighting**   The lighting system planned for a basketball court has bulbs that should be placed at least 30 ft above the floor for safety reasons. One bulb is located directly above the center of the court at a height of 30 ft. A ball is located at a position x ft from center court.

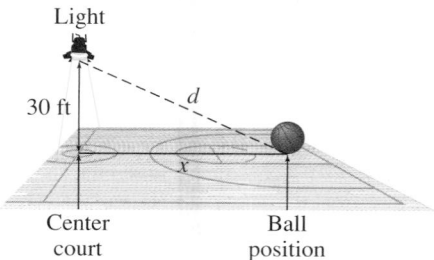

a. Express d, the distance between the ball and the light, as a function of x.
b. If a ball is 16 ft from center court, how far is it from the light to the ball's location?
c. If the limit of the light's effective range is 35 ft, how far can the ball be positioned from center court and still receive adequate light from this bulb? Approximate this distance to the nearest tenth of a foot.

76. Positioning an Effective Sound System   The sound system planned for an airport concourse has speakers that should be placed at least 18 ft above the floor and thus approximately 12 ft above the heads of most customers. A customer is standing at a position x ft from a reference point directly under one speaker that is 18 ft above the floor.

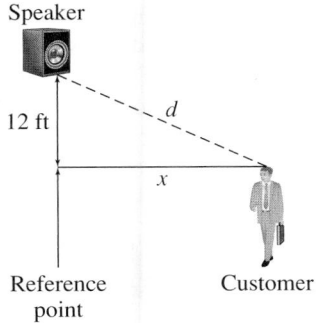

Speaker

12 ft

d

x

Reference point   Customer

a. Express d, the distance between the customer's ears and the speaker, as a function of x.

b. If a customer is standing 16 ft from the reference point, how far is it from the speaker to the customer's ears?

c. If the limit of the speaker's effective range is 20 ft, approximate to the nearest foot the distance a customer can walk from the reference point and stay within the effective range of the speaker.

77. Path of a Power Cable   The path of a power cable between booster stations located at A and C is shown in the following figure. The maximum length of cable between booster stations is 12 mi. Determine the distance x so that the path of the power cable from A to C through B will be 12 mi.

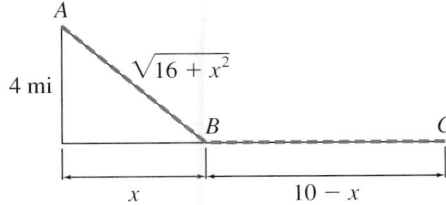

A

4 mi

$\sqrt{16 + x^2}$

B

C

x $10 - x$

78. Path of a Fiber-Optic Cable   The path of an underground fiber-optic cable between buildings located at A and C in a business district is shown in the following figure. The maximum length of cable allowed for connections of this type is 17 km. Determine the distance x so that the path of the fiber-optic cable from A to C through B will be 17 km.

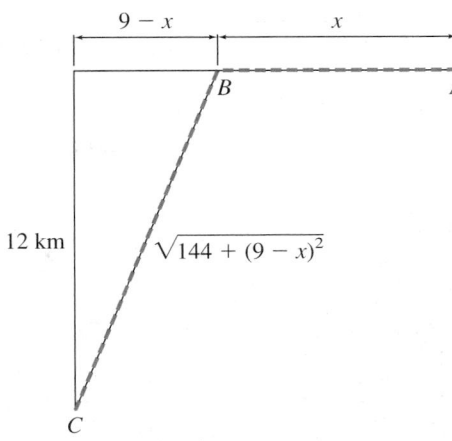

$9 - x$ x

B A

12 km $\sqrt{144 + (9 - x)^2}$

C

Group Discussion Questions

79. Discovery Question

a. Write the equation of all points 3 units from the origin. Describe the shape formed by these points.

b. Write the equation of all points 2 units from the point $(3, 5)$. Describe the shape formed by these points.

c. Write the equation of all points r units from the point (h, k). Describe the shape formed by these points.

80. Challenge Question   Solve each equation for x.

a. $y = \dfrac{\sqrt[3]{2x - z}}{3}$

b. $y = \sqrt{v^2 - w^2 + x}$

c. $5 = \sqrt{\dfrac{v + x}{v - x}}$

81. Challenge Question   The square shown in the figure was etched in stone, along with the single word "BEHOLD," by a Hindu mathematician, who meant that this figure offers visual proof of the Pythagorean theorem. Describe how this square can be used to prove the Pythagorean theorem. (*Hint:* Add the areas of the four triangles and the inner square.)

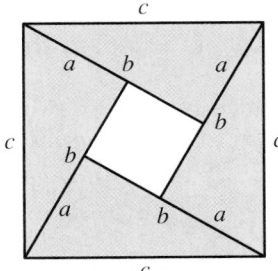

c

a b a

c b c

b

a b a

c

Section 8.7 More Applications of Quadratic Equations

Objective:

15. Use quadratic equations to solve word problems.

Most college graduates encounter mathematics in the context of verbally stated problems, problems that need to be translated into mathematics in order to be solved. It is rare that anyone will ask you to solve a quadratic equation. More likely someone will ask you "How do you . . ." or "What is the cheapest way to . . .". These are the real-world problems that will challenge your problem-solving skills.

To benefit from your algebra skills you must be able to work with word problems. Each sport or profession requires practice time on the basics in order to execute efficiently in a real performance. Even golfers like Tiger Woods practice chipping and putting to hone these skills for tournaments. Likewise, we include some basic drill exercises here as well as some applications that you may find more challenging. The basic strategy used to solve these word problems is summarized in the following box. It is the same strategy we used in Section 3.6 to solve word problems using linear equations. This time the equations we use are quadratic equations.

Strategy for Solving Word Problems

STEP 1. Read the problem carefully to determine what you are being asked to find.

STEP 2. Select a variable to represent each unknown quantity. Specify precisely what each variable represents.

STEP 3. If necessary, translate the problem into word equations. Then translate the word equations into algebraic equations.

STEP 4. Solve the equation(s) and answer the question asked by the problem.

STEP 5. Check the reasonableness of your answer.

After solving the equations that you form, always inspect the solutions to see if they are appropriate for the original problem. Check not only for extraneous values but also for answers that may not be meaningful in the application, such as negative lengths.

Example 1 is a skill developer whose primary purpose is to provide practice in reading word problems and translating the word equations into algebraic form.

■ EXAMPLE 1 Finding the Dimensions of Two Squares

The dimensions in centimeters for two squares are consecutive integers and their combined area is 61 cm^2. Find the dimensions of each square.

SOLUTION _____

Let x = the length in centimeters of each side of the smaller square
Let $x + 1$ = the length in centimeters of each side of the larger square   Consecutive integers differ by 1 unit.

VERBALLY

The sum of their areas is 61 cm^2.

NUMERICALLY

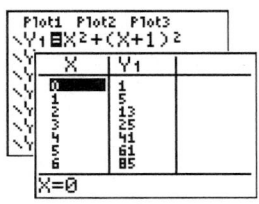

The table reveals that the value of 5 for x produces a y-value of 61. Because x is 5, $x + 1$ must be 6.

Answer: The dimensions of the squares are, respectively, 5 cm and 6 cm on each side.

ALGEBRAICALLY

$x^2 + (x + 1)^2 = 61$
$x^2 + x^2 + 2x + 1 = 61$

$2x^2 + 2x - 60 = 0$
$x^2 + x - 30 = 0$
$(x - 5)(x + 6) = 0$
$x - 5 = 0$ or $x + 6 = 0$
        $x = 5$ $x = -6$
$x + 1 = 6$ $x + 1 = -5$

Form a word equation. Then use the formula $A = s^2$, the formula for the area of a square, to translate this statement into an algebraic equation.

Write the quadratic equation in standard form and simplify by dividing both sides of the equation by 2.

Then solve this equation by factoring.

Be sure to find $x + 1$ as well as x. The negative values -6 and -5 are not meaningful dimensions.

Check:

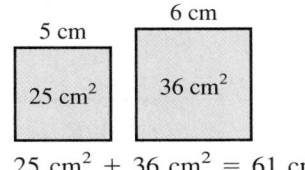

$25 \text{ cm}^2 + 36 \text{ cm}^2 = 61 \text{ cm}^2$

The danger in just guessing the solution to a word problem, even if you are correct and the answer checks, is that you can overlook other solutions. The solutions you overlook can actually be preferable if they are cheaper or better in some other way.

Sometimes it is easy to determine an answer to a word problem without taking time to write the equation. These answers may even check. What you risk by not using algebra to form an equation is overlooking other possible solutions. In a practical problem, this may mean an engineer or a business person has overlooked another solution which may be cheaper or has preferable attributes over the obvious solution. Example 2 has two pairs of solutions. It is common for some students to overlook one of these pairs.

■ **EXAMPLE 2**   Finding Consecutive Even Integers with a Known Product

Find two consecutive even integers whose product is 48.

SOLUTION _____

Let n = smaller of the two even integers
$n + 2$ = larger of the two even integers

Consecutive even integers (like 2, 4, 6, and 8) differ by 2 units.

VERBALLY

The smaller number times the larger number equals 48.

Form a word equation and translate it into an algebraic equation.

ALGEBRAICALLY

$$n(n + 2) = 48$$
$$n^2 + 2n = 48$$
$$n^2 + 2n - 48 = 0$$ Write the quadratic equation in standard form and then solve it by factoring.
$$(n - 6)(n + 8) = 0$$
$$n - 6 = 0 \quad \text{or} \quad n + 8 = 0$$
$$n = 6 \qquad\qquad n = -8$$
$$n + 2 = 8 \qquad n + 2 = -6$$ Be sure to find $n + 2$ as well as n.

Answer: The consecutive even integers are either -8 and -6 or 6 and 8.

Both pairs of even integers check because their product is 48. Be careful if you guess or use a table to check values. It is easy to overlook one pair of answers.

SELF-CHECK 8.7.1 ANSWER
The numbers are either -8 and -5 or 5 and 8.

SELF-CHECK 8.7.1

One number is 3 more than another and their product is 40. Find both numbers.

■ EXAMPLE 3   Determining the Height of Fireworks

A faulty fireworks rocket is launched vertically with an initial velocity of 96 ft/sec and then falls to the earth unexploded. Its height, h, in feet after t sec is given by $h(t) = -16t^2 + 96t$. During what time interval after launch will the height of the rocket exceed 90 ft? The figure shows the relationship of the height to the time. It does not show the path of the rocket, which goes straight up and then down.

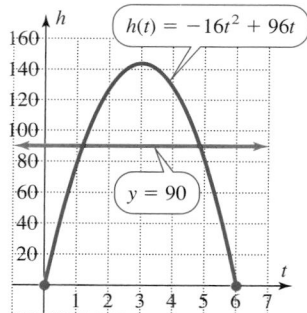

SOLUTION

$$h(t) = -16t^2 + 96t$$
$$90 = -16t^2 + 96t$$ First determine the times when the height is 90 ft.
$$16t^2 - 96t + 90 = 0$$ Write the quadratic equation in standard form and then divide both sides of the equation by 2.
$$8t^2 - 48t + 45 = 0$$ Solve this equation using the quadratic formula.

$$t = \frac{48 \pm \sqrt{(-48)^2 - 4(8)(45)}}{2(8)}$$

$$t = \frac{48 \pm \sqrt{864}}{16}$$ Simplify the radicand noting that $\sqrt{864} = \sqrt{(144)(6)} = 12\sqrt{6}$.

$$t = \frac{48 \pm \sqrt{(144)(6)}}{16}$$

$$t = \frac{48 \pm 12\sqrt{6}}{16}$$

$$t = \frac{12 \pm 3\sqrt{6}}{4}$$ Divide both the numerator and denominator by 4. Take care to divide each term in the numerator by 4.

$$t = \frac{12 - 3\sqrt{6}}{4} \quad \text{or} \quad t = \frac{12 + 3\sqrt{6}}{4}$$ These are the exact values of t.

$$t \approx 1.2 \qquad\qquad t \approx 4.8$$ We can use a calculator to approximate these values to the nearest tenth of a second. At these times the height of the rocket is 90 ft.

Answer: The height of the rocket will exceed 90 ft between 1.2 sec and 4.8 sec after launch.

Using the given figure, we note the parabola is above $y = 90$ for t values between 1.2 and 4.8. The height will be over 90 ft from 1.2 to 4.8 sec after launch.

■ EXAMPLE 4   Calculating an Interest Rate

The formula for computing the amount A of an investment of principal P invested at interest rate r for 1 year and compounded semiannually is $A = P\left(1 + \dfrac{r}{2}\right)^2$. Approximately what interest rate is necessary for \$800 to grow to \$848.72 in 1 year if the interest is compounded semiannually?

SOLUTION

$$A = P\left(1 + \frac{r}{2}\right)^2$$

$$848.72 = 800\left(1 + \frac{r}{2}\right)^2$$

Substitute the given values into the formula for compound interest. The resulting amount of the investment is $A = 848.72$ and the original principal is $P = 800$.

$$1.0609 = \left(1 + \frac{r}{2}\right)^2$$

Divide both sides of this quadratic equation by 800.

$$\pm 1.03 = 1 + \frac{r}{2}$$

Solve this quadratic equation by extraction of roots (or use the quadratic formula).

$$\pm 2.06 = 2 + r$$
$$r = -2 \pm 2.06$$
$$r = 0.06 \quad \text{or} \quad r = -4.06$$
$$r = 6\%$$

A negative interest rate is not meaningful in this problem. Only 6% is a meaningful answer.

Answer: At a rate of 6% compounded semiannually a principal of \$800 will grow to \$848.72 in 1 year.

Does this rate check and is it a reasonable answer?   ■

SELF-CHECK 8.7.2

1. Rework Example 3 to determine the time interval that the rocket is above 80 ft.
2. Rework Example 4 if the principal of \$800 grows to \$856.98 in 1 year.

SELF-CHECK 8.7.2 ANSWERS

1. The height of the fireworks rocket will exceed 80 ft between 1 and 5 sec after launch.
2. At a rate of 7% compounded semiannually a principal of \$800 will grow to \$856.98 in 1 year.

■ EXAMPLE 5   Determining a Profit Interval

The profit in dollars made by selling x boats per week is given by the quadratic function $P(x) = -x^2 + 275x - 6250$. Determine the number of boats that will generate a weekly profit for this manufacturer.

SOLUTION

Solve $P(x) > 0$.

$P(x) = 0$

$-x^2 + 275x - 6250 = 0$

$x = \dfrac{-275 \pm \sqrt{275^2 - 4(-1)(-6250)}}{2(-1)}$

$x = \dfrac{-275 \pm \sqrt{50625}}{-2}$

$x = \dfrac{-275 \pm 225}{-2}$

$x = \dfrac{-275 + 225}{-2}$ or $x = \dfrac{-275 - 225}{-2}$

$x = \dfrac{-50}{-2}$ $x = \dfrac{-500}{-2}$

$x = 25$ $x = 250$

The manufacturer will have a profit when $P(x) > 0$. To solve $P(x) > 0$, first solve $P(x) = 0$ to determine the break-even values—the values that result in a profit of $0.

Solve this quadratic equation by using the quadratic formula.

Simplify the radicand. Use a calculator to note that $\sqrt{50{,}625} = 225$.

For $x = 25$ and $x = 250$ the profit is 0. The manufacturer will break even.

When the graph of $y = P(x)$ is above the x-axis, $P(x) > 0$.

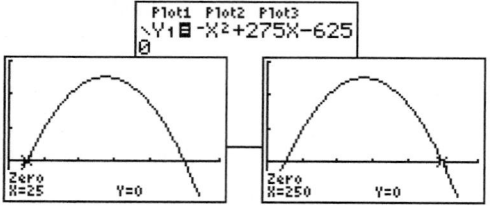

[0, 300, 50] by [−5000, 15000, 5000]

Using a graphics calculator we note that the parabola defined by $y_1 = -x^2 + 275x - 6250$ opens downward. It is above the x-axis for x-values between 25 and 250 units.

Answer: The manufacturer will make a weekly profit for sales between 25 and 250 boats per week.

The manufacturer makes a profit if $P(x) > 0$. The graph shows us that the x-values that make this happen are between 25 and 250.

The triangle is one of the most basic shapes in geometry and in the world around us. Thus it should not be surprising that right triangles and the Pythagorean theorem play a key role in solving many applied problems. Examples 6–8 all involve the Pythagorean theorem.

■ EXAMPLE 6   Applying the Pythagorean Theorem

The hypotenuse of a right triangle is 3 cm more than twice the length of the shortest side. The longer leg is 3 cm less than 3 times the length of the shortest side. Find the length of each side.

SOLUTION

Let $a =$ length of the shorter leg in centimeters
$3a - 3 =$ length of the longer leg in centimeters
$2a + 3 =$ length of the hypotenuse in centimeters

It is wise to start by making a sketch of the problem and labeling the sketch as illustrated in the figure.

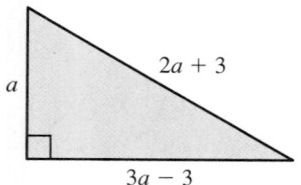

VERBALLY

$$\left(\begin{matrix}\text{Length of}\\\text{shorter leg}\end{matrix}\right)^2 + \left(\begin{matrix}\text{Length of}\\\text{longer leg}\end{matrix}\right)^2 = \left(\begin{matrix}\text{Length of}\\\text{hypotenuse}\end{matrix}\right)^2$$

The word equation is based on the Pythagorean theorem, $a^2 + b^2 = c^2$.

ALGEBRAICALLY

$$a^2 + (3a - 3)^2 = (2a + 3)^2$$
$$a^2 + 9a^2 - 18a + 9 = 4a^2 + 12a + 9$$
$$6a^2 - 30a = 0$$
$$a^2 - 5a = 0$$
$$(a - 5)a = 0$$
$$a = 5 \quad \text{or} \quad a = 0$$
$$3a - 3 = 12 \qquad \text{(not meaningful)}$$
$$2a + 3 = 13$$

Substitute the lengths of each side into the Pythagorean theorem for a, b, and c. Simplify both sides of the equation.

Combine like terms and write the quadratic equation in standard form.

Divide both sides of the equation by 6.

Solve this quadratic equation by factoring.

A length of 0 would not yield a triangle and is therefore not a meaningful answer for this problem. Be sure to find the lengths of all three sides.

Answer: The legs are 5 and 12 cm and the hypotenuse is 13 cm.

Do these values check?

SELF-CHECK 8.7.3

The length of a rectangle is 3 m more than 3 times the width. A diagonal of the rectangle is 1 m more than the length. Find the dimensions of the rectangle. (*Hint:* Use the Pythagorean theorem.)

SELF-CHECK 8.7.3 ANSWER

The width is 7 m and the length is 24 m.

Example 7 is based on the analysis of structural components for a building. The shape that results may not be exactly a right triangle. However, this mathematical model is close enough to the actual result to yield useful information.

■ EXAMPLE 7 Determining the Deflection of a Beam

A steel beam that is 20 ft long is placed in a hydraulic press for testing. Under extreme pressure the beam deflects upward so that the distance between the hydraulic rams decreases by 2 in. Assume the deflected shape is approximated by the figure shown. Determine the height of the bulge in the middle of the beam.

SOLUTION _____

Let $x =$ the height of the bulge in inches

VERBALLY

$$\left(\begin{matrix}\text{Length to the}\\\text{center point}\end{matrix}\right)^2 + \left(\begin{matrix}\text{Height of}\\\text{the bulge}\end{matrix}\right)^2 = \left(\begin{matrix}\text{Length of the left}\\\text{half of the beam}\end{matrix}\right)^2$$

First identify the unknown value with a variable.

Then note that the shape in the sketch is a right triangle, so we can apply the Pythagorean theorem.

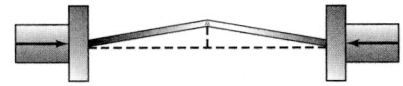

120 in

x

119 in

ALGEBRAICALLY

$$119^2 + x^2 = 120^2$$
$$14{,}161 + x^2 = 14{,}400$$
$$x^2 = 239$$
$$x = \pm\sqrt{239}$$

$x = \sqrt{239}$ or $x = -\sqrt{239}$
$x \approx 15.5$ $x \approx -15.5$

Answer: The bulge would be approximately 15.5 in.

Half the length of the beam is 10 ft or 120 in. Half the distance between the presses is 119 in, 1 in less than 120 in. Substitute these values into the verbal equation.

Solve this quadratic equation by extraction of roots. Then approximate these solutions.
The beam could deflect either up or down 15.46 in.

This large of a bulge is not tolerable in most building projects. Therefore engineers plan for expansion and contraction spaces in their designs. ∎

SELF-CHECK 8.7.4

In Example 6 determine the height of the bulge if the rams decrease the distance from 20 ft to 19 ft 11 in.

SELF-CHECK 8.7.4 ANSWER

The bulge would be approximately 10.9 in.

Example 8 uses both the Pythagorean theorem and the formula $D = R \cdot T$ for calculating the distance for a given rate and time.

▮ EXAMPLE 8   Applying the Pythagorean Theorem

Two airplanes depart simultaneously from an airport. One flies due south; the other flies due east at a rate 50 mi/h faster than that of the first airplane. After 2 h, radar indicates that the airplanes are 500 mi apart. What is the ground speed of each airplane?

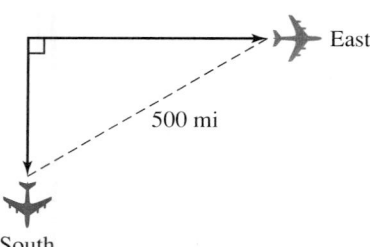

Make a sketch (as shown above) that models the problem.

SOLUTION

Let r = ground speed of the first plane in miles per hour
$r + 50$ = ground speed of the second plane in miles per hour

Represent each unknown quantity using a variable. To be consistent with the formula $D = R \cdot T$, we use the variable r to represent speed.

VERBALLY

$$\left(\begin{matrix}\text{Distance first}\\\text{plane travels}\end{matrix}\right)^2 + \left(\begin{matrix}\text{Distance second}\\\text{plane travels}\end{matrix}\right)^2 = \left(\begin{matrix}\text{Distance between}\\\text{the two planes}\end{matrix}\right)^2$$

The word equation is based on the Pythagorean theorem.

ALGEBRAICALLY

$$(2r)^2 + (2(r + 50))^2 = (500)^2$$
$$(2r)^2 + (2r + 100)^2 = (500)^2$$
$$4r^2 + (4r^2 + 400r + 10{,}000) = 250{,}000$$
$$8r^2 + 400r - 240{,}000 = 0$$
$$r^2 + 50r - 30{,}000 = 0$$
$$(r - 150)(r + 200) = 0$$

$r - 150 = 0$ or $r + 200 = 0$
   $r = 150$ $r = -200$ (not meaningful)
$r + 50 = 200$

Answer: The first plane is flying south at 150 mi/h, and the second is flying east at 200 mi/h.

Using $D = RT$ and a time of 2 h, the first plane travels $2r$ mi and the second travels $2(r + 50)$ mi.

Substitute the distances into the word equation and simplify.

Write the quadratic equation in standard form.
Divide both sides of the equation by 8.

Solve this quadratic equation by factoring.

A negative rate is not meaningful for this problem.

Find both r and $r + 50$.

Do these rates check and are they reasonable? ∎

USING THE LANGUAGE AND SYMBOLISM OF MATHEMATICS 8.7

1. The first step in the word problem strategy given in this book is to read the problem carefully to determine what you are being asked to _____.

2. The second step in the word problem strategy given in this book is to select a _____ to represent each unknown quantity.

3. The third step in the word problem strategy given in this book is to model the problem verbally with word equations and then translate these word equations into _____ equations.

4. The fourth step in the word problem strategy given in this book is to _____ the equation or system of equations and answer the question asked by the problem.

5. The fifth step in the word problem strategy given in this book is to check your answer to make sure the answer is _____.

6. In the formula $D = RT$, D represents distance, R represents _____, and T represents _____.

EXERCISES 8.7

In Exercises 1–8 solve each problem.

Numeric Word Problems

1. Find two consecutive integers whose product is 156.
2. Find two consecutive integers whose product is 240.
3. Find two consecutive even integers whose product is 288.
4. Find two consecutive odd integers whose product is 99.
5. The sum of the squares of two consecutive integers is 113. Find these integers.
6. The sum of the squares of three consecutive integers is 50. Find these integers.
7. One number is 4 more than 3 times another number. Find these numbers if their product is 175.
8. One number is 3 less than 4 times another number. Find these numbers if their product is 10.

Dimensions of a Rectangle

9. The length of a rectangle (see the figure) is 2 cm more than 3 times the width. Find the dimensions of this rectangle if the area is 21 cm^2.

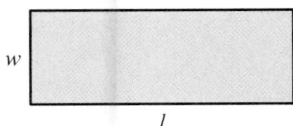

10. The length of a rectangle (see the figure above) is 10 m less than twice the width. Find the dimensions of this rectangle if the area is 72 m^2.

Dimensions of a Triangle

11. The base of a triangle (see the figure) is 3 m longer than the height. Find the base if the area of this triangle is 77 m^2.

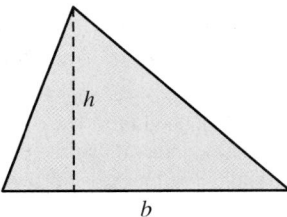

12. The base of the triangle shown in the figure is 5 cm longer than the height. Determine the height of the triangle if its area is 12 cm^2.

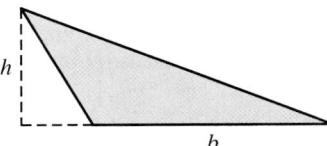

Dimensions of a Square

13. If each side of a square is increased by 5 cm, the total area of both the new square and the original square will be 200 cm^2. Approximate to the nearest tenth of a centimeter the length of each side of the original square.

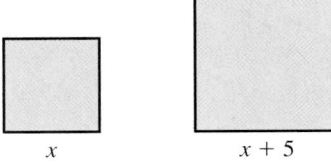

14. If each side of a square is increased by 2 cm, the total area of both the new square and the original square will be 400 cm^2. Approximate to the nearest tenth of a centimeter the length of each side of the original square.

15. Trajectory of a Projectile   The height in meters of a ball released from a ramp is given by the function $h(t) = -4.9t^2 + 29.4t + 34.3$, where t represents the time in seconds since the ball was released from the end of the ramp. Determine the time interval that the ball is above 50 m.

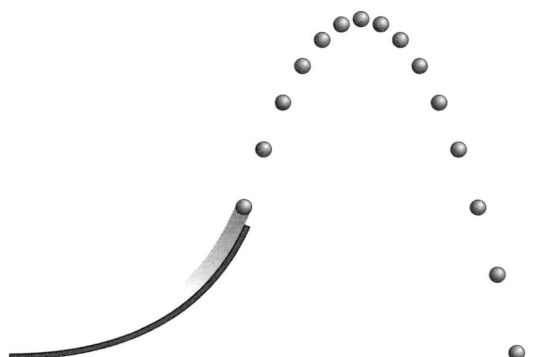

16. Height of a Golf Ball   The height in feet of a golf ball hit from an elevated tee box is given by $h(t) = -16t^2 + 60t + 25$ where t represents the time in seconds the ball is in flight.

    a. Determine the number of seconds before the golf ball hits the ground at a height level of 0 ft.

    b. Determine the time interval that the ball is above 25 ft.

17. Profit Polynomial   The business manager of a small windmill manufacturer projects that the profit in dollars from making x windmills per week will be $P(x) = -x^2 + 70x - 600$. How many windmills should be produced each week in order to generate a profit? (*Hint:* Only an integral number of units can be produced.)

18. Average Cost   The president of a company producing water pumps has gathered data suggesting that the average cost in dollars of producing x units per hour of a new style of pump will be $C(x) = x^2 - 22x + 166$. Determine the number of pumps per hour needed to be produced to keep the average cost below \$75 per pump.

19. Radius of a Circle   The area enclosed between two concentric circles is 56π cm^2. The radius of the larger circle is 1 cm less than twice the radius of the smaller circle. Determine the length of the shorter radius. (See the figure.)

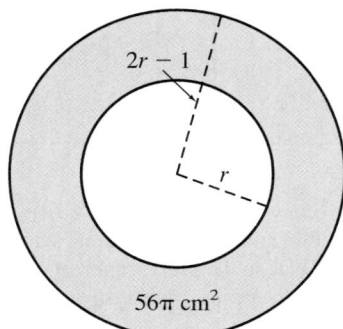

20. Dimensions of a Mat   A square mat used for athletic exercises has a uniform red border on all four sides. The rest of the mat is blue. The width of the blue square is three-fourths the width of the entire square. If the area colored red is 28 m^2, determine the length of each side of the mat. (See the figure.)

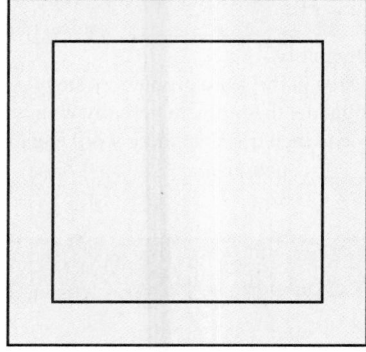

21. Deflection of a Beam   A 40-ft concrete beam is fixed between two rigid anchor points. If this beam expands by 0.5 in due to a 50° increase in temperature, determine the bulge in the middle of the beam. (Assume for simplicity that the beam deflects as shown in this diagram.) What do engineers do to avoid this problem?

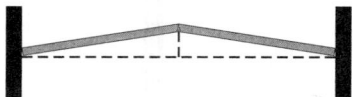

22. Deflection of a Beam   A 28-ft aluminum beam is tested between two hydraulic rams. If the distance between the rams is decreased by 0.25 in, determine the bulge in the middle of the beam. (Assume for simplicity that the beam deflects as shown in this diagram.)

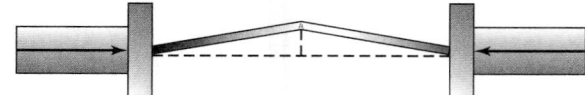

23. Screen Size   The size of a computer monitor is usually given as the length of a diagonal of the screen. A new computer comes with a 17-in monitor. Give the height of the screen if the width is 13 in.

24. Screen Size   A hand-held organizer offers a 3.5-in screen (diagonal length). Approximate its width to the nearest hundredth if its height is 3.0 in.

25. **Screen Size** A student is comparing two different 19-in computer monitors (see Exercise 23). One has a width of 15 in and the other has a width of 16 in. Determine the height of each monitor. Which screen offers more viewing area?

26. **Screen Size** A hand-held organizer is advertised to have a 4-in screen (diagonal length). The width of the screen is 2.25 in and the length of the screen is 3.25 in. Determine the actual length of the diagonal of the screen. Determine the relative error of the advertised length. (*Hint:* See Section 1.5.)

27. **Distance by Plane** Upon leaving an airport, an airplane flew due south and then due east. After it had flown 17 mi farther east than it had flown south, it was 25 mi from the airport. How far south had it flown? (See the figure.)

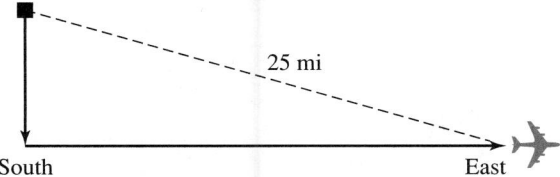

South   East

28. **Distance by Plane** Upon leaving an airport, an airplane flew due west and then due north. After it had flown 1 mi farther north than it had flown west, it was 29 mi from the airport. How far west had it flown? (See the figure.)

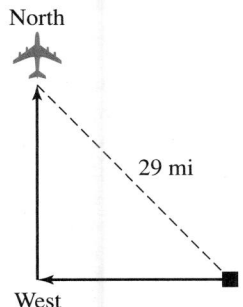

29. **Speed of an Airplane** Two airplanes depart simultaneously from an airport. One flies due south; the other flies due east at a rate 30 mi/h faster than that of the first airplane. After 3 h, radar indicates that the airplanes are 450 mi apart. What is the ground speed of each airplane? (See the figure.)

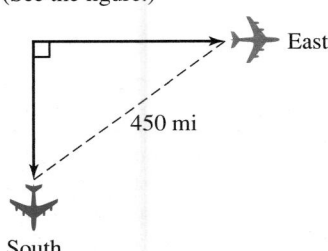

30. **Speed of an Airplane** Two airplanes depart simultaneously from an airport. One flies due south; the other flies due east at a rate 10 mi/h faster than that of the

first airplane. After 1 h, radar indicates that the airplanes are 290 mi apart. What is the ground speed of each airplane? (See the figure.)

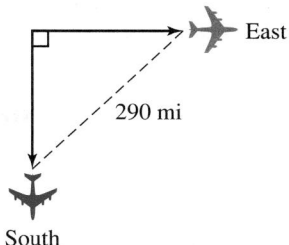

31. **Room Width** Examining the blueprints for a rectangular room that is 17 ft longer than it is wide, an electrician determines that a wire run diagonally across this room will be 53 ft long. What is the width of the room? (See the figure.)

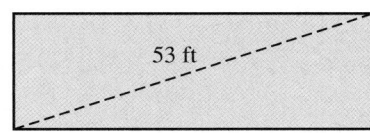

32. **Diameter of a Storage Bin** The length of the diagonal brace in the cylindrical storage bin shown in the figure is 17 m. If the height of the cylindrical portion of the bin is 7 m more than the diameter, determine the diameter.

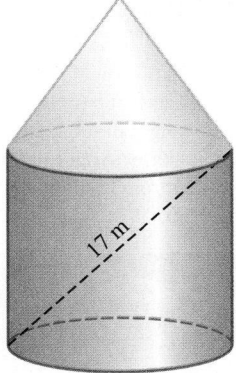

33. **Rope Length** The length of one piece of rope is 8 m less than twice the length of another piece of rope. Each rope is used to enclose a square region. The area of the region enclosed by the longer rope is 279 m^2 more than the area of the region enclosed by the shorter rope. Determine the length of the shorter rope. (See the figure.)

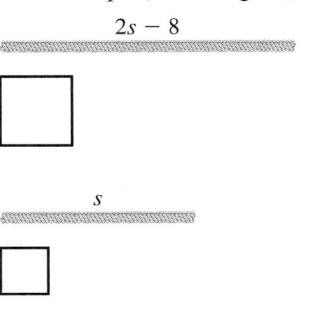

34. Dimensions of Poster Board   A rectangular piece of poster board is 6.0 cm longer than it is wide. A 1-cm strip is cut off each side. The area of the remaining poster board is 616 cm². Find the original dimensions.

35. Distances on a Baseball Diamond   The bases on a baseball diamond are placed at the corners of a square whose sides are 90 ft long. How much farther does a catcher have to throw the ball to get it from home plate to second base than from home plate to third base? (See the figure.)

36. Speed of a Baseball   If a catcher on a baseball team throws a baseball at 120 ft/sec (over 80 mi/h), approximately how long will it take his throw to go from home plate to second base? Approximately how long will it take his throw to go from home plate to third base? (*Hint:* See Exercise 35.)

37. Interest Rate   The formula for computing the amount A of an investment of principal P invested at interest rate r for 1 year and compounded semiannually is
$A = P\left(1 + \dfrac{r}{2}\right)^2$. Approximately what interest rate is necessary for $1000 to grow to $1095 in 1 year if the interest is compounded semiannually?

38. Interest Rate   The formula for computing the amount A of an investment of principal P invested at interest rate r for 1 year and compounded semiannually is
$A = P\left(1 + \dfrac{r}{2}\right)^2$. Approximately what interest rate is necessary for $1000 to grow to $1075 in 1 year if the interest is compounded semiannually?

39. Distance to the Horizon   The radius of the earth is approximately 4000 mi. Approximate to within 10 mi the distance from the horizon to a plane flying at an altitude of 4 mi. (See the figure.)

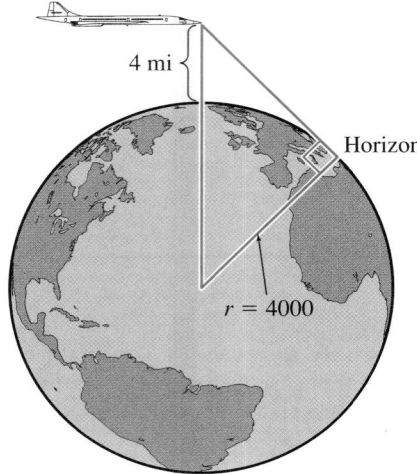

40. Distance to the Horizon   The radius of the earth is approximately 4000 mi. Approximate to within 10 mi the distance from the horizon to a plane flying at an altitude of 5 mi. (See the figure in Exercise 39.)

41. Metal Machining   A round stock of metal that is $18\sqrt{2}$ cm in diameter is milled into a square piece of stock. How long are the sides of the largest square that can be milled? (See the figure.)

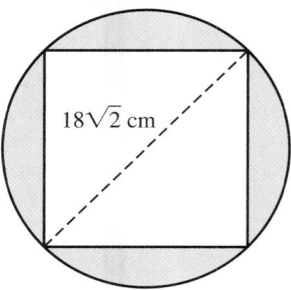

42. Metal Machining   A round stock of metal that is 30 cm in diameter is milled into a square piece of stock. How long are the sides of the largest square that can be milled? (See the figure in Exercise 41.)

Group Discussion Questions

43. Discovery Question
a. Determine the length of diagonal d_1 on the base of this box.
b. Determine the length of diagonal d_2 from the lower left front to the upper right rear of this box.

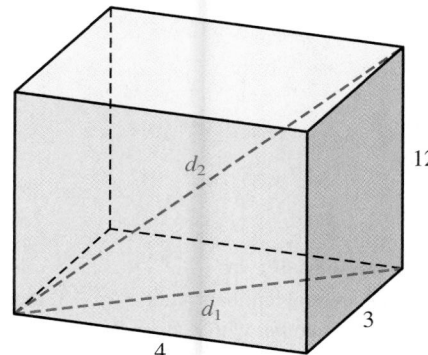

c. Calculate the distance in three dimensions from A to C.
d. Calculate the distance in three dimensions from A to D.

A: (x_1, y_1, z_1) B: (x_1, y_2, z_1)
C: (x_2, y_2, z_1) D: (x_2, y_2, z_2)

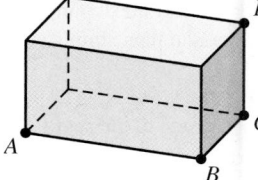 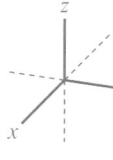

44. Discovery Question   The height of a ball in feet after t sec is given by $h(t) = -16t^2 + 90t + 8$.
a. When will the ball hit the ground?
b. When will the ball be above 100 ft?
c. When will the ball be above 120 ft?
d. When will the ball be above 130 ft?
e. What is the maximum height reached by this ball?
f. When will the ball reach its maximum height?

45. Challenge Question   Given the equation
$x^2 + (x + 2)^2 = 244$:
a. Write a numeric word problem that is modeled by this equation.
b. Write a word problem that models the areas of two squares. (See Exercise 13.)
c. Write a word problem that models the distance between two airplanes. (See Exercise 27.)

46. Discovery Question

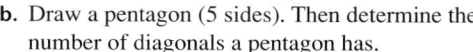

a. Draw a quadrilateral (4 sides). Then determine the number of diagonals a quadrilateral has.
b. Draw a pentagon (5 sides). Then determine the number of diagonals a pentagon has.
c. Draw a hexagon (6 sides). Then determine the number of diagonals a hexagon has.
d. Use this data and a graphics calculator with QuadReg to calculate the equation of the parabola of best fit for this data.
e. Use this formula to complete the table for polygons with 7, 8, 9, and 10 sides. Does this data check?
f. Use this formula to determine how many sides a polygon has if it has 65 diagonals.
g. Assume that each vertex represents a city. Discuss why a package delivery company or a communications company might be interested in the number of sides and diagonals a polygon has.

NUMBER OF SIDES x	NUMBER OF DIAGONALS y
4	
5	
6	
7	
8	
9	
10	
	65

KEY CONCEPTS   FOR CHAPTER 8

1. **Radical notation:**
   $\sqrt[n]{x}$ is read "the principal nth root of x." x is called the radicand, $\sqrt{}$ is called the radical symbol, and n is called the index or the order of the radical. If $\sqrt[n]{x}$ is used as a factor n times, the product is x.

2. **Principal nth root:**
   The principal nth root of the real number x is denoted by either $x^{1/n}$ or $\sqrt[n]{x}$.
   a. For $x > 0$: The principal root is positive for all natural numbers n.
   b. For $x < 0$: If n is odd, the principal root is negative. If n is even, there is no principal real nth root; it is imaginary.
   c. For $x = 0$: The principal root is 0.

3. **Rational exponents, $x^{m/n}$:**
   For a real number x and natural numbers m and n,
   $$x^{m/n} = \left(x^{1/n}\right)^m = (x^m)^{1/n} \quad \text{if } x^{1/n} \text{ is a real number}$$
   $$x^{-m/n} = \frac{1}{x^{m/n}} \quad \text{if } x \neq 0 \text{ and } x^{1/n} \text{ is a real number}$$
   If $x < 0$ and n is even, $x^{1/n}$ is not a real number and the equalities listed above are not true. For example, $\sqrt{-5}$ is not a real number, and $\sqrt{(-5)^2} \neq \left(\sqrt{-5}\right)^2$.

4. **$\sqrt[n]{x^n}$:**
   For any real number x and natural number n:
   $$\sqrt[n]{x^n} = |x| \quad \text{if } n \text{ is even}$$
   $$\sqrt[n]{x^n} = x \quad \text{if } n \text{ is odd}$$

5. **Properties of radicals:**
   If $\sqrt[n]{x}$ and $\sqrt[n]{y}$ are both real numbers, then:
   $$\sqrt[n]{xy} = \sqrt[n]{x}\sqrt[n]{y} \quad \text{and} \quad \sqrt[n]{\frac{x}{x}} = \frac{\sqrt[n]{x}}{\sqrt[n]{y}} \quad \text{for } y \neq 0$$

6. **Like radicals:**
   Like radicals have the same index and the same radicand; only the coefficients of like terms can differ.

7. **Conjugates:**
   The radical expressions $\sqrt{x} + \sqrt{y}$ and $\sqrt{x} - \sqrt{y}$ are called conjugates of each other.

8. **Operations with radical expressions:**
   The addition, subtraction, and multiplication of radical expressions are similar to the corresponding operations with polynomials. Division by a radical expression is accomplished by rationalizing the denominator.

9. **Rationalizing the denominator:**
   If an expression contains a radical in the denominator, the process of rewriting the expression so that there is no radical in the denominator is called rationalizing the denominator.

10. **The imaginary number i:**
    $$i = \sqrt{-1} \qquad \text{so } i^2 = -1$$
    $$-i = -\sqrt{-1} \qquad \text{so } (-i)^2 = i^2 = -1$$
    For any positive real number x, $\sqrt{-x} = i\sqrt{x}$.

The first four powers of i are:
$$i^1 = i, \, i^2 = -1, \, i^3 = -i, \, i^4 = 1$$

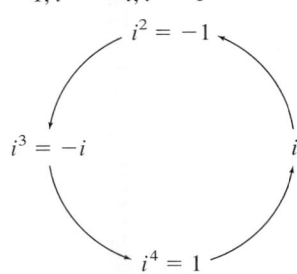

11. **Complex numbers:**
    ■ If a and b are real numbers and $i = \sqrt{-1}$, then $a + bi$ is a complex number with real term a and imaginary term bi.
    ■ $a + bi$ and $a - bi$ are complex conjugates.

12. **Operations with complex numbers:**
    The addition, subtraction, and multiplication of complex numbers are similar to the corresponding operations with binomials. Division by a complex number is accomplished by multiplying both the numerator and the denominator by the conjugate of the denominator and then simplifying the result.

13. **Subsets of the complex numbers:**
    The relationships of important subsets of the set of complex numbers are summarized in the tree diagram, where a and b are real numbers and $i = \sqrt{-1}$.

			Examples
		Pure imaginary numbers $a = 0$	$3i$
	Imaginary numbers $b \neq 0$	$a \neq 0$	$2 + 3i$
Complex numbers $a + bi$	**Real numbers** $b = 0$		$2, -\frac{3}{4}, \pi, \sqrt{3}$

14. **Standard form of a quadratic equation:**
    If x is a real variable and a, b, and c are real constants with $a \neq 0$, the standard form of a quadratic equation in x is:
    $$ax^2 + bx + c = 0$$

15. **Methods of solving quadratic equations:**
    ■ Graphically
    ■ Numerically
    ■ Factoring
    ■ Extraction of roots
    ■ Completing the square
    ■ The quadratic formula: $x = \dfrac{-b \pm \sqrt{b^2 - 4ac}}{2a}$

16. Nature of the solutions of a quadratic equation:
There are three possibilities for the solutions of $ax^2 + bx + c = 0$.

VALUE OF THE DISCRIMINANT	SOLUTIONS OF $ax^2 + bx + c = 0$	THE PARABOLA $y = ax^2 + bx + c$	GRAPHICAL EXAMPLE
$b^2 - 4ac > 0$	Two distinct real solutions	Two x-intercepts	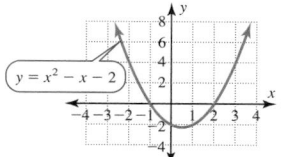
$b^2 - 4ac = 0$	A double real solution	One x-intercept with the vertex on the x-axis	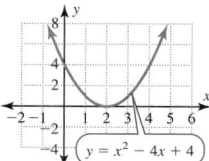
$b^2 - 4ac < 0$	Neither solution is real; both solutions are complex numbers with imaginary parts. These solutions will be complex conjugates.	No x-intercepts	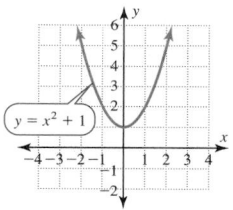

17. Solutions of a quadratic inequality:
- The solution of $ax^2 + bx + c > 0$ is the set of x-values for which the graph of $y = ax^2 + bx + c$ is above the x-axis.
- The solution of $ax^2 + bx + c < 0$ is the set of x-values for which the graph of $y = ax^2 + bx + c$ is below the x-axis.

18. Power theorem:
For any real numbers, x and y and natural number n:
- If $x = y$, then $x^n = y^n$.
- *Caution:* $x^n = y^n$ can be true when $x \neq y$. For example, $(-3)^2 = 3^2$ but $-3 \neq 3$.

19. Extraneous value:
Any value that occurs as a solution of the last equation of a solution process but is not a solution of the original equation is called an extraneous value.

20. Solving radical equations:
a. Isolate a radical term on one side of the equation.
b. Raise both sides to the nth power.

c. Solve the resulting equation. (If this equation contains a radical, repeat steps **a** and **b**.)
d. Check each possible solution in the original equation to determine whether it is a solution or an extraneous value.

21. Pythagorean theorem and converse:
ABC is a right triangle if, and only if, $a^2 + b^2 = c^2$.

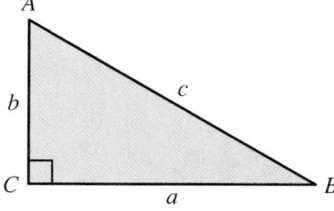

22. Formula for the distance between two points in a plane:
The distance between (x_1, y_1) and (x_2, y_2) is given by
$d = \sqrt{(x_2 - x_1)^2 + (y_2 - y_1)^2}$.

REVIEW EXERCISES   FOR CHAPTER 8

1. Write each of these radical expressions in exponential form.
 a. $\sqrt{x}$ **b.** $\sqrt[3]{x^2}$ **c.** $\dfrac{1}{\sqrt[3]{x}}$

2. Write each of these exponential expressions in radical form.
 a. $x^{1/4}$ **b.** $x^{-1/4}$ **c.** $x^{3/4}$

3. Represent each expression using both radical notation and exponential notation.
  a. Twice the principal square root of w.
  b. The principal square root of two w.
  c. The principal cube root of the quantity x plus 4.

In Exercises 4–9 simplify each expression without using a calculator. Use a calculator only to check your answer.

4. a. $49^{1/2}$ **b.** $-49^{1/2}$
  c. $49^{-1/2}$

5. a. $\left(\dfrac{27}{125}\right)^{2/3}$ **b.** $\left(\dfrac{-27}{125}\right)^{2/3}$
  c. $\left(\dfrac{27}{125}\right)^{-2/3}$

6. a. $1{,}000{,}000^{1/2}$ **b.** $1{,}000{,}000^{1/3}$
  c. $1{,}000{,}000^{1/6}$

7. a. $\sqrt{625}-576$ **b.** $\sqrt{625}-\sqrt{576}$
  c. $\sqrt{\dfrac{576}{625}}$

8. a. $\sqrt[5]{-32}$ **b.** $\sqrt[7]{-1}$
  c. $\sqrt[8]{0}$

9. a. $\sqrt{(-7)^2}$ **b.** $\sqrt[3]{(-7)^3}$
  c. $\sqrt[4]{(-5)^4}$

10. Use a calculator to approximate each expression to the nearest hundredth.
  a. $\sqrt{70}$ **b.** $\sqrt[3]{70}$
  c. $\sqrt[3]{-70}$

In Exercises 11–22 simplify each expression without using a calculator. Assume that all variables are positive real numbers.

11. a. $16^{5/8}16^{3/8}$ **b.** $\dfrac{16^{5/8}}{16^{1/8}}$
  c. $\left(16^{3/8}\right)^{2/3}$

12. a. $\left(8x^{-6}y^9\right)^{2/3}$ **b.** $\left(-32x^{5/3}y^{3/2}\right)^{2/5}$
  c. $\dfrac{\left(x^2y^2z^2\right)^{1/3}}{(xyz)^{-4/3}}$

13. a. $2x^{1/2}\left(3x^{1/2}-5x^{-1/2}\right)$ **b.** $\left(3x^{1/2}+5\right)\left(3x^{1/2}-5\right)$
  c. $\left(x^2+2xy+y^2\right)^{1/2}$

14. a. $12\sqrt{3}+8\sqrt{3}$ **b.** $12\sqrt{3}-8\sqrt{3}$
  c. $\left(5\sqrt{2}-7\sqrt{3}\right)-\left(\sqrt{2}-4\sqrt{3}\right)$

15. a. $6\sqrt[3]{5}-2\sqrt[3]{5}$ **b.** $\dfrac{6\sqrt[3]{5}}{2\sqrt[3]{5}}$
  c. $\left(6\sqrt[3]{5}\right)\left(2\sqrt[3]{5}\right)$

16. a. $\sqrt{20}$ **b.** $\sqrt[3]{24}$
  c. $3\sqrt{72}-2\sqrt{98}$

17. a. $5\sqrt{3x}+9\sqrt{3x}$ **b.** $2\sqrt{50v}-3\sqrt{8v}$
  c. $2\sqrt[3]{8v}-\sqrt[3]{125v}$

18. a. $\dfrac{\sqrt{175}}{\sqrt{7}}$ **b.** $\sqrt[3]{\dfrac{27x^3}{y^6}}$
  c. $\sqrt[3]{\dfrac{25}{27}}$

19. a. $3\sqrt{8}-7\sqrt{50}$ **b.** $3\sqrt{3}\left(2\sqrt{12}-9\sqrt{75}\right)$
  c. $\left(2\sqrt{2}-5\sqrt{3}\right)\left(2\sqrt{2}+5\sqrt{3}\right)$

20. a. $\left(3\sqrt{14}\right)\left(15\sqrt{2}\right)$ **b.** $\dfrac{3\sqrt{14}}{15\sqrt{2}}$
  c. $\dfrac{15\sqrt{2}}{3\sqrt{14}}$

21. a. $\left(\sqrt{2}-\sqrt{3}\right)^2$ **b.** $\dfrac{\sqrt{2}-\sqrt{3}}{\sqrt{2}+\sqrt{3}}$
  c. $\dfrac{12}{\sqrt{7}-\sqrt{3}}$

22. a. $\left(2\sqrt{3}\right)\left(3\sqrt{2}\right)\left(5\sqrt{6}\right)$ **b.** $\dfrac{\left(15\sqrt{2}\right)\left(2\sqrt{3}\right)}{10\sqrt{6}}$
  c. $\left(2\sqrt{3}\right)\left(3\sqrt{2}\right)-5\sqrt{6}$

In Exercises 23–27 simplify each expression without using a calculator and write the result in standard $a+bi$ form.

23. a. $\sqrt{-100}$ **b.** $\sqrt{-36}-\sqrt{-4}$
  c. $\sqrt{64}-\sqrt{-64}$

24. a. $(5-6i)+(3-2i)$ **b.** $(5-6i)-(3-2i)$
  c. $2(4-3i)-5(2+6i)$

25. a. $2i(5-6i)$ **b.** $(5-7i)(6+3i)$
  c. $(7-3i)^2$

26. a. $(5-2i)(5+2i)$ **b.** i^5
  c. i^6+i^7

27. a. $\dfrac{1+i}{1-i}$ **b.** $\dfrac{58}{2+5i}$
  c. $\dfrac{3+2i}{i}$

28. Use the calculator display shown in the figure to evaluate each of these expressions.
  a. i^{25}
  b. i^{-11}
  c. i^{32}

```
i^25
                -5E-13+i
i^(-11)
                1E-13+i
i^32
                1+2E-13i
■
```

In Exercises 29–34 solve each quadratic equation.

29. a. $(x+3)(x-7)=0$ **b.** $(2x+5)(3x-4)=0$
  c. $(x-19)^2=0$

30. a. $x^2=100$ **b.** $x^2=7$
  c. $x^2=-16$

31. $v^2=9v-20$ **32.** $y^2-4y+2=0$
33. $10m^2=21m+10$ **34.** $2x^2=6x-9$

In Exercises 35 and 36 use the given graph to solve each equation.

35. $x^2-5x-150=0$ **36.** $100x^2-40x-21=0$

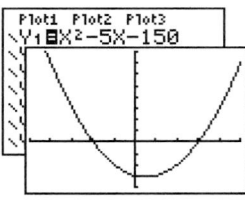

$[-25, 25, 5]$ by $[-200, 400, 50]$

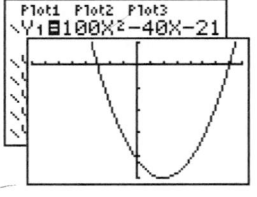

$[-.8, .8, .1]$ by $[-25, 5, 5]$

In Exercises 37 and 38 use the given table to solve each equation.

37. $x^2 + 18x + 77 = 0$ **38.** $50x^2 - 5x - 1 = 0$

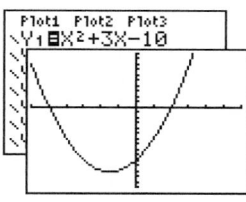

In Exercises 39–41 use the discriminant to determine the nature of the solutions of each quadratic equation.

39. $3x^2 = 30x - 75$

40. $4v^2 + 2v + 1 = 0$

41. $x^2 + \sqrt{11}x = 2$

In Exercises 42 and 43 use the given graph to solve each equation and inequality.

42. a. $x^2 + 3x - 10 = 0$
  b. $x^2 + 3x - 10 > 0$
  c. $x^2 + 3x - 10 < 0$

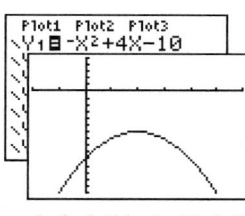

$[-6, 6, 1]$ by $[-15, 10, 1]$

43. a. $-x^2 + 4x - 10 > 0$
  b. $-x^2 + 4x - 10 < 0$

$[-2, 6, 1]$ by $[-15, 5, 1]$

In Exercises 44 and 45 solve each inequality.

44. $x^2 + 2x - 4 > 0$ **45.** $x^2 + 2x + 5 > 0$

In Exercises 46–49 solve each equation.

46. $\sqrt{x - 5} = 4$ **47.** $\sqrt{x + 12} - x = 0$

48. $\sqrt{2v - 1} + 2 = v$ **49.** $\sqrt[3]{4w + 5} = -3$

50. Calculate the distance between $(-4, 8)$ and $(1, -4)$.

51. The points $(-2, 4)$, $(4, 6)$, and $(2, 2)$ form a triangle. First calculate the perimeter of this triangle and then determine if it is a right triangle.

52. Dimensions of a Right Triangle   The length of the hypotenuse of the right triangle shown in the figure is 2 cm more than the length of the longer leg. If the longer leg is 7 cm longer than the shorter leg, determine the length of each side.

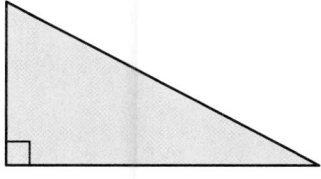

53. Compound Interest   The formula for computing the amount A of an investment of principal P at interest rate r for 1 year compounded semiannually is $A = P\left(1 + \dfrac{r}{2}\right)^2$. Approximately what interest rate is necessary for $1000 to grow to $1200 in 1 year if the interest is compounded semiannually?

54. Profit   The net income in dollars produced by selling x units of a product is given by $P(x) = x^2 + 45x - 200$. If $P(x) > 0$, there is a profit. Determine the values of x that will generate a profit.

55. Height of a Baseball   The height h in feet of a baseball t sec after being hit by a batter is given by $h(t) = -16t^2 + 80t + 3$. During what time interval after the baseball is hit will its height exceed 67 ft? (See the figure.)

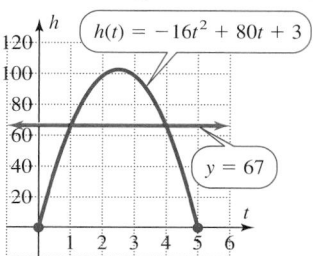

56. Find all the points with an x-coordinate of 8 that are 5 units from the point $(4, 4)$.

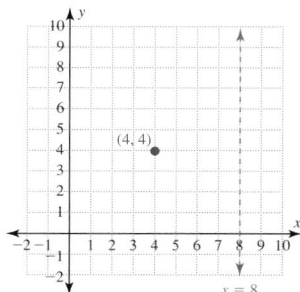

57. Deflection of a Beam   A 40-ft concrete beam is fixed between two rigid anchor points. If this beam expands by 0.4 in due to a 45° increase in temperature, determine the bulge in the middle of the beam. (Assume for simplicity that the beam deflects as shown in this diagram.) What can engineers do to avoid this problem?

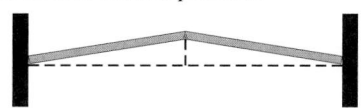

In Exercises 58–62 write a quadratic equation in standard form with the given solutions.

58. -1 and 9   **59.** $-\dfrac{2}{3}$ and $\dfrac{3}{5}$

60. $-\sqrt{3}$ and $\sqrt{3}$ **61.** $1 - 5i$ and $1 + 5i$

62. A double solution of $\dfrac{5}{7}$

MASTERY TEST   FOR CHAPTER 8

[8.1] **1.** Simplify each expression without using a calculator. Assume that $x > 0$.

   a. $81^{1/2}$ **b.** $(-1000)^{1/3}$

   c. $81^{3/4}$ **d.** $\left(\dfrac{x^{5/12}}{x^{1/12}}\right)^{3/7}$

[8.1] **2.** Simplify each expression without using a calculator.

   a. $\sqrt{169}$ **b.** $\sqrt[3]{-0.008}$

   c. $\sqrt{25} + \sqrt{144}$ **d.** $\sqrt{25 + 144}$

[8.2] **3.** Simplify each sum or difference without using a calculator. Assume that $x > 0$.

   a. $8\sqrt{7} - 3\sqrt{7}$

   b. $\left(5\sqrt{2} - 3\sqrt{5}\right) - \left(2\sqrt{2} - 7\sqrt{5}\right)$

   c. $4\sqrt[3]{7} - 11\sqrt[3]{7} + 6\sqrt[3]{7}$

   d. $13\sqrt{2x} - 5\sqrt{2x}$

[8.2] **4.** Simplify each expression without using a calculator.

   a. $\sqrt{40}$ **b.** $\sqrt[3]{40}$

   c. $3\sqrt{28} - 5\sqrt{63}$ **d.** $5\sqrt[3]{16} - \sqrt[3]{54}$

[8.3] **5.** Perform each multiplication and simplify the product without using a calculator. Assume that $x > 0$.

   a. $\left(3\sqrt{2}\right)\left(5\sqrt{2}\right)$ **b.** $2\sqrt{5}\left(3\sqrt{5} - 2\sqrt{10}\right)$

   c. $\sqrt[3]{16x^5}\sqrt[3]{-4x}$

   d. $\left(2\sqrt{3x} - \sqrt{5}\right)\left(2\sqrt{3x} + \sqrt{5}\right)$

[8.3] **6.** Perform each division by rationalizing the denominator and simplifying each quotient without using a calculator.

   a. $\dfrac{\sqrt{18}}{\sqrt{2}}$ **b.** $\dfrac{18}{\sqrt{6}}$

   c. $\dfrac{40}{\sqrt{7} - \sqrt{2}}$ **d.** $\dfrac{12}{2 - \sqrt{7}}$

[8.4] **7.** Write each complex number in standard $a + bi$ form without using a calculator.

   a. $\sqrt{-81}$ **b.** $\sqrt{16} - 25$

   c. $\sqrt{-16} - \sqrt{25}$ **d.** $i^2 + i^3$

[8.4] **8.** Perform each operation without using a calculator and write the result in standard $a + bi$ form.

   a. $2(4 - 5i) - 3(3 - 4i)$

   b. $(4 - 5i)(2 - 4i)$

   c. $\dfrac{4 - 5i}{2 - 4i}$

   d. $(3 - i)^2$

[8.5] **9.** Solve each quadratic equation by using the quadratic formula.

   a. $6x^2 - 19x + 10 = 0$

   b. $w^2 = 4 - 2w$

   c. $4x^2 + 49 = 28x$

   d. $3v^2 = 2v - 1$

[8.5] **10.** Use the discriminant to determine the nature of the solutions of each quadratic equation.

   a. $5x^2 + 5x + 1 = 0$

   b. $x^2 + 3x = 40$

   c. $7y^2 = 84y - 252$

   d. $-3w^2 = 2w + 1$

[8.5] **11.** Solve each of these inequalities.

   a. $x^2 - 3x - 10 > 0$

   b. $x^2 - 3x - 10 < 0$

   c. $3x^2 + 5x + 1 \geq 0$

   d. $3x^2 + 5x + 1 \leq 0$

[8.6] **12.** Solve each of these equations.

   a. $\sqrt{x - 3} = 11$

   b. $\sqrt[3]{2x - 17} = -3$

   c. $\sqrt{x + 4} = x + 11$

   d. $\sqrt{3x + 1} + 3 = x$

[8.6] **13.** Use the Pythagorean theorem to solve for x in each figure.

   a.

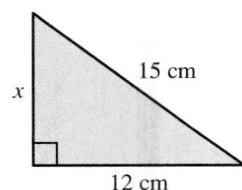

   b.

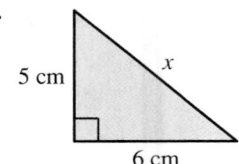

   c.

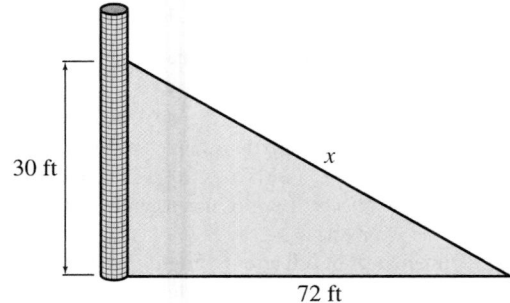

d.

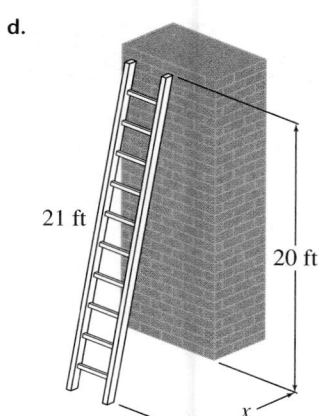

21 ft

20 ft

x

[8.6] **14.** Calculate the distance between each pair of points.
   a. $(3, 9)$ and $(-5, 9)$ **b.** $(-4, 7)$ and $(-4, -5)$
   c. $(1, -2)$ and $(7, 6)$ **d.** $(1, 1)$ and $(-4, 3)$

[8.7] **15. a. Length of a Rafter**   What length rafter is needed to span a horizontal distance of 24 ft if the roof must rise 7 ft?

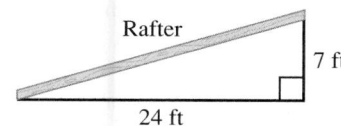

Rafter

7 ft

24 ft

b. The side of one square is 5 cm longer than the side of another square. Their total area is 193 cm^2. Determine the length of a side of each square.

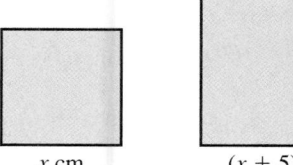

x cm   (x + 5) cm

c. $P(x) = -x^2 + 190x - 925$ is a profit polynomial that gives the profit in dollars made by selling x units of a product. Determine the values of x that will generate a profit.

d. An Application of the Pythagorean Theorem
Two airplanes depart simultaneously from an airport. One flies due south; the other flies due east at a rate 100 mi/h faster than that of the first airplane. After 2 h, radar indicates that the airplanes are 600 mi apart. What is the ground speed of each airplane?

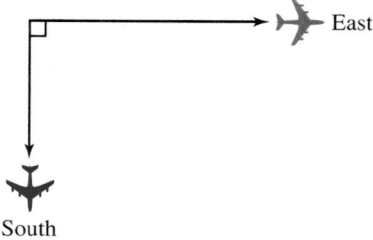

East

South

An Algebraic Model for Real Data

Supplies needed for the lab:

- A strong string at least 2 m long, a connector on the top to fasten to a ceiling, doorway frame, or stepladder, and a connector to attach a weight to the bottom of the string.
- A weight that is small but is relatively heavy, for example, a large steel nut.
- A metric ruler to measure the length of the string.
- Scissors to shorten the length of the string.
- A stopwatch.
- A graphics calculator with a Power Regression feature (PwrReg).

x LENGTH (CM)	y TIME (SEC)
200	
185	
170	
155	
140	
125	
110	
95	
80	
65	
50	

Lab: Recording the Time for the Period of a Pendulum

Suspend the top of the string from an elevated position such as a ceiling or the top of a doorframe. (Your physics lab may have a setup for this type of experiment.) Measure the length of string and attach a weight that is much heavier than the string. Using clock positions as a reference system, start with the top of the string in the 12:00 position and the bottom of the string in the 6:00 position. Then move the bottom of the string and the weight to approximately the 4:30 position. Simultaneously release the weight and start the stopwatch. Stop the stopwatch after this pendulum has completed 10 full periods. Divide this time by 10 to determine the time for one period. Then record the length of the string in centimeters and the time in seconds for one period. Shorten the string by about 15 cm and repeat this experiment about 11 times.

1. Record your data in a table similar to the one given.
2. Use this data to create a scatter diagram using an appropriate scale for each axis.

Creating an Algebraic Model for this Data

1. Using your graphics calculator, calculate the power function of best fit. This function will be of the form $f(x) = ax^b$.
2. Letting $y = T(x)$ represent the curve of best fit that you just calculated.
   a. Evaluate and interpret $T(165)$.
   b. Evaluate and interpret $T(70)$.
   c. Determine the value of x for which $T(x) = 1.0$. Interpret the meaning of this value.

9

RATIONAL

EXPRESSIONS

A fraction that contains at least one variable in the denominator is frequently called an **algebraic fraction.** A fraction that is the ratio of two polynomials is called a **rational expression.** The following are rational expressions:

$$\frac{4}{3}, \quad \frac{7x}{5y}, \quad \frac{x^2 - 1}{x^3 - 3}, \quad \text{and} \quad \frac{x^2 + 3xy + 2y^2}{x - y}$$

Problems that involve ratios, rates, and averages can all produce rational expressions. This chapter examines rational functions and rational expressions. We simplify rational expressions and add, subtract, multiply, and divide these expressions. We also solve equations with rational expressions and examine applications yielding rational expressions.

Section 9.1 Rational Functions and Reducing Rational Expressions

Objectives:
1. Determine the domain of a rational function.
2. Reduce a rational expression to lowest terms.

Monomials can have only positive integers as exponents.

Thus $\sqrt{x} = x^{1/2}$ and $\dfrac{1}{x} = x^{-1}$ are not monomials.

This section examines rational expressions. A **rational expression** is the ratio of two polynomials. Recall from Section 5.4 that a **monomial** is a real number, a variable, or a product of real numbers and variables. A **polynomial** is a monomial or a sum of monomials. Examples of polynomials are 7, x, $5x + 3$, $x^2 - 9x + 11$, and $x^2 - y^2$. Examples of expressions that are not polynomials are $\sqrt{x}, \dfrac{1}{x}$, and $\dfrac{x - y}{x + y}$.

Rational Function

ALGEBRAICALLY	VERBALLY	ALGEBRAIC EXAMPLE	GRAPHIC EXAMPLE
$f(x) = \dfrac{P(x)}{Q(x)}$ is a rational function if $P(x)$ and $Q(x)$ are polynomials and $Q(x) \neq 0$.	A rational function is defined as the ratio of two polynomials.	$f(x) = \dfrac{1}{x}$	

Examples of rational functions are:

$$f(x) = \frac{1}{x}, \quad f(x) = \frac{1}{2x - 10}, \quad g(x) = \frac{2x + 7}{x^2 - 9}, \quad \text{and} \quad h(x) = \frac{3x^2 - 7x + 19}{2x^2 - 5x - 3}.$$

The function $f(x) = \dfrac{x}{\sqrt{x} + 1}$ is not a rational function because the denominator $\sqrt{x} + 1$ is not a polynomial.

We begin an examination of rational expressions by determining the values for which rational expressions are defined and the values for which they are undefined. One way to determine this information is to examine the denominator of a rational function. The domain of a rational function must exclude values that would cause division by 0.

The domain of a rational function must exclude values that would cause division by 0.

We can determine the domain of a rational function algebraically by solving for the values that make the denominator 0 and excluding these values from the domain. In a numerical table there will be error messages at the excluded values due to division by 0. Graphically there will be breaks in the graph of a rational function at the excluded values in the domain.

Domain of a Rational Function

The domain of a rational function must exclude values that would cause division by 0.

ALGEBRAIC EXAMPLE	NUMERICAL EXAMPLE	GRAPHIC EXAMPLE	VERBAL EXAMPLE
The domain of $f(x) = \dfrac{1}{x}$ is $(-\infty, 0) \cup (0, +\infty)$.	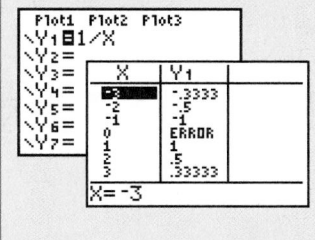	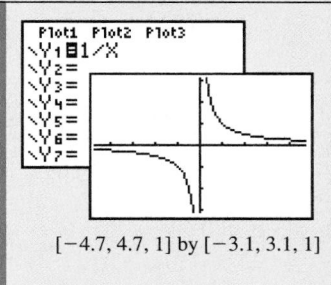 $[-4.7, 4.7, 1]$ by $[-3.1, 3.1, 1]$	Only 0 is excluded from the domain to prevent division by 0. Note that 0 causes an error message to appear in the table. Also note that there is a break in the graph at $x = 0$.

■ EXAMPLE 1   Using Multiple Perspectives to Determine the Domain of a Rational Function

The key to determining the domain of a rational function is to first determine the values that are not allowed because they would cause division by 0.

Determine the domain of $f(x) = \dfrac{x}{x^2 - 4}$ algebraically, numerically, and graphically.

SOLUTION

ALGEBRAICALLY

$$x^2 - 4 = 0$$
$$(x + 2)(x - 2) = 0$$
$$x + 2 = 0 \quad \text{or} \quad x - 2 = 0$$
$$x = -2 \qquad x = 2$$

The function is undefined at $x = -2$ and $x = 2$.

To determine the domain, first calculate the values that would cause division by 0. Do this by setting the denominator equal to 0 and solving for x.

The domain includes all real numbers except -2 and 2.

NUMERICALLY

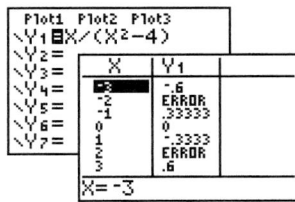

The error messages mean the function is undefined at $x = -2$ and $x = 2$.

The error messages in the table occur because of division by 0. This indicates the function is undefined at $x = -2$ and $x = 2$.

Since the denominator is second degree, these are the only two values that make the denominator 0.

The domain is the set of all real numbers except -2 and 2.

GRAPHICALLY

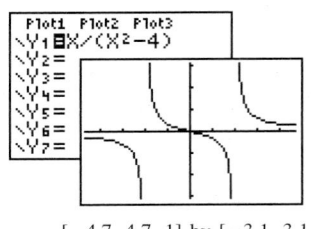

[−4.7, 4.7, 1] by [−3.1, 3.1, 1]

The graph of $y = \dfrac{x}{x^2 - 4}$ consists of three distinct branches. There are breaks in the graph at $x = -2$ and $x = 2$. There are no points on the graph for the input values of −2 and 2.

The graph has breaks at $x = -2$ and $x = 2$.

The domain contains all real numbers except −2 and 2.

Answer: Domain: $(-\infty, -2) \cup (-2, 2) \cup (2, \infty)$

SELF-CHECK 9.1.1

Given the function $f(x) = \dfrac{3x^2 - 7x + 19}{2x^2 - 5x - 3}$, evaluate:

1. $f(0)$
2. $f(1)$
3. $f(-1)$
4. $f\left(-\dfrac{1}{2}\right)$
5. $f(3)$
6. Determine the domain of this function.

If $\dfrac{P(x)}{Q(x)}$ is a rational expression, then we define the **excluded values** of this rational expression to be the values that cause division by 0. The excluded values for the expression $\dfrac{P(x)}{Q(x)}$ are the same values excluded from the domain of the function $f(x) = \dfrac{P(x)}{Q(x)}$.

■ **EXAMPLE 2** Determining the Excluded Values of a Rational Expression

Determine the excluded values of $\dfrac{2x^2 + 7x + 8}{3x^2 + 7x - 6}$.

SOLUTION

Find the values that make the denominator 0.

$3x^2 + 7x - 6 = 0$

$(3x - 2)(x + 3) = 0$

$3x - 2 = 0$ or $x + 3 = 0$

$x = \dfrac{2}{3}$ $x = -3$

To determine the excluded values set the denominator equal to 0 and solve for x. Solve this quadratic equation by factoring.

The given rational expression is defined for all real numbers except −3 and $\dfrac{2}{3}$.

Answer: The excluded values are −3 and $\dfrac{2}{3}$.

Some rational expressions do not have any excluded values. Example 3 has a denominator that is not 0 for any real values.

■ EXAMPLE 3   Determining the Excluded Values of a Rational Expression

Determine the excluded values of $\dfrac{x}{x^2 + 1}$ algebraically. Then check this answer graphically.

SOLUTION

ALGEBRAICALLY

Determine if there are any values that make the denominator 0.

$$x^2 + 1 = 0$$
$$x^2 = -1$$
$$x = \pm i$$

Thus there are no excluded values.

An excluded value is a real number that makes the denominator 0.

Set the denominator $x^2 + 1$ equal to 0 and solve for x.

Because the only values that make the denominator 0 are imaginary numbers, there are no excluded values.

GRAPHICAL CHECK

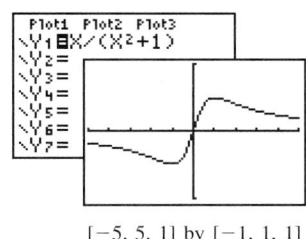

$[-5, 5, 1]$ by $[-1, 1, 1]$

Answer: No excluded values

There are no breaks in the graph of this function. Thus there are no excluded values.*

Note that the projection of this graph onto the x-axis includes the whole real axis. Thus the domain of this function includes all real numbers.

$\dfrac{x}{x^2 + 1}$ is defined for all real numbers.   ■

SELF-CHECK 9.1.2

Determine the excluded values of each expression.

1. $\dfrac{x - 5}{x - 7}$

2. $\dfrac{x - 7}{x - 5}$

3. $\dfrac{2x - 3}{x^2 - 36}$

4. $\dfrac{2x - 3}{x^2 + 36}$

SELF-CHECK 9.1.2 ANSWERS

1. 7 **2.** 5 **3.** −6 and 6
4. No excluded values

*Some breaks in the graph of a function may not be revealed by a graphics calculator. See Exercise 86 in Section 9.1 and Exercise 67 in Section 9.2.

To reduce a rational expression, the numerator and denominator must have a common factor.

The rules for simplifying rational expressions are identical to those for simplifying rational numbers in Chapter 1. A rational expression is in its **lowest terms** when the numerator and the denominator have no common factor other than -1 or 1. The basic principle for reducing rational expressions is as follows: If A, B, and C are real algebraic expressions and $B \neq 0$ and $C \neq 0$, then

$$\frac{AC}{BC} = \frac{A}{B}.$$

Note that $\dfrac{C}{C} = 1$, the multiplicative identity, for $C \neq 0$.

That is, we can divide both the numerator and the denominator of a rational expression by any nonzero factor.

■ **EXAMPLE 4**   Reducing a Rational Expression with a Monomial Denominator

Reduce $\dfrac{15x^3y^2}{25x^4y^2}$ to its lowest terms. Assume that x and y are not 0.

SOLUTION _____

$$\frac{15x^3y^2}{25x^4y^2} = \frac{\overset{1}{5x^3y^2}(3)}{\underset{1}{5x^3y^2}(5x)}$$ Factor the GCF out of the numerator and the denominator. Note that both x and y must be nonzero to avoid division by 0.

$$= \frac{3}{5x}$$ This is the same answer you would get using the properties of exponents given in Section 5.2. ■

Reducing a Rational Expression to Lowest Terms		
ALGEBRAICALLY	**VERBALLY**	**ALGEBRAIC EXAMPLE**
If A, B, and C are polynomials and $B \neq 0$ and $C \neq 0$, then $\dfrac{AC}{BC} = \dfrac{A}{B}.$	1. Factor both the numerator and the denominator of the rational expression. 2. Divide the numerator and the denominator by any common nonzero factors.	$\dfrac{3x - 6}{x^2 - 2x} = \dfrac{3(x - 2)}{x(x - 2)}$ $= \dfrac{3}{x} \quad$ for $x \neq 2$

■ **EXAMPLE 5**   Using Tables to Compare Two Rational Expressions

Reduce $\dfrac{x^2 + 3x}{x^2 + x - 6}$ and compare this result to the original expression using a table of values.

SOLUTION _____

ALGEBRAICALLY

$$\frac{x^2 + 3x}{x^2 + x - 6} = \frac{\overset{1}{x(x + 3)}}{(x - 2)\underset{1}{(x + 3)}}$$

Factor both the numerator and the denominator. Note that the excluded values are -3 and 2.

$$= \frac{x}{x - 2} \quad \text{for } x \neq 2 \text{ and } x \neq -3.$$

Divide both the numerator and denominator by the common factor $x + 3$. For $x \neq -3$, $x + 3$ is not 0.

NUMERICAL CHECK

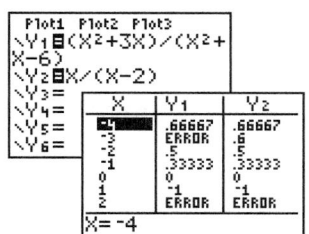

Let y_1 represent the original expression and y_2 represent the reduced expression.

The expressions are identical except for $x = -3$. The original expression has excluded values of -3 and 2. The reduced expression was obtained under these restrictions.

Answer: $\dfrac{x^2 + 3x}{x^2 + x - 6} = \dfrac{x}{x - 2}$ for $x \neq 2$ and $x \neq -3$

It is understood that this equality means that these expressions are identical for all values for which they are both defined. For $x \neq -3$ and $x \neq 2$ these expressions are identical.

SELF-CHECK 9.1.3

Reduce each rational expression to lowest terms. Assume that variables are restricted to values that prevent division by 0.

1. $\dfrac{22a^2b^3}{33a^3b^2}$ 2. $\dfrac{x^2 - 9}{4x - 12}$

SELF-CHECK 9.1.3 ANSWERS

1. $\dfrac{2b}{3a}$

2. $\dfrac{x + 3}{4}$

We assume that the simplified form of all rational expressions excludes values that cause division by 0. Then you will not need to write these restrictions with each of your answers.

■ **EXAMPLE 6** Reducing Rational Expressions

Reduce each rational expression to its lowest terms.

SOLUTIONS

(a) $\dfrac{x^2 - y^2}{5x - 5y}$

$\dfrac{x^2 - y^2}{5x - 5y} = \dfrac{(x + y)\overset{1}{\cancel{(x - y)}}}{5\underset{1}{\cancel{(x - y)}}}$

$= \dfrac{x + y}{5}$

Factor the numerator as the difference of two squares and factor the GCF, 5, out of the denominator. Then divide both by the common factor, $x - y$. Note that we must assume that $x \neq y$ to avoid division by 0.

(b) $\dfrac{x^2 - 4y^2}{x^2 - 5xy - 14y^2}$

$\dfrac{x^2 - 4y^2}{x^2 - 5xy - 14y^2} = \dfrac{\overset{1}{\cancel{(x + 2y)}}(x - 2y)}{\underset{1}{\cancel{(x + 2y)}}(x - 7y)}$

$= \dfrac{x - 2y}{x - 7y}$

Factor the numerator as a difference of two squares. Factor the trinomial in the denominator. Then divide both by the common factor, $x + 2y$, assuming $x \neq -2y$.

If two polynomials are opposites, their ratio is -1.

Two polynomials are **opposites** if every term of the first polynomial is the opposite of the corresponding term in the second polynomial. The ratio of opposites is -1.

■ **EXAMPLE 7** Reducing a Rational Expression When the Denominator Is the Opposite of the Numerator

Reduce each rational expression to its lowest terms.

SOLUTIONS

(a) $\dfrac{2x - 3y}{3y - 2x}$

$\dfrac{2x - 3y}{3y - 2x} = \dfrac{\overset{1}{\cancel{2x - 3y}}}{-\underset{1}{\cancel{(2x - 3y)}}}$

$= -1$

The denominator is the opposite of the numerator. The ratio of these opposites is -1.

(b) $\dfrac{-4a + 3b + c}{4a - 3b - c}$

$\dfrac{-4a + 3b + c}{4a - 3b - c} = \dfrac{\overset{1}{-\cancel{(4a - 3b - c)}}}{\underset{1}{\cancel{4a - 3b - c}}}$

$= -1$

This expression equals -1, because the numerator and the denominator are opposites.

SELF-CHECK 9.1.4

Reduce each rational expression to its lowest terms.

1. $\dfrac{5r - 10t}{r^2 - 4t^2}$

2. $\dfrac{5r - 15t}{3t - r}$

To reduce a rational expression to its lowest terms we divide both the numerator and denominator by a common nonzero factor. Sometimes students use the word *cancel* to describe this step. This can be dangerous because students often cancel where it is not applicable: you cannot cancel terms. A classic error would be to cancel the $3x$ terms in the numerator and the denominator in Example 8.

Warning: Reduce only by dividing by common factors— do *not* cancel terms.

EXAMPLE 8 Recognizing a Rational Expression Already in Reduced Form

Reduce $\dfrac{3x - 7}{3x + 11}$ to its lowest terms.

Do *not* cancel the $3x$ terms in the numerator and the denominator.

SOLUTION

$\dfrac{3x - 7}{3x + 11}$ is already in its reduced form because the greatest common factor of the numerator and the denominator is 1. ■

The fundamental principle of fractions, the property that $\dfrac{ac}{bc} = \dfrac{a}{b}$ for $b \neq 0$ and $c \neq 0$, is used both to reduce fractions to their lowest terms and to express fractions so that they have a common denominator. In Example 9 both the numerator and the denominator are multiplied by the same nonzero value, called the **building factor.** This step is important in adding rational expressions.

EXAMPLE 9 Converting a Rational Expression to a Given Denominator

Convert $\dfrac{x + 3}{x - 2}$ to an expression with a denominator of $x^2 - 3x + 2$.

SOLUTION

$\dfrac{x + 3}{x - 2} = \dfrac{?}{x^2 - 3x + 2}$ To determine the building factor, first factor $x^2 - 3x + 2$.

$\dfrac{x + 3}{x - 2} = \dfrac{?}{(x - 2)(x - 1)}$ The other factor needed to build the denominator is $x - 1$.

$\dfrac{x + 3}{x - 2} = \dfrac{x + 3}{x - 2} \cdot \dfrac{x - 1}{x - 1}$ Multiply $\dfrac{x + 3}{x - 2}$ by the multiplicative identity, 1, in the form $\dfrac{x - 1}{x - 1}$.

$\dfrac{x + 3}{x - 2} = \dfrac{x^2 + 2x - 3}{x^2 - 3x + 2}$ ■

SELF-CHECK 9.1.5

Determine the missing numerator in

$\dfrac{3x - 2}{2x - 3} = \dfrac{?}{10x^2 - 7x - 12}$.

USING THE LANGUAGE AND SYMBOLISM OF MATHEMATICS   9.1

1. A rational function is defined as the _____ of two polynomials.
2. Division by 0 is _____.
3. To determine the excluded values of a rational expression, we look at the values for which the _____ is equal to _____.

4. A rational expression is in _____ _____ when the numerator and denominator have no common factor other than -1 or 1.
5. If A, B, and C are polynomials and $B \neq 0$ and $C \neq 0$, then $\dfrac{AC}{BC} = $ _____.
6. If two polynomials are opposites, their ratio is _____.

EXERCISES   9.1

1. If $f(x) = \dfrac{2x+1}{2x-1}$, evaluate each expression.

    a. $f(0)$ **b.** $f(1)$ **c.** $f\left(\dfrac{1}{2}\right)$

2. If $f(x) = \dfrac{5x-11}{7x-14}$, evaluate each expression.

    a. $f(0)$ **b.** $f(1)$ **c.** $f(2)$

3. Given the rational function $f(x) = \dfrac{5}{x-4}$:

    a. What is the excluded value of this function?
    b. What is the domain of this function?

4. Given the rational function $f(x) = \dfrac{x+1}{x+6}$:

    a. What is the excluded value of this function?
    b. What is the domain of this function?

In Exercises 5–8 match the excluded values for the rational function $y = f(x)$ with the domain of this function.

EXCLUDED VALUES	DOMAIN
5. -3	**A.** $(-\infty, 0) \cup (0, \infty)$
6. 0	**B.** $(-\infty, 0) \cup (0, 3) \cup (3, \infty)$
7. -3 and 0	**C.** $(-\infty, -3) \cup (-3, \infty)$
8. 0 and 3	**D.** $(-\infty, -3) \cup (-3, 0) \cup (0, \infty)$

In Exercises 9–12 match the excluded values with the corresponding rational expression.

9. $f(x) = \dfrac{2x-1}{x+1}$ **A.** Excluded value 0.5

10. $f(x) = \dfrac{x+1}{2x-1}$ **B.** Excluded values 0.5 and 1

11. $f(x) = \dfrac{2x-1}{(x+1)(2x+1)}$ **C.** Excluded values -1 and -0.5

12. $f(x) = \dfrac{2x+1}{(x-1)(2x-1)}$ **D.** Excluded value -1

In Exercises 13 and 14 determine the excluded values for each rational expression.

13. **a.** $\dfrac{2x+1}{2x-1}$ **b.** $\dfrac{3x-7}{x^2-81}$

14. **a.** $\dfrac{5x-11}{7x-14}$ **b.** $\dfrac{x^2-9}{x^2-x-42}$

In Exercises 15–18 use the given table to determine the domain of each rational function.

15. $f(x) = \dfrac{x+4}{2x-3}$ 16. $f(x) = \dfrac{2x-3}{x+4}$

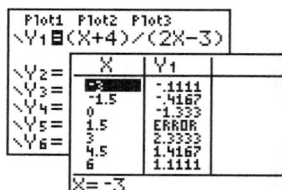 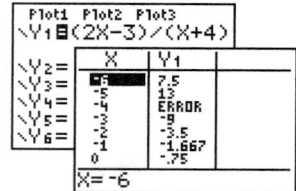

17. $f(x) = \dfrac{x^2+4}{x^2+x-6}$ 18. $f(x) = \dfrac{2x^2+3}{x^2-3x-4}$

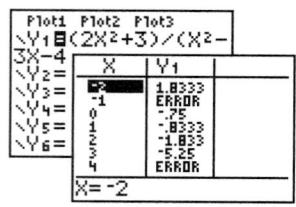

In Exercises 19–22 use the given graph to determine the domain of each rational function.

19. $f(x) = \dfrac{2x}{x-1}$ 20. $f(x) = \dfrac{-x}{x+1}$

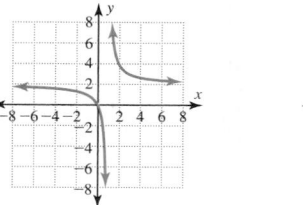 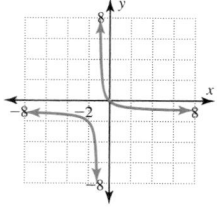

21. $f(x) = \dfrac{x - 1}{x^2 - x - 6}$

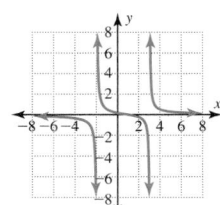

22. $f(x) = \dfrac{2 - x}{x^2 - 5x + 4}$

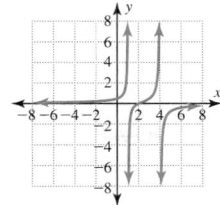

In Exercises 23–66 reduce each rational expression to its lowest terms.

23. $\dfrac{22a^2b^3}{33a^3b}$

24. $\dfrac{21x^7y^5}{35x^4y^8}$

25. $\dfrac{30x^2 - 45x}{5x}$

26. $\dfrac{46x^4 - 69x^3}{23x^2}$

27. $\dfrac{7x}{14x^2 - 21x}$

28. $\dfrac{6m^2}{9m^4 - 15m^3}$

29. $\dfrac{a^2b(2x - 3y)}{-ab^2(2x - 3y)}$

30. $\dfrac{ab(2x - 3y)^2}{ab(2x - 3y)^3}$

31. $\dfrac{7x - 8y}{8y - 7x}$

32. $\dfrac{3a - 5b}{5b - 3a}$

33. $\dfrac{ax - ay}{by - bx}$

34. $\dfrac{5x - 10}{8 - 4x}$

35. $\dfrac{(x - 2y)(x + 5y)}{(x + 2y)(x + 5y)}$

36. $\dfrac{(x - 4y)(x + 4y)}{(x - 4y)(x + 7y)}$

37. $\dfrac{x^2 - y^2}{3x + 3y}$

38. $\dfrac{v^2 - w^2}{7v - 7w}$

39. $\dfrac{25x^2 - 4}{14 - 35x}$

40. $\dfrac{4x^2 - 9}{9 - 6x}$

41. $\dfrac{3x^2 - 2xy}{3x^2 - 5xy + 2y^2}$

42. $\dfrac{x^2 - 2xy + y^2}{x^2 - y^2}$

43. $\dfrac{4x^2 + 12xy + 9y^2}{14x + 21y}$

44. $\dfrac{x^2 + 9x + 14}{x^2 + 5x - 14}$

45. $\dfrac{14ax + 21a}{35ay + 42a}$

46. $\dfrac{3ax - 3ay}{3ax + 3ay}$

47. $\dfrac{ax - y - z}{y + z - ax}$

48. $\dfrac{x^2 - x - 3}{x + 3 - x^2}$

49. $\dfrac{2a^2 - ab - b^2}{a^2 - b^2}$

50. $\dfrac{ax + bx - ay - by}{5x - 5y}$

51. $\dfrac{vx + vy - wx - wy}{v^2 - w^2}$

52. $\dfrac{2x^2 - 9x - 5}{4x^2 - 1}$

53. $\dfrac{x^2 - 25}{3x^2 + 14x - 5}$

54. $\dfrac{2x^2 + xy - 3y^2}{3x^2 - 5xy + 2y^2}$

55. $\dfrac{5a^2 + 4ab - b^2}{5a^2 - 6ab + b^2}$

56. $\dfrac{4a^2 - 4ab + b^2}{2a^2 + ab - b^2}$

57. $\dfrac{b^2 - 2b + 1 - a^2}{5ab - 5a + 5a^2}$

58. $\dfrac{a^2 - 4ab + 4b^2 - z^2}{5a - 10b + 5z}$

59. $\dfrac{x^3 - y^3}{-x^2 + 2xy - y^2}$

60. $\dfrac{x^4 - y^3}{y^3 - x^4}$

61. $\dfrac{12x^2 + 24xy + 12y^2}{16x^2 - 16y^2}$

62. $\dfrac{6y^2 + 11yz - 7z^2}{9y^2 + 42yz + 49z^2}$

63. $\dfrac{14x^2 - 9xy + y^2}{y^2 - 7xy}$

64. $\dfrac{(a + b)^2 + 7(a + b) + 6}{(a + b)^2 - 1}$

65. $\dfrac{a^2 + 2a + 1 + ab + b}{9a + 9b + 9}$

66. $\dfrac{(a - 2b)^2 - 9}{(a - 2b)^2 - (a - 2b) - 6}$

In Exercises 67–78 fill in the missing numerator or denominator.

67. $\dfrac{7}{12} = \dfrac{?}{24}$

68. $\dfrac{3}{5} = \dfrac{?}{20}$

69. $\dfrac{2x}{x - 5} = \dfrac{?}{7(x - 5)}$

70. $\dfrac{3x}{x + 4} = \dfrac{?}{5(x + 4)}$

71. $\dfrac{7x - 8y}{3a - 5b} = \dfrac{?}{10b - 6a}$

72. $\dfrac{3x - 2y}{b - 3a} = \dfrac{?}{9a - 3b}$

73. $\dfrac{10}{a + b} = \dfrac{?}{a^2 - b^2}$

74. $\dfrac{6}{x - y} = \dfrac{?}{x^2 - y^2}$

75. $\dfrac{5}{x - 6} = \dfrac{?}{x^2 - 7x + 6}$

76. $\dfrac{2x - y}{x + 3y} = \dfrac{?}{x^2 + 5xy + 6y^2}$

77. $\dfrac{2x - y}{x + 3y} = \dfrac{2x^2 + xy - y^2}{?}$

78. $\dfrac{x - 2y}{x + 3y} = \dfrac{x^2 - 4y^2}{?}$

79. Canoeing Time Upstream   The time it takes campers to paddle 12 mi upstream in a river that flows 2 mi/h is given by $T(x) = \dfrac{12}{x - 2}$, where x is the speed that the campers can paddle in still water. Evaluate and interpret each of these expressions.

a. $T(6)$ **b.** $T(8)$ **c.** $T(2)$

80. Canoeing Time Downstream   The time it takes campers
to paddle 12 mi downstream in a river that flows 3 mi/h
is given by $T(x) = \dfrac{12}{x + 3}$ where x is the speed the
campers can paddle in still water. Evaluate and interpret
each of these expressions.
   a. $T(0)$ **b.** $T(1)$ **c.** $T(3)$

81. Modeling Plant Breakdowns   During year t the
estimated number of breakdowns at one plant is given by
$N(t) = 2.5t + 4.5$. The total cost in thousands of dollars of
these breakdowns is estimated to be $T(t) = t^2 + 1.8t$.
   a. Write a function $A(t)$ that models the average cost of
these breakdowns during year t and express this
rational function in its lowest terms.
   b. Use this function to estimate the average cost of a
breakdown during the 9th year.

82. Modeling Average Cost   The number of units a plant
can produce by operating t h/day is given by $N(t) = 25t$.
The cost of operating the plant for t hours is given by
$C(t) = -25t^3 + 2500t^2 + 12,500t$.
   a. Write a function $A(t)$ that models the average cost per
unit when the plant is operated for t hours and express
this rational function in its lowest terms.
   b. Use this function to estimate the average cost per unit if
the plant operates 20 h/day.

Group Discussion Questions

83. Error Analysis   Examine each student's work and
correct any mistakes you find.
   a. Student A   **b.** Student B

$$\dfrac{x-3}{x+5} = \dfrac{-3}{5} \qquad \dfrac{x^2-9}{x^2-4x+1} = \dfrac{(x-3)(x+3)}{(x-3)(x-1)}$$
$$= \dfrac{x+3}{x-1}$$
$$= \dfrac{3}{-1}$$
$$= -3$$

84. Challenge Question   Reduce each rational expression to
its lowest terms. Assume that m is a natural number and
that all values of the variables that cause division by 0 are
excluded.
   a. $\dfrac{x^{2m}-25}{7x^m+35}$ **b.** $\dfrac{x^{m+1}-4x}{x^{2m}-16}$

85. Challenge Question
   a. Factor x^{-2} out of both the numerator and the
denominator of $\dfrac{2 + x^{-1}y - 15x^{-2}y^2}{1 + 6x^{-1}y + 9x^{-2}y^2}$ and then reduce
the expression to its lowest terms.
   b. Factor $x^{-1}y^{-1}$ out of both the numerator and the
denominator of $\dfrac{4xy^{-1} + 12 + 9x^{-1}y}{6xy^{-1} + 5 - 6x^{-1}y}$ and then reduce
the expression to its lowest terms.

86. Discovery Question
   a. Reduce $\dfrac{x^2-5x+6}{x-2}$ to its lowest terms.
   b. Determine the excluded value for $\dfrac{x^2-5x+6}{x-2}$.
   c. Graph $f(x) = \dfrac{x^2-5x+6}{x-2}$.
   d. Graph $g(x) = x - 3$.
   e. What is the domain of $f(x)$?
   f. What is the domain of $g(x)$?
   g. The graphs of $y = f(x)$ and $y = g(x)$ are different
(although you may not see the difference on a graphics
calculator). Describe the difference between these
graphs.

87. Challenge Question
   a. Have each person in your group create a different
rational function whose domain is all the real numbers
except for 2 and 5.
   b. Compare these functions algebraically.
   c. Use a graphics calculator to compare the graphs of
these functions.

88. Challenge Question
   a. Have each person in your group create a rational
function whose denominator is a second-degree
polynomial and whose domain is all real numbers.
   b. Compare these functions algebraically.
   c. Use a graphics calculator to compare the graphs of
these functions.

Section 9.2   Multiplication and Division of Rational Expressions

Objective: **3.** Multiply and divide rational expressions.

Since the variables in a rational expression represent real numbers, the rules and procedures for performing multiplication and division with rational expressions are the same
as those used in Chapter 1 for performing these operations with arithmetic fractions.
Multiplication and division will be covered before addition and subtraction so that you
can develop some of the skills needed to add rational expressions with different denominators. The rule for multiplying rational expressions is given in the following box.

Multiplying Rational Expressions

ALGEBRAICALLY	VERBALLY	ALGEBRAIC EXAMPLE
If A, B, C, and D are real polynomials and $B \neq 0$ and $D \neq 0$, then $$\frac{A}{B} \cdot \frac{C}{D} = \frac{AC}{BD}.$$	1. Factor the numerators and the denominators. Then write the product as a single fraction, indicating the product of the numerators and the product of the denominators. 2. Reduce this fraction by dividing the numerator and the denominator by any common nonzero factors.	$$\frac{x^2 - xy}{4a - 12b} \cdot \frac{3a - 9b}{5x - 5y}$$ $$= \frac{x(x - y)}{4(a - 3b)} \cdot \frac{3(a - 3b)}{5(x - y)}$$ $$= \frac{3x\overset{1}{\cancel{(x - y)}}\overset{1}{\cancel{(a - 3b)}}}{20\underset{1}{\cancel{(x - y)}}\underset{1}{\cancel{(a - 3b)}}}$$ $$= \frac{3x}{20}$$

■ EXAMPLE 1   Multiplying Rational Expressions

Find the product of $\dfrac{2x - 2y}{35}$ and $\dfrac{15}{3x - 3y}$. Assume that the variables are restricted to values that prevent division by 0.

SOLUTION

$$\frac{2x - 2y}{35} \cdot \frac{15}{3x - 3y} = \frac{2(x - y)}{35} \cdot \frac{15}{3(x - y)}$$

Factor the numerators and the denominators. (We usually do not factor the constant coefficients.)

$$= \frac{2\overset{1}{\cancel{(x - y)}}\overset{1}{\cancel{(3)}}\overset{1}{\cancel{(5)}}}{\underset{1}{\cancel{5}}(7)\underset{1}{\cancel{(3)}}\underset{1}{\cancel{(x - y)}}}$$

Then write the product as a single fraction, indicating the product of these factors.

Reduce by dividing out the common factors.

$$= \frac{2}{7}$$ ◼

In Example 2 note that each factor in the numerator also occurs in the denominator. Thus the reduced form has only a factor of 1 remaining in the numerator.

■ EXAMPLE 2   Multiplying Rational Expressions

Multiply $\dfrac{2x - y}{4x^2 - 4y^2} \cdot \dfrac{x^2 + 2xy + y^2}{2x^2 + xy - y^2}$. Assume that the variables are restricted to values that prevent division by 0.

SOLUTION

Reduce before multiplying out: Simplifying multiplication and division problems with rational expressions will be much easier if you divide both the numerator and the denominator by any common factors before multiplying any of the factors.

$$\frac{2x - y}{4x^2 - 4y^2} \cdot \frac{x^2 + 2xy + y^2}{2x^2 + xy - y^2} = \frac{2x - y}{4(x^2 - y^2)} \cdot \frac{(x + y)(x + y)}{(2x - y)(x + y)}$$

Factor the numerators and the denominators. Then write the product as a single fraction.

$$= \frac{(2x - y)(x + y)(x + y)}{4(x - y)(x + y)(2x - y)(x + y)}$$

Reduce this fraction by dividing by the common factors.

$$= \frac{1}{4(x - y)}$$

SELF-CHECK 9.2.1

Find each product. Assume that the variables are restricted to values that prevent division by 0.

1. $\dfrac{2x - 5y}{7xy} \cdot \dfrac{35x^2}{25y - 10x}$

2. $\dfrac{10x^2 - 6x}{7xy - 7y} \cdot \dfrac{5x^2 - 2x - 3}{25x^2 - 9}$

If two rational expressions are equal, then tables for these two expressions will be equal for all values for which both expressions are defined. Looking at a limited set of values in a table does not confirm that the expressions are equal for all values, but it does serve as a good check that is likely to catch potential errors from an algebraic computation. This is illustrated in Example 3.

Writing an equal symbol between two rational expressions means that these expressions are identical for all values for which both expressions are defined.

■ EXAMPLE 3   Using Tables to Compare Two Rational Expressions

Multiply $\dfrac{x^3 + x}{x^3 - 3x^2} \cdot \dfrac{x^2 - 4x + 3}{5x^2 + 5}$ and compare the product to the original expression using a table of values. Assume that the variables are restricted to values that prevent division by 0.

SOLUTION

ALGEBRAICALLY

$$\frac{x^3 + x}{x^3 - 3x^2} \cdot \frac{x^2 - 4x + 3}{5x^2 + 5} = \frac{x(x^2 + 1)}{x^2(x - 3)} \cdot \frac{(x - 1)(x - 3)}{5(x^2 + 1)}$$

Factor the numerators and the denominators.

$$= \frac{x(x^2 + 1)(x - 1)(x - 3)}{x^2(x - 3)(5)(x^2 + 1)}$$

Then write the product as a single fraction. Reduce this fraction by dividing the common factors.

$$= \frac{x - 1}{5x}$$

SELF-CHECK 9.2.1 ANSWERS

1. $-\dfrac{x}{y}$ 2. $\dfrac{2x}{7y}$

NUMERICAL CHECK

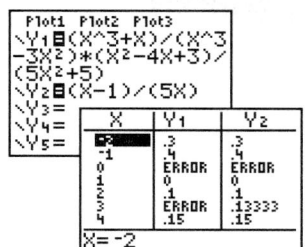

Let y_1 represent the original expression and y_2 represent the simplified product.

Both expressions are undefined for $x = 0$.

The expressions are identical except for $x = 3$.

The original expression has excluded values of 0 and 3. The reduced expression has an excluded value of 0.

It is understood that this equality means that these expressions are identical for all values for which they are both defined. For $x \neq 0$ and $x \neq 3$ these expressions are identical.

Answer: $\dfrac{x^3 + x}{x^3 - 3x^2} \cdot \dfrac{x^2 - 4x + 3}{5x^2 + 5} = \dfrac{x - 1}{5x}$

We now examine the division of rational expressions. The rule for dividing rational expressions is given in the following box.

Dividing Rational Expressions

ALGEBRAICALLY	VERBALLY	ALGEBRAIC EXAMPLE
If A, B, C, and D are real polynomials and $B \neq 0$, $C \neq 0$, and $D \neq 0$, then $$\frac{A}{B} \div \frac{C}{D} = \frac{A}{B} \cdot \frac{D}{C}$$ $$= \frac{AD}{BC}.$$	1. Rewrite the division problem as the product of the dividend and the reciprocal of the divisor. 2. Perform the multiplication using the rule for multiplying rational expressions.	$$\frac{12x}{10x - 35} \div \frac{15x}{4x - 14}$$ $$= \frac{12x}{5(2x - 7)} \cdot \frac{2(2x - 7)}{15x}$$ $$= \frac{\overset{4}{(12x)}(2)\overset{1}{(2x - 7)}}{5\underset{1}{(2x - 7)}\underset{5}{(15x)}}$$ $$= \frac{8}{25}$$

■ **EXAMPLE 4** Dividing Rational Expressions

Divide $\dfrac{10ab}{3y - x} \div \dfrac{5a^3}{7x - 21y}$. Assume that the variables are restricted to values that prevent division by 0.

SOLUTION _____

$$\frac{10ab}{3y - x} \div \frac{5a^3}{7x - 21y} = \frac{10ab}{3y - x} \cdot \frac{7x - 21y}{5a^3}$$

Rewrite the division problem as the product of the dividend and the reciprocal of the divisor.

$$= \frac{10ab}{(-1)(x - 3y)} \cdot \frac{7(x - 3y)}{5a^3}$$

Factor the binomials in the numerator and in the denominator. Note that -1 is factored out of $3y - x$.

$$= \frac{\overset{2b}{(10ab)}(7)\overset{1}{(x - 3y)}}{(-1)(x - 3y)(5a^3)}$$
$$\phantom{= \frac{}{(-1)}} \; {}_{1} \qquad {}_{a^2}$$

Write this product as a single fraction. Then reduce by dividing out the common factors.

$$= \frac{(2b)(7)}{(-1)a^2}$$

$$= -\frac{14b}{a^2}$$

■

SELF-CHECK 9.2.2

Divide $\dfrac{10x - 15}{x^2 - 5x} \div \dfrac{14x - 21}{10 - 2x}$. Assume that the variables are restricted to values that prevent division by 0.

■ **EXAMPLE 5** Dividing Rational Expressions

Divide $\dfrac{x^2 - 9}{x^2 - 25}$ by $\dfrac{x - 3}{x + 5}$. Assume that the variables are restricted to values that prevent division by 0.

SELF-CHECK 9.2.2 ANSWER

$-\dfrac{10}{7x}$

SOLUTION _____

$$\frac{x^2 - 9}{x^2 - 25} \div \frac{x - 3}{x + 5} = \frac{x^2 - 9}{x^2 - 25} \cdot \frac{x + 5}{x - 3}$$

Rewrite the division problem as the product of the dividend and the reciprocal of the divisor. Factor the numerator and denominator of the first fraction.

$$= \frac{(x + 3)(x - 3)}{(x + 5)(x - 5)} \cdot \frac{x + 5}{x - 3}$$

$$= \frac{(x + 3)(x - 3)(x + 5)}{(x + 5)(x - 5)(x - 3)}$$

Then reduce by dividing out the common factors.

$$= \frac{x + 3}{x - 5}$$

■

With practice you may be able to combine one or two of the steps illustrated in Example 5. Be sure your proficiency is well developed before you try to perform too many steps mentally. The steps that are skipped are the source of many student errors—especially errors in sign.

Examples 6 and 7 involve both multiplication and division. Compare these two examples to note the correct order of operations in each example.

■ EXAMPLE 6   Simplifying a Rational Expression Involving Both Multiplication and Division

Simplify $\dfrac{12x}{x + y} \div \left(\dfrac{5x - 5}{x^2 - y^2} \cdot \dfrac{3xy}{xy - y} \right)$. Assume that the variables are restricted to values that prevent division by 0.

SOLUTION _____

$$\frac{12x}{x + y} \div \left(\frac{5x - 5}{x^2 - y^2} \cdot \frac{3xy}{xy - y} \right) = \frac{12x}{x + y} \div \left[\frac{(5x - 5)(3xy)}{(x^2 - y^2)(xy - y)} \right]$$

$$= \frac{12x}{x + y} \cdot \frac{(x^2 - y^2)(xy - y)}{(5x - 5)(3xy)}$$

$$= \frac{12x}{x + y} \cdot \frac{(x - y)(x + y)(y)(x - 1)}{(5)(x - 1)(3xy)}$$

$$= \frac{\overset{4}{\cancel{(12x)}}(x - y)\overset{1}{\cancel{(x + y)}}\overset{1}{\cancel{(y)}}\overset{1}{\cancel{(x - 1)}}}{\cancel{(x + y)}(5)\cancel{(x - 1)}\underset{\underset{1}{y}}{\cancel{(3xy)}}}$$

$$= \frac{4(x - y)}{5}$$

The grouping symbols indicate that the multiplication inside the brackets has priority over the division. Multiply the numerators and the denominators within the brackets first. Then invert the divisor within the brackets and multiply. Factor the numerator and the denominator. Then reduce by dividing out the common factors.

■

■ EXAMPLE 7   Simplifying a Rational Expression Involving Both Multiplication and Division

Simplify $\dfrac{x^2 - 25}{x^2 - x - 12} \div \dfrac{x^2 - x - 20}{3x - 3} \cdot \dfrac{x^2 - 16}{x^2 + 4x - 5}$. Assume that the variables are restricted to values that prevent division by 0.

SOLUTION _____

$$\frac{x^2 - 25}{x^2 - x - 12} \div \frac{x^2 - x - 20}{3x - 3} \cdot \frac{x^2 - 16}{x^2 + 4x - 5}$$

$$= \frac{x^2 - 25}{x^2 - x - 12} \cdot \frac{3x - 3}{x^2 - x - 20} \cdot \frac{x^2 - 16}{x^2 + 4x - 5}$$

$$= \frac{(x + 5)(x - 5)}{(x + 3)(x - 4)} \cdot \frac{3(x - 1)}{(x - 5)(x + 4)} \cdot \frac{(x + 4)(x - 4)}{(x + 5)(x - 1)}$$

$$= \frac{\overset{1}{\cancel{(x + 5)}}\overset{1}{\cancel{(x - 5)}}(3)\overset{1}{\cancel{(x - 1)}}\overset{1}{\cancel{(x + 4)}}\overset{1}{\cancel{(x - 4)}}}{(x + 3)\underset{1}{\cancel{(x - 4)}}\underset{1}{\cancel{(x - 5)}}\underset{1}{\cancel{(x + 4)}}\underset{1}{\cancel{(x + 5)}}\underset{1}{\cancel{(x - 1)}}}$$

$$= \frac{3}{x + 3}$$

Following the correct order of operations from left to right, invert only the fraction in the middle.

Factor each numerator and denominator and then write this product as a single fraction.

Reduce by dividing out the common factors.

■

SELF-CHECK 9.2.3

Simplify $\dfrac{5x^2}{x^2 - 3x + 2} \cdot \dfrac{x^2 - 4x + 4}{x^2 - x} \div \dfrac{x^2 - 2x}{x^2 - 2x + 1}$. Assume that the variables are restricted to values that prevent division by 0.

USING THE LANGUAGE AND SYMBOLISM OF MATHEMATICS   9.2

1. If $A, B, C,$ and D are real polynomials and $B \neq 0$ and $D \neq 0$, then $\dfrac{A}{B} \cdot \dfrac{C}{D} =$ _____.

2. To multiply two rational expressions:
   a. Factor the numerators and the denominators. Then write the product as a single _____, indicating the product of the numerators and the product of the denominators.
   b. Reduce this fraction by dividing the numerator and the denominator by any common _____ factors.

3. If $A, B, C,$ and D are real polynomials and $B \neq 0, C \neq 0,$ and $D \neq 0$, then $\dfrac{A}{B} \div \dfrac{C}{D} =$ _____.

4. To divide two rational expressions, rewrite the division problem as the product of the dividend and the _____ of the divisor. Then perform the multiplication using the rule for multiplying rational expressions.

EXERCISES   9.2

In Exercises 1–46 perform the indicated operations and reduce each result to its lowest terms. Assume that the variables are restricted to values that prevent division by 0.

1. $\dfrac{14}{15} \cdot \dfrac{55}{42}$

2. $\dfrac{33}{10} \cdot \dfrac{28}{77}$

3. $\dfrac{9}{35} \div \dfrac{27}{55}$

4. $\dfrac{36}{49} \div \dfrac{9}{4}$

5. $\dfrac{7m}{9} \cdot \dfrac{6}{14m^2}$

6. $\dfrac{5m^3}{21} \cdot \dfrac{14}{15m^9}$

7. $\dfrac{66x^2}{30y} \div \dfrac{77x^8}{45y^3}$

8. $\dfrac{36x^5}{143} \div \dfrac{45x^7}{22}$

9. $\dfrac{x - 2y}{6} \cdot \dfrac{3}{2y - x}$

10. $\dfrac{15}{3a - 2b} \cdot \dfrac{2b - 3a}{12}$

11. $\dfrac{2x(x - 1)}{6(x - 3)} \cdot \dfrac{3(x + 3)}{x^2(x + 1)}$

12. $\dfrac{x^2(x + 5)}{12(x - 7)} \cdot \dfrac{36(x + 7)}{5x(x + 5)}$

13. $\dfrac{(x - 2)(x - 3)}{10(x - 2)} \cdot \dfrac{5(x - 3)}{(x + 3)(x - 3)}$

14. $\dfrac{3(x^2 - 4)}{14(x - 2)} \cdot \dfrac{7x}{11(x + 2)}$

15. $\dfrac{x^2 - y^2}{x^2 - 2xy + y^2} \div \dfrac{3x + 3y}{7x - 21}$

16. $\dfrac{4x^2 + 12x + 9}{5x^3 - 2x^2} \div \dfrac{14x^2 + 21x}{10x^2y^2}$

17. $\dfrac{x^2 - 3x + 2}{x^2 - 4x + 4} \div \dfrac{x^2 - 2x + 1}{3x^2 - 12}$

18. $\dfrac{a^2 - 9b^2}{4a^2 - b^2} \cdot \dfrac{4a^2 - 4ab + b^2}{2a^2 - 7ab + 3b^2}$

19. $\dfrac{7(c - y) - a(c - y)}{2c^2 - 7cy + 5y^2} \cdot \dfrac{2c - 5y}{c - y}$

20. $\dfrac{5(a + b) - x(a + b)}{2a^2 - 2b^2} \cdot \dfrac{6ax}{15 - 3x}$

21. $\dfrac{10y - 14x}{x - 1} \cdot \dfrac{x^2 - 2x + 1}{21x - 15y}$

22. $\dfrac{3w - 5z}{6w^2z} \cdot \dfrac{18w^2z + 30wz^2}{9w^2 - 25z^2}$

23. $\dfrac{9x^2 - 9xy + 9y^2}{5x^2y + 5xy^2} \cdot \dfrac{2xy}{3x^3 + 3y^3}$

24. $\dfrac{4x^2 - 1}{18xy} \div \dfrac{6x - 3}{16x^2 + 8x}$

25. $\dfrac{4x^2 - 4}{18xy} \cdot \dfrac{30x^2y}{5 - 5x^2}$

26. $\dfrac{2x^2 - 4xy + 2y^2}{6xy} \cdot \dfrac{25xy^2}{2xy - x^2 - y^2}$

27. $(3x^2 - 14x - 5) \cdot \dfrac{x^2 - 2x - 35}{3x^2 - 20x - 7}$

28. $(6x^2 - 19x + 10) \div \dfrac{6x^2 - 11x - 10}{9x^2 + 12x + 4}$

29. $\dfrac{20x^2 + 3x - 9}{21x - 35x^2} \div (12x^2 - 11x - 15)$

30. $\dfrac{4a^2 - ab - 5b^2}{ax + by + ay + bx} \div (8a - 10b)$

31. $\dfrac{-4}{7} \cdot \left(\dfrac{15}{6} \div \dfrac{-10}{14} \right)$

32. $-\dfrac{2}{9} \div \left(\dfrac{22}{12} \cdot \dfrac{-20}{33} \right)$

33. $\dfrac{18x^2}{5a} \cdot \dfrac{15ax}{81a^2} \cdot \dfrac{44ax}{24x^3}$

34. $\dfrac{14ab^2}{18x} \cdot \dfrac{27x^3}{15a^3b^3} \div \dfrac{7a^2x}{25ab}$

35. $\dfrac{x^2 + x - y^2 - y}{3x^2 - 3y^2} \div \dfrac{5x + 5y + 5}{7x^2y + 7xy^2}$

36. $\dfrac{3x^2 + 3y^2}{x^2 - y^2} \div \dfrac{-5x^2 - 5y^2}{x^2 - 2xy + y^2}$

37. $\dfrac{2x - 5}{2x^2 - x - 15} \cdot \dfrac{-2x^2 - x + 10}{3x + 4} \cdot \dfrac{-15x - 20}{2x^2 - 9x + 10}$

38. $\dfrac{x^4 - y^4}{-2x - 3y} \cdot \dfrac{2x + 3y}{7x^2 + 7y^2} \cdot \dfrac{12xy}{8x^2 + 8xy}$

39. $\dfrac{5a - b}{a^2 - 5ab + 4b^2} \div \left(\dfrac{6ab}{3a - 12b} \cdot \dfrac{b^2 - 5ab}{4a - 4b} \right)$

40. $\dfrac{x - y}{x^2 - y^2} \div \dfrac{6x}{x + y} \cdot \dfrac{15x^2}{4}$

41. $\dfrac{x^2 - xy}{5x^2y^2} \div \dfrac{4x - 4y}{3xy} \cdot \dfrac{20y}{11}$

42. $\dfrac{4v^2 + 11v + 6}{2v^2 - v - 10} \div \left(\dfrac{4v^2 - 21v - 18}{2v^2 - 3v - 5} \cdot \dfrac{4v^2 - 7v + 3}{4v^2 - 27v + 18} \right)$

43. $\dfrac{x(a - b + 2c) - 2y(a - b + 2c)}{x^2 - 4xy + 4y^2} \div \dfrac{2a^2 - 2ab + 4ac}{7xy^2 - 14y^3}$

44. $\dfrac{a(2x - y) - b(2x - y)}{a^2 - b^2} \div \dfrac{6x - 3y}{4a + 4b}$

45. $\dfrac{x^3 - y^3}{6x^2 - 6y^2} \div \dfrac{2x^2 + 2xy + 2y^2}{9xy}$

46. $\dfrac{x^3 - y^3}{18x^2y^3} \cdot \dfrac{9x^2y + 9xy^2}{x^2 - y^2} \div \dfrac{x^2 + xy + y^2}{36y^4}$

Estimation Skills and Calculator Skills`
In Exercises 47–50 mentally estimate the value of each expression for $x = 10.01$ and then use a calculator to approximate each value to the nearest hundredth.

PROBLEM	MENTAL ESTIMATE	CALCULATOR APPROXIMATION
47. $\dfrac{(x + 6)(x - 7)}{(x - 7)(x - 6)}$		
48. $\dfrac{(x - 2)(x + 5)}{(x - 2)(x - 5)}$		
49. $\dfrac{x^2 + 50}{x^2 - 50}$		
50. $\dfrac{x^2 - 80}{x - 6}$		

In Exercises 51–54 use the table of values for each function to match each rational expression with its reduced form.

51. $y_1 = \dfrac{3x^2 + 6x + 3}{(x + 1)^2}$

X	Y₁
-3	3
-2	3
-1	ERROR
0	3
1	3
2	3
3	3
X= -3	

52. $y_1 = \dfrac{-x^3 + 4x^2 - 4x}{(x - 2)^2}$

X	Y₁
-3	3
-2	2
-1	1
0	0
1	-1
2	ERROR
3	-3
X= -3	

53. $y_1 = \dfrac{x^3 + 4x^2 + 4x}{(x + 2)^2}$

X	Y₁
-3	-3
-2	ERROR
-1	-1
0	0
1	1
2	2
3	3
X= -3	

54. $y_1 = \dfrac{6x^2 - 6}{2 - 2x^2}$

X	Y₁
-3	-3
-2	-3
-1	ERROR
0	-3
1	ERROR
2	-3
3	-3
X= -3	

A. $y_2 = -3$

X	Y₂
-3	-3
-2	-3
-1	-3
0	-3
1	-3
2	-3
3	-3
X= -3	

B. $y_2 = x$

X	Y₂
-3	-3
-2	-2
-1	-1
0	0
1	1
2	2
3	3
X= -3	

C. $y_2 = -x$

X	Y₂
-3	3
-2	2
-1	1
0	0
1	-1
2	-2
3	-3
X= -3	

D. $y_2 = 3$

X	Y₂
-3	3
-2	3
-1	3
0	3
1	3
2	3
3	3
X= -3	

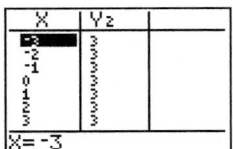

In Exercises 55–64 determine the unknown expression in each equation.

55. $12 \cdot (?) = 60$

56. $12 \div (?) = 4$

57. $(x^2y^3) \div (?) = xy$

58. $(x^2y^3) \cdot (?) = x^4y^4$

59. $(x^2 - 7x - 8) \div (?) = x + 1$

60. $(2x - 3) \cdot (?) = 4x^2 - 9$

61. $\left(\dfrac{x + 2}{x - 3}\right) \cdot (?) = \dfrac{x^2 - x - 6}{x^2 - 6x + 9}$

62. $\left(\dfrac{x + 5}{x - 5}\right) \cdot (?) = \dfrac{x^2 + 7x + 10}{x^2 - 3x - 10}$

63. $\left(\dfrac{2x^2 - 5x + 2}{3x^2 - 4x + 1}\right) \div (?) = \dfrac{2x - 1}{3x - 1}$

64. $\dfrac{x^2 - 2x - 15}{x^2 + x - 12} \div (?) = \dfrac{x + 3}{x - 3}$

Group Discussion Questions

65. Challenge Question   Perform the indicated operations and reduce each result to its lowest terms. Assume that m is a natural number and that all values of the variables that cause division by 0 are excluded.

a. $\dfrac{x^{2m+1}}{x^m - 1} \cdot \dfrac{3x^m - 3}{x}$

b. $\dfrac{x^{2m} - 1}{x^{2m} + x^m} \div \dfrac{x^m - 1}{x^{2m}}$

c. $\dfrac{x^{2m} - 1}{x^m - 1} \cdot \dfrac{3x^m + 3}{(x^m + 1)^2}$

d. $\dfrac{x^{3m} - x^m}{x^{2m} + x^m} \div \dfrac{x^{3m} - 1}{x^{2m} + x^m + 1}$

66. Challenge Question
Reduce $\dfrac{x}{x - 1} \cdot \dfrac{x - 1}{x - 2} \cdot \dfrac{x - 2}{x - 3} \cdots \dfrac{x - 8}{x - 9} \cdot \dfrac{x - 9}{x - 10}$ and list all the excluded values.

67. Error Analysis   A student graphed $f(x) = \dfrac{x^2 - x}{x - 1}$ on a TI-83 Plus calculator and obtained the graph shown here.

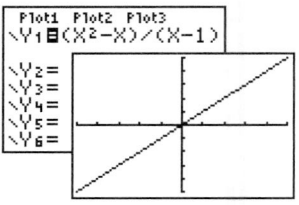

$[-5, 5, 1]$ by $[-5, 5, 1]$

Because no breaks appear in the graph of this function, the student concluded the domain of $f(x) = \dfrac{x^2 - x}{x - 1}$ is $\mathbb{R}$. This is incorrect.

a. What is the actual domain?

b. How do you think you would sketch this graph by hand to properly show the correct domain?

c. Graph this function on a TI-83 Plus calculator using the ⟨ZOOM⟩ ⟨4⟩ option. Then carefully examine this graph. What do you observe at the excluded value using this window?

d. Why do you think the excluded value is not revealed by a break in the given graph?

Section 9.3 Addition and Subtraction of Rational Expressions

Objective: **4.** Add and subtract rational expressions.

The rules for adding and subtracting rational expressions are the same rules given in Chapter 1 for adding and subtracting rational numbers. The rules for adding and subtracting rational expressions with the same denominator are given in the following box. The result should be reduced to its lowest terms.

Adding and Subtracting Rational Expressions with the Same Denominator

ALGEBRAICALLY	VERBALLY	ALGEBRAIC EXAMPLE
If A, B, and C are real polynomials and $C \neq 0$, then $$\frac{A}{C} + \frac{B}{C} = \frac{A + B}{C}$$	To add two rational expressions with the same denominator, add the numerators and use this common denominator.	$$\frac{7}{24x} + \frac{5}{24x} = \frac{7 + 5}{24x}$$ $$= \frac{12}{24x}$$ $$= \frac{1}{2x}$$
and $$\frac{A}{C} - \frac{B}{C} = \frac{A - B}{C}.$$	To subtract two rational expressions with the same denominator, subtract the numerators and use this common denominator.	$$\frac{7}{24x} - \frac{5}{24x} = \frac{7 - 5}{24x}$$ $$= \frac{2}{24x}$$ $$= \frac{1}{12x}$$

It is easy to make errors in the order of operations if you are not careful when adding or subtracting rational expressions. Example 1 illustrates how parentheses can be used to avoid these errors.

■ EXAMPLE 1   Subtracting Rational Expressions with the Same Denominator

Simplify $\dfrac{5x + 7}{2x + 2} - \dfrac{3x - 1}{2x + 2}$. Assume that the variables are restricted to values that prevent division by 0.

SOLUTION _____

$$\frac{5x + 7}{2x + 2} - \frac{3x - 1}{2x + 2} = \frac{(5x + 7) - (3x - 1)}{2x + 2}$$

Note: A common error is to write the numerator as $5x + 7 - 3x - 1$. To avoid this error, enclose the terms of the numerator in parentheses before combining like terms.

$$= \frac{5x + 7 - 3x + 1}{2x + 2}$$

$$= \frac{2x + 8}{2x + 2}$$

$$= \frac{\overset{1}{\cancel{2}}(x + 4)}{\underset{1}{\cancel{2}}(x + 1)}$$

Factor and then reduce to lowest terms.

$$= \frac{x + 4}{x + 1}$$

This sum is defined for all real numbers other than -1. The only excluded value is $x = -1$. ■

■ **EXAMPLE 2**   **Subtracting Rational Expressions with the Same Denominator**

Simplify $\dfrac{x^2}{x^2-1} - \dfrac{2x-1}{x^2-1}$. Assume that the variables are restricted to values that prevent division by 0.

SOLUTION _____

$$\dfrac{x^2}{x^2-1} - \dfrac{2x-1}{x^2-1} = \dfrac{x^2-(2x-1)}{x^2-1}$$

Use parentheses in the numerator to avoid an error in sign.

$$= \dfrac{x^2-2x+1}{x^2-1}$$

Simplify the numerator.

$$= \dfrac{\overset{1}{\cancel{(x-1)}}(x-1)}{\underset{1}{\cancel{(x-1)}}(x+1)}$$

Factor and then reduce to lowest terms. The excluded values are 1 and -1.

$$= \dfrac{x-1}{x+1}$$

This sum is defined for all real numbers other than 1 and -1. The only excluded values are $x = -1$ and $x = 1$. ■

Remember that $-\dfrac{a}{b} = \dfrac{-a}{b} = \dfrac{a}{-b}$. These alternative forms are often useful when you are working with two fractions whose denominators are opposites of each other.

■ **EXAMPLE 3**   **Adding Rational Expressions Whose Denominators Are Opposites**

Simplify $\dfrac{5x}{3x-6} + \dfrac{-4x}{6-3x}$. Assume that the variables are restricted to values that prevent division by 0.

SOLUTION _____

$$\dfrac{5x}{3x-6} + \dfrac{-4x}{6-3x} = \dfrac{5x}{3x-6} + \dfrac{-4x}{6-3x} \cdot \dfrac{-1}{-1}$$

Note that the denominators $3x-6$ and $6-3x$ are opposites. Convert $\dfrac{-4x}{6-3x}$ to the equivalent form

$$= \dfrac{5x}{3x-6} + \dfrac{4x}{3x-6}$$

$\left(\dfrac{-4x}{6-3x}\right) \cdot \left(\dfrac{-1}{-1}\right) = \dfrac{4x}{3x-6}$. Now that the

$$= \dfrac{5x+4x}{3x-6}$$

denominators are the same, you can add the numerators of these like terms.

$$= \dfrac{\overset{3}{\cancel{9}}x}{\underset{1}{\cancel{3}}(x-2)}$$

Then factor the denominator and reduce this result by dividing both the numerator and the denominator by 3.

$$= \dfrac{3x}{x-2}$$

This sum is defined for all real numbers other than 2. ■

SELF-CHECK 9.3.1

Simplify each expression. Assume that the variables are restricted to values that prevent division by 0.

1. $\dfrac{2x - 1}{x^2} + \dfrac{4x + 1}{x^2}$

2. $\dfrac{9x - 2}{2x + 1} - \dfrac{5x - 4}{2x + 1}$

3. $\dfrac{2y - 1}{7y - 3} - \dfrac{5y - 2}{3 - 7y}$

Fractions can be added or subtracted only after they are expressed in terms of a common denominator. Although any common denominator can be used, we can simplify our work considerably by using the least common denominator (LCD). Sometimes the LCD can be determined by inspection. In other cases it may be necessary to use the procedure given in the following box.

Finding the LCD of Two or More Rational Expressions

VERBALLY	ALGEBRAIC EXAMPLE
	Determine the LCD of $\dfrac{7}{6xy^2}$ and $\dfrac{2}{15x^2y}$.
1. Factor each denominator completely, including constant factors. Express repeated factors in exponential form.	$6xy^2 = 2 \cdot 3 \cdot x \cdot y^2$ $15x^2y = 3 \cdot 5 \cdot x^2 \cdot y$
2. Form the LCD by indicating the product of each factor to the highest power to which it occurs in any single factorization.	
3. Multiply the factors listed in Step 2.	$\text{LCD} = 2 \cdot 3 \cdot 5 \cdot x^2 \cdot y^2$ $\text{LCD} = 30x^2y^2$

The procedure for adding or subtracting rational expressions with different denominators is given in the following box. The common denominator in the algebraic example was calculated in the previous box.

Adding (Subtracting) Rational Expressions

VERBALLY	ALGEBRAIC EXAMPLE
1. Determine the LCD of all the terms.	$\dfrac{7}{6xy^2} + \dfrac{2}{15x^2y} = \dfrac{?}{30x^2y^2} + \dfrac{?}{30x^2y^2}$
2. Convert each term to an equivalent rational expression whose denominator is the LCD.	$= \dfrac{7}{6xy^2} \cdot \dfrac{5x}{5x} + \dfrac{2}{15x^2y} \cdot \dfrac{2y}{2y}$
3. Retain the LCD as the denominator and add (subtract) the numerators to form the sum (difference).	$= \dfrac{35x}{30x^2y^2} + \dfrac{4y}{30x^2y^2}$ $= \dfrac{35x + 4y}{30x^2y^2}$
4. Reduce the expression to its lowest terms.	

SELF-CHECK 9.3.1 ANSWERS

1. $\dfrac{6}{x}$ 2. 2   3. 1

■ **EXAMPLE 4** Subtracting Two Rational Expressions with Different Denominators

Simplify $\dfrac{5}{x+4} - \dfrac{4}{x-6}$ and compare the difference to the original expression using a table of values. Assume that the variables are restricted to values that prevent division by 0.

SOLUTION

ALGEBRAICALLY

$$\frac{5}{x+4} - \frac{4}{x-6} = \frac{5}{x+4} \cdot \frac{x-6}{x-6} - \frac{4}{x-6} \cdot \frac{x+4}{x+4}$$

The LCD is $(x+4)(x-6)$. Convert each fraction so that it has this LCD as its denominator.

$$= \frac{5(x-6)}{(x+4)(x-6)} - \frac{4(x+4)}{(x+4)(x-6)}$$

$$= \frac{5(x-6) - 4(x+4)}{(x+4)(x-6)}$$

Add like terms and simplify the numerator.

$$= \frac{5x - 30 - 4x - 16}{(x+4)(x-6)}$$

$$= \frac{x - 46}{(x+4)(x-6)}$$

For many applications it is more useful to leave the denominator in factored form.

NUMERICAL CHECK

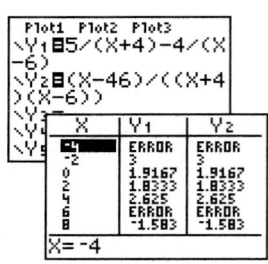

Let y_1 represent the original expression and y_2 represent the difference.

The denominator of a sum or difference is usually left in factored form because this form reveals the most information about the rational expression.

These tables are identical for all values of x. Both are undefined at -4 and 6.

Answer: $\dfrac{5}{x+4} - \dfrac{4}{x-6} = \dfrac{x-46}{(x+4)(x-6)}$

SELF-CHECK 9.3.2

Simplify $\dfrac{x}{x-5} + \dfrac{3}{x+2}$. Assume that the variables are restricted to values that prevent division by 0.

SELF-CHECK 9.3.2 ANSWER

$\dfrac{x^2 + 5x - 15}{(x-5)(x+2)}$

In Example 5 we must first factor the quadratic trinomials in order to determine the LCD of these two rational expressions.

■ EXAMPLE 5   Adding Rational Expressions with Different Denominators

Add $\dfrac{x}{6x^2 - 30x + 36} + \dfrac{7}{9x^2 + 9x - 108}$. Assume that the variables are restricted to values that prevent division by 0.

SOLUTION

$\dfrac{x}{6x^2 - 30x + 36} + \dfrac{7}{9x^2 + 9x - 108}$

$= \dfrac{x}{6(x - 2)(x - 3)} + \dfrac{7}{9(x - 3)(x + 4)}$

$= \dfrac{x}{6(x - 2)(x - 3)} \cdot \dfrac{3(x + 4)}{3(x + 4)} + \dfrac{7}{9(x - 3)(x + 4)} \cdot \dfrac{2(x - 2)}{2(x - 2)}$

$= \dfrac{3x(x + 4)}{18(x - 2)(x - 3)(x + 4)} + \dfrac{14(x - 2)}{18(x - 2)(x - 3)(x + 4)}$

$= \dfrac{3x(x + 4) + 14(x - 2)}{18(x - 2)(x - 3)(x + 4)}$

$= \dfrac{3x^2 + 12x + 14x - 28}{18(x - 2)(x - 3)(x + 4)}$

$= \dfrac{3x^2 + 26x - 28}{18(x - 2)(x - 3)(x + 4)}$

First factor each denominator. Then determine the LCD of these two fractions:
$2 \cdot 3(x - 2)(x - 3)$
$3^2(x - 3)(x + 4)$
$\text{LCD} = 2 \cdot 3^2(x - 2)(x - 3)(x + 4)$
$\text{LCD} = 18(x - 2)(x - 3)(x + 4)$
Convert each term to a fraction with this LCD as its denominator.

Leave the denominator in factored form and simplify the numerator by multiplying and then combining like terms.

The numerator is prime, so this expression is in reduced form.   ■

Some of the steps illustrated in Example 5 can be combined in order to decrease the amount of writing necessary in these problems. Although most of the steps are still shown in the following examples, some have been combined. Be sure your proficiency is well developed before you try to perform too many steps mentally. The steps that are skipped are the source of many student errors—especially errors in sign.

■ EXAMPLE 6   Combining Three Rational Expressions with Different Denominators

Simplify $\dfrac{2v + 11}{v^2 + v - 6} - \dfrac{2}{v + 3} + \dfrac{3}{2 - v}$. Assume that the variables are restricted to values that prevent division by 0.

SOLUTION

$\dfrac{2v + 11}{v^2 + v - 6} - \dfrac{2}{v + 3} + \dfrac{3}{2 - v}$

$= \dfrac{2v + 11}{(v + 3)(v - 2)} - \dfrac{2}{v + 3} + \dfrac{-3}{v - 2}$

Factor each denominator. Noting that the factor $v - 2$ in the first denominator is the opposite of the denominator $2 - v$ in the last term, change $\dfrac{3}{2 - v}$ to the equivalent form

$\left(\dfrac{3}{2 - v}\right)\left(\dfrac{-1}{-1}\right) = \dfrac{-3}{v - 2}.$

$$= \frac{2v + 11}{(v + 3)(v - 2)} - \frac{2}{v + 3} \cdot \frac{v - 2}{v - 2} + \frac{-3}{v - 2} \cdot \frac{v + 3}{v + 3}$$

Now convert each term so that it has the LCD, $(v + 3)(v - 2)$, as its denominator.

$$= \frac{2v + 11}{(v + 3)(v - 2)} - \frac{2(v - 2)}{(v + 3)(v - 2)} + \frac{-3(v + 3)}{(v + 3)(v - 2)}$$

$$= \frac{(2v + 11) - 2(v - 2) - 3(v + 3)}{(v + 3)(v - 2)}$$

$$= \frac{2v + 11 - 2v + 4 - 3v - 9}{(v + 3)(v - 2)}$$

Simplify the numerator by multiplying and then combining like terms.

$$= \frac{-3v + 6}{(v + 3)(v - 2)}$$

$$= \frac{\overset{1}{-3(v - 2)}}{(v + 3)\underset{1}{(v - 2)}}$$

Factor the numerator and reduce by dividing out the common factor, $v - 2$.

$$= \frac{-3}{v + 3}$$

$$= -\frac{3}{v + 3}$$

To multiply or divide rational expressions: first reduce as much as possible and then perform the multiplication or division.

To add or subtract rational expressions: first add or subtract the terms, using a common denominator, and then reduce afterward.

Note that when we multiply or divide rational expressions, we first reduce as much as possible and then perform the multiplication or division. On the other hand, when we add or subtract rational expressions, we add or subtract first, using a common denominator, and then reduce afterward.

SELF-CHECK 9.3.3

SELF-CHECK 9.3.3 ANSWER

$-\dfrac{2}{v - 1}$

Simplify $\dfrac{3v - 1}{v^2 - 1} - \dfrac{3}{v - 1} - \dfrac{2}{v + 1}$. Assume that the variables are restricted to values that prevent division by 0.

USING THE LANGUAGE AND SYMBOLISM OF MATHEMATICS 9.3

1. If A, B, and C are real polynomials and $C \neq 0$, then $\dfrac{A}{C} + \dfrac{B}{C} = $ _____ and $\dfrac{A}{C} - \dfrac{B}{C} = $ _____.
2. To add two rational expressions with the same denominator, _____ the numerators and use this common denominator.
3. To subtract two rational expressions with the same denominator, _____ the numerators and use this common denominator.
4. To find the least common denominator of two or more rational expressions:
   a. Factor each _____ completely, including constant factors. Express repeated factors in _____ form.
   b. Form the LCD by indicating the product of each factor to the _____ power to which it occurs in any single factorization.
   c. _____ the factors listed in the previous step.
5. To add or subtract two rational expressions:
   a. Determine the _____ of all the terms.
   b. Convert each term to an equivalent rational expression whose _____ is the LCD.
   c. Retaining the LCD as the denominator, add (subtract) the _____ to form the sum (difference).
   d. Reduce the expression to _____ _____.

EXERCISES 9.3

In Exercises 1–8 perform the indicated operations and reduce the results to lowest terms. Assume that the variables are restricted to values that prevent division by 0.

1. $\dfrac{5b + 13}{3b^2} + \dfrac{b - 4}{3b^2}$

2. $\dfrac{3x^2 + 1}{8x^3} - \dfrac{1 - 3x^2}{8x^3}$

3. $\dfrac{7}{x - 7} - \dfrac{x}{x - 7}$

4. $\dfrac{a - 1}{a + 7} + \dfrac{8}{a + 7}$

5. $\dfrac{3s + 7}{s^2 - 9} + \dfrac{s + 5}{s^2 - 9}$

6. $\dfrac{5t - 22}{t^2 - 5t + 6} + \dfrac{4t - 5}{t^2 - 5t + 6}$

7. $\dfrac{x + a}{x(a + b) + y(a + b)} - \dfrac{x - b}{x(a + b) + y(a + b)}$

8. $\dfrac{2y^2 + 1}{2y^2 - 5y - 12} - \dfrac{4 - y}{2y^2 - 5y - 12}$

In Exercises 9–14 fill in the blanks to complete each addition or subtraction.

9. $\dfrac{1}{15xy^2} + \dfrac{2}{27x^2y} = \dfrac{1}{15xy^2}\left(\dfrac{?}{9x}\right) + \dfrac{2}{27x^2y}\left(\dfrac{?}{5y}\right)$

$= \dfrac{9x}{135x^2y^2} + \dfrac{?}{135x^2y^2}$

$= \dfrac{?}{135x^2y^2}$

10. $\dfrac{3}{35x^3y} + \dfrac{2}{55x^2y^3} = \dfrac{3}{35x^3y}\left(\dfrac{?}{11y^2}\right) + \dfrac{2}{55x^2y^3}\left(\dfrac{?}{7x}\right)$

$= \dfrac{?}{385x^3y^3} + \dfrac{14x}{385x^3y^3}$

$= \dfrac{?}{385x^3y^3}$

11. $\dfrac{a + b}{21(3a - b)} - \dfrac{a - b}{14(3a - b)} = \dfrac{a + b}{21(3a - b)}\left(\dfrac{?}{?}\right) - \dfrac{a - b}{14(3a - b)}\left(\dfrac{?}{?}\right)$

$= \dfrac{?}{42(3a - b)} - \dfrac{?}{42(3a - b)}$

$= \dfrac{?}{42(3a - b)}$

12. $\dfrac{2x - y}{12(x + y)} - \dfrac{x - y}{15(x + y)} = \dfrac{2x - y}{12(x + y)}\left(\dfrac{?}{?}\right) - \dfrac{x - y}{15(x + y)}\left(\dfrac{?}{?}\right)$

$= \dfrac{?}{60(x + y)} - \dfrac{?}{60(x + y)}$

$= \dfrac{?}{60(x + y)}$

13. $\dfrac{2}{(x - 3)(x - 15)} + \dfrac{1}{(x - 3)(x + 3)} = \dfrac{2}{(x - 3)(x - 15)}\left(\dfrac{?}{x + 3}\right) + \dfrac{1}{(x - 3)(x + 3)}\left(\dfrac{?}{x - 15}\right)$

$= \dfrac{?}{(x - 3)(x - 15)(x + 3)} + \dfrac{?}{(x - 3)(x - 15)(x + 3)}$

$= \dfrac{?}{(x - 3)(x - 15)(x + 3)}$

$= \dfrac{?}{(x - 15)(x + 3)}$

14. $\dfrac{3}{(2x-1)(x+4)} - \dfrac{1}{(2x-1)(x+1)} = \dfrac{3}{(2x-1)(x+4)}\left(\dfrac{?}{x+1}\right) - \dfrac{1}{(2x-1)(x+1)}\left(\dfrac{?}{x+4}\right)$

$$= \dfrac{?}{(2x-1)(x+4)(x+1)} - \dfrac{?}{(2x-1)(x+4)(x+1)}$$

$$= \dfrac{?}{(2x-1)(x+4)(x+1)}$$

$$= \dfrac{?}{(x+4)(x+1)}$$

In Exercises 15–40 perform the indicated operations and reduce each result to its lowest terms. Assume that the variables are restricted to values that prevent division by 0.

15. $\dfrac{4}{9w} - \dfrac{7}{6w}$

16. $\dfrac{4}{5z} - \dfrac{1}{2z}$

17. $\dfrac{3v-1}{7v} - \dfrac{v-2}{14v}$

18. $\dfrac{12a-7}{35a} - \dfrac{a-1}{5a}$

19. $\dfrac{4}{b} + \dfrac{b}{b+4}$

20. $\dfrac{5}{c} - \dfrac{c}{c-7}$

21. $5 - \dfrac{1}{x}$

22. $7 + \dfrac{3}{y}$

23. $\dfrac{1}{x} - \dfrac{2}{x^2} + \dfrac{3}{x^3}$

24. $\dfrac{3}{y} + \dfrac{5}{y^2} - \dfrac{7}{y^3}$

25. $\dfrac{3}{x-2} - \dfrac{2}{x+3}$

26. $\dfrac{2}{x-5} - \dfrac{3}{x+6}$

27. $\dfrac{x}{x-1} - \dfrac{x}{x+1}$

28. $\dfrac{x}{x+2} - \dfrac{x}{x-2}$

29. $\dfrac{x+5}{x-4} - \dfrac{x+3}{x+2}$

30. $\dfrac{x-3}{x-2} - \dfrac{x-2}{x-3}$

31. $\dfrac{x-1}{x+2} + \dfrac{x+2}{x-1}$

32. $\dfrac{x+1}{x-3} - \dfrac{x-3}{x+1}$

33. $\dfrac{x}{77x-121y} - \dfrac{y}{49x-77y}$

34. $\dfrac{a}{6a-9b} - \dfrac{b}{4a-6b}$

35. $\dfrac{2}{(m+1)(m-2)} + \dfrac{3}{(m-2)(m+3)}$

36. $\dfrac{5}{(m-1)(m+2)} + \dfrac{2}{(m+2)(m-3)}$

37. $\dfrac{m+2}{m^2-6m+8} - \dfrac{8-3m}{m^2-5m+6}$

38. $\dfrac{3n+8}{n^2+6n+8} - \dfrac{4n+2}{n^2+n-12}$

39. $\dfrac{1}{a-3b} + \dfrac{b}{a^2-7ab+12b^2} + \dfrac{1}{a-4b}$

40. $\dfrac{5x}{5x-y} + \dfrac{2y^2}{25x^2-y^2} - \dfrac{5x}{5x+y}$

41. Total Area of Two Regions   Find the sum of the areas of the triangle and rectangle shown.

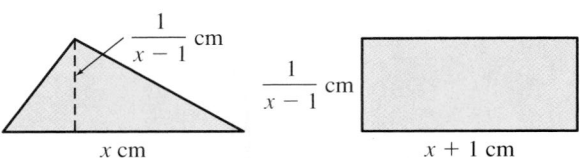

42. Total Area of Two Regions   Find the sum of the areas of the parallelogram and rectangle.

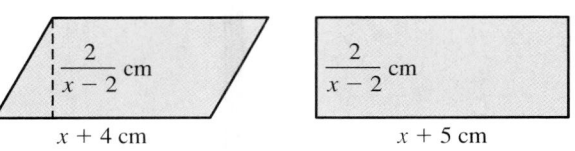

43. Total Work by Two Painters   One painter can paint a house in t h, while a second painter would take $t+5$ h to do the same house. In 8 h the fractional portion of the job done by each is $\dfrac{8}{t}$ and $\dfrac{8}{t+5}$. Write a simplified rational expression for the total portion of the job that is completed by the two painters working together for 8 h.

44. Total Work by Two Pipes   One pipe can fill a cooling tank in t h. A second pipe would take $t + 3$ h to fill the same tank. If both pipes are used, the fractional portion of the tank that can be filled by each pipe in 10 h is $\dfrac{10}{t}$ and $\dfrac{10}{t + 3}$. Write a simplified rational expression for the total portion of the tank that can be filled by the two pipes working together for 10 h.

45. Total Time of an Automobile Trip   An automobile traveled 90 mi on a gravel road in a time of $\dfrac{90}{r}$ h. Then the automobile traveled 165 mi on paved roads in a time of $\dfrac{165}{r + 10}$ h.

   a. Write a simplified rational expression for the total time of this trip.

   b. What is the total time of the 255-mi trip if the average speed of the car on the gravel road is 45 mi/h?

46. Total Time of a Delivery Truck   The route for a delivery truck includes both interstate highway and local two-way highways. The truck can travel 120 mi on the two-way highway in a time of $\dfrac{120}{r}$ h. It also can travel 180 mi on the interstate in a time of $\dfrac{180}{r + 15}$ h.

   a. Write a simplified rational expression for the total time of this trip.

   b. What is the total time of the 300-mi trip if the average speed of the truck on the two-way highway is 50 mi/h?

In Exercises 47–50 use the table of values for each function to match each expression with its sum.

47. $y_1 = \dfrac{5}{x - 1} + \dfrac{3}{x + 2}$

X	Y1
-3	-4.25
-2	ERROR
-1	.5
0	-3.5
1	ERROR
2	5.75
3	3.1

X= -3

48. $y_1 = \dfrac{5}{x - 1} - \dfrac{3}{x + 2}$

X	Y1
-3	1.75
-2	ERROR
-1	-5.5
0	-6.5
1	ERROR
2	4.25
3	1.9

X= -3

49. $y_1 = \dfrac{2x}{x - 1} - \dfrac{2x}{x + 2}$

X	Y1
-3	-4.5
-2	ERROR
-1	3
0	0
1	ERROR
2	3
3	1.8

X= -3

50. $y_1 = \dfrac{x}{x - 1} + \dfrac{x}{x + 2}$

X	Y1
-3	3.75
-2	ERROR
-1	.5
0	0
1	ERROR
2	2.5
3	2.1

X= -3

A. $y_2 = \dfrac{8x + 7}{(x - 1)(x + 2)}$

X	Y2
-3	-4.25
-2	ERROR
-1	.5
0	-3.5
1	ERROR
2	5.75
3	3.1

X= -3

B. $y_2 = \dfrac{x(2x + 1)}{(x - 1)(x + 2)}$

X	Y2
-3	3.75
-2	ERROR
-1	.5
0	0
1	ERROR
2	2.5
3	2.1

X= -3

C. $y_2 = \dfrac{2x + 13}{(x - 1)(x + 2)}$

X	Y2
-3	1.75
-2	ERROR
-1	-5.5
0	-6.5
1	ERROR
2	4.25
3	1.9

X= -3

D. $y_2 = \dfrac{6x}{(x - 1)(x + 2)}$

X	Y2
-3	-4.5
-2	ERROR
-1	3
0	0
1	ERROR
2	3
3	1.8

X= -3

In Exercises 51–68 perform the indicated operations and reduce each result to its lowest terms. Assume that the variables are restricted to values that prevent division by 0.

51. $\dfrac{x^2 + 5}{x^2 - 3x + 4} - 2$

52. $\dfrac{2x^2 - 5x + 3}{3x^2 + 7x + 9} + 2$

53. $\dfrac{5}{2x + 2} + \dfrac{x + 5}{2x^2 - 2} - \dfrac{3}{x - 1}$

54. $\dfrac{p + 6}{p^2 - 4} - \dfrac{4}{p + 2} - \dfrac{2}{p - 2}$

55. $\dfrac{1}{x - 5} + \dfrac{1}{x + 5} - \dfrac{10}{x^2 - 25}$

56. $\dfrac{42}{x^2 - 49} + \dfrac{3}{x + 7} + \dfrac{3}{x - 7}$

57. $\dfrac{3}{s - 5t} + \dfrac{7}{s - 2t} - \dfrac{9t}{s^2 - 7st + 10t^2}$

58. $\dfrac{y + 8}{y^2 - 2y - 8} + \dfrac{1}{y + 2} - \dfrac{4}{4 - y}$

59. $\dfrac{2z + 11}{z^2 + z - 6} + \dfrac{2}{z + 3} - \dfrac{3}{z - 2}$

60. $\dfrac{3}{2w - 1} + \dfrac{4}{2w + 1} - \dfrac{10w - 3}{4w^2 - 1}$

61. $\dfrac{v + w}{vw} + \dfrac{w}{v^2 - vw} - \dfrac{1}{w}$

62. $\dfrac{5v}{v^3 - 5v^2 + 6v} - \dfrac{4}{v - 2} - \dfrac{3v}{3 - v}$

63. $\dfrac{9w + 2}{3w^2 - 2w - 8} - \dfrac{7}{4 - w - 3w^2}$

64. $\dfrac{2x + y}{(x - y)(x - 2y)} - \dfrac{x + 4y}{(x - 3y)(x - y)} - \dfrac{x - 7y}{(x - 3y)(x - 2y)}$

65. $\dfrac{2w - 7}{w^2 - 5w + 6} - \dfrac{2 - 4w}{w^2 - w - 6} + \dfrac{5w + 2}{4 - w^2}$

66. $\dfrac{2x^2 - 5y^2}{x^3 - y^3} + \dfrac{x - y}{x^2 + xy + y^2} + \dfrac{1}{x - y}$

67. $\dfrac{4m}{1 - m^2} + \dfrac{2}{m + 1} - 2$

68. $\dfrac{y - 5}{x^2 + 5x + xy + 5y} + \dfrac{1}{x + y} - \dfrac{2}{x + 5}$

69. The sum of a rational expression and $\dfrac{1}{x + 5}$ is $\dfrac{4x}{2x^2 + 5x - 25}$. Determine this rational expression.

70. The difference formed by a rational expression minus $\dfrac{1}{x + 5}$ is $\dfrac{27}{4x^2 + 13x - 35}$. Determine this rational expression.

Group Discussion Questions

71. Error Analysis   Discuss the error made by the student whose work is shown here.

$$\dfrac{x}{x + 1} - \dfrac{1}{x + 1} \cdot \dfrac{2x}{x - 1} = \dfrac{x - 1}{x + 1} \cdot \dfrac{2x}{x - 1}$$

$$= \dfrac{2x}{x + 1}$$

72. Challenge Question   Simplify each expression. Assume that m is a natural number and that all values of the variables that cause division by 0 are excluded.

a. $\dfrac{x^m}{x^{2m} - 4} - \dfrac{2}{x^{2m} - 4}$

b. $\dfrac{2x^m}{x^{2m} - 6x^m + 9} - \dfrac{6}{x^{2m} - 6x^m + 9}$

c. $\dfrac{3x^m}{x^{2m} - x^m - 20} - \dfrac{15}{x^{2m} - x^m - 20}$

73. Challenge Question   Assume that m is an even integer. Simplify $\dfrac{1}{m + 1} + \dfrac{2}{m + 1} + \dfrac{3}{m + 1} + \cdots + \dfrac{m}{m + 1}$. Then test your result by evaluating both the original expression and your result for $m = 4$, 6, and 8.

Section 9.4   Combining Operations and Simplifying Complex Rational Expressions

Objectives: **5.** Simplify rational expressions in which the order of operations must be determined.
6. Simplify complex fractions.

This section shows how to combine the operations already covered in the earlier sections of this chapter and provides further exercises involving rational expressions. The correct order of operations, from Section 1.6, is as follows:

A Mathematical Note

Evelyn Boyd Granville completed her Ph.D. in functional analysis in 1949 and became one of the first African-American women to earn a doctorate in mathematics. She later worked on both the Mercury and Apollo space programs.

Order of Operations

STEP 1 Start with the expression within the innermost pair of grouping symbols.

STEP 2 Perform all exponentiations.

STEP 3 Perform all multiplications and divisions as they appear from left to right.

STEP 4 Perform all additions and subtractions as they appear from left to right.

■ EXAMPLE 1   Using the Correct Order of Operations When Simplifying Rational Expressions

Simplify $\dfrac{2x}{5} + \dfrac{x^3}{15} \cdot \dfrac{9}{x^2}$. Assume that the variables are restricted to values that prevent division by 0.

SOLUTION _____

$$\frac{2x}{5} + \frac{x^3}{15} \cdot \frac{9}{x^2} = \frac{2x}{5} + \frac{\overset{3x}{\cancel{9x^3}}}{\underset{5}{\cancel{15x^2}}}$$ Multiplication has a higher priority than addition, so perform the indicated multiplication first. Reduce the product to its lowest terms by dividing both the numerator and the denominator by $3x^2$.

$$= \frac{2x}{5} + \frac{3x}{5}$$

$$= \frac{2x + 3x}{5}$$ Then perform the indicated addition and reduce this sum to lowest terms.

$$= \frac{\overset{1}{\cancel{5x}}}{\underset{1}{\cancel{5}}}$$

$$= x$$ ■

The expressions in Examples 1 and 2 look very much alike. However, the pair of parentheses in Example 2 completely changes the order of operations used in Example 1.

■ EXAMPLE 2   Using the Correct Order of Operations When Simplifying Rational Expressions

Simplify $\left(\dfrac{2x}{5} + \dfrac{x^3}{15}\right) \cdot \dfrac{9}{x^2}$. Assume that the variables are restricted to values that prevent division by 0.

SOLUTION _____

$$\left(\frac{2x}{5} + \frac{x^3}{15}\right) \cdot \frac{9}{x^2} = \left(\frac{2x}{5} \cdot \frac{3}{3} + \frac{x^3}{15}\right) \cdot \frac{9}{x^2}$$ First add the terms inside the parentheses using a common denominator, 15.

$$= \left(\frac{6x}{15} + \frac{x^3}{15}\right) \cdot \frac{9}{x^2}$$

$$= \left(\frac{6x + x^3}{15}\right) \cdot \frac{9}{x^2}$$ Then multiply these factors. Reduce the product to its lowest terms.

$$= \frac{\overset{1}{\cancel{x}}(x^2 + 6)}{\underset{5}{\cancel{15}}} \cdot \frac{\overset{3}{\cancel{9}}}{\underset{x}{\cancel{x^2}}}$$

$$= \frac{3(x^2 + 6)}{5x}$$

$$= \frac{3x^2 + 18}{5x}$$

SELF-CHECK 9.4.1

Simplify each expression. Assume that x is restricted to values that prevent division by 0.

1. $\left(\dfrac{x}{3} - \dfrac{2}{x^2}\right) \cdot \dfrac{x^3}{4}$

2. $\dfrac{x}{3} - \dfrac{2}{x^2} \cdot \dfrac{x^3}{4}$

Example 3, which involves division of one rational expression by another, is used as a lead-in to the topic of complex rational expressions.

■ **EXAMPLE 3**   Using the Correct Order of Operations When Simplifying Rational Expressions

Simplify $\left(\dfrac{2}{x} - \dfrac{1}{3}\right) \div \left(\dfrac{2}{x} + \dfrac{1}{3}\right)$ and compare the result to the original expression using a table of values. Assume that the variables are restricted to values that prevent division by 0.

SOLUTION

ALGEBRAICALLY

$\left(\dfrac{2}{x} - \dfrac{1}{3}\right) \div \left(\dfrac{2}{x} + \dfrac{1}{3}\right) = \left[\dfrac{2(3) - 1(x)}{3x}\right] \div \left[\dfrac{2(3) + 1(x)}{3x}\right]$ First simplify the terms inside each set of parentheses. The LCD inside each set of parentheses is $3x$.

$= \left(\dfrac{6 - x}{3x}\right) \div \left(\dfrac{6 + x}{3x}\right)$ Simplify each numerator. Note that 0 is an excluded value to prevent division by 0.

$= \dfrac{6 - x}{\overset{}{\underset{1}{\cancel{3x}}}} \cdot \dfrac{\overset{1}{\cancel{3x}}}{6 + x}$ To divide, invert the divisor and then multiply. Note that -6 is also an excluded value.
Reduce by dividing out the common factor of $3x$.

$= \dfrac{6 - x}{6 + x}$

$= \dfrac{-(x - 6)}{x + 6}$ To rewrite the numerator so that x has a positive coefficient, factor a -1 out of the numerator.

$= -\dfrac{x - 6}{x + 6}$

NUMERICAL CHECK

Let y_1 represent the original expression and y_2 represent the result.

```
Plot1 Plot2 Plot3
\Y1■(2/X-1/3)/(2
/X+1/3)
\Y2■-(X-6)/(X+6)
\Y3
\Y4
\Y5
```

X	Y1	Y2
-6	ERROR	ERROR
-5	11	11
-4	5	5
-3	3	3
-2	2	2
-1	1.4	1.4
0	ERROR	1

X= -6

One or both of the functions generates an error message at the excluded values of -6 and 0. The table indicates that these expressions are identical for all other real numbers. This supports the fact that these two expressions are equal.

Answer: $\left(\dfrac{2}{x} - \dfrac{1}{3}\right) \div \left(\dfrac{2}{x} + \dfrac{1}{3}\right) = -\dfrac{x - 6}{x + 6}$

SELF-CHECK 9.4.2

Simplify $\left(\dfrac{2}{x} - \dfrac{1}{y}\right) \div \left(\dfrac{6}{x} - \dfrac{3}{y}\right)$. Assume that the variables are restricted to values that prevent division by 0.

Complex Rational Expressions

The original expression in Example 3 can be written as either

$$\left(\dfrac{2}{x} - \dfrac{1}{3}\right) \div \left(\dfrac{2}{x} + \dfrac{1}{3}\right) \quad \text{or} \quad \dfrac{\dfrac{2}{x} - \dfrac{1}{3}}{\dfrac{2}{x} + \dfrac{1}{3}}.$$

Make sure the main fraction bar is longer than the other fraction bars to clearly denote the meaning of the expression.

The second form is called a complex rational expression. A **complex rational expression** is a rational expression whose numerator or denominator, or both, is also a rational expression. Each of the following fractions contains more than one fraction bar and is therefore a complex fraction:

$$\dfrac{7 + \dfrac{1}{x}}{y}, \qquad \dfrac{8}{19 + \dfrac{1}{2}}, \qquad \text{and} \qquad \dfrac{\dfrac{m - n}{2m + 3}}{\dfrac{m^2 - 5n}{m^2 - 2mn + n^2}}$$

■ EXAMPLE 4   Simplifying Two Complex Fractions

Simplify each complex fraction.

SOLUTIONS

(a) $\dfrac{2}{\dfrac{3}{4}}$
$\qquad \dfrac{2}{\dfrac{3}{4}} = 2 \div \dfrac{3}{4}$
$\qquad\qquad$ In part (a) the main fraction bar indicates that the numerator is 2 and the denominator is $\dfrac{3}{4}$.

$\qquad\qquad = \dfrac{2}{1} \cdot \dfrac{4}{3}$
$\qquad\qquad$ To divide by $\dfrac{3}{4}$, invert the divisor and multiply by $\dfrac{4}{3}$.

$\qquad\qquad = \dfrac{8}{3}$

(b) $\dfrac{\dfrac{2}{3}}{4}$
$\qquad \dfrac{\dfrac{2}{3}}{4} = \dfrac{2}{3} \div 4$
$\qquad\qquad$ In part (b) the main fraction bar indicates that the numerator is $\dfrac{2}{3}$ and the denominator is 4.

$\qquad\qquad = \dfrac{\overset{1}{\cancel{2}}}{3} \cdot \dfrac{1}{\underset{2}{\cancel{4}}}$
$\qquad\qquad$ To divide by 4, invert the divisor and multiply by $\dfrac{1}{4}$.

$\qquad\qquad = \dfrac{1}{6}$
$\qquad\qquad$ Note that the expressions in parts (a) and (b) have different values.

SELF-CHECK 9.4.2 ANSWER

$\dfrac{1}{3}$

SELF-CHECK 9.4.3

Simplify each complex fraction.

1. $\dfrac{\frac{3}{4}}{\frac{5}{6}}$ 2. $\dfrac{\frac{3}{4}}{5}$ 3. $\dfrac{3}{\frac{4}{5}}$

Rewriting Complex Fractions Using the Division Symbol

There are two useful methods of simplifying complex fractions. Both are worth learning, since some problems can be worked more easily by one method than the other. Because the fraction bar represents division we can rewrite a complex fraction using the division symbol. This is the first method we use to simplify complex fractions.

■ **EXAMPLE 5**   **Simplifying a Complex Rational Expression**

Simplify $\dfrac{\dfrac{a^2 - b^2}{5a + 5b}}{15a}$ by rewriting the expression as a division problem. Assume that the variables are restricted to values that prevent division by 0.

SOLUTION _____

$$\dfrac{\dfrac{a^2 - b^2}{5a + 5b}}{15a} = \left(\dfrac{a^2 - b^2}{1}\right) \div \left(\dfrac{5a + 5b}{15a}\right)$$

Rewrite this complex rational expression as a division problem. A denominator of 1 is understood for $a^2 - b^2$.

$$= \left(\dfrac{a^2 - b^2}{1}\right) \cdot \left(\dfrac{15a}{5a + 5b}\right)$$

Then rewrite the division problem as the product of the dividend and the reciprocal of the divisor.

$$= \dfrac{\overset{1}{(a + b)}(a - b)}{1} \cdot \dfrac{\overset{3}{15a}}{\underset{1}{5}\underset{1}{(a + b)}}$$

Factor and reduce by dividing out the common factors.

$$= \dfrac{3a(a - b)}{1}$$

$$= 3a^2 - 3ab$$ ■

The use of the equal symbol in these examples means that the expressions are equal for all values, except the excluded values.

In Example 6 we first simplify the denominator of the main fraction and then replace the main fraction bar by a division symbol.

SELF-CHECK 9.4.3 ANSWERS

1. $\dfrac{9}{10}$ 2. $\dfrac{3}{20}$ 3. $\dfrac{15}{4}$

◼ EXAMPLE 6   Simplifying a Complex Rational Expression

Simplify $\dfrac{\dfrac{x^2 + b^2}{x^2 - b^2}}{\dfrac{x - b}{x + b} + \dfrac{x + b}{x - b}}$ by rewriting the expression as a division problem. Assume

that the variables are restricted to values that prevent division by 0.

SOLUTION

$$\frac{\dfrac{x^2 + b^2}{x^2 - b^2}}{\left(\dfrac{x - b}{x + b}\right) + \left(\dfrac{x + b}{x - b}\right)} = \frac{\dfrac{x^2 + b^2}{x^2 - b^2}}{\dfrac{(x - b)^2 + (x + b)^2}{(x + b)(x - b)}}$$

First add the two terms in the denominator of the main fraction.

$$= \frac{x^2 + b^2}{x^2 - b^2} \div \frac{(x^2 - 2bx + b^2) + (x^2 + 2bx + b^2)}{(x + b)(x - b)}$$

Then rewrite this complex rational expression as a division problem.

$$= \frac{x^2 + b^2}{x^2 - b^2} \div \frac{2x^2 + 2b^2}{(x + b)(x - b)}$$

Simplify the numerator of the divisor.

$$= \frac{\overset{1}{x^2 + b^2}}{\underset{1}{(x + b)(x - b)}} \cdot \frac{\overset{1}{(x + b)(x - b)}}{\underset{1}{2(x^2 + b^2)}}$$

Rewrite the division problem by inverting the divisor and multiplying.

$$= \frac{1}{2}$$

Factor and reduce by dividing out the common factors. ◼

SELF-CHECK 9.4.4

Simplify each expression. Assume that the variables are restricted to values that prevent division by 0.

1. $\dfrac{\dfrac{x^2 - y^2}{15xy^2}}{\dfrac{x - y}{6x^2y}}$

2. $\dfrac{\dfrac{3}{m} - 1}{m^2 - 9}$

SELF-CHECK 9.4.4 ANSWERS

1. $\dfrac{2x^2 + 2xy}{5y}$

2. $-\dfrac{1}{m(m + 3)}$

Simplifying Complex Fractions by Multiplying by the LCD

Multiplying both the numerator and the denominator by the same value is equivalent to multiplying the fraction by the multiplicative identity, 1.

Another method for simplifying complex rational expressions is illustrated in Example 7. To use this method, multiply both the numerator and the denominator of the complex fraction by the LCD of all the fractions that occur in the numerator and the denominator of the complex fraction. The effect of this is to remove all the individual fractions that occur in the numerator and the denominator.

■ EXAMPLE 7   Simplifying a Complex Rational Expression

Simplify $\dfrac{1 + \dfrac{2}{x} + \dfrac{1}{x^2}}{1 - \dfrac{1}{x^2}}$ by multiplying both the numerator and the denominator by the LCD of all terms. Assume that the variables are restricted to values that prevent division by 0.

SOLUTION

$$\dfrac{1 + \dfrac{2}{x} + \dfrac{1}{x^2}}{1 - \dfrac{1}{x^2}} = \dfrac{1 + \dfrac{2}{x} + \dfrac{1}{x^2}}{1 - \dfrac{1}{x^2}} \cdot \dfrac{x^2}{x^2}$$ By inspection, the LCD of all the fractions in the numerator and the denominator is x^2.

$$= \dfrac{\left(1 + \dfrac{2}{x} + \dfrac{1}{x^2}\right)x^2}{\left(1 - \dfrac{1}{x^2}\right)x^2}$$ Multiply both the numerator and the denominator by this LCD.

$$= \dfrac{(1)x^2 + \left(\dfrac{2}{x}\right)x^2 + \left(\dfrac{1}{x^2}\right)x^2}{(1)x^2 - \left(\dfrac{1}{x^2}\right)x^2}$$ Using the distributive property, multiply each term by x^2.

$$= \dfrac{x^2 + 2x + 1}{x^2 - 1}$$ Simplify the numerator and the denominator.

$$= \dfrac{(x + 1)(x + 1)}{(x + 1)(x - 1)}$$ Factor the numerator and denominator.

$$= \dfrac{x + 1}{x - 1}$$ Then reduce the fraction to its lowest terms.

SELF-CHECK 9.4.5

Simplify each complex fraction. Assume that x is restricted to values that prevent division by 0.

1. $\dfrac{\dfrac{5}{4} - \dfrac{2}{3}}{\dfrac{1}{2} + 6}$

2. $\dfrac{1 - \dfrac{4}{x} + \dfrac{3}{x^2}}{1 - \dfrac{1}{x} - \dfrac{6}{x^2}}$

Because $a^{-n} = \dfrac{1}{a^n}$, fractions that involve negative exponents can be expressed as complex fractions. One way to simplify some expressions involving negative exponents is to first rewrite these expressions in terms of positive exponents. Then you can apply the methods just described for complex fractions.

■ EXAMPLE 8   Simplifying a Fraction with Negative Exponents

Simplify $\dfrac{x^{-1} - x^{-2}}{x^{-1} + x^{-2}}$. Assume that the variables are restricted to values that prevent division by 0.

SOLUTION

$$\frac{x^{-1} - x^{-2}}{x^{-1} + x^{-2}} = \frac{\dfrac{1}{x} - \dfrac{1}{x^2}}{\dfrac{1}{x} + \dfrac{1}{x^2}}$$

Rewrite this expression as a complex fraction, using the definition of negative exponents.

$$= \frac{\dfrac{1}{x} - \dfrac{1}{x^2}}{\dfrac{1}{x} + \dfrac{1}{x^2}} \cdot \frac{x^2}{x^2}$$

Multiply both the numerator and the denominator by x^2, the LCD of all fractions that occur in the numerator and the denominator of the complex fraction.

$$= \frac{\left(\dfrac{1}{x}\right)x^2 - \left(\dfrac{1}{x^2}\right)x^2}{\left(\dfrac{1}{x}\right)x^2 + \left(\dfrac{1}{x^2}\right)x^2}$$

Use the distributive property to multiply each term by x^2.

$$= \frac{x - 1}{x + 1}$$

Simplify the numerator and the denominator.

In Example 8 the LCD of all the fractions within the numerator and the denominator of the complex fraction is x^2. Thus to simplify this complex fraction, we multiplied both the numerator and the denominator by x^2. This is exactly what we do in Example 9, but without going through the work of writing the expression as a complex fraction. The alternative strategy illustrated in Example 9 is particularly appropriate when negative exponents are applied only to monomials. The key is to select the appropriate expression to multiply times both the numerator and the denominator of the given fraction. The appropriate LCD to use as a factor can be determined by inspecting all the negative exponents in the original expression.

From definitions in Chapter 5 we know that for any nonzero real number x and exponent n, $x^0 = 1$ and $x^{-n} = \dfrac{1}{x^n}$.

■ EXAMPLE 9 Using an Alternative Method for Simplifying a Fraction with Negative Exponents

Simplify $\dfrac{x^{-1} - x^{-2}}{x^{-1} + x^{-2}}$. Assume that the variables are restricted to values that prevent division by 0.

SOLUTION

$$\frac{x^{-1} - x^{-2}}{x^{-1} + x^{-2}} = \frac{x^{-1} - x^{-2}}{x^{-1} + x^{-2}} \cdot \frac{x^2}{x^2}$$

Multiply both the numerator and the denominator by x^2. Note that this is the lowest power of x that will eliminate all of the negative exponents on x in both the numerator and the denominator.

$$= \frac{(x^{-1})x^2 - (x^{-2})x^2}{(x^{-1})x^2 + (x^{-2})x^2}$$

Use the distributive property to multiply each term by x^2.

$$= \frac{x - x^0}{x + x^0}$$

$$= \frac{x - 1}{x + 1}$$

Then simplify by replacing x^0 with 1. Note that this result is the same as the result in Example 8. ■

SELF-CHECK 9.4.6

Assume that x is restricted to values that prevent division by 0.

1. Determine the lowest power of x to multiply times both the numerator and the denominator in order to eliminate all of the negative exponents in

$$\frac{x^{-1} - 4x^{-2} - 21x^{-3}}{x^{-1} + 6x^{-2} + 9x^{-3}}.$$

2. Simplify the expression in part **1.**

SELF-CHECK 9.4.6 ANSWERS

1. x^3 **2.** $\dfrac{x - 7}{x + 3}$

USING THE LANGUAGE AND SYMBOLISM OF MATHEMATICS   9.4

1. When simplifying a rational expression, the correct order of operations is essential. This order is:
  a. Start with the expression within the innermost pair of

  _____ _____.
  b. Perform all _____.
  c. Perform all _____ and _____ as they appear from left to right.
  d. Perform all _____ and _____ as they appear from left to right.

2. A complex rational expression is a rational expression whose _____ or _____, or both, is also a rational expression.

3. Dividing by a fraction is equivalent to multiplying by the _____ of that fraction.

4. Two rational expressions must have the same LCD when combined using the operations of _____ or _____.

EXERCISES   9.4

In Exercises 1–40 simplify each expression. Assume that the variables are restricted to values that prevent division by 0.

1. a. $\dfrac{2x}{3} + \dfrac{x^2}{15} \cdot \dfrac{5}{x}$

  b. $\left(\dfrac{2x}{3} + \dfrac{x^2}{15}\right) \cdot \dfrac{5}{x}$

2. a. $\dfrac{3x}{4} - \dfrac{x^2}{12} \cdot \dfrac{3}{x}$

  b. $\left(\dfrac{3x}{4} - \dfrac{x^2}{12}\right) \cdot \dfrac{3}{x}$

3. a. $-\dfrac{2y}{3} + \dfrac{y^2}{6} \div \dfrac{y}{9}$

  b. $\left(-\dfrac{2y}{3} + \dfrac{y^2}{6}\right) \div \dfrac{y}{9}$

4. a. $\dfrac{3y}{5} - \dfrac{2y^2}{3} \div \dfrac{y}{10}$

  b. $\left(\dfrac{3y}{5} - \dfrac{2y^2}{3}\right) \div \dfrac{y}{10}$

5. a. $\dfrac{v}{v-6} - \dfrac{3}{v+2} \cdot \dfrac{2v+4}{v^2-6v}$

b. $\left(\dfrac{v}{v-6} - \dfrac{3}{v+2}\right) \cdot \dfrac{2v+4}{v^2-6v}$

6. a. $\dfrac{v}{v+10} + \dfrac{5v}{v-3} \cdot \dfrac{2v-6}{v^2+10v}$

b. $\left(\dfrac{v}{v+10} + \dfrac{5v}{v-3}\right) \cdot \dfrac{2v-6}{v^2+10v}$

7. a. $\dfrac{\dfrac{4}{6}}{\dfrac{5}{8}}$ **b.** $\dfrac{4}{\dfrac{6}{5}}$

8. a. $\dfrac{\dfrac{3}{4}}{\dfrac{5}{6}}$ **b.** $\dfrac{\dfrac{4}{6}}{5}$

9. $\dfrac{1+\dfrac{1}{5}}{1-\dfrac{1}{5}}$ **10.** $\dfrac{\dfrac{1}{2}+\dfrac{2}{3}}{\dfrac{3}{2}+2}$

11. $\dfrac{\dfrac{3}{5}-\dfrac{5}{3}}{\dfrac{1}{3}+\dfrac{1}{5}}$ **12.** $\dfrac{\dfrac{7}{10}-\dfrac{3}{4}}{\dfrac{3}{8}+\dfrac{1}{3}}$

13. $\dfrac{2-3x}{2x-3} + \dfrac{8-2x}{3x-6} \div \dfrac{4x-6}{3x-6}$

14. $\dfrac{6z+3}{3z-1} - \dfrac{5}{9z^2-1} \div \dfrac{1}{3z+1}$

15. $\left(\dfrac{2}{v+3} + v\right)\left(v - \dfrac{3}{v+2}\right)$

16. $\left(v - \dfrac{6v+35}{v+4}\right)\left(v - \dfrac{44}{v-7}\right)$

17. $\left(x - 2 - \dfrac{3}{x}\right) \div \left(1 + \dfrac{1}{x}\right)$

18. $\left(\dfrac{w+3}{2w+1} - 2w\right) \div \left(\dfrac{w-1}{2w+1} + 2w\right)$

19. $1 - \dfrac{4}{x} - \left(1 - \dfrac{2}{x}\right)^2$

20. $4 - \dfrac{12}{x} - \left(2 - \dfrac{3}{x}\right)^2$

21. $\left(1 + \dfrac{3}{5x}\right)^2 - \left(1 - \dfrac{3}{5x}\right)^2$

22. $\left(1 + \dfrac{2}{3x}\right)^2 - \left(1 - \dfrac{2}{3x}\right)^2$

23. $\dfrac{\dfrac{12x^2}{5y}}{\dfrac{16x^2}{15y^2}}$ **24.** $\dfrac{\dfrac{18a^3}{25x}}{\dfrac{24a^3}{35x^2}}$

25. $\dfrac{\dfrac{w-z}{x^2y^2}}{\dfrac{w^2-z^2}{2xy}}$ **26.** $\dfrac{\dfrac{a^2-b^2}{6a^2b^3}}{\dfrac{a+b}{9a^3b}}$

27. $\dfrac{\dfrac{x^2-9x+14}{34x^4}}{\dfrac{5x^2-20}{17x^5}}$ **28.** $\dfrac{\dfrac{5s^2+5st+5t^2}{8s^3-8t^3}}{\dfrac{7s^2+28st+7t^2}{14s^2-14t^2}}$

29. $\dfrac{2-\dfrac{1}{x}}{4-\dfrac{1}{x^2}}$ **30.** $\dfrac{\dfrac{1}{x^2}-49}{7-\dfrac{1}{x}}$

31. $\dfrac{vw}{\dfrac{1}{v}+\dfrac{1}{w}}$ **32.** $\dfrac{\dfrac{1}{v}-\dfrac{1}{w}}{vw}$

33. $\dfrac{\dfrac{3}{x^2}-\dfrac{3}{x}+3}{18+\dfrac{18}{x^3}}$ **34.** $\dfrac{\dfrac{1}{x}-\dfrac{8}{x^2}+\dfrac{15}{x^3}}{1-\dfrac{5}{x}}$

35. $\dfrac{3+\dfrac{9}{x}}{\dfrac{15}{x^3}+\dfrac{8}{x^2}+\dfrac{1}{x}}$ **36.** $\dfrac{x+2-\dfrac{6}{2x+3}}{x+\dfrac{8x}{2x-1}}$

37. $\dfrac{\dfrac{w-a}{w+a}-\dfrac{w+a}{w-a}}{\dfrac{w^2+a^2}{w^2-a^2}}$ **38.** $\dfrac{w-3+\dfrac{2}{w-6}}{w-2-\dfrac{22}{w+7}}$

39. $\dfrac{\dfrac{12}{v^2}+\dfrac{1}{v}-1}{\dfrac{24}{v^2}-\dfrac{2}{v}-1}$ **40.** $\dfrac{\dfrac{12}{v^2}-\dfrac{1}{v}-1}{\dfrac{8}{v^2}+\dfrac{6}{v}+1}$

Estimation Skills and Calculator Skills
In Exercises 41–44 mentally estimate the value of
each expression and then use a calculator to
approximate each value to the nearest thousandth.

PROBLEM	MENTAL ESTIMATE	CALCULATOR APPROXIMATION
41. $\dfrac{\frac{211}{429}}{\frac{107}{325}}$		
42. $\dfrac{\frac{211}{107}}{\frac{899}{298}}$		
43. $\dfrac{\frac{401}{99}}{\frac{499}{101}}$		
44. $\dfrac{\frac{1022}{350}}{\frac{1017}{204}}$		

In Exercises 45–60 simplify each expression. Assume that
the variables are restricted to values that prevent division
by 0.

45. $\dfrac{a - b}{ab} + \dfrac{a^3 + b^3}{a^3 b^3} \cdot \dfrac{a^2 b^2}{4a^2 - 4ab + 4b^2}$

46. $\dfrac{a + 2}{a + 3} - \dfrac{a^3 - 8}{a^2 - 9} \div \dfrac{a^2 + 2a + 4}{a - 3}$

47. $\dfrac{x^{-2}}{x^{-2} + y^{-2}}$

48. $\dfrac{x^{-2} - y^{-2}}{y^{-2}}$

49. $\dfrac{v^{-1} + w^{-1}}{v^{-1} - w^{-1}}$

50. $\dfrac{m^2 - n^2}{m^{-1} + n^{-1}}$

51. $\dfrac{m^3 - n^3}{m^{-1} - n^{-1}}$

52. $\dfrac{x^3 + y^3}{x^{-1} + y^{-1}}$

53. $\dfrac{a^{-2} - b^{-2}}{a^{-1} - b^{-1}}$

54. $\dfrac{b - a^{-1}}{a - b^{-1}}$

55. $\dfrac{xy^{-1} + x^{-1}y}{x^{-2} + y^{-2}}$

56. $\dfrac{x^{-1}y^{-2} + x^{-2}y^{-1}}{x^{-3} + y^{-3}}$

57. $\dfrac{\dfrac{1}{a} - \dfrac{1}{b}}{\dfrac{ab}{1 - \dfrac{a}{b}}}{1 + \dfrac{a}{b}}$

58. $\dfrac{\dfrac{\frac{1}{x} + \frac{1}{y}}{\frac{1}{x}}}{\dfrac{y}{x} + 1}$

59. $\dfrac{\dfrac{1 - \frac{2}{x}}{1 - \frac{4}{x^2}}}{\dfrac{1 + \frac{1}{x}}{1 + \frac{1}{x^3}}}$

60. $\dfrac{\dfrac{1 + \frac{x}{y}}{1 - \frac{x}{y}}}{\dfrac{x - \frac{y^2}{x}}{x + y}}$

In Exercises 61–64 use the table of values for each
expression to match the expression with its simplified form.

61. $\dfrac{3}{x} + \dfrac{1}{4} \div \dfrac{3}{x} - \dfrac{1}{4}$

62. $\left(\dfrac{3}{x} + \dfrac{1}{4}\right) \div \left(\dfrac{3}{x} - \dfrac{1}{4}\right)$

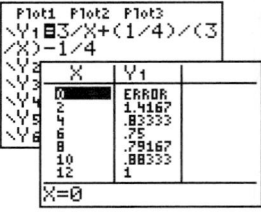 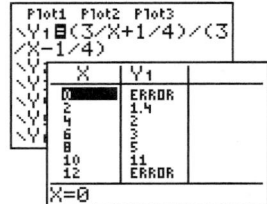

63. $\dfrac{3}{x} + \dfrac{1}{4} \div \left(\dfrac{3}{x} - \dfrac{1}{4}\right)$

64. $\left(\dfrac{3}{x} + \dfrac{1}{4}\right) \div \dfrac{3}{x} - \dfrac{1}{4}$

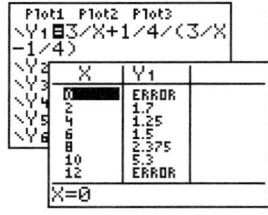

A. $\dfrac{x + 9}{12}$

B. $\dfrac{x^2 - 3x + 36}{12x}$

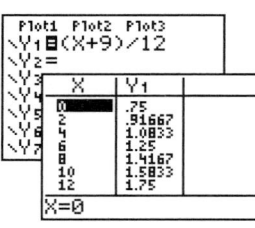 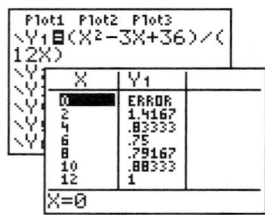

C. $-\dfrac{x + 12}{x - 12}$

D. $-\dfrac{x^2 - 3x + 36}{x(x - 12)}$

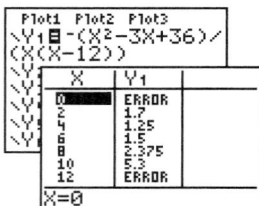

65. Area of a Trapezoid   The formula for the area of a trapezoid is $A = \dfrac{h}{2}(a + b)$. Use this formula to find an expression that represents the area of the trapezoid shown.

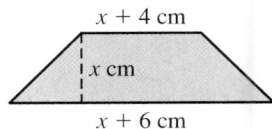

66. Total Area of Two Regions   Determine an expression that represents the total area of the rectangle and triangle shown.

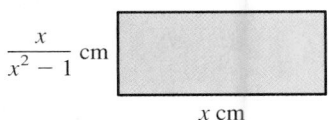

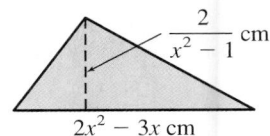

67. Average Cost   The number of units of patio chairs that a factory can make by operating t h/week is given by $N(t) = 5t - 10$. The cost of operating the factory t h/week is given by $C(t) = 8400 - \dfrac{16{,}800}{t}$. Determine an expression for $A(t)$ that represents the average cost per unit when the plant operates t h/week.

68. Average Cost   The number of units of patio tables a factory can make by operating t h/week is given by $N(t) = 2t - 4$. The cost of operating the factory t h/week is given by $C(t) = 5040 - \dfrac{10{,}080}{t}$. Determine an expression for $A(t)$ that represents the average cost per unit when the plant operates t h/week.

69. Lens Formula   The relationship between the focal length f of a lens, the distance of an object d_o from the lens, and the distance of an image d_i from the lens is

$$\frac{1}{f} = \frac{1}{d_o} + \frac{1}{d_i}. \text{ So } f = \frac{1}{\dfrac{1}{d_o} + \dfrac{1}{d_i}}.$$

  a. Give a simplified rational expression for the focal length f.

b. Determine f when the object distance is 20 ft and the image distance is 0.5 ft. Round to the nearest thousandth.

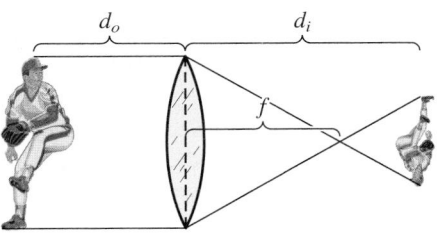

70. Electrical Resistance   The resistance R in a parallel circuit with two individual resistors r_1 and r_2 can be calculated using the formula $\dfrac{1}{R} = \dfrac{1}{r_1} + \dfrac{1}{r_2}$. So

$$R = \frac{1}{\dfrac{1}{r_1} + \dfrac{1}{r_2}}.$$

  a. Give a simplified rational expression for the total resistance R.

  b. Determine R when $r_1 = 8$ ohms and $r_2 = 56$ ohms.

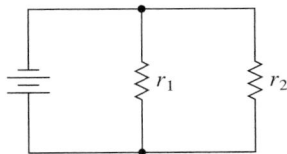

Group Discussion Questions

71. Challenge Question   Simplify each expression. Assume that m and n are natural numbers and that x and y are restricted to values that prevent division by 0.

  a. $\dfrac{x^{-m} + y^{-n}}{x^{-m}y^{-n}}$

  b. $\dfrac{x^{-m} - y^{-n}}{x^m - y^n}$

  c. $\dfrac{x^{-m} - x^{-2m}}{x^{-m} + x^{-2m}}$

  d. $\dfrac{x^{-m} + 3x^{-2m}}{x^{-m} - 2x^{-2m}}$

72. Error Analysis   A student simplified the expression $\dfrac{x^{-1} - y^{-1}}{x^{-2} - y^{-2}}$ as follows: $\dfrac{x^{-1} - y^{-1}}{x^{-2} - y^{-2}} = \dfrac{x^2 - y^2}{x - y}$

$$= \frac{(x + y)(x - y)}{x - y} = x + y. \text{ What is the source of the}$$

error?

Section 9.5 Solving Equations Containing Rational Expressions

Objective: 7. Solve equations containing rational expressions.

Equivalent equations are equations with exactly the same set of solutions. Multiplying both sides of an equation by any nonzero number produces an equivalent equation. Multiplying both sides of an equation by 0 produces $0 = 0$.

In Section 2.4 we solved fractional equations by multiplying both sides of the equation by the LCD. This method produces an equation equivalent to the original one as long as we do not multiply by 0. Multiplying by an expression that is equal to 0 can produce an equation that is *not* equivalent to the original equation and thus can produce an extraneous value. An extraneous value is not a solution of the original equation.

When we solve an equation with a variable in the denominator, we must check each possible solution. We do this to verify that our answer does not include a value that would cause division by 0.

Solving an Equation Containing Rational Expressions

VERBALLY	ALGEBRAIC EXAMPLE
1. Multiply both sides of the equation by the LCD. 2. Solve the resulting equation. 3. Check each solution to determine whether it is an excluded value and therefore extraneous.	Solve: $\dfrac{21}{x} - \dfrac{15}{x} = 2$ The only excluded value is 0. $x\left(\dfrac{21}{x} - \dfrac{15}{x}\right) = x(2)$ $x\left(\dfrac{21}{x}\right) - x\left(\dfrac{15}{x}\right) = 2x$ $21 - 15 = 2x$ $6 = 2x$ $3 = x$ 3 is not an excluded value, so it will check. Check: $\dfrac{21}{3} - \dfrac{15}{3} \overset{?}{=} 2$ $7 - 5 \overset{?}{=} 2$ $2 \overset{?}{=} 2$ checks Answer: $x = 3$

The solution of an equation cannot include any excluded values.

■ EXAMPLE 1   Solving an Equation Containing Rational Expressions

Solve $\dfrac{3}{x-7} + 5 = \dfrac{8}{x-7}$ algebraically and numerically using a table of values.

SOLUTION

$$\frac{3}{x-7} + 5 = \frac{8}{x-7}$$

Note that $x = 7$ is the only excluded value.

$$(x-7)\left(\frac{3}{x-7} + 5\right) = (x-7)\left(\frac{8}{x-7}\right)$$

Multiply both sides of the equation by the nonzero LCD, $x - 7$. The LCD is nonzero since 7 is an excluded value.

$$(x-7)\left(\frac{3}{x-7}\right) + (x-7)(5) = (x-7)\left(\frac{8}{x-7}\right)$$

Use the distributive property to multiply each term by $x - 7$.
Then combine like terms.

$$3 + 5x - 35 = 8$$
$$5x = 40$$
$$x = 8$$

Since $x = 8$ is not an excluded value, this solution will check.

NUMERICALLY

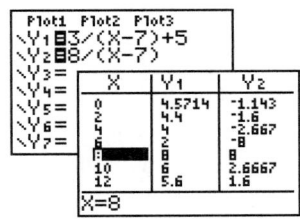

Let y_1 represent the left side of the equation and y_2 represent the right side of the equation.

Caution: A table can locate or confirm a solution of an equation. However, unless we already know how many solutions the equation has, we will not know if the solution shown in the table is the only solution.

For $x = 8$, y_1 and y_2 are equal. This confirms that the algebraic solution checks. *Caution:* Unless we already know how many solutions this equation has, this table would not confirm that $x = 8$ is the only solution.

Answer: $x = 8$

The solution process in Example 2 produces an excluded value for the variable. This value cannot be a solution; it is extraneous.

■ EXAMPLE 2   Solving an Equation with an Extraneous Value

Solve $\dfrac{v}{v-3} = 4 - \dfrac{3}{3-v}$.

SOLUTION

$$\frac{v}{v-3} = 4 - \frac{3}{3-v}$$

Note that the only excluded value is $v = 3$. Noting that the denominators are opposites, change $-\dfrac{3}{3-v}$ to $\dfrac{3}{v-3}$.

$$\frac{v}{v-3} = 4 + \frac{3}{v-3}$$

$$(v-3)\left(\frac{v}{v-3}\right) = (v-3)\left(4 + \frac{3}{v-3}\right)$$

Multiply both sides of the equation by the nonzero LCD, $v - 3$. The LCD is nonzero since 3 is an excluded value.

$$(v-3)\left(\frac{v}{v-3}\right) = (v-3)(4) + (v-3)\left(\frac{3}{v-3}\right)$$

Use the distributive property to multiply each term by $v - 3$.

$$v = 4v - 12 + 3$$
$$-3v = -9$$
$$v = 3$$

Solve the resulting equation for v.

$v = 3$ is the excluded value noted previously, so this value is extraneous.

This value causes division by 0 in the original equation, so there is no solution.

Answer: No solution

■

Solve each equation.

1. $\dfrac{x - 12}{x - 4} = \dfrac{2x}{4 - x} + 2$

2. $\dfrac{2y}{y + 2} = \dfrac{y}{y + 2} - 3$

SELF-CHECK 9.5.1 ANSWERS

1. No solution

2. $y = -\dfrac{3}{2}$

The solution process in Example 3 produces two values. One of these is an excluded value and cannot be a solution, while the other solution will check.

■ **EXAMPLE 3** Solving an Equation with an Extraneous Value

Solve $\dfrac{y + 1}{y + 2} + \dfrac{y}{y - 2} + 1 = \dfrac{8}{y^2 - 4}$.

SOLUTION _____

$$\frac{y + 1}{y + 2} + \frac{y}{y - 2} + 1 = \frac{8}{y^2 - 4}$$

Factor the denominator of the right side of the equation in order to determine the LCD, $(y + 2)(y - 2)$, and the excluded values, -2 and 2.

$$\frac{y + 1}{y + 2} + \frac{y}{y - 2} + 1 = \frac{8}{(y + 2)(y - 2)}$$

$$(y + 2)(y - 2)\left[\frac{y + 1}{y + 2} + \frac{y}{y - 2} + 1\right] =$$

$$\cancel{(y + 2)(y - 2)}\left[\frac{8}{\cancel{(y + 2)(y - 2)}}\right]$$

Multiply both sides of the equation by the nonzero LCD, $(y + 2)(y - 2)$. The LCD is nonzero since -2 and 2 are excluded values.

$$\cancel{(y + 2)}(y - 2)\left(\frac{y + 1}{\cancel{y + 2}}\right) + (y + 2)\cancel{(y - 2)}\left(\frac{y}{\cancel{y - 2}}\right) + (y + 2)(y - 2)(1) = 8$$

Use the distributive property to multiply each term by $(y + 2)(y - 2)$.

$$(y - 2)(y + 1) + (y + 2)(y) + (y + 2)(y - 2)(1) = 8$$

$$y^2 - y - 2 + y^2 + 2y + y^2 - 4 = 8$$

$$3y^2 + y - 14 = 0$$

$$(3y + 7)(y - 2) = 0$$

Write the resulting quadratic equation in standard form and then solve it by factoring.

$$3y + 7 = 0 \quad \text{or} \quad y - 2 = 0$$

$$y = -\frac{7}{3} \qquad\qquad y = 2$$

Because $y = 2$ is an excluded value it cannot be a solution.

The value $y = 2$ causes division by 0 in the original equation and is therefore an extraneous value.

The value $y = -\dfrac{7}{3}$ is not an excluded value. Therefore it should check.

Does $y = -\dfrac{7}{3}$ check?

Answer: $y = -\dfrac{7}{3}$

■

If the steps in the solution of an equation produce a contradiction, then the original equation has no solution. Example 4 shows how to interpret the answer when the steps in the solution of an equation produce an identity.

■ EXAMPLE 4 Solving an Equation That Simplifies to an Identity

Solve $\dfrac{3}{x-4} - \dfrac{2}{x-2} = \dfrac{x+2}{x^2-6x+8}$.

SOLUTION

$$\frac{3}{x-4} - \frac{2}{x-2} = \frac{x+2}{x^2-6x+8}$$

Factor the denominator of the right side of the equation in order to determine the LCD, $(x-4)(x-2)$, and the excluded values, 2 and 4.

$$\frac{3}{x-4} - \frac{2}{x-2} = \frac{x+2}{(x-4)(x-2)}$$

$$(x-4)(x-2)\left[\frac{3}{x-4} - \frac{2}{x-2}\right] = (x-4)(x-2)\left[\frac{x+2}{(x-4)(x-2)}\right]$$

Multiply both sides of the equation by the nonzero LCD, $(x-4)(x-2)$. The LCD is nonzero since 2 and 4 are excluded values. Use the distributive property to multiply each term by $(x-4)(x-2)$.

$$(x-4)(x-2)\left(\frac{3}{x-4}\right) - (x-4)(x-2)\left(\frac{2}{x-2}\right) = x+2$$

$$3(x-2) - 2(x-4) = x+2$$
$$3x - 6 - 2x + 8 = x+2$$
$$x + 2 = x + 2$$
$$0 = 0 \ (\text{an identity})$$

Simplify and then solve the equation.

This simplified equation is an identity. Thus all real numbers are solutions of the original equation except for the values excluded to avoid division by 0.

Answer: All real numbers except 2 and 4

In interval notation the answer is $(-\infty, 2) \cup (2, 4) \cup (4, +\infty)$.

SELF-CHECK 9.5.2 ANSWERS

1. All real numbers except 0 and 1
2. No solution
3. $x = 2$

SELF-CHECK 9.5.2

Solve each equation.

1. $\dfrac{2}{x} + \dfrac{1}{x-1} = \dfrac{3x-2}{x^2-x}$ 2. $\dfrac{2}{x} + \dfrac{1}{x-1} = \dfrac{3x+2}{x^2-x}$ 3. $\dfrac{2}{x} + \dfrac{1}{x-1} = \dfrac{4}{x^2-x}$

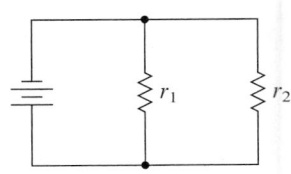

The resistance R in a parallel circuit with two individual resistors r_1 and r_2 can be calculated using the formula $\dfrac{1}{R} = \dfrac{1}{r_1} + \dfrac{1}{r_2}$. We will use this formula to determine appropriate values for r_1 and r_2.

■ **EXAMPLE 5**   Calculating Resistances in a Parallel Circuit

The resistance planned by an electrical technician for a parallel electrical circuit with two resistors is 4 ohms. The resistance of r_2 must be 15 ohms greater than that of r_1. Determine the resistance of each resistor.

SOLUTION

Let r = the resistance in ohms of the first resistor
$r + 15$ = the resistance in ohms of the second resistor

The resistance of the second resistor is 15 ohms greater than that of the first resistor.

$$\frac{1}{r_1} + \frac{1}{r_2} = \frac{1}{R}$$

$$\frac{1}{r} + \frac{1}{r + 15} = \frac{1}{4}$$

Substitute the total resistance of 4 ohms and the identified variables into the resistance formula.

$$4r(r + 15)\left(\frac{1}{r} + \frac{1}{r + 15}\right) = 4r(r + 15)\left(\frac{1}{4}\right)$$

Multiply both sides of the equation by the LCD, $4r(r + 15)$.

$$4r(r + 15)\left(\frac{1}{r}\right) + 4r(r + 15)\left(\frac{1}{r + 15}\right) = r(r + 15)$$

Use the distributive property to multiply each term by $4r(r + 15)$.

$$4(r + 15) + 4r = r^2 + 15r$$
$$4r + 60 + 4r = r^2 + 15r$$
$$0 = r^2 + 7r - 60 \quad \text{or} \quad r^2 + 7r - 60 = 0$$
$$(r - 5)(r + 12) = 0$$
$$r - 5 = 0 \quad \text{or} \quad r + 12 = 0$$
$$r = 5 \qquad\qquad r = -12$$
$$r + 15 = 20$$

Simplify and write this quadratic equation in standard form. Then solve this equation by factoring.

The value of 5 ohms seems reasonable for r (we will check it). However, the resistance of -12 ohms is not reasonable because resistance cannot be negative.

Check:

$$\frac{1}{5} + \frac{1}{20} \overset{?}{=} \frac{1}{4}$$
$$0.25 \overset{?}{=} 0.25 \text{ checks.}$$

Answer: The resistance of the first resistor is 5 ohms, and the resistance of the second resistor is 20 ohms.

In Section 2.7 we practiced rewriting formulas to solve for a specified variable. Example 6 involves rewriting a formula that includes a rational expression. To solve for b, we must isolate this variable in the numerator on one side of the equation.

■ EXAMPLE 6   Solving for a Specified Variable

Solve each equation for b.

(a) $\dfrac{2}{a} = \dfrac{3}{b} - \dfrac{1}{ab}$ (b) $\dfrac{1}{a} + \dfrac{1}{c} = \dfrac{1}{b}$

SOLUTIONS

(a) $\dfrac{2}{a} = \dfrac{3}{b} - \dfrac{1}{ab}$

$ab\left(\dfrac{2}{a}\right) = ab\left(\dfrac{3}{b} - \dfrac{1}{ab}\right)$ To get b in the numerator, multiply both sides of the equation by the LCD, ab.

$ab\left(\dfrac{2}{a}\right) = ab\left(\dfrac{3}{b}\right) - ab\left(\dfrac{1}{ab}\right)$ Use the distributive property to multiply each term by ab.

$2b = 3a - 1$ Simplify and then divide both sides of the equation by 2 to isolate b on the left side with a coefficient of 1.

Answer: $b = \dfrac{3a - 1}{2}$

(b) $\dfrac{1}{a} + \dfrac{1}{c} = \dfrac{1}{b}$

$abc\left(\dfrac{1}{a} + \dfrac{1}{c}\right) = abc\left(\dfrac{1}{b}\right)$ To get b in the numerator, multiply both sides of the equation by the LCD, abc.

$abc\left(\dfrac{1}{a}\right) + abc\left(\dfrac{1}{c}\right) = abc\left(\dfrac{1}{b}\right)$ Use the distributive property to multiply each term by abc.

$bc + ab = ac$ Simplify and then use the distributive property to factor b out of each term on the left side of the equation.

$b(c + a) = ac$

$\dfrac{b(c + a)}{c + a} = \dfrac{ac}{c + a}$ Divide both sides of the equation by $c + a$ to isolate b on the left side of the equation.

Answer: $b = \dfrac{ac}{a + c}$ Simplify the left side of the equation and on the right side write the terms in the denominator in alphabetical order.   ■

SELF-CHECK 9.5.3 ANSWERS

1. $x = \dfrac{vw}{z}$

2. $x = \dfrac{wy}{v}$

3. $x = \dfrac{2v - 3yz}{z}$

4. $x = \dfrac{wy}{w - y}$

SELF-CHECK 9.5.3

Solve each equation for x.

1. $z = \dfrac{vw}{x}$

2. $\dfrac{w}{x} = \dfrac{v}{y}$

3. $\dfrac{2v}{x + 3y} = z$

4. $\dfrac{1}{x} + \dfrac{1}{w} = \dfrac{1}{y}$

USING THE LANGUAGE AND SYMBOLISM OF MATHEMATICS 9.5

1. Equivalent equations have exactly the _____ set of solutions.
2. To produce equivalent equations we can multiply both sides of the equation by the same number as long as we do not multiply by _____.
3. To solve an equation containing rational expressions:
   a. Multiply both sides of the equation by the _____.

b. _____ the resulting equation.
   c. Check each solution to determine whether it is an excluded value and therefore _____.
4. An excluded value for an equation containing rational expressions is one that causes division by _____.

EXERCISES 9.5

In Exercises 1 and 2 determine the values excluded from the domain of the variable because they would cause division by 0.

1. a. $\dfrac{3}{m-3} + 5 = \dfrac{2}{m-2}$

b. $\dfrac{3y}{(2y+3)(3y-2)} - \dfrac{7}{2y+3} = \dfrac{y-1}{3y-2}$

c. $\dfrac{4y-5}{2y^2+5y-3} = \dfrac{5y-4}{6y^2-y-1}$

2. a. $\dfrac{2}{n+2} + 3 = \dfrac{2}{n-5}$

b. $\dfrac{y+2}{4y^2-13y+3} = \dfrac{y}{4y-1} - \dfrac{5}{y-3}$

c. $\dfrac{7y}{6y^2-5y-6} = \dfrac{9y-1}{6y^2-11y-10}$

In Exercises 3–6 each equation has exactly one solution. Use the given table to determine this solution.

3. $\dfrac{x}{x+3} = \dfrac{1}{x+3} + \dfrac{1}{3}$

4. $\dfrac{3}{x-1} = \dfrac{1}{x-1} - 1$

5. $\dfrac{x}{x-3} + \dfrac{x-1}{x+3} = \dfrac{6x}{x^2-9}$

6. $\dfrac{x-20}{x^2-5x-6} + \dfrac{12}{2x^2-13x+6} = \dfrac{7x-2}{2x^2+x-1}$

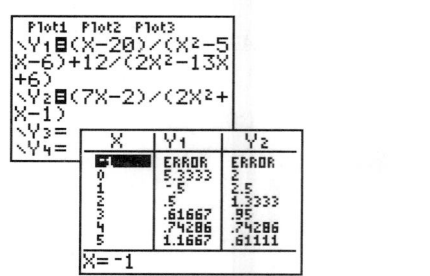

In Exercises 7–40 solve each equation.

7. $\dfrac{3}{z-1} + 2 = \dfrac{5}{z-1}$

8. $\dfrac{5}{z+3} - 2 = \dfrac{4}{z+3}$

9. $\dfrac{6w-1}{2w-1} - 5 = \dfrac{2w-3}{1-2w}$

10. $\dfrac{m}{m-2} - 5 = \dfrac{2}{m-2}$

11. $\dfrac{-3}{p+2} = \dfrac{-8}{p-3}$

12. $\dfrac{7}{p-4} = \dfrac{2}{p+1}$

13. $\dfrac{7}{3n-1} = \dfrac{2}{n+2}$

14. $\dfrac{10}{r-3} = \dfrac{34}{2r+1}$

15. $\dfrac{4}{k+2} = \dfrac{1}{3k+6} + \dfrac{11}{9}$

16. $\dfrac{5}{4k+1} = \dfrac{3}{8k+2} + 1$

17. $\dfrac{3y}{(y+4)(y-2)} = \dfrac{5}{y-2} + \dfrac{2}{y+4}$

18. $\dfrac{y^2 + 18}{(2y + 3)(y - 3)} = \dfrac{5}{y - 3} - \dfrac{1}{2y + 3}$

19. $1 - \dfrac{14}{y^2 + 4y + 4} = \dfrac{7y}{y^2 + 4y + 4}$

20. $1 + \dfrac{6w}{w^2 - 6w + 9} = \dfrac{18}{w^2 - 6w + 9}$

21. $\dfrac{4}{x - 5} + \dfrac{5}{x - 2} = \dfrac{x + 6}{3x - 6}$

22. $\dfrac{w + 1}{w} + \dfrac{14}{w - 7} = \dfrac{3w - 7}{w^2 - 7w}$

23. $\dfrac{1}{(t - 1)^2} - 3 = \dfrac{2}{1 - t}$

24. $\dfrac{8}{(t + 4)^2} + \dfrac{2}{t + 4} = 3$

25. $\dfrac{z}{(z - 2)(z + 1)} - \dfrac{z}{(z + 1)(z + 3)} = \dfrac{3z}{(z + 3)(z - 2)}$

26. $\dfrac{4}{(z + 1)(z - 1)} = \dfrac{6 - z}{(z + 1)(z - 2)} - \dfrac{8}{(z - 2)(z - 1)}$

27. $\dfrac{2v - 5}{3v^2 - v - 14} + \dfrac{7}{3v - 7} = \dfrac{8}{v + 2}$

28. $\dfrac{v^2 - 2v + 2}{6v^2 + 23v - 4} + \dfrac{2}{v + 4} = \dfrac{v}{6v - 1}$

29. $\dfrac{m + 4}{6m^2 + 5m - 6} = \dfrac{m}{3m - 2} - \dfrac{m}{2m + 3}$

30. $\dfrac{2m + 17}{2m^2 + 11m + 14} + \dfrac{m - 2}{m + 2} = \dfrac{m - 3}{2m + 7}$

31. $\dfrac{x + 1}{3x^2 - 4x + 1} - \dfrac{x + 1}{2x^2 + x - 3} = \dfrac{2}{6x^2 + 7x - 3}$

32. $\dfrac{4y}{6y^2 - 7y - 3} + \dfrac{2}{3y^2 - 2y - 1} = \dfrac{y + 2}{2y^2 - 5y + 3}$

33. $\dfrac{2}{m + 2} - \dfrac{1}{m + 1} = \dfrac{1}{m}$

34. $\dfrac{m^2 - 1}{2m + 1} = \dfrac{1 - m}{3}$

35. $\dfrac{z - 2}{2z^2 - 5z + 3} + \dfrac{3}{3z^2 - 2z - 1} = \dfrac{3z}{6z^2 - 7z - 3}$

36. $\dfrac{n - 3}{n^2 + 5n + 4} + \dfrac{n - 2}{n^2 + 3n + 2} = \dfrac{n^2 - 12}{(n + 1)(n + 2)(n + 4)}$

37. $\dfrac{1}{n^2 - 5n + 6} - \dfrac{1}{n^2 - n - 2} + \dfrac{3}{n^2 - 2n - 3} = 0$

38. $\dfrac{12}{2w^2 - 13w + 6} - \dfrac{7w - 2}{2w^2 + w - 1} + \dfrac{w - 20}{w^2 - 5w - 6} = 0$

39. $\dfrac{x^2}{x^2 - x - 2} = \dfrac{2x}{x^2 + x - 6}$

40. $\dfrac{6v + 6}{2v^2 + 7v - 4} = \dfrac{3v}{v^2 + 2v - 8} - \dfrac{5v - 7}{2v^2 - 5v + 2}$

In Exercises 41–46 simplify the expression in part **a** and solve the equation in part **b**.

Simplify	Solve
41. a. $\dfrac{1}{p - 1} - \dfrac{3}{p + 1}$	**b.** $\dfrac{1}{p - 1} = \dfrac{3}{p + 1}$
42. a. $\dfrac{m - 1}{m + 1} - \dfrac{m - 3}{m - 2}$	**b.** $\dfrac{m - 1}{m + 1} = \dfrac{m - 3}{m - 2}$
43. a. $\dfrac{x - 1}{x + 1} - 1 - \dfrac{x - 6}{x - 2}$	**b.** $\dfrac{x - 1}{x + 1} - 1 = \dfrac{x - 6}{x - 2}$
44. a. $\dfrac{x^2}{x + 2} + \dfrac{x - 1}{x - 3}$	**b.** $\dfrac{x^2}{x + 2} + \dfrac{x - 1}{x - 3} = 0$

45. a. $\dfrac{2x - 8}{6x^2 + x - 2} - \dfrac{4}{3x + 2} + \dfrac{2}{2x - 1}$

b. $\dfrac{2x - 8}{6x^2 + x - 2} = \dfrac{4}{3x + 2} - \dfrac{2}{2x - 1}$

46. a. $\dfrac{w^2 - w - 3}{2w^2 - 9w + 9} + \dfrac{1}{3 - w} - \dfrac{w}{2w - 3}$

b. $\dfrac{w^2 - w - 3}{2w^2 - 9w + 9} + \dfrac{1}{3 - w} = \dfrac{w}{2w - 3}$

47. Resistance in a Parallel Circuit   The resistance planned by an electrical technician for a parallel electrical circuit with two resistors is 5 ohms. The resistance of r_2 must be 5 times the resistance of r_1. Determine the resistance of each resistor. The resistance R in a parallel circuit with two individual resistors r_1 and r_2 can be calculated using the formula $\dfrac{1}{R} = \dfrac{1}{r_1} + \dfrac{1}{r_2}$.

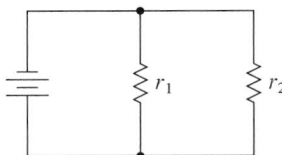

48. Resistance in a Parallel Circuit   The resistance planned by an electrical technician for a parallel electrical circuit with two resistors is 3 ohms. The resistance of r_2 must be 8 ohms greater than that of r_1. Determine the resistance of each resistor. (See Exercise 47.)

In Exercises 49–60 solve each equation for the variable specified.

49. $\dfrac{a}{b} = \dfrac{c}{d}$ for b

50. $a = \dfrac{bc}{d}$ for d

51. $\dfrac{a}{b - 1} = \dfrac{c}{d + 1}$ for b

52. $\dfrac{a}{b - 1} = \dfrac{c}{d + 1}$ for d

53. $I = \dfrac{k}{d}$ for d

54. $F = \dfrac{k}{r}$ for r

55. $\dfrac{1}{R} = \dfrac{1}{r_1} + \dfrac{1}{r_2}$ for R

56. $\dfrac{1}{R} = \dfrac{1}{r_1} + \dfrac{1}{r_2}$ for r_1

57. $h = \dfrac{2A}{B + b}$ for B

58. $h = \dfrac{2A}{B + b}$ for b

59. $\dfrac{1}{x} = \dfrac{1}{y} - \dfrac{1}{z}$ for z

60. $I = \dfrac{E}{r_1 + r_2}$ for r_1

In Exercises 61–68 solve each equation.

61. a. $\dfrac{5v + 4}{2v^2 + v - 15} + \dfrac{3}{5 - 2v} - \dfrac{1}{v + 3} = 0$

b. $\dfrac{5v + 3}{2v^2 + v - 15} + \dfrac{3}{5 - 2v} - \dfrac{1}{v + 3} = 0$

62. a. $\dfrac{w}{3w^2 - 11w + 10} + \dfrac{5}{3w - 5} = \dfrac{2}{w - 2}$

b. $\dfrac{w + 1}{3w^2 - 11w + 10} + \dfrac{5}{3w - 5} = \dfrac{2}{w - 2}$

63. $\dfrac{x^2}{2x^2 + 9x - 5} + \dfrac{2x}{x^2 + 2x - 15} = \dfrac{4x}{(2x - 1)(x + 5)(x - 3)}$

64. $\dfrac{x}{x - 3} + \dfrac{1}{x - 2} - \dfrac{1}{x + 2} = \dfrac{x - 12}{x^3 - 3x^2 - 4x + 12}$

65. $\dfrac{z - 2}{4z^2 - 29z + 30} - \dfrac{z + 2}{5z^2 - 27z - 18} = \dfrac{z + 1}{20z^2 - 13z - 15}$

66. $\dfrac{3n - 7}{n^2 - 5n + 6} + \dfrac{2n + 8}{9 - n^2} - \dfrac{n + 2}{n^2 + n - 6} = 0$

67. $\dfrac{z - 1}{z^2 - 2z - 3} + \dfrac{z + 1}{z^2 - 4z + 3} = \dfrac{z + 8}{z^2 - 1} + \dfrac{20}{(z - 1)(z + 1)(z - 3)}$

68. $\dfrac{3w - 8}{w^2 - 5w + 6} + \dfrac{w + 2}{w^2 - 6w + 8} = \dfrac{5 - 2w}{w^2 - 7w + 12} + \dfrac{12}{(w - 4)(w - 3)(w - 2)}$

Group Discussion Questions

69. Error Analysis   Determine the error in the following argument:

Let $x = 1$
Then $3x = 2x + 1$
$3x - 3 = 2x - 2$
$3(x - 1) = 2(x - 1)$
$\dfrac{3(x - 1)}{x - 1} = \dfrac{2(x - 1)}{x - 1}$
Thus $3 = 2$

70. Challenge Question
a. Complete this equation so that the solution is all real numbers except for -1 and 2.

$$\dfrac{5}{x - 2} + \dfrac{2}{x + 1} = \dfrac{?}{x^2 - x - 2}$$

b. Complete this equation so that the solution is all real numbers except $-\dfrac{2}{3}$ and $\dfrac{3}{2}$.

$$\dfrac{?}{6x^2 - 5x - 6} = \dfrac{4}{3x + 2} - \dfrac{1}{2x - 3}$$

71. Challenge Question   Complete each equation so that the equation is a contradiction with no solution.

a. $\dfrac{5}{x - 2} + \dfrac{2}{x + 1} = \dfrac{7x + ?}{x^2 - x - 2}$

b. $\dfrac{5x + ?}{6x^2 - 5x - 6} = \dfrac{4}{3x + 2} - \dfrac{1}{2x - 3}$

Section 9.6 Applications Yielding Equations with Fractions

Objectives: **8.** Solve problems involving inverse variation.
   9. Solve applied problems that yield equations with fractions.

The problem-solving skills that we have developed in earlier chapters are now used to solve applications that yield equations containing rational expressions. The problems selected here illustrate some of the variety that you may encounter outside the classroom. These problems also include some problems that are skill builders to gradually develop your expertise for other applications.

In Section 2.6 we examined problems that involved direct variation. If y varies directly as x with a constant of variation k, then $y = kx$. We now examine problems involving inverse variation.

Inverse Variation

If x and y are real variables, k is a real constant, $x \neq 0$, and $k \neq 0$, then:

VERBALLY	ALGEBRAICALLY	NUMERICAL EXAMPLE	GRAPHICAL EXAMPLE FOR $x > 0$
y varies inversely as x with a constant of variation k.	$y = \dfrac{k}{x}$ *Example:* $y = \dfrac{120}{x}$	$\begin{array}{c\|c} x & y = \dfrac{120}{x} \\ \hline 1 & 120 \\ 2 & 60 \\ 3 & 40 \\ 4 & 30 \\ 5 & 24 \end{array}$	$y = \dfrac{120}{x}$, for $x > 0$

◼ EXAMPLE 1 Translating Statements of Variation

SOLUTIONS

(a) Translate $V = \dfrac{k}{P}$ into a verbal statement of variation.

The volume V of a fixed amount of gas (at a given temperature) varies inversely with the pressure P exerted on the gas.

This is a statement of Boyle's law from chemistry.

(b) Translate $T = \dfrac{D}{R}$ into a verbal statement of variation.

The time T for a trip varies directly as the distance D and inversely as the rate of travel R.

This equation can also be written as $D = RT$.

(c) The average overhead cost per item $A(t)$ produced varies inversely with the number of items produced $N(t)$. Translate this statement of variation into an algebraic equation.

$$A(t) = \dfrac{C}{N(t)}$$

The constant of variation, C, is the fixed overhead cost.

$N(t)$ is the number of units produced.

$A(t)$ is the average overhead cost per unit. ◼

Example 2 works through a numeric word problem involving direct and inverse variation. This allows us to compare these two types of variation before the more involved application shown in Example 3.

■ EXAMPLE 2   Solving Problems with Direct and Inverse Variation

The variable y equals 48 when x equals 4. Find y when $x = 6$ if:
(a) y varies directly as x **(b)** y varies inversely as x

SOLUTIONS

(a) VERBALLY y varies directly as x

ALGEBRAICALLY	$y = kx$	Translate the word equation into algebraic form using k for the constant of variation. Substitute in the given values of 48 for y and 4 for x. Then solve for the constant k.
	$48 = k(4)$	
	$12 = k$	

$$y = 12x$$
$$y = 12(6)$$
$$y = 72$$

Use this constant to write the equation of variation. Then substitute in the value of 6 for x and solve for y.

(b) VERBALLY y varies inversely as x

ALGEBRAICALLY $y = \dfrac{k}{x}$

Translate the word equation into algebraic form using k for the constant of variation.

$$48 = \frac{k}{4}$$

Substitute in the given values of 48 for y and 4 for x.

$$192 = k$$

Then solve for the constant k.

$$y = \frac{192}{x}$$

Use this constant to write the equation of variation.

$$y = \frac{192}{6}$$

Then substitute in the value of 6 for x and solve for y.

$$y = 32$$

SELF-CHECK 9.6.1 ANSWERS

1. $y = 4$ 2. $y = 9$

SELF-CHECK 9.6.1

1. y varies directly as x. $y = 6$ when $x = 3$. Find y when x is 2.
2. y varies inversely as x. $y = 6$ when $x = 3$. Find y when $x = 2$.

■ **EXAMPLE 3**   Using Inverse Variation

The illumination provided by a car's headlight varies inversely as the square of the distance from the headlight. If a headlight produces 15 footcandles at a distance of 20 ft, what will be the illumination at 40 ft?

SOLUTION _____

Let I = the illumination of the headlight in footcandles

d = the distance in feet of an object from the headlight

k = the constant of variation for this headlight

Identify each unknown with a variable. Be sure to include units in this identification.

VERBALLY   The illumination varies inversely as the square of the distance.

Precisely state an equation that can be translated into algebraic form.

ALGEBRAICALLY

$$I = \frac{k}{d^2}$$

Translate the word equation into algebraic form using the identified variables.

$$15 = \frac{k}{20^2}$$

Substitute in the given values for I and d and solve for the constant k.

$$15 = \frac{k}{400}$$

$$15(400) = k$$

$$k = 6000$$

$$I = \frac{6000}{d^2}$$

Using the constant of variation for this headlight, find I when the distance is 40 ft.

$$I = \frac{6000}{40^2}$$

$$I = 3.75$$

NUMERICALLY

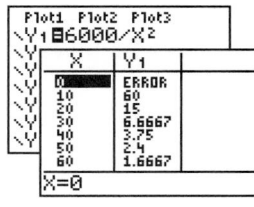

GRAPHICALLY

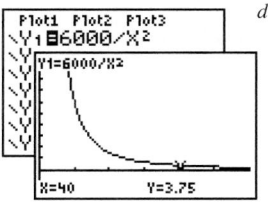

[0, 60, 10] by [−20, 100, 10]

For a constant of variation of 6000, both the table and the graph confirm $I = 3.75$ when $d = 40$.

Answer: The illumination at 40 ft will be 3.75 footcandles.   ■

Variables related by direct and inverse variation have quite distinctive behaviors. If x and y vary directly, then increasing magnitudes of x produce increasing magnitudes of y. If x and y vary inversely, then increasing magnitudes of x produce decreasing magnitudes of y. This is illustrated in the following box.

Numbers with a large magnitude or absolute value are located relatively far from the origin, while numbers with a small magnitude or absolute value are located relatively close to the origin.

Comparison of Direct and Inverse Variation

If x and y are real variables, $x \neq 0$, and k is a real constant:

VERBALLY	ALGEBRAICALLY	NUMERICAL EXAMPLE	GRAPHICAL EXAMPLE
y varies directly as x.	$y = kx$ Example: $y = 2x$		 $[-10, 10, 1]$ by $[-10, 10, 1]$ As the magnitude of x increases, the magnitude of y increases linearly.

VERBALLY	ALGEBRAICALLY	NUMERICAL EXAMPLE	GRAPHICAL EXAMPLE
y varies inversely as x.	$y = \dfrac{k}{x}$ Example: $y = \dfrac{12}{x}$		 $[-10, 10, 1]$ by $[-10, 10, 1]$ As the magnitude of x increases, the magnitude of y decreases.

It is important to keep this distinction between direct and inverse variation in mind as you analyze real-world variables. This is illustrated by Example 4.

■ EXAMPLE 4 Selecting Direct or Inverse Variation

Determine for each relationship whether the variation between the two variables is more likely to be direct variation or inverse variation.

SOLUTIONS _____

(a) The average weight of a group of people of the same age varies _____ as the height of this group of people.

Directly — Taller people are overall bigger people and also weigh more. Thus, these variables are more likely to vary directly.

(b) The average vertical jumping height of a group of adults of approximately the same size varies _____ as the age of this group of people.

Inversely — After maturity and past age 25, the athletic ability of people declines as they age. Their jumping height will decrease as their age increases. Thus, these variables are more likely to vary inversely. ■

The formula $D = RT$ states that the distance traveled equals the product of the rate of travel times the time traveled.

Many problems involving rates and times involve inverse variation. For example, the formula for distance, $D = RT$, can be rewritten as $R = \dfrac{D}{T}$. This form states that the rate of travel varies directly as the distance traveled and inversely as the time traveled. As the time you have available for a trip increases, you can travel at a slower rate.

The formula $W = RT$ states that the work done is the product of the rate of work times the time worked.

Likewise, the formula for work, $W = RT$, can be rewritten as $R = \dfrac{W}{T}$. For a fixed amount of work the rate of work varies inversely as the time required to do this work. For less time you must work faster—for more time you can work slower.

Example 5 involves shared work: two people working together to complete a job. This problem assumes that when the painters work together there is no gain or loss of efficiency of either painter. Each will continue to work at the same rate as when they were working alone. Another key point in this example is that information about the time it takes to do a job is also information about the rate of work for this job. This is true because of the inverse relationship between the time and the rate of work.

■ EXAMPLE 5   Determining the Time It Takes to Paint a Sign

Working alone, painter A can paint a sign in 6 h less time than it would take painter B working alone. When A and B work together, the job takes only 4 h. How many hours would it take each painter working alone to paint the sign?

This problem assumes that when the painters work together there is no gain or loss of efficiency for either painter.

SOLUTION

Let $t =$ time in hours for painter B to paint the sign working alone
$t - 6 =$ time in hours for painter A to paint the sign working alone

$\dfrac{1}{t - 6} =$ rate of work for painter A

$\dfrac{1}{t} =$ rate of work for painter B

First identify the time it takes for each painter to paint the sign working alone.

The rate of work varies inversely as the time; that is, $R = \dfrac{W}{T}$. The work equals 1 sign painted; thus this equation becomes $R = \dfrac{1}{T}$.

VERBALLY

Work A does		Work B does		Total work of painting 1 sign
	+		=	

(Rate of A)(Time of A) + (Rate of B)(Time of B) = Total work

The word equation is based on the mixture principle when the painters share the work and work together.

ALGEBRAICALLY

$$\left(\frac{1}{t-6}\right)(4) + \left(\frac{1}{t}\right)(4) = 1$$

Substitute the rates of each painter and the time of 4 h into the work equation $W = RT$. The total work done is 1 painted sign.

$$\frac{4}{t-6} + \frac{4}{t} = 1$$

$$t(t-6)\frac{4}{t-6} + t(t-6)\frac{4}{t} = t(t-6)(1)$$

Multiply both sides of the equation by the LCD, $t(t-6)$.

$$4t + 4(t-6) = t^2 - 6t$$
$$4t + 4t - 24 = t^2 - 6t$$
$$8t - 24 = t^2 - 6t$$
$$0 = t^2 - 14t + 24$$
$$0 = (t-12)(t-2)$$

Simplify this quadratic equation and write it in standard form.

$$t - 12 = 0 \quad \text{or} \quad t - 2 = 0$$
$$t = 12 \qquad\qquad t = 2$$
$$t - 6 = 6 \qquad\quad t - 6 = -4$$

Then solve this equation by factoring. While 2 h may seem meaningful for painter B, the value of -4 h is not meaningful for painter A.

Check: The rates for each painter are $\frac{1}{6}$ sign/h

This value is not appropriate.

for A and $\frac{1}{12}$ sign/h for B. In 4 h they paint a

Answer: Working alone, painter A could paint the sign in 6 h and painter B could paint the sign in 12 h.

total of $\frac{1}{6}(4) + \frac{1}{12}(4) = \frac{4}{6} + \frac{4}{12}$

$$= \frac{2}{3} + \frac{1}{3}$$

$$= 1 \text{ sign.}$$

SELF-CHECK 9.6.2

Working alone a welder can weld a metal framework 5 h faster than an apprentice can. How long will it take the apprentice to do the job alone if the welder and the apprentice can do it in 6 h when they share the work?

SELF-CHECK 9.6.2 ANSWER

The apprentice can do the job alone in 15 h. (No other values check in this problem.)

Example 6 has many similarities to the $W = RT$ problem in Example 5. However, in this example the formula is $D = RT$ and the rate involved is the rate of travel instead of the rate of work.

■ EXAMPLE 6 Determining the Speed of a Boat in Still Water

Two tugboats that have the same speed in still water travel in opposite directions in a river with a constant current of 8 mi/h. The tugboats departed at the same time from a refueling station; and after a period of time, one has traveled 30 mi downstream and the other has traveled 6 mi upstream. Determine the rate of each boat in still water.

SOLUTION

Let r = rate of each boat in still water in miles per hour
    $r + 8$ = rate of boat going downstream in miles per hour
    $r - 8$ = rate of boat going upstream in miles per hour

$$\frac{30}{r+8} = \text{time of boat going downstream in hours}$$

$$\frac{6}{r-8} = \text{time of boat going upstream in hours}$$

First identify the quantities being sought with a variable—the rates of the two boats. It is wise to include the units of measurement when you identify your variables. This is especially true if there are different units within the same problem (e.g., both minutes and hours).

The formula $D = RT$ also can be written as $T = \dfrac{D}{R}$. Substitute the given distances and the rates $r + 8$ and $r - 8$ into this formula to identify the time for each boat.

VERBALLY

| Time of boat going downstream | = | Time of boat going upstream |

Although the times are unknown, we do know that they are the same because the boats left at the same time.

ALGEBRAICALLY

$$\frac{30}{r + 8} = \frac{6}{r - 8}$$

Substitute the identified times into the word equation.

$$(r + 8)(r - 8)\frac{30}{r + 8} = (r + 8)(r - 8)\frac{6}{r - 8}$$

Multiply both sides of the equation by the LCD, $(r + 8)(r - 8)$.

$$30(r - 8) = 6(r + 8)$$
$$30r - 240 = 6r + 48$$
$$24r = 288$$
$$r = 12$$

Then simplify this equation and solve for r.

Check: $r + 8 = 20$
$r - 8 = 4$
Going downstream at 20 mi/h, the tugboat will take $\frac{30}{20} = 1.5$ h to

go 30 mi. Going upstream at 4 mi/h, the tugboat will take

$\frac{6}{4} = 1.5$ h to go 6 mi. These times are equal.

Answer: The rate of each tugboat in still water is 12 mi/h.   ∎

SELF-CHECK 9.6.3

Rework Example 6 assuming that the rate of the current is 5 mi/h.

The formula $I = PRT$, states that the interest on an investment equals the product of the principal invested times the interest rate times the time of the investment.

Example 7 also involves a rate problem. For this application the rate is the interest rate charged for borrowed money. The formula $I = PRT$ can be rewritten as $P = \frac{I}{RT}$. For a fixed time of 1 year, this formula can be simplified to $P = \frac{I}{R}$ as is done in Example 7. Recall that P represents the principal, R represents the interest rate, T represents the time, and I represents the interest.

■ EXAMPLE 7   Determining Two Interest Rates

One question to ask when trying to form the word equation for a problem is "What things are equal or the same?"

A local bank pays interest on both checking accounts and savings accounts. The interest rate for a savings account is 1% higher than that for a checking account. A customer calculates that a deposit in a checking account would earn yearly interest of $80, whereas this same deposit would earn yearly interest of $100 in a savings account. What is the interest rate on each account?

SOLUTION

Let
r = rate of interest on the checking account
$r + 0.01$ = rate of interest on the savings account
$\dfrac{80}{r}$ = principal in checking account

$\dfrac{100}{r + 0.01}$ = principal in savings account

First identify the quantities being sought with a variable—the interest rates of the two accounts.

$I = PRT$; for $T = 1$ we can write $P = \dfrac{I}{R}$. Substitute the given interest for each account and the rates identified to label the principal in each account.

SELF-CHECK 9.6.3 ANSWER

The rate of each tugboat in still water is 7.5 mi/h.

VERBALLY

Principal in a checking account	=	Principal in a savings account

The principal is the same for both accounts.

ALGEBRAICALLY

$$\frac{80}{r} = \frac{100}{r + 0.01}$$

Substitute the identified values into the word equation.

$$r(r + 0.01)\left(\frac{80}{r}\right) = r(r + 0.01)\left(\frac{100}{r + 0.01}\right)$$

Multiply both sides of the equation by the LCD, $r(r + 0.01)$.

$$80(r + 0.01) = 100r$$
$$80r + 0.8 = 100r$$
$$0.8 = 20r$$
$$0.04 = r$$
$$r = 0.04$$
$$r + 0.01 = 0.05$$

Simplify, and then solve for r.

The checking account rate is 4%.
The savings account rate is 5%.

Answer: The interest rate on the checking account is 4% and on the savings account is 5%.

Do these values check?

SELF-CHECK 9.6.4

Rework Example 7 assuming that the interest rate for the savings account is 0.5% higher than that for the checking account.

The next problem that we examine is a numeric word problem involving reciprocals. These problems do not have the context of a real-world problem, but exercises like this can help us to develop our problem-solving skills.

■ EXAMPLE 8   Solving a Numeric Word Problem

The sum of the reciprocals of two consecutive even integers is $\frac{13}{84}$. Find these integers.

SOLUTION

Let n = smaller integer

$\dfrac{1}{n}$ = reciprocal of the smaller integer

Identify each number and its reciprocal.

$n + 2$ = larger integer

$\dfrac{1}{n + 2}$ = reciprocal of the larger integer

Consecutive even integers differ by 2.

VERBALLY

Reciprocal of first integer	+	Reciprocal of second integer	=	$\dfrac{13}{84}$

Write the word equation.

ALGEBRAICALLY

$$\frac{1}{n} + \frac{1}{n+2} = \frac{13}{84}$$

Substitute the identified values into the word equation.

$$84n(n+2)\left[\frac{1}{n} + \frac{1}{n+2}\right] = 84n(n+2)\left(\frac{13}{84}\right)$$

Multiply both sides of the equation by the LCD, $84n(n+2)$.

$$84(n+2) + 84n = 13n(n+2)$$
$$84n + 168 + 84n = 13n^2 + 26n$$
$$13n^2 - 142n - 168 = 0$$
$$(n-12)(13n+14) = 0$$

Simplify and write the quadratic equation in standard form.

Factor the left side of the equation.

$$n - 12 = 0 \quad \text{or} \quad 13n + 14 = 0$$

Set each factor equal to 0.

$$n = 12 \qquad\qquad n = -\frac{14}{13}$$

$$n + 2 = 14$$

Then solve for the smaller integer, n, and the larger integer, $n + 2$.

This value is not an integer.

Answer: The integers are 12 and 14.

Check: $\dfrac{1}{12} + \dfrac{1}{14} = \dfrac{7}{84} + \dfrac{6}{84} = \dfrac{13}{84}$. ■

USING THE LANGUAGE AND SYMBOLISM OF MATHEMATICS   9.6

1. If a varies directly as b, then $a =$ _____ .
2. If a varies inversely as b, then $a =$ _____ .
3. If a varies directly as b and b increases in magnitude, then a will _____ in magnitude.
4. If a varies directly as b and b decreases in magnitude, then a will _____ in magnitude.
5. If a varies inversely as b and b increases in magnitude, then a will _____ in magnitude.
6. If a varies inversely as b and b decreases in magnitude, then a will _____ in magnitude.

7. In the formula $D = RT$, D represents distance, R represents _____ , and T represents _____ .
8. In the formula $I = PRT$, I represents interest, P represents _____ , R represents _____ , and T represents _____ .
9. In the formula $W = RT$, W represents work, R represents _____ , and T represents _____ .

EXERCISES   9.6

In Exercises 1–10 write an equation for each statement of variation. Use k as the constant of variation.

1. **a.** m varies directly as n
    b. m varies inversely as p
    c. m varies directly as n and inversely as p
2. **a.** w varies directly as z
    b. w varies inversely as v
    c. w varies directly as z and inversely as v
3. **a.** v varies directly as the square root of w
    b. v varies inversely as the square of x
    c. v varies directly as the square root of w and inversely as the square of x

4. **a.** y varies directly as x cubed
    b. y varies inversely as the cube root of z
    c. y varies directly as x cubed and inversely as the cube root of z
5. **Electricity from a Windmill**   The number of kilowatts W of electricity that can be produced by a windmill varies directly as the cube of the speed v of the wind.
6. **Travel Time**   The time t required to drive between two towns varies inversely as the rate r.
7. **Ohm's Law**   The electrical current I varies directly as the voltage V.

8. **Wind Resistance** The force of the wind resistance R on a moving automobile varies directly as the square of the velocity v of the automobile.

9. **Weight of an Astronaut** The weight w of an astronaut varies inversely as the square of the distance d from the center of the earth.

10. **Period of a Pendulum** The period T of a pendulum (time for one complete swing of the pendulum) varies directly as the square root of the length L of the pendulum.

In Exercises 11–14 use the given statement of variation to solve each problem.

11. **a.** y varies directly as x, and $y = 24$ when $x = 8$. Find y when $x = 10$.
    b. y varies inversely as x, and $y = 24$ when $x = 8$. Find y when $x = 10$.

12. **a.** a varies directly as b, and $a = 12$ when $b = 3$. Find a when $b = 2$.
    b. a varies inversely as b, and $a = 12$ when $b = 3$. Find a when $b = 2$.

13. a varies directly as b and inversely as c, and $a = 3$ when $b = 9$ and $c = 12$. Find a when $b = 15$ and $c = 6$.

14. a varies directly as b and inversely as c, and $a = 6$ when $b = 2$ and $c = 24$. Find a when $b = 5$ and $c = 60$.

In Exercises 15–18 match each table with the corresponding statement of variation.

15. y varies directly as x with a positive constant of variation.
16. y varies directly as x with a negative constant of variation.
17. y varies inversely as x with a positive constant of variation.
18. y varies inversely as x with a negative constant of variation.

A.

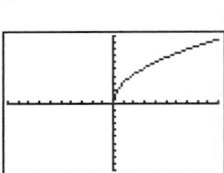

B.

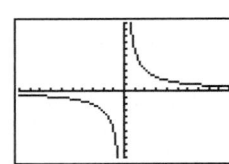

C.

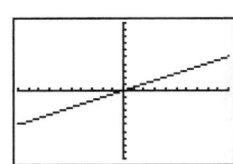

D.

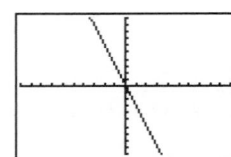

In Exercises 19–23 match each graph with the corresponding statement of variation. All graphs are displayed with the window $[-10, 10, 1]$ by $[-10, 10, 1]$.

19. y varies directly as x with a positive constant of variation.
20. y varies directly as x with a negative constant of variation.
21. y varies directly as the square of x.
22. y varies directly as the square root of x.
23. y varies inversely as x.

A.

B.

C.

D.

E.

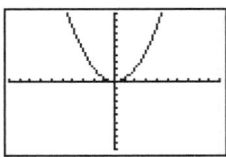

24. **Stopping Distance of a Car** The distance required for a car to make an emergency stop varies directly as the square of the car's speed. A car traveling at 50 mi/h requires 140 ft to stop. Under the same conditions, how many feet will be required for the same car traveling at 70 mi/h?

25. Resistance of a Wire   The electrical resistance of a wire varies inversely as the square of the diameter of the wire. The resistance of a wire with a diameter of 4 mm is 4.5 ohms. Find the resistance of another wire of the same length and made from similar materials with a diameter of 6 mm.

26. Illumination   The illumination I received at an object from a light source varies inversely as the square of the distance from the source. If the illumination 5 m from a light source is 4 lumens, find the illumination 10 m from this same light source.

In Exercises 27–62 solve each problem.

27. What number can be added to both the numerator and the denominator of $\frac{17}{25}$ to produce a fraction equal to $\frac{5}{7}$?

28. What number can be subtracted from both the numerator and the denominator of $\frac{17}{25}$ to produce a fraction equal to $\frac{3}{5}$?

29. The sum of the reciprocals of two consecutive integers is $\frac{11}{30}$. Find these integers.

30. The sum of the reciprocals of two consecutive even integers is $\frac{7}{24}$. Find these integers.

31. The sum of the reciprocals of two consecutive odd integers is 16 times the reciprocal of their product. Find these integers.

32. The sum of the reciprocals of two consecutive even integers is 10 times the reciprocal of their product. Find these integers.

33. The denominator of a fraction is an integer that is 4 more than the square of the numerator. If the fraction is reduced, it equals $\frac{3}{20}$. Find this numerator.

34. The denominator of a fraction is an integer that is 3 more than the square of the numerator. If the fraction is reduced, it equals $\frac{1}{4}$. Find this numerator.

35. The sum of a number and its reciprocal is $\frac{13}{6}$. Find this number.

36. The difference of a number minus its reciprocal is $\frac{9}{20}$. Find this number.

37. Ratio of Gauge Readings   The ratio of two readings from a gauge is $\frac{4}{5}$. The first reading is 3 units above normal, and the second reading is 5 units above normal. What is the normal reading?

38. Ratio of Temperatures   The ratio of two temperature readings is $\frac{7}{8}$. The first reading is 7° below normal, whereas the second reading is 2° above normal. What is the normal reading?

39. Wire Length   A wire 16 m long is cut into two pieces whose lengths have a ratio of 3 to 1. Find the length of each piece.

40. Rope Length   A rope 20 m long is cut into two pieces whose lengths have a ratio of 4 to 1. Find the length of each piece.

41. Electrical Resistance   The resistance of the parallel circuit with two resistors shown in the figure is 40 ohms. If one resistor has twice the resistance of the other, what is the resistance of each?

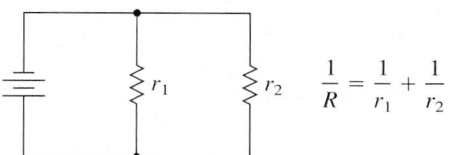

42. Electrical Resistance   The resistance of one resistor in a parallel circuit is 3 ohms greater than that of the second resistor in the circuit. Find the resistance of each resistor if the resistance of the circuit is $5\frac{1}{7}$ ohms.

43. Complementary Angles   The ratio of the measures of two complementary angles is 3 to 2. Find the measure of each angle.

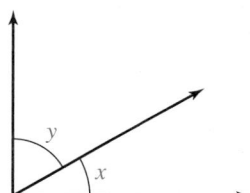

44. Supplementary Angles   The ratio of the measures of two supplementary angles is 4 to 1. Find the measure of each angle.

45. Investment Ratio   An investor invests $12,000 in a combination of secure funds and high-risk funds. The ratio of dollars invested in secure funds to dollars invested in high-risk funds is 7 to 3. Find the amount invested in secure funds.

46. Investment Ratio   By charter a mutual fund must invest its funds in bonds and stocks in a 5 to 3 ratio, respectively. If the mutual fund has $16,000,000, how much is invested in bonds?

47. Ratio of Assets to Liabilities   The assets of a small automobile dealership are approximated by $500x + 10{,}000$, and its liabilities are approximated by $100x + 14{,}000$, where x represents the number of vehicles sold. One measure of the strength of this business is the ratio of its assets to its liabilities. For the month of January the ratio was 5:4. How many vehicles were sold in January?

48. Ratio of Assets to Liabilities The assets of a heating and air-conditioning business are approximated by $300x + 5000$, and its liabilities are approximated by $200x + 6000$, where x represents the number of furnaces sold. One measure of the strength of this business is the ratio of its assets to its liabilities. For the month of February the ratio was $\frac{7}{6}$. How many furnaces were sold in February?

49. Boats in a Flowing River A paddlewheel riverboat travels between two towns 24 km apart. The current on the river is 3 km/h. This causes the boat to take 2 hours more for the trip upstream than for the trip downstream. What is the rate of the boat in still water?

50. Boats in a Flowing River Two boats having the same speed in still water depart simultaneously from a dock, traveling in opposite directions on a river that has a current of 7 km/h. After a period of time one boat is 30 km downstream and the other boat is 9 km upstream. What is the speed of each boat in still water?

51. Airplanes Traveling in an Air Stream Two planes departed from the same airport at the same time, flying in opposite directions. After a period of time, the slower plane has traveled 1080 mi and the faster plane has traveled 1170 mi. The faster plane is traveling 40 mi/h faster than the slower plane. Determine the rate of each plane. (*Hint:* What is the same for each plane?)

52. Airplanes Traveling in an Air Stream Two planes having the same airspeed depart simultaneously from an airport, flying in opposite directions. One plane flies directly into a 40-mi/h wind, and the other flies in the same direction as this wind. After a period of time, one plane has traveled 540 mi and the other has traveled 660 mi. What is the airspeed of each plane?

53. Time for Pipes to Fill a Tank One pipe can fill a tank in 15 h, and a larger pipe can fill the same tank in 10 h. If both pipes are used simultaneously, how many hours will it take to fill this tank?

54. Time for Pipes to Fill a Tank Working together, two pipes can fill one tank in 4 h. Working alone, the smaller pipe would take 6 h longer than the larger pipe to fill the tank. How many hours would it take the larger pipe to fill the tank working alone?

55. Time for Assembly Line to Fill an Order Working alone, one assembly line can complete an order in 36 h. A newer type of assembly line in this same factory can complete the same order in 18 h. If both assembly lines are used simultaneously, how many hours will it take to complete this order?

56. Time for Assembly Line to Fill an Order Working together, two assembly lines can complete an order in 18 h. Working alone, the older assembly line would take 15 h longer than the newer assembly line to complete the order. How many hours would it take the newer assembly line to complete the order working alone?

57. Search Time for Planes Search plane A can search an area for a crash victim in 50 h. Planes A and B can jointly search the area in 30 h. How many hours would it take plane B to search the area alone?

58. Generating a Mailing List Printer A takes twice as long as printer B to generate a mailing list. If a particular list can be generated in 9 h when both printers are used, how many hours would it take each printer to generate the list working separately?

59. Interest Rates on Two Accounts A 1-year investment will earn $120 in interest in a savings account or $200 in interest in a money market account that pays 2% more than the savings account. Determine each interest rate.

60. Interest Rates on Two Bonds Two bonds pay yearly interest. The longer-term bond pays 1.5% more than the shorter-term bond. The yearly interest on an investment would be $275 if invested in the first bond or $350 if invested in the second bond. Determine each interest rate.

61. Comparison of Mortgage Rates A prospective homeowner is looking for a mortgage for a new house. For a 30-year mortgage the interest for the first month would be $850. For a 15-year mortgage the interest for the first month would be $700. The rate is 1.5% less for the 15-year mortgage than for the 30-year mortgage. Determine each interest rate.

62. Comparison of Mortgage Rates A prospective homeowner is looking for a mortgage for a new house. For a 30-year mortgage the interest for the first month would be $720. For a 15-year mortgage the interest for the first month would be $630. The rate is 1% less for the 15-year mortgage than for the 30-year mortgage. Determine each interest rate.

Group Discussion Questions

63. Challenge Question   For each of the following pairs of variables determine whether it is more reasonable that these variables vary directly or vary inversely.

   a. The volume of paint to be painted on a wall and the time to brush this paint on the wall.

   b. The area covered by a bucket of paint and the thickness with which the paint is applied.

   c. The resistance of a pipe to the flow of water and the square of the radius of the pipe.

   d. The radius of a household water pipe and the volume of water flowing through this pipe.

64. Challenge Question

   a. Write a numerical word problem that is modeled by
   $$\frac{4}{t} + \frac{4}{t+6} = 1.$$

   b. Write a word problem involving the work done by two farm tractors that is modeled by the equation
   $$\frac{4}{t} + \frac{4}{t+6} = 1.$$

   c. Write a word problem involving two investments that is modeled by the equation $\frac{600}{r} = \frac{400}{r - 0.02}$.

   d. Write a word problem involving the distance traveled by two planes that is modeled by the equation
   $$\frac{600}{r+50} = \frac{400}{r-50}.$$

65. Discovery Question

   a. Describe two variables that you believe should vary directly.

   b. Collect 10 pairs of data involving these variables and construct a scatter diagram for these points.

   c. Does this data point support your claim?

66. Discovery Question

   a. Describe two variables that you believe should vary inversely.

   b. Collect 10 pairs of data involving these variables and construct a scatter diagram for these points.

   c. Does this data point support your claim?

KEY CONCEPTS   FOR CHAPTER 9

1. **Rational Expression:**   A rational expression is the ratio of two polynomials.
2. **Excluded Value:**   A value of a variable that must be excluded to prevent division by 0 in an expression is called an excluded value.
3. **Domain of a Rational Function:**   The domain of a rational function consists of all real numbers except those that are excluded to prevent division by 0.
4. **Reducing Rational Expressions:**   If A, B, and C are real algebraic expressions and $B \neq 0$ and $C \neq 0$, then $\dfrac{AC}{BC} = \dfrac{A}{B}$. Note $\dfrac{C}{C} = 1$, the multiplicative identity.

   A rational expression is in its lowest terms when the numerator and the denominator have no common factor other than -1 or 1.
5. **Operations with Rational Expressions:**   If A, B, C, and D are real polynomials, then
   - $\dfrac{A}{C} + \dfrac{B}{C} = \dfrac{A + B}{C}$ for $C \neq 0$
   - $\dfrac{A}{C} - \dfrac{B}{C} = \dfrac{A - B}{C}$ for $C \neq 0$
   - $\dfrac{A}{B} \cdot \dfrac{C}{D} = \dfrac{AC}{BD}$ for $B \neq 0$ and $D \neq 0$
   - $\dfrac{A}{B} \div \dfrac{C}{D} = \dfrac{A}{B} \cdot \dfrac{D}{C} = \dfrac{AD}{BC}$ for $B \neq 0$, $C \neq 0$, and $D \neq 0$
6. **Signs of Rational Expressions:**   If A and B are rational expressions,
   - $-\dfrac{A}{B} = \dfrac{-A}{B} = \dfrac{A}{-B}$ for $B \neq 0$
   - $\dfrac{1}{B - A} = -\dfrac{1}{A - B}$ for $A \neq B$
   - $\dfrac{A - B}{B - A} = -1$ for $A \neq B$
7. **Least Common Denominator (LCD):**   The least common denominator of two or more fractions is the product formed by using each factor the greatest number of times it occurs in any of the denominators.
8. **Order of Operations:**
   Step 1 Start with the expression within the innermost pair of grouping symbols.
   Step 2 Perform all exponentiations.
   Step 3 Perform all multiplications and divisions as they appear from left to right.
   Step 4 Perform all additions and subtractions as they appear from left to right.

9. **Complex Rational Expression:**   A complex rational expression is a rational expression whose numerator or denominator (or both) is also a rational expression.
10. **Simplifying Complex Fractions:**
    - Complex fractions can be simplified by rewriting the complex fraction as a division problem. Divide the main numerator by the main denominator.
    - Complex fractions can be simplified by multiplying both the numerator and the denominator by the LCD of all the fractions that occur in the numerator and denominator of the complex fraction.
11. **Extraneous Value:**   An extraneous value is a value produced in the solution process that does not check in the original equation.
12. **Solving an Equation Containing Rational Expressions:**
    Step 1 Multiply both sides of the equation by the LCD.
    Step 2 Solve the resulting equation.
    Step 3 Check the solution to determine whether it is an excluded value and therefore extraneous.
13. **Variation:**
    - **Direct Variation:**   If x and y are variables and k is a constant, then stating "y varies directly as x with a constant of variation k" means $y = kx$.
    - **Inverse Variation:**   If x and y are variables and k is a constant, then stating "y varies inversely as x with a constant of variation k" means $y = \dfrac{k}{x}$ for $k \neq 0$ and $x \neq 0$.
14. **Rate of Work:**   If one job can be done in time t, then the rate of work is $\dfrac{1}{t}$ of the job per unit of time.
15. **Strategy for Solving Word Problems:**
    Step 1 Read the problem carefully to determine what you are being asked to find.
    Step 2 Select a variable to represent each unknown quantity. Specify precisely what each variable represents.
    Step 3 If necessary, translate the problem into word equations. Then translate the word equations into algebraic equations.
    Step 4 Solve the equation(s) and answer the question asked by the problem.
    Step 5 Check the reasonableness of your answer.

REVIEW EXERCISES   FOR CHAPTER 9

In Exercises 1–4 evaluate each expression for $f(x) = \dfrac{2x - 9}{x^2 - 1}$.

1. $f(0)$ **2.** $f(1)$ **3.** $f(4)$ **4.** $f(10)$

5. Use the given table to determine the domain of

$$f(x) = \frac{3x + 1}{x^2 - x - 6}.$$

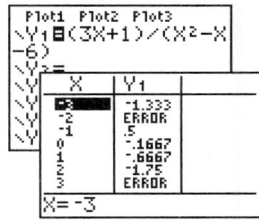

6. Use the given graph to determine the domain of

$$f(x) = \frac{x + 3}{x^2 + 2x - 15}.$$

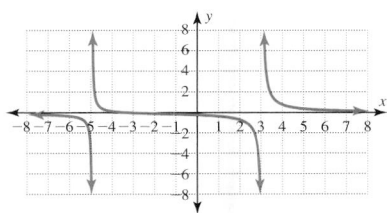

In Exercises 7 and 8 determine the domain of each function.

7. $f(x) = \dfrac{(5x + 1)(x - 3)}{(x + 5)(3x - 1)}$ **8.** $f(x) = \dfrac{x^2 - 11}{x^2 + 13}$

In Exercises 9–11 determine the excluded values for each rational expression.

9. $\dfrac{7x - 9}{4x - 2}$ **10.** $\dfrac{5x^2 - 7x + 11}{x^2 - 36}$

11. $\dfrac{6x^2 - 23x + 21}{20x^2 - 23x - 7}$

In Exercises 12–20 reduce each expression to its lowest terms.

12. $\dfrac{36x^2y}{12xy^3}$ **13.** $\dfrac{6a - 18b}{12b - 4a}$

14. $\dfrac{15x^2 - 15}{25x + 25}$ **15.** $\dfrac{x^2 + x - 30}{2x^2 + 11x - 6}$

16. $\dfrac{cx - cy}{ax - ay + bx - by}$ **17.** $\dfrac{3 - 7m}{14m^2 - 6m}$

18. $\dfrac{10x^2 + 29xy + 10y^2}{6x^2 + 13xy - 5y^2}$ **19.** $\dfrac{9x^2 - 24xy + 16y^2}{12x^2 - 25xy + 12y^2}$

20. $\dfrac{x^2 - 4y^2}{x^3 - 8y^3}$

In Exercises 21 and 22 find the missing numerators.

21. $\dfrac{2x + 1}{x - 3} = \dfrac{?}{x^2 - 6x + 9}$

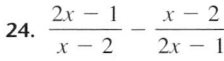

22. $\dfrac{y - 3}{y + 4} = \dfrac{?}{2y^2 + 9y + 4}$

In Exercises 23–26 match the table of values for the given expression with the table of values given in A–D representing the reduced expression.

23. $\dfrac{2x - 1}{x - 2} + \dfrac{x - 2}{2x - 1}$ **24.** $\dfrac{2x - 1}{x - 2} - \dfrac{x - 2}{2x - 1}$

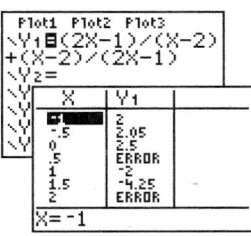   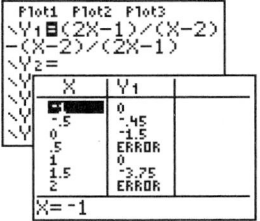

25. $\dfrac{2x - 1}{x - 2} \cdot \dfrac{x - 2}{2x - 1}$ **26.** $\dfrac{2x - 1}{x - 2} \div \dfrac{x - 2}{2x - 1}$

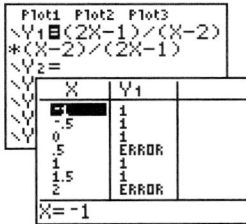   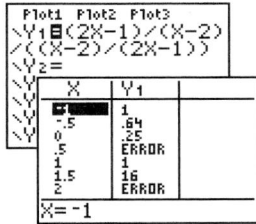

A. 1 **B.** $\dfrac{4x^2 - 4x + 1}{(x - 2)^2}$

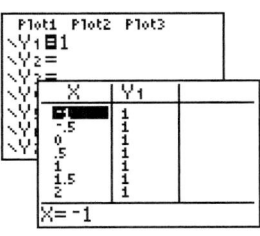   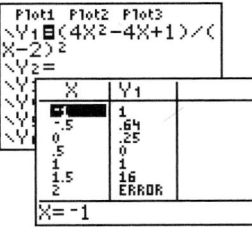

C. $\dfrac{3x^2 - 3}{(x - 2)(2x - 1)}$ **D.** $\dfrac{5x^2 - 8x + 5}{(x - 2)(2x - 1)}$

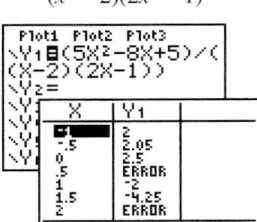

In Exercises 27–48 simplify each expression as completely as possible.

27. $\dfrac{36x^2 - 24x}{6x}$

28. $\dfrac{6xy}{2x - y} \cdot \dfrac{4x^2 - y^2}{3x^2}$

29. $\dfrac{9t^2 - 4}{16st} \div \dfrac{3t + 2}{16st^2 + 8st}$

30. $\dfrac{v^2 + 9vw + 8w^2}{v^2 - w^2} \div \dfrac{v^2 + 7vw - 8w^2}{v^2 + 5vw - 6w^2}$

31. $\dfrac{3x}{6x^2 + x - 1} - \dfrac{1}{6x^2 + x - 1}$

32. $\dfrac{1}{w + 1} - \dfrac{w}{w - 2} + \dfrac{w^2 + 2}{w^2 - w - 2}$

33. $\dfrac{3v}{3v^2 - 5v + 2} - \dfrac{2v}{2v^2 - v - 1}$

34. $\dfrac{6x}{2x - 3} - \dfrac{9x + 18}{4x^2 - 9} \cdot \dfrac{2x^2 - x - 6}{x^2 - 4}$

35. $\dfrac{2w + 4}{w^2 + 4w - 12} + \dfrac{w^2 - 169}{2w^2 - 13w + 21} \div \dfrac{w^2 - 15w + 26}{2w - 7}$

36. $\dfrac{6v^2 - 25v + 4}{6v^2 + 5v - 6} \div \dfrac{2v^2 - 3v - 20}{4v^2 + 16v + 15}$

37. $\left(\dfrac{y}{4} - \dfrac{4}{y}\right)\left(y - \dfrac{y^2}{y + 4}\right)$

38. $\dfrac{5z + 5}{6z^2 + 13z + 6} - \dfrac{1 - 4z}{3z^2 - 7z - 6} - \dfrac{3z}{2z^2 - 3z - 9}$

39. $\dfrac{\dfrac{1}{a} + \dfrac{1}{a + 1}}{\dfrac{1}{a} + \dfrac{1}{a^2}}$

40. $\dfrac{x + \dfrac{44}{x + 5} - 10}{x + \dfrac{33}{x + 5} - 9}$

41. $\left(\dfrac{3v^2 + 11v + 6}{3v^2 + 5v + 2}\right)^2 \div \dfrac{v^2 - 9}{v^2 + 2v + 1}$

42. $\dfrac{x^{-2} - x^{-1}}{x^{-2} + x^{-1}}$

43. $\dfrac{x + y^{-1}}{x - y^{-1}}$

44. $\dfrac{36w^{-2} + 23w^{-1} - 8}{24 - 5w^{-1} - 36w^{-2}}$

45. $\left(m - \dfrac{15}{m + 2}\right) \div \left(m - 1 - \dfrac{10}{m + 2}\right)$

46. $\dfrac{6y}{3y + 2} + \dfrac{20y + 4}{15y^2 + 7y - 2} \div \dfrac{5y^2 - 24y - 5}{5y^2 - 26y + 5}$

47. $\dfrac{2y}{27y^3 - 1} + \dfrac{y - 1}{27y^3 - 1}$

48. $\dfrac{z^2 - 2z + 1}{z^5 - z^4} \cdot \dfrac{2z^4}{z^3 - 1} + \dfrac{2z^2 + 2z}{z^2 + z + 1}$

In Exercises 49–52 simplify the expression in the first column and solve the equation in the second column.

Simplify	Solve
49. a. $\dfrac{1}{m + 2} - \dfrac{3}{m + 4}$	**b.** $\dfrac{1}{m + 2} = \dfrac{3}{m + 4}$
50. a. $\dfrac{x - 2}{x + 1} - \dfrac{x - 4}{x - 6}$	**b.** $\dfrac{x - 2}{x + 1} = \dfrac{x - 4}{x - 6}$
51. a. $\dfrac{x}{x - 3} + \dfrac{x + 4}{x - 2}$	**b.** $\dfrac{x}{x - 3} = \dfrac{x + 4}{x - 2}$
52. a. $\dfrac{2x}{x^2 - 1} + \dfrac{2}{x^2 - 1}$	**b.** $\dfrac{2x}{x^2 - 1} = \dfrac{2}{x^2 - 1}$

In Exercises 53–60 solve each equation.

53. $\dfrac{7}{2x} = \dfrac{2}{x} - \dfrac{3}{2}$

54. $\dfrac{15}{w^2 + 5w} + \dfrac{w + 4}{w + 5} = \dfrac{w + 3}{w}$

55. $\dfrac{1}{y^2 + 5y + 6} - \dfrac{2}{y + 3} = \dfrac{7}{y + 2}$

56. $1 - \dfrac{14}{y^2 + 4y + 4} = \dfrac{7y}{y^2 + 4y + 4}$

57. $\dfrac{w}{w^2 - 9w + 20} + \dfrac{14}{w^2 - 3w - 4} = \dfrac{18}{w^2 - 4w - 5}$

58. $\dfrac{4}{v^2 + 7v + 10} - \dfrac{3}{v + 2} + \dfrac{8}{v + 5} = 0$

59. $\dfrac{5z + 11}{2z^2 + 7z - 4} = \dfrac{1}{z + 4} - \dfrac{3}{1 - 2z}$

60. $\dfrac{1}{2z^2 - 9z - 5} - \dfrac{1}{2z^2 - 6z - 20} = \dfrac{3}{(2z + 1)(2z + 4)(z - 5)}$

Estimation Skills and Calculator Skills
In Exercises 61–64 mentally estimate the value of each expression for $f(x) = \dfrac{(x + 5)(x - 8)}{x^2 - 4}$ and then use a calculator to approximate each value to the nearest hundredth.

PROBLEM	MENTAL ESTIMATE	CALCULATOR APPROXIMATION
61. $f(1.01)$		
62. $f(7.99)$		
63. $f(-4.99)$		
64. $f(-0.99)$		

65. Perimeter and Area of a Trapezoid

    a. Determine the perimeter of the trapezoid shown in the figure.

    b. Determine the area of this trapezoid using the formula $A = \dfrac{h}{2}(a + b)$.

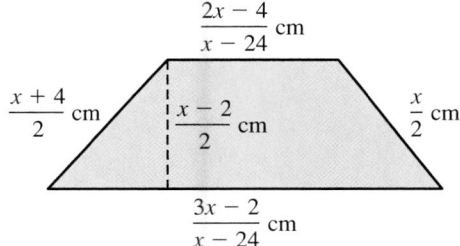

66. Area of a Region   Determine the area of the region shown in the figure.

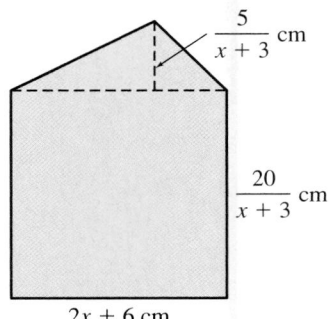

67. Volume of a Box   Determine the volume of the box shown in the figure.

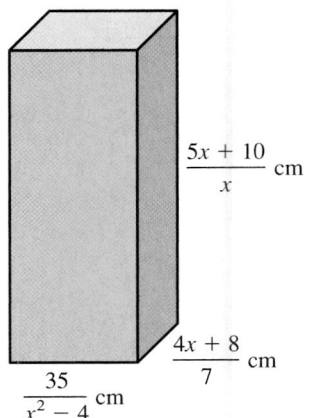

68. Average Cost   The number of units a plant can produce by operating t h/day is given by $N(t) = 30t$. The cost in dollars of operating the plant for t h is given by $C(t) = -6t^3 + 600t^2 + 3000t$.

    a. Determine $A(t)$ the average cost per unit when the plant is operated for t h.

    b. Evaluate and interpret $N(10)$.

    c. Evaluate and interpret $C(10)$.

    d. Evaluate and interpret $A(10)$.

Numeric Word Problems

In Exercises 69–77 solve each word problem.

69. The denominator of a fraction is 6 more than the numerator, and the fraction equals -1. Find the numerator.

70. The ratio of one number to another is $\dfrac{5}{2}$. If the first number is 5 less than 3 times the second, find both numbers.

71. The sum of the reciprocals of two consecutive odd integers is 12 times the reciprocal of their product. Find these integers.

72. Travel Time   The amount of time required for a bus to travel from St. Louis, Missouri, to Chicago, Illinois, varies inversely with the average speed of the bus. The trip takes 5 h if the bus averages 60 mi/h. How long will the trip take at an average speed of 50 mi/h? What is the significance of the constant of variation k?

73. Electrical Resistance   The resistance of the parallel circuit with two resistors shown in the figure is 3 ohms. If one resistor has 3 times the resistance of the other, what is the resistance of each?

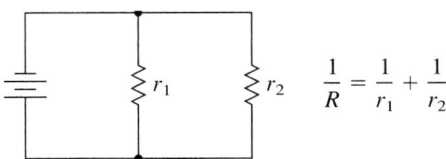

74. Filling a Bathtub   A hot water faucet takes 7 min longer than the cold water faucet to fill a bathtub. If both faucets were turned on, the tub would be half full in 6 min. How many minutes would it take to fill an empty tub with cold water if the hot water faucet were turned off?

75. Average Cost   A furniture company can produce x units of a desk for $54 per desk, plus an initial start-up investment of $20,000. To compete on the market, it must be able to achieve an average cost of only $74 per unit. How many units must the company produce to reach an average cost of $74 per unit?

76. Dimensions of a Metal Sheet A rectangular piece of metal is 6 cm longer than it is wide. A 0.5-cm strip is cut off of each side, leaving an area thirteen-sixteenths of the original area. Find the original dimensions. (See the figure.)

w cm

$w + 6$ cm

77. Speed of Boats Two boats having the same speed in still water depart from a dock at the same time. The boats travel in opposite directions in a river that has a current of 6 mi/h. After a period of time, one boat is 54 mi downstream, and the other boat is 30 mi upstream. What is the speed of each boat in still water?

MASTERY TEST FOR CHAPTER 9

[9.1] **1.** Determine the domain of each rational function.

a. $f(x) = \dfrac{x - 5}{2x - 12}$ b. $f(x) = \dfrac{x + 5}{(x - 1)(x + 6)}$

c. $f(x) = \dfrac{x + 7}{x^2 - 64}$ d. $f(x) = \dfrac{2x + 19}{x^2 + 1}$

[9.1] **2.** Reduce each rational expression to its lowest terms.

a. $\dfrac{6x}{12x^2 - 18x}$

b. $\dfrac{2x - 6}{x^2 - 9}$

c. $\dfrac{2x^2 - 7x - 15}{x^2 - 25}$

d. $\dfrac{4x^2 - 12xy + 9y^2}{3ay - 6by - 2ax + 4bx}$

[9.2] **3.** Perform each multiplication or division and reduce the result to its lowest terms.

a. $\dfrac{ax - bx}{x^2} \cdot \dfrac{5x}{4a - 4b}$

b. $\dfrac{x^2 - 1}{x + 1} \div \dfrac{x^2 - 3x + 2}{x - 2}$

c. $\dfrac{v^2 - v - 20}{v^2 - 16} \div \dfrac{3v - 15}{2v - 8}$

d. $\dfrac{x^2 - 4}{x^2 - 4x - 21} \cdot \dfrac{x^2 - 2x - 35}{x^2 - 7x + 10}$

[9.3] **4.** Perform each addition or subtraction and reduce each result to its lowest terms.

a. $\dfrac{3x - 4}{2x - 3} + \dfrac{x - 2}{2x - 3}$

b. $\dfrac{3x + 7}{x^2 + x - 12} - \dfrac{2x + 3}{x^2 + x - 12}$

c. $\dfrac{7}{w^2 + w - 12} + \dfrac{2}{w^2 - 8w + 15}$

d. $\dfrac{1}{x + y} + \dfrac{3}{x - y} + \dfrac{2y}{x^2 - y^2}$

[9.4] **5.** Simplify each expression.

a. $\dfrac{4x}{5} + \dfrac{x^2}{15} \cdot \dfrac{3}{x}$

b. $\left(\dfrac{4x}{5} + \dfrac{x^2}{15}\right) \cdot \dfrac{3}{x}$

c. $\dfrac{x - y}{x + y} + \dfrac{3x - 21y}{x^2 - y^2} \div \dfrac{15x^2 - 105xy}{10x^2y - 10xy^2}$

d. $\left(\dfrac{w}{w - 2} + \dfrac{1}{w - 3}\right) \div \dfrac{w^2 - 2w - 2}{w^2 - w - 2}$

[9.4] **6.** Simplify each expression.

a. $\dfrac{\dfrac{4}{5}}{\dfrac{6}{}}$

b. $\dfrac{\dfrac{x}{3} - 2 + \dfrac{3}{x}}{1 - \dfrac{3}{x}}$

c. $\dfrac{36x^{-2} - 3x^{-1} - 18}{12x^{-2} - 25x^{-1} + 12}$

d. $\dfrac{\dfrac{v}{v - 4} - \dfrac{2v}{v + 3}}{\dfrac{v^2 - 4v}{2v + 6}}$

[9.5] **7.** Solve each equation.

 a. $\dfrac{z - 4}{z - 2} = \dfrac{1}{z - 2}$

 b. $\dfrac{z - 2}{z - 1} + \dfrac{z - 3}{2z - 5} = 1$

 c. $\dfrac{z + 11}{z^2 - 5z + 4} = \dfrac{5}{z - 4} + \dfrac{3}{1 - z}$

 d. Solve $\dfrac{x + 4}{x - 5} = y$ for x.

[9.6] **8. a.** If y varies inversely as x, and y is 12 when x is 6, find y when x is 4.

 b. The volume of a gas varies inversely with the pressure when temperature is held constant. If the pressure exerted by 4 liters of a gas is 6 newtons/cm^2, what will the pressure be if the volume is 3 liters?

[9.6] **9.** Solve each of the following problems.

 a. Ratio of Economic Indicators The ratio of two monthly economic indicators is $\dfrac{2}{3}$. For the first month the indicator is 2 units below normal, and for the second month the indicator is 2 units above normal. What is the normal number of units for this indicator?

 b. Work by Two Machines When members of a construction crew use two end loaders at once, they can move a pile of sand in 6 h. If they use only the larger end loader, they can do the job in 5 h less time than it would take if they used the smaller machine. How many hours would it take to do the job using the larger machine?

 c. Interest Rates on Two Bonds Two bonds pay yearly interest. The longer-term bond pays 2.5% more than the shorter-term bond. The yearly interest on an investment would be $680 if invested in the first bond or $480 if invested in the second bond. Determine each interest rate.

 d. Airplanes Traveling in an Air Stream Two planes departed from the same airport at the same time, flying in opposite directions. After a period of time, the slower plane has traveled 950 mi and the faster plane has traveled 1050 mi. The faster plane is traveling 50 mi/h faster than the slower plane. Determine the rate of each plane. (*Hint:* What is the same for each plane?)

GROUP PROJECT **FOR CHAPTER 9**

Exploring the Asymptotic Behavior of a Rational Function

A line is an **asymptote** of a graph if this graph gets closer and closer to the line for points farther and farther from the origin.

The graph of $y = \dfrac{1}{x}$, which is shown in the figure, has both a horizontal and a vertical asymptote. The line $x = 0$ is a vertical asymptote, and the line $y = 0$ is a horizontal asymptote.

Exploration: Use a graphics calculator to examine the function

$$f(x) = \frac{3(x + 1)}{x - 2}$$

graphically, numerically, and verbally. Then try to generalize your observations so you can analyze other rational functions algebraically.

Vertical Asymptote

1.

GRAPHICALLY	NUMERICALLY	VERBALLY	ALGEBRAICALLY
After examining several graphics calculator windows, sketch a graph $y = \dfrac{3(x + 1)}{x - 2}$ on a full sheet of graph paper. Is there a vertical line that you can sketch that is a vertical asymptote for this function?	**a.** Use a calculator to complete the y-values in table **A.** The x-values in this table get closer and closer to 2 while staying less than 2.	**a.** Describe the trend in the y-values in table **A** as the x-values in this table get closer and closer to 2.	Write the equation of the vertical asymptote for this graph.
	b. Use a calculator to complete the y-values in table **B.** The x-values in this table get closer and closer to 2 while remaining more than 2.	**b.** Describe the trend in the y-values in table **B** as the x-values in this table get closer and closer to 2.	

2. Generalize from your observations of $y = \dfrac{3(x + 1)}{x - 2}$ to predict what the vertical asymptote of the graph of $y = \dfrac{2(x - 3)}{x + 1}$ will be. What is the equation of this vertical asymptote? Describe the behavior of the y-values as x gets closer and closer to -1. Use a graphics calculator to test your generalizations.

Horizontal Asymptote

1.

GRAPHICALLY	NUMERICALLY	VERBALLY	ALGEBRAICALLY
After examining several graphics calculator windows, sketch a graph $y = \dfrac{3(x + 1)}{x - 2}$ on a full sheet of graph paper. Is there a horizontal line that you can sketch that is a horizontal asymptote for this function?	**a.** Use a calculator to complete the y-values in table **C**. The positive x-values in this table get larger and larger in magnitude. **b.** Use a calculator to complete the y-values in table **D**. The negative x-values in this table get larger and larger in magnitude.	**a.** Describe the trend in the y-values in table **C** as the x-values in this table get larger and larger in magnitude. **b.** Describe the trend in the y-values in table **D** as the x-values in this table get larger and larger in magnitude.	Write the equation of the horizontal asymptote for this graph.

2. Generalize from your observations of $y = \dfrac{3(x + 1)}{x - 2}$ to predict what the horizontal asymptote of the graph of $y = \dfrac{2(x - 3)}{x + 1}$ will be. What is the equation of this horizontal asymptote? Describe the behavior of the y-values as x gets larger and larger in magnitude. Use a graphics calculator to test your generalizations.

Tables

A. x	y	B. x	y	C. x	y	D. x	y
0		4		5		-5	
1		3		10		-10	
1.5		2.5		50		-50	
1.9		2.1		100		-100	
1.99		2.01		1,000		$-1,000$	
1.999		2.001		5,000		$-5,000$	
1.9999		2.0001		10,000		$-10,000$	

10

EXPONENTIAL AND LOGARITHMIC FUNCTIONS

Exponential and logarithmic functions are used to solve many types of growth and decay problems. These functions are used by bankers to compute compound interest, by sociologists to predict population growth, and by archaeologists to compute the age of ancient objects through carbon-14 dating. One specific example that uses exponential growth is computing the safe shelf life of products such as milk. By calculating how long it takes bacteria to reach an unacceptable level, we can determine how many days this perishable product is safe.

This chapter examines applications of both exponential and logarithmic functions. Because exponential and logarithmic functions are inverses of each other, we also examine inverse functions in this chapter. We start by examining exponential functions and geometric sequences.

Section 10.1   Geometric Sequences and Exponential Functions

Objectives: **1.** Identify a geometric sequence.
2. Graph and use exponential functions.

An arithmetic sequence has a constant change from term to term (a common difference). This common difference is often denoted by d.

The following tables and graphs illustrate two distinctive types of sequences. The first sequence is an arithmetic sequence with a common difference $d = 2$. As we noted in Section 2.1, arithmetic sequences have a constant change from term to term and produce a linear pattern when graphed. The second sequence is an example of a geometric sequence. Geometric sequences differ significantly from arithmetic sequences. Following these two examples, we examine geometric sequences.

The graph of this arithmetic sequence contains only the discrete points given by the sequence: $a_1 = 1, a_2 = 3$, $a_3 = 5, a_4 = 7, a_5 = 9, a_6 = 11$. The dashed line is shown here to emphasize that these points do form a linear pattern.

VERBALLY	NUMERICAL EXAMPLE	GRAPHICAL EXAMPLE
Arithmetic sequence	1, 3, 5, 7, 9, 11	

x	y
1	1
2	3
3	5
4	7
5	9
6	11

VERBALLY	NUMERICAL EXAMPLE	GRAPHICAL EXAMPLE
Geometric sequence	1, 2, 4, 8, 16, 32	

x	y
1	1
2	2
3	4
4	8
5	16
6	32

Geometric Sequences

The sequence 1, 2, 4, 8, 16, 32 has a constant ratio from term to term. Each term is twice the preceding term. This is an example of a geometric sequence. A **geometric sequence** is a sequence with a constant ratio from term to term. That is, the ratio of any two consecutive terms is constant. This constant is called the **common ratio** and is usually denoted by r. The common ratio for the sequence 1, 2, 4, 8, 16, 32 is 2 $\left(\dfrac{2}{1} = \dfrac{4}{2} = \dfrac{8}{4} = \dfrac{16}{8} = \dfrac{32}{16} = 2\right)$.

> A geometric sequence has a constant ratio from term to term (a common ratio). This common ratio is often denoted by r.

■ **EXAMPLE 1**   **Writing the Terms of Arithmetic and Geometric Sequences**

Write the first five terms of each sequence.

SOLUTIONS ————————————————

(a) An arithmetic sequence with $a_1 = 5$ and a common difference of 4.

$a_1 = 5$
$a_2 = 5 + 4 = 9$
$a_3 = 9 + 4 = 13$
$a_4 = 13 + 4 = 17$
$a_5 = 17 + 4 = 21$

Each term of this arithmetic sequence is obtained by adding the common difference of 4 to the previous term.

(b) A geometric sequence with $a_1 = 5$ and a common ratio of 4.

$a_1 = 5$
$a_2 = 5(4) = 20$
$a_3 = 20(4) = 80$
$a_4 = 80(4) = 320$
$a_5 = 320(4) = 1280$

Each term of this geometric sequence is obtained by multiplying the common ratio of 4 times the previous term.

■

In Example 2 we examine each sequence to determine if the sequence is geometric. If a sequence is geometric, the ratio of consecutive terms will be constant.

■ **EXAMPLE 2**   Identifying Geometric Sequences

Determine which of these sequences are geometric sequences. For those that are geometric, determine the common ratio r.
(a) 2, 3, 4.5, 6.75, 10.125, 15.1875
(b) 32, 16, 8, 4, 2, 1
(c) 2, 3.5, 5, 6.5, 8, 9.5

SOLUTIONS _____

(a) NUMERICALLY: RATIO OF CONSECUTIVE TERMS

$$\frac{3}{2} = 1.5$$

$$\frac{4.5}{3} = 1.5$$

$$\frac{6.75}{4.5} = 1.5$$

$$\frac{10.125}{6.75} = 1.5$$

$$\frac{15.1875}{10.125} = 1.5$$

$$r = 1.5$$

VERBALLY

Because there is a common ratio $r = 1.5$, this is a geometric sequence. The graph of these points exhibits the shape that we define as geometric growth.

GRAPHICALLY

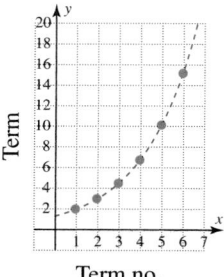
Term no.

(b) NUMERICALLY: RATIO OF CONSECUTIVE TERMS

$$\frac{16}{32} = 0.5$$

$$\frac{8}{16} = 0.5$$

$$\frac{4}{8} = 0.5$$

$$\frac{2}{4} = 0.5$$

$$\frac{1}{2} = 0.5$$

$$r = 0.5$$

VERBALLY

Because there is a common ratio $r = 0.5$, this is a geometric sequence. The graph of these points exhibits the shape that we define as geometric decay.

GRAPHICALLY

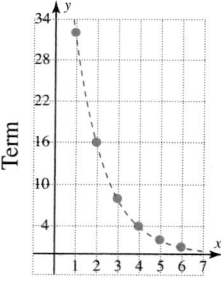
Term no.

(c) NUMERICALLY: RATIO OF CONSECUTIVE TERMS

$$\frac{3.5}{2} = 1.75$$

$$\frac{5}{3.5} \approx 1.43$$

$$1.75 \neq 1.43$$

VERBALLY

This is not a geometric sequence because the ratio from term to term is not constant. In fact, this is an arithmetic sequence with a common difference $d = 1.5$. Note that the points exhibit a linear pattern.

GRAPHICALLY

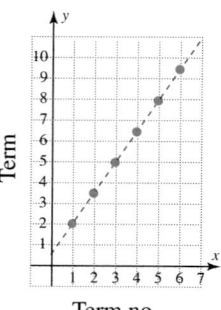
Term no.

SELF-CHECK 10.1.1

1. Write an arithmetic sequence of seven terms with $a_1 = 2$ and a common difference $d = 3$.
2. Write a geometric sequence of seven terms with $a_1 = 2$ and a common ratio $r = 3$.
3. Write a geometric sequence of seven terms with $a_1 = 2$ and a common ratio $r = -3$.

If the common ratio r of a geometric sequence is greater than 1, then we say the sequence exhibits **geometric growth.** Each successive term grows by a factor of r. If the common ratio r of a geometric sequence is positive and less than 1, then we say the sequence exhibits **geometric decay.** Each successive term will decrease or decay by a factor of r.

Geometric Growth and Decay

GROWTH	DECAY
If $r > 1$, then a geometric sequence exhibits geometric growth.	If $0 < r < 1$, then a geometric sequence exhibits geometric decay.

$r = 2$

$r = 0.5$

■ EXAMPLE 3   Determining a Sequence that Models the Height of a Bouncing Ball

A golf ball is dropped onto a concrete cart path from a height of 10 ft. Each bounce is 0.6 as high as the height of the previous bounce. Determine the height of the first four bounces.

SOLUTION

SELF-CHECK 10.1.1 ANSWERS

1. 2, 5, 8, 11, 14, 17, 20
2. 2, 6, 18, 54, 162, 486, 1458
3. 2, −6, 18, −54, 162, −486, 1458

$a_0 = 10$
$a_1 = 0.6(10) = 6$
$a_2 = 0.6(6) = 3.6$
$a_3 = 0.6(3.6) = 2.16$
$a_4 = 0.6(2.16) = 1.296$

The sequence exhibits geometric decay. The common ratio $r = 0.6 < 1$. The height of each bounce is decreasing geometrically by a factor of 0.6. a_0 denotes the initial position of 10 ft.

Answer: The heights of the first four bounces are 6, 3.6, 2.16, and 1.296 ft.   ■

Exponential Functions

The terms of a geometric sequence can be obtained by repeated multiplication by the common ratio r. If a_1 is the first term, then we can write the first six terms as:

$$a_1, \; a_1 r, \; a_1 r^2, \; a_1 r^3, \; a_1 r^4, \; a_1 r^5$$

Since exponentiation was originally developed to represent repeated multiplication like that exhibited in the given sequence, we now examine exponential functions. In the work that follows, the exponents can be any real number, including irrational values.

Exponential Function $f(x) = b^x$

ALGEBRAICALLY	VERBALLY	ALGEBRAIC EXAMPLE	GRAPHICAL EXAMPLE
If $b > 0$ and $b \neq 1$, then $f(x) = b^x$ is an exponential function with base b.	An exponential function has a constant for the base and the variable is in the exponent.	$y = 2^x$	

■ EXAMPLE 4   Identifying Exponential Functions

Determine which of these equations define exponential functions.

SOLUTIONS

(a) $f(x) = 3^x$ An exponential function   This function has a base of 3.

(b) $f(x) = x^3$ Not an exponential function   The exponent is the constant 3, not a variable. This is called a cubic polynomial function.

(c) $f(x) = \left(\dfrac{4}{5}\right)^x$ An exponential function   This function has a base of $\dfrac{4}{5}$.

(d) $f(x) = (-5)^x$ Not an exponential function   The base of this exponential expression is -5. Since the base is negative, this equation does not represent an exponential function.

(e) $y = 1^x$ Not an exponential function $y = 1^x$ simplifies to $y = 1$, since 1 to any power equals 1. Thus it is a constant function. ■

The domain of an exponential function $f(x) = b^x$ is understood to be the set of all real numbers. You may be able to evaluate some selected values such as 2^4 by hand, but you need a calculator to approximate values like 2^π.

◼ **EXAMPLE 5**   Evaluating an Exponential Function

Given $f(x) = 2^x$, evaluate each expression.

SOLUTIONS _____

(a) $f(4)$
(b) $f(-2)$
(c) $f(\pi)$

```
2^4
            16
2^(-2)
           .25
2^π
     8.824977827
∎
```

Answers: (a) $f(4) = 2^4 = 16$

(b) $f(-2) = 2^{-2} = \dfrac{1}{2^2} = \dfrac{1}{4} = 0.25$

(c) $f(\pi) = 2^\pi \approx 8.825$ ◼

SELF-CHECK 10.1.3 ANSWERS

1. $f(2) = 3^2 = 9$

2. $f(-1) = 3^{-1} = \dfrac{1}{3} \approx 0.333$

3. $f(\sqrt{2}) = 3^{\sqrt{2}} \approx 4.729$

SELF-CHECK 10.1.3

Given $f(x) = 3^x$, evaluate each expression.

1. $f(2)$
2. $f(-1)$
3. $f(\sqrt{2})$

A function whose graph rises to the right is called an **increasing function.** Similarly, a function whose graph falls to the right is called a **decreasing function.** A line with a positive slope represents an increasing function, and a line with a negative slope represents a decreasing function.

If the base b of an exponential function is greater than 1, as in $f(x) = 2^x$, then the graph of the function rises rapidly to the right as the base is used repeatedly as a factor. Thus an exponential function $f(x) = b^x$ with $b > 1$ is an increasing function, called an **exponential growth function.** If $0 < b < 1$, then the graph of $f(x) = b^x$ declines rapidly to the right and is called an **exponential decay function.** In the following box the exponential growth function is $f(x) = 2^x$ and the exponential decay function is $f(x) = \left(\dfrac{1}{2}\right)^x$.

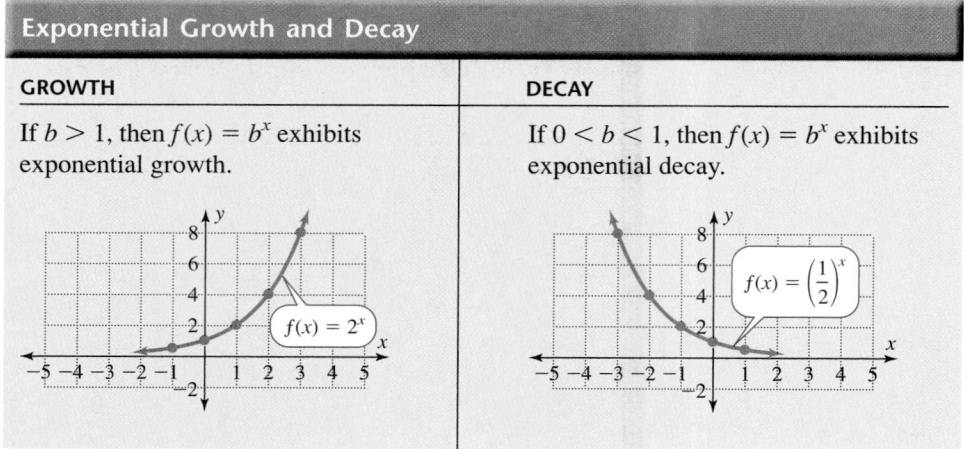

Exponential Growth and Decay

GROWTH	DECAY
If $b > 1$, then $f(x) = b^x$ exhibits exponential growth.	If $0 < b < 1$, then $f(x) = b^x$ exhibits exponential decay.

The exponential decay function $f(x) = \left(\dfrac{1}{2}\right)^x$ can also be written as $f(x) = 2^{-x}$.

Recall that the domain of a function is represented by the projection of its graph onto the x-axis and the range is represented by the projection of its graph onto the y-axis. Note that this is illustrated by the growth and decay functions given in the box.

These graphs of $f(x) = 2^x$ and $f(x) = \left(\dfrac{1}{2}\right)^x$ reveal several properties and features of exponential functions. We now list some of these properties.

Properties of the Graphs of Exponential Functions

For the exponential function $f(x) = b^x$ with $b > 0$ and $b \neq 1$:

1. The domain of f is the set of all real numbers $\mathbb{R}$.
2. The range of f is the set of all positive real numbers.
3. There is no x-intercept.
4. The y-intercept is $(0, 1)$.
5. a. The growth function is asymptotic to the negative portion of the x-axis (it approaches but does not touch the x-axis).
   b. The decay function is asymptotic to the positive portion of the x-axis.
6. a. The growth function rises (or grows) from left to right.
   b. The decay function falls (or decays) from left to right.

Newspapers and magazines often incorrectly use the term "exponential growth" simply to indicate rapid growth. If this term is used correctly, the data would need to fit an exponential function.

   Some financial advisors and insurance salespeople promoting annuities and other investments refer to the "magic of compound interest." The "magic" they are referring to is exponential growth—where growth is always a factor of the previous base. Thus as the terms grow, so does the change from term to term. Example 6 involves exponential growth with a base $b = 1.0925$.

■ EXAMPLE 6 Modeling Compound Interest

The formula $A = P(1 + r)^t$ can be used to compute the total amount of money A that accumulates when a principal P is invested at a yearly interest rate r and left to compound annually for t years. Use this growth formula to:
(a) Write a function for A for an investment of $5000 at 9.25% for t years.
(b) Evaluate this function for $t = 8$ years.

SOLUTIONS

(a) $A = P(1 + r)^t$
  $A(t) = 5000(1 + 0.0925)^t$
  $A(t) = 5000(1.0925)^t$

Substitute the given values into the compound interest formula: $P = 5000$ and $r = 0.0925$. This is an exponential function of the form $f(x) = ab^x$ with a base of $b = 1.0925$.

(b) $A(8) = 5000(1.0925)^8$
  $A(8) \approx 5000(2.029418267)$
  $A(8) \approx \$10,147$

To determine the value of this investment after 8 years evaluate $A(t)$ for $t = 8$.

The value of this investment after 8 years is approximately $10,147.

This calculator approximation has been rounded to the nearest dollar.

SELF-CHECK 10.1.4

1. Graph $y = 3^x$ and $y = \left(\dfrac{1}{3}\right)^x$ on the same rectangular coordinate system.

2. Use the formula $A = P(1 + r)^t$ and a calculator to determine the value of $450 invested at 8.75% and left to compound annually for 7 years.

A Mathematical Note

Swiss-born Leonhard Euler (1701–1783) was hired by Catherine the Great of Russia to write the elementary mathematics textbooks for Russian schools. He wrote prolifically on various mathematical topics. From Euler's textbook *Introductio in Analysin Infinitorum* came many symbols, such as i for $\sqrt{-1}$, π for the ratio of the circumference of a circle to its diameter, and e for the base of natural logarithms.

Problems involving growth and decay often involve the irrational number e that is approximately equal to 2.718281828. Both π and e are fundamental constants that show up over and over in our study of the world and universe around us.

```
π     3.141592654
e     2.718281828
■
```

These calculator approximations are just that—approximations. Because π and e are irrational their decimal representations never repeat or terminate.

CALCULATOR PERSPECTIVE 10.1.1   Using the e^x Feature

To evaluate the expression $1000e^{-0.01245}$ from Example 7 on a TI-83 Plus calculator, enter the following keystrokes:

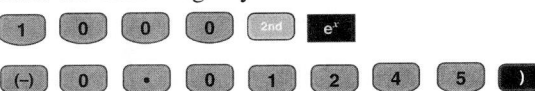

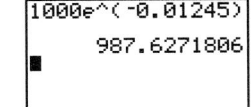

```
1000e^(-0.01245)
         987.6271806
■
```

The e^x feature is the secondary function of the LN key.

SELF-CHECK 10.1.4 ANSWERS

1.
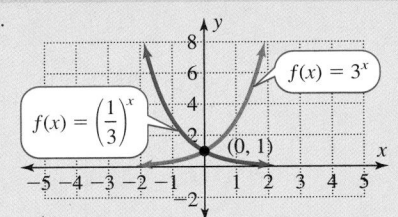

$f(x) = \left(\dfrac{1}{3}\right)^x$ $f(x) = 3^x$ (0, 1)

2. $A(7) \approx \$809.50$; after 7 years the investment is worth $809.50.

Example 7 involves e and a function that models radioactive decay.

■ EXAMPLE 7 Determining the Amount of Radioactive Decay

The amount of 1000 grams of radioactive carbon-14 remaining after t years is given by the function $A(t) = 1000e^{-0.0001245t}$. How much of a 1000-gram sample will remain a century later?

SOLUTION _____

$$A(t) = 1000e^{-0.0001245t}$$
$$A(100) = 1000e^{-0.0001245(100)}$$
$$A(100) = 1000e^{-0.01245}$$
$$A(100) \approx 987.6$$

A century is 100 years. Substitute this value of t into the given function.

Approximate this value using a calculator.

Answer: After 100 years, approximately 988 grams of carbon-14 remain.

SELF-CHECK 10.1.5

1. Approximate e^2 to the nearest thousandth.
2. Approximate e^{-2} to the nearest thousandth.
3. Use the formula in Example 7 to determine the amount of carbon-14 remaining after 1000 years.

The equations in Example 8 can all be solved easily without the use of a calculator because they involve familiar powers of 1, 2, 3, 4, or other small integers. The properties we will use are given in the following box.

Properties of Exponential Functions

ALGEBRAICALLY	VERBALLY
For real exponents x and y and bases $a > 0$ and $b > 0$:	
1. For $b \neq 1$, $b^x = b^y$ if and only if $x = y$.	1. The exponents must be equal because the expressions are equal and have the same base.
2. For $x \neq 0$, $a^x = b^x$ if and only if $a = b$.	2. The bases must be equal because the expressions are equal and have the same exponents.

■ EXAMPLE 8   Solving Exponential Equations

Solve each equation.

SOLUTIONS

(a) $3^{x-4} = 9$

$3^{x-4} = 9$	Substitute 3^2 for 9 in order to express both members in
$3^{x-4} = 3^2$	terms of the common base 3.
$x - 4 = 2$	The exponents are equal since the bases are the same.

Answer: $x = 6$

(b) $4^x = \dfrac{1}{8}$

$4^x = \dfrac{1}{8}$	Express both 4 and $\dfrac{1}{8}$ in terms of the common base 2.
$\left(2^2\right)^x = 2^{-3}$	Use the power rule for exponents to simplify the left
$2^{2x} = 2^{-3}$	side of this equation.
$2x = -3$	The exponents are equal since the bases are the same.

Answer:$x = -\dfrac{3}{2}$

(c) $(b + 2)^4 = 625$

$(b + 2)^4 = 625$	
$(b + 2)^4 = 5^4$	Substitute 5^4 for 625.
$b + 2 = 5$	The bases are equal since the exponents are the same.

Answer: $b = 3$

SELF-CHECK 10.1.6

Solve each equation.

1. $2^{x+7} = 16$
2. $27^x = 9$
3. $b^{-3} = \dfrac{1}{64}$

SELF-CHECK 10.1.6 ANSWERS

1. $x = -3$
2. $x = \dfrac{2}{3}$
3. $b = 4$

We solve more complicated exponential equations in Section 10.6 after we develop some skills with logarithms.

USING THE LANGUAGE AND SYMBOLISM OF MATHEMATICS   10.1

1. An arithmetic sequence is a sequence with a constant _____ from term to term.

2. A geometric sequence is a sequence with a constant _____ from term to term.

3. If the common ratio r of a geometric sequence is greater than 1, then we say the sequence exhibits geometric _____.

4. If the common ratio r of a geometric sequence is between 0 and 1, then we say the sequence exhibits geometric _____.

5. An exponential function is of the form $f(x) =$ _____ with a constant for the _____ and a variable in the _____.

6. An exponential function $f(x) = b^x$ with $b > 1$ is called an exponential _____ function.

7. An exponential function $f(x) = b^x$ with $0 < b < 1$ is called an exponential _____ function.

8. The graph of an exponential function $f(x) = b^x$ is asymptotic to a portion of the _____-axis.

9. The irrational number $e \approx$ _____ (to the nearest thousandth).

10. For real exponents x and y and base $b > 0$ and $b \neq 1$, $b^x = b^y$ if and only if _____ = _____.

11. For a real exponent $x \neq 0$ and bases $a > 0$ and $b > 0$, $a^x = b^x$ if and only if _____ = _____.

EXERCISES   10.1

In Exercises 1–4 determine which sequences are geometric. If a sequence is geometric, determine the common ratio r.

1. **a.** 1, 5, 25, 125, 625, 3125
    b. 3125, 625, 125, 25, 5, 1
    c. 1, −5, 25, −125, 625, −3125
    d. 3125, −625, 125, −25, 5, −1
2. **a.** 4, 12, 36, 108, 324, 972
    b. 972, 324, 108, 36, 12, 4
    c. −4, 12, −36, 108, −324, 972
    d. 972, −324, 108, −36, 12, −4
3. **a.** 36, 24, 16, $\dfrac{32}{3}$, $\dfrac{64}{9}$
    b. 36, 24, 12, 0, −12
    c. 4, 8, 12, 16, 20
    d. 4, 8, 16, 32, 64
4. **a.** 2, 10, 40, 200, 1000
    b. 2, $2\sqrt{2}$, 4, $4\sqrt{2}$, 8
    c. 1, 10, 100, 1000, 10,000
    d. 100, 90, 80, 70, 60
5. **a.** Write an arithmetic sequence of five terms with a first term $a_1 = 10$ and a common difference $d = 2$.
    b. Write an arithmetic sequence of five terms with a first term $a_1 = 10$ and a common difference $d = -2$.
    c. Write a geometric sequence of five terms with a first term $a_1 = 10$ and a common ratio $r = 2$.
    d. Write a geometric sequence of five terms with a first term $a_1 = 10$ and a common ratio $r = -2$.
6. **a.** Write an arithmetic sequence of five terms with a first term of $a_1 = 12$ and a common difference $d = \dfrac{1}{2}$.
    b. Write an arithmetic sequence of five terms with a first term of $a_1 = 12$ and a common difference $d = -\dfrac{1}{2}$.
    c. Write a geometric sequence of five terms with a first term of $a_1 = 12$ and a common ratio $r = \dfrac{1}{2}$.
    d. Write a geometric sequence of five terms with a first term of $a_1 = 12$ and a common ratio $r = -\dfrac{1}{2}$.

In Exercises 7–10 match each graph with the corresponding description.

7. A linear growth function (slope positive)
8. An exponential growth function
9. A linear decay function (slope negative)
10. An exponential decay function

A.   **B.**

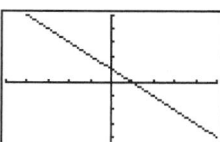

C.   **D.**

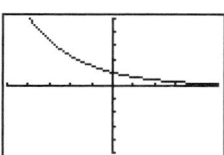   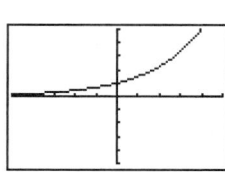

In Exercises 11–14 complete the next three terms of each sequence: (a) so that the sequence will be arithmetic and (b) so that the sequence will be geometric.

11. 3, 6, ___ , ___ , ___
12. 5, 15, ___ , ___ , ___
13. 2, 2, ___ , ___ , ___
14. −4, −4, ___ , ___ , ___
15. Given $f(x) = 4^x$, mentally evaluate each expression.
    a. $f(1)$ **b.** $f(3)$ **c.** $f(0)$
    d. $f(-2)$ **e.** $f\left(\dfrac{1}{2}\right)$
16. Given $f(x) = 8^x$, mentally evaluate each expression.
    a. $f(1)$ **b.** $f(2)$ **c.** $f(0)$
    d. $f(-2)$ **e.** $f\left(\dfrac{1}{3}\right)$

Estimation Skills and Calculator Skills
In Exercises 17–20 mentally estimate the value of each expression and then use a calculator to approximate each value to the nearest hundredth. Use the function $f(x) = (3.9)^x$.

PROBLEM	MENTAL ESTIMATE	CALCULATOR APPROXIMATION
17. $f(2)$		
18. $f(0.01)$		
19. $f(-1)$		
20. $f(0.5)$		

21. Given $f(x) = 2.7^x$, use a calculator to approximate each expression to the nearest thousandth.
    a. $f(8)$ **b.** $f(-2.3)$ **c.** $f(\pi)$
22. Given $f(x) = 0.83^x$, use a calculator to approximate each expression to the nearest thousandth.
    a. $f(3)$ **b.** $f(-3)$ **c.** $f(\pi)$
23. Given $f(x) = e^x$, use a calculator to approximate each expression to the nearest thousandth.
    a. $f\left(\sqrt{8}\right)$ **b.** $f(-1.6)$ **c.** $f(\pi)$
24. Given $f(x) = e^{-0.2x}$, use a calculator to approximate each expression to the nearest thousandth.
    a. $f\left(\sqrt{7}\right)$ **b.** $f(-2.4)$ **c.** $f(\pi)$

In Exercises 25–28 use the given function to calculate by hand a table of values. Use this table of values to sketch the function. You then may check your work using a calculator.

25. $f(x) = 4^x$

26. $f(x) = \left(\dfrac{1}{4}\right)^x$

27. $f(x) = 5^x$

28. $f(x) = \left(\dfrac{1}{5}\right)^x$

In Exercises 29 and 30 solve each equation without using a calculator.

29. a. $2x = 64$ **b.** $x^2 = 64$ **c.** $2^x = 64$

30. a. $3x = 1$ **b.** $x^3 = 1$ **c.** $3^x = 1$

In Exercises 31–62 solve each equation without using a calculator.

31. $4^x = 16$

32. $2^y = 32$

33. $5^{v+1} = 125$

34. $7^{v+5} = 49$

35. $\left(\dfrac{2}{3}\right)^w = \dfrac{8}{27}$

36. $\left(\dfrac{3}{4}\right)^w = \dfrac{9}{16}$

37. $2^m = \dfrac{1}{2}$

38. $5^m = \dfrac{1}{5}$

39. $5^n = 1$

40. $2^n = 1$

41. $25^x = 5$

42. $27^{2y} = 3$

43. $49^x = 7$

44. $8^{5y} = 2$

45. $7^x = \dfrac{1}{49}$

46. $5^x = \dfrac{1}{25}$

47. $\left(\dfrac{2}{5}\right)^w = \dfrac{125}{8}$

48. $\left(\dfrac{2}{7}\right)^w = \dfrac{49}{4}$

49. $32^{3n-5} = 2$

50. $64^{2n+9} = 2$

51. $3^v = \sqrt{3}$

52. $7v = \sqrt[3]{7}$

53. $10^x = 10{,}000$

54. $10^x = 100{,}000$

55. $10^y = 0.001$

56. $10^y = 0.00001$

57. $5^{x+7} = 25$

58. $7^{x-3} = 49$

59. $7^{3x} = \sqrt{7}$

60. $11^{2x-1} = \sqrt[3]{11}$

61. $\left(\dfrac{3}{7}\right)^x = \dfrac{49}{9}$

62. $\left(\dfrac{5}{4}\right)^{-x} = \dfrac{64}{125}$

63. Stock Prices A share of stock on the NASDAQ is priced at $100. Determine the price per share at the end of each of the next 4 years under each of the following assumptions.
   a. Arithmetic growth of $10 per year.
   b. Arithmetic decay of $10 per year.
   c. Geometric growth of 10% per year.
   d. Geometric decay of 10% per year.

64. Stock Prices A share of stock on the NASDAQ is priced at $100. Determine the price per share at the end of each of the next 4 years under each of the following assumptions.
   a. Arithmetic growth of $15 per year.
   b. Arithmetic decay of $15 per year.
   c. Geometric growth of 15% per year.
   d. Geometric decay of 15% per year.

65. Height of a Bouncing Ball A golf ball is dropped from 36 m onto a synthetic surface that causes it to rebound to one-half its previous height on each bounce. Determine the height of the first four bounces.

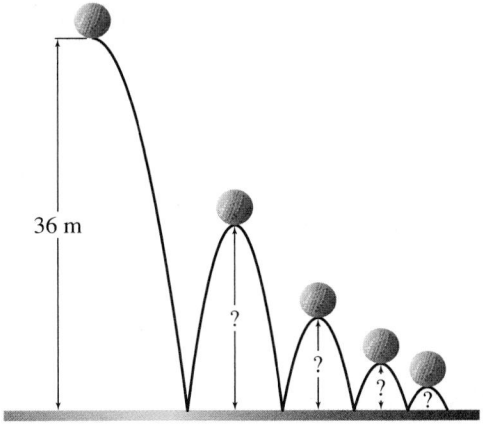

36 m

66. Radioactive Decay A nuclear chemist starts an experiment with 100 grams of a radioactive material. At the end of each time period only one-half the amount present at the start of the period is left. Determine the amount of this material left at the end of each of the first five time periods.

In Exercises 67 and 68 use the compound interest formula $A = P(1 + r)^t$ with A representing the total amount of money that accumulates when a principal P is invested at an annual interest rate r and left to compound for a time of t years.

67. Compound Interest Write A as a function of t when $1000 is invested at 6% for t years. Then evaluate this function for $t = 7$ years.

68. Compound Interest Write A as a function of t when $4000 is invested at 5.5% for t years. Then evaluate this function for $t = 10$ years.

69. Radioactive Decay The strontium-90 in a nuclear reactor decays continuously. If 100 mg is present initially, the amount present after t years is given by $A(t) = 100e^{-0.0248t}$. Approximate to the nearest tenth of a milligram the amount left after 10 years.

70. Radioactive Decay The strontium-90 in a nuclear reactor decays continuously. If 100 mg is present initially, the amount present after t years is given by $A(t) = 100e^{-0.0248t}$. Approximate to the nearest tenth of a milligram the amount left after 50 years.

Group Discussion Questions

71. Challenge Question $f(x) = b^x$ is an exponential function if $b > 0$ and $b \neq 1$.
   a. Discuss the nature of the function $f(x) = b^x$ if $b = 0$.
   b. Discuss the nature of the function $f(x) = b^x$ if $b = 1$.

72. **Challenge Question** The potential of geometric growth is demonstrated by this classic problem of a blacksmith shoeing a horse. A blacksmith attaches each horseshoe with eight nails. The blacksmith offers to charge by the nail for shoeing all four hooves. The cost for the first nail would be 1 cent, for the second 2 cents, for the third 4 cents, and so on. At this rate, complete the following table.

Nail	1	2	4	8	10	16	20	30	32
Cost of Nail	0.01	0.02							

73. **Challenge Question** An employee is offered two salary options. Option I starts at $20,000 per year with a $3000 raise after each of the first 9 years. Option II starts at $20,000 per year with a 10% raise after each of the first 9 years. Complete the following two tables.

OPTION I

YEAR	SALARY ($)	RAISE ($)	RAISE (%)
1	20,000		
2	23,000	3,000	15
3	26,000	3,000	13
4	29,000		
5			
6			
7			
8			
9			
10			

OPTION II

YEAR	SALARY ($)	RAISE ($)	RAISE (%)
1	20,000		
2	22,000	2,000	10
3	24,200	2,200	10
4	26,620		
5			
6			
7			
8			
9			
10			

Discuss the advantages and disadvantages of each option using the terms arithmetic growth and geometric growth where these terms are appropriate. Which option is better for a person who plans to quit this job after 5 years, after 10 years, after 20 years?

Section 10.2 Inverse Functions

Objectives:

3. Write the inverse of a function.
4. Graph the inverse of a function.

A function pairs each input value with exactly one output value.

Loan amount → Function → Monthly payment

Monthly payment → Inverse → Loan amount

A function (see Section 6.1) is a correspondence between two sets of values; a function matches each input value with exactly one output value. Because a function clearly pairs two sets of values, the information available from this pairing often can be used both ways—from x to y and from y to x.

For example, we can write a formula for a function that for a fixed time and interest rate will pair each car loan amount with the required monthly payment. By reversing this function we can use the monthly payment we can afford to determine the loan amount that is affordable. This reversing of a function produces what we call the **inverse of a function.**

Inverse of a Function

If f is a function that matches each input value x with an output value y, then the inverse of f reverses this correspondence to match this y-value with the x-value.

In Section 6.1 we examined several representations for functions including: mapping notation, ordered pairs, tables, graphs, and function notation. We now examine the inverse of a function using each of these representations. Note that in each case the inverse undoes what f does.

◼ **EXAMPLE 1**   Determining the Inverse of a Function

Determine the inverse of each function.

SOLUTIONS

	Function f			*The Inverse of f*		

(a)

x		y
7	⟶	1
8	⟶	3
9	⟶	5

x		y
1	⟶	7
3	⟶	8
5	⟶	9

In the mapping notation the inverse of f reverses the arrows and the order of the input-output pairs.

(b)

x	y
4	-1
7	2
$-\pi$	5
0	6

x	y
-1	4
2	7
5	$-\pi$
6	0

In the table the inverse of f reverses the x-y pairing.

(c) $\{(0, 1), (2, 7), (-5, 4)\}$

$\{(1, 0), (7, 2), (4, -5)\}$

The inverse of f reverses the order of each (x, y) pair.

(d)

LOAN AMOUNT ($)	MONTHLY PAYMENT ($)
5,000	122.06
10,000	244.13
15,000	366.19
20,000	488.26

MONTHLY PAYMENT ($)	LOAN AMOUNT ($)
122.06	5,000
244.13	10,000
366.19	15,000
488.26	20,000

The given function pairs each loan amount with the payment required to pay off an 8% loan in 4 years. The inverse pairs a payment with the loan amount this payment will pay off.   ◼

Note that all the samples in Example 1 illustrate that the inverse of f reverses the roles of the domain and range of the function f. In Example 1(c) the domain of $f = \{(0, 1), (2, 7), (-5, 4)\}$ is the set $\{0, 2, -5\}$ and the range is the set $\{1, 7, 4\}$. The inverse of f is $\{(1, 0), (7, 2), (4, -5)\}$. Its domain is the set $\{1, 7, 4\}$, and its range is the set $\{0, 2, -5\}$.

SELF-CHECK 10.2.1 ANSWERS

1. $D = \{8, 4, -2\}$,
   $R = \{6, 11, 0\}$

2.

x		y
6	⟶	8
11	⟶	4
0	⟶	-2

3. $D = \{6, 11, 0\}$,
   $R = \{8, 4, -2\}$

SELF-CHECK 10.2.1

For the function f:

x		y
8	⟶	6
4	⟶	11
-2	⟶	0

1. Give the domain and range of f.
2. Give the inverse of f.
3. Give the domain and range of the inverse of f.

We now investigate the conditions under which the inverse of a function f is also a function. Remember that a function must pair each input value with exactly one output value.

The inverse of the function $f = \{(3, 4), (5, 4)\}$ is given by $\{(4, 3), (4, 5)\}$. In this case the inverse of f does not match each input with exactly one output; 4 is paired with both 3 and 5. Examining $f = \{(3, 4), (5, 4)\}$, note that f has two input values with an output of 4. This causes the inverse of f to have two outputs for the input of 4. Thus the inverse of f is not a function.

However, if different inputs of f have different outputs, then the inverse of f also will be a function. We call functions with this property **one-to-one functions.**

One-to-One Function

A function is one-to-one if different input values produce different output values.

The inverse of a function f is an inverse function if f is one-to-one. We denote this inverse by f^{-1}.

If a function f is one-to-one, then the inverse of f will also be a function. In this case we denote the inverse of f by f^{-1} and call f^{-1} an **inverse function.**

■ EXAMPLE 2   Identifying One-to-One Functions

Determine whether the function f is one-to-one and whether the inverse of f is an inverse function.

SOLUTIONS

Caution: Treat f^{-1} as a single symbol that undoes what the function f does. This is *not* exponential notation and does not represent $\frac{1}{f}$. The context of the notation should make the intent of the notation clear.

(a) $f = \{(-5, 6), (2, 7), (8, 9), (9, 3)\}$ This function is one-to-one, and thus $f^{-1} = \{(6, -5), (7, 2), (9, 8), (3, 9)\}$ is an inverse function.

(b) $f = \{(-3, 9), (0, 0), (3, 9)\}$ This function is not one-to-one because both -3 and 3 are paired with 9. Thus the inverse of f, $\{(9, -3), (0, 0), (9, 3)\}$, is not a function. ■

SELF-CHECK 10.2.2

Determine whether f is one-to-one and whether the inverse of f is an inverse function.

1. $f = \{(3, \pi), (e, 4), (8, -1)\}$
2. $f = \{(5, 9), (7, 6), (-3, 9)\}$

SELF-CHECK 10.2.2 ANSWERS

1. f is one-to-one; the inverse of f, $\{(\pi, 3), (4, e), (-1, 8)\}$, is an inverse function.
2. f is not one-to-one; the inverse of f, $\{(9, 5), (6, 7), (9, -3)\}$, is the inverse of a function but is not an inverse function.

The vertical-line test covered in Section 6.1 is a quick method for determining whether a graph represents a function. This test allows us to determine visually whether each x input value is paired with exactly one y output value. Similarly, the horizontal-line test is a quick method for determining whether the graph of a function represents a one-to-one function.

The Horizontal-Line Test

A graph of a function represents a one-to-one function if it is impossible to have any horizontal line intersect the graph at more than one point.

■ EXAMPLE 3   Using the Horizontal-Line Test

Use the horizontal-line test to determine whether each graph represents a one-to-one function.

Use the vertical-line test to determine whether a graph represents a function. Then use the horizontal-line test to determine whether the graph of a function represents a one-to-one function.

SOLUTIONS

(a)

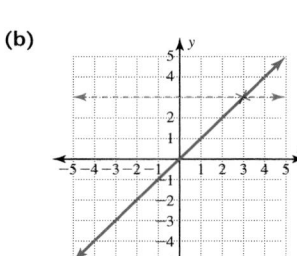

Not a one-to-one function

The horizontal line passes through more than one point of the graph of this function.

(b)

A one-to-one function

Any horizontal line will pass through exactly one point of the graph of this function.

SELF-CHECK 10.2.3

Use a graphics calculator to graph each function and to determine if the function is one-to-one.

1. $f(x) = \sqrt{4x - 1}$
2. $f(x) = |x - 1| - 2$

A key concept to understand about f^{-1} is that f^{-1} undoes what the function f does. This is illustrated in Example 4.

■ EXAMPLE 4   Describing Inverse Functions

Describe each function and its inverse verbally and algebraically. Give a numerical example from the domain and range of each function.

SOLUTIONS

	VERBALLY	**ALGEBRAICALLY**	**NUMERICAL EXAMPLE**
(a) $f(x) = 2x$	f doubles x	$f(x) = 2x$	$8 \xrightarrow{\ f\ } 16$
	f^{-1} takes one-half of x	$f^{-1}(x) = \dfrac{x}{2}$	$16 \xrightarrow{\ f^{-1}\ } 8$
(b) $f(x) = \sqrt[3]{x}$	f takes the cube root of x	$f(x) = \sqrt[3]{x}$	$8 \xrightarrow{\ f\ } 2$
	f^{-1} cubes x	$f^{-1}(x) = x^3$	$2 \xrightarrow{\ f^{-1}\ } 8$

> ### SELF-CHECK 10.2.4
>
> Describe f^{-1} verbally and write both f and f^{-1} using function notation.
>
> **1.** f triples x.
> **2.** f increases x by 5.

If f is a one-to-one function defined by an equation, then the inverse function f^{-1} also is given indirectly by this equation. In Example 4 we noted this when we used simple functions to describe how f^{-1} undoes what f does. Another example of this is:

VERBALLY	**ALGEBRAICALLY**
$f : 5 \longrightarrow \boxed{\begin{array}{c}\text{First}\\ \text{double}\end{array}} \xrightarrow{\ 10\ } \boxed{\begin{array}{c}\text{Then}\\ \text{add 1}\end{array}} \longrightarrow 11$	$f(x) = 2x + 1$
$f^{-1} : 11 \longrightarrow \boxed{\begin{array}{c}\text{First}\\ \text{subtract 1}\end{array}} \xrightarrow{\ 10\ } \boxed{\begin{array}{c}\text{Then take half}\\ \text{of this value}\end{array}} \longrightarrow 5$	$f^{-1}(x) = \dfrac{x - 1}{2}$

It is customary to use x to represent the input value for both f and f^{-1}. Thus the formulas for both f and f^{-1} often are expressed using the variable x. Because the inverse function f^{-1} reverses the order of each ordered pair (x, y), we interchange the roles of x and y in the formula for f in order to find a formula for f^{-1} in terms of x. The following box describes this procedure and reexamines the function $f(x) = 2x + 1$ given previously.

Finding an Equation for an Inverse Function

To find the inverse of a one-to-one function $y = f(x)$:

VERBALLY	ALGEBRAIC EXAMPLE	
STEP 1 Replace $f(x)$ by y. Write this as a function of y in terms of x.	Function:	$f(x) = 2x + 1$ $y = 2x + 1$
STEP 2 To form the inverse, replace each x with y and each y with x.	Inverse function:	$x = 2y + 1$
STEP 3 If possible, solve the resulting equation for y.		$x - 1 = 2y$ $\dfrac{x - 1}{2} = y$
STEP 4 The inverse can then be written in function notation by replacing y with $f^{-1}(x)$.		$f^{-1}(x) = \dfrac{x - 1}{2}$

■ EXAMPLE 5   Determining an Inverse Function

Determine the inverse of $f(x) = 3x - 5$ algebraically and compare f and f^{-1} both verbally and graphically.

SOLUTION

ALGEBRAICALLY

Function:
$$f(x) = 3x - 5$$
$$y = 3x - 5 \qquad \text{First replace } f(x) \text{ by } y \text{ to express } y \text{ as a function of } x.$$

Inverse function:
$$x = 3y - 5 \qquad \text{Then interchange } x \text{ and } y \text{ in the equation to form the inverse.}$$
$$x + 5 = 3y \qquad \text{Solve for } y \text{ by adding 5 to both sides of the equation and then dividing both sides by 3.}$$
$$y = \frac{x + 5}{3}$$
$$f^{-1}(x) = \frac{x + 5}{3} \qquad \text{Rewrite the inverse using function notation.}$$

VERBALLY

$f(x) = 3x - 5$ $f : x \rightarrow$ $\boxed{\text{First triple } x}$ $\xrightarrow{\ 3x\ }$ $\boxed{\text{Then subtract 5}}$ $\rightarrow 3x - 5$

$f^{-1}(x) = \dfrac{x + 5}{3}$ $f^{-1} : x \rightarrow$ $\boxed{\text{First add 5}}$ $\xrightarrow{\ x + 5\ }$ $\boxed{\text{Then divide by 3}}$ $\rightarrow \dfrac{x + 5}{3}$

NUMERICAL EXAMPLE

$f(2) = 1 :$ $2 \rightarrow$ $\boxed{\text{First triple 2}}$ $\xrightarrow{\ 6\ }$ $\boxed{\text{Then subtract 5}}$ $\rightarrow 1$

$f^{-1}(1) = 2 :$ $1 \rightarrow$ $\boxed{\text{First add 5}}$ $\xrightarrow{\ 6\ }$ $\boxed{\text{Then divide by 3}}$ $\rightarrow 2$

GRAPHICALLY

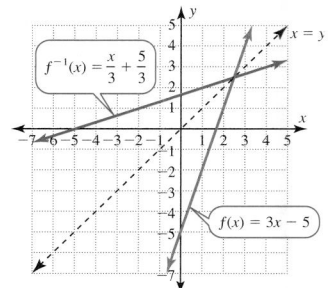

The graphs of both f and f^{-1} are straight lines. Inspection of the slope-intercept form reveals that f has slope 3 and y-intercept -5 and f^{-1} has slope $\dfrac{1}{3}$ and y-intercept $\dfrac{5}{3}$.

Note that these graphs are symmetric about the line $x = y$.

Determine the inverse of $f(x) = \dfrac{2x - 1}{3}$.

As illustrated by Example 5, the graphs of f and f^{-1} are symmetric about the line $x = y$. Note in the figure that the points (a, b) and (b, a) are mirror images about the graph of $x = y$. Thus f and f^{-1} always will be symmetric about the line $x = y$. Example 6 further illustrates this symmetry.

Symmetry of (a, b) and (b, a) about $x = y$

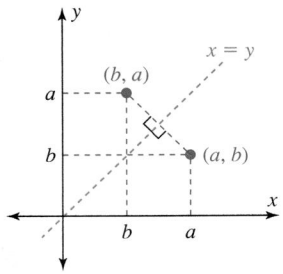

■ **EXAMPLE 6**   Graphing the Inverse of a Function

The graph of the function $f = \{(2, 1), (4, 2), (6, 3)\}$ is shown in the figure. Graph the inverse of this function.

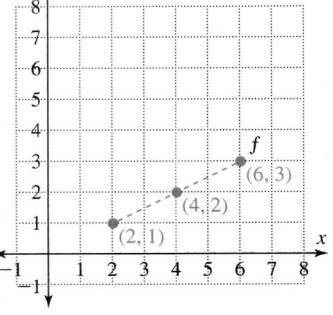

SOLUTION _____

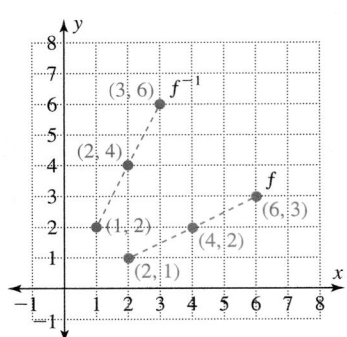

The inverse of $f = \{(2, 1), (4, 2), (6, 3)\}$ is $f^{-1} = \{(1, 2), (2, 4), (3, 6)\}$. In fact, the inverse of all the points on the dashed line segment through the points of f is the set of all the points on the dashed line segment through the points of f^{-1}. You can check some of these points yourself. ■

In Example 7 we take advantage of the symmetry of f and f^{-1} about the line $x = y$. Using this symmetry we can graph f^{-1} directly from the graph of f without knowing the equation for f^{-1}.

■ EXAMPLE 7   Using Symmetry to Graph f^{-1}

Graph $f(x) = 2^x$ and use symmetry about the line $x = y$ to graph $y = f^{-1}(x)$.

SOLUTION

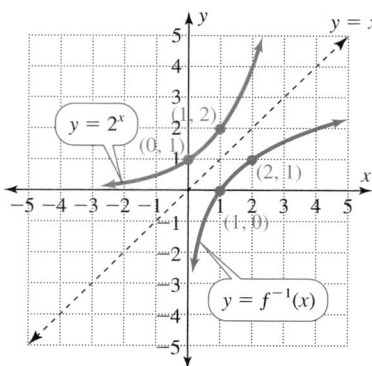

First graph the exponential function $f(x) = 2^x$ and sketch in the dashed line $x = y$. Then use symmetry to create a symmetric image for $f(x) = 2^x$ about $x = y$. This symmetric image is the graph of $y = f^{-1}(x)$.

Note that the symmetric image of $(0, 1)$ is $(1, 0)$ and of $(1, 2)$ is $(2, 1)$.

Some graphics calculators have the ability to draw the inverse of a function. This is illustrated in Calculator Perspective 10.2.1.

CALCULATOR PERSPECTIVE 10.2.1	Drawing the Inverse of a Function

To draw the inverse of the function $f(x) = 3x - 5$ from Example 5 on a TI-83 Plus calculator, enter the following keystrokes:

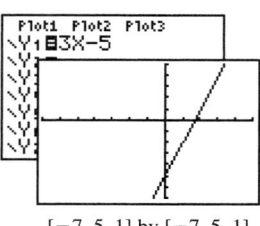

Y= 3 X,T,θ,n − 5
GRAPH

$[-7, 5, 1]$ by $[-7, 5, 1]$

The **DRAW** feature is the secondary function of the **PRGM** key.

2nd DRAW 8 VARS ▶ ENTER
ENTER ENTER

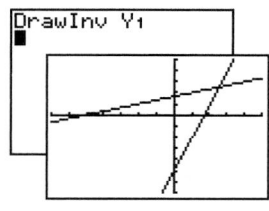

$[-7, 5, 1]$ by $[-7, 5, 1]$

SELF-CHECK 10.2.6

Use a graphics calculator to check Example 7. First graph $f(x) = 2^x$. Then use DrawInv to graph $y = f^{-1}(x)$.

SELF-CHECK 10.2.6 ANSWER

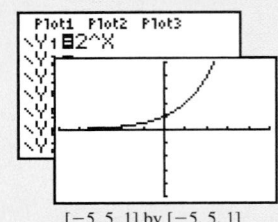

$[-5, 5, 1]$ by $[-5, 5, 1]$

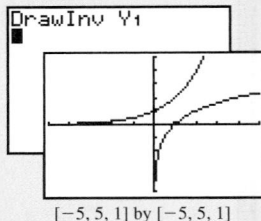

$[-5, 5, 1]$ by $[-5, 5, 1]$

USING THE LANGUAGE AND SYMBOLISM OF MATHEMATICS   10.2

1. If f is a function that matches each input value x with an output value y, then the inverse of f reverses this correspondence to match this _____-value with the _____-value.
2. A function is one-to-one if different input values produce _____ output values.
3. A graph of a function represents a _____-_____-_____ function if it is impossible to

have any horizontal line intersect the graph at more than one point.
4. If f is a one-to-one function, then the inverse of f is a function denoted by _____.
5. The graphs of f and f^{-1} are symmetric about the line _____.

EXERCISES   10.2

In Exercises 1–10 write the inverse of each function using ordered-pair notation.

1. $\{(1, 4), (3, 11), (8, 2)\}$
2. $\{(2, -1), (0, 3), (-4, 6)\}$
3. $\{(-3, 2), (-1, 2), (0, 2), (2, 2)\}$
4. $\{(-0.7, \pi), (-0.5, \pi), (1.3, \pi)\}$
5. $\{(a, b), (c, d)\}$
6. $\{(w, x), (y, z)\}$

7.
x		y
4	→	7
e	→	9
−3	→	−2

8.
x		y
−8	→	1
−9	→	3
6	→	−4

9.
x	y
9	0
4	3
6	−2
−4	5

10.
x	y
−2	3
−1	5
0	7
1	9

In Exercises 11 and 12 first express the function f, whose graph is given, in ordered-pair notation. Then write f^{-1} using ordered-pair notation.

11.

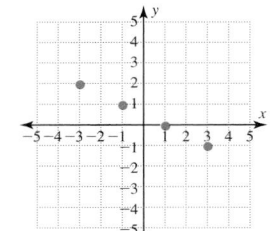

12.

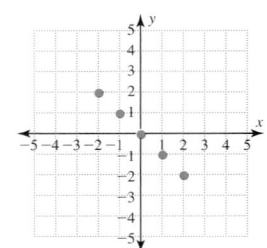

In Exercises 13–16 state the domain and range of the given function f and its inverse f^{-1}.

13.

x		y
0	→	−3
1	→	−2
2	→	−1
3	→	0
4	→	1

14.

x		y
−1	→	−6
0	→	−4
1	→	−2
2	→	0
3	→	2

15.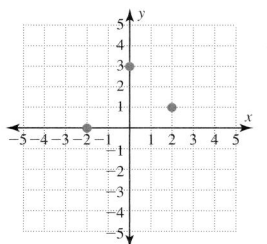

16.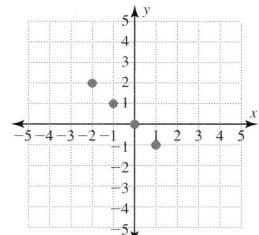

In Exercises 17–20 determine whether each function is one-to-one.

17. $\{(-\pi, 4), (3, \pi), (7, -9)\}$
18. $\{(-8, 9), (7, 13), (8, 9)\}$
19. $\{(7.5, 7), (7.8, 7), (8.3, 8)\}$
20. $\{(-5, 2), (-4, -1), (3, 7), (8, 8)\}$

In Exercises 21–30 use the horizontal line test to determine whether each graph represents a one-to-one function.

21.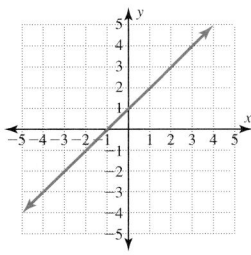

22.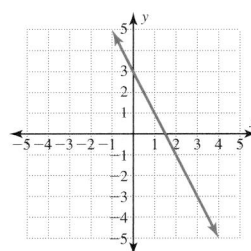

23.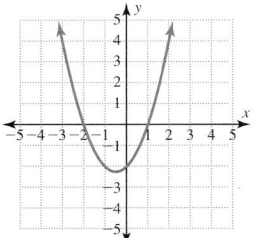

24.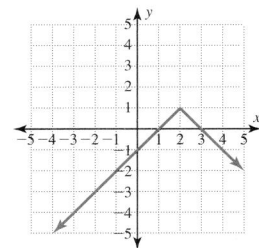

25.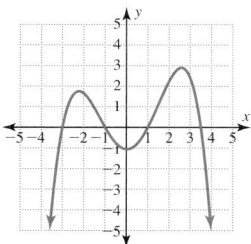

26.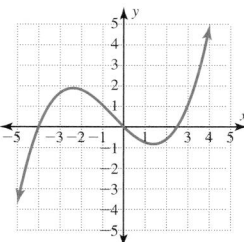

27.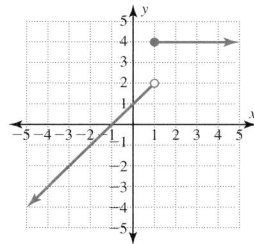

28.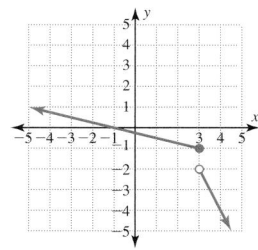

29.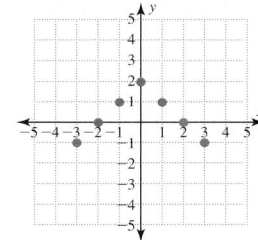

30.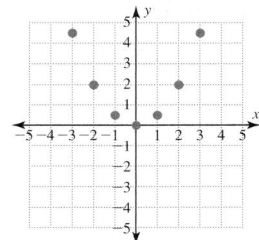

In Exercises 31–40 sketch the inverse of each function using the symmetry of f and f^{-1} about the line $x = y$. (*Hint:* For Exercises 33–40 pick a few key points on f and f^{-1} and then sketch the rest of f^{-1}.)

31.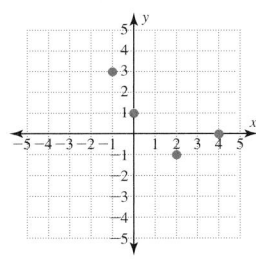

32.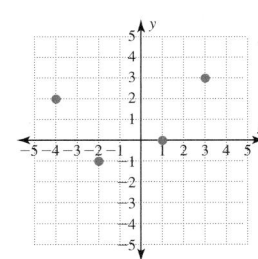

33.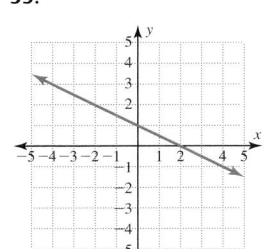

34.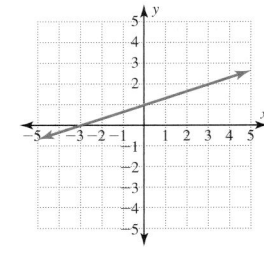

35.

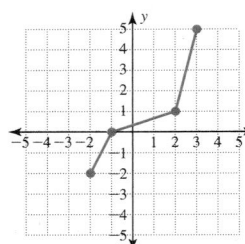

36.

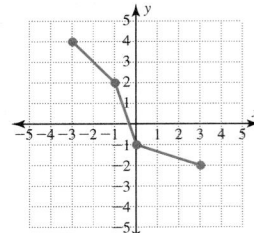

37.

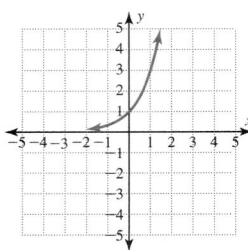

38.

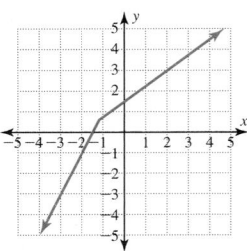

39.

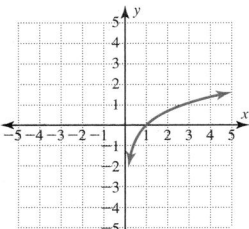

40.

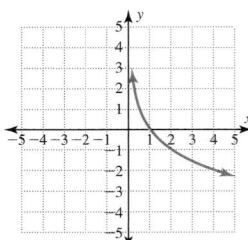

In Exercises 41–44 use a graphics calculator with the DrawInv feature to graph $y = f(x)$ and to draw $f^{-1}(x)$.

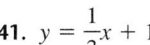

41. $y = \dfrac{1}{3}x + 1$

42. $y = 3x - 3$

43. $y = \left(\dfrac{1}{2}\right)^x$

44. $y = 3^x$

In Exercises 45–50 the function f is described verbally and algebraically. Describe f^{-1} both verbally and algebraically.

VERBALLY	ALGEBRAICALLY
45. f decreases x by 2	$f(x) = x - 2$
46. f increases x by 7	$f(x) = x + 7$
47. f quadruples x	$f(x) = 4x$
48. f takes one-fifth of x	$f(x) = \dfrac{x}{5}$
49. f doubles x and then subtracts 3	$f(x) = 2x - 3$
50. f takes one-half of x and then adds 5	$f(x) = \dfrac{1}{2}x + 5$

In Exercises 51–58 write the inverse of each function using function notation.

51. $f(x) = 5x + 2$

52. $f(x) = 2x - 5$

53. $f(x) = \dfrac{1}{3}x - 7$

54. $f(x) = -\dfrac{1}{4}x + 3$

55. $f(x) = \sqrt[3]{x} + 2$

56. $f(x) = \sqrt[3]{x + 2}$

57. $f(x) = -x$

58. $f(x) = \dfrac{1}{x}$

In Exercises 59–64 use $f(x) = 4x - 3$ and $f^{-1}(x) = \dfrac{x + 3}{4}$ to evaluate each expression.

59. a. $f(5)$ **b.** $f^{-1}(17)$
60. a. $f(0)$ **b.** $f^{-1}(-3)$
61. a. $f(2)$ **b.** $f^{-1}(5)$
62. a. $f\left(\dfrac{3}{4}\right)$ **b.** $f^{-1}(0)$
63. a. $f(1)$ **b.** $f^{-1}(1)$
64. a. $f(10)$ **b.** $f^{-1}(37)$

In Exercises 65–70 each table defines a function. Form a table that defines the inverse of each function. State conditions for which you think the inverse function might be more useful than the original function.

65. Shoe Sizes

U.S. SIZE	EUROPEAN SIZE
7	39
8	40.5
9	42
10	43
11	44

66. Environmental Application

YEAR	U.S. LAND USED FOR RECREATION (IN MILLIONS OF ACRES)
1977	66
1982	71
1987	84
1992	87
1997	96

U.S. Department of Agriculture
http://www.usda.gov/

67. Car Loan at 7.5% for 4 Years

LOAN AMOUNT ($)	MONTHLY PAYMENT ($)
5,000	120.89
10,000	241.79
15,000	362.68
20,000	483.58
25,000	604.47

68. Car Loan at 5% for 4 Years

LOAN AMOUNT ($)	MONTHLY PAYMENT ($)
5,000	115.15
10,000	230.29
15,000	345.44
20,000	460.59
25,000	575.73

69. Temperature Conversions

TEMPERATURE, °F	TEMPERATURE, °C
0	−17.8
20	−6.7
40	4.4
60	15.6
80	26.7
100	37.8

70. Tuition at Gulf Coast Community College

SEMESTER HOURS	TUITION ($)
1	50
2	100
3	150
4	200
5	250
10	500
15	800

71. Production Cost   The cost of producing x units of a product is given by the function $C(x) = 12x + 350$.
  a. What does the variable x represent in $C(x)$?
  b. What does the output of $C(x)$ represent?
  c. Determine the formula for $C^{-1}(x)$.
  d. What does the variable x represent in $C^{-1}(x)$?
  e. What does the output of $C^{-1}(x)$ represent?
  f. Determine the cost of producing 100 units.
  g. Determine the number of units that can be produced for $1934.

72. Sales Commission   A car dealership hires new salespeople and calculates the monthly salary for each person for the first 6 months using the formula $S(x) = 500x + 1000$, where x represents the number of cars this person sold for the month.
  a. Evaluate and interpret $S(0)$.
  b. Evaluate and interpret $S(8)$.
  c. Determine the formula for $S^{-1}(x)$.
  d. What does the variable x represent in $S^{-1}(x)$?
  e. Evaluate and interpret $S^{-1}(2500)$.
  f. Evaluate and interpret $S^{-1}(6000)$.

Group Discussion Questions
73. Discovery Question
  a. Make two identical copies of the graph of a function $y = f(x)$ on two sheets of clear plastic. Also, dash in the line $x = y$ on each sheet of plastic.
  b. Place one sheet of plastic on an overhead projector.
  c. Then flip the second sheet of plastic over and turn the sheet 90° so that the upside-down sheet has its positive y-axis align with the positive x-axis on the other sheet.
  d. Try this for three or four functions and compare these results to using the DrawInv feature on a graphics calculator.
  e. Describe to your instructor what you have observed and why you think this is so.
74. Challenge Question
  a. Is $f(x) = x^2$ a one-to-one function?
  b. If we restrict $f(x) = x^2$ to the domain $D = [0, +\infty)$, is this restricted function one-to-one?
  c. Graph the restricted function in part (b) and its inverse.
  d. Write the equation of the inverse of the restricted function in part (b).
  e. If we restrict $f(x) = x^2$ to the domain $D = (-\infty, 0]$, is this restricted function one-to-one?
  f. Graph the restricted function in part (e) and its inverse.
  g. Write the equation of the inverse of the restricted function in part (e).
75. Calculator Discovery
  a. Use a graphics calculator to graph $y_1 = e^x$.
  b. Use a graphics calculator to graph $y_2 = \ln x$. (*Hint:* Use the LN key.)
  c. Compare these two graphs and make a conjecture on the relationship between the functions $y_1 = e^x$ and $y_2 = \ln x$.

Section 10.3 Logarithmic Functions

Objectives: 5. Interpret and use logarithmic notation.

6. Simplify logarithmic expressions using the basic identities.

The following graphs of the exponential functions illustrate that an exponential function $f(x) = b^x$ is a one-to-one function for both exponential growth and exponential decay. Thus each exponential function has an inverse function.

Also note that the projection of the graph of an exponential function onto the x-axis is the entire real axis. Thus the domain of $f(x) = b^x$ is $\mathbb{R}$, the set of all real numbers. The projection of each graph onto the y-axis is the interval $(0, +\infty)$. Thus the range of $f(x) = b^x$ is $(0, +\infty)$.

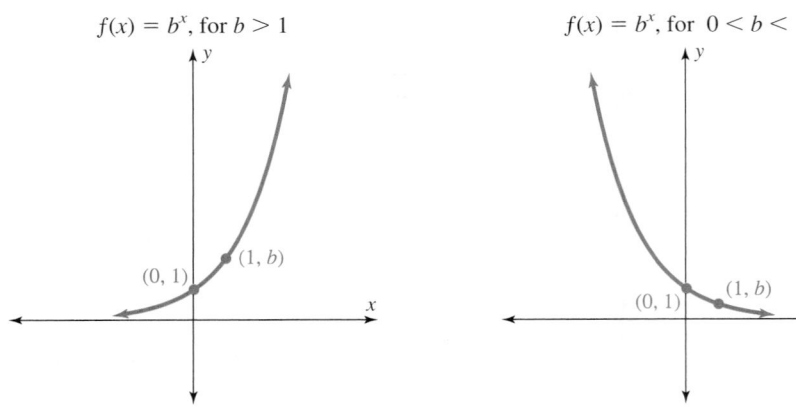

The domain of an exponential function $f(x) = b^x$ is $\mathbb{R}$ and the range is $(0, +\infty)$.

Domain: $\mathbb{R}$ Range: $(0, +\infty)$
The points $(0, 1)$ and $(1, b)$ are on the graph of $f(x) = b^x$.

The symmetry of f and f^{-1} about the line $x = y$ is used to sketch the inverse of $f(x) = b^x$ in the following graphs. Note that the domain of each inverse function is $(0, +\infty)$ and the range is $\mathbb{R}$.

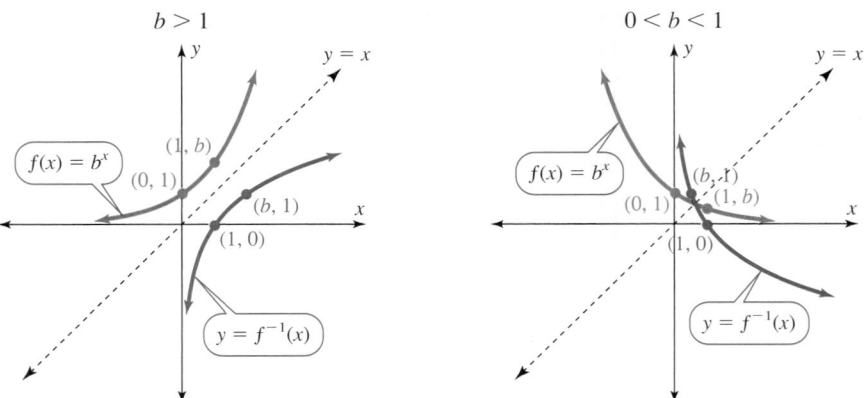

The domain of the inverse of $f(x) = b^x$ is $(0, +\infty)$ and the range is $\mathbb{R}$.

Domain of $y = f^{-1}(x)$: $(0, +\infty)$ Range: $\mathbb{R}$
The points $(1, 0)$ and $(b, 1)$ are on the graph of $y = f^{-1}(x)$.

The inverse of $f(x) = b^x$ cannot be rewritten by using the basic operations to solve $x = b^y$ for y. Thus we introduce a new notation to denote the inverse of $f(x) = b^x$. This inverse is denoted by $f^{-1}(x) = \log_b x$, which is read "the log base b of x." Logarithmic functions are defined in the following box.

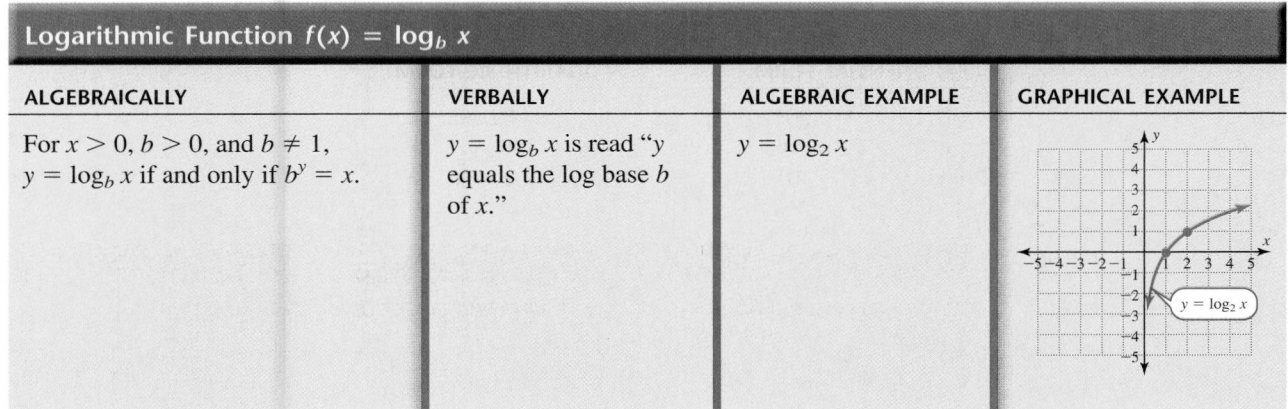

Logarithmic Function $f(x) = \log_b x$

ALGEBRAICALLY	VERBALLY	ALGEBRAIC EXAMPLE	GRAPHICAL EXAMPLE
For $x > 0$, $b > 0$, and $b \neq 1$, $y = \log_b x$ if and only if $b^y = x$.	$y = \log_b x$ is read "y equals the log base b of x."	$y = \log_2 x$	

A function and its inverse both contain the same information but from different perspectives—the ordered pairs in the original function are reversed to obtain the ordered pairs in the inverse. As you work with $y = b^x$ and $y = \log_b x$ remember that these functions are inverses of each other and contain the same information.

Think: Logarithms are exponents.

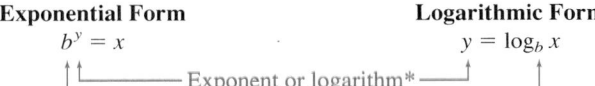

Exponential Form
$$b^y = x$$

Logarithmic Form
$$y = \log_b x$$

Exponent or logarithm*
Base

The relationship between exponential and logarithmic forms is examined further in Examples 1 and 2.

■ EXAMPLE 1   Translating Logarithmic Equations to Exponential Form

Write each logarithmic equation in verbal form and translate this equation to the equivalent exponential form.

SOLUTIONS

LOGARITHMIC FORM	VERBALLY	EXPONENTIAL FORM
(a) $\log_5 25 = 2$	The log base 5 of 25 is 2.	$5^2 = 25$
(b) $\log_2 8 = 3$	The log base 2 of 8 is 3.	$2^3 = 8$
(c) $\log_3\left(\dfrac{1}{3}\right) = -1$	The log base 3 of $\dfrac{1}{3}$ is -1.	$3^{-1} = \dfrac{1}{3}$
(d) $\log_7 \sqrt{7} = \dfrac{1}{2}$	The log base 7 of $\sqrt{7}$ is $\dfrac{1}{2}$.	$7^{1/2} = \sqrt{7}$

■ **EXAMPLE 2** Translating Exponential Equations to Logarithmic Form

Translate each exponential equation to logarithmic form.

SOLUTIONS

EXPONENTIAL FORM	LOGARITHMIC FORM
(a) $10^{-3} = 0.001$	$\log_{10} 0.001 = -3$
(b) $e^2 \approx 7.389$	$\log_e 7.389 \approx 2$
(c) $A = (1 + r)^x$	$\log_{(1+r)} A = x$

SELF-CHECK 10.3.1

Write each logarithmic equation in exponential form and each exponential equation in logarithmic form.

1. $\log_7 49 = 2$ **2.** $\log_6 \sqrt[5]{6} = \dfrac{1}{5}$

3. $3^{-4} = \dfrac{1}{81}$ **4.** $6^0 = 1$

To evaluate $\log_b x$, think "What exponent on b is needed to obtain x?" Problems involving familiar powers of small integers can often be solved easily without the use of a calculator. For problems involving more complicated values, a calculator is generally used.

■ **EXAMPLE 3** Evaluating Logarithms by Inspection

Determine the value of each logarithm by inspection.

SOLUTIONS

(a) $\log_5 125$	$\log_5 125 = 3$	In exponential form, $5^3 = 125$.
(b) $\log_2 \dfrac{1}{8}$	$\log_2 \dfrac{1}{8} = -3$	In exponential form, $2^{-3} = \dfrac{1}{8}$.
(c) $\log_{49} 7$	$\log_{49} 7 = \dfrac{1}{2}$	In exponential form, $49^{1/2} = 7$.
(d) $\log_5 0$	$\log_5 0$ is undefined.	The input of a logarithmic function cannot be 0.
(e) $\log_5(-25)$	$\log_5(-25)$ is undefined.	The input of a logarithmic function cannot be negative.

SELF-CHECK 10.3.1 ANSWERS

1. $7^2 = 49$
2. $6^{1/5} = \sqrt[5]{6}$
3. $\log_3 \dfrac{1}{81} = -4$
4. $\log_6 1 = 0$

Four properties that result from the definition of a logarithm are given in the following box. These important properties are used often in solving the logarithmic equations that arise in the solution of some word problems.

Properties of Logarithmic Functions

For $b > 0$ and $b \neq 1$,

LOGARITHMIC FORM	EXPONENTIAL FORM	NUMERICAL EXAMPLE	
1. $\log_b 1 = 0$	$b^0 = 1$	$\log_5 1 = 0$	$5^0 = 1$
2. $\log_b b = 1$	$b^1 = b$	$\log_5 5 = 1$	$5^1 = 5$
3. $\log_b \dfrac{1}{b} = -1$	$b^{-1} = \dfrac{1}{b}$	$\log_5 \dfrac{1}{5} = -1$	$5^{-1} = \dfrac{1}{5}$
4. $\log_b b^x = x$	$b^x = b^x$	$\log_5 5^4 = 4$	$5^4 = 5^4$

We use these properties to evaluate the logarithms in Example 4. Other properties of logarithms are examined in the following sections of this chapter.

■ EXAMPLE 4   Using Properties to Evaluate Logarithms

Determine the value of each logarithm by inspection.

SOLUTIONS

(a) $\log_{19} 1$ $\log_{19} 1 = 0$ In exponential form $19^0 = 1$.

(b) $\log_7 7$ $\log_7 7 = 1$ In exponential form $7^1 = 7$.

(c) $\log_8 \dfrac{1}{8}$ $\log_8 \dfrac{1}{8} = -1$ In exponential form $8^{-1} = \dfrac{1}{8}$.

(d) $\log_8 8^{2y}$ $\log_8 8^{2y} = 2y$ In exponential form $8^{2y} = 8^{2y}$. ■

SELF-CHECK 10.3.2

Determine the value of each logarithm.

1. $\log_2 64$ 2. $\log_4 64$

3. $\log_8 64$ 4. $\log_7 \dfrac{1}{7}$

5. $\log_e 1$ 6. $\log_e e$

7. $\log_e \dfrac{1}{e}$ 8. $\log_e e^8$

SELF-CHECK 10.3.2 ANSWERS

1. 6   2. 3   3. 2   4. -1
5. 0   6. 1   7. -1 8. 8

The equations in Example 5 can be solved without use of a calculator because the numbers involve well-known powers. We will examine equations of this type in more detail in Section 10.6.

■ **EXAMPLE 5** Solving Equations

Solve each equation for x.

SOLUTIONS

(a) $\log_3(x - 1) = 2$

$\log_3(x - 1) = 2$

$x - 1 = 3^2$ First rewrite this equation in exponential form.

$x - 1 = 9$ Substitute 9 for 3^2 and then solve for x.

$x = 10$

(b) $\log_x 9 = -2$

$\log_x 9 = -2$

$x^{-2} = 9$ First rewrite this equation in exponential form.

$x^{-2} = 3^2$ Now work toward expressing each side in terms of the same exponent.

$x^{-2} = \left(\dfrac{1}{3}\right)^{-2}$

$x = \dfrac{1}{3}$ The bases are equal since the exponents are equal.

(c) $\log_4 8 = x$

$\log_4 8 = x$

$4^x = 8$ First rewrite this equation in exponential form.

$\left(2^2\right)^x = 2^3$ Then express each side in terms of the common base of 2. Use the power rule for exponents to rewrite the

$2^{2x} = 2^3$ left side of the equation.

$2x = 3$ Then equate the exponents since the bases are the same.

$x = \dfrac{3}{2}$

■

SELF-CHECK 10.3.3

Solve each equation for x.

1. $\log_2 x = 3$
2. $\log_{11} \sqrt{11} = x$
3. $\log_x 125 = 3$

SELF-CHECK 10.3.3 ANSWERS

1. $x = 8$ 2. $x = \dfrac{1}{2}$

3. $x = 5$

Linear growth exhibits a constant rate of growth. By contrast, exponential growth exhibits a rate of growth that eventually becomes very very rapid. On the other extreme, logarithmic growth eventually exhibits very very slow growth. In Example 6 we use a graphics calculator to graph the logarithmic growth function $f(x) = \log_e x$.

■ EXAMPLE 6   Graphing a Logarithmic Growth Function

To graph $f(x) = \log_e x$, first use a graphics calculator to graph the corresponding exponential function $y = e^x$. Then use the DrawInv feature to sketch the graph of the $f(x) = \log_e x$.

SOLUTION

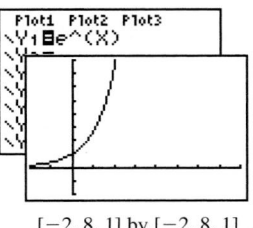

[−2, 8, 1] by [−2, 8, 1]

Because $f(x) = \log_e x$ and $y_1 = e^x$ are inverses of each other, we can use the graph of y_1 and DrawInv to sketch the graph of $f(x) = \log_e x$.

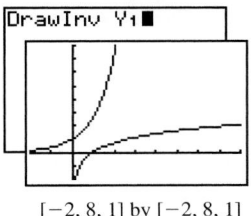

[−2, 8, 1] by [−2, 8, 1]

For larger values of x, note the extremely slow growth rate for $f(x) = \log_e x$.

The following graphs compare exponential growth to logarithmic growth. We examine each of the patterns again with the applications in Section 10.7.

Exponential growth functions eventually grow very rapidly while logarithmic growth functions eventually grow very slowly.

Exponential Growth

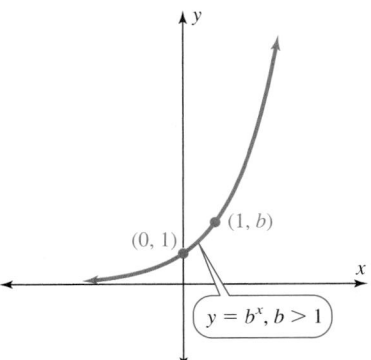

Features: For large values of x, extremely rapid growth.

x-intercept: None
y-intercept: (0, 1)
Key point: (1, b)
Asymptotic to negative x-axis

Logarithmic Growth

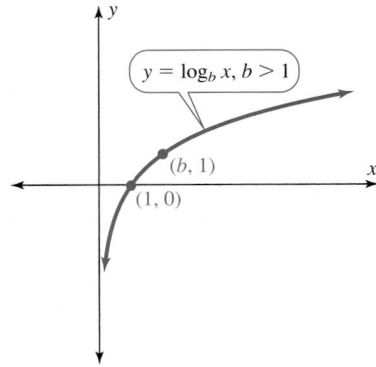

Features: For large values of x, extremely slow growth.

x-intercept: (1, 0)
y-intercept: None
Key point: (b, 1)
Asymptotic to negative y-axis

It is important to keep this distinction between the rapid growth of exponential functions and the slow growth of logarithmic functions in mind as you analyze real-world problems. This is illustrated by Example 7.

■ **EXAMPLE 7**   Selecting an Exponential or Logarithmic Growth Model

Determine for each relationship whether the growth over time is more likely to be an exponential growth function or a logarithmic growth function.

SOLUTIONS

(a) A tadpole fetus grows _____ over a period of time.

Exponentially

At the beginning the first cell doubles to form two cells; then these cells grow and split to form four cells. This growing and splitting continues to form millions of cells and a live creature within a few weeks.

(b) The productivity gain of an employee on an assembly line grows _____ .

Logarithmically

Employees normally make their greatest gains of efficiency from their novice state to becoming seasoned employees. After they have experienced all of the unique features of this job, there is little new to learn and any new gains of productivity are minimal.

■

USING THE LANGUAGE AND SYMBOLISM OF MATHEMATICS   10.3

1. For $x > 0$, $b > 0$, and $b \neq 1$ the inverse of the function $f(x) = b^x$ is written as $f^{-1}(x) =$ _____ .

2. The $\log_b x = y$ is read "the log base _____ of _____ equals _____ ."

3. For $b > 0$ and $b \neq 1$, the $\log_b 0$ is _____ .

4. For $b > 0$ and $b \neq 1$, the $\log_b 1 =$ _____ .

5. For $b > 0$ and $b \neq 1$, the $\log_b b =$ _____ .

6. For $b > 0$ and $b \neq 1$, the $\log_b \dfrac{1}{b} =$ _____ .

7. For $b > 0$ and $b \neq 1$, the $\log_b b^x =$ _____ .

8. For $b > 1$ the graph of $f(x) = \log_b x$ is asymptotic to the negative portion of the _____ -axis.

EXERCISES   10.3

In Exercises 1–16 use the given information to complete each entry in the table as illustrated by the example in the first row.

LOGARITHMIC FORM	VERBAL DESCRIPTION	EXPONENTIAL FORM
Example: $\log_7 49 = 2$	The log base 7 of 49 is 2.	$7^2 = 49$
1. $\log_5 125 = 3$		
2. $\log_2 16 = 4$		
3. $\log_3 \sqrt{3} = \dfrac{1}{2}$		
4. $\log_5 \sqrt[3]{5} = \dfrac{1}{3}$		
5. $\log_5 \dfrac{1}{5} = -1$		
6. $\log_2 \dfrac{1}{4} = -2$		
7.	The log base 16 of 4 is $\dfrac{1}{2}$.	
8.	The log base 16 of $\dfrac{1}{16}$ is -1.	
9.		$8^{2/3} = 4$
10.		$4^{3/2} = 8$
11.		$3^{-2} = \dfrac{1}{9}$
12.		$\left(\dfrac{1}{3}\right)^{-4} = 81$
13.		$m^p = n$
14. $\log_x y = z$		
15.	The log base n of m is k.	
16. $\log_{10} 0.0001 = -4$		

In Exercises 17–46 determine the value of each logarithm by inspection.

17. $\log_{10} 100$

18. $\log_{10} 1000$

19. $\log_{10} 0.001$

20. $\log_{10} 0.01$

21. $\log_{10} \sqrt[3]{10}$

22. $\log_{10} \sqrt{10}$

23. $\log_2 32$

24. $\log_{11} 121$

25. $\log_3 \dfrac{1}{3}$

26. $\log_7 \dfrac{1}{7}$

27. $\log_{18} 1$

28. $\log_{25} 1$

29. $\log_{3/4} \dfrac{4}{3}$

30. $\log_{2/7} \dfrac{7}{2}$

31. $\log_3 81$

32. $\log_9 81$

33. $\log_2 64$

34. $\log_{64} 64$

35. $\log_3 27$

36. $\log_{27} 3$

37. $\log_{16} 2$

38. $\log_2 16$

39. $\log_{25} 125$

40. $\log_9 27$

41. $\log_{19} 19$

42. $\log_\pi \pi$

43. $\log_5 5^{2.7}$

44. $\log_3 3^{4.1}$

45. $\log_{3/4} \dfrac{16}{9}$

46. $\log_{5/7} \dfrac{49}{25}$

In Exercises 47–50 determine which expressions are defined and which are not defined. Do not evaluate these expressions.

47. **a.** $\log_7 10$ **b.** $\log_7 0$

48. **a.** $\log_0 7$ **b.** $\log_7 7$

49. **a.** $\log_5(-5)$ **b.** $\log_5 1$

50. **a.** $\log_{-3} 3$ **b.** $\log_{0.3} 3$

In Exercises 51–68 solve each equation without using a calculator.

51. $\log_6 36 = x$

52. $\log_{11} 121 = x$

53. $\log_6 x = 1$

54. $\log_{11} x = 1$

55. $\log_6 x = -1$

56. $\log_{11} x = -1$

57. $\log_6 x = \dfrac{1}{2}$

58. $\log_{11} x = \dfrac{1}{3}$

59. $\log_x 4 = 2$

60. $\log_x 4 = \dfrac{1}{2}$

61. $\log_x 125 = 3$

62. $\log_x 8 = 3$

63. $\log_b \sqrt[5]{b} = x$

64. $\log_b \sqrt[5]{b^2} = x$

65. $\log_3 1 = 2x + 1$

66. $\log_3 3 = 4x - 1$

67. $\log_2(3x - 4) = 3$

68. $\log_5(3x + 1) = 2$

In Exercises 69–72 match each graph (A–D) with the function that defines this graph. All graphs are displayed with the window $[-5, 5, 1]$ by $[-5, 5, 1]$.

69. $f(x) = \left(\dfrac{1}{3}\right)^x$

70. $f(x) = 2^x$

71. $f(x) = \log_3 x$

72. $f(x) = \log_{1/3} x$

A.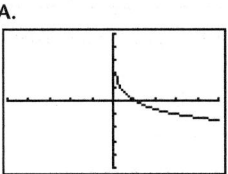

B.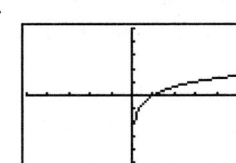

C.

D.

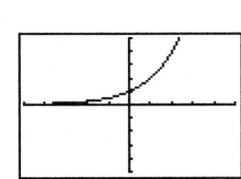

73. Use the graph of $f(x) = \log_2 x$ shown in the figure to determine the domain and range of this function.

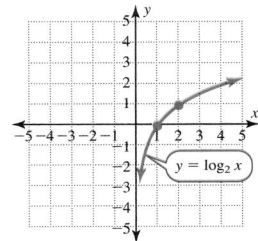

74. Use the graph of $f(x) = \log_4 x$ shown in the figure to determine the domain and range of this function.

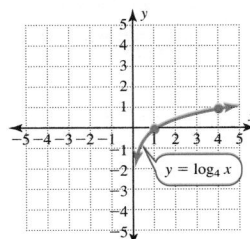

In Exercises 75 and 76 first use a graphics calculator to graph the corresponding exponential function. Then use the DrawInv to sketch the graph of the logarithmic function.

75. $f(x) = \log_{1.5} x$

76. $f(x) = \log_{2.5} x$

Group Discussion Questions

77. **Challenge Question**   Determine for each relationship whether the growth over time is more likely to be an exponential growth function or a logarithmic growth function.

   a. The value of an investment in a U.S. Treasury note over a 30-year period

   b. The amount of learning by an individual over their 85-year lifetime

   c. The population of locusts from the start of a growing season until the end of the growing season

   d. The heights obtained by a pole-vaulter over a 20-year career

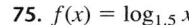

78. Discovery Question Examine tables of values for each of these growth functions and compare their growth rates for a range of input values. Which function grows the fastest for values near 2?
a. $f(x) = 10x$
b. $f(x) = 10^x$
c. $f(x) = \log_{10} x$

79. Discovery Question Examine tables of values for each of these decay functions and compare their decay rates for a range of input values. Which function decays the fastest for values near 1? Which function decays the fastest for values near 100?
a. $f(x) = -4x$
b. $f(x) = \left(\dfrac{1}{4}\right)^x$
c. $f(x) = \log_{1/4} x$

Section 10.4 Evaluating Logarithms

Objective: 7. Evaluate common and natural logarithms with a calculator.

Although $\log_b x$ is defined for any base $b > 0$ and $b \neq 1$, the only two bases commonly used are base 10 and base e. Since our number system is based upon powers of 10, the most convenient base for many computations is base 10. Logarithms to the base 10 are called **common logarithms.** The abbreviated form $\log x$ is used to denote $\log_{10} x$.

Mathematicians and scientists use base e extensively because many formulas are stated most easily using this base. The number e is irrational:

Remember $e \approx 2.718$.

$$e \approx 2.718\ 281\ 828\ 459\ 045\ \ldots$$

You should remember that e is approximately equal to 2.718. Logarithms to the base e are called **natural logarithms.** The abbreviated form $\ln x$ is used to denote $\log_e x$.

Common and Natural Logarithms

COMMON LOGARITHMS	VERBALLY	NUMERICAL EXAMPLE
$\log x$ means $\log_{10} x$	$\log x$ is read "the (common) log of x" or "the log base 10 of x."	$\log 100 = 2$ because $10^2 = 100$
NATURAL LOGARITHMS $\ln x$ means $\log_e x$	$\ln x$ is read "the natural log of x" or "the log base e of x."	$\ln 100 \approx 4.605$ because $e^{4.605} \approx 100$

■ EXAMPLE 1 Evaluating Logarithms by Inspection

Determine the value of each expression by inspection.

SOLUTIONS

(a) $\log 10{,}000$ $\log 10{,}000 = 4$ In exponential form $10^4 = 10{,}000$.

(b) $\log 0.1$ $\log 0.1 = -1$ In exponential form $10^{-1} = \dfrac{1}{10} = 0.1$.

(c) $\ln e^5$ $\ln e^5 = 5$ In exponential form $e^5 = e^5$.

(d) $\ln e^{-4}$ $\ln e^{-4} = -4$ In exponential form $e^{-4} = e^{-4}$. ■

Only relatively simple logarithms such as log 100 = 2 can be determined by inspection. Thus common and natural logarithms have historically been determined through use of tables, slide rules, and other devices. Today calculators are usually used to determine these values. The examples in this section assume calculator usage.

Logarithmic functions also are available in many computer languages; however, these functions may not mean what you expect. Note the following warning.

Warning for Computer Users

Some computer languages provide only the natural logarithmic function, in which case this function may be called LOG instead of ln, even though the base is *e*. If you are unsure of the meaning of LOG on a particular computer, test a known value such as LOG 10 to determine which base is being used. We know log 10 = 1 but ln 10 ≈ 2.30.

■ EXAMPLE 2   Evaluating Common Logarithms

Use a calculator to approximate each expression to the nearest thousandth. Check each answer using the equivalent exponential form.

SOLUTIONS

The 10^x feature is the secondary function of the **LOG** key.

(a) log 748

log 748 ≈ 2.874

Check: $10^{2.874} \approx 748$

```
log(748)
        2.873901598
10^(Ans)
              748
```

(b) log 0.0034

log 0.0034 ≈ −2.469

Check: $10^{-2.469} \approx 0.0034$

```
log(0.0034)
        -2.468521083
10^(Ans)
              .0034
```

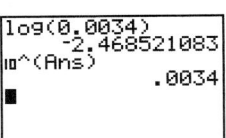

Although the expressions in Examples 2 and 3 look very similar, the values are quite distinct. Example 3 involves natural logarithms with base *e*.

SELF-CHECK 10.4.1

Use a calculator to approximate each expression to the nearest ten-thousandth.

1. log 0.47912
2. log 512.8

■ **EXAMPLE 3**   Evaluating Natural Logarithms

Use a calculator to approximate each expression to the nearest thousandth. Check each answer using the equivalent exponential form.

SOLUTIONS

(a) ln 748

ln 748 $\approx$ 6.617

Check: $e^{6.617} \approx 748$

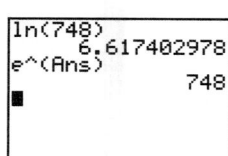

(b) ln 0.0034

ln 0.0034 $\approx$ −5.684

Check: $e^{-5.684} \approx 0.0034$

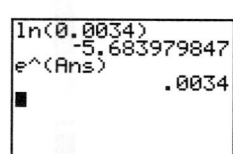

We often use scientific notation (see Calculator Perspective 5.3.1 in Section 5.3) to write numbers whose magnitude is either very large or very small. We use scientific notation to enter the logarithmic expressions in Example 4.

■ **EXAMPLE 4**   Using Scientific Notation with Logarithmic Expressions

Use scientific notation to enter the argument of each logarithm. Use a calculator to approximate each expression to the nearest ten-thousandth.

SOLUTIONS

(a) ln 894,000,000,000

ln 894,000,000,000
$= \ln(8.94 \times 10^{11})$
≈ 27.5190

Check: $e^{27.5190} \approx 8.94 \times 10^{11}$

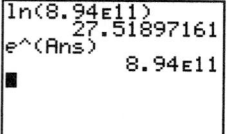

(b) log 0.000 000 001 234

log 0.000 000 001 234
$= \log(1.234 \times 10^{-9})$
≈ -8.9087

Check: $10^{-8.9087} \approx 1.234 \times 10^{-9}$

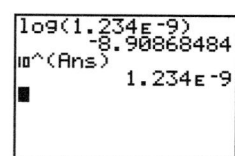

SELF-CHECK 10.4.2

Use a calculator to approximate each expression to the nearest thousandth.
1. ln 0.47912
2. ln 512.8
3. $\log(5.81 \times 10^4)$
4. $\ln(3.96 \times 10^{-5})$

SELF-CHECK 10.4.2 ANSWERS

1. −0.736   2. 6.240
3. 4.764   4. −10.137

To avoid order-of-operations errors, it is wise to use sufficient parentheses to clearly denote the correct interpretation of logarithmic expressions. Note the similar appearance but distinct meanings of the expressions in Example 5. It is better to use extra parentheses than not to use enough to denote the meaning that you intend.

■ EXAMPLE 5   Using a Calculator to Evaluate Logarithmic Expressions

Rewrite each expression using parentheses and then use a calculator to approximate each expression to the nearest ten-thousandth.

SOLUTIONS ───────────────────────

Note that all three of these expressions have different orders of operation and different values.

(a) $\dfrac{\log 12}{5}$ $\qquad$ $\dfrac{\log 12}{5} = \big(\log(12)\big) \div 5 \approx 0.2158$

(b) $\log \dfrac{12}{5}$ $\qquad$ $\log \dfrac{12}{5} = \log\left(\dfrac{12}{5}\right) \approx 0.3802$

(c) $\dfrac{\log 12}{\log 5}$ $\qquad$ $\dfrac{\log 12}{\log 5} = \dfrac{\log(12)}{\log(5)} \approx 1.5440$

```
(log(12))/5
         .2158362492
log(12/5)
         .3802112417
log(12)/log(5)
         1.543959311
■
```

SELF-CHECK 10.4.3

Use a calculator to approximate the value of each expression to the nearest hundredth.

1. $\dfrac{5}{\log 7}$ $\qquad$ 2. $\log \dfrac{5}{7}$ $\qquad$ 3. $\dfrac{\log 5}{\log 7}$

We use the fact that logarithmic functions and exponential functions are inverses of each other to solve for x in Example 6.

■ EXAMPLE 6   Solving Logarithmic Equations

Use a calculator to approximate the value of x to the nearest thousandth.

SOLUTIONS ───────────────────────

(a) $\log x = 0.69897$ $\qquad$ $\log x = 0.69897$

$x = 10^{0.69897}$

$x \approx 5.000$

First rewrite the equation in exponential form with a base of 10. Then approximate x using a calculator.

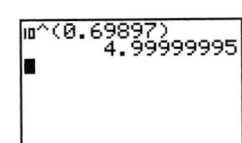

```
10^(0.69897)
         4.99999995
■
```

(b) $\ln x = -1.045$ $\ln x = -1.045$ First rewrite the equation
$$x = e^{-1.045}$$ $\log_e x = -1.045$ in exponential
$$x \approx 0.352$$ form with a base of e. Then
approximate x using a calculator.

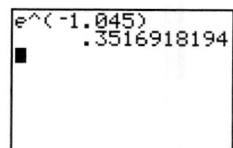

SELF-CHECK 10.4.4

Use a calculator to approximate x to the nearest thousandth.
1. $x = 10^{1.4583}$
2. $x = e^{1.4583}$
3. $\log x = 1.4583$
4. $\ln x = 1.4583$

It is important to know that the values that we calculate are reasonable and not in error because of some careless keystroke. Example 7 illustrates two estimation problems. We will examine a formula that will allow us to calculate logarithms to any base in Section 10.5.

■ EXAMPLE 7   Estimation Skills

Mentally estimate the value of each of these expressions.

SOLUTIONS

(a) $2^{3.98765}$

$2^{3.98765} \approx 2^4$
$2^{3.98765} \approx 16$

Since 3.98765 is only slightly less than 4, $2^{3.98765}$ is slightly less than 2^4, which is 16.

(b) $\log_7 50.0017$

$\log_7 50.0017 \approx \log_7 49$
$\log_7 50.0017 \approx 2$

Since 50.0017 is slightly larger than 49, $\log_7 50.0017$ is slightly larger than $\log_7 49$. $\log_7 49 = 2$ because $7^2 = 49$.

USING THE LANGUAGE AND SYMBOLISM OF MATHEMATICS   10.4

1. The irrational number $e \approx$ _____ (to the nearest thousandth).
2. The common logarithm of x, denoted $\log x$, means _____.
3. The natural logarithm of x, denoted $\ln x$, means _____.

EXERCISES   10.4

In Exercises 1–8 determine the value of the expression in part (a) by inspection. Then use a calculator to approximate the value of the expressions in parts (b) and (c) to the nearest thousandth.

1. a. $\log_6 36$ b. $\log 36$ c. $\ln 36$
2. a. $\log_3 27$ b. $\log 27$ c. $\ln 27$
3. a. $\log_3 \dfrac{1}{9}$ b. $\log \dfrac{1}{9}$ c. $\ln \dfrac{1}{9}$
4. a. $\log_5 \dfrac{1}{125}$ b. $\log \dfrac{1}{125}$ c. $\ln \dfrac{1}{125}$
5. a. $\log_8 1$ b. $\log 1$ c. $\ln 1$
6. a. $\log 1000$ b. $\log 900$ c. $\ln 900$
7. a. $\ln e^3$ b. $\log e^3$ c. $\ln 100$
8. a. $\ln e^{-3}$ b. $\log e^{-3}$ c. $\ln 10^{-3}$

In Exercises 9–12 determine the value of each expression by inspection.

9.   a. $\log 100{,}000$ b. $\log 0.0001$ c. $\log 10^9$
10. a. $\log 1{,}000{,}000$ b. $\log 0.01$ c. $\log 10^{-6}$
11. a. $\ln e^5$ b. $\ln e^{-5}$ c. $\ln \sqrt{e}$
12. a. $\ln e^7$ b. $\ln e^{-7}$ c. $\ln \sqrt[3]{e}$

In Exercises 13–42 use a calculator to approximate the value of each expression to the nearest thousandth.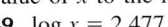

13. a. $\log 47$ b. $10^{1.672}$
14. a. $\log 2.47$ b. $10^{0.393}$
15. a. $\ln 47$ b. $e^{3.850}$
16. a. $\ln 2.47$ b. $e^{0.904}$
17. a. $\log 10$ b. $\ln 10$
18. a. $\log 100$ b. $\ln 100$
19. a. $\log e$ b. $\ln e$
20. a. $\log e^3$ b. $\ln e^3$
21. a. $\log 113$ b. $10^{2.053}$
22. a. $\ln 113$ b. $e^{4.727}$
23. a. $\ln 0.00621$ b. $e^{-5.082}$
24. a. $\log 0.00621$ b. $10^{-2.207}$
25. a. $\log \pi$ b. $\ln \pi$
26. a. $\log \sqrt{7}$ b. $\ln \sqrt{7}$
27. a. $\log 0.00009$ b. $\ln 0.00009$
28. a. $\log(3.87 \times 10^{-7})$ b. $\ln(3.87 \times 10^{-7})$
29. a. $\log(5.93 \times 10^{-12})$ b. $\ln(5.93 \times 10^{-12})$
30. a. $\log(4.07 \times 10^5)$ b. $\ln(4.07 \times 10^5)$
31. a. $\log(6.93 \times 10^{11})$ b. $\ln(6.93 \times 10^{11})$
32. a. $\log(\ln 9)$ b. $\ln(\log 9)$
33. a. $\log(\ln 13)$ b. $\ln(\log 13)$
34. a. $(\log 7.52)^2$ b. $\log(7.52^2)$
35. a. $(\ln 0.84)^2$ b. $\ln(0.84^2)$
36. a. $\ln \dfrac{11}{4}$ b. $\dfrac{\ln 11}{\ln 4}$

37. a. $\log \dfrac{17}{5}$ b. $\dfrac{\log 17}{\log 5}$
38. a. $(\log 11)(\log 4)$ b. $\log(11 \cdot 4)$
39. a. $(\ln 5)(\ln 17)$ b. $\ln(5 \cdot 17)$
40. a. $\ln 5 + \ln 17$ b. $\ln(5 + 17)$
41. a. $\log 11 - \log 4$ b. $\log(11 - 4)$
42. a. $\ln 13 - \ln 5$ b. $\ln(13 - 5)$

In Exercises 43–48 identify whether each expression is defined. Then use a calculator to approximate to the nearest thousandth the value of each expression that is defined.

43. a. $-\ln 8$ b. $\ln(-8)$
44. a. $-\log 11$ b. $\log(-11)$
45. a. $0 \cdot \log 23$ b. $\log(23 \cdot 0)$
46. a. $\ln 17 - \ln 17$ b. $\ln(17 - 17)$
47. a. $\ln 17 - \ln 19$ b. $\ln(17 - 19)$
48. a. $\dfrac{\log 41}{\log 1}$ b. $\dfrac{\log 41}{\log 10}$

In Exercises 49–56 use a calculator to approximate the value of x to the nearest thousandth.

49. $\log x = 2.477$ 50. $\log x = 2.778$
51. $\log x = -0.301$ 52. $\log x = -0.699$
53. $\ln x = 2.079$ 54. $\ln x = 2.773$
55. $\ln x = -2.996$ 56. $\ln x = -4.711$

Estimation Skills

In Exercises 57–60 mentally estimate the value of each expression to the nearest integer.

PROBLEM	MENTAL ESTIMATE
57. $\log_8 63$	
58. $\log_2 17$	
59. $2.99^{4.01}$	
60. $5.01^{1.99}$	

Group Discussion Questions

61. **Discovery Question**
    a. Evaluate each of these expressions:
       $10^{\log 33}, \; e^{\ln 19}, \; 2^{\log_2 8}, \; 3^{\log_3 81}$
    b. Based on your observations, make a conjecture stated as an algebraic equation.
    c. Test your conjecture and describe it to your instructor.

62. **Discovery Question**
    a. Evaluate each of these expressions:
       $\log(5 \cdot 6), \; \log 5 + \log 6, \; \ln(7 \cdot 11), \; \ln 7 + \ln 11$
    b. Based on your observations, make a conjecture stated as an algebraic equation.
    c. Test your conjecture and describe it to your instructor.

63. Discovery Question

a. Evaluate each of these expressions:

$\log \dfrac{15}{5}$, $\log 15 - \log 5$, $\ln \dfrac{24}{6}$, $\ln 24 - \ln 6$

b. Based on your observations, make a conjecture stated as an algebraic equation.

c. Test your conjecture and describe it to your instructor.

64. Discovery Question

a. Evaluate each of these expressions:

$\log 5^7$, $7 \log 5$, $\log \pi^3$, $3 \log \pi$, $\ln 10^9$, $9 \ln 10$

b. Based on your observations, make a conjecture stated as an algebraic equation.

c. Test your conjecture and describe it to your instructor.

Section 10.5 Properties of Logarithms

Objectives:
8. Use the product rule, the quotient rule, and the power rule for logarithms.
9. Use the change-of-base formula to evaluate logarithms.

The properties of logarithms follow directly from the definition of a logarithmic function as the inverse of an exponential function. Thus for every exponential property there is a corresponding logarithmic property. The logarithmic properties provide a means of simplifying some algebraic expressions and make it easier to solve exponential equations.

The logarithmic and exponential properties are stated here side by side so that you can compare them. (The capital letters X and Y are used in the logarithmic form to show that these variables are not the same as x and y in the exponential form.)

In exponential form, we add exponents; in logarithmic form, we add logarithms.
In exponential form, we subtract exponents; in logarithmic form, we subtract logarithms.
In exponential form, we multiply exponents; in logarithmic form, we multiply the power p times the log of X.

	EXPONENTIAL FORM	**LOGARITHMIC FORM**
PRODUCT RULE:	$b^x b^y = b^{x+y}$	$\log_b XY = \log_b X + \log_b Y$
QUOTIENT RULE:	$\dfrac{b^x}{b^y} = b^{x-y}$	$\log_b \dfrac{X}{Y} = \log_b X - \log_b Y$
POWER RULE:	$(b^x)^p = b^{xp}$	$\log_b X^p = p \log_b X$

The proof of the product rule is presented here. The quotient rule and the power rule can be proven in a similar fashion; their proofs are left to you in Exercises 82 and 83 at the end of this section.

Proof of the Product Rule for Logarithms Let $X = b^m$ and $Y = b^n$

$\log_b X = m$ and $\log_b Y = n$ Rewrite these statements in logarithmic form.

Now examine XY.

$XY = b^m b^n$	Substitute b^m for X and b^n for Y.
$XY = b^{m+n}$	Use the product rule for exponents.
$\log_b XY = m + n$	Rewrite this statement in logarithmic form.
$\log_b XY = \log_b X + \log_b Y$	Substitute $\log_b X$ for m and $\log_b Y$ for n.

PROPERTIES OF LOGARITHMS

For $x, y > 0$, $b > 0$, and $b \neq 1$,

	ALGEBRAICALLY	VERBALLY	NUMERICAL EXAMPLE
PRODUCT RULE:	$\log_b xy = \log_b x + \log_b y$	The log of a product is the sum of the logs.	$\log_2 32 = \log_2(4 \cdot 8) = \log_2 4 + \log_2 8$ $5 = 2 + 3$
QUOTIENT RULE:	$\log_b \dfrac{x}{y} = \log_b x - \log_b y$	The log of a quotient is the difference of the logs.	$\log_2 4 = \log_2 \dfrac{32}{8} = \log_2 32 - \log_2 8$ $2 = 5 - 3$
POWER RULE:	$\log_b x^p = p \log_b x$	The log of the pth power of x is p times the log of x.	$\log_2 64 = \log_2 4^3 = 3 \log_2 4$ $6 = 3 \cdot 2$

■ EXAMPLE 1 Using a Calculator to Examine the Properties of Logarithms

Use the properties of logarithms to rewrite each expression. Then use a calculator to verify each equation.

SOLUTIONS

(a) $\ln 17^5$

$\ln 17^5 = 5 \ln 17$
$14.166 \approx 14.166$

Use the power rule:
$\log_b x^p = p \log_b x$

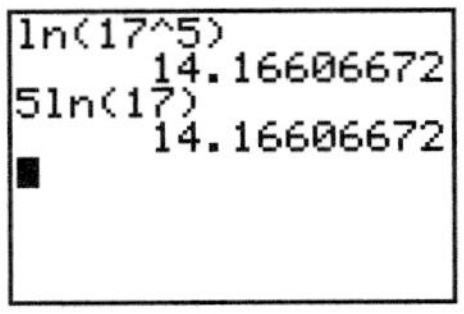

```
ln(17^5)
          14.16606672
5ln(17)
          14.16606672
■
```

(b) $\ln (15 \cdot 91)$

$\ln (15 \cdot 91) = \ln 15 + \ln 91$
$7.219 \approx 7.219$

Use the product rule:
$\log_b xy = \log_b x + \log_b y$

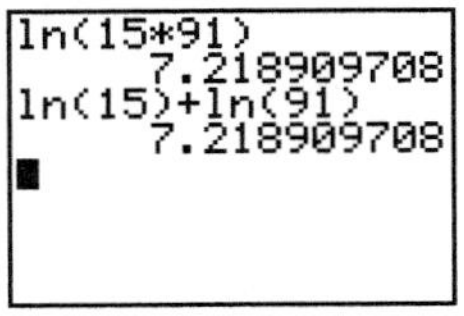

```
ln(15*91)
          7.218909708
ln(15)+ln(91)
          7.218909708
■
```

(c) $\ln \dfrac{86}{37}$

$\ln \dfrac{86}{37} = \ln 86 - \ln 37$
$0.843 \approx 0.843$

Use the quotient rule:
$\log_b \dfrac{x}{y} = \log_b x - \log_b y$

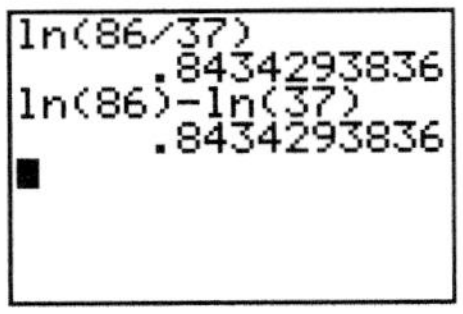

```
ln(86/37)
          .8434293836
ln(86)-ln(37)
          .8434293836
■
```

The properties of logarithms are used both to rewrite logarithmic expressions in a simpler form and to combine logarithms. These properties also are used to solve logarithmic equations and to simplify expressions in calculus.

The input value of $y = f(x)$ also is called the **argument** of the function. In the logarithmic function $y = \log(x + 2)$, we refer to $x + 2$ as the argument of the logarithmic function. Likewise, the argument of $y = \ln \dfrac{x}{x - 1}$ is $\dfrac{x}{x - 1}$. This terminology is often used for clarification when taking a logarithm of an expression other than just x.

■ EXAMPLE 2 Using the Properties of Logarithms to Form Simpler Expressions

Use the properties of logarithms to write these expressions in terms of logarithms of simpler expressions. Assume that the argument of each logarithm is a positive real number.

SOLUTIONS

(a) $\log xyz$ — $\log xyz = \log x + \log y + \log z$ — Product rule (generalized to three factors): Add the logarithms.

(b) $\ln x^7$ — $\ln x^7 = 7 \ln x$ — Power rule: Multiply the power 7 times the logarithm.

(c) $\ln x^4 y^5$ — $\ln x^4 y^5 = \ln x^4 + \ln y^5$ — Product rule: Add the logarithms.
$= 4 \ln x + 5 \ln y$ — Power rule: Multiply each power times the logarithm.

(d) $\log_5 \dfrac{x + 3}{x - 2}$ — $\log_5 \dfrac{x + 3}{x - 2} = \log_5(x + 3) - \log_5(x - 2)$ — Quotient rule: Subtract the logarithms.

(e) $\ln \sqrt{\dfrac{x}{y}}$ — $\ln \sqrt{\dfrac{x}{y}} = \ln \left(\dfrac{x}{y}\right)^{1/2}$
$= \dfrac{1}{2} \ln \dfrac{x}{y}$ — Power rule: Multiply the power $\dfrac{1}{2}$ times the logarithm.
$= \dfrac{1}{2} (\ln x - \ln y)$ — Quotient rule: Subtract the logarithms. ■

SELF-CHECK 10.5.1

Express each logarithm in terms of logarithms of simpler expressions. Assume that the argument of each logarithm is a positive real number.

1. $\log_{11} x^3 y^2$
2. $\log_6 \dfrac{(x + 1)^5}{(y - 2)^3}$

SELF-CHECK 10.5.1 ANSWERS

1. $3 \log_{11} x + 2 \log_{11} y$
2. $5 \log_6(x + 1) - 3 \log_6(y - 2)$

Example 3 is essentially the reverse of Example 2. The properties of logarithms are used to combine logarithmic terms.

■ **EXAMPLE 3**   Using the Properties of Logarithms to Combine Expressions

Combine the logarithmic terms in each expression into a single logarithmic expression with a coefficient of 1. Assume that the argument of each logarithm is a positive real number.

SOLUTIONS

(a) $2 \ln x + 3 \ln y - 4 \ln z$

$$\begin{aligned} 2 \ln x + 3 \ln y - 4 \ln z &= \ln x^2 + \ln y^3 - \ln z^4 & \text{Power rule} \\ &= \ln x^2 y^3 - \ln z^4 & \text{Product rule} \\ &= \ln \frac{x^2 y^3}{z^4} & \text{Quotient rule} \end{aligned}$$

(b) $5 \log x + \dfrac{1}{3} \log y$

$$\begin{aligned} 5 \log x + \frac{1}{3} \log y &= \log x^5 + \log y^{1/3} & \text{Power rule} \\ &= \log x^5 + \log \sqrt[3]{y} \\ &= \log x^5 \sqrt[3]{y} & \text{Product rule} \end{aligned}$$

■

SELF-CHECK 10.5.2

Combine the logarithmic terms in each expression into a single logarithmic expression with a coefficient of 1. Assume that the argument of each logarithm is a positive real number.

1. $\log_{12}(x + 3) + \log_{12}(2x - 9)$

2. $3 \log_b x + 2 \log_b y - \dfrac{1}{2} \log_b z$

Example 4 can serve as a good source of practice in observing patterns and using the properties of logarithms. In Example 5 we recalculate these values using the change-of-base formula for logarithms. However, the purpose of Example 4 is to give practice using the product, quotient, and power rules for logarithms.

■ **EXAMPLE 4**   Logarithmic Properties

Use $\log_5 2 \approx 0.431$, $\log_5 3 \approx 0.683$, and the properties of logarithms to approximate the value of each of these logarithmic expressions.

SOLUTIONS

(a) $\log_5 6$

$$\begin{aligned} \log_5 6 &= \log_5(2 \cdot 3) & \text{Product rule} \\ &= \log_5 2 + \log_5 3 & \text{Substitute the given values for } \log_5 2 \text{ and } \log_5 3. \\ &\approx 0.431 + 0.683 \\ &\approx 1.114 \end{aligned}$$

(b) $\log_5 9$ $\log_5 9 = \log_5 3^2$

$= 2 \log_5 3$ Power rule

$\approx 2(0.683)$ Substitute the given value of $\log_5 3$.

≈ 1.366

(c) $\log_5 0.4$ $\log_5 0.4 = \log_5 \dfrac{2}{5}$ Quotient rule

$= \log_5 2 - \log_5 5$ The value of $\log_5 2$ is given, and $\log_5 5$ can be

$\approx 0.431 - 1$ determined by inspection.

≈ -0.569

SELF-CHECK 10.5.3

Use the values given in Example 4 to approximate the value of each expression.

1. $\log_5 10$

2. $\log_5 8$

3. $\log_5 1.5$

Calculators have special keys for both $\log x$ and $\ln x$. However, no other logarithmic keys are needed. The change-of-base formula for logarithms can be used to convert any logarithmic expression to either base 10 or base e.

Although any base b can be used in the formula $\log_a x = \dfrac{\log_b x}{\log_b a}$, we usually use either $\dfrac{\log x}{\log a}$ or $\dfrac{\ln x}{\ln a}$. One reason we use these bases is that they are available on most graphics calculators.

Change-of-Base Formula

For $a, b > 0$, $a \neq 1$, and $b \neq 1$:

FOR LOGARITHMS

$\log_a x = \dfrac{\log_b x}{\log_b a}$ for $x > 0$

FOR EXPONENTS

$a^x = b^{x \log_b a}$

EXAMPLES

$\log_2 x = \dfrac{\log x}{\log 2}$ or $\log_2 x = \dfrac{\ln x}{\ln 2}$

$2^x = e^{(\ln 2)x}$

Proof of the Change-of-Base Formula for Logarithms

Let $\log_a x = y$, then

$x = a^y$ Rewrite this equation in exponential form.

$\log_b x = \log_b a^y$ Take the log to base b of both sides of this equation.

$\log_b x = y \log_b a$ Use the power rule for logarithms.

$y = \dfrac{\log_b x}{\log_b a}$ Solve for y.

$\log_a x = \dfrac{\log_b x}{\log_b a}$ Substitute $\log_a x$ for y.

The most useful forms of this identity are:

$\log_a x = \dfrac{\log x}{\log a}$ and $\log_a x = \dfrac{\ln x}{\ln a}$.

The proof of the change-of-base formula for exponents is similar to the proof given for logarithms.

■ EXAMPLE 5   Using the Change-of-Base Formula for Logarithms

Use the change-of-base formula for logarithms and a calculator to approximate the value of each expression to the nearest thousandth.

SOLUTIONS _____

(a) $\log_5 6$

$$\log_5 6 = \frac{\ln 6}{\ln 5}$$
$$\approx 1.113$$

Use the change-of-base formula $\log_a x = \dfrac{\ln x}{\ln a}$ to convert to natural logs. (Converting to common logs would produce the same results.)

(b) $\log_5 9$

$$\log_5 9 = \frac{\ln 9}{\ln 5}$$
$$\approx 1.365$$

```
ln(6)/ln(5)
         1.113282753
ln(9)/ln(5)
         1.365212389
ln(0.4)/ln(5)
         -.5693234419
■
```

(c) $\log_5 0.4$

$$\log_5 0.4 = \frac{\ln 0.4}{\ln 5}$$
$$\approx -0.569$$

Note that these results confirm the results calculated in Example 4.   ■

SELF-CHECK 10.5.4

1. Estimate $\log_3 80$ to the nearest integer.
2. Use the change-of-base formula for logarithms and a calculator to approximate $\log_3 80$ to the nearest thousandth.

The change-of-base formula for exponents can be used to convert any exponential growth or decay function to base e. In scientific applications base e is used almost exclusively.

■ EXAMPLE 6   Using the Change-of-Base Formula for Exponents

Use the change-of-base formula for exponents to write $f(x) = 3^x$ as an exponential function with base e.

SOLUTION _____

SELF-CHECK 10.5.4 ANSWERS

1. $\log_3 80 \approx \log_3 81 = 4$

2. $\log_3 80 = \dfrac{\ln 80}{\ln 3} \approx 3.989$

ALGEBRAICALLY

$y = 3^x$ — Use the change-of-base formula $a^x = e^{(\ln a)x}$ to convert to base e.

$y = e^{(\ln 3)x}$ — Then approximate $\ln 3$ with a calculator.

$y \approx e^{1.099x}$

NUMERICAL CHECK

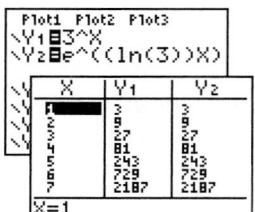

Let y_1 represent the original expression 3^x. Let y_2 represent the new expression $e^{(\ln 3)x}$.

Because y_1 and y_2 have identical values, these two functions are equal. A comparison of their graphs would reveal that their graphs are identical.

Example 7 proves an identity involving logarithms and exponents.

■ EXAMPLE 7 Verifying an Identity Involving Logarithms

Verify that $b^{\log_b x} = x$ for any positive real number x.

SOLUTION

Let $y = \log_b x$.

Then $b^y = x$ and   Rewrite the logarithmic expression in exponential form.

$b^{\log_b x} = x$ Substitute $\log_b x$ for y.

Another Property of Logarithms

For any positive real number x, $b > 0$, and $b \neq 1$:

ALGEBRAICALLY	NUMERICAL EXAMPLE
$b^{\log_b x} = x$	$23^{\log_{23} 19} = 19$

■ EXAMPLE 8 Using the Properties of Logarithms

Determine the value of $e^{\ln 22}$ and check this result with a calculator.

SOLUTION

$e^{\ln 22} = 22$

This value can be determined by inspection using the formula $b^{\log_b x} = x$; $e^{\ln 22} = e^{\log_e 22} = 22$. The calculator check confirms that the value of this expression is 22.

CALCULATOR CHECK

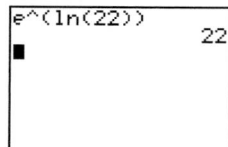

This property of logarithms is used in Example 9 to evaluate some expressions by inspection.

■ EXAMPLE 9   Using the Properties of Logarithms

Determine the value of each expression by inspection.

SOLUTIONS

(a) $7^{\log_7 8}$ $7^{\log_7 8} = 8$ Each of these expressions can be determined by inspection, using the identity $b^{\log_b x} = x$.

(b) $3^{\log_3 0.179}$ $3^{\log_3 0.179} = 0.179$

(c) $a^{\log_a 231}$ $a^{\log_a 231} = 231$ for $a > 0, a \neq 1$

SELF-CHECK 10.5.5

Determine the value of each expression by inspection.

1. $e^{\ln 31}$
2. $e^{\ln 0.07}$
3. $10^{\log 86}$
4. $10^{\log 0.041}$

SELF-CHECK 10.5.5 ANSWERS

1. 31
2. 0.07
3. 86
4. 0.041

USING THE LANGUAGE AND SYMBOLISM OF MATHEMATICS   10.5

1. By the product rule for exponents, $b^x b^y = $ _____ .
2. By the product rule for logarithms, $\log_b xy = $ _____ .
3. By the quotient rule for exponents, $\dfrac{b^x}{b^y} = $ _____ .
4. By the quotient rule for logarithms, $\log_b \dfrac{x}{y} = $ _____ .
5. By the power rule for exponents, $(b^x)^p = $ _____ .
6. By the power rule for logarithms, $\log_b x^p = $ _____ .

7. The change-of-base formula for logarithms is $\log_a x = $ _____ .
8. The change-of-base formula for exponents is $a^x = $ _____ .
9. By the change-of-base formula for exponents, $e^{(\ln 7)x} = $ _____ .
10. If $x > 0, b > 0$, and $b \neq 1$, then $b^{\log_b x} = $ _____ .

EXERCISES   10.5

In Exercises 1–20 use the properties of logarithms to express each logarithm in terms of logarithms of simpler expressions. Assume that the argument of each logarithm is a positive real number.

1. $\log abc$
2. $\log xyz$
3. $\ln \dfrac{x}{11}$
4. $\ln \dfrac{y}{7}$
5. $\log xy^5$
6. $\log x^3 y^4$
7. $\ln x^2 y^3 z^4$
8. $\ln (2x + 3)(x + 7)$
9. $\ln \dfrac{2x + 3}{x + 7}$
10. $\ln \dfrac{5x + 8}{3x + 4}$
11. $\log \sqrt{4x + 7}$
12. $\log \sqrt[3]{6x + 1}$
13. $\ln \dfrac{\sqrt{x + 4}}{(y + 5)^2}$
14. $\ln \dfrac{(x + 9)^3}{\sqrt{y + 1}}$

15. $\log \sqrt{\dfrac{xy}{z - 8}}$
16. $\log \sqrt[4]{\dfrac{x^2 y}{z^3}}$
17. $\log \dfrac{x^2(2y + 3)^3}{z^4}$
18. $\log \dfrac{(x + 1)^3(y - 2)^2}{(z + 4)^5}$
19. $\ln \left(\dfrac{xy^2}{z^3} \right)$
20. $\ln \left(\dfrac{x^2 y^3}{z^5} \right)^3$

In Exercises 21–30 combine the logarithmic terms into a single logarithmic expression with a coefficient of 1. Assume that the argument of each logarithm is a positive real number.

21. $2 \log x + 5 \log y$
22. $7 \ln x + 3 \ln y$
23. $3 \ln x + 7 \ln y - \ln z$
24. $4 \log x + 9 \log y - 5 \log z$
25. $\dfrac{1}{2} \log(x + 1) - \log(2x + 3)$

26. $\ln(3x + 8) - \dfrac{1}{2}\ln(5x + 1)$

27. $\dfrac{1}{3}[\ln(2x + 7) + \ln(7x + 1)]$

28. $\dfrac{1}{4}[\log(x + 3) - \log(2x + 9)]$

29. $2\log_5 x + \dfrac{2}{3}\log_5 y$

30. $\dfrac{3}{5}\log_7 x - 2\log_7 y$

In Exercises 31–40 use the properties of logarithms to determine the value of each expression. Assume that x and y are positive.

31. $17^{\log_{17} 34}$ **32.** $83^{\log_{83} 51}$

33. $1.93^{\log_{1.93} 0.53}$ **34.** $4.6^{\log_{4.6} 0.068}$

35. $10^{\log 37}$ **36.** $10^{\log 0.0093}$

37. $e^{\ln 0.045}$ **38.** $e^{\ln 453}$

39. $e^{\ln x}$ **40.** $10^{\log y}$

In Exercises 41–52 use $\log_b 2 \approx 0.3562$, $\log_b 5 \approx 0.8271$, and the properties of logarithms to determine the value of each logarithmic expression.

41. $\log_b 10$ **42.** $\log_b 4$

43. $\log_b 25$ **44.** $\log_b \sqrt{5}$

45. $\log_b \sqrt[3]{2}$ **46.** $\log_b 5b$

47. $\log_b 2b$ **48.** $\log_b 2.5$

49. $\log_b 0.4$ **50.** $\log_b \dfrac{b}{5}$

51. $\log_b \dfrac{b}{2}$ **52.** $\log_b \sqrt{10b}$

In Exercises 53–62 match each expression with an equal expression from choices A–F. Note choice F can be the correct response to more than one exercise. Assume that the argument of each logarithm is a positive real number.

53. $\log \dfrac{x}{y}$

54. $\dfrac{\log x}{\log y}$

55. $\log xy$

56. $(\log x)(\log y)$

57. $\log(x - y)$

58. $\log(x + y)$

59. $\log x^y$

60. $\log y^x$

61. $10^{\log x}$

62. $e^{\ln x}$

A. x
B. $\log x + \log y$
C. $\log x - \log y$
D. $x \log y$
E. $y \log x$
F. This expression cannot be simplified.

In Exercises 63–70 use the change-of-base formula for logarithms and a calculator to approximate each expression to the nearest ten-thousandth. Check your answers by using the exponential form of the expression.

63. $\log_5 37.1$ **64.** $\log_7 81.8$

65. $\log_{13} 7.08$ **66.** $\log_{11} 4.31$

67. $\log_{6.8} 0.085$ **68.** $\log_{7.2} 0.0032$

69. $\log_{0.49} 3.86$ **70.** $\log_{0.61} 18.4$

Estimation Skills and Calculator Skills
In Exercises 71–74 mentally estimate the value of each expression to the nearest integer and then use a calculator to approximate each value to the nearest thousandth.

PROBLEM	MENTAL ESTIMATE	CALCULATOR APPROXIMATION
71. $\log_4 63$		
72. $\log_2 33$		
73. $1.99^{5.02}$		
74. $10.01^{1.98}$		

In Exercises 75 and 76 use the change-of-base formula for exponents to rewrite each exponential function as a function with base e.

75. a. $f(x) = 5^x$ **b.** $f(x) = \left(\dfrac{1}{5}\right)^x$

76. a. $f(x) = 10^x$ **b.** $f(x) = (0.1)^x$

In Exercises 77–79 the first column contains a student's answer to a calculus problem. The second column contains the book's answer. Use the properties of logarithms to show that these expressions are equal.

STUDENT'S ANSWER	BOOK'S ANSWER
77. $\ln\left(\dfrac{1}{2}\right)$	$-\ln 2$
78. $\log 24^x - \log 60^x - \log 0.2^{2x}$	x
79. $\ln y = kt + c_1$	$y = c_2 e^{kt}$

(*Hint:* c_1 and c_2 are different constants.)

Group Discussion Questions
80. Challenge Question   Every exponential growth or decay function can be expressed in terms of base e.
  a. Select an exponential growth function and write it in terms of base e.
  b. Select an exponential decay function and write it in terms of base e.
  c. Describe how one can tell by inspection whether an exponential function base e represents growth or decay.
  d. Prepare a justification for your answer in part **(c)** to present to your instructor.
81. Challenge Question   Write $\ln x + t = 0$ as an exponential equation that does not contain logarithms.
82. Challenge Question   Prove the quotient rule for logarithms. (*Hint:* See the book's proof on page 720 for the product rule.)
83. Challenge Question   Prove the power rule for logarithms.

Section 10.6 Exponential and Logarithmic Equations

Objective: **10.** Solve exponential and logarithmic equations.

An **exponential equation** contains a variable in the exponent. A **logarithmic equation** contains a variable in the argument of a logarithm. We solve both types of equations in this section.

Only relatively simple exponential equations, such as the one in Example 1, result in nice integer solutions. For other equations we often use a calculator to approximate the solutions.

Later in this section we also examine some identities involving exponents and logarithms.

■ EXAMPLE 1 Solving an Exponential Equation

Solve $2^{3x+2} = 32$.

SOLUTION

$$2^{3x+2} = 32$$
$$2^{3x+2} = 2^5 \qquad \text{Express both sides of the equation in terms of base 2.}$$
$$3x + 2 = 5 \qquad \text{The exponents are equal since the bases are the same.}$$
$$3x = 3$$
$$x = 1$$

Answer: $x = 1$ ■

If both sides of an exponential equation cannot be expressed easily in terms of a common base, then we solve the equation by first taking logarithms of both sides of the equation. If two expressions are equal, then their logarithms are equal.

■ EXAMPLE 2 Solving an Exponential Equation

Algebraically determine the exact solution of $5^x = 30$. Also approximate this solution to the nearest ten-thousandth and determine this solution graphically.

SOLUTION

ALGEBRAICALLY

$$5^x = 30 \qquad \text{Take the common log of both sides of the equation.}$$
$$\log 5^x = \log 30 \qquad \text{Use the power rule for logarithms. Then divide both sides of the}$$
$$x \log 5 = \log 30 \qquad \text{equation by } \log 5. \text{ This is the exact solution. } \left(\textit{Caution:}\ \dfrac{\log 30}{\log 5} \text{ is not}\right.$$
$$x = \frac{\log 30}{\log 5}$$
$$x \approx 2.1133 \qquad \text{the same as } \log \dfrac{30}{5}.\right) \text{ Then approximate this value with a calculator.}$$

GRAPHICALLY

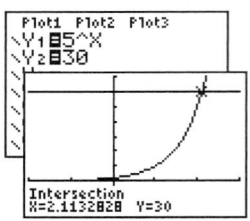

[−2, 3, 1] by [−10, 35, 5]

Let $y_1 = 5x$ and $y_2 = 30$ and use the intersect feature on a calculator to determine the point of intersection.

The approximate x-coordinate of the point of intersection is $x \approx 2.1133$.

Answer: $x = \dfrac{\log 30}{\log 5} \approx 2.1133$ *Check:* $5^{2.1133} \approx 30.000833 \approx 30$

SELF-CHECK 10.6.1

Graphically approximate the solution of $2^x = 7$ to the nearest thousandth.

The procedure used in Examples 1 and 2 is summarized in the following box.

Solving Exponential Equations Algebraically

VERBALLY	ALGEBRAIC EXAMPLE
STEP 1 **a.** If it is obvious that both sides of the equation are powers of the same base, express each side of the equation as a power of this base and then equate the exponents.	**a.** $2^x = 8$ $2^x = 2^3$ $x = 3$
b. Otherwise, take the logarithm of both sides of the equation and use the power rule to form an equation that does not contain variable exponents. **STEP 2** Solve the equation formed in step 1.	**b.** $2^x = 7$ $\log 2^x = \log 7$ $x \log 2 = \log 7$ $x = \dfrac{\log 7}{\log 2}$

SELF-CHECK 10.6.1 ANSWER

$x \approx 2.807$

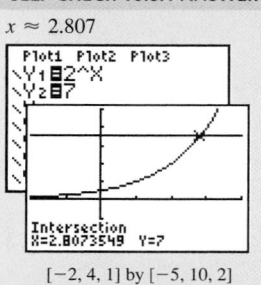

[−2, 4, 1] by [−5, 10, 2]

When we take the logarithm of both sides of an equation, we usually use either common logs or natural logs. We do this because these functions are available on calculators. In Example 3 we use natural logs.

To solve exponential equations we can take either the common log of both sides of the equation or the natural log of both sides of the equation.

■ EXAMPLE 3   Solving an Exponential Equation

Solve $6^{z+3} = 8^{2z-1}$ and approximate this solution to the nearest ten-thousandth.

SOLUTION

$$6^{z+3} = 8^{2z-1}$$

Because 6 and 8 are not powers of the same base, take the natural log of both sides of the equation.

$$\ln 6^{z+3} = \ln 8^{2z-1}$$

Use the power rule for logarithms.

$$(z+3)\ln 6 = (2z-1)\ln 8$$

Multiply using the distributive property and then rewrite the equation so that the variable terms are on the left side of the equation.

$$z \ln 6 + 3 \ln 6 = 2z \ln 8 - \ln 8$$

$$z \ln 6 - 2z \ln 8 = -\ln 8 - 3 \ln 6$$

$$z(\ln 6 - 2 \ln 8) = -(\ln 8 + 3 \ln 6)$$

Factor out z and then divide both sides of the equation by the coefficient of z. This is the exact solution.

$$z = -\frac{\ln 8 + 3 \ln 6}{\ln 6 - 2 \ln 8}$$

$$z \approx 3.1492736$$

Use a calculator to determine a numerical approximation of the exact answer obtained in the previous step.

Answer: $z \approx 3.1493$

Does this value check? ■

SELF-CHECK 10.6.2

Solve $3^{2z+1} = 5^z$ to the nearest thousandth.

As Examples 2 and 3 illustrate, either common logs or natural logs can be used to solve exponential equations. If the exponential equation involves base e, then we usually use natural logarithms.

■ EXAMPLE 4   Solving an Exponential Equation

Solve $3e^{x^2+2} = 49.287$ and then approximate this solution to the nearest ten-thousandth.

SOLUTION

$$3e^{x^2+2} = 49.287$$

Divide both sides of the equation by 3.

$$e^{x^2+2} = 16.429$$

Rewrite this exponential equation in logarithmic form with base e.

$$x^2 + 2 = \ln 16.429$$

Subtract 2 from both sides of the equation.

$$x^2 = \ln 16.429 - 2$$

Find the exact roots of this quadratic equation by extraction of roots.

$$x = \pm\sqrt{\ln 16.429 - 2}$$

$$x \approx \pm 0.8938948853$$

Use a calculator to approximate this value.

Answer: $x \approx -0.8939$ or $x \approx 0.8939$

Do these values check? ■

Since exponential and logarithmic functions are inverses of each other, it is not surprising that we can use logarithms to solve some exponential equations. Likewise, we can solve some logarithmic equations by first rewriting them in exponential form.

■ **EXAMPLE 5** Solving a Logarithmic Equation with One Logarithmic Term

Solve $\log(3x + 7) = 2$ algebraically, numerically, and graphically.

SOLUTION

ALGEBRAICALLY

$\log(3x + 7) = 2$

$3x + 7 = 10^2$ Rewrite this logarithmic equation in exponential form with base 10.

$3x + 7 = 100$

$3x = 93$ Then solve for x.

$x = 31$

NUMERICALLY

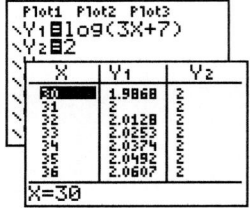

Let $y_1 = \log(3x + 7)$ and $y_2 = 2$ and then examine these functions both by tables and by graphs.

For $x = 31$, $y_1 = y_2$.

GRAPHICALLY

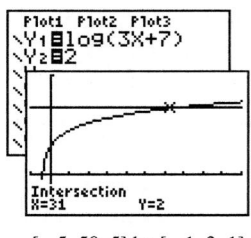

$[-5, 50, 5]$ by $[-1, 3, 1]$

The x-coordinate of the point of intersection is $x = 31$.

Answer: $x = 31$ *Check:* $\log[3(31) + 7] \overset{?}{=} 2$

$\log 100 \overset{?}{=} 2$

$2 \overset{?}{=} 2$ checks.   ■

Logarithmic functions are one-to-one functions.

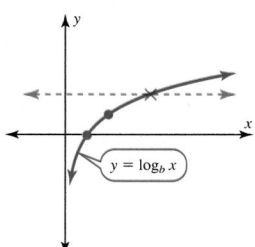

Logarithmic functions are one-to-one functions. (The horizontal-line test can be applied to the graph of $y = \log_b x$ to show that this function is one-to-one.) Thus if x, $y > 0$, and $\log_b x = \log_b y$, then $x = y$. The fact that the arguments are equal when logarithms are equal allows us to replace a logarithmic equation with one that does not involve logarithms. Because real logarithms are defined only for positive arguments, each possible solution must be checked to make sure that it is not an extraneous value.

■ **EXAMPLE 6**   Solving a Logarithmic Equation with Two
                     Logarithmic Terms

Solve $\ln y = \ln(4y + 6)$.

SOLUTION _____

$\ln y = \ln(4y + 6)$	Since the logarithms are equal, the arguments are
$y = 4y + 6$	equal.
$-3y = 6$	
$y = -2$	Now solve for y.

Check: $\ln y = \ln(4y + 6)$
$\ln(-2) \overset{?}{=} \ln[4(-2) + 6]$
$\ln(-2)$ is not a real number.   Logarithms of negative argument values are not
real numbers.
Answer: There is no solution. -2 is an extraneous value.   ■

SELF-CHECK 10.6.3 ANSWERS
1. $t = 11$
2. No solution; $v = 3$ is
   extraneous.

SELF-CHECK 10.6.3

Solve these logarithmic equations.
1. $\log(t - 1) = 1$
2. $\log(v - 5) = \log(1 - v)$

Solving Logarithmic Equations Algebraically

VERBALLY	ALGEBRAIC EXAMPLE
STEP 1 a. If possible, rewrite the logarithmic equation in exponential form. b. Otherwise, use the properties of logarithms to write each side of the equation as a single logarithmic term with the same base. Then form a new equation by equating the arguments of these logarithms.	**a.** $\log(x + 1) = 2$ $x + 1 = 10^2$ $x = 99$ *Check:* $\log(99 + 1) \overset{?}{=} 2$ $\log(100) \overset{?}{=} 2$ $2 \overset{?}{=} 2$ checks.
STEP 2 Solve the equation formed in step 1. **STEP 3** Check all possible solutions for extraneous values; logarithms of zero or negative arguments are undefined.	**b.** $\ln 2x = \ln(x + 1)$ $2x = x + 1$ $x = 1$ *Check:* $\ln 2(1) \overset{?}{=} \ln(1 + 1)$ $\ln 2 \overset{?}{=} \ln 2$ checks.

If a logarithmic equation has more than one logarithmic term on one side of the
equation, then we may need to use the properties of logarithms to rewrite these terms as
a single logarithm.

■ **EXAMPLE 7** Solving a Logarithmic Equation with Three Logarithmic Terms

Solve $\ln(3 - w) + \ln(1 - w) = \ln(11 - 6w)$.

SOLUTION _____

$\ln(3 - w) + \ln(1 - w) = \ln(11 - 6w)$

$\ln[(3 - w)(1 - w)] = \ln(11 - 6w)$

$(3 - w)(1 - w) = 11 - 6w$

$3 - 4w + w^2 = 11 - 6w$

$w^2 + 2w - 8 = 0$

$(w + 4)(w - 2) = 0$

$w + 4 = 0 \qquad$ or $\quad w - 2 = 0$

$w = -4 \qquad\qquad\qquad w = 2$

Express the left side of the equation as a single logarithm, using the product rule for logarithms. Since the natural logarithms are equal, the arguments are equal.

Combine like terms and write the quadratic equation in standard form. Then factor and solve for w.

Be sure to check possible solutions to a logarithmic equation to make sure the values are not extraneous.

Check: For $w = -4$,

$\ln[3 - (-4)] + \ln[1 - (-4)] \overset{?}{=} \ln[11 - 6(-4)]$

$\ln 7 + \ln 5 \overset{?}{=} \ln(11 + 24)$

$\ln 35 \overset{?}{=} \ln 35$ checks.

Check: For $w = 2$,

$\ln(3 - 2) + \ln(1 - 2) \overset{?}{=} \ln[11 - 6(2)]$

$\ln 1 + \ln(-1) \overset{?}{=} \ln(-1)$

$\ln(-1)$ is undefined.

Thus $w = 2$ is an extraneous value.

Answer: $w = -4$ ∎

> ### SELF-CHECK 10.6.4
>
> Solve for z.
> $\ln(2z + 5) + \ln z = \ln 3$.

■ **EXAMPLE 8** Solving a Logarithmic Equation with Two Logarithmic Terms

Solve $\log(2v + 1) - \log(v - 4) = 1$.

SOLUTION _____

$\log(2v + 1) - \log(v - 4) = 1$

$\log\left(\dfrac{2v + 1}{v - 4}\right) = 1$

$\dfrac{2v + 1}{v - 4} = 10^1$

$2v + 1 = 10(v - 4)$

$2v + 1 = 10v - 40$

$-8v = -41$

$v = \dfrac{41}{8}$

Express the left side of the equation as a single logarithm, using the quotient rule for logarithms.

Rewrite this equation in exponential form, using a base of 10.

Then solve for v.

SELF-CHECK 10.6.4 ANSWER

$z = \dfrac{1}{2}$

Check: $\log\left[2\left(\dfrac{41}{8}\right) + 1\right] - \log\left(\dfrac{41}{8} - 4\right) \overset{?}{=} 1$

$$\log\left(\dfrac{45}{4}\right) - \log\left(\dfrac{9}{8}\right) \overset{?}{=} 1$$

$$\log\left(\dfrac{45}{4} \div \dfrac{9}{8}\right) \overset{?}{=} 1$$

$$\log 10 \overset{?}{=} 1$$

$$1 \overset{?}{=} 1 \text{ checks.}$$

Answer: $v = \dfrac{41}{8}$

The exponential and logarithmic equations that we have examined in this section have been either contradictions with no solution or conditional equations with one or two solutions. We now examine some exponential and logarithmic equations that are identities.

■ EXAMPLE 9   Verifying Logarithmic and Exponential Identities

Verify the following identities.

SOLUTIONS

(a) $\log_7 14^x - \log_7 2^x = x$

$\log_7 14^x - \log_7 2^x = x \log_7 14 - x \log_7 2$
$\log_7 14^x - \log_7 2^x = x(\log_7 14 - \log_7 2)$

Rewrite the left side of the equation using the power rule for logarithms. Then factor out x.

$\log_7 14^x - \log_7 2^x = x \log_7 \dfrac{14}{2}$

Simplify this expression using the quotient rule for logarithms.

$\log_7 14^x - \log_7 2^x = x \log_7 7$
$\log_7 14^x - \log_7 2^x = x(1)$
$\log_7 14^x - \log_7 2^x = x$

Thus we have verified that the left side of the equation does equal the right side of the equation.

(b) $e^{-\ln x} = \dfrac{1}{x}$ for $x > 0$

$e^{-\ln x} = e^{\ln(x^{-1})}$
$e^{-\ln x} = x^{-1}$

$e^{-\ln x} = \dfrac{1}{x}$

Rewrite the left side of the equation using the power rule for logarithms: $p \ln x = \ln x^p$ with $p = -1$. Simplify, using the identity $b^{\log_b y} = y$, with $b = e$ and $y = x^{-1}$.

Replace x^{-1} with $\dfrac{1}{x}$ to obtain the right side of this identity.

USING THE LANGUAGE AND SYMBOLISM OF MATHEMATICS   10.6

1. An _____ equation contains a variable in the exponent.

2. A _____ equation contains a variable in the argument of a logarithm.

3. If $x > 0$, $y > 0$, and $\log_b x = \log_b y$, then _____ = _____.

EXERCISES **10.6**

In Exercises 1–24 solve each equation without using a calculator.

1. $3^{w-5} = 27$

2. $7^{2w-1} = \sqrt{7}$

3. $\left(\dfrac{2}{3}\right)^{x^2} = \dfrac{16}{81}$

4. $5^{x^2-6} = \dfrac{1}{25}$

5. $\log(y - 5) = 1$

6. $\log(2y + 1) = 1$

7. $\log(3n - 5) = 2$

8. $\log(5n - 4) = 0$

9. $\ln(3m - 7) = \ln(2m + 9)$

10. $\log(5y + 6) = \log(2y - 9)$

11. $\log(4w + 3) = \log(8w + 5)$

12. $\ln(5y - 7) = \ln(2y + 1)$

13. $\ln(3 - x) = \ln(1 - 2x)$

14. $\log(7 - 5x) = \log(4 - 8x)$

15. $\log(t + 3) + \log(t - 1) = \log 5$

16. $\ln(7t + 3) - \ln(t + 1) = \ln(6t + 2)$

17. $\ln(v^2 - 9) - \ln(v + 3) = \ln 7$

18. $\ln(2v + 6) - \ln(v + 1) = \ln(v + 3)$

19. $\ln(5x - 7) - \ln(2x + 3) = \ln 3$

20. $\log(7x + 13) - \log(4x + 13) = \log 5$

21. $\log(1 - y) + \log(4 - y) = \log(18 - 10y)$

22. $\ln(2 - y) + \ln(1 - y) = \ln(32 - 4y)$

23. $\log(w - 3) - \log(w^2 + 9w - 32) = -1$

24. $\log(w^2 + 1) - \log(w - 2) = 1$

In Exercises 25 and 26 use the given tables to solve each equation.

25. $4^x = 32$

26. $32^x = 64$

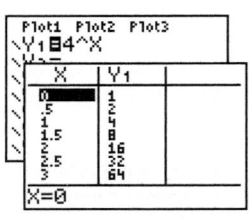

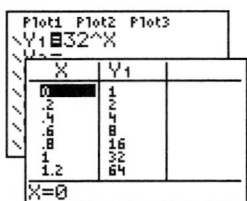

In Exercises 27 and 28 use the given graphs to solve each equation.

27. $8^{\frac{x+2}{3}} = 16$

28. $27^{\frac{2x-4}{3}} = 9$

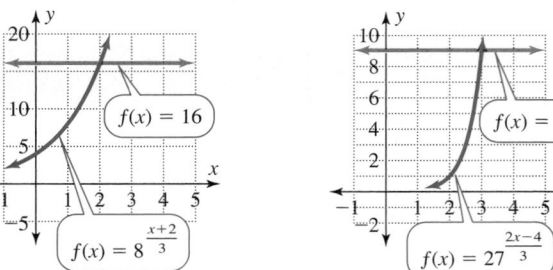

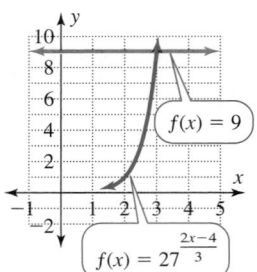

In Exercises 29–44 solve each equation and then use a calculator to approximate the solution to the nearest thousandth.

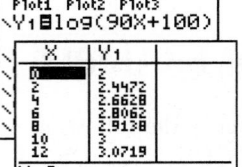

29. $4^v = 15$

30. $5^v = 18$

31. $3^{-w+7} = 22$

32. $6^{-w+3} = 81$

33. $9.2^{2t+1} = 11.3^t$

34. $8.7^{3t-1} = 10.8^{2t}$

35. $7.6^{-2z} = 5.3^{2z-1}$

36. $8.1^{-3z} = 6.5^{1-2z}$

37. $e^{3x} = 78.9$

38. $e^{3x+5} = 15.9$

39. $10^{2y+1} = 51.3$

40. $10^{3y-4} = 73.8$

41. $0.83^{v^2} = 0.68$

42. $0.045^{v^2} = 0.0039$

43. $3.7e^{x^2+1} = 689.7$

44. $2.5e^{x^2+4} = 193.2$

In Exercises 45 and 46 use the given tables to solve each equation.

45. $\log(5x - 9.99) = -2$

46. $\log(90x + 100) = 3$

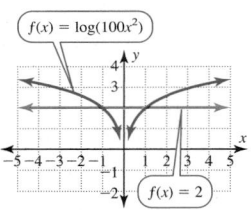

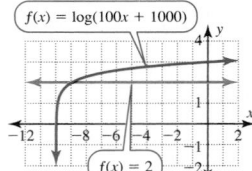

In Exercises 47 and 48 use the given graphs to solve each equation.

47. $\log(100x^2) = 2$

48. $\log(100x + 1000) = 2$

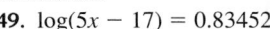

In Exercises 49–58 solve each equation and then use a calculator to approximate each solution to the nearest thousandth.

49. $\log(5x - 17) = 0.83452$

50. $\ln(11x - 3) = 1.44567$

51. $\ln(\ln y) = 1$

52. $\log(\log y) = 0.48913$

53. $\ln(v - 4) + \ln(v - 3) = \ln(5 - v)$

54. $\log(3x + 1) + \log(x + 2) = \log x$

55. $\ln(11 - 5x) - \ln(x - 2) = \ln(x - 6)$

56. $\ln(3 - x) - \ln(x - 2) = \ln(x + 1)$

57. $(\ln x)^2 = \ln x^2$

58. $(\log x)^2 = \log x^2$

In Exercises 59–62 use a graphics calculator to approximate the solution(s) of each equation to the nearest thousandth.

59. $\log(398x + 5) = x + 0.5$

60. $\ln(18x + 3) = 3x - 5$

61. $0.1e^x = 8 - 2x$

62. $e^{0.5x} = 4 - x^2$

In Exercises 63–70 verify each identity, assuming that x is a positive real number.

63. $10^{-\log x} = \dfrac{1}{x}$

64. $100^{\log x} = x^2$

65. $e^{-x \ln 3} = \left(\dfrac{1}{3}\right)^x$

66. $e^{(\ln x)/2} = \sqrt{x}$

67. $\log 60^x - \log 6^x = x$

68. $\log 5^x + \log 2^x = x$

69. $\ln\left(\dfrac{4}{5}\right)^x + \ln\left(\dfrac{5}{3}\right)^x + \ln\left(\dfrac{3}{4}\right)^x = 0$

70. $\ln\left(\dfrac{2}{3}\right)^x + \ln\left(\dfrac{5}{2}\right)^x - \ln\left(\dfrac{5}{3}\right)^x = 0$

71. Depreciation   The value V of an industrial lathe after t years of depreciation is given by the formula $V = 35{,}000e^{-0.2t} + 1000$. Approximately how many years will it take for the value to depreciate to \$10,000?

72. Depreciation   The value V of an irrigation system after t years of depreciation is given by the formula $V = 59{,}000e^{-0.2t} + 3000$. Approximately how many years will it take for the value to depreciate to \$7000?

Group Discussion Questions

73. Challenge Question   Which is larger, 2001^{2002} or 2002^{2001}?

74. Error Analysis   The steps used to solve a certain logarithmic equation produced one possible solution that was a negative number. One student claimed this number could not check in the equation, since it was negative. Discuss the logic of this student's claim.

75. Discovery Question
   a. Create an exponential equation whose only solution is 5.
   b. Create an exponential equation whose only solutions are -2 and 5.

76. Discovery Question
   a. Create a logarithmic equation with $x = -2$ as an extraneous value and that has no solution.
   b. Create a logarithmic equation with $x = -2$ as an extraneous value and with a solution of 5.
   c. Create a logarithmic equation with solutions of -2 and 5.

Section 10.7   Applications of Exponential and Logarithmic Equations

Objective: **11.** Use exponential and logarithmic equations to solve applied problems.

Exponential and logarithmic equations can be used to describe an increase or decrease in an amount of money, the growth or decline of populations, and the decay of radioactive elements. Logarithmic scales are often used to measure natural phenomena. For example, decibels measure the intensity of sounds, and the Richter scale measures the intensity of earthquakes. In this section we examine all of these applications.

   Two important formulas that we apply in the following examples are the formulas for periodic growth and continuous growth. Growth that occurs at discrete intervals (such as yearly, monthly, or weekly) is called **periodic growth;** growth that occurs continually at each instant is called **continuous growth.** The formula for continuous growth is a good model for periodic growth when the number of growth periods is relatively large. For example, the population of rabbits over a period of years can be accurately estimated using the continuous-growth formula, even though female rabbits do not have offspring continuously.

Growth and Decay Formulas

PERIODIC GROWTH FORMULA	VERBALLY	ALGEBRAIC EXAMPLE
$A = P\left(1 + \dfrac{r}{n}\right)^{nt}$	The amount A that is produced by an original amount P growing at an annual rate r with periodic compounding n times a year for t years.	$A = 1000\left(1 + \dfrac{0.07}{4}\right)^{(4)(10)}$ This is the amount resulting from investing \$1000 at 7% compounded quarterly for 10 years.
CONTINUOUS GROWTH AND DECAY FORMULA $A = Pe^{rt}$	The amount A that is produced by an original amount P growing (decaying) continuously at a rate r for a time t. If $r > 0$, there is growth. If $r < 0$, there is decay.	$A = 1000e^{(0.07)(10)}$ This is the amount resulting from investing \$1000 at 7% compounded continuously for 10 years.

■ EXAMPLE 1   Periodic Growth of an Investment

How many years will it take an investment of \$1000 to double in value if interest on the investment is compounded semiannually at a rate of 11%?

It is customary that most statements of interest rates are given as annual rates. In Example 1 the rate of 11% means an annual rate of 11%.

SOLUTION

$$A = P\left(1 + \frac{r}{n}\right)^{nt}$$

$$2000 = 1000\left(1 + \frac{0.11}{2}\right)^{2t}$$
Substitute $A = 2000$, $P = 1000$, $r = 0.11$, and $n = 2$ into the periodic-growth formula and then simplify.

$$2 = (1.055)^{2t}$$

$$\ln 2 = \ln(1.055)^{2t}$$
Take the natural log of both sides of the equation.

$$\ln 2 = 2t \ln 1.055$$
Use the power rule for logarithms.

$$t = \frac{\ln 2}{2 \ln 1.055}$$
Solve for t by dividing both sides of the equation by 2 ln 1.055.

$$t \approx 6.4730785$$
Use a calculator to approximate this value.

Answer: The investment will double in value in approximately 6.5 years.   ■

SELF-CHECK 10.7.1

1. Determine the value of \$1000 invested at 7% interest compounded quarterly for 10 years.
2. Compute the number of years it will take an investment of \$1 to triple in value if the interest is compounded annually at 8%.

SELF-CHECK 10.7.1 ANSWERS

1. \$2001.60
2. Approximately 14.3 years

A comparison of periodic growth to continuous growth is given in Table 10.7.1. This table shows various compounding periods and the amount to which \$1 accumulates in each situation after 1 year. The interest formula, with $P = 1$ and $r = 1.00$, is

$$A = 1\left(1 + \frac{1}{n}\right)^{n(1)} = \left(1 + \frac{1}{n}\right)^{n}.$$

Note that compounding monthly produces a big gain over simple interest. However, further gains are relatively small, and as n increases, the expression $\left(1 + \frac{1}{n}\right)^{n}$ approaches the irrational number e.

$A = P\left(1 + \frac{r}{n}\right)^{nt}$ with $P = 1$, $r = 1$, and $t = 1$.

Table 10.7.1 Periodic Compounding and Continuous Compounding for 1 Year

PERIODIC COMPOUNDING	n	$A = \left(1 + \frac{1}{n}\right)^{n}$ (to six places)	A (to the nearest penny)
Annually	1	2.000000	$2.00
Semiannually	2	2.250000	2.25
Quarterly	4	2.441406	2.44
Monthly	12	2.613035	2.61
Weekly	52	2.692597	2.69
Daily	365	2.714567	2.71
Hourly	8,760	2.718127	2.72
Every minute	525,600	2.718279	2.72
Every second	31,536,000	2.718282	2.72
.	.	.	
.	.	.	
.	.	.	
Continuously	Infinite	e	
CONTINUOUS COMPOUNDING		$A = PE^{RT} = E$	

$A = Pe^{rt}$ with $P = 1$, $r = 1$, and $t = 1$.

The formula for continuous compounding provides a reasonable approximation of periodic compounding when there are many growth periods.

A key point to note is that continuous compounding provides a reasonable approximation for some periodic events; this is especially true when there are many growth periods. Because the formula $A = Pe^{rt}$ is easy to use and can provide a good approximation of periodic growth, it is used widely in growth and decay problems. Examples 2 through 4 illustrate some of the possibilities.

■ EXAMPLE 2   Comparing Periodic Growth to Continuous Growth

Compute the value of $1000 invested at 6% for 5 years under two conditions:
(a) the interest is compounded daily, and **(b)** the interest is compounded continuously.

SOLUTIONS _____

(a) $A = P\left(1 + \frac{r}{n}\right)^{nt}$

$A = 1000\left(1 + \frac{0.06}{365}\right)^{(365)(5)}$

$A \approx 1000(1.000164384)^{1825}$

$A \approx \$1349.83$

Use the formula for periodic growth. Substitute in the given values. Use 365 days for each year (assume that there is no change due to leap year).

Use a calculator to approximate this value.

(b) $A = Pe^{rt}$

$A = 1000e^{(0.06)(5)}$

$A = 1000e^{0.30}$

$A \approx \$1349.86$

Use the formula for continuous growth. Substitute in the given values.

Use a calculator to approximate this value. ■

Note that the results in Example 2 differ by only 3 cents over a 5-year period. Thus continuous growth provided a good model for growth compounded daily.

SELF-CHECK 10.7.2

Compute the value of $5000 invested at 7% for 8 years under two conditions:

1. The interest is compounded monthly.
2. The interest is compounded continuously.

■ EXAMPLE 3   Continuous Growth of Bacteria

A new culture of bacteria grows continuously at the rate of 20% per day. If a culture of 10,000 bacteria isolated in a laboratory is allowed to multiply, how many bacteria will there be at the end of 1 week?

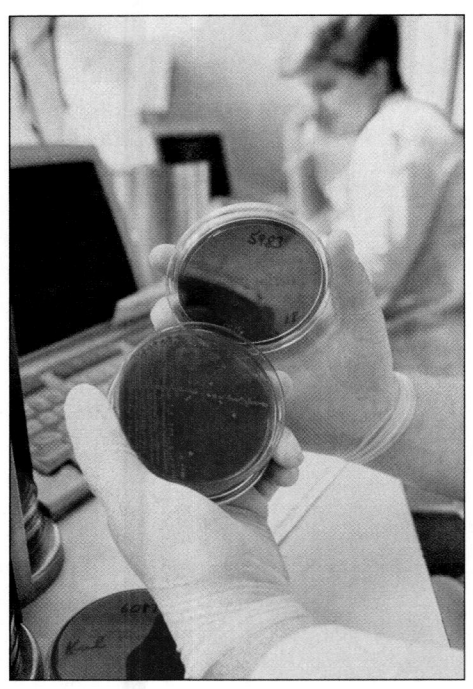

SOLUTION _____

$A = Pe^{rt}$ Substitute $P = 10,000$, $r = 0.20$, and $t = 7$ (1 week = 7 days)
$A = 10,000e^{0.20(7)}$ into the continuous growth formula, and then simplify.
$A = 10,000e^{1.4}$
$A \approx 40,552$ Use a calculator to approximate this value.

Answer: At the end of the week there will be approximately 41,000 bacteria.

SELF-CHECK 10.7.2 ANSWERS

1. $8739.13
2. $8753.36

For physical phenomena such as radioactive decay, exponential decay models have been experimentally tested and provide an excellent mathematical model for predicting the future or describing the past. This is illustrated in Example 4.

■ EXAMPLE 4   Continuous Decay of Carbon-14

Carbon-14 decays continuously at the rate of 0.01245% per year. When living tissue dies, it no longer absorbs carbon-14; any carbon-14 present decays and is not replaced. An archaeologist has determined that only 20% of the carbon-14 originally in a plant specimen remains. Estimate the age of this specimen.

SOLUTION

$$A = Pe^{rt}$$

Substitute the given values into the continuous-decay formula. The 0.01245% decay rate means that $r = -0.0001245$, and 20% of the original amount P left means that $A = 0.20P$.

$$0.20P = Pe^{-0.0001245t}$$

$$0.20 = e^{-0.0001245t}$$

Divide both sides of the equation by P.

$$\ln 0.20 = -0.0001245t$$

Rewrite this statement in logarithmic form.

$$t = \frac{\ln 0.20}{-0.0001245}$$

Solve this linear equation for t.

$$t \approx 12{,}927.21215$$

Use a calculator to approximate this value.

Answer: This specimen is approximately 13,000 years old.   ■

SELF-CHECK 10.7.3

1. The population of an island is growing continuously at the rate of 5% per year. Estimate to the nearest thousand the population of the island in 10 years, given that the population is now 100,000.
2. The population of a species of whales is estimated to be decreasing continuously at a rate of 5% per year ($r = -0.05$). If this rate of decrease continues, in how many years will the population have declined from its current level of 20,000 to 8000?

Logarithmic growth functions grow very slowly. For example, log 100,000 = 5, while log 1,000,000 = 6. The increase from 100,000 to 1,000,000 causes an increase in the logarithm of only 1 unit from 5 to 6. Scientists take advantage of this property to describe phenomena whose numerical measure covers a wide range of scores. By converting these measures to a logarithmic scale the numbers become smaller and more comprehensible. Two examples of this are the Richter scale and the decibel scale. These scales are illustrated in Examples 5 and 6.

Seismologists use the Richter scale to measure the magnitude of earthquakes. The equation $R = \log \dfrac{A}{a}$ compares the amplitude A of the shock wave of an earthquake to the amplitude a of a reference shock wave of minimal intensity.

■ EXAMPLE 5   Using the Richter Scale

In the Denver, Colorado, area in the 1960s an earthquake with an amplitude 40,000 times the reference amplitude occurred after a liquid was injected under pressure into a well more than 2 mi deep. Calculate the magnitude of this earthquake on the Richter scale.

SOLUTION

$$R = \log\left(\frac{A}{a}\right)$$

$$R = \log\left(\frac{40,000a}{a}\right) \quad \text{Substitute the given amplitude into the Richter scale equation.}$$

$$R = \log 40,000 \quad \text{Simplify the fraction.}$$

$$R \approx 4.6020600 \quad \text{Use a calculator to approximate this value.}$$

Answer: This earthquake measured approximately 4.6 on the Richter scale.   ■

The human ear can hear a vast range of sound intensities, so it is more practical to use a logarithmic scale than to use an absolute scale to measure the intensity of sound. The unit of measurement on this scale is the **decibel.** The number of decibels D of a sound is given by the formula $D = 10 \log \dfrac{I}{I_o}$, which compares the intensity I of the sound to the reference intensity I_o, which is at the threshold of hearing ($I_o \approx 10^{-16}$ watts/cm^2).

■ EXAMPLE 6   Determining the Decibel Level of Music

If a store in a mall plays background music at an intensity of 10^{-14} watts/cm^2, determine the number of decibels of the music you hear.

SOLUTION

$$D = 10 \log \frac{I}{I_o}$$

$$D = 10 \log \frac{10^{-14}}{10^{-16}} \quad \text{Substitute } 10^{-14} \text{ for } I \text{ and } 10^{-16} \text{ for } I_o \text{ into the decibel equation.}$$

$$D = 10 \log 10^2 \quad \text{Simplify and then solve for } D \text{ by inspection.}$$

$$D = 10(2)$$

$$D = 20$$

Answer: 20 decibels   ■

SELF-CHECK 10.7.4

Compute the number of decibels of a child's cry if its intensity is 10^{-7} watts/cm^2.

Chemists use pH to measure the hydrogen ion concentration in a solution. Distilled water has a pH of 7, acids have a pH of less than 7, and bases have a pH of more than 7. The formula for the pH of a solution is pH $= -\log$ H$^+$, where H$^+$ measures the concentration of hydrogen ions in moles per liter.

■ EXAMPLE 7   Determining the pH of a Beer

Determine the pH of a light beer if its H$^+$ is measured at 6.3×10^{-5} moles/liter.

SOLUTION _____

pH $= -\log$ H$^+$
pH $= -\log(6.3 \times 10^{-5})$ Substitute H$^+ = 6.3 \times 10^{-5}$ into the pH equation.
pH ≈ 4.2007 Approximate this value with a calculator.

Answer: pH ≈ 4.2 ■

USING THE LANGUAGE AND SYMBOLISM OF MATHEMATICS   10.7

1. Growth that occurs at discrete intervals is called
   _____ growth.
2. Growth that occurs continually at each instant is called
   _____ growth.
3. The periodic growth formula is $A =$ _____ .

4. The continuous growth formula is $A =$ _____ .
5. Seismologists use the Richter scale to measure the magnitude of _____ .
6. The decibel is a unit of measure for the intensity of _____ .

EXERCISES   10.7

Compound Interest   In Exercises 1–8 use the formula for periodic growth to solve each problem. All interest rates are stated as annual rates of interest.
1. Find the value of $150 invested at 9% with interest compounded quarterly for 5 years.
2. Find the value of $210 invested at 7% with interest compounded monthly for 9 years.
3. How many years will it take an investment to double in value if interest is compounded annually at 8%?
4. How many years will it take an investment to double in value if interest is compounded semiannually at 10%?
5. How many years will it take a savings account to triple in value if interest is compounded monthly at 6%?
6. How many years will it take a zero-coupon bond to triple in value if interest is compounded monthly at 7.5%?
7. If an investment on which interest is compounded monthly doubles in value in 8 years, what is the annual rate of interest?
8. If an investment on which interest is compounded quarterly doubles in value in 9 years, what is the annual rate of interest?

In Exercises 9–22 use the formula for continuous growth and decay to solve each problem.
9. **Rate of Inflation**   If prices will double in 10 years at the current rate of inflation, what is the current rate of inflation? Assume that the effect of inflation is continuous.
10. **Rate of Inflation**   If prices will double in 9 years at the current rate of inflation, what is the current rate of inflation? Assume that the effect of inflation is continuous.
11. **Continuous Compound Interest**   How many years will it take an investment to double in value if interest is compounded continuously at 7%?
12. **Continuous Compound Interest**   How many years will it take an investment to double in value if interest is compounded continuously at 9%?
13. **Carbon-14 Dating**   Carbon-14 decays continuously at the rate of 0.01245% per year. An archaeologist has determined that only 5% of the original carbon-14 from a plant specimen remains. Estimate the age of this specimen.

14. Carbon-14 Dating Carbon-14 decays continuously at the rate of 0.01245% per year. An archaeologist has determined that only 10% of the original carbon-14 from a plant specimen remains. Estimate the age of this specimen.

15. Modeling the Safe Storage Time for Milk The safe level of psychrotrophic bacteria in a gallon of milk is 100 units. In a refrigerator set at 38°F the number of units of these bacteria in a gallon of skim milk is approximated by the exponential function $B(t) = 4.0e^{0.24t}$, where t is the time in days. Evaluate and interpret each expression.
   a. $B(0)$ **b.** $B(5)$ **c.** $B(10)$ **d.** $B(t) = 100, t = ?$

16. Modeling the Safe Storage Time for Milk The safe level of psychrotrophic bacteria in a gallon of milk is 100 units. In a refrigerator set at 38°F the number of units of these bacteria in a gallon of whole milk is approximated by the exponential function $B(t) = 4.0e^{0.26t}$, where t is the time in days. Evaluate and interpret each expression.
   a. $B(0)$ **b.** $B(5)$ **c.** $B(10)$ **d.** $B(t) = 100, t = ?$

17. Radioactive Decay The radioactive material used to power a satellite decays at a rate that decreases the available power by 0.05% per day. When the power supply reaches $\frac{1}{100}$ of its original level, the satellite is no longer functional. How many days should the power supply last?

18. Radioactive Decay The radioactive material used to power a satellite decays at a rate that decreases the available power by 0.03% per day. When the power supply reaches $\frac{1}{100}$ of its original level, the satellite is no longer functional. How many days should the power supply last?

19. Population Decline The population of a species of whales is estimated to be decreasing at a rate of 4% per year ($r = -0.04$). The current population is approximately 15,000. First estimate the population 10 years from now and then determine how many years from now the population will have declined to 5000.

20. Population Decline If the number of white owls in Illinois has decreased from 750 to 500 in 10 years, what is the annual rate of decrease?

21. Population Growth If the population of a town has grown from 1200 to 1800 in 3 years, what is the annual rate of increase?

22. Population Growth The human population in a remote area has doubled in the last 18 years. What is the annual rate of increase?

Magnitude of an Earthquake In Exercises 23–26 use the formula $R = \log \dfrac{A}{a}$ to solve each problem.

23. The amplitude of the September 19, 1985, earthquake in Mexico City was 63,100,000 times the reference amplitude. Calculate the magnitude of this earthquake on the Richter scale.

24. The amplitude of the September 20, 1965, earthquake in Mexico City was 20,000,000 times the reference amplitude. Calculate the magnitude of this earthquake on the Richter scale.

25. The damage done by an earthquake is related to its magnitude, the population of the area affected, and the quality of the buildings in this area. The August 17, 1999, earthquake that hit Izmit, Turkey, is rated by some (The Learning Channel, "Ultimate Natural Disasters") as the number seven natural disaster to ever occur. At 7.4 on the Richter scale it is the greatest earthquake to hit a major city since the 1906 San Francisco earthquake, which was estimated to be 8.25. Calculate how many times more intense the 1906 San Francisco earthquake was than the 1999 Izmit earthquake.

26. Calculate how many times more intense an earthquake with a Richter scale reading of 8.6 is than an earthquake with a Richter scale reading of 8.3.

Noise Levels In Exercises 27–30 use the formula $D = 10 \log \dfrac{I}{I_o}$ to solve each problem ($I_o = 10^{-16}$ watts/cm^2).

27. Find the number of decibels of a whisper if its intensity is 3×10^{-14} watts/cm^2.

28. Find the number of decibels of city traffic if its intensity is 8.9×10^{-7} watts/cm^2.

29. The noise level in a bar measures 85 decibels. Calculate the intensity of this noise.

30. The decibel reading near a jet aircraft is 105 decibels. Calculate the intensity of this noise.

pH Measurements In Exercises 31–34 use the formula $pH = -\log H^+$ to solve each problem.

31. Determine the pH of grape juice that has an H^+ concentration of 0.000109 moles/liter.

32. Determine the pH of saccharin, a sugar substitute, which has an H^+ concentration of 4.58×10^{-7} moles/liter.

33. A leading shampoo has a pH of 9.13. What is the H^+ concentration in moles per liter?

34. Blood is buffered (kept constant) at a pH of 7.35. What is the H^+ concentration in moles per liter?

The monthly payment P required to pay off a loan of amount A at an annual interest rate R in n years is given by the formula

$$P = \frac{A\left(\dfrac{R}{12}\right)}{1 - \left(1 + \dfrac{R}{12}\right)^{-12n}}.$$

In Exercises 35 and 36 use this formula to calculate the monthly payment.

35. Loan Payments   Determine the monthly payment necessary to pay off a $47,400 home loan that is financed for 30 years at 9.875%.

36. Loan Payments   Determine the monthly payment necessary to pay off a $47,400 home loan that is financed for 30 years at 10.5%.

The number of monthly payments of amount P required to completely pay off a loan of amount A borrowed at interest rate R is given by the formula

$$n = -\frac{\log\left(1 - \dfrac{AR}{12P}\right)}{\log\left(1 + \dfrac{R}{12}\right)}.$$

In Exercises 37 and 38 use this formula to calculate the number of monthly payments.

37. Car Payments   Determine the number of monthly car payments of $253.59 required to pay off a $7668.00 car loan when the interest rate is 11.7%.

38. Car Payments   Determine the number of monthly car payments of $200.80 required to pay off a $7668.00 car loan when the interest rate is 11.7%.

Group Discussion Questions

39. Writing Mathematically   Which type of function—exponential or logarithmic—would provide a better model for the number of skills a dog acquires during the first 10 years of its life? Explain your answer.

40. Challenge Question   Suppose that a bacteria doubles in volume every day. After 30 days, the population of these bacteria has grown to fill a laboratory jar. On what day was the jar half-full of these bacteria?

KEY CONCEPTS   FOR CHAPTER 10

1. Geometric Sequence:

- A geometric sequence is a sequence with a constant ratio for any two consecutive terms. This common ratio often is denoted by r.
- If the common ratio $r > 1$, the geometric sequence exhibits geometric growth.
- If the common ratio $0 < r < 1$, the geometric sequence exhibits geometric decay.
- The graph of the terms of a geometric sequence consists of discrete points that lie on the graph of an exponential function.

2. Exponential Function, $f(x) = b^x$:

- If $b > 0$ and $b \neq 1$, then $f(x) = b^x$ is an exponential function with base b.
- If $b > 1$, then $f(x) = b^x$ is called an exponential growth function.
- If $0 < b < 1$, then $f(x) = b^x$ is called an exponential decay function.
- For $b \neq 1$, $b^x = b^y$ if and only if $x = y$.
- For $a, b > 0$, and $x \neq 0$, $a^x = b^x$ if and only if $a = b$.

3. Exponential Growth and Decay: Exponential Growth Function

If $b > 1$, then $f(x) = b^x$ exhibits exponential growth.

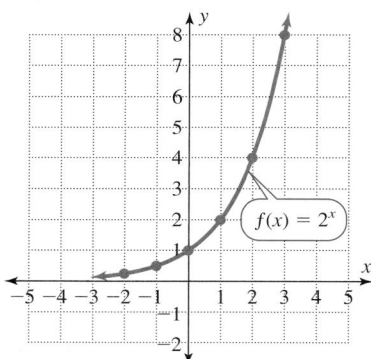

Exponential Decay Function

If $0 < b < 1$, then $f(x) = b^x$ exhibits exponential decay.

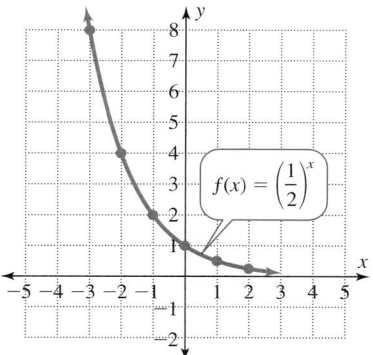

4. Properties of the Graphs of Exponential Functions:

For the exponential function $f(x) = b^x$ with $b > 0$ and $b \neq 1$:

- The domain of f is the set of all real numbers, $\mathbb{R}$.
- The range of f is the set of all positive real numbers.
- There is no x-intercept.
- The y-intercept is $(0, 1)$.
- The growth function is asymptotic to the negative portion of the x-axis (it approaches but does not touch the x-axis).
- The decay function is asymptotic to the positive portion of the x-axis.
- The growth function rises (or grows) from left to right.
- The decay function falls (or decays) from left to right.

5. The Irrational Constant e:

$e \approx 2.718\ 281\ 828\ldots$

Remember $e \approx 2.718$.

6. A One-to-One Function:

- A function is one-to-one if different input values produce different output values.
- The vertical-line test can be used to determine whether a graph represents a function.
- The horizontal-line test can be used to determine whether the graph of a function represents a one-to-one function.

7. Horizontal-Line Test:

- The graph of a function represents a one-to-one function if it is impossible to have any horizontal line intersect the graph at more than one point.

8. Inverse of a Function:

- If f is a function that matches each input value x with an output value y, then the inverse of f reverses this correspondence to match this y-value with the x-value.
- The inverse of a one-to-one function f is also a function. In this case we call f^{-1} an inverse function.
- The graphs of f and f^{-1} are symmetric about the line $x = y$.
- The DrawInv feature on a graphics calculator can be used to sketch the inverse of $y_1 = f(x)$.

9. Finding an Equation for the Inverse of a Function:

Step 1. Replace $f(x)$ by y. Write this as a function of y in terms of x.

Step 2. To form the inverse, replace each x with y and each y with x.

Step 3. If possible, solve the resulting equation for y.

Step 4. The inverse then can be written in function notation by replacing y with $f^{-1}(x)$.

10. Logarithmic and Exponential Functions:

- Logarithmic and exponential functions are inverses of each other.

11. Logarithmic Function, $f(x) = \log_b x$:

- For $x > 0$, $b > 0$, and $b \neq 1$, $y = \log_b x$ if and only if $b^y = x$.
- Common logarithms: $\log x$ means $\log_{10} x$.

- Natural logarithms: $\ln x$ means $\log_e x$.
- $y = \log x$ is equivalent to $x = 10^y$.
- $y = \ln x$ is equivalent to $x = e^y$.

12. Properties of Logarithms:

For $x > 0$, $y > 0$, $b > 0$, and $b \neq 1$,

- $\log_b 1 = 0$
- $\log_b b = 1$
- $\log_b \dfrac{1}{b} = -1$
- $\log_b b^x = x$
- $b^{\log_b x} = x$
- Product rule: $\log_b xy = \log_b x + \log_b y$
- Quotient rule: $\log_b \dfrac{x}{y} = \log_b x - \log_b y$
- Power rule: $\log_b x^p = p \log_b x$

13. Logarithmic Growth Function:

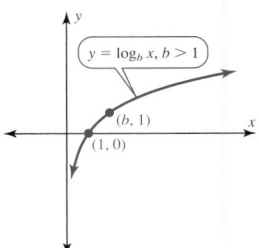

Features: For large values of x, extremely slow growth.

x-intercept: $(1, 0)$
y-intercept: None
Key point: $(b, 1)$
Asymptotic to negative y-axis

14. Properties of the Graphs of Logarithmic Functions, $f(x) = \log_b x$:

For the logarithmic function $f(x) = \log_b x$ with $x > 0$, $b > 0$, and $b \neq 1$:

- The domain of f is the set of all positive real numbers, $(0, +\infty)$.
- The range of f is the set of all real numbers, $\mathbb{R}$.
- The x-intercept is $(1, 0)$.
- There is no y-intercept.
- The growth function is asymptotic to the negative portion of the y-axis.

- The decay function is asymptotic to the positive portion of the y-axis.
- The growth function rises (or grows) from left to right.
- The decay function falls (or decays) from left to right.

15. Change-of-Base Formulas:

For $a, b > 0$ and $a, b \neq 1$,

- $\log_a x = \dfrac{\log_b x}{\log_b a}$ for $x > 0$
- $a^x = b^{x \log_b a}$

16. Solving Exponential Equations:

a. If it is obvious that both sides of an equation are powers of the same base, express each side as a power of this base and then equate the exponents. Otherwise, take the logarithm of both sides of the equation and use the power rule to form an equation that does not contain variable exponents.

b. Then solve the equation formed.

17. Solving Logarithmic Equations:

a. If possible, rewrite the logarithmic equation in exponential form. Otherwise, use the properties of logarithms to write the two sides of the equation as two single logarithms with the same base. Form a new equation by equating the arguments of these logarithms.

b. Then solve the equation formed.

c. Check all possible solutions for extraneous values, since logarithms of zero and negative arguments are undefined.

18. Growth and Decay Formulas:

- Periodic-growth formula:
$$A = P\left(1 + \frac{r}{n}\right)^{nt}$$

- Continuous-growth formula:
$$A = Pe^{rt}$$

REVIEW EXERCISES   **FOR CHAPTER 10**

In Exercises 1 and 2 determine whether each sequence is arithmetic, geometric, both, or neither. If the sequence is arithmetic, write the common difference d. If the sequence is geometric, write the common ratio r.

1. a. $2, 4, 6, 8, 10, \ldots$
   b. $2, 4, 8, 16, 32, \ldots$
   c. $2, 4, 6, 8, 10, 16, 26, \ldots$

2. a. $5, 5, 5, 5, 5, \ldots$
   b. $7, 3, -1, -5, -9, \ldots$
   c. $5, -5, 5, -5, 5, \ldots$

3. Dilution of a Mixture   A tank contains 100 gallons of a cleaning solvent. Then 10 gallons are drained and replaced with pure water. The contents are thoroughly mixed. Then the process is repeated by draining 10 gallons and replacing this mixture with pure water. Determine the volume of solvent in the tank after each of the first of five such drainings.

100, 90, 81, _____, _____, _____

4. Use $f(x) = 9^x$ to evaluate each expression.
   a. $f(0)$ **b.** $f(-1)$
   c. $f(2)$ **d.** $f\left(\dfrac{1}{2}\right)$

5. Use $f(x) = \pi^x$ and a calculator to approximate each expression to the nearest thousandth.
   a. $f(-1)$ **b.** $f(2)$
   c. $f(\pi)$ **d.** $f(e)$

In Exercises 6–8 match each graph with the most appropriate description.
6. Not a function of x.
7. A function of x but not a one-to-one function.
8. A one-to-one function of x.

A.

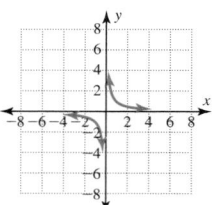

B.

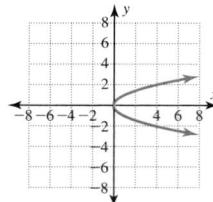

C.

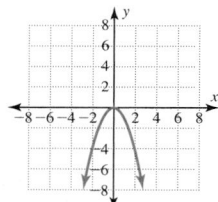

9. Graph each function and its inverse on the same coordinate system.
   a. $f(x) = \left(\dfrac{4}{3}\right)^x$
   b. $f(x) = \left(\dfrac{3}{4}\right)^x$
   c. $f(x) = \ln x$

In Exercises 10 and 11 the function f is described verbally and algebraically. Describe f^{-1} both verbally and algebraically.

Verbally	Algebraically
10. $f(x)$ triples x and then subtracts 2.	$f(x) = 3x - 2$
11. $f(x)$ takes one-fourth of x and then adds 6.	$f(x) = \dfrac{x}{4} + 6$

In Exercises 12 and 13 write the inverse of each function using ordered-pair notation.
12. $f = \{(-4, -5), (-3, -3), (0, 3), (2, 7), (3, 9)\}$
13.

x	-2	-1	0	1	2	3
y	0	2	4	6	8	10

In Exercises 14 and 15 write the equation of the inverse of each function.
14. $f(x) = \dfrac{1}{3}x - 4$
15. $f(x) = 3^x$
16. **Product Costs** The cost of producing x pizzas at a small pizzeria is given by the function $C(x) = 2x + 250$.
   a. What does the variable x represent in $C(x)$?
   b. Determine the formula for $C^{-1}(x)$.
   c. What does the variable x represent in $C^{-1}(x)$?
   d. What does the output of $C^{-1}(x)$ represent?

e. Determine the cost of producing 100 pizzas.
f. Determine the number of pizzas that can be produced for $398.

In Exercises 17 and 18 graph the inverse of the given function.
17.

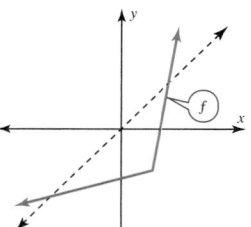

18.

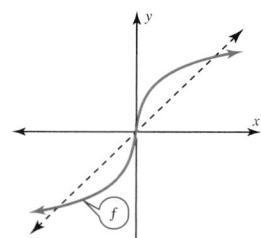

In Exercises 19 and 20 write each logarithmic equation in exponential form.
19. a. $\log_6 \sqrt{6} = \dfrac{1}{2}$ b. $\log_{17} 1 = 0$ c. $\log_8 \dfrac{1}{64} = -2$
20. a. $\log_b a = c$ b. $\ln a = c$ c. $\log c = d$

In Exercises 21 and 22 write each exponential equation in logarithmic form.
21. a. $7^3 = 343$ b. $19^{1/3} = \sqrt[3]{19}$ c. $\left(\dfrac{4}{7}\right)^{-2} = \dfrac{49}{16}$
22. a. $e^{-1} = \dfrac{1}{e}$ b. $10^{-4} = 0.0001$ c. $8^x = y$

In Exercises 23 and 24 evaluate each expression by inspection.
23. a. $\log_{12} 144$ b. $\log_{12} \sqrt{12}$ c. $\log_{12} \dfrac{1}{12}$
24. a. $\log_{12} 1$ b. $\log_{12} 12^{0.23}$ c. $\log_{12} \dfrac{1}{\sqrt[3]{12}}$

In Exercises 25–32 use a calculator to approximate each expression to the nearest thousandth.
25. a. $(\sqrt{7})^\pi$ b. $e^{7.83}$ c. $10^{-0.8107}$
26. a. $\pi^{\sqrt{7}}$ b. $e^{-2.4897}$ c. $10^{4.83}$
27. a. $\log 113.58$ b. $\ln 113.58$ c. $\log_5 113.58$
28. a. $\log(8.1 \times 10^{-4})$ b. $\ln(8.1 \times 10^{-4})$ c. $\log_5(8.1 \times 10^{-4})$
29. a. $\dfrac{\ln 23}{\ln 19}$ b. $\ln \dfrac{23}{19}$ c. $\dfrac{\ln 23}{19}$
30. a. $\ln(6 + 5)$ b. $\ln(6 \cdot 5)$ c. $\ln 6 + \ln 5$
31. a. $\log_7 50$ b. $\log_8 9$ c. $\log_2 9$
32. a. $\log_4 63$ b. $\log_{12} 150$ c. $\log_3 8$

In Exercises 33–66 solve each equation without using a calculator. Indicate which expressions are undefined.
33. $11^x = \dfrac{1}{121}$
34. $125^x = 25$
35. $2^y = \sqrt[3]{4}$
36. $\left(\dfrac{4}{9}\right)^y = \dfrac{3}{2}$
37. $2^{4w-1} = 8$
38. $9^{2v+1} = 27$
39. $2^{x^2-1} = 8$
40. $9^{x^2} = 3^{x+1}$

41. $\log_8 64 = z$

42. $\log_{16} 64 = x$

43. $\log_2 x = 2$

44. $\log_{13} x = \dfrac{1}{2}$

45. $\log_3 w = -2$

46. $\log_3(-2) = w$

47. $\log_{-2} 3 = y$

48. $\log_t 169 = 2$

49. $\log_t 8 = -3$

50. $5^{\log_5 11} = x$

51. $\log_5 5^{17} = x$

52. $\log_7 x = 1$

53. $\ln 0 = x$

54. $\log(3n - 4) = \log(2n - 1)$

55. $\log_3 20 + \log_3 7 = \log_3 y$

56. $\log(3x + 1) = 2$

57. $\log_3(x^2 - 19) = 4$

58. $\ln(w + 2) + \ln w = \ln 3$

59. $\ln(1 - w) + \ln(1 - 2w) = \ln(7 - 4w)$

60. $\log(5 - 2v) - \log(1 - v) = \log(3 - 2v)$

61. $\ln(5v + 3) = \ln(3v + 9)$

62. $\log_5 27 - \log_5 2 = \log_5 y$

63. $\log(2 - 6v) - \log(2 - v) = \log(1 - v)$

64. $\ln(5 - x) + \ln(x + 1) = \ln(3x - 1)$

65. $\log x + \log(x - 9) = 1$

66. $\log_2 x + \log_2(x - 2) = 3$

In Exercises 67–70 express each logarithm in terms of logarithms of simpler expressions. Assume that the argument of each logarithm is a positive real number.

67. $\log x^3 y^5$

68. $\ln \dfrac{7x - 9}{2x + 3}$

69. $\ln \dfrac{\sqrt{2x + 1}}{5x + 9}$

70. $\log \sqrt{\dfrac{x^2 y^3}{z}}$

In Exercises 71–74 combine the logarithmic terms into a single logarithmic expression with a coefficient of 1. Assume that the argument of each logarithm is a positive real number.

71. $2 \ln x + 3 \ln y$

72. $5 \ln x - 4 \ln y$

73. $\ln(x^2 - 3x - 4) - \ln(x - 4)$

74. $\dfrac{1}{2}(\ln x - \ln y)$

In Exercises 75 and 76 approximate the solution to each equation to the nearest thousandth.

75. $\log(5x - 2) + \log(x - 1) = \log 10$

76. $\ln(2w + 3) + \ln(w + 1) = \ln(w + 2)$

77. Convert $y = 5^x$ to an exponential function with base e.

78. Verify that $\log 50^x + \log 6^x - \log 3^x = 2x$ is an identity.

79. Verify that $1000^{\log x} = x^3$ is an identity.

Estimation and Calculator Skills

In Exercises 80–83 mentally estimate the value of each expression to the nearest integer and then use a calculator to approximate the value to the nearest thousandth.

PROBLEM	MENTAL ESTIMATE	CALCULATOR APPROXIMATION
80. $\log 990$		
81. $\ln 3$		
82. $\log_3 30$		
83. $\log_5 120$		

84. Given $f(x) = 5x - 2$, evaluate each expression.

   a. $f(3)$ b. $f^{-1}(13)$ c. $\dfrac{1}{f(3)}$

In Exercises 85–89 match each graph with the most appropriate description. All graphs are displayed using the window $[-5, 5, 1]$ by $[-5, 5, 1]$.

85. A linear growth function

86. A logarithmic growth function

87. An exponential growth function

88. An exponential decay function

89. A constant function

A.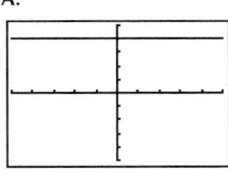

B.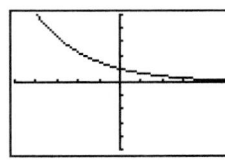

C.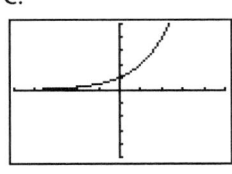

D.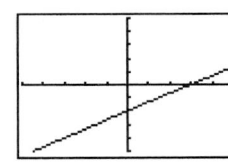

E.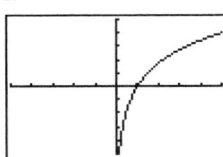

90. **Compound Interest**   A $2000 investment is invested at 8% for 5 years. Determine the value of the investment under each condition.

   a. The interest is compounded yearly.

   b. The interest is compounded monthly.

   c. The interest is compounded continuously.

91. **Radioactive Decay**   The radioactive material used to power a satellite decays at a rate that decreases the available power by 0.045% per day. When the power supply reaches $\dfrac{1}{100}$ of its original level, the satellite is no longer functional. How many days should the power supply last?

92. **Doubling Period for an Investment**   How many years will it take a savings bond to double in value if interest is compounded annually at 7.5%?

93. **Rate of Interest**   If an investment on which interest is compounded continuously doubles in value in 8 years, what is the rate of interest?

94. Richter Scale   The 2001 Kodiak Island, Alaska, earthquake had an amplitude A that was 6,310,000 times the reference amplitude a. Use the formula $R = \log \dfrac{A}{a}$ to calculate the magnitude of this earthquake on the Richter scale. (Source: **http://neic.usgs.gov/**)

95. Modeling the Safe Storage Time for Milk   The safe level of psychrotrophic bacteria in a gallon of milk is 100 units. In a refrigerator set at 40°F the number of units of these bacteria in a gallon of whole milk is approximated by the exponential function $B(t) = 3.0e^{0.29t}$, where t is the time in days. Evaluate and interpret each expression.
   a. $B(0)$
   b. $B(5)$
   c. $B(10)$
   d. $B(t) = 100$, $t = ?$

MASTERY TEST   FOR CHAPTER 10

[10.1] **1.** Determine whether each sequence is a geometric sequence. If the sequence is geometric, write the common ratio r.
   a. 1, 5, 25, 125, 625   **b.** 1, 5, 9, 13, 17
   c. 48, 24, 12, 6, 3   **d.** 0.9, 0.09, 0.009, 0.0009, 0.00009

[10.1] **2.** Graph each function.
   a. $f(x) = \left(\dfrac{5}{2}\right)^x$ **b.** $f(x) = \left(\dfrac{2}{5}\right)^x$

   Evaluate each expression given $f(x) = 16^x$.
   c. $f(-1)$ **d.** $f\left(\dfrac{1}{2}\right)$

[10.2] **3.** Write the inverse of each function. For parts **a, c,** and **d** use ordered-pair notation.
   a. $\{(-1, 4), (8, 9), (-7, 11)\}$
   b. Write the equation for the inverse of $f(x) = 3x - 6$.

   c.

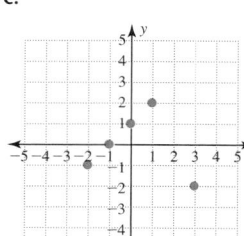

   d.

x	y
$\dfrac{1}{2}$	2
$\dfrac{1}{3}$	3
6	$\dfrac{1}{6}$
1	1

[10.2] **4.** Graph each function and its inverse on the same coordinate system.
   a. $f = \{(-3, 2), (-2, 1), (-1, 3), (2, 4)\}$
   b. $f(x) = 3x + 2$

   c. $f(x) = \left(\dfrac{1}{2}\right)^x$
   d. $f(x) = x^2$ for $x \geq 0$

[10.3] **5.** Translate the following logarithmic equations to exponential form.
   a. $\log_5 \sqrt[3]{25} = \dfrac{2}{3}$
   b. $\log_5 \dfrac{1}{125} = -3$
   c. $\log_b (y + 1) = x$
   d. $\log_b x = y + 1$

[10.3] **6.** Determine the value of each logarithm by inspection.
   a. $\log_8 1$ **b.** $\log_8 8$
   c. $\log_8 \dfrac{1}{8}$ **d.** $\log_8 8^{17}$

[10.4] **7.** Use a calculator to approximate each expression to the nearest ten-thousandth.
   a. $\log 19.1$ **b.** $\ln 19.1$
   c. $\log(4.876 \times 10^{12})$ **d.** $\ln(3.04 \times 10^{-5})$

[10.5] **8.** Use the properties of logarithms to express each logarithm in terms of logarithms of simpler expressions. Assume that x and y are positive real numbers.
   a. $\log x^4 y^5$
   b. $\ln \dfrac{x^3}{\sqrt{y}}$

Use the properties of logarithms to combine these logarithmic terms into a single logarithmic expression with a coefficient of 1. Assume that the argument of each logarithm is a positive real number.

c. $2 \ln(7x + 9) - \ln x$

d. $\dfrac{1}{2} \log(x + 3) + \log x$

[10.5] **9.** Use the change-of-base formula for logarithms and a calculator to approximate each expression to the nearest thousandth.

   a. $\log_4 11$ **b.** $\log_\pi \sqrt{7}$
   c. $\log_2 100$ **d.** $\log_\pi 5\pi$

[10.6] **10.** Solve each equation without using a calculator.

   a. $3^x = \dfrac{1}{81}$

   b. $16^w = 64$

   c. $\log_2 x = 4$

   d. $\log_8 2 = y$

   e. $\log(1 - 4t) - \log(5 + t) = \log 3$

   f. $\ln(1 - z) + \ln(2 - z) = \ln(17 - z)$

   Using a calculator, approximate the solution of each equation to the nearest thousandth.

   g. $3^{4y+1} = 17.83$

   h. $\ln(x + 1) + \ln(3x - 1) = \ln(6x)$

[10.7] **11. a.** Assume that the population of a new space colony is growing continuously at a rate of 5% per year. How many years will it take the population to grow from 500 to 3000?

   b. How many years will it take an investment to double in value if interest is compounded monthly at 8.25%?

GROUP PROJECT	FOR CHAPTER 10

An Algebraic Model for Real Data

Supplies Needed for the Lab:

- A golf ball or other hard ball
- A meterstick or a metric tape measure
- A chair or a short stepladder
- Masking tape and a marking pen
- Hard, smooth floor area next to a wall
- A graphics calculator with an exponential regression feature (ExpReg)

Lab: Recording the Height of a Bouncing Golf Ball

Move to a location with a relatively high ceiling and a smooth concrete or other hard floor. (Tile with cracks can cause problems with the bounce of the golf ball.) Ask the tallest member of the class to get on a chair or small stepladder. Run a strip of masking tape from the base of the floor to the top of the student's reach. Stick this masking tape to the wall in a straight line. (Starting from a height of 250 to 300 cm will make the data collection easier.) Using your meterstick and the pen, label the heights on the masking tape every 10 cm starting with 0 cm at floor level.

Placing the bottom of the golf ball level with your starting point at the top of the tape (250 to 300 cm high) and about 5 cm away from the wall, have one student drop the golf ball and have a second student note the approximate point that marks the height of the golf ball at the top of its first bounce. It would be wise to have a third student catch the golf ball at the beginning of the second bounce. Repeat this experiment now that you know approximately where to look and have the second student mark the height of this bounce using his or her finger. Place a pen mark on the tape at this height. Using the reference marks on the tape and the meterstick, record the height of this bounce to the nearest centimeter.

Use the height just recorded as the new release point and repeat the experiment. You may need to repeat drops from the same height before you are comfortable that you have recorded accurately the height of the first bounce. Repeat this experiment until you have recorded the results for 8 to 10 bounces.

1. Record your data in a table similar to the one shown.
2. Use this data to create a scatter diagram using an appropriate scale for each axis.

x BOUNCE #	y HEIGHT (CM)
0	250
1	
2	
3	
4	
5	
6	
7	
8	
9	
10	

Creating an Algebraic Model for This Data

1. Using your graphics calculator, calculate the exponential function of best fit. This function will be of the form $f(x) = ab^x$.
2. Letting $y = H(x)$ represent the curve of best fit that you just calculated,
   a. Evaluate and interpret $H(7)$. How does this value compare to the corresponding value in the table?
   b. Determine the value of x for which $H(x) = 100$. Interpret the meaning of this value.
3. In the function $f(x) = ab^x$ that models your data, what is the value of a? Interpret the meaning of this value in this application.
4. In the function $f(x) = ab^x$ that models your data, what is the value of b? Interpret the meaning of this value in this application.

11

A PREVIEW
OF COLLEGE
ALGEBRA

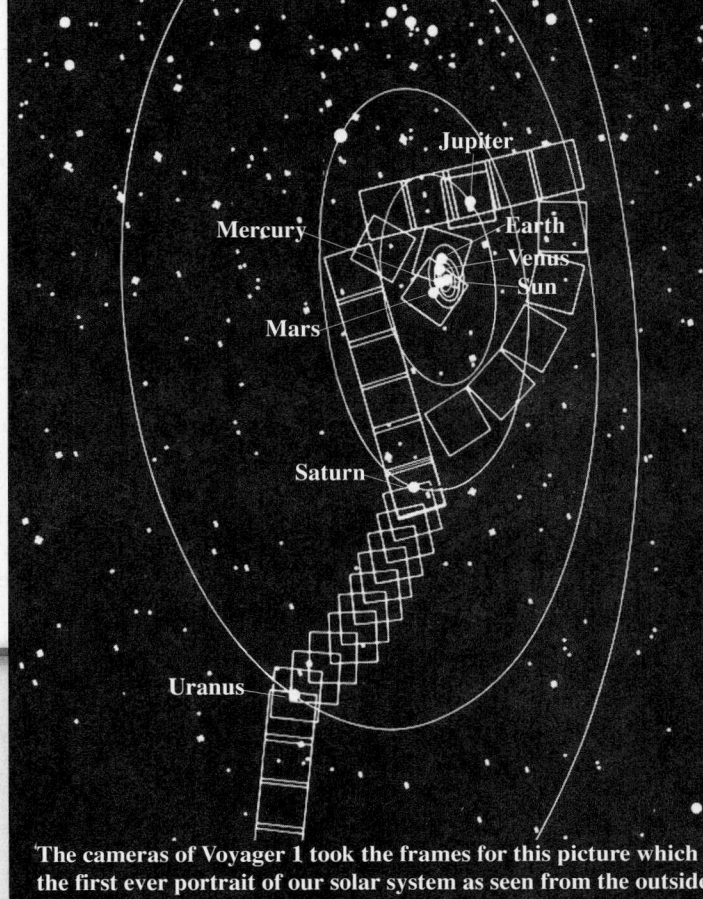

The cameras of Voyager 1 took the frames for this picture which [is?] the first ever portrait of our solar system as seen from the outside.

In this chapter we extend some of the topics presented in earlier chapters. This chapter can serve as a resource for those taking college algebra or other more advanced courses. For those already taking more advanced courses, it can serve as a source of additional explanation and further examples. The organization of this chapter permits you to select the topics independently to meet your needs.

Section 11.1   Algebra of Functions

Objectives: 1. Add, subtract, multiply, and divide two functions.

2. Form the composition of two functions.

Problems—such as those in business—are often broken down into simpler components for analysis. For example, in order to determine the profit made by producing and selling an item, both the revenue and the cost must be known. Separate divisions of a business might be asked to find the revenue function and the cost function; the profit function would be found then by properly combining these two functions. We shall examine five ways to combine functions: sum, difference, product, quotient, and composition of functions.

We have added two real numbers to obtain another real number. We have added two polynomials to obtain another polynomial. Now we are adding two functions to obtain a new function.

The **sum of two functions** f and g, denoted by $f + g$, is defined as

$$(f + g)(x) = f(x) + g(x)$$

for all values of x that are in the domain of both f and g. Note that if either $f(x)$ or $g(x)$ is undefined, then $(f + g)(x)$ is also undefined.

■ EXAMPLE 1   Determining the Sum of Two Functions

Find the sum of $f(x) = x^2$ and $g(x) = 2$ algebraically, numerically, and graphically.

SOLUTION _____

ALGEBRAICALLY

$$(f + g)(x) = f(x) + g(x)$$
$$(f + g)(x) = x^2 + 2$$

The new function called $f + g$ is determined by adding $f(x)$ and $g(x)$. Substitute x^2 for $f(x)$ and 2 for $g(x)$.

NUMERICALLY

x	$f(x) = x^2$	x	$g(x) = 2$	x	$(f + g)(x) = x^2 + 2$
-2	4	-2	2	-2	6
-1	1	-1	2	-1	3
0	0	0	2	0	2
1	1	1	2	1	3
2	4	2	2	2	6

This table contains only a few values from the domain of input values from $\mathbb{R}$. These values illustrate that $(f + g)(x)$ can be determined by adding $f(x) + g(x)$ for each input x, $4 + 2 = 6, 1 + 2 = 3, 0 + 2 = 2, 1 + 2 = 3$, and $4 + 2 = 6$.

GRAPHICALLY

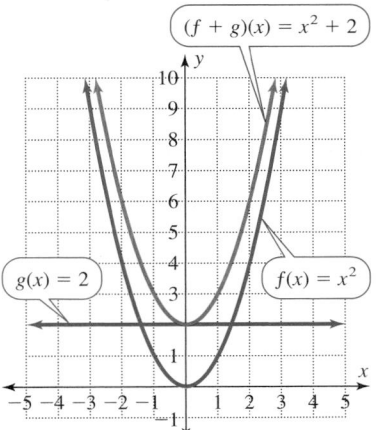

$(f + g)(x) = x^2 + 2$

$g(x) = 2$

$f(x) = x^2$

The graph of $f(x) = x^2$ is a parabola opening upward with its vertex at $(0, 0)$.

The graph of $g(x) = 2$ is the horizontal line representing a constant output of 2.

The new function $f + g$ adds all output values. Since $g(x) = 2$, all output values of f are translated up 2 units when $g(x)$ is added to $f(x)$.

SELF-CHECK 11.1.1

Given $f(x) = x^2$ and $g(x) = -4x + 4$, determine $f + g$.

If two functions are equal, their input and output values are identical. If the domain of input values is the set of all real numbers, then we cannot list all the input-output pairs. However, a table of values can serve to check that two formulas yield the same values. The following calculator perspective uses a table of values to check the result of $f + g$ in Example 1.

CALCULATOR PERSPECTIVE 11.1.1   **Verifying the Sum of Two Functions**

To verify the sum of the functions $f(x) = x^2$ and $g(x) = 2$ from Example 1 on a TI-83 Plus calculator, enter the following keystrokes:

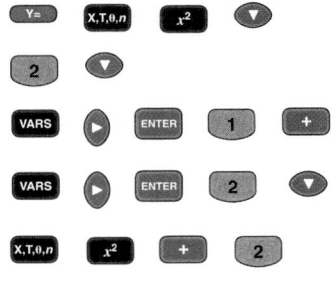

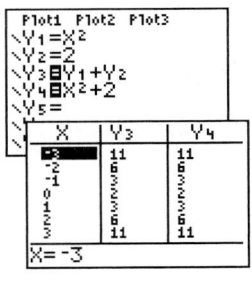

SELF-CHECK 11.1.1 ANSWER

$(f + g)(x) = x^2 - 4x + 4$

Note: Y_3 is formed by adding the functions Y_1 and Y_2. Y_4 is formed by adding x^2 and 2. Y_3 and Y_4 are equal for each input value of x. Notice that Y_1 and Y_2 have been deselected on the ⬛Y= screen, so that they do not appear in the table.

The **difference of two functions** f and g, denoted by $f - g$, is defined as

$$(f - g)(x) = f(x) - g(x)$$

for all values of x that are in the domain of both f and g.

■ EXAMPLE 2 Using the Difference of Two Functions to Model Profit

Suppose that the weekly revenue function for u units of a product sold is $R(u) = 5u^2 - 7u$ dollars and the cost function for u units is $C(u) = 8u + 23$. Assume 0 is the fewest number of units that can be produced and 100 is the greatest number that can be marketed. Assuming profit can be determined by subtracting the cost from the revenue, find the profit function P and determine $P(4)$, the profit made by selling four units.

SOLUTION _____

VERBALLY

Profit = revenue − cost   First write a word equation to model this problem.

ALGEBRAICALLY

$P(u) = R(u) - C(u)$
$P(u) = (5u^2 - 7u) - (8u + 23)$ Substitute the given expressions for $R(u)$ and $C(u)$.
$P(u) = 5u^2 - 7u - 8u - 23$ The domain is $0 \le u \le 100$. (For some types of units, u may
                                   have to be an integer.)

$P(u) = 5u^2 - 15u - 23$ This is the profit function.

The profit on four units is
$P(4) = 5(4)^2 - 15(4) - 23$ Substitute 4 for x in the profit function P.
$P(4) = -3$

Thus $3 will be lost if only four units are sold.   ■

SELF-CHECK 11.1.2

Given $f(x) = 3x^2 - x + 5$ and $g(x) = 2x^2 + 4x + 7$, determine these two functions:

1. $(f + g)(x)$
2. $(f - g)(x)$

SELF-CHECK 11.1.2 ANSWERS

1. $(f + g)(x) =$
   $5x^2 + 3x + 12$
2. $(f - g)(x) =$
   $x^2 - 5x - 2$

The operations of multiplication and division are defined similarly to addition except that the quotient $\left(\dfrac{f}{g}\right)$ is not defined if $g(x) = 0$.

Operations on Functions

	NOTATION	DEFINITION*	EXAMPLES: FOR $f(x) = x^2 - 4$ and $g(x) = x - 2$:
Sum	$f + g$	$(f + g)(x) = f(x) + g(x)$	$(f + g)(x) = x^2 + x - 6$ for all real numbers
Difference	$f - g$	$(f - g)(x) = f(x) - g(x)$	$(f - g)(x) = x^2 - x - 2$ for all real numbers
Product	$f \cdot g$	$(f \cdot g)(x) = f(x)g(x)$	$(f \cdot g)(x) = x^3 - 2x^2 - 4x + 8$ for all real numbers
Quotient	$\dfrac{f}{g}$	$\left(\dfrac{f}{g}\right)(x) = \dfrac{f(x)}{g(x)}$	$\left(\dfrac{f}{g}\right)(x) = x + 2$ for all reals except 2

*The domain of all these functions except $\dfrac{f}{g}$ is the set of values in both the domain of f and the domain of g. For $\dfrac{f}{g}$, we also must have $g(x) \neq 0$.

■ EXAMPLE 3 Determining the Product and Quotient of Two Functions

Given $f(x) = x^2 + 5x$ and $g(x) = \dfrac{x + 5}{x}$, find the following.

SOLUTIONS

(a) $(f \cdot g)(x)$

$(f \cdot g)(x) = f(x) \cdot g(x)$

$(f \cdot g)(x) = (x^2 + 5x)\left(\dfrac{x + 5}{x}\right)$

$(f \cdot g)(x) = (x + 5)^2$

(b) The domain of $(f \cdot g)(x)$

f is defined for $\mathbb{R}$
g is defined for $x \neq 0$
$f \cdot g$ is defined for $x \neq 0$

$(f \cdot g)(x)$ is defined only for values for which both f and g are defined.

(c) $\left(\dfrac{f}{g}\right)(x)$

$\left(\dfrac{f}{g}\right)(x) = \dfrac{f(x)}{g(x)}$

$\left(\dfrac{f}{g}\right)(x) = \dfrac{x^2 + 5x}{\dfrac{x + 5}{x}}$

$\left(\dfrac{f}{g}\right)(x) = \dfrac{x(x + 5)}{1} \cdot \dfrac{x}{x + 5}$

$\left(\dfrac{f}{g}\right)(x) = x^2$

To simplify this complex rational expression, invert the divisor and multiply.

(d) The domain of $\left(\dfrac{f}{g}\right)(x)$

f is defined for $\mathbb{R}$

g is defined for $x \neq 0$
$g(x) = 0$ for $x = -5$
$\dfrac{f}{g}$ is defined for $x \neq 0$ and $x \neq -5$

$\left(\dfrac{f}{g}\right)(x)$ is defined for values for which both f and g are defined and $g(x) \neq 0$.

SELF-CHECK 11.1.3

SELF-CHECK 11.1.3 ANSWERS

1. $(f \cdot g)(x) = x^4 - 1$
2. $\left(\dfrac{f}{g}\right)(x) = \dfrac{x^2 - 1}{x^2 + 1}$

Given $f(x) = x^2 - 1$ and $g(x) = x^2 + 1$, determine these two functions:

1. $(f \cdot g)(x)$

2. $\left(\dfrac{f}{g}\right)(x)$

Functions are defined not only by their formulas but also by the set of input values contained in the domains of the functions. We often allow the domain of a function to be implied by the formula. In this case, the domain is understood to be all real numbers for which the formula is defined and produces real-number outputs. When we combine functions to produce new functions, we must take care that the formulas are only used for input values that are allowed for the new function. This is illustrated by Example 4. Two functions f and g are **equal** if the domain of f equals the domain of g and $f(x) = g(x)$ for each x in their common domain.

■ **EXAMPLE 4** **Comparing Functions to Determine If They Are Equal**

Given $f(x) = x$, $g(x) = \dfrac{x^3 - 4x}{x^2 - 4}$, and $h(x) = \dfrac{x^3 + 4x}{x^2 + 4}$, determine whether:

SOLUTIONS

(a) $f = g$
$\quad f(x) = x \quad\quad\quad\quad$ for all real x

$\quad g(x) = \dfrac{x^3 - 4x}{x^2 - 4} \quad x \neq \pm 2 \quad\quad$ Both -2 and 2 result in division by 0.

$\quad g(x) = \dfrac{x(x^2 - 4)}{x^2 - 4} \quad$ for $x \neq \pm 2$

$\quad g(x) = x \quad\quad\quad$ for $x \neq \pm 2$

$f \neq g$ because f is defined for $x = -2$ and $x = 2$ but g is not. $f(x)$ and $g(x)$ have the same values for all real numbers except $x = -2$ and $x = 2$.

(b) $f = h$
$\quad f(x) = x \quad\quad\quad\quad$ for all real x

$\quad h(x) = \dfrac{x^3 + 4x}{x^2 + 4} \quad$ for all real $x \quad\quad$ There are no real numbers for which the denominator of h is 0.

$\quad h(x) = \dfrac{x(x^2 + 4)}{x^2 + 4} \quad$ for all real x

$\quad h(x) = x \quad\quad\quad$ for all real x

$f = h$ because the formulas are equal for all real numbers.

As shown in the following figure, the graphs of the three functions f, g, and h in Example 4 are nearly identical. The only difference is that the graph of g has "holes" at $x = -2$ and $x = 2$ because these values are not in its domain.

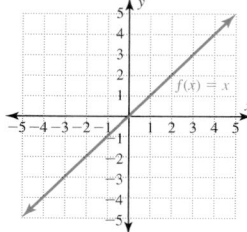

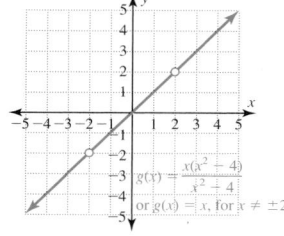

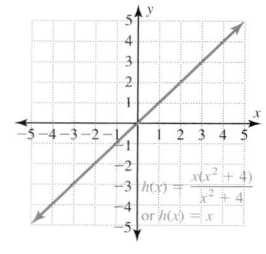

Functions, especially in formula form, are a powerful means of describing the relationship between two quantities. We can further amplify this power by "chaining" two functions together. This way of combining two functions is called **composition.**

Caution: The symbol $\circ$ denoting the composition of two functions, $f \circ g$, should not be confused with the symbol for the multiplication of two functions, $f \cdot g$.

Composite Function $f \circ g$

ALGEBRAICALLY	VERBALLY	ALGEBRAIC EXAMPLE
$(f \circ g)(x) = f[g(x)]$ The domain of $f \circ g$ is the set of x values from the domain of g for which $g(x)$ is in the domain of f.	$f \circ g$ denotes the composition of function f with function g. $f \circ g$ is read "f composed with g."	For $f(x) = x + 1$ and $g(x) = 2x$: $(f \circ g)(x) = f[g(x)]$ $\qquad\qquad = f(2x)$ $\qquad\qquad = 2x + 1$

Functions can be examined using mapping notation, ordered-pair notation, tables, graphs, and function notation. In each case, to evaluate $(f \circ g)(x)$ we first evaluate $g(x)$ and then apply f to this result.

■ EXAMPLE 5   Determining the Composition of Two Functions

Find $f \circ g$ for the given functions f and g.

SOLUTION _____

g	f	**VERBALLY**	**MAPPING NOTATION** $f \circ g$	**NUMERICAL EXAMPLE** $f \circ g$
$6 \to 3$	$3 \to 9$	g maps 6 to 3; then f maps 3 to 9.	$6 \xrightarrow{g} 3 \xrightarrow{f} 9$	$6 \to 9$
$5 \to 4$	$4 \to 7$	g maps 5 to 4; then f maps 4 to 7.	$5 \xrightarrow{g} 4 \xrightarrow{f} 7$	$5 \to 7$
$-1 \to 0$	$0 \to 2$	g maps -1 to 0; then f maps 0 to 2.	$-1 \xrightarrow{g} 0 \xrightarrow{f} 2$	$-1 \to 2$ ■

■ EXAMPLE 6   Determining the Composition of Two Functions

Given $f(x) = x^2$ and $g(x) = 3x$, evaluate these expressions.

SOLUTIONS _____

(a) $(f \circ g)(4)$

$(f \circ g)(4) = f[g(4)] = f[3(4)]$
$\qquad\qquad\quad = f(12)$
$\qquad\qquad\quad = (12)^2$
$\qquad\qquad\quad = 144$

First apply the formula for $g(x)$. Evaluate $g(4) = 3(4)$. Then apply the formula for $f(x)$ to 12. Evaluate $f(12) = 12^2$.

(b) $(g \circ f)(4)$

$(g \circ f)(4) = g[f(4)] = g[4^2]$
$\qquad\qquad\quad = g(16)$
$\qquad\qquad\quad = 3(16)$
$\qquad\qquad\quad = 48$

First apply the formula for $f(x)$. Evaluate $f(4) = 4^2$. Then apply the formula for $g(x)$ to 16. Evaluate $g(16) = 3(16)$.

(c) $(f \circ g)(x)$ $(f \circ g)(x) = f[g(x)] = f(3x)$ First apply the formula for g to x. Then apply the
$= (3x)^2$ formula for f to $3x$.
$= 9x^2$

(d) $(g \circ f)(x)$ $(g \circ f)(x) = g[f(x)] = g(x^2)$ First apply the formula for f to x. Then apply the
$= 3(x^2)$ formula for g to x^2.
$= 3x^2$

Note that in this example
$f \circ g \neq g \circ f$.

In Example 6, $f \circ g \neq g \circ f$. Although $f \circ g$ can equal $g \circ f$ in special cases, in general the order that we perform composition of functions is important.

SELF-CHECK 11.1.4

Given $f(x) = 3x - 1$ and $g(x) = \dfrac{x + 1}{3}$, evaluate these expressions.

1. $(f + g)(2)$ **2.** $(f - g)(2)$

3. $(f \cdot g)(-1)$ **4.** $\left(\dfrac{f}{g}\right)(1)$

5. $(f \circ g)(2)$ **6.** $(g \circ f)(2)$

The functions $f(x) = 3x - 1$ and $g(x) = \dfrac{x + 1}{3}$ from Self-Check 11.1.4 are inverses of each other. Note that:

$$(f \circ g)(2) = 2: \ 2 \xrightarrow{\ g\ } 1 \xrightarrow{\ f\ } 2 \ \text{ and } \ (g \circ f)(2) = 2: \ 2 \xrightarrow{\ f\ } 5 \xrightarrow{\ g\ } 2.$$

For any input value x,

$$(f \circ g)(x) = f(g(x)) = f\left(\frac{x + 1}{3}\right) = 3\left(\frac{x + 1}{3}\right) - 1 = (x + 1) - 1 = x.$$

The inverse of the function f reverses the order of each ordered pair (x, y) of f.

Likewise $(g \circ f)(x) = x$. Thus the composition of functions gives us another way to characterize the inverse of a function.

Inverse of a Function

ALGEBRAICALLY	EXAMPLE
The functions f and f^{-1} are inverses of each other if and only if $(f \circ f^{-1})(x) = x$ for each input value of f^{-1}	$f(x) = 3x - 1$ and $f^{-1}(x) = \dfrac{x + 1}{3}$ are inverses because: $(f \circ f^{-1})(x) = f\left(\dfrac{x + 1}{3}\right)$ $(f \circ f^{-1})(x) = 3\left(\dfrac{x + 1}{3}\right) - 1$ $(f \circ f^{-1})(x) = x$
and $(f^{-1} \circ f)(x) = x$ for each input value of f	and $(f^{-1} \circ f)(x) = f^{-1}(3x - 1)$ $(f^{-1} \circ f)(x) = \dfrac{(3x - 1) + 1}{3}$ $(f^{-1} \circ f)(x) = x$

SELF-CHECK 11.1.4 ANSWERS

1. 6
2. 4
3. 0
4. 3
5. 2
6. 2

CALCULATOR PERSPECTIVE 11.1.2   Composing a Function with Its Inverse

To numerically check that $(f \circ g)(x) = x$ for the functions $f(x) = 3x - 4$ and $g(x) = \dfrac{x + 4}{3}$, on a TI-83 Plus calculator, enter the following keystrokes:

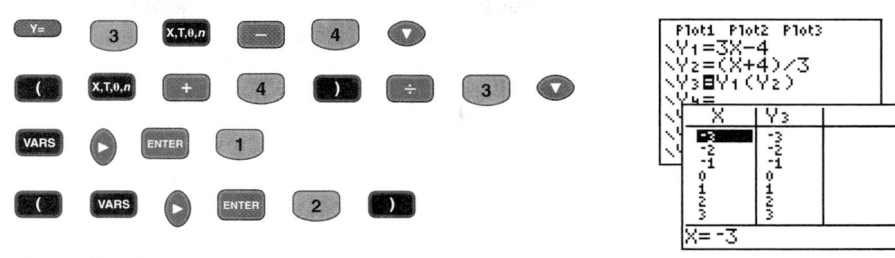

Note: $Y_1(Y_2)$ represents $f(g(x))$. Observe that Y_3 has the same values as x, thus confirming that $(f \circ g)(x) = x$.

Example 7 illustrates how a problem can be broken down into pieces with individual functions as mathematical models. Then the overall relationship can be modeled by a composite function.

■ **EXAMPLE 7** Composing Cost and Production Functions

The quantity of items a factory can produce weekly is a function of the number of hours it operates. For one company this is given by $q(t) = 40t$ for $0 \le t \le 168$. The dollar cost of manufacturing these items is a function of the quantity produced; in this case, $C(q) = q^2 - 40q + 750$ for $q \ge 0$. Evaluate and interpret the following expressions.

SOLUTIONS

(a) $q(8)$

$q(t) = 40t$
$q(8) = 40(8)$ Substitute 8 into the formula for $q(t)$.
$q(8) = 320$
320 units can be produced in 8 h.

(b) $C(320)$

$C(q) = q^2 - 40q + 750$
$C(320) = (320)^2 - 40(320) + 750$ Substitute 320 into the formula for $C(q)$.
$C(320) = 90{,}350$
$90{,}350 is the cost of manufacturing 320 units.

(c) $(C \circ q)(8)$

$(C \circ q)(8) = C[q(8)]$
$(C \circ q)(8) = C(320)$ Substitute for $q(8)$ from part **a.**
$(C \circ q)(8) = \$90{,}350$ Substitute for $C(320)$ from part **b.**
$90{,}350 is the cost of 8 h of production.

(d) $(C \circ q)(t)$

$(C \circ q)(t) = C[q(t)]$
$(C \circ q)(t) = C[40t]$ Substitute $40t$ for $q(t)$. Then evaluate C for $40t$.
$(C \circ q)(t) = (40t)^2 - 40(40t) + 750$
$(C \circ q)(t) = 1600t^2 - 1600t + 750$
This is the cost of t h of production.

Example 8 illustrates the analysis of a geometric problem using a composite function.

■ EXAMPLE 8   Composing Area and Length Functions

A piece of wire 20 m long is cut into two pieces. The length of the shorter piece is s meters, and the length of the longer piece is L meters. The longer piece is then bent into the shape of a square of area A m^2.

(a) Express the length L as a function of s.
(b) Express the area A as a function of L.
(c) Express the area A as a function of s.

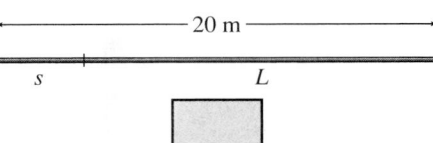

SOLUTIONS

(a) $\left(\begin{matrix}\text{Length of the} \\ \text{longer piece}\end{matrix}\right) = \left(\begin{matrix}\text{Total} \\ \text{length}\end{matrix}\right) - \left(\begin{matrix}\text{Length of the} \\ \text{shorter piece}\end{matrix}\right)$ First write a word equation that describes the relationship between the length of the longer piece and the length of the shorter piece.

$\qquad\qquad L(s) \quad = \quad 20 \quad - \quad s$ This equation expresses the length as a function of s.

(b) Area = square of the length of one side

$\qquad A(L) = \left(\dfrac{L}{4}\right)^2$ The length of one side of the square is one-fourth the perimeter.

(c) $(A \circ L)(s) = A[L(s)]$ The composition $A \circ L$ expresses A as a function of s.

$\quad (A \circ L)(s) = A(20 - s)$ Substitute $20 - s$ for $L(s)$ from part **a.**

$\quad (A \circ L)(s) = \left(\dfrac{20 - s}{4}\right)^2$ Evaluate A using the formula from part **b.**   ■

An important skill in calculus is the ability to take a given function and decompose it into simpler components.

■ EXAMPLE 9   Decomposing Functions into Simpler Components

Express each of these functions in terms of $f(x) = 2x - 3$ and $g(x) = \sqrt{x}$.

SOLUTIONS

(a) $h(x) = \sqrt{2x - 3}$ $h(x) = \sqrt{2x - 3} = \sqrt{f(x)}$ First substitute $f(x)$ for $2x - 3$. Then replace the square root function with the g function.
$\qquad\qquad\qquad\quad = g(f(x))$
$\qquad\qquad\qquad\quad = (g \circ f)(x)$

(b) $h(x) = 2\sqrt{x} - 3$ $h(x) = 2\sqrt{x} - 3 = 2g(x) - 3$ First substitute $g(x)$ for $\sqrt{x}$. Then rewrite the expression using the definition of the f function.
$\qquad\qquad\qquad\quad = f(g(x))$
$\qquad\qquad\qquad\quad = (f \circ g)(x)$ ■

SELF-CHECK 11.1.5 ANSWER

$h(x) = (g \circ f)(x)$

SELF-CHECK 11.1.5
Express $h(x) = \dfrac{1}{x - 6}$ in terms of $f(x) = x - 6$ and $g(x) = \dfrac{1}{x}$.

USING THE LANGUAGE AND SYMBOLISM OF MATHEMATICS   11.1

1. The sum of two functions f and g, denoted by $f + g$, is defined as $(f + g)(x) = $ _____ for all values of x that are in the domain of both f and g.
2. The difference of two functions f and g, denoted by $f - g$, is defined as $(f - g)(x) = $ _____ for all values of x that are in the domain of both f and g.
3. The product of two functions f and g, denoted by $f \cdot g$, is defined as $(f \cdot g)(x) = $ _____ for all values of x that are in the domain of both f and g.
4. The quotient of two functions f and g, denoted by $\dfrac{f}{g}$, is defined as $\left(\dfrac{f}{g}\right)(x) = $ _____ for all values of x that

are in the domain of both f and g, provided $g(x) \neq$ _____.

5. The composition of the function f with the function g, denoted by $f \circ g$, is defined by $(f \circ g)(x) = $ _____.
6. The domain of $f \circ g$ is the set of x values from the domain of g for which _____ is in the domain of _____.
7. The functions f and f^{-1} are inverses of each other if and only if $(f \circ f^{-1})(x) = $ _____ for each input value from the domain of _____ and $(f^{-1} \circ f)(x) = $ _____ for each input value from the domain of _____.

EXERCISES   11.1

In Exercises 1–4 evaluate each expression, given $f(x) = x^2 - 1$ and $g(x) = 2x + 5$.

1. a. $(f + g)(2)$
   b. $(f - g)(2)$
   c. $(f \cdot g)(2)$
   d. $\left(\dfrac{f}{g}\right)(2)$

2. a. $(f + g)(-3)$
   b. $(f - g)(-3)$
   c. $(f \cdot g)(-3)$
   d. $\left(\dfrac{f}{g}\right)(-3)$

3. a. $(f \circ g)(2)$
   b. $(g \circ f)(2)$
   c. $(f \circ f)(2)$
   d. $(g \circ g)(2)$

4. a. $(f \circ g)(-3)$
   b. $(g \circ f)(-3)$
   c. $(f \circ f)(-3)$
   d. $(g \circ g)(-3)$

In Exercises 5–10 use the given tables for f and g to form a table of values for each function.

x	f(x)	x	g(x)
0	-2	0	-1
3	0	3	5
8	7	8	9

5. $f + g$
6. $f - g$
7. $g - f$
8. $f \cdot g$
9. $g \cdot f$
10. $\dfrac{f}{g}$

In Exercises 11–14 use $f = \{(-2, 3), (1, 5), (4, 7)\}$ and $g = \{(-2, 4), (1, -1), (4, 6)\}$ to form a set of ordered pairs for each function.

11. $f - g$
12. $f + g$
13. $\dfrac{f}{g}$
14. $f \cdot g$

In Exercises 15–18 use the given graphs for f and g to graph each function. (*Hint:* You first may want to write each function as a set of ordered pairs.)

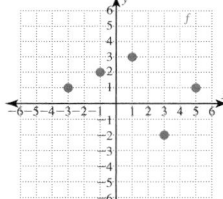 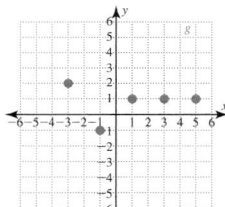

15. $f + g$
16. $f - g$
17. $f \cdot g$
18. $\dfrac{f}{g}$

In Exercises 19–22 determine $f + g, f - g, f \cdot g, \dfrac{f}{g},$ and $f \circ g$ for the given functions. State the domain of the resulting function.

19. $f(x) = 2x^2 - x - 3$
    $g(x) = 2x - 3$
20. $f(x) = 6x^2 - x - 15$
    $g(x) = 3x - 5$
21. $f(x) = \dfrac{x}{x + 1}$
    $g(x) = \dfrac{1}{x}$
22. $f(x) = 5x - 7$
    $g(x) = 3$

In Exercises 23–28 determine if the functions f and g are equal. If $f \neq g$, state the reason.

23. $f(x) = x - 3$
    $g(x) = \dfrac{x^2 - 5x + 6}{x - 2}$

24. $f(x) = 1 - \dfrac{1}{x}$

$g(x) = \dfrac{x - 1}{x}$

25. $f = \{(-1, 1), (0, 0), (1, 1), (2, 4)\}$

$g(x) = x^2$

26. $f(x) = x$

$g(x) = \dfrac{x^3 + x}{x^2 + 1}$

27. $f = \{(-7, 7), (0, 0), (8, 8)\}$

$g(x) = |x|$

28. $f(x) = 5$

$g(x) = \dfrac{5x^2 + 15}{x^2 + 3}$

29. Use the given tables for f and g to complete the tables for $f \circ g$.

$x \rightarrow g(x)$	$x \rightarrow f(x)$	$x \xrightarrow{g} g(x) \xrightarrow{f} f(g(x))$	$x \rightarrow f(g(x))$
$3 \rightarrow 5$	$5 \rightarrow 9$	$3 \longrightarrow 5 \longrightarrow 9$	$3 \rightarrow 9$
$4 \rightarrow 2$	$2 \rightarrow 8$	$4 \longrightarrow 2 \longrightarrow ?$	$? \rightarrow ?$
$7 \rightarrow 1$	$1 \rightarrow 0$	$7 \longrightarrow ? \longrightarrow ?$	$? \rightarrow ?$
$8 \rightarrow 4$	$4 \rightarrow 6$	$8 \longrightarrow ? \longrightarrow ?$	$? \rightarrow ?$

30. Use $f = \{(1, 3), (4, 5), (6, 2)\}$ and

$g = \{(3, 6), (5, 1), (2, 4)\}$ to form a set of ordered pairs for

a. $f \circ g$

b. $g \circ f$

31. Use the given graphs of f and g to graph $f \circ g$. (*Hint:* You first may want to write each function as a set of ordered pairs.)

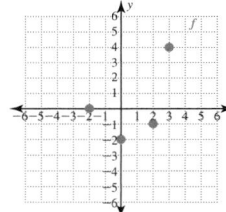 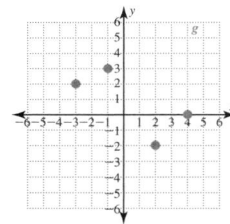

In Exercises 32–37 determine $f \circ g$ and $g \circ f$.

32. $f(x) = 3x - 4$

$g(x) = 4x + 3$

33. $f(x) = x^2 - 5x + 3$

$g(x) = 4x - 2$

34. $f(x) = |x|$

$g(x) = x - 8$

35. $f(x) = \sqrt{x}$

$g(x) = x + 5$

36. $f(x) = \dfrac{1}{x}$

$g(x) = x^3 + 1$

37. $f(x) = \dfrac{1}{x + 2}$

$g(x) = x^2 - 4$

In Exercises 38 and 39 use the graphs of f and g to graph $f + g$.

38. 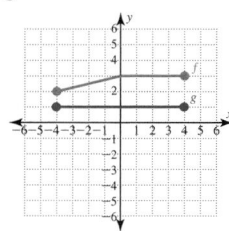 **39.**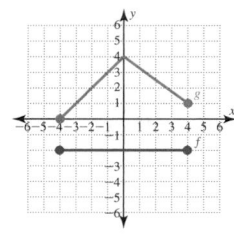

40. Profit as the Difference of Revenue Minus Cost The weekly revenue function for u units of a product sold is $R(u) = 4u^2 - 3u$ dollars, and the cost function for u units is $C(u) = 10u + 25$. Assume 0 is the least number of units that can be produced and 100 is the greatest number that can be marketed. Find the profit function P, and find $P(10)$, the profit made by selling 10 units.

41. Determining a Formula for Total Cost and Average Cost The fixed monthly cost F (rent, insurance, etc.) of a manufacturer is $5000. The variable cost (labor, materials, etc.) for producing u units is given by $V(u) = u^2 + 5u$ for $0 \le u \le 1200$. The total cost of producing u units is $C(u) = F(u) + V(u)$. Determine:

a. $F(500)$

b. $V(500)$

c. $C(500)$

d. The average cost of producing 500 units

e. A formula for $C(u)$

f. A formula for $A(u)$, the average cost of producing u units

42. Determining a Formula for Total Cost and Average Cost A manufacturer produces circuit boards for the electronics industry. The fixed cost F associated with this production is $3000 per week, and the variable cost V is $10 per board. The circuit boards produce revenue of $12 each. Determine:

a. $V(b)$, the variable cost of producing b boards per week

b. $F(b)$, the fixed cost of producing b boards per week

c. $C(b)$, the total cost of producing b boards per week

d. $A(b)$, the average cost of producing b boards per week

e. $R(b)$, the revenue from selling b boards per week

f. $P(b)$, the profit from selling b boards per week

g. $P(1000)$

h. $P(1500)$

i. $P(2000)$

j. The break-even value (the value producing a profit of $0)

43. Determining the Product of a Price Function and a Demand Function The number of items demanded by consumers is a function of the number of months that the product has been advertised. The price per item is varied each month as part of the marketing strategy. The number demanded during the mth month is $N(m) = 36m - m^2$, and the price per item during the mth month is $P(m) = 5m + 45$. The revenue for the mth month $R(m)$ is the product of the price per item and the number of items demanded. Determine:

a. $N(7)$
b. $P(7)$
c. $R(7)$
d. A formula for $R(m)$

44. Composing Cost and Production Functions The number of sofas a factory can produce weekly is a function of the number of hours t it operates. This function is $S(t) = 5t$ for $0 \le t \le 168$. The cost of manufacturing s sofas is given by $C(s) = s^2 - 6s + 500$ for $s \ge 0$. Evaluate and interpret:

a. $S(10)$
b. $C(50)$
c. $(C \circ S)(10)$
d. $(C \circ S)(t)$
e. $(C \circ S)(40)$
f. $(C \circ S)(100)$

45. Composing Cost and Markup Functions The weekly cost C of making d doses of a vaccine is $C(d) = 0.30d + 400$. The company charges 150% of cost to its wholesaler for this drug; that is, $R(C) = 1.5C$. Evaluate and interpret:

a. $C(5000)$
b. $R(1900)$
c. $(R \circ C)(5000)$
d. $(R \circ C)(d)$

46. Composing an Area and a Radius Function A circular concrete pad was poured to serve as the base for a grain bin. This pad was inscribed in a square plot as shown.

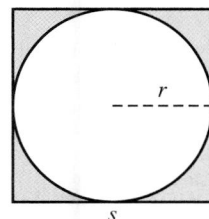

a. Express the radius r of this circle as a function of the length s of a side of the square.
b. Express the area A of this circle as a function of the radius r.
c. Determine $(A \circ r)(s)$ and interpret this result.

47. Composing an Area and a Width Function A border of uniform width x is trimmed from all sides of the square poster board as shown.

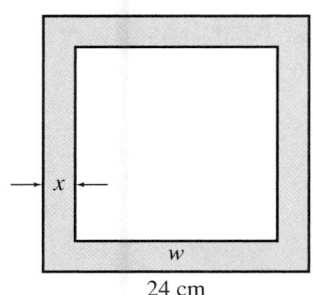

a. Express the area A of the border that is removed as a function of the remaining width w.
b. Express the remaining width w as a function of x.
c. Determine $(A \circ w)(x)$ and interpret this function.

48. Composing an Area and a Width Function A metal box with an open top can be formed by cutting squares of sides x cm from each corner of a square piece of sheet metal of width 44 cm as shown.

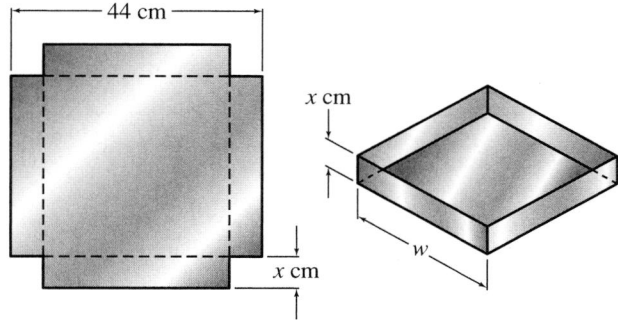

a. Express the area of the base of this box as a function of its width w.
b. Express w as a function of x.
c. Determine $(A \circ w)(x)$ and interpret this function.
d. Express the volume V of the box as a function of x.

In Exercises 49 and 50 determine both $(f \circ f^{-1})(x)$ and $(f^{-1} \circ f)(x)$.

49. $f(x) = 2x - 3$

$f^{-1}(x) = \dfrac{x + 3}{2}$

50. $f(x) = \dfrac{2x - 1}{3}$

$f^{-1}(x) = \dfrac{3x + 1}{2}$

51. Use $f(x) = 4x + 5$ to determine:
a. $f^{-1}(x)$
b. $(f \circ f^{-1})(x)$
c. $(f^{-1} \circ f)(x)$

In Exercises 52–55 express each function in terms of $f(x) = x^2 + 1$ and $g(x) = \sqrt{x}$.

52. $h(x) = x^2 + \sqrt{x} + 1$
53. $h(x) = x + 1, x \ge 0$
54. $h(x) = \sqrt{x^2 + 1}$
55. $h(x) = \dfrac{\sqrt{x}}{x^2 + 1}$

In Exercises 56–59 decompose each function into functions f and g such that $h(x) = (f \circ g)(x)$. (Answers may vary.)

56. $h(x) = \sqrt[3]{x^2 + 4}$
57. $h(x) = \dfrac{1}{3x^2 - 7x + 9}$
58. $h(x) = (x + 2)^2 + 3(x + 2) + 5$
59. $h(x) = x^3 - 2 + \dfrac{1}{x^3 - 2}$

In Exercises 60 and 61 use a graphics calculator to obtain the graph of $y = (f \circ g)(x)$.

60. $f(x) = \sqrt{x}, g(x) = 2x + 3$
61. $f(x) = 3x - 1, g(x) = x^2$

62. Discovery Question For $f(x) = \log x$ and $g(x) = 10^x$, determine $(f \circ g)(x)$. What can you observe from this result?

63. Discovery Question

a. If f and f^{-1} are inverses, can $f(x) = f^{-1}(x)$ for any value of x? If so, describe the location of these points on the plane.

b. If f and f^{-1} are inverses, can $f(x) = f^{-1}(x)$ for all values of x? If so, give an example. Can you give more than one example?

64. Discovery Question

a. For $f(x) = \dfrac{3x - 2}{7x - 3}$, determine f^{-1}.

b. Compare f and f^{-1}.

c. What does this imply about the graph of $y = f(x)$?

Section 11.2 Sequences, Series, and Summation Notation

Objectives:
3. Calculate the terms of arithmetic and geometric sequences.
4. Use summation notation and evaluate the series associated with a finite sequence.
5. Evaluate an infinite geometric series.

Sequences

Objects and natural phenomena often form regular and interesting patterns. The study of these objects and phenomena generally involves data collected sequentially or in a systematic manner. Here are some examples of occurrences of sequences:

- Business: To enumerate the payments necessary to repay a loan
- Biology: To describe the growth pattern of a living organism
- Calculator design: To specify the sequence of terms used to calculate functions such as the exponential function e^x and the trigonometric function $\cos x$
- Calculus: To give the areas of a sequence of rectangles used to approximate the area of a region

We first examined sequences in Section 1.4, arithmetic sequences in Section 2.1, and geometric sequences in Section 10.1. We now will give a more formal definition of a sequence. A **sequence** is a function whose domain is a set of consecutive natural numbers. For example, the sequence 2, 5, 8, 11, 14, 17 can be viewed as the function $\{(1, 2), (2, 5), (3, 8), (4, 11), (5, 14), (6, 17)\}$ with input values $\{1, 2, 3, 4, 5, 6\}$ and output values $\{2, 5, 8, 11, 14, 17\}$.

ALTERNATIVE NOTATIONS FOR THE SEQUENCE 2, 5, 8, 11, 14, 17

MAPPING NOTATION	FUNCTION NOTATION	SUBSCRIPT NOTATION
$1 \rightarrow 2$	$f(1) = 2$	$a_1 = 2$
$2 \rightarrow 5$	$f(2) = 5$	$a_2 = 5$
$3 \rightarrow 8$	$f(3) = 8$	$a_3 = 8$
$4 \rightarrow 11$	$f(4) = 11$	$a_4 = 11$
$5 \rightarrow 14$	$f(5) = 14$	$a_5 = 14$
$6 \rightarrow 17$	$f(6) = 17$	$a_6 = 17$

The notation consisting of three dots, which is used to represent the sequence $a_1, a_2, \ldots, a_n$ is called ellipsis notation. This notation is used to indicate that terms in the sequence are missing in the listing but the pattern shown is continued.

A **finite sequence** has a last term. A sequence that continues without end is called an **infinite sequence**. A finite sequence with n terms can be denoted by $a_1, a_2, \ldots, a_n$, and an infinite sequence can be denoted by $a_1, a_2, \ldots, a_n, \ldots$

Arithmetic and Geometric Sequences

We now will review the definitions and descriptions of arithmetic and geometric sequences.

Arithmetic and Geometric Sequences

Arithmetic:

ALGEBRAICALLY	VERBALLY	NUMERICAL EXAMPLE	GRAPHICAL EXAMPLE
$a_n - a_{n-1} = d$ or $a_n = a_{n-1} + d$	An arithmetic sequence has a constant change d from term to term. The graph will form a set of discrete points lying on a straight line.	$-3, -1, 1, 3, 5, 7$	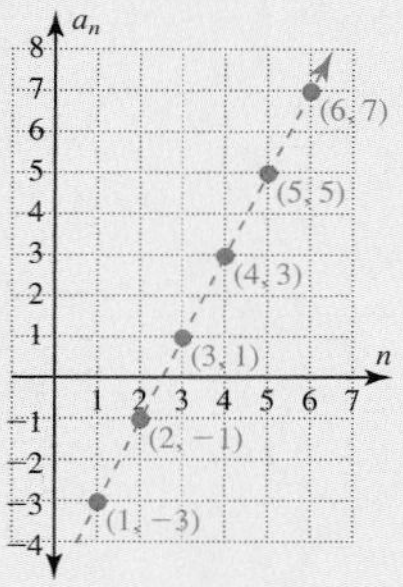

Geometric:

ALGEBRAICALLY	VERBALLY	NUMERICAL EXAMPLE	GRAPHICAL EXAMPLE
$\dfrac{a_n}{a_{n-1}} = r$ or $a_n = ra_{n-1}$	A geometric sequence has a constant ratio r from term to term. The graph will form a set of discrete points lying on an exponential curve.	$\dfrac{1}{4}, \dfrac{1}{2}, 1, 2, 4, 8$	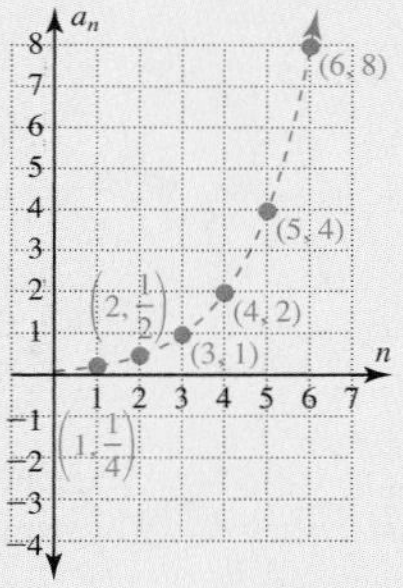

■ EXAMPLE 1   Identifying Arithmetic and Geometric Sequences

Determine whether each sequence is arithmetic, geometric, both, or neither. If the sequence is arithmetic, write the common difference d. If the sequence is geometric, write the common ratio r.

SOLUTIONS

(a) $17, 14, 11, 8, 5, \ldots$ Arithmetic sequence with $d = -3$ $14 - 17 = -3; 11 - 14 = -3;$
$8 - 11 = -3; 5 - 8 = -3$

$d = -3$

Not a geometric sequence $\dfrac{14}{17} \neq \dfrac{11}{14}$

(b) $3, -6, 12, -24, 48, \ldots$ Not an arithmetic sequence

$$\begin{aligned} -6 - 3 &= -9 \\ 12 - (-6) &= 18 \\ -9 &\neq 18 \end{aligned}$$

Geometric sequence with $r = -2$

$$-\frac{6}{3} = -2; \frac{12}{-6} = -2;$$
$$\frac{-24}{12} = -2; \frac{48}{-24} = -2$$
$$r = -2$$

(c) $3, 3, 3, 3, 3, \ldots$ Arithmetic sequence with $d = 0$ $3 - 3 = 0$

Geometric sequence with $r = 1$ $\dfrac{3}{3} = 1$

(d) $1, 1, 2, 3, 5, 8, \ldots$ Not an arithmetic sequence

$$\begin{aligned} 1 - 1 &= 0 \\ 2 - 1 &= 1 \\ 0 &\neq 1 \end{aligned}$$

Not a geometric sequence

$$\frac{1}{1} = 1$$
$$\frac{2}{1} = 2$$
$$1 \neq 2$$

SELF-CHECK 11.2.1

Complete the finite sequence $3, 6, \underline{\hspace{1cm}}, \underline{\hspace{1cm}}, \underline{\hspace{1cm}}$, so that the sequence will be:

1. An arithmetic sequence.
2. A geometric sequence.

Recursive Definitions

Since a_n can represent any term of the sequence, it is called the **general term.** The general term of a sequence sometimes is defined in terms of one or more of the preceding terms. A sequence defined in this manner is said to be **defined recursively.** Examples of recursive definitions for parts (a) and (b) of Example 1 are

Recursively defined sequences are used extensively by computer programmers.

(a) $a_n = a_{n-1} - 3$ with $a_1 = 17$ $17, 14, 11, 8, 5, \ldots$
(b) $a_n = -2a_{n-1}$ with $a_1 = 3$ $3, -6, 12, -24, 48, \ldots$

Because recursive definitions relate terms to preceding terms, there must be an initial term or condition given so that the sequence can get started.

This initial or "seed" value is frequently used by computer programmers to construct looping structures within their programs. Example 2 reexamines the sequence in Example 1(d). This recursive definition requires two initial conditions.

The sequence 1, 1, 2, 3, 5, 8, 13, ... is known as a *Fibonacci sequence,* in honor of the Italian mathematician Leonard Fibonacci (1170–1250). Fibonacci is known as the greatest mathematician of the thirteenth century. His sequence appears in many surprising ways in nature, including the arrangement of seeds in some flowers, the layout of leaves on the stems of some plants, and the spirals on some shells. There are so many applications that in 1963 the Fibonacci Association was founded and began to publish *The Fibonacci Quarterly.* In its first 3 years the association published nearly 1000 pages of research.

■ EXAMPLE 2   Using a Recursive Definition

Write the first five terms of the sequence defined recursively by the formulas $a_1 = 1$, $a_2 = 1$, and $a_n = a_{n-2} + a_{n-1}$.

SOLUTION

$a_1 = 1$ The first two terms are given; for $n = 3$, a_n is a_3, a_{n-2} is a_1, and a_{n-1} is a_2.
$a_2 = 1$
$a_3 = a_1 + a_2$
$\quad = 1 + 1$ Substitute 1 for a_1 and 1 for a_2.
$a_3 = 2$
$a_4 = a_2 + a_3$ For $n = 4$, a_n is a_4, a_{n-2} is a_2, and a_{n-1} is a_3.
$\quad = 1 + 2$ Substitute 1 for a_2 and 2 for a_3.
$a_4 = 3$
$a_5 = a_3 + a_4$ For $n = 5$, a_n is a_5, a_{n-2} is a_3, and a_{n-1} is a_4.
$\quad = 2 + 3$ Substitute 2 for a_3 and 3 for a_4.
$a_5 = 5$

Answer: 1, 1, 2, 3, 5   ■

SELF-CHECK 11.2.2

Use the recursive definition $a_1 = 1$, $a_2 = 1$, and $a_n = a_{n-2} + a_{n-1}$ to write the next four terms of 1, 1, 2, 3, 5, _____, _____, _____, _____.

| CALCULATOR PERSPECTIVE 11.2.1 | Using the seq Feature to Display the Terms of a Sequence |

To display the first five terms of the sequence defined by $a_n = 2n + 3$ on a TI-83 Plus calculator, enter the following keystrokes:

The **LIST** feature is the secondary function of the **STAT** key.

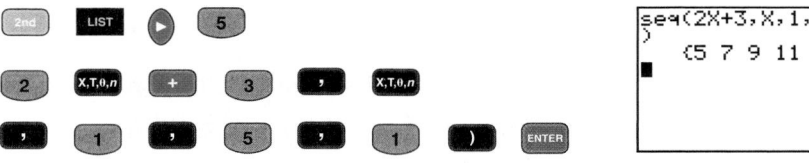

Note: The **seq** feature has 5 parameters: seq(*expression, variable, first value, last value, increment*). In this case the variable used is *x*, the first value is 1, the last value is 5 (for the fifth term), and the increment is 1. If no increment is given, the default is to increase by 1.   ■

Formulas for the *n*th Term of an Arithmetic and a Geometric Sequence

Recursive formulas are used extensively in computer science and in mathematics. However, these formulas do require us to compute a sequence term by term. For example, to compute the 100th term of $a_n = a_{n-1} - 3$ we would first need the 99th term. It is also very useful to have a form that allows us to directly compute a_n from *n*. We will now examine formulas for a_n for arithmetic and geometric sequences.

SELF-CHECK 11.2.2 ANSWERS

1, 1, 2, 3, 5, 8, 13, 21, 34

ARITHMETIC SEQUENCE	GEOMETRIC SEQUENCE	
$a_n = a_{n-1} + d$	$a_n = ra_{n-1}$	Start with the recursive formula for a_n and note the pattern that develops. Arithmetic sequences *add* a common difference for each new term, and geometric sequences *multiply* by a common ratio for each new term.
a_1	a_1	
$a_2 = a_1 + d$	$a_2 = a_1 r$	
$a_3 = a_2 + d = (a_1 + d) + d = a_1 + 2d$	$a_3 = a_2 r = (a_1 r)r = a_1 r^2$	
$a_4 = a_3 + d = (a_1 + 2d) + d = a_1 + 3d$	$a_4 = a_3 r = (a_1 r^2)r = a_1 r^3$	
$\vdots$	$\vdots$	
$a_n = a_{n-1} + d = a_1 + (n-1)d$	$a_n = a_{n-1}r = a_1 r^{n-1}$	
$\mathbf{a_n = a_1 + (n-1)d}$	$\mathbf{a_n = a_1 r^{n-1}}$	

■ EXAMPLE 3   Calculating a_n for an Arithmetic and a Geometric Sequence

Use the formulas for a_n for an arithmetic sequence and a geometric sequence to calculate a_{10} for each sequence.

SOLUTIONS

(a) Determine a_{10} for an arithmetic sequence with $a_1 = 8$ and $d = 7$.

$a_n = a_1 + (n-1)d$ Substitute the given values into the formula for a_n for an arithmetic
$a_{10} = 8 + (10-1)(7)$ sequence.
$a_{10} = 8 + 9(7)$
$a_{10} = 8 + 63$
$a_{10} = 71$

(b) Determine a_{10} for a geometric sequence with $a_1 = 8$ and $r = 3$.

$a_n = a_1 r^{n-1}$ Substitute the given values into the formula for a_n for a geometric
$a_{10} = 8(3)^{10-1}$ sequence.
$a_{10} = 8(3)^9$
$a_{10} = 8(19,683)$ Use a calculator to evaluate this expression.
$a_{10} = 157,464$

(c) Determine a_{10} for a geometric sequence with $a_1 = 5$ and $r = -2$.

$a_n = a_1 r^{n-1}$ Because r is negative the terms in this geometric sequence will alternate
$a_{10} = 5(-2)^{10-1}$ in sign. All even-numbered terms will be negative.
$a_{10} = 5(-2)^9$
$a_{10} = 5(-512)$
$a_{10} = -2560$

The formula $a_n = a_1 + (n-1)d$ relates a_n, a_1, n, and d. Therefore, we can find any of these four variables when the other three are known. A similar statement can be made for the formula $a_n = a_1 r^{n-1}$. This is illustrated in Example 4.

■ EXAMPLE 4   Determining the Common Difference and the Common Ratio

Use the formulas for a_n for an arithmetic sequence and a geometric sequence to calculate d and r as requested in these problems.

SOLUTIONS

(a) Find d in an arithmetic sequence with $a_1 = -87$ and $a_{57} = 529$.

$a_n = a_1 + (n-1)d$ Substitute the given values into the formula for a_n for an arithmetic
$529 = -87 + (57-1)d$ sequence.
$616 = 56d$
$d = 11$ Then solve for d.

(b) Find r in a geometric sequence with $a_1 = 6$ and $a_3 = 24$.

$$a_n = a_1 r^{n-1}$$
$$24 = (6)(r)^{3-1}$$
$$4 = r^2$$

$$r = 2 \quad \text{or} \quad r = -2$$

Substitute the given values into the formula for a_n for a geometric sequence.

Then solve this quadratic equation for both possible values of r.

Check: The geometric sequences $6, 12, 24, 48, \ldots$ and $6, -12, 24, -48, \ldots$ both satisfy the given conditions. ■

In Example 5 we use the formulas for a_n to calculate the number of terms in an arithmetic sequence and a geometric sequence.

■ **EXAMPLE 5** **Determining the Number of Terms in a Sequence**

Use the formulas for a_n for an arithmetic sequence and a geometric sequence to calculate the number of terms in each sequence.

SOLUTIONS _____

(a) $-20, -13, -6, \ldots, 281$

$$d = -13 - (-20) = 7$$
$$a_n = a_1 + (n-1)d$$
$$281 = -20 + (n-1)7$$
$$301 = 7(n-1)$$
$$n - 1 = 43$$
$$n = 44$$

This arithmetic sequence has 44 terms.

First note that this is an arithmetic sequence with a common difference $d = 7$ and a first term of -20. Substitute $a_1 = -20$, $a_n = 281$, and $d = 7$ into the formula for a_n for an arithmetic sequence. Then solve for n.

(b) $3125, 1250, 500, \ldots, 32$

$$r = \frac{1250}{3125} = \frac{500}{1250} = 0.4$$
$$a_n = a_1 r^{n-1}$$
$$32 = (3125)(0.4)^{n-1}$$
$$(0.4)^{n-1} = 0.01024$$
$$\log 0.4^{n-1} = \log 0.01024$$
$$(n-1)\log 0.4 = \log 0.01024$$
$$n - 1 = \frac{\log 0.01024}{\log 0.4}$$
$$n - 1 \approx \frac{-1.9897}{-0.3979}$$
$$n - 1 \approx 5$$
$$n \approx 6$$

This geometric sequence has 6 terms.

First note that this is a geometric sequence with a common ratio of $r = 0.4$.

Substitute $a_1 = 3125$, $a_n = 32$, and $r = 0.4$ into the formula for a_n for a geometric sequence. Divide both sides of the equation by 3125. Take the common log of both members. Simplify by using the power rule for logarithms. Divide both sides of the equation by $\log 0.4$.

Approximate the right side with a calculator.

You can check this answer by writing the first six terms of this geometric sequence. ■

Series

The sum of the terms of a sequence is called a **series.** If $a_1, a_2, \ldots, a_n$ is a finite sequence, then the indicated sum $a_1 + a_2 + \cdots + a_n$ is the series associated with this sequence. For arithmetic and geometric series there are formulas that serve as shortcuts to evaluating the series. We will use Example 6 to illustrate the meaning of a sum, and then we will examine the shortcuts.

SELF-CHECK 11.2.3

Use the formulas $a_n = a_1 + (n-1)d$ and $a_n = a_1 r^{n-1}$ to determine:

1. a_{21} for an arithmetic sequence with $a_1 = 4$ and $d = 5$.
2. d for an arithmetic sequence with $a_1 = 10$ and $a_{81} = 250$.
3. The number of terms in the arithmetic sequence with $a_1 = -12$, $a_n = 288$, and $d = 4$.
4. a_9 for a geometric sequence with $a_1 = 4$ and $r = 5$.
5. r for a geometric sequence with $a_1 = 5$ and $a_4 = 1080$.
6. The number of terms in the geometric sequence with $a_1 = \dfrac{32}{3125}$, $a_n = \dfrac{625}{16}$, and $r = \dfrac{5}{2}$.

■ EXAMPLE 6   Evaluating an Arithmetic Series and a Geometric Series

Find the value of the six-term series associated with these arithmetic and geometric sequences.

SOLUTIONS _____

(a) $a_n = 3n$

$a_1 + a_2 + a_3 + a_4 + a_5 + a_6$
$= 3(1) + 3(2) + 3(3) + 3(4) + 3(5) + 3(6)$
$= 3 + 6 + 9 + 12 + 15 + 18$
$= 63$

The series is the sum of the first six terms. Substitute the first six natural numbers into the formula $a_n = 3n$ to determine the first six terms and then add the terms of this arithmetic sequence.

(b) $a_n = 3^n$

$a_1 + a_2 + a_3 + a_4 + a_5 + a_6$
$= 3^1 + 3^2 + 3^3 + 3^4 + 3^5 + 3^6$
$= 3 + 9 + 27 + 81 + 243 + 729$
$= 1092$

Calculate each of the six terms of this geometric sequence and then determine the sum.

SELF-CHECK 11.2.3 ANSWERS

1. $a_{21} = 104$
2. $d = 3$
3. $n = 76$
4. $a_9 = 1,562,500$
5. $r = 6$
6. $n = 10$

Summation Notation

A convenient way of denoting a series is to use **summation notation,** in which the Greek letter Σ (*sigma,* which corresponds to the letter S for "sum") indicates the summation.

Summation Notation

ALGEBRAICALLY	VERBALLY	ALGEBRAIC EXAMPLE
Last value of the index — $\displaystyle\sum_{i=1}^{n} a_i = a_1 + a_2 + \cdots + a_{n-1} + a_n$ — Formula for general term; Index variable — Initial value of the index	The sum of a sub i from i equals 1 to i equals n.	$\displaystyle\sum_{i=1}^{5} a_i = a_1 + a_2 + a_3 + a_4 + a_5$

Generally the index variable is denoted by i, j, or k. The index variable is always replaced with successive integers from the initial value through the last value. Example 7 illustrates that the initial value can be a value other than 1. For example, in $\sum_{i=5}^{8} a_i$, i is replaced with 5, 6, 7, and then 8 to yield $\sum_{i=5}^{8} a_i = a_5 + a_6 + a_7 + a_8$.

■ EXAMPLE 7   Using Summation Notation to Evaluate an Arithmetic Series and a Geometric Series

Evaluate each series.

SOLUTIONS

(a) $\displaystyle\sum_{i=1}^{4} 5i$ $\displaystyle\sum_{i=1}^{4} 5i = 5(1) + 5(2) + 5(3) + 5(4)$ Replace i with 1, 2, 3, and then 4 and indicate the sum of these terms.

$= 5 + 10 + 15 + 20$ Evaluate each term of this arithmetic series and then add these terms.

$= 50$

(b) $\displaystyle\sum_{k=3}^{7} 2^k$ $\displaystyle\sum_{k=3}^{7} 2^k = 2^3 + 2^4 + 2^5 + 2^6 + 2^7$ Replace k with 3, 4, 5, 6, and then 7 and indicate the sum of these terms.

$= 8 + 16 + 32 + 64 + 128$ Evaluate each term of this geometric series and then add these terms.

$= 248$

SELF-CHECK 11.2.4

Evaluate each series.

SELF-CHECK 11.2.4 ANSWERS

1. 60   **2.** 10   **3.** 19,500

1. $\displaystyle\sum_{i=1}^{5} 4i$ 2. $\displaystyle\sum_{j=1}^{4} (j^2 - 2j)$ 3. $\displaystyle\sum_{k=3}^{6} 5^k$

CALCULATOR PERSPECTIVE 11.2.2   Computing the Sum of a Sequence

To evaluate the series $\displaystyle\sum_{i=1}^{4} 5i = 5 + 10 + 15 + 20$ from Example 7(a) on a TI-83 Plus calculator, enter the following keystrokes:

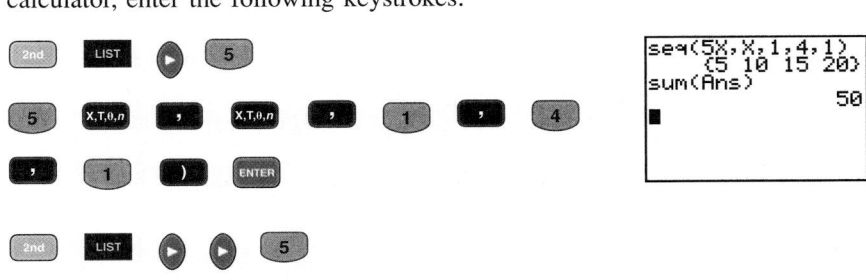

The [ANS] feature is the secondary function of the [(-)] key.

```
seq(5X,X,1,4,1)
      {5 10 15 20}
sum(Ans)
                50
■
```

For a series with many terms, such as $\sum_{i=1}^{100} i$, it is useful to have a shortcut formula that allows us to calculate the sum without actually doing all of the adding. We will now develop formulas for an arithmetic series and a geometric series.

Development of a Formula for an Arithmetic Series

$$S_n = a_1 + (a_1 + d) + \cdots + [a_1 + (n-2)d] + [a_1 + (n-1)d]$$
$$\underline{S_n = [a_1 + (n-1)d] + [a_1 + (n-2)d] + \cdots + (a_1 + d) + a_1}$$

$$2S_n = [a_1 + a_1 + (n-1)d] + [a_1 + a_1 + (n-1)d] + \cdots$$
$$\qquad + [a_1 + a_1 + (n-1)d] + [a_1 + a_1 + (n-1)d]$$
$$2S_n = (a_1 + a_n) + (a_1 + a_n) + \cdots + (a_1 + a_n) + (a_1 + a_n)$$
$$2S_n = n(a_1 + a_n)$$
$$S_n = \frac{n(a_1 + a_n)}{2}$$
$$S_n = \frac{n[2a_1 + (n-1)d]}{2}$$

Sum the terms from a_1 to a_n.
Sum the terms from a_n to a_1.
Add corresponding terms.

Substitute a_n for $a_1 + (n-1)d$.
Note that $(a_1 + a_n)$ is added n times.
Solve for S_n by dividing both sides by 2.

Substitute $a_1 + (n-1)d$ for a_n for an alternative form of this formula.

Development of a Formula for a Geometric Series

$$S_n = a_1 + a_1r + a_1r^2 + \qquad \cdots + a_1r^{n-2} + a_1r^{n-1}$$
$$\underline{rS_n = \qquad a_1r + a_1r^2 + a_1r^3 + \cdots + a_1r^{n-2} + a_1r^{n-1} + a_1r^n}$$

$$S_n - rS_n = a_1 + \quad 0 + \quad 0 + \quad \cdots + \quad 0 + \quad 0 - a_1r^n$$

$$S_n(1-r) = a_1(1-r^n)$$
$$S_n = \frac{a_1(1-r^n)}{1-r} \quad \text{for } r \neq 1$$
$$S_n = \frac{a_1 - a_1r^n}{1-r}$$
$$S_n = \frac{a_1 - ra_n}{1-r}$$

To obtain the second equation, multiply both sides of the first equation by r and shift terms to the right to align similar terms. Subtract the second equation from the first equation.

Factor both sides.

Divide both members by $1-r$. If $r \neq 1$, $1 - r \neq 0$.

Substitute a_n for a_1r^{n-1} to obtain an alternative form of this formula.

Example 8 examines an application for both arithmetic and geometric series.

Formulas for Arithmetic and Geometric Series $S_n = \sum_{i=1}^{n} a_i$

	ALGEBRAICALLY	ALGEBRAIC EXAMPLE
Arithmetic series:	$S_n = \frac{n}{2}(a_1 + a_n)$ or $S_n = \frac{n}{2}[2a_1 + (n-1)d]$	$S_n = \sum_{i=1}^{4} 5i = 5 + 10 + 15 + 20$ $S_4 = \frac{4}{2}(5 + 20)$ $S_4 = 2(25)$ $S_4 = 50$
Geometric series:	$S_n = \frac{a_1(1-r^n)}{1-r}$ or $S_n = \frac{a_1 - ra_n}{1-r}$	$S_n = \sum_{k=1}^{5} 2^k = 2 + 4 + 8 + 16 + 32$ $S_5 = \frac{2(1-2^5)}{1-2}$ $S_5 = \frac{2(-31)}{-1}$ $S_5 = 62$

■ EXAMPLE 8   Using Arithmetic and Geometric Series to Model Applications

(a) Rolls of carpet are stacked in a warehouse, with 20 rolls on the first level, 19 on
the second level, and so on. The top level has only 1 roll. How many rolls are in
this stack? (See the figure.)

(b) If you could arrange to be paid $1000 at the end of January, $2000 at the end of
February, $4000 at the end of March, and so on, what is the total amount you
would be paid for the year?

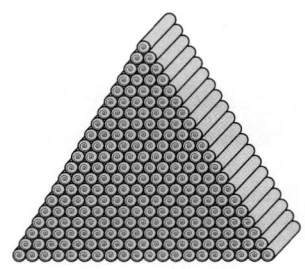

SOLUTIONS

(a) The number of rolls on the various levels form an arithmetic sequence
with $a_1 = 20$, $n = 20$, and $a_{20} = 1$.

ALGEBRAICALLY

$$S_n = \frac{n(a_1 + a_n)}{2}$$

$$S_{20} = \frac{20(20 + 1)}{2}$$

$$S_{20} = 210$$

NUMERICALLY

$$S_n = a_1 + a_2 + \cdots + a_n$$
$$S_{20} = a_1 + a_2 + \cdots + a_{20}$$
$$S_{20} = 20 + 19 + 18 + \cdots + 3 + 2 + 1$$
$$S_{20} = 210$$

There is a constant difference of 1 roll between
consecutive levels in this stack. Thus the
numbers of rolls form an arithmetic sequence,
and the total number of rolls is determined by
adding the terms of this arithmetic series.

Answer: There are 210 rolls in this stack.

(b) The payments at the end of each month form a geometric sequence with
$a_1 = 1000$, $r = 2$, and $n = 12$.

The total amount for the year is determined by
adding the 12 terms of this geometric
sequence.

ALGEBRAICALLY

$$S_n = \frac{a_1(1 - r^n)}{1 - r}$$

$$S_{12} = \frac{1000(1 - 2^{12})}{1 - 2}$$

$$S_{12} = \frac{1000(1 - 4096)}{-1}$$

$$S_{12} = \frac{-4,095,000}{-1}$$

$$S_{12} = 4,095,000$$

NUMERICALLY

$$S_n = a_1 + a_2 + \cdots + a_n$$
$$S_{12} = a_1 + a_2 + \cdots + a_{12}$$
$$S_{12} = 1000 + 2000 + 4000$$
$$\quad + 8000 + 16,000 + 32,000$$
$$\quad + 64,000 + 128,000 + 256,000$$
$$\quad + 512,000 + 1,024,000$$
$$\quad + 2,048,000$$
$$S_{12} = 4,095,000$$

Simplify and calculate S_{12}.

Although this answer may seem unreasonable,
it is the doubling pay scheme that makes the
total unreasonable—not the arithmetic.

Answer: The total amount for the year would be $4,095,000.   ■

Compare the answers obtained algebraically in Example 8 with the answers obtained numerically by actually adding the terms in the series. Which method do you prefer? Which method would you prefer if there were a thousand terms?

SELF-CHECK 11.2.5

Use the formulas for arithmetic and geometric series to determine these sums.

1. $\displaystyle\sum_{i=1}^{100} i$

2. $\displaystyle\sum_{i=1}^{20} 2^i$

If $|r| < 1$, then the absolute values of the terms of the geometric sequence are decreasing. For example, the geometric sequence $\dfrac{1}{2}, \dfrac{1}{4}, \dfrac{1}{8}, \dfrac{1}{16}, \ldots, \dfrac{1}{2^n}, \ldots$ is a decreasing, infinite geometric sequence with $r = \dfrac{1}{2}$. We already have a formula for finding the sum of a finite geometric sequence. Is the infinite sum $\displaystyle\sum_{i=1}^{\infty}\left(\dfrac{1}{2}\right)^i$ meaningful? Although we could never actually add an infinite number of terms, we can adopt a new meaning for this sum. If the sum $\displaystyle\sum_{i=1}^{n} a_i$ approaches some limiting value S as n becomes large, then we will call this value the **infinite sum.** Symbolically, $\displaystyle\sum_{i=1}^{\infty} a_i = S$ if $\displaystyle\sum_{i=1}^{n} a_i$ approaches S as n increases.

A general formula for the infinite sum can be obtained by examining the formula for S_n.

$$S_n = \frac{a_1\left(1 - r^n\right)}{1 - r}$$

If $|r| < 1$, then $|r|^n$ approaches 0 as n becomes larger. Thus $S_n = \dfrac{a_1\left(1 - r^n\right)}{1 - r}$ approaches $\dfrac{a_1(1 - 0)}{1 - r}$; that is, S_n approaches $\dfrac{a_1}{1 - r}$. The infinite sum S is the limiting value, so $S = \dfrac{a_1}{1 - r}$. If $|r| \geq 1$, the terms of a geometric series do not approach 0 and $\displaystyle\sum_{i=1}^{n} a_i$ does not approach any limit as n becomes large. In this case, we do not assign a value to $\displaystyle\sum_{i=1}^{\infty} a_i$.

Infinite Geometric Series $S = \sum\limits_{i=1}^{\infty} a_i$	
ALGEBRAICALLY	**ALGEBRAIC EXAMPLE**
If $\|r\| < 1$, $S = \dfrac{a_1}{1 - r}$ If $\|r\| \geq 1$, this sum does not exist.	$S = 0.333 \ldots$ $S = 0.3 + 0.03 + 0.003 + \cdots$ with $a_1 = 0.3$ and $r = 0.1$ $S = \dfrac{a_1}{1 - r}$ $S = \dfrac{0.3}{1 - 0.1}$ $S = \dfrac{0.3}{0.9}$ $S = \dfrac{1}{3}$

■ EXAMPLE 9   Writing a Repeating Decimal in Fractional Form

Write $0.272727 \ldots$ as a fraction.

SOLUTION

$$0.272727 \ldots = 0.27 + 0.0027 + 0.000027 + \cdots$$
$$= 27(0.01) + 27(0.01)^2 + 27(0.01)^3 + \cdots$$
$$= \sum_{i=1}^{\infty} 27(0.01)^i$$

This series is an infinite geometric series with $a_1 = 0.27$ and $r = 0.01$.

$$S = \frac{a_1}{1 - r}$$

This is the formula for the sum of an infinite geometric series.

$$S = \frac{0.27}{1 - 0.01}$$

Substitute 0.27 for a_1 and 0.01 for r.

$$S = \frac{0.27}{0.99}$$

Simplify and express S in fractional form.

$$S = \frac{3}{11}$$

Answer: $0.272727 \ldots = \dfrac{3}{11}$

You can check this answer by dividing 3 by 11.   ■

SELF-CHECK 11.2.6 ANSWER

$0.545454 \ldots = \dfrac{6}{11}$

SELF-CHECK 11.2.6

Write $0.545454 \ldots$ as a fraction.

USING THE LANGUAGE AND SYMBOLISM OF MATHEMATICS   11.2

1. A sequence is a function whose domain is a set of consecutive _____ numbers.
2. A sequence that has a last term is called a _____ sequence.
3. A sequence that continues without end is called an _____ sequence.
4. A sequence with $a_n = a_{n-1} + d$ is an _____ sequence.
5. A sequence with $a_n = ra_{n-1}$ is a _____ sequence.
6. The sequence 1, 1, 2, 3, 5, 8, 13, . . . is an example of a _____ sequence.
7. Since a_n can represent any term of a sequence, it is called the _____ term.
8. A sequence where the general term is defined in terms of one or more of the preceding terms is said to be defined _____ .

9. The formula for a_n for an arithmetic sequence is $a_n = $ _____ .
10. The formula for a_n for a geometric sequence is $a_n = $ _____ .
11. In the _____ notation $\sum\limits_{i=1}^{n} a_i$, the index variable is _____ . The initial value of the index variable in this example is _____ , and the last value of the index variable is _____ .
12. The formula for an arithmetic series is $S_n = $ _____ .
13. The formula for a geometric series is $S_n = $ _____ .
14. The formula for an infinite geometric series is $S = $ _____ if $|r| < 1$.

EXERCISES   11.2

In Exercises 1–4 determine whether each sequence is arithmetic, geometric, both, or neither. If the sequence is arithmetic, write the common difference d. If the sequence is geometric, write the common ratio r.

1. **a.** 1, 6, 36, 216, 1296, 7776
   b. 1, 6, 11, 16, 21, 26
   c. 2, 3, 5, 8, 12, 13
   d. 6, 6, 6, 6, 6, 6
2. **a.** 144, 124, 104, 84, 64, 44
   b. 144, 72, 36, 18, 9, 4.5
   c. 9, −9, 9, −9, 9, −9
   d. 9, −8, 7, −6, 5, −4
3. **a.** $a_n = 3n - 5$
   b. $a_n = n^2$
   c. $a_n = 4^n$
   d. $a_n = 7$
4. **a.** $a_1 = 1$ and $a_n = 5a_{n-1}$ for $n > 1$
   b. $a_1 = 2$ and $a_n = (a_{n-1})^2$ for $n > 1$
   c. $a_1 = 3$ and $a_n = a_{n-1} + 5$ for $n > 1$
   d. $a_1 = 4$ and $a_n = 4$ for $n > 1$

In Exercises 5 and 6 write the first six terms of each sequence.

5. **a.** $a_n = 5n + 3$
   b. $a_n = n^2 - n + 1$
   c. $a_n = 16\left(\dfrac{1}{2}\right)^n$
   d. $a_n = 5(-2)^n$
6. **a.** $a_1 = 4$ and $a_n = a_{n-1} + 5$ for $n > 1$
   b. $a_1 = 4$ and $a_n = 5a_{n-1}$ for $n > 1$
   c. $a_1 = 4$ and $a_n = -a_{n-1}$ for $n > 1$
   d. $a_1 = 1, a_2 = 2$, and $a_n = 2a_{n-2} + a_{n-1}$ for $n > 1$

In Exercises 7 and 8 write the first six terms of each arithmetic sequence.

7. **a.** 18, 14, _____, _____, _____, _____
   b. 18, _____, 14, _____, _____, _____
   c. $a_1 = 7, d = -2$
   d. $a_1 = -9, d = 3$
8. **a.** $a_1 = -8, a_6 = 12$
   b. $a_n = 4n - 1$
   c. $a_n = 11 - 2n$
   d. $a_1 = 1.2$ and $a_n = a_1 + 0.4$ for $n > 1$

In Exercises 9 and 10 write the first five terms of each geometric sequence.

9. **a.** 1, 4, _____, _____, _____
   b. 1, _____, 4, _____, _____
   c. $a_1 = 16, r = \dfrac{1}{2}$
   d. $a_1 = 5, r = -3$
10. **a.** $a_1 = 3, a_4 = 24$
    b. $a_n = 9\left(\dfrac{1}{3}\right)^n$
    c. $a_n = 96\left(-\dfrac{1}{2}\right)^n$
    d. $a_1 = 10$ and $a_n = 2a_{n-1}$ for $n > 1$
11. Use the given information to calculate the indicated term of the arithmetic sequence.
    a. $a_1 = 6, d = 5, a_{83} = ?$
    b. $a_1 = 17, a_2 = 15, a_{51} = ?$
    c. $a_{80} = 25, a_{81} = 33, a_{82} = ?$
    d. $a_1 = 2, a_3 = 12, a_{101} = ?$

12. Use the given information to calculate the indicated term of the geometric sequence.
   a. $a_1 = \dfrac{1}{256}, r = 2, a_{10} = ?$
   b. $a_1 = 4, a_2 = 12, a_8 = ?$
   c. $a_{50} = 8, a_{51} = 40, a_{52} = ?$
   d. $a_1 = 5, a_3 = 20, a_{10} = ?$

In Exercises 13–20 use the information given for the arithmetic sequence to find the quantities indicated.

13. $18, 14, 10, \ldots, -62; n = ?$
14. $48, 55, \ldots, 496; n = ?$
15. $a_{44} = 216, d = 12, a_1 = ?$
16. $a_{113} = -109, d = -2, a_1 = ?$
17. $a_{11} = 4, a_{31} = 14, d = ?$
18. $a_{47} = 23, a_{62} = 28, d = ?$
19. $a_1 = 5, a_n = 14, d = \dfrac{1}{5}, n = ?$
20. $a_1 = -12, a_n = 6, d = \dfrac{1}{2}, n = ?$

In Exercises 21–28 use the information given for the geometric sequences to find the quantities indicated.

21. $243, 81, \ldots, \dfrac{1}{3}; n = ?$
22. $1024, 512, \ldots, 1; n = ?$
23. $a_5 = 24, r = 2, a_1 = ?$
24. $a_6 = 64, r = 4, a_1 = ?$
25. $a_9 = 32, a_{11} = 288, r = ?$
26. $a_{45} = 17, a_{47} = 425, r = ?$
27. $a_n = 5a_{n-1}, a_1 = \dfrac{1}{3125}, a_9 = ?$
28. $a_n = 0.1a_{n-1}, a_1 = 7000, a_8 = ?$

In Exercises 29 and 30 write the terms of each series and then add these terms.

29. a. $\displaystyle\sum_{i=1}^{6} (2i + 3)$ b. $\displaystyle\sum_{i=1}^{5} \left(i^2 - 1\right)$

    c. $\displaystyle\sum_{j=1}^{6} 2^j$ d. $\displaystyle\sum_{k=1}^{10} 5$

30. a. $\displaystyle\sum_{k=1}^{5} \dfrac{k+3}{5}$ b. $\displaystyle\sum_{i=4}^{8} \left(i^2 + 1\right)$

    c. $\displaystyle\sum_{j=4}^{7} 3^j$ d. $\displaystyle\sum_{k=10}^{16} 7$

In Exercises 31–40 use the information given to evaluate each arithmetic series.

31. $a_1 = 2, a_{40} = 80, S_{40} = ?$
32. $a_1 = 3, a_{51} = 153, S_{51} = ?$
33. $a_1 = \dfrac{1}{2}, a_{12} = \dfrac{1}{3}, S_{12} = ?$
34. $a_1 = 0.36, a_{18} = 0.64, S_{18} = ?$
35. $a_1 = 10, d = 4, S_{66} = ?$
36. $a_1 = 11, d = -3, S_{11} = ?$
37. $\displaystyle\sum_{i=1}^{61} (2i + 3)$

38. $\displaystyle\sum_{k=1}^{47} (3k - 2)$

39. $\displaystyle\sum_{k=1}^{24} \dfrac{k+3}{5}$

40. $\displaystyle\sum_{j=1}^{40} \dfrac{j-5}{3}$

In Exercises 41–52 use the information given to evaluate each geometric series.

41. $a_1 = 3, r = 2, S_7 = ?$
42. $a_1 = 2, r = 3, S_6 = ?$
43. $a_1 = 0.2, r = 0.1, S_5 = ?$
44. $a_1 = 0.5, r = 0.1, S_6 = ?$
45. $a_1 = 48, r = -\dfrac{1}{2}, S_8 = ?$
46. $a_1 = 729, r = -\dfrac{1}{3}, S_7 = ?$
47. $a_1 = 1, a_n = 3.71293, r = 1.3, S_n = ?$
48. $64 + 32 + \cdots + \dfrac{1}{8}$
49. $\displaystyle\sum_{i=1}^{7} 8(0.1)^i$
50. $\displaystyle\sum_{k=1}^{6} 7(0.1)^k$
51. $a_1 = 40, a_n = \dfrac{1}{5}a_{n-1}$ for $n > 1, S_6 = ?$
52. $a_1 = 81, a_n = \dfrac{1}{3}a_{n-1}$ for $n > 1, S_6 = ?$

In Exercises 53–64 use the information given to evaluate the infinite geometric series.

53. $a_1 = 5, r = \dfrac{2}{3}$
54. $a_1 = 16, r = \dfrac{1}{5}$
55. $a_1 = 14, r = -\dfrac{3}{4}$
56. $a_1 = 7, r = -\dfrac{2}{5}$
57. $a_1 = 0.12, r = \dfrac{1}{100}$
58. $a_1 = 0.9, r = \dfrac{1}{10}$
59. $\displaystyle\sum_{k=1}^{\infty} \left(\dfrac{4}{9}\right)^k$
60. $\displaystyle\sum_{j=1}^{\infty} \left(\dfrac{3}{7}\right)^j$
61. $6 - 4 + \dfrac{8}{3} - \dfrac{16}{9} + \cdots$
62. $16 - 12 + 9 - 6.75 + \cdots$
63. $a_1 = 24, a_n = \dfrac{3}{8}a_{n-1}$ for $n > 1$
64. $a_1 = -36, a_n = \left(-\dfrac{5}{9}\right)a_{n-1}$ for $n > 1$

In Exercises 65 and 66 write each repeating decimal as a fraction.

65. a. 0.444 . . .   **b.** 0.212121 . . .
  c. 0.409409409 . . .   **d.** 2.5555 . . .
66. a. 0.555 . . .   **b.** 0.363636 . . .
  c. 0.495495495 . . .   **d.** 8.3333 . . .

In Exercises 67–72 use the information given for the arithmetic sequences to find the quantities indicated.

67. $a_{77} = 19, d = -11, S_{77} = ?$
68. $S_n = 240, a_1 = 4, a_n = 16, n = ?$
69. $S_{30} = 1560, a_{30} = 93, a_1 = ?$
70. $S_{17} = 527, a_1 = 15, a_{17} = ?$
71. $S_{40} = 680, a_1 = 11, d = ?$
72. $S_{25} = 60, a_1 = 17, d = ?$

In Exercises 73–78 use the information given for the geometric sequences to find the quantities indicated.

73. $r = 2, S_{10} = 6138, a_1 = ?$
74. $a_1 = 12, r = 5, S_n = 46{,}872, n = ?$
75. $a_1 = 6, a_n = 24{,}576, S_n = 32{,}766, r = ?$
76. $\sum_{i=1}^{\infty} a_i = 27, a_1 = 12, r = ?$
77. $\sum_{k=1}^{\infty} a_k = 21, r = \frac{2}{9}, a_1 = ?$
78. $\sum_{k=1}^{\infty} a_i = 2, r = \frac{6}{7}, a_1 = ?$

79. Stacks of Logs   Logs are stacked so that each layer after the first has 1 less log than the previous layer. If the bottom layer has 24 logs and the top layer has 8 logs, how many logs are in the stack? (See the figure.)

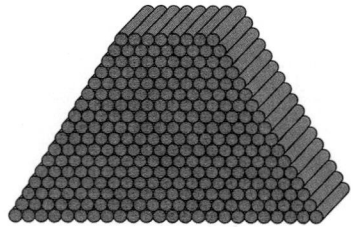

80. Rolls of Insulation   Rolls of insulation are stacked so that each layer after the first has 8 fewer rolls than the previous layer. How many layers will a lumberyard need to use in order to stack 120 rolls if 40 rolls are placed on the bottom layer? (See the figure.)

81. Increased Productivity   The productivity gain from installing a new robot welder on a machinery assembly line is estimated to be $4000 the first month of operation, $4500 the second month, and $5000 the third month. If this trend continues, what will be the total productivity gain for the first 12 months of operation?

82. Seats in a Theater   A theater has 20 rows of seats, with 100 seats in the back row. Each row has 2 fewer seats than the row immediately behind it. How many seats are in the theater?

83. Chain Letter   A chain-letter scam requires that each participant persuade four other people to participate. If one person starts this venture as a first-generation participant, determine how many people will have been involved by the time the eighth generation has signed on but not yet contacted anyone.

84. Bouncing Ball   A ball dropped from a height of 36 m rebounds to six-tenths its previous height on each bounce. How far has it traveled when it reaches the apex of its eighth bounce? Give your answer to the nearest tenth of a meter. (*Hint:* You may wish to consider the distances going up separately from the distances going down.)

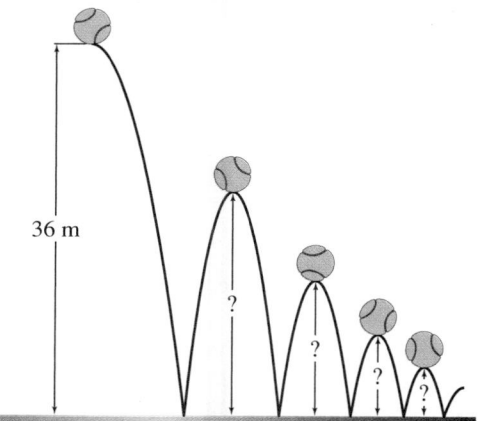

36 m

85. Vacuum Pump   With each cycle a vacuum pump removes one-third of the air in a glass vessel. What percent of the air has been removed after eight cycles?

86. Arc of a Swing   A child's swing moves through a 3-m arc. On each swing it travels only two-thirds the distance it traveled on the previous arc. How far does the swing travel before coming to rest?

87. Multiplier Effect City planners estimate that a new manufacturing plant located in their area will contribute $600,000 in salaries to the local economy. They estimate that those who earn the salaries will spend three-fourths of this money within the community. The merchants, service providers, and others who receive this $450,000 from the salary earners in turn will spend three-fourths of it in the community, and so on. Taking into account the multiplier effect, find the total amount of spending within the local economy that will be generated by the salaries from this new plant.

Group Discussion Questions

88. Challenge Question If 20 people in a room shake hands with each other exactly once, how many handshakes will take place?

89. Discovery Question The trichotomy property of real numbers states that exactly one of the following three statements must be true in each case. Determine which statement is true.

a. i. $0.333 \ldots < \dfrac{1}{3}$ ii. $0.333 \ldots = \dfrac{1}{3}$ iii. $0.333 \ldots > \dfrac{1}{3}$

b. i. $0.666 \ldots < \dfrac{2}{3}$ ii. $0.666 \ldots = \dfrac{2}{3}$ iii. $0.666 \ldots > \dfrac{2}{3}$

c. i. $0.888 \ldots < \dfrac{8}{9}$ ii. $0.888 \ldots = \dfrac{8}{9}$ iii. $0.888 \ldots > \dfrac{8}{9}$

d. What is the meaning of the three dots in $0.999 \ldots$?

e. Which of the following is true?
  i. $0.999 \ldots < 1$ ii. $0.999 \ldots = 1$ iii. $0.999 \ldots > 1$

f. Are there any real numbers between $0.999 \ldots$ and 1? If so, give an example.

g. Multiply both sides of $\dfrac{1}{3} = 0.333 \ldots$ by 3. What does this prove?

h. Is the statement, "The number one-half on the number line can be represented by $\dfrac{1}{2}, \dfrac{2}{4}$, 50%, 0.5, and $0.499 \ldots$" a true statement? If not, why is it false?

90. Risks and Choices One gimmick advertised in a newspaper was a "surefire" secret formula for becoming a millionaire. The secret discovered by those unwise enough to pay for it was as follows: "On the first day of the month save 1¢, on the second day save 2¢, on the third 4¢, etc." How many days would be required to save a total of at least $1,000,000? At this rate, what amount would be saved on the last day?

Section 11.3 Solving Systems of Linear Equations Using Augmented Matrices

Objective:

6. Use augmented matrices to solve systems of two linear equations with two variables.

In Section 3.5 we used the addition method to solve systems of two linear equations in two variables. We use this work to help us introduce another representation for systems of equations called augmented matrices. This representation is very systematic and simplifies the solution of larger systems of linear equations. This method is also well suited to solution by a computer or calculator.

A **matrix** is a rectangular array of numbers arranged into rows and columns. Each **row** consists of entries arranged horizontally. Each **column** consists of entries arranged vertically. The **dimension** of a matrix is given by stating first the number of rows and then the number of columns. The dimension of the following matrix is 2×3, read "two by three," because it has two rows and three columns.

The columns on a building are vertical pillars. The columns of a matrix consist of entries arranged vertically.

MATRIX	DIMENSION
$\begin{bmatrix} 5 & 6 & -1 \\ 4 & 3 & 1 \end{bmatrix}$	2×3 2 rows by 3 columns

The entries in an **augmented matrix** for a system of linear equations consist of the coefficients and constants in the equations. To form the augmented matrix for a system, we first align the similar variables on the left side of each equation and the constants on the right side. Then we form each row in the matrix from the coefficients and the constant of the corresponding equation. A 0 should be written in any position that corresponds to a missing variable in an equation.

SYSTEM OF LINEAR EQUATIONS	AUGMENTED MATRIX

$$\begin{cases} 4x + y = 3 \\ 3x - 2y = 16 \end{cases}$$

$$\begin{bmatrix} 4 & 1 & 3 \\ 3 & -2 & 16 \end{bmatrix}$$

Brackets enclose the matrix

Coefficients of x ——

Constants

Coefficients of y ——

Optional vertical bar to separate coefficients from the constant terms

■ **EXAMPLE 1**   Writing Augmented Matrices for Systems of Linear Equations

Write an augmented matrix for each system of linear equations.

SOLUTIONS

(a) $2x - 5y = 11$
$3x + 4y = 5$

$$\begin{bmatrix} 2 & -5 & 11 \\ 3 & 4 & 5 \end{bmatrix}$$

The x-coefficients 2 and 3 are in the first column. The y-coefficients -5 and 4 are in the second column. The constants 11 and 5 are in the third column.

(b) $2x + 3y = 29$
$5x - 4y + 8 = 0$

$2x + 3y = 29$
$5x - 4y = -8$

First write each equation with the constant on the right side of the equation. Then form the augmented matrix.

$$\begin{bmatrix} 2 & 3 & 29 \\ 5 & -4 & -8 \end{bmatrix}$$

(c) $2x - 4 = 0$
$6x + y = 4$

$2x + 0y = 4$
$6x + y = 4$

Write each equation in the form $Ax + By = C$ using zero coefficients as needed. Then form the augmented matrix.

$$\begin{bmatrix} 2 & 0 & 4 \\ 6 & 1 & 4 \end{bmatrix}$$

In order to solve systems of linear equations using augmented matrices, we need to be able to represent these systems using matrices and also be able to write the system of linear equations represented by a given augmented matrix.

■ **EXAMPLE 2**   Writing Systems of Linear Equations Represented by Augmented Matrices

Write a system of linear equations that is represented by each augmented matrix.

SOLUTIONS

(a) $\begin{bmatrix} 4 & 5 & 17 \\ 5 & -3 & 12 \end{bmatrix}$ $4x + 5y = 17$
$5x - 3y = 12$

Each row of the matrix represents an equation with an x-coefficient, a y-coefficient, and a constant on the right side of the equation.

(b) $\begin{bmatrix} 8 & 3 & | & -1 \\ 0 & 2 & | & 10 \end{bmatrix}$ $\begin{aligned} 8x + 3y &= -1 \\ 0x + 2y &= 10 \end{aligned}$ When the coefficient of a variable is 0, we do not have to write this variable in the equation. For example $0x + 2y = 10$ can be written as $2y = 10$.

(c) $\begin{bmatrix} 1 & 0 & | & 7 \\ 0 & 1 & | & -8 \end{bmatrix}$ $\begin{aligned} x + 0y &= 7 \\ 0x + y &= -8 \end{aligned}$ The last pair of equations can be written as
$x = 7$
$y = -8$
Thus the solution of this system is the ordered pair $(7, -8)$.

CALCULATOR PERSPECTIVE 11.3.1 **Entering a Matrix into a Calculator**

To enter the matrix from Example 2(a) into a TI-83 Plus calculator, enter the following keystrokes:

The [MATRX] feature is the secondary function of the x^{-1} key.

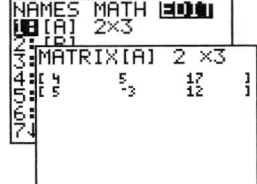

Enter the dimension of the matrix row by column and then enter each element of the matrix. Press [ENTER] after each value.

Press [2nd] [QUIT] after you have finished entering the values.

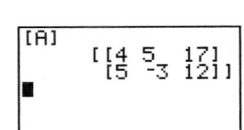

This will bring a matrix name into the home screen to use in other calculations. Pressing [ENTER] will display the actual matrix elements.

Note: This matrix has 2 rows and 3 columns so its dimension is 2×3.

The augmented matrix $\begin{bmatrix} 1 & 0 & | & 7 \\ 0 & 1 & | & -8 \end{bmatrix}$ in Example 2(c) is an excellent example of the type of matrices that we want to produce. It follows immediately from this matrix that the solution of the corresponding system of linear equations has $x = 7$ and $y = -8$. In ordered-pair notation the answer is $(7, -8)$. In general, the form $\begin{bmatrix} 1 & 0 & | & k_1 \\ 0 & 1 & | & k_2 \end{bmatrix}$ will represent a system of linear equations whose solution is the ordered pair (k_1, k_2). The material that follows will show how to produce this form.

SELF-CHECK 11.3.1

1. Write the augmented matrix for the system $\begin{cases} 5x + y = -3 \\ 3x + 2y = 8 \end{cases}$.

2. Write the system of linear equations represented by the augmented matrix $\begin{bmatrix} 4 & -3 & | & 1 \\ 2 & 9 & | & 4 \end{bmatrix}$.

3. What is the solution of the system of linear equations represented by $\begin{bmatrix} 1 & 0 & | & 5 \\ 0 & 1 & | & 6 \end{bmatrix}$?

SELF-CHECK 11.3.1 ANSWERS

1. $\begin{bmatrix} 5 & 1 & | & -3 \\ 3 & 2 & | & 8 \end{bmatrix}$

2. $\begin{cases} 4x - 3y = 1 \\ 2x + 9y = 4 \end{cases}$

3. $(5, 6)$

Equivalent systems of equations have the same solution set. The strategy for solving a system of linear equations is to transform the system into an equivalent system composed of simpler equations. In Chapter 3 we used the properties of equality to justify the transformations we used to solve systems of linear equations. We will review these transformations and give the corresponding elementary row operations on a matrix.

TRANSFORMATIONS RESULTING IN EQUIVALENT SYSTEMS	ELEMENTARY ROW OPERATIONS ON AUGMENTED MATRICES
1. Any two equations in a system may be interchanged.	**1.** Any two rows in the matrix may be interchanged.
2. Both sides of any equation in a system may be multiplied by a nonzero constant.	**2.** Any row in the matrix may be multiplied by a nonzero constant.
3. Any equation in a system may be replaced by the sum of itself and a constant multiple of another equation in the system.	**3.** Any row in the matrix may be replaced by the sum of itself and a constant multiple of another row.

The strategy that we use to solve a system of linear equations is to transform the system to an equivalent system composed of simpler equations.

We use the elementary row operations on an augmented matrix just as if the rows were the equations they represent. This is illustrated by the parallel development in Example 3.

■ **EXAMPLE 3** Solving a System Using the Addition Method and the Augmented Matrix Method

The goal we are working toward is to transform the augmented matrix to the form $\begin{bmatrix} 1 & 0 & | & k_1 \\ 0 & 1 & | & k_2 \end{bmatrix}$.

Solve the system $\begin{cases} 3x + 2y = -2 \\ x - 3y = 14 \end{cases}$.

SOLUTION _____

ADDITION METHOD

$\begin{cases} 3x + 2y = -2 \\ x - 3y = 14 \end{cases}$

↓

$\begin{cases} x - 3y = 14 \\ 3x + 2y = -2 \end{cases}$

↓

$\begin{cases} x - 3y = 14 \\ 11y = -44 \end{cases}$

↓

$\begin{cases} x - 3y = 14 \\ y = -4 \end{cases}$

↓

$\begin{cases} x = 2 \\ y = -4 \end{cases}$

Answer: (2, −4)

AUGMENTED MATRIX METHOD

$\begin{bmatrix} 3 & 2 & | & -2 \\ 1 & -3 & | & 14 \end{bmatrix}$

↓ $r_1 \leftrightarrow r_2$

$\begin{bmatrix} 1 & -3 & | & 14 \\ 3 & 2 & | & -2 \end{bmatrix}$

↓ $r_2' = r_2 - 3r_1$

$\begin{bmatrix} 1 & -3 & | & 14 \\ 0 & 11 & | & -44 \end{bmatrix}$

↓ $r_2' = \frac{1}{11}r_2$

$\begin{bmatrix} 1 & -3 & | & 14 \\ 0 & 1 & | & -4 \end{bmatrix}$

↓ $r_1' = r_1 + 3r_2$

$\begin{bmatrix} 1 & 0 & | & 2 \\ 0 & 1 & | & -4 \end{bmatrix}$

This notation means that the first row is interchanged with the second row.

Replace the second row with itself minus 3 times the first row.

Multiply the second row by $\frac{1}{11}$.

Replace the first row with itself plus 3 times the second row.
This form gives the answer.

Does this answer check?

Note that we use arrows, not equal symbols, to denote the flow from one matrix to the next. The matrices are not equal because entries in them have been changed. However, if we use the elementary row operations, they do represent equivalent systems of equations and will help us to determine the solution of the system of linear equations.

The recommended first step in transforming an augmented matrix to the reduced form $\begin{bmatrix} 1 & 0 & | & k_1 \\ 0 & 1 & | & k_2 \end{bmatrix}$ is to get a 1 to occur in the row one, column one position.

In Example 4 we illustrate three ways to accomplish this. We suggest that you master the notation that is used to describe each step. This will help you better understand this topic and also prepare you to use calculator and computer commands to perform these elementary row operations.

EXAMPLE 4   Using the Notation for Elementary Row Operations

Use the elementary row operations to transform $\begin{bmatrix} 3 & 1 & | & 2 \\ 1 & 2 & | & 9 \end{bmatrix}$ into the form

$\begin{bmatrix} 1 & \_ & | & \_ \\ \_ & \_ & | & \_ \end{bmatrix}$.

SOLUTIONS

(a) By interchanging rows one and two.

$\begin{bmatrix} 3 & 1 & | & 2 \\ 1 & 2 & | & 9 \end{bmatrix} \xrightarrow{r_1 \leftrightarrow r_2} \begin{bmatrix} 1 & 2 & | & 9 \\ 3 & 1 & | & 2 \end{bmatrix}$

The notation $r_1 \leftrightarrow r_2$ denotes that rows one and two have been interchanged.

(b) By multiplying row one by $\frac{1}{3}$.

$\begin{bmatrix} 3 & 1 & | & 2 \\ 1 & 2 & | & 9 \end{bmatrix} \xrightarrow{r_1' = \frac{r_1}{3}} \begin{bmatrix} 1 & \frac{1}{3} & | & \frac{2}{3} \\ 1 & 2 & | & 9 \end{bmatrix}$

r_1' denotes the new row one obtained by multiplying r_1 by $\frac{1}{3}$.

(c) By replacing row one with the sum of row one and -2 times row two.

$\begin{bmatrix} 3 & 1 & | & 2 \\ 1 & 2 & | & 9 \end{bmatrix} \xrightarrow{r_1' = r_1 - 2r_2} \begin{bmatrix} 1 & -3 & | & -16 \\ 1 & 2 & | & 9 \end{bmatrix}$

r_1' denotes the new row one obtained by adding row one and -2 times row two. ■

The method used in Example 4(a) may be the easiest to apply, but it only works when there is a coefficient of 1 in another row to shift to this first row. The method used in part (c) is often used to avoid the fractions that can result from the method in part (b).

SELF-CHECK 11.3.2

Perform the indicated elementary row operations to produce new matrices.

1. $\begin{bmatrix} 4 & 1 & | & 11 \\ 3 & 2 & | & 12 \end{bmatrix} \xrightarrow{r_1' = r_1 - r_2} \begin{bmatrix} \_ & \_ & | & \_ \\ 3 & 2 & | & 12 \end{bmatrix}$

2. $\begin{bmatrix} 1 & 1 & | & 9 \\ 2 & 3 & | & 23 \end{bmatrix} \xrightarrow{r_2' = r_2 - 2r_1} \begin{bmatrix} 1 & 1 & | & 9 \\ \_ & \_ & | & \_ \end{bmatrix}$

3. $\begin{bmatrix} 1 & 3 & | & 7 \\ 0 & 2 & | & 6 \end{bmatrix} \xrightarrow{r_2' = \frac{r_2}{2}} \begin{bmatrix} 1 & 3 & | & 7 \\ \_ & \_ & | & \_ \end{bmatrix}$

SELF-CHECK 11.3.2 ANSWERS

1. $\begin{bmatrix} 1 & -1 & | & -1 \\ 3 & 2 & | & 12 \end{bmatrix}$

2. $\begin{bmatrix} 1 & 1 & | & 9 \\ 0 & 1 & | & 5 \end{bmatrix}$

3. $\begin{bmatrix} 1 & 3 & | & 7 \\ 0 & 1 & | & 3 \end{bmatrix}$

In Example 5 we use the elementary row operations to solve a system of linear equations.

EXAMPLE 5 Solving a System of Linear Equations Using an Augmented Matrix

Use an augmented matrix and elementary row operations to solve $\left\{ \begin{array}{l} 2x + 3y = 4 \\ x + 4y = -3 \end{array} \right\}$.

SOLUTION _____

$\left\{ \begin{array}{l} 2x + 3y = 4 \\ x + 4y = -3 \end{array} \right\}$ $\qquad$ $\begin{bmatrix} 2 & 3 & | & 4 \\ 1 & 4 & | & -3 \end{bmatrix}$ $\qquad$ First form the augmented matrix.

$\xrightarrow{r_1 \leftrightarrow r_2}$ $\begin{bmatrix} 1 & 4 & | & -3 \\ 2 & 3 & | & 4 \end{bmatrix}$ To work toward the reduced form $\begin{bmatrix} 1 & 0 & | & k_1 \\ 0 & 1 & | & k_2 \end{bmatrix}$ interchange rows one and two to place a 1 in row one, column one.

$\xrightarrow{r_2' = r_2 - 2r_1}$ $\begin{bmatrix} 1 & 4 & | & -3 \\ 0 & -5 & | & 10 \end{bmatrix}$ Next add $-2r_1$ to r_2 to produce a 0 in row two, column one.

$\xrightarrow{r_2' = \frac{r_2}{-5}}$ $\begin{bmatrix} 1 & 4 & | & -3 \\ 0 & 1 & | & -2 \end{bmatrix}$ Then divide row two by -5 to produce a 1 in row two, column two.

$\xrightarrow{r_1' = r_1 - 4r_2}$ $\begin{bmatrix} 1 & 0 & | & 5 \\ 0 & 1 & | & -2 \end{bmatrix}$ To complete the transformation to the form $\begin{bmatrix} 1 & 0 & | & k_1 \\ 0 & 1 & | & k_2 \end{bmatrix}$ produce a 0 in row one, column two by adding $-4r_2$ to r_1.

$\begin{array}{l} x + 0y = 5 \\ 0x + y = -2 \end{array}$ Write the equivalent system of equations for the reduced form.

Answer: $(5, -2)$ $\qquad$ Does this answer check?

Note that in Example 5 we first worked on column one and then on column two. This column-by-column strategy is recommended for transforming all augmented matrices to reduced form. This strategy also is used in Example 6.

EXAMPLE 6 Solving a System of Linear Equations Using an Augmented Matrix

Use an augmented matrix and elementary row operations to solve $\left\{ \begin{array}{l} 4x + 3y = 3 \\ 2x + 9y = 4 \end{array} \right\}$.

SOLUTION _____

$\left\{ \begin{array}{l} 4x + 3y = 3 \\ 2x + 9y = 4 \end{array} \right\}$ $\qquad$ $\begin{bmatrix} 4 & 3 & | & 3 \\ 2 & 9 & | & 4 \end{bmatrix}$ First form the augmented matrix.

$\xrightarrow{r_1' = \frac{r_1}{4}}$ $\begin{bmatrix} 1 & \frac{3}{4} & | & \frac{3}{4} \\ 2 & 9 & | & 4 \end{bmatrix}$ Then work on putting column one in the form $\begin{bmatrix} 1 & - & | & - \\ 0 & - & | & - \end{bmatrix}$. $\frac{r_1}{4}$ means to multiply row one by $\frac{1}{4}$.

$$\xrightarrow{r_2' = r_2 - 2r_1} \begin{bmatrix} 1 & \dfrac{3}{4} & \Big| & \dfrac{3}{4} \\ 0 & \dfrac{15}{2} & \Big| & \dfrac{5}{2} \end{bmatrix}$$

$r_2 - 2r_1$ means to subtract twice row one from row two.

$$\xrightarrow{r_2' = \frac{2}{15}r_2} \begin{bmatrix} 1 & \dfrac{3}{4} & \Big| & \dfrac{3}{4} \\ 0 & 1 & \Big| & \dfrac{1}{3} \end{bmatrix}$$

Next work on column two to put the matrix in the reduced form $\begin{bmatrix} 1 & 0 & | & k_1 \\ 0 & 1 & | & k_2 \end{bmatrix}$.

$\dfrac{2}{15}r_2$ means to multiply row two by $\dfrac{2}{15}$.

$$\xrightarrow{r_1' = r_1 - \frac{3}{4}r_2} \begin{bmatrix} 1 & 0 & \Big| & \dfrac{1}{2} \\ 0 & 1 & \Big| & \dfrac{1}{3} \end{bmatrix}$$

$r_1 - \dfrac{3}{4}r_2$ means to subtract $\dfrac{3}{4}r_2$ from row one.

$$x + 0y = \frac{1}{2}$$
$$0x + y = \frac{1}{3}$$

Write the equivalent system of equations for the reduced form.

Does this answer check?

Answer: $\left(\dfrac{1}{2}, \dfrac{1}{3}\right)$

SELF-CHECK 11.3.3

Write the solution of a system of linear equations represented by

1. $\begin{bmatrix} 1 & 0 & | & -4 \\ 0 & 1 & | & 8 \end{bmatrix}$

2. $\begin{bmatrix} 1 & 0 & \Big| & -\dfrac{3}{5} \\ 0 & 1 & \Big| & \dfrac{2}{3} \end{bmatrix}$

Solve using augmented matrices.

3. $\begin{cases} x + 5y = -3 \\ 2x + 9y = -7 \end{cases}$

The reduced form $\begin{bmatrix} 1 & 0 & | & k_1 \\ 0 & 1 & | & k_2 \end{bmatrix}$ always yields a unique solution for a consistent system of independent equations. For an inconsistent system with no solution or a consistent system of dependent equations with an infinite number of solutions the reduced form is different. We now examine these two possibilities. Solving an inconsistent system by the addition method produces an equation that is a contradiction. Note the matrix equivalent of this in Example 7.

SELF-CHECK 11.3.3 ANSWERS
1. $(-4, 8)$
2. $\left(-\dfrac{3}{5}, \dfrac{2}{3}\right)$
3. $(-8, 1)$

■ EXAMPLE 7   Solving an Inconsistent System Using an Augmented Matrix

Use an augmented matrix and elementary row operations to solve $\begin{cases} x + 2y = 3 \\ 5x + 10y = 11 \end{cases}$.

SOLUTION

$$\begin{bmatrix} 1 & 2 & | & 3 \\ 5 & 10 & | & 11 \end{bmatrix}$$

First form the augmented matrix.

$$\xrightarrow{r_2' = r_2 - 5r_1} \begin{bmatrix} 1 & 2 & | & 3 \\ 0 & 0 & | & -4 \end{bmatrix}$$

Then work on putting column one in the form $\begin{bmatrix} 1 & - & | & - \\ 0 & - & | & - \end{bmatrix}$.

$$x + 2y = 3$$
$$0 + 0 = -4$$

There is no need to proceed further because the last row in the matrix corresponds to an equation that is a contradiction. Thus the original system of equations is inconsistent and has no solution.

Answer: No solution

We encourage you to compare Example 7 to Example 6 in Section 3.5 where the same problem was worked by the addition method. Example 8 examines a consistent system of dependent equations.

■ EXAMPLE 8 Solving a Consistent System of Dependent Equations Using an Augmented Matrix

Use an augmented matrix and elementary row operations to solve $\begin{cases} -4x + 10y = -2 \\ 6x - 15y = 3 \end{cases}$.

SOLUTION

$$\begin{cases} -4x + 10y = -2 \\ 6x - 15y = 3 \end{cases} \qquad \begin{bmatrix} -4 & 10 & | & -2 \\ 6 & -15 & | & 3 \end{bmatrix}$$

First form the augmented matrix.

$$\xrightarrow{r_1' = -\frac{1}{4}r_1} \begin{bmatrix} 1 & -\dfrac{5}{2} & | & \dfrac{1}{2} \\ 6 & -15 & | & 3 \end{bmatrix}$$

Then work on putting column one in the form $\begin{bmatrix} 1 & - & | & - \\ 0 & - & | & - \end{bmatrix}$.

$$\xrightarrow{r_2' = r_2 - 6r_1} \begin{bmatrix} 1 & -\dfrac{5}{2} & | & \dfrac{1}{2} \\ 0 & 0 & | & 0 \end{bmatrix}$$

The last row corresponds to an equation that is an identity. It cannot be used to simplify column two further.

$$x - \frac{5}{2}y = \frac{1}{2}$$

This is a dependent system with an infinite number of solutions.

$$0x + 0y = 0$$

$$x = \frac{5}{2}y + \frac{1}{2}$$

Solve the first equation for x in terms of y.

Answer: An infinite number of solutions all having the form $\left(\dfrac{5}{2}y + \dfrac{1}{2}, y\right)$.

All the solutions are points on the same line, points that can be written in this form.

The **general solution** of a dependent system of equations describes all solutions of the system and is given by indicating the relationship between the coordinates of the solutions. The **particular solutions** obtained from the general solution contain only constant coordinates.

The general solution in Example 8 is $\left(\dfrac{5}{2}y + \dfrac{1}{2}, y\right)$. By arbitrarily selecting y-values, we can produce as many particular solutions as we wish. For $y = 0$ and $y = 1$, we obtain two particular solutions $\left(\dfrac{1}{2}, 0\right)$ and $(3, 1)$.

SELF-CHECK 11.3.4 ANSWERS

1. No solution, an inconsistent system.
2. A dependent system with a general solution $(y + 4, y)$ and three particular solutions $(4, 0)$, $(5, 1)$, and $(6, 2)$.

SELF-CHECK 11.3.4

Write the solution of the system of linear equations represented by

1. $\begin{bmatrix} 1 & 0 & | & 7 \\ 0 & 0 & | & 6 \end{bmatrix}$
2. $\begin{bmatrix} 1 & -1 & | & 4 \\ 0 & 0 & | & 0 \end{bmatrix}$

USING THE LANGUAGE AND SYMBOLISM OF MATHEMATICS   11.3

1. A matrix is a _____ array of numbers.
2. The entries in an augmented matrix for a system of linear equations consist of the _____ and _____ in the equations.
3. _____ systems of equations have the same solution set.
4. The notation $r_1 \leftrightarrow r_2$ denotes that rows one and two for a matrix are _____.
5. The notation $r_1' = \dfrac{1}{2}r_1$ denotes that row _____ is being replaced by multiplying the current row _____ by _____.

6. The notation to represent that row two is being replaced by the sum of the current row two plus twice row one is _____.
7. The _____ solution of a dependent system of equations describes all solutions of the system and is given by indicating the relationship between the coordinates of the solutions.
8. The _____ solutions obtained from a general solution contain only constant coordinates.

EXERCISES   11.3

In Exercises 1–6 write an augmented matrix for each system of linear equations.

1. $\begin{cases} 3x + y = 0 \\ 2x - y = -5 \end{cases}$
2. $\begin{cases} 5x - y = 3 \\ 4x + 3y = 29 \end{cases}$
3. $\begin{cases} 4x = 12 \\ 3x + 2y = 1 \end{cases}$
4. $\begin{cases} 2x + 7y = 3 \\ -3y = 3 \end{cases}$
5. $\begin{cases} x = 5 \\ y = -6 \end{cases}$
6. $\begin{cases} x = -4 \\ y = 11 \end{cases}$

In Exercises 7–12 write a system of linear equations in x and y that is represented by each augmented matrix.

7. $\begin{bmatrix} 2 & 3 & | & 2 \\ 4 & -3 & | & 1 \end{bmatrix}$
8. $\begin{bmatrix} 5 & -1 & | & 0 \\ 3 & 2 & | & 13 \end{bmatrix}$
9. $\begin{bmatrix} 2 & 1 & | & 1 \\ 1 & 3 & | & 0 \end{bmatrix}$
10. $\begin{bmatrix} 2 & 3 & | & 5 \\ 6 & -4 & | & 2 \end{bmatrix}$
11. $\begin{bmatrix} 1 & 0 & | & 7 \\ 0 & 1 & | & -8 \end{bmatrix}$
12. $\begin{bmatrix} 1 & 0 & | & \dfrac{2}{3} \\ 0 & 1 & | & -\dfrac{4}{5} \end{bmatrix}$

In Exercises 13–20 use the given elementary row operations to complete each matrix.

13. $\begin{bmatrix} 2 & 1 & | & 1 \\ 1 & 3 & | & 0 \end{bmatrix} \xrightarrow{r_1 \leftrightarrow r_2} \begin{bmatrix} \_ & \_ & | & \_ \\ \_ & \_ & | & \_ \end{bmatrix}$

14. $\begin{bmatrix} -6 & 3 & | & 4 \\ 1 & 2 & | & 1 \end{bmatrix} \xrightarrow{r_1 \leftrightarrow r_2} \begin{bmatrix} \_ & \_ & | & \_ \\ \_ & \_ & | & \_ \end{bmatrix}$

15. $\begin{bmatrix} 1 & -3 & | & -2 \\ 2 & -5 & | & 4 \end{bmatrix} \xrightarrow{r_2' = r_2 - 2r_1} \begin{bmatrix} 1 & -3 & | & -2 \\ \_ & \_ & | & \_ \end{bmatrix}$

16. $\begin{bmatrix} 1 & -1 & | & 1 \\ 2 & -3 & | & -2 \end{bmatrix} \xrightarrow{r_2' = r_2 - 2r_1} \begin{bmatrix} 1 & -1 & | & 1 \\ \_ & \_ & | & \_ \end{bmatrix}$

17. $\begin{bmatrix} 3 & -1 & | & 3 \\ 6 & 4 & | & -6 \end{bmatrix} \xrightarrow{r_1' = \frac{1}{3}r_1} \begin{bmatrix} \_ & \_ & | & \_ \\ 6 & 4 & | & -6 \end{bmatrix}$

18. $\begin{bmatrix} 2 & 3 & | & 1 \\ 3 & -3 & | & -21 \end{bmatrix} \xrightarrow{r_1' = \frac{1}{2}r_1} \begin{bmatrix} \_ & \_ & | & \_ \\ 3 & -3 & | & -21 \end{bmatrix}$

19. $\begin{bmatrix} 1 & 5 & | & 16 \\ 0 & 1 & | & 3 \end{bmatrix} \xrightarrow{r_1' = r_1 - 5r_2} \begin{bmatrix} \_ & \_ & | & \_ \\ 0 & 1 & | & 3 \end{bmatrix}$

20. $\begin{bmatrix} 1 & -4 & | & 10 \\ 0 & 1 & | & -2 \end{bmatrix} \xrightarrow{\ r_1' = r_1 + 4r_2\ } \begin{bmatrix} \_ & \_ & | & \_ \\ 0 & 1 & | & -2 \end{bmatrix}$

In Exercises 21–26 write the solution for the system of linear equations represented by each augmented matrix. If the matrix represents a consistent system of dependent equations, write the general solution and three particular solutions.

21. $\begin{bmatrix} 1 & 0 & | & -5 \\ 0 & 1 & | & 9 \end{bmatrix}$

22. $\begin{bmatrix} 1 & 0 & | & 4 \\ 0 & 1 & | & 6 \end{bmatrix}$

23. $\begin{bmatrix} 1 & 4 & | & 3 \\ 0 & 0 & | & 7 \end{bmatrix}$

24. $\begin{bmatrix} 1 & -3 & | & 5 \\ 0 & 0 & | & -2 \end{bmatrix}$

25. $\begin{bmatrix} 1 & 3 & | & 5 \\ 0 & 0 & | & 0 \end{bmatrix}$

26. $\begin{bmatrix} 1 & -2 & | & 3 \\ 0 & 0 & | & 0 \end{bmatrix}$

In Exercises 27–34 label the elementary row operation used to transform the first matrix to the second. Use the notation developed in this section.

27. $\begin{bmatrix} 2 & 6 & | & 8 \\ 3 & 7 & | & 10 \end{bmatrix} \xrightarrow{\ r_1' = ?\ } \begin{bmatrix} 1 & 3 & | & 4 \\ 3 & 7 & | & 10 \end{bmatrix}$

28. $\begin{bmatrix} 2 & 5 & | & 9 \\ 3 & 2 & | & 8 \end{bmatrix} \xrightarrow{\ ?\ } \begin{bmatrix} 1 & \frac{5}{2} & | & \frac{9}{2} \\ 3 & 2 & | & 8 \end{bmatrix}$

29. $\begin{bmatrix} 1 & 5 & | & 8 \\ 3 & 2 & | & -2 \end{bmatrix} \xrightarrow{\ ?\ } \begin{bmatrix} 1 & 5 & | & 8 \\ 0 & -13 & | & -26 \end{bmatrix}$

30. $\begin{bmatrix} 1 & 3 & | & 8 \\ 2 & -3 & | & 7 \end{bmatrix} \xrightarrow{\ ?\ } \begin{bmatrix} 1 & 3 & | & 8 \\ 0 & -9 & | & -9 \end{bmatrix}$

31. $\begin{bmatrix} 1 & -2 & | & 8 \\ 0 & 3 & | & -9 \end{bmatrix} \xrightarrow{\ ?\ } \begin{bmatrix} 1 & -2 & | & 8 \\ 0 & 1 & | & -3 \end{bmatrix}$

32. $\begin{bmatrix} 1 & 6 & | & 2 \\ 0 & 3 & | & 1 \end{bmatrix} \xrightarrow{\ ?\ } \begin{bmatrix} 1 & 6 & | & 2 \\ 0 & 1 & | & \frac{1}{3} \end{bmatrix}$

33. $\begin{bmatrix} 1 & 2 & | & 6 \\ 0 & 1 & | & 1 \end{bmatrix} \xrightarrow{\ ?\ } \begin{bmatrix} 1 & 0 & | & 4 \\ 0 & 1 & | & 1 \end{bmatrix}$

34. $\begin{bmatrix} 1 & 3 & | & 2 \\ 0 & 1 & | & \frac{1}{3} \end{bmatrix} \xrightarrow{\ ?\ } \begin{bmatrix} 1 & 0 & | & 1 \\ 0 & 1 & | & \frac{1}{3} \end{bmatrix}$

In Exercises 35–50 use an augmented matrix and elementary row operations to solve each system of linear equations.

35. $x + 3y = 5$
$2x + y = -5$

36. $x - 2y = 9$
$3x + 4y = 7$

37. $x + 3y = 1$
$3x + 7y = 7$

38. $x + 2y = 7$
$4x + 3y = 3$

39. $2x + 5y = -4$
$4x + 3y = 6$

40. $2x - 3y = 17$
$4x + y = 13$

41. $4x - 9y = 5$
$3x + 12y = 10$

42. $8x + 3y = -39$
$7x - 2y = -11$

43. $3x - y = 2$
$2x + y = 6$

44. $3x + 4y = 0$
$2x - 3y = 16$

45. $6x + 4y = 11$
$10x + 6y = 17$

46. $5x - y = 7$
$2x + 4y = 5$

47. $3x + 4y = 7$
$6x + 8y = 10$

48. $4x + 3y = 2$
$16x + 12y = 7$

49. $2x - y = 5$
$4x - 2y = 10$

50. $3x - 6y = 12$
$4x - 8y = 16$

In Exercises 51–60 write a system of linear equations using the variables x and y and use this system to solve the problem.

51. **Numeric Word Problem**   Find two numbers whose sum is 160 and whose difference is 4.

52. **Numeric Word Problem**   Find two numbers whose sum is 260 if one number is 3 times the other number.

53. **Complementary Angles**   The two angles shown are complementary, and one angle is 32° larger than the other. Determine the number of degrees in each angle.

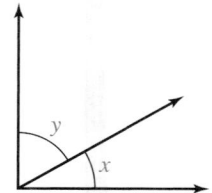

54. **Supplementary Angles**   The two angles shown are supplementary, and one angle is 74° larger than the other. Determine the number of degrees in each angle.

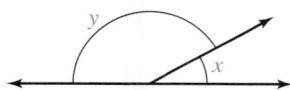

55. **Fixed and Variable Costs**   A seamstress makes custom costumes for operas. One month the total of fixed and variable costs for making 20 costumes was $3200. The next month the total of the fixed and variable costs for making 30 costumes was $4300. Determine the fixed cost and the variable cost per costume.

56. **Rates of Two Bicyclists**   Two bicyclists depart at the same time from a common location traveling in opposite directions. One averages 5 km/h more than the other. After 2 h, they are 130 km apart. Determine the speed of each bicyclist.

57. **Rate of a River Current**   A small boat can go 30 km downstream in an hour, but only 14 km upstream in an hour. Determine the rate of the boat and the rate of the current.

58. **Mixture of Disinfectant**   A hospital needs 100 liters of a 15% solution of disinfectant. How many liters of a 40% solution and a 5% solution should be mixed to obtain this 15% solution?

59. **Mixture of Fruit Drinks**   A fruit concentrate is 15% water. How many liters of pure water and how many liters of concentrate should be mixed to produce 100 liters of mixture that is 83% water?

60. **Basketball Scoring**   During one game for the Los Angeles Lakers, Kobe Bryant scored 39 points on 17 field goals. How many of these field goals were 2-pointers and how many were 3-pointers?

Group Discussion Questions

61. Challenge Question   Solve $\begin{cases} a_1x + b_1y = c_1 \\ a_2x + b_2y = c_2 \end{cases}$ for (x, y)

in terms of a_1, a_2, b_1, b_2, c_1, and c_2. Assume that
$a_1b_2 - a_2b_1 \neq 0$.

62. Discovery Question

  a. Extend the augmented matrix notation given for
systems of two linear equations with two variables to
write an augmented matrix for this system.

$$\begin{cases} x + 2y + 3z = 5 \\ 2x + y - z = -1 \\ 3x + y + z = 4 \end{cases}$$

  b. Write a system of linear equations that is represented
by this augmented matrix. Use the variables x, y, and z.

$$\left[\begin{array}{ccc|c} 2 & -3 & 3 & 2 \\ 4 & 0 & 2 & -1 \\ -2 & 4 & -3 & -2 \end{array}\right]$$

  c. Write a system of linear equations that is represented by
this augmented matrix. Use the variables w, x, y, and z.

$$\left[\begin{array}{cccc|c} 1 & -1 & 3 & -4 & 1 \\ -2 & 4 & -1 & 1 & -4 \\ 3 & 2 & -4 & 5 & 7 \\ 2 & -1 & -1 & 4 & 7 \end{array}\right]$$

Section 11.4   Systems of Linear Equations in Three Variables

Objective: **7.** Solve a system of three linear equations in three variables.

A first-degree equation in two variables of the form $Ax + By = C$ is called a linear equation because its graph is a straight line if A and B are not both 0. Similarly, a first-degree equation in three variables of the form $Ax + By + Cz = D$ also is called a linear equation. However, this name is misleading because if A, B, and C are not all 0, the graph of $Ax + By + Cz = D$ is not a line but a plane in three-dimensional space.

The graph of a three-dimensional space on two-dimensional paper is limited in its portrayal of the third dimension. Nonetheless, we can give the viewer a feeling for planes in a three-dimensional space by orienting the x-, y-, and z-axes as shown in the figure. This graph illustrates the plane $2x + 3y + 4z = 12$, whose x-intercept is $(6, 0, 0)$, whose y-intercept is $(0, 4, 0)$, and whose z-intercept is $(0, 0, 3)$. Drawing lines to connect these intercepts gives the view of the plane in the region where all coordinates are positive.

A system of 3 linear equations in 3 variables is referred to as a 3 × 3 (three-by-three) system. A **solution** of a system of equations with three variables is an ordered triple that is a solution of each equation in the system.

The plane defined by
$2x + 3y + 4z = 12$

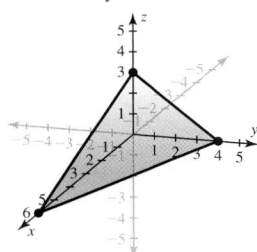

■ EXAMPLE 1   Determining Whether an Ordered Triple Is a Solution of a System of Linear Equations

Determine whether $(2, -3, 5)$ is a solution of $\begin{cases} x + y - z = -6 \\ 2x - y + z = 12 \\ 3x - 2y - 2z = 3 \end{cases}$.

SOLUTIONS

FIRST EQUATION

$$x + y - z = -6$$
$$2 + (-3) - 5 \overset{?}{=} -6$$
$$-6 \overset{?}{=} -6 \text{ is true.}$$

SECOND EQUATION

$$2x - y + z = 12$$
$$2(2) - (-3) + 5 \overset{?}{=} 12$$
$$4 + 3 + 5 \overset{?}{=} 12$$
$$12 \overset{?}{=} 12 \text{ is true.}$$

THIRD EQUATION

$$3x - 2y - 2z = 3$$
$$3(2) - 2(-3) - 2(5) \overset{?}{=} 3$$
$$6 + 6 - 10 \overset{?}{=} 3$$
$$2 \overset{?}{=} 3 \text{ is false.}$$

Answer: $(2, -3, 5)$ is not a solution of this system.   To be a solution of this system, the point must satisfy all three equations. ■

Determine whether $(2, -3.25, 4.75)$ is a solution of the system in Example 1.

The graph of each linear equation $Ax + By + Cz = D$ is a plane in three-dimensional space unless A, B, and C are 0.

A system of three linear equations in three variables can be viewed geometrically as the intersection of a set of three planes. These planes may intersect in one point, no points, or an infinite number of points. The illustrations in the following box show some of the ways we can obtain these solutions. Can you sketch other ways of obtaining these solution sets?

Types of Solution Sets for Linear Systems with Three Equations

The linear system $\begin{cases} A_1x + B_1y + C_1z = D_1 \\ A_2x + B_2y + C_2z = D_2 \\ A_3x + B_3y + C_3z = D_3 \end{cases}$ can have:

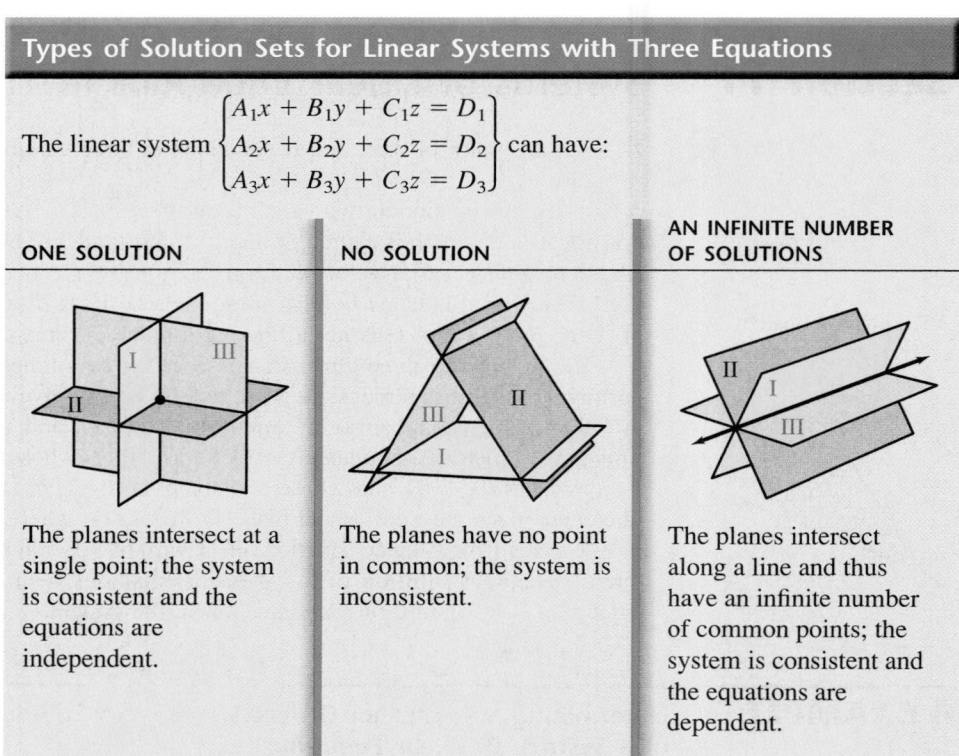

ONE SOLUTION	NO SOLUTION	AN INFINITE NUMBER OF SOLUTIONS
The planes intersect at a single point; the system is consistent and the equations are independent.	The planes have no point in common; the system is inconsistent.	The planes intersect along a line and thus have an infinite number of common points; the system is consistent and the equations are dependent.

The figures in this box can give us an intuitive understanding of the possible solutions to these systems; however, it is not practical to actually solve these systems graphically. Thus we rely entirely on algebraic methods. In this section we use the augmented matrix method to solve systems of three linear equations in three variables.

In Example 2, the entries in the first column of the matrix are the coefficients of x, the entries in the second column are the coefficients of y, the entries in the third column are the coefficients of z, and the entries in the fourth column are the constants.

■ EXAMPLE 2 Writing Augmented Matrices for Systems of Linear Equations

Write an augmented matrix for each system of linear equations.

SOLUTIONS

(a) $\begin{cases} x + y - z = -6 \\ 2x - y + z = 12 \\ 3x - 2y - 2z = 3 \end{cases}$ $\begin{bmatrix} 1 & 1 & -1 & | & -6 \\ 2 & -1 & 1 & | & 12 \\ 3 & -2 & -2 & | & 3 \end{bmatrix}$ The augmented matrix contains the coefficients and the constants from each equation.

(b) $\begin{cases} 2x + y = 7 \\ y - z = 2 \\ x + z = 2 \end{cases}$ $\begin{bmatrix} 2 & 1 & 0 & | & 7 \\ 0 & 1 & -1 & | & 2 \\ 1 & 0 & 1 & | & 2 \end{bmatrix}$ Use zero coefficients as needed for any missing terms.

■

SELF-CHECK 11.4.2

1. Form the augmented matrix for this system of linear equations:
$$\begin{cases} x + 2y + z = -2 \\ 2x + 4y - 3z = 0 \\ -x + 2y = -3 \end{cases}$$

2. Write the system of equations represented by the augmented matrix:
$$\begin{bmatrix} 1 & -1 & 0 & | & 2 \\ 2 & 3 & -1 & | & 4 \\ 0 & 3 & -4 & | & 2 \end{bmatrix}$$

The reduced form for the augmented matrix associated with a consistent system of three independent linear equations with three variables is $\begin{bmatrix} 1 & 0 & 0 & | & k_1 \\ 0 & 1 & 0 & | & k_2 \\ 0 & 0 & 1 & | & k_3 \end{bmatrix}$. The solution of this system is (k_1, k_2, k_3). Similar forms can be obtained for dependent or inconsistent systems. The reduced form defined here is also called **reduced echelon form.**

Properties of the Reduced Row-Echelon Form of a Matrix

1. The first nonzero entry in a row is a 1. All other entries in the column containing the leading 1 are 0s.
2. All nonzero rows are above any rows containing only 0s.
3. The first nonzero entry in a row is to the left of the first nonzero entry in the following row.

■ **EXAMPLE 3** Using Elementary Row Operations

Use the elementary row operations to transform the matrix $\begin{bmatrix} 5 & 6 & -4 & | & -8 \\ 2 & 4 & -1 & | & 1 \\ 1 & 1 & -3 & | & 0 \end{bmatrix}$ into the

form $\begin{bmatrix} 1 & \_ & \_ & | & \_ \\ 0 & \_ & \_ & | & \_ \\ 0 & \_ & \_ & | & \_ \end{bmatrix}$.

SOLUTION _____

$$\begin{bmatrix} 5 & 6 & -4 & | & -8 \\ 2 & 4 & -1 & | & 1 \\ 1 & 1 & -3 & | & 0 \end{bmatrix} \xrightarrow{r_1 \leftrightarrow r_3} \begin{bmatrix} 1 & 1 & -3 & | & 0 \\ 2 & 4 & -1 & | & 1 \\ 5 & 6 & -4 & | & -8 \end{bmatrix}$$

Place a 1 in the upper-left position by interchanging the first and third rows.

$$\xrightarrow{r_2' = r_2 - 2r_1} \begin{bmatrix} 1 & 1 & -3 & | & 0 \\ 0 & 2 & 5 & | & 1 \\ 5 & 6 & -4 & | & -8 \end{bmatrix}$$

Introduce 0s into column one, rows two and three.

$$\xrightarrow{r_3' = r_3 - 5r_1} \begin{bmatrix} 1 & 1 & -3 & | & 0 \\ 0 & 2 & 5 & | & 1 \\ 0 & 1 & 11 & | & -8 \end{bmatrix}$$ ∎

The last two steps in Example 3 can be combined. This is illustrated in Example 4.

■ EXAMPLE 4   Solving a System of Linear Equations Using an Augmented Matrix

Solve $\begin{cases} 2a + 3b - 2c = -8 \\ a - b + 2c = 25 \\ 4a + 6b - c = -7 \end{cases}$.

SOLUTION _____

The first step is to form the augmented matrix:

$$\begin{bmatrix} 2 & 3 & -2 & | & -8 \\ 1 & -1 & 2 & | & 25 \\ 4 & 6 & -1 & | & -7 \end{bmatrix} \xrightarrow{r_1 \leftrightarrow r_2} \begin{bmatrix} 1 & -1 & 2 & | & 25 \\ 2 & 3 & -2 & | & -8 \\ 4 & 6 & -1 & | & -7 \end{bmatrix}$$

Transform the first column into the form
$$\begin{bmatrix} 1 & — & — & | & — \\ 0 & — & — & | & — \\ 0 & — & — & | & — \end{bmatrix}.$$

$$\xrightarrow[r_3' = r_3 - 4r_1]{r_2' = r_2 - 2r_1} \begin{bmatrix} 1 & -1 & 2 & | & 25 \\ 0 & 5 & -6 & | & -58 \\ 0 & 10 & -9 & | & -107 \end{bmatrix}$$

$$\xrightarrow{r_2' = \frac{1}{5}r_2} \begin{bmatrix} 1 & -1 & 2 & | & 25 \\ 0 & 1 & -\frac{6}{5} & | & -\frac{58}{5} \\ 0 & 10 & -9 & | & -107 \end{bmatrix}$$

Transform the second column into the form
$$\begin{bmatrix} 1 & 0 & — & | & — \\ 0 & 1 & — & | & — \\ 0 & 0 & — & | & — \end{bmatrix}.$$

$$\xrightarrow[r_3' = r_3 - 10r_2]{r_1' = r_1 + r_2} \begin{bmatrix} 1 & 0 & \frac{4}{5} & | & \frac{67}{5} \\ 0 & 1 & -\frac{6}{5} & | & -\frac{58}{5} \\ 0 & 0 & 3 & | & 9 \end{bmatrix}$$

$$\xrightarrow{r_3' = \frac{1}{3}r_3} \begin{bmatrix} 1 & 0 & \frac{4}{5} & | & \frac{67}{5} \\ 0 & 1 & -\frac{6}{5} & | & -\frac{58}{5} \\ 0 & 0 & 1 & | & 3 \end{bmatrix}$$

Transform the third column into the form
$$\begin{bmatrix} 1 & 0 & 0 & | & — \\ 0 & 1 & 0 & | & — \\ 0 & 0 & 1 & | & — \end{bmatrix}.$$

$$\xrightarrow[\;\; r_2{}' = r_2 + \frac{6}{5}r_3 \;\;]{r_1{}' = r_1 - \frac{4}{5}r_3} \begin{bmatrix} 1 & 0 & 0 & 11 \\ 0 & 1 & 0 & -8 \\ 0 & 0 & 1 & 3 \end{bmatrix}$$

The answer is displayed in the last column of the reduced form.

Answer: $(11, -8, 3)$ Does this answer check?

The overall strategy used in Example 4 can be summarized as "work from left to right and produce the leading 1s before you produce the 0s." This strategy is formalized in the following box.

Transforming an Augmented Matrix into Reduced Echelon Form

STEP 1 $\begin{bmatrix} 1 & \cdots \\ 0 & \\ 0 & \\ \vdots & \vdots \\ 0 & \cdots \end{bmatrix}$ **Transform the first column** into this form by using the elementary row operations to
a. Produce a 1 in the top position.
b. Use the 1 in row one to produce 0s in the other positions of column one.

STEP 2 $\begin{bmatrix} 1 & 0 & \cdots \\ 0 & 1 & \\ 0 & 0 & \\ \vdots & & \vdots \\ 0 & 0 & \cdots \end{bmatrix}$ **Transform the next column,** if possible, into this form by using the elementary operations to
a. Produce a 1 in the next row.
b. Use the 1 in this row to produce 0s in the other positions of this column.
If it is not possible to produce a 1 in the next row, proceed to the next column.

STEP 3   Repeat step 2 column by column, always producing the 1 in the next row, until you arrive at the reduced form.

For emphasis, remember your goal is to produce the leading 1s before you produce the 0s. You may use shortcuts in this process whenever they are appropriate. This procedure is also appropriate for dependent and inconsistent systems.

SELF-CHECK 11.4.3

Produce a 1 in row two, column two of the matrix $\begin{bmatrix} 1 & 1 & -3 & 0 \\ 0 & 2 & 5 & 1 \\ 0 & 1 & 11 & -8 \end{bmatrix}$ by interchanging the second and third rows. Then introduce two 0s into the second column of this matrix.

SELF-CHECK 11.4.3 ANSWER

$$\begin{bmatrix} 1 & 1 & -3 & 0 \\ 0 & 2 & 5 & 1 \\ 0 & 1 & 11 & -8 \end{bmatrix} \xrightarrow{\; r_2 \leftrightarrow r_3 \;} \begin{bmatrix} 1 & 1 & -3 & 0 \\ 0 & 1 & 11 & -8 \\ 0 & 2 & 5 & 1 \end{bmatrix} \xrightarrow[\; r_3{}' = r_3 - 2r_2 \;]{r_1{}' = r_1 - r_2} \begin{bmatrix} 1 & 0 & -14 & 8 \\ 0 & 1 & 11 & -8 \\ 0 & 0 & -17 & 17 \end{bmatrix}$$

■ **EXAMPLE 5** Solving an Inconsistent System Using an Augmented Matrix

Solve $\begin{cases} r + 8s + 2t = 20 \\ 11s + t = 28 \\ -22s - 2t = -55 \end{cases}$.

SOLUTION

$$\begin{bmatrix} 1 & 8 & 2 & | & 20 \\ 0 & 11 & 1 & | & 28 \\ 0 & -22 & -2 & | & -55 \end{bmatrix} \xrightarrow{r_3' = r_3 + 2r_2} \begin{bmatrix} 1 & 8 & 2 & | & 20 \\ 0 & 11 & 1 & | & 28 \\ 0 & 0 & 0 & | & 1 \end{bmatrix}$$

Although this matrix is not in reduced form, the last row indicates that the system is inconsistent with no solution. The last row represents the equation $0r + 0s + 0t = 1$, which is a contradiction.

Answer: No solution. ■

Example 6 produces the general solution for a consistent system of dependent linear equations.

■ **EXAMPLE 6** Solving a Consistent System of Dependent Equations Using an Augmented Matrix

Solve $\begin{cases} x_1 + 4x_2 + 5x_3 = -2 \\ x_2 + 2x_3 = -1 \\ -5x_2 - 10x_3 = 5 \end{cases}$.

SOLUTION

$$\begin{bmatrix} 1 & 4 & 5 & | & -2 \\ 0 & 1 & 2 & | & -1 \\ 0 & -5 & -10 & | & 5 \end{bmatrix} \xrightarrow[r_3' = r_3 + 5r_2]{r_1' = r_1 - 4r_2} \begin{bmatrix} 1 & 0 & -3 & | & 2 \\ 0 & 1 & 2 & | & -1 \\ 0 & 0 & 0 & | & 0 \end{bmatrix}$$

The new matrix is in reduced form. A row of 0s in an $n \times n$ consistent system indicates a dependent system with infinitely many solutions.

$$\longrightarrow \begin{cases} x_1 + 0x_2 - 3x_3 = 2 \\ 0x_1 + x_2 + 2x_3 = -1 \\ 0x_1 + 0x_2 + 0x_3 = 0 \end{cases}$$

This is the system represented by the reduced matrix. The system is dependent, since the last equation is an identity satisfied by all values of $x_1, x_2,$ and x_3.

Thus

$$x_1 = 3x_3 + 2$$
$$x_2 = -2x_3 - 1$$

The general solution is acquired by solving the first two equations for x_1 and x_2 in terms of x_3.

Answer: The general solution is $(3x_3 + 2, -2x_3 - 1, x_3)$; three particular solutions are $(8, -5, 2), (2, -1, 0),$ and $(-7, 5, -3)$.

These particular solutions were found by letting x_3 be 2, 0, and -3, respectively. ■

Let A be the augmented matrix of an $n \times n$ system of equations.
1. If the reduced form of A has a row of the form $[0\ 0\ 0\ \ldots\ 0\ k]$, where $k \neq 0$,
   then the system is inconsistent and has no solution.
2. If the system is consistent and the reduced form of A has a row of the form
   $[0\ 0\ 0\ \ldots\ 0\ 0]$ (all zeros), then the equations in the system are dependent and
   the system has infinitely many solutions.

SELF-CHECK 11.4.4

Write the solution for each system of linear equations represented by the
following reduced augmented matrices.

1. $\begin{bmatrix} 1 & 0 & 0 & | & -1 \\ 0 & 1 & 0 & | & 1 \\ 0 & 0 & 1 & | & 3 \end{bmatrix}$
2. $\begin{bmatrix} 1 & 0 & 1 & | & 5 \\ 0 & 1 & -1 & | & 2 \\ 0 & 0 & 0 & | & 0 \end{bmatrix}$
3. $\begin{bmatrix} 1 & 0 & 0 & | & -2 \\ 0 & 1 & 2 & | & 1 \\ 0 & 0 & 0 & | & 1 \end{bmatrix}$

SELF-CHECK 11.4.4 ANSWERS

1. $(-1, 1, 3)$
2. General solution
   $(5 - z, z + 2, z)$
3. No solution

The matrix method is also a convenient method for solving systems of linear equations that do not have the same number of variables as equations.

■ EXAMPLE 7 Solving a System of Equations with More Variables Than Equations

Find a general solution and three particular solutions for the 2×3
system $\begin{cases} 2x - 3y + 17z = 12 \\ 8x + 2y - 2z = 20 \end{cases}$.

SOLUTION

$\begin{bmatrix} 2 & -3 & 17 & | & 12 \\ 8 & 2 & -2 & | & 20 \end{bmatrix} \xrightarrow{r_1' = \frac{1}{2}r_1} \begin{bmatrix} 1 & -\frac{3}{2} & \frac{17}{2} & | & 6 \\ 8 & 2 & -2 & | & 20 \end{bmatrix}$

Use the elementary row operations to transform the matrix to reduced row-echelon form.

$\xrightarrow{r_2' = r_2 - 8r_1} \begin{bmatrix} 1 & -\frac{3}{2} & \frac{17}{2} & | & 6 \\ 0 & 14 & -70 & | & -28 \end{bmatrix}$

$\xrightarrow{r_2' = \frac{1}{14}r_2} \begin{bmatrix} 1 & -\frac{3}{2} & \frac{17}{2} & | & 6 \\ 0 & 1 & -5 & | & -2 \end{bmatrix}$

$\xrightarrow{r_1' = r_1 + \frac{3}{2}r_2} \begin{bmatrix} 1 & 0 & 1 & | & 3 \\ 0 & 1 & -5 & | & -2 \end{bmatrix}$

$$\begin{cases} x + z = 3 \\ y - 5z = -2 \end{cases}$$

This is the system represented by the reduced matrix.

$$x = 3 - z$$
$$y = 5z - 2$$

The general solution is obtained by solving these equations for x and y in terms of z.

Answer: The general solution is $(3 - z, 5z - 2, z)$; particular solutions are $(3, -2, 0)$, $(4, -7, -1)$, and $(1, 8, 2)$.

The particular solutions were found by letting z be 0, -1, and 2, respectively.

CALCULATOR PERSPECTIVE 11.4.1	Using rref (Reduced Row-Echelon Form) to Solve a System of Linear Equations

To solve the system in Example 4 on a TI-83 Plus calculator, enter the following keystrokes:

Note: The fourth column of this matrix has been entered but is not visible on this display window. The arrow keys can be used to scroll through all of the entries.

[2nd] [MATRX] [▶] [▶] [ENTER]

Then enter the dimension of the matrix and the elements of the matrix.

[2nd] [QUIT]

[2nd] [MATRX] [▶] [▲] (5 times to choice B:)

[ENTER] [2nd] [MATRX] [1] [)] [ENTER]

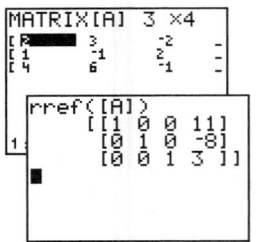

Note: The rref feature skips all the intermediate steps and proceeds immediately to the reduced row-echelon form. The solution to the system is $(11, -8, 3)$.

The graphical method and tables of values can be used to solve some systems of linear equations with two variables. Example 8 is a 4×4 system with four equations and four variables. Graphs and tables are not appropriate tools for these larger systems because it is difficult to graphically depict more than three dimensions. For these larger systems, augmented matrices are often used with calculators or computers.

■ EXAMPLE 8 Solving a System of Four Linear Equations Using an Augmented Matrix

Use an augmented matrix and a calculator to solve $\begin{cases} w + x + 2y - z = 2 \\ 2w + 4x - 6y + z = -4 \\ w - 4x + 5y - 3z = 11 \\ 3w + 3x - y - 2z = -5 \end{cases}$.

SOLUTION _____

$$\begin{bmatrix} 1 & 1 & 2 & -1 & | & 2 \\ 2 & 4 & -6 & 1 & | & -4 \\ 1 & -4 & 5 & -3 & | & 11 \\ 3 & 3 & -1 & -2 & | & -5 \end{bmatrix}$$

First form the augmented matrix. Note that it is a 4×5 matrix. Enter this as matrix [A] on a calculator.

$$\longrightarrow \begin{bmatrix} 1 & 0 & 0 & 0 & | & 10 \\ 0 & 1 & 0 & 0 & | & -4 \\ 0 & 0 & 1 & 0 & | & 3 \\ 0 & 0 & 0 & 1 & | & 10 \end{bmatrix}$$

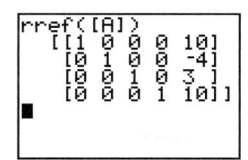

Then use the rref feature to transform this matrix to reduced row-echelon form.

Obtain the answer from this reduced form.

Answer: $(10, -4, 3, 10)$

Does this answer check? ■

USING THE LANGUAGE AND SYMBOLISM OF MATHEMATICS   11.4

1. The graph of $Ax + By + Cz = D$, if A, B, and C are not all 0, is a _____ in three-dimensional space.

2. A _____ of a system of linear equations with three variables is an ordered triple, which satisfies each equation in the system.

3. A 3×3 system of linear equations contains _____ linear equations with _____ variables.

4. The reduced form for an augmented matrix is also called reduced _____ form.

5. In reduced row-echelon form, a row of the form $[0\ 0\ 0 \ldots 0\ k]$ where $k \neq 0$ represents an _____ system of linear equations.

EXERCISES   11.4

In Exercises 1–4 write an augmented matrix for each system of linear equations.

1. $\begin{cases} x + y + z = 2 \\ -x + y - 2z = 1 \\ x + y - z = 0 \end{cases}$

2. $\begin{cases} x - 3y + 3z = 10 \\ x + y - 4z = -17 \\ -x + y + z = 6 \end{cases}$

3. $\begin{cases} x + 2y - z = 19 \\ 2x - y = -1 \\ 3x - 2y + 4z = -32 \end{cases}$

4. $\begin{cases} 6x - 3y = 2 \\ 3y + 2z = 1 \\ 2x - z = 5 \end{cases}$

In Exercises 5–10 write a system of linear equations that is represented by each augmented matrix.

5. $\begin{bmatrix} 1 & -2 & 5 & | & 0 \\ 2 & 4 & -3 & | & 8 \\ 3 & 5 & 7 & | & 11 \end{bmatrix}$

6. $\begin{bmatrix} 1 & 2 & 3 & | & 7 \\ 2 & -1 & 2 & | & 5 \\ 1 & -3 & -2 & | & -13 \end{bmatrix}$

7. $\begin{bmatrix} 1 & 0 & 2 & | & 5 \\ 1 & 1 & 0 & | & 0 \\ 0 & 1 & 1 & | & 4 \end{bmatrix}$

8. $\begin{bmatrix} 1 & 1 & 0 & | & 0 \\ 1 & 0 & 2 & | & 5 \\ 0 & 1 & 1 & | & 4 \end{bmatrix}$

9. $\begin{bmatrix} 1 & 0 & 0 & | & 7 \\ 0 & 1 & 0 & | & -5 \\ 0 & 0 & 1 & | & 8 \end{bmatrix}$

10. $\begin{bmatrix} 1 & 0 & 0 & | & 2 \\ 0 & 1 & 0 & | & 6 \\ 0 & 0 & 1 & | & -3 \end{bmatrix}$

In Exercises 11–22 use the given elementary row operations to complete each matrix.

11. $\begin{bmatrix} 2 & 1 & -2 & | & -11 \\ 1 & 2 & 3 & | & 16 \\ 3 & 2 & 1 & | & 3 \end{bmatrix} \xrightarrow{r_1 \leftrightarrow r_2} \begin{bmatrix} \_ & \_ & \_ & | & \_ \\ \_ & \_ & \_ & | & \_ \\ 3 & 2 & 1 & | & 3 \end{bmatrix}$

12. $\begin{bmatrix} 2 & 3 & 4 & | & 5 \\ 1 & -1 & 3 & | & 6 \\ 3 & -2 & 2 & | & 10 \end{bmatrix} \xrightarrow{r_1 \leftrightarrow r_2} \begin{bmatrix} \_ & \_ & \_ & | & \_ \\ \_ & \_ & \_ & | & \_ \\ 3 & -2 & 2 & | & 10 \end{bmatrix}$

13. $\begin{bmatrix} 2 & 4 & 8 & | & 6 \\ 3 & 5 & 7 & | & 1 \\ 4 & 9 & 2 & | & 8 \end{bmatrix} \xrightarrow{r_1' = \frac{1}{2}r_1} \begin{bmatrix} \_ & \_ & \_ & | & \_ \\ 3 & 5 & 7 & | & 1 \\ 4 & 9 & 2 & | & 8 \end{bmatrix}$

14. $\begin{bmatrix} 1 & 2 & 5 & | & 11 \\ 0 & 5 & 6 & | & 4 \\ 0 & 8 & 2 & | & 5 \end{bmatrix} \xrightarrow{r_2' = \frac{1}{5}r_2} \begin{bmatrix} 1 & 2 & 5 & | & 11 \\ \_ & \_ & \_ & | & \_ \\ 0 & 8 & 2 & | & 5 \end{bmatrix}$

15. $\begin{bmatrix} 1 & 3 & 5 & | & 11 \\ 2 & 7 & 9 & | & 13 \\ 4 & 8 & 3 & | & 7 \end{bmatrix} \xrightarrow{r_2' = r_2 - 2r_1} \begin{bmatrix} 1 & 3 & 5 & | & 11 \\ \_ & \_ & \_ & | & \_ \\ 4 & 8 & 3 & | & 7 \end{bmatrix}$

16. $\begin{bmatrix} 1 & 3 & 5 & | & 11 \\ 2 & 7 & 9 & | & 13 \\ 4 & 8 & 3 & | & 7 \end{bmatrix} \xrightarrow{r_3' = r_3 - 4r_1} \begin{bmatrix} 1 & 3 & 5 & | & 11 \\ 2 & 7 & 9 & | & 13 \\ \_ & \_ & \_ & | & \_ \end{bmatrix}$

17. $\begin{bmatrix} 1 & 2 & 2 & | & 3 \\ 2 & 1 & 3 & | & -1 \\ 3 & -1 & 4 & | & -8 \end{bmatrix} \xrightarrow[r_3' = r_3 - 3r_1]{r_2' = r_2 - 2r_1} \begin{bmatrix} 1 & 2 & 2 & | & 3 \\ 0 & \_ & \_ & | & \_ \\ 0 & \_ & \_ & | & \_ \end{bmatrix}$

18. $\begin{bmatrix} 1 & 3 & 5 & | & 51 \\ -3 & 4 & 6 & | & 70 \\ 4 & -1 & 2 & | & 11 \end{bmatrix} \xrightarrow[r_3' = r_3 - 4r_1]{r_2' = r_2 + 3r_1} \begin{bmatrix} 1 & 3 & 5 & | & 51 \\ 0 & \_ & \_ & | & \_ \\ 0 & \_ & \_ & | & \_ \end{bmatrix}$

19. $\begin{bmatrix} 1 & 3 & -2 & | & 19 \\ 0 & 2 & 4 & | & -10 \\ 0 & 4 & 1 & | & 8 \end{bmatrix} \xrightarrow{r_2' = \frac{1}{2}r_2} \begin{bmatrix} 1 & 3 & -2 & | & 19 \\ \_ & \_ & \_ & | & \_ \\ 0 & 4 & 1 & | & 8 \end{bmatrix}$

20. $\begin{bmatrix} 1 & 0 & 2 & | & -1 \\ 0 & 1 & 3 & | & -8 \\ 0 & 0 & 4 & | & -12 \end{bmatrix} \xrightarrow{r_3' = \frac{1}{4}r_3} \begin{bmatrix} 1 & 0 & 2 & | & -1 \\ 0 & 1 & 3 & | & -8 \\ \_ & \_ & \_ & | & \_ \end{bmatrix}$

21. $\begin{bmatrix} 1 & 2 & 4 & | & 9 \\ 0 & 1 & 2 & | & 3 \\ 0 & -3 & 5 & | & 13 \end{bmatrix} \xrightarrow[r_3' = r_3 + 3r_2]{r_1' = r_1 - 2r_2} \begin{bmatrix} 1 & 0 & \_ & | & \_ \\ 0 & 1 & 2 & | & 3 \\ 0 & 0 & \_ & | & \_ \end{bmatrix}$

22. $\begin{bmatrix} 1 & 0 & -5 & | & -2 \\ 0 & 1 & 3 & | & 0 \\ 0 & 0 & 1 & | & 1 \end{bmatrix} \xrightarrow[r_2' = r_2 - 3r_3]{r_1' = r_1 + 5r_3} \begin{bmatrix} 1 & 0 & \_ & | & \_ \\ 0 & 1 & \_ & | & \_ \\ 0 & 0 & 1 & | & 1 \end{bmatrix}$

In Exercises 23–28 write the solution for the system of linear equations represented by each augmented matrix. If the matrix represents a dependent system, write the general solution and three particular solutions.

23. $\begin{bmatrix} 1 & 0 & 0 & | & -2 \\ 0 & 1 & 0 & | & 7 \\ 0 & 0 & 1 & | & 3 \end{bmatrix}$ 24. $\begin{bmatrix} 1 & 0 & 0 & | & 5 \\ 0 & 1 & 0 & | & -9 \\ 0 & 0 & 1 & | & 11 \end{bmatrix}$

25. $\begin{bmatrix} 1 & 0 & 7 & | & 3 \\ 0 & 1 & 2 & | & -9 \\ 0 & 0 & 0 & | & -2 \end{bmatrix}$ 26. $\begin{bmatrix} 1 & 0 & 3 & | & 8 \\ 0 & 1 & -3 & | & 4 \\ 0 & 0 & 0 & | & 7 \end{bmatrix}$

27. $\begin{bmatrix} 1 & 0 & 3 & | & 2 \\ 0 & 1 & 2 & | & -5 \\ 0 & 0 & 0 & | & 0 \end{bmatrix}$ 28. $\begin{bmatrix} 1 & 0 & -4 & | & 2 \\ 0 & 1 & 5 & | & 3 \\ 0 & 0 & 0 & | & 0 \end{bmatrix}$

In Exercises 29–34 label the elementary row operation used to transform the first matrix to the second. Use the notation introduced in Section 11.2.

29. $\begin{bmatrix} 2 & 4 & 6 & | & -6 \\ 3 & 1 & -1 & | & 6 \\ 1 & 2 & -4 & | & 11 \end{bmatrix} \xrightarrow{r_1' = ?} \begin{bmatrix} 1 & 2 & 3 & | & -3 \\ 3 & 1 & -1 & | & 6 \\ 1 & 2 & -4 & | & 11 \end{bmatrix}$

30. $\begin{bmatrix} 1 & -2 & 5 & | & 25 \\ 0 & 3 & -6 & | & -18 \\ 0 & 4 & -3 & | & -9 \end{bmatrix} \xrightarrow{r_2' = ?} \begin{bmatrix} 1 & -2 & 5 & | & 25 \\ 0 & 1 & -2 & | & -6 \\ 0 & 4 & -3 & | & -9 \end{bmatrix}$

31. $\begin{bmatrix} 1 & 3 & -2 & | & -3 \\ -3 & 1 & 1 & | & -16 \\ 4 & 2 & -1 & | & 15 \end{bmatrix} \longrightarrow \begin{bmatrix} 1 & 3 & -2 & | & -3 \\ 0 & 10 & -5 & | & -25 \\ 4 & 2 & -1 & | & 15 \end{bmatrix}$

32. $\begin{bmatrix} 1 & 3 & -2 & | & -3 \\ 0 & 10 & -5 & | & -25 \\ 4 & 2 & -1 & | & 15 \end{bmatrix} \longrightarrow \begin{bmatrix} 1 & 3 & -2 & | & -3 \\ 0 & 10 & -5 & | & -25 \\ 0 & -10 & 7 & | & 27 \end{bmatrix}$

33. $\begin{bmatrix} 1 & -3 & 2 & | & 10 \\ 0 & 1 & 5 & | & -12 \\ 0 & -2 & -1 & | & 1 \end{bmatrix} \longrightarrow \begin{bmatrix} 1 & 0 & 17 & | & -26 \\ 0 & 1 & 5 & | & -12 \\ 0 & -2 & -1 & | & 1 \end{bmatrix}$

34. $\begin{bmatrix} 1 & 0 & 17 & | & -26 \\ 0 & 1 & 5 & | & -12 \\ 0 & -2 & -1 & | & 1 \end{bmatrix} \longrightarrow \begin{bmatrix} 1 & 0 & 17 & | & -26 \\ 0 & 1 & 5 & | & -12 \\ 0 & 0 & 9 & | & -23 \end{bmatrix}$

In Exercises 35–44 use an augmented matrix and elementary row operations to solve each system of linear equations.

35. $2x + y = 7$
$y - z = 2$
$x + z = 2$

36. $x + y = -2$
$-y + z = 2$
$x - z = -1$

37. $x + 2y + z = 11$
$-x - y + 2z = 1$
$2x - y + z = 4$

38. $3x - y + 2z = 4$
$2x + 2y - z = 10$
$x - y + 3z = -4$

39. $x_1 + 2x_2 - 2x_3 = -7$
$-2x_1 + 3x_2 - 17x_3 = -14$
$4x_1 + 2x_2 + 10x_3 = -4$

40. $x_1 - 2x_2 + 7x_3 = 3$
$2x_1 + 2x_2 - 3x_3 = -5$
$x_1 - 11x_2 + 24x_3 = 11$

41. $5a + b - 2c = -3$
$2a + 4b + c = -3$
$-3a + 5b - 6c = -21$

42. $6a - 3b + 3c = 3$
$3a + 3b - c = 5$
$5a + 2b - 2c = 4$

43. $r + 2s - 5t = 4$
$3r - s + 2t = 3$
$r + 9s - 22t = 10$

44. $r - 3s + 2t = 1$
$4r - 2s + t = 2$
$2r + 4s - 2t = 0$

In Exercises 45 and 46 find a general solution and two particular solutions for each system.

45. $2a - b - 3c = -5$
$3a + b - 2c = -10$

46. $3a - b + 5c = 4$
$2a + 2b - 2c = 0$

47. **Numeric Word Problem**   The sum of three numbers is 108. The largest number is 16 less than the sum of the other two numbers. The sum of the largest and the smallest is twice the other number. Find the three numbers.

48. **Numeric Word Problem**   The largest of three numbers is 7 times the second number. The second number is 7 times the smallest number. The sum of the numbers is 285. Find the three numbers.

49. **Dimensions of a Triangle**   The perimeter of this triangle is 168 cm. The length of the longest side is twice that of the shortest side. The sum of the lengths of the shortest side and the longest side is 48 cm more than the length of side b. Find the length of each side. (See the figure.)

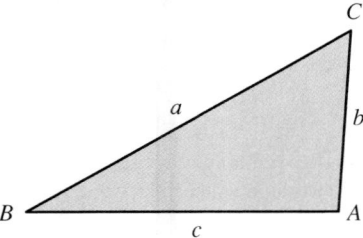

50. **Dimensions of a Triangle**   Triangle ABC has sides a, b, and c with side a the longest side and side b the shortest side. The length of the longest side of a triangle is 12 cm less than the sum of the lengths of the other two sides. The length of the shortest side is 10 cm more than one-half the length of side c. Find the length of each side if the perimeter is 188 cm.

51. **Angles of a Triangle**   Triangle ABC has angles A, B, and C with angle A the largest angle and angle B the smallest angle. Angle C is twice as large as the smallest angle. The largest angle is 9° larger than the sum of the other two angles. Find the number of degrees in each angle. (*Hint:* The sum of the angles of a triangle is 180°.)

52. **Angles of a Triangle**   Triangle ABC has angles A, B, and C with angle A the largest angle and angle B the smallest angle. The smallest angle of this triangle is 78° less than the largest angle. Angle C is 3 times as large as the smallest angle. How many degrees are in each angle?

53. **Mixture of Foods**   A zookeeper mixes three foods, the contents of which are described in the following table. How many grams of each food are needed to produce a

mixture with 133 grams of fat, 494 grams of protein, and 1700 grams of carbohydrates?

	A	B	C
Fat (%)	6	4	5
Protein (%)	15	18	20
Carbohydrates (%)	45	65	70

54. Use of Farmland A farmer must decide how many acres of each of three crops to plant during this growing season. The farmer must pay a certain amount for seed and devote a certain amount of labor and water to each acre of crop planted, as shown in the following table.

	A	B	C
Seed cost ($)	120	85	80
Hours of labor	4	12	8
Gallons of water	500	900	700

The amount of money available to pay for seed is $26,350. The farmer's family can devote 2520 h to tending the crops, and the farmer has access to 210,000 gallons of water for irrigation. How many acres of each crop would use *all* of these resources?

In Exercises 55 and 56 match each graph with the linear equation that defines this plane.
55. $2x + y + 3z = 6$
56. $3x + 2y + 2z = 6$

A.

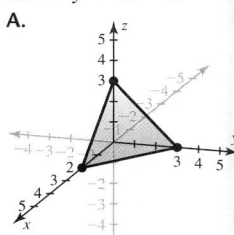

B.

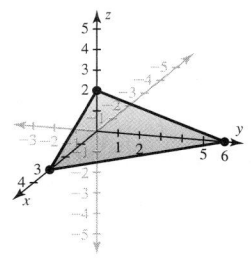

In Exercises 57 and 58 graph the plane defined by each linear equation.
57. $4x + 2y + 3z = 12$
58. $2x + 3y + 6z = 6$

Group Discussion Questions
59. Discovery Question

a. Solve $\begin{cases} 3a - 5b = 1 \\ a + 2b = 4 \end{cases}$.

b. Use the solution in part **a** to solve this nonlinear

system: $\begin{cases} \dfrac{3}{x} - \dfrac{5}{y} = 1 \\ \dfrac{1}{x} + \dfrac{2}{y} = 4 \end{cases}$

c. Use the solution in part **a** to solve this nonlinear

system: $\begin{cases} 3 \log x - 5 \log y = 1 \\ \log x + 2 \log y = 4 \end{cases}$

60. Discovery Question

a. Solve $\begin{cases} a - 2b - c = 5 \\ 2a - b + 3c = 11 \\ -3a - 2b + 2c = 7 \end{cases}$.

b. Use the solution in part **a** to solve this nonlinear

system: $\begin{cases} \dfrac{1}{x} - \dfrac{2}{y} - \dfrac{1}{z} = 5 \\ \dfrac{2}{x} - \dfrac{1}{y} + \dfrac{3}{z} = 11 \\ -\dfrac{3}{x} - \dfrac{2}{y} + \dfrac{2}{z} = 7 \end{cases}$

61. Challenge Question Find the constants a, b, and c such that $(1, -3, 5)$ is a solution of the linear system:
$\begin{cases} ax + by + cz = 5 \\ ax - by - cz = -1 \\ 2ax + 3by + 4cz = 13 \end{cases}$

62. Challenge Question Solve the following system for (x, y, z) in terms of the nonzero constants a, b, and c:
$\begin{cases} ax + by + cz = 0 \\ 2ax - by + cz = 14 \\ -ax + by + 2cz = -21 \end{cases}$

Section 11.5 Conic Sections

Objectives: **8.** Use the distance and midpoint formulas.
9. Graph parabolas, circles, ellipses, and hyperbolas.
10. Write the equations of parabolas, circles, ellipses, and hyperbolas in standard form.

Parabolas, circles, ellipses, and hyperbolas can be formed by cutting a cone or a pair of cones with a plane. Therefore these figures collectively are referred to as conic sections. Conic sections originally were studied from a geometric viewpoint. (See Fig. 11.5.1.)

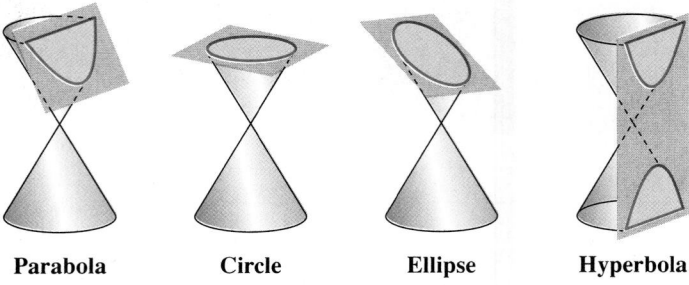

| Parabola | Circle | Ellipse | Hyperbola |

Figure 11.5.1 Conic sections.

Two formulas that are useful for examining relationships between points are the distance formula (Section 8.6) and the midpoint formula. Note that the midpoint is found by taking the average of the x-coordinates and the average of the y-coordinates.

Midpoint Formula:

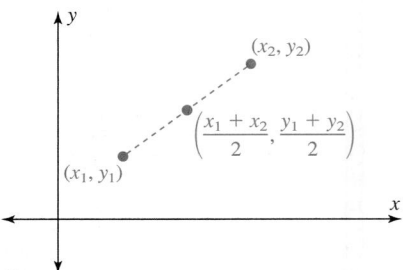

The distance and midpoint formulas are summarized in the following box.

Distance and Midpoint Formulas

If (x_1, y_1) and (x_2, y_2) are two points, then:

ALGEBRAICALLY	NUMERICAL EXAMPLE
Distance formula: $d = \sqrt{(x_2 - x_1)^2 + (y_2 - y_1)^2}$	For $(-2, 7)$ and $(4, -1)$: Distance between the points: $d = \sqrt{(4 - (-2))^2 + (-1 - 7)^2}$ $d = \sqrt{36 + 64} = \sqrt{100}$ $d = 10$
Midpoint formula: $(x, y) = \left(\dfrac{x_1 + x_2}{2}, \dfrac{y_1 + y_2}{2} \right)$	Midpoint between the points: $(x, y) = \left(\dfrac{-2 + 4}{2}, \dfrac{7 + (-1)}{2} \right) = \left(\dfrac{2}{2}, \dfrac{6}{2} \right)$ $(x, y) = (1, 3)$

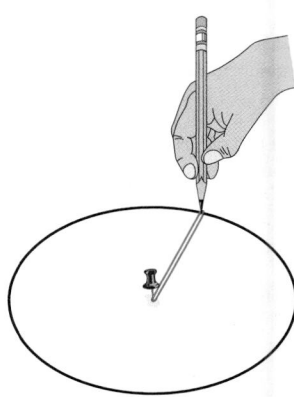

Circle

Using the distance formula, we now develop the equation of a circle with center (h, k) and radius r. A **circle** is the set of all points in a plane that are a constant distance from a fixed point. (See Fig. 11.5.2.) The fixed point is called the **center** of the circle, and the distance from the center to the points on the circle is called the length of a **radius.** A **diameter** is a line segment from one point on a circle through the center to another point on the circle. The length of a diameter is twice the length of a radius.

The distance r from any point (x, y) on the circle to the center (h, k) is given by

$$r = \sqrt{(x - h)^2 + (y - k)^2}$$

Squaring both sides of this equation gives an equation satisfied by the points on the circle:

$$(x - h)^2 + (y - k)^2 = r^2$$

Figure 11.5.2 To draw a circle, fix a loop of string with a tack and draw the circle as illustrated.

Standard Form of the Equation of a Circle

ALGEBRAICALLY	GRAPHICALLY	ALGEBRAIC EXAMPLE
The equation of a circle with center (h, k) and radius r is $(x - h)^2 + (y - k)^2 = r^2$.		The equation of a circle with center $(-1, 3)$ and radius 5 is $(x - (-1))^2 + (y - 3)^2 = 5^2$ $(x + 1)^2 + (y - 3)^2 = 25$

■ **EXAMPLE 1**   Determining the Equation of a Circle

Determine the equation of each graphed circle.

SOLUTIONS

(a)

Center $(h, k) = (4, -3)$

Radius $r = \sqrt{(6 - 4)^2 + (-3 - (-3))^2}$

$r = \sqrt{2^2 + 0^2} = \sqrt{4}$

$r = 2$

Circle: $(x - h)^2 + (y - k)^2 = r^2$

$(x - 4)^2 + (y - (-3))^2 = 2^2$

$(x - 4)^2 + (y + 3)^2 = 4$

Determine the center by inspection and use the distance formula to calculate the length of the radius. The segment $(4, -3)$ to $(6, -3)$ is a radius of this circle.

Substitute the center and radius into the standard form for the equation of a circle.

(b)

Center $(h, k) = \left(\dfrac{1 + 9}{2}, \dfrac{1 + 1}{2}\right)$

$(h, k) = (5, 1)$

Radius $r = \sqrt{(9 - 5)^2 + (1 - 1)^2}$

$r = \sqrt{4^2 + 0^2} = \sqrt{16}$

$r = 4$

Circle: $(x - h)^2 + (y - k)^2 = r^2$

$(x - 5)^2 + (y - 1)^2 = 4^2$

$(x - 5)^2 + (y - 1)^2 = 16$

Use the midpoint formula to calculate the center of the circle with a diameter from $(1, 1)$ to $(9, 1)$.

Then use the center, one of the points on the circle, and the distance formula to calculate the length of the radius.

Substitute the center and radius into the standard form for the equation of a circle.

■

SELF-CHECK 11.5.1

A circle has a diameter from $(-3, 4)$ to $(3, -4)$.

1. Determine the center of this circle.

2. Determine the length of this diameter.

3. Determine the length of a radius.

4. Determine the equation of this circle.

SELF-CHECK 11.5.1 ANSWERS

1. $(0, 0)$

2. 10

3. 5

4. $x^2 + y^2 = 25$

The standard form of the circle $(x - 4)^2 + (y + 3)^2 = 4$ can be expanded to give the general form $x^2 + y^2 - 8x + 6y + 21 = 0$. If we are given the equation of a circle in general form, we can use the process of completing the square to rewrite the equation in standard form so that the center and radius will be obvious.

■ **EXAMPLE 2** Writing the Equation of a Circle in Standard Form

Determine the center and the radius of the circle defined by the equation $3x^2 + 3y^2 + 30x - 66y + 330 = 0$.

SOLUTION _____

$$3x^2 + 3y^2 + 30x - 66y + 330 = 0$$
$$x^2 + y^2 + 10x - 22y + 110 = 0$$
$$(x^2 + 10x) + (y^2 - 22y) = -110$$
$$(x^2 + 10x + 25) + (y^2 - 22y + 121) = -110 + 25 + 121$$
$$(x + 5)^2 + (y - 11)^2 = 36$$
$$(x - (-5))^2 + (y - 11)^2 = 6^2$$

Divide both sides of the equation by 3.
Regroup terms.
Complete the square.
Write in standard form.

Answer: The circle has center $(-5, 11)$ and radius 6. ■

Equations of the form $x^2 + y^2 = c$, for $c > 0$, define circles centered at the origin. The following box also describes the cases where $c = 0$ or $c < 0$.

Equations of the Form $x^2 + y^2 = c$

As shown in the figure,
1. $x^2 + y^2 = 4$ is a circle with center $(0, 0)$ and radius 2.
2. $x^2 + y^2 = 1$ is a circle with center $(0, 0)$ and radius 1.
3. $x^2 + y^2 = \dfrac{1}{4}$ is a circle with center $(0, 0)$ and radius $\dfrac{1}{2}$.
4. $x^2 + y^2 = 0$ is the degenerate case of a circle—a single point, $(0, 0)$.
5. $x^2 + y^2 = -1$ has no real solutions and thus no points to graph. Both x^2 and y^2 are nonnegative, so $x^2 + y^2$ also must be nonnegative.

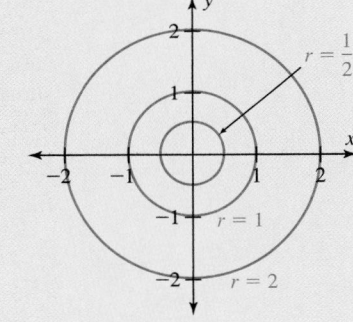

SELF-CHECK 11.5.2

Write $4x^2 + 4y^2 - 24x + 32y + 99 = 0$ in standard form and identify the center and the radius.

By the vertical-line test, circles are not functions. Thus graphing calculators do not graph circles with y as a function of x. Nonetheless, there are ways to get the graph of a circle to appear on a graphics calculator display. One way is to split a circle into two semicircles—an upper semicircle and a lower semicircle, both of which are functions. This is illustrated with the circle $x^2 + y^2 = 16$. Solving for y, we obtain $y = \pm\sqrt{16 - x^2}$. The upper semicircle is defined by $y_1 = +\sqrt{16 - x^2}$, and the lower semicircle is defined by $y_2 = -\sqrt{16 - x^2}$. These semicircles are both graphed in Calculator Perspective 11.5.1.

CALCULATOR PERSPECTIVE 11.5.1	Graphing a Circle Using Two Functions

To graph the circle $x^2 + y^2 = 16$ on a TI-83 Plus calculator, graph the two semicircles, $y_1 = \sqrt{16 - x^2}$ and $y_2 = -\sqrt{16 - x^2}$ by entering the following keystrokes:

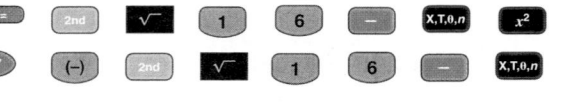

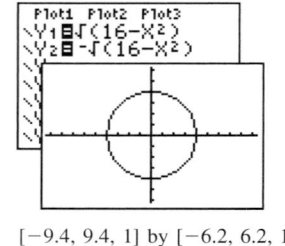

If a circle looks distorted on your calculator display, try using the ZOOM SQUARE feature to adjust for the fact that the display window is wider than it is tall.

$[-9.4, 9.4, 1]$ by $[-6.2, 6.2, 1]$

Ellipse

An **ellipse** is the set of all points in a plane, the sum of whose distances from two fixed points is constant. (See Fig. 11.5.3.) The two fixed points $F_1(-c, 0)$ and $F_2(c, 0)$ are called **foci.** The **major axis** of the ellipse passes through the foci. The **minor axis** is shorter than the major axis and is perpendicular to it at the center. The ends of the major axis are called the **vertices,** and the ends of the minor axis are called the **covertices.** If the ellipse is centered at the origin, then the equation of the ellipse as shown in Fig. 11.5.4 is $\dfrac{x^2}{a^2} + \dfrac{y^2}{b^2} = 1$ where $a > b$. This formula is derived in the group exercises.

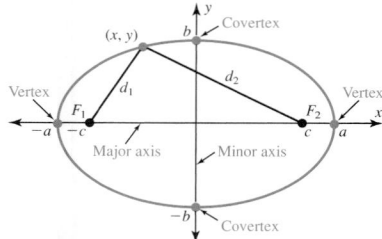

Figure 11.5.3 To draw an ellipse, loop a piece of string around two tacks and draw the ellipse as illustrated.

Figure 11.5.4 $d_1 + d_2 = 2a$.

■ **EXAMPLE 3**   Graphing an Ellipse Centered at the Origin

Graph $\dfrac{x^2}{49} + \dfrac{y^2}{16} = 1$.

SOLUTION _____

Plot the x-intercepts $(-7, 0)$ and $(7, 0)$.
Plot the y-intercepts $(0, -4)$ and $(0, 4)$.
Then use the known shape to complete the ellipse.

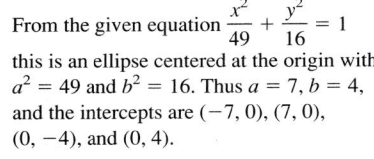

From the given equation $\dfrac{x^2}{49} + \dfrac{y^2}{16} = 1$ this is an ellipse centered at the origin with $a^2 = 49$ and $b^2 = 16$. Thus $a = 7$, $b = 4$, and the intercepts are $(-7, 0)$, $(7, 0)$, $(0, -4)$, and $(0, 4)$.

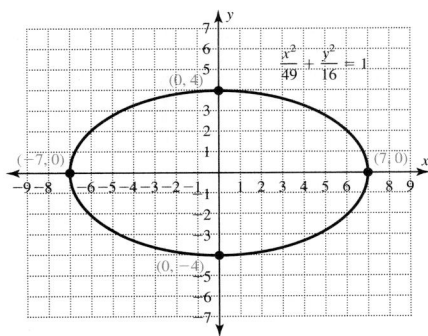

We now use translations (Section 6.3) to examine ellipses that are not centered at the origin. The graph of $\dfrac{(x - h)^2}{a^2} + \dfrac{(y - k)^2}{b^2} = 1$ is identical to the graph of $\dfrac{x^2}{a^2} + \dfrac{y^2}{b^2} = 1$, except that it has been translated so that the center is at (h, k) instead of $(0, 0)$.

Standard Form of the Equation of an Ellipse

Algebraically

The equation of an ellipse with center (h, k), major axis of length $2a$, and minor axis of length $2b$ is:

HORIZONTAL MAJOR AXIS	ALGEBRAIC EXAMPLE
$$\frac{(x - h)^2}{a^2} + \frac{(y - k)^2}{b^2} = 1$$	$$\frac{(x - 2)^2}{49} + \frac{(y + 3)^2}{16} = 1$$

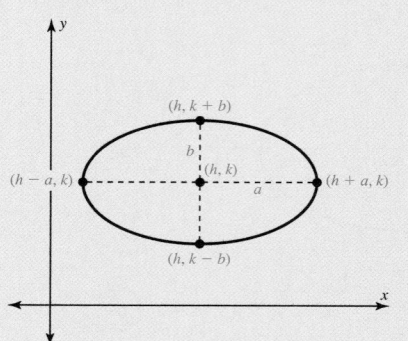

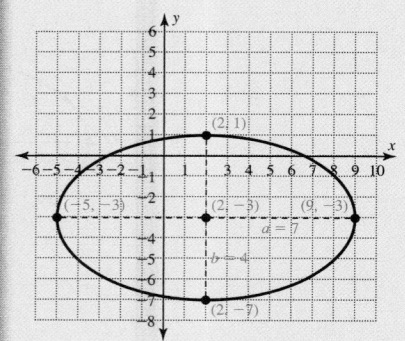

VERTICAL MAJOR AXIS	
$$\frac{(x - h)^2}{b^2} + \frac{(y - k)^2}{a^2} = 1$$	$$\frac{(x + 4)^2}{16} + \frac{(y - 5)^2}{36} = 1$$

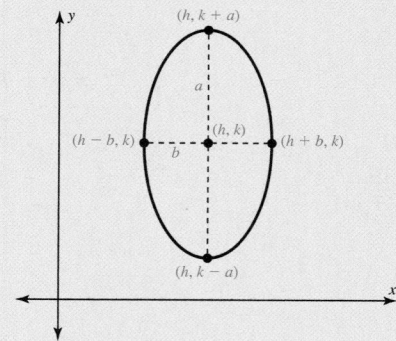

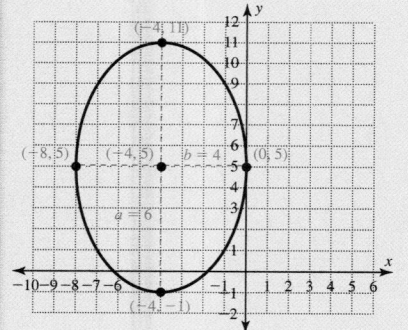

Note: $a > b > 0$.

■ EXAMPLE 4   Writing the Equation of an Ellipse in Standard Form

Determine the equation of the graphed ellipse.

SOLUTION _____

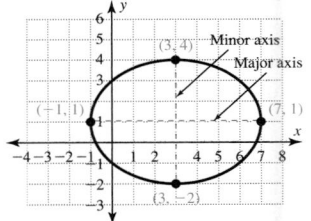

$$(h, k) = \left(\frac{-1 + 7}{2}, \frac{1 + 1}{2} \right)$$

$$(h, k) = (3, 1)$$

The center of the ellipse is at the midpoint of the major axis.

$$2a = \sqrt{7 - (-1)^2 + (1 - 1)^2}$$
$$2a = \sqrt{8^2 + 0^2}$$
$$2a = \sqrt{64}$$
$$2a = 8$$
$$a = 4$$

The length of the major axis is $2a$ and the length of the minor axis is $2b$. Use the distance formula to determine these values.

$$2b = \sqrt{(3 - 3)^2 + (4 - (-2))^2}$$
$$2b = \sqrt{0^2 + 6^2}$$
$$2b = \sqrt{36}$$
$$2b = 6$$
$$b = 3$$

$$\frac{(x - h)^2}{a^2} + \frac{(y - k)^2}{b^2} = 1$$

The horizontal axis is the major axis; thus we use the standard form of an ellipse with a horizontal major axis.

$$\frac{(x - 3)^2}{4^2} + \frac{(y - 1)^2}{3^2} = 1$$

Substitute a, b, h, and k into the standard form.

$$\frac{(x - 3)^2}{16} + \frac{(y - 1)^2}{9} = 1$$

SELF-CHECK 11.5.3

Write the equation of each ellipse in standard form and graph the ellipse.
1. $9x^2 + 36y^2 = 324$
2. $9x^2 + 16y^2 - 72x - 96y + 144 = 0$

SELF-CHECK 11.5.3 ANSWERS

1. $\dfrac{x^2}{36} + \dfrac{y^2}{9} = 1$

2. $\dfrac{(x - 4)^2}{16} + \dfrac{(y - 3)^2}{9} = 1$

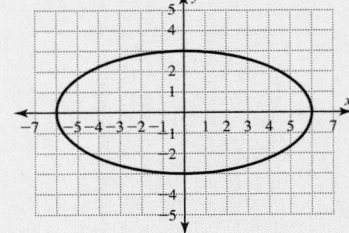

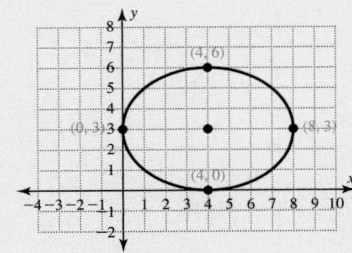

Hyperbola

A **hyperbola** is the set of all points in a plane whose distances from two fixed points have a constant difference. (See Fig. 11.5.5.) The fixed points are the **foci** of the hyperbola. If the hyperbola is centered at the origin and opens horizontally as in Fig. 11.5.5, then the equation of the hyperbola is $\dfrac{x^2}{a^2} - \dfrac{y^2}{b^2} = 1$. This hyperbola is asymptotic to the lines $y = -\dfrac{b}{a}x$ and $y = \dfrac{b}{a}x$. As $|x|$ becomes larger, the hyperbola gets closer and closer to these lines. The asymptotes pass through the corners of the rectangle formed by (a, b), $(-a, b)$, $(-a, -b)$, and $(a, -b)$. This rectangle, which is shown in Fig. 11.5.6, is called the **fundamental rectangle** and is used to sketch quickly the linear asymptotes.

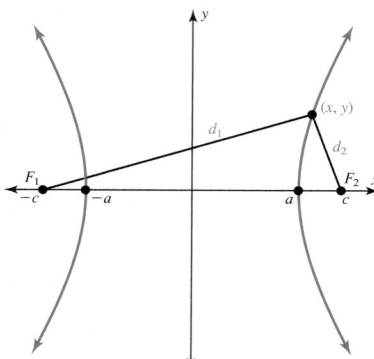

Figure 11.5.5 $|d_1 - d_2| = 2a$.

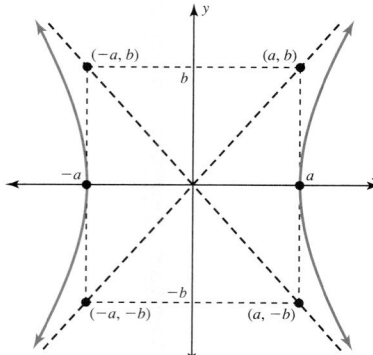

Figure 11.5.6 $\dfrac{x^2}{a^2} - \dfrac{y^2}{b^2} = 1$.

■ EXAMPLE 5   Graphing a Hyperbola Centered at the Origin

Graph $\dfrac{x^2}{16} - \dfrac{y^2}{9} = 1$.

SOLUTION _____

Plot the x-intercepts $(-4, 0)$ and $(4, 0)$. Sketch the fundamental rectangle with corners $(4, 3)$, $(-4, 3)$, $(-4, -3)$, and $(4, -3)$. Draw the asymptotes through the corners of this rectangle. Plot the x-intercepts and then sketch the hyperbola that opens horizontally using the asymptotes as guidelines.

From the given equation $\dfrac{x^2}{16} - \dfrac{y^2}{9} = 1$, this is a hyperbola centered at the origin with $a^2 = 16$ and $b^2 = 9$. Thus $a = 4$ and $b = 3$.

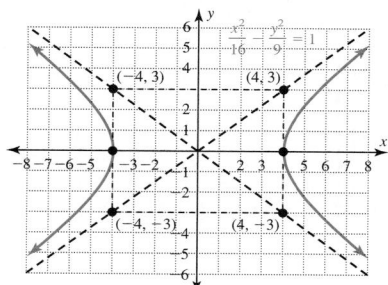

We now use translations to examine hyperbolas whose center is (h, k). The following box covers both hyperbolas that open horizontally and those that open vertically.

Standard Form of the Equation of a Hyperbola

Algebraically

The equation of a hyperbola with center (h, k) is

OPENING HORIZONTALLY	ALGEBRAIC EXAMPLE
$$\frac{(x - h)^2}{a^2} - \frac{(y - k)^2}{b^2} = 1$$	$$\frac{(x + 1)^2}{16} - \frac{(y - 2)^2}{9} = 1$$
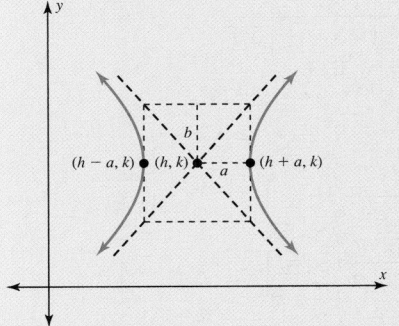	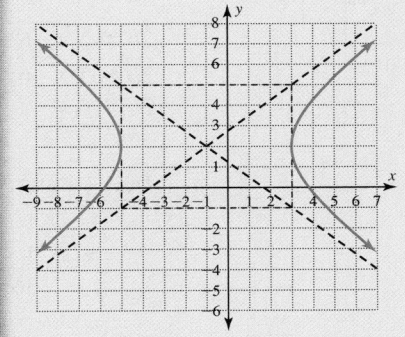

Vertices: $(h - a, k)$ and $(h + a, k)$.
The fundamental rectangle has base $2a$ and height $2b$.

OPENING VERTICALLY	
$$\frac{(y - k)^2}{a^2} - \frac{(x - h)^2}{b^2} = 1$$	$$\frac{(y - 2)^2}{9} - \frac{(x + 1)^2}{16} = 1$$
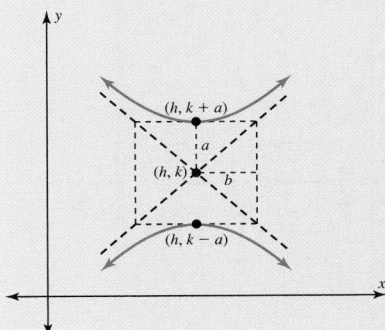	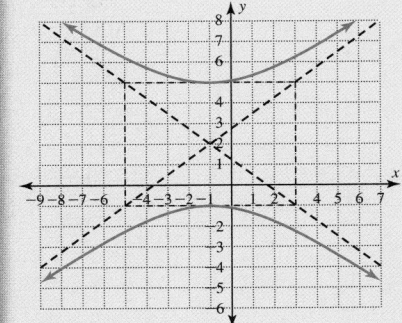

Vertices: $(h, k - a)$ and $(h, k + a)$.
The fundamental rectangle has base $2b$ and height $2a$.

■ EXAMPLE 6   Writing the Equation of a Hyperbola in Standard Form

Determine the equation of the hyperbola graphed in the figure.

SOLUTION

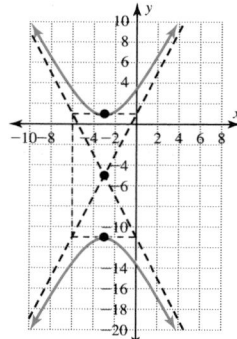

$(h, k) = (-3, -5)$

$$\frac{(y - k)^2}{a^2} - \frac{(x - h)^2}{b^2} = 1$$

The center of the hyperbola was determined by inspection.
Select the form for a hyperbola that opens vertically.

$2a = \sqrt{(-3 - (-3))^2 + (1 - (-11))^2}$

$2a = \sqrt{0 + 12^2} = \sqrt{144}$

$2a = 12$

$a = 6$

Use points on the fundamental rectangle to compute the values of a and b.

$2b = \sqrt{(-6 - 0)^2 + (-5 - (-5))^2}$

$2b = \sqrt{(-6)^2 + 0} = \sqrt{36}$

$2b = 6$

$b = 3$

$$\frac{(y - (-5))^2}{6^2} - \frac{(x - (-3))^2}{3^2} = 1$$

Substitute $(-3, -5)$ for (h, k) and 6 for a and 3 for b into the standard form for this hyperbola.

$$\frac{(y + 5)^2}{36} - \frac{(x + 3)^2}{9} = 1$$

■

SELF-CHECK 11.5.4

Graph each hyperbola.

1. $\dfrac{x^2}{25} - \dfrac{y^2}{4} = 1$

2. $\dfrac{(y - 2)^2}{4} - \dfrac{(x - 3)^2}{25} = 1$

SELF-CHECK 11.5.4 ANSWERS

1.

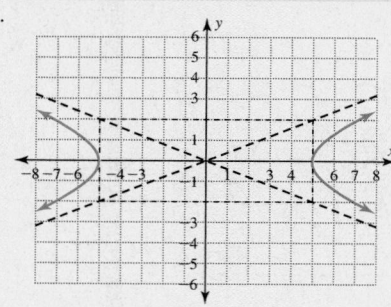

2.

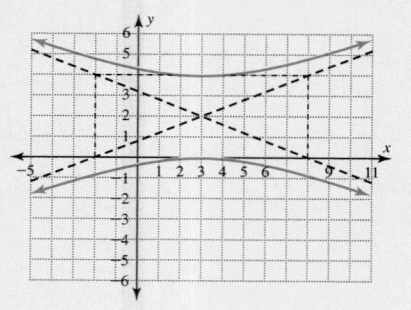

Parabolas

We already have graphed parabolas in several sections of this book (including Section 6.3). For completeness of this introduction to conic sections, we now note that a parabola can be formed by cutting a cone with a plane. A **parabola** is the set of all points in a plane the same distance from a fixed line L (the **directrix**) and from a fixed point F (the

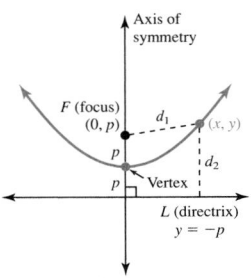

Figure 11.5.7 $d_1 = d_2$

focus). See Fig. 11.5.7. Note that the axis of symmetry passes through the vertex, and it is perpendicular to the directrix. If the vertex of the parabola in Fig. 11.5.7 is at the origin, then the equation of this parabola can be written in the form $y = \dfrac{1}{4p}x^2$. If the vertex is translated to the point (h, k), then the equation becomes $y - k = \dfrac{1}{4p}(x - h)^2$.

The equation $y + 2 = \dfrac{1}{4\left(\dfrac{1}{4}\right)}(x - 3)^2$ can also be rewritten in the form $y = (x - 3)^2 - 2$. This is the form that we used extensively in Chapter 6.

■ EXAMPLE 7   Graphing a Parabola

Graph $y = (x - 3)^2 - 2$.

SOLUTION

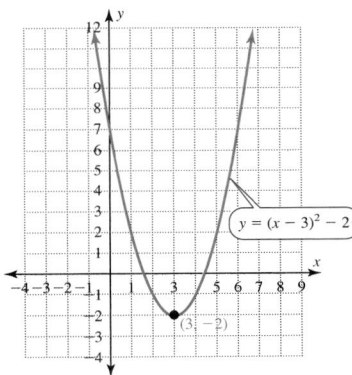

This parabola can be obtained by translating the graph of $y = x^2$ right 3 units and down 2 units. The vertex of this parabola is $(3, -2)$ and the y-intercept is $(0, 7)$.

■

SELF-CHECK 11.5.5

Graph $y = -(x + 2)^2 + 1$.

SELF-CHECK 11.5.5 ANSWER

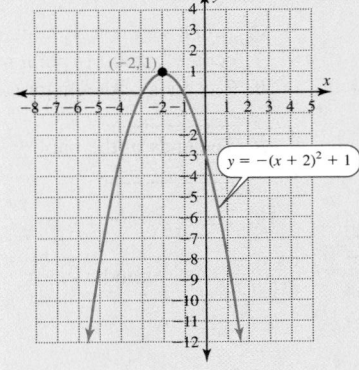

Example 8 involves a circle and a parabola. We now use the graphs of these conic sections to solve the corresponding system of equations and inequalities.

■ **EXAMPLE 8**   Solving a Nonlinear System of Equations and a Nonlinear System of Inequalities

Use graphs to solve: **(a)** $\begin{cases} x^2 + y^2 = 25 \\ y = x^2 - 5 \end{cases}$ and **(b)** $\begin{cases} x^2 + y^2 \le 25 \\ y \ge x^2 - 5 \end{cases}$.

SOLUTIONS _____

(a)

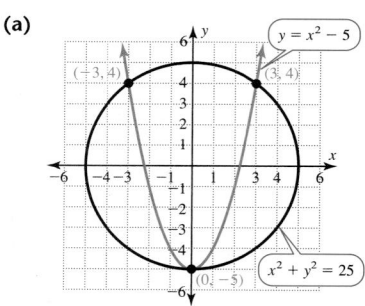

$x^2 + y^2 = 25$ is a circle centered at the origin with radius $r = 5$.

$y = x^2 - 5$ is a parabola opening upward with its vertex at $(0, -5)$.

$(-3, 4)$, $(0, -5)$, and $(3, 4)$ all satisfy both equations.

Do all these points check?

(b)

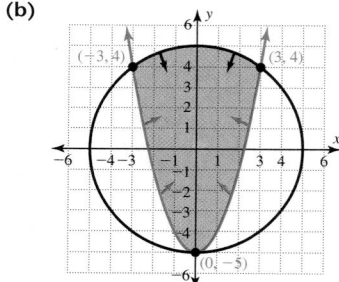

The points satisfying $x^2 + y^2 \le 25$ lie on and inside the circle. Use $(0, 0)$ as a test value. The points satisfying $y \ge x^2 - 5$ lie on and above the parabola. Use $(0, 0)$ as a test value.

The shaded points satisfy both inequalities.

■

USING THE LANGUAGE AND SYMBOLISM OF MATHEMATICS   11.5

1. Parabolas, circles, ellipses, and hyperbolas are collectively referred to as _____ _____.

2. A _____ is the set of all points in a plane that are a constant distance from a fixed point.

3. An _____ is the set of all points in a plane the sum of whose distances from two fixed points is constant.

4. A _____ is the set of all points in a plane whose distances from two fixed points have a constant difference.

5. A _____ is the set of all points in a plane the same distance from a fixed line and a fixed point.

6. A line segment from the center of a circle to a point on the circle is a _____.

7. A line segment from one point on a circle through the center to another point on the circle is a _____.

8. The ends of the major axis of an ellipse are called _____.

9. The ends of the minor axis of an ellipse are called _____.

10. The hyperbola $\dfrac{x^2}{a^2} - \dfrac{y^2}{b^2} = 1$ is _____ to the lines $y = -\dfrac{b}{a}x$ and $y = \dfrac{b}{a}x$.

EXERCISES 11.5

In Exercises 1–4 calculate the distance between each pair of points and the midpoint between these points.

1. $(-3, 2)$ and $(1, -1)$ **2.** $(-6, 2)$ and $(6, 3)$

3. $(0, 1)$ and $(1, 2)$ **4.** $(a + 1, b)$ and $(a - 3, b + 3)$

5. Calculate the length of the radius of a circle with center at $(0, 0)$ and with the point $(7, 24)$ on the circle.

6. Calculate the length of the diameter of a circle with endpoints on the circle at $(-8, 0)$ and $(15, 0)$.

7. A circle has a diameter with endpoints at $(-1, 5)$ and $(5, -3)$. Determine the center of this circle.

In Exercises 8–12 match each graph with the corresponding equation.

8. $y = x^2 + 4$

9. $x^2 + y^2 = 4$

10. $\dfrac{x^2}{1} + \dfrac{y^2}{4} = 1$

11. $\dfrac{x^2}{1} - \dfrac{y^2}{4} = 1$

12. $y = x + 4$

A.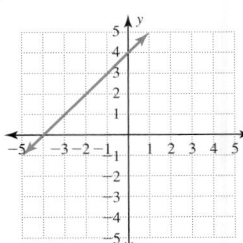

B.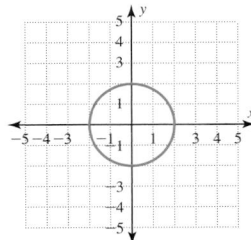

C.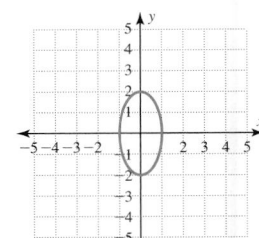

D.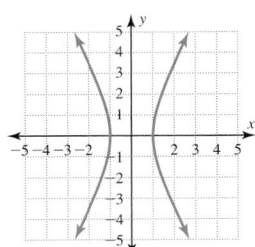

E.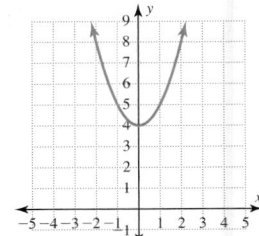

In Exercises 13–17 write in standard form the equation of the circle satisfying the given conditions.

13. Center $(0, 0)$, radius 10

14. Center $(5, -4)$, radius 4

15. Center $(2, 6)$, radius $\sqrt{2}$

16. Center $(1, 1)$, radius 0.1

17. Center $\left(0, \dfrac{1}{2}\right)$, radius $\dfrac{1}{2}$

In Exercises 18–24 determine the center and the length of the radius of the circle defined by each equation.

18. $x^2 + y^2 = 144$

19. $(x + 5)^2 + (y - 4)^2 = 64$

20. $(x - 4)^2 + (y + 5)^2 = 36$

21. $x^2 + y^2 - 6x = 0$

22. $x^2 + y^2 + 12y - 13 = 0$

23. $x^2 + y^2 - 2x + 10y + 22 = 0$

24. $4x^2 + 4y^2 - 8x - 40y + 103 = 0$

In Exercises 25–28 determine the center and the lengths of the major and minor axes of the ellipse defined by each equation and then graph the ellipse.

25. $\dfrac{x^2}{36} + \dfrac{y^2}{16} = 1$

26. $\dfrac{(x - 1)^2}{9} + \dfrac{(y + 2)^2}{49} = 1$

27. $\dfrac{(y - 3)^2}{25} + \dfrac{(x + 4)^2}{16} = 1$

28. $36x^2 + 49y^2 = 1764$

In Exercises 29–36 write in standard form the equation of the ellipse satisfying the given conditions.

29. Center $(0, 0)$, x-intercepts $(-9, 0)$ and $(9, 0)$, y-intercepts $(0, -5)$ and $(0, 5)$

30. Center $(0, 0)$, x-intercepts $(-5, 0)$ and $(5, 0)$, y-intercepts $(0, -3)$ and $(0, 3)$

31. Center $(3, 4)$, horizontal major axis of length 4, vertical minor axis of length 2

32. Center $(-5, 2)$, horizontal major axis of length 6, vertical minor axis of length 2

33. Center $(6, -2)$, vertical major axis of length 10, horizontal minor axis of length 6

34. Center $(-3, -4)$, vertical major axis of length 8, horizontal minor axis of length $\dfrac{1}{2}$

35. Center $(-3, -4)$, $a = 6$, $b = 2$, major axis vertical

36. Center $(2, 5)$, $a = 8$, $b = 4$, major axis horizontal

In Exercises 37–42 determine the center, the values of a and b, and the direction that the hyperbola opens, and then sketch the hyperbola.

37. $\dfrac{x^2}{25} - \dfrac{y^2}{81} = 1$

38. $\dfrac{y^2}{36} - \dfrac{x^2}{16} = 1$

39. $25x^2 = 100y^2 + 100$

40. $\dfrac{(x - 4)^2}{49} - \dfrac{(y - 6)^2}{9} = 1$

41. $\dfrac{(y - 6)^2}{9} - \dfrac{(x - 4)^2}{49} = 1$

42. $16x^2 + 96x = 9y^2 - 126y + 153$

In Exercises 43–46 write the standard form of the equation of the hyperbola satisfying the given conditions.

43. The center is $(0, 0)$, the hyperbola opens vertically, and the fundamental rectangle has height 10 and width 8.

44. The center is $(0, 0)$, the hyperbola opens horizontally, and the fundamental rectangle has height 6 and width 12.

45. The hyperbola has vertices $(-3, 0)$ and $(3, 0)$, and the height of the fundamental rectangle is 14.

46. The hyperbola has vertices $(0, -5)$ and $(0, 5)$, and the width of the fundamental rectangle is 16.

In Exercises 47–50 graph each parabola.

47. a. $y = x^2$
   b. $y = x^2 - 5$
   c. $y = (x + 2)^2 - 5$

48. a. $y = -x^2$
   b. $y = -x^2 + 3$
   c. $y = -(x - 1)^2 + 3$

49. a. $y = x^2$
   b. $y = \dfrac{1}{2}x^2$
   c. $y = -2x^2$

50. a. $y = x^2$
   b. $y = \dfrac{1}{3}x^2$
   c. $y = -3x^2$

In Exercises 51–56 use a graphics calculator to graph each pair of functions.

51. $y_1 = \sqrt{25 - x^2}$, $y_2 = -y_1$
   (*Hint:* Compare to $x^2 + y^2 = 25$.)

52. $y_1 = \sqrt{16 - (x + 2)^2}$, $y_2 = -y_1$
   (*Hint:* Compare to $(x + 2)^2 + y^2 = 16$.)

53. $y_1 = \dfrac{1}{3}\sqrt{36 - 4x^2}$, $y_2 = -y_1$
   (*Hint:* Compare to $\dfrac{x^2}{9} + \dfrac{y^2}{4} = 1$.)

54. $y_1 = \dfrac{1}{3}\sqrt{36 + 4x^2}$, $y_2 = -y_1$
   (*Hint:* Compare to $\dfrac{y^2}{4} - \dfrac{x^2}{9} = 1$.)

55. Split $x^2 + y^2 = 36$ into y_1 and y_2, which represent upper and lower semicircles. Then graph y_1 and y_2.

56. Split $\dfrac{x^2}{25} + \dfrac{y^2}{16} = 1$ into y_1 and y_2, which represent upper and lower semiellipses. Then graph y_1 and y_2.

In Exercises 57–60 use the given graphs to solve each system of equations and each system of inequalities.

57. a. $\begin{cases} y = x^2 - 2x - 1 \\ 2x + y = 3 \end{cases}$
   b. $\begin{cases} y \geq x^2 - 2x - 1 \\ y \leq -2x + 3 \end{cases}$

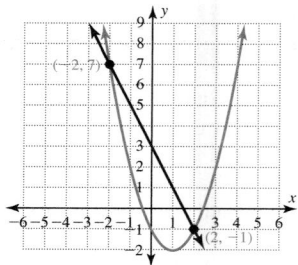

58. a. $\begin{cases} x + y = -1 \\ (x - 4)^2 + (y + 1)^2 = 40 \end{cases}$
   b. $\begin{cases} y \leq -x - 1 \\ (x - 4)^2 + (y + 1)^2 \leq 40 \end{cases}$

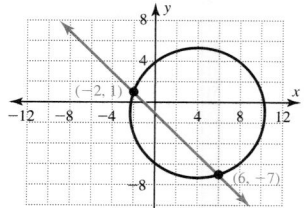

59. a. $\begin{cases} x^2 + y^2 = 8 \\ y = x \end{cases}$
   b. $\begin{cases} x^2 + y^2 \leq 8 \\ y \geq x \end{cases}$

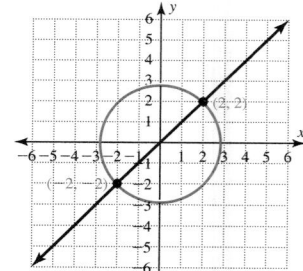

60. a. $\begin{cases} y = e^x \\ y = -x + 1 \end{cases}$
   b. $\begin{cases} y \geq e^x \\ y \leq -x + 1 \end{cases}$

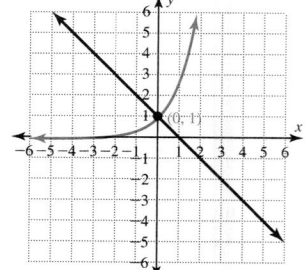

Group Discussion Questions

61. Discovery Question   A line and a circle can intersect at 0, 1, or 2 points. Illustrate each possibility with a sketch.

62. Discovery Question   A circle and an ellipse can intersect at 0, 1, 2, 3, or 4 points. Illustrate each possibility with a sketch.

63. Discovery Question   A circle and a hyperbola can intersect at 0, 1, 2, 3, or 4 points. Illustrate each possibility with a sketch.

64. Challenge Question   Use the following figure to complete each part of this question.

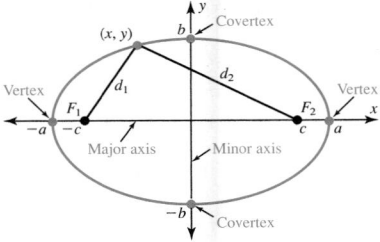

a. Place (x, y) at a convenient point on the figure and present a justification for $d_1 + d_2 = 2a$.

b. Place (x, y) at a convenient point on the figure and present a justification for $a^2 = b^2 + c^2$.

c. Use the distance formula to represent d_1 and d_2. Then substitute these values into $d_1 + d_2 = 2a$. Show that this equation can be simplified to $\dfrac{x^2}{a^2} + \dfrac{y^2}{b^2} = 1$.

KEY CONCEPTS   FOR CHAPTER 11

1. Operations on Functions:

If f and g are functions, then for all input values in both the domain of f and g:

- Sum $(f + g)(x) = f(x) + g(x)$
- Difference $(f - g)(x) = f(x) - g(x)$
- Product $(f \cdot g)(x) = f(x)g(x)$
- Quotient $\left(\dfrac{f}{x}\right)(x) = \dfrac{f(x)}{g(x)}$ for $g(x) \neq 0$

2. Composite Function, $f \circ g$:

- If f and g are functions, then $(f \circ g)(x) = f[g(x)]$ for all input values x in the domain of g for which $g(x)$ is an input value in the domain of f.
- If f and f^{-1} are inverses of each other, then $(f \circ f^{-1})(x) = x$ for each input value of f^{-1} and $(f^{-1} \circ f)(x)$ for each input value of f.

3. Sequences:

- A sequence is a function whose domain is a set of consecutive natural numbers.
- A finite sequence has a last term.
- An infinite sequence continues without end.
- A sequence is arithmetic if $a_n = a_{n-1} + d$; d is called the common difference.
- The points of an arithmetic sequence lie on a line.
- A sequence is geometric if $a_n = ra_{n-1}$; r is called the common ratio.
- The points of a geometric sequence lie on an exponential curve.

4. General Term of a Sequence:

- A formula for a_n is a formula for the general term of a sequence.
- A formula defining a_n in terms of one or more of the preceding terms is called a recursive definition for a_n.
- The Fibonacci sequence defined by $a_1 = 1$, $a_2 = 1$, and $a_n = a_{n-2} + a_{n-1}$ is an example of a recursive definition.
- A formula for the general term of an arithmetic sequence is $a_n = a_1 + (n - 1)d$.
- A formula for the general term of a geometric sequence is $a_n = a_1 r^{n-1}$.

5. Series:

- A series is a sum of the terms of a sequence.
- One notation used to denote a series is summation notation: $\displaystyle\sum_{i=1}^{n} a_i = a_1 + a_2 + \cdots + a_{n-1} + a_n$.
- The sum of the first n terms of an arithmetic sequence is $S_n = \dfrac{n}{2}(a_1 + a_n)$.
- The sum of the first n terms of a geometric sequence is $S_n = \dfrac{a_1(1 - r^n)}{1 - r}$.
- If $|r| < 1$, the infinite geometric series $S = \displaystyle\sum_{i=1}^{\infty} a_i$ is $S = \dfrac{a_1}{1 - r}$. If $|r| \geq 1$, this sum does not exist.

6. Augmented Matrix:

- A matrix is a rectangular array of numbers consisting of rows and columns.
- A row consists of entries arranged horizontally.
- A column consists of entries arranged vertically.
- The dimension of a matrix is given by first stating the number of rows and then the number of columns.
- An augmented matrix for a system of linear equations contains rows consisting of the coefficients and the constant for each equation.
- The augmented matrix for $\begin{cases} a_1 x + b_1 y + c_1 z = d_1 \\ a_2 x + b_2 y + c_2 z = d_2 \\ a_3 x + b_3 y + c_3 z = d_3 \end{cases}$ is

$$\begin{bmatrix} a_1 & b_1 & c_1 & | & d_1 \\ a_2 & b_2 & c_2 & | & d_2 \\ a_3 & b_3 & c_3 & | & d_3 \end{bmatrix}.$$

7. Elementary Row Operations on Augmented Matrices:

- Any two rows in the matrix may be interchanged.
- Any row in the matrix may be multiplied by a nonzero constant.
- Any row in the matrix may be replaced by the sum of itself and a constant multiple of another row.

8. Equation of a Plane:

- The graph of a linear equation of the form $Ax + By + Cz = D$ is a plane in three-dimensional space.

9. Reduced Row-Echelon Form of a Matrix:

- The first nonzero entry in a row is 1. All other entries in the column containing the leading 1 are 0s.
- All nonzero rows are above any rows containing only 0s.
- The first nonzero entry in a row is to the left of the first nonzero entry in the following row.

10. Distance and Midpoint Formulas:

For two points (x_1, y_1) and (x_2, y_2):

- Distance: $d = \sqrt{(x_2 - x_1)^2 + (y_2 - y_1)^2}$
- Midpoint: $(x, y) = \left(\dfrac{x_1 + x_2}{2}, \dfrac{y_1 + y_2}{2}\right)$

11. Standard Forms for Conic Sections:

- Circle: $(x - h)^2 + (y - k)^2 = r^2$
- Ellipse:

$$\dfrac{(x - h)^2}{a^2} + \dfrac{(y - k)^2}{b^2} = 1 \text{ with horizontal major axis}$$

$$\dfrac{(x - h)^2}{b^2} + \dfrac{(y - k)^2}{a^2} = 1 \text{ with vertical major axis}$$

- Hyperbola:

$$\dfrac{(x - h)^2}{a^2} - \dfrac{(y - k)^2}{b^2} = 1 \text{ opens horizontally}$$

$$\dfrac{(y - k)^2}{a^2} - \dfrac{(x - h)^2}{b^2} = 1 \text{ opens vertically}$$

- Parabola: $y - k = \dfrac{1}{4p}(x - h)^2$

In Exercises 1–3 evaluate each expression given
$f(x) = 3x - 7$ and $g(x) = 2x^2 + 1$.

1. a. $f(10)$
   b. $g(10)$
   c. $(f + g)(10)$
   d. $(f - g)(10)$

2. a. $(f \cdot g)(2)$
   b. $\dfrac{f}{g}(2)$
   c. $\dfrac{g}{f}(2)$
   d. $(g - f)(2)$

3. a. $(f \circ g)(2)$
   b. $(g \circ f)(2)$
   c. $(f \circ f)(2)$
   d. $(g \circ g)(2)$

4. Use the given tables for f and g to form a table of values
   for each function.

x	f(x)
0	−7
1	−4
2	−1
3	2

x	g(x)
0	1
1	3
2	9
3	19

   a. $f + g$ **b.** $f - g$ **c.** $f \cdot g$ **d.** $\dfrac{f}{g}$

5. Use the given graphs for f and g to form a set of ordered
   pairs for each function.

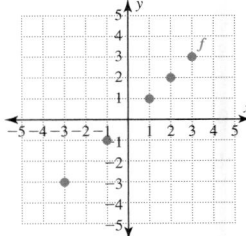

   a. $f + g$ **b.** $f - g$ **c.** $f \cdot g$ **d.** $\dfrac{f}{g}$

6. Use the given tables for f and g to complete a table for
   $f \circ g$.

x	f(x)
0	9
2	7
4	5
6	11

x	g(x)
−1	4
2	6
3	2
5	0

7. Use the given graphs of f and g to graph $f + g$.

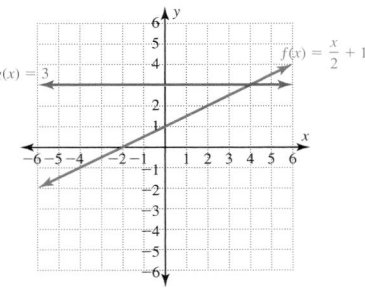

In Exercises 8 and 9 determine each function, given
$f(x) = 4x - 2$ and $g(x) = 1 - 2x$. Give the domain for each
function.

8. a. $f + g$ **b.** $f \cdot g$ **c.** $f \circ g$

9. a. $f - g$ **b.** $\dfrac{f}{g}$ **c.** $g \circ f$

10. Given $f(x) = 2x$ and $g(x) = \dfrac{2x^2 - 6x}{x - 3}$, state the reason
   that $f \neq g$.

11. Given $f(x) = 4x + 3$, determine f^{-1}. Then determine
   $(f \circ f^{-1})(x)$.

**12. Determining a Formula for Total Cost and Average
   Cost**   The fixed weekly costs for a sandwich shop are
   \$400; $F(x) = 400$. The variable dollar cost for x
   sandwiches is $V(x) = 1.5x$. Determine:
   a. $F(100)$ **b.** $V(100)$
   c. $C(x)$; the total cost of making x sandwiches in a week.
   d. $A(x)$; the average cost per sandwich when x sandwiches
       are made in a week.

13. Composing Cost and Production Functions   The
   number of desks a factory can produce weekly is a
   function of the number of hours t it operates. This function
   is $N(t) = 3t$ for $0 \le t \le 168$. The cost of manufacturing
   N desks is given by $C(N) = \left(150 - \dfrac{N}{4}\right)N$.

   a. Evaluate and interpret $N(40)$.
   b. Evaluate and interpret $C(120)$.
   c. Evaluate and interpret $(C \circ N)(40)$.
   d. Determine $(C \circ N)(t)$.
   e. Can you explain the logic of the domain $0 \le t \le 168$?

14. Write the first six terms of an arithmetic sequence that satisfies the given conditions.

    a. $a_1 = 5, d = 4$ **b.** $a_1 = 5, d = -4$

    c. $a_1 = 5, a_2 = 8$ **d.** $a_1 = 5, a_6 = 20$

15. Write the first six terms of a geometric sequence that satisfies the given conditions.

    a. $a_1 = 3, r = 2$ **b.** $a_1 = 3, r = -2$

    c. $a_1 = 3, a_2 = 12$ **d.** $a_1 = 1, a_6 = 243$

16. Write the first five terms of each sequence.

    a. $a_n = 2n + 7$ **b.** $a_n = n^2 + 3$

    c. $a_n = 64\left(\dfrac{1}{2}\right)^n$

    d. $a_1 = 3$ and $a_n = a_{n-1} + 7$ for $n > 1$

17. An arithmetic sequence has $a_1 = 11$ and $d = 3$. Find a_{101}.

18. A geometric sequence has $a_1 = \dfrac{1}{64}$ and $r = 2$. Find a_{11}.

19. An arithmetic sequence has $a_1 = -20$, $a_n = 300$, and $d = 8$. Find n.

20. A geometric sequence has $a_1 = \dfrac{1}{15{,}625}$, $a_n = 78{,}125$, and $r = 5$. Find n.

21. Write the terms of each series and then add these terms.

    a. $\displaystyle\sum_{i=1}^{6}(3i - 2)$ **b.** $\displaystyle\sum_{k=3}^{7}(k^2 - 2k)$

    c. $\displaystyle\sum_{j=1}^{5}j^3$ **d.** $\displaystyle\sum_{i=1}^{6}10$

22. An arithmetic sequence has $a_1 = 23$ and $a_{50} = 366$. Find S_{50}, the sum of the first 50 terms.

23. Write $0.121212\ldots$ as a fraction.

24. Evaluate $\displaystyle\sum_{k=1}^{\infty}\left(\dfrac{3}{5}\right)^k$.

25. Rungs of a Ladder The lengths of the rungs of a wooden ladder form an arithmetic sequence. There are 17 rungs ranging in length from 80 to 46 cm. Find the total length of these 17 rungs.

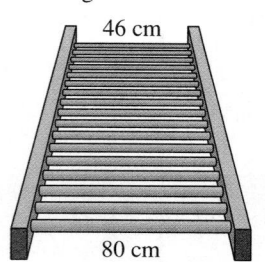

46 cm

80 cm

26. Perimeter of an Art Design An art design is formed by drawing a square with sides of 20 cm and then connecting the midpoints of the sides to form a second square. (See the following figure.) If this process is continued infinitely, determine the perimeter of the third square and the total perimeter of all the squares.

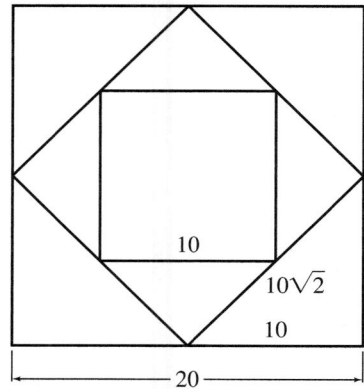

10

$10\sqrt{2}$

10

20

27. Write an augmented matrix for each system of linear equations.

    a. $\begin{cases} 2x - 5y = 17 \\ 3x + 4y = 14 \end{cases}$ **b.** $\begin{cases} 3x - 4y + 2z = -11 \\ 2x + 2y + 3z = -3 \\ 4x - y + 5z = -13 \end{cases}$

28. Write the solution for the system of linear equations represented by each augmented matrix.

    a. $\begin{bmatrix} 1 & 0 & | & -2 \\ 0 & 1 & | & 6 \end{bmatrix}$ **b.** $\begin{bmatrix} 1 & 0 & 0 & | & 4 \\ 0 & 1 & 0 & | & 5 \\ 0 & 0 & 1 & | & 8 \end{bmatrix}$

29. Write the general solution and three particular solutions for the system of linear equations represented by this augmented matrix: $\begin{bmatrix} 1 & 0 & 3 & | & 5 \\ 0 & 1 & -2 & | & 4 \\ 0 & 0 & 0 & | & 0 \end{bmatrix}$

In Exercises 30–33 use the given elementary row operations to complete each matrix.

30. $\begin{bmatrix} 3 & 5 & | & 13 \\ 1 & 4 & | & 2 \end{bmatrix} \xrightarrow{r_1 \leftrightarrow r_2} \begin{bmatrix} \_ & \_ & | & \_ \\ \_ & \_ & | & \_ \end{bmatrix}$

31. $\begin{bmatrix} 2 & 4 & | & -10 \\ 3 & 2 & | & -3 \end{bmatrix} \xrightarrow{r_1' = \frac{1}{2}r_1} \begin{bmatrix} \_ & \_ & | & \_ \\ 3 & 2 & | & -3 \end{bmatrix}$

32. $\begin{bmatrix} 1 & 2 & | & 14 \\ 3 & -1 & | & 13 \end{bmatrix} \xrightarrow{r_2' = r_2 - 3r_1} \begin{bmatrix} 1 & 2 & | & 14 \\ \_ & \_ & | & \_ \end{bmatrix}$

33. $\begin{bmatrix} 1 & 3 & -2 & | & -6 \\ 0 & 1 & 2 & | & 3 \\ 0 & 2 & 3 & | & 4 \end{bmatrix} \xrightarrow[r_3' = r_3 - 2r_2]{r_1' = r_1 - 3r_2} \begin{bmatrix} \_ & \_ & \_ & | & \_ \\ 0 & 1 & 2 & | & 3 \\ \_ & \_ & \_ & | & \_ \end{bmatrix}$

In Exercises 34–40 use an augmented matrix and elementary row operations to solve each system of linear equations.

34. $x + 2y = 17$
$2x + 5y = 41$

35. $3x - 2y = -16$
$x + 3y = 13$

36. $3x + 4y = -1$
$2x - 3y = 5$

37. $x - 5y = 8$
$-3x + 15y = 10$

38. $x + 2y - z = -2$
$2x - y + z = 4$
$3x + 2y + 2z = 3$

39. $x - 3z = 10$
$2x + y = 6$
$2y + z = 5$

40. $2x - 3y + 4z = 11$
$3x + 2y - 4z = -4$
$4x + y - 3z = -1$

41. Graph the plane defined by the linear equation $5x + 2y + 2.5z = 10$.

42. Weights of Pallets of Bricks and Blocks   A truck has an empty weight of 30,000 lb. On the first trip the truck delivered two pallets of bricks and three pallets of concrete blocks. The gross weight on this delivery was 48,000 lb. On the second trip the truck delivered four pallets of bricks and one pallet of concrete blocks. The gross weight on this delivery was 49,600 lb. Use this information to determine the weight of a pallet of bricks and the weight of a pallet of concrete blocks.

43. Price of Calculators   A department store chain with three stores retails calculators of types A, B, and C. The table shows the number sold of each type of calculator and the total income from these sales at each store. Find the price of each type of calculator.

	A	B	C	TOTAL SALES ($)
Store 1	1	2	5	156
Store 2	2	4	8	294
Store 3	3	3	7	321

44. Determine the distance from $(-4, 7)$ to $(1, -5)$.

45. Determine the midpoint between $(-5, 2)$ and $(7, 12)$.

In Exercises 46–48, refer to the following figure.

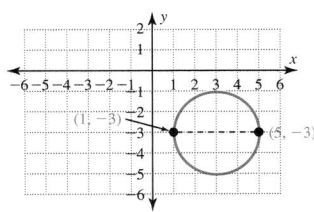

46. Determine the center of the graphed circle.

47. Determine the length of a radius of the graphed circle.

48. Determine the equation of the graphed circle.

In Exercises 49–54 match each graph with the corresponding equation.

49. $y = 2x - 4$

50. $y = 2(x - 1)^2$

51. $(x - 1)^2 + (y + 2)^2 = 4$

52. $\dfrac{(x - 1)^2}{9} + \dfrac{(y + 2)^2}{4} = 1$

53. $\dfrac{(x - 1)^2}{9} - \dfrac{(y + 2)^2}{4} = 1$

54. $y = \sqrt{4 - x^2}$

A. 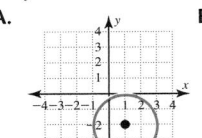   **B.** 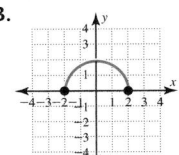   **C.**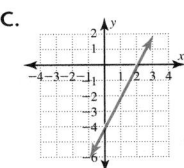

D. 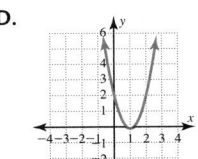   **E.** 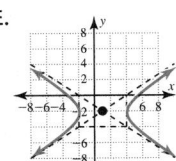   **F.**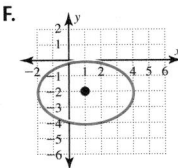

In Exercises 55–64 graph each equation.

55. $y = \dfrac{1}{2}x^2 - 3$

56. $y = \dfrac{1}{2}x - 3$

57. $y = -x^2 + 4$

58. $x^2 + y^2 = 36$

59. $(x + 3)^2 + (y - 2)^2 = 9$

60. $y = \sqrt{25 - x^2}$

61. $y = -\sqrt{25 - x^2}$

62. $\dfrac{x^2}{49} + \dfrac{y^2}{16} = 1$

63. $\dfrac{(x - 1)^2}{4} + \dfrac{(y + 3)^2}{9} = 1$

64. $\dfrac{x^2}{36} - \dfrac{y^2}{9} = 1$

In Exercises 65 and 66, refer to the following figure.

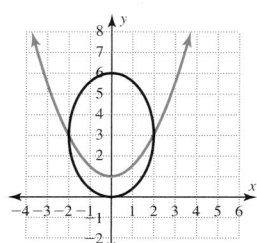

65. Use the graphs shown in the figure to solve this system of equations:

$$\begin{cases} y = \dfrac{1}{2}x^2 + 1 \\ \dfrac{x^2}{4} + \dfrac{(y-3)^2}{9} = 1 \end{cases}$$

66. Use the graphs shown in the figure to graph the solution of this system of inequalities:

$$\begin{cases} y \geq \dfrac{1}{2}x^2 + 1 \\ \dfrac{x^2}{4} + \dfrac{(y-3)^2}{9} \leq 1 \end{cases}$$

MASTERY TEST   FOR CHAPTER 11

[11.1] 1. Given $f(x) = 3x - 5$ and $g(x) = x + 2$, determine each function.
   a. $f + g$ **b.** $f - g$
   c. $f \cdot g$ **d.** $\dfrac{f}{g}$

[11.1] 2. Given $f(x) = x^2 + 1$ and $g(x) = 2x - 1$, determine
   a. $(f \circ g)(5)$ **b.** $(g \circ f)(5)$
   c. $(f \circ g)(x)$ **d.** $(g \circ f)(x)$

[11.2] 3. a. Write the first six terms of an arithmetic sequence with $a_1 = 4$ and $d = 3$.
   b. Write the first six terms of a geometric sequence with $a_1 = 4$ and $r = 3$.
   c. Write the first six terms of an arithmetic sequence with $a_1 = 2$ and $a_2 = 8$.
   d. Write the first six terms of a geometric sequence with $a_1 = 2$ and $a_2 = 8$.

[11.2] 4. Evaluate each series.
   a. The sum of the first five terms of an arithmetic sequence with $a_1 = 2$ and $d = 5$.
   b. The sum of the first five terms of a geometric sequence with $a_1 = 2$ and $r = 5$.
   c. The sum of the first 100 terms of an arithmetic sequence with $a_1 = 10$ and $a_{100} = 21$.
   d. The sum of the first 21 terms of a geometric sequence with $a_1 = 6$ and $r = -2$.
   e. $\displaystyle\sum_{i=1}^{6} (3i^2 + i)$

[11.2] 5. a. Evaluate $\displaystyle\sum_{i=1}^{10} 2\left(\frac{1}{2}\right)^i$
   b. Evaluate $\displaystyle\sum_{i=1}^{\infty} 2\left(\frac{1}{2}\right)^i$
   c. Write $0.888\ldots$ as a fraction.
   d. Write $0.545454\ldots$ as a fraction.

[11.3] 6. a. Write the solution for the system of linear equations represented by $\begin{bmatrix} 1 & 0 & | & -2 \\ 0 & 1 & | & 7 \end{bmatrix}$.
   b. Write the solution for the system of linear equations represented by $\begin{bmatrix} 1 & 0 & | & -2 \\ 0 & 0 & | & 7 \end{bmatrix}$.

   c. Write the general solution and three particular solutions for the system of linear equations represented by $\begin{bmatrix} 1 & 2 & | & 6 \\ 0 & 0 & | & 0 \end{bmatrix}$.
   d. Use augmented matrices to solve
$$\begin{cases} 2x + 4y = -6 \\ 3x - 5y = 68 \end{cases}.$$

[11.4] 7. Use an augmented matrix to solve each system of linear equations.
   a. $\begin{cases} x + 3y + 2z = 13 \\ 3x + 3y - 2z = 13 \\ 6x + 2y - 5z = 13 \end{cases}$
   b. $\begin{cases} x + 3y - 2z = 2 \\ 2x - y + z = -1 \\ -5x + 6y - 5z = 5 \end{cases}$
   c. $\begin{cases} x + y + z = 1 \\ -2x + y + z = -2 \\ 3x + 6y + 6z = 5 \end{cases}$

[11.5] 8. a. Calculate the distance between $(-9, 30)$ and $(5, -18)$.
   b. Determine the midpoint between $(-9, 30)$ and $(5, -18)$.
   c. The endpoints of the diameter of a circle are at $(3, 11)$ and $(-2, -1)$. Calculate the length of this diameter.
   d. The endpoints of the diameter of a circle are at $(3, 11)$ and $(-2, -1)$. Determine the center of this circle.

[11.5] 9. Graph each of these conic sections.
   a. $y = (x + 3)^2 - 4$
   b. $(x - 3)^2 + (y + 4)^2 = 16$
   c. $\dfrac{(x-3)^2}{25} + \dfrac{(y+4)^2}{16} = 1$
   d. $\dfrac{x^2}{25} - \dfrac{y^2}{16} = 1$

[11.5] **10.** Write the equation for each of these conic sections.

a.

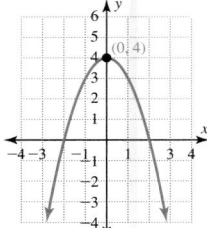

b.

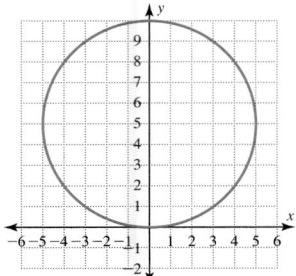

c.

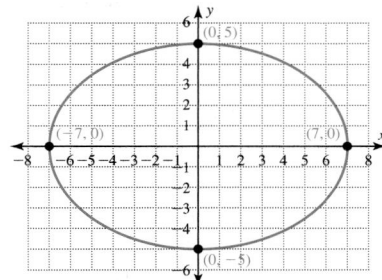

d.

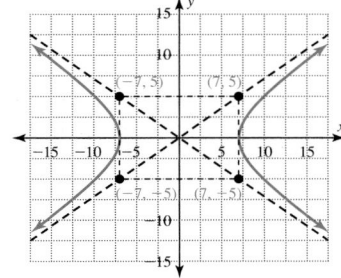

APPENDIX A

Answers to Selected Exercises

Chapter 1

Using the Language and Symbolism of Mathematics   1.1
1. Variable **2.** Additive inverse **3.** 0 **4.** a **5.** 7; left **6.** Absolute value; distance **7.** Greater; equal **8.** Less; left **9.** Equal
10. Approximately equal **11.** Principal square root **12.** Rational
13. Rational **14.** Irrational **15.** Pi **16.** Infinity **17.** Is not **18.** Is
19. $(-\infty, \infty)$ **20.** Larger; $[-3, 2]$ **21.** (a, b)

Exercises 1.1
1. a. -9 **b.** 13 **c.** $\frac{3}{5}$ **d.** -7.23 **e.** $\sqrt{13}$ **f.** 0 **3. a.** 7 **b.** -7 **c.** 7
5. a. 0 **b.** 0 **c.** 0 **d.** 0 **e.** 0 **f.** 0 **7. a.** -7 **b.** 8 **c.** 21 **d.** 23 **9. a.** 29
b. 29 **c.** -29 **d.** -29 **11. a.** 25 **b.** 9 **c.** 9 **d.** -9 **13. a.** 14 **b.** 4
c. 4 **d.** -14 **15. a.** 3 **b.** 4 **c.** 7 **d.** 5 **17. a.** 13 **b.** 9 **19.** (b) **21.** (c)
23. (d) **25.** Integer estimate: 7; inequality: $<$; approximation: 7.389
27. Integer estimate: 1; inequality: $>$; approximation: 0.858
29. Integer estimate: 14; inequality: $<$; approximation: 14.272 **31. a.**
$x = 7$ **b.** $x = 3$ **33. a.** $>$ **b.** $<$ **c.** $<$ **d.** $<$ **35. a.** $<$ **b.** $>$ **c.** $=$
d. $=$ **37. a.** $>$ **b.** $>$ **c.** $<$ **d.** $<$
39. a. 15 **b.** 0, 15 **c.** $-11, -\sqrt{9}, 0, 15$ **d.** $-11, -4.8, -\sqrt{9}, 0, 1\frac{3}{5}$,
15 **e.** $\sqrt{5}$ **41.** $-3 < x \le 3$; x is greater than -3 and less than or
equal to 3; $(-3, 3]$

43. x is greater than 1; ; $(1, \infty)$

45. $x \ge -3$; 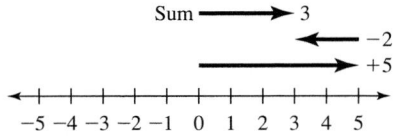 ; $[-3, \infty)$

47. $x \le 4$; x is less than or equal to 4;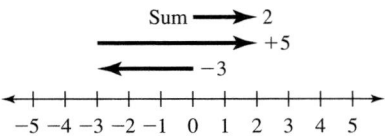

49. Integer, rational number **51.** Natural number, whole number,
integer, rational number **53.** Irrational number **55.** Rational number
57.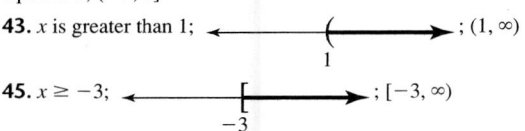

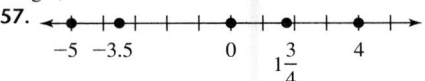

59.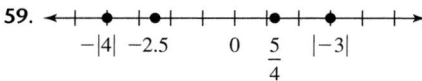

61. a. Rational **b.** Irrational **63. a.** Rational **b.** Irrational
65. a. Rational **b.** Rational **67.** $|x| = y$ **69.** $3.14 < \pi < 3.15$
71. $-x \le -2$ **73.** Answers may vary. **a.** 5 **b.** -5 **75.** Answers may
vary. **a.** 0 **b.** $-5, -3$ **c.** $-\frac{1}{2}, -\frac{7}{3}$ **d.** $\frac{3}{2}, \frac{5}{4}$ **77.** b **79.** a

Using the Language and Symbolism of Mathematics   1.2
1. Terms; addends; sum **2.** Commutative **3.** Associative **4.** Both
5. Larger **6.** Commutative; associative **7.** $\frac{w + y}{x}$; 0 **8.** Evaluate
9. $x + 5$ **10.** $x + y + z$

Exercises 1.2
1. a. 13 **b.** -3 **c.** 3 **d.** -13 **3. a.** 0 **b.** -16 **c.** 9 **d.** -43 **5. a.** -21
b. -19 **7. a.** -18 **b.** -17 **9.** -3 **11.** -1 **13. a.** -2 **b.** 1
15. a. -2 **b.** 6 **17. a.** 34 **b.** 0 **c.** 4 **19. a.** $\frac{4}{11}$ **b.** $\frac{7}{20}$ **c.** $-\frac{13}{12}$
21. a. 0 **b.** -0.3 **c.** 1.29
23. Verbally: the sum of 5 and -2 is 3; graphically:

Sum ⟶ 3
⟵ -2
⟶ $+5$

$-5 \ -4 \ -3 \ -2 \ -1 \quad 0 \quad 1 \quad 2 \quad 3 \quad 4 \quad 5$

25. Verbally: the sum of -3 and 5 is 2; graphically:

Sum ⟶ 2
⟶ $+5$
⟵ -3

$-5 \ -4 \ -3 \ -2 \ -1 \quad 0 \quad 1 \quad 2 \quad 3 \quad 4 \quad 5$

27. a. 3 **b.** -3 **c.** 3 **29. a.** 4 **b.** 2 **c.** -2 **31. a.** -2 **b.** 2 **c.** -10
33. a. 10 **b.** 4 **c.** -4 **35.** 21.6 cm **37.** 70 cm **39.** Commutative
41. $(11 + 12) + 13$ **43.** 32 h **45.** $-\$485,000,000$ **47.** 22%

49. 22% **51.** 26% **53.** $-2°$F **55.** 76.6 m **57.** 8 cm **59.** Q1: +$12,000; Q2: +$2000; Q3: −$2000; Q4: −$12,000 **61.** +; 5.8 **63.** −; $-\frac{47}{20}$ **65.** +; 3.97 **67.** b. **69.** b. **71.** b. **73.** d. **75.** $\frac{2}{5}$ **77.** 0.375 **79. a.** $a + b = b + a$ **b.** $x + (-x) = 0$ **81. a.** $\pi \neq 3.14$ **b.** $\pi \approx 3.14$ **83.** $r + 50$ **85.** $n + 1$

Using the Language and Symbolism of Mathematics   1.3

1. Minuend; subtrahend; difference **2.** $x - 5$ **3.** $w - 17$ **4.** $\frac{w - y}{x}$; 0 **5.** $x_2 - x_1$ **6.** $y_2 - y_1$ **7.** "a sub 1"; "a sub n" **8.** Sequence **9.** Solution **10.** Satisfy **11.** Checking **12.** Solution set

Exercises 1.3

1. a. -6 **b.** 20 **c.** -20 **d.** 6 **3. a.** -7 **b.** 7 **c.** -37 **d.** 37 **5. a.** 0 **b.** $-\frac{7}{3}$ **c.** $-\frac{1}{12}$ **d.** $\frac{7}{30}$ **7. a.** -15 **b.** -57 **c.** -7 **9. a.** -34 **b.** -12 **c.** 7 **11. a.** 49 **b.** -23 **13. a.** -9 **b.** 23 **15. a.** 2 **b.** 6 **17. a.** 3 **b.** -3 **19. a.** 6 **b.** -14 **21. a.** 0 **b.** -2.7 **c.** 3.8 **23.** Verbally: the difference of three minus one is two; graphically:

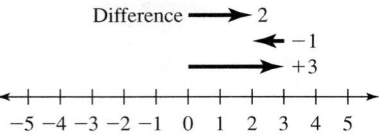

25. Verbally: the difference of -4 minus 2 is -6; graphically:

27. a. 0 **b.** 10 **c.** 0 **29. a.** -3 **b.** 13 **c.** 2 **31. a.** 6 **b.** 6 **c.** 0 **33. a.** 4 is not a solution; 5 is a solution **b.** 4 is a solution; 5 is not a solution **35. a.** 4 is a solution; 5 is not a solution **b.** 4 is not a solution; 5 is a solution **37. a.** $-3, -2, -1, 0, 1$ **b.** $2, 1, 0, -1, -2$ **39.** 70 ft **41.** $624 **43.** $-$37,370,000 **45.** 17% **47.** $-7°$ **49. a.** $+6°$ **b.** $-6°$ **51. a.** $+9°$ **b.** $-9°$ **53. a.** $-4°$ **b.** $+4°$ **55.** Q1 change: $+0.04$; Q2 change: -0.04; Q3 change: $+0.09$; Q4 change: $+0.10$; **57.** $+5.5$ yd **59.** $97.79 **61.** 79.449 m **63.** sign of difference: −; difference: -175.93 **65.** sign of difference: +; difference: $\frac{47}{24}$ **67.** sign of difference: +; difference: ≈ 4.303 **69.** b **71.** b **73.** a **75. a.** $x - 8 = z$ **b.** $z - y = x$ **77. a.** $|x| - |-y| = 11$ **b.** $|x - y| = 4$ **79.** $r - 40$ **81.** $P - 2000$ **83.** $n - 2$

Using the Language and Symbolism of Mathematics   1.4

1. Factors; product **2.** PRT **3.** $0.14C$ **4.** Commutative **5.** Associative **6.** 0 **7.** Positive **8.** Negative **9.** 60 **10.** $\frac{wy}{xz}$; 0; 0 **11.** Reciprocals; inverses **12.** 1 **13.** 0 **14.** 1

Exercises 1.4

1. a. -77 **b.** -77 **c.** 77 **d.** -77 **3. a.** -60 **b.** 60 **c.** -60 **d.** 60 **5. a.** 8 **b.** 0 **c.** 30 **d.** -1000 **7. a.** -123.4 **b.** $-123,400$ **c.** -1.234 **d.** $-1,234,000$ **9. a.** $\frac{3}{10}$ **b.** $-\frac{2}{15}$ **c.** $\frac{15}{88}$ **d.** $\frac{10}{39}$ **11. a.** 20 **b.** 20 **c.** 6 **d.** 6 **13. a.** 1 **b.** 1 **c.** 1 **d.** -5 **15. a.** 1 **b.** 1 **c.** -1 **d.** 1.7 **17. a.** -300 **b.** -300 **c.** 300 **19. a.** 900 **b.** -1200 **c.** -1800 **21. a.** -2 is a solution; 2 is not a solution **b.** -2 is not a solution; 2 is a solution **23. a.** -2 is not a solution; 2 is a solution **b.** -2 is not a solution; 2 is not a solution **25. a.** $2, 4, 6, 8, 10$ **b.** $1, 4, 9, 16, 25$

27. 13, 40 **29.** $-3, -40$ **31.** π, 0 **33. a.** $8x$ **b.** $7y = 11$ **35.** $3P$ **37. a.** Sum **b.** Difference **c.** Product **39. a.** Commutative **b.** Associative **41. a.** Multiplication **b.** Addition **43.** 180 cm^2 **45.** 288 cm^2 **47.** 3360 in^3 **49.** 14.82 cm **51.** $280 **53.** $90.63 **55.** 75 ml **57.** 0.675 liters **59.** 1260 mi **61.** 13 m **63.** 15.7 cm **65.** 20 g **67.** $31.25 **69.** $304.56 **71.** $64; $32; $75; $150; Total: $321 **73.** Q1: $16,800; Q2: $19,200; Q3: $18,200; Q4: $16,000; **75.** +; 125.28 **77.** +; $\frac{28}{65}$ **79.** −; -5.477 **81.** a **83.** c

Using the Language and Symbolism of Mathematics   1.5

1. Dividend; divisor; quotient **2.** $\frac{D}{R} = T$ **3.** $m{:}w = 3{:}5$ **4.** Positive **5.** Negative **6.** Multiplicative; reciprocal **7.** $\frac{wz}{xy}$; x; y; 0 **8.** Ratio **9.** Undefined **10.** Range **11.** Mean **12.** Relative **13.** Units

Exercises 1.5

1. a. -8 **b.** 8 **c.** -8 **d.** 0 **3. a.** -96 **b.** -24 **c.** 144 **d.** 16 **5. a.** 1230 **b.** $-12,300$ **c.** -0.123 **d.** 12.3 **7. a.** $-\frac{2}{3}$ **b.** -4 **c.** $\frac{1}{9}$ **d.** $\frac{3}{5}$ **9. a.** 0 **b.** Undefined **c.** Undefined **d.** Undefined **11. a.** 4 **b.** -4 **c.** -3 **d.** -2 **13. a.** 3 **b.** -3 **c.** -3 **d.** 3 **15. a.** -6 is not a solution; 6 is a solution **b.** -6 is a solution; 6 is not a solution **17. a.** 120, 60, 40, 30, 24 **b.** $-\frac{1}{2}, -1, -\frac{3}{2}, -2, -\frac{5}{2}$ **19.** 18; 20 **21.** $-9, -13$ **23.** 6, 12 **25.** 1:4 **27.** 3:1 **29.** 1:5 **31.** 8:5 **33.** 64-oz bottle **35.** 82 **37.** Actual perimeter: 155.8 cm; estimated perimeter: 160 cm; percent error: 2.7% **39.** Actual perimeter: 50.9 cm; estimated perimeter: 52 cm; percent error: 2.2% **41.** Range: $108.31; mean: $156.77 **43.** 290 mi/h **45.** 24 cm/min **47.** 300 lightbulbs/h **49.** $\frac{1}{5}$ pool/h **51.** −; -3.28 **53.** +; $\frac{9}{20}$ **55.** +; 5 **57.** −; -1.475 **59.** c **61.** c **63.** b **65.** $82.99; $69.87 **67.** Q1: 0.12; Q2: $0.10; Q3: $0.10; Q4: 0.075 **69.** 14.5 cm **71. a.** $a \div 5$ **b.** $\frac{d}{t} = 7$ **73. a.** $\frac{x}{y} = 8$ **b.** $xy = 8$ **75. a.** $5 > \frac{x}{3}$ **b.** $-2 \leq \frac{8}{w}$ **77.** -1

Using the Language and Symbolism of Mathematics   1.6

1. x^7 **2.** 5th power **3.** Factor **4.** w; 4 **5.** Multiplication **6.** Addition **7.** Exponentiation **8.** $-x$ **9.** x **10.** Distributive, multiplication, addition **11.** Coefficient **12.** Distributive **13.** Combining

Exercises 1.6

1. a. 5^4 **b.** $(-4)^3$ **c.** y^5 **d.** $(3z)^6$ **3. a.** 8 **b.** 9 **c.** 9 **d.** -9 **5. a.** 0 **b.** 1 **c.** 1 **d.** -1 **7. a.** -100 **b.** 100 **c.** 0.001 **d.** -1000 **9. a.** $\frac{1}{8}$ **b.** $\frac{9}{25}$ **c.** $\frac{1}{9}$ **d.** $\frac{17}{19}$ **11. a.** 21 **b.** 56 **13. a.** 2 **b.** 70 **15. a.** 100 **b.** 52 **17. a.** 25 **b.** 1 **19. a.** -19 **b.** 0 **21. a.** -45 **b.** 9 **23. a.** -27 **b.** 26 **25. a.** 0 **b.** 12 **27. a.** -117 **b.** -27 **29. a.** -1000 **b.** 52 **31. a.** 17 **b.** -17 **33. a.** $\frac{1}{16}$ **b.** $\frac{17}{32}$ **35. a.** 24 **b.** 208 **37. a.** 16 **b.** 160 **39.** 158 **41.** 56 **43.** $\frac{1}{2}$ **45. a.** $7x + 35$ **b.** $9(x + 4)$ **47. a.** $-6x + 14$ **b.** $-5(a + b)$ **49. a.** $10x - 15y$ **b.** $11(2x - 3y)$ **51. a.** $-b + 3$ **b.** $b - 5$ **53. a.** $7x$ **b.** $-3x$ **c.** $7\sqrt{3}$ **55. a.** $6w$ **b.** $9w - 3z$ **57. a.** $6a - 2b$ **b.** $-2a + 8b$ **59.** $6\sqrt{3} - 2\sqrt{5}$ **61. a.** -2 is a solution; 5 is not a solution **b.** -2 is a solution; 5 is a solution **63.** -2 is not a solution; 5 is a solution **65. a.** $>$ **b.** $<$ **67.** 81; 80.407 **69.** 27; 27.465 **71.** 941,192 cm^3 **73.** 180,000 ft^3 **75.** 21,205.75 ft^3 **77. a.** $2(4x + 5y)$ **b.** $9(6x - 11) = 1$ **79. a.** $c = \sqrt{a^2 + b^2}$ **b.** $d = \sqrt{x^2 + 5}$ **81. a.** $x^2 + y^2$ **b.** $(x + y)^2$

Review Exercises for Chapter 1

1. a. -12 **b.** -20 **c.** -20 **d.** -12 **2. a.** -64 **b.** 64 **c.** -4 **d.** 4
3. a. -7 **b.** 0 **c.** 0 **d.** Undefined **4. a.** 18 **b.** 30 **c.** -144 **d.** -4
5. a. 9.01 **b.** 8.99 **c.** 0.09 **d.** 900 **6. a.** 995.5 **b.** -1004.5 **c.** -4500
d. -0.0045 **7. a.** -6 **b.** -24 **c.** 10 **d.** -8 **8. a.** -24 **b.** 6 **c.** -24
d. -10 **9. a.** -27 **b.** -3 **c.** -27 **d.** -27 **10. a.** -18 **b.** 27
c. -50 **d.** 70 **11. a.** 9 **b.** 5.25 **c.** 1.2 **d.** 18 **12. a.** 9 **b.** -9 **c.** -27
d. -27 **13. a.** 49 **b.** 25 **c.** 1 **d.** -7 **14. a.** 25 **b.** 32 **c.** 25 **d.** -32
15. a. 8 **b.** -8 **c.** -8 **d.** 14 **16. a.** 14 **b.** 10 **c.** 4 **d.** 2 **17. a.** 24
b. -120 **c.** 0 **d.** 720 **18. a.** 0 **b.** 0 **c.** -1 **d.** 1 **19. a.** $\dfrac{17}{12}$ **b.** $-\dfrac{1}{12}$
c. $\dfrac{1}{2}$ **d.** $\dfrac{8}{9}$ **20. a.** $-\dfrac{7}{75}$ **b.** $-\dfrac{133}{75}$ **c.** $\dfrac{98}{125}$ **d.** $-\dfrac{10}{9}$

21. a. -44 **b.** -71 **c.** -38 **d.** -26 **22.** 22 **23.** 79 **24.** -2 **25.** -7

26. $-$; -533 **27.** $+$; 133 **28.** $+$; 31 **29.** $+$; 41,181 **30.** $+$; 128

31. $-$; -315 **32.** $-$; -557 **33.** $-$; -5 **34.** 16; 16.049 **35.** 800;

799.4636 **36.** 3000; 2950.388 **37.** 6; 6.03 **38.** 125; 125.751501

39. 7; 7.1 **40.** $x > -2$; $(-2, \infty)$

41. $x \le 4$;

42. ; $(-\infty, 3)$

43. $-3 < x < 0$;

44. ; $[-7, -1]$

45. $4 \le x < 10$; $[4, 10)$ **46.** $-x = 11$ **47.** $|x| \le 7$ **48.** $\sqrt{26} > 5$

49. $x + 5 = 4$ **50.** $x - 3 < y$ **51.** $-3y = -1$ **52.** $\dfrac{x}{3} = 12$

53. $\dfrac{x}{y} = \dfrac{3}{4}$ **54.** $2(x + 2) = 9$ **55.** $5(3x - 4) = 13$ **56.** 7, 17, 27, 37, 47

57. 25, 20, 15, 10, 5 **58.** B **59.** A **60.** D **61.** E **62.** C **63.** 2
64. -4 **65.** 2 is not a solution; 3 is a solution **66.** 2 is a solution; 3 is
not a solution **67.** Perimeter: 54 m; area: 162 m² **68.** Perimeter error:
0.4 m; relative error: 0.7%; **69.** $2.35 per CD **70.** C **71.** A **72.** E
73. B **74.** D **75.** 20-ounce bottle **76.** 17:18 **77.** $1039.20

78. a. 40 liters **b.** 30 gal **79. a.** 200 golf balls/h **b.** $\dfrac{1}{4}$ vat/h

Mastery Test for Chapter 1

1. a. 2 **b.** 0 **c.** -1.5 **d.** -25 **2. a.** 23 **b.** 23 **c.** 0 **d.** 46
3. a. $(-2, 3]$ **b.** $[-1, \infty)$ **c.** $(-\infty, 4)$ **d.** $[5, 9)$ **4. a.** 5; 5.10
b. 10; 9.95 **c.** 3; 2.83 **d.** 4; 4.00 **5. a.** B **b.** C **c.** A **d.** E **e.** D
6. a. 6 **b.** -28 **c.** -6 **d.** $\dfrac{1}{2}$ **e.** $-\dfrac{1}{4}$ **f.** $-\dfrac{31}{40}$ **g.** 0 **h.** -12
7. a. Commutative; addition **b.** Associative; addition **c.** $x + (y + 5)$
d. $5(y + x)$ **8. a.** -6 **b.** 6 **c.** -4 **d.** 6 **9. a.** 7 **b.** -23 **c.** -7 **d.** 23
e. $\dfrac{2}{3}$ **f.** -1 **g.** $\dfrac{5}{6}$ **h.** $\dfrac{7}{30}$ **10. a.** 3, 4, 5, 6, 7 **b.** $-1, 0, 1, 2, 3$
c. 3, 5, 7, 9, 11 **d.** 11, 10, 9, 8, 7 **11. a.** Solution **b.** Not a solution
c. Solution **d.** Not a solution **12. a.** -30 **b.** 30 **c.** -30 **d.** 0 **e.** $\dfrac{1}{15}$
f. $\dfrac{1}{3}$ **g.** $-\dfrac{3}{7}$ **h.** -24 **13. a.** Cummutative; multiplication

b. Associative; multiplication **c.** $5(xy)$ **d.** $(a + b)x$ **14.** 60 m²
15. a. -3 **b.** 3 **c.** -3 **d.** 0 **e.** Undefined **f.** $-\dfrac{3}{2}$ **g.** $\dfrac{5}{4}$ **h.** $-\dfrac{66}{245}$
16. a. $\dfrac{1}{5}$ **b.** 12-ounce bottle **17. a.** 25 **b.** 32 **c.** 25 **d.** -25 **e.** 0
f. $\dfrac{9}{49}$ **g.** 219 **h.** -1 **18. a.** 9 **b.** 27 **c.** 24 **d.** $-\dfrac{2}{33}$ **e.** 64 **f.** 34
g. -59 **h.** 2 **19. a.** Distributive; multiplication; addition
b. $33x - 44$ **c.** $5(3x + 4)$ **d.** $-14x + 35$ **e.** $8(2x - 3)$

Chapter 2

Using the Language and Symbolism of Mathematics   2.1
1. Cartesian **2.** x; y **3.** Origin; 0; 0 **4.** 0 **5.** Counterclockwise **6.** III
7. Scatter **8.** Arithmetic **9.** Difference **10.** Linear

Exercises 2.1
1. A: $(-1, 5)$, II; B: $(5, -2)$, IV; C: $(3, 2)$, I; D: $(-4, -1)$, III
3.

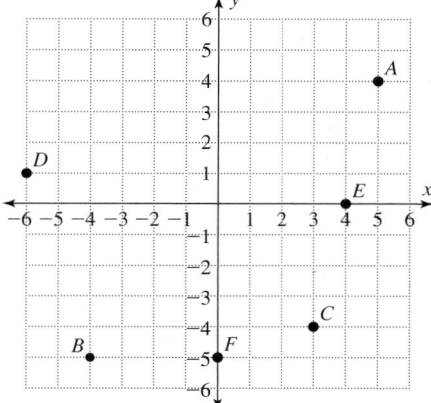

5. a. II **b.** III **c.** IV **d.** I **7. a.** y-axis **b.** x-axis **c.** x-axis **d.** y-axis
9.

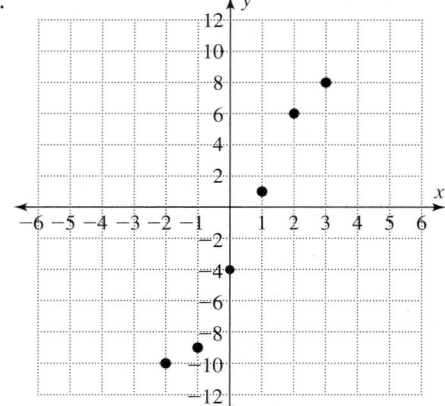

The points do not lie on one line, but they all lie close to a line.

11.

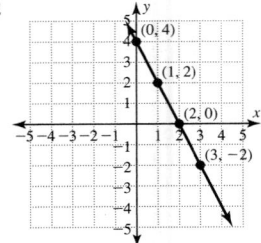

The points all lie on a line; the sequence is arithmetic; $d = 2$

13.

The points all lie on a line; the sequence is arithmetic; $d = -3$
15. 4, 9, 14, 19, 24   **17.** $-4, -9, -14, -19, -24$ **19. a.** 7   **b.** 10
c. 19   **d.** 37   **21. a.** 4   **b.** -2 **c.** -20 **d.** -236 **23.** 1, 1.5, 2, 2.5, 3
25. 4, 1, -2, -5, -8 **27.** Arithmetic; $d = 4$ **29.** Arithmetic;
$d = -10$ **31.** Not arithmetic   **33. a.** Arithmetic; $d = 2$ **b.** Not
arithmetic   **35.** Arithmetic; $d = 2$ **37.** Not arithmetic   **39.** Arithmetic;
$d = -2$ **41.** Arithmetic; $d = 0$ **43.** $-4, -3, -2, -1, 0, 1, 2$;
arithmetic   **45.** 4, 4, 4, 4, 4; arithmetic   **47.** $-9, -6, -1, 6$; not
arithmetic   **49. a.** $a_0 = 0$; nothing is spent yet   **b.** $a_6 = 2550$; a total of
$2550 has been spent on the loan   **c.** $a_{12} = 5100$; a total of $5100 has
been spent on the loan   **51. a.** $a_1 = 2850$; a total of $2850 has been
spent on the lease   **b.** $a_{18} = 10,500$; a total of $10,500 has been spent
on the lease   **c.** $a_{24} = 13,200$; a total of $13,200 has been spent on the
lease   **53.** $>; >$ **55.** $<; <$ **57.** I; III   **59.** Up   **61.** Yes; $d = 2$
assuming these integers are listed in increasing order   **63.** Area:
20 units^2; perimeter: 18 units   **65.** Yes; $d = 1$ **67.** No

Using the Language and Symbolism of Mathematics   2.2
1. Function   **2.** f; x **3.** x; output   **4.** Straight line   **5.** Linear   **6.** 5; 9
7. Modeling

Exercises 2.2
1. 7   **3.** 4   **5.** 4   **7.** -56 **9.** 28   **11.** 3   **13.** 3   **15.** 6   **17.** 0.5, 1, 1.5, 2, 2.5

19. 4, 2, 0, -2

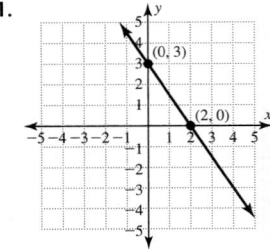

21.

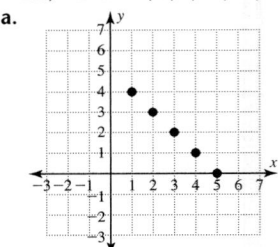

23. a. $-3, -1, 1, 3, 5, 7, 9$ **b.** 3, 4, 5, 6, 7, 8, 9   **c.** 3, 23, 43, 63, 83,
103, 123   **25. a.** $-4, -3, -2, -1, 0, 1, 2$ **b.** $-5, -4.5, -4, -3.5,$
$-3, -2.5, -2$ **c.** $-2, 3, 8, 13, 18, 23, 28$
27. a.

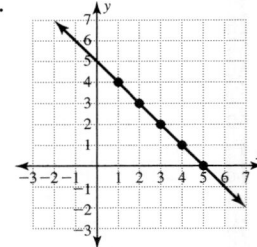

b.

29. $-3, -1, 1, 3, 5$ **31.** $-2, 0, 3, 5$
33. 2, 1, 0, -1, -2

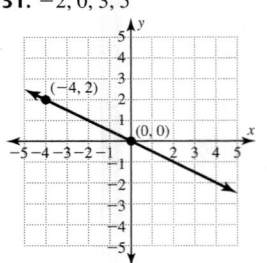

35. $-8, -5.5, -3, -0.5, 2$

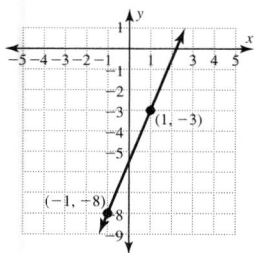

37. a. $f(x) = 0.15x$ **b.** $f(x) = 0.85x$ **c.** $37.40, 51.00, 57.80, 76.50
39. a. $f(x) = x + 5$ **b.** $f(x) = x - 5$ **41. a.** $f(x) = 2x + 64$
b. $f(x) = 32x$ **43. a.** $f(x) = 180 - x$ **45. a.** $f(x) = 15x + 500$
b. 500, 2000, 3500, 5000, 6500, 8000 **c.** $f(250) = 4250$; the cost of
producing 250 units is $4250

Using the Language and Symbolism of Mathematics   2.3
1. x-intercept **2.** y-intercept **3.** True **4.** Infinite **5.** Solution
6. Intersection **7.** y-coordinate **8.** x-coordinate **9.** y **10.** x

Exercises 2.3
1. Solution **3.** Solution **5.** Solution **7.** Solution **9.** Not a solution
11. Not a solution **13.** Solution **15.** Solution **17.** The points A and D
are solutions **19.** The point C is a solution

21.

x	-2	0	2	3
y	5	3	1	0

x-intercept: (3, 0)
y-intercept (0, 3)

23.

x	-2	0	2	8
y	-5	-4	-3	0

x-intercept: (8, 0)
y-intercept (0, -4)

25.

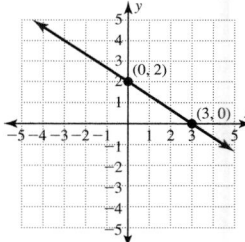

27.

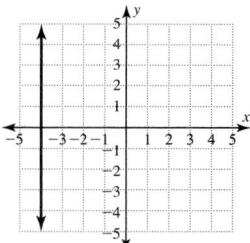

29.

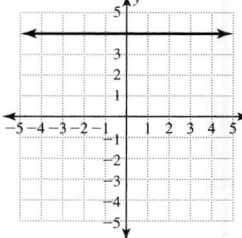

31.

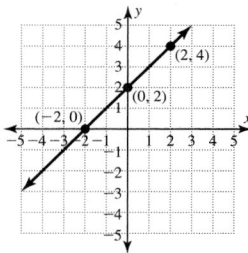

33.

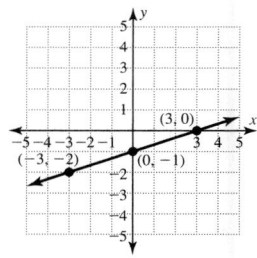

35.

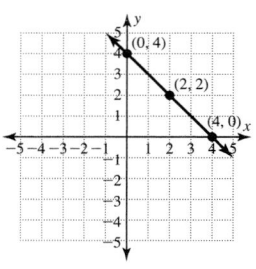

37.

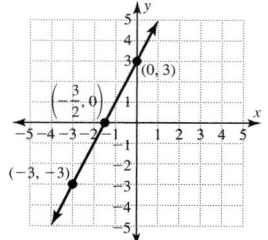

39.

x	y
-2	-15
-1	-12
0	-9
1	-6
2	-3

x-intercept: (3, 0)
y-intercept: (0, -9)

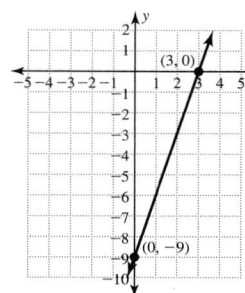

41.

x	y
-2	2
-1	1.5
0	1
1	0.5
2	0

x-intercept: (2, 0)
y-intercept: (0, 1)

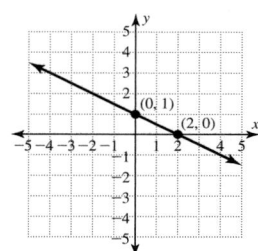

43. $(-2, 3)$ **45.** $(-1, -2)$ **47.** $(3, 2)$ **49.** $(1, -3)$ **51.** $(2, 3)$
53. x-intercept: (80, 0); y-intercept: (0, -500): The x-intercept tells us
that a profit of $0.00 occurs when 80 units are produced. (This is the
break-even point.) The y-intercept tells us that we lose $500.00 when
no items are produced. $500 is the overhead cost.

55.

x	f(x)
1	790.83
2	1124.16
3	1457.49
4	1790.82
5	2124.15
6	2457.48
7	2790.81
8	3124.14
9	3457.47
10	3790.80
11	4124.13
12	4457.46

57. Each option will have the same cost of $11.00 when the
production is 4 units **59.** Each option will have the same cost of
$550.00 at the end of the second month

Using the Language and Symbolism of Mathematics   2.4
1. Linear **2.** Linear **3.** Linear **4.** One **5.** Equivalent **6.** Addition
7. $b - c$ **8.** Conditional **9.** Identity **10.** Contradiction

Exercises 2.4

1. b **3.** $x = 24$ **5.** $v = -4$ **7.** $y = -1$ **9.** $z = -11$ **11.** $y = 4$
13. $x = 16$ **15.** $y = 63$ **17.** $v = 2$ **19.** $y = 0$ **21.** $m = 26$
23. $n = 99$ **25.** $t = 0$ **27.** $v = 0$ **29.** $x = 7$ **31.** $v = 25$ **33.** $w = 22$
35. $m = -5$ **37.** $n = 9$ **39.** $y = -12$ **41. a.** Conditional; the
solution is $x = 0$ **b.** Contradiction; no solution **c.** Identity; every real
number is a solution **43. a.** Contradiction; no solution **b.** Identity;
every real number is a solution **c.** Conditional; the solution is $v = 4$
45. a. $9x - 5$ **b.** $x = -7$ **47. a.** $x - 7$ **b.** $x = 7$ **49. a.** $5.8x + 0.6$
b. $x = 4$ **51.** $x = 5$ **53.** $x = 2.3$ **55.** $x = -1$ **57.** $x = 2$ **59.** $x = 3$
61. 2; 1.905 **63.** 470; 470.177 **65.** $3m + 7 = 2m - 8$; $m = -15$
67. $12 - 9m = 2 - 10m$; $m = -10$ **69.** $2(3m - 9) = 5(m + 13)$;
$m = 83$ **71.** $a = 7$ **73.** $x = 253$

Using the Language and Symbolism of Mathematics   2.5

1. Equivalent **2.** Multiplication **3.** $\dfrac{b}{c}$ **4.** 6 **5.** Conditional **6.** 0

7. Distributive **8.** Least common denominator **9.** 4 **10.** -4

Exercises 2.5

1. a. $x = 6$ **b.** $x = \dfrac{1}{6}$ **c.** $x = -7$ **d.** $x = 7$ **3. a.** $t = -\dfrac{3}{2}$ **b.** $t = -\dfrac{2}{3}$

c. $t = 96$ **d.** $t = \dfrac{3}{2}$ **5. a.** $m = -72$ **b.** $m = 8$ **7. a.** $t = -4$

b. $t = -1$ **9. a.** $y = -5$ **b.** $y = 3$ **11.** $z = -3$ **13.** $m = -\dfrac{1}{9}$

15. $w = 0$ **17.** $v = -15$ **19.** $x = 0$ **21.** $y = -5$ **23.** $a = 110$
25. $x = -7$ **27.** $v = -16$ **29.** No solution **31.** $w = -10$ **33.** $x = 4$

35. $x = -\dfrac{3}{2}$ **37. a.** Contradiction; no solution **b.** Identity; every real

number is a solution **c.** Conditional; solution: $x = 0$ **39. a.** $x - 2$

b. $x = 2$ **41. a.** $21x - 19$ **b.** $x = \dfrac{19}{21}$ **43.** $s \approx 25$ cm **45.** $75.80

47. -4; -4.2 **49.** 4; 4.1 **51.** $x = 1.5$ **53.** $x = 0.7$

55. $5(x + 2) = 7(x - 3)$; $x = \dfrac{31}{2}$ **57.** $2(3v - 2) = 4(v + 9)$; $v = 20$

59. $2(3m - 5) = 3(m + 6) - 4$; $m = 8$ **61.** $\dfrac{1}{3}(2x + 5) = 4x + 2$;

$x = -\dfrac{1}{10}$ **63.** $x = 6$ **65.** $a = \dfrac{1}{2}$ **67.** $x = 2$

Using the Language and Symbolism of Mathematics   2.6
1. Proportion **2.** Terms **3.** b; c **4.** a; d **5.** Directly **6.** Linear
7. Arithmetic **8.** Line

Exercises 2.6
1. $x = 12$ **3.** $x = 10$ **5.** $x = 3$ **7.** $y = 11$ **9.** $y = 16$ **11.** $a = 2$
13. $b = 33$ **15.** $x = 3$ **17.** $z = -8$ **19.** $x = 9$ **21.** $v = 3$ **23.** $d = kt$

25. v varies directly as w **27.** $v = 10$ **29.** $g = 4$ **31.** $k = \dfrac{4}{5}$

33. Algebraically: $y = \dfrac{1}{2}x$

Numerically:

x	y
1	0.5
2	1
3	1.5
4	2
5	2.5

Graphically:

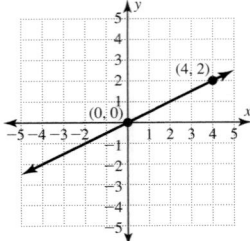

35. Numerically:

x	y
1	-0.5
2	-1
3	-1.5
4	-2
5	-2.5

Graphically:

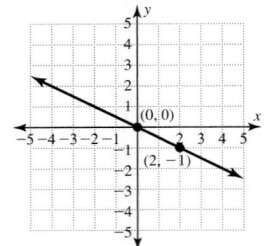

Verbally: y varies directly as x with a constant of variation $-\dfrac{1}{2}$

37. $4\dfrac{1}{2}$ cups **39.** $a = 18$ ft **41.** 10 mm **43.** 50 bulbs

45. 1416 bricks **47.** 1.64 Australian dollars for one U.S. dollar
49. 8.75 m × 7.5 m **51.** 13.5 lb **53.** 9 **55.** 22.4 ft **57.** 17.5 cm

59. 17.5 ft **61.** $53\dfrac{1}{3}$ ft **63.** The ERA would be 9. **65.** 24 doses

67. 1.5 **69.** 0.5

Using the Language and Symbolism of Mathematics   2.7
1. Specified variable **2.** x-intercept **3.** y-intercept **4.** Distributive

Exercises 2.7
1. $y = -2x + 7$ **3.** $y = 3x - 2$ **5.** $y = 2x - 3$ **7.** $y = 2x + 4$
9. $y = -3x - 2$ **11.** $y = -4x + 4$ **13.** $y = 6x - 8$ **15.** x-intercept:

$(5, 0)$; y-intercept: $(0, 3)$ **17.** x-intercept: $\left(-\dfrac{15}{2}, 0\right)$; y-intercept: $(0, 5)$

19. $l = \dfrac{A}{w}$ **21.** $r = \dfrac{C}{2\pi}$ **23.** $V_1 = \dfrac{V_2 T_1}{T_2}$ **25.** $w = \dfrac{P - 2l}{2}$

27. $E = V + F - 2$ **29.** $b = \dfrac{2A}{h} - a$ **31.** $F = \dfrac{9}{5}C + 32$

33. $a = l - (n - 1)d$ **35.** $n = \dfrac{l - a}{d} + 1$ **37.** $a = S(1 - r)$

39. $m = \dfrac{y - b}{x}$ **41.** $h = \dfrac{3V}{\pi r^2}$

43. Algebraically: $y = 2x - 3$;

Numerically:

x	y
-3	-9
-2	-7
-1	-5
0	-3
1	-1
2	1
3	3

Graphically:

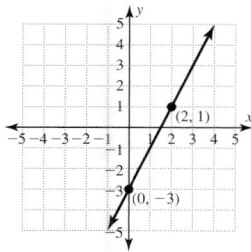

45. Algebraically: $y = -\dfrac{1}{2}x - 2$;

Numerically:

x	y
-3	$-.5$
-2	-1
-1	-1.5
0	-2
1	-2.5
2	-3
3	-3.5

Graphically:

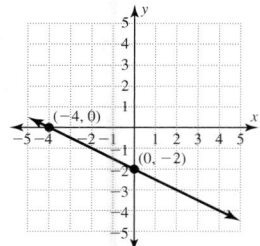

47. $P = \dfrac{T}{R}$;

T	P
50	625
100	1250
150	1875
200	2500
250	3125
300	3750
350	4375

49. $w = \dfrac{P - 2l}{2}$;

l	w
50	150
60	140
70	130
80	120
90	110
100	100
110	90

Review Exercises for Chapter 2

1. A: $(4, 0)$; B: $(2, 4)$, I; C: $(-6, 2)$, II; D: $(0, -2)$

2.

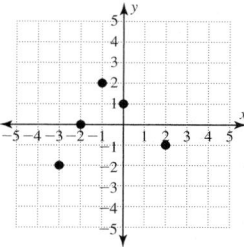

3. $1, -1, -3, -5, -7$

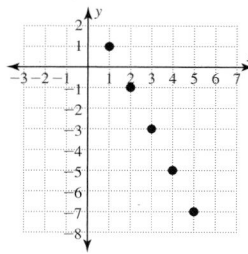

4. $2, 2.5, 3, 3.5, 4$

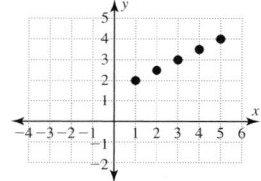

5. Arithmetic; $d = 3$ **6.** Not arithmetic **7.** Arithmetic; $d = -2$ **8.** Not arithmetic **9.** Not arithmetic **10.** Arithmetic; $d = 2$ **11.** Neither are solutions **12.** $(-3, 2)$ is a solution; $(3, -2)$ is not a solution **13.** $(-3, 2)$ is a solution; $(3, -2)$ is not a solution **14.** Neither are solutions **15.** x-intercept: $(3, 0)$; y-intercept: $(0, -2)$ **16.** x-intercept: $(-5, 0)$; y-intercept: $(0, 5)$ **17.** x-intercept: $(-2.75, 0)$; y-intercept: $(0, 11)$ **18.** B **19.** C **20.** D **21.** A **22.** $(1, 2)$ **23.** $(2, -3)$ **24.** $x = 2.5$ **25.** $x = 24$ **26.** $x = 15$ **27.** $v = 2$ **28.** $m = 0$ **29.** $x = -3$ **30.** $m = 2$ **31.** $n = -36$ **32.** $y = -98$ **33.** $w = -4$ **34.** $t = 11{,}700$ **35.** $b = 0$ **36.** $x = 7$ **37.** $z = \dfrac{3}{16}$ **38.** $r = 0$ **39.** $x = -4$ **40.** $x = 27$ **41.** $y = 3$ **42.** $y = 0$ **43.** $w = -31.6$ **44.** $t = -24$ **45.** $m = -9$ **46.** $n = -11$ **47.** $y = -1295$ **48.** $x = -13.3198$ **49.** All real numbers **50.** $a = 0$ **51.** $x = 1.5$ **52.** $x = -2.4$ **53.** Contradiction; no solution **54.** Contradiction; no solution **55.** Identity; all real numbers **56.** Conditional; $v = 0$ **57. a.** -81 **b.** -11 **c.** 52 **d.** $7\pi - 11 \approx 10.991$ **58. a.** -74 **b.** -14 **c.** 106 **d.** 1986 **59.** $x = v + w - y$ **60.** $x = \dfrac{vw}{y}$ **61.** $x = \dfrac{5y + 7z}{3}$ **62.** $x = -\dfrac{1}{8}y + \dfrac{19}{4}$ **63. a.** $8x + 10$ **b.** $x = -1$ **64. a.** $-2x - 2$ **b.** $x = \dfrac{1}{4}$ **65. a.** $2x - 1$ **b.** $x = \dfrac{1}{2}$ **66. a.** $-x - 7$ **b.** $x = -7$ **67.** $x = 10$; $x = 10.1$ **68.** $x = 20$; $x = 19.95$ **69.** $5m + 4 = 49$; $m = 9$ **70.** $\dfrac{1}{3}(m + 7) = 12$; $m = 29$ **71.** $2(m + 2) = -(m - 3)$; $m = -\dfrac{1}{3}$

72. $m + (m + 1) + (m + 2) = 93$; $m = 30$ **73. a.** $f(x) = 0.25x + 500$ **b.** 500; 750; 1000; 1250; 1500; 1750 **c.** $f(2500) = 1125$; it costs \$1125 to sell 2500 snow cones. **74. a.** $f(x) = 0.40x$ **b.** 0, 400, 800, 1200, 1600, 2000 **c.** $f(2500) = 1000$; the revenue generated by selling 2500 snow cones is \$1000. **75.** $w = 5$ **76.** $b = 3$ **77.** 210 mi **78.** 20 defective display panels **79.** $a = 4.5$ cm; $b = 13.5$ cm; $c = 7.5$ cm

80. 4.4 m **81.** $v = 10$ **82.** $k = \dfrac{4}{5}$ **83. a.** $y = 280x + 680$

b.

x	0	1	2	3	4	5	6
y	680	960	1240	1520	1800	2080	2360

c.

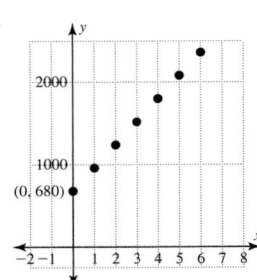

d. The sequence is arithmetic

Mastery Test for Chapter 2

1. A: $(-3, 1)$, II; B: $(1, -3)$, IV; C: $(4, 2)$, I; D: $(-2, -2)$, III
2.

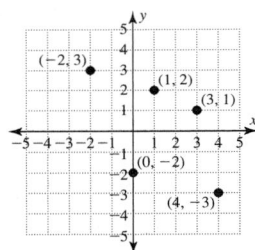

3. a. Arithmetic; $d = 2$ **b.** Not arithmetic **c.** Arithmetic; $d = 0$
d. Arithmetic; $d = -3$ **4. a.** -7 **b.** -18 **c.** 70 **d.** 103
5. a.

x	y
1	2
2	3
3	4
4	5
5	6

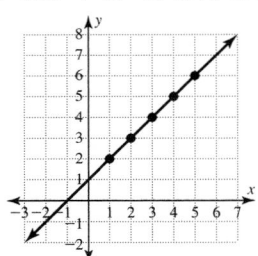

b.

x	y
1	1
2	0
3	-1
4	-2
5	-3

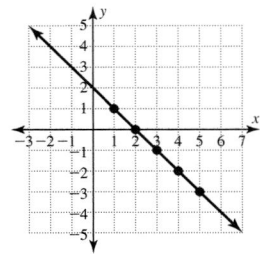

c.

x	y
1	-3
2	-1
3	1
4	3
5	5

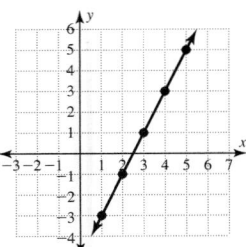

d.

x	y
1	2
2	0
3	-2
4	-4
5	-6

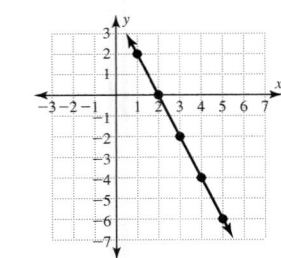

6. a. $f(x) = 375x + 800$
b.

x	0	6	12	18	24	30
f(x)	800	3050	5300	7550	9800	12,050

c. $f(25) = 10{,}175$; the total amount paid at the end of the 25th month is \$10,175 **7. a.** Solution **b.** Not a solution **c.** Solution **d.** Solution
8. a. $(-2, 0)$; $(0, 3)$ **b.** $(-3, 0)$; $(0, -1)$ **c.** no x-intercept; $(0, 2)$
d. $(-1, 0)$; no y-intercept **9. a.** $(2, 1)$ **b.** $(-1, 3)$ **10. a.** $x = 4$ **b.** No solution **c.** $x = 0$ **d.** All real numbers **11. a.** $x = 2$

b. $x = -2$ **12. a.** $x = 3$ **b.** $y = -\dfrac{5}{9}$ **c.** $v = \dfrac{3}{2}$ **d.** $w = 13$

13. a. $m = -3$ **b.** $x = 11$ **c.** 187.5 km **14. a.** $y = 12$ **b.** 9.5 pesos for each U.S. dollar

15.

x	$y = 4x - 1$
-2	-9
-1	-5
0	-1
1	3
2	7

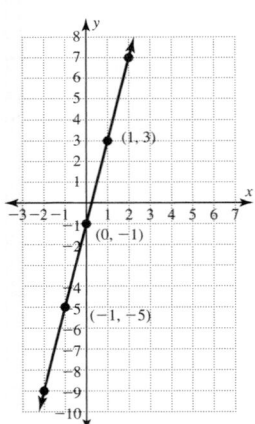

16. a. $(3, 0)$; $(0, -6)$ **b.** No x-intercept; $(0, 4)$ **c.** $(5, 0)$; $(0, -2)$
d. $(-2, 0)$; $(0, 1)$

Chapter 3
Using the Language and Symbolism of Mathematics 3.1

1. y **2.** x **3.** m **4.** $m = \dfrac{y_2 - y_1}{x_2 - x_1}$ **5.** 1 **6.** 0 **7.** Undefined **8.** Increases
9. Decreases **10.** Parallel **11.** Perpendicular **12.** Slope **13.** Grade

Exercises 3.1

1. a. $m = 2$ **b.** $m = -\dfrac{2}{3}$ **c.** $m = \dfrac{8}{5}$ **3. a.** $m = 0$ **b.** m is undefined

c. $m = \dfrac{4}{3}$ **5. a.** $m = -\dfrac{3}{2}$ **b.** $m = \dfrac{3}{10}$ **c.** $m = -1.5$ **7. a.** $m = \dfrac{1}{3}$

b. $m = -2$ **c.** m is undefined **d.** $m = 0$ **9. a.** $m = -\dfrac{3}{5}$ **b.** $m = \dfrac{2}{7}$

c. $m = \dfrac{5}{8}$ **d.** $m = 2$ **11. a.** m is undefined **b.** $m = 0$ **c.** $m = -1$

d. $m = 0$ **13.** Slope: $m = -\dfrac{8}{5}$ **15.** Change in y: 2 **17.** Change in

y: $\dfrac{2}{3}$ **19.** Change in y: 4 **21.** Change in x: 3 **23.** Change in x: -9

25. Change in y: 0

27. Answers may vary. Point: (4, 5)

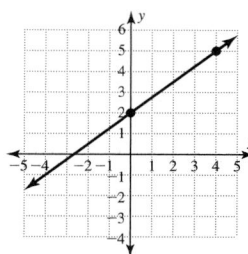

29. Answers may vary. Point: (1, −5)

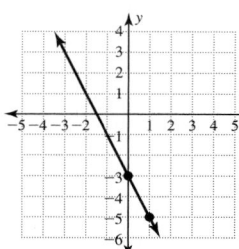

31.

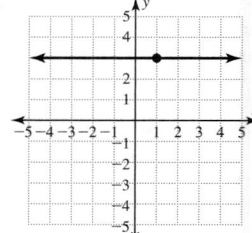

33.

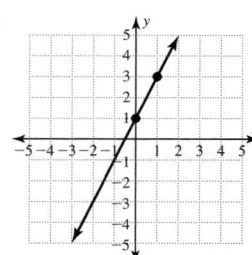

35.

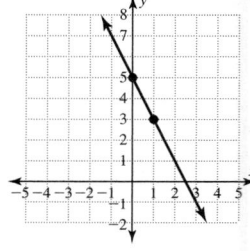

37. $m = 3$; y-intercept: (0, −5) **39.** $m = -\dfrac{1}{2}$; y-intercept: (0, 3)

41. $m = 20$; the cost per shirt is $20 **43.** $\dfrac{1}{2}$; 0.494 **45.** $-\dfrac{1}{3}$; −0.333

47. $m_2 = \dfrac{3}{7}$; $m_3 = -\dfrac{7}{3}$ **49.** $m_2 = 0$; m_3 is undefined

51. $m_1 = \dfrac{8}{5}$; $m_3 = -\dfrac{5}{8}$ **53.** Parallel **55.** Perpendicular **57.** Neither

59. Perpendicular **61.** Parallel

63. a. $6250, $7500, $8750, $10,000

b.

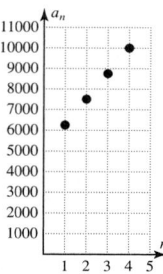

c. $m = 1250$; monthly payment is $1250 **d.** $a_0 = 5000$; down

payment is $5000 **65.** $m = \dfrac{1}{7}$ **67.** 90 m **69.** $h = 3.5$ ft **71.** $m = 0$

73. Positive **75.** Negative **77.** Positive **79.** $m = 2$

Using the Language and Symbolism of Mathematics   3.2

1. $y = mx + b$ **2.** Same **3.** Perpendicular **4.** Vertical; 5, 0
5. Horizontal; 0, −5 **6.** $Ax + By = C$ **7.** 1 **8.** Point; slope; point;
slope

Exercises 3.2

1. a. $m = 2$; y-intercept: (0, 5) **b.** $m = -\dfrac{3}{11}$; y-intercept: $\left(0, -\dfrac{4}{5}\right)$

c. $m = 6$; y-intercept: (0, 0) **3. a.** $m = 0$; y-intercept: (0, 2) **b.** m is

undefined; no y-intercept **c.** $m = -1$; y-intercept: (0, 0) **5.** $y = 4x + 7$

7. $y = -\dfrac{2}{11}x + 5$ **9.** $y = -6$

11.

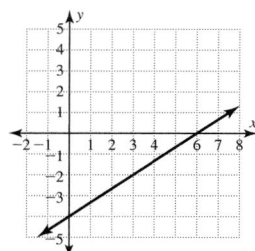

13.

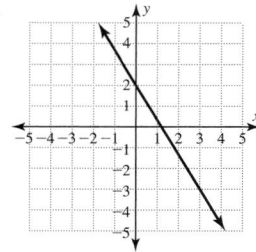

15.

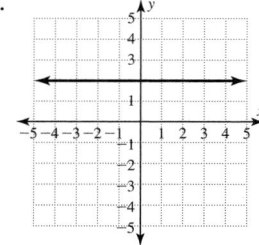

17. $f(x) = \frac{2}{5}x + 1$ **19.** $f(x) = -\frac{1}{4}x - 1$ **21.** $f(x) = 0.05x + 35$;
The cost is $0.05 per minute plus a flat fee of $35 **23.** $y = 7x + 1$

25. $y = -\frac{1}{2}x + 8$ **27.** $f(x) = 50x + 75$; The cost is $50 per hour plus

a flat fee of $75 per visit **29.** $y = 5$ **31.** $x = -4$ **33.** Neither
35. Perpendicular **37.** Parallel **39.** Neither **41. a.** $m = 5$; point:

$(3, 4)$ **b.** $m = -7$; point: $(2, -5)$ **c.** $m = \frac{1}{2}$; point: $(-6, 0)$

43. 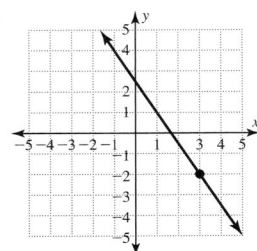   **45.**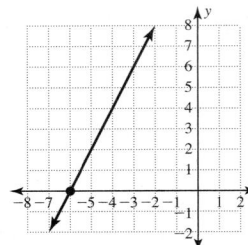

47. 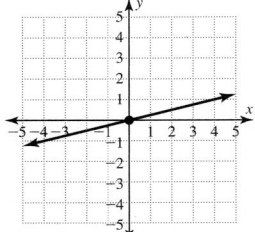   **49.**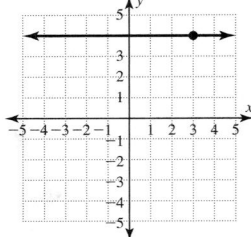

51. $y = -4x + 11$ **53.** $y = \frac{2}{3}x + \frac{23}{3}$ **55.** $x = 0$ **57.** $y = \frac{1}{4}x + 3$

59. $y = 2x + 6$ **61.** $x = 4$ **63.** $y = 8$ **65.** $y = \frac{8}{3}x - 1$

67. $f(x) = 0.8x + 3.25$; The cost is $0.80 per mile plus a flat fee of
$3.25 per ride **69.** Slope-intercept form: $y = 4x + 7$; general form:

$4x - y = -7$ **71.** Slope-intercept form: $y = -\frac{2}{3}x - \frac{2}{3}$; general form:

$2x + 3y = -2$

73. Slope-intercept form: $y = -\frac{1}{2}x + 2$;

Table:

x	y
-3	$3\frac{1}{2}$
-2	3
-1	$2\frac{1}{2}$
0	2
1	$1\frac{1}{2}$
2	1
3	$\frac{1}{2}$

Graph: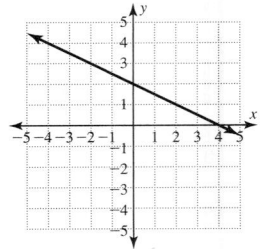

75. Slope-intercept form: $y = 3$;

Table:

x	y
-3	3
-2	3
-1	3
0	3
1	3
2	3
3	3

Graph: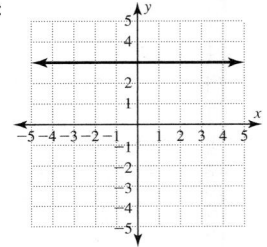

77. $y = 350x + 1250$; the slope represents the monthly payment and
the y-intercept represents the down payment **79.** $y = 0.15x + 40$; the
slope represents the commission rate and the y-intercept represents the
fixed daily salary **81.** $y = -0.045x + 91.8$ (where x is the calendar
year and y is the dividend in dollars per share); estimate for 1995 is
$y = -0.045(1995) + 91.8 \approx \$2.03/\text{share}$

Using the Language and Symbolism of Mathematics   3.3
1. Solution **2.** Intersection **3.** Inconsistent; parallel **4.** One; intersect;
one **5.** Infinite number; coincide **6. a.** Infinitely many **b.** Are
parallel; zero **c.** Are not; one

Exercises 3.3
1. $(2, 2)$ **3.** $(2, -6)$ **5.** Solution **7.** Solution **9.** Solution **11.** Not a
solution **13.** $(300, 90)$; each company charges $90 when 300 mi are
driven **15.** $(6, 0)$ **17.** $(-6, 4)$ **19.** $(-3, 2)$ **21.** $(6, -2)$ **23.** $(4, 1)$

25. $\left(-\frac{9}{2}, -5\right)$ **27.** No solution **29.** Infinite number of solutions

31. One solution; consistent system of independent equations **33.** No
solution; inconsistent system **35.** Infinitely many solutions; consistent
system of dependent equations **37.** One solution; consistent system of
independent equations **39.** No solution; inconsistent system

41. $\left(\frac{2}{7}, \frac{9}{7}\right)$ **43.** $\left(\frac{13}{15}, -\frac{9}{55}\right)$ **45.** $(5, -8)$ **47.** $(-0.5, 4)$ **49.** $(1, 450)$;

at 3 P.M. each plane is 450 mi from O'Hare Airport. **51.** Consistent
system of independent equations **53.** Inconsistent system

55. Numerically:

x	$y_1 = 2x + 7$	$y_2 = 13 - x$
-3	1	16
-2	3	15
-1	5	14
0	7	13
1	9	12
2	11	11
3	13	10

Graphically:

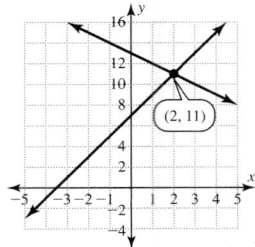

The two numbers are 2 and 11.
57. $43.75°, 136.25°$ **59.** The daily costs both will be $105 when 10 items are produced **61.** $x = 3; y = 5$ **63.** $y = \frac{1}{2}x + 3; y = -4x + 3$
65. $2x - 3y = -6; 4x - 3y = 0$; solution $(3, 4)$

Using the Language and Symbolism of Mathematics   3.4
1. Equal **2.** Ordered pair **3.** False **4.** True **5.** A conditional equation
6. An identity **7.** A contradiction **8.** Consistent; independent; $(3, 4)$
9. Inconsistent; independent; no **10.** Consistent; dependent; infinite

Exercises 3.4
1. $(4, 5)$ **3.** $(2, 0)$ **5.** $(-34, 8)$ **7.** $\left(\frac{4}{7}, \frac{5}{7}\right)$ **9.** $(0, -6)$ **11.** $(-2, -1)$
13. $\left(3, -\frac{3}{5}\right)$ **15.** $\left(-\frac{13}{3}, -7\right)$ **17.** $(-1, 2)$ **19.** $(-2, -3)$ **21.** $(12, 9)$
23. $(0, -6)$ **25.** $(3, 2)$ **27.** $(-4, -2)$ **29.** $(12, 15)$ **31.** Infinitely many solutions; consistent system of dependent equations **33.** No solution; inconsistent system **35.** $(0, 0)$ **37.** $(3, 4)$ **39.** $(2, -1)$
41. $\left(\frac{1}{2}, -\frac{1}{3}\right)$ **43.** B; 75 and 45 **45.** C; 30 and 0
47. Numerically:

x	$y_1 = 8 - x$	$y_2 = x - 2$
2	6	0
3	5	1
4	4	2
5	3	3
6	2	4
7	1	5
8	0	6

Graphically:

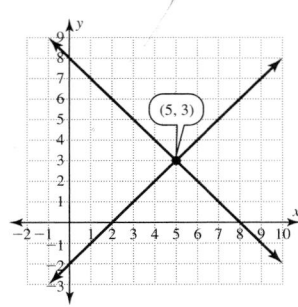

The two numbers are 5 and 3.

49. $x + y = 90; y = x + 18; 36°$ and $54°$ **51.** $\frac{x + y}{2} = 76$;

$x - y = 28; 90$ and 62 **53.** $\frac{3x + y}{4} = 66; x - y = 8$; Tiger scored a

68 in each of his first three rounds and he scored a 60 in the last round
55. $y = 10 + 0.20x; y = 20 + 0.10x$; the cost for both plans is $30 when 100 checks are written **57.** Company A: $y = 0.5x + 1$; company
B: $y = 0.2x + 2$; solution: $\left(\frac{10}{3}, \frac{8}{3}\right)$; each company charges $2.67
when approximately 3.33 mi are driven **59.** Investment A:
$y = 180x + 1500$; investment B: $y = 140x + 2000$; solution:
$(12.5, 3750)$; after 12.5 years, each investment plan is worth $3750.00
61. $B = -6$ **63.** $C = 6$ **65.** $(4.65, 2.75)$ **67.** $(-11.3, -1.1)$

Using the Language and Symbolism of Mathematics   3.5
1. Equal **2.** Elimination; eliminate **3.** Opposites **4.** Inconsistent; no
5. Consistent; dependent; infinite

Exercises 3.5
1. $(2, 2)$ **3.** $(-8, 7)$ **5.** $\left(\frac{1}{6}, -1\right)$ **7.** $\left(-2, \frac{3}{2}\right)$ **9.** $(3, -5)$ **11.** $(4, 5)$
13. $(1, -1)$ **15.** $(-9, 11)$ **17.** $(19, 0)$ **19.** $(0, 0)$ **21.** $(-3, 5)$
23. $(6, 6)$ **25.** No solution; inconsistent system **27.** Infinitely many
solutions; consistent system of dependent equations **29.** $\left(\frac{4}{3}, -\frac{1}{6}\right)$
31. $(5, 9)$ **33.** $(2, 2)$ **35.** $\left(\frac{1}{2}, -\frac{1}{2}\right)$ **37.** No solution; inconsistent
system **39.** $(0, 3)$ **41.** C; 2 and -5 **43.** D; -2 and 6
45. $x + y = 88; x - y = 28$; 58 and 30 **47.** $x + y = 102; y = 2x$;
34 and 68 **49.** $\frac{4x + y}{5} = 25; y - x = 15$; he scored 22 points in each
of the first 4 games and 37 points in the last game **51.** 4-year colleges:
$y = -0.5x + 1030.5$; community college: $y = 1.5x - 2959.5$;
solution: $(1995, 33)$; in 1995 the percent of those attending 4-year
colleges and community colleges was 33% **53.** $(10, 20)$ **55.** $(4.4, 5.5)$

Using the Language and Symbolism of Mathematics   3.6
1. Find **2.** Variable **3.** Algebraic **4.** Solve **5.** Reasonable **6.** Mixture
7. Rate; time **8.** Principal; rate; time

Exercises 3.6
1. 72 and 28 **3.** 48 cm by 54 cm **5.** The base is 32 cm and the two
equal sides measure 20 cm each **7.** 7 and 4 **9.** 9 and 13 **11.** 11.5
units and 30.5 units **13.** The fixed cost is $30.00 and the pay-per-view
movies are $3.50 each **15.** 47 union employees and 3 union stewards
17. 332 adult tickets and 456 student tickets **19.** $31°$ and $59°$ **21.** $35°$
and $145°$ **23.** 540 right-handed desks and 60 left-handed desks

25. The fixed cost is $1000 a month and the variable cost is $125 per chair **27.** $9500 at 7% and $2500 at 9% **29.** $6500 in bonds and $3500 in stocks **31.** 84 and 94 km/h **33.** 85 and 100 km/h **35.** In still water, the boat travels 15 km/h. The current is going 9 km/h **37.** Mix 60 liters of the 5% solution and 20 liters of the 33% solution **39.** 220 g **41.** 8.4 liters

Review Exercises for Chapter 3

1. 2 **2.** $-\dfrac{3}{5}$ **3.** $\dfrac{4}{7}$ **4.** -1 **5.** 0 **6.** $\dfrac{2}{5}$ **7.** Undefined **8.** 0

9. Undefined **10.** 0.5 **11. a.** 4 **b.** 5 **c.** -4 **d.** $\dfrac{4}{5}$ **e.** $-\dfrac{4}{5}$

12. a. 3800, 4600, 5400, 6200

b.
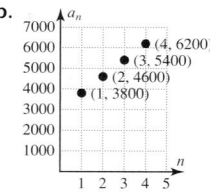

c. $m = 800$; monthly payment is $800 **d.** $a_0 = 3000$; down payment is $3000 **13.** $\dfrac{1}{6}$ **14. a.** Parallel **b.** Perpendicular **c.** Neither

15. $2x - y = 2$ **16.** $10x - 15y = 3$ **17.** $x + 2y = 12$ **18.** $2x - 3y = 6$ **19.** $y = 9$ **20.** $x = 4$ **21.** $3x - y = -5$ **22.** $(3, -1)$ **23.** No solution **24.** Infinitely many solutions **25.** $(6, 5)$ **26.** Infinitely many solutions **27.** No solution **28.** $(5, 11)$

29. $\left(\dfrac{18}{11}, \dfrac{1}{22}\right)$ **30.** $(-148, 225)$ **31.** $(25, 60)$ **32.** No solution

33. $\left(-\dfrac{2}{7}, \dfrac{3}{5}\right)$ **34.** Infinitely many solutions **35.** $m \approx 1$; $m = 0.9$

36. $m \approx -1$; $m = -1.1$ **37.** C; 3 and 6 **38.** A; 10 and 1 **39.** B; 11 and 2 **40.** Both options cost $3500 when 10,000 pages are printed **41.** 46° and 134° **42.** $6000 **43.** 9 cm **44.** 96 km/h and 104 km/h **45.** 45 liters of the $6 chemical and 55 liters of the $9 chemical **46. a.** $y = 0.95x + 6$ **b.** The slope, $m = 0.95$ represents the cost per CD. The y-intercept (0.6) represents the initiation fee $6 **c.** $y = 1.95x$ **d.** The slope, $m = 1.95$, represents the cost per CD; the y-intercept $(0, 0)$ represents the initiation fee $0 **e.** (6, 11.7) **f.** When 6 CDs are purchased, each plan costs $11.70 **47. a.** $y = 30x + 20$ **b.** The slope, $m = 30$, represents the charge per hour; the y-intercept represents a $20 flat fee **c.** $y = 26x + 30$ **d.** The slope, $m = 26$, represents charge per hour; the y-intercept represents a $30 flat fee **e.** (2.5, 95) **f.** The total charge for each shop is $95 when the repairs require 2.5 h

Mastery Test for Chapter 3

1. a. 2 **b.** -3 **c.** 0 **d.** Undefined **2. a.** Perpendicular **b.** Parallel **c.** Perpendicular **d.** Neither **3. a.** $y = -2x + 6$ **b.** $y = 4x + 7$ **c.** $y = \dfrac{2}{3}x + 5$ **d.** $y = -\dfrac{5}{3}x - 2$ **4. a.** $y = 3$ **b.** $x = -4$ **c.** $x = -4$ **d.** $y = 3$

5. a.
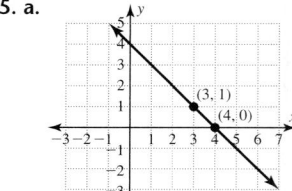

b.
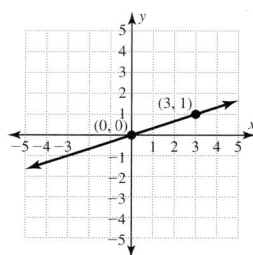

c.
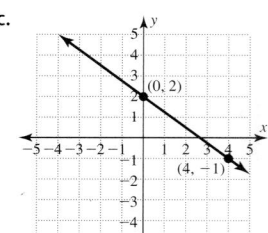

d.
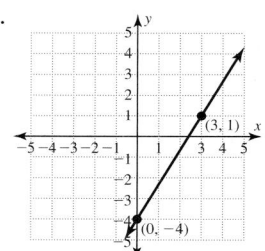

6. a. $(0, 2)$ **b.** $(2, 5)$ **c.** $(-2, -1)$ **d.** $(12, 5)$ **7. a.** B **b.** C **c.** A **8. a.** $(-2, 7)$ **b.** $\left(\dfrac{1}{14}, -\dfrac{5}{7}\right)$ **c.** $\left(2, -\dfrac{1}{3}\right)$ **d.** $\left(\dfrac{37}{10}, -\dfrac{13}{20}\right)$ **9. a.** $(-2, 1)$ **b.** $(-1, 1)$ **c.** Infinitely many solutions **d.** No solution **10. a.** 13°, 77° **b.** $7500.00 in bonds, $2500.00 in the savings account **c.** 15 ml 30% solution, 10 ml 80% solution **d.** 475 mi/h, 225 mi/h

Chapter 4

Using the Language and Symbolism of Mathematics 4.1
1. Linear **2.** One **3.** Equivalent **4.** Conditional **5.** Solution **6.** Addition; subtraction **7.** Addition; subtraction **8.** $\geq$ **9.** $\leq$ **10.** $\leq$ **11.** $\leq$ **12.** $\geq$

Exercises 4.1
1. b **3. a.** $(-\infty, -4]$ **b.** $(5, \infty)$ **c.** $(-\infty, 3)$ **d.** $[-1, \infty)$ **5. a.** Not a solution **b.** Solution **c.** Not a solution **d.** Solution **7. a.** Solution **b.** Not a solution **c.** Solution **d.** Not a solution **9. a.** $x = 1$ **b.** $(-\infty, 1)$ **c.** $(1, \infty)$ **11. a.** $x = -3$ **b.** $(-\infty, -3]$ **c.** $[-3, \infty)$ **13. a.** $x = 500$; when 500 h of overtime are worked, the spending limit of $15,000 is met **b.** $(500, \infty)$; when more than 500 h of overtime are worked, the spending limit of $15,000 is exceeded **c.** $[0, 500)$; when less than 500 h of overtime are worked, the spending limit of $15,000 is not met **15. a.** $x = 5$ **b.** $(-\infty, 5]$ **c.** $[5, \infty)$ **17. a.** $x = 0.3$ **b.** $(-\infty, 0.3)$ **c.** $(0.3, \infty)$ **19. a.** $x = 8$; each firm charges the same amount when 8 boxes of paper are ordered **b.** $[0, 8)$; Ace Office Supply charges less for orders of less than 8 boxes of paper **c.** $(8, 24]$; Ace Office Supply charges more for orders of more than 8 boxes of paper **21.** $x + 7 > 11$; $x + 7 - 7 > 11 - 7$; $x > 4$; solution: $(4, \infty)$ **23.** $2x + 3 \leq x + 4$; $2x - x + 3 \leq x - x + 4$; $x + 3 \leq 4$; $x + 3 - 3 \leq 4 - 3$; $x \leq 1$; solution: $(-\infty, 1]$ **25.** $[3, \infty)$; all values of x greater than or equal to 3 satisfy this inequality **27.** $(-\infty, 2)$; all values of x less than 2 satisfy this inequality **29.** $(-\infty, 4]$ **31.** $(-\infty, 0)$ **33.** $(-\infty, 7]$ **35.** $(-\infty, -1)$ **37.** $(-\infty, 1)$ **39.** $(-\infty, 2]$ **41.** $(-\infty, -2)$ **43.** $[40, \infty)$ **45.** $(1, \infty)$; $(1.112, \infty)$ **47.** $(-\infty, -22)$; $(-\infty, -21.93)$ **49.** $(8, \infty)$ **51.** $(-\infty, -1]$ **53.** $[-20, \infty)$ **55.** $(1.5, \infty)$ **57.** No solution **59.** $[2, \infty)$ **61.** $5 - 2x \leq 7 - 3x$; $(-\infty, 2]$ **63.** $2w + 3 \geq 11$; $w + 11$; $[8, \infty)$ **65.** $3(m + 5) > 2m + 12$; $(-3, \infty)$ **67.** 8 **69.** From 0 ft to 60 ft **71.** The company will lose money when less than 450 mice are produced; the company will have a profit if more than 450 mice are produced **73.** $a + b > c$; $a + c > b$; and $b + c > a$

Using the Language and Symbolism of Mathematics   4.2
1. Multiplication-division **2.** Multiplication-division **3.** $>$ **4.** $<$
5. Least common denominator **6.** Distributive

Exercises 4.2
1. a. $>$ **b.** $>$ **c.** $>$ **d.** $<$ **3. a.** $x = -1$ **b.** $(-\infty, -1)$ **c.** $(-1, \infty)$
5. a. $x = -2$ **b.** $[-2, \infty)$ **c.** $(-\infty, -2]$ **7. a.** $x = 2$; each company charges the same amount when the machines are rented for 2 h **b.** $(2, \infty)$; the Dependable Rental Company charges less when the machines are rented for more than 2 h **c.** $[0, 2)$; the Dependable Rental Company charges more when the machines are rented for less than 2 h **9. a.** $x = 3$ **b.** $(-\infty, 3)$ **c.** $(3, \infty)$ **11. a.** $x = -5$ **b.** $[-5, \infty)$
c. $(-\infty, -5]$ **13. a.** $x = 30$; each store charges the same amount when 30 shirts are ordered **b.** $[0, 30)$; Mel's Sports charges less for orders of less than 30 shirts **c.** $(30, 60]$; Mel's Sports charges more for orders of more than 30 shirts **15.** $-2x < 4$; $\dfrac{-2x}{-2} > \dfrac{4}{-2}$; $x > -2$; $(-2, \infty)$

17. $-3x + 4 \geq 7$; $-3x + 4 - 4 \geq 7 - 4$; $-3x \geq 3$; $\dfrac{-3x}{-3} \leq \dfrac{3}{-3}$;
$x \leq -1$; $(-\infty, -1]$ **19.** $(-2, \infty)$; all values of x greater than -2 satisfy the inequality **21.** $(-\infty, 4]$; all values of x less than or equal to 4 satisfy the inequality **23.** $[2, \infty)$ **25.** $(-\infty, -8)$ **27.** $(16, \infty)$
29. $(-\infty, 0]$ **31.** $(-\infty, 2.3)$ **33.** $(-\infty, -2)$ **35.** $(-\infty, -5]$
37. $(-\infty, 0]$ **39.** $(2, \infty)$ **41.** $(-\infty, 3]$ **43.** $(-5, \infty)$ **45.** $(-\infty, -5)$
47. $[-2, \infty)$ **49.** $(1, \infty)$ **51.** $\left[\dfrac{2}{3}, \infty\right)$ **53.** $[-3, \infty)$ **55.** $\left(-\dfrac{31}{3}, \infty\right)$
57. $(1, \infty)$ **59.** $[-8, \infty)$ **61.** $(-\infty, 7]$; $(-\infty, 7.1]$ **63.** $(-\infty, 8)$;
$(-\infty, 8.9)$ **65.** $2(x + 5) < 6$; $(-\infty, -2)$ **67.** $4(y - 7) \geq 6(y + 3)$;
$(-\infty, -23]$ **69.** At least 5°C **71.** At most $60,000 **73.** The company loses money when less than 200 bricks are produced; the company has a profit when more than 200 bricks are produced

Using the Language and Symbolism of Mathematics   4.3
1. Conditional **2.** Unconditional **3.** Contradiction **4.** Compound; and
5. Intersection **6.** Union

Exercises 4.3
1. C **3.** B **5.** $-5 < x \leq 1$ **7.** $0 < x < 6$ **9.** $150 \leq x \leq 450$
11. $x \leq 1$ or $x \geq 5$ **13.** $155 < x < 190$ **15.** $[0, 150] \cup [220, 270]$
17. $A \cap B = [5, 7]$; $A \cup B = (-4, 11)$ **19.** $A \cap B = [5, 11)$;
$A \cup B = (-4, \infty)$ **21.** $A \cap B = [-4, -2]$; $A \cup B = (-\infty, \infty)$
23. Interval notation: $[-1, 6)$

Graph:

Verbally: x is greater than or equal to negative one and less than six
25. a. $-2 < x \leq 5$ **b.** $0 \leq x \leq 4$ **27.** $[-1, 2)$ **29.** $(-1, 4)$
31. $[50, 100]$; the strength requirements are met when from 50 to 100 fiber strands are used **33.** $(1, 6)$ **35.** $(-2, 3]$ **37.** Contradiction; no solution **39.** Unconditional; the set of all real numbers
41. Conditional; $(-\infty, 2]$ **43.** $[-5, 1)$; all real numbers greater than or equal to negative five and less than one **45.** $[-1, 2]$; all real numbers greater than or equal to one and less than or equal to two
47. $(-6, -4)$ **49.** $[-2, 1)$ **51.** $[0, 4]$ **53.** $[0, 4)$ **55.** $(2, 8]$
57. $(-1, 2)$ **59.** $[2, 4]$ **61.** $(-1, 1]$ **63.** $\left(-\infty, \dfrac{11}{7}\right]$ **65.** $(-1, 2)$
67. $[-5, 1]$ **69.** $(2, 3)$; $(1.9, 3.1)$ **71.** $(-3, 2]$; $(-2.9, 2.1]$
73. No solution **75.** $\left(\dfrac{1}{3}, \dfrac{8}{5}\right)$ **77.** $(-\infty, -3) \cup (-2, \infty)$
79. More than 5°C and less than 35°C **81.** At least 4 m but no more than 19 m

Using the Language and Symbolism of Mathematics   4.4
1. Distance **2. a.** $<$ **b.** $=$ **c.** $>$ **3.** $|x - a| = d$ **4.** Tolerance

Exercises 4.4
1. a. $|x| < 5$ **b.** $|x| \geq 2$ **c.** $|x - 1| = 3$ **d.** $|x| < 4$ **3. a.** $|x - 4| \leq 2$
b. $|x - 1| \geq 2$ **c.** $|x + 7| < 3$ **d.** $|x| \leq 5$ **5. a.** $|x| \leq 3$ **b.** $|x| > 3$
c. $|x| < 7$ **7. a.** $x = -6$ or $x = 6$ **b.** $x = 0$ **c.** $x = -3$ or $x = 7$
d. $x = -7$ or $x = 3$ **9. a.** $x = 6$ or $x = -2$ **b.** $(-2, 6)$
c. $(-\infty, -2) \cup (6, \infty)$ **11. a.** $x = 6$ or $x = 2$ **b.** $(2, 6)$
c. $(-\infty, 2] \cup [6, \infty)$ **13. a.** $x = -4$ or $x = 0$ **b.** $[-4, 0]$
c. $(-\infty, -4] \cup [0, \infty)$ **15. a.** $x = -5$ or $x = 15$ **b.** $(-5, 15)$
c. $(-\infty, -5] \cup [15, \infty)$ **17.** $(-2, 0)$; values greater than -2 and less than 0 **19.** $(-1, 2)$; values greater than -1 and less than 2
21. $(-\infty, -2] \cup \left[-\dfrac{2}{3}, \infty\right)$ **23.** $[-4, 4]$ **25.** $(-4, 1)$ **27.** $\left(-\dfrac{3}{2}, -\dfrac{1}{2}\right)$
29. $(-\infty, -4) \cup (3, \infty)$ **31.** $(-\infty, -97] \cup [99, \infty)$ **33.** $\left(-4, \dfrac{10}{3}\right)$
35. $x = \dfrac{1}{3}$ or $x = -1$ **37.** $x = 3$ or $x = \dfrac{1}{3}$ **39.** $x = 4$ or $x = -\dfrac{2}{5}$
41. $x = 5$ or $x = -5$; $x = 4.9$ or $x = -4.9$ **43.** $x = 7$ or $x = -23$;
$x = 7.1$ or $x = -22.96$ **45.** $|x + 3| < 9$ **47.** $|x - 11| \leq 15$
49. $|x - 2| > 4$ **51.** $|x - 15| \leq 0.001$; $[14.999, 15.001]$
53. $|x - 16| \leq 0.25$; $[15.75, 16.25]$ **55.** $|x - 25| \leq 2$; $[23, 27]$

Using the Language and Symbolism of Mathematics   4.5
1. Ordered pair; true **2.** Half; planes **3.** Solid **4.** Dashed **5.** Contains
6. Not contain **7.** Upper **8.** Lower

Exercises 4.5
1. a. Solution **b.** Solution **c.** Not a solution **d.** Not a solution
3. a. Not a solution **b.** Solution **c.** Not a solution **d.** Solution
5. A and B are solutions; C and D are not solutions **7.** C and B are solutions; A and D are not solutions **9.** a **11.** A and C

13. a.

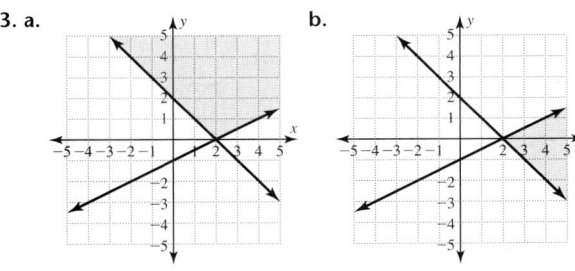

c.

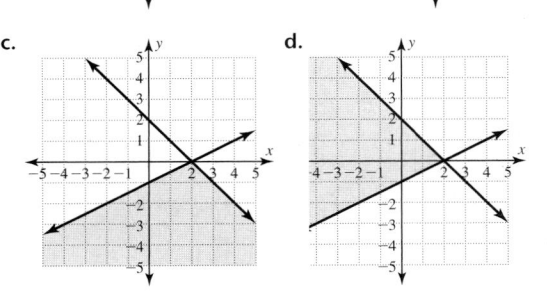

15. a. 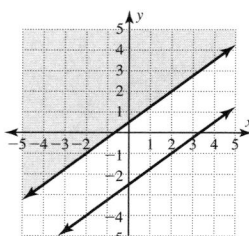   **b.** No solution

c. 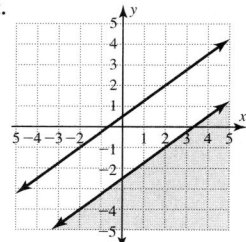   **d.**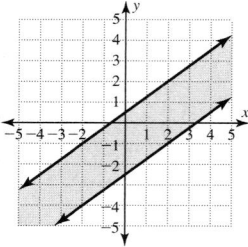

17. 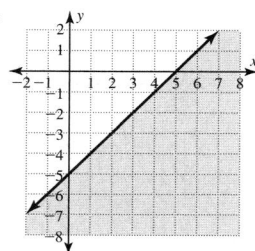   **19.**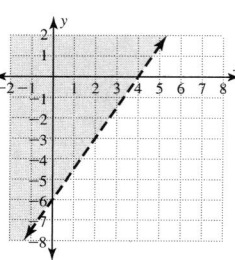

21. 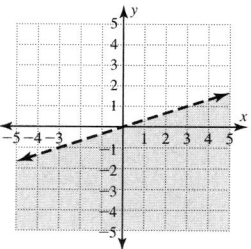   **23.**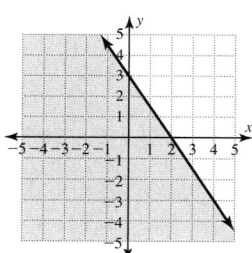

25. 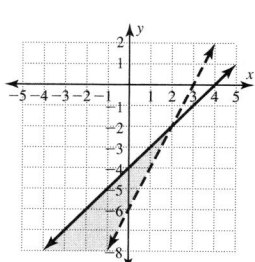   **27.**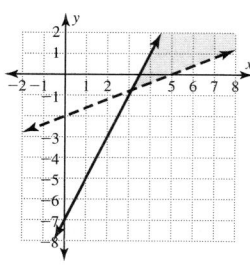

29. 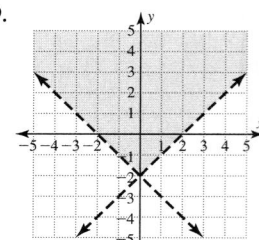   **31.**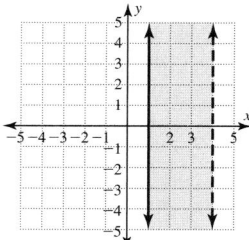

33. 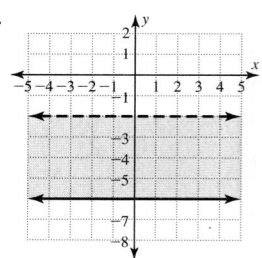   **35.**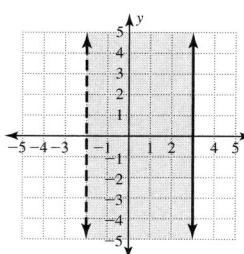

37. No solution   **39.**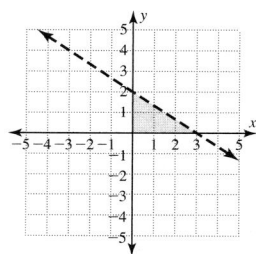

41. 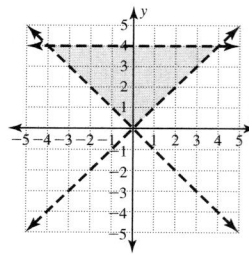   **43.**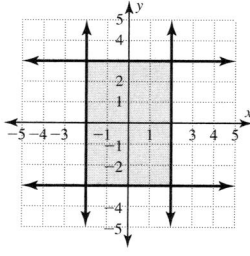

45.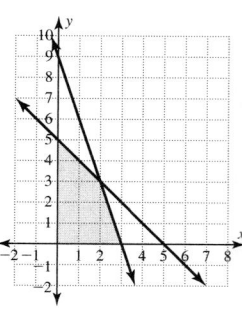

47. $y \le x - 4$; $y \le \dfrac{x}{2} - 3$ **49.** $x \ge y + 2$ **51.** $y \le x + 4$ **53.** $x > 0$; $y > 0$; $x + y \le 10$ **55.** $x \ge 0$; $y \ge 0$; $x + 2y \le 5$

57. x > 0; y > 0; x + y ≤ 150

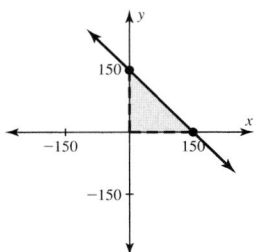

59. x ≥ 0; y ≥ 0; 40x + 50y ≤ 4400

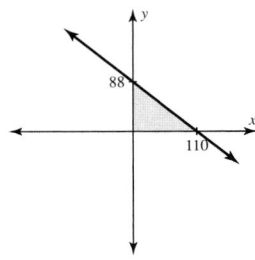

Review Answers for Chapter 4

1. a. Not a solution **b.** Solution **c.** Solution **d.** Not a solution
2. C and D **3. a.** Not a solution **b.** Solution **c.** Not a solution
d. Solution **4. a.** x = 3 **b.** (3, ∞) **c.** (−∞, 3) **5. a.** x = −1
b. (−∞, −1) **c.** (−1, ∞) **6. a.** x = 1998; in 1998 the percents
generated by coal and wind were equal **b.** [1990, 1998); the
community used less electricity generated by wind than coal between
1990 and 1998 **c.** (1998, 2000]; the community used more electricity
generated by wind than coal after 1998 **7. a.** x = 2; when 2 h are
required, the cost of each shop is $85 **b.** (0, 2); when less than 2 h are
required, repair shop A charges less **c.** (2, 4]; when more than 2 h are
required, repair shop B charges less **8.** (Verbal answers vary.)
a. x > 3; (3, ∞);

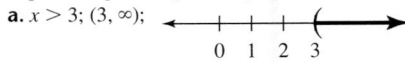

b. x is at most 4; (−∞, 4];

c. x is at least 2; x ≥ 2;

d. x is less than 5; x < 5; (−∞, 5)
9. a. 2 < x < 6; (2, 6);

b. x is greater than −3 and less than or equal to 4; (−3, 4];

c. x is greater than or equal to −2 and less than 0; −2 ≤ x < 0;

d. x is greater than or equal to −1 and less than or equal to 3;
−1 ≤ x ≤ 3; [−1, 3] **10.** [−1, 3) **11.** [−1, 5] **12. a.** x + 7 ≤ 11

b. 2(x − 1) > 13 **c.** 3x − 5 ≥ 7 **d.** 4x + 9 ≤ 21 **13.** (−2, ∞); all
values greater than −2 **14.** (−∞, 4]; all values less than or equal to 4
15. [−3, ∞); all values greater than or equal to −3 **16.** (−∞, 3); all
values less than 3 **17.** [1, ∞); all values greater than or equal to 1
18. (0, 4); all values between 0 and 4 **19.** (−∞, −2) **20.** (1, ∞)
21. (−∞, −5] **22.** [−4, ∞) **23.** (−∞, −3) **24.** [0, ∞)
25. $\left(-\frac{1}{2}, \infty\right)$ **26.** (−∞, 3) **27.** [2, ∞) **28.** (−∞, −3] **29.** (−3, ∞)
30. $\left(-\infty, -\frac{4}{7}\right)$ **31.** (−∞, −3) **32.** (−∞, −13] **33.** (−∞, ∞)
34. $\left(-\infty, \frac{25}{6}\right]$ **35.** (−∞, −22) **36.** (−24, ∞) **37.** (2, 6] **38.** [0, 25)
39. (−140, −98) **40.** [4, 14] **41.** [−2, 2] **42.** (−1, 5) **43.** x = 7,
x = −1 **44.** x = 4 or x = 1 **45.** $\left(-\infty, -\frac{6}{5}\right) \cup (2, \infty)$ **46.** (2, 8)
47. Conditional; (0, ∞) **48.** Conditional inequality; (−∞, 0]
49. Unconditional; (−∞, ∞) **50.** Contradiction; no solution
51. A ∩ B = (2, 4); A ∪ B = (1, 7) **52.** A ∩ B = [3, 5); A ∪ B =
[2, 6] **53.** A ∩ B = [−3, 2]; A ∪ B = (−∞, ∞) **54.** A ∩ B = [−2, 5);
A ∪ B = (−∞, 6) **55.** −3 < x ≤ 4 **56.** 0 ≤ x < π **57.** (−∞, 1)
58. $(-\infty, -3) \cup \left(-\frac{4}{3}, \infty\right)$

59.   **60.**

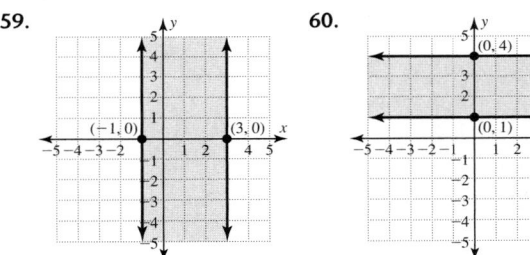

61.   **62.**

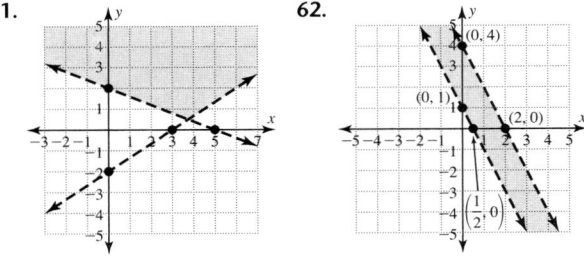

63.   **64.**

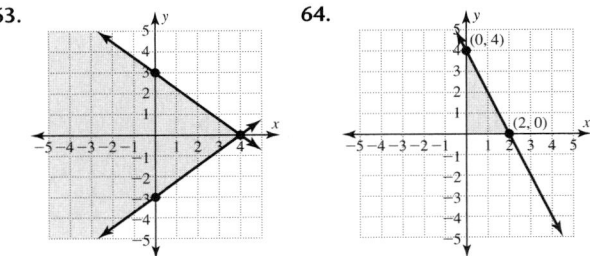

65. |x| < 5 **66.** |x| ≤ 4 **67.** |x − 4| < 6 **68.** |x − 8| ≤ 12
69. 4(x + 3) > 8; (−1, ∞) **70.** 3(x − 17) ≤ 2(x + 11); (−∞, 73]
71. 7 ≤ 3x − 2 < 19; [3, 7) **72.** He must score at least 18 points
73. At least $510 **74.** Between 10 cm and 20 cm **75.** 160 < x < 200
76. |x − 7.2| ≤ 0.25; [6.95, 7.45] **77.** There is a loss if less than 50
posters are ordered; there is a profit if 50 or more posters are ordered

78. $(-\infty; 10)$; $(-\infty, 10.05)$ **79.** $(-\infty, -2]$; $(-\infty, -2.01]$ **80.** $(1, 3]$; $(1.1, 2.99]$

Mastery Test for Chapter 4
1. a. Solution **b.** Solution **c.** Not a solution **d.** Not a solution
2. a. $(-\infty, 3)$ **b.** $[3, \infty)$ **c.** $(-\infty, 4]$ **d.** $(4, \infty)$ **3. a.** $(-\infty, 5]$
b. $(-\infty, -2)$ **c.** $(-11, \infty)$ **d.** $[4, \infty)$ **4. a.** $[-15, \infty)$ **b.** $(245, \infty)$
c. $(-\infty, -9)$ **d.** $\left[-\dfrac{3}{2}, \infty\right)$ **5. a.** Conditional; $(-\infty, 0)$
b. Unconditional; $(-\infty, \infty)$ **c.** Contradiction; no solution
d. Conditional; $(0, \infty)$ **6. a.** $[-8, 5)$ **b.** $[-24, 54)$ **c.** $[-1, 4]$
d. $(-\infty, -4] \cup [3, \infty)$ **7. a.** $x = 8, x = -8$ **b.** $x = 22, x = -27$
c. $(-8, 11)$ **d.** $(-\infty, 1) \cup (9, \infty)$ **8. a.** $(0, 3)$ **b.** $(-\infty, 0) \cup (3, \infty)$
c. $[-4, 0]$ **d.** $(-\infty, -4] \cup [0, \infty)$

9. a.
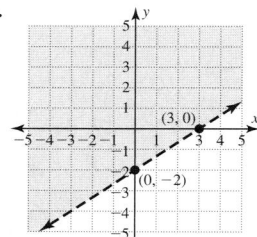
b.
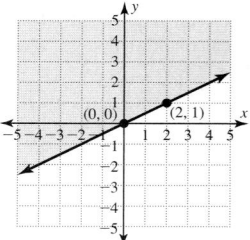

c.
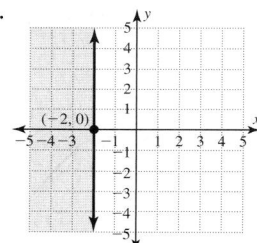
d.
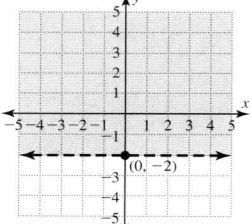

10. a.
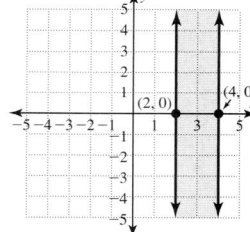
b.
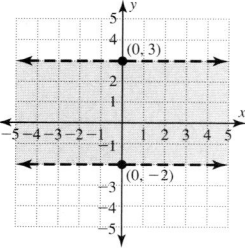

c.
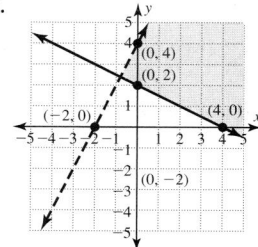
d.
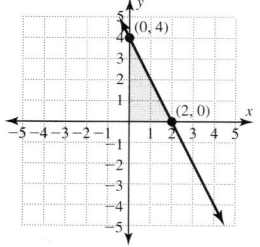

Chapter 5

Using the Language and Symbolism of Mathematics 5.1
1. Base; exponent **2.** Factors **3.** Terms **4.** x **5.** $-x$ **6.** xy **7.** y
8. Exponential; expanded **9.** x^{m+n} **10.** x^{mn} **11.** $x^m y^m$ **12.** $\dfrac{x^m}{y^m}$

Exercises 5.1
a. x^5 **b.** xy^4 **c.** $(xy)^3$ **3. a.** $(a + b)^3$ **b.** $a^3 + b^3$ **c.** $a + b^3$
5. a. $m \cdot m \cdot m \cdot m \cdot m \cdot m$ **b.** $(-m) \cdot (-m) \cdot (-m) \cdot (-m) \cdot (-m) \cdot (-m)$
c. $-m \cdot m \cdot m \cdot m \cdot m \cdot m$ **7. a.** $(m + n)(m + n)$ **b.** $m \cdot m + n \cdot n$
c. $m + n \cdot n$ **9. a.** $\dfrac{a \cdot a}{b \cdot b \cdot b}$ **b.** $\left(\dfrac{a}{b}\right) \cdot \left(\dfrac{a}{b}\right) \cdot \left(\dfrac{a}{b}\right) \cdot \left(\dfrac{a}{b}\right)$
c. $\dfrac{a}{b \cdot b \cdot b \cdot b}$ **11. a.** 36 **b.** 36 **c.** -36 **d.** -1 **13. a.** -1 **b.** 1
c. -1 **d.** 1000 **15. a.** 169 **b.** 289

17.

x	y_1	y_2	
-3	-243	-243	$x^2 \cdot x^3 = x^5$
-2	-32	-32	
-1	-1	-1	
0	0	0	
1	1	1	
2	32	32	
3	243	243	

19.

x	y_1	y_2	
-3	729	729	$(x^3)^2 = x^6$
-2	64	64	
-1	1	1	
0	0	0	
1	1	1	
2	64	64	
3	729	729	

21. a. x^{20} **b.** $-x^6$ **23. a.** $6m^5$ **b.** $-20y^5$ **25. a.** y^{14} **b.** y^{20}
27. a. $x^3 y^3$ **b.** $32n^5$ **29. a.** n^{12} **b.** $-n^{12}$ **31. a.** $25a^2b^2c^2$ **b.** $9b^2c^2d^2$
33. a. $x^3 y^6$ **b.** $x^8 y^{12}$ **35. a.** $\dfrac{9}{16}$ **b.** $\dfrac{81}{10{,}000}$ **37. a.** $\dfrac{64}{w^3}$ **b.** $\dfrac{25}{v^2}$
39. a. $\dfrac{x^{11}}{y^{11}}$ **b.** $\dfrac{v^9}{w^9}$ **41. a.** $\dfrac{4x^2}{9y^2}$ **b.** $\dfrac{27v^3}{1000w^3}$ **43. a.** -4 **b.** 4 **45. a.** 36
b. 18 **47. a.** 35 **b.** 125 **49. a.** 180 **b.** 900 **51. a.** -19 **b.** -1
53. a. 54 **b.** 40 **55. a.** $3v$ **b.** v^3 **57. a.** $10m^3$ **b.** $24m^6$ **59.** -49;
-48.7204 **61.** 4; 4.0804 **63.** 28; 29.0301 **65.** \$2.16
67. a. $P(20) = 80$; a wind speed of 20 mi/h generates 80 kW of
electricity **b.** $P(27) = 196.83$; a wind speed of 27 mi/h generates
196.83 kW of electricity **69.** 8 times **71.** $x^6 y^3$
73. $16x^4 y^8$ **75.** $200x^5$ **77.** $\dfrac{27x^3}{8v^6}$

Using the Language and Symbolism of Mathematics 5.2

1. Addends; terms **2.** Factors **3.** Quotient **4.** x^{m-n}; 0; $\dfrac{1}{x^{n-m}}$; subtract
5. 1; 0 **6.** Undefined

Exercises 5.2
1. a. -5 **b.** 0 **c.** 1 **d.** 0 **e.** Undefined **3. a.** 1000 **b.** 81 **c.** $\dfrac{1}{49}$ **d.** $\dfrac{1}{8}$
e. 1 **5. a.** 1 **b.** -1 **c.** 1 **d.** 1 **e.** 2 **7. a.** 2 **b.** 1 **c.** 0 **d.** 1 **e.** 1
9. a. 1 **b.** 0 **c.** 1 **d.** 3 **e.** -1 **11. a.** 1 **b.** Undefined **13. a.** -3
b. Undefined

15.

x	y_1	y_2
-3	9	9
-2	4	4
-1	1	1
0	ERROR	0
1	1	1
2	4	4
3	9	9

$\dfrac{x^6}{x^4} = x^2$ for all $x \neq 0$

17.

x	y_1	y_2
-3	-0.3333	-0.3333
-2	-0.5	-0.5
-1	-1	-1
0	ERROR	ERROR
1	1	1
2	0.5	0.5
3	0.3333	0.3333

$\dfrac{x^4}{x^5} = \dfrac{1}{x}$ for all $x \neq 0$

19. x^{10} **21.** m^{24} **23.** 1 **25.** $-x^4$ **27.** $\dfrac{1}{t^4}$ **29.** $\dfrac{1}{v^5}$ **31.** $-\dfrac{1}{a^3}$ **33.** $\dfrac{a^3}{b^4}$

35. $\dfrac{2m^3}{3}$ **37.** $\dfrac{6}{11v^3}$ **39.** $\dfrac{3b}{4}$ **41.** $\dfrac{4n^5}{3m^2}$ **43.** $-\dfrac{3w^3}{5v^2}$ **45.** x^6 **47.** v^{12}

49. $\dfrac{1}{m^{12}}$ **51. a.** $36x$ **b.** 19 **53. a.** $3x^2 - 3x$ **b.** x **55.** $\dfrac{1}{8}$; 0.126

57. 200; 198.59 **59. a.** 1 **b.** 4 **c.** 1 **d.** 0 **e.** 1 **61. a.** 4 **b.** $\dfrac{1}{4}$ **c.** 1

d. -1 **63.** 11 **65.** $36x^{10}$ **67.** $8000a^{27}$ **69.** $32x^{15}$ **71.** $144m^{18}$

73. $200x^{21}$ **75.** $6x^6$ **77.** $3a^4b^5$ **79.** $\dfrac{x^4}{y}$ **81.** \$7,518.30 **83.** \$5,105.13

Using the Language and Symbolism of Mathematics   5.3
1. 0; $\dfrac{1}{x^n}$ **2. a.** x^{m+n} **b.** x^{mn} **c.** x^{m-n} **3.** 10 **4.** 10 **5.** Significant
6. All **7.** Precision

Exercises 5.3

1.

x	y_1	y_2
0	ERROR	ERROR
0.5	4	4
1	1	1
1.5	0.44444	0.44444
2	0.25	0.25
2.5	0.16	0.16
3	0.11111	0.11111

$x^{-2} = \dfrac{1}{x^2}$

3.

x	y_1	y_2
0	ERROR	ERROR
0.5	8	8
1	1	1
1.5	0.2963	0.2963
2	0.125	0.125
2.5	0.064	0.064
3	0.03704	0.03704

$\dfrac{x^4}{x^7} = x^{-3}$

5. a. $\dfrac{1}{9}$ **b.** $\dfrac{1}{32}$ **c.** -32 **d.** -25 **7. a.** $\dfrac{5}{6}$ **b.** $\dfrac{25}{36}$ **c.** $\dfrac{6}{5}$ **d.** 1 **9. a.** $\dfrac{1}{100}$

b. $\dfrac{1}{1000}$ **c.** $\dfrac{1}{10,000}$ **d.** $-\dfrac{1}{100}$ **11. a.** $\dfrac{1}{7}$ **b.** $\dfrac{7}{10}$ **c.** $-\dfrac{9}{5}$ **d.** 1

13. a. $\dfrac{10}{7}$ **b.** 7 **c.** $\dfrac{100}{49}$ **d.** 2 **15. a.** $\dfrac{y}{x}$ **b.** $\dfrac{y^2}{x^2}$ **c.** 1 **d.** $\dfrac{1}{x^2y}$ **17. a.** $\dfrac{1}{9x^2}$

b. $\dfrac{3}{x^2}$ **c.** $9x^2$ **19. a.** $\dfrac{1}{m+n}$ **b.** $\dfrac{1}{m}+\dfrac{1}{n}$ **c.** $m+\dfrac{1}{n}$ **21. a.** 45,800

b. 0.000458 **c.** 4,580,000 **d.** 0.00000458 **23. a.** 8100 **b.** 0.0081
c. -8100 **d.** -0.00000081 **25. a.** 9.7×10^3 **b.** 9.7×10^7
c. 9.7×10^{-1} **d.** 9.7×10^{-4} **27. a.** 3.5×10^{10} **b.** 3.5×10^{-9}
c. -3.5×10^3 **d.** -3.5×10^{-2} **29.** 4,493,000,000
31. 0.000000000001 **33.** 1.4×10^7 **35.** 2.0×10^{-8}
37. a. 6,000,000 **b.** 250 times as fast **c.** 312,500 times as much
memory **39.** 0.7943; 0.7953 **41.** 0.05034; 0.05088 **43.** 0.00002700;
0.00002628 **45.** 0.0001764 **47. a.** 0.367 **b.** 51,300,000 **49. a.** v^9
b. v^{15} **c.** $\dfrac{1}{v^{15}}$ **51. a.** $\dfrac{1}{x^4}$ **b.** 1 **c.** $\dfrac{9}{x^{14}}$ **53. a.** $\dfrac{3}{x}$ **b.** $12x$ **c.** $\dfrac{x^{12}}{16}$

55. a. $\dfrac{9x^2}{25y^2}$ **b.** $\dfrac{25y^2}{9x^2}$ **c.** $\dfrac{a^8}{b^6}$ **57.** $-\dfrac{1}{20x}$ **59.** $\dfrac{72y^8}{x}$ **61.** $4x^{10}$ **63.** $\dfrac{m^{12}}{n^{12}}$

65. $\dfrac{3y^6}{2x^7}$ **67.** 13; 25; $\dfrac{5}{6}$; $\dfrac{1}{5}$ **69.** 25; 49; $-\dfrac{7}{12}$; $-\dfrac{1}{7}$

Using the Language and Symbolism of Mathematics   5.4
1. Factors **2.** Addition; subtraction **3.** 0, 1, 2, 3 . . . **4.** Monomial
5. Polynomial **6.** Coefficient **7. a.** Monomial **b.** Binomial
c. Trinomial **8.** Sum **9.** Highest **10. a.** 0 **b.** 0 **11.** Alphabetical;
descending **12. a.** 1 **b.** -1 **c.** 1

Exercises 5.4
1. a. -12 is a monomial; the coefficient is -12 and the degree is 0
b. -12x is a monomial; the coefficient is -12 and the degree is 1
c. $-x^{12}$ is a monomial; the coefficient is -1 and the degree is 12

d. x^{-12} is not a monomial **3. a.** $\dfrac{2}{5}x^4$ is a monomial; the coefficient is

$\dfrac{2}{5}$ and the degree is 4 **b.** $\dfrac{5x^4}{2}$ is a monomial; the coefficient is $\dfrac{5}{2}$ and

the degree is 4 **c.** $-\dfrac{5x^4}{2}$ is a monomial; the coefficient is $-\dfrac{5}{2}$ and the

degree is 4 **d.** $\dfrac{2}{5x^4}$ is not a monomial **5. a.** $4xy^2z^3$ is a monomial; the

coefficient is 4 and the degree is 6 **b.** $-5x^2y^2$ is a monomial; the

coefficient is -5 and the degree is 4 **c.** $\dfrac{5x^2}{y^2}$ is not a monomial

d. $\dfrac{x^3y^4}{6}$ is a monomial; the coefficient is $\dfrac{1}{6}$ and the degree is 7

7. a. $3x^2 - 7x$ is a binomial of degree 2 **b.** $x^3 - 8x^2 + 9$ is a trinomial
of degree 3 **c.** $-9x^4$ is a monomial of degree 4 **d.** -9 is a monomial
of degree 0 **9. a.** $11xy$ is a monomial of degree 2 **b.** $11x + y$ is a

binomial of degree 1 **c.** $\dfrac{11x}{y}$ is not a polynomial **d.** $x^3 + y^2 - 11$ is a

trinomial of degree 3 **11. a.** $8x^2yz^3$ **b.** $x^2 - 5x - 9$
c. $-4x^3 - 2x^2 + 9x + 7$ **13. a.** Equal **b.** Not equal **15. a.** $21x$
b. $7x^2$ **17. a.** $7x^2 - 12x$ **b.** $-x^2 - 2x$ **19. a.** $9x^2 - x + 6$
b. $x^2 - 13x + 12$ **21.** $14x - 2$ **23.** $-9v + 14$ **25.** $9a^2 - a + 6$
27. $6x^2 - x + 5$ **29.** $3w^2 - 7w + 10$ **31.** $-m^4 + 3m^3 - 2m + 4$
33. $-4n^4 + 9n^3 + n^2 + n - 12$
35. $-4y^6 + y^5 - 6y^4 + 11y^2 + 4y - 18$ **37.** $-5x^2 - 20x - 3$
39. $-13a + 13$ **41.** $12x - 12$ **43.** $6x + 1$ **45.** $11x^2 + 4xy - 8y^2$
47. $-20x^2 + 7xy + 11y^2$ **49.** $9x^3 + 6x^2 - 12x + 30$
51. $-8x^4 - 3x^3y + 3x^2y^2 + xy^3 + 2y^4$ **53. a.** $P(0) = -2$ **b.** $P(2) = 24$ **c.** $P(10) = 528$ **d.** $P(8) = 342$ **55.** $P(0) = -40$; \$40 is lost if no
units are sold **57.** $P(2) = 0$; a breakeven point is when 2 units are sold

59. $P(20) = 0$; a breakeven point when 20 units are sold

61.

x	y_1
0	-1.1
0.2	-2.672
0.4	-3.668
0.6	-4.088
0.8	-3.932
1	-3.2
1.2	-1.892

63. $9.02x^2 + 4.02x$ **65.** $11.18x^2 - 1.97x - 5.98$ **67.** $3x + y + 13$

69. $x + y + 13$ **71.** $\frac{1}{2}xy$ **73.** $2x^3 - 5x + 11$ **75.** $3w^6 + 5w^4$

77. $x^2 - y^2$ **79.** $2w + 7$ **81.** 370,000 tons

Using the Language and Symbolism of Mathematics 5.5
1. a. 2 **b.** 3 **c.** 5 **2.** Distributive **3.** Distributive **4.** Distributive
5 a. 2 **b.** 2 **c.** 3 **6. a.** 1 **b.** 1 **c.** 2 **7.** Equal **8.** of x

Exercises 5.5
1. $-10x^2$ **3.** $63v^5$ **5.** $4a^9b^5$ **7.** $24x^3 - 20x^2$ **9.** $-32v^3 - 12v^2$
11. $14x^3 - 18x^2 - 8x$ **13.** $-44x^4 + 55x^3 + 88x^2$
15. $10x^3y - 5x^2y^2 - 5xy^3$ **17.** $6a^4b + 12a^3b^2 - 3a^2b^3$
19. $y^2 + 5y + 6$ **21.** $6v^2 + 7v - 20$ **23.** $54x^2 + 63x + 15$
25. $24m^2 - 34m + 7$ **27.** $2x^2 + xy - y^2$ **29.** $10x^2 + 6xy - 28y^2$
31. $x^3 - x^2 - 10x - 8$ **33.** $6x^3 - 8x^2 - 18x + 20$
35. $2m^3 + m^2 - 11m - 10$ **37.** $x^3 + y^3$ **39.** $5x^3 + 5x^2 - 30x$
41. $2x^3 + 13x^2 + 17x - 12$ **43.** $(8x^2 + 10x - 3)$ cm²
45. $(x^2 + x - 2)$ cm² **47.** $(x^2 + x + 1)$ cm² **49.** $(4x^2 - 1)$ cm² **51.** Yes
53. a. $36x^2 + 12x$ **b.** $4x(9x + 3)$ **55. a.** $6x^4y + 21x^3y^2$
b. $3x^3y(2x + 7y)$ **57. a.** $2m^3n - 2m^2n^2 + 10mn^3$
b. $2mn(m^2 - mn + 5n^2)$ **59. a.** $x^2 - x - 6$ **b.** $(x + 2)(x - 3)$
61. a. $15x^2 - 26x + 8$ **b.** $(3x - 4)(5x - 2)$ **63.** $R(x) = -2x^2 + 280x$;
$R(50) = \$9000$ **65.** $R(x) = -0.5x^2 + 1000x$; $R(50) = \$48{,}750$
67. $-6x^4 + 7x^2 + 20$ **69.** $-x^4 + 2x^3 - 2x^2 + x + 20$
71. $v^2 + 10v + 25$ **73.** $x^3 + 6x^2 + 12x + 8$
75. $-5xy(4x - y) = -20x^2y + 5xy^2$ **77.** $(x + y)^2 = x^2 + 2xy + y^2$
79. $(x^3 + 5x^2 + 6x)$ cm³ **81.** $3x + 4$; $3x + 4$; $3x + 4$ **83.** $2x$; y; $2x + y$

Using the Language and Symbolism of Mathematics 5.6
1. Quotient **2.** $x - 3$; quotient **3.** Product **4.** $(x - 3)$; $(x + 2)$;
$x^2 - x - 6$ **5.** Standard **6.** Zero; divisor **7.** 0

Exercises 5.6
1. $3x^2$ **3.** $-5a^2b$ **5.** $3a - 4$ **7.** $3a - 5 + \frac{2}{a}$ **9.** $8m^3 - 4m^2 + 5m + 3$
11. $5x^3 - 6x^2y^3 + 11y^2$ **13.** $6v^3 + 3v^2 - 9v - 2$
15. $35v^3w - 7v^2w - 28vw^2 = 7vw(5v^2 - 7v - 4w)$
17. $100x^{20} - 50x^{12} + 30x^9 = 10x^9(10x^{11} - 5x^3 + 3)$
19. $x^2 + 9x + 14 = (x + 2)(x + 7)$ **21.** $v^2 + 2v - 24 = (v - 4)(v + 6)$
23. $6w^2 + w - 12 = (2w + 3)(3w - 4)$ **25.** $20m^2 - 43m + 14 =$
$(5m - 2)(4m - 7)$ **27.** $63y^2 - 130y + 63 = (7y - 9)(9y - 7)$
29. $4m^2 + 3m + 19 + \dfrac{64}{m - 3}$ **31.** $8b^2 + 7b$ **33.** $21v^2 - 7v$
35. $3b^2 - 2b + 9$ **37.** $9m - 10n$ **39.** $7x^3 - 9x^2 + 6x + 2$
41. $x - 7$ **43.** $3a + 2$ **45.** $3v^2 + 2v + 4$ **47.** $x^2 + 2x + 4$
49. $4y^2 + 8y + 9$ **51.** $(2x - 1)$ cm **53.** $(6x + 2)$ cm **55.** $(2x + 3)$ cm

57.

x	y_1	y_2
0	3	3
0.5	2.5	2.5
1	2	2
1.5	1.5	1.5
2	1	1
2.5	ERROR	0.5
3	0	0

$P_1(x) = P_2(x)$ for all values of $x \neq 2.5$

59. $5k + 4$ columns **61.** $x^3 - 14x - 15$ **63.** $3x + 5$ **65. a.** 4 **b.** 6
c. 2 **67.** $2x^2 + 7x + 10$ **69.** $t - 2$ **71.** $t^2 - 4t + 16$

Using the Language and Symbolism of Mathematics 5.7
1. Two **2.** Three **3.** Multiplication **4.** Factoring **5. i.** c **ii.** d **iii.** a **iv.** b

Exercises 5.7
1. $49a^2 - 1$ **3.** $z^2 - 100$ **5.** $4w^2 - 9$ **7.** $81a^2 - 16b^2$
9. $16x^2 - 121y^2$ **11.** $x^4 - 4$ **13.** $16m^2 + 8m + 1$ **15.** $n^2 - 18n + 81$
17. $9t^2 + 12t + 4$ **19.** $25v^2 - 80v + 64$ **21.** $16x^2 + 40xy + 25y^2$
23. $36a^2 - 132ab + 121b^2$ **25.** $a^2 + 2abc + b^2c^2$ **27.** $x^4 - 6x^2 + 9$
29. $x^4 + 2x^2y + y^2$ **31.** $(7x + 1)(7x - 1)$ **33.** $(a + 10)(a - 10)$
35. $(9x + 5)(9x - 5)$ **37.** $(2a + 3b)(2a - 3b)$ **39.** $(x^2 + 2)(x^2 - 2)$
41. $(3x + 1)^2$ **43.** $(5y - 1)^2$ **45.** $(m - 13)^2$ **47.** $(2w + 3)^2$
49. $(3x + 5y)^2$ **51.** $(x^2 - 6)^2$ **53.** $(11m + 4n)^2$

55.

x	y_1	y_2
-3	35	35
-2	15	15
-1	3	3
0	-1	-1
1	3	3
2	15	15
3	35	35

$P_1(x) = P_2(x)$

57. $45x^2 - 20 = 5(9x^2) + 5(-4) = 5(9x^2 - 4) = 5(3x + 2)(3x - 2)$
59. $98x^3 - 18x = 2x(49x^2) + 2x(-9) = 2x(49x^2 - 9) =$
$2x(7x + 3)(7x - 3)$ **61.** $(x - 1)^2 - 4 = [(x - 1) + 2][(x - 1) - 2] =$
$(x + 1)(x - 3)$ **63.** $14v + 98$ **65.** $4xy$ **67.** $2vw - 2w^2$ **69.** 0
71. $(3x + 7y)^2 = 9x^2 + 42xy + 49y^2$ **73.** $(x + 8)^2 - (x - 6)^2 =$
$28x + 28$ **75.** $(a^2 + 1)(a - 1)(a + 1) = a^4 - 1$

Review Exercises for Chapter 5
1. a. xy^3 **b.** $(xy)^3$ **c.** $x^2 + y^2$ **d.** $(x + y)^2$ **2. a.** $x \cdot x + y \cdot y$
b. $(x + y)(x + y)$ **c.** $-x \cdot x \cdot x \cdot x$ **d.** $(-x)(-x)(-x)(-x)$ **3. a.** 9 **b.** 8
c. -9 **d.** $\frac{1}{9}$ **4. a.** 1 **b.** 0 **c.** 2 **d.** 1 **5. a.** 5 **b.** $\frac{6}{5}$ **c.** $\frac{3}{2}$ **d.** $\frac{1}{6}$ **6. a.** -1
b. 1 **c.** $\frac{1}{6}$ **d.** $-\frac{1}{6}$ **7. a.** $2x^3$ **b.** x^6 **c.** 1 **d.** 0 **8. a.** 100 **b.** 100 **c.** $\frac{1}{100}$
d. $-\frac{1}{100}$ **9. a.** 3 **b.** 2 **c.** 1 **10. a.** 9 **b.** 1 **c.** 3 **11. a.** x^7 **b.** x^3 **c.** x^{10}
12. a. $48x^8y^6$ **b.** $3x^2y^2$ **c.** $16x^6y^4$ **13. a.** -9 **b.** 9 **14. a.** 48 **b.** 144
15. a. 25 **b.** 49 **16. a.** $\frac{7}{12}$ **b.** $\frac{1}{7}$ **17. a.** 2 **b.** 1 **18. a.** $\frac{3}{16}$ **b.** $\frac{1}{144}$

19. $30x^3y^7z^{11}$ **20.** $25x^4y^6z^8$ **21.** $\dfrac{3ac^2}{5b^3}$ **22.** $\dfrac{n^4}{9m^6}$ **23.** $60x^{16}$ **24.** $9x^3$

25. $-200x^{21}$ **26.** $56y^2$ **27.** $2w^2$ **28.** $\dfrac{m^{24}}{625}$ **29.** $\dfrac{6}{x^7}$ **30.** 1 **31.** a^3b^4

32. $\dfrac{y^7z^{10}}{x^8}$ **33.** 0.0000000000001 watts **34.** 299,790,000 m/s

35. 5,230,000 **36.** 9.57842×10^9 **37.** 4.37×10^{-3} **38.** b

39. Monomial of degree 0 **40.** Sixth-degree binomial **41.** Third-degree polynomial with 4 terms **42.** Sixth-degree trinomial
43. $7x^5 + 11x^4 + 9x^3 - 3x^2 - 8x + 4$ **44.** $-7x^5$ **45.** $x^2 - 3$
46. $11x^2 - 3x + 2$ **47.** $5x^3 - 5x^2 - 8x + 4$
48. $5x^4 - 8x^3 + 13x^2 + 9x + 5$ **49.** $3x^5 + 3x^4 + 16x^3 + x^2 + 7x - 11$
50. $2x^2 - 11x - 12$ **51.** $35x^5 - 45x^4 + 15x^3 + 5x^2$ **52.** $35v^2 + 2v - 1$
53. $25y^2 - 70y + 49$ **54.** $81y^2 + 90y + 25$ **55.** $9a^2 - 25b^2$

56. $6m^3 - 8m^2 - 9m + 35$ **57.** $-\dfrac{4b^3}{a}$ **58.** $5m^2n - 7m^2n^2 - mn^3$

59. $v - 2$ **60.** $7w + 3$ **61.** $a + b$ **62.** $24xy + 32y^2$ **63.** $24xy$
64. $x - 5$ **65.** $(x + 4)(x^2 - 4x + 16) = x^3 + 64$

66. $\left(x + 6 + \dfrac{7}{x - 2}\right)(x - 2) = (x + 6)(x - 2) + 7 =$

$x^2 + 4x - 12 + 7 = x^2 + 4x - 5$ **67. a.** $3x^2 - 12x$ **b.** $3x(x - 4)$
68. a. $10x^2 + 15x$ **b.** $5x(2x + 3)$ **69. a.** $2m^3n - 6m^2n^2 - 10mn^3$
b. $2mn(m^2 - 3mn - 5n^2)$ **70. a.** $2x^3y - 6x^2y^2 + 2xy^3$
b. $2xy(x^2 - 3xy + y^2)$ **71. a.** $2x^2 + 5xy - 12y^2$ **b.** $(2x - 3y)(x + 4y)$
72. a. $21x^2 + 41x + 18$ **b.** $(3x + 2)(7x + 9)$ **73. a.** $25x^2 - 64$
b. $(7x - 11)(7x + 11)$ **74. a.** $4x^2 - 169y^2$ **b.** $(3x + 4y)(3x - 4y)$
75. a. $36x^2 - 84x + 49$ **b.** $(3x - 10)^2$ **76. a.** $4x^2 + 36xy + 81y^2$
b. $(5x + 4y)^2$ **77.** $2x^2 + 5x - 4$ **78.** $a + 3b - 4$ **79.** $x - 5$ **80.** $x + 7$
81. $x - 11$ **82.** $x - 2$ **83.** $(6x^2 - 7x - 3)$ ft^2 **84.** $(x + 1)$ cm
85. a. $R(x) = x(350 - 4x) = -4x^2 + 350x$ **b.** $R(0) = 0$; when the price is \$0 per unit, the revenue is \$0 **c.** $R(40) = 7600$; when the price is \$40 per unit, the revenue is \$7600 **d.** $R(80) = 2400$; when the price is \$80 per unit, the revenue is \$2400 **86.** \$17,834.78
87. $a = d$ and $b = c$

Mastery Test for Chapter 5
1. a. x^{13} **b.** $40y^7$ **c.** $-v^{15}$ **d.** m^4n^3 **2. a.** w^{40} **b.** $32y^{20}$ **c.** v^8w^{12}

d. $-\dfrac{27m^3}{8n^3}$ **3. a.** z^6 **b.** $2x^6$ **c.** -1 **d.** $-2a^2b^4$ **4. a.** 1 **b.** 1 **c.** 1 **d.** 8

5. a. x^5y^{11} **b.** x^3y^6 **c.** $\dfrac{4}{3x^5}$ **d.** $36x^{12}$ **6. a.** $\dfrac{1}{3}$ **b.** $\dfrac{1}{49}$ **c.** $\dfrac{9}{4}$ **d.** 9

7. a. 357,000; 0.0000735 **b.** 5.09×10^{-4}; 9.305×10^7 **8. a.** Third-degree binomial **b.** Monomial of degree 0 **c.** Second-degree trinomial
d. Fifth-degree polynomial with 4 terms **9. a.** $7x - y$
b. $-3x^2 + 6x + 18$ **c.** $9x^4 - 9x^3 + 19x^2 + 9x + 8$ **d.** $3x^3 - 12x^2 - 16x + 6$ **10. a.** $55x^6y^{10}$ **b.** $-15x^5 + 6x^4 - 21x^3 + 27x^2$ **c.** $x^3 + 8x^2 + 16x + 5$ **d.** $x^3 - 2x^2y - 2xy^2 - 3y^3$ **11. a.** $3a^2$ **b.** $3m^2n - 4n^3$ **c.** $x + 11$ **d.** $2x^2 + 6x + 8$ **12. a.** $x - 7$ **b.** $4x + 5$ **c.** $x^2 + 4x + 16$
d. $3x - 2$ **13. a.** $25x^2 - 121$ **b.** $x^2 - 81y^2$ **c.** $4x^2 + 28x + 49$
d. $9x^2 - 60xy + 100y^2$ **14. a.** $(9m + 4)(9m - 4)$
b. $(5x + 3y)(5x - 3y)$ **c.** $(7x + 3)^2$ **d.** $(4x - 11y)^2$

Chapter 6
Using the Language and Symbolism of Mathematics 6.1
1. One; output **2.** Domain; x **3.** Range; y **4.** x **5.** y **6.** x **7.** y
8. Vertical **9.** Function **10.** $f, x; x$; output

Exercises 6.1
1. a. Not a function **b.** Function; $D = \{-3, -1, 1, 2\}$; $R = \{1, 2, 3\}$
c. Function; $D = \{0, 1, 2, 3\}$; $R = \{-1, 0, 1, 3\}$ **3. a.** Function;
$D = \{-1, 1, 2\}$; $R = \{0, 1\}$ **b.** Not a function **c.** Function
$D = \{-3, -2, -1, 0, 1\}$; $R = \{-1, 0, 1, 2, 3\}$ **5. a.** Function **b.** Not a function **c.** Function **7. a.** Function **b.** Not a function **c.** Not a function **9. a.** Function **b.** Function **c.** Not a function **11. a.** -27

b. -7 **c.** 3 **13. a.** 0 **b.** 3 **c.** $\dfrac{1}{2}$ **15. a.** Output value **b.** Input value

c. 3 **d.** 8 **17. a.** Input value **b.** Output value **c.** 0 **d.** 3 **19. a.** 8
b. -2 **c.** -5 **21. a.** 0 **b.** -1 **c.** -3 **23. a.** 4 **b.** 0 **c.** -3 **25. a.** -2
b. 0 **c.** 2 **27. a.** $D = \{-3, -2, -1, 0, 1, 2, 3\}$; $R = \{2\}$
b. $D = (-\pi, \pi]$; $R = [-1, 1]$ **c.** $D = [-2, \infty)$; $R = (-\infty, 2]$
29. a. $D = [-2, 2]$; $R = [0, 2]$ **b.** $D = [-2, 4)$; $R = [-1, 4]$
c. $D = (-\infty, \infty)$; $R = [-2, 2]$

31. a.

D		R
-5	$\to$	4
-3	$\to$	2
-2	$\to$	0
0	$\to$	-2
1	$\to$	-3
4	$\to$	-4

b. $\{(-5, 4), (-3, 2), (-2, 0), (0, -2), (1, -3), (4, -4)\}$
c.

33. a.

D		R
-3	$\to$	1
-2	$\to$	-1
-1	$\to$	3
1	$\to$	-2
3	$\to$	2

b. $\{(-3, 1), (-2, -1), (-1, 3), (1, -2), (3, 2)\}$

c.

x	-3	-2	-1	1	3
y	1	-1	3	-2	2

35. a.

x	-1	1	2	4
y	1	3	-1	-2

b. $\{(-1, 1), (1, 3), (2, -1), (4, -2)\}$
c.

37. a.

x	-5	-3	1	2	3
y	4	4	4	4	4

b.

D		R
-5	$\to$	4
-3	$\to$	4
1	$\to$	4
2	$\to$	4
3	$\to$	4

c.

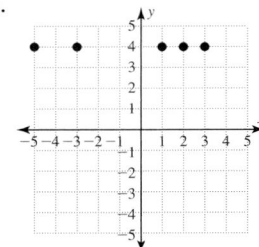

39. a. The number of students answering that problem incorrectly
b. 3 **c.** 1 **d.** 3 **41. a.** Pressure **b.** 16 in^3 **c.** 15 lb/in^2
43. a.

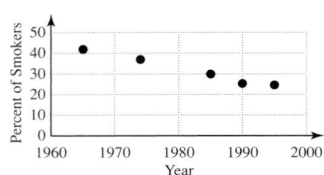

b. 29.9% **c.** In 1990 25.3% of U.S. adults smoked.

45. a.

Year	Tuition
1970	1000
1980	3000
1990	5000
2000	7000

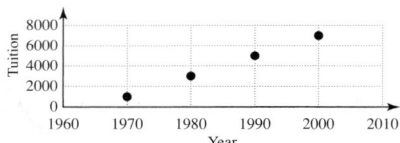

b. $9000

47.

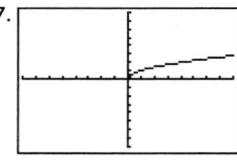

$D = [0, \infty); R = [0, \infty)$

49.

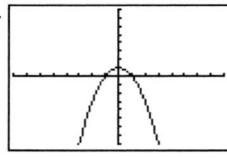

$D = (-\infty, \infty); R = (-\infty, 1]$

51.

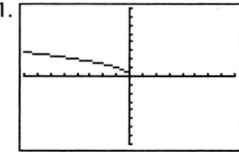

$D = (-\infty, 0]; R = [0, \infty)$

53. D **55.** B

57.

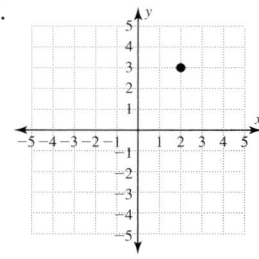

59.

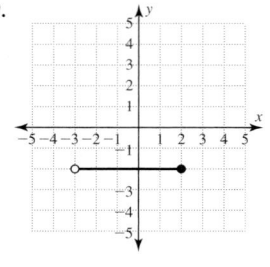

61.

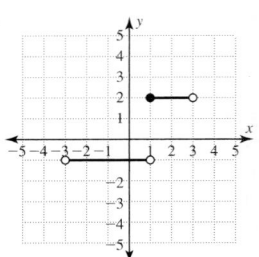

63.

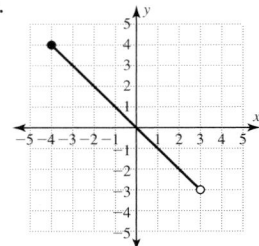

65.

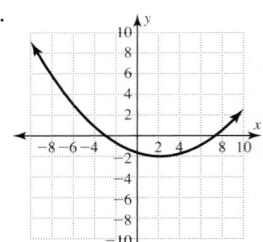

67.

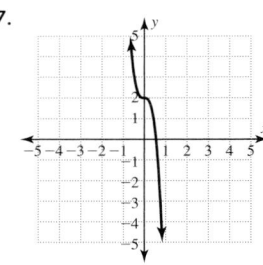

69. a.

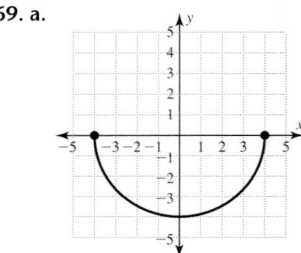

b.

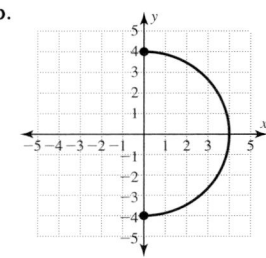

Using the Language and Symbolism of Mathematics 6.2
1. Line **2.** Slope; intercept; slope; y-intercept **3.** Point; slope

4. general **5.** $m = \dfrac{y_2 - y_1}{x_2 - x_1}$ **6.** Same **7.** Multiplicative inverses

8. Vertical; (7, 0) **9.** Horizontal; (0, −7) **10.** Horizontal **11.** Vertical
12. V **13.** Vertex

Exercises 6.2
1. a. −2 **b.** 1 **3. a.** $-\dfrac{2}{3}$ **b.** $\dfrac{5}{2}$ **5. a.** 0 **b.** Undefined **7. a.** −1 **b.** −2

9. a. $\dfrac{1}{2}$ **b.** $-\dfrac{2}{5}$ **11. a.** 0 **b.** Undefined **13. a.** $\dfrac{3}{5}$ **b.** $-\dfrac{4}{3}$ **15. a.** 3

b. −3 **17. a.** $-\dfrac{3}{4}$ **b.** $\dfrac{2}{5}$ **19. a.** $\dfrac{3}{8}$ **b.** $-\dfrac{2}{5}$ **21. a.** Undefined **b.** 0

23. a. −8 **b.** $-\dfrac{11}{12}$ **c.** −5 **d.** −3

25.

x	−3	−2	−1	0	1	2	3
y	−11	−9	−7	−5	−3	−1	1

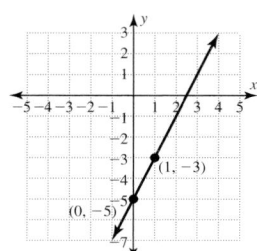

27.

x	−3	−2	−1	0	1	2	3
y	6	5	4	3	2	1	0

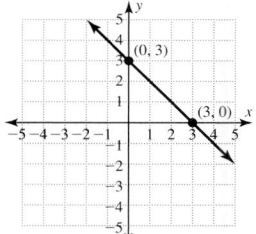

29. 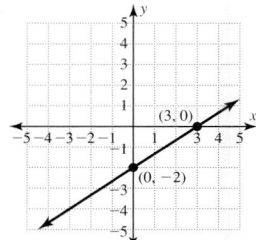   **31.**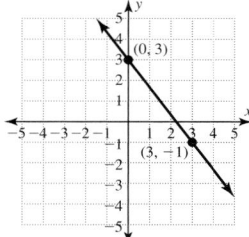

33. 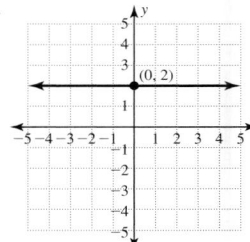   **35.**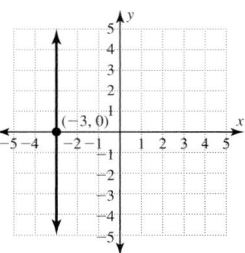

37. 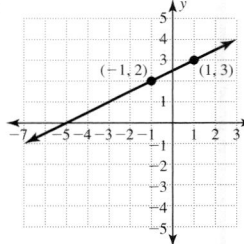   **39.**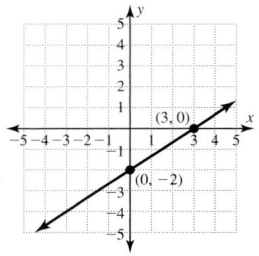

41. $y = 3x - 5$ **43.** $y = -\dfrac{2}{7}x + 1$ **45.** $y = \dfrac{3}{7}x + \dfrac{27}{7}$

47. $y = -2x - 1$ **49.** $y = 3$ **51.** $x = 2$ **53.** $y = 3x - 1$

55. $y = -\dfrac{1}{3}x + \dfrac{7}{3}$ **57. a.** 32 **b.** 5 **c.** 2 **d.** 13

59.

x	−3	−2	−1	0	1	2	3
y	1	0	1	2	3	4	5

$D = (-\infty, \infty)$; $R = [0, \infty)$; vertex: $(-2, 0)$

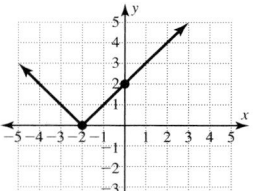

61.

x	−3	−2	−1	0	1	2	3
y	0	−1	−2	−3	−2	−1	0

$D = (-\infty, \infty)$; $R = [-3, \infty)$; vertex: $(0, -3)$

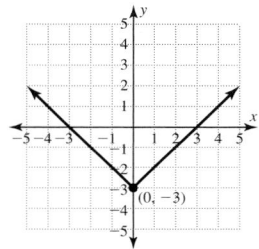

63. C **65.** D **67.** $D = (-\infty, \infty)$; $R = (-\infty, \infty)$ **69.** $D = (-\infty, \infty)$; $R = \{3\}$ **71.** $D = (-\infty, \infty)$; $R = [0, \infty)$; vertex: $(3, 0)$ **73.** $D = (-\infty, \infty)$; $R = (-\infty, 2]$; vertex: $(1, 2)$ **75. a.** $T(0) = -21.6$; a 10-mi/h wind makes 0° feel like $-21.6°$ **b.** $T(5) = -15.6$; a 10-mi/h wind makes 5° feel like $-15.6°$ **c.** $T(20) = 2.4$; a 10-mi/h wind makes 20° feel like $2.4°$ **77. a.** $P(2) = \dfrac{1}{36}$; the probability of rolling a sum of 2 on a pair of dice is $\dfrac{1}{36}$ **b.** $P(7) = \dfrac{1}{6}$; the probability of rolling a sum of 7 on a pair of dice is $\dfrac{1}{6}$ **c.** $P(12) = \dfrac{1}{36}$; the probability of rolling a sum of 12 on a pair of dice is $\dfrac{1}{36}$ **79. a.** $f(x) = 500 - 50x$ **b.** $f(7) = 150$; the value of the tools is $150 after 7 years

c.

x	0	1	2	3	4	5	6
f(x)	500	450	400	350	300	250	200

81. $f(x) = 0.60x + 2$; there is a flat fee of $2 per collect call and an additional charge of $0.60 per minute **83. a.** $y = -\dfrac{1}{7}x + 30$ **b.** Yes

c. $y = 7x - 220$

Using the Language and Symbolism of Mathematics   6.3
1. Parabola **2.** Quadratic **3.** Vertex **4.** Cubic **5.** Real **6.** Zero; negative

Exercises 6.3
1. a. 26 **b.** 10 **c.** 2 **d.** 10 **3. a.** 5 **b.** 3 **c.** 2 **d.** 0 **5. a.** −1004 **b.** −5 **c.** −4 **d.** 121 **7. a.** 2 **b.** 1 **c.** −2 **d.** 3

9.

x	−3	−2	−1	0	1	2	3
y	7	2	−1	−2	−1	2	7

$D = (-\infty, \infty)$; $R = [-2, \infty)$; vertex: $(0, -2)$

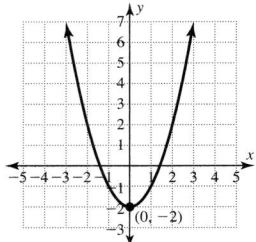

11.

x	−3	−2	−1	0	1	2	3
y	−4	−1	0	−1	−4	−9	−16

$D = (-\infty, \infty)$; $R = (-\infty, 0]$; vertex: $(-1, 0)$

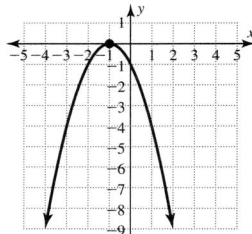

13.

x	−3	−2	−1	0	1	2	3
y	−1	0	1	8	27	64	125

$D = (-\infty, \infty)$; $R = (-\infty, \infty)$

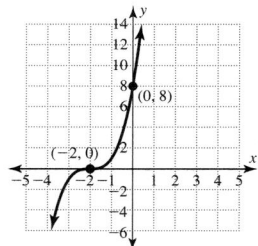

15.

x	−3	−2	−1	0	1	2	3
y	29	10	3	2	1	−6	−25

$D = (-\infty, \infty)$; $R = (-\infty, \infty)$

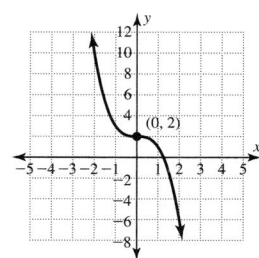

17.

x	0	1	4	9	16	25
y	−2	−1	0	1	2	3

$D = [0, \infty)$; $R = [-2, \infty)$

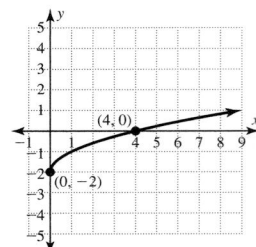

19.

x	−8	−1	$-\frac{1}{8}$	0	$\frac{1}{8}$	1	8
y	0	1	$\frac{3}{2}$	2	$\frac{5}{2}$	3	4

$D = (-\infty, \infty)$; $R = (-\infty, \infty)$

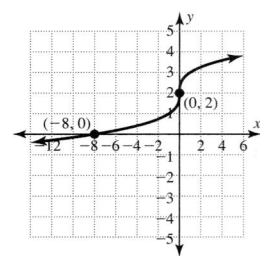

21. G **23.** B **25.** F **27.** D **29.** H **31.** B **33.** E **35.** A **37.** D
39. $D = (-\infty, \infty)$; $R = \{5\}$ **41.** $D = (-\infty, \infty)$; $R = [-5, \infty)$
43. $D = (-\infty, \infty)$; $R = [1, \infty)$ **45.** $D = (-\infty, \infty)$; $R = (-\infty, 1]$
47. $D = [-3, \infty)$; $R = (-\infty, 0]$ **49.** $D = (-\infty, \infty)$; $R = (-\infty, \infty)$
51. $D = (-\infty, \infty)$; $R = (-\infty, \infty)$ **53.** x-intercepts: $(5, 0)$, $(-1, 0)$;
y-intercept: $(0, -1)$ **55.** x-intercepts: $(3, 0)$, $(-1, 0)$; y-intercept: $(0, 3)$
57. x-intercepts: $(1, 0)$, $(0, 0)$, $(-1, 0)$; y-intercept: $(0, 0)$
59. x-intercepts: $(3, 0)$, $(1, 0)$, $(-2, 0)$; y-intercept: $(0, -3)$
61. a. $h(0) = 5000$; the initial height of the parachutist is 5000 ft
b. $h(5) = 4600$; the parachutist is 4600 ft high after 5 s
c. $h(10) = 3400$; the parachutist is 3400 ft high after 10 s
63. a. $C(0) = 1200$; it costs \$1200 to run a power line straight across
the river **b.** $C(2) \approx 2683.28$; it costs $\approx$ \$2683.28 to run the power
line to the other side of the river 2 mi downstream **c.** $C(5) \approx 6118.82$;
it costs $\approx$ \$6118.82 to run the power line to the other side of the river
5 mi downstream **65. a.** $V(1) = 144$; the volume of the box is
144 cm^3 when the height is 1 cm **b.** $V(3) = 168$; the volume of the
box is 168 cm^3 when the height is 3 cm **c.** $V(4) = 96$; the volume of
the box is 96 cm^3 when the height is 4 cm **67.** $D = (-\infty, \infty)$
69. $D = (-\infty, \infty)$ **71.** $D = [-4, \infty)$ **73.** $D = (-\infty, \infty)$
75. $D = (-\infty, \infty)$ **77.** $D = (-\infty, \infty)$

Using the Language and Symbolism of Mathematics 6.4
1. Shape **2.** Vertical; vertical **3.** Horizontal; horizontal **4. a.** Vertical;
up **b.** Vertical; down **c.** Horizontal; left **d.** Horizontal; right
5. a. Horizontal; left; vertical; up **b.** Horizontal; right; vertical; down
6. Parabola; vertex

Exercises 6.4
1. C **3.** D **5.** D **7.** B **9.** C **11.** D **13.** B **15.** D

17.

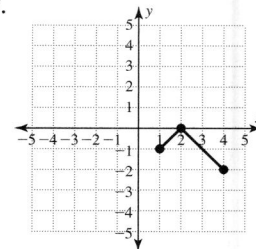

19.

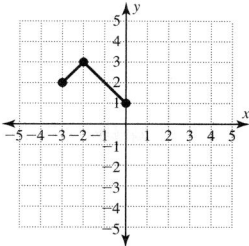

21.

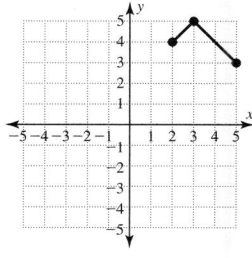

23. $y = \dfrac{x}{2} + 3$ or $y = \dfrac{x + 6}{2}$; $(0, 3)$

25.

x	$y = f(x) + 10$
-3	10
-2	18
-1	16
0	10
1	6
2	10

27. $y_2 = f(x) - 7$

29.

x	$y_2 = f(x - 4)$
4	-3
5	-1
6	1
7	3
8	5
9	7

31. $y_2 = f(x - 3)$ **33.** Vertex: $(6, 0)$ **35.** Vertex: $(0, 11)$ **37.** Vertex: $(-5, -8)$ **39.** Vertex: $(0, -13)$ **41.** Vertex: $(-15, 0)$ **43.** Vertex: $(7, -6)$ **45.** E **47.** D **49.** A **51.** $D = [3, 8)$; $R = [2, 4)$ **53.** $D = [0, 5)$; $R = [-5, -3)$ **55.** $D = [-4, 1)$; $R = [7, 9)$ **57.** D **59.** A **61.** E **63.** H **65. a.** 2, 4, 6, 8, 10 **b.** 5, 7, 9, 11, 13 **c.** 8, 10, 12, 14, 16 **d.** $-2, -4, -6, -8, -10$

67.

t Kelvin	$V = 4(t - 273)$
473	800
423	600
373	400
323	200
273	0

69.

x Item #	New Price
1	25.95
2	34.79
3	13.98
4	27.78
5	20.95

Using the Language and Symbolism of Mathematics   6.5
1. Same; same **2.** $(x, -y)$ **3.** Stretching **4.** Shrinking **5.** Reflection; x **6.** Same; stretching

Exercises 6.5
1. G **3.** B **5.** E **7.** D **9.** B **11.** A **13.** D **15.** A **17.** D **19.** A **21.** B **23.** A

25.

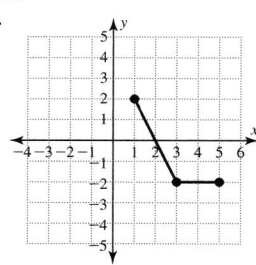

27.

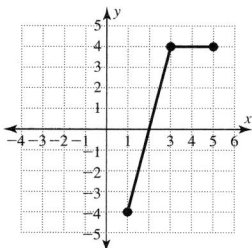

29.

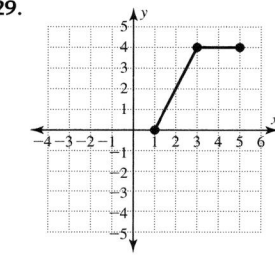

31.

x	$y = -f(x)$
-2	-18
-1	-9
0	-3
1	-1
2	-3

33.

x	$y = 2f(x)$
-2	36
-1	18
0	6
1	2
2	6

35. $y_2 = \dfrac{1}{2}f(x)$ **37.** $y_2 = -\dfrac{1}{4}f(x)$ **39.** $y_2 = f(x) - 8$

41. $y_2 = f(x + 3)$ **43.** $(0, 0)$ **45.** $(0, -7)$ **47.** $(8, -7)$ **49.** $(8, 7)$

51. a.

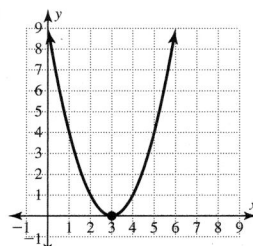

The graph of $f(x) = (x - 3)^2$ is obtained by shifting $f(x) = x^2$ three units to the right

b.

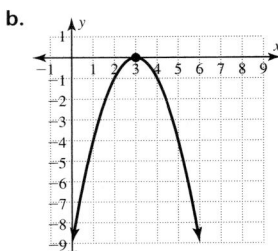

The graph of $f(x) = -(x - 3)^2$ is obtained by reflecting the graph of $f(x) = (x - 3)^2$ about the x-axis

c.

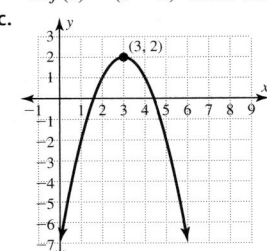

The graph of $f(x) = -(x - 3)^2 + 2$ is obtained by shifting the graph of $f(x) = -(x - 3)^2$ up two units

53. B **55.** D **57.** F **59.** $R = [4, 8)$ **61.** $R = [4, 12)$

63. $R = (-6, -2]$ **65. a.** 1, 8, 27, 64, 125 **b.** $-1, -8, -27, -64,$ -125 **c.** 2, 16, 54, 128, 250 **d.** $-1, 2, 7, 14, 23$ **67. a.** $A_n = n$

b. $V_n = 50n$

69.

T Years	I Interest
5	3,000
10	6,000
15	9,000
20	12,000

Using the Language and Symbolism of Mathematics 6.6
1. Scatter **2.** Line **3.** Parabola **4.** Curve **5.** Regression **6.** Slope; intercept; slope; $(0, b)$

Exercises 6.6
1. C **3.** F **5.** D

7. A.

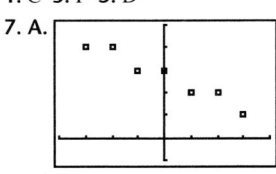

9. B.

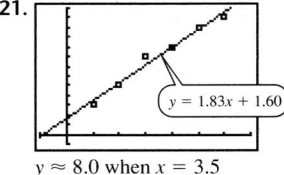

11. $y = \dfrac{4}{7}x + 3$ **13.** $y = -3x - 6$ **15.** $y = -\dfrac{2}{3}x - \dfrac{8}{3}$

17. $y = -2x - 1$ **19.** $y = 0.04x$

21.

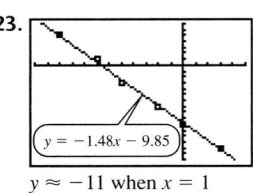

$y \approx 8.0$ when $x = 3.5$

23.

$y \approx -11$ when $x = 1$

25.

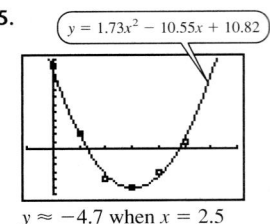

$y = 1.73x^2 - 10.55x + 10.82$

$y \approx -4.7$ when $x = 2.5$

27.

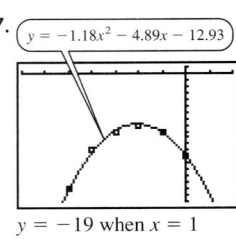

$y = -1.18x^2 - 4.89x - 12.93$

$y = -19$ when $x = 1$

29.

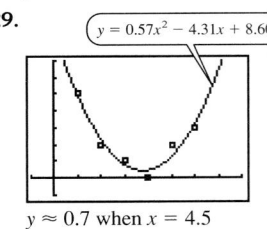

$y = 0.57x^2 - 4.31x + 8.60$

$y \approx 0.7$ when $x = 4.5$

31.

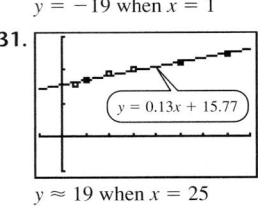

$y = 0.13x + 15.77$

$y \approx 19$ when $x = 25$

33.

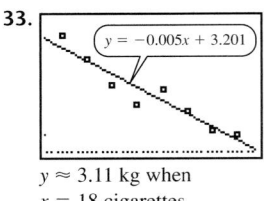

$y = -0.005x + 3.201$

$y \approx 3.11$ kg when $x = 18$ cigarettes

35.

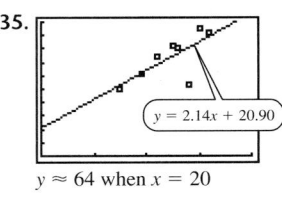

$y = 2.14x + 20.90$

$y \approx 64$ when $x = 20$

37.

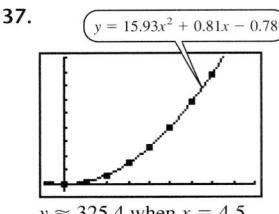

$y = 15.93x^2 + 0.81x - 0.78$

$y \approx 325.4$ when $x = 4.5$

Using the Language and Symbolism of Mathematics 6.7
1. Greatest common **2.** Coefficient; exponent **3.** Distributive **4.** Zero
5. Zero **6.** $(x - 5)$; $(5, 0)$ **7.** -4; $(-4, 0)$ **8.** 0; $(x - 8)$ **9.** Real **10.** n

Exercises 6.7
1. $2x - 3$ **3.** $5a + 3b$ **5.** $2x^2 - 3xy + 4y^2$ **7.** $2x + y$ **9.** $7x - 9$
11. $7x(2x + 11)$ **13.** $4xy(2x - 5y)$ **15.** $2x(3x^2 - 4x + 5)$
17. $(5x - 2)(x + 3)$ **19.** $(x + 2y)(6x - 7y)$ **21.** $(x + 8)(x - 4)$
23. $(x - 5)(x - 25)$ **25.** $(x + 4)(x - 2)(x - 4)$ **27.** $(x + 5)(x - 6)$
29. $(x - 2)(x - 5)$ **31.** $x(x + 2)(x - 3)$ **33.** Zeros: -5 and 9;
x-intercepts: $(-5, 0)$, $(9, 0)$ **35.** Zeros: -3 and 17; x-intercepts:
$(-3, 0)$, $(17, 0)$ **37.** Zeros: $-3, -1,$ and 8; x-intercepts: $(-3, 0)$,
$(-1, 0)$, $(8, 0)$

39.

x	-12	-9	-6	-3	0	3	6
y	45	0	-27	-36	-27	0	45

$x^2 + 6x - 27 = (x + 9)(x - 3)$

41.

x	0	5	10	15	20	25	30
y	300	150	50	0	0	50	150

$x^2 - 35x + 300 = (x - 15)(x - 20)$

43.

x	0	1	2	3	4	5	6
y	0	15	16	9	0	-5	0

$x^3 - 10x^2 + 24x = x(x - 4)(x - 6)$

45. $(x - 7)(x - 4)$ **47.** $(x - 5)(x + 25)$ **49.** $(x + 9)(x - 1)(x - 5)$
51. $(x + 2)(x + 1)(x - 1)(x - 2)$ **53.** Zeros: 7 and -9; x-intercepts:
$(7, 0), (-9, 0)$ **55.** Factored form: $(x + 8)(x - 10)$; zeros: -8 and 10
57. $(x + 8y)(x - 9y)$ **59.** $(x - 11y)(x - 13y)$

Review Exercises for Chapter 6

1. a. Function **b.** Function **c.** Not a function **d.** Function **2. a.** Not a
function **b.** Function **c.** Function **d.** Not a function
3. a. $D = \{5, 6, 7\}$; $R = \{8\}$ **b.** $D = (-1, 3]$; $R = [0, 2]$
c. $D = \{-3, -2, -1, 0, 1\}$; $R = \{11, 9, 4, 1, 6\}$ **d.** $D = \{2, \infty\}$;
$R = \{3, \infty\}$ **4. a.** 3 **b.** 7 **c.** 12 **d.** 103 **5. a.** 4 **b.** 2 **c.** 8 **d.** 10
6. a. -2 **b.** -1 **c.** 1 **d.** 2

7. a.

D		R
-10	$\to$	4
-5	$\to$	3
0	$\to$	2
5	$\to$	1
10	$\to$	0
15	$\to$	-1
20	$\to$	-2

b. $\{(-10, 4), (-5, 3), (0, 2), (5, 1), (10, 0), (15, -1), (20, -2)\}$

c. 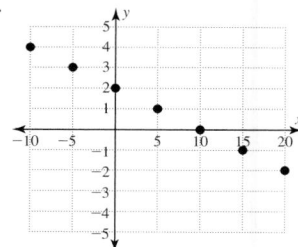   **d.** $y = -\dfrac{1}{5}x + 2$

8. a.

D		R
-4	$\to$	-1
-2	$\to$	0
0	$\to$	1
2	$\to$	2
4	$\to$	3

b.

x	y
-4	-1
-2	0
0	1
2	2
4	3

c. $\{(-4, -1), (-2, 0), (0, 1), (2, 2), (4, 3)\}$ **d.** $y = \dfrac{1}{2}x + 1$

9. a. 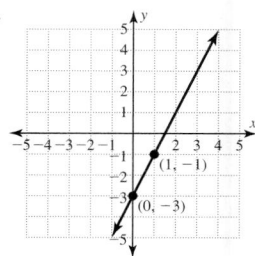   **b.**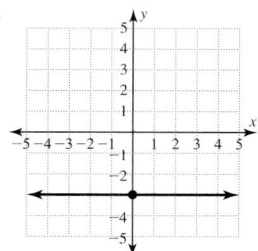

c. 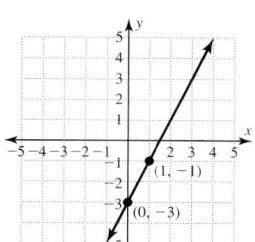   **d.**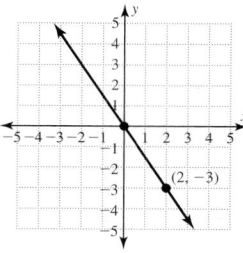

10. a. $(-\infty, \infty)$ **b.** $(-\infty, \infty)$ **c.** $[5, \infty)$ **d.** $(-\infty, \infty)$ **11. a.** B **b.** C **c.** D
d. A **12. a.** $D = (-\infty, \infty)$; $R = (-\infty, 3]$ **b.** $D = (-\infty, 4]$; $R = [0, \infty)$
c. $D = (-\infty, \infty)$; $R = (-\infty, \infty)$ **d.** $D = (-\infty, \infty)$; $R = \{1\}$
13. a. $h(0) = 1296$; the initial height of the wrench is 1296 ft
b. $h(2) = 1232$; after 2 sec the wrench is 1232 ft high **c.** $h(5) = 896$;
after 5 sec the wrench is 896 ft high **d.** $h(9) = 0$; the wrench hits the
ground after 9 sec **14. a.** B **b.** A **c.** C **15. a.** A **b.** C **c.** B **16. a.** B
b. A **c.** C

17. a. 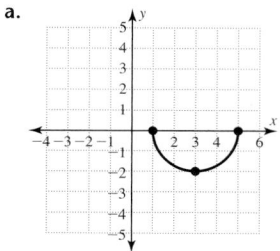   **b.**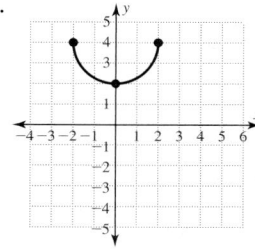

c.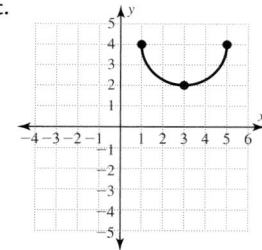

18. a. $(0, 7)$ **b.** $(9, 0)$ **c.** $(-11, -12)$

19. a.

x	$y = f(x) + 5$
-2	10
-1	14
0	8
1	1
2	5

b.

x	$y = f(x + 5)$
-7	5
-6	9
-5	3
-4	-4
-3	0

20. a. D **b.** A **c.** B **d.** C **21. a.** D **b.** B **c.** C **d.** A **22. a.** B **b.** C
c. D **d.** A **23. a.** C **b.** D **c.** A **d.** B

24. a.

x	$y = -f(x)$
0	0
1	-8
2	-4
3	-12
4	-24

b.

x	$y = \dfrac{1}{4}f(x)$
0	0
1	2
2	1
3	3
4	6

c.

x	y = 4f(x)
0	0
1	32
2	16
3	48
4	96

d.

x	y = −3f(x)
0	0
1	−24
2	−12
3	−36
4	−72

25. a. D **b.** A **c.** B **d.** C **26. a.** 1, 4, 9, 16, 25 **b.** 2, 5, 10, 17, 26
c. 2, 8, 18, 32, 50 **d.** −1, −4, −9, −16, −25 **27. a.** (0, 0) **b.** (−4, 0)

c. (0, 4) **d.** (0, 0) **28.** $y = 3x + 5$ **29.** $y = \frac{2}{3}x + 5$ **30.** $y = -3x + 11$

31. $y = \frac{3}{2}x - 4$ **32.** $x = 2$ **33.** $y = 5x$ **34.** $y = 2.50x + 3.86$

35. $y = -3.19x + 4.49$ **36.** $y = 1.15x^2 + 2.34x + 4.55$
37. $y = -1.08x^2 - 0.18x + 8.05$ **38.** $y = 0.080x + 0.057$; a mass of
12 kg would stretch the spring approximately 1.0 cm
39. $y = 7.92x^2 + 45.27x - 27.98$; a pendulum of length approximately
94 cm will have a period of 2 s **40.** Associative **41.** Distributive
42. Distributive **43.** $5x - 6$ **44.** $5a + 4b$ **45.** $3x^2 - 5x + 7$
46. $9x + 7$ **47.** $4x - 11$ **48.** $6x + 5y$ **49.** $11x(2x + 5)$
50. $7xy(x - 2y)$ **51.** $3(3x^2 - 4x + 9)$ **52.** $2ab(3a^2 + 5ab + 11b^2)$
53. $(2x - 9)(3x + 17)$ **54.** $(4x - 5y)(7x - y)$ **55.** $(x + 6)(x - 4)$
56. $(x + 5)(x - 20)$ **57.** $(x - 3)(x - 5)(x - 9)$
58. $(x + 5)(x + 1)(x - 1)(x - 5)$ **59.** $(x + 5)(x + 1)$
60. $(x + 6)(x - 3)$ **61.** $x(x + 5)(x - 2)$

62.

x	0	3	6	9	12	15	18
y	36	0	−18	−18	0	36	90

$(x - 3)(x - 12)$

63.

x	0	4	8	12	16	20	24
y	48	0	−16	0	48	128	240

$(x - 4)(x - 12)$

64. $(x + 8)(x - 7)$ **65.** $x(x + 4)(x - 4)$ **66.** $(x - 2)(x - 1)(x + 3) \cdot$
$(x + 5)$ **67.** −11 and 4; $(-11, 0), (4, 0)$ **68.** $(x + 5)(x - 13)$; $(-5, 0)$,
$(13, 0)$ **69.** $(x + 9)(x - 12)$; −9 and 12 **70.** $(x + 3)(x - 3)(x - 7)$;
$(-3, 0), (3, 0), (7, 0)$ **71.** $(x + 11y)(x - 13y)$ **72.** $(x + 6y)(x - 14y)$

Mastery Test for Chapter 6
1. a. Function **b.** Not a function **c.** Function **d.** Function
e. Not a function **f.** Function **2. a.** $D = \{1, 2, 7, 8\}$; $R = \{8, 4, 11, 13\}$
b. $D = (-\infty, \infty)$; $R = \{7\}$ **c.** $D = [3, \infty)$; $R = [0, \infty)$
d. $D = \{-3, -2, -1, 0\}$; $R = \{0, 1, 2, 7\}$ **e.** $D = \{-3, -2, -1, 0, 1\}$;
$R = \{-2, -1, 0, 1, 2\}$ **f.** $D = [-2, \infty)$; $R = (-\infty, 2]$ **3. a.** −12 **b.** 2
c. −2 **d.** −2

4. a. 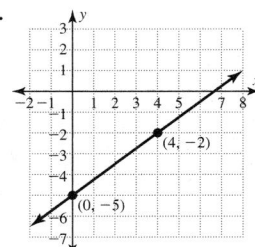 **b.**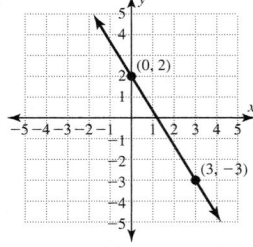

c. 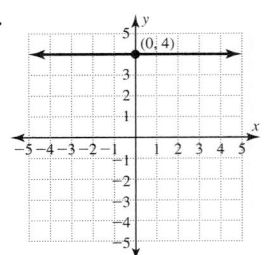 **d.**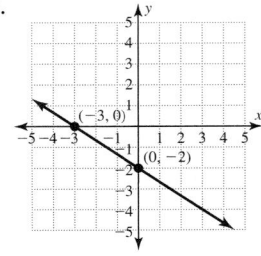

5. $f(x) = 0.4x + 1.25$; there is a flat fee of $1.25 per collect call and an
additional charge of $0.40 per minute

6. a.

x	−3	−2	−1	0	1	2	3
y	8	7	6	5	4	3	4

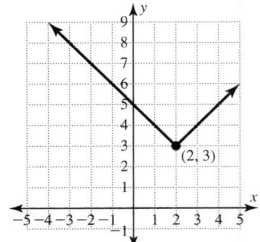

Vertex: (2, 3); $D = (-\infty, \infty)$; $R = [3, \infty)$

b.

x	−3	−2	−1	0	1	2	3
y	−3	−4	−5	−4	−3	−2	−1

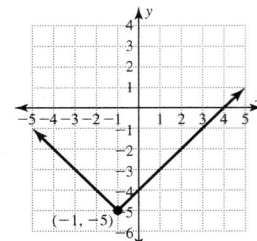

Vertex: (−1, −5); $D = (-\infty, \infty]$; $R = [-5, \infty)$
7. a. C **b.** E **c.** A **d.** F **e.** D **f.** B **8. a.** $D = (-\infty, \infty)$
b. $D = (-\infty, \infty)$ **c.** $D = (-\infty, \infty)$ **d.** $D = [6, \infty)$ **e.** $D = (-\infty, \infty)$
f. $D = (-\infty, \infty)$

9. a. A **b.** B

c.

x	f(x) − 2
0	−7
1	−6
2	1
3	20
4	57

d.

x	f(x) + 2
0	−3
1	−2
2	5
3	24
4	61

10. a. A **b.** B

c.

x	f(x − 3)
3	−5
4	−4
5	3
6	22
7	59

d.

x	f(x + 3)
−3	−5
−2	−4
−1	3
0	22
1	59

11. a.

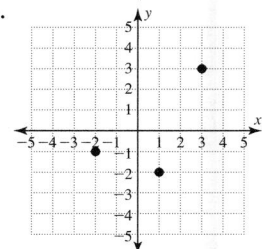

b.

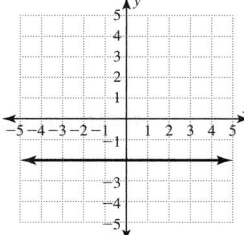

c.

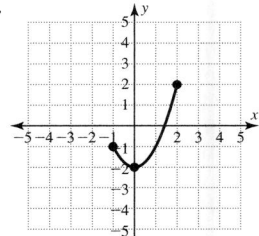

d.

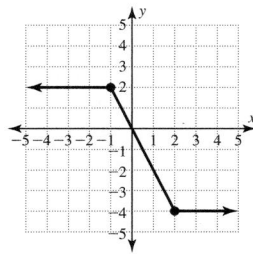

12. a. C **b.** D **c.** A **d.** B **13.** $y = 11.71x - 45.00$
14. $y = 0.560x^2 - 0.036x + 5.190$ **15. a.** $9x^2(2x - 11)$
b. $2xy^2(2x^2 - 3xy + 5y^2)$ **c.** $(7x - 2)(3x + 4)$ **d.** $(2x + 3y)(x - 5y)$
16. a. $(x + 8)(x + 6)$ **b.** $x(x + 2)(x - 3)$ **17. a.** $(x + 8)(x - 5)$
b. $(x + 3)(x + 2)(x - 1)$

Chapter 7
Using the Language and Symbolism of Mathematics   7.1
1. Prime **2.** Quadratic **3. a.** Quadratic **b.** Linear **c.** Constant
4. Distributive; greatest common factor **5.** Product; sum **6.** Same;
opposite

Exercises 7.1
1. $x(x - 5) - 7(x - 5) = (x - 7)(x - 5)$
3. $x^2 - 8x + 10x - 80 = x(x - 8) + 10(x - 8) = (x + 10)(x - 8)$
5. $(5x)(x^2 + 5x + 4) = (5x)(x^2 + x + 4x + 4) =$
$(5x)[x(x + 1) + 4(x + 1)] = (5x)(x + 4)(x + 1)$
7. $x^2 + 4xy - 9xy - 36y^2 = x(x + 4y) - 9y(x + 4y) =$
$(x - 9y)(x + 4y)$ **9. a.** $(x + 3)(x + 4)$ **b.** $(x + 6)(x + 2)$ **c.** Prime
d. $(x + 1)(x + 12)$ **11. a.** $(x - 7)(x + 4)$ **b.** $(x + 2)(x - 14)$ **c.** Prime
d. $(x + 1)(x - 28)$ **13. a.** $(x - 6)(x + 6)$ **b.** $(x + 12)(x - 3)$
c. $(x - 2)(x + 18)$ **d.** Prime **15. a.** $(x + 10)(x + 1)$
b. $(x + 2)(x + 5)$ **17. a.** $(x + 8)(x - 3)$ **b.** $(x + 6)(x - 4)$
19. $(a - 11)(a + 1)$ **21.** $(y - 17)(y - 1)$ **23.** $(x - 1)(x - 7)$
25. $(y - 11)(y + 9)$ **27.** $(p - 4)(p - 25)$ **29.** $(t + 8)(t + 6)$
31. $(n + 9)(n - 4)$ **33.** $(x + 2y)(x + 3y)$ **35.** $(x + 3y)(x - 2y)$
37. $(x - 5y)(x - 5y)$ **39.** $(a + 2b)(a - 18b)$ **41.** $(x + 3)(x + 30)$
43. Prime **45.** $(m - 14n)(m - 7n)$ **47.** Prime **49.** Prime
51. $5x(x + 2)(x + 1)$ **53.** $4ax(x - 2)(x - 3)$ **55.** $10a(z + 25)(z + 4)$
57. $2ab(x + 8)(x - 6)$ **59.** $(a + b)(x - 4)(x - 2)$
61. $(-1)(x + 5)(x - 7)$ **63.** $(-1)(m + 6)(m - 3)$
65. $(-1)(x + 3y)(x - 15y)$ **67.** $(x + 5)(x - 15)$ **69.** $(x - 8)(x + 8)$
71. Factored form: $(x + 5)(x - 3)$; zeros of $P(x)$: -5 and 3;
x-intercepts: $(-5, 0), (3, 0)$ **73.** Factored form: $(x + 6)(x - 9)$; zeros
of $P(x)$: -6 and 9; x-intercepts: $(-6, 0), (9, 0)$

Using the Language and Symbolism of Mathematics   7.2
1. Greatest common factor **2.** b **3.** Same **4.** Opposite

Exercises 7.2
1. $3a(a + 5) + 2(a + 5) = (3a + 2)(a + 5)$
3. $3m^2 - 12m + 2m - 8 = 3m(m - 4) + 2(m - 4) = (3m + 2)(m - 4)$

5. $(4a)(12m^2 - 4m - 21) = (4a)(12m^2 - 18m + 14m - 21) =$
$(4a)[6m(2m - 3) + 7(2m - 3)] = (4a)(6m + 7)(2m - 3)$
7. $3x(5x - 4y) - 5y(5x - 4y) = (3x - 5y)(5x - 4y)$
9. a. $(m + 2)(3m + 4)$ **b.** $(3m + 8)(m + 1)$ **c.** $(m + 4)(3m + 2)$
d. $(m + 8)(3m + 1)$ **11. a.** $(n + 10)(4n - 1)$ **b.** $(n + 4)(2n - 5)$
c. $(n + 2)(5n - 4)$ **d.** Prime **13. a.** Prime **b.** $(4x - 3y)(x - 3y)$
c. $(x - 2y)(2x - 9y)$ **d.** $(6x - y)(x - 6y)$ **15. a.** $(x - 1)(3x - 5)$
b. $(5x - 1)(x - 3)$ **c.** $(15x - 1)(x - 1)$ **d.** Prime
17. a. $(10y + 1)(y - 7)$ **b.** $(14y + 1)(y - 5)$ **c.** $(y + 1)(2y - 35)$
d. $(14y + 1)(5y - 1)$ **19.** Prime **21.** $(w - 1)(4w - 5)$
23. $(3z + 2)(3z - 5)$ **25.** $(2b + 5)(5b + 2)$ **27.** $(3m + 5n)(4m - 7n)$
29. $(2y + 7)(9y - 4)$ **31.** $(11y + 3)(5y - 4)$ **33.** $(4x + 5y)(5x + 3y)$
35. $(7x - 2y)(2x - 5y)$ **37.** $(3m + 2n)(11m - 3n)$
39. $(2v + 3w)(3v + 2w)$ **41.** $-(6x - 1)(2x - 3)$
43. $-(7y + 2)(5y - 2)$ **45.** $-(3z + 8)(4z + 5)$
47. $-2a(19x + 3)(x - 1)$ **49.** $5x^3(4x - 3)(3x - 5)$
51. $3mn(m + n)(5m - 7n)$ **53.** $7av(5v + 6w)(6v - 5w)$
55. $(a + b)(5x - 2)(7x - 3)$

Using the Language and Symbolism of Mathematics   7.3
1. Perfect square trinomial **2.** Difference of two squares **3.** Sum of
two squares **4.** Difference of two cubes

Exercises 7.3
1. $a^2 + 2(3)(a) + (3)^2 = (a + 3)^2$ **3.** $(2w)^2 - 2(2w)(3) + 3^2 =$
$(2w - 3)^2$ **5.** $(m - 1)^2$ **7.** $(6v + 1)^2$ **9.** $(5x - 2y)^2$ **11.** $(11m + 4n)^2$
13. $-(x - 8y)^2$ **15.** $(4x - 3)^2$ **17.** $2x(x + 4)^2$ **19.** $(w + 7)(w - 7)$
21. $(3v + 1)(3v - 1)$ **23.** $(9m + 5)(9m - 5)$ **25.** $(2a + 3b)(2a - 3b)$
27. $(4v + 11w)(4v - 11w)$ **29.** $(6 + m)(6 - m) = -(m + 6)(m - 6)$
31. $5(2x + 3y)(2x - 3y)$ **33.** $(5x + 8y)(5x - 8y)$ **35.** Prime
37. $16(m^2 + 4n^2)$ **39.** $(10x + 9y)(10x - 9y)$ **41.** $(10x - 9y)^2$
43. $4a(5v - 3w)(5v + 3w)$ **45.** $(x + 3)(x^2 - 3x + 9)$
47. $(m - 5)(m^2 + 5m + 25)$ **49.** $(4a - b)(16a^2 + 4ab + b^2)$
51. $(5x + 2y)(25x^2 - 10xy + 4y^2)$ **53.** $(x_1 + x_2)(x_1 - x_2)$
55. $\pi h(r_1 + r_2)(r_1 - r_2)$ **57.** $(x + 10)(x^2 - 10x + 100)$
59. $(x^2 - 3)(x^2 + 3)$ **61.** $(5y^2 + 1)(5y^2 - 1)$ **63.** $(3y - 1)(3y + 1) \cdot$
$(9y^2 + 1)$ **65.** $-3(a - b)(a + b)$ **67.** $(a + b)(x + 3)^2$
69. $(a - b)(a + b)(x - y)(x + y)$ **71.** $(z - 10)(z^2 + 10z + 100)$
73. $(7v + w)(49v^2 - 7vw + w^2)$ **75.** $7.46(x^2 - 12x + 36) =$
$7.46(x - 6)^2$ **77.** $3.64(v^2 - 25) = 3.64(v - 5)(v + 5)$

Using the Language and Symbolism of Mathematics   7.4
1. Grouping **2.** Greatest common factor **3. a.** D **b.** E **c.** A **d.** C **e.** B

Exercises 7.4
1. $x(x - y) + 5(x - y) = (x + 5)(x - y)$ **3.** $x^2 - (y^2 - 2y + 1) =$
$x^2 - (y - 1)^2 = [x + (y - 1)][x - (y - 1)] = (x + y - 1)(x - y + 1)$
5. $(c + d)(a + b)$ **7.** $(5c + 3)(a - 2b)$ **9.** $(b - d)(a + c)$
11. $(v - 7)(v - w)$ **13.** $(2a + 4b + 3)(2a - 4b + 3)$
15. $(n + 5)(3m - k)$ **17.** $(x - y)(x^2 + xy + y^2 - 1)$
19. $(a + 1)(a + b + 1)$ **21.** $(z^3 + w^2)(a + b)$
23. $(3b + a - 4)(3b - a - 4)$ **25.** $(y + 1)(ay + a - 1)$
27. $(8y + 3z)(8y - 3z)$ **29.** Prime **31.** $3(x - 1)(4x - 5)$
33. $(7a - 2)^2$ **35.** $(a - b)(x + y)$ **37.** Prime **39.** Prime
41. $(2x^5 + 3y^3)^2$ **43.** $12xy(x - y)(x + y)$ **45.** $(c + d)(x + y)$
47. $(x + 1)(ax + by)$ **49.** $(x^3 + 2y)^2$ **51.** $4b(x - 2)(x^2 + 2x + 4)$
53. $5(x^2 - 11)$ **55.** $(5x - y - 4z)(5x - y + 6z)$ **57.** $(7b + 9c)^2$
59. $-71a(x - 1)(x + 1)(x^2 + 1)$ **61.** $(2x^2 - y)(4x^4 + 2x^2y + y^2)$
63. $(x + y + 4z)(x + y - 4z)$ **65.** $2(x + y)(x - y + 1)$
67. $8a(x - 9y^2)(x + 9y^2)$ **69.** $-6ax(x - 2)(x + 2)$ **71.** $x(x + 2y)^2$
73. $(a - 2b)(x + 4)(x - 6)$ **75.** $(-3, 0), (2, 0), (4, 0); -3, 2, 4;$
$(x + 3)(x - 2)(x - 4)$ **77.** $\left(-\frac{3}{2}, 0\right), (0, 0), \left(\frac{3}{2}, 0\right); \ -\frac{3}{2}, 0, \frac{3}{2};$
$x(2x + 3)(2x - 3)$

Using the Language and Symbolism of Mathematics 7.5
1. Quadratic equation **2.** Quadratic; linear; constant **3.** $(x - 3)$;
$(x + 6)$ **4.** Least common denominator **5.** Two; double; multiplicity
6. Double **7.** $(c, 0)$ **8.** $(x - c)$ **9.** $=$ **10.** $>$ **11.** $<$

Exercises 7.5
1. a. D **b.** C **c.** A **d.** B **3. a.** $2x^2 - 7x + 3 = 0$; $a = 2, b = -7, c = 3$
b. $8x^2 - 3x = 0$; $a = 8, b = -3, c = 0$ **c.** $7x^2 + 5 = 0$; $a = 7, b = 0$,
$c = 5$ **d.** $2x^2 - x - 4 = 0$; $a = 2, b = -1, c = -4$ **5.** $8, -17$
7. $\frac{5}{2}, -\frac{1}{3}$ **9.** $1, -2, \frac{7}{2}$ **11.** $-11, 11$ **13.** $-2, -1$ **15.** $-3, 6$
17. $0, -\frac{1}{3}$ **19.** $-\frac{3}{2}, 5$ **21.** $-\frac{5}{2}, -\frac{2}{3}$ **23.** 5 **25.** $\frac{7}{3}$ **27.** $-\frac{3}{7}, \frac{1}{2}$
29. $-\frac{5}{3}, \frac{5}{3}$ **31.** $-5, 2$ **33.** $-3, 6$ **35.** $-2, 6$ **37.** $4, 7$ **39.** $-1, 4$
41. $-1, \frac{5}{2}$ **43.** $-\frac{1}{3}$ **45.** $0, -3, 8$ **47.** $0, -\frac{3}{2}, \frac{2}{3}$ **49.** $0, -\frac{3}{2}, \frac{1}{7}$
51. a. $x^2 - 9 = 0$ **b.** $7x^2 + 11x - 6 = 0$ **c.** $x^2 - 6x = 0$
d. $x^2 + 4x + 4 = 0$ **53. a.** $x^3 - 3x^2 + 2x = 0$ **b.** $x^3 - x = 0$
55. a. $\frac{3}{5}, 2$ **b.** $5m^2 - 13m + 6$ **57. a.** $3, -\frac{2}{5}$ **b.** $20x^2 - 52x - 24$
59. $(3x + 5)(4x - 3)$ **61. a.** $-3, 2$ **b.** $(-3, 2)$ **c.** $(-\infty, -3) \cup (2, \infty)$
63. a. $\frac{5}{2}$ **b.** No solution **c.** $\left(-\infty, \frac{5}{2}\right) \cup \left(\frac{5}{2}, \infty\right)$
65. $(5x - 1)(10x - 7)$ **67. a.** $-9, 7$ **b.** $(-9, 7)$ **c.** $(-\infty, -9) \cup (7, \infty)$
69. a. $1, 3, 5$ **b.** $(-\infty, 1) \cup (3, 5)$ **c.** $(1, 3) \cup (5, \infty)$ **71.** 3 cm × 11 cm
73. 8 m **75.** 15 cm **77.** $(-3, 0), (7, 0)$; $(x + 3)(x - 7)$; $-3, 7$; $-3, 7$
79. $(6, 0)$; $(x - 6)^2$; 6; 6 **81.** $2, -7$; $(-7, 2)$; $(-\infty, -7] \cup [2, \infty)$
83. $\frac{2}{3}, -7$; $(-\infty, -7) \cup \left(\frac{2}{3}, \infty\right)$; $\left[-7, \frac{2}{3}\right]$

Review Exercises for Chapter 7
1. a. $(x - 50)(x + 2)$ **b.** $(x - 20)(x + 5)$ **c.** $(x + 10)(x - 10)$
d. Prime **2. a.** $(x - 2)(x + 24)$ **b.** $(x - 3)(x + 16)$ **c.** $(x - 6)(x + 8)$
d. Prime **3. a.** $(2x + 1)(50x + 1)$ **b.** $(2x + 1)(2x + 25)$
c. $(x + 5y)(5x + 4y)$ **d.** $(5x + 2y)(2x + 5y)$ **4. a.** $(3x - 1)(16x - 1)$
b. $(x - 8)(3x - 2)$ **c.** $(x - 24y)(2x - y)$ **d.** $(4x - 3y)(4x - y)$
5. $4(x - 9)$ **6.** $2x(x - 5)$ **7.** $12ax(x - 2)$ **8.** $(x + 2)(x - 2)$
9. $(2x - 1)(2x + 1)$ **10.** $(x + 2)^2$ **11.** $7(x - 2)(x + 2)$
12. $(m + 8)(m - 8)$ **13.** $(x - 9)(x - 2)$ **14.** $(m + 3)(m + 7)$
15. $(a + b)(x + y)$ **16.** $(a - 3)(x + 2y)$ **17.** $(x - 7y)(x + 6y)$
18. $(v + w + 1)(v - w + 1)$ **19.** $(3y + 5)^2$ **20.** $(x - 4)(x + 4)$
21. $(2x - y)(4x^2 + 2xy + y^2)$ **22.** $2x(x - 2)(x^2 + 2x + 4)$
23. $11(x - 2y)(x + 2y)$ **24.** $11(x + y)(x - y + 3)$
25. $a(2x - 3y)(2x + 3y)$ **26.** $(5a + 3b)(2x - 3y)$
27. $2(x + 3y)(2x - 5y)$ **28.** $(2x - 5y)(3x + 4y)$
29. $xy(x - 4)(x - 1)$ **30.** $(x + 10y)(6x + y)$ **31.** $(2x + 3y)^2$
32. $(a + b)(x + 5)(x - 5)$ **33.** $(x + 5y)(a + b)^2$
34. $(2x - 3y)(a + b)(a - b)$ **35.** Prime **36.** $10(x + 3)(2x - 1)$
37. $(v - 1)^2(v^2 + v + 1)$ **38.** $(v - 1)(v + 1)(v^2 + 1)$ **39.** Prime
40. $(8x + 1)(8x - 1)$ **41.** $(4x + 1)(16x^2 - 4x + 1)$
42. $(4x - 1)(16x^2 + 4x + 1)$ **43.** $5, -1$ **44.** $\frac{3}{2}, -\frac{2}{3}$ **45.** $0, 7, \frac{2}{7}$
46. $7, -3$ **47.** $\frac{1}{5}, -\frac{3}{2}$ **48.** $3, -\frac{7}{6}$ **49.** $5, -3$ **50.** $3, -7$ **51.** $0, 6, -6$
52. $0, 5$ **53.** $-2, -4$ **54.** $\frac{5}{2}, 2$ **55.** $6, -2$ **56.** $-\frac{9}{2}$ **57.** $x^2 - 49 = 0$
58. $x^2 + 9x - 22 = 0$ **59.** $10x^2 - x - 3 = 0$ **60.** $x^2 - 8x = 0$
61. $x^2 + 4x + 4 = 0$ **62.** $16x^2 - 24x + 9 = 0$
63. $x^3 - 12x^2 + 35x = 0$ **64.** $x^3 - 7x^2 + 16x - 12 = 0$
65. a. $2\pi(r_1 - r_2)$ **b.** $\pi(L_1 - L_2)(L_1 + L_2)$

66. $\frac{4}{3}\pi(r_1 - r_2)(r_1^2 + r_1 r_2 + r_2^2)$ **67.** $-18, 12$ **68.** $-3.5, 2.5$
69. a. $-20, 25$ **b.** $(-20, 25)$ **c.** $(-\infty, -20) \cup (25, \infty)$
70. a. $-1, 2, 3$ **b.** $(-\infty, -1) \cup (2, 3)$ **c.** $(-1, 2) \cup (3, \infty)$
71. $(x - 7)(x + 11)$; $+7, -11$; $7, -11$; $(7, 0), (-11, 0)$
72. $(5x + 4)(7x - 3)$; $-\frac{4}{5}, \frac{3}{7}$; $-\frac{4}{5}, \frac{3}{7}$; $\left(-\frac{4}{5}, 0\right), \left(\frac{3}{7}, 0\right)$
73. $(x - 35)(x + 15)$; $35, -15$; $35, -15$; $(35, 0), (-15, 0)$
74. $(4x - 9)^2$; $\frac{9}{4}, \frac{9}{4}$; $\left(\frac{9}{4}, 0\right)$ **75.** $h = 4$ cm **76.** Either 10 cm or 20 cm

Mastery Test for Chapter 7
1. a. $(w - 9)(w + 5)$ **b.** $(w + 9)(w + 5)$ **c.** $(v - 6)(v - 4)$
d. $(v + 9)(v - 4)$ **2. a.** $(x - 4y)(x + 3y)$ **b.** $(x - 12y)(x - y)$
c. $(a + 6b)(a + 24b)$ **d.** $(a + 8b)(a - 6b)$ **3. a.** $(2x - 7)(3x + 2)$
b. $(2x - 1)(3x - 14)$ **c.** $(x + 14)(6x + 1)$ **d.** $(x + 2)(10x - 3)$
4. a. $(3x + 4y)(4x + 3y)$ **b.** $(3x - 4y)(4x + 3y)$
c. $(x - 12y)(12x - y)$ **d.** $(5x - y)(8x + 3y)$ **5. a.** $(x + 7y)^2$
b. $(x - 8y)^2$ **c.** $(3x + 10y)^2$ **d.** $(5x - 11y)^2$ **6. a.** $(x - 2y)(x + 2y)$
b. $(20a - b)(20a + b)$ **c.** $(4v - 7w)(4v + 7w)$ **d.** $(6x + 5y)(6x - 5y)$
7. a. $(4v - 1)(16v^2 + 4v + 1)$ **b.** $(v + 5)(v^2 - 5v + 25)$
c. $(2x + 5y)(4x^2 - 10xy + 25y^2)$ **d.** $(3a - 10b)(9a^2 + 30ab + 100b^2)$
8. a. $(a + b)(2x + 3)$ **b.** $(7a - 3b)(2x - 5y)$ **c.** $(a - 2b)(a + 2b + 1)$
d. $(x + 5y - 2)(x + 5y + 2)$ **9. a.** $5(x - 7)(x + 7)$
b. $19(x - 1)(x^2 + x + 1)$ **c.** $2a(x + 5)^2$ **d.** $-5a(x - 3y)(x - y)$

10. a. $-\frac{1}{2}; 3$ **b.** $11; -9$ **c.** $-3; 2; \frac{9}{4}$ **d.** $-\frac{1}{6}; -3$

11. a. $(x + 5)(x - 7)$; $-5, 7$; $-5, 7$ **b.** $(2x + 1)(8x - 3)$; $-\frac{1}{2}, \frac{3}{8}$;
$-\frac{1}{2}, \frac{3}{8}$ **c.** $(4x - 5)^2$; $\frac{5}{4}, \frac{5}{4}$ **d.** $x(x - 1)(x - 6)$; $0, 1, 6$; $0, 1, 6$

Chapter 8
Using the Language and Symbolism of Mathematics 8.1
1. Radicand; radical; index; order **2.** $\sqrt[5]{x}$ **3.** $x^{3/4}$ **4.** $x^{1/n}$; $\sqrt[n]{x}$
5. Even; negative

Exercises 8.1
1. a. $\sqrt[3]{w} = w^{1/3}$ **b.** $\sqrt[4]{x} = x^{1/4}$ **c.** $\sqrt[7]{v} = v^{1/7}$ **3. a.** $\sqrt{4} = 4^{1/2} = 2$
b. $\sqrt[3]{1000} = 1000^{1/3} = 10$ **c.** $\sqrt[5]{32} = 32^{1/5} = 2$ **5. a.** $16^{1/2} = 4$
b. $16^{1/4} = 2$ **c.** $0^{1/5} = 0$ **7. a.** $(-8)^{1/3} = -2$ **b.** $\left(\frac{-1}{32}\right)^{1/5} = \frac{-1}{2}$
c. $0.01^{1/2} = 0.1$ **9. a.** $\sqrt{36} = 6$ **b.** $\frac{1}{\sqrt{36}} = \frac{1}{6}$ **c.** $-\sqrt{36} = -6$
11. a. $\sqrt[3]{27} = 3$ **b.** $\sqrt[3]{-27} = -3$ **c.** $\frac{1}{\sqrt[3]{27}} = \frac{1}{3}$ **13. a.** $\sqrt{0.09} = 0.3$
b. $-\sqrt{0.09} = -0.3$ **c.** $\frac{1}{\sqrt{400}} = \frac{1}{20}$ **15. a.** $\sqrt[3]{0.008} = 0.2$
b. $\sqrt[3]{-0.008} = -0.2$ **c.** $\frac{1}{\sqrt[3]{64}} = \frac{1}{4}$ **17. a.** $\sqrt[4]{16} = 2$ **b.** $\frac{1}{\sqrt[4]{16}} = \frac{1}{2}$
c. $-\sqrt[4]{16} = -4$ **19 a.** $\sqrt[3]{\left(\frac{8}{125}\right)^2} = \frac{4}{25}$ **b.** $\left(\sqrt[3]{\frac{9}{4}}\right)^3 = \frac{27}{8}$
c. $\left(\sqrt[3]{\frac{125}{8}}\right)^2 = \frac{25}{4}$ **21. a.** $\sqrt{25} + \sqrt{144} = 17$ **b.** $\sqrt{25 + 144} = 13$
c. $\sqrt{25 - 16} = 3$ **23. a.** $\sqrt[3]{8} + \sqrt[3]{1} = 3$ **b.** $\sqrt[3]{26 + 1} = 3$
c. $\sqrt[3]{1001 - 1} = 10$ **25. a.** $\sqrt{0.000001} = 0.001$
b. $\sqrt[3]{0.000001} = 0.01$ **c.** $\sqrt[6]{0.000001} = 0.1$ **27. a.** 5.916 **b.** 3.271
c. 1.809 **29. a.** -2.759 **b.** -1.545 **c.** -4.583 **31.** b **33.** 25 **35.** 4
37. 11 **39.** 9 **41.** $x^{5/6}$ **43.** $x^{1/6}$ **45.** $z^{3/14}$ **47.** w **49.** v^2w^3 **51.** $\frac{64}{v^{3/5}}$

53. $\dfrac{27}{8n}$ **55.** $\dfrac{9x}{5y}$ **57.** $x-1$ **59.** $2y-3$ **61.** $6w^2-15w-27$
63. $a-9$ **65.** $b^{6/5}-c^{10/3}$ **67.** $b^{6/5}-2b^{3/5}c^{5/3}+c^{10/3}$
69. $\dfrac{1}{x}+2+x$ **71.** $y+8$ **73.** 32 **75. a.** A **b.** B

Using the Language and Symbolism of Mathematics 8.2

1. Radicands; indices **2.** $\sqrt[n]{x}\,\sqrt[n]{y}$ **3.** $\dfrac{\sqrt[n]{x}}{\sqrt[n]{y}}$ **4.** $|x|$ **5.** x

Exercises 8.2

1. d **3.** 12 **5.** −6 **7.** $30\sqrt{2}$ **9.** $-4\sqrt{7}$ **11.** $-5\sqrt[3]{6}$ **13.** $10\sqrt[4]{13}$
15. $6\sqrt{5}-4\sqrt{7}$ **17.** $-3\sqrt{7x}$ **19.** 0 **21.** $5\sqrt{3}$ **23.** $3\sqrt{7}$ **25.** $2\sqrt[3]{3}$
27. $2\sqrt[3]{3}$ **29.** $5\sqrt{7}$ **31.** $\sqrt{3}$ **33.** $-13\sqrt{2v}$ **35.** $-2\sqrt{7w}$
37. $-3\sqrt[3]{3}$ **39.** $-74t\sqrt[3]{3}$ **41.** $-2\sqrt[3]{5z^2}$ **43.** $\dfrac{5\sqrt{5}}{6}$ **45.** $\dfrac{9\sqrt{11x}}{35}$
47. $7.9\sqrt{2}+0.8\sqrt{3}$ **49.** 0 **51. a.** 5x **b.** $5|x|$ **53. a.** 2x **b.** 2x
55. a. $1000x^3$ **b.** $1000|x|^3$ **57. a.** x^6 **b.** x^6 **59.** D **61.** A
63. a. Integer estimate: 4 **b.** Inequality: $4>\sqrt[3]{63}$
c. Approximation: 3.979 **65. a.** Integer estimate: 2 **b.** Inequality:
$2<\dfrac{1+\sqrt{9.05}}{2}$ **c.** Approximation: 2.004 **67. a.** Integer estimate: −5
b. Inequality: $-5<\dfrac{-4-\sqrt{35.97}}{2}$ **c.** Approximation: −4.999
69. c **71.** 46.3 cm

Using the Language and Symbolism of Mathematics 8.3

1. a. $\sqrt[n]{xy}$ **b.** $\sqrt[n]{\dfrac{x}{y}}$ **2.** Same order **3.** Even **4.** Conjugates
5. Rationalizing

Exercises 8.3

1. $2\sqrt{3}$ **3.** $10-\sqrt{2}$ **5.** $8\sqrt{15}$ **7.** $84\sqrt{35}$ **9.** $\sqrt{6}+\sqrt{15}$
11. $30\sqrt{3}-105\sqrt{7}$ **13.** 240 **15.** 4w **17.** $24z\sqrt{2}$ **19.** $6x-15\sqrt{x}$
21. $3x+3\sqrt{3xy}-4y$ **23.** $14-4\sqrt{6}$ **25.** $a+10\sqrt{3ab}+75b$
27. $v-11$ **29.** 7 **31.** $-3v$ **33.** $-4-\sqrt[3]{20}$ **35.** $y\sqrt[3]{x^2y}$
37. $12\sqrt{7}-8\sqrt{5}-12$ **39.** $10\sqrt{3}+7\sqrt{2}-6$ **41.** 3 **43.** 14
45. $x-3y$ **47.** $-v+2$ **49.** $\dfrac{\sqrt{3}}{2}$ **51.** $\dfrac{\sqrt{6}}{3}$ **53.** $\dfrac{\sqrt{10}}{4}$ **55.** $\sqrt{5}$
57. $3\sqrt{6}$ **59.** $\dfrac{13\sqrt{10}}{5}$ **61.** $\dfrac{5\sqrt{3x}}{x}$ **63.** $\dfrac{\sqrt{7}-1}{2}$ **65.** $15+3\sqrt{13}$
67. $-3\sqrt{7}-3\sqrt{2}$ **69.** $\dfrac{a+\sqrt{ab}}{a-b}$ **71.** $4\sqrt[3]{9}$ **73.** $\dfrac{\sqrt[3]{6}}{2}$ **75.** $\dfrac{\sqrt[3]{3vw}}{3w}$
77. $\dfrac{5(\sqrt{3x}-\sqrt{2y})}{3x-2y}$ **79.** Solution **81. a.** 10 **b.** $x+y$
83. Integer estimate: 20; inequality: $20<\dfrac{100}{\sqrt{24}}$; approximation:
20.412 **85.** Integer estimate: 10; inequality: $10>\dfrac{50}{\sqrt{37}-\sqrt{0.98}}$;
approximation: 9.818

Using the Language and Symbolism of Mathematics 8.4

1. $\sqrt{-1}$ **2.** $a; bi$ **3.** Sometimes **4.** Always **5.** $a+bi$ **6.** Four
7. Conjugate

Exercises 8.4

1. a. −6 **b.** 6i **c.** −6i **d.** 6 **3. a.** 4 + 3i **b.** 3 + 4i **c.** −3 − 4i
d. −4 + 3i **5. a.** 5i **b.** −5i **c.** −6 **d.** 6 **7. a.** −10 **b.** 10 **c.** −10i
d. 10 **9. a.** $\dfrac{5}{3}i$ **b.** $\dfrac{5}{3}i$ **c.** $-\dfrac{5}{3}i$ **d.** $\dfrac{5}{3}$ **11. a.** $a=18, b=-5$
b. $a=-6, b=-2$ **c.** $a=4, b=-11$ **d.** $a=24, b=0$
13. $9-i$ **15.** $3+i$ **17.** $60+24i$ **19.** 7 **21.** −6 **23.** $10+6i$

25. $8-3i$ **27.** 53 **29.** $34+27i$ **31.** $24+10i$ **33.** $7-24i$
35. $56+33i$ **37.** $-3+i\sqrt{6}$ **39.** $13i\sqrt{3}$ **41.** $-1-2i$ **43.** 29
45. 169 **47.** $2-2i$ **49.** $\dfrac{15}{17}-\dfrac{8}{17}i$ **51.** $7+6i$ **53.** i **55.** −1
57. $-1-2i, -1+2i$ **59.** $-30-29i$ **61.** $-11+2i$ **63.** 1
65. $4+10i$ **67.** $i^{301}=i$ **69.** $i^{440}=1$ **71.** Solution **73.** Solution
75. a. $3+i$ **b.** $-3+i$ **c.** $\dfrac{3}{10}+\dfrac{1}{10}i$ **77. a.** The number is real
b. 2i and −2i **c.** 3i and 5i **d.** 0

Using the Language and Symbolism of Mathematics 8.5

1. Plus or minus **2.** Extraction of roots **3.** x-intercept; zero
4. Completing the square **5.** Quadratic formula **6.** Discriminant
7. Two; two **8.** Two; tangent **9.** No; no

Exercises 8.5

1. B **3.** A **5.** 9, −9 **7.** $3\sqrt{2}, -3\sqrt{2}$ **9.** 3, −1 **11.** 4i, −4i
13. $-\dfrac{1}{2}+\dfrac{3}{2}i, -\dfrac{1}{2}-\dfrac{3}{2}i$ **15.** $-\dfrac{1}{4}, \dfrac{1}{2}$ **17.** $2\sqrt{2}-2\sqrt{2}$ **19.** $0, \dfrac{6}{5}$
21. $\dfrac{3}{2}\pm\dfrac{1}{2}i$ **23.** $3\pm 2i$ **25.** $-\dfrac{2}{3}$ **27.** $\dfrac{1-\sqrt{6}}{5}, \dfrac{1+\sqrt{6}}{5}$
29. a. −1, 5 **b.** (−1, 5) **c.** $(-\infty, -1)\cup(5, \infty)$
31. a. $1-\sqrt{7}, 1+\sqrt{7}$ **b.** $(-\infty, 1-\sqrt{7})\cup(1+\sqrt{7}, \infty)$
c. $(1-\sqrt{7}, 1+\sqrt{7})$ **33. a.** No real solutions **b.** $(-\infty, \infty)$
35. a. $-5, \dfrac{1}{2}$ **b.** $\left(-5, \dfrac{1}{2}\right)$ **c.** $\left(-\infty, -5\right)\cup\left(\dfrac{1}{2}, \infty\right)$
37. a. $-\sqrt{7}, \sqrt{7}$ **b.** $(-\sqrt{7}, \sqrt{7})$ **c.** $(-\infty, -\sqrt{7})\cup(\sqrt{7}, \infty)$
39. a. $i\sqrt{7}, -i\sqrt{7}$ **b.** No real solutions **c.** $(-\infty, \infty)$
41. $(x+5)^2=3$ **43.** $(x-7)^2=1$ **45.** $(x-9)^2=90$ **47. a.** 4 **b.** 1
c. 1 **d.** 1 **e.** $-1\pm\sqrt{5}$ **49.** 0, 4 **51.** 1, −5 **53.** $-1-i, -1+i$
55. $-\dfrac{1}{4}, \dfrac{7}{4}$ **57.** C **59.** A **61.** 24; the equation has two distinct real
solutions **63.** −11, 9; (−11, 0), (9, 0); $(x+11)(x-9)$; −11, 9
65. $-\sqrt{13}, \sqrt{13}$; $(-\sqrt{13}, 0), (\sqrt{13}, 0)$; $(x-\sqrt{13})(x+\sqrt{13})$;
$-\sqrt{13}, \sqrt{13}$ **67.** $2x^2-9x-5=0$ **69.** $x^2-7=0$ **71.** $x^2+16=0$
73. $x^2-4x+1=0$ **75.** Two possible solutions: $-3+\sqrt{17}$ and
$3+\sqrt{17}$ or $-3-\sqrt{17}$ and $3-\sqrt{17}$ **77.** $(-2+2\sqrt{3})$ cm by
$(2+2\sqrt{3})$ cm **79.** 0, 1, 4

Using the Language and Symbolism of Mathematics 8.6

1. Radicand **2.** $x^n=y^n$ **3.** Extraneous **4.** Legs, hypotenuse
5. $a^2+b^2=c^2$ **6.** $\sqrt{(x_2-x_1)^2+(y_2-y_1)^2}$

Exercises 8.6

1. 5 **3.** −1 **5.** 10 **7.** 6 **9.** 13 **11.** 393 **13.** No solution **15.** $-\dfrac{9}{2}$
17. 3 **19.** −4, 8 **21.** 4 **23.** 2 **25.** $\dfrac{1}{3}$ **27.** 4 **29.** 0 **31.** 13 cm
33. 9 cm **35.** $\sqrt{113}$ cm **37.** 5 **39.** 17 **41.** $\sqrt{2}$ **43.** 3
45. $7\sqrt{2}+\sqrt{58}$; the triangle is a right triangle
47. $2\sqrt{10}+5\sqrt{2}+\sqrt{106}$; the triangle is not a right triangle
49. 42.5 m **51.** (5, −4) and (5, 8) **53.** 4 **55.** 9 **57.** $1+\sqrt{7}$
59. $\dfrac{-1+\sqrt{5}}{2}$ **61.** 0, 2 **63.** No solution **65.** $450 **67.** 64
69. $S(0)=0$; the strength of a beam with no volume is 0 newtons
71. $V=1$; a square box beam with a strength of 750 newtons has a
volume of 1 cm³ **73. a.** $f(x)=\sqrt{289-x^2}$ **b.** 15 ft
75. a. $d=\sqrt{x^2+900}$ **b.** 34 ft **c.** 18.0 ft **77.** $x=3$ mi

Using the Language and Symbolism of Mathematics 8.7

1. Find **2.** Variable **3.** Algebraic **4.** Solve **5.** Reasonable **6.** Rate;
time

Exercises 8.7

1. -13 and -12 or 12 and 13 **3.** -18 and -16 or 16 and 18
5. -8 and -7 or 7 and 8 **7.** $-\dfrac{25}{3}$ and -21 or 7 and 25 **9.** $\dfrac{7}{3}$ cm by
9 cm **11.** 14 m **13.** 7.2 cm **15.** Between approximately 0.6 sec and
5.4 sec **17.** Make more than 10, but less than 60 windmills **19.** 5 cm
21. 10.96 in; engineers use expansion joints **23.** Approximately
10.95 in **25.** 11.7 in and 10.2 in; the monitor that measures 15 in by
11.7 in offers more viewing area **27.** 7 mi **29.** 90 mi/h and 120 mi/h
31. 28 ft **33.** 44 m **35.** 37.3 ft **37.** 9.28% **39.** 180 mi **41.** 18 cm

Review Exercises for Chapter 8

1. a. $x^{1/2}$ **b.** $x^{2/3}$ **c.** $x^{-1/3}$ **2. a.** $\sqrt[4]{x}$ **b.** $\dfrac{1}{\sqrt[4]{x}}$ **c.** $\sqrt[4]{x^3}$

3. a. $2\sqrt{w} = 2w^{1/2}$ **b.** $\sqrt{2w} = (2w)^{1/2}$ **c.** $\sqrt[3]{x+4} = (x+4)^{1/3}$

4. a. 7 **b.** -7 **c.** $\dfrac{1}{7}$ **5. a.** $\dfrac{9}{25}$ **b.** $\dfrac{9}{25}$ **c.** $\dfrac{25}{9}$ **6. a.** 1000 **b.** 100

c. 10 **7. a.** 7 **b.** 1 **c.** $\dfrac{24}{25}$ **8. a.** -2 **b.** -1 **c.** 0 **9. a.** 7 **b.** -7

c. 5 **10. a.** 8.37 **b.** 4.12 **c.** -2.34 **11. a.** 16 **b.** 4 **c.** 2 **12. a.** $\dfrac{4y^6}{x^4}$

b. $4x^{2/3}y^{3/5}$ **c.** $x^2y^2z^2$ **13. a.** $6x - 10$ **b.** $9x - 25$ **c.** $x + y$
14. a. $20\sqrt{3}$ **b.** $4\sqrt{3}$ **c.** $4\sqrt{2} - 3\sqrt{3}$ **15. a.** $4\sqrt[3]{5}$ **b.** 3 **c.** $12\sqrt[3]{25}$
16. a. $2\sqrt{5}$ **b.** $2\sqrt[3]{3}$ **c.** $4\sqrt{2}$ **17. a.** $14\sqrt{3x}$ **b.** $4\sqrt{2v}$ **c.** $-\sqrt[3]{v}$
18. a. 5 **b.** $\dfrac{3x}{y^2}$ **c.** $\dfrac{\sqrt[3]{25}}{3}$ **19. a.** $-29\sqrt{2}$ **b.** -369 **c.** -67

20. a. $90\sqrt{7}$ **b.** $\dfrac{\sqrt{7}}{5}$ **c.** $\dfrac{5\sqrt{7}}{7}$ **21. a.** $5 - 2\sqrt{6}$ **b.** $2\sqrt{6} - 5$

c. $3\sqrt{7} + 3\sqrt{3}$ **22. a.** 180 **b.** 3 **c.** $\sqrt{6}$ **23. a.** $10i$ **b.** $4i$ **c.** $8 - 8i$
24. a. $8 - 8i$ **b.** $2 - 4i$ **c.** $-2 - 36i$ **25. a.** $12 + 10i$ **b.** $51 - 27i$
c. $40 - 42i$ **26. a.** 29 **b.** i **c.** $-1 - i$ **27. a.** i **b.** $4 - 10i$ **c.** $2 - 3i$

28. a. i **b.** i **c.** 1 **29. a.** $-3, 7$ **b.** $-\dfrac{5}{2}, \dfrac{4}{3}$ **c.** 19 **30. a.** $-10, 10$

b. $-\sqrt{7}, \sqrt{7}$ **c.** $-4i, 4i$ **31.** 4, 5 **32.** $2 - \sqrt{2}, 2 + \sqrt{2}$

33. $-\dfrac{2}{5}, \dfrac{5}{2}$ **34.** $\dfrac{3}{2} + \dfrac{3}{2}i, \dfrac{3}{2} - \dfrac{3}{2}i$ **35.** $-10, 15$ **36.** $-0.3, 0.7$

37. $-11, -7$ **38.** $x = -0.1, x = 0.2$ **39.** 0; one double real solution
40. -12; the solutions are complex conjugates **41.** 19; two distinct
real solutions **42. a.** $-5, 2$ **b.** $(-\infty, -5) \cup (2, \infty)$ **c.** $(-5, 2)$
43. a. No solution **b.** $(-\infty, \infty)$
44. $(-\infty, -1 - \sqrt{5}) \cup (-1 + \sqrt{5}, \infty)$ **45.** $(-\infty, \infty)$ **46.** 21 **47.** 4
48. 5 **49.** -8 **50.** 13 **51.** $4\sqrt{5} + 2\sqrt{10}$; a right triangle **52.** 8 cm,
15 cm, 17 cm **53.** 19.09% **54.** A profit is generated if more than 4
units are sold **55.** Between 1 and 4 s **56.** (8, 7), (8, 1) **57.** 9.8 in;
engineers could use expansion joints to avoid this problem
58. $x^2 - 8x - 9 = 0$ **59.** $15x^2 + x - 6 = 0$ **60.** $x^2 - 3 = 0$
61. $x^2 - 2x + 26 = 0$ **62.** $49x^2 - 70x + 25 = 0$

Mastery Test for Chapter 8

1. a. 9 **b.** -10 **c.** 27 **d.** $x^{1/7}$ **2. a.** 13 **b.** -0.2 **c.** 17 **d.** 13
3. a. $5\sqrt{7}$ **b.** $3\sqrt{2} + 4\sqrt{5}$ **c.** $-\sqrt[3]{7}$ **d.** $8\sqrt{2x}$ **4. a.** $2\sqrt{10}$ **b.** $2\sqrt[3]{5}$
c. $-9\sqrt{7}$ **d.** $7\sqrt[3]{2}$ **5. a.** 30 **b.** $30 - 20\sqrt{2}$ **c.** $-4x^2$ **d.** $12x - 5$
6. a. 3 **b.** $3\sqrt{6}$ **c.** $8(\sqrt{7} + \sqrt{2})$ **d.** $-8 - 4\sqrt{7}$ **7. a.** $9i$ **b.** $3i$
c. $-5 + 4i$ **d.** $-1 - i$ **8. a.** $-1 + 2i$ **b.** $-12 - 26i$ **c.** $\dfrac{7}{5} + \dfrac{3}{10}i$

d. $8 - 6i$ **9. a.** $\dfrac{5}{2}, \dfrac{2}{3}$ **b.** $-1 + \sqrt{5}; -1 - \sqrt{5}$ **c.** $\dfrac{7}{2}$

d. $\dfrac{1}{3} + \dfrac{\sqrt{2}}{3}i; \dfrac{1}{3} - \dfrac{\sqrt{2}}{3}i$ **10. a.** 5; two distinct real solutions **b.** 169;
two distinct real solutions **c.** 0; one double real solution **d.** -8; two
complex conjugate solutions **11. a.** $(-\infty, -2) \cup (5, \infty)$ **b.** $(-2, 5)$

c. $\left(-\infty, \dfrac{-5 - \sqrt{13}}{6}\right] \cup \left[\dfrac{-5 + \sqrt{13}}{6}, \infty\right)$

d. $\left[\dfrac{-5 - \sqrt{13}}{6}, \dfrac{-5 + \sqrt{13}}{6}\right]$ **12. a.** 124 **b.** -5 **c.** No solution

d. 8 **13. a.** 9 cm **b.** $\sqrt{61}$ cm ≈ 7.8 cm **c.** 78 ft **d.** $\sqrt{41}$ ft ≈ 6.4 ft
14. a. 8 **b.** 12 **c.** 10 **d.** $\sqrt{29}$ **15. a.** 25 ft **b.** 7 cm and 12 cm **c.** A
profit is generated if between 5 and 185 units are sold **d.** The
approximate ground speed of the southbound airplane is 156 mi/h and
of the eastbound plane is 256 mi/h.

Chapter 9

Using the Language and Symbolism of Mathematics   9.1

1. Quotient **2.** Undefined **3.** Denominator; zero **4.** Lowest terms

5. $\dfrac{A}{B}$ **6.** -1

Exercises 9.1

1. a. -1 **b.** 3 **c.** Undefined **3. a.** 4 **b.** $(-\infty, 4) \cup (4, \infty)$ **5.** C **7.** D

9. D **11.** C **13. a.** $\dfrac{1}{2}$ **b.** $-9, 9$ **15.** $\left(-\infty, \dfrac{3}{2}\right) \cup \left(\dfrac{3}{2}, \infty\right)$

17. $(-\infty, -3) \cup (-3, 2) \cup (2, \infty)$ **19.** $(-\infty, 1) \cup (1, \infty)$

21. $(-\infty, -2) \cup (-2, 3) \cup (3, \infty)$ **23.** $\dfrac{2b^2}{3a}$ **25.** $6x - 9$ **27.** $\dfrac{1}{2x - 3}$

29. $-\dfrac{a}{b}$ **31.** -1 **33.** $-\dfrac{a}{b}$ **35.** $\dfrac{x - 2y}{x + 2y}$ **37.** $\dfrac{x - y}{3}$ **39.** $\dfrac{-5x - 2}{7}$

41. $\dfrac{x}{x - y}$ **43.** $\dfrac{2x + 3y}{7}$ **45.** $\dfrac{2x + 3}{5y + 6}$ **47.** -1 **49.** $\dfrac{2a + b}{a + b}$

51. $\dfrac{x + y}{v + w}$ **53.** $\dfrac{x - 5}{3x - 1}$ **55.** $\dfrac{a + b}{a - b}$ **57.** $\dfrac{b - a - 1}{5a}$

59. $\dfrac{x^2 + xy + y^2}{y - x}$ **61.** $\dfrac{3(x + y)}{4(x - y)}$ **63.** $\dfrac{y - 2x}{y}$ **65.** $\dfrac{a + 1}{9}$ **67.** 14

69. $14x$ **71.** $16y - 14x$ **73.** $10a - 10b$ **75.** $5x - 5$ **77.** $x^2 + 4xy + 3y^2$
79. a. $T(6) = 3$; when paddling at 6 mi/h, it takes 3 h to go 12 mi
upstream **b.** $T(8) = 2$; when paddling at 8 mi/h, it takes 2 h to go 12
mi upstream **c.** $T(2)$ is undefined; the camper would never make it
upstream when paddling at 2 mi/h since the rate of the current is 2 mi/h

81. a. $A(t) = \dfrac{2t}{5}$ **b.** \$3600

Using the Language and Symbolism of Mathematics   9.2

1. $\dfrac{AC}{BD}$ **2. a.** Fraction **b.** Nonzero **3.** $\dfrac{AD}{BC}$ **4.** Reciprocal

Exercises 9.2

1. $\dfrac{11}{9}$ **3.** $\dfrac{11}{21}$ **5.** $\dfrac{1}{3m}$ **7.** $\dfrac{9y^2}{7x^6}$ **9.** $-\dfrac{1}{2}$ **11.** $\dfrac{(x - 1)(x + 3)}{x(x - 3)(x + 1)}$

13. $\dfrac{x - 3}{2(x + 3)}$ **15.** $\dfrac{7(x - 3)}{3(x - y)}$ **17.** $\dfrac{3(x + 2)}{x - 1}$ **19.** $\dfrac{a - 7}{y - c}$ **21.** $\dfrac{-2x + 2}{3}$

23. $\dfrac{6}{5(x + y)^2}$ **25.** $\dfrac{4x}{3}$ **27.** $(x + 5)(x - 5)$ **29.** $\dfrac{-1}{7x(3x - 5)}$

31. 2 **33.** $\dfrac{11x}{9a}$ **35.** $\dfrac{7xy}{15}$ **37.** $\dfrac{5}{x - 3}$ **39.** $\dfrac{-2}{ab^2}$ **41.** $\dfrac{3}{11}$ **43.** $\dfrac{7y^2}{2a}$

45. $\dfrac{3xy}{4(x + y)}$ **47.** 4; 3.99 **49.** 3; 2.99 **51.** D **53.** B **55.** 5

57. xy^2 **59.** $x - 8$ **61.** $\dfrac{x - 3}{x - 3}$ **63.** $\dfrac{x - 2}{x - 1}$

Using the Language and Symbolism of Mathematics   9.3

1. $\dfrac{A + B}{C}; \dfrac{A - B}{C}$ **2.** Add **3.** Subtract **4. a.** Denominator;
exponential **b.** Highest **c.** Multiply **5. a.** LCD **b.** Denominator
c. Numerators **d.** Lowest terms

Exercises 9.3

1. $\dfrac{2b + 3}{b^2}$ **3.** -1 **5.** $\dfrac{4}{s - 3}$ **7.** $\dfrac{1}{x + y}$

9. $\frac{1}{15xy^2}\left(\frac{9x}{9x}\right) + \frac{2}{27x^2y}\left(\frac{5y}{5y}\right) = \frac{9x}{135x^2y^2} + \frac{10y}{135x^2y^2} = \frac{9x+10y}{135x^2y^2}$

11. $\frac{a+b}{21(3a-b)}\left(\frac{2}{2}\right) - \frac{a-b}{14(3a-b)}\left(\frac{3}{3}\right) =$

$\frac{2a+2b}{42(3a-b)} - \frac{3a-3b}{42(3a-b)} = \frac{-a+5b}{42(3a-b)}$

13. $\frac{2}{(x-3)(x-15)}\left(\frac{x+3}{x+3}\right) +$

$\frac{1}{(x-3)(x+3)}\left(\frac{x-15}{x-15}\right) = \frac{2x+6}{(x-3)(x-15)(x+3)} +$

$\frac{x-15}{(x-3)(x+3)(x-15)} = \frac{3x-9}{(x-3)(x-15)(x+3)} =$

$\frac{3}{(x-15)(x+3)}$ 15. $\frac{-13}{18w}$ 17. $\frac{5}{14}$ 19. $\frac{b^2+4b+16}{b(b+4)}$

21. $\frac{5x-1}{x}$ 23. $\frac{x^2-2x+3}{x^3}$ 25. $\frac{x+13}{(x-2)(x+3)}$

27. $\frac{2x}{(x-1)(x+1)}$ 29. $\frac{8x+22}{(x-4)(x+2)}$ 31. $\frac{2x^2+2x+5}{(x+2)(x-1)}$

33. $\frac{1}{77}$ 35. $\frac{5m+9}{(m+1)(m-2)(m+3)}$ 37. $\frac{4m-13}{(m-3)(m-4)}$

39. $\frac{2}{a-4b}$ 41. $\frac{3x+2}{2(x-1)}$ cm² 43. $\frac{16t+40}{t(t+5)}$ 45. a. $\frac{255r+900}{r(r+10)}$

b. 5 h 47. A 49. D 51. $\frac{-x^2+6x-3}{x^2-3x+4}$ 53. $\frac{-3}{(x-1)(x+1)}$

55. $\frac{2}{x+5}$ 57. $\frac{10}{s-2t}$ 59. $\frac{1}{z+3}$ 61. $\frac{1}{v-w}$ 63. $\frac{3w-4}{(w-1)(w-2)}$

65. $\frac{1}{w-3}$ 67. $\frac{-2m}{m-1}$ 69. $\frac{2x+5}{(x+5)(2x-5)}$

Using the Language and Symbolism of Mathematics   9.4
1. a. Grouping symbols b. Exponentiations c. Multiplications; divisions d. Additions; subtractions 2. Numerator; denominator
3. Reciprocal 4. Addition; subtraction

Exercises 9.4

1. a. x b. $\frac{x+10}{3}$ 3. a. $\frac{5y}{6}$ b. $\frac{3y-12}{2}$ 5. a. $\frac{v^2-6}{v(v-6)}$

b. $\frac{2(v^2-v+18)}{v(v-6)^2}$ 7. a. $\frac{16}{15}$ b. $\frac{10}{3}$ 9. $\frac{3}{2}$ 11. -2 13. -2

15. v^2-1 17. $x-3$ 19. $-\frac{4}{x^2}$ 21. $\frac{12}{5x}$ 23. $\frac{9y}{4}$ 25. $\frac{2}{xy(w+z)}$

27. $\frac{x(x-7)}{10(x+2)}$ 29. $\frac{x}{2x+1}$ 31. $\frac{v^2w^2}{v+w}$ 33. $\frac{x}{6(x+1)}$ 35. $\frac{3x^2}{x+5}$

37. $\frac{-4aw}{w^2+a^2}$ 39. $\frac{v+3}{v+6}$ 41. $\frac{3}{2} = 1.5; 1.494$ 43. $\frac{4}{5} = 0.8; 0.820$

45. $\frac{5a-3b}{4ab}$ 47. $\frac{y^2}{x^2+y^2}$ 49. $\frac{w+v}{w-v}$ 51. $-mn(m^2+mn+n^2)$

53. $\frac{a+b}{ab}$ 55. xy 57. $\frac{a+b}{a^2b^2}$ 59. $\frac{x^2-x+1}{x(x+2)}$ 61. B 63. D

65. $x(x+5)$ cm² 67. $A(t) = \frac{1680}{t}$ 69. a. $f = \frac{d_o d_i}{d_o+d_i}$ b. $f \approx 0.488$ ft

Using the Language and Symbolism of Mathematics   9.5
1. Same 2. Zero 3. a. LCD b. Solve c. Extraneous 4. Zero

Exercises 9.5

1. a. 2 and 3 b. $-\frac{3}{2}$ and $\frac{2}{3}$ c. $-3, -\frac{1}{3}$, and $\frac{1}{2}$ 3. 3 5. 0.5 7. 2

9. No solution 11. -5 13. -16 15. 1 17. No solution 19. 5

21. 3, 23 23. $\frac{2}{3}$, 2 25. 0, $\frac{2}{3}$ 27. $\frac{13}{3}$ 29. 2 31. $-2, 3$ 33. $-\frac{2}{3}$

35. $\frac{11}{4}$ 37. $\frac{2}{3}$ 39. $-2, 0, 1$ 41. a. $\frac{-2(p-2)}{(p-1)(p+1)}$ b. 2

43. a. $\frac{-(x+2)(x-5)}{(x+1)(x-2)}$ b. $-2, 5$ 45. a. 0 b. All real numbers

except $-\frac{2}{3}$ and $\frac{1}{2}$ 47. $r_1 = 6$ ohms; $r_2 = 30$ ohms 49. $b = \frac{ad}{c}$

51. $b = \frac{ad+a+c}{c}$ 53. $d = \frac{k}{I}$ 55. $R = \frac{r_1 r_2}{r_1+r_2}$ 57. $B = \frac{2A-bh}{h}$

59. $z = \frac{xy}{x-y}$ 61. a. All real numbers except -3 and $\frac{5}{2}$

b. No solution 63. $-3, 0, 2$ 65. 2 67. 2

Using the Language and Symbolism of Mathematics   9.6

1. kb 2. $\frac{k}{b}$ 3. Increase 4. Decrease 5. Decrease 6. Increase

7. Rate; time 8. Principal; rate; time 9. Rate; time

Exercises 9.6

1. a. $m = kn$ b. $m = \frac{k}{p}$ c. $m = \frac{kn}{p}$ 3. a. $v = k\sqrt{w}$ b. $v = \frac{k}{x^2}$

c. $v = \frac{k\sqrt{w}}{x^2}$ 5. $W = kv^3$ 7. $I = kV$ 9. $w = \frac{k}{d^2}$ 11. a. 30

b. 19.2 13. 10 15. D 17. A 19. C 21. E 23. B 25. 2 ohms 27. 3

29. 5, 6 31. 7, 9 33. 6 35. $\frac{2}{3}$ or $\frac{3}{2}$ 37. 5 39. 12 m, 4 m

41. 60 ohms and 120 ohms 43. 54° and 36° 45. $8400 47. 20 vehicles 49. 9 km/h 51. 480 mi/h and 520 mi/h 53. 6 h 55. 12 h
57. 75 h 59. 3% and 5% 61. 8.5% and 7%

Review Exercises for Chapter 9

1. 9 2. Undefined 3. $-\frac{1}{15}$ 4. $\frac{1}{9}$ 5. $(-\infty, -2) \cup (-2, 3) \cup (3, \infty)$

6. $(-\infty, -5) \cup (-5, 3) \cup (3, \infty)$ 7. $(-\infty, -5) \cup \left(-5, \frac{1}{3}\right) \cup \left(\frac{1}{3}, \infty\right)$

8. $(-\infty, \infty)$ 9. $\frac{1}{2}$ 10. $-6, 6$ 11. $-\frac{1}{4}, \frac{7}{5}$ 12. $\frac{3x}{y^2}$ 13. $-\frac{3}{2}$

14. $\frac{3(x-1)}{5}$ 15. $\frac{x-5}{2x-1}$ 16. $\frac{c}{a+b}$ 17. $-\frac{1}{2m}$ 18. $\frac{5x+2y}{3x-y}$

19. $\frac{3x-4y}{4x-3y}$ 20. $\frac{x+2y}{x^2+2xy+4y^2}$ 21. $(x-3)(2x+1)$

22. $(y-3)(2y+1)$ 23. D 24. C 25. A 26. B 27. $6x-4$

28. $\frac{2y(2x+y)}{x}$ 29. $\frac{(2t+1)(3t-2)}{2}$ 30. $\frac{v+6w}{v-w}$ 31. $\frac{1}{2x+1}$

32. 0 33. $\frac{7v}{(v-1)(2v+1)(3v-2)}$ 34. 3

35. $\frac{3w^2+17w+66}{(w-3)(w-2)(w+6)}$ 36. $\frac{6v-1}{3v-2}$ 37. $y-4$ 38. $\frac{2}{3z+2}$

39. $\frac{a(2a+1)}{(a+1)^2}$ 40. $\frac{x+1}{x+2}$ 41. $\frac{v+3}{v-3}$ 42. $\frac{-(x-1)}{x+1}$ 43. $\frac{xy+1}{xy-1}$

44. $-\frac{w-4}{3w-4}$ 45. $\frac{m+5}{m+4}$ 46. 2 47. $\frac{1}{9y^2+3y+1}$ 48. 2

49. a. $-\frac{2m+2}{(m+2)(m+4)}$ b. $m = -1$ 50. a. $\frac{-5x+16}{(x+1)(x-6)}$

b. $x = \frac{16}{5}$ 51. a. $\frac{2x^2-x-12}{(x-3)(x-2)}$ b. $x = 4$ 52. a. $\frac{2}{x-1}$

b. No solution 53. -1 54. No solution 55. $-\frac{8}{3}$ 56. 5 57. 1, 2

58. -1 59. All real numbers except $\frac{1}{2}$ and -4 60. All real numbers

except $-\frac{1}{2}$, -2, and 5 61. 14; 14.10 62. 0; $-0.002 \approx -0.00$

63. 0; −0.01 **64.** 12; 11.94 **65. a.** $\dfrac{x^2 - 17x - 54}{x - 24}$ cm

b. $\dfrac{5x^2 - 16x + 12}{4(x - 24)}$ cm² **66.** 45 cm²

67. $\dfrac{100(x + 2)}{x(x - 2)}$ cm³ **68. a.** $A(t) = \dfrac{-t^2 + 100t + 500}{5}$

b. $N(10) = 300$; the plant can produce 300 units in 10 h
c. $C(10) = 84,000$; the cost of operating the plant for 10 h is $84,000
d. $A(10) = 280$; the average cost per unit is $280 when the plant is
operated for 10 h **69.** −3 **70.** 25 and 10 **71.** 5 and 7 **72.** 6 h; the
constant of variation, $k = 300$, is the distance in miles between
St. Louis and Chicago **73.** 4 ohms and 12 ohms **74.** 21 min
75. 1000 desks **76.** 8 cm × 14 cm **77.** 21 mi/h

Mastery Test for Chapter 9
1. a. $(-\infty, 6) \cup (6, \infty)$ **b.** $(-\infty, -6) \cup (-6, 1) \cup (1, \infty)$

c. $(-\infty, -8) \cup (-8, 8) \cup (8, \infty)$ **d.** $(-\infty, \infty)$ **2. a.** $\dfrac{1}{2x - 3}$ **b.** $\dfrac{2}{x + 3}$

c. $\dfrac{2x + 3}{x + 5}$ **d.** $-\dfrac{2x - 3y}{a - 2b}$ **3. a.** $\dfrac{5}{4}$ **b.** 1 **c.** $\dfrac{2}{3}$ **d.** $\dfrac{(x + 2)(x + 5)}{(x - 5)(x + 3)}$

4. a. 2 **b.** $\dfrac{1}{x - 3}$ **c.** $\dfrac{9}{(w - 5)(w + 4)}$ **d.** $\dfrac{4}{x - y}$ **5. a.** x **b.** $\dfrac{x + 12}{5}$

c. 1 **d.** $\dfrac{w + 1}{w - 3}$ **6. a.** $\dfrac{24}{5}$ **b.** $\dfrac{x - 3}{3}$ **c.** $-\dfrac{3(2x + 3)}{4x - 3}$ **d.** $-\dfrac{2(v - 11)}{(v - 4)^2}$

7. a. 5 **b.** 2; 4 **c.** No solution **d.** $x = \dfrac{5y + 4}{y - 1}$ **8. a.** 18 **b.** 8 N/cm²

9. a. 10 **b.** 10 h **c.** 6%, 8.5% **d.** 475 mi/h, 525 mi/h

Chapter 10
Using the Language and Symbolism of Mathematics   10.1
1. Difference **2.** Ratio **3.** Growth **4.** Decay **5.** b^x; base; exponent
6. growth **7.** decay **8.** x **9.** 2.718 **10.** x; y **11.** a; b

Exercises 10.1

1. a. Geometric; $r = 5$ **b.** Geometric; $r = \dfrac{1}{5}$ **c.** Geometric; $r = -5$

d. Geometric; $r = -\dfrac{1}{5}$ **3. a.** Geometric; $r = \dfrac{2}{3}$ **b.** Not geometric

c. Not geometric **d.** Geometric; $r = 2$ **5. a.** 10, 12, 14, 16, 18
b. 10, 8, 6, 4, 2 **c.** 10, 20, 40, 80, 160 **d.** 10, −20, 40, −80, 160 **7.** B
9. A **11. a.** 3, 6, 9, 12, 15 **b.** 3, 6, 12, 24, 48 **13. a.** 2, 2, 2, 2, 2

b. 2, 2, 2, 2, 2 **15. a.** 4 **b.** 64 **c.** 1 **d.** $\dfrac{1}{16}$ **e.** 2 **17.** 16; 15.21

19. $\dfrac{1}{4} = 0.25$; 0.26 **21. a.** 2824.295 **b.** 0.102 **c.** 22.655

23. a. 16.919 **b.** 0.202 **c.** 23.141
25.   **27.**

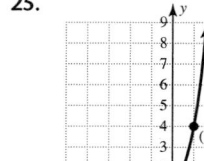

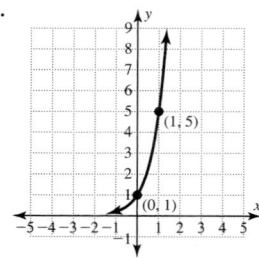

29. a. 32 **b.** −8, 8 **c.** 6 **31.** 2 **33.** 2 **35.** 3 **37.** −1 **39.** 0 **41.** $\dfrac{1}{2}$

43. $\dfrac{1}{2}$ **45.** −2 **47.** −3 **49.** $\dfrac{26}{15}$ **51.** $\dfrac{1}{2}$ **53.** 4 **55.** −3 **57.** −5 **59.** $\dfrac{1}{6}$

61. −2 **63. a.** $110, $120, $130, $140 **b.** $90, $80, $70, $60
c. $110, $121, $133.10, $146.41 **d.** $90, $81, $72.90, $65.61

65. 18 m, 9 m, 4.5 m, 2.25 m **67.** $A(t) = 1000(1.06)^t$; $A \approx $1503.63
69. 78.0 mg

Using the Language and Symbolism of Mathematics   10.2
1. y; x **2.** Different **3.** One-to-one **4.** f^{-1} **5.** $x = y$

Exercises 10.2
1. $\{(4, 1), (11, 3), (2, 8)\}$ **3.** $\{(2, -3), (2, -1), (2, 0), (2, 2)\}$
5. $\{(b, a), (d, c)\}$ **7.** $\{(7, 4), (9, e), (-2, -3)\}$ **9.** $\{(0, 9), (3, 4), (-2, 6), (5, -4)\}$ **11.** $f = \{(-3, 2), (-1, 1), (1, 0), (3, -1)\}$;
$f^{-1} = \{(2, -3), (1, -1), (0, 1), (-1, 3)\}$ **13.** Domain of $f = \{0, 1, 2, 3, 4\}$;
range of $f = \{-3, -2, -1, 0, 1\}$; domain of $f^{-1} = \{-3, -2, -1, 0, 1\}$;
range of $f^{-1} = \{0, 1, 2, 3, 4\}$ **15.** Domain of $f = \{-2, 0, 2\}$; range of
$f = \{0, 3, 1\}$; domain of $f^{-1} = \{0, 3, 1\}$; range of $f^{-1} = \{-2, 0, 2\}$
17. One-to-one **19.** Not one-to-one **21.** One-to-one **23.** Not one-to-
one **25.** Not one-to-one **27.** Not one-to-one **29.** Not one-to-one
31.   **33.**

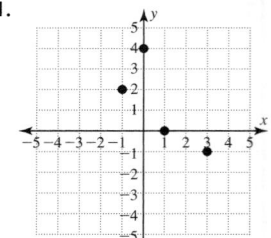

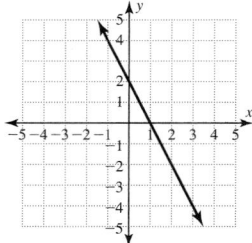

35.   **37.**

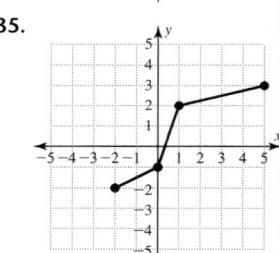

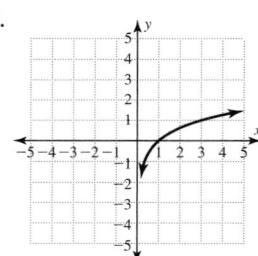

39.   **41.**

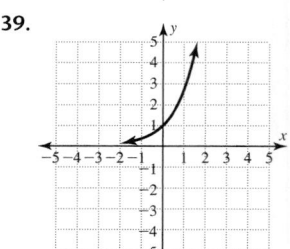

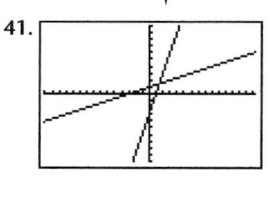

43.

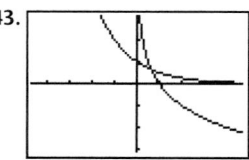

45. $f^{-1}(x)$ increases x by 2; $f^{-1}(x) = x + 2$ **47.** $f^{-1}(x)$ takes one-
fourth of x; $f^{-1}(x) = \dfrac{x}{4}$ **49.** $f^{-1}(x)$ increases x by three and then

divides this quantity by two; $f^{-1}(x) = \dfrac{x + 3}{2}$ **51.** $f^{-1}(x) = \dfrac{x - 2}{5}$
53. $f^{-1}(x) = 3(x + 7)$ **55.** $f^{-1}(x) = (x - 2)^3$ **57.** $f^{-1}(x) = -x$
59. a. 17 **b.** 5 **61. a.** 5 **b.** 2 **63. a.** 1 **b.** 1 **65.** The inverse function
is useful in converting a European shoe size to the equivalent U.S. size

European Size	U.S. Size
39	7
40.5	8
42	9
43	10
44	11

67. The inverse function is useful in determining the cost of the car one can afford on a given monthly budget

Monthly Payment	Loan Amount
120.89	5,000
241.79	10,000
362.68	15,000
483.58	20,000
604.47	25,000

69. The inverse function is useful in converting Celsius to Fahrenheit

Celsius	Fahrenheit
-17.8	0
-6.7	20
4.4	40
15.6	60
26.7	80
37.8	100

71. a. The number of units produced **b.** The cost of producing x units
c. $C^{-1}(x) = \dfrac{x - 350}{12}$ **d.** The cost **e.** The number of units produced
f. $C(100) = \$1550$ **g.** $C^{-1}(1934) = 132$ units

Using the Language and Symbolism of Mathematics   10.3
1. $\log_b x$ **2.** $b; x; y$ **3.** Undefined **4.** 0 **5.** 1 **6.** -1 **7.** x **8.** y

Exercises 10.3
1. Verbal description: the log base 5 of 125 is 3; exponential form:
$5^3 = 125$ **3.** Verbal description: the log base 3 of $\sqrt{3}$ is $\dfrac{1}{2}$; exponential
form: $3^{1/2} = \sqrt{3}$ **5.** Verbal description: the log base 5 of $\dfrac{1}{5}$ is -1;
exponential form: $5^{-1} = \dfrac{1}{5}$ **7.** Logarithmic form: $\log_{16} 4 = \dfrac{1}{2}$;
exponential form: $16^{1/2} = 4$ **9.** Logarithmic form: $\log_8 4 = \dfrac{2}{3}$;
verbal description: the log base 8 of 4 is $\dfrac{2}{3}$ **11.** Logarithmic form:
$\log_3 \dfrac{1}{9} = -2$; verbal description: the log base 3 of $\dfrac{1}{9}$ is -2
13. Logarithmic form: $\log_m n = p$; verbal description: the log base m
of n is p **15.** Logarithmic form: $\log_n m = k$; exponential form:
$n^k = m$ **17.** 2 **19.** -3 **21.** $\dfrac{1}{3}$ **23.** 5 **25.** -1 **27.** 0 **29.** -1 **31.** 4
33. 6 **35.** 3 **37.** $\dfrac{1}{4}$ **39.** $\dfrac{3}{2}$ **41.** 1 **43.** 2.7 **45.** -2 **47. a.** Defined
b. Not defined **49. a.** Not defined **b.** Defined **51.** 2 **53.** 6 **55.** $\dfrac{1}{6}$
57. $\sqrt{6}$ **59.** 2 **61.** 5 **63.** $\dfrac{1}{5}$ **65.** $-\dfrac{1}{2}$ **67.** 4 **69.** C **71.** B
73. D: $(0, \infty)$; R: $(-\infty, \infty)$

75.

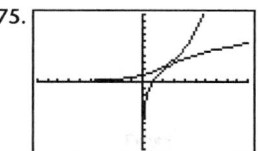

Using the Language and Symbolism of Mathematics   10.4
1. 2.718 **2.** $\log_{10} x$ **3.** $\log_e x$

Exercises 10.4
1. a. 2 **b.** 1.556 **c.** 3.584 **3. a.** -2 **b.** -0.954 **c.** -2.197 **5. a.** 0
b. 0 **c.** 0 **7. a.** 3 **b.** 1.303 **c.** 4.605 **9. a.** 5 **b.** -4 **c.** 9 **11. a.** 5
b. -5 **c.** $\dfrac{1}{2}$ **13. a.** 1.672 **b.** 46.989 **15. a.** 3.850 **b.** 46.993 **17. a.** 1
b. 2.303 **19. a.** 0.434 **b.** 1 **21. a.** 2.053 **b.** 112.980 **23. a.** -5.082
b. 0.006 **25. a.** 0.497 **b.** 1.145 **27. a.** -4.046 **b.** -9.316
29. a. -11.227 **b.** -25.851 **31. a.** 11.841 **b.** 27.264 **33. a.** 0.409
b. 0.108 **35. a.** 0.030 **b.** -0.349 **37. a.** 0.531 **b.** 1.760
39. a. 4.560 **b.** 4.443 **41. a.** 0.439 **b.** 0.845 **43. a.** -2.079
b. Undefined **45. a.** 0 **b.** Undefined **47. a.** -0.111 **b.** Undefined
49. 299.916 **51.** 0.500 **53.** 7.996 **55.** 0.050 **57.** 2 **59.** 81

Using the Language and Symbolism of Mathematics   10.5
1. b^{x+y} **2.** $\log_b x + \log_b y$ **3.** b^{x-y} **4.** $\log_b x - \log_b y$ **5.** b^{xp}
6. $p \log_b x$ **7.** $\dfrac{\log_b x}{\log_b a}$ **8.** $b^{x \log_b a}$ **9.** 7^x **10.** x

Exercises 10.5
1. $\log a + \log b + \log c$ **3.** $\ln x - \ln 11$ **5.** $\log x + 5 \log y$
7. $2 \ln x + 3 \ln y + 4 \ln z$ **9.** $\ln(2x + 3) - \ln(x + 7)$
11. $\dfrac{1}{2} \log(4x + 7)$ **13.** $\dfrac{1}{2} \ln(x + 4) - 2 \ln(y + 5)$
15. $\dfrac{1}{2}[\log x + \log y - \log (z - 8)]$ **17.** $2 \log x + 3 \log (2y + 3) -$
$4 \log z$ **19.** $\ln x + 2 \ln y - 3 \ln z$ **21.** $\log (x^2 y^5)$ **23.** $\ln \dfrac{x^3 y^7}{z}$
25. $\log \dfrac{\sqrt{x + 1}}{2x + 3}$ **27.** $\ln \sqrt[3]{(2x + 7)(7x + 1)}$ **29.** $\log_5 x^2 \sqrt[3]{y^2}$ **31.** 34
33. 0.53 **35.** 37 **37.** 0.045 **39.** x **41.** 1.1833 **43.** 1.6542 **45.** 0.1187
47. 1.3562 **49.** -0.4709 **51.** 0.6438 **53.** C **55.** B **57.** F **59.** E
61. A **63.** 2.2453 **65.** 0.7631 **67.** -1.2860 **69.** -1.8934 **71.** 3;
2.989 **73.** 32; 31.640 **75. a.** $f(x) = e^{x \ln 5}$ **b.** $f(x) = e^{-x \ln 5}$
77. $\ln \dfrac{1}{2} = \ln 1 - \ln 2 = 0 - \ln 2 = -\ln 2$ **79.** $\ln y = kt + c_1$ is
equivalent to $y = e^{kt + c_1} = e^{kt} e^{c_1} = c_2 e^{kt}$ (where $c_2 = e^{c_1}$)

Using the Language and Symbolism of Mathematics   10.6
1. Exponential **2.** Logarithmic **3.** $x; y$

Exercises 10.6
1. 8 **3.** $-2, 2$ **5.** 15 **7.** 35 **9.** 16 **11.** $-\dfrac{1}{2}$ **13.** -2 **15.** 2 **17.** 10
19. No solution **21.** -7 **23.** No solution **25.** 2.5 **27.** 2
29. $\dfrac{\ln 15}{\ln 4} \approx 1.953$ **31.** $7 - \dfrac{\ln 22}{\ln 3} \approx 4.186$ **33.** $\dfrac{\ln 9.2}{\ln 11.3 - 2 \ln 9.2} \approx$
-1.102 **35.** $\dfrac{\ln 5.3}{2 \ln 5.3 + 2 \ln 7.6} \approx 0.226$ **37.** $\dfrac{\ln 78.9}{2} \approx 1.456$
39. $\dfrac{\log 51.3 - 1}{2} \approx 0.355$ **41.** $\pm \sqrt{\dfrac{\ln 0.68}{\ln 0.83}} \approx \pm 1.439$
43. $\pm \sqrt{\ln\left(\dfrac{689.7}{3.7}\right) - 1} \approx \pm 2.056$ **45.** 2 **47.** $-1, 1$

49. $\dfrac{10^{0.83452} + 17}{5} \approx 4.766$ **51.** $e^e \approx 15.154$ **53.** $3 + \sqrt{2} \approx 4.414$

55. No solution **57.** $1, e^2 \approx 7.389$ **59.** $-0.005, 2.500$ **61.** 2.998

63. $10^{-\log x} = 10^{\log x^{-1}} = x^{-1} = \dfrac{1}{x}$

65. $e^{-x \ln 3} = e^{\ln 3^{-x}} = 3^{-x} = \dfrac{1}{3^x} = \left(\dfrac{1}{3}\right)^x$

67. $\log 60^x - \log 6^x = \log \dfrac{60^x}{6^x} = \log \left(\dfrac{60}{6}\right)^x = \log 10^x = x$

69. $\ln \left(\dfrac{4}{5}\right)^x + \ln \left(\dfrac{5}{3}\right)^x + \ln \left(\dfrac{3}{4}\right)^x = x \ln \dfrac{4}{5} + x \ln \dfrac{5}{3} + x \ln \dfrac{3}{4} =$

$x\left(\ln \dfrac{4}{5} + \ln \dfrac{5}{3} + \ln \dfrac{3}{4}\right) = x \ln \left(\dfrac{4}{5}\right)\left(\dfrac{5}{3}\right)\left(\dfrac{3}{4}\right) = x \ln 1 = x(0) = 0$

71. 6.8 years

Using the Language and Symbolism of Mathematics 10.7

1. Periodic **2.** Continuous **3.** $P\left(1 + \dfrac{r}{n}\right)^{nt}$ **4.** Pe^{rt} **5.** Earthquakes

6. Sound

Exercises 10.7

1. \$234.08 **3.** 9 years **5.** 18.4 years **7.** 8.7% **9.** 6.9% **11.** 9.9 years
13. 24,000 years **15. a.** $B(0) = 4$; there are 4 units of bacteria present
initially **b.** $B(5) \approx 13.3$; after 5 days there are approximately 13.3
units of bacteria present **c.** $B(10) \approx 44$; after 10 days there are
approximately 44 units of bacteria present **d.** $B(13.4) \approx 100$; there
will be 100 units of bacteria present after approximately 13.4 days
17. 9210 days **19.** There will be approximately 10,000 (round
10,055) whales left in 10 years; it will take approximately 27.5 years
for the whale population to reach 5000 **21.** 13.5% **23.** 7.8
25. 7 times **27.** 24.8 decibels **29.** 3.2×10^{-8} W/cm^2 **31.** 3.96
33. 7.4×10^{-10} mol/l **35.** \$411.60 **37.** 36 payments

Review Exercises for Chapter 10

1. a. Arithmetic; $d = 2$ **b.** Geometric; $r = 2$ **c.** Neither
2. a. Geometric; $r = 1$; Arithmetic; $d = 0$ **b.** Arithmetic; $d = -4$
c. Geometric; $r = -1$ **3.** 100, 90, 81, 72.9, 65.61, and 59.049 gal
4. a. 1 **b.** $\dfrac{1}{9}$ **c.** 81 **d.** 3 **5. a.** 0.318 **b.** 9.870 **c.** 36.462 **d.** 22.459
6. B **7.** C **8.** A
9. a.

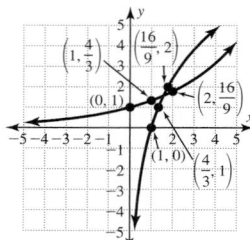

b.

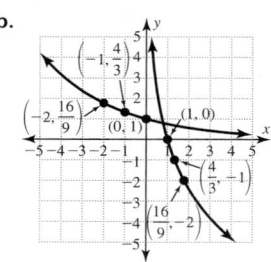

c.

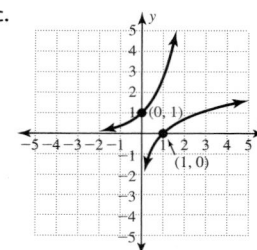

10. f^{-1} adds 2 and then divides the quantity by 3; $f^{-1}(x) = \dfrac{x + 2}{3}$

11. f^{-1} subtracts 6 from x and then multiplies the quantity by 4;
$f^{-1}(x) = 4(x - 6)$ **12.** $f^{-1} = \{(-5, -4), (-3, -3), (3, 0), (7, 2), (9, 3)\}$
13. $f^{-1} = \{(0, -2), (2, -1), (4, 0), (6, 1), (8, 2), (10, 3)\}$
14. $f^{-1}(x) = 3(x + 4)$ **15.** $f^{-1}(x) = \log_3 x$ **16. a.** x represents the
number of pizzas **b.** $C^{-1}(x) = \dfrac{x - 250}{2}$ **c.** The cost of producing
the pizzas **d.** The number of pizzas **e.** \$450 **f.** 74 pizzas
17.

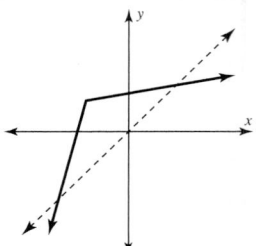

18.

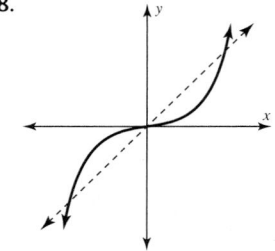

19. a. $6^{1/2} = \sqrt{6}$ **b.** $17^0 = 1$ **c.** $8^{-2} = \dfrac{1}{64}$ **20. a.** $a = b^c$

b. $a = e^c$ **c.** $c = 10^d$ **21. a.** $\log_7 343 = 3$ **b.** $\log_{19} \sqrt[3]{19} = \dfrac{1}{3}$

c. $\log_{4/7} \dfrac{49}{16} = -2$ **22. a.** $\ln \dfrac{1}{e} = -1$ **b.** $\log 0.0001 = -4$

c. $\log_8 y = x$ **23. a.** 2 **b.** $\dfrac{1}{2}$ **c.** -1 **24. a.** 0 **b.** 0.23 **c.** $-\dfrac{1}{3}$

25. a. 21.256 **b.** 2514.929 **c.** 0.155 **26. a.** 20.670 **b.** 0.083
c. 67,608.298 **27. a.** 2.055 **b.** 4.733 **c.** 2.940 **28.** -3.092
b. -7.118 **c.** -4.423 **29. a.** 1.065 **b.** 0.191 **c.** 0.165 **30. a.** 2.398
b. 3.401 **c.** 3.401 **31. a.** 2.010 **b.** 1.057 **c.** 3.170 **32. a.** 2.989
b. 2.016 **c.** 1.893 **33.** -2 **34.** $\dfrac{2}{3}$ **35.** $\dfrac{2}{3}$ **36.** $-\dfrac{1}{2}$ **37.** 1 **38.** $\dfrac{1}{4}$

39. $2; -2$ **40.** $-\dfrac{1}{2}, 1$ **41.** 2 **42.** $\dfrac{3}{2}$ **43.** 4 **44.** $\sqrt{13}$ **45.** $\dfrac{1}{9}$ **46.** No

solution **47.** No solution **48.** 13 **49.** $\dfrac{1}{2}$ **50.** 11 **51.** 17 **52.** 7

53. No solution **54.** 3 **55.** 140 **56.** 33 **57.** 10; -10 **58.** 1 **59.** -2

60. $-\dfrac{1}{2}$ **61.** 3 **62.** $\dfrac{27}{2}$ **63.** $-3, 0$ **64.** 3 **65.** 10 **66.** 4 **67.** $3 \log x +$

$5 \log y$ **68.** $\ln (7x - 9) - \ln (2x + 3)$ **69.** $\dfrac{1}{2} \ln (2x + 1) - \ln (5x + 9)$

70. $\dfrac{1}{2}(2 \log x + 3 \log y - \log z)$ **71.** $\ln x^2 y^3$ **72.** $\ln \dfrac{x^5}{y^4}$

73. $\ln (x + 1)$ **74.** $\ln \sqrt{\dfrac{x}{y}}$ **75.** 2.146 **76.** -0.293 **77.** $e^{x \ln 5}$

78. $\log 50^x + \log 6^x - \log 3^x = x \log 50 + x \log 6 - x \log 3 =$

$x(\log 50 + \log 6 - \log 3) = x\left(\log \dfrac{50 \cdot 6}{3}\right) = x \log 100 = x(2) = 2x$

79. $1000^{\log x} = (10^3)^{\log x} = 10^{3 \log x} = 10^{\log x^3} = x^3$ **80.** 3; 2.996

81. 1; 1.099 **82.** 3; 3.096 **83.** 3; 2.975 **84. a.** 13 **b.** 3 **c.** $\dfrac{1}{13}$ **85.** D

86. E **87.** C **88.** B **89.** A **90. a.** \$2938.66 **b.** \$2979.69
c. \$2983.65 **91.** $\approx 10,200$ days **92.** 9.6 years **93.** 8.7% **94.** 6.8
95. a. $B(0) = 3$; the initial amount of bacteria is 3 units **b.** $B(5) \approx$
12.8; after 5 days there are approximately 12.8 units of bacteria
c. $B(10) \approx 54.5$; after 10 days there are approximately 54.5 units of
bacteria **d.** $t \approx 12.1$ days; the amount of bacteria reaches 100 units
after approximately 12.1 days

Mastery Test for Chapter 10

1. a. Geometric; $r = 5$ **b.** Not geometric **c.** Geometric; $r = \dfrac{1}{2}$

d. Geometric; $r = 0.1$

2. a.

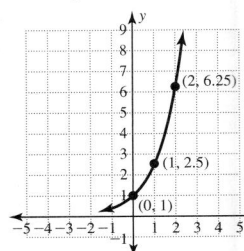

b.

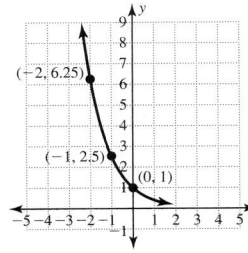

c. $\dfrac{1}{16}$ **d.** 4 **3. a.** $\{(4, -1), (9, 8), (11, -7)\}$ **b.** $f^{-1}(x) = \dfrac{x + 6}{3}$

c. $\{(-1, -2), (0, -1), (1, 0), (2, 1), (-2, 3)\}$

d. $\left\{\left(2, \dfrac{1}{2}\right), \left(3, \dfrac{1}{3}\right), \left(\dfrac{1}{6}, 6\right), (1, 1)\right\}$

4. a.

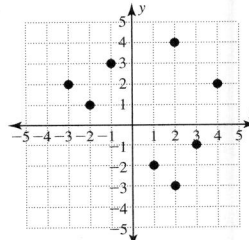

b.

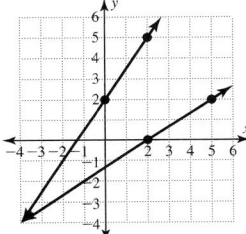

c.

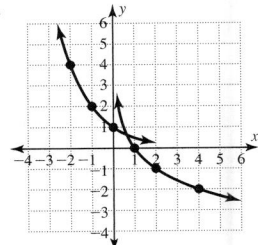

d.

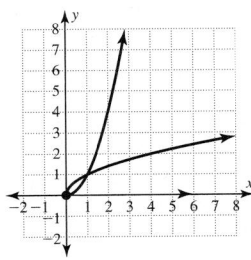

5. a. $5^{2/3} = \sqrt[3]{25}$ **b.** $5^{-3} = \dfrac{1}{125}$ **c.** $b^x = y + 1$ **d.** $b^{y+1} = x$ **6. a.** 0

b. 1 **c.** -1 **d.** 17 **7. a.** 1.2810 **b.** 2.9497 **c.** 12.6881 **d.** -10.4011

8. a. $4 \log x + 5 \log y$ **b.** $3 \ln x - \dfrac{1}{2} \ln y$ **c.** $\ln \dfrac{(7x + 9)^2}{x}$

d. $\log x\sqrt{x + 3}$ **9. a.** 1.730 **b.** 0.850 **c.** 6.644 **d.** 2.406

10. a. -4 **b.** $\dfrac{3}{2}$ **c.** 16 **d.** $\dfrac{1}{3}$ **e.** -2 **f.** -3 **g.** 0.406 **h.** 1.549

11. a. 35.8 years **b.** 8.4 years

Chapter 11
Using the Language and Symbolism of Mathematics   11.1

1. $f(x) + g(x)$ **2.** $f(x) - g(x)$ **3.** $f(x) \cdot g(x)$ **4.** $\dfrac{f(x)}{g(x)}$; 0 **5.** $f(g(x))$

6. $g(x); f$ **7.** $x; f^{-1}(x); x; f(x)$

Exercises 11.1

1. a. 12 **b.** -6 **c.** 27 **d.** $\dfrac{1}{3}$ **3. a.** 80 **b.** 11 **c.** 8 **d.** 23

5.

x	$(f + g)(x)$
0	-3
3	5
8	16

7.

x	$(g - f)(x)$
0	1
3	5
8	2

9.

x	$(g \cdot f)(x)$
0	2
3	0
8	63

11. $f - g = \{(-2, -1), (1, 6), (4, 1)\}$ **13.** $\dfrac{f}{g} = \left\{\left(-2, \dfrac{3}{4}\right), (1, -5), \left(4, \dfrac{7}{6}\right)\right\}$

15.

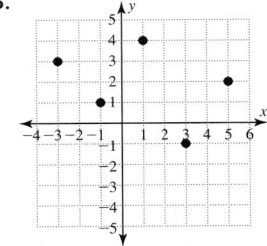

17.

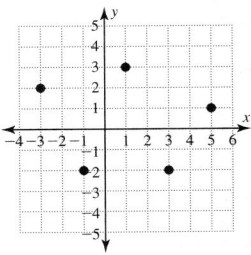

19. $(f + g)(x) = 2x^2 + x - 6, D = (-\infty, \infty); (f - g)(x) = 2x^2 - 3x,$
$D = (-\infty, \infty); (f \cdot g)(x) = 4x^3 - 8x^2 - 3x + 9, D = (-\infty, \infty);$
$\left(\dfrac{f}{g}\right)(x) = x + 1, D = \left(-\infty, \dfrac{3}{2}\right) \cup \left(\dfrac{3}{2}, \infty\right);$
$(f \circ g)(x) = 8x^2 - 26x + 18, D = (-\infty, \infty)$

21. $(f + g)(x) = \dfrac{x^2 + x + 1}{x^2 + x}, D = (-\infty, 1) \cup (-1, 0) \cup (0, \infty);$
$(f - g)(x) = \dfrac{x^2 - x - 1}{x^2 + x}, D = (-\infty, -1) \cup (-1, 0) \cup (0, \infty);$
$(f \cdot g)(x) = \dfrac{1}{x + 1}, D = (-\infty, -1) \cup (-1, 0) \cup (0, \infty);$
$\left(\dfrac{f}{g}\right)(x) = \dfrac{x^2}{x + 1}, D = (-\infty, -1) \cup (-1, 0) \cup (0, \infty);$
$(f \circ g)(x) = \dfrac{1}{x + 1}, D = (-\infty, -1) \cup (-1, 0) \cup (0, \infty);$

23. Not equal; $f(2)$ is defined and $g(2)$ is not defined **25.** Not equal; the two functions have different domains **27.** Not equal; the two functions have different domains

29.

$x \rightarrow g(x) \rightarrow f(g(x))$	$x \rightarrow f(g(x))$
$3 \rightarrow 5 \rightarrow 9$	$3 \rightarrow 9$
$4 \rightarrow 2 \rightarrow 8$	$4 \rightarrow 8$
$7 \rightarrow 1 \rightarrow 0$	$7 \rightarrow 0$
$8 \rightarrow 4 \rightarrow 6$	$8 \rightarrow 6$

31.

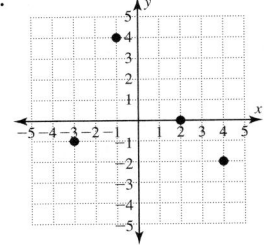

33. $(f \circ g)(x) = 16x^2 - 36x + 17; (g \circ f)(x) = 4x^2 - 20x + 10$
35. $(f \circ g)(x) = \sqrt{x + 5}; (g \circ f)(x) = \sqrt{x} + 5$

37. $(f \circ g)(x) = \dfrac{1}{x^2 - 2}$; $(g \circ f)(x) = \dfrac{1}{(x+2)^2} - 4 =$
$-\dfrac{(2x+3)(2x+5)}{(x+2)^2}$

39.

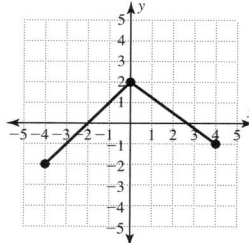

41. a. $5000 **b.** $252,500 **c.** $257,500 **d.** $515 per unit
e. $C(u) = u^2 + 5u + 5000$ **f.** $A(u) = \dfrac{u^2 + 5u + 5000}{u}$

43. a. 203 items **b.** $80 **c.** $16,240
d. $R(m) = -5m^3 + 135m^2 + 1620m$ **45. a.** $C(5000) = 1900$; the cost
of making 5000 doses is $1900 **b.** $R(1900) = 2850$; the revenue
derived from vaccine costing $1900 is $2850
c. $(R \circ C)(5000) = 2850$; the revenue derived from making 5000 doses
of the vaccine is $2850 **d.** $(R \circ C)(d) = 0.45d + 600$; the revenue
derived from making d doses of vaccine **47. a.** $A(w) = 576 - w^2$
b. $w(x) = 24 - 2x$ **c.** $(A \circ w)(x) = 96x - 4x^2$; $(A \circ w)(x)$ is the area of
the board when x cm is trimmed from all sides

49. $(f \circ f^{-1})(x) = x$; $(f^{-1} \circ f)(x) = x$ **51. a.** $f^{-1}(x) = \dfrac{x - 5}{4}$
b. $(f \circ f^{-1})(x) = x$ **c.** $(f^{-1} \circ f)(x) = x$ **53.** $h(x) = f(g(x))$
55. $h(x) = \dfrac{g(x)}{f(x)}$ **57.** $f(x) = \dfrac{1}{x}$; $g(x) = 3x^2 - 7x + 9$

59. $f(x) = x + \dfrac{1}{x}$; $g(x) = x^3 - 2$ **61.**

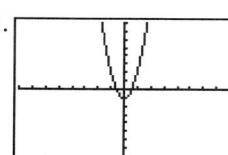

Using the Language and Symbolism of Mathematics 11.2
1. Natural **2.** Finite **3.** Infinite **4.** Arithmetic **5.** Geometric
6. Fibonacci **7.** General **8.** Recursively **9.** $a_1 + (n-1)d$ **10.** $a_1 r^{n-1}$

11. Summation; i; 1; n **12.** $\dfrac{n}{2}(a_1 + a_n)$ or $\dfrac{n}{2}[2a_1 + (n-1)d]$

13. $\dfrac{a_1(1 - r^n)}{1 - r}$ or $\dfrac{a_1 - ra_n}{1 - r}$ **14.** $\dfrac{a_1}{1 - r}$

Exercises 11.2
1. a. Geometric; $r = 6$ **b.** Arithmetic; $d = 5$ **c.** Neither
d. Geometric; $r = 1$; arithmetic; $d = 0$ **3. a.** Arithmetic; $d = 3$
b. Neither **c.** Geometric; $r = 4$ **d.** Geometric; $r = 1$; arithmetic;
$d = 0$ **5. a.** 8, 13, 18, 23, 28, 33 **b.** 1, 3, 7, 13, 21, 31 **c.** 8, 4, 2, 1, $\dfrac{1}{2}, \dfrac{1}{4}$
d. $-10, 20, -40, 80, -160, 320$ **7. a.** 18, 14, 10, 6, 2, -2 **b.** 18, 16,
14, 12, 10, 8 **c.** 7, 5, 3, 1, -1, -3 **d.** $-9, -6, -3, 0, 3, 6$ **9. a.** 1, 4,
16, 64, 256 **b.** 1, 2, 4, 8, 16 or 1, -2, 4, -8, 16 **c.** 16, 8, 4, 2, 1 **d.** 5,
$-15, 45, -135, 405$ **11. a.** 416 **b.** -83 **c.** 41 **d.** 502 **13.** 21

15. -300 **17.** $\dfrac{1}{2}$ **19.** 46 **21.** 7 **23.** $\dfrac{3}{2}$ **25.** -3 or 3 **27.** 125

29. a. $5 + 7 + 9 + 11 + 13 + 15 = 60$ **b.** $0 + 3 + 8 + 15 + 24 = 50$
c. $2 + 4 + 8 + 16 + 32 + 64 = 126$
d. $5 + 5 + 5 + 5 + 5 + 5 + 5 + 5 + 5 + 5 = 50$

31. 1640 **33.** 5 **35.** 9240 **37.** 3965 **39.** $\dfrac{372}{5}$ **41.** 381 **43.** 0.22222

45. $\dfrac{255}{8}$ **47.** 12.75603 **49.** 0.8888888 **51.** $\dfrac{31248}{625}$ **53.** 15 **55.** 8

57. $\dfrac{4}{33}$ **59.** $\dfrac{4}{5}$ **61.** $\dfrac{18}{5}$ **63.** $\dfrac{192}{5}$ **65. a.** $\dfrac{4}{9}$ **b.** $\dfrac{7}{33}$ **c.** $\dfrac{409}{999}$ **d.** $\dfrac{23}{9}$

67. 33,649 **69.** 11 **71.** $\dfrac{4}{13}$ **73.** 6 **75.** 4 **77.** $\dfrac{49}{3}$ **79.** 272 logs

81. $81,000 **83.** 21,845 people **85.** 96% **87.** $1,800,000

Using the Language and Symbolism of Mathematics 11.3
1. Rectangular **2.** Coefficients; constants **3.** Equivalent
4. Interchanged **5.** One; one; $\dfrac{1}{2}$ **6.** $r_2' = r_2 + 2r_1$ **7.** General
8. Particular

Exercises 11.3

1. $\begin{bmatrix} 3 & 1 & | & 0 \\ 2 & -1 & | & -5 \end{bmatrix}$ **3.** $\begin{bmatrix} 4 & 0 & | & 12 \\ 3 & 2 & | & 1 \end{bmatrix}$ **5.** $\begin{bmatrix} 1 & 0 & | & 5 \\ 0 & 1 & | & -6 \end{bmatrix}$

7. $2x + 3y = 2$; $4x - 3y = 1$ **9.** $2x + y = 1$; $x + 3y = 0$ **11.** $x = 7$;

$y = -8$ **13.** $\begin{bmatrix} 1 & 3 & | & 0 \\ 2 & 1 & | & 1 \end{bmatrix}$ **15.** $\begin{bmatrix} 1 & -3 & | & -2 \\ 0 & 1 & | & 8 \end{bmatrix}$ **17.** $\begin{bmatrix} 1 & -\dfrac{1}{3} & | & 1 \\ 6 & 4 & | & -6 \end{bmatrix}$

19. $\begin{bmatrix} 1 & 0 & | & 1 \\ 0 & 1 & | & 3 \end{bmatrix}$ **21.** $(-5, 9)$ **23.** No solution **25.** $(5 - 3y, y)$; $(5, 0)$,

$(2, 1)$, $(8, -1)$ **27.** $r_1' = \dfrac{1}{2}r_1$ **29.** $r_2' = r_2 - 3r_1$ **31.** $r_2' = \dfrac{1}{3}r_2$

33. $r_2' = r_1 - 2r_2$ **35.** $(-4, 3)$ **37.** $(7, -2)$ **39.** $(3, -2)$

41. $\left(2, \dfrac{1}{3}\right)$ **43.** $\left(\dfrac{8}{5}, \dfrac{14}{5}\right)$ **45.** $\left(\dfrac{1}{2}, 2\right)$ **47.** No solution

49. $\left(\dfrac{5}{2} + \dfrac{1}{2}y, y\right)$ **51.** $x + y = 160$; $x - y = 4$; 82 and 78

53. $x + y = 90$; $x - y = 32$; 61° and 29° **55.** $20x + y = 3200$;
$30x + y = 4300$; the fixed cost is $1000 and the variable cost per
costume is $110 **57.** $x + y = 30$; $x - y = 14$; the rate of the boat is
22 km/h; the rate of the current is 8 km/h **59.** $x + y = 100$;
$0.15x + y = 83$; 20 l of the concentrate and 80 l of water

Using the Language and Symbolism of Mathematics 11.4
1. Plane **2.** Solution **3.** Three; three **4.** Echelon **5.** Inconsistent

Exercises 11.4

1. $\begin{bmatrix} 1 & 1 & 1 & | & 2 \\ -1 & 1 & -2 & | & 1 \\ 1 & 1 & -1 & | & 0 \end{bmatrix}$ **3.** $\begin{bmatrix} 1 & 2 & -1 & | & 19 \\ 2 & -1 & 0 & | & -1 \\ 3 & -2 & 4 & | & -32 \end{bmatrix}$

5. $x - 2y + 5z = 0$; $2x + 4y - 3z = 8$; $3x + 5y + 7z = 11$
7. $x + 2z = 5$; $x + y = 0$; $y + z = 4$ **9.** $x = 7$; $y = -5$; $z = 8$

11. $\begin{bmatrix} 1 & 2 & 3 & | & 16 \\ 2 & 1 & -2 & | & -11 \\ 3 & 2 & 1 & | & 3 \end{bmatrix}$ **13.** $\begin{bmatrix} 1 & 2 & 4 & | & 3 \\ 3 & 5 & 7 & | & 1 \\ 4 & 9 & 2 & | & 8 \end{bmatrix}$

15. $\begin{bmatrix} 1 & 3 & 5 & | & 11 \\ 0 & 1 & -1 & | & -9 \\ 4 & 8 & 3 & | & 7 \end{bmatrix}$ **17.** $\begin{bmatrix} 1 & 2 & 2 & | & 3 \\ 0 & -3 & -1 & | & -7 \\ 0 & -7 & -2 & | & -17 \end{bmatrix}$

19. $\begin{bmatrix} 1 & 3 & -2 & | & 19 \\ 0 & 1 & 2 & | & -5 \\ 0 & 4 & 1 & | & 8 \end{bmatrix}$ **21.** $\begin{bmatrix} 1 & 0 & 0 & | & 3 \\ 0 & 1 & 2 & | & 3 \\ 0 & 0 & 11 & | & 22 \end{bmatrix}$

23. $(-2, 7, 3)$ **25.** No solution **27.** $(2 - 3z, -5 - 2z, z)$, $(2, -5, 0)$,

$(-1, -7, 1)$, $(-4, -9, 2)$ **29.** $r_1' = \dfrac{1}{2}r_1$ **31.** $r_2' = r_2 + 3r_1$

33. $r_1' = r_1 + 3r_2$ **35.** $(3, 1, -1)$ **37.** $(2, 3, 3)$

39. $(1 - 4x_3, -4 + 3x_3, x_3)$ **41.** $\left(\dfrac{1}{2}, -\dfrac{3}{2}, 2\right)$ **43.** No solution

45. $(-3 + c, -1 - c, c); (-3, -1, 0), (-1, -3, 2)$

47. 26, 36, 46 **49.** 72 cm, 60 cm, 36 cm **51.** 94.5°, 28.5°, and 57°
53. 600 g of A, 800 g of B, and 1300 g of C **55.** B

57.

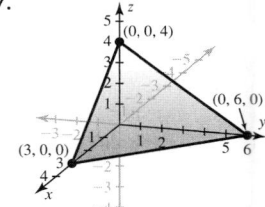

Using the Language and Symbolism of Mathematics   11.5
1. Conic sections **2.** Circle **3.** Ellipse **4.** Hyperbola **5.** Parabola
6. Radius **7.** Diameter **8.** Vertices **9.** Covertices **10.** Asymptotic

Exercises 11.5
1. 5; $(-1, 0.5)$ **3.** $\sqrt{2}$, $(0.5, 1.5)$ **5.** 25 **7.** $(2, 1)$ **9.** B **11.** D
13. $x^2 + y^2 = 100$ **15.** $(x - 2)^2 + (y - 6)^2 = 2$
17. $x^2 + \left(y - \dfrac{1}{2}\right)^2 = \dfrac{1}{4}$ **19.** Center: $(-5, 4)$; radius: 8 **21.** Center:
$(3, 0)$; radius: 3 **23.** Center: $(1, -5)$; radius: 2 **25.** Center: $(0, 0)$;
Length of major axis: 12; Length of minor axis: 8

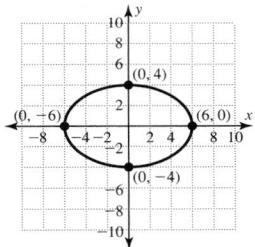

27. Center: $(-4, 3)$; Length of major axis: 10; Length of minor axis: 8

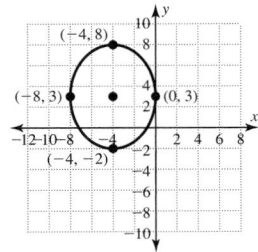

29. $\dfrac{x^2}{81} + \dfrac{y^2}{25} = 1$ **31.** $\dfrac{(x - 3)^2}{4} + \dfrac{(y - 4)^2}{1} = 1$
33. $\dfrac{(x - 6)^2}{9} + \dfrac{(y + 2)^2}{25} = 1$ **35.** $\dfrac{(x + 3)^2}{4} + \dfrac{(y + 4)^2}{36} = 1$
37. Center: $(0, 0)$

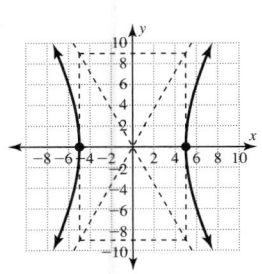

39. Center: $(0, 0)$

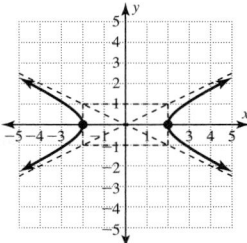

41. Center: $(4, 6)$

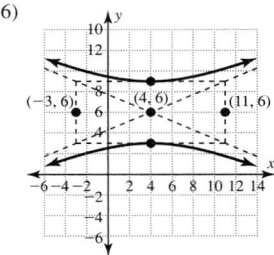

43. $\dfrac{y^2}{25} - \dfrac{x^2}{16} = 1$ **45.** $\dfrac{x^2}{9} - \dfrac{y^2}{49} = 1$
47. a.   **b.**

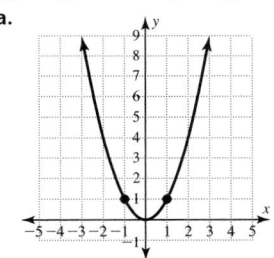

c.

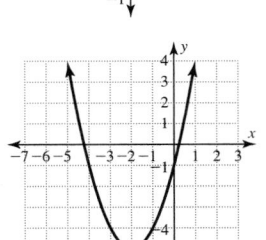

49. a.   **b.**

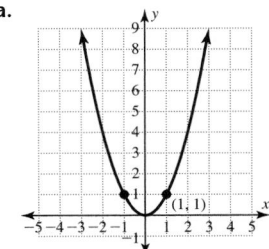
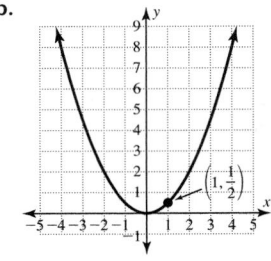

c.

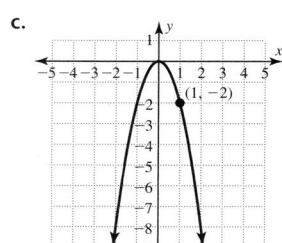

51. 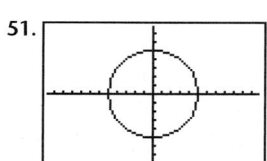 53.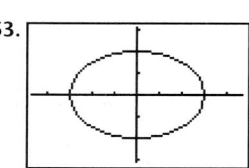

55. $y_1 = \sqrt{36 - x^2}; y_2 = -\sqrt{36 - x^2}$

57. **a.** $(-2, 7); (2, -1)$ **b.**

59. **a.** $(2, 2); (-2, -2)$ **b.**

Review Exercises for Chapter 11

1. **a.** 23 **b.** 201 **c.** 224 **d.** -178 2. **a.** -9 **b.** $-\dfrac{1}{9}$ **c.** -9 **d.** 10

3. **a.** 20 **b.** 3 **c.** -10 **d.** 163

4. **a.**

x	f + g
0	-6
1	-1
2	8
3	21

b.

x	f - g
0	-8
1	-7
2	-10
3	-17

c.

x	f · g
0	-7
1	-12
2	-9
3	38

d.

x	$\dfrac{f}{g}$
0	-7
1	$-\dfrac{4}{3}$
2	$-\dfrac{1}{9}$
3	$\dfrac{2}{19}$

5. **a.** $f + g = \{(-3, -4), (-1, -2), (1, 0), (2, 1), (3, 2)\}$
b. $f - g = \{(-3, -2), (-1, 0), (1, 2), (2, 3), (3, 4)\}$
c. $f \cdot g = \{(-3, 3), (-1, 1), (1, -1), (2, -2), (3, -3)\}$
d. $\dfrac{f}{g} = \{(-3, 3), (-1, 1), (1, -1), (2, -2), (3, -3)\}$

6.

x	f ∘ g
-1	5
2	11
3	7
5	9

7.

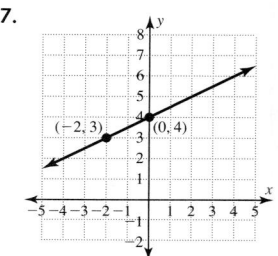

8. **a.** $(f + g)(x) = 2x - 1; D = (-\infty, \infty)$ **b.** $(f \cdot g)(x) = -8x^2 + 8x - 2; D = (-\infty, \infty)$ **c.** $(f \circ g)(x) = 2 - 8x; D = (-\infty, \infty)$

9. **a.** $(f - g)(x) = 6x - 3; D = (-\infty, \infty)$ **b.** $\left(\dfrac{f}{g}\right)(x) = -2;$
$D = \left(-\infty, \dfrac{1}{2}\right) \cup \left(\dfrac{1}{2}, \infty\right)$ **c.** $(g \circ f)(x) = 5 - 8x; D = (-\infty, \infty)$

10. $f(x)$ is defined when $x = 3$, and $g(x)$ is not defined when $x = 3$

11. $f^{-1}(x) = \dfrac{x - 3}{4}; (f \circ f^{-1})(x) = x$ 12. **a.** $F(100) = 400$

b. $V(100) = 150$ **c.** $C(x) = 1.5x + 400$ **d.** $A(x) = \dfrac{1.5x + 400}{x}$

13. **a.** $N(40) = 120$; the factory can produce 120 desks in 40 h
b. $C(120) = 14,400$; the cost of producing 120 desks is $14,400
c. $(C \circ N)(40) = 14,400$; the cost of operating the factory for 40 h is

$14,400 **d.** $(C \circ N)(t) = \left(150 - \dfrac{3t}{4}\right)(3t)$ **e.** There are only 168 h in a

week 14. **a.** 5, 9, 13, 17, 21, 25 **b.** 5, 1, -3, -7, -11, -15 **c.** 5, 8,
11, 14, 17, 20 **d.** 5, 8, 11, 14, 17, 20 15. **a.** 3, 6, 12, 24, 48, 96 **b.** 3,
-6, 12, -24, 48, -96 **c.** 3, 12, 48, 192, 768, 3072 **d.** 1, 3, 9, 27, 81,
243 16. **a.** 9, 11, 13, 15, 17 **b.** 4, 7, 12, 19, 28 **c.** 32, 16, 8, 4, 2 **d.** 3,
10, 17, 24, 31 17. 311 18. 16 19. 41 20. 14 21. **a.** $1 + 4 + 7 +$
$10 + 13 + 16 = 51$ **b.** $3 + 8 + 15 + 24 + 35 = 85$ **c.** $1 + 8 +$
$27 + 64 + 125 = 225$ **d.** $10 + 10 + 10 + 10 + 10 + 10 = 60$

22. 9725 23. $\dfrac{4}{33}$ 24. $\dfrac{3}{2}$ 25. 1071 cm 26. 40 cm; $(160 + 80\sqrt{2})$ cm

27. a. $\begin{bmatrix} 2 & -5 & | & 17 \\ 3 & 4 & | & 14 \end{bmatrix}$ **b.** $\begin{bmatrix} 3 & -4 & 2 & | & -11 \\ 2 & 2 & 3 & | & -3 \\ 4 & -1 & 5 & | & -13 \end{bmatrix}$ **28. a.** $(-2, 6)$

b. $(4, 5, 8)$ **29.** $(5 - 3z, 4 + 2z, z)$; $(5, 4, 0)$; $(2, 6, 1)$; $(-10, 14, 5)$

30. $\begin{bmatrix} 1 & 4 & | & 2 \\ 3 & 5 & | & 13 \end{bmatrix}$ **31.** $\begin{bmatrix} 1 & 2 & | & -5 \\ 3 & 2 & | & -3 \end{bmatrix}$ **32.** $\begin{bmatrix} 1 & 2 & | & 14 \\ 0 & -7 & | & -29 \end{bmatrix}$

33. $\begin{bmatrix} 1 & 0 & -8 & | & -15 \\ 0 & 1 & 2 & | & 3 \\ 0 & 0 & -1 & | & -2 \end{bmatrix}$ **34.** $(3, 7)$ **35.** $(-2, 5)$ **36.** $(1, -1)$

37. No solution **38.** $(1, -1, 1)$ **39.** $(1, 4, -3)$ **40.** $(2, 3, 4)$

41.

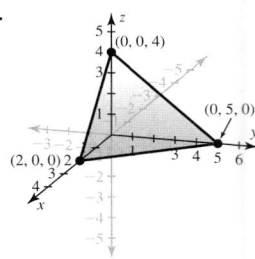

42. 4080 lb; 3280 lb **43.** $61 for A, $25 for B, $9 for C **44.** 13
45. $(1, 7)$ **46.** $(3, -3)$ **47.** 2 **48.** $(x - 3)^2 + (y + 3)^2 = 4$ **49.** C
50. D **51.** A **52.** F **53.** E **54.** B

55.

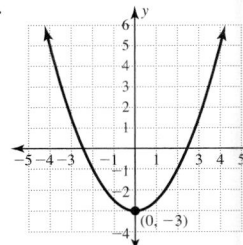

56.

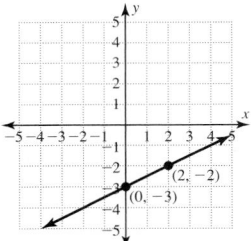

57.

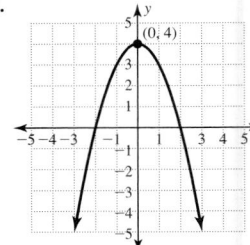

58.

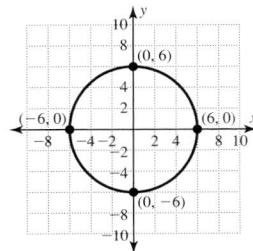

59.

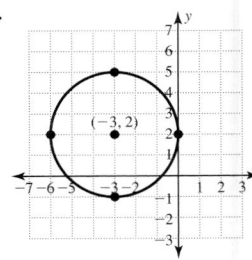

60.

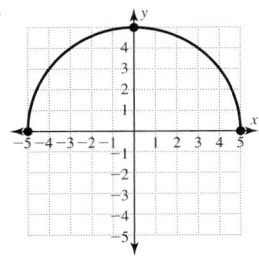

61.

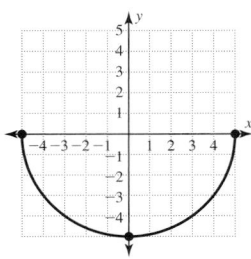

62.

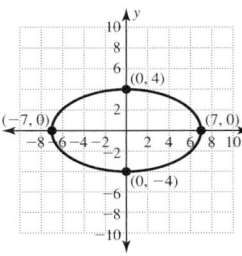

63.

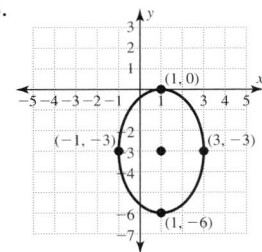

64.

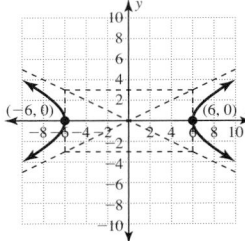

65. $(-2, 3)$; $(2, 3)$

66.

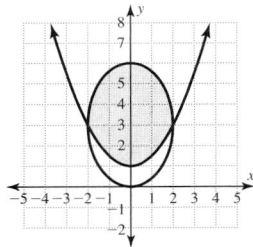

Mastery Test for Chapter 11
1. a. $(f + g)(x) = 4x - 3$ **b.** $(f - g)(x) = 2x - 7$ **c.** $(f \cdot g)(x) = 3x^2 + x - 10$ **d.** $\left(\dfrac{f}{g}\right)(x) = \dfrac{3x - 5}{x + 2}$ **2. a.** 82 **b.** 51 **c.** $4x^2 - 4x + 2$
d. $2x^2 + 1$ **3. a.** 4, 7, 10, 13, 16, 19 **b.** 4, 12, 36, 108, 324, 972
c. 2, 8, 14, 20, 26, 32 **d.** 2, 8, 32, 128, 512, 2048 **4. a.** 60 **b.** 1562
c. 1550 **d.** 4,194,306 **e.** 294 **5. a.** $\dfrac{1023}{512}$ **b.** 2 **c.** $\dfrac{8}{9}$ **d.** $\dfrac{6}{11}$
6. a. $(-2, 7)$ **b.** No solution **c.** $(6 - 2y, y)$; $(6, 0)$, $(0, 3)$, $(4, 1)$
d. $(11, -7)$ **7. a.** $(2, 3, 1)$**b.** $\left(-\dfrac{1}{7} - \dfrac{1}{7}z, \dfrac{5}{7} + \dfrac{5}{7}z, z\right)$ **c.** No solution
8. a. 50 **b.** $(-2, 6)$ **c.** 13 **d.** $\left(\dfrac{1}{2}, 5\right)$
9. a.

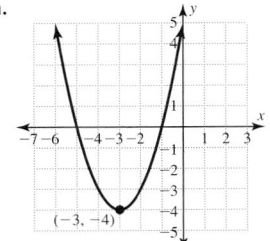

b.

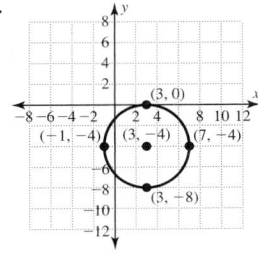

c.

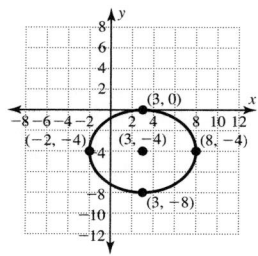

d.

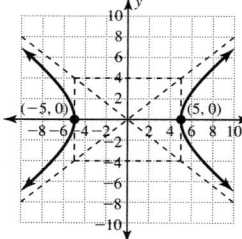

10. a. $y = -x^2 + 4$ **b.** $x^2 + (y - 5)^2 = 25$ **c.** $\dfrac{x^2}{49} + \dfrac{y^2}{25} = 1$

d. $\dfrac{x^2}{49} - \dfrac{y^2}{25} = 1$

CREDITS

INDEX

Strategy for Solving Word Problems

Step 1. Read the problem carefully to determine what you are asked to find.

Step 2. Select a variable to represent each unknown quantity. Specify precisely what each variable represents.

Step 3. If necessary, translate the problem into word equations. Then translate the word equations into a system of algebraic equations.

Step 4. Solve the equation or the system of equations, and answer the question asked by the problem.

Step 5. Check the reasonableness of your answer.

Statements of Variation

Direct variation: $y = kx$, y varies directly as x

Inverse variation: $y = \dfrac{k}{x}$, y varies inversely as x

Absolute Value Equations and Inequalities

For $a > 0$,

$|x| = a$ means $x = -a$ or $x = a$

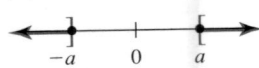

$-a \quad 0 \quad a$

$|x| \le a$ means $-a \le x \le a$

$-a \quad 0 \quad a$

$|x| \ge a$ means $x \le -a$ or $x \ge a$

$-a \quad 0 \quad a$

Properties of Radicals

If $\sqrt[n]{x}$ and $\sqrt[n]{y}$ are both real numbers, then:

$$\sqrt[n]{xy} = \sqrt[n]{x}\,\sqrt[n]{y} \qquad \text{and}$$

$$\sqrt[n]{\dfrac{x}{y}} = \dfrac{\sqrt[n]{x}}{\sqrt[n]{y}} \qquad \text{for } y \ne 0$$

$$\sqrt[n]{x^m} = \left(\sqrt[n]{x}\right)^m = x^{m/n}$$

For any real number x and natural number n:

$\sqrt[n]{x^n} = |x|$ if n is even

$\sqrt[n]{x^n} = x$ if n is odd

Properties of Logarithms

For $x > 0$, $y > 0$, $b > 0$, and $b \ne 1$, $\log_b x = m$ is equivalent to $b^m = x$.

Product rule: $\log_b xy = \log_b x + \log_b y$

Quotient rule: $\log_b \dfrac{x}{y} = \log_b x - \log_b y$

Power rule: $\log_b x^p = p \log_b x$

Special identities:

$\log_b 1 = 0$

$\log_b \dfrac{1}{b} = -1$

$b^{\log_b x} = x$

$\log_b b = 1$

$\log_b b^x = x$

Sequences and Series

a_n is read as "a sub n."

Arithmetic sequences: $d = a_n - a_{n-1}$

$$a_n = a_1 + (n-1)d$$

$$S_n = \dfrac{n}{2}(a_1 + a_n)$$

Geometric sequences: $r = \dfrac{a_n}{a_{n-1}}$

$$a_n = a_1 r^{n-1}$$

$$S_n = \dfrac{a_1(1 - r^n)}{1 - r}$$

Infinite geometric sequences: $S = \dfrac{a_1}{1 - r}$, $|r| < 1$.

Summation notation:

$$\sum_{i=1}^{n} a_i = a_1 + a_2 + \ldots + a_{n-1} + a_n$$

Methods of Solving Quadratic Equations

$ax^2 + bx + c = 0$ with $a \ne 0$.

Graphically
Numerically
Factoring
Extraction of roots
Completing the square

The quadratic formula: $x = \dfrac{-b \pm \sqrt{b^2 - 4ac}}{2a}$

Slope

The slope of a line through (x_1, y_1) and (x_2, y_2) is given by

$$m = \frac{\text{Change in } y}{\text{Change in } x} = \frac{y_2 - y_1}{x_2 - x_1} \qquad \text{for } x_1 \neq x_2.$$

Positive slope: The line goes upward to the right.
Negative slope: The line goes downward to the right.
Zero slope: The line is horizontal.
Undefined slope: The line is vertical.

Forms of Linear Equations

Slope-intercept form: $y = mx + b$ or $f(x) = mx + b$
                    with slope m and y-intercept
                    $(0, b)$
Vertical line: $x = a$ for a real constant a
Horizontal line: $y = b$ for a real constant b
General form: $Ax + By = C$
Point-slope form: $y - y_1 = m(x - x_1)$ through (x_1, y_1)
                    with slope m

Properties of Exponents

For real numbers x and y and real exponents m and n,
Product rule: $x^m \cdot x^n = x^{m+n}$
Power rule: $\left(x^m\right)^n = x^{mn}$
Product to a power: $(xy)^m = x^m y^m$
Quotient to a power: $\left(\dfrac{x}{y}\right)^m = \dfrac{x^m}{y^m} \qquad$ for $y \neq 0$
Quotient rule: $\dfrac{x^m}{x^n} = x^{m-n} \qquad$ for $x \neq 0$
Negative power: $\left(\dfrac{x}{y}\right)^{-n} = \left(\dfrac{y}{x}\right)^n \qquad$ for $x \neq 0, y \neq 0$
Special identities:
   $x^0 = 1$ for $x \neq 0 \qquad x^1 = x \qquad x^{-1} = \dfrac{1}{x}$ for $x \neq 0$

Factoring Special Forms

Difference of two squares: $x^2 - y^2 = (x + y)(x - y)$
Perfect square trinomial: $x^2 + 2xy + y^2 = (x + y)^2$
Perfect square trinomial: $x^2 - 2xy + y^2 = (x - y)^2$
Difference of two cubes:
   $x^3 - y^3 = (x - y)(x^2 + xy + y^2)$
Sum of two cubes:
   $x^3 + y^3 = (x + y)(x^2 - xy + y^2)$

Systems of Two Linear Equations

One Solution

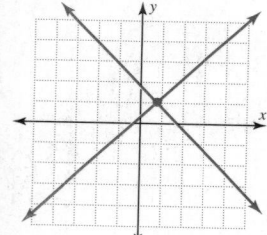

A **consistent system of independent equations:** The solution process will produce unique x- and y-coordinates.

No Solution

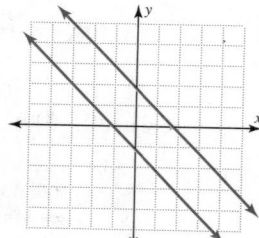

An inconsistent system: The solution process will produce a contradiction.

An Infinite Number of Solutions

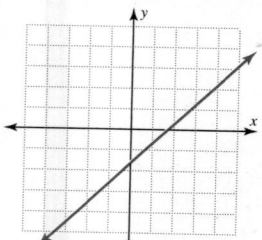

A **consistent system of dependent equations:** The solution process will produce an identity.

Equivalent Statements about Linear Factors of a Polynomial

For a real constant c and a real polynomial $P(x)$, the following statements are equivalent:

Algebraically

$x - c$ is a factor of $P(x)$.
$x = c$ is a solution of $P(x) = 0$.

Numerically

$P(c) = 0$; that is,
c is a zero of $P(x)$.

Graphically

$(c, 0)$ is an x-intercept
of the graph of $y = P(x)$.